Trigonometry

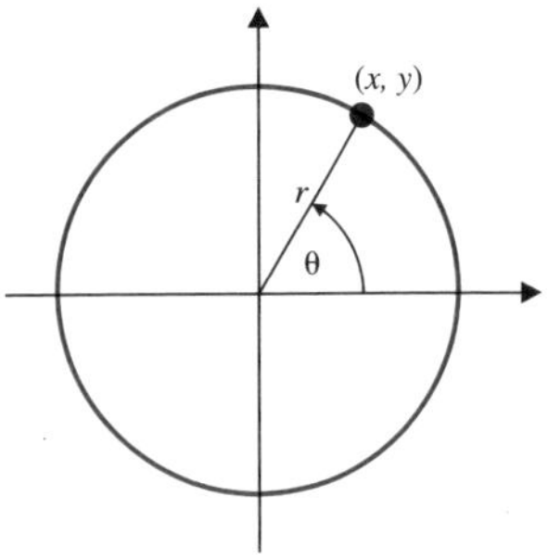

$\sin\theta = \dfrac{y}{r}$

$\cos\theta = \dfrac{x}{r}$

$\tan\theta = \dfrac{y}{x}$

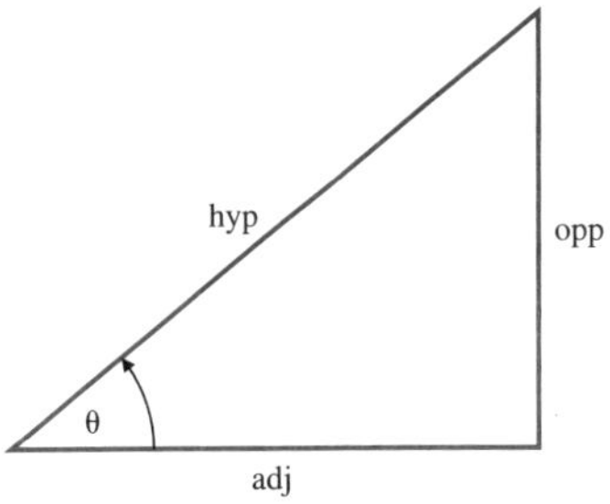

$\sin\theta = \dfrac{\text{opp}}{\text{hyp}}$

$\cos\theta = \dfrac{\text{adj}}{\text{hyp}}$

$\tan\theta = \dfrac{\text{opp}}{\text{adj}}$

Reciprocals

$\cot\theta = \dfrac{1}{\tan\theta}$ $\quad \sec\theta = \dfrac{1}{\cos\theta}$ $\quad \csc\theta = \dfrac{1}{\sin\theta}$

Definitions

$\cot\theta = \dfrac{\cos\theta}{\sin\theta}$ $\quad \sec\theta = \dfrac{1}{\cos\theta}$ $\quad \csc\theta = \dfrac{1}{\sin\theta}$

Pythagorean

$\sin^2\theta + \cos^2\theta = 1$ $\quad \tan^2\theta + 1 = \sec^2\theta$ $\quad 1 + \cot^2\theta = \csc^2\theta$

Cofunction

$\sin\left(\frac{\pi}{2} - \theta\right) = \cos\theta$ $\quad \cos\left(\frac{\pi}{2} - \theta\right) = \sin\theta$ $\quad \tan\left(\frac{\pi}{2} - \theta\right) = \cot\theta$

Even/Odd

$\sin(-\theta) = -\sin\theta$ $\quad \cos(-\theta) = \cos\theta$ $\quad \tan(-\theta) = -\tan\theta$

Double-Angle

$\sin 2\theta = 2\sin\theta\cos\theta$ $\quad \cos 2\theta = \cos^2\theta - \sin^2\theta$ $\quad \cos 2\theta = 1 - 2\sin^2\theta$

Half-Angle

$\sin^2\theta = \dfrac{1 - \cos 2\theta}{2}$ $\quad \cos^2\theta = \dfrac{1 + \cos 2\theta}{2}$

Addition

$\sin(a + b) = \sin a\cos b + \cos a\sin b$ $\quad \cos(a + b) = \cos a\cos b - \sin a\sin b$

Subtraction

$\sin(a - b) = \sin a\cos b - \cos a\sin b$ $\quad \cos(a - b) = \cos a\cos b + \sin a\sin b$

Sum

$\sin u + \sin v = 2\sin\dfrac{u+v}{2}\cos\dfrac{u-v}{2}$

$\cos u + \cos v = 2\cos\dfrac{u+v}{2}\cos\dfrac{u-v}{2}$

Product

$\sin u\sin v = \frac{1}{2}[\cos(u - v) - \cos(u + v)]$

$\cos u\cos v = \frac{1}{2}[\cos(u - v) + \cos(u + v)]$

$\sin u\cos v = \frac{1}{2}[\sin(u + v) + \sin(u - v)]$

$\cos u\sin v = \frac{1}{2}[\sin(u + v) - \sin(u - v)]$

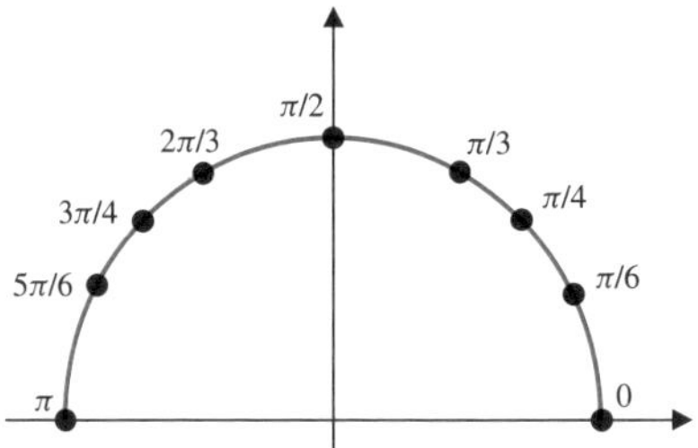

Radians

$\sin(0) = 0$	$\cos(0) = 1$
$\sin\left(\frac{\pi}{6}\right) = \frac{1}{2}$	$\cos\left(\frac{\pi}{6}\right) = \frac{\sqrt{3}}{2}$
$\sin\left(\frac{\pi}{4}\right) = \frac{\sqrt{2}}{2}$	$\cos\left(\frac{\pi}{4}\right) = \frac{\sqrt{2}}{2}$
$\sin\left(\frac{\pi}{3}\right) = \frac{\sqrt{3}}{2}$	$\cos\left(\frac{\pi}{3}\right) = \frac{1}{2}$
$\sin\left(\frac{\pi}{2}\right) = 1$	$\cos\left(\frac{\pi}{2}\right) = 0$
$\sin\left(\frac{2\pi}{3}\right) = \frac{\sqrt{3}}{2}$	$\cos\left(\frac{2\pi}{3}\right) = -\frac{1}{2}$
$\sin\left(\frac{3\pi}{4}\right) = \frac{\sqrt{2}}{2}$	$\cos\left(\frac{3\pi}{4}\right) = -\frac{\sqrt{2}}{2}$
$\sin\left(\frac{5\pi}{6}\right) = \frac{1}{2}$	$\cos\left(\frac{5\pi}{6}\right) = -\frac{\sqrt{3}}{2}$
$\sin(\pi) = 0$	$\cos(\pi) = -1$
$\sin(2\pi) = 0$	$\cos(2\pi) = 1$

42

Derivative Formulas

General Rules

$$\frac{d}{dx}[f(x)+g(x)] = f'(x)+g'(x)$$

$$\frac{d}{dx}[f(x)-g(x)] = f'(x)-g'(x)$$

$$\frac{d}{dx}[cf(x)] = cf'(x)$$

$$\frac{d}{dx}[f(g(x))] = f'(g(x))g'(x)$$

$$\frac{d}{dx}[f(x)g(x)] = f'(x)g(x)+f(x)g'(x)$$

$$\frac{d}{dx}\left[\frac{f(x)}{g(x)}\right] = \frac{f'(x)g(x)-f(x)g'(x)}{[g(x)]^2}$$

Power Rules

$$\frac{d}{dx}(x^n) = nx^{n-1}$$

$$\frac{d}{dx}(c) = 0$$

$$\frac{d}{dx}(cx) = c$$

$$\frac{d}{dx}(\sqrt{x}) = \frac{1}{2\sqrt{x}}$$

Exponential

$$\frac{d}{dx}[e^x] = e^x$$

$$\frac{d}{dx}[a^x] = a^x \ln a$$

$$\frac{d}{dx}\left[e^{u(x)}\right] = e^{u(x)}u'(x)$$

$$\frac{d}{dx}\left[e^{rx}\right] = r\,e^{rx}$$

Trigonometric

$$\frac{d}{dx}(\sin x) = \cos x$$

$$\frac{d}{dx}(\cos x) = -\sin x$$

$$\frac{d}{dx}(\tan x) = \sec^2 x$$

$$\frac{d}{dx}(\cot x) = -\csc^2 x$$

$$\frac{d}{dx}(\sec x) = \sec x \tan x$$

$$\frac{d}{dx}(\csc x) = -\csc x \cot x$$

Inverse Trigonometric

$$\frac{d}{dx}(\sin^{-1} x) = \frac{1}{\sqrt{1-x^2}}$$

$$\frac{d}{dx}(\cos^{-1} x) = -\frac{1}{\sqrt{1-x^2}}$$

$$\frac{d}{dx}(\tan^{-1} x) = \frac{1}{1+x^2}$$

$$\frac{d}{dx}(\cot^{-1} x) = -\frac{1}{1+x^2}$$

$$\frac{d}{dx}(\sec^{-1} x) = \frac{1}{|x|\sqrt{x^2-1}}$$

$$\frac{d}{dx}(\csc^{-1} x) = -\frac{1}{|x|\sqrt{x^2-1}}$$

Hyperbolic

$$\frac{d}{dx}(\sinh x) = \cosh x$$

$$\frac{d}{dx}(\cosh x) = \sinh x$$

$$\frac{d}{dx}(\tanh x) = \operatorname{sech}^2 x$$

$$\frac{d}{dx}(\coth x) = -\operatorname{csch}^2 x$$

$$\frac{d}{dx}(\operatorname{sech} x) = -\operatorname{sech} x \tanh x$$

$$\frac{d}{dx}(\operatorname{csch} x) = -\operatorname{csch} x \coth x$$

Inverse Hyperbolic

$$\frac{d}{dx}(\sinh^{-1} x) = \frac{1}{\sqrt{1+x^2}}$$

$$\frac{d}{dx}(\cosh^{-1} x) = \frac{1}{\sqrt{x^2-1}}$$

$$\frac{d}{dx}(\tanh^{-1} x) = \frac{1}{1-x^2}$$

$$\frac{d}{dx}(\coth^{-1} x) = \frac{1}{1-x^2}$$

$$\frac{d}{dx}(\operatorname{sech}^{-1} x) = -\frac{1}{x\sqrt{1-x^2}}$$

$$\frac{d}{dx}(\operatorname{csch}^{-1} x) = -\frac{1}{|x|\sqrt{x^2+1}}$$

Multivariable

Calculus

EARLY TRANSCENDENTAL FUNCTIONS

Third Edition

ROBERT T. SMITH

Millersville University of Pennsylvania

ROLAND B. MINTON

Roanoke College

McGraw Hill Higher Education

Boston Burr Ridge, IL Dubuque, IA Madison, WI New York San Francisco St. Louis
Bangkok Bogotá Caracas Kuala Lumpur Lisbon London Madrid Mexico City
Milan Montreal New Delhi Santiago Seoul Singapore Sydney Taipei Toronto

The McGraw·Hill Companies

Higher Education

CALCULUS: EARLY TRANSCENDENTAL FUNCTIONS, MULTIVARIABLE, THIRD EDITION

Published by McGraw-Hill, a business unit of The McGraw-Hill Companies, Inc., 1221 Avenue of the Americas, New York, NY 10020.

This book is printed on acid-free paper.

4 5 6 7 8 9 0 VNH/VNH 0 9 8

ISBN-13 978–0–07–287029–9
ISBN-10 0–07–287029–X

Publisher: *Elizabeth J. Haefele*
Senior Sponsoring Editor: *Elizabeth Covello*
Director of Development: *David Dietz*
Senior Developmental Editor: *Randy Welch*
Senior Marketing Manager: *Dawn R. Bercier*
Lead Project Manager: *Peggy J. Selle*
Senior Production Supervisor: *Laura Fuller*
Senior Media Project Manager: *Sandra M. Schnee*
Lead Media Producer: *Jeff Huettman*
Senior Designer: *David W. Hash*
Cover/Interior Designer: *Kaye Farmer*
Cover Photo: ©*PictureArts/CORBIS*
Senior Photo Research Coordinator: *John C. Leland*
Photo Research: *Emily Tietz*
Supplement Producer: *Melissa M. Leick*
Compositor: *The GTS Companies/York, PA Campus*
Typeface: *10/12 Times Roman*
Printer: *Von Hoffmann Corporation*

The credits section for this book begins on page C-1 and is considered an extension of the copyright page.

Library of Congress Cataloging-in-Publication Data

Smith, Robert T. (Robert Thomas), 1955–
Calculus : early transcendental functions, multivariable / Robert T. Smith, Roland B. Minton.—3rd ed.
p. cm.
Includes bibliographical references and index.
ISBN 978–0–07–287029–9— ISBN 0–07–287029–X (hard copy : alk. paper)
1. Calculus. I. Minton, Roland B., 1956–. II. Title.

QA303.2.S65 2007
515—dc22

2005030701
CIP

www.mhhe.com

DEDICATION

To Pam, Katie and Michael
To Jan, Kelly and Greg
And our parents—
Thanks for your love and inspiration.

About the Authors

Robert T. Smith is Professor of Mathematics and Chair of the Department of Mathematics at Millersville University of Pennsylvania, where he has taught since 1987. Prior to that, he was on the faculty at Virginia Tech. He earned his Ph.D. in mathematics from the University of Delaware in 1982.

Professor Smith's mathematical interests are in the application of mathematics to problems in engineering and the physical sciences. He has published a number of research articles on the applications of partial differential equations as well as on computational problems in x-ray tomography. He is a member of the American Mathematical Society, the Mathematical Association of America, and the Society for Industrial and Applied Mathematics.

Professor Smith lives in Lancaster, Pennsylvania, with his wife Pam, his daughter Katie and his son Michael. When time permits, he enjoys playing volleyball, tennis, and softball. In his spare time, he coaches youth league soccer. His present extracurricular goal is to learn the game of golf well enough not to come in last in his annual mathematicians/statisticians tournament.

Roland B. Minton is Professor of Mathematics at Roanoke College, where he has taught since 1986. Prior to that, he was on the faculty at Virginia Tech. He earned his Ph.D. from Clemson University in 1982. He is the recipient of the 1998 Roanoke College Exemplary Teaching Award and the 2005 Virginia Outstanding Faculty Award.

Professor Minton has supervised numerous student research projects in such topics as sports science, complexity theory, and fractals. He has published several articles on the use of technology and sports examples in mathematics, in addition to a technical monograph on control theory. He has received grants for teacher training from the State Council for Higher Education in Virginia. He is a member of the Mathematical Association of America, the American Mathematical Society, and other mathematical societies.

Professor Minton lives in Salem, Virginia, with his wife Jan and occasionally with his daughter Kelly and son Greg when they are home from college. He enjoys playing golf and tennis when time permits and watching sports on television even when time doesn't permit. Jan also teaches mathematics at Roanoke College and is very active in mathematics education.

In addition to *Calculus: Early Transcendental Functions,* Professors Smith and Minton are also coauthors of *Calculus: Concepts and Connections* © 2006, and three earlier books for McGraw-Hill Higher Education. The second edition of *Calculus* has been translated into Spanish and is used in several Spanish-speaking countries.

Brief Table of Contents

Table of Contents

CHAPTER 12 •• Functions of Several Variables and Partial Differentiation 919

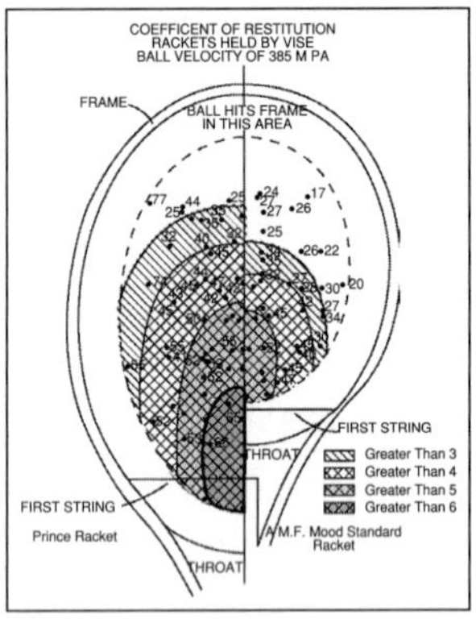

CHAPTER 13 •• Multiple Integrals 1029

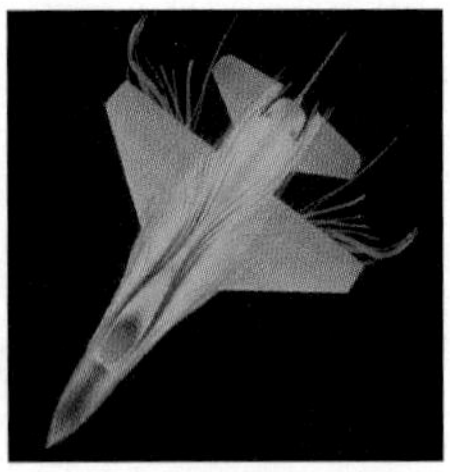

CHAPTER 14 •• Vector Calculus 1115

New Features

BEYOND FORMULAS

TODAY IN MATHEMATICS

Preface

The wide-ranging debate brought about by the calculus reform movement has had a significant impact on calculus textbooks. In response to many of the questions and concerns surrounding this debate, we have written a modern calculus textbook, intended for students majoring in mathematics, physics, chemistry, engineering, and related fields.

Our intention is that students should be able to read our book, rather than merely use it as an encyclopedia filled with the facts of calculus. We have written in a conversational style that reviewers have compared to listening to a good lecture. Our sense of what works well with students has been honed by teaching mathematics for more than a combined 50 years at a variety of colleges and universities, both public and private, ranging from a small liberal arts college to large engineering schools.

In an effort to ensure that this textbook successfully addresses our concerns about the effective teaching of calculus we have continually asked instructors around the world for their opinions on the calculus curriculum, the strengths and weaknesses of current textbooks, and the strengths and weaknesses of our own text. In preparing this third edition, as with the previous editions, we enjoyed the benefit of countless insightful comments from a talented panel of reviewers that was selected to help us with this project.

OUR PHILOSOPHY

We agree with many of the ideas that have come out of the calculus reform movement. In particular, we believe in the **Rule of Four:** that concepts should be presented **graphically, numerically, algebraically** and **verbally,** whenever these are appropriate. In fact, we would add **physically** to this list, since the modeling of physical problems is an important skill that students need to develop. We also believe that, while the calculus curriculum has been in need of reform, we should not throw out those things that already work well. Our book thus represents an updated approach to the traditional topics of calculus. We follow a mainstream order of presentation, while integrating technology and thought-provoking exercises throughout.

One of the thrusts of the calculus reform movement has been to place greater emphasis on problem solving and to present students with more realistic applications as well as open-ended problems. We have incorporated meaningful writing exercises and extended, open-ended problems into **every problem set.** You will also find a **much wider range of applications** than in most traditional texts. We make frequent use of applications from students' experience both to **motivate the development of new topics** and to illustrate concepts we have already presented. In particular, we have included numerous examples from a wide range of fields to give students a familiar context in which to think of various concepts and their applications.

We believe that a conceptual development of the calculus must motivate the text. Although we have **integrated technology throughout,** we have not allowed the technology to drive the book. Our goal is to use the available technology to help students reach a conceptual understanding of the calculus as it is used today.

MOTIVATION AND UNDERSTANDING

Perhaps the most important task when preparing a calculus text is the actual *writing* of it. We have endeavored to write this text in a manner that combines an appropriate level of informality with an honest discussion regarding the difficulties that students commonly face in their study of calculus. In addition to the concepts and applications of calculus, we have also included many frank discussions about what is practical and impractical, and what is difficult and not so difficult to students in the course.

Our primary objectives are to find better ways to motivate students and facilitate their understanding. To accomplish this, we go beyond the standard textbook presentation and tell students **why** they are learning something, **how** they will use it, and **why** it is important. As a result students master problem-solving skills while also **learning how to think mathematically,** an important goal for most instructors teaching the calculus course.

This edition of our text incorporates an early introduction to all transcendental functions. Our students have seen these functions before they ever set foot in a calculus classroom, so we would like to take advantage of their familiarity. We introduce the calculus of these functions in Chapter 2, along with the other rules of differentiation. We have found that this early introduction allows for more varied examples and exercises in the applications of differentiation (including graphing), integration, and applications of integration.

In our view, techniques of integration remain of great importance. Our emphasis is on helping students develop the ability to carefully distinguish among similar-looking integrals and identify the appropriate technique of integration to apply to each integral. The attention to detail and mathematical sophistication required by this process are invaluable skills. We do not attempt to be encyclopedic about techniques of integration, especially given the widespread use of computer algebra systems. Accordingly, in section 6.5, we include a discussion of integration tables and the use of computer algebra systems for performing symbolic integration.

In addition to a focus on the central concepts of calculus, we have included several sections that are not typically found in other calculus texts, as well as expanded coverage of specific topics. This provides instructors with the flexibility to tailor their courses to the interests and abilities of each class.

- For instance, in section 1.7, we explore **loss-of significance errors.** Here, we discuss how computers and calculators perform arithmetic operations and how these can cause errors, in the context of numerical approximation of limits.
- In section 3.9, we present a diverse group of applications of differentiation, including **chemical reaction rates and heart rates.**
- Separable differential equations and logistic growth are discussed in section 7.2, followed by direction fields and Euler's method for first-order ordinary differential equations in section 7.3.
- In Chapter 8, we follow our discussion of power series and Taylor's Theorem with a section on **Fourier series.**
- In sections 9.1–9.3 we provide **expanded coverage of parametric equations.**
- In section 10.4 we include a discussion of **Magnus force.**

CALCULUS AND TECHNOLOGY

It is our conviction that graphing calculators and computer algebra systems must not be used indiscriminately. The focus must always remain on the calculus. We have ensured that each of our exercise sets offers an extensive array of problems that should be worked by hand. We also believe, however, that calculus study supplemented with an intelligent use of technology gives students an extremely powerful arsenal of problem-solving skills. Many passages in the text provide guidance on how to judiciously use—and not abuse—graphing calculators and computers. We also provide ample opportunity for students to practice using these tools. Exercises that are most easily solved with the aid of a graphing calculator or a computer algebra system are easily identified with a icon.

IMPROVEMENTS IN THE THIRD EDITION

Building upon the success of the Second Edition of *Calculus,* we have made the following revisions to produce an even better Third Edition:

Organization

- **All transcendental functions are introduced early,** and their calculus is covered with the calculus of algebraic functions, to accommodate instructors who prefer this approach.
- **Differential equations** receive substantially more coverage in Chapter 7 and in the **all-new** Chapter 15.

Presentation

- A **thorough rewrite** of the book resulted in a **more concise and direct presentation** of all concepts and techniques.
- The **multivariable chapters** were thoroughly revised in response to user feedback to provide a **more cogent and refined** presentation of this material.
- The entire text was redesigned for a **more open, clean appearance** to aid students in locating and focusing on essential information.

Exercises

- **More challenging exercises** appear throughout the book, and ***Exploratory Exercises* conclude every section** to encourage students to synthesize what they've learned.
- **Technology icons** now appear next to all exercises requiring the use of a computer algebra system.

Aesthetics and Relevance of Mathematics

- *NEW **Beyond Formulas*** boxes appear in every chapter to encourage students to **think mathematically** and go beyond routine answer calculation.
- *NEW **Today in Mathematics*** boxes appear in every chapter showing students that mathematics is a dynamic discipline with many discoveries continually being made by **people inspired by the beauty of the subject.**

- *NEW* The ***Index of Applications*** shows students of diverse majors the **immediate relevance** of what they are studying.

SUPPLEMENTS

INSTRUCTOR'S SOLUTIONS MANUAL (ISBN 978-0-07-327655-7)

An invaluable, timesaving resource, the Instructor's Solutions Manual contains comprehensive, worked-out solutions to the odd- and even-numbered exercises in the text.

STUDENT SOLUTIONS MANUAL (ISBN 978-0-07-286967-5)

The Student Solutions Manual is a helpful reference that contains comprehensive, worked-out solutions to the odd-numbered exercises in the text.

INSTRUCTOR'S TESTING AND RESOURCE CD-ROM (ISBN 978-0-07-286962-0)

Brownstone Diploma® testing software, available on CD-ROM, offers instructors a quick and easy way to create customized exams and view student results. Instructors may use the software to sort questions by section, difficulty level, and type; add questions and edit existing questions; create multiple versions of questions using algorithmically-randomized variables; prepare multiple-choice quizzes; and construct a grade book.

MathZone www.mathzone.com

McGraw-Hill's MathZone is a cutting-edge, customizable web-based system that offers a complete solution to instructors' online homework, quizzing and testing needs. MathZone guides students through step-by-step solutions to practice problems and facilitates student assessment through the use of algorithmically-generated test questions. Student activity within the MathZone site is **automatically graded** and accessible to instructors in an integrated, exportable grade book.

MathZone also provides a wide variety of **interactive student tutorials,** including **new applets for every section** in the book to give students interactive practice on important concepts and procedures; algorithmic practice problems; **e-Professor,** a collection of step-by-step animated instructions for solving exercises from the text; **Calculus Concepts Videos;** and **NetTutor,** a live, personalized tutoring service offered via the Internet.

CALCULUS CONCEPTS VIDEOS (978-0-07-312476-6)

Students will see **essential concepts** explained and brought to life through **dynamic animations** in this new video series available on DVD and on the Smith/Minton MathZone site. The **twenty-five key concepts,** chosen after consultation with calculus instructors across the country, are the most commonly taught topics that students need help with and that also lend themselves most readily to on-camera demonstration.

ALEKS PREP FOR CALCULUS

ALEKS (**A**ssessment and **LE**arning in **K**nowledge **S**paces) is an artificial intelligence-based system for mathematics learning, available online 24/7. Using unique adaptive questioning, ALEKS accurately assesses what topics each student knows and then determines exactly what each student is ready to learn next. ALEKS interacts with the students much as a skilled human tutor would, moving between explanation and practice as needed, correcting and analyzing errors, defining terms and changing topics on request, and helping them master the course content more quickly and easily. **New ALEKS 3.0** now links to text-specific videos, multimedia tutorials, and textbook pages in PDF format. ALEKS also offers a robust classroom management system that allows instructors to monitor and direct student progress toward mastery of curricular goals. See www.highed.aleks.com.

ACKNOWLEDGMENTS

A project of this magnitude requires the collaboration of an incredible number of talented and dedicated individuals. Our editorial staff worked tirelessly to provide us with countless surveys, focus group reports, and reviews, giving us the best possible read on the current state of calculus instruction. First and foremost, we want to express our appreciation to our sponsoring editor Liz Covello and our developmental editor Randy Welch for their encouragement and support to keep us on track throughout this project. They challenged us to make this a better book. We also wish to thank our publisher Liz Haefele, and director of development David Dietz for their ongoing strong support.

We are indebted to the McGraw-Hill production team, especially project manager Peggy Selle and design coordinator David Hash, for (among other things) producing a beautifully designed text. Cindy Trimble and Santo D'Agostino provided us with numerous suggestions for clarifying and improving the exercise sets and ensuring the text's accuracy. Our marketing manager Dawn Bercier has been instrumental in helping to convey the story of this book to a wider audience, and media producer Jeff Huettman created an innovative suite of media supplements.

Our work on this project benefited tremendously from the insightful comments we received from many reviewers, survey respondents and symposium attendees. We wish to thank the following individuals whose contributions helped to shape this book:

REVIEWERS OF THE THIRD EDITION

Kent Aeschliman, *Oakland Community College*
Stephen Agard, *University of Minnesota*
Charles Akemann, *University of California, Santa Barbara*
Tuncay Aktosun, *University of Texas–Arlington*
Gerardo Aladro, *Florida International University*
Dennis Bila, *Washtenaw Community College*
Ron Blei, *University of Connecticut*
Joseph Borzellino, *California Polytechnic State University*
Timmy Bremer, *Broome Community College*
Qingying Bu, *University of Mississippi*
Katherine Byler, *California State University–Fresno*
Fengxin Chen, *University of Texas at San Antonio*
Youn-Min Chou, *University of Texas at San Antonio*
Leo G. Chouinard, *University of Nebraska–Lincoln*
Si Kit Chung, *The University of Hong Kong*
Donald Cole, *University of Mississippi*
David Collingwood, *University of Washington*
Tristan Denley, *University of Mississippi*
Jin Feng, *University of Massachusetts, Amherst*
Carl FitzGerald, *University of California, San Diego*
John Gilbert, *University of Texas*
Rajiv Gupta, *University of British Columbia*
Guershon Harel, *University of California, San Diego*
Richard Hobbs, *Mission College*
Shun-Chieh Hsieh, *Chang Jung Christian University*
Josefina Barnachea Janier, *University Teknologi Petronas*
Jakub Jasinski, *University of Scranton*
George W. Johnson, *University of South Carolina*
Nassereldeen Ahmed Kabbashi, *International Islamic University*

G. P. Kapoor, *Indian Institute of Technology Kanpur*
Jacob Kogan, *University of Maryland, Baltimore*
Carole King Krueger, *University of Texas at Arlington*
Kenneth Kutler, *Brigham Young University*
Hong-Jian Lai, *West Virginia University*
John Lawlor, *University of Vermont*
Richard Le Borne, *Tennessee Technological University*
Glenn Ledder, *University of Nebraska–Lincoln*
Sungwook Lee, *University of Southern Mississippi*
Steffen Lempp, *University of Wisconsin–Madison*
Barbara MacCluer, *University of Virginia*
William Margulies, *California State University, Long Beach*
Mary B. Martin, *Middle Tennessee State University*
James Meek, *University of Arkansas*
Carrie Muir, *University of Colorado*
Michael M. Neumann, *Mississippi State University*
Sam Obeid, *University of North Texas*
Iuliana Oprea, *Colorado State University*
Anthony Peressini, *University of Illinois*
Greg Perkins, *Hartnell College*
Tan Ban Pin, *National University of Singapore*
Linda Powers, *Virginia Polytechnic Institute and State University*
Mohammad A. Rammaha, *University of Nebraska–Lincoln*
Richard Rebarber, *University of Nebraska–Lincoln*
Edgar Reyes, *Southeastern Louisiana University*
Mark Smith, *University of Illinois*
Donald Solomon, *University of Wisconsin–Milwaukee*
Rustem Suncheleev, *Universiti Putra Malaysia*
Anthony Thomas, *University of Wisconsin–Platteville*
Anthony Vance, *Austin Community College*
P. Veeramani, *Indian Institute of Technology Madras*
Anke Walz, *Kutztown University of Pennsylvania*
Scott Wilde, *Baylor University*
James Wilson, *Iowa State University*
Raymond Wong, *University of California, Santa Barbara*
Bernardine R. Wong Cheng Kiat, *University of Malaya*
Teri Woodington, *Colorado School of Mines*
Gordon Woodward, *University of Nebraska–Lincoln*
Haidong Wu, *University of Mississippi*
Adil Yaqub, *University of California, Santa Barbara*
Hong-Ming Yin, *Washington State University, Pullman*
Paul Yun, *El Camino College*
Jennifer Zhao, *University of Michigan at Dearborn*

TECHNOLOGY BOARD OF ADVISORS FOR THE THIRD EDITION

Roxanne Byrne, *University of Colorado–Denver*
Judith Downey, *University of Nebraska–Omaha*
Linda Duchrow, *Regis University*
Mihail Frumosu, *Boston University*
Mary Legner, *Riverside Community College*
Mike Martin, *Johnson County Community College*
Kim Rescorla, *Eastern Michigan University*

With many thanks to our previous editions' reviewer panels:

Alisher S. Abdullayev, *American River College*
Edward Aboufadel, *Grand Valley State University*
Shair Ahmad, *University of Texas at San Antonio*
Tom Akers, *University of Missouri–Rolla*
Tuncay Aktosun, University of Texas–Arlington
Gerardo Aladro, *Florida International University*
Ariyadasa Aluthge, *Marshall University*
David Anderson, *University of Tennessee*
Michael R. Anderson, *West Virginia State College*
Wilma Anderson, *University of Nebraska–Omaha*

Tamas Antal, *Ohio State University*
Seth Armstrong, *Arkansas State University*
Leon Arriola, *Western New Mexico University*
Nuh Aydin, *Ohio State University*
Prem N. Bajaj, *Wichita State University*
Robert Bakula, *Ohio State University*
Robert Beezer, *University of Puget Sound*
Rachel Belinsky, *Morris Brown College*
Neil Berger, *University of Illinois*
Chris Black, *Seattle University*
Karen Bolinger, *Clarion University of Pennsylvania*
Mike Bonnano, *Suffolk Community College*
Robert Brabenec, *Wheaton College*
George Bradley, *Duquesne University*
Dave Bregenzer, *Utah State University*
C. Allen Brown, *Wabash Valley College*
Linda K. Buchanan, *Howard College*
James Caggiano, *Arkansas State University*
Jorge Alberto Calvo, *North Dakota State University*
James T. Campbell, *University of Memphis*
Jianguo Cao, *University of Notre Dame*
Florin Catrina, *Utah State University*
Deanna M. Caveny, *College of Charleston*
Maurice J. Chabot, *University of Southern Maine*
Wai Yuen Chan, *University of Science and Arts of Oklahoma*
Mei-Chu Chang, *University of California–Riverside*
Benito Chen, *University of Wyoming*
Karin Chess, *Owensboro Community College*
Moody Chu, *North Carolina State University*
Raymond Clapsadle, *University of Memphis*
Dominic P. Clemence, *North Carolina Agricultural and Technical State University*
Barbara Cortzen, *DePaul University*
Julane B. Crabtree, *Johnson County Community College*
Ellen Cunningham, *Saint Mary-of-the-Woods College*
Daniel J. Curtin, *Northern Kentucky University*
Sujay Datta, *Northern Michigan University*
Gregory Davis, *University of Wisconsin–Green Bay*
Joe Diestel, *Kent State University*
Shusen Ding, *Seattle University*
Michael Dorff, *University of Missouri–Rolla*
Michael M. Dougherty, *Penn State Berks*
Judith Downey, *University of Nebraska at Omaha*
Tevian Dray, *Oregon State University*
Dan Drucker, *University of Puget Sound*
Bennett Eisenberg, *Lehigh University*
Alan Elcrat, *Wichita State University*
Sherif T. El-Helaly, *Catholic University of America*
Eugene Enneking, *Portland State University*
David L. Fama, *Germanna Community College*
Judith Hanks Fethe, *Pellissippi State Technical Community College*
Earl D. Fife, *Calvin College*
Jose D. Flores, *University of South Dakota*
Teresa Floyd, *Mississippi College*
William P. Francis, *Michigan Technological University*
Michael Frantz, *University of LaVerne*
Chris Gardiner, *Eastern Michigan University*
Charles H. Giffen, *University of Virginia*
Kalpana Godbole, *Michigan Technological University*
Michael Green, *Metropolitan State University*
Harvey Greenwald, *California Polytechnic State University*
Ronald Grimmer, *Southern Illinois University*
Laxmi N. Gupta, *Rochester Institute of Technology*
Joel K. Haack, *University of Northern Iowa*
H. Allen Hamilton, *Delaware State University*
John Hansen, *Iowa Central Community College*
John Harding, *New Mexico State University*
Mel Hausner, *New York University*
John Haverhals, *Bradley University*
Johnny Henderson, *Auburn University*
Sue Henderson, *Georgia Perimeter College*
Guy T. Hogan, *Norfolk State University*
Robert Horvath, *El Camino College*
Jack Howard, *Clovis Community College*
Cornelia Wang Hsu, *Morgan State University*
Shirley Huffman, *Southwest Missouri State University*
Gail Kaufmann, *Tufts University*
Hadi Kharaghani, *University of Lethbridge (Alberta)*
Masato Kimura, *College of William and Mary*
Robert Knott, *University of Evansville*
Hristo V. Kojouharov, *Arizona State University*
Emanuel Kondopirakis, *Cooper Union*
Kathryn Kozak, *Coconino County Community College*

Kevin Kreider, *University of Akron*
Tor A. Kwembe, *Chicago State University*
Joseph Lakey, *New Mexico State University*
Melvin D. Lax, *California State University–Long Beach*
James W. Lea, *Middle Tennessee State University*
John Lee, *University of Kentucky*
William L. Lepowsky, *Laney College*
Fengshan Liu, *Delaware State University*
Yung-Chen Lu, *Ohio State University*
Stephen A. MacDonald, *University of Southern Maine*
John Maginnis, *Kansas State University*
Michael Maller, *Queens College*
Nicholas A. Martin, *Shepherd College*
Paul A. Martin, *University of Wisconsin Colleges*
Alex Martin McAllister, *Centre College*
Daniel McCallum, *University of Arkansas at Little Rock*
Philip McCartney, *Northern Kentucky University*
Michael J. McConnell, *Clarion University of Pennsylvania*
Chris McCord, *University of Cincinnati*
David McKay, *California State University, Long Beach*
Aaron Melman, *University of San Francisco*
Gordon Melrose, *Old Dominion University*
Richard Mercer, *Wright State University*
Scott Metcalf, *Eastern Kentucky University*
Remigijus Mikulevicius, *University of Southern California*
Allan D. Mills, *Tennessee Technological University*
Jeff Mock, *Diablo Valley College*
Mike Montano, *Riverside Community College*
Laura Moore-Mueller, *Green River Community College*
Shahrooz Moosavizadeh, *Norfolk State University*
Kandasamy Muthevel, *University of Wisconsin–Oshkosh*
Kouhestani Nader, *Prairie View A & M University*
Sergey Nikitin, *Arizona State University*
Terry A. Nyman, *University of Wisconsin–Fox Valley*
Altay Özgener, *Elizabethtown Community College*
Christina Pereyra, *University of New Mexico*
Bent E. Petersen, *Oregon State University*
Cyril Petras, *Lord Fairfax Community College*
Donna Pierce, *Washington State University*
Jim Polito, *North Harris College*
Yiu Tong Poon, *Iowa State University*
Linda Powers, *Virginia Tech*
Evelyn Pupplo-Cody, *Marshall University*
Anthony Quas, *University of Memphis*
Doraiswamy Ramachandran, *California State University–Sacramento*
William C. Ramaley, *Fort Lewis College*
W. Ramasinghage, *Ohio State University*
M. Rama Mohana Rao, *University of Texas at San Antonio*
Nandita Rath, *Arkansas Tech University*
S. Barbara Reynolds, *Cardinal Stritch University*
Joe Rody, *Arizona State University*
Errol Rowe, *North Carolina Agricultural and Technical State University*
Harry M. Schey, *Rochester Institute of Technology*
Charles Seebeck, *Michigan State University*
George L. Selitto, *Iona College*
Shagi-Di Shih, *University of Wyoming*
Mehrdad Simkani, *University of Michigan–Flint*
Eugenia A. Skirta, *University of Toledo*
Rod Smart, *University of Wisconsin–Madison*
Alex Smith, *University of Wisconsin–Eau Claire*
Scott Smith, *Columbia College*
Frederick Solomon, *Warren Wilson College*
V. K. Srinivasan, *University of Texas at El Paso*
Mary Jane Sterling, *Bradley University*
Adam Stinchcombe, *Adirondack Community College*
Jerry Stonewater, *Miami University of Ohio*
Jeff Stuart, *University of Southern Mississippi*
D'Loye Swift, *Nunez Community College*
Randall J. Swift, *Western Kentucky University*
Lawrence Sze, *California Polytechnic State University*
Wanda Szpunar-Lojasiewicz, *Rochester Institute of Technology*
Fereja Tahir, *Eastern Kentucky University*
J. W. Thomas, *Colorado State University*
Juan Tolosa, *Richard Stockton College of New Jersey*
Michael M. Tom, *Louisiana State University*
William K. Tomhave, *Concordia College*

Stefania Tracogna, *Arizona State University*
Jay Treiman, *Western Michigan University*
Patricia Treloar, *University of Mississippi*
Thomas C. Upson, *Rochester Institute of Technology*
Richard G. Vinson, *University of South Alabama*
David Voss, *Western Illinois University*
Mu-Tao Wang, *Stanford University*
Paul Weichsel, *University of Illinois*
Richard A. Weida, *Lycoming College*
Michael Weiner, *Penn State Altoona*
Alan Wilson, *Kaskaskia College*
Michael Wilson, *University of Vermont*
Jim Wolper, *Idaho State University*
Jiahong Wu, *University of Texas at Austin*
DaGang Yang, *Tulane University*
Marvin Zeman, *Southern Illinois University*
Xiao-Dong Zhang, *Florida Atlantic University*
Jianqiang Zhao, *University of Pennsylvania*

In addition, a number of our colleagues graciously gave their time and energy to help create or improve portions of the manuscript. We would especially like to thank Richard Grant, Bill Ergle, Jack Steehler, Ben Huddle, Chris Lee and Jan Minton of Roanoke College for sharing their expertise in calculus and related applications; Tom Burns for help with an industrial application; Gregory Minton and James Albrecht for suggesting several brilliant problems; Dorothee Blum of Millersville University for helping to class-test an early version of the manuscript; Bruce Ikenaga of Millersville University for generously sharing his expertise in TeX and Corel Draw and Pam Vercellone-Smith, for lending us her expertise in many of the biological applications. We also wish to thank Dorothee Blum, Bob Buchanan, Roxana Costinescu, Chuck Denlinger, Bruce Ikenaga, Zhoude Shao and Ron Umble of Millersville University for offering numerous helpful suggestions for improvement. In addition, we would like to thank all of our students throughout the years, who have (sometimes unknowingly) field-tested innumerable ideas, some of which worked and the rest of which will not be found in this book.

Ultimately, this book is for our families. We simply could not have written a book of this magnitude without their strong support. We thank them for their love and inspiration throughout our growth as textbook authors. Their understanding, in both the technical and the personal sense, was essential. They provide us with the reason why we do all of the things we do. So, it is fitting that we especially thank our wives, Pam Vercellone-Smith and Jan Minton; our children, Katie and Michael Smith and Kelly and Greg Minton; and our parents, Anne Smith and Paul and Mary Frances Minton.

Robert T. Smith
Lancaster, Pennsylvania

Roland B. Minton
Salem, Virginia

A COMMITMENT TO ACCURACY

You have a right to expect an accurate textbook, and McGraw-Hill invests considerable time and effort to make sure that we deliver one. Listed below are the many steps we take to make sure this happens.

OUR ACCURACY VERIFICATION PROCESS

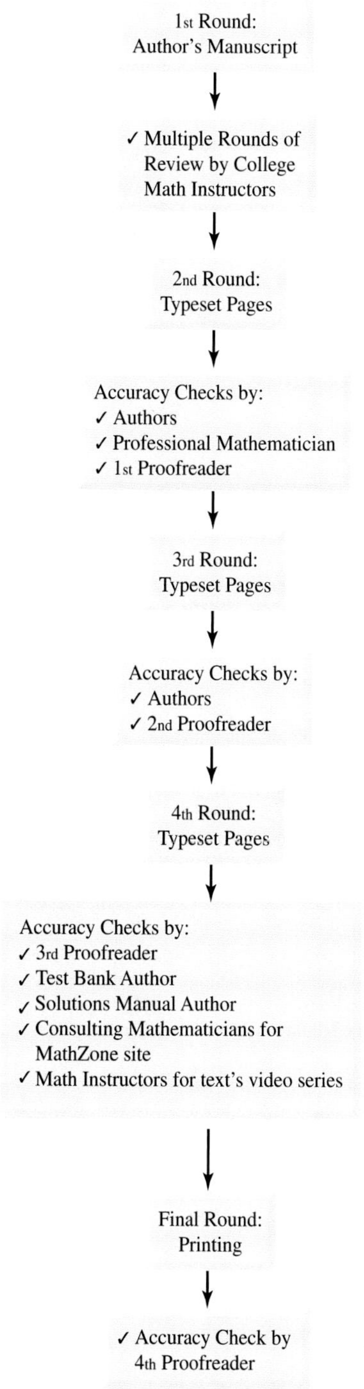
1st Round:
Author's Manuscript
✓ Multiple Rounds of Review by College Math Instructors
2nd Round:
Typeset Pages
Accuracy Checks by:
✓ Authors
✓ Professional Mathematician
✓ 1st Proofreader
3rd Round:
Typeset Pages
Accuracy Checks by:
✓ Authors
✓ 2nd Proofreader
4th Round:
Typeset Pages
Accuracy Checks by:
✓ 3rd Proofreader
✓ Test Bank Author
✓ Solutions Manual Author
✓ Consulting Mathematicians for MathZone site
✓ Math Instructors for text's video series
Final Round:
Printing
✓ Accuracy Check by
4th Proofreader

First Round

Step 1: Numerous **college math instructors** review the manuscript and report on any errors that they may find, and the authors make these corrections in their final manuscript.

Second Round

Step 2: Once the manuscript has been typeset, the **authors** check their manuscript against the first page proofs to ensure that all illustrations, graphs, examples, exercises, solutions, and answers have been correctly laid out on the pages, and that all notation is correctly used.

Step 3: An outside, **professional mathematician** works through every example and exercise in the page proofs to verify the accuracy of the answers.

Step 4: A **proofreader** adds a triple layer of accuracy assurance in the first pages by hunting for errors, then a second, corrected round of page proofs is produced.

Third Round

Step 5: The **author team** reviews the second round of page proofs for two reasons: 1) to make certain that any previous corrections were properly made, and 2) to look for any errors they might have missed on the first round.

Step 6: A **second proofreader** is added to the project to examine the new round of page proofs to double check the author team's work and to lend a fresh, critical eye to the book before the third round of paging.

Fourth Round

Step 7: A **third proofreader** inspects the third round of page proofs to verify that all previous corrections have been properly made and that there are no new or remaining errors.

Step 8: Meanwhile, in partnership with **independent mathematicians,** the text accuracy is verified from a variety of fresh perspectives:

- The **test bank author** checks for consistency and accuracy as they prepare the computerized test item file.
- The **solutions manual author** works every single exercise and verifies their answers, reporting any errors to the publisher.
- A **consulting group of mathematicians,** who write material for the text's MathZone site, notifies the publisher of any errors they encounter in the page proofs.
- A video production company employing **expert math instructors** for the text's videos will alert the publisher of any errors they might find in the page proofs.

Final Round

Step 9: The **project manager,** who has overseen the book from the beginning, performs a **fourth proofread** of the textbook during the printing process, providing a final accuracy review.

⇒ What results is a mathematics textbook that is as accurate and error-free as is humanly possible, and our authors and publishing staff are confident that our many layers of quality assurance have produced textbooks that are the leaders of the industry for their integrity and correctness.

Guided Tour

TOOLS FOR LEARNING

Real-World Emphasis

Each chapter opens with a real-world application that illustrates the usefulness of the concepts being developed and motivates student interest.

Parametric Equations and Polar Coordinates

CHAPTER

9

You are all familiar with sonic booms, those loud crashes of noise caused by aircraft flying faster than the speed of sound. You may have even heard a sonic boom, but you have probably never *seen* a sonic boom. The remarkable photograph here shows water vapor outlining the surface of a shock wave created by an F-18 jet flying supersonically. (Note that there is also a small cone of water vapor trailing the back of the cockpit of the jet.)

You may be surprised at the apparently conical shape assumed by the shock waves. A mathematical analysis of the shock waves verifies that the shape is indeed conical. (You will have an opportunity to explore this in the exercises in section 9.1.) To visualize how sound waves propagate, imagine an exploding firecracker. If you think of this in two dimensions, you'll recognize that the sound waves propagate in a series of ever-expanding concentric circles that reach everyone standing a given distance away from the firecracker at the same time.

In this chapter, we extend the concepts of calculus to curves described by parametric equations and polar coordinates. For instance, in order to study the motion of an object such as an airplane in two dimensions, we would need to describe the object's position (x, y) as a function of the parameter t (time). That is, we write the position in the form $(x, y) = (x(t), y(t))$, where $x(t)$ and $y(t)$ are functions to which our existing techniques of calculus can be applied. The equations $x = x(t)$ and $y = y(t)$ are called parametric equations. Additionally, we'll explore how to use polar coordinates to represent curves, not as a set of points (x, y), but rather, by specifying the points by the distance from the origin to the point and an angle corresponding to the direction from the origin to the point. Polar coordinates are especially convenient for describing circles, such as those that occur in propagating sound waves.

These alternative descriptions of curves bring us a great deal of needed flexibility in attacking many problems. Often, even very complicated looking curves have a simple description in terms of parametric equations or polar coordinates. We explore a variety of interesting curves in this chapter and see how to extend the methods of calculus to such curves.

715

Definitions, Theorems and Proofs

All formal definitions and theorems are clearly boxed within the text for easy visual reference. Proofs are clearly labeled. Proofs of some results are found in Appendix A.

DEFINITION 2.2

The **derivative** of $f(x)$ is the function $f'(x)$ given by

$$f'(x) = \lim_{h \to 0} \frac{f(x+h) - f(x)}{h}, \tag{2.3}$$

provided the limit exists. The process of computing a derivative is called **differentiation.**

Further, f is differentiable on an interval I if it is differentiable at every point in I.

THEOREM 2.1

If $f(x)$ is differentiable at $x = a$, then $f(x)$ is continuous at $x = a$.

PROOF

For f to be continuous at $x = a$, we need only show that $\lim_{x \to a} f(x) = f(a)$. We consider

$$\lim_{x \to a} [f(x) - f(a)] = \lim_{x \to a} \left[\frac{f(x) - f(a)}{x - a}(x - a) \right] \quad \text{Multiply and divide by } (x - a).$$

$$= \lim_{x \to a} \left[\frac{f(x) - f(a)}{x - a} \right] \lim_{x \to a}(x - a) \quad \text{By Theorem 3.1 (iii) from section 1.3.}$$

$$= f'(a)(0) = 0, \quad \text{Since } f \text{ is differentiable at } x = a.$$

where we have used the alternative definition of derivative (2.2) discussed earlier. By Theorem 3.1 in section 1.3, it now follows that

$$0 = \lim_{x \to a}[f(x) - f(a)] = \lim_{x \to a} f(x) - \lim_{x \to a} f(a)$$

$$= \lim_{x \to a} f(x) - f(a),$$

which gives us the result. ■

Examples

Each chapter contains a large number of worked examples, ranging from the simple and concrete to more complex and abstract.

Use of Graphs and Tables

Being able to visualize a problem is an invaluable aid in understanding the concept presented. To this purpose, we have integrated more than 1500 computer-generated graphs throughout the text.

Point	**(0, 0)**	$\left(-\frac{1}{6}, -\frac{1}{3}\right)$
$f_{xx} = 4$	4	4
$f_{yy} = -6y$	0	−2
$f_{xy} = -2$	−2	−2
$D(a, b)$	−4	4

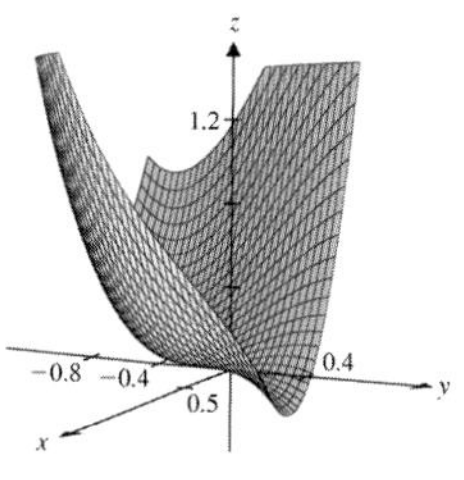

FIGURE 12.42
$z = 2x^2 - y^3 - 2xy$

EXAMPLE 7.2 Using the Discriminant to Find Local Extrema

Locate and classify all critical points for $f(x, y) = 2x^2 - y^3 - 2xy$.

Solution We first compute the first partial derivatives: $f_x = 4x - 2y$ and $f_y = -3y^2 - 2x$. Since both f_x and f_y are defined for all (x, y), the critical points are solutions of the two equations:

$$f_x = 4x - 2y = 0$$

and

$$f_y = -3y^2 - 2x = 0.$$

Solving the first equation for y, we get $y = 2x$. Substituting this into the second equation, we have

$$0 = -3(4x^2) - 2x = -12x^2 - 2x$$

$$= -2x(6x + 1),$$

so that $x = 0$ or $x = -\frac{1}{6}$. The corresponding y-values are $y = 0$ and $y = -\frac{1}{3}$. The only two critical points are then $(0, 0)$ and $\left(-\frac{1}{6}, -\frac{1}{3}\right)$. To classify these points, we first compute the second partial derivatives: $f_{xx} = 4$, $f_{yy} = -6y$ and $f_{xy} = -2$ and then test the discriminant. We have

$$D(0, 0) = (4)(0) - (-2)^2 = -4$$

and

$$D\left(-\tfrac{1}{6}, -\tfrac{1}{3}\right) = (4)(2) - (-2)^2 = 4.$$

From Theorem 7.2, we conclude that there is a saddle point of f at $(0, 0)$, since $D(0, 0) < 0$. Further, there is a local minimum at $\left(-\frac{1}{6}, -\frac{1}{3}\right)$ since $D\left(-\frac{1}{6}, -\frac{1}{3}\right) > 0$ and $f_{xx}\left(-\frac{1}{6}, -\frac{1}{3}\right) > 0$. The surface is shown in Figure 12.42. ■

As we see in example 7.3, the second derivatives test does not always help us to classify a critical point.

COMMENTARY AND GUIDANCE

Beyond Formulas

Beyond Formulas boxes appear in every chapter to encourage students to think mathematically and go beyond routine answer calculation.

BEYOND FORMULAS

The Mean Value Theorem is subtle, but its implications are far-reaching. Although the illustration in Figure 2.49 makes the result seem obvious, the consequences of the Mean Value Theorem, such as example 9.4, are powerful and not at all obvious. For example, most of the rest of the calculus developed in this book depends on the Mean Value Theorem either directly or indirectly. A thorough understanding of the theory of calculus can lead you to important conclusions, particularly when the problems are beyond what your intuition alone can handle. What other theorems have you learned that continue to provide insight beyond their original context?

TODAY IN MATHEMATICS

Cathleen Morawetz (1923–) A Canadian mathematician whose work on transonic flows and wave scattering greatly influenced the design of air foils. Both of her parents were mathematically trained, but her mathematics

Today in Mathematics

Today in Mathematics boxes appear in every chapter, showing students that mathematics is a dynamic discipline with many discoveries continually being made by people inspired by the beauty of the subject.

HISTORICAL NOTES

Johannes Kepler (1571–1630) German astronomer and mathematician whose discoveries revolutionized Western science. Kepler's remarkable mathematical ability and energy produced connections among many areas of research. A study of observations

Historical Notes

These biographical features provide background information on prominent mathematicians and their contributions to the development of calculus and put the subject matter into perspective.

CONCEPTUAL UNDERSTANDING THROUGH PRACTICE

Applications

The text provides numerous and varied applications that relate calculus to the real world, many of which are unique to the Smith/Minton series. Worked examples and exercises are frequently developed with an applied focus in order to motivate the presentation of new topics, further illustrate familiar topics, and connect the conceptual development of calculus with students' everyday experiences.

EXAMPLE 1.6 Steering an Aircraft in a Head Wind and a Crosswind

An airplane has an airspeed of 400 mph. Suppose that the wind velocity is given by the vector $\mathbf{w} = \langle 20, 30 \rangle$. In what direction should the airplane head in order to fly due west (i.e., in the direction of the unit vector $-\mathbf{i} = \langle -1, 0 \rangle$)?

Solution We illustrate the velocity vectors for the airplane and the wind in Figure 10.14. We let the airplane's velocity vector be $\mathbf{v} = \langle x, y \rangle$. The effective velocity of the plane is then $\mathbf{v} + \mathbf{w}$, which we set equal to $\langle c, 0 \rangle$, for some negative constant c. Since

$$\mathbf{v} + \mathbf{w} = \langle x + 20, y + 30 \rangle = \langle c, 0 \rangle,$$

FIGURE 10.14
Forces on an airplane.

we must have $x + 20 = c$ and $y + 30 = 0$, so that $y = -30$. Further, since the plane's airspeed is 400 mph, we must have $\|\mathbf{v}\| = \sqrt{x^2 + y^2} = \sqrt{x^2 + 900} = 400$. Squaring this gives us $x^2 + 900 = 160{,}000$, so that $x = -\sqrt{159{,}100}$. (We take the negative square root so that the plane heads westward.) Consequently, the plane should head in the direction of $\mathbf{v} = \langle -\sqrt{159{,}100}, -30 \rangle$, which points left and down, or southwest, at an angle of $\tan^{-1}(30/\sqrt{159{,}100}) \approx 4°$ below due west. ■

33. A baseball is hit from a height of 3 feet with initial speed 120 feet per second and at an angle of 30 degrees above the horizontal. Find a vector-valued function describing the position of the ball t seconds after it is hit. To be a home run, the ball must clear a wall that is 385 feet away and 6 feet tall. Determine whether this is a home run.

34. Repeat exercise 33 if the ball is launched with an initial angle of 31 degrees.

Balanced Exercise Sets

This text contains thousands of exercises, found at the end of each section and chapter. Each problem set has been carefully designed to provide a balance of routine, moderate and challenging exercises. The authors have taken great care to create original and imaginative exercises that provide an appropriate review of the topics covered in each section and chapter.

Special types of exercises:

Technology Icon

Exercises in the section exercise sets and chapter review exercises that can most easily be solved using a graphing calculator or computer are clearly identified with this icon.

42.

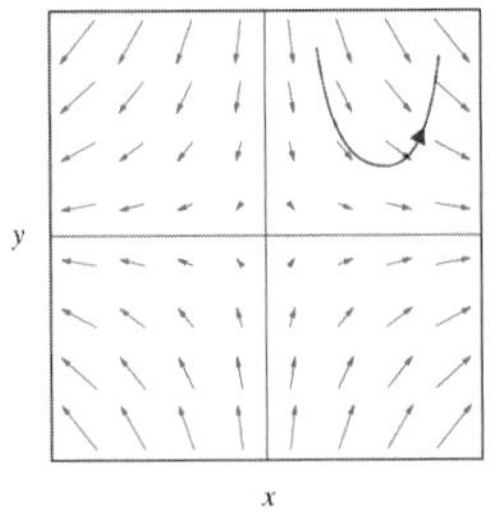

In exercises 43–52, use the formulas $m = \int_C \rho ds$, $\bar{x} = \frac{1}{m}\int_C x\rho ds$, $\bar{y} = \frac{1}{m}\int_C y\rho ds$, $I = \int_C w^2 \rho ds$.

43. Compute the mass m of a rod with density $\rho(x, y) = x$ in the shape of $y = x^2$, $0 \le x \le 3$.

44. Compute the mass m of a rod with density $\rho(x, y) = y$ in the shape of $y = 4 - x^2$, $0 \le x \le 2$.

45. Compute the center of mass $(\bar{x}, \bar{y})$ of the rod of exercise 43.

46. Compute the center of mass $(\bar{x}, \bar{y})$ of the rod of exercise 44.

47. Compute the moment of inertia I for rotating the rod of exercise 43 about the y-axis. Here, w is the distance from the point (x, y) to the y-axis.

48. Compute the moment of inertia I for rotating the rod of exercise 44 about the x-axis. Here, w is the distance from the point (x, y) to the x-axis.

56. Above the portion of $y = x^2$ from $(0, 0, 0)$ to $(2, 4, 0)$ up to the surface $z = x^2 + y^2$

57. Above the line segment from $(2, 0, 0)$ to $(-2, 0, 0)$ up to the surface $z = 4 - x^2 - y^2$

58. Above the line segment from $(1, 1, 0)$ to $(-1, 1, 0)$ up to the surface $z = \sqrt{x^2 + y^2}$

59. Above the unit square $x \in [0, 1]$, $y \in [0, 1]$ up to the plane $z = 4 - x - y$

60. Above the ellipse $x^2 + 4y^2 = 4$ up to the plane $z = 4 - x$

In exercises 61 and 62, estimate the line integrals (a) $\int_C f\,ds$, (b) $\int_C f\,dx$ and (c) $\int_C f\,dy$.

61.

(x, y)	$(0, 0)$	$(1, 0)$	$(1, 1)$	$(1.5, 1.5)$
$f(x, y)$	2	3	3.6	4.4

(x, y)	$(2, 2)$	$(3, 2)$	$(4, 1)$
$f(x, y)$	5	4	4

62.

(x, y)	$(0, 0)$	$(1, -1)$	$(2, 0)$	$(3, 1)$
$f(x, y)$	1	0	-1.2	0.4

(x, y)	$(4, 0)$	$(3, -1)$	$(2, -2)$
$f(x, y)$	1.5	2.4	2

Writing Exercises

Each section exercise set begins with writing exercises that encourage students to carefully consider important mathematical concepts and ideas in new contexts, and to express their findings in their own words. The writing exercises may also be used as springboards for class discussion.

EXERCISES 5.5

WRITING EXERCISES

1. In example 5.6, the assumption that air resistance can be ignored is obviously invalid. Discuss the validity of this assumption in examples 5.1 and 5.3.

2. In the discussion preceding example 5.3, we showed that Michael Jordan (and any other human) spends half of his air-time in the top one-fourth of the height. Compare his velocities at various points in the jump to explain why relatively more time is spent at the top than at the bottom.

3. In example 5.4, we derived separate equations for the horizontal and vertical components of position. To discover one consequence of this separation, consider the following situation. Two people are standing next to each other with arms raised to the same height. One person fires a bullet horizontally from a gun. At the same time, the other person drops a bullet. Explain why the bullets will hit the ground at the same time.

4. For the falling raindrop in example 5.6, a more accurate model would be $y''(t) = -32 + f(t)$, where $f(t)$ represents the force due to air resistance (divided by the mass). If $v(t)$ is the downward velocity of the raindrop, explain why this equation is equivalent to $v'(t) = 32 - f(t)$. Explain in physical terms why the larger $v(t)$ is, the larger $f(t)$ is. Thus, a model such as $f(t) = v(t)$ or $f(t) = [v(t)]^2$ would be reasonable. (In most situations, it turns out that $[v(t)]^2$ matches the experimental data better.)

8. The Washington Monument is 555 feet, $5\frac{1}{8}$ inches high. In a famous experiment, a baseball was dropped from the top of the monument to see if a player could catch it. How fast would the ball be going?

9. A certain not-so-wily coyote discovers that he just stepped off the edge of a cliff. Four seconds later, he hits the ground in a puff of dust. How high was the cliff?

10. A large boulder dislodged by the falling coyote in exercise 9 falls for 3 seconds before landing on the coyote. How far did the boulder fall? What was its velocity when it flattened the coyote?

11. The coyote's next scheme involves launching himself into the air with an Acme catapult. If the coyote is propelled vertically from the ground with initial velocity 64 ft/s, find an equation for the height of the coyote at any time t. Find his maximum height, the amount of time spent in the air and his velocity when he smacks back into the catapult.

12. On the rebound, the coyote in exercise 11 is propelled to a height of 256 feet. What is the initial velocity required to reach this height?

13. One of the authors has a vertical "jump" of 20 inches. What is the initial velocity required to jump this high? How does this compare to Michael Jordan's velocity, found in example 5.3?

Exploratory Exercises

Each section exercise set concludes with a series of in-depth exploratory exercises designed to challenge students' knowledge.

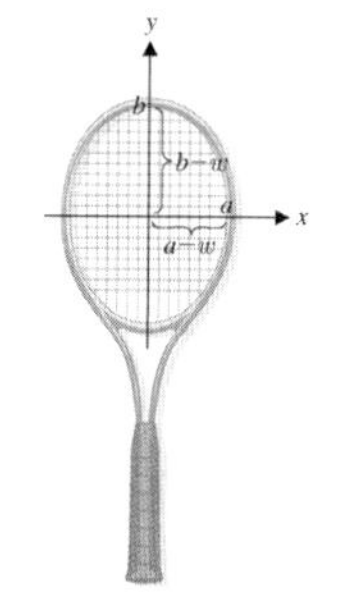

49. For tennis rackets, a large second moment (see exercises 47 and 48) means less twisting of the racket on off-center shots. Compare the second moment of a wooden racket ($a = 9$, $b = 12$, $w = 0.5$), a midsize racket ($a = 10$, $b = 13$, $w = 0.5$) and an oversized racket ($a = 11$, $b = 14$, $w = 0.5$).

50. Let M be the second moment found in exercise 48. Show that $\frac{dM}{da} > 0$ and conclude that larger rackets have larger second moments. Also, show that $\frac{dM}{dw} > 0$ and interpret this result.

EXPLORATORY EXERCISES

1. As equipment has improved, heights cleared in the pole vault have increased. A crude estimate of the maximum pole vault possible can be derived from conservation of energy principles. Assume that the maximum speed a pole-vaulter could run carrying a long pole is 25 mph. Convert this speed to ft/s. The kinetic energy of this vaulter would be $\frac{1}{2}mv^2$. (Leave m as an unknown for the time being.) This initial kinetic energy would equal the potential energy at the top of the vault minus whatever energy is absorbed by the pole (which we will ignore). Set the potential energy, $32mh$, equal to the kinetic energy and solve for h. This represents the maximum amount the vaulter's center of mass could be raised. Add 3 feet for the height of the vaulter's center of mass and you have an estimate of the maximum vault possible. Compare this to Sergei Bubka's 1994 world record vault of $20'1\frac{3}{4}''$.

2. An object will remain on a table as long as the center of mass of the object lies over the table. For example, a board of length 1 will balance if half the board hangs over the edge of the table. Show that two homogeneous boards of length 1 will balance if $\frac{1}{4}$ of the first board hangs over the edge of the table and $\frac{1}{2}$ of the second board hangs over the edge of the first board. Show that three boards of length 1 will balance if $\frac{1}{6}$ of the first board hangs over the edge of the table, $\frac{1}{4}$ of the second board hangs over the edge of the first board and $\frac{1}{2}$ of the third board hangs over the edge of the second board. Generalize this to a procedure for balancing n boards. How many boards are needed so that the last board hangs completely over the edge of the table?

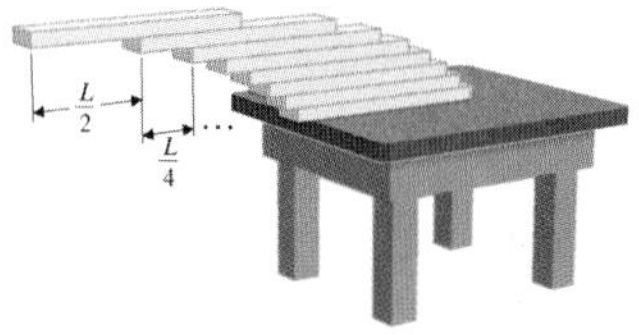

Review Exercises

Review Exercises sets are provided as an overview of the chapter and include Writing Exercises, True or False, and Exploratory Exercises.

Review Exercises

WRITING EXERCISES

The following list includes terms that are defined and theorems that are stated in this chapter. For each term or theorem, (1) give a precise definition or statement, (2) state in general terms what it means and (3) describe the types of problems with which it is associated.

Volume by slicing	Volume by disks	Volume by washers
Volume by shells	Arc length	Surface area
Newton's second law	Work	Impulse
Center of mass	Probability density function	Mean

TRUE OR FALSE

State whether each statement is true or false and briefly explain why. If the statement is false, try to "fix it" by modifying the given statement to a new statement that is true.

1. The area between f and g is given by $\int_a^b [f(x) - g(x)]\,dx$.

2. The method of disks is a special case of volume by slicing.

7. The area of the region bounded by $y = x^2$, $y = 2 - x$ and $y = 0$

8. The area of the region bounded by $y = x^2$, $y = 0$ and $x = 2$

9. A town has a population of 10,000 with a birthrate of $10 + 2t$ people per year and a death rate of $4 + t$ people per year. Compute the town's population after 6 years.

10. From the given data, estimate the area between the curves for $0 \le x \le 2$.

x	0.0	0.2	0.4	0.6	0.8	1.0
$f(x)$	3.2	3.6	3.8	3.7	3.2	3.4
$g(x)$	1.2	1.5	1.6	2.2	2.0	2.4

x	1.2	1.4	1.6	1.8	2.0
$f(x)$	3.0	2.8	2.4	2.9	3.4
$g(x)$	2.2	2.1	2.3	2.8	2.4

11. Find the volume of the solid with cross-sectional area $A(x) = \pi(3 + x)^2$ for $0 \le x \le 2$.

EXPLORATORY EXERCISES

1. As indicated in section 5.5, general formulas can be derived for many important quantities in projectile motion. For an object launched from the ground at angle θ_0 with initial speed v_0 ft/s, find the horizontal range R ft and use the trig identity $\sin(2\theta_0) = 2\sin\theta_0\cos\theta_0$ to show that $R = \frac{v_0^2 \sin(2\theta_0)}{32}$. Conclude that the maximum range is achieved with angle $\theta_0 = \pi/4$ (45°).

2. To follow up on exploratory exercise 1, suppose that the ground makes an angle of $A°$ with the horizontal. If $A > 0$ (i.e., the projectile is being launched uphill), explain why the maximum range would be achieved with an angle larger than 45°. If $A < 0$ (launching downhill), explain why the maximum range would be achieved with an angle less than 45°. To determine the exact value of the optimal angle, first argue that the ground can be represented by the line $y = (\tan A)x$. Show that the projectile reaches the ground at time $t = v_0 \frac{\sin\theta_0 - \tan A\cos\theta_0}{16}$. Compute $x(t)$ for this value of t and use a trig identity to replace the quantity $\sin\theta_0\cos A - \sin A\cos\theta_0$ with $\sin(\theta_0 - A)$. Then use another trig identity to replace $\cos\theta_0 \sin(\theta_0 - A)$ with $\sin(2\theta_0 - A) - \sin A$. At this stage, the only term involving θ_0 will be $\sin(2\theta_0 - A)$. To maximize the range, maximize this term by taking $\theta_0 = \frac{\pi}{4} + \frac{1}{2}A$.

Applications Index

Biology/Natural World

Chemistry

Economics

Engineering

Health/Physiology

Physics

Sports/Entertainment

Vectors and the Geometry of Space

CHAPTER

10

The Bristol Motor Speedway is one of the most popular racetracks on the NASCAR circuit. Races there are considered the most intense of the year, thanks to a combination of high speed, tight quarters and high-energy crowds. With 43 drivers going over 120 miles per hour on a half-mile oval for two and a half hours with 160,000 people cheering them on, the experience is both thrilling and exhausting for all involved.

Auto racing at all levels is a highly technological competition. A Formula One race car seems to have more in common with an airplane than with a standard car. At speeds of 100–200 mph, air resistance forces on the car can reach hurricane levels. The racing engineer's task is to design the car aerodynamically so that these forces help keep the car on the road. The large wings on the back of cars create a downward force that improves the traction of the car. So-called "ground effect" downforces have an even greater impact. Here, the entire underside of the car is shaped like an upside-down airplane wing and air drawn underneath the car generates a tremendous downward force. Such designs are so successful that the downward forces exceed three times the weight of the car. This means that theoretically the car could race upside down!

Stock car racers have additional challenges, since intricate rules severely limit the extent to which the cars can be modified. So, how do stock cars safely speed around the Bristol Motor Speedway? The track is an oval only 0.533 mile in length and racers regularly exceed 120 miles per hour, completing a lap in just over 15 seconds. These speeds would be unsafe if the track were not specially designed for high-speed racing. In particular, the Bristol track is steeply banked, with a 16-degree bank on straightaways and a spectacular 36-degree bank in the corners.

As you will see in the exercises in section 10.3, the banking of a road changes the role of gravity. In effect, part of the weight of the car is diverted into a force that helps the car make its turn safely. In this chapter, we introduce vectors and develop the calculations needed to resolve vectors into components. This is a fundamental tool for engineers designing race cars and racetracks.

This chapter represents a crossroads from the primarily two-dimensional world of first-year calculus to the three-dimensional world of many important scientific and engineering problems. The rest of the calculus we develop in this book builds directly on the basic ideas developed here.

10.1 VECTORS IN THE PLANE

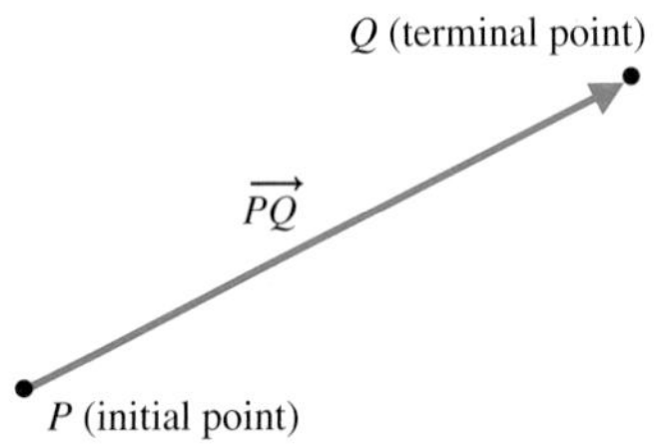

FIGURE 10.1
Directed line segment

We have already considered very simple models of velocity and acceleration in one dimension, but these are *not* generally one-dimensional quantities. In particular, to describe the velocity of a moving object, we must specify both its speed *and* the direction in which it's moving. In fact, velocity, acceleration and force each have both a *size* (e.g., speed) and a *direction*. We represent such a quantity graphically as a directed line segment, that is, a line segment with a specific direction (i.e., an arrow).

We denote the directed line segment extending from the point P (the **initial point**) to the point Q (the **terminal point**) as $\overrightarrow{PQ}$ (see Figure 10.1). We refer to the length of $\overrightarrow{PQ}$ as its **magnitude,** denoted $\|\overrightarrow{PQ}\|$. Mathematically, we consider all directed line segments with the same magnitude and direction to be equivalent, regardless of the location of their initial point and we use the term **vector** to describe any quantity that has both a magnitude and a direction. We should emphasize that the location of the initial point is not relevant; only the magnitude and direction matter. In other words, if $\overrightarrow{PQ}$ is the directed line segment from the initial point P to the terminal point Q, then the corresponding vector $\mathbf{v}$ represents $\overrightarrow{PQ}$ as well as every other directed line segment having the same magnitude and direction as $\overrightarrow{PQ}$. In Figure 10.2, we indicate three vectors that are all considered to be equivalent, even though their initial points are different. In this case, we write

$$\mathbf{a} = \mathbf{b} = \mathbf{c}.$$

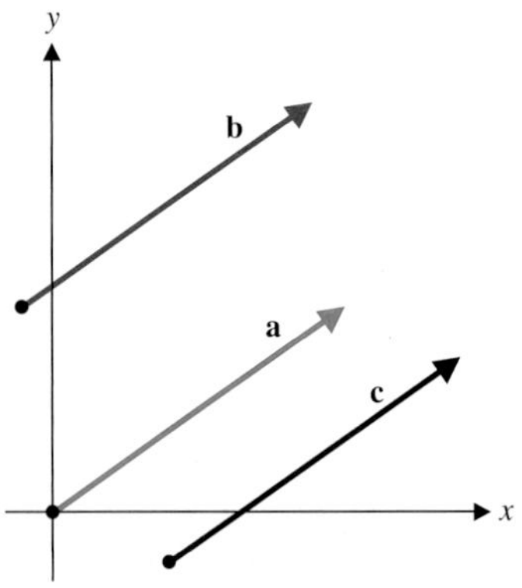

FIGURE 10.2
Equivalent vectors

When considering vectors, it is often helpful to think of them as representing some specific physical quantity. For instance, when you see the vector $\overrightarrow{PQ}$, you might imagine moving an object from the initial point P to the terminal point Q. In this case, the magnitude of the vector would represent the distance the object is moved and the direction of the vector would point from the starting position to the final position.

In this text, we usually denote vectors by boldface characters such as **a**, **b** and **c**, as seen in Figure 10.2. Since you will not be able to write in boldface, you should use the arrow notation (e.g., $\overrightarrow{a}$). When discussing vectors, we refer to real numbers as **scalars.** It is *very important* that you begin now to carefully distinguish between vector and scalar quantities. This will save you immense frustration both now and as you progress through the remainder of this text.

Look carefully at the three vectors shown in Figure 10.3a. If you think of the vector $\overrightarrow{AB}$ as representing the displacement of a particle from the point A to the point B, notice that the end result of displacing the particle from A to B (corresponding to the vector $\overrightarrow{AB}$), followed by displacing the particle from B to C (corresponding to the vector $\overrightarrow{BC}$) is the same as displacing the particle directly from A to C, which corresponds to the vector $\overrightarrow{AC}$

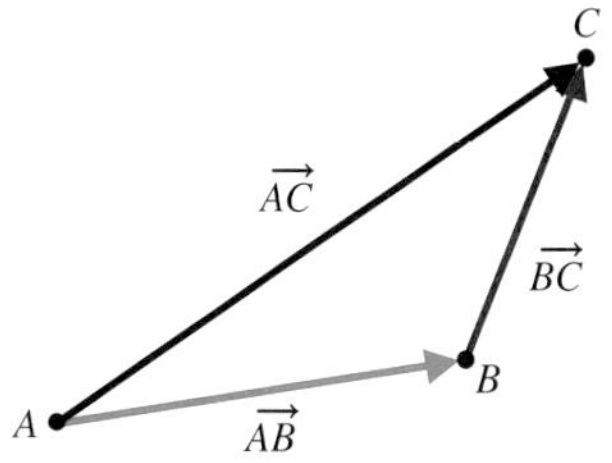

FIGURE 10.3a
Resultant vector

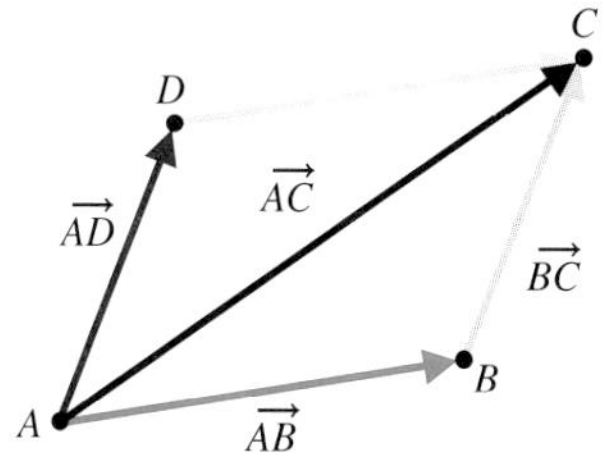

FIGURE 10.3b
Sum of two vectors

(called the **resultant vector**). We call $\overrightarrow{AC}$ the **sum** of $\overrightarrow{AB}$ and $\overrightarrow{BC}$ and write

$$\overrightarrow{AC} = \overrightarrow{AB} + \overrightarrow{BC}.$$

Given two vectors that we want to add, we locate their initial points at the same point, translate the initial point of one to the terminal point of the other and complete the parallelogram, as indicated in Figure 10.3b. The vector lying along the diagonal, with initial point at A and terminal point at C is the sum

$$\overrightarrow{AC} = \overrightarrow{AB} + \overrightarrow{AD}.$$

A second basic arithmetic operation for vectors is **scalar multiplication.** If we multiply a vector $\mathbf{u}$ by a scalar (a real number) $c > 0$, the resulting vector will have the same direction as $\mathbf{u}$, but will have magnitude $c\|\mathbf{u}\|$. On the other hand, multiplying a vector $\mathbf{u}$ by a scalar $c < 0$ will result in a vector with opposite direction from $\mathbf{u}$ and magnitude $|c|\|\mathbf{u}\|$ (see Figure 10.4).

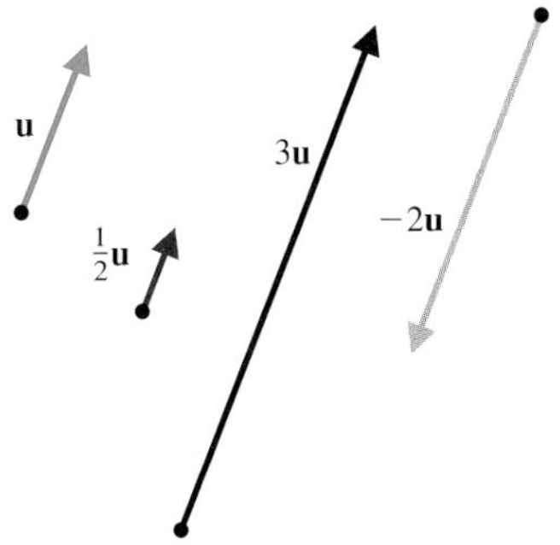

FIGURE 10.4
Scalar multiplication

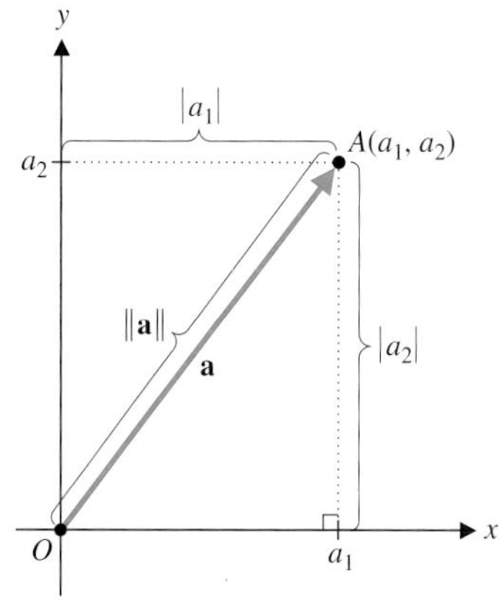

FIGURE 10.5
Position vector $\mathbf{a} = \langle a_1, a_2 \rangle$

Since the location of the initial point is irrelevant, we typically draw vectors with their initial point located at the origin. Such a vector is called a **position vector.** Notice that the terminal point of a position vector will completely determine the vector, so that specifying the terminal point will also specify the vector. For the position vector $\mathbf{a}$ with initial point at the origin and terminal point at the point $A(a_1, a_2)$ (see Figure 10.5), we denote the vector by

$$\mathbf{a} = \overrightarrow{OA} = \langle a_1, a_2 \rangle.$$

We call a_1 and a_2 the **components** of the vector $\mathbf{a}$; a_1 is the **first component** and a_2 is the **second component.** Be careful to distinguish between the *point* (a_1, a_2) and the position vector $\langle a_1, a_2 \rangle$. Note from Figure 10.5 that the magnitude of the position vector $\mathbf{a}$ follows

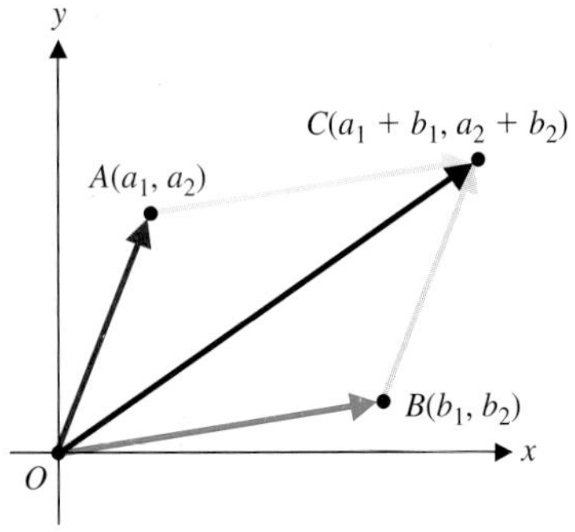

FIGURE 10.6
Adding position vectors

directly from the Pythagorean Theorem. We have

$$\|\mathbf{a}\| = \sqrt{a_1^2 + a_2^2}. \quad \text{Magnitude of a vector} \quad (1.1)$$

Notice that it follows from the definition that for two position vectors $\mathbf{a} = \langle a_1, a_2 \rangle$ and $\mathbf{b} = \langle b_1, b_2 \rangle$, $\mathbf{a} = \mathbf{b}$ if and only if their terminal points are the same, that is if $a_1 = b_1$ and $a_2 = b_2$. In other words, two position vectors are equal only when their corresponding components are equal.

To add two position vectors, $\overrightarrow{OA} = \langle a_1, a_2 \rangle$ and $\overrightarrow{OB} = \langle b_1, b_2 \rangle$, we draw the position vectors in Figure 10.6 and complete the parallelogram, as before. From Figure 10.6, we have

$$\overrightarrow{OA} + \overrightarrow{OB} = \overrightarrow{OC}.$$

Writing down the position vectors in their component form, we take this as our definition of vector addition:

$$\langle a_1, a_2 \rangle + \langle b_1, b_2 \rangle = \langle a_1 + b_1, a_2 + b_2 \rangle. \quad \text{Vector addition} \quad (1.2)$$

So, to add two vectors, we simply add the corresponding components. For this reason, we say that addition of vectors is done **componentwise.** Similarly, we *define* subtraction of vectors componentwise, so that

$$\langle a_1, a_2 \rangle - \langle b_1, b_2 \rangle = \langle a_1 - b_1, a_2 - b_2 \rangle. \quad \text{Vector subtraction} \quad (1.3)$$

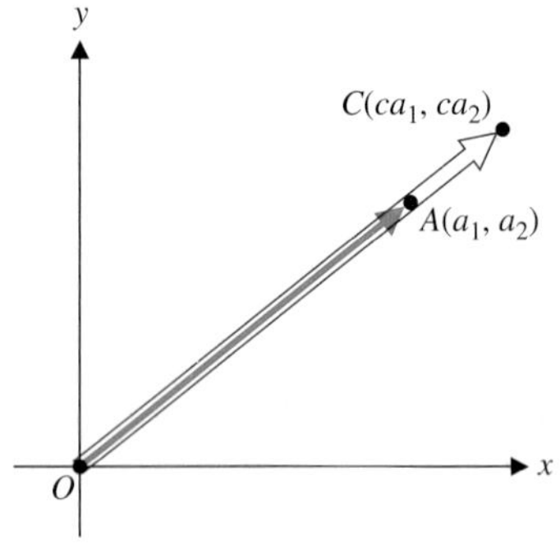

FIGURE 10.7a
Scalar multiplication ($c > 1$)

We give a geometric interpretation of subtraction later in this section.

Recall that if we multiply a vector $\mathbf{a}$ by a scalar c, the result is a vector in the same direction as $\mathbf{a}$ (for $c > 0$) or the opposite direction as $\mathbf{a}$ (for $c < 0$), in each case with magnitude $|c|\|\mathbf{a}\|$. We indicate the case of a position vector $\mathbf{a} = \langle a_1, a_2 \rangle$ and scalar multiple $c > 1$ in Figure 10.7a and for $0 < c < 1$ in Figure 10.7b. The situation for $c < 0$ is illustrated in Figures 10.7c and 10.7d.

For the case where $c > 0$, notice that a vector in the same direction as $\mathbf{a}$, but with magnitude $|c|\|\mathbf{a}\|$, is the position vector $\langle ca_1, ca_2 \rangle$, since

$$\begin{aligned}\|\langle ca_1, ca_2 \rangle\| &= \sqrt{(ca_1)^2 + (ca_2)^2} = \sqrt{c^2 a_1^2 + c^2 a_2^2} \\ &= |c|\sqrt{a_1^2 + a_2^2} = |c|\|\mathbf{a}\|.\end{aligned}$$

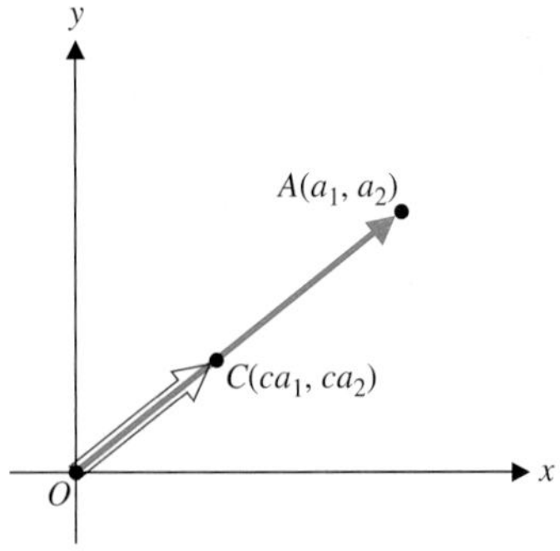

FIGURE 10.7b
Scalar multiplication ($0 < c < 1$)

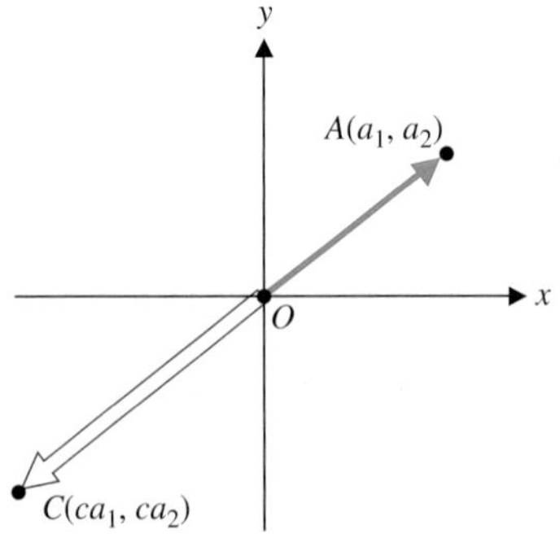

FIGURE 10.7c
Scalar multiplication ($c < -1$)

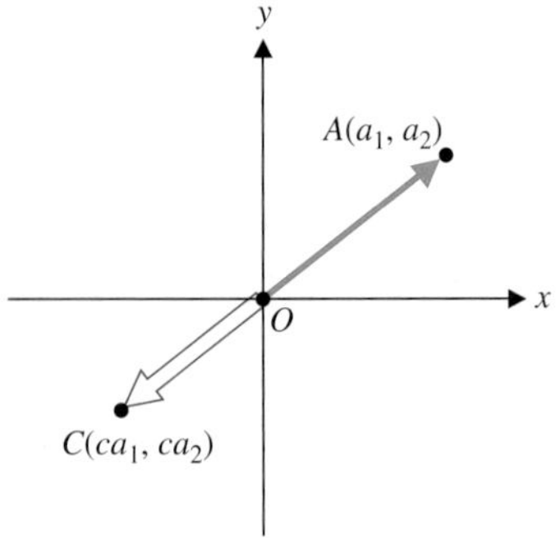

FIGURE 10.7d
Scalar multiplication ($-1 < c < 0$)

Similarly, if $c < 0$, you can show that $\langle ca_1, ca_2 \rangle$ is a vector in the **opposite** direction from **a**, with magnitude $|c|\,\|\mathbf{a}\|$. For this reason, we define scalar multiplication of position vectors by

Scalar multiplication

$$c\langle a_1, a_2 \rangle = \langle ca_1, ca_2 \rangle, \tag{1.4}$$

for any scalar c. Further, notice that this says that

$$\|c\mathbf{a}\| = |c|\,\|\mathbf{a}\|. \tag{1.5}$$

EXAMPLE 1.1 Vector Arithmetic

For vectors $\mathbf{a} = \langle 2, 1 \rangle$ and $\mathbf{b} = \langle 3, -2 \rangle$, compute (a) $\mathbf{a} + \mathbf{b}$, (b) $2\mathbf{a}$, (c) $2\mathbf{a} + 3\mathbf{b}$, (d) $2\mathbf{a} - 3\mathbf{b}$ and (e) $\|2\mathbf{a} - 3\mathbf{b}\|$.

Solution (a) From (1.2), we have

$$\mathbf{a} + \mathbf{b} = \langle 2, 1 \rangle + \langle 3, -2 \rangle = \langle 2 + 3, 1 - 2 \rangle = \langle 5, -1 \rangle.$$

(b) From (1.4), we have

$$2\mathbf{a} = 2\langle 2, 1 \rangle = \langle 2 \cdot 2, 2 \cdot 1 \rangle = \langle 4, 2 \rangle.$$

(c) From (1.2) and (1.4), we have

$$2\mathbf{a} + 3\mathbf{b} = 2\langle 2, 1 \rangle + 3\langle 3, -2 \rangle = \langle 4, 2 \rangle + \langle 9, -6 \rangle = \langle 13, -4 \rangle.$$

(d) From (1.3) and (1.4), we have

$$2\mathbf{a} - 3\mathbf{b} = 2\langle 2, 1 \rangle - 3\langle 3, -2 \rangle = \langle 4, 2 \rangle - \langle 9, -6 \rangle = \langle -5, 8 \rangle.$$

(e) Finally, from (1.1), we have

$$\|2\mathbf{a} - 3\mathbf{b}\| = \|\langle -5, 8 \rangle\| = \sqrt{25 + 64} = \sqrt{89}. \blacksquare$$

Observe that if we multiply any vector (with any direction) by the scalar $c = 0$, we get a vector with zero length, the **zero vector:**

$$\mathbf{0} = \langle 0, 0 \rangle.$$

Further, notice that this is the *only* vector with zero length. (Why is that?) The zero vector also has no particular direction. Finally, we define the **additive inverse** $-\mathbf{a}$ of a vector **a** in the expected way:

$$-\mathbf{a} = -\langle a_1, a_2 \rangle = (-1)\langle a_1, a_2 \rangle = \langle -a_1, -a_2 \rangle.$$

Notice that this says that the vector $-\mathbf{a}$ is a vector with the **opposite** direction as **a** and since

$$\|-\mathbf{a}\| = \|(-1)\langle a_1, a_2 \rangle\| = |-1|\,\|\mathbf{a}\| = \|\mathbf{a}\|,$$

$-\mathbf{a}$ has the same length as **a**.

DEFINITION 1.1

Two vectors having the same or opposite direction are called **parallel.**

It then follows that two (nonzero) position vectors **a** and **b** are parallel if and only if $\mathbf{b} = c\mathbf{a}$, for some scalar c. In this event, we say that **b** is a *scalar multiple* of **a**.

EXAMPLE 1.2 Determining When Two Vectors Are Parallel

Determine whether the given pair of vectors is parallel: (a) $\mathbf{a} = \langle 2, 3\rangle$ and $\mathbf{b} = \langle 4, 5\rangle$, (b) $\mathbf{a} = \langle 2, 3\rangle$ and $\mathbf{b} = \langle -4, -6\rangle$.

Solution (a) Notice that from (1.4), we have that if $\mathbf{b} = c\mathbf{a}$, then

$$\langle 4, 5\rangle = c\langle 2, 3\rangle = \langle 2c, 3c\rangle.$$

For this to hold, the corresponding components of the two vectors must be equal. That is, $4 = 2c$ (so that $c = 2$) *and* $5 = 3c$ (so that $c = 5/3$). This is a contradiction and so, **a** and **b** are not parallel.

(b) Again, from (1.4), we have

$$\langle -4, -6\rangle = c\langle 2, 3\rangle = \langle 2c, 3c\rangle.$$

In this case, we have $-4 = 2c$ (so that $c = -2$) and $-6 = 3c$ (which again leads us to $c = -2$). This says that $-2\mathbf{a} = \langle -4, -6\rangle = \mathbf{b}$ and so, $\langle 2, 3\rangle$ and $\langle -4, 6\rangle$ are parallel. ■

We denote the set of all position vectors in two-dimensional space by

$$V_2 = \{\langle x, y\rangle | x, y \in \mathbb{R}\}.$$

You can easily show that the rules of algebra given in Theorem 1.1 hold for vectors in V_2.

THEOREM 1.1

For any vectors **a**, **b** and **c** in V_2, and any scalars d and e in $\mathbb{R}$, the following hold:

(i)	$\mathbf{a} + \mathbf{b} = \mathbf{b} + \mathbf{a}$	(commutativity)
(ii)	$\mathbf{a} + (\mathbf{b} + \mathbf{c}) = (\mathbf{a} + \mathbf{b}) + \mathbf{c}$	(associativity)
(iii)	$\mathbf{a} + \mathbf{0} = \mathbf{a}$	(zero vector)
(iv)	$\mathbf{a} + (-\mathbf{a}) = \mathbf{0}$	(additive inverse)
(v)	$d(\mathbf{a} + \mathbf{b}) = d\mathbf{a} + d\mathbf{b}$	(distributive law)
(vi)	$(d + e)\mathbf{a} = d\mathbf{a} + e\mathbf{a}$	(distributive law)
(vii)	$(1)\mathbf{a} = \mathbf{a}$	(multiplication by 1) and
(viii)	$(0)\mathbf{a} = \mathbf{0}$	(multiplication by 0).

PROOF

We prove the first of these and leave the rest as exercises. By definition,

$$\begin{aligned}\mathbf{a} + \mathbf{b} &= \langle a_1, a_2\rangle + \langle b_1, b_2\rangle = \langle a_1 + b_1, a_2 + b_2\rangle \\ &= \langle b_1 + a_1, b_2 + a_2\rangle = \mathbf{b} + \mathbf{a}. \quad ■\end{aligned}$$

Since addition of real numbers is commutative.

Notice that using the commutativity and associativity of vector addition, we have

$$\mathbf{b} + (\mathbf{a} - \mathbf{b}) = (\mathbf{a} - \mathbf{b}) + \mathbf{b} = \mathbf{a} + (-\mathbf{b} + \mathbf{b}) = \mathbf{a} + \mathbf{0} = \mathbf{a}.$$

From our graphical interpretation of vector addition, we get Figure 10.8. Notice that this now gives us a geometric interpretation of vector subtraction.

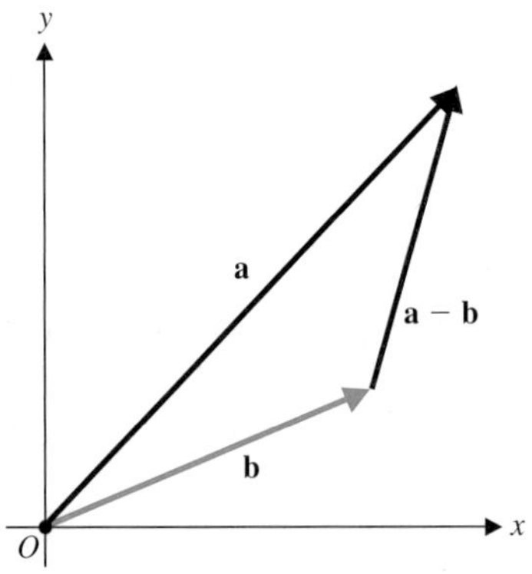

FIGURE 10.8
$\mathbf{b} + (\mathbf{a} - \mathbf{b}) = \mathbf{a}$

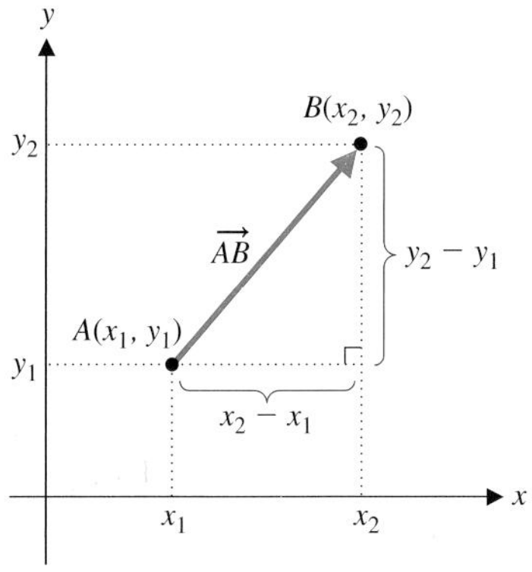

FIGURE 10.9
Vector from A to B

For any two points $A(x_1, y_1)$ and $B(x_2, y_2)$, observe from Figure 10.9 that the vector $\overrightarrow{AB}$ corresponds to the position vector $\langle x_2 - x_1, y_2 - y_1 \rangle$.

EXAMPLE 1.3 Finding a Position Vector

Find the vector with (a) initial point at $A(2, 3)$ and terminal point at $B(3, -1)$ and (b) initial point at B and terminal point at A.

Solution (a) We show this graphically in Figure 10.10a. Notice that

$$\overrightarrow{AB} = \langle 3 - 2, -1 - 3 \rangle = \langle 1, -4 \rangle.$$

(b) Similarly, the vector with initial point at $B(3, -1)$ and terminal point at $A(2, 3)$ is given by

$$\overrightarrow{BA} = \langle 2 - 3, 3 - (-1) \rangle = \langle 2 - 3, 3 + 1 \rangle = \langle -1, 4 \rangle.$$

We indicate this graphically in Figure 10.10b. ■

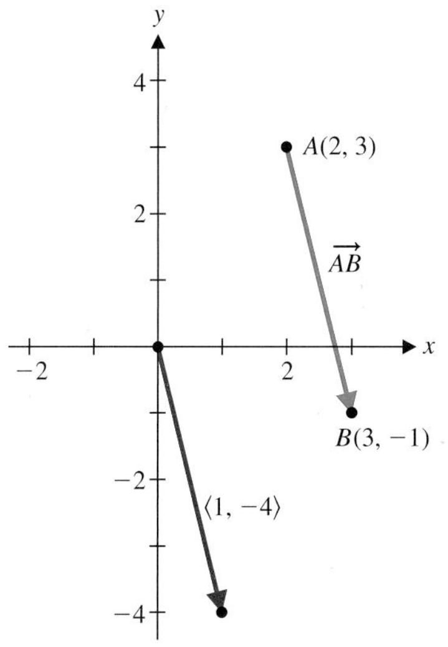

FIGURE 10.10a
$\overrightarrow{AB} = \langle 1, -4 \rangle$

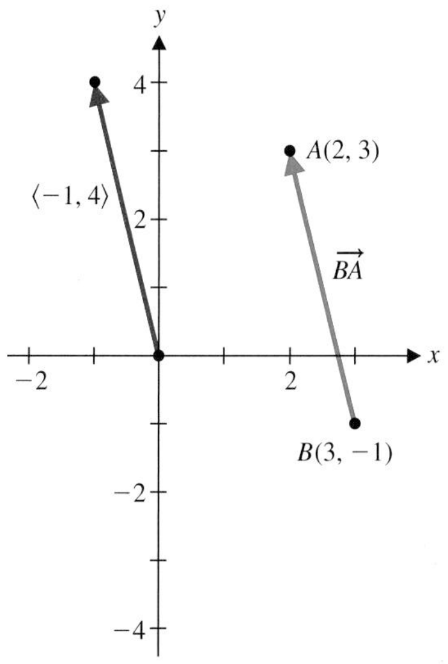

FIGURE 10.10b
$\overrightarrow{BA} = \langle -1, 4 \rangle$

We often find it convenient to write vectors in terms of some standard vectors. We define the **standard basis vectors i** and **j** by

$$\mathbf{i} = \langle 1, 0 \rangle \quad \text{and} \quad \mathbf{j} = \langle 0, 1 \rangle$$

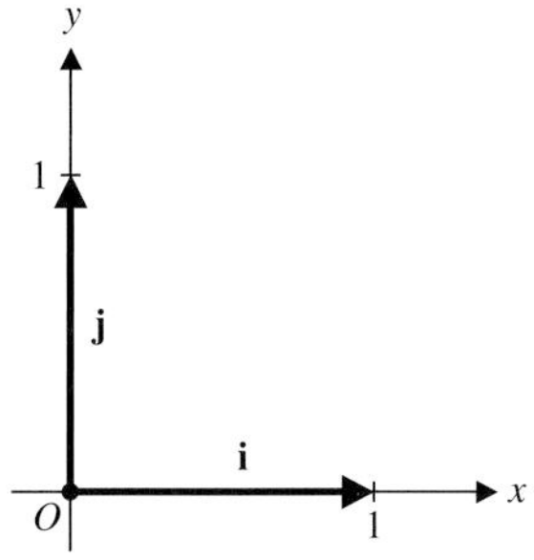

FIGURE 10.11
Standard basis

(see Figure 10.11). Notice that $\|\mathbf{i}\| = \|\mathbf{j}\| = 1$. Any vector **a** with $\|\mathbf{a}\| = 1$ is called a **unit vector.** So, **i** and **j** are unit vectors.

Finally, we say that **i** and **j** form a **basis** for V_2, since we can write any vector $\mathbf{a} \in V_2$ uniquely in terms of **i** and **j**, as follows:

$$\mathbf{a} = \langle a_1, a_2 \rangle = a_1\mathbf{i} + a_2\mathbf{j}.$$

We call a_1 and a_2 the **horizontal** and **vertical components** of **a**, respectively.

For any nonzero vector, we can always find a unit vector with the same direction, as in Theorem 1.2.

THEOREM 1.2 (Unit Vector)

For any nonzero position vector $\mathbf{a} = \langle a_1, a_2 \rangle$, a unit vector having the same direction as $\mathbf{a}$ is given by

$$\mathbf{u} = \frac{1}{\|\mathbf{a}\|}\mathbf{a}.$$

The process of dividing a nonzero vector by its magnitude is sometimes called **normalization.** (A vector's magnitude is sometimes called its **norm.**) As we'll see, some problems are simplified by using normalized vectors.

PROOF

First, notice that since $\mathbf{a} \neq \mathbf{0}$, $\|\mathbf{a}\| > 0$ and so, $\mathbf{u}$ is a *positive* scalar multiple of $\mathbf{a}$. This says that $\mathbf{u}$ and $\mathbf{a}$ have the same direction. To see that $\mathbf{u}$ is a unit vector, notice that since $\dfrac{1}{\|\mathbf{a}\|}$ is a positive scalar, we have from (1.5) that

$$\|\mathbf{u}\| = \left\| \frac{1}{\|\mathbf{a}\|}\mathbf{a} \right\| = \frac{1}{\|\mathbf{a}\|}\|\mathbf{a}\| = 1. \quad \blacksquare$$

EXAMPLE 1.4 Finding a Unit Vector

Find a unit vector in the same direction as $\mathbf{a} = \langle 3, -4 \rangle$.

Solution First, note that

$$\|\mathbf{a}\| = \|\langle 3, -4 \rangle\| = \sqrt{3^2 + (-4)^2} = \sqrt{25} = 5.$$

A unit vector in the same direction as $\mathbf{a}$ is then

$$\mathbf{u} = \frac{1}{\|\mathbf{a}\|}\mathbf{a} = \frac{1}{5}\langle 3, -4 \rangle = \left\langle \frac{3}{5}, -\frac{4}{5} \right\rangle. \quad \blacksquare$$

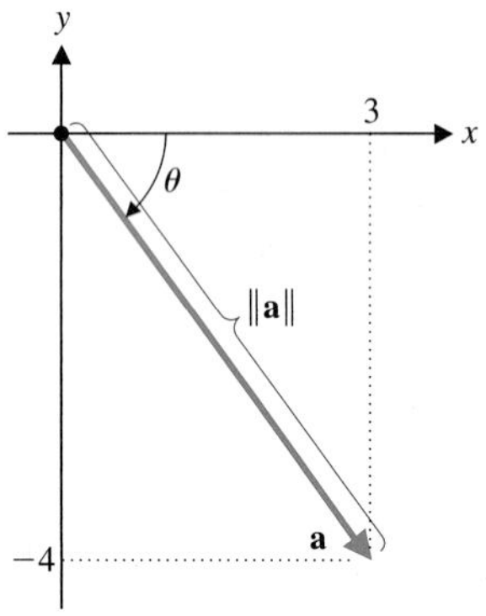

FIGURE 10.12
Polar form of a vector

It is often convenient to write a vector explicitly in terms of its magnitude and direction. For instance, in example 1.4, we found that the magnitude of $\mathbf{a} = \langle 3, -4 \rangle$ is $\|\mathbf{a}\| = 5$, while its direction is indicated by the unit vector $\langle \frac{3}{5}, -\frac{4}{5} \rangle$. Notice that we can now write $\mathbf{a} = 5\langle \frac{3}{5}, -\frac{4}{5} \rangle$. Graphically, we can represent $\mathbf{a}$ as a position vector (see Figure 10.12). Notice also that if θ is the angle between the positive x-axis and $\mathbf{a}$, then

$$\mathbf{a} = 5\langle \cos\theta, \sin\theta \rangle,$$

where $\theta = \tan^{-1}\left(-\frac{4}{3}\right) \approx -0.93$. This representation is called the **polar form** of the vector $\mathbf{a}$. Note that this corresponds to writing the rectangular point $(3, -4)$ as the polar point (r, θ), where $r = \|\mathbf{a}\|$.

We close this section with two applications of vector arithmetic. Whenever two or more forces are acting on an object, the net force acting on the object (often referred to as the **resultant force**) is simply the sum of all of the force vectors. That is, the net effect of two or more forces acting on an object is the same as a single force (given by the sum) applied to the object.

EXAMPLE 1.5 Finding the Net Force Acting on a Sky Diver

At a certain point during a jump, there are two principal forces acting on a sky diver: gravity exerting a force of 180 pounds straight down and air resistance exerting a force of 180 pounds up and 30 pounds to the right. What is the net force acting on the sky diver?

Solution We write the gravity force vector as $\mathbf{g} = \langle 0, -180 \rangle$ and the air resistance force vector as $\mathbf{r} = \langle 30, 180 \rangle$. The net force on the sky diver is the sum of the two forces, $\mathbf{g} + \mathbf{r} = \langle 30, 0 \rangle$. We illustrate the forces in Figure 10.13. Notice that at this point, the vertical forces are balanced, producing a "free-fall" vertically, so that the sky diver is neither accelerating nor decelerating vertically. The net force is purely horizontal, combating the horizontal motion of the sky diver after jumping from the plane. ■

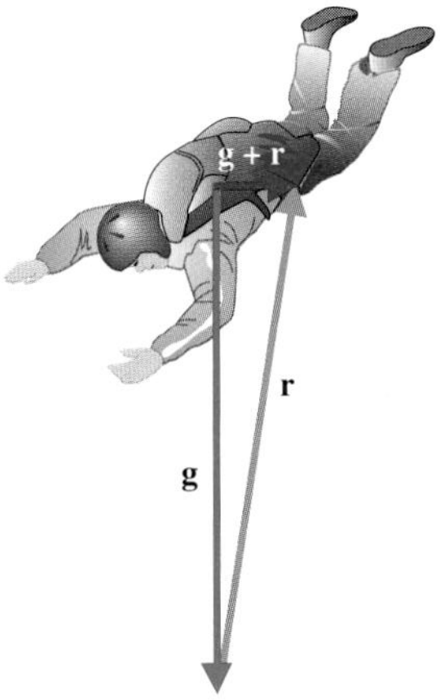

FIGURE 10.13
Forces on a sky diver

When flying an airplane, it's important to consider the velocity of the air in which you are flying. Observe that the effect of the velocity of the air can be quite significant. Think about it this way: if a plane flies at 200 mph (its airspeed) and the air in which the plane is moving is itself moving at 35 mph in the same direction (i.e., there is a 35 mph tailwind), then the effective speed of the plane is 235 mph. Conversely, if the same 35 mph wind is moving in exactly the opposite direction (i.e., there is a 35 mph headwind), then the plane's effective speed is only 165 mph. If the wind is blowing in a direction that's not parallel to the plane's direction of travel, we need to add the velocity vectors corresponding to the plane's airspeed and the wind to get the effective velocity. We illustrate this in example 1.6.

EXAMPLE 1.6 Steering an Aircraft in a Headwind and a Crosswind

An airplane has an airspeed of 400 mph. Suppose that the wind velocity is given by the vector $\mathbf{w} = \langle 20, 30 \rangle$. In what direction should the airplane head in order to fly due west (i.e., in the direction of the unit vector $-\mathbf{i} = \langle -1, 0 \rangle$)?

Solution We illustrate the velocity vectors for the airplane and the wind in Figure 10.14. We let the airplane's velocity vector be $\mathbf{v} = \langle x, y \rangle$. The effective velocity of the plane is then $\mathbf{v} + \mathbf{w}$, which we set equal to $\langle c, 0 \rangle$, for some negative constant c. Since

$$\mathbf{v} + \mathbf{w} = \langle x + 20, y + 30 \rangle = \langle c, 0 \rangle,$$

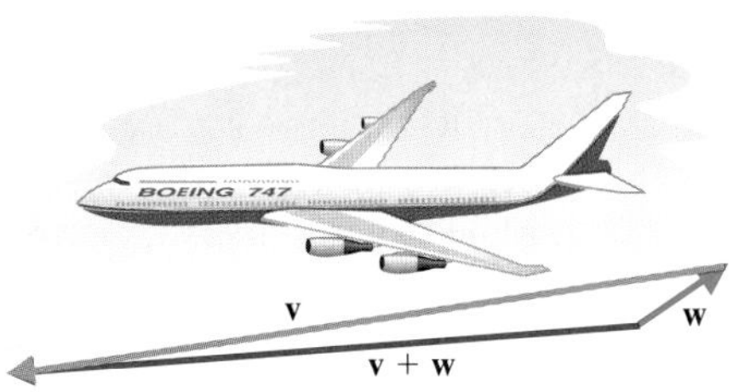

FIGURE 10.14
Forces on an airplane

we must have $x + 20 = c$ and $y + 30 = 0$, so that $y = -30$. Further, since the plane's airspeed is 400 mph, we must have $400 = \|\mathbf{v}\| = \sqrt{x^2 + y^2} = \sqrt{x^2 + 900}$. Squaring this gives us $x^2 + 900 = 160{,}000$, so that $x = -\sqrt{159{,}100}$. (We take the negative square root so that the plane heads westward.) Consequently, the plane should head in the direction of $\mathbf{v} = \langle -\sqrt{159{,}100}, -30 \rangle$, which points left and down, or southwest, at an angle of $\tan^{-1}(30/\sqrt{159{,}100}) \approx 4°$ below due west. ■

BEYOND FORMULAS

It is important that you understand vectors in both symbolic and graphical terms. Much of the notation introduced in this section is used to simplify calculations. However, the visualization of vectors as directed line segments is often the key to identifying which calculation is appropriate. For example, notice in example 1.6 that Figure 10.14 leads directly to an equation. The equation is more easily solved than the corresponding trigonometric problem implied by Figure 10.14. What are some of the ways in which symbolic and graphical representations reinforce each other in one-variable calculus?

EXERCISES 10.1

WRITING EXERCISES

1. Discuss whether each of the following is a vector or a scalar quantity: force, area, weight, height, temperature, wind velocity.
2. Some athletes are blessed with "good acceleration." In calculus, we define acceleration as the rate of change of velocity. Keeping in mind that the velocity vector has magnitude (i.e., speed) and direction, discuss why the ability to accelerate rapidly is beneficial.
3. The location of the initial point of a vector is irrelevant. Using the example of a velocity vector, explain why we want to focus on the magnitude of the vector and its direction, but not on the initial point.
4. Describe the changes that occur when a vector is multiplied by a scalar $c \neq 0$. In your discussion, consider both positive and negative scalars, discuss changes both in the components of the vector and in its graphical representation, and consider the specific case of a velocity vector.

In exercises 1 and 2, sketch the vectors 2a, −3b, a + b and 2a − 3b.

1.

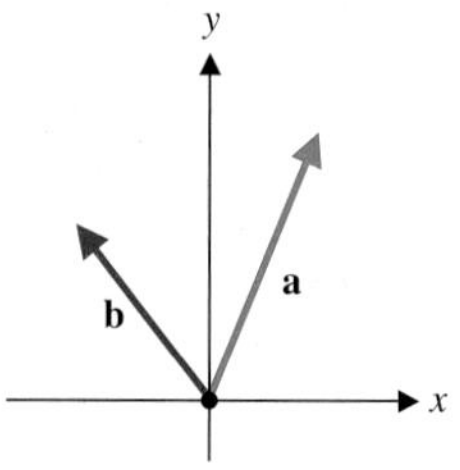

2.

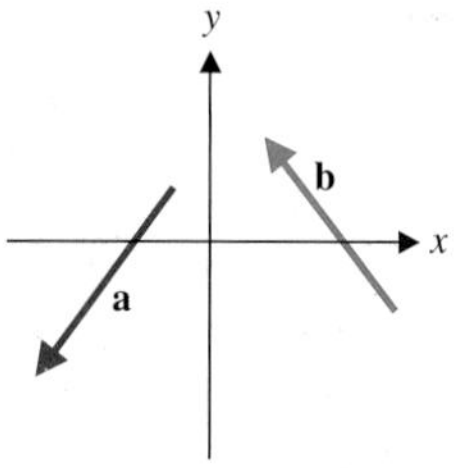

In exercises 3–6, compute, a + b, a − 2b, 3a and ‖5b − 2a‖.

3. $\mathbf{a} = \langle 2, 4\rangle, \mathbf{b} = \langle 3, -1\rangle$
4. $\mathbf{a} = \langle 3, -2\rangle, \mathbf{b} = \langle 2, 0\rangle$
5. $\mathbf{a} = \mathbf{i} + 2\mathbf{j}, \mathbf{b} = 3\mathbf{i} - \mathbf{j}$
6. $\mathbf{a} = -2\mathbf{i} + \mathbf{j}, \mathbf{b} = 3\mathbf{i}$
7. For exercises 3 and 4, illustrate the sum $\mathbf{a} + \mathbf{b}$ graphically.
8. For exercises 5 and 6, illustrate the difference $\mathbf{a} - \mathbf{b}$ graphically.

In exercises 9–14, determine whether the vectors a and b are parallel.

9. $\mathbf{a} = \langle 2, 1\rangle, \mathbf{b} = \langle -4, -2\rangle$
10. $\mathbf{a} = \langle 1, -2\rangle, \mathbf{b} = \langle 2, 1\rangle$
11. $\mathbf{a} = \langle -2, 3\rangle, \mathbf{b} = \langle 4, 6\rangle$
12. $\mathbf{a} = \langle 1, -2\rangle, \mathbf{b} = \langle -4, 8\rangle$
13. $\mathbf{a} = \mathbf{i} + 2\mathbf{j}, \mathbf{b} = 3\mathbf{i} + 6\mathbf{j}$
14. $\mathbf{a} = -2\mathbf{i} + \mathbf{j}, \mathbf{b} = 4\mathbf{i} + 2\mathbf{j}$

In exercises 15–18, find the vector with initial point *A* and terminal point *B*.

15. $A = (2, 3), B = (5, 4)$
16. $A = (4, 3), B = (1, 0)$
17. $A = (-1, 2), B = (1, -1)$
18. $A = (1, 1), B = (-2, 4)$

In exercises 19–24, (a) find a unit vector in the same direction as the given vector and (b) write the given vector in polar form.

19. $\langle 4, -3\rangle$
20. $\langle 3, 6\rangle$
21. $2\mathbf{i} - 4\mathbf{j}$
22. $4\mathbf{i}$
23. from (2, 1) to (5, 2)
24. from (5, −1) to (2, 3)

In exercises 25–30, find a vector with the given magnitude in the same direction as the given vector.

25. magnitude 3, $\mathbf{v} = 3\mathbf{i} + 4\mathbf{j}$
26. magnitude 4, $\mathbf{v} = 2\mathbf{i} - \mathbf{j}$
27. magnitude 29, $\mathbf{v} = \langle 2, 5\rangle$
28. magnitude 10, $\mathbf{v} = \langle 3, 1\rangle$
29. magnitude 4, $\mathbf{v} = \langle 3, 0\rangle$
30. magnitude 5, $\mathbf{v} = \langle 0, -2\rangle$
31. Suppose that there are two forces acting on a sky diver: gravity at 150 pounds down and air resistance at 140 pounds up and 20 pounds to the right. What is the net force acting on the sky diver?

32. Suppose that there are two forces acting on a sky diver: gravity at 200 pounds down and air resistance at 180 pounds up and 40 pounds to the right. What is the net force acting on the sky diver?

33. Suppose that there are two forces acting on a sky diver: gravity at 200 pounds down and air resistance. If the net force is 10 pounds down and 30 pounds to the right, what is the force of air resistance acting on the sky diver?

34. Suppose that there are two forces acting on a sky diver: gravity at 180 pounds down and air resistance. If the net force is 20 pounds down and 20 pounds to the left, what is the force of air resistance acting on the sky diver?

35. In the accompanying figure, two ropes are attached to a large crate. Suppose that rope A exerts a force of $\langle -164, 115\rangle$ pounds on the crate and rope B exerts a force of $\langle 177, 177\rangle$ pounds on the crate. If the crate weighs 275 pounds, what is the net force acting on the crate? Based on your answer, which way will the crate move?

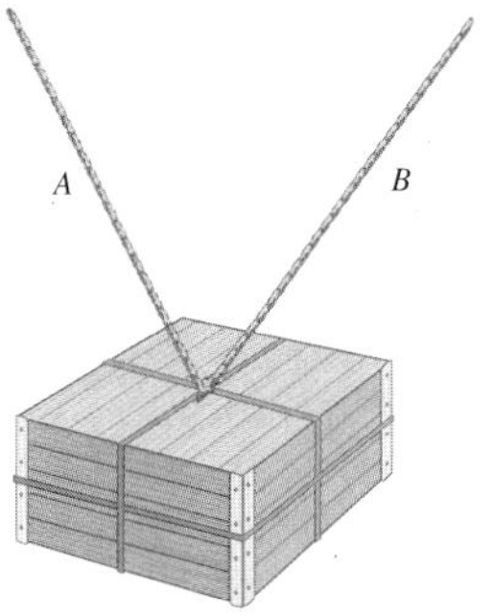

36. Repeat exercise 35 with forces of $\langle -131, 92\rangle$ pounds from rope A and $\langle 92, 92\rangle$ from rope B.

37. The thrust of an airplane's engines produces a speed of 300 mph in still air. The wind velocity is given by $\langle 30, -20\rangle$. In what direction should the airplane head to fly due west?

38. The thrust of an airplane's engines produces a speed of 600 mph in still air. The wind velocity is given by $\langle -30, 60\rangle$. In what direction should the airplane head to fly due west?

39. The thrust of an airplane's engines produces a speed of 400 mph in still air. The wind velocity is given by $\langle -20, 30\rangle$. In what direction should the airplane head to fly due north?

40. The thrust of an airplane's engines produces a speed of 300 mph in still air. The wind velocity is given by $\langle 50, 0\rangle$. In what direction should the airplane head to fly due north?

41. A paperboy is riding at 10 ft/s on a bicycle and tosses a paper over his left shoulder at 50 ft/s. If the porch is 50 ft off the road, how far up the street should the paperboy release the paper to hit the porch?

42. A papergirl is riding at 12 ft/s on a bicycle and tosses a paper over her left shoulder at 48 ft/s. If the porch is 40 ft off the road, how far up the street should the papergirl release the paper to hit the porch?

43. The water from a fire hose exerts a force of 200 pounds on the person holding the hose. The nozzle of the hose weighs 20 pounds. What force is required to hold the hose horizontal? At what angle to the horizontal is this force applied?

44. Repeat exercise 43 for holding the hose at a 45° angle to the horizontal.

45. A person is paddling a kayak in a river with a current of 1 ft/s. The kayaker is aimed at the far shore, perpendicular to the current. The kayak's speed in still water would be 4 ft/s. Find the kayak's actual speed and the angle between the kayak's direction and the far shore.

46. For the kayak in exercise 45, find the direction that the kayaker would need to paddle to go straight across the river. How does this angle compare to the angle found in exercise 45?

47. If vector $\mathbf{a}$ has magnitude $\|\mathbf{a}\| = 3$ and vector $\mathbf{b}$ has magnitude $\|\mathbf{b}\| = 4$, what is the largest possible magnitude for the vector $\mathbf{a}+\mathbf{b}$? What is the smallest possible magnitude for the vector $\mathbf{a}+\mathbf{b}$? What will be the magnitude of $\mathbf{a}+\mathbf{b}$ if $\mathbf{a}$ and $\mathbf{b}$ are perpendicular?

48. Use vectors to show that the points $(1, 2)$, $(3, 1)$, $(4, 3)$ and $(2, 4)$ form a parallelogram.

49. Prove the associativity property of Theorem 1.1.

50. Prove the distributive laws of Theorem 1.1.

51. For vectors $\mathbf{a} = \langle 2, 3\rangle$ and $\mathbf{b} = \langle 1, 4\rangle$, compare $\|\mathbf{a}+\mathbf{b}\|$ and $\|\mathbf{a}\| + \|\mathbf{b}\|$. Repeat this comparison for two other choices of $\mathbf{a}$ and $\mathbf{b}$. Use the sketch in Figure 10.6 to explain why $\|\mathbf{a}+\mathbf{b}\| \le \|\mathbf{a}\| + \|\mathbf{b}\|$ for any vectors $\mathbf{a}$ and $\mathbf{b}$.

52. To prove that $\|\mathbf{a}+\mathbf{b}\| \le \|\mathbf{a}\| + \|\mathbf{b}\|$ for $\mathbf{a} = \langle a_1, a_2\rangle$ and $\mathbf{b} = \langle b_1, b_2\rangle$, start by showing that $2a_1a_2b_1b_2 \le a_1^2b_2^2 + a_2^2b_1^2$. [Hint: Compute $(a_1b_2 - a_2b_1)^2$.] Then, show that $a_1b_1 + a_2b_2 \le \sqrt{a_1^2+a_2^2}\sqrt{b_1^2+b_2^2}$. (Hint: Square both sides and use the previous result.) Finally, compute $\|\mathbf{a}+\mathbf{b}\|^2 - (\|\mathbf{a}\| + \|\mathbf{b}\|)^2$ and use the previous inequality to show that this is less than or equal to 0.

53. In exercises 51 and 52, you explored the inequality $\|\mathbf{a}+\mathbf{b}\| \le \|\mathbf{a}\| + \|\mathbf{b}\|$. Use the geometric interpretation of Figure 10.6 to conjecture the circumstances under which $\|\mathbf{a}+\mathbf{b}\| = \|\mathbf{a}\| + \|\mathbf{b}\|$. Similarly, use a geometric interpretation to determine circumstances under which $\|\mathbf{a}+\mathbf{b}\|^2 = \|\mathbf{a}\|^2 + \|\mathbf{b}\|^2$. In general, what is the relationship between $\|\mathbf{a}+\mathbf{b}\|^2$ and $\|\mathbf{a}\|^2 + \|\mathbf{b}\|^2$ (i.e., which is larger)?

EXPLORATORY EXERCISES

1. The figure shows a foot striking the ground, exerting a force of $\mathbf{F}$ pounds at an angle of θ from the vertical. The force is resolved into vertical and horizontal components $\mathbf{F}_v$ and $\mathbf{F}_h$, respectively. The friction force between floor and foot is $\mathbf{F}_f$, where $\|\mathbf{F}_f\| = \mu\|\mathbf{F}_v\|$ for a positive constant μ known as the **coefficient of friction.** Explain why the foot will slip if $\|\mathbf{F}_h\| > \|\mathbf{F}_f\|$ and show that this happens if and only if

$\tan\theta > \mu$. Compare the angles θ at which slipping occurs for coefficients $\mu = 0.6$, $\mu = 0.4$ and $\mu = 0.2$.

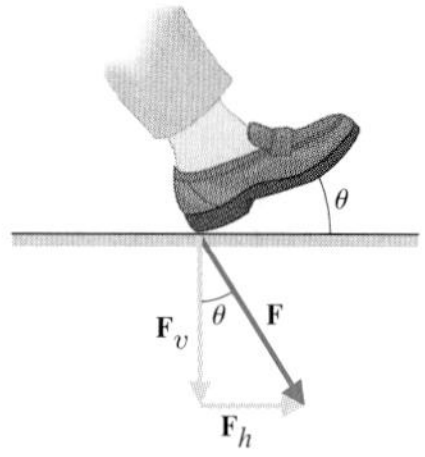

2. The vectors **i** and **j** are not the only basis vectors that can be used. In fact, any two nonzero and nonparallel vectors can be used as basis vectors for two-dimensional space. To see this, define $\mathbf{a} = \langle 1, 1\rangle$ and $\mathbf{b} = \langle 1, -1\rangle$. To write the vector $\langle 5, 1\rangle$ in terms of these vectors, we want constants c_1 and c_2 such that $\langle 5, 1\rangle = c_1\mathbf{a} + c_2\mathbf{b}$. Show that this requires that $c_1 + c_2 = 5$ and $c_1 - c_2 = 1$, and then solve for c_1 and c_2. Show that any vector $\langle x, y\rangle$ can be represented uniquely in terms of **a** and **b**. Determine as completely as possible the set of all vectors **v** such that **a** and **v** form a basis.

10.2 VECTORS IN SPACE

HISTORICAL NOTES

William Rowan Hamilton (1805–1865)
Irish mathematician who first defined and developed the theory of vectors. Hamilton was an outstanding student who was appointed Professor of Astronomy at Trinity College while still an undergraduate. After publishing several papers in the field of optics, Hamilton developed an innovative and highly influential approach to dynamics. He then became obsessed with the development of his theory of "quaternions" in which he also defined vectors. Hamilton thought that quaternions would revolutionize mathematical physics, but vectors have proved to be his most important contribution to mathematics.

We now extend several ideas from the two-dimensional Euclidean space, $\mathbb{R}^2$ to the three-dimensional Euclidean space, $\mathbb{R}^3$. We specify each point in three dimensions by an ordered triple (a, b, c), where the coordinates a, b and c represent the (signed) distance from the origin along each of three coordinates axes (x, y and z), as indicated in Figure 10.15a. This orientation of the axes is an example of a **right-handed** coordinate system. That is, if you align the fingers of your right hand along the positive x-axis and then curl them toward the positive y-axis, your thumb will point in the direction of the positive z-axis (see Figure 10.15b).

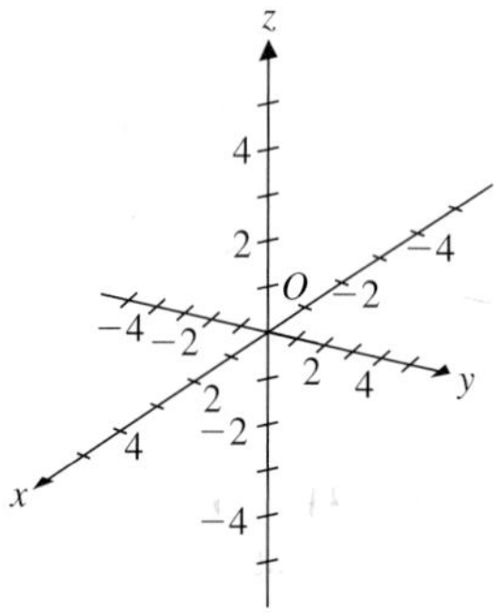

FIGURE 10.15a
Coordinate axes in $\mathbb{R}^3$

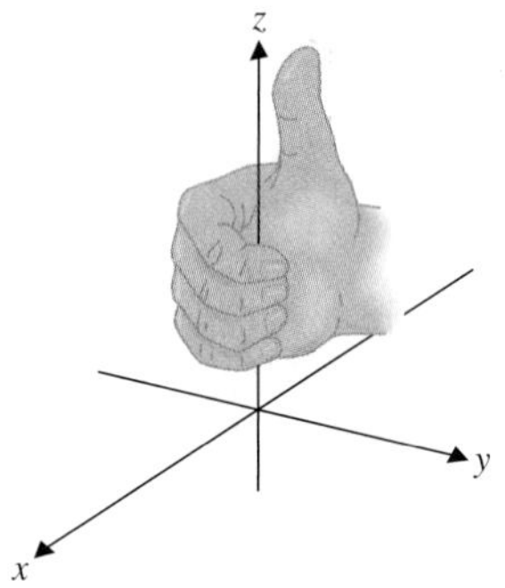

FIGURE 10.15b
Right-handed system

To locate the point $(a, b, c) \in \mathbb{R}^3$, where a, b and c are all positive, first move along the x-axis a distance of a units from the origin. This will put you at the point $(a, 0, 0)$. Continuing from this point, move parallel to the y-axis a distance of b units from $(a, 0, 0)$. This leaves you at the point $(a, b, 0)$. Finally, continuing from this point, move c units parallel to the z-axis. This is the location of the point (a, b, c) (see Figure 10.16).

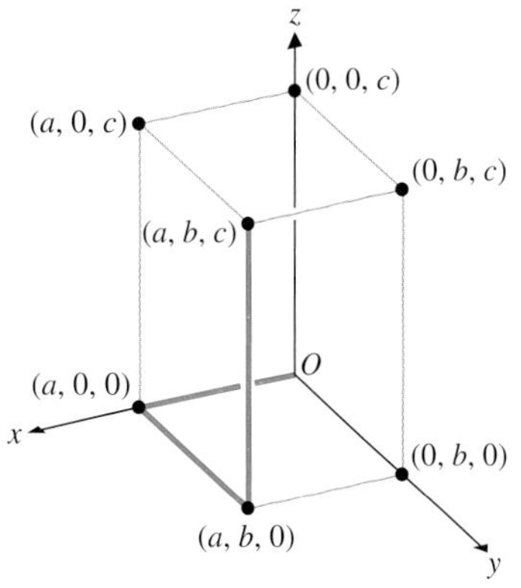

FIGURE 10.16
Locating the point (a, b, c)

EXAMPLE 2.1 Plotting Points in Three Dimensions

Plot the points $(1, 2, 3)$, $(3, -2, 4)$ and $(-1, 3, -2)$.

Solution Working as indicated above, we see the points plotted in Figures 10.17a, 10.17b and 10.17c, respectively.

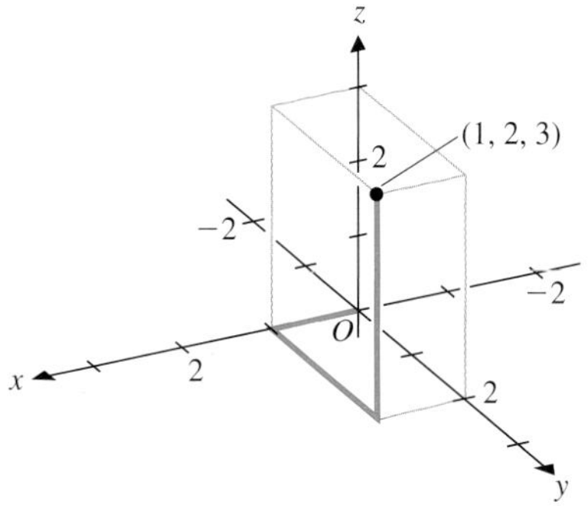

FIGURE 10.17a
The point $(1, 2, 3)$

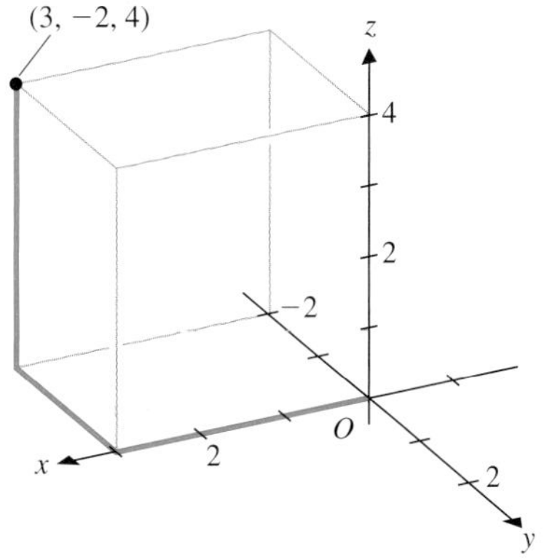

FIGURE 10.17b
The point $(3, -2, 4)$

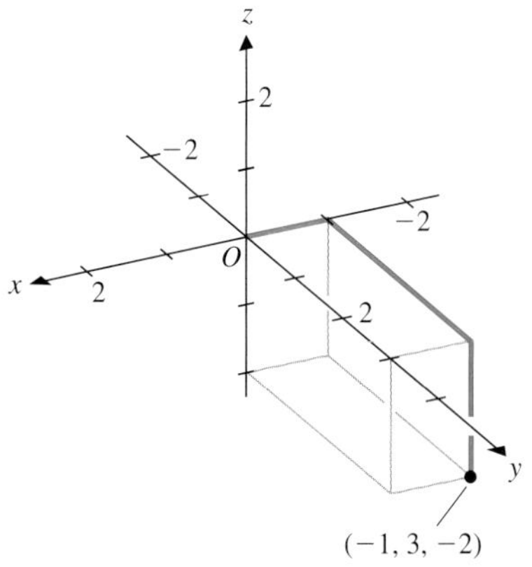

FIGURE 10.17c
The point $(-1, 3, -2)$

Recall that in $\mathbb{R}^2$, the coordinate axes divide the xy-plane into four quadrants. In a similar fashion, the three coordinate planes in $\mathbb{R}^3$ (the xy-plane, the yz-plane and the xz-plane) divide space into **eight octants** (see Figure 10.18 on the following page). The **first octant** is the one with $x > 0$, $y > 0$ and $z > 0$. We do not usually distinguish among the other seven octants.

We can compute the distance between two points in $\mathbb{R}^3$ by thinking of this as essentially a two-dimensional problem. For any two points $P_1(x_1, y_1, z_1)$ and $P_2(x_2, y_2, z_2)$ in $\mathbb{R}^3$, first locate the point $P_3(x_2, y_2, z_1)$ and observe that the three points are the vertices of a right triangle, with the right angle at the point P_3 (see Figure 10.19). The Pythagorean Theorem then says that the distance between P_1 and P_2, denoted $d\{P_1, P_2\}$, satisfies

$$d\{P_1, P_2\}^2 = d\{P_1, P_3\}^2 + d\{P_2, P_3\}^2. \tag{2.1}$$

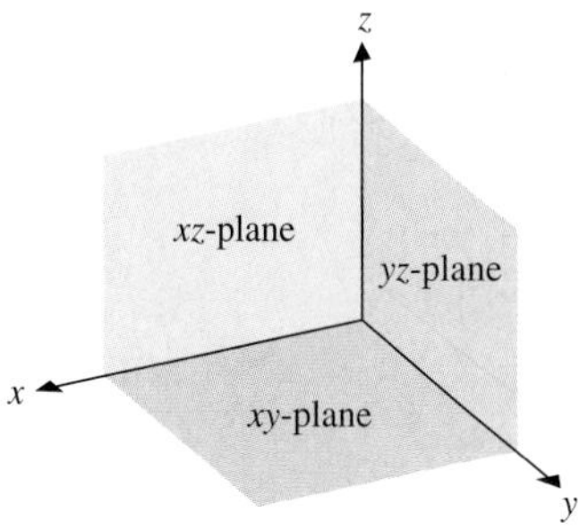

FIGURE 10.18
The coordinate planes

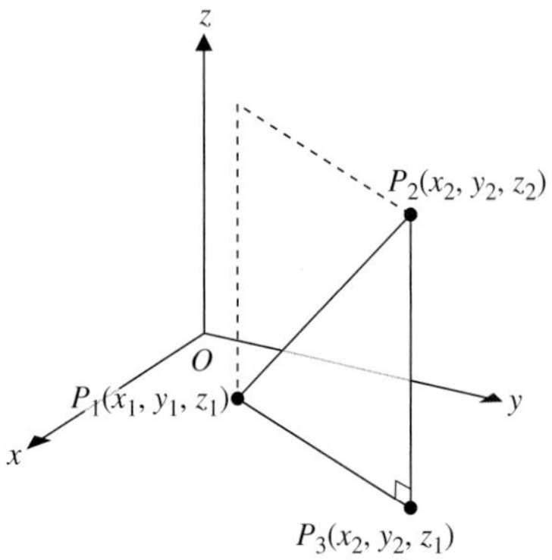

FIGURE 10.19
Distance in $\mathbb{R}^3$

Notice that P_2 lies directly above P_3 (or below, if $z_2 < z_1$), so that

$$d\{P_2, P_3\} = d\{(x_2, y_2, z_2), (x_2, y_2, z_1)\} = |z_2 - z_1|.$$

Since P_1 and P_3 both lie in the plane $z = z_1$, we can ignore the third coordinates of these points (since they're the same!) and use the usual two-dimensional distance formula:

$$d\{P_1, P_3\} = d\{(x_1, y_1, z_1), (x_2, y_2, z_1)\} = \sqrt{(x_2 - x_1)^2 + (y_2 - y_1)^2}.$$

From (2.1), we now have

$$\begin{aligned} d\{P_1, P_2\}^2 &= d\{P_1, P_3\}^2 + d\{P_2, P_3\}^2 \\ &= \left[\sqrt{(x_2 - x_1)^2 + (y_2 - y_1)^2}\right]^2 + |z_2 - z_1|^2 \\ &= (x_2 - x_1)^2 + (y_2 - y_1)^2 + (z_2 - z_1)^2. \end{aligned}$$

Taking the square root of both sides gives us the **distance formula** for $\mathbb{R}^3$:

Distance in $\mathbb{R}^3$

$$\boxed{d\{(x_1, y_1, z_1), (x_2, y_2, z_2)\} = \sqrt{(x_2 - x_1)^2 + (y_2 - y_1)^2 + (z_2 - z_1)^2},} \tag{2.2}$$

which is a straightforward generalization of the familiar formula for the distance between two points in the plane.

EXAMPLE 2.2 Computing Distance in $\mathbb{R}^3$

Find the distance between the points (1, −3, 5) and (5, 2, −3).

Solution From (2.2), we have

$$\begin{aligned} d\{(1, -3, 5), (5, 2, -3)\} &= \sqrt{(5 - 1)^2 + [2 - (-3)]^2 + (-3 - 5)^2} \\ &= \sqrt{4^2 + 5^2 + (-8)^2} = \sqrt{105}. \ \blacksquare \end{aligned}$$

○ Vectors in $\mathbb{R}^3$

As in two dimensions, vectors in three-dimensional space have both direction and magnitude. We again visualize vectors as directed line segments joining two points. A vector **v** is represented by any directed line segment with the appropriate magnitude and direction. The position vector **a** with terminal point at $A(a_1, a_2, a_3)$ (and initial point at the origin) is denoted by $\langle a_1, a_2, a_3\rangle$ and is shown in Figure 10.20a.

We denote the set of all three-dimensional position vectors by

$$V_3 = \{\langle x, y, z\rangle \mid x, y, z \in \mathbb{R}\}.$$

The **magnitude** of the position vector $\mathbf{a} = \langle a_1, a_2, a_3\rangle$ follows directly from the distance formula (2.2). We have

Magnitude of a vector

$$\|\mathbf{a}\| = \|\langle a_1, a_2, a_3\rangle\| = \sqrt{a_1^2 + a_2^2 + a_3^2}. \tag{2.3}$$

Note from Figure 10.20b that the vector with initial point at $P(a_1, a_2, a_3)$ and terminal point at $Q(b_1, b_2, b_3)$ corresponds to the position vector

$$\overrightarrow{PQ} = \langle b_1 - a_1, b_2 - a_2, b_3 - a_3\rangle.$$

We define vector addition in V_3 just as we did in V_2, by drawing a parallelogram, as in Figure 10.20c.

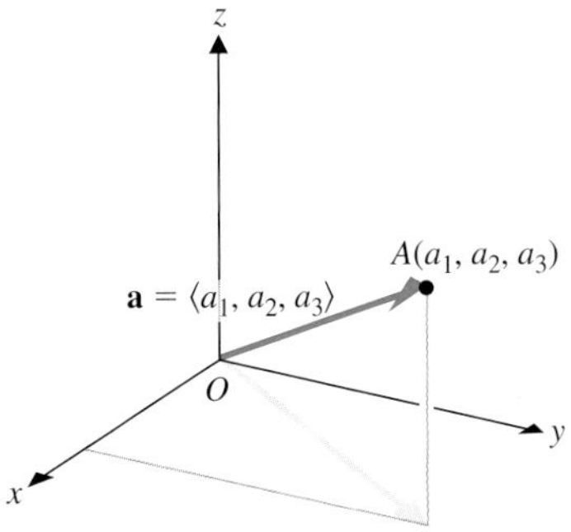

FIGURE 10.20a
Position vector in $\mathbb{R}^3$

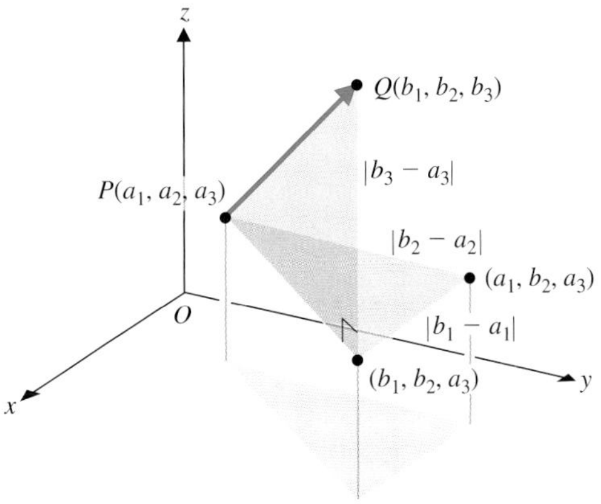

FIGURE 10.20b
Vector from P to Q

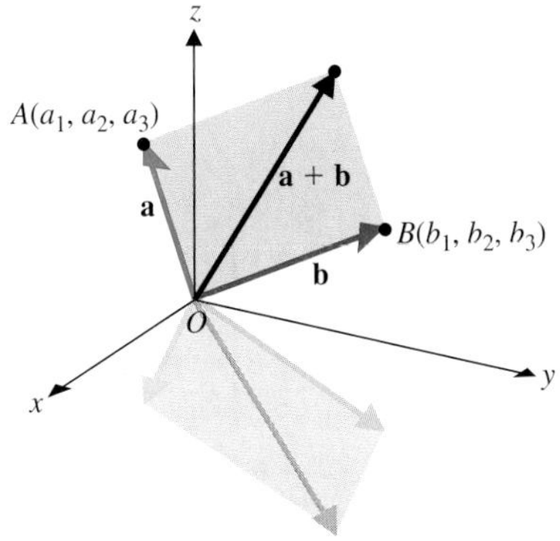

FIGURE 10.20c
Vector addition

Notice that for vectors $\mathbf{a} = \langle a_1, a_2, a_3\rangle$ and $\mathbf{b} = \langle b_1, b_2, b_3\rangle$, we have

Vector addition

$$\mathbf{a} + \mathbf{b} = \langle a_1, a_2, a_3\rangle + \langle b_1, b_2, b_3\rangle = \langle a_1 + b_1, a_2 + b_2, a_3 + b_3\rangle.$$

That is, as in V_2, addition of vectors in V_3 is done componentwise. Similarly, subtraction is done componentwise:

Vector subtraction

$$\mathbf{a} - \mathbf{b} = \langle a_1, a_2, a_3\rangle - \langle b_1, b_2, b_3\rangle = \langle a_1 - b_1, a_2 - b_2, a_3 - b_3\rangle.$$

Again as in V_2, for any scalar $c \in \mathbb{R}$, $c\mathbf{a}$ is a vector in the same direction as $\mathbf{a}$ when $c > 0$ and the opposite direction as $\mathbf{a}$ when $c < 0$. We have

Scalar multiplication

$$c\mathbf{a} = c\langle a_1, a_2, a_3\rangle = \langle ca_1, ca_2, ca_3\rangle.$$

Further, it's easy to show using (2.3), that

$$\|c\mathbf{a}\| = |c|\,\|\mathbf{a}\|.$$

We define the **zero vector 0** to be the vector in V_3 of length 0:

$$\mathbf{0} = \langle 0, 0, 0\rangle.$$

As in two dimensions, the zero vector has no particular direction. As we did in V_2, we define the **additive inverse** of a vector $\mathbf{a} \in V_3$ to be

$$-\mathbf{a} = -\langle a_1, a_2, a_3\rangle = \langle -a_1, -a_2, -a_3\rangle.$$

The rules of algebra established for vectors in V_2 hold verbatim in V_3, as seen in Theorem 2.1.

THEOREM 2.1

For any vectors $\mathbf{a}$, $\mathbf{b}$ and $\mathbf{c}$ in V_3, and any scalars d and e in $\mathbb{R}$, the following hold:

(i)	$\mathbf{a} + \mathbf{b} = \mathbf{b} + \mathbf{a}$	(commutativity)
(ii)	$\mathbf{a} + (\mathbf{b} + \mathbf{c}) = (\mathbf{a} + \mathbf{b}) + \mathbf{c}$	(associativity)
(iii)	$\mathbf{a} + \mathbf{0} = \mathbf{a}$	(zero vector)
(iv)	$\mathbf{a} + (-\mathbf{a}) = \mathbf{0}$	(additive inverse)
(v)	$d(\mathbf{a} + \mathbf{b}) = d\mathbf{a} + d\mathbf{b}$	(distributive law)
(vi)	$(d + e)\,\mathbf{a} = d\mathbf{a} + e\mathbf{a}$	(distributive law)
(vii)	$(1)\,\mathbf{a} = \mathbf{a}$	(multiplication by 1) and
(viii)	$(0)\,\mathbf{a} = \mathbf{0}$	(multiplication by 0).

We leave the proof of Theorem 2.1 as an exercise.

Since V_3 is three-dimensional, the standard basis consists of three unit vectors, each lying along one of the three coordinate axes. We define these as a straightforward generalization of the standard basis for V_2 by

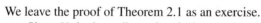

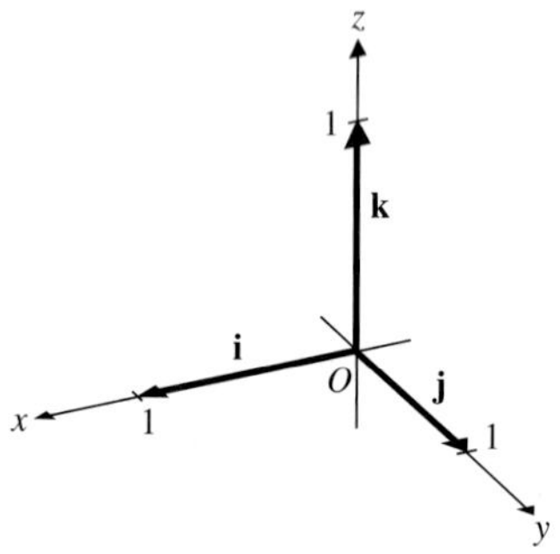

FIGURE 10.21
Standard basis for V_3

$$\mathbf{i} = \langle 1, 0, 0\rangle, \quad \mathbf{j} = \langle 0, 1, 0\rangle \quad \text{and} \quad \mathbf{k} = \langle 0, 0, 1\rangle,$$

as pictured in Figure 10.21. As in V_2, these basis vectors are unit vectors, since $\|\mathbf{i}\| = \|\mathbf{j}\| = \|\mathbf{k}\| = 1$. Also as in V_2, it is sometimes convenient to write position vectors in V_3 in terms of the standard basis. This is easily accomplished, as for any $\mathbf{a} \in V_3$, we can write

$$\mathbf{a} = \langle a_1, a_2, a_3\rangle = a_1\mathbf{i} + a_2\mathbf{j} + a_3\mathbf{k}.$$

If you're getting that déjà vu feeling that you've done all of this before, you're not imagining it. Vectors in V_3 follow all of the same rules as vectors in V_2. As a final note, observe that for any $\mathbf{a} = \langle a_1, a_2, a_3 \rangle \neq \mathbf{0}$, a unit vector in the same direction as $\mathbf{a}$ is given by

Unit vector

$$\mathbf{u} = \frac{1}{\|\mathbf{a}\|}\mathbf{a}. \tag{2.4}$$

The proof of this result is identical to the proof of the corresponding result for vectors in V_2, found in Theorem 1.2. Once again, it is often convenient to normalize a vector (i.e., produce a vector in the same direction, but with length 1).

EXAMPLE 2.3 Finding a Unit Vector

Find a unit vector in the same direction as $\langle 1, -2, 3 \rangle$ and write $\langle 1, -2, 3 \rangle$ as the product of its magnitude and a unit vector.

Solution First, we find the magnitude of the vector:

$$\|\langle 1, -2, 3 \rangle\| = \sqrt{1^2 + (-2)^2 + 3^2} = \sqrt{14}.$$

From (2.4), we have that a unit vector having the same direction as $\langle 1, -2, 3 \rangle$ is given by

$$\mathbf{u} = \frac{1}{\sqrt{14}}\langle 1, -2, 3 \rangle = \left\langle \frac{1}{\sqrt{14}}, \frac{-2}{\sqrt{14}}, \frac{3}{\sqrt{14}} \right\rangle.$$

Further,
$$\langle 1, -2, 3 \rangle = \sqrt{14}\left\langle \frac{1}{\sqrt{14}}, \frac{-2}{\sqrt{14}}, \frac{3}{\sqrt{14}} \right\rangle.$$

Of course, going from two dimensions to three dimensions gives us a much richer geometry, with more interesting examples. For instance, we define a **sphere** to be the set of all points whose distance from a fixed point (the **center**) is constant.

EXAMPLE 2.4 Finding the Equation of a Sphere

Find the equation of the sphere of radius r centered at the point (a, b, c).

Solution The sphere consists of all points (x, y, z) whose distance from (a, b, c) is r, as illustrated in Figure 10.22. This says that

$$\sqrt{(x-a)^2 + (y-b)^2 + (z-c)^2} = d\{(x, y, z), (a, b, c)\} = r.$$

Squaring both sides gives us

$$(x-a)^2 + (y-b)^2 + (z-c)^2 = r^2,$$

the standard form of the **equation of a sphere.**

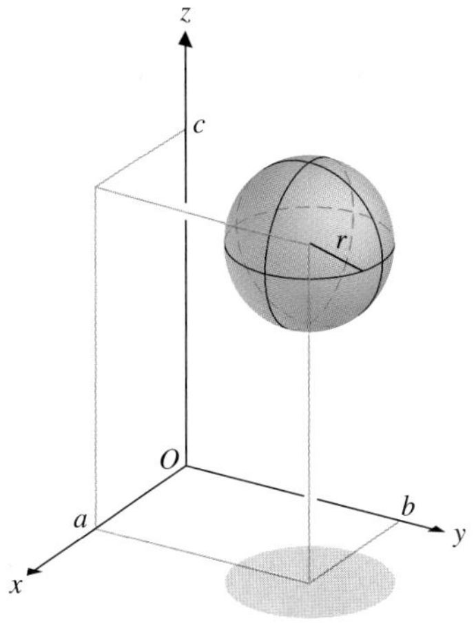

FIGURE 10.22
Sphere of radius r centered at (a, b, c)

You will occasionally need to recognize when a given equation represents a common geometric shape, as in example 2.5.

EXAMPLE 2.5 Finding the Center and Radius of a Sphere

Find the geometric shape described by the equation:

$$0 = x^2 + y^2 + z^2 - 4x + 8y - 10z + 36.$$

Solution Completing the squares in each variable, we have

$$\begin{aligned}0 &= (x^2 - 4x + 4) - 4 + (y^2 + 8y + 16) - 16 + (z^2 - 10z + 25) - 25 + 36\\ &= (x - 2)^2 + (y + 4)^2 + (z - 5)^2 - 9.\end{aligned}$$

Adding 9 to both sides gives us

$$3^2 = (x - 2)^2 + (y + 4)^2 + (z - 5)^2,$$

which is the equation of a sphere of radius 3 centered at the point (2, −4, 5). ■

EXERCISES 10.2

WRITING EXERCISES

1. Visualize the circle $x^2 + y^2 = 1$. With three-dimensional axes oriented as in Figure 10.15a, describe how to sketch this circle in the plane $z = 0$. Then, describe how to sketch the parabola $y = x^2$ in the plane $z = 0$. In general, explain how to translate a two-dimensional curve into a three-dimensional sketch.
2. It is difficult, if not impossible, for most people to visualize what points in four dimensions would look like. Nevertheless, it is easy to generalize the distance formula to four dimensions. Describe what the distance formula looks like in general dimension n, for $n \geq 4$.
3. It is very important to be able to quickly and accurately visualize three-dimensional relationships. In three dimensions, describe how many lines are perpendicular to the unit vector **i**. Describe all lines that are perpendicular to **i** and that pass through the origin. In three dimensions, describe how many planes are perpendicular to the unit vector **i**. Describe all planes that are perpendicular to **i** and that contain the origin.
4. In three dimensions, describe all planes that contain a given vector **a**. Describe all planes that contain two given vectors **a** and **b** (where **a** and **b** are not parallel). Describe all planes that contain a given vector **a** and pass through the origin. Describe all planes that contain two given (nonparallel) vectors **a** and **b** and pass through the origin.

In exercises 1 and 2, plot the indicated points.

1. (a) (2, 1, 5) (b) (3, 1, −2) (c) (−1, 2, −4)
2. (a) (−2, 1, 2) (b) (2, −3, −1) (c) (3, −2, 2)

In exercises 3 and 4, sketch the third axis to make xyz a right-handed system.

3. 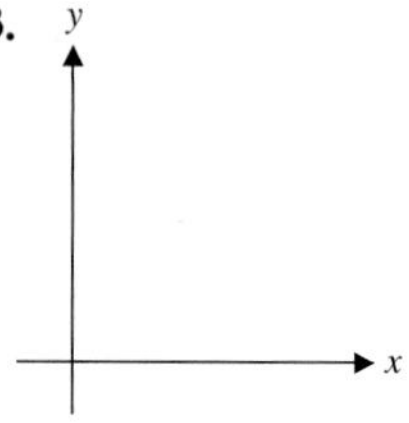

4. 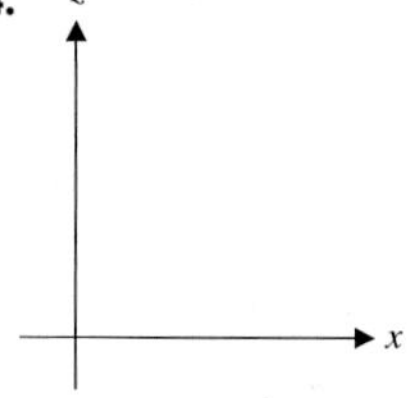

In exercises 5–8, find the distance between the given points.

5. (2, 1, 2), (5, 5, 2)
6. (1, 2, 0), (7, 10, 0)
7. (−1, 0, 2), (1, 2, 3)
8. (3, 1, 0), (1, 3, −4)

In exercises 9–12, compute a + b, a − 3b and ‖4a + 2b‖.

9. $\mathbf{a} = \langle 2, 1, -2 \rangle$, $\mathbf{b} = \langle 1, 3, 0 \rangle$
10. $\mathbf{a} = \langle -1, 0, 2 \rangle$, $\mathbf{b} = \langle 4, 3, 2 \rangle$
11. $\mathbf{a} = 3\mathbf{i} - \mathbf{j} + 4\mathbf{k}$, $\mathbf{b} = 5\mathbf{i} + \mathbf{j}$
12. $\mathbf{a} = \mathbf{i} - 4\mathbf{j} - 2\mathbf{k}$, $\mathbf{b} = \mathbf{i} - 3\mathbf{j} + 4\mathbf{k}$

In exercises 13–18, (a) find two unit vectors parallel to the given vector and (b) write the given vector as the product of its magnitude and a unit vector.

13. $\langle 3, 1, 2 \rangle$
14. $\langle 2, -4, 6 \rangle$
15. $2\mathbf{i} - \mathbf{j} + 2\mathbf{k}$
16. $4\mathbf{i} - 2\mathbf{j} + 4\mathbf{k}$
17. From (1, 2, 3) to (3, 2, 1)
18. From (1, 4, 1) to (3, 2, 2)

In exercises 19–22, find a vector with the given magnitude and in the same direction as the given vector.

19. Magnitude 6, $\mathbf{v} = \langle 2, 2, -1 \rangle$
20. Magnitude 10, $\mathbf{v} = \langle 3, 0, -4 \rangle$
21. Magnitude 4, $\mathbf{v} = 2\mathbf{i} - \mathbf{j} + 3\mathbf{k}$
22. Magnitude 3, $\mathbf{v} = 3\mathbf{i} + 3\mathbf{j} - \mathbf{k}$

In exercises 23–26, find an equation of the sphere with radius r and center (a, b, c).

23. $r = 2$, $(a, b, c) = (3, 1, 4)$
24. $r = 3$, $(a, b, c) = (2, 0, 1)$
25. $r = \sqrt{5}$, $(a, b, c) = (\pi, 1, -3)$
26. $r = \sqrt{7}$, $(a, b, c) = (1, 3, 4)$

In exercises 27–30, identify the geometric shape described by the given equation.

27. $(x-1)^2 + y^2 + (z+2)^2 = 4$

28. $x^2 + (y-1)^2 + (z-4)^2 = 2$

29. $x^2 - 2x + y^2 + z^2 - 4z = 0$

30. $x^2 + x + y^2 - y + z^2 = \frac{7}{2}$

In exercises 31–34, identify the plane as parallel to the xy-plane, xz-plane or yz-plane and sketch a graph.

31. $y = 4$

32. $x = -2$

33. $z = -1$

34. $z = 3$

In exercises 35–38, give an equation (e.g., $z = 0$) for the indicated figure.

35. xz-plane

36. xy-plane

37. yz-plane

38. x-axis

39. Prove the commutative property of Theorem 2.1.

40. Prove the associative property of Theorem 2.1.

41. Prove the distributive properties of Theorem 2.1.

42. Prove the multiplicative properties of Theorem 2.1.

43. Find the displacement vectors $\overrightarrow{PQ}$ and $\overrightarrow{QR}$ and determine whether the points $P = (2, 3, 1)$, $Q = (4, 2, 2)$ and $R = (8, 0, 4)$ are colinear (on the same line).

44. Find the displacement vectors $\overrightarrow{PQ}$ and $\overrightarrow{QR}$ and determine whether the points $P = (2, 3, 1)$, $Q = (0, 4, 2)$ and $R = (4, 1, 4)$ are colinear (on the same line).

45. Use vectors to determine whether the points $(0, 1, 1)$, $(2, 4, 2)$ and $(3, 1, 4)$ form an equilateral triangle.

46. Use vectors to determine whether the points $(2, 1, 0)$, $(4, 1, 2)$ and $(4, 3, 0)$ form an equilateral triangle.

47. Use vectors and the Pythagorean Theorem to determine whether the points $(3, 1, -2)$, $(1, 0, 1)$ and $(4, 2, -1)$ form a right triangle.

48. Use vectors and the Pythagorean Theorem to determine whether the points $(1, -2, 1)$, $(4, 3, 2)$ and $(7, 1, 3)$ form a right triangle.

49. Use vectors to determine whether the points $(2, 1, 0)$, $(5, -1, 2)$, $(0, 3, 3)$ and $(3, 1, 5)$ form a square.

50. Use vectors to determine whether the points $(1, -2, 1)$, $(-2, -1, 2)$, $(2, 0, 2)$ and $(-1, 1, 3)$ form a square.

51. In the accompanying figure, two ropes are attached to a 500-pound crate. Rope A exerts a force of $\langle 10, -130, 200\rangle$ pounds on the crate, and rope B exerts a force of $\langle -20, 180, 160\rangle$ pounds on the crate. If no further ropes are added, find the net force on the crate and the direction it will move. If a third rope C is added to balance the crate, what force must this rope exert on the crate?

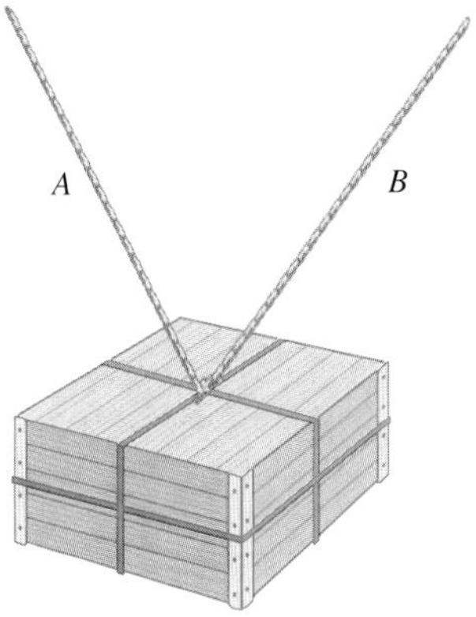

52. For the crate in exercise 51, suppose that the crate weighs only 300 pounds and the goal is to move the crate up and to the right with a constant force of $\langle 0, 30, 20\rangle$ pounds. If a third rope is added to accomplish this, what force must the rope exert on the crate?

53. The thrust of an airplane's engine produces a speed of 600 mph in still air. The plane is aimed in the direction of $\langle 2, 2, 1\rangle$ and the wind velocity is $\langle 10, -20, 0\rangle$ mph. Find the velocity vector of the plane with respect to the ground and find the speed.

54. The thrust of an airplane's engine produces a speed of 700 mph in still air. The plane is aimed in the direction of $\langle 6, -3, 2\rangle$ but its velocity with respect to the ground is $\langle 580, -330, 160\rangle$ mph. Find the wind velocity.

In exercises 55–62, you are asked to work with vectors of dimension higher than three. Use rules analogous to those introduced for two and three dimensions.

55. $\langle 2, 3, 1, 5\rangle + 2\langle 1, -2, 3, 1\rangle$

56. $2\langle 3, -2, 1, 0\rangle - \langle 2, 1, -2, 1\rangle$

57. $\langle 3, -2, 4, 1, 0, 2\rangle - 3\langle 1, 2, -2, 0, 3, 1\rangle$

58. $\langle 2, 1, 3, -2, 4, 1, 0, 2\rangle + 2\langle 3, 1, 1, 2, -2, 0, 3, 1\rangle$

59. $\|\mathbf{a}\|$ for $\mathbf{a} = \langle 3, 1, -2, 4, 1\rangle$

60. $\|\mathbf{a}\|$ for $\mathbf{a} = \langle 1, 0, -3, -2, 4, 1\rangle$

61. $\|\mathbf{a} + \mathbf{b}\|$ for $\mathbf{a} = \langle 1, -2, 4, 1\rangle$ and $\mathbf{b} = \langle -1, 4, 2, -4\rangle$

62. $\|\mathbf{a} - 2\mathbf{b}\|$ for $\mathbf{a} = \langle 2, 1, -2, 4, 1\rangle$ and $\mathbf{b} = \langle 3, -1, 4, 2, -4\rangle$

63. Take four unit circles and place them tangent to the *x*- and *y*-axes as shown. Find the radius of the inscribed circle (shown in the accompanying figure in red).

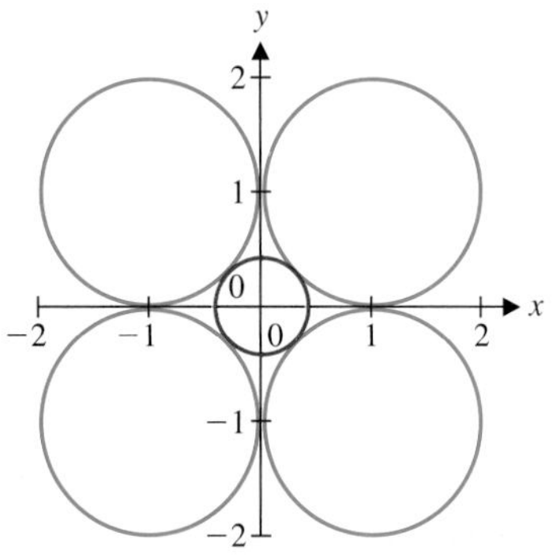

64. Extend exercise 63 to three dimensions by finding the radius of the sphere inscribed by eight unit spheres that are tangent to the coordinate planes.

65. Generalize the results of exercises 63 and 64 to *n* dimensions. Show that for $n \geq 10$, the inscribed hypersphere is actually

not contained in the "box" $-2 \leq x \leq 2$, $-2 \leq y \leq 2$ and so on, that contains all of the individual hyperspheres.

⊕ EXPLORATORY EXERCISES

1. Find an equation describing all points equidistant from $A = (0, 1, 0)$ and $B = (2, 4, 4)$ and sketch a graph. Based on your graph, describe the relationship between the displacement vector $\overrightarrow{AB} = \langle 2, 3, 4 \rangle$ and your graph. Simplify your equation for the three-dimensional surface until 2, 3 and 4 appear as coefficients of *x*, *y* and z. Use what you have learned to quickly write down an equation for the set of all points equidistant from $A = (0, 1, 0)$ and $C = (5, 2, 3)$.

2. In this exercise, you will try to identify the three-dimensional surface defined by the equation $a(x - 1) + b(y - 2) + c(z - 3) = 0$ for nonzero constants *a*, *b* and *c*. First, show that $(1, 2, 3)$ is one point on the surface. Then, show that any point that is equidistant from the points $(1 + a, 2 + b, 3 + c)$ and $(1 - a, 2 - b, 3 - c)$ is on the surface. Use this geometric fact to identify the surface.

10.3 THE DOT PRODUCT

In sections 10.1 and 10.2, we defined vectors in $\mathbb{R}^2$ and $\mathbb{R}^3$ and examined many of the properties of vectors, including how to add and subtract two vectors. It turns out that two different kinds of products involving vectors have proved to be useful: the dot product (or scalar product) and the cross product (or vector product). We introduce the first of these two products in this section.

DEFINITION 3.1

Scalar → the answer is a real number

The **dot product** of two vectors $\mathbf{a} = \langle a_1, a_2, a_3 \rangle$ and $\mathbf{b} = \langle b_1, b_2, b_3 \rangle$ in V_3 is defined by

$$\mathbf{a} \cdot \mathbf{b} = \langle a_1, a_2, a_3 \rangle \cdot \langle b_1, b_2, b_3 \rangle = a_1 b_1 + a_2 b_2 + a_3 b_3. \tag{3.1}$$

Likewise, the dot product of two vectors in V_2 is defined by

$$\mathbf{a} \cdot \mathbf{b} = \langle a_1, a_2 \rangle \cdot \langle b_1, b_2 \rangle = a_1 b_1 + a_2 b_2.$$

Be sure to notice that the dot product of two vectors is a *scalar* (i.e., a number, not a vector). For this reason, the dot product is also called the **scalar product.**

EXAMPLE 3.1   Computing a Dot Product in $\mathbb{R}^3$

Compute the dot product $\mathbf{a} \cdot \mathbf{b}$ for $\mathbf{a} = \langle 1, 2, 3 \rangle$ and $\mathbf{b} = \langle 5, -3, 4 \rangle$.

Solution We have

$$\mathbf{a} \cdot \mathbf{b} = \langle 1, 2, 3 \rangle \cdot \langle 5, -3, 4 \rangle = (1)(5) + (2)(-3) + (3)(4) = 11. \ \blacksquare$$

Certainly, dot products are very simple to compute, whether a vector is written in component form or written in terms of the standard basis vectors, as in example 3.2.

EXAMPLE 3.2 Computing a Dot Product in $\mathbb{R}^2$

Find the dot product of the two vectors $\mathbf{a} = 2\mathbf{i} - 5\mathbf{j}$ and $\mathbf{b} = 3\mathbf{i} + 6\mathbf{j}$.

Solution We have

$$\mathbf{a} \cdot \mathbf{b} = (2)(3) + (-5)(6) = 6 - 30 = -24. \blacksquare$$

The dot product in V_2 or V_3 satisfies the following simple properties.

REMARK 3.1

Since vectors in V_2 can be thought of as special cases of vectors in V_3 (where the third component is zero), all of the results we prove for vectors in V_3 hold equally for vectors in V_2.

THEOREM 3.1

For vectors **a**, **b** and **c** and any scalar d, the following hold:

(i) $\mathbf{a} \cdot \mathbf{b} = \mathbf{b} \cdot \mathbf{a}$ (commutativity)
(ii) $\mathbf{a} \cdot (\mathbf{b} + \mathbf{c}) = \mathbf{a} \cdot \mathbf{b} + \mathbf{a} \cdot \mathbf{c}$ (distributive law)
(iii) $(d\mathbf{a}) \cdot \mathbf{b} = d(\mathbf{a} \cdot \mathbf{b}) = \mathbf{a} \cdot (d\mathbf{b})$
(iv) $\mathbf{0} \cdot \mathbf{a} = 0$ and
(v) $\mathbf{a} \cdot \mathbf{a} = \|\mathbf{a}\|^2$.

PROOF

We prove (i) and (v) for $\mathbf{a}, \mathbf{b} \in V_3$. The remaining parts are left as exercises.

(i) For $\mathbf{a} = \langle a_1, a_2, a_3 \rangle$ and $\mathbf{b} = \langle b_1, b_2, b_3 \rangle$, we have from (3.1) that

$$\begin{aligned}\mathbf{a} \cdot \mathbf{b} &= \langle a_1, a_2, a_3 \rangle \cdot \langle b_1, b_2, b_3 \rangle = a_1 b_1 + a_2 b_2 + a_3 b_3 \\ &= b_1 a_1 + b_2 a_2 + b_3 a_3 = \mathbf{b} \cdot \mathbf{a},\end{aligned}$$

since multiplication of real numbers is commutative.

(v) For $\mathbf{a} = \langle a_1, a_2, a_3 \rangle$, we have

$$\mathbf{a} \cdot \mathbf{a} = \langle a_1, a_2, a_3 \rangle \cdot \langle a_1, a_2, a_3 \rangle = a_1^2 + a_2^2 + a_3^2 = \|\mathbf{a}\|^2. \blacksquare$$

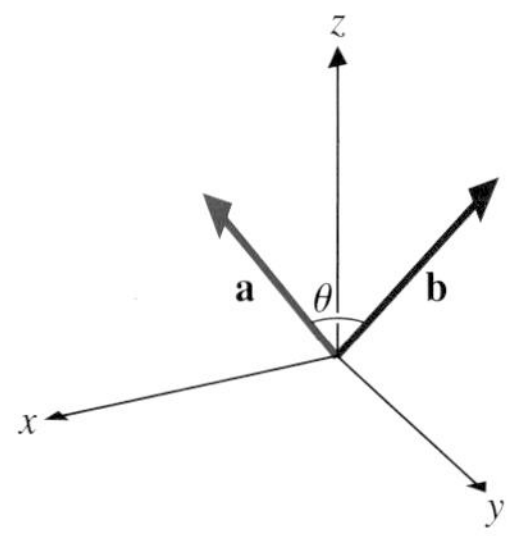

FIGURE 10.23a
The angle between two vectors

Notice that properties (i)–(iv) of Theorem 3.1 are also properties of multiplication of real numbers. This is why we use the word *product* in dot product. However, there are some properties of multiplication of real numbers not shared by the dot product. For instance, we will see that $\mathbf{a} \cdot \mathbf{b} = 0$ does not imply that either $\mathbf{a} = \mathbf{0}$ or $\mathbf{b} = \mathbf{0}$.

For two *nonzero* vectors **a** and **b** in V_3, we define the **angle** θ $(0 \le \theta \le \pi)$ **between the vectors** to be the smaller angle between **a** and **b**, formed by placing their initial points at the same point, as illustrated in Figure 10.23a.

Notice that if **a** and **b** have the *same* direction, then $\theta = 0$. If **a** and **b** have *opposite* directions, then $\theta = \pi$. We say that **a** and **b** are **orthogonal** (or **perpendicular**) if $\theta = \frac{\pi}{2}$. We consider the zero vector **0** to be orthogonal to every vector. The general case is stated in Theorem 3.2.

THEOREM 3.2

Let θ be the angle between nonzero vectors **a** and **b**. Then,

$$\mathbf{a} \cdot \mathbf{b} = \|\mathbf{a}\| \|\mathbf{b}\| \cos \theta. \qquad (3.2)$$

PROOF

We must prove the theorem for three separate cases.

(i) If **a** and **b** have the *same direction,* then $\mathbf{b} = c\mathbf{a}$, for some scalar $c > 0$ and the angle between **a** and **b** is $\theta = 0$. This says that

$$\mathbf{a} \cdot \mathbf{b} = \mathbf{a} \cdot (c\mathbf{a}) = c\mathbf{a} \cdot \mathbf{a} = c\|\mathbf{a}\|^2.$$

Further,

$$\|\mathbf{a}\|\,\|\mathbf{b}\| \cos\theta = \|\mathbf{a}\|\,|c|\,\|\mathbf{a}\| \cos 0 = c\|\mathbf{a}\|^2 = \mathbf{a} \cdot \mathbf{b},$$

since for $c > 0$, we have $|c| = c$.

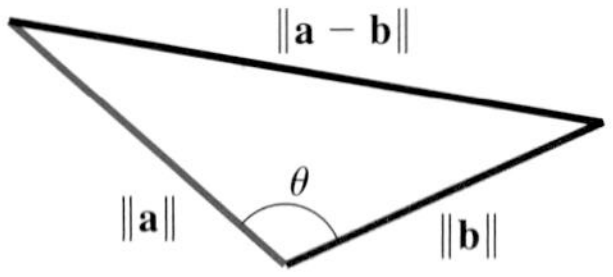

FIGURE 10.23b
The angle between two vectors

(ii) If **a** and **b** have the opposite direction, the proof is nearly identical to case (i) above and we leave the details as an exercise.

(iii) If **a** and **b** are not parallel, then we have that $0 < \theta < \pi$, as shown in Figure 10.23b. Recall that the Law of Cosines allows us to relate the lengths of the sides of triangles like the one in Figure 10.23b. We have

$$\|\mathbf{a} - \mathbf{b}\|^2 = \|\mathbf{a}\|^2 + \|\mathbf{b}\|^2 - 2\|\mathbf{a}\|\,\|\mathbf{b}\| \cos\theta. \qquad (3.3)$$

Now, observe that

$$\begin{aligned}
\|\mathbf{a} - \mathbf{b}\|^2 &= \|\langle a_1 - b_1, a_2 - b_2, a_3 - b_3\rangle\|^2 \\
&= (a_1 - b_1)^2 + (a_2 - b_2)^2 + (a_3 - b_3)^2 \\
&= \left(a_1^2 - 2a_1b_1 + b_1^2\right) + \left(a_2^2 - 2a_2b_2 + b_2^2\right) + \left(a_3^2 - 2a_3b_3 + b_3^2\right) \\
&= \left(a_1^2 + a_2^2 + a_3^2\right) + \left(b_1^2 + b_2^2 + b_3^2\right) - 2(a_1b_1 + a_2b_2 + a_3b_3) \\
&= \|\mathbf{a}\|^2 + \|\mathbf{b}\|^2 - 2\mathbf{a} \cdot \mathbf{b} \qquad (3.4)
\end{aligned}$$

Equating the right-hand sides of (3.3) and (3.4), we get (3.2), as desired. ■

We can use (3.2) to find the angle between two vectors, as in example 3.3.

EXAMPLE 3.3 Finding the Angle between Two Vectors

Find the angle between the vectors $\mathbf{a} = \langle 2, 1, -3\rangle$ and $\mathbf{b} = \langle 1, 5, 6\rangle$.

Solution From (3.2), we have

$$\cos\theta = \frac{\mathbf{a} \cdot \mathbf{b}}{\|\mathbf{a}\|\,\|\mathbf{b}\|} = \frac{-11}{\sqrt{14}\sqrt{62}}.$$

It follows that $\theta = \cos^{-1}\left(\frac{-11}{\sqrt{14}\sqrt{62}}\right) \approx 1.953$ (radians)

(or about 112°), since $0 \le \theta \le \pi$ and the inverse cosine function returns an angle in this range. ■

The following result is an immediate and important consequence of Theorem 3.2.

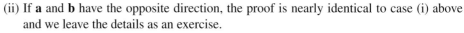

COROLLARY 3.1

Two vectors **a** and **b** are orthogonal if and only if $\mathbf{a} \cdot \mathbf{b} = 0$.

PROOF

First, observe that if either **a** or **b** is the zero vector, then $\mathbf{a} \cdot \mathbf{b} = 0$ and **a** and **b** are orthogonal, as the zero vector is considered orthogonal to every vector. If **a** and **b** are nonzero vectors and if θ is the angle between **a** and **b**, we have from Theorem 3.2 that

$$\|\mathbf{a}\|\|\mathbf{b}\| \cos\theta = \mathbf{a} \cdot \mathbf{b} = 0$$

if and only if $\cos\theta = 0$ (since neither **a** nor **b** is the zero vector). This occurs if and only if $\theta = \frac{\pi}{2}$, which is equivalent to having **a** and **b** orthogonal and so, the result follows. ■

EXAMPLE 3.4 Determining Whether Two Vectors Are Orthogonal

Determine whether the following pairs of vectors are orthogonal: (a) $\mathbf{a} = \langle 1, 3, -5\rangle$ and $\mathbf{b} = \langle 2, 3, 10\rangle$ and (b) $\mathbf{a} = \langle 4, 2, -1\rangle$ and $\mathbf{b} = \langle 2, 3, 14\rangle$.

Solution For (a), we have:

$$\mathbf{a} \cdot \mathbf{b} = 2 + 9 - 50 = -39 \neq 0,$$

so that **a** and **b** are *not* orthogonal.

For (b), we have

$$\mathbf{a} \cdot \mathbf{b} = 8 + 6 - 14 = 0,$$

so that **a** and **b** are orthogonal, in this case. ■

The following two results provide us with some powerful tools for comparing the magnitudes of vectors.

THEOREM 3.3 (Cauchy-Schwartz Inequality)

For any vectors **a** and **b**,

$$|\mathbf{a} \cdot \mathbf{b}| \leq \|\mathbf{a}\|\|\mathbf{b}\|. \tag{3.5}$$

PROOF

If either **a** or **b** is the zero vector, notice that (3.5) simply says that $0 \leq 0$, which is certainly true. On the other hand, if neither **a** nor **b** is the zero vector, we have from (3.2) that

$$|\mathbf{a} \cdot \mathbf{b}| = \|\mathbf{a}\|\|\mathbf{b}\||\cos\theta| \leq \|\mathbf{a}\|\|\mathbf{b}\|,$$

since $|\cos\theta| \leq 1$ for all values of θ. ■

One benefit of the Cauchy-Schwartz Inequality is that it allows us to prove the following very useful result. If you were going to learn only one inequality in your lifetime, this is probably the one you would want to learn.

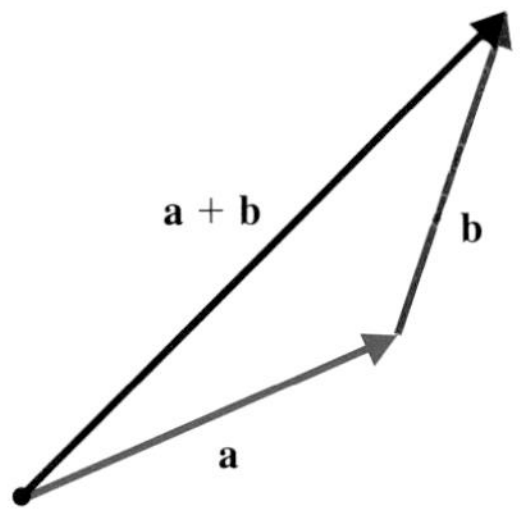

FIGURE 10.24
The Triangle Inequality

THEOREM 3.4 (The Triangle Inequality)

For any vectors **a** and **b**,

$$\|\mathbf{a}+\mathbf{b}\| \le \|\mathbf{a}\| + \|\mathbf{b}\|. \tag{3.6}$$

Before we prove the theorem, consider the triangle formed by the vectors **a**, **b** and **a** + **b**, shown in Figure 10.24. Notice that the Triangle Inequality says that the length of the vector **a** + **b** never exceeds the sum of the individual lengths of **a** and **b**.

PROOF

From Theorem 3.1 (i), (ii) and (v), we have

$$\begin{aligned}\|\mathbf{a}+\mathbf{b}\|^2 &= (\mathbf{a}+\mathbf{b})\cdot(\mathbf{a}+\mathbf{b}) = \mathbf{a}\cdot\mathbf{a}+\mathbf{a}\cdot\mathbf{b}+\mathbf{b}\cdot\mathbf{a}+\mathbf{b}\cdot\mathbf{b}\\ &= \|\mathbf{a}\|^2 + 2\mathbf{a}\cdot\mathbf{b} + \|\mathbf{b}\|^2.\end{aligned}$$

From the Cauchy-Schwartz Inequality (3.5), we have $\mathbf{a}\cdot\mathbf{b} \le |\mathbf{a}\cdot\mathbf{b}| \le \|\mathbf{a}\|\|\mathbf{b}\|$ and so, we have

$$\begin{aligned}\|\mathbf{a}+\mathbf{b}\|^2 &= \|\mathbf{a}\|^2 + 2\mathbf{a}\cdot\mathbf{b} + \|\mathbf{b}\|^2\\ &\le \|\mathbf{a}\|^2 + 2\|\mathbf{a}\|\|\mathbf{b}\| + \|\mathbf{b}\|^2 = (\|\mathbf{a}\| + \|\mathbf{b}\|)^2.\end{aligned}$$

Taking square roots gives us (3.6). ■

TODAY IN MATHEMATICS

Lene Hau (1959–)
A Danish mathematician and physicist known for her experiments to slow down and stop light. Although neither of her parents had a background in science or mathematics, she says that as a student, "I loved mathematics. I would rather do mathematics than go to the movies in those days. But after awhile, I discovered quantum mechanics and I've been hooked ever since." Hau credits a culture of scientific achievement with her success. "I was lucky to be a Dane. Denmark has a long scientific tradition that included the great Niels Bohr. . . . In Denmark, physics is widely respected by laymen as well as scientists and laymen contribute to physics."

○ Components and Projections

Think about the case where a vector represents a force. Often, it's impractical to exert a force in the direction you'd like. For instance, in pulling a child's wagon, we exert a force in the direction determined by the position of the handle, instead of in the direction of motion. (See Figure 10.25.) An important question is whether there is a force of smaller magnitude that can be exerted in a different direction and still produce the same effect on the wagon. Notice that it is the horizontal portion of the force that most directly contributes to the motion of the wagon. (The vertical portion of the force only acts to reduce friction.) We now consider how to compute such a component of a force.

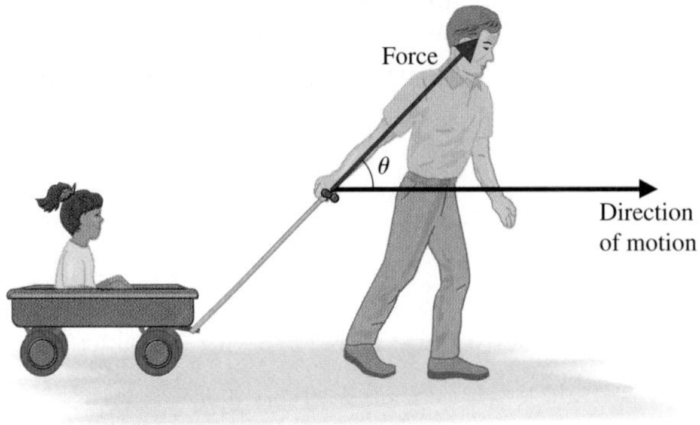

FIGURE 10.25
Pulling a wagon

For any two nonzero position vectors **a** and **b**, let the angle between the vectors be θ. If we drop a perpendicular line segment from the terminal point of **a** to the line containing the vector **b**, then from elementary trigonometry, the base of the triangle (in the case where $0 < \theta < \frac{\pi}{2}$) has length given by $\|\mathbf{a}\| \cos\theta$ (see Figure 10.26a).

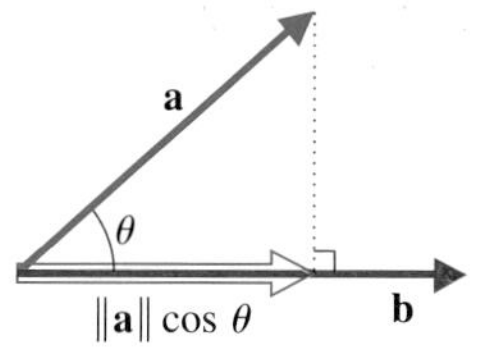

FIGURE 10.26a
$\text{comp}_{\mathbf{b}}\, \mathbf{a}$, for $0 < \theta < \frac{\pi}{2}$

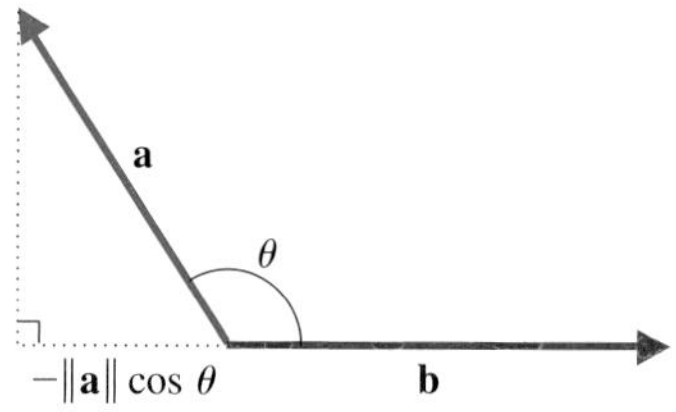

FIGURE 10.26b
$\text{comp}_{\mathbf{b}}\, \mathbf{a}$, for $\frac{\pi}{2} < \theta < \pi$

On the other hand, notice that if $\frac{\pi}{2} < \theta < \pi$, the length of the base is given by $-\|\mathbf{a}\| \cos\theta$ (see Figure 10.26b). In either case, we refer to $\|\mathbf{a}\| \cos\theta$ as the **component** of **a** along **b**, denoted $\text{comp}_{\mathbf{b}}\, \mathbf{a}$. Using (3.2), observe that we can rewrite this as

$$\begin{aligned} \text{comp}_{\mathbf{b}}\, \mathbf{a} &= \|\mathbf{a}\| \cos\theta = \frac{\|\mathbf{a}\|\,\|\mathbf{b}\|}{\|\mathbf{b}\|} \cos\theta \\ &= \frac{1}{\|\mathbf{b}\|} \|\mathbf{a}\|\,\|\mathbf{b}\| \cos\theta = \frac{1}{\|\mathbf{b}\|} \mathbf{a} \cdot \mathbf{b} \end{aligned}$$

Component of **a** along **b**

or

$$\boxed{\text{comp}_{\mathbf{b}}\, \mathbf{a} = \frac{\mathbf{a} \cdot \mathbf{b}}{\|\mathbf{b}\|}.} \tag{3.7}$$

Notice that $\text{comp}_{\mathbf{b}}\, \mathbf{a}$ is a scalar and that we divide the dot product in (3.7) by $\|\mathbf{b}\|$ and not by $\|\mathbf{a}\|$. One way to keep this straight is to recognize that the components in Figures 10.26a and 10.26b depend on how long **a** is but not on how long **b** is. We can view (3.7) as the dot product of the vector **a** and a unit vector in the direction of **b**, given by $\dfrac{\mathbf{b}}{\|\mathbf{b}\|}$.

Once again, consider the case where the vector **a** represents a force. Rather than the component of **a** along **b**, we are often interested in finding a force vector parallel to **b** having the same component along **b** as **a**. We call this vector the **projection** of **a** onto **b**, denoted $\textbf{proj}_{\mathbf{b}}\, \mathbf{a}$, as indicated in Figures 10.27a and 10.27b. Since the projection has magnitude $|\text{comp}_{\mathbf{b}}\, \mathbf{a}|$ and points in the direction of **b**, for $0 < \theta < \frac{\pi}{2}$ and opposite **b**, for $\frac{\pi}{2} < \theta < \pi$,

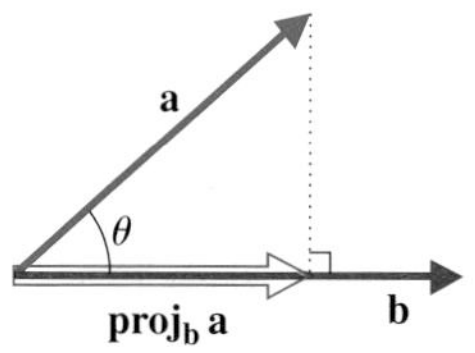

FIGURE 10.27a
$\textbf{proj}_{\mathbf{b}}\, \mathbf{a}$, for $0 < \theta < \frac{\pi}{2}$

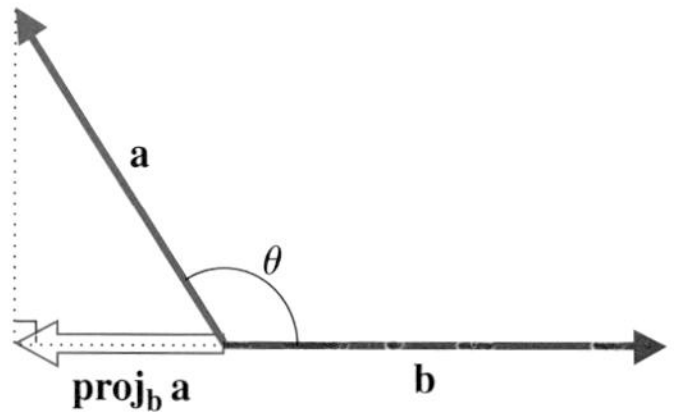

FIGURE 10.27b
$\textbf{proj}_{\mathbf{b}}\, \mathbf{a}$, for $\frac{\pi}{2} < \theta < \pi$

we have from (3.7) that

$$\mathbf{proj_b\, a} = (\text{comp}_\mathbf{b}\,\mathbf{a})\frac{\mathbf{b}}{\|\mathbf{b}\|} = \left(\frac{\mathbf{a}\cdot\mathbf{b}}{\|\mathbf{b}\|}\right)\frac{\mathbf{b}}{\|\mathbf{b}\|},$$

Projection of **a** onto **b**

or

$$\boxed{\mathbf{proj_b\, a} = \frac{\mathbf{a}\cdot\mathbf{b}}{\|\mathbf{b}\|^2}\mathbf{b},} \tag{3.8}$$

where $\frac{\mathbf{b}}{\|\mathbf{b}\|}$ represents a unit vector in the direction of **b**.

In example 3.5, we illustrate the process of finding components and projections.

CAUTION

Be careful to distinguish between the *projection* of **a** onto **b** (a vector) and the *component* of **a** along **b** (a scalar). It is very common to confuse the two.

EXAMPLE 3.5 Finding Components and Projections

For $\mathbf{a} = \langle 2, 3\rangle$ and $\mathbf{b} = \langle -1, 5\rangle$, find the component of **a** along **b** and the projection of **a** onto **b**.

Solution From (3.7), we have

$$\text{comp}_\mathbf{b}\,\mathbf{a} = \frac{\mathbf{a}\cdot\mathbf{b}}{\|\mathbf{b}\|} = \frac{\langle 2,3\rangle\cdot\langle -1,5\rangle}{\|\langle -1,5\rangle\|} = \frac{-2+15}{\sqrt{1+5^2}} = \frac{13}{\sqrt{26}}.$$

Similarly, from (3.8), we have

$$\mathbf{proj_b\, a} = \left(\frac{\mathbf{a}\cdot\mathbf{b}}{\|\mathbf{b}\|}\right)\frac{\mathbf{b}}{\|\mathbf{b}\|} = \left(\frac{13}{\sqrt{26}}\right)\frac{\langle -1,5\rangle}{\sqrt{26}}$$
$$= \frac{13}{26}\langle -1,5\rangle = \frac{1}{2}\langle -1,5\rangle = \left\langle -\frac{1}{2}, \frac{5}{2}\right\rangle.$$

We leave it as an exercise to show that, in general, $\text{comp}_\mathbf{b}\,\mathbf{a} \neq \text{comp}_\mathbf{a}\,\mathbf{b}$ and $\mathbf{proj_b\, a} \neq \mathbf{proj_a b}$. One reason for needing to consider components of a vector in a given direction is to compute work, as we see in example 3.6.

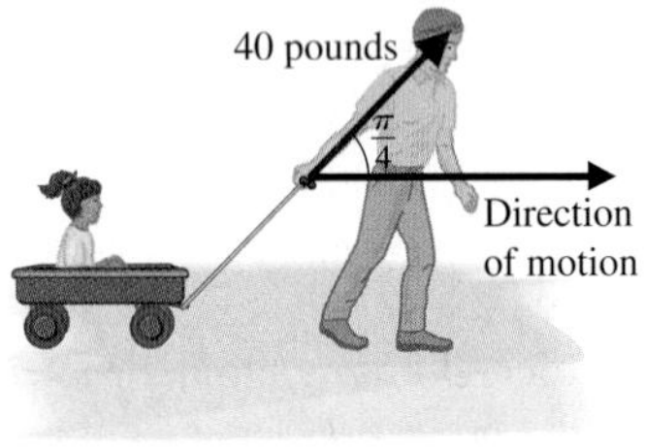

FIGURE 10.28
Pulling a wagon

EXAMPLE 3.6 Calculating Work

You exert a constant force of 40 pounds in the direction of the handle of the wagon pictured in Figure 10.28. If the handle makes an angle of $\frac{\pi}{4}$ with the horizontal and you pull the wagon along a flat surface for 1 mile (5280 feet), find the work done.

Solution First, recall from our discussion in Chapter 5 that if we apply a constant force F for a distance d, the work done is given by $W = Fd$. Unfortunately, the force exerted in the direction of motion is not given. Since the magnitude of the force is 40, the force vector must be

$$\mathbf{F} = 40\left\langle \cos\frac{\pi}{4}, \sin\frac{\pi}{4}\right\rangle = 40\left\langle \frac{\sqrt{2}}{2}, \frac{\sqrt{2}}{2}\right\rangle = \langle 20\sqrt{2}, 20\sqrt{2}\rangle.$$

The force exerted in the direction of motion is simply the component of the force along the vector **i** (that is, the horizontal component of **F**) or $20\sqrt{2}$. The work done is then

$$W = Fd = 20\sqrt{2}(5280) \approx 149{,}341 \text{ foot-pounds}.$$

More generally, if a constant force **F** moves an object from point P to point Q, we refer to the vector $\mathbf{d} = \overrightarrow{PQ}$ as the **displacement vector.** The work done is the product of the

component of **F** along **d** and the distance:

$$W = \text{comp}_{\mathbf{d}}\,\mathbf{F}\|\mathbf{d}\|$$
$$= \frac{\mathbf{F}\cdot\mathbf{d}}{\|\mathbf{d}\|}\cdot\|\mathbf{d}\| = \mathbf{F}\cdot\mathbf{d}.$$

Here, this gives us

$$W = \langle 20\sqrt{2}, 20\sqrt{2}\rangle \cdot \langle 5280, 0\rangle = 20\sqrt{2}(5280), \text{ as before.} \blacksquare$$

BEYOND FORMULAS

You can think of the dot product as a shortcut for computing components and projections. The dot product test for perpendicular vectors follows directly from this interpretation. In general, components and projections are used to isolate a particular portion of a large problem for detailed analysis. This sort of reductionism is central to much of modern science.

EXERCISES 10.3

WRITING EXERCISES

1. Explain in words why the Triangle Inequality is true.

2. The dot product is called a "product" because the properties listed in Theorem 3.1 are true for multiplication of real numbers. Two other properties of multiplication of real numbers involve factoring: (1) if $ab = ac\,(a \neq 0)$ then $b = c$ and (2) if $ab = 0$ then $a = 0$ or $b = 0$. Discuss the extent to which these properties are true for the dot product.

3. On several occasions you have been asked to find unit vectors. To understand the importance of unit vectors, first identify the simplification in formulas for finding the angle between vectors and for finding the component of a vector, if the vectors are unit vectors. There is also a theoretical benefit to using unit vectors. Compare the number of vectors in a particular direction to the number of unit vectors in that direction. (For this reason, unit vectors are sometimes called **direction vectors.**)

4. It is important to understand why work is computed using only the component of force in the direction of motion. To take a simple example, suppose you are pushing on a door to try to close it. If you are pushing on the edge of the door straight at the door hinges, are you accomplishing anything useful? In this case, the work done would be zero. If you change the angle at which you push very slightly, what happens? Discuss what happens as you change that angle more and more (up to 90°). As the angle increases, discuss how the component of force in the direction of motion changes and how the work done changes.

In exercises 1–6, compute a · b.

1. $\mathbf{a} = \langle 3, 1\rangle, \mathbf{b} = \langle 2, 4\rangle$
2. $\mathbf{a} = 3\mathbf{i} + \mathbf{j}, \mathbf{b} = -2\mathbf{i} + 3\mathbf{j}$
3. $\mathbf{a} = \langle 2, -1, 3\rangle, \mathbf{b} = \langle 0, 2, 4\rangle$
4. $\mathbf{a} = \langle 3, 2, 0\rangle, \mathbf{b} = \langle -2, 4, 3\rangle$
5. $\mathbf{a} = 2\mathbf{i} - \mathbf{k}, \mathbf{b} = 4\mathbf{j} - \mathbf{k}$
6. $\mathbf{a} = 3\mathbf{i} + 3\mathbf{k}, \mathbf{b} = -2\mathbf{i} + \mathbf{j}$

In exercises 7–10, compute the angle between the vectors.

7. $\mathbf{a} = 3\mathbf{i} - 2\mathbf{j}, \mathbf{b} = \mathbf{i} + \mathbf{j}$
8. $\mathbf{a} = \langle 2, 0, -2\rangle, \mathbf{b} = \langle 0, -2, 4\rangle$
9. $\mathbf{a} = 3\mathbf{i} + \mathbf{j} - 4\mathbf{k}, \mathbf{b} = -2\mathbf{i} + 2\mathbf{j} + \mathbf{k}$
10. $\mathbf{a} = \mathbf{i} + 3\mathbf{j} - 2\mathbf{k}, \mathbf{b} = 2\mathbf{i} - 3\mathbf{k}$

In exercises 11–14, determine whether the vectors are orthogonal.

11. $\mathbf{a} = \langle 2, -1\rangle, \mathbf{b} = \langle 2, 4\rangle$
12. $\mathbf{a} = \langle 4, -1, 1\rangle, \mathbf{b} = \langle 2, 4, 4\rangle$
13. $\mathbf{a} = 6\mathbf{i} + 2\mathbf{j}, \mathbf{b} = -\mathbf{i} + 3\mathbf{j}$
14. $\mathbf{a} = 3\mathbf{i}, \mathbf{b} = 6\mathbf{j} - 2\mathbf{k}$

In exercises 15–18, find a vector perpendicular to the given vector.

15. $\langle 2, -1\rangle$

16. $\langle 4, -1, 1\rangle$

17. $6\mathbf{i} + 2\mathbf{j} - \mathbf{k}$

18. $2\mathbf{i} - 3\mathbf{k}$

In exercises 19–24, find $\text{comp}_\mathbf{b}\,\mathbf{a}$ and $\text{proj}_\mathbf{b}\,\mathbf{a}$.

19. $\mathbf{a} = \langle 2, 1\rangle, \mathbf{b} = \langle 3, 4\rangle$

20. $\mathbf{a} = 3\mathbf{i} + \mathbf{j}, \mathbf{b} = 4\mathbf{i} - 3\mathbf{j}$

21. $\mathbf{a} = \langle 2, -1, 3\rangle, \mathbf{b} = \langle 1, 2, 2\rangle$

22. $\mathbf{a} = \langle 1, 4, 5\rangle, \mathbf{b} = \langle -2, 1, 2\rangle$

23. $\mathbf{a} = \langle 2, 0, -2\rangle, \mathbf{b} = \langle 0, -3, 4\rangle$

24. $\mathbf{a} = \langle 3, 2, 0\rangle, \mathbf{b} = \langle -2, 2, 1\rangle$

25. Repeat example 3.6 with an angle of $\frac{\pi}{3}$ with the horizontal.

26. Repeat example 3.6 with an angle of $\frac{\pi}{6}$ with the horizontal.

27. Explain why the answers to exercises 25 and 26 aren't the same, even though the force exerted is the same. In this setting, explain why a larger amount of work corresponds to a more efficient use of the force.

28. Find the force needed in exercise 25 to produce the same amount of work as in example 3.6.

29. A constant force of $\langle 30, 20\rangle$ pounds moves an object in a straight line from the point (0, 0) to the point (24, 10). Compute the work done.

30. A constant force of $\langle 60, -30\rangle$ pounds moves an object in a straight line from the point (0, 0) to the point (10, −10). Compute the work done.

31. Label each statement as true or false. If it is true, briefly explain why; if it is false, give a counterexample.

(a) If $\mathbf{a}\cdot\mathbf{b} = \mathbf{a}\cdot\mathbf{c}$, then $\mathbf{b} = \mathbf{c}$.

(b) If $\mathbf{b} = \mathbf{c}$, then $\mathbf{a}\cdot\mathbf{b} = \mathbf{a}\cdot\mathbf{c}$.

(c) $\mathbf{a}\cdot\mathbf{a} = \|\mathbf{a}\|^2$.

(d) If $\|\mathbf{a}\| > \|\mathbf{b}\|$ then $\mathbf{a}\cdot\mathbf{c} > \mathbf{b}\cdot\mathbf{c}$.

(e) If $\|\mathbf{a}\| = \|\mathbf{b}\|$ then $\mathbf{a} = \mathbf{b}$.

32. To compute $\mathbf{a}\cdot\mathbf{b}$, where $\mathbf{a} = \langle 2, 5\rangle$ and $\mathbf{b} = \dfrac{\langle 4, 1\rangle}{\sqrt{17}}$, you can first compute $\langle 2, 5\rangle\cdot\langle 4, 1\rangle$ and then divide the result (13) by $\sqrt{17}$. Which property of Theorem 3.1 is being used?

33. By the Cauchy-Schwartz Inequality, $|\mathbf{a}\cdot\mathbf{b}| \le \|\mathbf{a}\|\|\mathbf{b}\|$. What relationship must exist between $\mathbf{a}$ and $\mathbf{b}$ to have $|\mathbf{a}\cdot\mathbf{b}| = \|\mathbf{a}\|\|\mathbf{b}\|$?

34. By the Triangle Inequality, $\|\mathbf{a}+\mathbf{b}\| \le \|\mathbf{a}\| + \|\mathbf{b}\|$. What relationship must exist between $\mathbf{a}$ and $\mathbf{b}$ to have $\|\mathbf{a}+\mathbf{b}\| = \|\mathbf{a}\| + \|\mathbf{b}\|$?

35. Use the Triangle Inequality to prove that $\|\mathbf{a}-\mathbf{b}\| \ge \|\mathbf{a}\| - \|\mathbf{b}\|$.

36. Prove parts (ii) and (iii) of Theorem 3.1.

37. For vectors $\mathbf{a}$ and $\mathbf{b}$, use the Cauchy-Schwartz Inequality to find the maximum value of $\mathbf{a}\cdot\mathbf{b}$ if $\|\mathbf{a}\| = 3$ and $\|\mathbf{b}\| = 5$.

38. Find a formula for $\mathbf{a}$ in terms of $\mathbf{b}$ where $\|\mathbf{a}\| = 3$, $\|\mathbf{b}\| = 5$ and $\mathbf{a}\cdot\mathbf{b}$ is maximum.

39. Use the Cauchy-Schwartz Inequality in n dimensions to show that $\left(\sum_{k=1}^{n} |a_k b_k|\right)^2 \le \left(\sum_{k=1}^{n} a_k^2\right)\left(\sum_{k=1}^{n} b_k^2\right)$. If both $\sum_{k=1}^{\infty} a_k^2$ and $\sum_{k=1}^{\infty} b_k^2$ converge, what can be concluded? Apply the result to $a_k = \frac{1}{k}$ and $b_k = \frac{1}{k^2}$.

40. Show that $\sum_{k=1}^{n} |a_k b_k| \le \frac{1}{2}\sum_{k=1}^{n} a_k^2 + \frac{1}{2}\sum_{k=1}^{n} b_k^2$. If both $\sum_{k=1}^{\infty} a_k^2$ and $\sum_{k=1}^{\infty} b_k^2$ converge, what can be concluded? Apply the result to $a_k = \frac{1}{k}$ and $b_k = \frac{1}{k^2}$. Is this bound better or worse than the bound found in exercise 39?

41. Use the Cauchy-Schwartz Inequality in n dimensions to show that $\sum_{k=1}^{n} |a_k| \le \left(\sum_{k=1}^{n} |a_k|^{2/3}\right)^{1/2}\left(\sum_{k=1}^{n} |a_k|^{4/3}\right)^{1/2}$.

42. Use the Cauchy-Schwartz Inequality in n dimensions to show that $\sum_{k=1}^{n} |a_k| \le \sqrt{n}\left(\sum_{k=1}^{n} a_k^2\right)^{1/2}$.

43. If $p_1, p_2, \ldots, p_n$ are nonnegative numbers that sum to 1, show that $\sum_{k=1}^{n} p_k^2 \ge \dfrac{1}{n}$.

44. Among all sets of nonnegative numbers $p_1, p_2, \ldots, p_n$ that sum to 1, find the choice of $p_1, p_2, \ldots, p_n$ that minimizes $\sum_{k=1}^{n} p_k^2$.

45. Show that $\sum_{k=1}^{n} a_k^2 b_k^2 \le \left(\sum_{k=1}^{n} a_k^2\right)\left(\sum_{k=1}^{n} b_k^2\right)$ and then $\left(\sum_{k=1}^{n} a_k b_k c_k\right)^2 \le \left(\sum_{k=1}^{n} a_k^2\right)\left(\sum_{k=1}^{n} b_k^2\right)\left(\sum_{k=1}^{n} c_k^2\right)$.

46. Show that $\sqrt{\dfrac{x+y}{x+y+z}} + \sqrt{\dfrac{y+z}{x+y+z}} + \sqrt{\dfrac{x+z}{x+y+z}} \le \sqrt{6}$.

47. In a methane molecule (CH_4), a carbon atom is surrounded by four hydrogen atoms. Assume that the hydrogen atoms are at (0, 0, 0), (1, 1, 0), (1, 0, 1) and (0, 1, 1) and the carbon atom is at $\left(\frac{1}{2}, \frac{1}{2}, \frac{1}{2}\right)$. Compute the **bond angle,** the angle from hydrogen atom to carbon atom to hydrogen atom.

48. Consider the parallelogram with vertices at (0, 0), (2, 0), (3, 2) and (1, 2). Find the angle at which the diagonals intersect.

49. Prove that $\text{comp}_\mathbf{c}(\mathbf{a}+\mathbf{b}) = \text{comp}_\mathbf{c}\,\mathbf{a} + \text{comp}_\mathbf{c}\,\mathbf{b}$ for any nonzero vectors $\mathbf{a}$, $\mathbf{b}$ and $\mathbf{c}$.

50. The **orthogonal projection** of vector **a** along vector **b** is defined as $\mathbf{orth_b\, a} = \mathbf{a} - \mathbf{proj_b\, a}$. Sketch a picture showing vectors **a**, **b**, $\mathbf{proj_b\, a}$ and $\mathbf{orth_b\, a}$, and explain what is orthogonal about $\mathbf{orth_b\, a}$.

51. Suppose that a beam of an oil rig is installed in a direction parallel to $\langle 10, 1, 5\rangle$. If a wave exerts a force of $\langle 0, -200, 0\rangle$ newtons, find the component of this force along the beam.

52. Repeat exercise 51 with a force of $\langle 13, -190, -61\rangle$ newtons. The forces here and in exercise 51 have nearly identical magnitudes. Explain why the force components are different.

53. A car makes a turn on a banked road. If the road is banked at 10°, show that a vector parallel to the road is $\langle \cos 10^\circ, \sin 10^\circ\rangle$. If the car has weight 2000 pounds, find the component of the weight vector along the road vector. This component of weight provides a force that helps the car turn.

54. Find the component of the weight vector along the road vector for a 2500-pound car on a 15° bank.

55. The racetrack at Bristol, Tennessee, is famous for being short with the steepest banked curves on the NASCAR circuit. The track is an oval of length 0.533 mile and the corners are banked at 36°. Circular motion at a constant speed v requires a centripetal force of $F = \frac{mv^2}{r}$, where r is the radius of the circle and m is the mass of the car. For a track banked at angle A, the weight of the car provides a centripetal force of $mg \sin A$, where g is the gravitational constant. Setting the two equal gives $\frac{v^2}{r} = g \sin A$. Assuming that the Bristol track is circular (it's not really) and using $g = 32$ ft/s^2, find the speed supported by the Bristol bank. Cars actually complete laps at over 120 mph. Discuss where the additional force for this higher speed might come from.

56. For a car driving on a 36° bank, compute the ratio of the component of weight along the road to the component of weight into the road. Discuss why it might be dangerous if this ratio is very small.

57. A small store sells CD players and DVD players. Suppose 32 CD players are sold at \$25 apiece and 12 DVD players are sold at \$125 apiece. The vector $\mathbf{a} = \langle 32, 12\rangle$ can be called the sales vector and $\mathbf{b} = \langle 25, 125\rangle$ the price vector. Interpret the meaning of $\mathbf{a}\cdot\mathbf{b}$.

58. Suppose that a company makes n products. The production vector $\mathbf{a} = \langle a_1, a_2, \ldots, a_n\rangle$ records how many of each product are manufactured and the cost vector $\mathbf{b} = \langle b_1, b_2, \ldots, b_n\rangle$ records how much each product costs to manufacture. Interpret the meaning of $\mathbf{a}\cdot\mathbf{b}$.

59. Parametric equations for one object are $x_1 = a\cos t$ and $y_1 = b\sin t$. The object travels along the ellipse $\frac{x^2}{a^2} + \frac{y^2}{b^2} = 1$. The parametric equations for a second object are $x_2 = a\cos(t + \frac{\pi}{2})$ and $y_2 = b\sin(t + \frac{\pi}{2})$. This object travels along the same ellipse but is $\frac{\pi}{2}$ time units ahead. If $a = b$, use the trigonometric identity $\cos u \cos v + \sin u \sin v = \cos(u - v)$ to show that the position vectors of the two objects are orthogonal. However, if $a \neq b$, the position vectors are not orthogonal.

60. Show that the object with parametric equations $x_3 = b\cos(t + \frac{\pi}{2})$ and $y_3 = a\sin(t + \frac{\pi}{2})$ has position vector that is orthogonal to the first object of exercise 59.

61. In the diagram, a crate of weight w pounds is placed on a ramp inclined at angle θ above the horizontal. The vector **v** along the ramp is given by $\mathbf{v} = \langle \cos\theta, \sin\theta\rangle$ and the **normal** vector by $\mathbf{n} = \langle -\sin\theta, \cos\theta\rangle$. Show that **v** and **n** are perpendicular. Find the component of $\mathbf{w} = \langle 0, -w\rangle$ along **v** and the component of **w** along **n**.

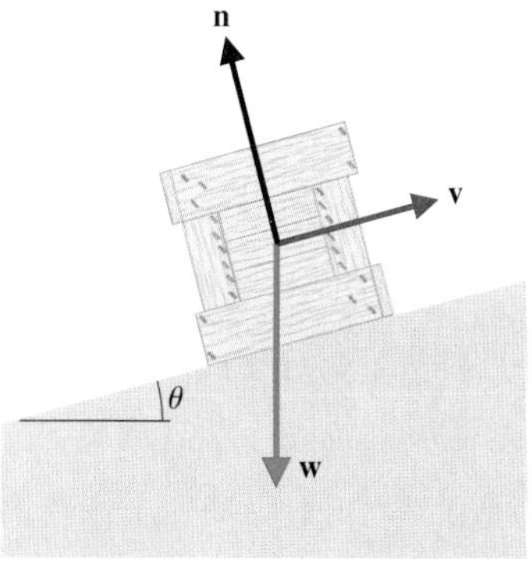

62. If the coefficient of static friction between the crate and ramp in exercise 61 equals μ_s, physics tells us that the crate will slide down the ramp if the component of **w** along **v** is greater than the product of μ_s and the component of **w** along **n**. Show that this occurs if the angle θ is steep enough that $\theta > \tan^{-1} \mu_s$.

63. A weight of 500 pounds is supported by two ropes that exert forces of $\mathbf{a} = \langle -100, 200 \rangle$ pounds and $\mathbf{b} = \langle 100, 300 \rangle$ pounds. Find the angle θ between the ropes.

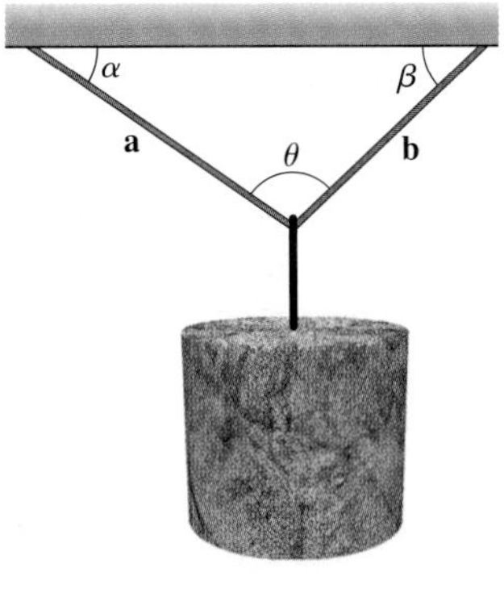

64. In the diagram for exercise 63, find the angles α and β.

65. Suppose a small business sells three products. In a given month, if 3000 units of product A are sold, 2000 units of product B are sold and 4000 units of product C are sold, then the **sales vector** for that month is defined by $\mathbf{s} = \langle 3000, 2000, 4000 \rangle$. If the prices of products A, B and C are \$20, \$15 and \$25, respectively, then the **price vector** is defined by $\mathbf{p} = \langle 20, 15, 25 \rangle$. Compute $\mathbf{s} \cdot \mathbf{p}$ and discuss how it relates to monthly revenue.

66. Suppose that in a particular county, ice cream sales (in thousands of gallons) for a year is given by the vector $\mathbf{s} = \langle 3, 5, 12, 40, 60, 100, 120, 160, 110, 50, 10, 2 \rangle$. That is, 3000 gallons were sold in January, 5000 gallons were sold in February, and so on. In the same county, suppose that murders for the year are given by the vector $\mathbf{m} = \langle 2, 0, 1, 6, 4, 8, 10, 13, 8, 2, 0, 6 \rangle$. Show that the average monthly ice cream sales is $\bar{s} = 56{,}000$ gallons and that the average monthly number of murders is $\bar{m} = 5$. Compute the vectors **a** and **b**, where the components of **a** equal the components of **s** with the mean 56 subtracted (so that $\mathbf{a} = \langle -53, -51, -44, \ldots \rangle$) and the components of **b** equal the components of **m** with the mean 5 subtracted. The correlation between ice cream sales and murders is defined as $\rho = \dfrac{\mathbf{a} \cdot \mathbf{b}}{\|\mathbf{a}\| \, \|\mathbf{b}\|}$. Often, a positive correlation is incorrectly interpreted as meaning that **a** "causes" **b**. (In fact, correlation should *never* be used to infer a cause-and-effect relationship.) Explain why such a conclusion would be invalid in this case.

67. Use the Cauchy-Schwartz Inequality to show that if $a_k \geq 0$, then $\displaystyle\sum_{k=1}^{n} \frac{\sqrt{a_k}}{k^p} \leq \sqrt{\sum_{k=1}^{n} a_k} \sqrt{\sum_{k=1}^{n} \frac{1}{k^{2p}}}$.

68. Show that if $a_k \geq 0$, $p > \frac{1}{2}$ and $\displaystyle\sum_{k=1}^{\infty} a_k$ converges, then $\displaystyle\sum_{k=1}^{\infty} \frac{\sqrt{a_k}}{k^p}$ converges.

69. For the Mandelbrot set and associated Julia sets, functions of the form $f(x) = x^2 - c$ are analyzed for various constants c. The iterates of the function increase if $|x^2 - c| > |x|$. Show that this is true if $|x| > \frac{1}{2} + \sqrt{\frac{1}{4} + c}$.

70. Show that the vector analog of exercise 69 is also true. For vectors $\mathbf{x}$, $\mathbf{x}_2$ and $\mathbf{c}$, if $\|\mathbf{x}\| > \frac{1}{2} + \sqrt{\frac{1}{4} + \|\mathbf{c}\|}$ and $\|\mathbf{x}_2\| = \|\mathbf{x}\|^2$, then $\|\mathbf{x}_2 - \mathbf{c}\| > \|\mathbf{x}\|$.

EXPLORATORY EXERCISES

1. One of the basic problems throughout calculus is computing distances. In this exercise, we will find the distance between a point (x_1, y_1) and a line $ax + by + d = 0$. To start with a concrete example, take the point (5, 6) and the line $2x + 3y + 4 = 0$. First, show that the intercepts of the line are the points $\left(-\frac{4}{2}, 0\right)$ and $\left(0, -\frac{4}{3}\right)$. Show that the vector $\mathbf{b} = \langle 3, -2 \rangle$ is parallel to the displacement vector between these points and hence, also to the line. Sketch a picture showing the point (5, 6), the line, the vector $\langle 3, -2 \rangle$ and the displacement vector **v** from $(-2, 0)$ to (5, 6). Explain why the magnitude of the vector $\mathbf{v} - \mathbf{proj_b}\, \mathbf{v}$ equals the desired distance between point and line. Compute this distance. Show that, in general, the distance between the point (x_1, y_1) and the line $ax + by + d = 0$ equals $\dfrac{|ax_1 + by_1 + d|}{\sqrt{a^2 + b^2}}$.

2. In the accompanying figure, the circle $x^2 + y^2 = r^2$ is shown. In this exercise, we will compute the time required for an object to travel the length of a chord from the top of the circle to another point on the circle at an angle of θ from the vertical, assuming that gravity (acting downward) is the only force.

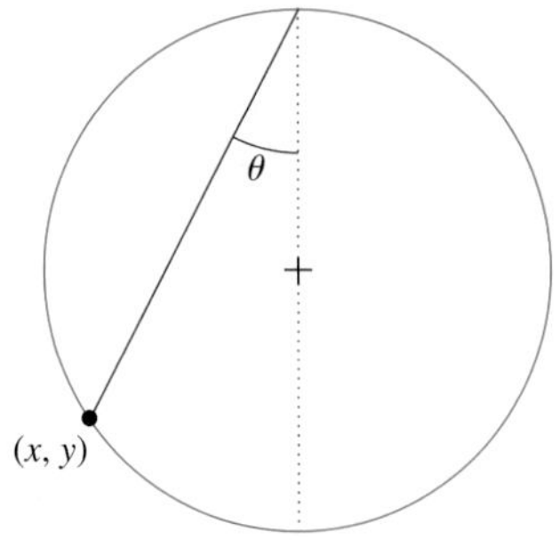

From our study of projectile motion in section 5.5, recall that an object traveling with a constant acceleration a covers a distance d in time $\sqrt{\frac{2d}{a}}$. Show that the component of gravity in the direction of the chord is $a = g\cos\theta$. If the chord ends at the point (x, y), show that the length of the chord is $d = \sqrt{2r^2 - 2ry}$. Also, show that $\cos\theta = \frac{r - y}{d}$. Putting this all together, compute the time it takes to travel the chord. Explain why it's surprising that the answer does not depend on the value of θ. Note that as θ increases, the distance d decreases but the effectiveness of gravity decreases. Discuss the balance between these two factors.

3. This exercise develops a basic principle used to create wire-frame and other 3-D computer graphics. In the drawing, an artist traces the image of an object onto a pane of glass. Explain why the trace will be distorted unless the artist keeps the pane of glass perpendicular to the line of sight. The trace is thus a projection of the object onto the pane of glass. To make this precise, suppose that the artist is at the point $(100, 0, 0)$ and the point $P_1 = (2, 1, 3)$ is part of the object being traced. Find the projection $\mathbf{p}_1$ of the position vector $\langle 2, 1, 3\rangle$ along the artist's position vector $\langle 100, 0, 0\rangle$. Then find the vector $\mathbf{q}_1$ such that $\langle 2, 1, 3\rangle = \mathbf{p}_1 + \mathbf{q}_1$. Which of the vectors $\mathbf{p}_1$ and $\mathbf{q}_1$ does the artist actually see and which one is hidden? Repeat this with the point $P_2 = (-2, 1, 3)$ and find vectors $\mathbf{p}_2$ and $\mathbf{q}_2$ such that $\langle -2, 1, 3\rangle = \mathbf{p}_2 + \mathbf{q}_2$. The artist would plot both points P_1 and P_2 at the same point on the pane of glass. Identify which of the vectors $\mathbf{p}_1, \mathbf{q}_1, \mathbf{p}_2$ and $\mathbf{q}_2$ correspond to this point. From the artist's perspective, one of the points P_1 or P_2 is hidden behind the other. Identify which point is hidden and explain how the information in the vectors $\mathbf{p}_1, \mathbf{q}_1, \mathbf{p}_2$ and $\mathbf{q}_2$ can be used to determine which point is hidden.

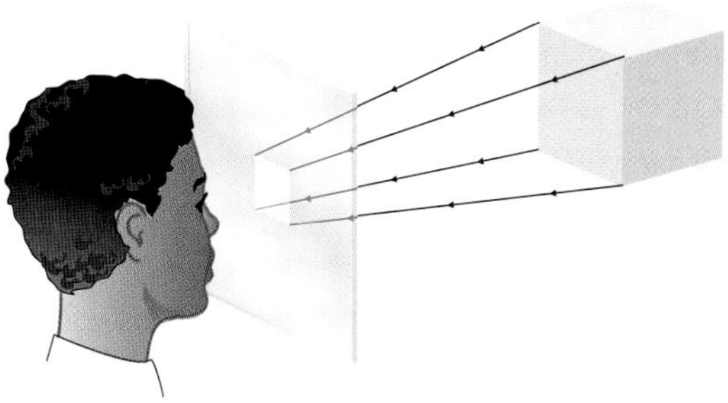

4. Take a cube and spin it around a diagonal.

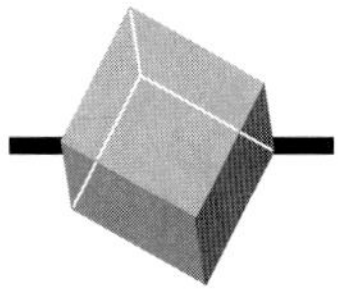

If you spin it rapidly, you will see a curved outline appear in the middle. (See the figure below.) How does a cube become curved? This exercise answers that question. Suppose that the cube is a unit cube with $0 \le x \le 1, 0 \le y \le 1$ and $0 \le z \le 1$, and we rotate about the diagonal from $(0, 0, 0)$ to $(1, 1, 1)$. What we see on spinning the cube is the combination of points on the cube at their maximum distance from the diagonal. The points on the edge of the cube have the maximum distance, so we focus on them. If (x, y, z) is a point on an edge of the cube, define h to be the component of the vector $\langle x, y, z\rangle$ along the diagonal $\langle 1, 1, 1\rangle$. The distance d from (x, y, z) to the diagonal is then $d = \sqrt{\|\langle x, y, z\rangle\|^2 - h^2}$, as in the diagram below. The curve is produced by the edge from $(0, 0, 1)$ to $(0, 1, 1)$. Parametric equations for this segment are $x = 0$, $y = t$ and $z = 1$, for $0 \le t \le 1$. For the vector $\langle 0, t, 1\rangle$, compute h and then d. Graph $d(t)$. You should see a curve similar to the middle of the outline shown below. Show that this curve is actually part of a hyperbola. Then find the outline created by other sides of the cube. Which ones produce curves and which produce straight lines?

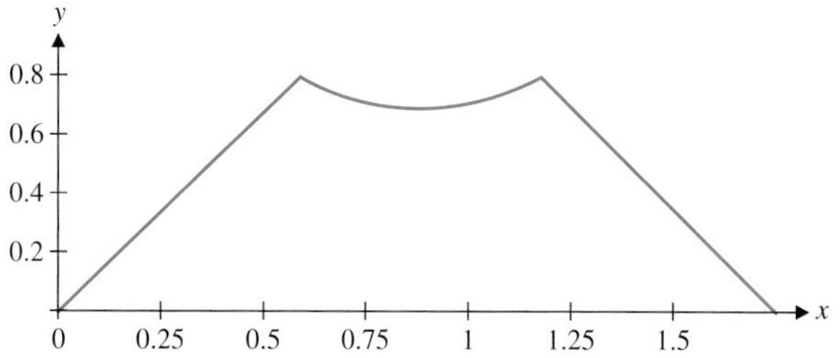

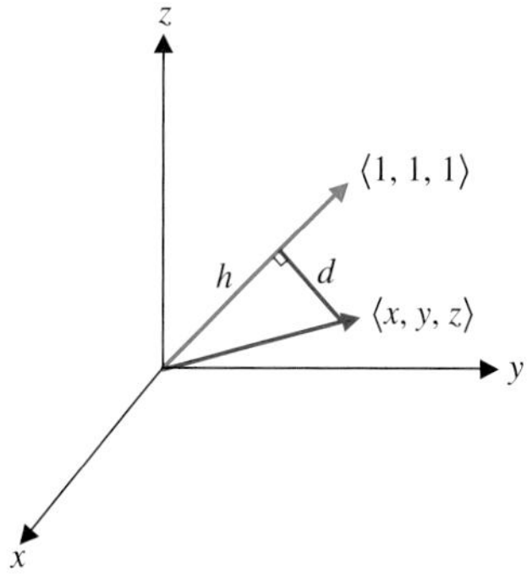

10.4 THE CROSS PRODUCT

In this section, we define a second type of product of vectors, the *cross product* or *vector product*. While the dot product of two vectors is a scalar, the cross product of two vectors is another vector. The cross product has many important applications, from physics and engineering mechanics to space travel. Before we define the cross product, we need a few definitions.

DEFINITION 4.1

The **determinant** of a 2×2 matrix of real numbers is defined by

$$\underbrace{\begin{vmatrix} a_1 & a_2 \\ b_1 & b_2 \end{vmatrix}}_{2 \times 2 \text{ matrix}} = a_1 b_2 - a_2 b_1. \tag{4.1}$$

EXAMPLE 4.1 Computing a 2×2 Determinant

Evaluate the determinant $\begin{vmatrix} 1 & 2 \\ 3 & 4 \end{vmatrix}$.

Solution From (4.1), we have

$$\begin{vmatrix} 1 & 2 \\ 3 & 4 \end{vmatrix} = (1)(4) - (2)(3) = -2.$$

■

DEFINITION 4.2

The **determinant** of a 3×3 matrix of real numbers is defined as a combination of three 2×2 determinants, as follows:

$$\underbrace{\begin{vmatrix} a_1 & a_2 & a_3 \\ b_1 & b_2 & b_3 \\ c_1 & c_2 & c_3 \end{vmatrix}}_{3 \times 3 \text{ matrix}} = a_1 \begin{vmatrix} b_2 & b_3 \\ c_2 & c_3 \end{vmatrix} - a_2 \begin{vmatrix} b_1 & b_3 \\ c_1 & c_3 \end{vmatrix} + a_3 \begin{vmatrix} b_1 & b_2 \\ c_1 & c_2 \end{vmatrix}. \tag{4.2}$$

Equation (4.2) is referred to as an **expansion** of the determinant **along the first row.** Notice that the multipliers of each of the 2×2 determinants are the entries of the first row of the 3×3 matrix. Each 2×2 determinant is the determinant you get if you eliminate the row and column in which the corresponding multiplier lies. That is, for the *first* term, the multiplier is a_1 and the 2×2 determinant is found by eliminating the first row and *first* column from the 3×3 matrix:

$$\begin{vmatrix} a_1 & a_2 & a_3 \\ b_1 & b_2 & b_3 \\ c_1 & c_2 & c_3 \end{vmatrix} = \begin{vmatrix} b_2 & b_3 \\ c_2 & c_3 \end{vmatrix}.$$

Likewise, the *second* 2×2 determinant is found by eliminating the first row and the *second* column from the 3×3 determinant:

$$\begin{vmatrix} a_1 & a_2 & a_3 \\ b_1 & b_2 & b_3 \\ c_1 & c_2 & c_3 \end{vmatrix} = \begin{vmatrix} b_1 & b_3 \\ c_1 & c_3 \end{vmatrix}.$$

Be certain to notice the minus sign in front of this term. Finally, the *third* determinant is found by eliminating the first row and the *third* column from the 3×3 determinant:

$$\begin{vmatrix} a_1 & a_2 & a_3 \\ b_1 & b_2 & b_3 \\ c_1 & c_2 & c_3 \end{vmatrix} = \begin{vmatrix} b_1 & b_2 \\ c_1 & c_2 \end{vmatrix}.$$

EXAMPLE 4.2 Evaluating a 3 × 3 Determinant

Evaluate the determinant $\begin{vmatrix} 1 & 2 & 4 \\ -3 & 3 & 1 \\ 3 & -2 & 5 \end{vmatrix}$.

Solution Expanding along the first row, we have:

$$\begin{vmatrix} 1 & 2 & 4 \\ -3 & 3 & 1 \\ 3 & -2 & 5 \end{vmatrix} = (1)\begin{vmatrix} 3 & 1 \\ -2 & 5 \end{vmatrix} - (2)\begin{vmatrix} -3 & 1 \\ 3 & 5 \end{vmatrix} + (4)\begin{vmatrix} -3 & 3 \\ 3 & -2 \end{vmatrix}$$

$$= (1)[(3)(5) - (1)(-2)] - (2)[(-3)(5) - (1)(3)] + (4)[(-3)(-2) - (3)(3)]$$

$$= 41. \blacksquare$$

We use determinant notation as a convenient device for defining the cross product, as follows.

DEFINITION 4.3

For two vectors $\mathbf{a} = \langle a_1, a_2, a_3 \rangle$ and $\mathbf{b} = \langle b_1, b_2, b_3 \rangle$ in V_3, we define the **cross product** (or **vector product**) of **a** and **b** to be

$$\mathbf{a} \times \mathbf{b} = \begin{vmatrix} \mathbf{i} & \mathbf{j} & \mathbf{k} \\ a_1 & a_2 & a_3 \\ b_1 & b_2 & b_3 \end{vmatrix} = \begin{vmatrix} a_2 & a_3 \\ b_2 & b_3 \end{vmatrix} \mathbf{i} - \begin{vmatrix} a_1 & a_3 \\ b_1 & b_3 \end{vmatrix} \mathbf{j} + \begin{vmatrix} a_1 & a_2 \\ b_1 & b_2 \end{vmatrix} \mathbf{k}. \tag{4.3}$$

Notice that $\mathbf{a} \times \mathbf{b}$ is also a vector in V_3. To compute $\mathbf{a} \times \mathbf{b}$, you must write the components of **a** in the second row and the components of **b** in the third row; *the order is important!* Also note that while we've used the determinant notation, the 3 × 3 determinant indicated in (4.3) is not really a determinant, in the sense in which we defined them, since the entries in the first row are vectors instead of scalars. Nonetheless, we find this slight abuse of notation convenient for computing cross products and we use it routinely.

EXAMPLE 4.3 Computing a Cross Product

Compute $\langle 1, 2, 3 \rangle \times \langle 4, 5, 6 \rangle$.

Solution From (4.3), we have

$$\langle 1, 2, 3 \rangle \times \langle 4, 5, 6 \rangle = \begin{vmatrix} \mathbf{i} & \mathbf{j} & \mathbf{k} \\ 1 & 2 & 3 \\ 4 & 5 & 6 \end{vmatrix} = \begin{vmatrix} 2 & 3 \\ 5 & 6 \end{vmatrix} \mathbf{i} - \begin{vmatrix} 1 & 3 \\ 4 & 6 \end{vmatrix} \mathbf{j} + \begin{vmatrix} 1 & 2 \\ 4 & 5 \end{vmatrix} \mathbf{k}$$

$$= -3\mathbf{i} + 6\mathbf{j} - 3\mathbf{k} = \langle -3, 6, -3 \rangle. \blacksquare$$

REMARK 4.1

The cross product is defined only for vectors in V_3. There is no corresponding operation for vectors in V_2.

THEOREM 4.1

For any vector $\mathbf{a} \in V_3$, $\mathbf{a} \times \mathbf{a} = \mathbf{0}$ and $\mathbf{a} \times \mathbf{0} = \mathbf{0}$.

PROOF

We prove the first of these two results. The second, we leave as an exercise. For $\mathbf{a} = \langle a_1, a_2, a_3 \rangle$, we have from (4.3) that

$$\mathbf{a} \times \mathbf{a} = \begin{vmatrix} \mathbf{i} & \mathbf{j} & \mathbf{k} \\ a_1 & a_2 & a_3 \\ a_1 & a_2 & a_3 \end{vmatrix} = \begin{vmatrix} a_2 & a_3 \\ a_2 & a_3 \end{vmatrix} \mathbf{i} - \begin{vmatrix} a_1 & a_3 \\ a_1 & a_3 \end{vmatrix} \mathbf{j} + \begin{vmatrix} a_1 & a_2 \\ a_1 & a_2 \end{vmatrix} \mathbf{k}$$

$$= (a_2a_3 - a_3a_2)\mathbf{i} - (a_1a_3 - a_3a_1)\mathbf{j} + (a_1a_2 - a_2a_1)\mathbf{k} = \mathbf{0}. \blacksquare$$

Let's take a brief look back at the result of example 4.3. There, we saw that

$$\langle 1, 2, 3 \rangle \times \langle 4, 5, 6 \rangle = \langle -3, 6, -3 \rangle.$$

There is something rather interesting to observe here. Note that

$$\langle 1, 2, 3 \rangle \cdot \langle -3, 6, -3 \rangle = 0$$

and

$$\langle 4, 5, 6 \rangle \cdot \langle -3, 6, -3 \rangle = 0.$$

That is, both $\langle 1, 2, 3 \rangle$ and $\langle 4, 5, 6 \rangle$ are orthogonal to their cross product. As it turns out, this is true in general, as we see in Theorem 4.2.

HISTORICAL NOTES

Josiah Willard Gibbs (1839–1903)
American physicist and mathematician who introduced and named the dot product and the cross product. A graduate of Yale, Gibbs published important papers in thermodynamics, statistical mechanics and the electromagnetic theory of light. Gibbs used vectors to determine the orbit of a comet from only three observations. Originally produced as printed notes for his students, Gibbs' vector system greatly simplified the original system developed by Hamilton. Gibbs was well liked but not famous in his lifetime. One biographer wrote of Gibbs that, "The greatness of his intellectual achievements will never overshadow the beauty and dignity of his life."

THEOREM 4.2

For any vectors **a** and **b** in V_3, $\mathbf{a} \times \mathbf{b}$ is orthogonal to both **a** and **b**.

PROOF

Recall that two vectors are orthogonal if and only if their dot product is zero. Now, using (4.3), we have

$$\begin{aligned} \mathbf{a} \cdot (\mathbf{a} \times \mathbf{b}) &= \langle a_1, a_2, a_3 \rangle \cdot \left[\begin{vmatrix} a_2 & a_3 \\ b_2 & b_3 \end{vmatrix} \mathbf{i} - \begin{vmatrix} a_1 & a_3 \\ b_1 & b_3 \end{vmatrix} \mathbf{j} + \begin{vmatrix} a_1 & a_2 \\ b_1 & b_2 \end{vmatrix} \mathbf{k} \right] \\ &= a_1 \begin{vmatrix} a_2 & a_3 \\ b_2 & b_3 \end{vmatrix} - a_2 \begin{vmatrix} a_1 & a_3 \\ b_1 & b_3 \end{vmatrix} + a_3 \begin{vmatrix} a_1 & a_2 \\ b_1 & b_2 \end{vmatrix} \\ &= a_1[a_2b_3 - a_3b_2] - a_2[a_1b_3 - a_3b_1] + a_3[a_1b_2 - a_2b_1] \\ &= a_1a_2b_3 - a_1a_3b_2 - a_1a_2b_3 + a_2a_3b_1 + a_1a_3b_2 - a_2a_3b_1 \\ &= 0, \end{aligned}$$

so that **a** and $(\mathbf{a} \times \mathbf{b})$ are orthogonal. We leave it as an exercise to show that $\mathbf{b} \cdot (\mathbf{a} \times \mathbf{b}) = 0$, also. $\blacksquare$

Notice that since $\mathbf{a} \times \mathbf{b}$ is orthogonal to both **a** and **b**, it is also orthogonal to every vector lying in the plane containing **a** and **b**. (We also say that $\mathbf{a} \times \mathbf{b}$ is orthogonal to the plane, in this case.) But, given a plane, out of which side of the plane does $\mathbf{a} \times \mathbf{b}$ point? We can get an idea by computing some simple cross products.

Notice that

$$\mathbf{i} \times \mathbf{j} = \begin{vmatrix} \mathbf{i} & \mathbf{j} & \mathbf{k} \\ 1 & 0 & 0 \\ 0 & 1 & 0 \end{vmatrix} = \begin{vmatrix} 0 & 0 \\ 1 & 0 \end{vmatrix} \mathbf{i} - \begin{vmatrix} 1 & 0 \\ 0 & 0 \end{vmatrix} \mathbf{j} + \begin{vmatrix} 1 & 0 \\ 0 & 1 \end{vmatrix} \mathbf{k} = \mathbf{k}.$$

Likewise,

$$\mathbf{j} \times \mathbf{k} = \mathbf{i}.$$

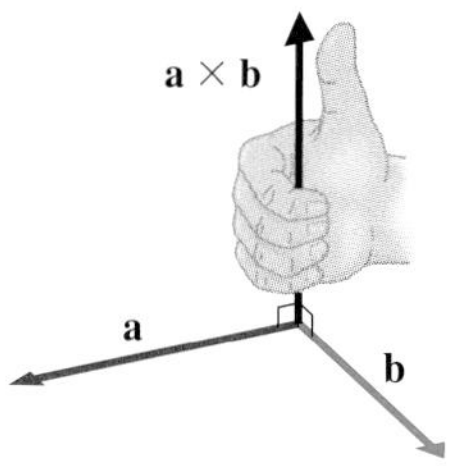

FIGURE 10.29a
$\mathbf{a} \times \mathbf{b}$

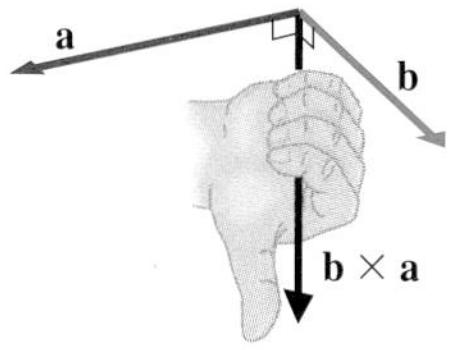

FIGURE 10.29b
$\mathbf{b} \times \mathbf{a}$

These are illustrations of the **right-hand rule:** If you align the fingers of your *right* hand along the vector $\mathbf{a}$ and bend your fingers around in the direction of rotation from $\mathbf{a}$ toward $\mathbf{b}$ (through an angle of less than 180°), your thumb will point in the direction of $\mathbf{a} \times \mathbf{b}$ (see Figure 10.29a). Now, following the right-hand rule, $\mathbf{b} \times \mathbf{a}$ will point in the direction opposite $\mathbf{a} \times \mathbf{b}$ (see Figure 10.29b). In particular, notice that

$$\mathbf{j} \times \mathbf{i} = \begin{vmatrix} \mathbf{i} & \mathbf{j} & \mathbf{k} \\ 0 & 1 & 0 \\ 1 & 0 & 0 \end{vmatrix} = -\mathbf{k}.$$

We leave it as an exercise to show that

$$\mathbf{j} \times \mathbf{k} = \mathbf{i}, \qquad \mathbf{k} \times \mathbf{j} = -\mathbf{i},$$
$$\mathbf{k} \times \mathbf{i} = \mathbf{j} \quad \text{and} \quad \mathbf{i} \times \mathbf{k} = -\mathbf{j}.$$

Take the time to think through the right-hand rule for each of these cross products.

There are several other unusual things to observe here. Notice that

$$\mathbf{i} \times \mathbf{j} = \mathbf{k} \neq -\mathbf{k} = \mathbf{j} \times \mathbf{i},$$

which says that the cross product is *not* commutative. Further, notice that

$$(\mathbf{i} \times \mathbf{j}) \times \mathbf{j} = \mathbf{k} \times \mathbf{j} = -\mathbf{i},$$

while

$$\mathbf{i} \times (\mathbf{j} \times \mathbf{j}) = \mathbf{i} \times \mathbf{0} = \mathbf{0},$$

so that the cross product is also *not* associative. That is, in general,

$$(\mathbf{a} \times \mathbf{b}) \times \mathbf{c} \neq \mathbf{a} \times (\mathbf{b} \times \mathbf{c}).$$

Since the cross product does not follow several of the rules you might expect a product to satisfy, you might ask what rules the cross product *does* satisfy. We summarize these in Theorem 4.3.

THEOREM 4.3

For any vectors $\mathbf{a}$, $\mathbf{b}$ and $\mathbf{c}$ in V_3 and any scalar d, the following hold:

(i) $\mathbf{a} \times \mathbf{b} = -(\mathbf{b} \times \mathbf{a})$ (anticommutativity)
(ii) $(d\mathbf{a}) \times \mathbf{b} = d(\mathbf{a} \times \mathbf{b}) = \mathbf{a} \times (d\mathbf{b})$
(iii) $\mathbf{a} \times (\mathbf{b} + \mathbf{c}) = \mathbf{a} \times \mathbf{b} + \mathbf{a} \times \mathbf{c}$ (distributive law)
(iv) $(\mathbf{a} + \mathbf{b}) \times \mathbf{c} = \mathbf{a} \times \mathbf{c} + \mathbf{b} \times \mathbf{c}$ (distributive law)
(v) $\mathbf{a} \cdot (\mathbf{b} \times \mathbf{c}) = (\mathbf{a} \times \mathbf{b}) \cdot \mathbf{c}$ (scalar triple product) and
(vi) $\mathbf{a} \times (\mathbf{b} \times \mathbf{c}) = (\mathbf{a} \cdot \mathbf{c})\mathbf{b} - (\mathbf{a} \cdot \mathbf{b})\mathbf{c}$ (vector triple product).

PROOF

We prove parts (i) and (iii) only. The remaining parts are left as exercises.

(i) For $\mathbf{a} = \langle a_1, a_2, a_3 \rangle$ and $\mathbf{b} = \langle b_1, b_2, b_3 \rangle$, we have from (4.3) that

$$\mathbf{a} \times \mathbf{b} = \begin{vmatrix} \mathbf{i} & \mathbf{j} & \mathbf{k} \\ a_1 & a_2 & a_3 \\ b_1 & b_2 & b_3 \end{vmatrix} = \begin{vmatrix} a_2 & a_3 \\ b_2 & b_3 \end{vmatrix} \mathbf{i} - \begin{vmatrix} a_1 & a_3 \\ b_1 & b_3 \end{vmatrix} \mathbf{j} + \begin{vmatrix} a_1 & a_2 \\ b_1 & b_2 \end{vmatrix} \mathbf{k}$$

$$= -\begin{vmatrix} b_2 & b_3 \\ a_2 & a_3 \end{vmatrix} \mathbf{i} + \begin{vmatrix} b_1 & b_3 \\ a_1 & a_3 \end{vmatrix} \mathbf{j} - \begin{vmatrix} b_1 & b_2 \\ a_1 & a_2 \end{vmatrix} \mathbf{k} = -(\mathbf{b} \times \mathbf{a}),$$

since swapping two rows in a 2×2 matrix (or in a 3×3 matrix, for that matter) changes the sign of its determinant.

(iii) For $\mathbf{c} = \langle c_1, c_2, c_3 \rangle$, we have

$$\mathbf{b} + \mathbf{c} = \langle b_1 + c_1, b_2 + c_2, b_3 + c_3 \rangle$$

and so,
$$\mathbf{a} \times (\mathbf{b} + \mathbf{c}) = \begin{vmatrix} \mathbf{i} & \mathbf{j} & \mathbf{k} \\ a_1 & a_2 & a_3 \\ b_1 + c_1 & b_2 + c_2 & b_3 + c_3 \end{vmatrix}.$$

Looking only at the $\mathbf{i}$ component of this, we have

$$\begin{aligned} \begin{vmatrix} a_2 & a_3 \\ b_2 + c_2 & b_3 + c_3 \end{vmatrix} &= a_2(b_3 + c_3) - a_3(b_2 + c_2) \\ &= (a_2 b_3 - a_3 b_2) + (a_2 c_3 - a_3 c_2) \\ &= \begin{vmatrix} a_2 & a_3 \\ b_2 & b_3 \end{vmatrix} + \begin{vmatrix} a_2 & a_3 \\ c_2 & c_3 \end{vmatrix}, \end{aligned}$$

which you should note is also the $\mathbf{i}$ component of $\mathbf{a} \times \mathbf{b} + \mathbf{a} \times \mathbf{c}$. Similarly, you can show that the $\mathbf{j}$ and $\mathbf{k}$ components also match, which establishes the result. ■

Always keep in mind that vectors are specified by two things: magnitude and direction. We have now shown that $\mathbf{a} \times \mathbf{b}$ is orthogonal to both $\mathbf{a}$ and $\mathbf{b}$. In Theorem 4.4, we make a general (and quite useful) statement about $\|\mathbf{a} \times \mathbf{b}\|$.

THEOREM 4.4

For nonzero vectors $\mathbf{a}$ and $\mathbf{b}$ in V_3, if θ is the angle between $\mathbf{a}$ and $\mathbf{b}$ $(0 \leq \theta \leq \pi)$, then

$$\|\mathbf{a} \times \mathbf{b}\| = \|\mathbf{a}\| \|\mathbf{b}\| \sin \theta. \tag{4.4}$$

PROOF

From (4.3), we get

$$\begin{aligned} \|\mathbf{a} \times \mathbf{b}\|^2 &= [a_2 b_3 - a_3 b_2]^2 + [a_1 b_3 - a_3 b_1]^2 + [a_1 b_2 - a_2 b_1]^2 \\ &= a_2^2 b_3^2 - 2a_2 a_3 b_2 b_3 + a_3^2 b_2^2 + a_1^2 b_3^2 - 2a_1 a_3 b_1 b_3 + a_3^2 b_1^2 \\ &\quad + a_1^2 b_2^2 - 2a_1 a_2 b_1 b_2 + a_2^2 b_1^2 \\ &= \left(a_1^2 + a_2^2 + a_3^2\right)\left(b_1^2 + b_2^2 + b_3^2\right) - (a_1 b_1 + a_2 b_2 + a_3 b_3)^2 \\ &= \|\mathbf{a}\|^2 \|\mathbf{b}\|^2 - (\mathbf{a} \cdot \mathbf{b})^2 \\ &= \|\mathbf{a}\|^2 \|\mathbf{b}\|^2 - \|\mathbf{a}\|^2 \|\mathbf{b}\|^2 \cos^2 \theta \qquad \text{From Theorem 3.2} \\ &= \|\mathbf{a}\|^2 \|\mathbf{b}\|^2 (1 - \cos^2 \theta) \\ &= \|\mathbf{a}\|^2 \|\mathbf{b}\|^2 \sin^2 \theta. \end{aligned}$$

Taking square roots, we get

$$\|\mathbf{a} \times \mathbf{b}\| = \|\mathbf{a}\| \|\mathbf{b}\| \sin \theta,$$

since $\sin \theta \geq 0$, for $0 \leq \theta \leq \pi$. ■

The following characterization of parallel vectors is an immediate consequence of Theorem 4.4.

COROLLARY 4.1

Two nonzero vectors $\mathbf{a}, \mathbf{b} \in V_3$ are parallel if and only if $\mathbf{a} \times \mathbf{b} = \mathbf{0}$.

PROOF

Recall that $\mathbf{a}$ and $\mathbf{b}$ are parallel if and only if the angle θ between them is either 0 or π. In either case, $\sin\theta = 0$ and so, by Theorem 4.4,

$$\|\mathbf{a} \times \mathbf{b}\| = \|\mathbf{a}\|\|\mathbf{b}\| \sin\theta = \|\mathbf{a}\|\|\mathbf{b}\|(0) = 0.$$

The result then follows from the fact that the only vector with zero magnitude is the zero vector. ■

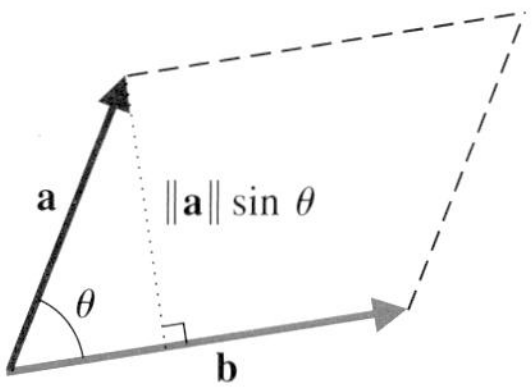

FIGURE 10.30
Parallelogram

Theorem 4.4 also provides us with the following interesting geometric interpretation of the cross product. For any two nonzero vectors $\mathbf{a}$ and $\mathbf{b}$, as long as $\mathbf{a}$ and $\mathbf{b}$ are not parallel, they form two adjacent sides of a parallelogram, as seen in Figure 10.30. Notice that the area of the parallelogram is given by the product of the base and the altitude. We have

$$\begin{aligned} \text{Area} &= (\text{base})(\text{altitude}) \\ &= \|\mathbf{b}\|\|\mathbf{a}\| \sin\theta = \|\mathbf{a} \times \mathbf{b}\|, \end{aligned} \tag{4.5}$$

from Theorem 4.4. That is, the magnitude of the cross product of two vectors gives the area of the parallelogram with two adjacent sides formed by the vectors.

EXAMPLE 4.4 Finding the Area of a Parallelogram Using the Cross Product

Find the area of the parallelogram with two adjacent sides formed by the vectors $\mathbf{a} = \langle 1, 2, 3 \rangle$ and $\mathbf{b} = \langle 4, 5, 6 \rangle$.

Solution First notice that

$$\mathbf{a} \times \mathbf{b} = \begin{vmatrix} \mathbf{i} & \mathbf{j} & \mathbf{k} \\ 1 & 2 & 3 \\ 4 & 5 & 6 \end{vmatrix} = \mathbf{i}\begin{vmatrix} 2 & 3 \\ 5 & 6 \end{vmatrix} - \mathbf{j}\begin{vmatrix} 1 & 3 \\ 4 & 6 \end{vmatrix} + \mathbf{k}\begin{vmatrix} 1 & 2 \\ 4 & 5 \end{vmatrix} = \langle -3, 6, -3 \rangle.$$

From (4.5), the area of the parallelogram is given by

$$\|\mathbf{a} \times \mathbf{b}\| = \|\langle -3, 6, -3 \rangle\| = \sqrt{54} \approx 7.348. \quad ■$$

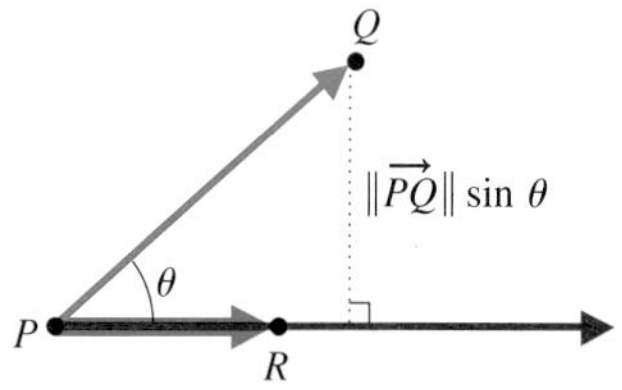

FIGURE 10.31
Distance from a point to a line

We can also use Theorem 4.4 to find the distance from a point to a line in $\mathbb{R}^3$, as follows. Let d represent the distance from the point Q to the line through the points P and R. From elementary trigonometry, we have that

$$d = \|\overrightarrow{PQ}\| \sin\theta,$$

where θ is the angle between $\overrightarrow{PQ}$ and $\overrightarrow{PR}$ (see Figure 10.31). From (4.4), we have

$$\|\overrightarrow{PQ} \times \overrightarrow{PR}\| = \|\overrightarrow{PQ}\|\|\overrightarrow{PR}\| \sin\theta = \|\overrightarrow{PR}\|(d).$$

Solving this for d, we get
$$d = \frac{\|\overrightarrow{PQ} \times \overrightarrow{PR}\|}{\|\overrightarrow{PR}\|}. \tag{4.6}$$

EXAMPLE 4.5 Finding the Distance from a Point to a Line

Find the distance from the point $Q(1, 2, 1)$ to the line through the points $P(2, 1, -3)$ and $R(2, -1, 3)$.

Solution First, the position vectors corresponding to $\overrightarrow{PQ}$ and $\overrightarrow{PR}$ are

$$\overrightarrow{PQ} = \langle -1, 1, 4\rangle \quad \text{and} \quad \overrightarrow{PR} = \langle 0, -2, 6\rangle,$$

and
$$\langle -1, 1, 4\rangle \times \langle 0, -2, 6\rangle = \begin{vmatrix} \mathbf{i} & \mathbf{j} & \mathbf{k} \\ -1 & 1 & 4 \\ 0 & -2 & 6 \end{vmatrix} = \langle 14, 6, 2\rangle.$$

We then have from (4.6) that

$$d = \frac{\|\overrightarrow{PQ} \times \overrightarrow{PR}\|}{\|\overrightarrow{PR}\|} = \frac{\|\langle 14, 6, 2\rangle\|}{\|\langle 0, -2, 6\rangle\|} = \frac{\sqrt{236}}{\sqrt{40}} \approx 2.429. \blacksquare$$

For any three noncoplanar vectors **a**, **b** and **c** (i.e., three vectors that do not lie in a single plane), consider the parallelepiped formed using the vectors as three adjacent edges (see Figure 10.32). Recall that the volume of such a solid is given by

$$\text{Volume} = (\text{Area of base})(\text{altitude}).$$

Further, since two adjacent sides of the base are formed by the vectors **a** and **b**, we know that the area of the base is given by $\|\mathbf{a} \times \mathbf{b}\|$. Referring to Figure 10.32, notice that the altitude is given by

$$|\text{comp}_{\mathbf{a}\times\mathbf{b}}\,\mathbf{c}| = \frac{|\mathbf{c}\cdot(\mathbf{a}\times\mathbf{b})|}{\|\mathbf{a}\times\mathbf{b}\|},$$

from (3.7). The volume of the parallelepiped is then

$$\text{Volume} = \|\mathbf{a}\times\mathbf{b}\|\frac{|\mathbf{c}\cdot(\mathbf{a}\times\mathbf{b})|}{\|\mathbf{a}\times\mathbf{b}\|} = |\mathbf{c}\cdot(\mathbf{a}\times\mathbf{b})|.$$

The scalar $\mathbf{c}\cdot(\mathbf{a}\times\mathbf{b})$ is called the **scalar triple product** of the vectors **a**, **b** and **c**. As you can see from the following, you can evaluate the scalar triple product by computing a single

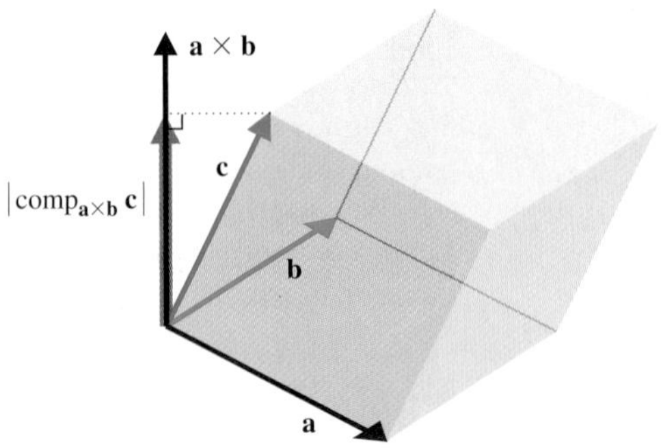

FIGURE 10.32
Parallelepiped formed by the vectors **a**, **b** and **c**

determinant. Note that for $\mathbf{a} = \langle a_1, a_2, a_3 \rangle$, $\mathbf{b} = \langle b_1, b_2, b_3 \rangle$ and $\mathbf{c} = \langle c_1, c_2, c_3 \rangle$, we have

$$\begin{aligned}
\mathbf{c} \cdot (\mathbf{a} \times \mathbf{b}) &= \mathbf{c} \cdot \begin{vmatrix} \mathbf{i} & \mathbf{j} & \mathbf{k} \\ a_1 & a_2 & a_3 \\ b_1 & b_2 & b_3 \end{vmatrix} \\
&= \langle c_1, c_2, c_3 \rangle \cdot \left(\mathbf{i} \begin{vmatrix} a_2 & a_3 \\ b_2 & b_3 \end{vmatrix} - \mathbf{j} \begin{vmatrix} a_1 & a_3 \\ b_1 & b_3 \end{vmatrix} + \mathbf{k} \begin{vmatrix} a_1 & a_2 \\ b_1 & b_2 \end{vmatrix} \right) \\
&= c_1 \begin{vmatrix} a_2 & a_3 \\ b_2 & b_3 \end{vmatrix} - c_2 \begin{vmatrix} a_1 & a_3 \\ b_1 & b_3 \end{vmatrix} + c_3 \begin{vmatrix} a_1 & a_2 \\ b_1 & b_2 \end{vmatrix} \\
&= \begin{vmatrix} c_1 & c_2 & c_3 \\ a_1 & a_2 & a_3 \\ b_1 & b_2 & b_3 \end{vmatrix}.
\end{aligned} \tag{4.7}$$

EXAMPLE 4.6 Finding the Volume of a Parallelepiped Using the Cross Product

Find the volume of the parallelepiped with three adjacent edges formed by the vectors $\mathbf{a} = \langle 1, 2, 3 \rangle$, $\mathbf{b} = \langle 4, 5, 6 \rangle$ and $\mathbf{c} = \langle 7, 8, 0 \rangle$.

Solution First, note that Volume $= |\mathbf{c} \cdot (\mathbf{a} \times \mathbf{b})|$. From (4.7), we have that

$$\begin{aligned}
\mathbf{c} \cdot (\mathbf{a} \times \mathbf{b}) &= \begin{vmatrix} 7 & 8 & 0 \\ 1 & 2 & 3 \\ 4 & 5 & 6 \end{vmatrix} = 7 \begin{vmatrix} 2 & 3 \\ 5 & 6 \end{vmatrix} - 8 \begin{vmatrix} 1 & 3 \\ 4 & 6 \end{vmatrix} + 0 \begin{vmatrix} 1 & 2 \\ 4 & 5 \end{vmatrix} \\
&= 7(-3) - 8(-6) = 27.
\end{aligned}$$

So, the volume of the parallelepiped is Volume $= |\mathbf{c} \cdot (\mathbf{a} \times \mathbf{b})| = |27| = 27$. ■

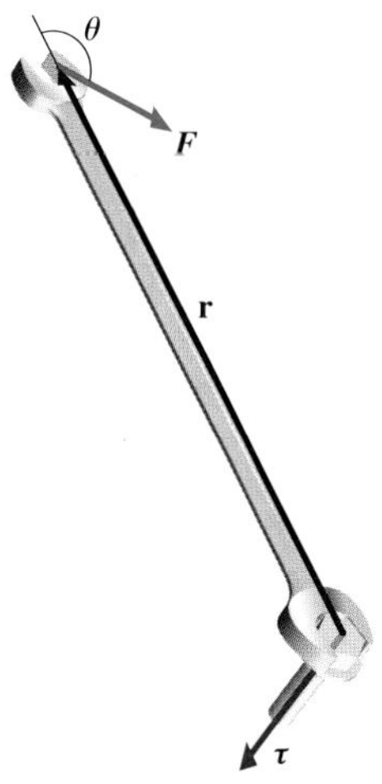

FIGURE 10.33
Torque, τ

Consider the action of a wrench on a bolt, as shown in Figure 10.33. In order to tighten the bolt, we apply a force $\mathbf{F}$ at the end of the handle, in the direction indicated in the figure. This force creates a **torque** τ acting along the axis of the bolt, drawing it in tight. Notice that the torque acts in the direction perpendicular to both $\mathbf{F}$ and the position vector $\mathbf{r}$ for the handle as indicated in Figure 10.33. In fact, using the right-hand rule, the torque acts in the same direction as $\mathbf{r} \times \mathbf{F}$ and physicists define the torque vector to be

$$\tau = \mathbf{r} \times \mathbf{F}.$$

In particular, this says that

$$\|\tau\| = \|\mathbf{r} \times \mathbf{F}\| = \|\mathbf{r}\| \|\mathbf{F}\| \sin\theta, \tag{4.8}$$

from (4.4). There are several observations we can make from this. First, this says that the farther away from the axis of the bolt we apply the force (i.e., the larger $\|\mathbf{r}\|$ is), the greater the magnitude of the torque. So, a longer wrench produces a greater torque, for a given amount of force applied. Second, notice that $\sin\theta$ is maximized when $\theta = \frac{\pi}{2}$, so that from (4.8) the magnitude of the torque is maximized when $\theta = \frac{\pi}{2}$ (when the force vector $\mathbf{F}$ is orthogonal to the position vector $\mathbf{r}$). If you've ever spent any time using a wrench, this should fit well with your experience.

EXAMPLE 4.7 Finding the Torque Applied by a Wrench

If you apply a force of magnitude 25 pounds at the end of a 15-inch-long wrench, at an angle of $\frac{\pi}{3}$ to the wrench, find the magnitude of the torque applied to the bolt. What is the maximum torque that a force of 25 pounds applied at that point can produce?

Solution From (4.8), we have

$$\|\tau\| = \|\mathbf{r}\|\|\mathbf{F}\|\sin\theta = \left(\frac{15}{12}\right)25\sin\frac{\pi}{3}$$

$$= \left(\frac{15}{12}\right)25\frac{\sqrt{3}}{2} \approx 27.1 \text{ foot-pounds.}$$

Further, the maximum torque is obtained when the angle between the wrench and the force vector is $\frac{\pi}{2}$. This would give us a maximum torque of

$$\|\tau\| = \|\mathbf{r}\|\|\mathbf{F}\|\sin\theta = \left(\frac{15}{12}\right)25\,(1) = 31.25 \text{ foot-pounds.} \quad\blacksquare$$

FIGURE 10.34
Spinning ball

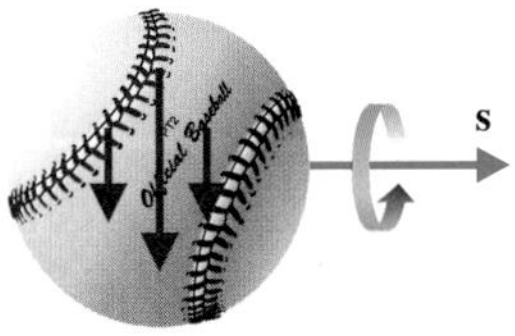

FIGURE 10.35a
Backspin

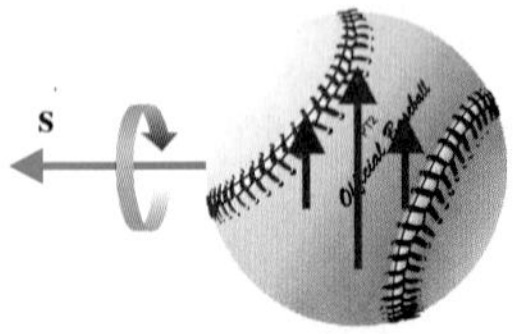

FIGURE 10.35b
Topspin

In many sports, the action is at least partially influenced by the motion of a spinning ball. For instance, in baseball, batters must contend with pitchers' curveballs and in golf, players try to control their slice. In tennis, players hit shots with topspin, while in basketball, players improve their shooting by using backspin. The list goes on and on. These are all examples of the **Magnus force,** which we describe below.

Suppose that a ball is spinning with angular velocity ω, measured in radians per second (i.e., ω is the rate of change of the rotational angle). The ball spins about an axis, as shown in Figure 10.34. We define the spin vector $\mathbf{s}$ to have magnitude ω and direction parallel to the spin axis. We use a right-hand rule to distinguish between the two directions parallel to the spin axis: curl the fingers of your right hand around the ball in the direction of the spin, and your thumb will point in the correct direction. Two examples are shown in Figures 10.35a and 10.35b. The motion of the ball disturbs the air through which it travels, creating a Magnus force $\mathbf{F}_m$ acting on the ball. For a ball moving with velocity $\mathbf{v}$ and spin vector $\mathbf{s}$, $\mathbf{F}_m$ is given by

$$\mathbf{F}_m = c(\mathbf{s} \times \mathbf{v}),$$

for some positive constant c. Suppose the balls in Figure 10.35a and Figure 10.35b are moving into the page and away from you. Using the usual sports terminology, the first ball has backspin and the second ball has topspin. Using the right-hand rule, we see that the Magnus force acting on the first ball acts in the upward direction, as shown in Figure 10.36a. This says that backspin (for example, on a basketball or golf shot) produces an upward force that helps the ball land more softly than a ball with no spin. Similarly, the Magnus force acting on the second ball acts in the downward direction (see Figure 10.36b), so that topspin (for example, on a tennis shot or baseball hit) produces a downward force that causes the ball to drop to the ground more quickly than a ball with no spin.

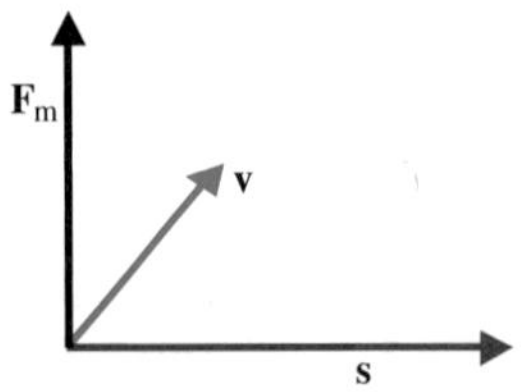

FIGURE 10.36a
Magnus force for a ball with
backspin

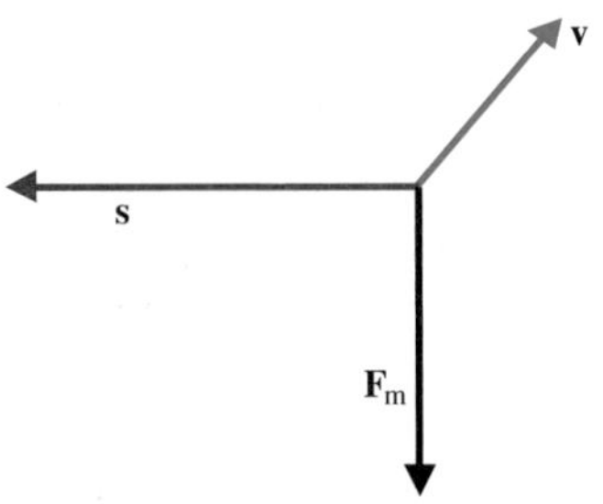

FIGURE 10.36b
Magnus force for a ball with
topspin

FIGURE 10.37a
Right-hand curveball

FIGURE 10.37b
Right-hand golf shot

EXAMPLE 4.8 Finding the Direction of a Magnus Force

The balls shown in Figures 10.37a and 10.37b are moving into the page and away from you with spin as indicated. The first ball represents a right-handed baseball pitcher's curveball, while the second ball represents a right-handed golfer's shot. Determine the direction of the Magnus force and discuss the effects on the ball.

Solution For the first ball, notice that the spin vector points up and to the left, so that $\mathbf{s} \times \mathbf{v}$ points down and to the left as shown in Figure 10.38a. Such a ball will curve to the left and drop faster than a ball that is not spinning, making it more difficult to hit. For the second ball, the spin vector points down and to the right, so $\mathbf{s} \times \mathbf{v}$ points up and to the right. Such a ball will move to the right (a "slice") and stay in the air longer than a ball that is not spinning (see Figure 10.38b).

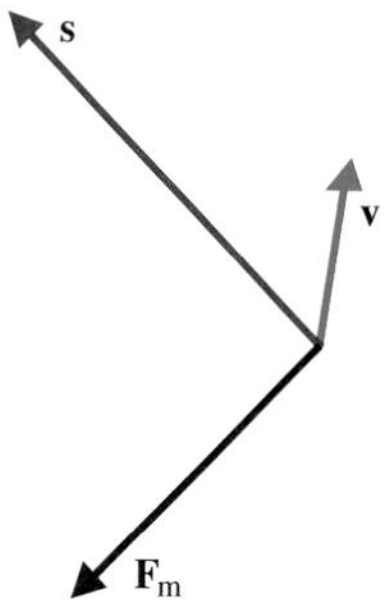

FIGURE 10.38a
Magnus force for a right-handed curveball

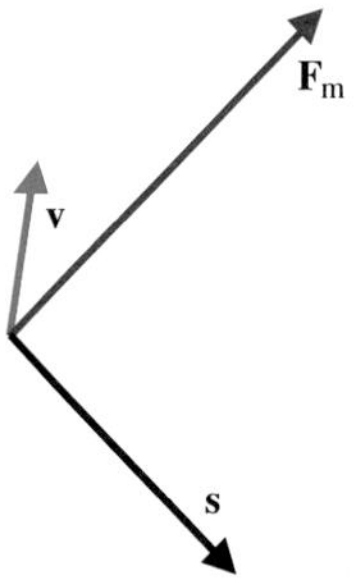

FIGURE 10.38b
Magnus force for a right-handed golf shot

EXERCISES 10.4

WRITING EXERCISES

1. In this chapter, we have developed several tests for geometric relationships. Briefly describe how to test whether two vectors are (a) parallel; (b) perpendicular. Briefly describe how to test whether (c) three points are colinear; (d) four points are coplanar.

2. The flip side of the problems in exercise 1 is to construct vectors with desired properties. Briefly describe how to construct a vector (a) parallel to a given vector; (b) perpendicular to a given vector. Given a vector, describe how to construct two other vectors such that the three vectors are mutually perpendicular.

3. Recall that torque is defined as $\boldsymbol{\tau} = \mathbf{r} \times \mathbf{F}$, where $\mathbf{F}$ is the force applied to the end of the handle and $\mathbf{r}$ is the position vector for the end of the handle. In example 4.7, how would the torque change if the force $\mathbf{F}$ were replaced with the force $-\mathbf{F}$? Answer both in mathematical terms and in physical terms.

4. Explain in geometric terms why $\mathbf{k} \times \mathbf{i} = \mathbf{j}$ and $\mathbf{k} \times \mathbf{j} = -\mathbf{i}$.

In exercises 1–4, compute the given determinant.

1. $\begin{vmatrix} 2 & 0 & -1 \\ 1 & 1 & 0 \\ -2 & -1 & 1 \end{vmatrix}$

2. $\begin{vmatrix} 0 & 2 & -1 \\ 1 & -1 & 2 \\ 1 & 1 & 2 \end{vmatrix}$

3. $\begin{vmatrix} 2 & 3 & -1 \\ 0 & 1 & 0 \\ -2 & -1 & 3 \end{vmatrix}$

4. $\begin{vmatrix} -2 & 2 & -1 \\ 0 & 3 & -2 \\ 0 & 1 & 2 \end{vmatrix}$

In exercises 5–10, compute the cross product $\mathbf{a} \times \mathbf{b}$.

5. $\mathbf{a} = \langle 1, 2, -1 \rangle$, $\mathbf{b} = \langle 1, 0, 2 \rangle$

6. $\mathbf{a} = \langle 3, 0, -1 \rangle$, $\mathbf{b} = \langle 1, 2, 2 \rangle$

7. $\mathbf{a} = \langle 0, 1, 4 \rangle$, $\mathbf{b} = \langle -1, 2, -1 \rangle$

8. $\mathbf{a} = \langle 2, -2, 0\rangle, \mathbf{b} = \langle 3, 0, 1\rangle$

9. $\mathbf{a} = 2\mathbf{i} - \mathbf{k}, \mathbf{b} = 4\mathbf{j} + \mathbf{k}$

10. $\mathbf{a} = -2\mathbf{i} + \mathbf{j} - 3\mathbf{k}, \mathbf{b} = 2\mathbf{j} - \mathbf{k}$

In exercises 11–16, find two unit vectors orthogonal to the two given vectors.

11. $\mathbf{a} = \langle 1, 0, 4\rangle, \mathbf{b} = \langle 1, -4, 2\rangle$

12. $\mathbf{a} = \langle 2, -2, 1\rangle, \mathbf{b} = \langle 0, 0, -2\rangle$

13. $\mathbf{a} = \langle 2, -1, 0\rangle, \mathbf{b} = \langle 1, 0, 3\rangle$

14. $\mathbf{a} = \langle 0, 2, 1\rangle, \mathbf{b} = \langle 1, 0, -1\rangle$

15. $\mathbf{a} = 3\mathbf{i} - \mathbf{j}, \mathbf{b} = 4\mathbf{j} + \mathbf{k}$

16. $\mathbf{a} = -2\mathbf{i} + 3\mathbf{j} - 3\mathbf{k}, \mathbf{b} = 2\mathbf{i} - \mathbf{k}$

In exercises 17–20, use the cross product to determine the angle between the vectors, assuming that $0 \le \theta \le \frac{\pi}{2}$.

17. $\mathbf{a} = \langle 1, 0, 4\rangle, \mathbf{b} = \langle 2, 0, 1\rangle$

18. $\mathbf{a} = \langle 2, 2, 1\rangle, \mathbf{b} = \langle 0, 0, 2\rangle$

19. $\mathbf{a} = 3\mathbf{i} + \mathbf{k}, \mathbf{b} = 4\mathbf{j} + \mathbf{k}$

20. $\mathbf{a} = \mathbf{i} + 3\mathbf{j} + 3\mathbf{k}, \mathbf{b} = 2\mathbf{i} + \mathbf{j}$

In exercises 21–24, find the distance from the point Q to the given line.

21. $Q = (1, 2, 0)$, line through $(0, 1, 2)$ and $(3, 1, 1)$

22. $Q = (2, 0, 1)$, line through $(1, -2, 2)$ and $(3, 0, 2)$

23. $Q = (3, -2, 1)$, line through $(2, 1, -1)$ and $(1, 1, 1)$

24. $Q = (1, 3, 1)$, line through $(1, 3, -2)$ and $(1, 0, -2)$

25. If you apply a force of magnitude 20 pounds at the end of an 8-inch-long wrench at an angle of $\frac{\pi}{4}$ to the wrench, find the magnitude of the torque applied to the bolt.

26. If you apply a force of magnitude 40 pounds at the end of an 18-inch-long wrench at an angle of $\frac{\pi}{3}$ to the wrench, find the magnitude of the torque applied to the bolt.

27. If you apply a force of magnitude 30 pounds at the end of an 8-inch-long wrench at an angle of $\frac{\pi}{6}$ to the wrench, find the magnitude of the torque applied to the bolt.

28. If you apply a force of magnitude 30 pounds at the end of an 8-inch-long wrench at an angle of $\frac{\pi}{3}$ to the wrench, find the magnitude of the torque applied to the bolt.

In exercises 29–32, assume that the balls are moving into the page (and away from you) with the indicated spin. Determine the direction of the Magnus force.

29. (a)

(b)

30. (a)

(b)

31. (a)

(b)

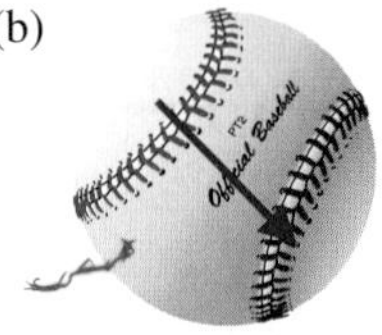

32. (a)

(b)

In exercises 33–40, a sports situation is described, with the typical ball spin shown in the indicated exercise. Discuss the effects on the ball and how the game is affected.

33. Baseball overhand fastball, spin in exercise 29(a)

34. Baseball right-handed curveball, spin in exercise 31(a)

35. Tennis topspin groundstroke, spin in exercise 32(a)

36. Tennis left-handed slice serve, spin in exercise 30(b)

37. Football spiral pass, spin in exercise 32(b)

38. Soccer left-footed "curl" kick, spin in exercise 29(b)

39. Golf "pure" hit, spin in exercise 29(a)

40. Golf right-handed "hook" shot, spin in exercise 31(b)

In exercises 41–46, label each statement as true or false. If it is true, briefly explain why. If it is false, give a counterexample.

41. If $\mathbf{a} \times \mathbf{b} = \mathbf{a} \times \mathbf{c}$, then $\mathbf{b} = \mathbf{c}$.

42. $\mathbf{a} \times \mathbf{b} = -\mathbf{b} \times \mathbf{a}$

43. $\mathbf{a} \times \mathbf{a} = \|\mathbf{a}\|^2$

44. $\mathbf{a} \cdot (\mathbf{b} \times \mathbf{c}) = (\mathbf{a} \cdot \mathbf{b}) \times \mathbf{c}$

45. If the force is doubled, the torque doubles.

46. If the spin rate is doubled, the Magnus force is doubled.

In exercises 47–52, find the indicated area or volume.

47. Area of the parallelogram with two adjacent sides formed by $\langle 2, 3\rangle$ and $\langle 1, 4\rangle$

48. Area of the parallelogram with two adjacent sides formed by $\langle -2, 1\rangle$ and $\langle 1, -3\rangle$

49. Area of the triangle with vertices $(0, 0, 0)$, $(2, 3, -1)$ and $(3, -1, 4)$

50. Area of the triangle with vertices $(0, 0, 0)$, $(0, -2, 1)$ and $(1, -3, 0)$

51. Volume of the parallelepiped with three adjacent edges formed by $\langle 2, 1, 0\rangle$, $\langle -1, 2, 0\rangle$ and $\langle 1, 1, 2\rangle$

52. Volume of the parallelepiped with three adjacent edges formed by $\langle 0, -1, 0\rangle$, $\langle 0, 2, -1\rangle$ and $\langle 1, 0, 2\rangle$

In exercises 53–58, use geometry to identify the cross product (do not compute!).

53. $\mathbf{i} \times (\mathbf{j} \times \mathbf{k})$

54. $\mathbf{j} \times (\mathbf{j} \times \mathbf{k})$

55. $\mathbf{j} \times (\mathbf{j} \times \mathbf{i})$

56. $(\mathbf{j} \times \mathbf{i}) \times \mathbf{k}$

57. $\mathbf{i} \times (3\mathbf{k})$

58. $\mathbf{k} \times (2\mathbf{i})$

In exercises 59–62, use the parallelepiped volume formula to determine whether the vectors are coplanar.

59. $\langle 2, 3, 1\rangle$, $\langle 1, 0, 2\rangle$ and $\langle 0, 3, -3\rangle$

60. $\langle 1, -3, 1\rangle$, $\langle 2, -1, 0\rangle$ and $\langle 0, -5, 2\rangle$

61. $\langle 1, 0, -2\rangle$, $\langle 3, 0, 1\rangle$ and $\langle 2, 1, 0\rangle$

62. $\langle 1, 1, 2\rangle$, $\langle 0, -1, 0\rangle$ and $\langle 3, 2, 4\rangle$

63. Show that $\|\mathbf{a} \times \mathbf{b}\|^2 = \|\mathbf{a}\|^2\|\mathbf{b}\|^2 - (\mathbf{a} \cdot \mathbf{b})^2$.

64. Prove parts (ii), (iv), (v) and (vi) of Theorem 4.3.

65. In each of the situations shown here, $\|\mathbf{a}\| = 3$ and $\|\mathbf{b}\| = 4$. In which case is $\|\mathbf{a} \times \mathbf{b}\|$ larger? What is the maximum possible value for $\|\mathbf{a} \times \mathbf{b}\|$?

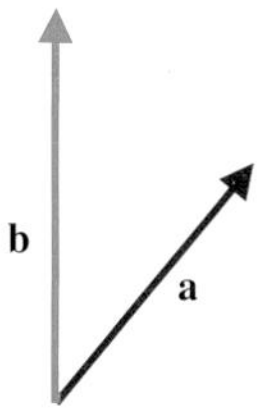

FIGURE A

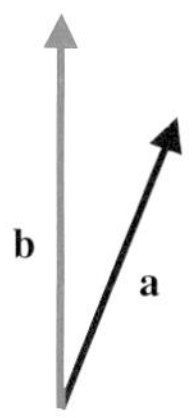

FIGURE B

66. Show that $(\mathbf{a} - \mathbf{b}) \times (\mathbf{a} + \mathbf{b}) = 2(\mathbf{a} \times \mathbf{b})$.

67. Show that $(\mathbf{a} \times \mathbf{b}) \cdot (\mathbf{c} \times \mathbf{d}) = \begin{vmatrix} \mathbf{a} \cdot \mathbf{c} & \mathbf{b} \cdot \mathbf{c} \\ \mathbf{a} \cdot \mathbf{d} & \mathbf{b} \cdot \mathbf{d} \end{vmatrix}$.

EXPLORATORY EXERCISES

1. Use the torque formula $\boldsymbol{\tau} = \mathbf{r} \times \mathbf{F}$ to explain the positioning of doorknobs. In particular, explain why the knob is placed as far as possible from the hinges and at a height that makes it possible for most people to push or pull on the door at a right angle to the door.

2. In the diagram, a foot applies a force $\mathbf{F}$ vertically to a bicycle pedal. Compute the torque on the sprocket in terms of θ and $\mathbf{F}$. Determine the angle θ at which the torque is maximized. When helping a young person to learn to ride a bicycle, most people rotate the sprocket so that the pedal sticks straight out to the front. Explain why this is helpful.

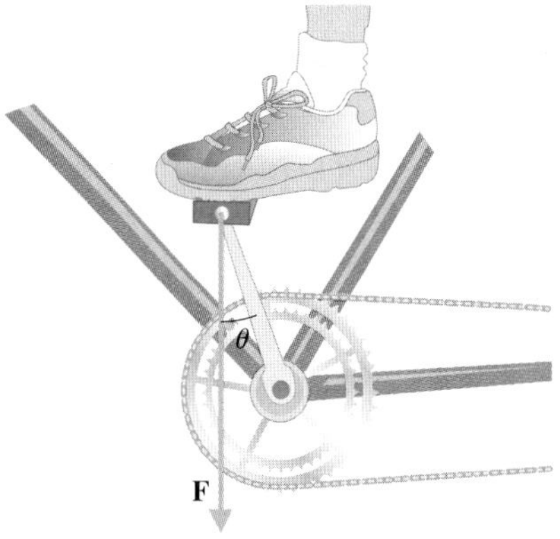

3. Devise a test that quickly determines whether $\|\mathbf{a} \times \mathbf{b}\| < |\mathbf{a} \cdot \mathbf{b}|$, $\|\mathbf{a} \times \mathbf{b}\| > |\mathbf{a} \cdot \mathbf{b}|$ or $\|\mathbf{a} \times \mathbf{b}\| = |\mathbf{a} \cdot \mathbf{b}|$. Apply your test to the following vectors: (a) $\langle 2, 1, 1\rangle$ and $\langle 3, 1, 2\rangle$ and (b) $\langle 2, 1, -1\rangle$ and $\langle -1, -2, 1\rangle$. For randomly chosen vectors, which of the three cases is the most likely?

4. In this exercise, we explore the equation of motion for a general projectile in three dimensions. Newton's second law remains $\mathbf{F} = m\mathbf{a}$, but now force and acceleration are vectors. Three forces that could affect the motion of the projectile are gravity, air drag and the Magnus force. Orient the axes such that positive z is up, positive x is right and positive y is straight ahead. The force due to gravity is weight, given by $\mathbf{F}_g = \langle 0, 0, -mg\rangle$. Air drag has magnitude proportional to the square of speed and direction opposite that of velocity. Show that if $\mathbf{v}$ is the velocity vector, then $\mathbf{F}_d = -\|\mathbf{v}\|\mathbf{v}$ satisfies both properties. That is, $\|\mathbf{F}_d\|^2 = \|\mathbf{v}\|^2$ and the angle between $\mathbf{F}_d$ and $\mathbf{v}$ is π. Finally, the Magnus force is proportional to $\mathbf{s} \times \mathbf{v}$, where $\mathbf{s}$ is the spin vector. The full model is then

$$\frac{d\mathbf{v}}{dt} = \langle 0, 0, -g\rangle - c_d\|\mathbf{v}\|\mathbf{v} + c_m(\mathbf{s} \times \mathbf{v}),$$

for positive constants c_d and c_m. With $\mathbf{v} = \langle v_x, v_y, v_z\rangle$ and $\mathbf{s} = \langle s_x, s_y, s_z\rangle$, expand this equation into separate differential equations for v_x, v_y and v_z. We can't solve these equations, but we can get some information by considering signs. For a golf drive, the spin produced could be pure backspin, in which case the spin vector is $\mathbf{s} = \langle \omega, 0, 0\rangle$ for some large $\omega > 0$. (A golf shot can have spins of 4000 rpm.) The initial velocity of a good shot would be straight ahead with some loft, $\mathbf{v}(0) = \langle 0, b, c\rangle$ for positive constants b and c. At the beginning of the flight, show that $v_y' < 0$ and thus, v_y decreases. If the ball spends approximately the same amount of time going up as coming down, conclude that the ball will travel further downrange going up than coming down. In fact, the ball does *not* spend equal amounts of time going up and coming down. By examining the sign of v_z' going up ($v_z > 0$) versus coming down ($v_z < 0$), determine whether the ball spends more time going up or coming

down. Next, consider the case of a ball with some sidespin, so that $s_x > 0$ and $s_y > 0$. By examining the sign of v_x', determine whether this ball will curve to the right or left. Examine the other equations and determine what other effects this sidespin may have.

10.5 LINES AND PLANES IN SPACE

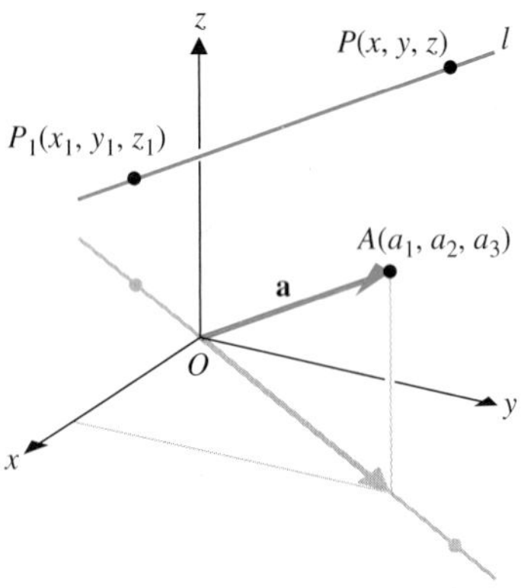

FIGURE 10.39
Line in space

Normally, you specify a line in the xy-plane by selecting any two points on the line or a single point on the line and its direction, as indicated by the *slope* of the line. In three dimensions, specifying two points on a line will still determine the line. An alternative is to specify a single point on the line and its *direction*. In three dimensions, direction should make you think about vectors right away.

Let's look for the line that passes through the point $P_1(x_1, y_1, z_1)$ and that is parallel to the position vector $\mathbf{a} = \langle a_1, a_2, a_3 \rangle$ (see Figure 10.39). For any other point $P(x, y, z)$ on the line, observe that the vector $\overrightarrow{P_1 P}$ will be parallel to $\mathbf{a}$. Further, two vectors are parallel if and only if one is a scalar multiple of the other, so that

$$\overrightarrow{P_1 P} = t\mathbf{a}, \tag{5.1}$$

for some scalar t. The line then consists of all points $P(x, y, z)$ for which (5.1) holds. Since

$$\overrightarrow{P_1 P} = \langle x - x_1, y - y_1, z - z_1 \rangle,$$

we have from (5.1) that

$$\langle x - x_1, y - y_1, z - z_1 \rangle = t\mathbf{a} = t \langle a_1, a_2, a_3 \rangle.$$

Finally, since two vectors are equal if and only if all of their components are equal, we have

Parametric equations of a line

set of parametric equations

$$x = ta_1 + x_1 \qquad y = ta_2 + y_1 \qquad z = ta_3 + z_1$$

$$\boxed{x - x_1 = a_1 t, \quad y - y_1 = a_2 t \quad \text{and} \quad z - z_1 = a_3 t.} \tag{5.2}$$

We call (5.2) **parametric equations** for the line, where t is the **parameter.** As in the two-dimensional case, a line in space can be represented by many different sets of parametric equations. Provided none of a_1, a_2 or a_3 are zero, we can solve for the parameter in each of the three equations, to obtain

Symmetric equations of a line

$$\boxed{\frac{x - x_1}{a_1} = \frac{y - y_1}{a_2} = \frac{z - z_1}{a_3}.} \tag{5.3}$$

We refer to (5.3) as **symmetric equations** of the line.

EXAMPLE 5.1   Finding Equations of a Line Given a Point and a Vector

Find an equation of the line through the point $(1, 5, 2)$ and parallel to the vector $\langle 4, 3, 7 \rangle$. Also, determine where the line intersects the yz-plane.

Solution   From (5.2), parametric equations for the line are

$$x - 1 = 4t, \quad y - 5 = 3t \quad \text{and} \quad z - 2 = 7t.$$

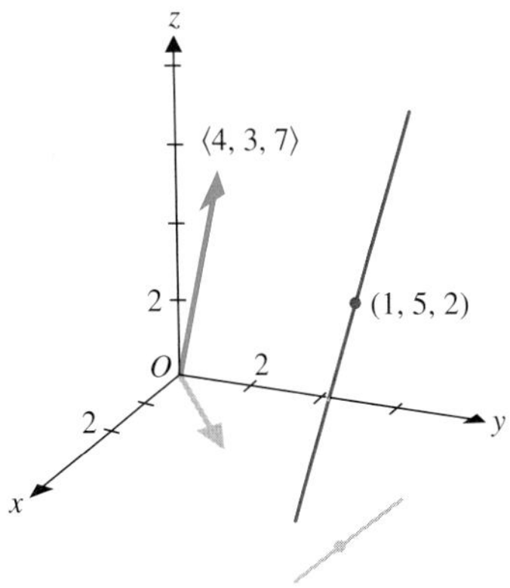

FIGURE 10.40
The line $x = 1 + 4t$, $y = 5 + 3t$, $z = 2 + 7t$

From (5.3), symmetric equations of the line are

$$\frac{x-1}{4} = \frac{y-5}{3} = \frac{z-2}{7}. \tag{5.4}$$

We show the graph of the line in Figure 10.40. Note that the line intersects the yz-plane where $x = 0$. Setting $x = 0$ in (5.4), we solve for y and z to obtain

$$y = \frac{17}{4} \quad \text{and} \quad z = \frac{1}{4}.$$

Alternatively, observe that we could solve $x - 1 = 4t$ for t (again where $x = 0$) and substitute this into the parametric equations for y and z. So, the line intersects the yz-plane at the point $\left(0, \frac{17}{4}, \frac{1}{4}\right)$. ■

Given two points, we can easily find the equations of the line passing through them, as in example 5.2.

EXAMPLE 5.2 Finding Equations of a Line Given Two Points

Find an equation of the line passing through the points $P(1, 2, -1)$ and $Q(5, -3, 4)$.

Solution First, a vector that is parallel to the line is

$$\overrightarrow{PQ} = \langle 5-1, -3-2, 4-(-1)\rangle = \langle 4, -5, 5\rangle.$$

Picking either point will give us equations for the line. Here, we use P, so that parametric equations for the line are

$$x - 1 = 4t, \quad y - 2 = -5t \quad \text{and} \quad z + 1 = 5t.$$

Similarly, symmetric equations of the line are

$$\frac{x-1}{4} = \frac{y-2}{-5} = \frac{z+1}{5}.$$

We show the graph of the line in Figure 10.41. ■

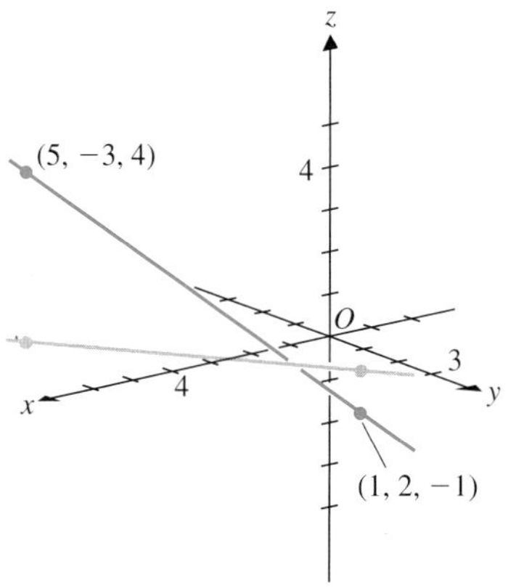

FIGURE 10.41
The line $\frac{x-1}{4} = \frac{y-2}{-5} = \frac{z+1}{5}$

Since we have specified a line by choosing a point on the line and a vector with the same direction, Definition 5.1 should be no surprise.

DEFINITION 5.1

Let l_1 and l_2 be two lines in $\mathbb{R}^3$, with parallel vectors **a** and **b**, respectively, and let θ be the angle between **a** and **b**.

(i) The lines l_1 and l_2 are **parallel** whenever **a** and **b** are parallel.
(ii) If l_1 and l_2 intersect, then
  (a) the angle between l_1 and l_2 is θ and
  (b) the lines l_1 and l_2 are **orthogonal** whenever **a** and **b** are orthogonal.

In two dimensions, two lines are either parallel or they intersect. This is not true in three dimensions, as we see in example 5.3.

EXAMPLE 5.3 Showing Two Lines Are Not Parallel but Do Not Intersect

Show that the lines

$$l_1 : x - 2 = -t, \quad y - 1 = 2t \quad \text{and} \quad z - 5 = 2t$$

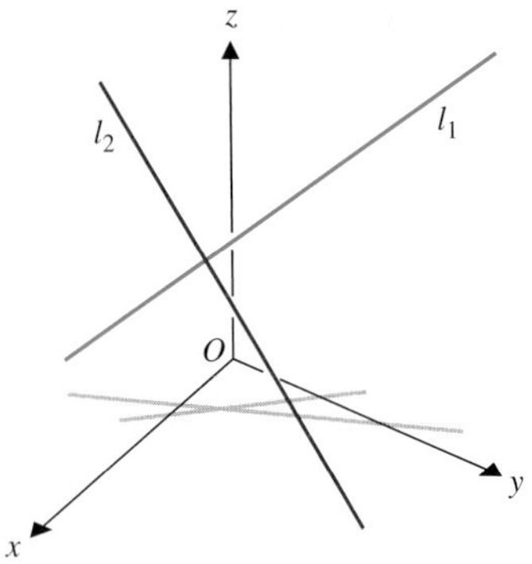

FIGURE 10.42
Skew lines

and
$$l_2 : x - 1 = s, \quad y - 2 = -s \quad \text{and} \quad z - 1 = 3s$$

are not parallel, yet do not intersect.

Solution Notice immediately that we have used different letters (t and s) as parameters for the two lines. In this setting, the parameter is a dummy variable, so the letter used is not significant. However, solving the first parametric equation of each line for the parameter in terms of x, we get

$$t = 2 - x \quad \text{and} \quad s = x - 1,$$

respectively. This says that the parameter represents something different in each line; so we must use different letters. Notice from the graph in Figure 10.42 that the lines are most certainly not parallel, but it is unclear whether or not they intersect. (Remember, the graph is a two-dimensional rendering of lines in three dimensions and so, while the two-dimensional lines drawn do intersect, it's unclear whether or not the three-dimensional lines that they represent intersect.)

You can read from the parametric equations that a vector parallel to l_1 is $\mathbf{a}_1 = \langle -1, 2, 2 \rangle$, while a vector parallel to l_2 is $\mathbf{a}_2 = \langle 1, -1, 3 \rangle$. Since $\mathbf{a}_1$ is not a scalar multiple of $\mathbf{a}_2$, the vectors are not parallel and so, the lines l_1 and l_2 are not parallel. The lines intersect if there's a choice of the parameters s and t that produces the same point, that is, that produces the same values for all of x, y and z. Setting the x-values equal, we get

$$2 - t = 1 + s,$$

so that $s = 1 - t$. Setting the y-values equal and setting $s = 1 - t$, we get

$$1 + 2t = 2 - s = 2 - (1 - t) = 1 + t.$$

Solving this for t yields $t = 0$, which further implies that $s = 1$. Setting the z-components equal gives

$$5 + 2t = 3s + 1,$$

but this is not satisfied when $t = 0$ and $s = 1$. So, l_1 and l_2 are not parallel, yet do not intersect. ■

DEFINITION 5.2

Nonparallel, nonintersecting lines are called **skew** lines.

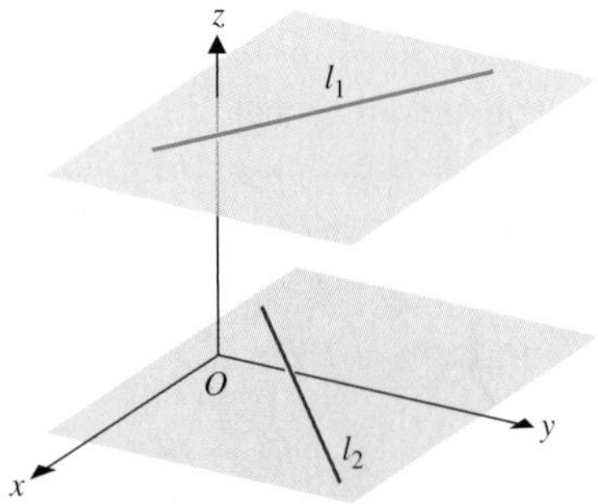

FIGURE 10.43
Skew lines

Note that it's fairly easy to visualize skew lines. Draw two planes that are parallel and draw a line in each plane (so that it lies completely in the plane). As long as the two lines are not parallel, these are skew lines (see Figure 10.43).

○ Planes in $\mathbb{R}^3$

Think about what information you might need to specify a plane in space. As a simple example, observe that the yz-plane is a set of points in space such that every vector connecting two points in the set is orthogonal to $\mathbf{i}$. However, every plane parallel to the yz-plane satisfies this criterion (see Figure 10.44). In order to select the one that corresponds to the yz-plane, you need to specify a point through which it passes (any one will do).

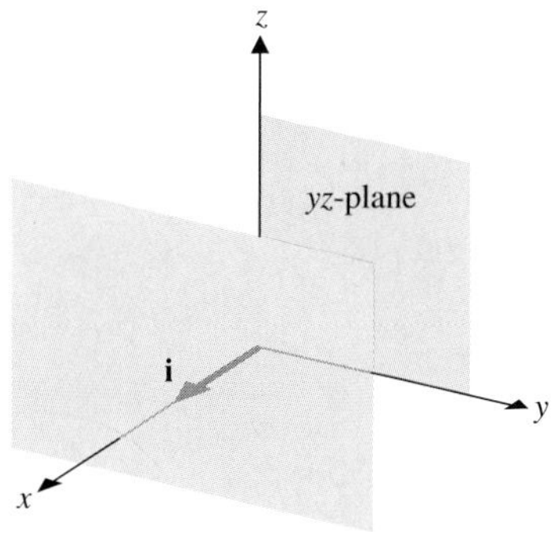

FIGURE 10.44
Parallel planes

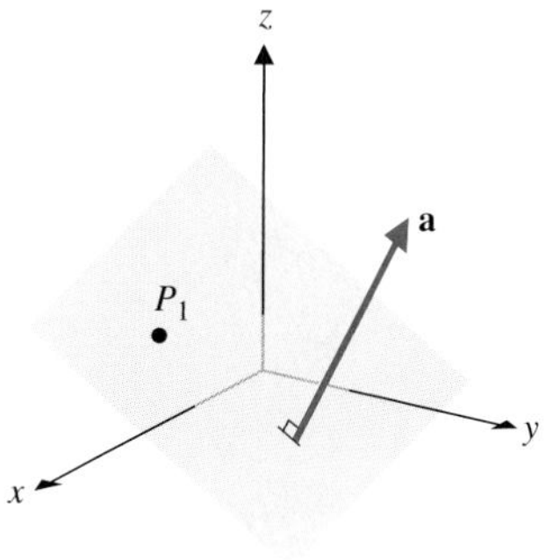

FIGURE 10.45
Plane in $\mathbb{R}^3$

In general, a plane in space is determined by specifying a vector $\mathbf{a} = \langle a_1, a_2, a_3\rangle$ that is **normal** to the plane (i.e., orthogonal to every vector lying in the plane) and a point $P_1(x_1, y_1, z_1)$ lying in the plane (see Figure 10.45). In order to find an equation of the plane, let $P(x, y, z)$ represent any point in the plane. Then, since P and P_1 are both points in the plane, the vector $\overrightarrow{P_1P} = \langle x - x_1, y - y_1, z - z_1\rangle$ lies in the plane and so, must be orthogonal to $\mathbf{a}$. By Corollary 3.1, we have that

$$0 = \mathbf{a} \cdot \overrightarrow{P_1P} = \langle a_1, a_2, a_3\rangle \cdot \langle x - x_1, y - y_1, z - z_1\rangle$$

Equation of a plane

or

$$\boxed{0 = a_1(x - x_1) + a_2(y - y_1) + a_3(z - z_1).} \tag{5.5}$$

Equation (5.5) is an equation for the plane passing through the point (x_1, y_1, z_1) with normal vector $\langle a_1, a_2, a_3\rangle$. It's a simple matter to use this to find the equation of any particular plane. We illustrate this in example 5.4.

EXAMPLE 5.4 The Equation and Graph of a Plane Given a Point and a Normal Vector

Find an equation of the plane containing the point (1, 2, 3) with normal vector $\langle 4, 5, 6\rangle$, and sketch the plane.

Solution From (5.5), we have the equation

$$0 = 4(x - 1) + 5(y - 2) + 6(z - 3). \tag{5.6}$$

To draw the plane, we locate three points lying in the plane. In this case, the simplest way to do this is to look at the intersections of the plane with each of the coordinate axes. When $y = z = 0$, we get from (5.6) that

$$0 = 4(x - 1) + 5(0 - 2) + 6(0 - 3) = 4x - 4 - 10 - 18,$$

so that $4x = 32$ or $x = 8$. The intersection of the plane with the x-axis is then the point (8, 0, 0). Similarly, you can find the intersections of the plane with the y- and z-axes: $\left(0, \frac{32}{5}, 0\right)$ and $\left(0, 0, \frac{16}{3}\right)$, respectively. Using these three points, we can draw the plane

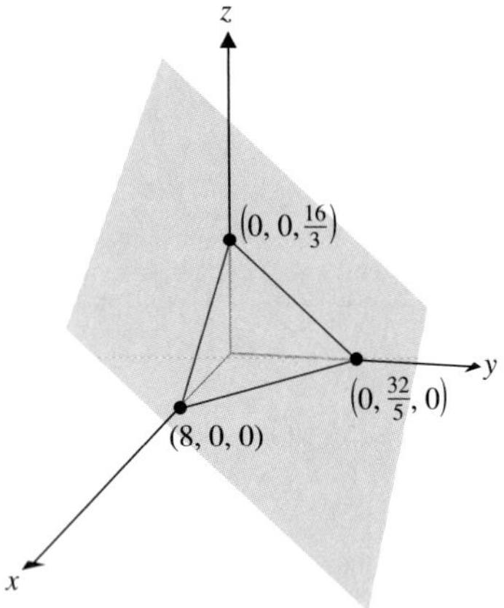

FIGURE 10.46
The plane through $(8, 0, 0)$, $\left(0, \frac{32}{5}, 0\right)$ and $\left(0, 0, \frac{16}{3}\right)$

For line → Parametric equation

For Plane → linear equation

seen in Figure 10.46. We start by drawing the triangle with vertices at the three points; the plane we want is the one containing this triangle. Notice that since the plane intersects all three of the coordinate axes, the portion of the plane in the first octant is the indicated triangle and its interior. ■

Note that if we expand out the expression in (5.5), we get

$$\begin{aligned} 0 &= a_1(x - x_1) + a_2(y - y_1) + a_3(z - z_1) \\ &= a_1x + a_2y + a_3z + \underbrace{(-a_1x_1 - a_2y_1 - a_3z_1)}_{\text{constant}}. \end{aligned}$$

We refer to this last equation as a **linear equation** in the three variables x, y and z. In particular, this says that every linear equation of the form

$$0 = ax + by + cz + d,$$

where a, b, c and d are constants, is the equation of a plane with normal vector $\langle a, b, c\rangle$.

We observed earlier that three points determine a plane. But, how can you find an equation of a plane given only three points? If you are to use (5.5), you'll first need to find a normal vector. We can easily resolve this, as in example 5.5.

EXAMPLE 5.5 Finding the Equation of a Plane Given Three Points

Find the plane containing the three points $P(1, 2, 2)$, $Q(2, -1, 4)$ and $R(3, 5, -2)$.

Solution First, we'll need to find a vector normal to the plane. Notice that two vectors lying in the plane are

$$\overrightarrow{PQ} = \langle 1, -3, 2\rangle \quad \text{and} \quad \overrightarrow{QR} = \langle 1, 6, -6\rangle.$$

Consequently, a vector orthogonal to both of $\overrightarrow{PQ}$ and $\overrightarrow{QR}$ is the cross product

$$\overrightarrow{PQ} \times \overrightarrow{QR} = \begin{vmatrix} \mathbf{i} & \mathbf{j} & \mathbf{k} \\ 1 & -3 & 2 \\ 1 & 6 & -6 \end{vmatrix} = \langle 6, 8, 9\rangle.$$

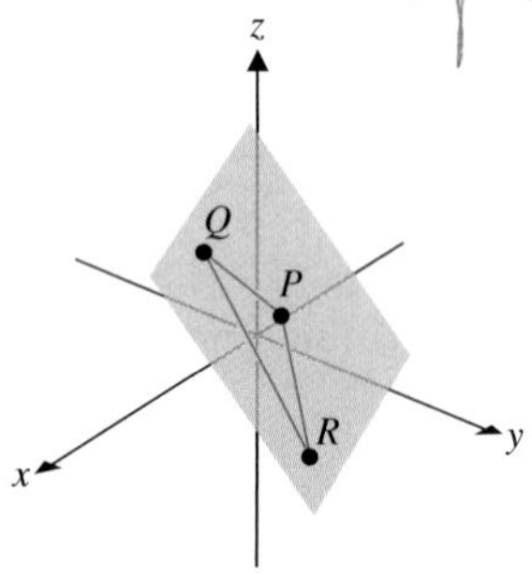

FIGURE 10.47
Plane containing three points

Since $\overrightarrow{PQ}$ and $\overrightarrow{QR}$ are not parallel, $\overrightarrow{PQ} \times \overrightarrow{QR}$ must be orthogonal to the plane, as well. (Why is that?) From (5.5), an equation for the plane is then

$$0 = 6(x - 1) + 8(y - 2) + 9(z - 2).$$

In Figure 10.47, we show the triangle with vertices at the three points. The plane in question is the one containing the indicated triangle. ■

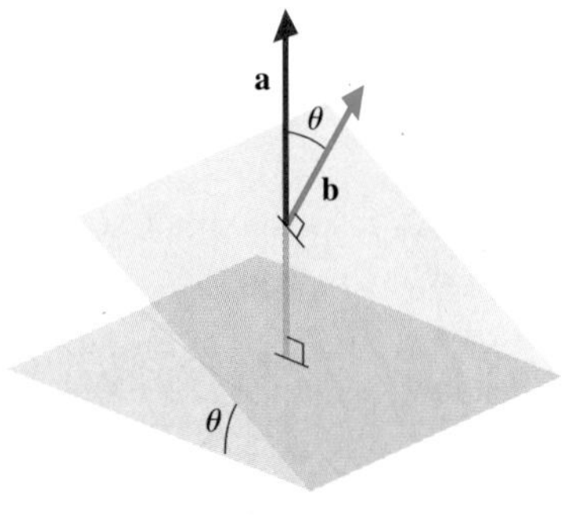

FIGURE 10.48
Angle between planes

In three dimensions, two planes are either parallel or they intersect in a straight line. (Think about this some.) Suppose that two planes having normal vectors **a** and **b**, respectively, intersect. Then the angle between the planes is the same as the angle between **a** and **b** (see Figure 10.48). With this in mind, we say that the two planes are **parallel** whenever their normal vectors are parallel and the planes are **orthogonal** whenever their normal vectors are orthogonal.

EXAMPLE 5.6 The Equation of a Plane Given a Point and a Parallel Plane

Find an equation for the plane through the point $(1, 4, -5)$ and parallel to the plane defined by $2x - 5y + 7z = 12$.

Solution First, notice that a normal vector to the given plane is $\langle 2, -5, 7\rangle$. Since the two planes are to be parallel, this vector is also normal to the new plane. From (5.5), we can write down the equation of the plane:

$$0 = 2(x - 1) - 5(y - 4) + 7(z + 5).$$ ■

It's particularly easy to see that some planes are parallel to the coordinate planes.

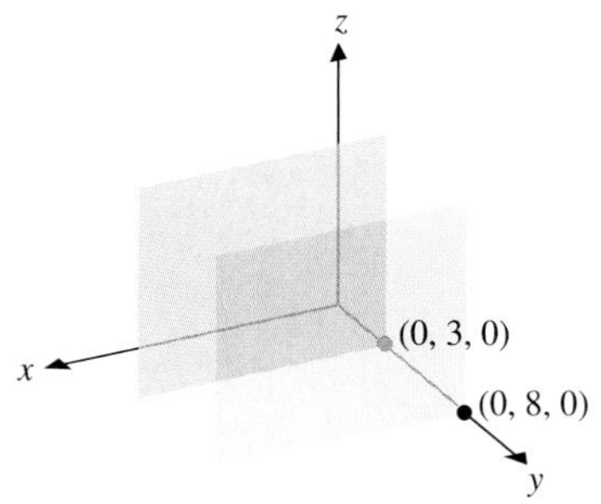

FIGURE 10.49
The planes $y = 3$ and $y = 8$

EXAMPLE 5.7 Drawing Some Simple Planes

Draw the planes $y = 3$ and $y = 8$.

Solution First, notice that both equations represent planes with the same normal vector, $\langle 0, 1, 0\rangle = \mathbf{j}$. This says that the planes are both parallel to the xz-plane, the first one passing through the point $(0, 3, 0)$ and the second one passing through $(0, 8, 0)$, as seen in Figure 10.49. ■

You should recognize that the intersection of two nonparallel planes will be a line. (Think about this some!) In example 5.8, we see how to find an equation of the line of intersection.

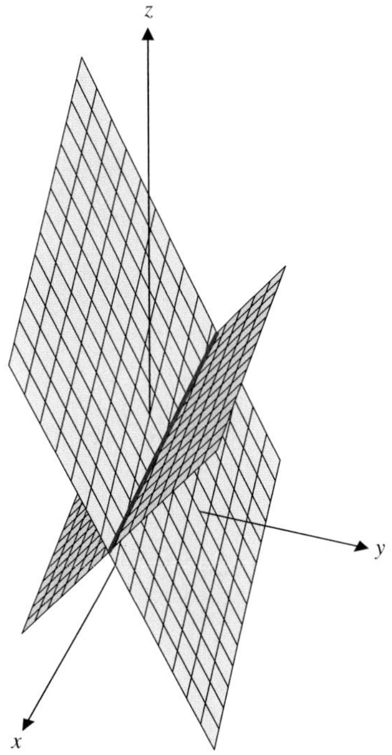

FIGURE 10.50
Intersection of planes

EXAMPLE 5.8 Finding the Intersection of Two Planes

Find the line of intersection of the planes $x + 2y + z = 3$ and $x - 4y + 3z = 5$.

Solution Solving both equations for x, we get

$$x = 3 - 2y - z \quad \text{and} \quad x = 5 + 4y - 3z. \tag{5.7}$$

Setting these expressions for x equal gives us

$$3 - 2y - z = 5 + 4y - 3z.$$

Solving this for z gives us

$$2z = 6y + 2 \quad \text{or} \quad z = 3y + 1.$$

Returning to either equation in (5.7), we can solve for x (also in terms of y). We have

$$x = 3 - 2y - z = 3 - 2y - (3y + 1) = -5y + 2.$$

Taking y as the parameter (i.e., letting $y = t$), we obtain parametric equations for the line of intersection:

$$x = -5t + 2, \quad y = t \quad \text{and} \quad z = 3t + 1.$$

You can see the line of intersection in the computer-generated graph of the two planes seen in Figure 10.50. ■

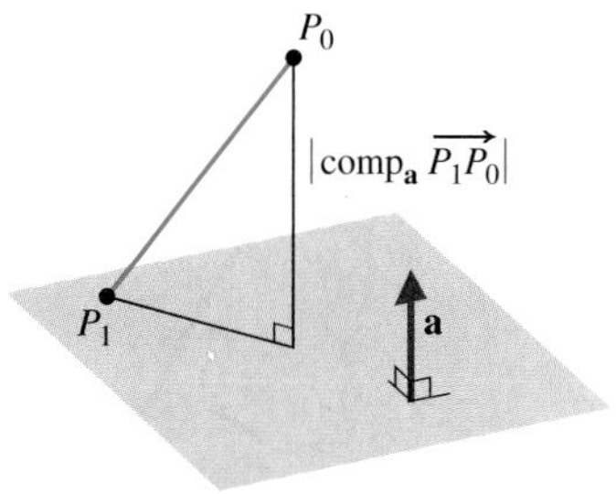

FIGURE 10.51
Distance from a point to a plane

Observe that the distance from the plane $ax + by + cz + d = 0$ to a point $P_0(x_0, y_0, z_0)$ not on the plane is measured along a line segment connecting the point to the plane that is orthogonal to the plane (see Figure 10.51). To compute this distance, pick any point $P_1(x_1, y_1, z_1)$ lying in the plane and let $\mathbf{a} = \langle a, b, c\rangle$ denote a vector normal to the plane. From Figure 10.51 notice that the distance from P_0 to the plane is simply $|\text{comp}_{\mathbf{a}} \overrightarrow{P_1P_0}|$, where

$$\overrightarrow{P_1P_0} = \langle x_0 - x_1, y_0 - y_1, z_0 - z_1\rangle.$$

From (3.7), the distance is

$$\left|\text{comp}_{\mathbf{a}} \overrightarrow{P_1 P_0}\right| = \left|\overrightarrow{P_1 P_0} \cdot \frac{\mathbf{a}}{\|\mathbf{a}\|}\right|$$

$$= \left|\langle x_0 - x_1, y_0 - y_1, z_0 - z_1\rangle \cdot \frac{\langle a, b, c\rangle}{\|\langle a, b, c\rangle\|}\right|$$

$$= \frac{|a(x_0 - x_1) + b(y_0 - y_1) + c(z_0 - z_1)|}{\sqrt{a^2 + b^2 + c^2}}$$

$$= \frac{|ax_0 + by_0 + cz_0 - (ax_1 + by_1 + cz_1)|}{\sqrt{a^2 + b^2 + c^2}}$$

$$= \frac{|ax_0 + by_0 + cz_0 + d|}{\sqrt{a^2 + b^2 + c^2}}, \tag{5.8}$$

since (x_1, y_1, z_1) lies in the plane and $ax + by + cz = -d$, for every point (x, y, z) in the plane.

EXAMPLE 5.9 Finding the Distance between Parallel Planes

Find the distance between the parallel planes:

$$P_1\colon 2x - 3y + z = 6$$

and

$$P_2\colon 4x - 6y + 2z = 8.$$

Solution First, observe that the planes are parallel, since their normal vectors $\langle 2, -3, 1\rangle$ and $\langle 4, -6, 2\rangle$ are parallel. Further, since the planes are parallel, the distance from the plane P_1 to every point in the plane P_2 is the same. So, pick any point in P_2, say $(0, 0, 4)$. (This is certainly convenient.) The distance d from the point $(0, 0, 4)$ to the plane P_1 is then given by (5.8) to be

$$d = \frac{|(2)(0) - (3)(0) + (1)(4) - 6|}{\sqrt{2^2 + 3^2 + 1^2}} = \frac{2}{\sqrt{14}}. \blacksquare$$

BEYOND FORMULAS

Both lines and planes are defined in this section in terms of a point and a vector. To avoid confusing the equations of lines and planes, focus on understanding the derivation of each equation. Parametric equations of a line simply express the line in terms of a starting point and a direction of motion. The equation of a plane is simply an expanded version of the dot product equation for the normal vector being perpendicular to the plane. Hopefully, you have discovered that a formula is easier to memorize if you understand the logic behind the result.

EXERCISES 10.5

WRITING EXERCISES

1. Explain how to shift back and forth between the parametric and symmetric equations of a line. Describe one situation in which you would prefer to have parametric equations to work with and one situation in which symmetric equations would be more convenient.

2. Lines and planes can both be specified with a point and a vector. Discuss the differences in the vectors used, and explain why the normal vector of the plane specifies an entire plane, while the direction vector of the line merely specifies a line.

3. Notice that if $c = 0$ in the general equation $ax + by + cz + d = 0$ of a plane, you have an equation that would describe a line in the xy-plane. Describe how this line relates to the plane.

4. Our hint about visualizing skew lines was to place the lines in parallel planes. Discuss whether every pair of skew lines must necessarily lie in parallel planes. (Hint: Discuss how the cross product of the direction vectors of the lines would relate to the parallel planes.)

In exercises 1–10, find (a) parametric equations and (b) symmetric equations of the line.

1. The line through $(1, 2, -3)$ and parallel to $\langle 2, -1, 4\rangle$

2. The line through $(3, -2, 4)$ and parallel to $\langle 3, 2, -1\rangle$

3. The line through $(2, 1, 3)$ and $(4, 0, 4)$

4. The line through $(0, 2, 1)$ and $(2, 0, 2)$

5. The line through $(1, 4, 1)$ and parallel to the line $x = 2 - 3t$, $y = 4$, $z = 6 + t$

6. The line through $(-1, 0, 0)$ and parallel to the line $\dfrac{x+1}{-2} = \dfrac{y}{3} = z - 2$

7. The line through $(2, 0, 1)$ and perpendicular to both $\langle 1, 0, 2\rangle$ and $\langle 0, 2, 1\rangle$

8. The line through $(-3, 1, 0)$ and perpendicular to both $\langle 0, -3, 1\rangle$ and $\langle 4, 2, -1\rangle$

9. The line through $(1, 2, -1)$ and normal to the plane $2x - y + 3z = 12$

10. The line through $(0, -2, 1)$ and normal to the plane $y + 3z = 4$

In exercises 11–16, state whether the lines are parallel or perpendicular and find the angle between the lines.

11. $\begin{cases} x = 1 - 3t \\ y = 2 + 4t \\ z = -6 + t \end{cases}$ and $\begin{cases} x = 1 + 2s \\ y = 2 - 2s \\ z = -6 + s \end{cases}$

12. $\begin{cases} x = 4 - 2t \\ y = 3t \\ z = -1 + 2t \end{cases}$ and $\begin{cases} x = 4 + s \\ y = -2s \\ z = -1 + 3s \end{cases}$

13. $\begin{cases} x = 1 + 2t \\ y = 3 \\ z = -1 + t \end{cases}$ and $\begin{cases} x = 2 - s \\ y = 10 + 5s \\ z = 3 + 2s \end{cases}$

14. $\begin{cases} x = 1 - 2t \\ y = 2t \\ z = 5 - t \end{cases}$ and $\begin{cases} x = 3 + 2s \\ y = -2 - 2s \\ z = 6 + s \end{cases}$

15. $\begin{cases} x = -1 + 2t \\ y = 3 + 4t \\ z = -6t \end{cases}$ and $\begin{cases} x = 3 - s \\ y = 1 - 2s \\ z = 3s \end{cases}$

16. $\begin{cases} x = 3 - t \\ y = 4 \\ z = -2 + 2t \end{cases}$ and $\begin{cases} x = 1 + 2s \\ y = 7 - 3s \\ z = -3 + s \end{cases}$

In exercises 17–20, determine whether the lines are parallel, skew or intersect.

17. $\begin{cases} x = 4 + t \\ y = 2 \\ z = 3 + 2t \end{cases}$ and $\begin{cases} x = 2 + 2s \\ y = 2s \\ z = -1 + 4s \end{cases}$

18. $\begin{cases} x = 3 + t \\ y = 3 + 3t \\ z = 4 - t \end{cases}$ and $\begin{cases} x = 2 - s \\ y = 1 - 2s \\ z = 6 + 2s \end{cases}$

19. $\begin{cases} x = 1 + 2t \\ y = 3 \\ z = -1 - 4t \end{cases}$ and $\begin{cases} x = 2 - s \\ y = 2 \\ z = 3 + 2s \end{cases}$

20. $\begin{cases} x = 1 - 2t \\ y = 2t \\ z = 5 - t \end{cases}$ and $\begin{cases} x = 3 + 2s \\ y = -2 \\ z = 3 + 2s \end{cases}$

In exercises 21–30, find an equation of the given plane.

21. The plane containing the point $(1, 3, 2)$ with normal vector $\langle 2, -1, 5\rangle$

22. The plane containing the point $(-2, 1, 0)$ with normal vector $\langle -3, 0, 2\rangle$

23. The plane containing the points $(2, 0, 3)$, $(1, 1, 0)$ and $(3, 2, -1)$

24. The plane containing the points $(1, -2, 1)$, $(2, -1, 0)$ and $(3, -2, 2)$

25. The plane containing the points $(-2, 2, 0)$, $(-2, 3, 2)$ and $(1, 2, 2)$

26. The plane containing the point $(3, -2, 1)$ and parallel to the plane $x + 3y - 4z = 2$

27. The plane containing the point $(0, -2, -1)$ and parallel to the plane $-2x + 4y = 3$

28. The plane containing the point $(3, 1, 0)$ and parallel to the plane $-3x - 3y + 2z = 4$

29. The plane containing the point $(1, 2, 1)$ and perpendicular to the planes $x + y = 2$ and $2x + y - z = 1$

30. The plane containing the point $(3, 0, -1)$ and perpendicular to the planes $x + 2y - z = 2$ and $2x - z = 1$

In exercises 31–40, sketch the given plane.

31. $x + y + z = 4$

32. $2x - y + 4z = 4$

33. $3x + 6y - z = 6$

34. $2x + y + 3z = 6$

35. $x = 4$

36. $y = 3$

37. $z = 2$

38. $x + y = 1$

39. $2x - z = 2$

40. $y = x + 2$

In exercises 41–44, find the intersection of the planes.

41. $2x - y - z = 4$ and $3x - 2y + z = 0$

42. $3x + y - z = 2$ and $2x - 3y + z = -1$

43. $3x + 4y = 1$ and $x + y - z = 3$

44. $x - 2y + z = 2$ and $x + 3y - 2z = 0$

In exercises 45–50, find the distance between the given objects.

45. The point $(2, 0, 1)$ and the plane $2x - y + 2z = 4$

46. The point $(1, 3, 0)$ and the plane $3x + y - 5z = 2$

47. The point $(2, -1, -1)$ and the plane $x - y + z = 4$

48. The point $(0, -1, 1)$ and the plane $2x - 3y = 2$

49. The planes $2x - y - z = 1$ and $2x - y - z = 4$

50. The planes $x + 3y - 2z = 3$ and $x + 3y - 2z = 1$

51. Show that the distance between planes $ax + by + cz = d_1$ and $ax + by + cz = d_2$ is given by $\dfrac{|d_2 - d_1|}{\sqrt{a^2 + b^2 + c^2}}$.

52. Suppose that $\langle 2, 1, 3\rangle$ is a normal vector for a plane containing the point $(2, -3, 4)$. Show that an equation of the plane is $2x + y + 3z = 13$. Explain why another normal vector for this plane is $\langle -4, -2, -6\rangle$. Use this normal vector to find an equation of the plane and show that the equation reduces to the same equation, $2x + y + 3z = 13$.

53. Find an equation of the plane containing the lines $\begin{cases} x = 4 + t \\ y = 2 \\ z = 3 + 2t \end{cases}$ and $\begin{cases} x = 2 + 2s \\ y = 2s \\ z = -1 + 4s \end{cases}$.

54. Find an equation of the plane containing the lines $\begin{cases} x = 1 - t \\ y = 2 + 3t \\ z = 2t \end{cases}$ and $\begin{cases} x = 1 - s \\ y = 5 \\ z = 4 - 2s \end{cases}$.

In exercises 55–62, state whether the statement is true or false (not always true).

55. Two planes either are parallel or intersect.

56. The intersection of two planes is a line.

57. The intersection of three planes is a point.

58. Lines that lie in parallel planes are always skew.

59. The set of all lines perpendicular to a given line forms a plane.

60. There is one line perpendicular to a given plane.

61. The set of all points equidistant from two given points forms a plane.

62. The set of all points equidistant from two given planes forms a plane.

In exercises 63–66, determine whether the given lines or planes are the same.

63. $x = 3 - 2t, y = 3t, z = t - 2$ and $x = 1 + 4t, y = 3 - 6t, z = -1 - 2t$

64. $x = 1 + 4t, y = 2 - 2t, z = 2 + 6t$ and $x = 9 - 2t, y = -2 + t, z = 8 - 3t$

65. $2(x - 1) - (y + 2) + (z - 3) = 0$ and $4x - 2y + 2z = 2$

66. $3(x + 1) + 2(y - 2) - 3(z + 1) = 0$ and $6(x - 2) + 4(y + 1) - 6z = 0$

67. Suppose two airplanes fly paths described by the parametric equations $P_1: \begin{cases} x = 3 \\ y = 6 - 2t \\ z = 3t + 1 \end{cases}$ and $P_2: \begin{cases} x = 1 + 2s \\ y = 3 + s \\ z = 2 + 2s \end{cases}$. Describe the shape of the flight paths. If $t = s$ represents time, determine whether the paths intersect. Determine if the planes collide.

EXPLORATORY EXERCISES

1. Compare the equations that we have developed for the distance between a (two-dimensional) point and a line and for a (three-dimensional) point and a plane. Based on these equations, hypothesize a formula for the distance between the (four-dimensional) point (x_1, y_1, z_1, w_1) and the hyperplane $ax + by + cz + dw + e = 0$.

2. In this exercise, we will explore the geometrical object determined by the parametric equations $\begin{cases} x = 2s + 3t \\ y = 3s + 2t \\ z = s + t \end{cases}$. Given that there are two parameters, what dimension do you expect the object to have? Given that the individual parametric equations are linear, what do you expect the object to be? Show that the points $(0, 0, 0)$, $(2, 3, 1)$ and $(3, 2, 1)$ are on the object. Find an equation of the plane containing these three points. Substitute in the equations for x, y and z and show that the object lies in the plane. Argue that the object is, in fact, the entire plane.

10.6 SURFACES IN SPACE

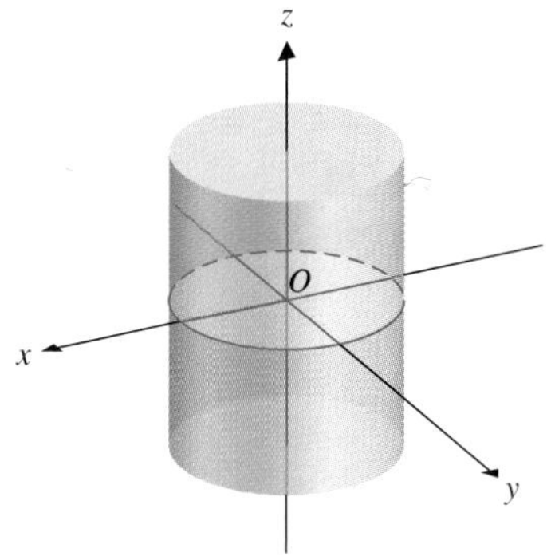

FIGURE 10.52
Right circular cylinder

Now that we have discussed lines and planes in $\mathbb{R}^3$, we continue our graphical development by drawing more complicated objects in three dimensions. Don't expect a general theory like we developed for two-dimensional graphs. Drawing curves and surfaces in three dimensions by hand or correctly interpreting computer-generated graphics is something of an art. After all, you must draw a two-dimensional image that somehow represents an object in three dimensions. Our goal here is not to produce artists, but rather to leave you with the ability to deal with a small group of surfaces in three dimensions. In numerous exercises throughout the rest of the book, taking a few extra minutes to draw a better graph will often result in a huge savings of time and effort.

○ Cylindrical Surfaces

We begin with a simple type of three-dimensional surface. When you see the word *cylinder,* you probably think of a right circular cylinder. For instance, consider the graph of the equation $x^2 + y^2 = 9$ in *three* dimensions. While the graph of $x^2 + y^2 = 9$ in *two* dimensions is the circle of radius 3, centered at the origin, what is its graph in *three* dimensions? Consider the intersection of the surface with the plane $z = k$, for some constant k. Since the equation has no z's in it, the intersection with every such plane (called the **trace** of the surface in the plane $z = k$) is the same: a circle of radius 3, centered at the origin. Think about it: whatever this three-dimensional surface is, its intersection with every plane parallel to the xy-plane is a circle of radius 3, centered at the origin. This describes a right circular cylinder, in this case one of radius 3, whose axis is the z-axis (see Figure 10.52).

More generally, the term **cylinder** is used to refer to any surface whose traces in every plane parallel to a given plane are the same. With this definition, many surfaces qualify as cylinders.

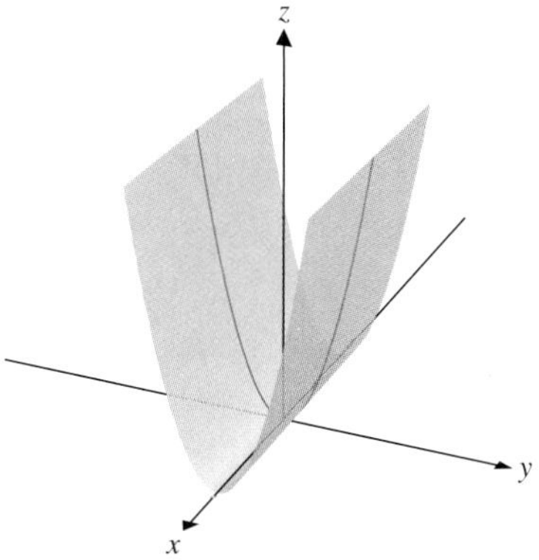

FIGURE 10.53a
$z = y^2$

EXAMPLE 6.1 Sketching a Surface

Draw a graph of the surface $z = y^2$ in $\mathbb{R}^3$.

Solution Since there are no x's in the equation, the trace of the graph in the plane $x = k$ is the same for every k. This is then a cylinder whose trace in every plane parallel to the yz-plane is the parabola $z = y^2$. To draw this, we first draw the trace in the yz-plane and then make several copies of the trace, locating the vertices at various points along the x-axis. Finally, we connect the traces with lines parallel to the x-axis to give the drawing its three-dimensional look (see Figure 10.53a). A computer-generated wireframe graph of the same surface is seen in Figure 10.53b. Notice that the wireframe consists of numerous traces for fixed values of x or y. ■

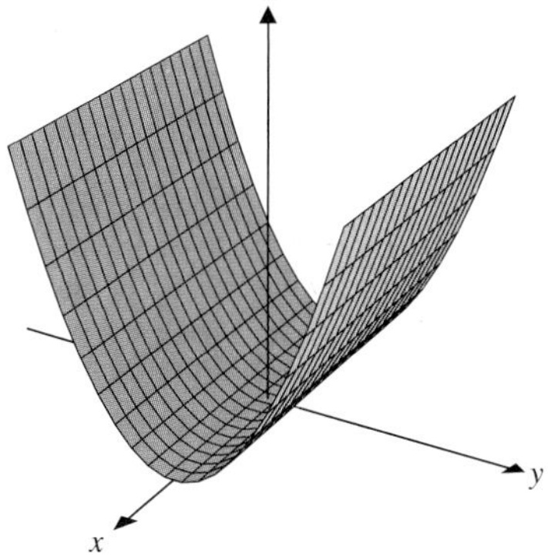

FIGURE 10.53b
Wireframe of $z = y^2$

EXAMPLE 6.2 Sketching an Unusual Cylinder

Draw a graph of the surface $z = \sin x$ in $\mathbb{R}^3$.

Solution Once again, one of the variables is missing; in this case, there are no y's. Consequently, traces of the surface in any plane parallel to the xz-plane are the same; they all look like the two-dimensional graph of $z = \sin x$. We draw one of these in the xz-plane and then make copies in planes parallel to the xz-plane, finally connecting the traces with lines parallel to the y-axis (see Figure 10.54a). In Figure 10.54b, we show a computer-generated wireframe plot of the same surface. In this case, the cylinder looks like a plane with ripples in it.

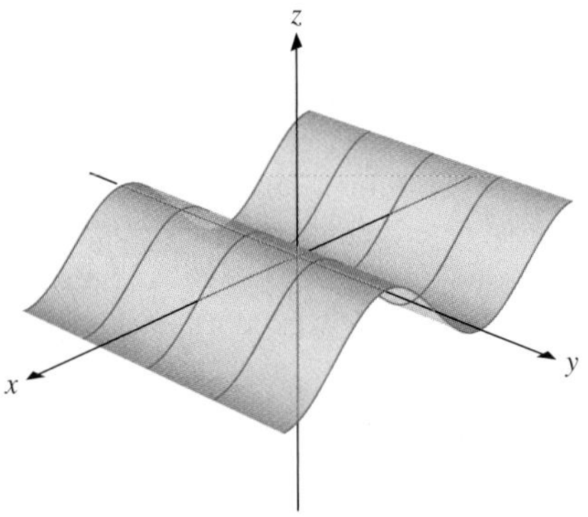

FIGURE 10.54a
The surface $z = \sin x$

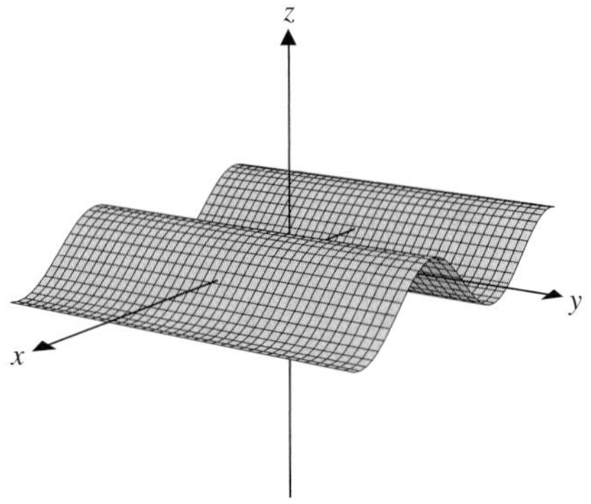

FIGURE 10.54b
Wireframe: $z = \sin x$

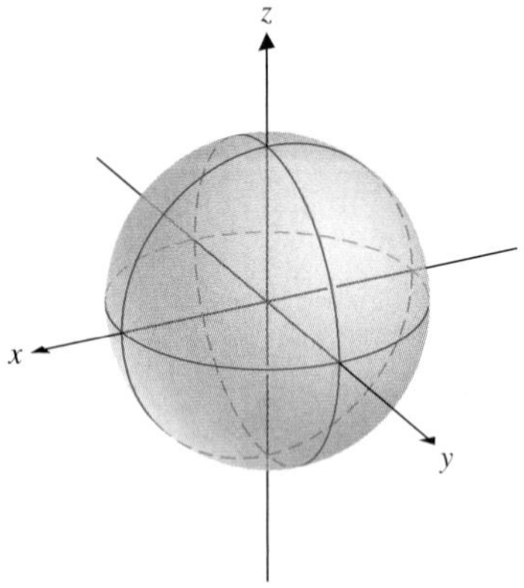

FIGURE 10.55
Sphere

Quadric Surfaces

The graph of the equation

$$ax^2 + by^2 + cz^2 + dxy + eyz + fxz + gx + hy + jz + k = 0$$

in three-dimensional space (where $a, b, c, d, e, f, g, h, j$ and k are all constants and at least one of a, b, c, d, e or f is nonzero) is referred to as a **quadric surface.**

The most familiar quadric surface is the **sphere:**

$$(x - a)^2 + (y - b)^2 + (z - c)^2 = r^2$$

of radius r centered at the point (a, b, c). To draw the sphere centered at $(0, 0, 0)$, first draw a circle of radius r, centered at the origin in the yz-plane. Then, to give the surface its three-dimensional look, draw circles of radius r centered at the origin, in both the xz- and xy-planes, as in Figure 10.55. Note that due to the perspective, these circles will look like ellipses and will be only partially visible. (We indicate the hidden parts of the circles with dashed lines.)

A generalization of the sphere is the **ellipsoid:**

$$\frac{(x-a)^2}{d^2} + \frac{(y-b)^2}{e^2} + \frac{(z-c)^2}{f^2} = 1.$$

(Notice that when $d = e = f$, the surface is a sphere.)

EXAMPLE 6.3 Sketching an Ellipsoid

Graph the ellipsoid

$$\frac{x^2}{1} + \frac{y^2}{4} + \frac{z^2}{9} = 1.$$

Solution First draw the traces in the three coordinate planes. (In general, you may need to look at the traces in planes parallel to the three coordinate planes, but the traces in the three coordinate planes will suffice, here.) In the yz-plane, $x = 0$, which gives us the ellipse

$$\frac{y^2}{4} + \frac{z^2}{9} = 1,$$

shown in Figure 10.56a. Next, add to Figure 10.56a the traces in the xy- and xz-planes. These are

$$\frac{x^2}{1} + \frac{y^2}{4} = 1 \quad \text{and} \quad \frac{x^2}{1} + \frac{z^2}{9} = 1,$$

respectively, which are both ellipses (see Figure 10.56b).

CASs have the capability of plotting functions of several variables in three dimensions. Many graphing calculators with three-dimensional plotting capabilities produce three-dimensional plots only when given z as a function of x and y. For the problem at hand, notice that we can solve for z and plot the two functions $z = 3\sqrt{1 - x^2 - \frac{y^2}{4}}$ and $z = -3\sqrt{1 - x^2 - \frac{y^2}{4}}$, to obtain the graph of the surface. Observe that the wireframe graph in Figure 10.56c (on the following page) is not particularly smooth and appears to have some gaps. To correctly interpret such a graph, you must mentally fill in the gaps. This requires an understanding of how the graph should look, which we obtained drawing Figure 10.56b.

As an alternative, many CASs enable you to graph the equation $x^2 + \frac{y^2}{4} + \frac{z^2}{9} = 1$ using **implicit plot** mode. In this mode, the CAS numerically solves the equation for the value of z corresponding to each one of a large number of sample values of x and y and plots the resulting points. The graph obtained in Figure 10.56d is an improvement over Figure 10.56c, but doesn't show the elliptical traces that we used to construct Figure 10.56b.

The best option, when available, is often a **parametric plot.** In three dimensions, this involves writing each of the three variables x, y and z in terms of two parameters, with the resulting surface plotted by plotting points corresponding to a sample of values of the two parameters. (A more extensive discussion of the mathematics of parametric surfaces is given in section 11.6.) As we develop in the exercises, parametric equations for the ellipsoid are $x = \sin s \cos t$, $y = 2 \sin s \sin t$ and $z = 3 \cos s$, with the parameters taken to be in the intervals $0 \le s \le 2\pi$ and $0 \le t \le 2\pi$. Notice how Figure 10.56e shows a nice smooth plot and clearly shows the elliptical traces.

TODAY IN MATHEMATICS

Grigori Perelman (1966–)
A Russian mathematician whose work on a famous problem known as Poincaré's conjecture is still (as of 2005) undergoing mathematical scrutiny to verify its correctness. Perelman showed his mathematical prowess as a high school student, winning a gold medal at the 1982 International Mathematical Olympiad. Eight years of isolated work produced what he believes proves a deep result known as Thurston's Geometrization Conjecture, which is more general than the 100-year-old Poincaré conjecture. Perelman's techniques have already enabled other researchers to make breakthroughs on less famous problems and are important advances in our understanding of the geometry of three-dimensional space.

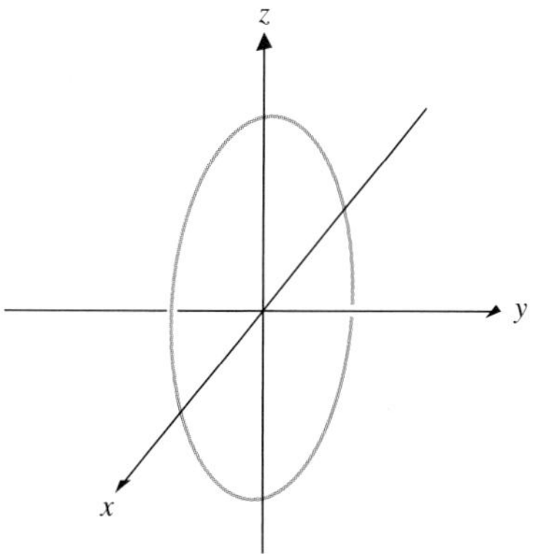

FIGURE 10.56a
Ellipse in yz-plane

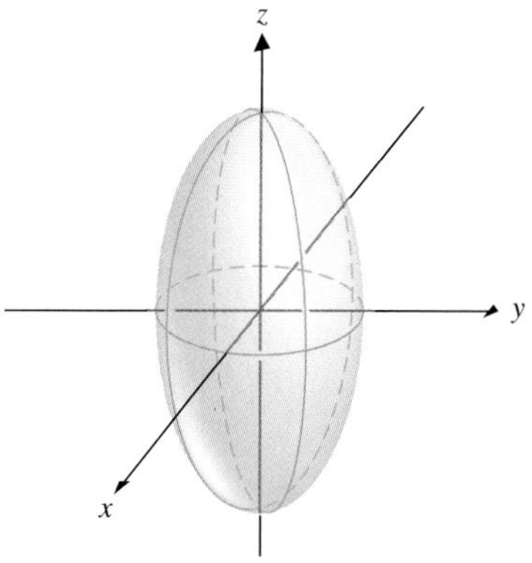

FIGURE 10.56b
Ellipsoid

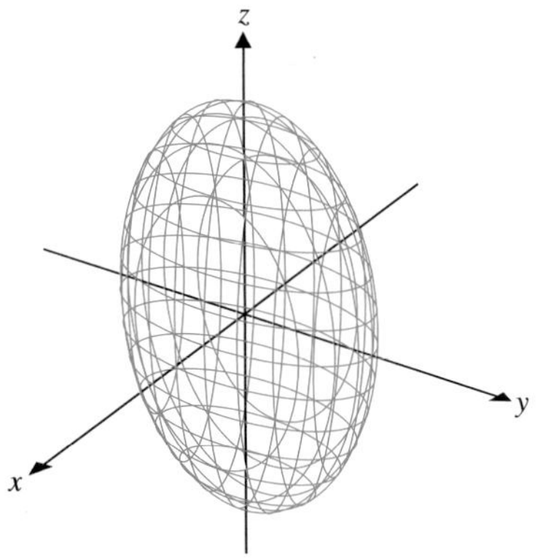

FIGURE 10.56c
Wireframe ellipsoid

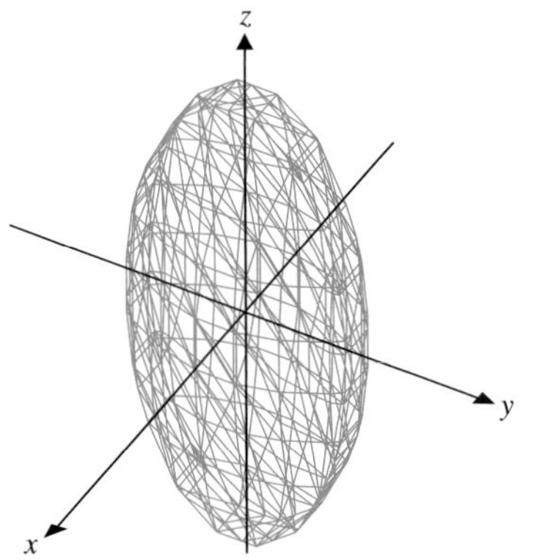

FIGURE 10.56d
Implicit wireframe plot

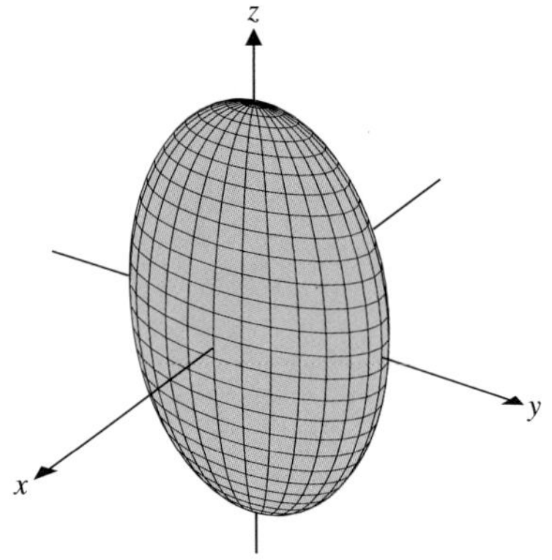

FIGURE 10.56e
Parametric plot

EXAMPLE 6.4 Sketching a Paraboloid

Draw a graph of the quadric surface

$$x^2 + y^2 = z.$$

Solution To get an idea of what the graph looks like, first draw its traces in the three coordinate planes. In the yz-plane, we have $x = 0$ and so, $y^2 = z$ (a parabola). In the xz-plane, we have $y = 0$ and so, $x^2 = z$ (a parabola). In the xy-plane, we have $z = 0$ and so, $x^2 + y^2 = 0$ (a point—the origin). We sketch the traces in Figure 10.57a. Finally, since the trace in the xy-plane is just a point, we consider the traces in the planes $z = k$ (for $k > 0$). Notice that these are the circles $x^2 + y^2 = k$, where for larger values of z (i.e., larger values of k), we get circles of larger radius. We sketch the surface in Figure 10.57b. Such surfaces are called **paraboloids** and since the traces in planes parallel to the xy-plane are circles, this is called a **circular paraboloid.**

Graphing utilities with three-dimensional capabilities generally produce a graph like Figure 10.57c for $z = x^2 + y^2$. Notice that the parabolic traces are visible, but not the circular cross sections we drew in Figure 10.57b. The four peaks visible in Figure 10.57c are due to the rectangular domain used for the plot (in this case, $-5 \le x \le 5$ and $-5 \le y \le 5$).

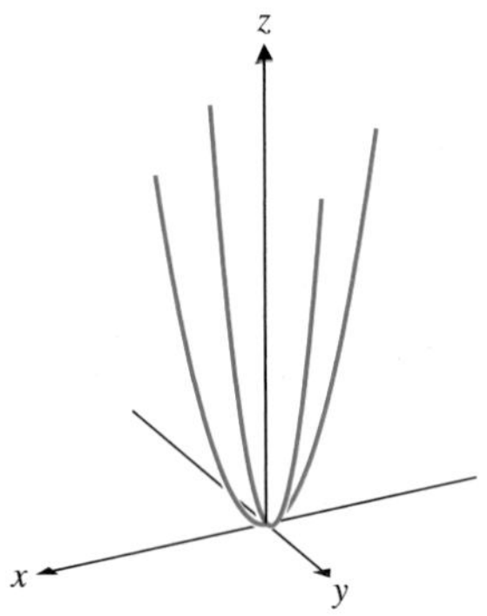

FIGURE 10.57a
Traces

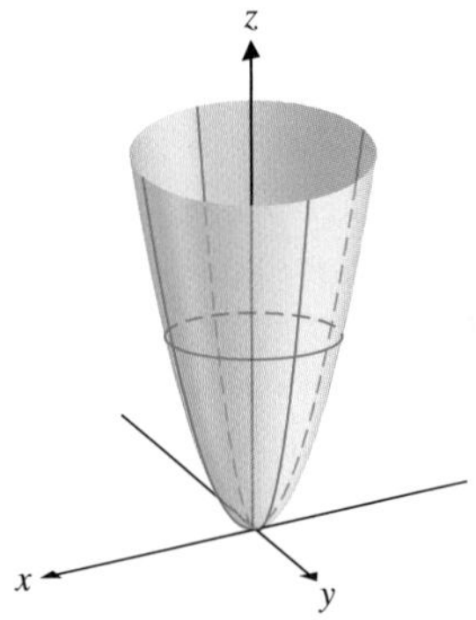

FIGURE 10.57b
Paraboloid

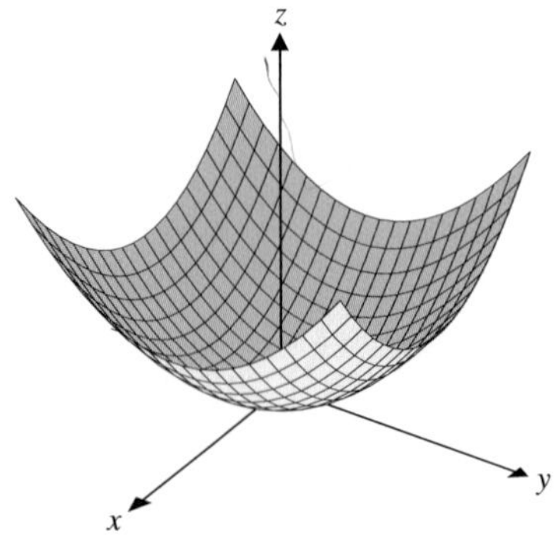

FIGURE 10.57c
Wireframe paraboloid

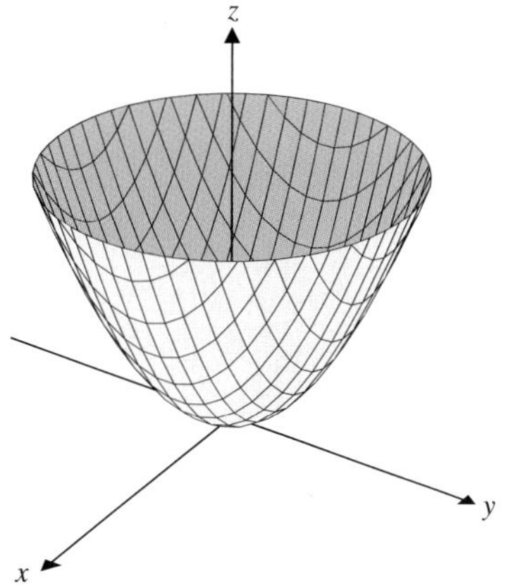

FIGURE 10.57d
Wireframe paraboloid for $0 \le z \le 15$

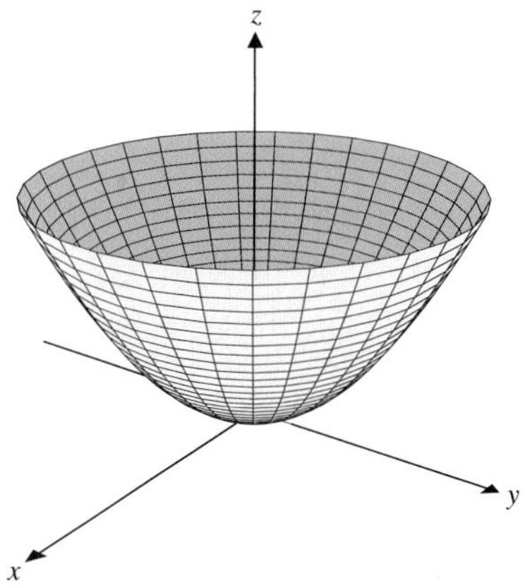

FIGURE 10.57e
Parametric plot paraboloid

We can improve this by restricting the range of the z-values. With $0 \le z \le 15$, you can clearly see the circular cross section in the plane $z = 15$ in Figure 10.57d.

As in example 6.3, a parametric surface plot is even better. Here, we have $x = s\cos t$, $y = s\sin t$ and $z = s^2$, with $0 \le s \le 5$ and $0 \le t \le 2\pi$. Figure 10.57e clearly shows the circular cross sections in the planes $z = k$, for $k > 0$. ■

Notice that in each of the last several examples, we have had to use some thought to produce computer-generated graphs that adequately show the important features of the given quadric surface. We want to encourage you to use your graphing calculator or CAS for drawing three-dimensional plots, because computer graphics are powerful tools for visualization and problem solving. However, be aware that you will need a basic understanding of the geometry of quadric surfaces to effectively produce and interpret computer-generated graphs.

EXAMPLE 6.5 Sketching an Elliptic Cone

Draw a graph of the quadric surface

$$x^2 + \frac{y^2}{4} = z^2.$$

Solution While this equation may look a lot like that of an ellipsoid, there is a significant difference. (Look where the z^2 term is!) Again, we start by looking at the traces in the coordinate planes. For the yz-plane, we have $x = 0$ and so, $\frac{y^2}{4} = z^2$ or $y^2 = 4z^2$, so that $y = \pm 2z$. That is, the trace is a pair of lines: $y = 2z$ and $y = -2z$. We show these in Figure 10.58a. Likewise, the trace in the xz-plane is a pair of lines: $x = \pm z$. The trace in the xy-plane is simply the origin. (Why?) Finally, the traces in the planes $z = k$ $(k \ne 0)$, parallel to the xy-plane, are the ellipses $x^2 + \frac{y^2}{4} = k^2$. Adding these to the drawing gives us the double-cone seen in Figure 10.58b (on the following page).

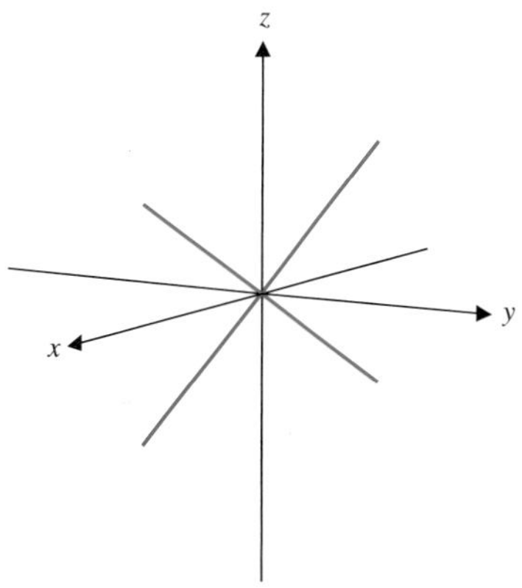

FIGURE 10.58a
Trace in yz-plane

Since the traces in planes parallel to the xy-plane are ellipses, we refer to this as an **elliptic cone.** One way to plot this with a CAS is to graph the two functions $z = \sqrt{x^2 + \frac{y^2}{4}}$ and $z = -\sqrt{x^2 + \frac{y^2}{4}}$. In Figure 10.58c, we restrict the z-range to

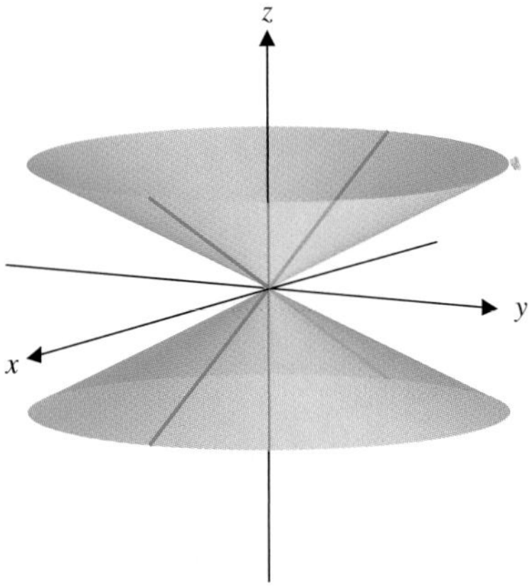

FIGURE 10.58b
Elliptic cone

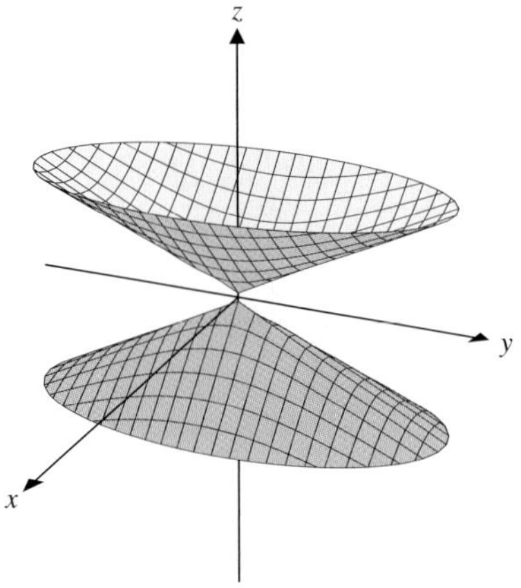

FIGURE 10.58c
Wireframe cone

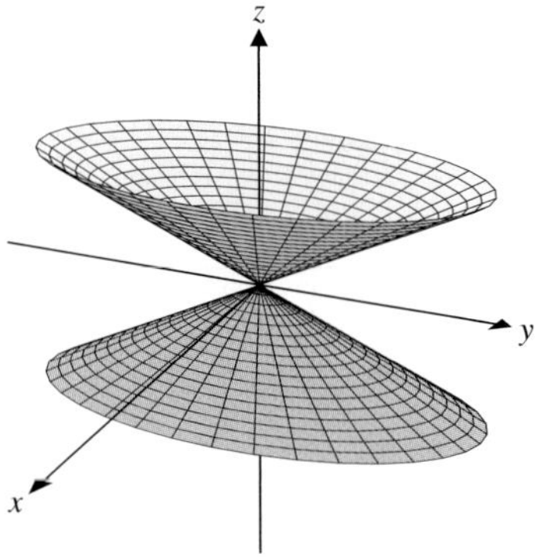

FIGURE 10.58d
Parametric plot

$-10 \le z \le 10$ to show the elliptical cross sections. Notice that this plot shows a gap between the two halves of the cone. If you have drawn Figure 10.58b yourself, this plotting deficiency won't fool you. Alternatively, the parametric plot shown in Figure 10.58d, with $x = \sqrt{s^2}\cos t$, $y = 2\sqrt{s^2}\sin t$ and $z = s$, with $-5 \le s \le 5$ and $0 \le t \le 2\pi$, shows the full cone with its elliptical and linear traces. ■

EXAMPLE 6.6 Sketching a Hyperboloid of One Sheet

Draw a graph of the quadric surface

$$\frac{x^2}{4} + y^2 - \frac{z^2}{2} = 1.$$

Solution The traces in the coordinate plane are as follows:

$$yz\text{-plane } (x = 0)\text{: } y^2 - \frac{z^2}{2} = 1 \text{ (hyperbola)}$$

(see Figure 10.59a),

$$xy\text{-plane } (z = 0)\text{: } \frac{x^2}{4} + y^2 = 1 \text{ (ellipse)}$$

and

$$xz\text{-plane } (y = 0)\text{: } \frac{x^2}{4} - \frac{z^2}{2} = 1 \text{ (hyperbola).}$$

Further, notice that the trace of the surface in each plane $z = k$ (parallel to the xy-plane) is also an ellipse:

$$\frac{x^2}{4} + y^2 = \frac{k^2}{2} + 1.$$

Finally, observe that the larger k is, the larger the axes of the ellipses are. Adding this information to Figure 10.59a, we draw the surface seen in Figure 10.59b. We call this surface a **hyperboloid of one sheet.**

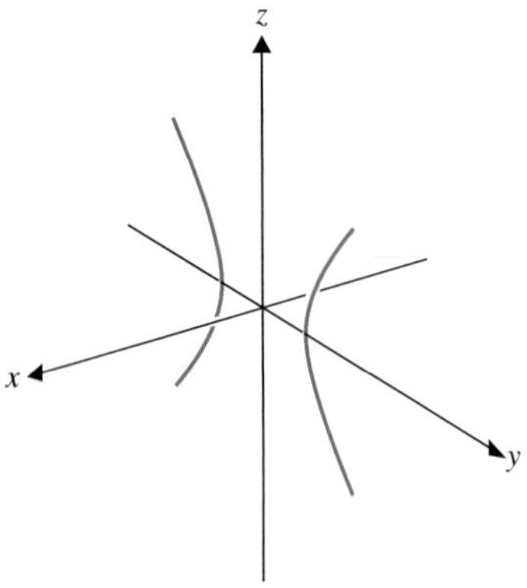

FIGURE 10.59a
Trace in yz-plane

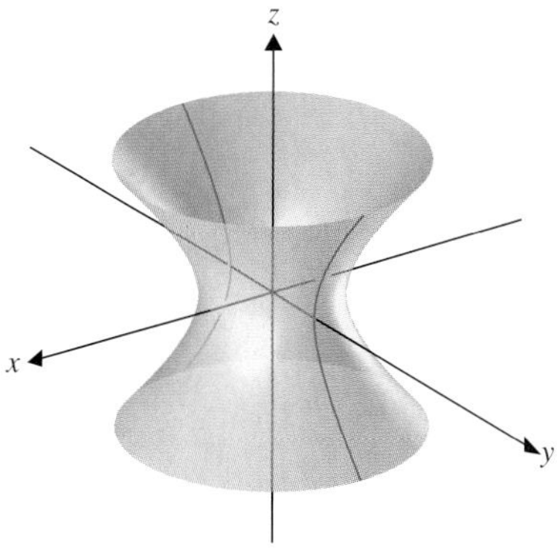

FIGURE 10.59b
Hyperboloid of one sheet

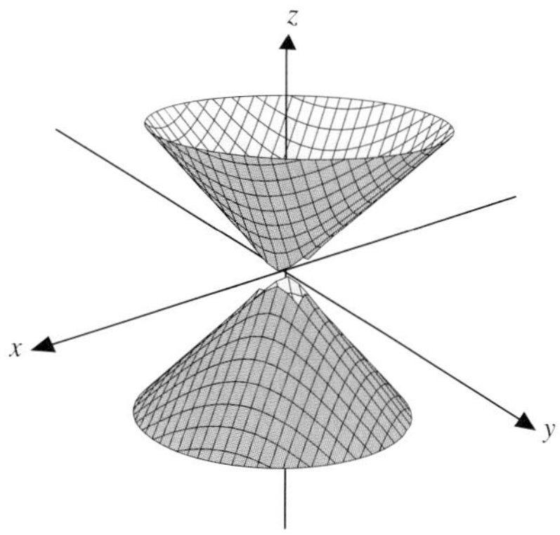

FIGURE 10.59c
Wireframe hyperboloid

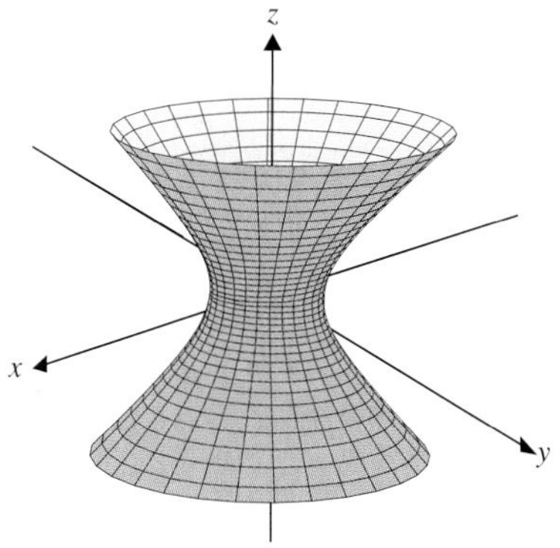

FIGURE 10.59d
Parametric plot

To plot this with a CAS, you could graph the two functions $z = \sqrt{2\left(\frac{x^2}{4} + y^2 - 1\right)}$ and $z = -\sqrt{2\left(\frac{x^2}{4} + y^2 - 1\right)}$. (See Figure 10.59c, where we have restricted the z-range to $-10 \le z \le 10$, to show the elliptical cross sections.) Notice that this plot looks more like a cone than the hyperboloid in Figure 10.59b. If you have drawn Figure 10.59b yourself, this plotting problem won't fool you.

Alternatively, the parametric plot seen in Figure 10.59d, with $x = 2\cos s \cosh t$, $y = \sin s \cosh t$ and $z = \sqrt{2}\sinh t$, with $0 \le s \le 2\pi$ and $-5 \le t \le 5$, shows the full hyperboloid with its elliptical and hyperbolic traces. ■

EXAMPLE 6.7 Sketching a Hyperboloid of Two Sheets

Draw a graph of the quadric surface

$$\frac{x^2}{4} - y^2 - \frac{z^2}{2} = 1.$$

Solution Notice that this is the same equation as in example 6.6, except for the sign of the y-term. As we have done before, we first look at the traces in the three coordinate planes. The trace in the yz-plane ($x = 0$) is defined by

$$-y^2 - \frac{z^2}{2} = 1.$$

Since it is clearly impossible for two negative numbers to add up to something positive, this is a contradiction and there is no trace in the yz-plane. That is, the surface does not intersect the yz-plane. The traces in the other two coordinate planes are as follows:

$$xy\text{-plane } (z = 0): \frac{x^2}{4} - y^2 = 1 \text{ (hyperbola)}$$

and

$$xz\text{-plane } (y = 0): \frac{x^2}{4} - \frac{z^2}{2} = 1 \text{ (hyperbola).}$$

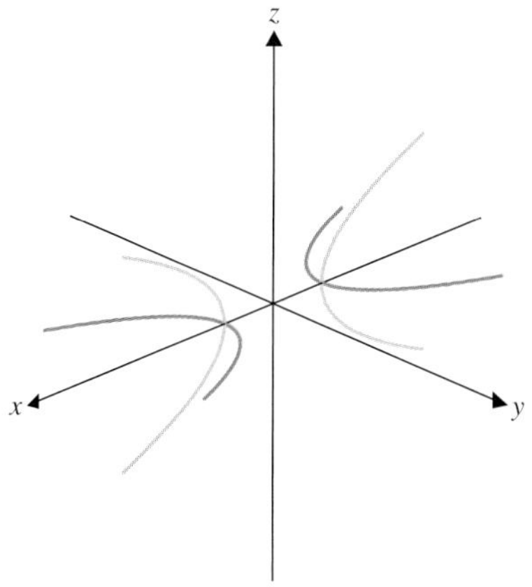

FIGURE 10.60a
Traces in xy- and xz-planes

We show these traces in Figure 10.60a. Finally, notice that for $x = k$, we have that

$$y^2 + \frac{z^2}{2} = \frac{k^2}{4} - 1,$$

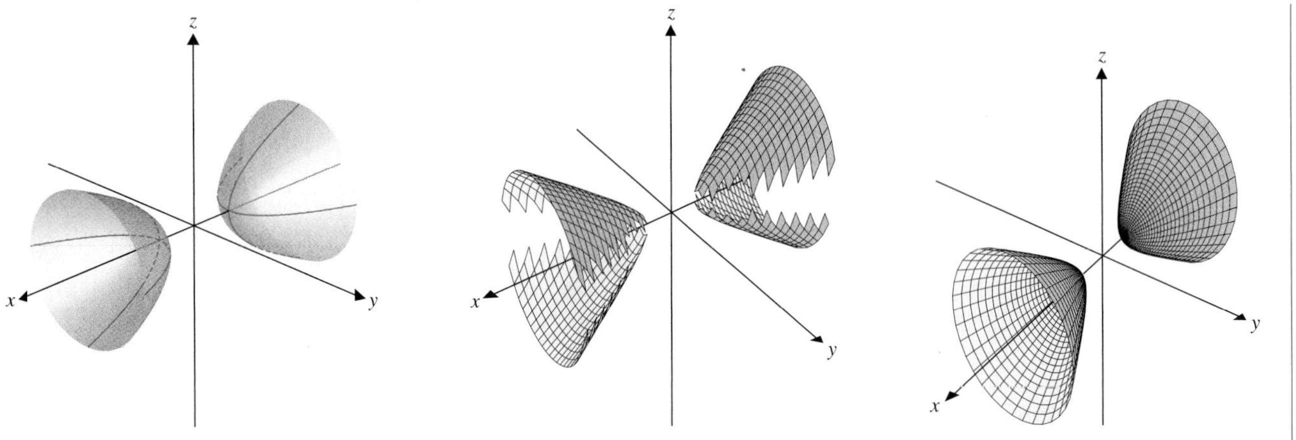

FIGURE 10.60b
Hyperboloid of two sheets

FIGURE 10.60c
Wireframe hyperboloid

FIGURE 10.60d
Parametric plot

so that the traces in the plane $x = k$ are ellipses for $k^2 > 4$. It is important to notice here that if $k^2 < 4$, the equation $y^2 + \frac{z^2}{9} = \frac{k^2}{4} - 1$ has no solution. (Why is that?) So, for $-2 < k < 2$, the surface has no trace at all in the plane $x = k$, leaving a gap that separates the hyperbola into two *sheets*. Putting this all together, we have the surface seen in Figure 10.60b. We call this surface a **hyperboloid of two sheets.**

We can plot this on a CAS by graphing the two functions $z = \sqrt{2\left(\frac{x^2}{4} - y^2 - 1\right)}$ and $z = -\sqrt{2\left(\frac{x^2}{4} - y^2 - 1\right)}$. (See Figure 10.60c, where we have restricted the z-range to $-10 \le z \le 10$, to show the elliptical cross sections.) Notice that this plot shows large gaps between the two halves of the hyperboloid. If you have drawn Figure 10.60b yourself, this plotting deficiency won't fool you.

Alternatively, the parametric plot with $x = 2\cosh s$, $y = \sinh s \cos t$ and $z = \sqrt{2}\sinh s \sin t$, for $-4 \le s \le 4$ and $0 \le t \le 2\pi$, produces the left half of the hyperboloid with its elliptical and hyperbolic traces. The right half of the hyperboloid has parametric equations $x = -2\cosh s$, $y = \sinh s \cos t$ and $z = \sqrt{2}\sinh s \sin t$, with $-4 \le s \le 4$ and $0 \le t \le 2\pi$. We show both halves in Figure 10.60d. ■

As our final example, we offer one of the more interesting quadric surfaces. It is also one of the more difficult surfaces to sketch.

EXAMPLE 6.8 Sketching a Hyperbolic Paraboloid

Sketch the graph of the quadric surface defined by the equation

$$z = 2y^2 - x^2.$$

Solution We first consider the traces in planes parallel to each of the coordinate planes:

$$\text{parallel to } xy\text{-plane } (z = k)\text{: } 2y^2 - x^2 = k \text{ (hyperbola, for } k \ne 0),$$
$$\text{parallel to } xz\text{-plane } (y = k)\text{: } z = -x^2 + 2k^2 \text{ (parabola opening down)}$$

and

$$\text{parallel to } yz\text{-plane } (x = k)\text{: } z = 2y^2 - k^2 \text{ (parabola opening up).}$$

We begin by drawing the traces in the xz- and yz-planes, as seen in Figure 10.61a. Since the trace in the xy-plane is the degenerate hyperbola $2y^2 = x^2$ (two lines: $x = \pm\sqrt{2}y$), we instead draw the trace in several of the planes $z = k$. Notice that for $k > 0$, these are hyperbolas opening toward the positive and negative y-direction and for $k < 0$, these are hyperbolas opening toward the positive and negative x-direction. We indicate one of these for $k > 0$ and one for $k < 0$ in Figure 10.61b, where we show a sketch of the surface. We refer to this surface as a **hyperbolic paraboloid.** More than anything else, the surface resembles a saddle. In fact, we refer to the origin as a **saddle point** for this graph. (We'll discuss the significance of saddle points in Chapter 12.)

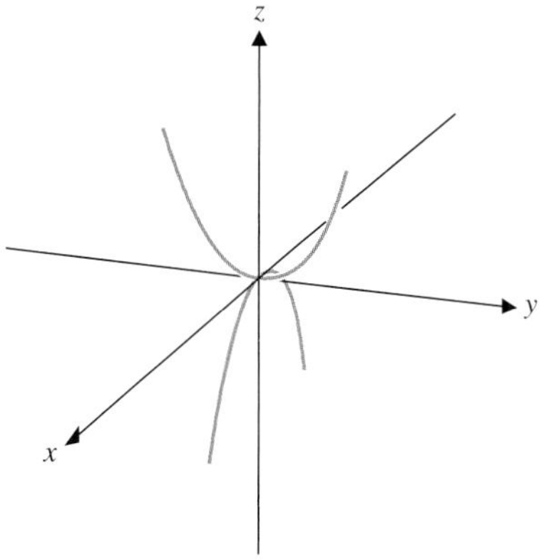

FIGURE 10.61a
Traces in the xz- and yz-planes

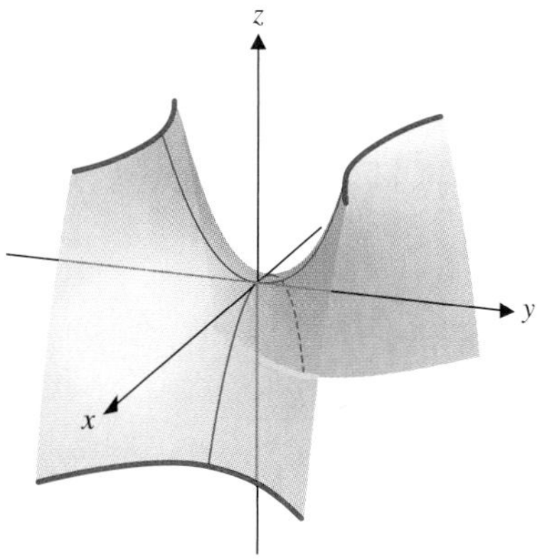

FIGURE 10.61b
The surface $z = 2y^2 - x^2$

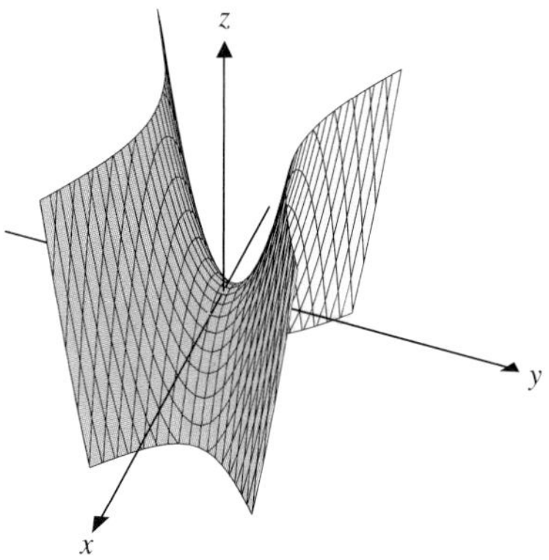

FIGURE 10.61c
Wireframe plot of $z = 2y^2 - x^2$

A wireframe graph of $z = 2y^2 - x^2$ is shown in Figure 10.61c (with $-5 \le x \le 5$ and $-5 \le y \le 5$ and where we limited the z-range to $-8 \le z \le 12$). Note that only the parabolic cross sections are drawn, but the graph shows all the features of Figure 10.61b. Plotting this surface parametrically is fairly tedious (requiring four different sets of equations) and doesn't improve the graph noticeably. ■

○ An Application

You may have noticed the large number of paraboloids around you. For instance, radiotelescopes and even home television satellite dishes have the shape of a portion of a paraboloid. Reflecting telescopes have parabolic mirrors that again, are a portion of a paraboloid. There is a very good reason for this. It turns out that in all of these cases, light waves and radio waves striking *any* point on the parabolic dish or mirror are reflected toward *one* point, the focus of each parabolic cross section through the vertex of the paraboloid. This remarkable fact means that all light waves and radio waves end up being concentrated at just one point. In the case of a radiotelescope, placing a small receiver just in front of the focus can take a very faint signal and increase its effective strength immensely (see Figure 10.62). The same principle is used in optical telescopes to concentrate the light from a faint source (e.g., a distant star). In this case, a small mirror is mounted in a line from the parabolic mirror to the focus. The small mirror then reflects the concentrated light to an eyepiece for viewing (see Figure 10.63).

FIGURE 10.62
Radiotelescope

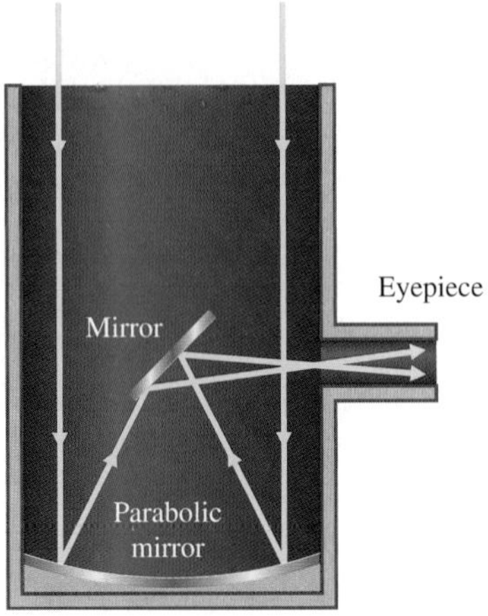

FIGURE 10.63
Reflecting telescope

The following table summarizes the graphs of quadric surfaces.

Name	*Generic Equation*	*Graph*
Ellipsoid	$\frac{x^2}{a^2} + \frac{y^2}{b^2} + \frac{z^2}{c^2} = 1$	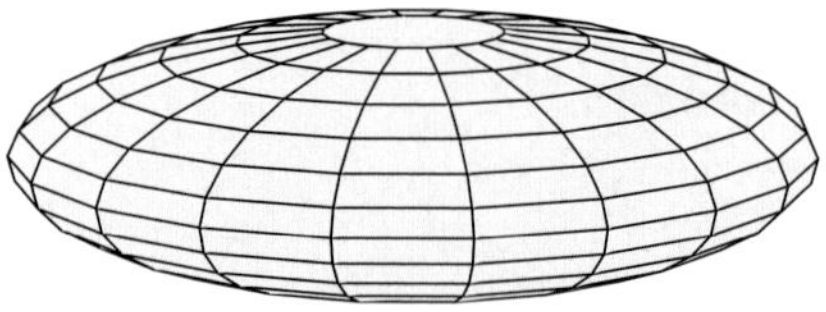
Elliptic paraboloid	$z = ax^2 + by^2 + c$ $(a, b > 0)$	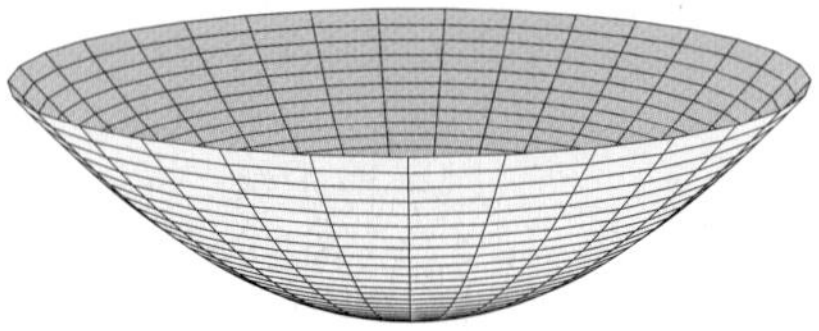
Hyperbolic paraboloid	$z = ax^2 - by^2 + c$ $(a, b > 0)$	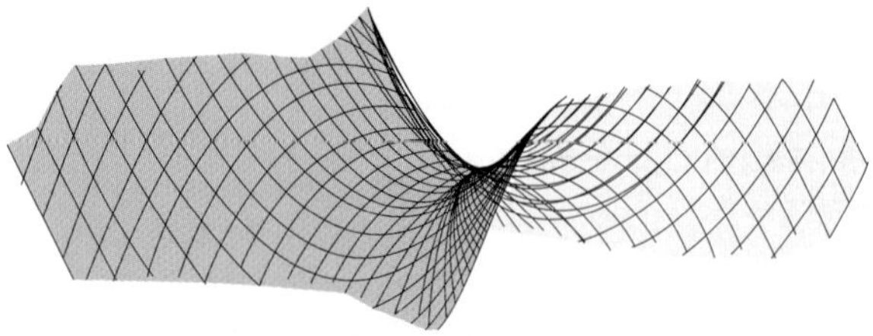
Cone	$z^2 = ax^2 + by^2$ $(a, b > 0)$	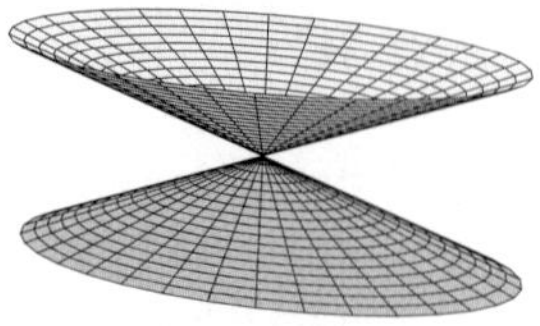

Continued

Name	*Generic Equation*	*Graph*
Hyperboloid of one sheet	$ax^2 + by^2 - cz^2 = 1$ $(a, b, c > 0)$	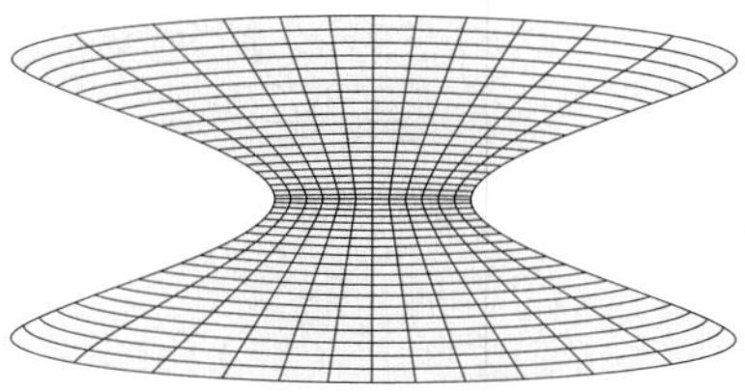
Hyperboloid of two sheets	$ax^2 - by^2 - cz^2 = 1$ $(a, b, c > 0)$	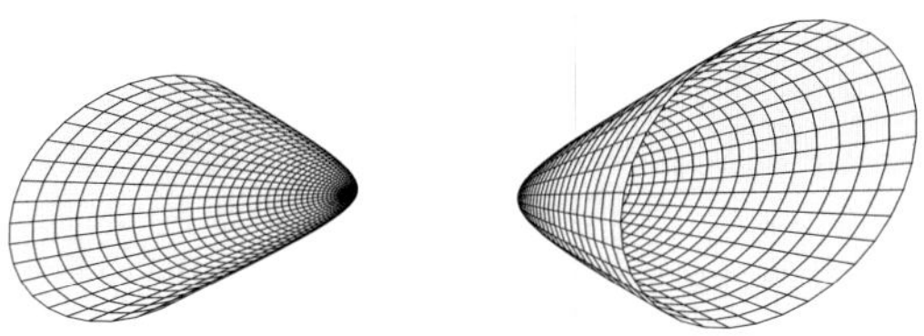

EXERCISES 10.6

WRITING EXERCISES

1. In the text, different hints were given for graphing cylinders as opposed to quadric surfaces. Explain how to tell from the equation whether you have a cylinder, a quadric surface, a plane or some other surface.
2. The first step in graphing a quadric surface is identifying traces. Given the traces, explain how to tell whether you have an ellipsoid, elliptical cone, paraboloid or hyperboloid. (Hint: For a paraboloid, how many traces are parabolas?)
3. Suppose you have identified that a given equation represents a hyperboloid. Explain how to determine whether the hyperboloid has one sheet or two sheets.
4. Circular paraboloids have a bowl-like shape. However, the paraboloids $z = x^2 + y^2$, $z = 4 - x^2 - y^2$, $y = x^2 + z^2$ and $x = y^2 + z^2$ all open up in different directions. Explain why these paraboloids are different and how to determine in which direction a paraboloid opens.

In exercises 1–40, sketch the appropriate traces, and then sketch and identify the surface.

1. $z = x^2$
2. $z = 4 - y^2$
3. $x^2 + \frac{y^2}{9} + \frac{z^2}{4} = 1$
4. $\frac{x^2}{4} + \frac{y^2}{4} + \frac{z^2}{9} = 1$
5. $z = 4x^2 + 4y^2$
6. $z = x^2 + 4y^2$
7. $z^2 = 4x^2 + y^2$
8. $z^2 = \frac{x^2}{4} + \frac{y^2}{9}$
9. $z = x^2 - y^2$
10. $z = y^2 - x^2$
11. $x^2 - y^2 + z^2 = 1$
12. $x^2 + \frac{y^2}{4} - z^2 = 1$
13. $x^2 - \frac{y^2}{9} - z^2 = 1$
14. $x^2 - y^2 - \frac{z^2}{4} = 1$
15. $z = \cos x$
16. $z = \sqrt{x^2 + 4y^2}$
17. $z = 4 - x^2 - y^2$
18. $x = y^2 + z^2$
19. $z = x^3$
20. $z = 4 - y^2$
21. $z = \sqrt{x^2 + y^2}$
22. $z = \sin y$
23. $y = x^2$
24. $x = 2 - y^2$
25. $y = x^2 + z^2$
26. $z = 9 - x^2 - y^2$
27. $x^2 + 4y^2 + 16z^2 = 16$
28. $2x - z = 4$
29. $4x^2 - y^2 - z = 0$
30. $-x^2 - y^2 + 9z^2 = 9$
31. $4x^2 + y^2 - z^2 = 4$
32. $x^2 - y^2 + 9z^2 = 9$
33. $-4x^2 + y^2 - z^2 = 4$
34. $x^2 - 4y^2 + z = 0$
35. $x + y = 1$
36. $9x^2 + y^2 + 9z^2 = 9$
37. $x^2 + y^2 = 4$
38. $9x^2 + z^2 = 9$
39. $x^2 + y^2 - z = 4$
40. $x + y^2 + z^2 = 2$

In exercises 41–44, sketch the given traces on a single three-dimensional coordinate system.

41. $z = x^2 + y^2; x = 0, x = 1, x = 2$
42. $z = x^2 + y^2; y = 0, y = 1, y = 2$
43. $z = x^2 - y^2; x = 0, x = 1, x = 2$
44. $z = x^2 - y^2; y = 0, y = 1, y = 2$
45. Hyperbolic paraboloids are sometimes called "saddle" graphs. The architect of the Saddle Dome in the Canadian city of

Calgary used this shape to create an attractive and symbolically meaningful structure.

One issue in using this shape is water drainage from the roof. If the Saddle Dome roof is described by $z = x^2 - y^2$, $-1 \le x \le 1, -1 \le y \le 1$, in which direction would the water drain? First, consider traces for which y is constant. Show that the trace has a minimum at $x = 0$. Identify the plane $x = 0$ in the picture. Next, show that the trace at $x = 0$ has an absolute maximum at $y = 0$. Use this information to identify the two primary points at which the water would drain.

46. Cooling towers for nuclear reactors are often constructed as hyperboloids of one sheet because of the structural stability of that surface. (See the accompanying photo.) Suppose all horizontal cross sections are circular, with a minimum radius of 200 feet occurring at a height of 600 feet. The tower is to be 800 feet tall with a maximum cross-sectional radius of 300 feet. Find an equation for the structure.

47. If $x = a \sin s \cos t$, $y = b \sin s \sin t$ and $z = c \cos s$, show that (x, y, z) lies on the ellipsoid $\frac{x^2}{a^2} + \frac{y^2}{b^2} + \frac{z^2}{c^2} = 1$.

48. If $x = as \cos t$, $y = bs \sin t$ and $z = s^2$, show that (x, y, z) lies on the paraboloid $z = \frac{x^2}{a^2} + \frac{y^2}{b^2}$.

49. If $x = as \cos t$, $y = bs \sin t$ and $z = s$, show that (x, y, z) lies on the cone $z^2 = \frac{x^2}{a^2} + \frac{y^2}{b^2}$.

50. If $x = a \cos s \cosh t$, $y = b \sin s \cosh t$ and $z = c \sinh t$, show that (x, y, z) lies on the hyperboloid of one sheet $\frac{x^2}{a^2} + \frac{y^2}{b^2} - \frac{z^2}{c^2} = 1$.

51. If $a > 0$ and $x = a \cosh s$, $y = b \sinh s \cos t$ and $z = c \sinh s \sin t$, show that (x, y, z) lies on the right half of the hyperboloid of two sheets $\frac{x^2}{a^2} - \frac{y^2}{b^2} - \frac{z^2}{c^2} = 1$.

52. If $a < 0$ and $x = a \cosh s$, $y = b \sinh s \cos t$ and $z = c \sinh s \sin t$, show that (x, y, z) lies on the left half of the hyperboloid of two sheets $\frac{x^2}{a^2} - \frac{y^2}{b^2} - \frac{z^2}{c^2} = 1$.

53. Find parametric equations as in exercises 47–52 for the surfaces in exercises 3, 5 and 7. Use a CAS to graph the parametric surfaces.

54. Find parametric equations as in exercises 47–52 for the surfaces in exercises 11 and 13. Use a CAS to graph the parametric surfaces.

55. Find parametric equations for the surface in exercise 17.

56. Find parametric equations for the surface in exercise 33.

57. You can improve the appearance of a wireframe graph by carefully choosing the viewing window. We commented on the curved edge in Figure 10.57c. Graph this function with domain $-5 \le x \le 5$ and $-5 \le y \le 5$, but limit the z-range to $-1 \le z \le 20$. Does this look more like Figure 10.57b?

EXPLORATORY EXERCISES

1. Golf club manufacturers use ellipsoids (called **inertia ellipsoids**) to visualize important characteristics of golf clubs. A three-dimensional coordinate system is set up as shown in the figure. The (second) moments of inertia are then computed for the clubhead about each coordinate axis. The inertia ellipsoid is defined as $I_{xx}x^2 + I_{yy}y^2 + I_{zz}z^2 + 2I_{xy}xy + 2I_{yz}yz + 2I_{xz}xz = 1$. The graph of this ellipsoid provides important information to the club designer. For comparison purposes, a homogeneous spherical shell would have a perfect sphere as its inertia ellipsoid. In *Science and Golf II,* the data given here are provided for a 6-iron and driver, respectively. Graph the ellipsoids and compare the shapes. (Recall that the larger the moment of inertia of an object, the harder it is to rotate.)

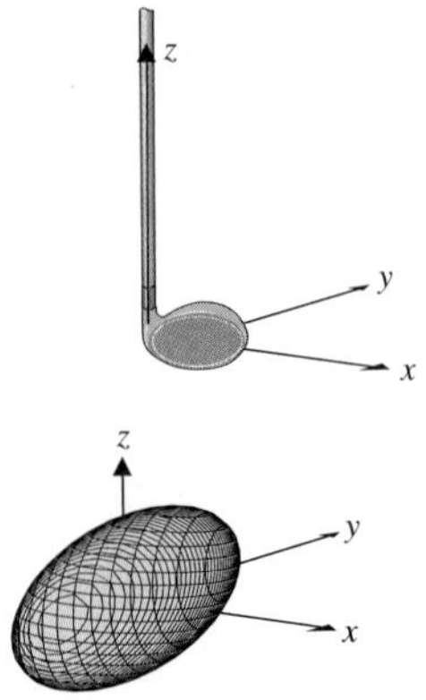

For the 6-iron, $89.4x^2 + 195.8y^2 + 124.9z^2 - 48.6xy - 111.8xz + 0.4yz = 1{,}000{,}000$ and for the driver, $119.3x^2 + 243.9y^2 + 139.4z^2 - 1.2xy - 71.4xz - 25.8yz = 1{,}000{,}000$.

2. Sketch the graphs of $x^2 + cy^2 - z^2 = 1$ for a variety of positive and negative constants c. If your CAS allows you to animate a sequence of graphs, set up an animation that shows a sequence of hyperboloids of one sheet morphing into hyperboloids of two sheets.

Review Exercises

WRITING EXERCISES

The following list includes terms that are defined and theorems that are stated in this chapter. For each term or theorem, (1) give a precise definition or statement, (2) state in general terms what it means and (3) describe the types of problems with which it is associated.

Vector	Scalar	Magnitude
Position vector	Unit vector	Displacement vector
First octant	Sphere	Dot product
Angle between vectors	Triangle Inequality	Component
Projection	Cross product	Torque
Magnus force	Parametric equations of line	Symmetric equations of line
Parallel planes	Orthogonal planes	Skew lines
Traces	Cylinder	Ellipsoid
Circular paraboloid	Hyperbolic paraboloid	Cone
Hyperboloid of one sheet	Hyperboloid of two sheets	Saddle

TRUE OR FALSE

State whether each statement is true or false and briefly explain why. If the statement is false, try to "fix it" by modifying the given statement to a new statement that is true.

1. Two vectors are parallel if one vector equals the other divided by a constant.
2. For a given vector, there is one unit vector parallel to it.
3. A sphere is the set of all points at a given distance from a fixed point.
4. The dot product $\mathbf{a} \cdot \mathbf{b} = 0$ implies that either $\mathbf{a} = \mathbf{0}$ or $\mathbf{b} = \mathbf{0}$.
5. If $\mathbf{a} \cdot \mathbf{b} > 0$, then the angle between $\mathbf{a}$ and $\mathbf{b}$ is less than $\frac{\pi}{2}$.
6. $(\mathbf{a} \cdot \mathbf{b}) \cdot \mathbf{c} = \mathbf{a} \cdot (\mathbf{b} \cdot \mathbf{c})$ for all vectors $\mathbf{a}$, $\mathbf{b}$ and $\mathbf{c}$.
7. $(\mathbf{a} \times \mathbf{b}) \times \mathbf{c} = \mathbf{a} \times (\mathbf{b} \times \mathbf{c})$ for all vectors $\mathbf{a}$, $\mathbf{b}$ and $\mathbf{c}$.
8. $\mathbf{a} \times \mathbf{b}$ is the unique vector perpendicular to the plane containing $\mathbf{a}$ and $\mathbf{b}$.
9. The cross product can be used to determine the angle between vectors.
10. Two planes are parallel if and only if their normal vectors are parallel.
11. The distance between parallel planes equals the distance between any two points in the planes.
12. The equation of a hyperboloid of two sheets has two minus signs in it.
13. In an equation of a quadric surface, if one variable is linear and the other two are squared, then the surface is a paraboloid wrapping around the axis corresponding to the linear variable.

In exercises 1–4, compute $\mathbf{a} + \mathbf{b}$, $4\mathbf{b}$ and $\|2\mathbf{b} - \mathbf{a}\|$.

1. $\mathbf{a} = \langle -2, 3 \rangle$, $\mathbf{b} = \langle 1, 0 \rangle$
2. $\mathbf{a} = \langle -1, -2 \rangle$, $\mathbf{b} = \langle 2, 3 \rangle$
3. $\mathbf{a} = 10\mathbf{i} + 2\mathbf{j} - 2\mathbf{k}$, $\mathbf{b} = -4\mathbf{i} + 3\mathbf{j} + 2\mathbf{k}$
4. $\mathbf{a} = -\mathbf{i} - \mathbf{j} + 2\mathbf{k}$, $\mathbf{b} = -\mathbf{i} + \mathbf{j} - 2\mathbf{k}$

In exercises 5–8, determine whether a and b are parallel, orthogonal or neither.

5. $\mathbf{a} = \langle 2, 3 \rangle$, $\mathbf{b} = \langle 4, 5 \rangle$
6. $\mathbf{a} = \mathbf{i} - 2\mathbf{j}$, $\mathbf{b} = 2\mathbf{i} - \mathbf{j}$
7. $\mathbf{a} = \langle -2, 3, 1 \rangle$, $\mathbf{b} = \langle 4, -6, -2 \rangle$
8. $\mathbf{a} = 2\mathbf{i} - \mathbf{j} + 2\mathbf{k}$, $\mathbf{b} = 4\mathbf{i} - 2\mathbf{j} + \mathbf{k}$

In exercises 9 and 10, find the displacement vector $\overrightarrow{PQ}$.

9. $P = (3, 1, -2)$, $Q = (2, -1, 1)$
10. $P = (3, 1)$, $Q = (1, 4)$

Review Exercises

In exercises 11–16, find a unit vector in the same direction as the given vector.

11. $\langle 3, 6\rangle$

12. $\langle -2, 3\rangle$

13. $10\mathbf{i} + 2\mathbf{j} - 2\mathbf{k}$

14. $-\mathbf{i} - \mathbf{j} + 2\mathbf{k}$

15. from $(4, 1, 2)$ to $(1, 1, 6)$

16. from $(2, -1, 0)$ to $(0, 3, -2)$

In exercises 17 and 18, find the distance between the given points.

17. $(0, -2, 2), (3, 4, 1)$

18. $(3, 1, 0), (1, 4, 1)$

In exercises 19 and 20, find a vector with the given magnitude and in the same direction as the given vector.

19. magnitude 2, $\mathbf{v} = 2\mathbf{i} - 2\mathbf{j} + 2\mathbf{k}$

20. magnitude $\frac{1}{2}$, $\mathbf{v} = -\mathbf{i} - \mathbf{j} + \mathbf{k}$

21. The thrust of an airplane's engine produces a speed of 500 mph in still air. The wind velocity is given by $\langle 20, -80\rangle$. In what direction should the plane head to fly due east?

22. Two ropes are attached to a crate. The ropes exert forces of $\langle -160, 120\rangle$ and $\langle 160, 160\rangle$, respectively. If the crate weighs 300 pounds, what is the net force on the crate?

In exercises 23 and 24, find an equation of the sphere with radius r and center (a, b, c).

23. $r = 6, (a, b, c) = (0, -2, 0)$

24. $r = \sqrt{3}, (a, b, c) = (-3, 1, 2)$

In exercises 25–28, compute $\mathbf{a} \cdot \mathbf{b}$.

25. $\mathbf{a} = \langle 2, -1\rangle, \mathbf{b} = \langle 2, 4\rangle$

26. $\mathbf{a} = \mathbf{i} - 2\mathbf{j}, \mathbf{b} = 4\mathbf{i} + 2\mathbf{j}$

27. $\mathbf{a} = 3\mathbf{i} + \mathbf{j} - 4\mathbf{k}, \mathbf{b} = -2\mathbf{i} + 2\mathbf{j} + \mathbf{k}$

28. $\mathbf{a} = \mathbf{i} + 3\mathbf{j} - 2\mathbf{k}, \mathbf{b} = 2\mathbf{i} - 3\mathbf{k}$

In exercises 29 and 30, find the angle between the vectors.

29. $\langle 3, 2, 1\rangle$ and $\langle -1, 1, 2\rangle$

30. $\langle 3, 4\rangle$ and $\langle 2, -1\rangle$

In exercises 31 and 32, find $\text{comp}_{\mathbf{b}}\, \mathbf{a}$ and $\text{proj}_{\mathbf{b}}\, \mathbf{a}$.

31. $\mathbf{a} = 3\mathbf{i} + \mathbf{j} - 4\mathbf{k}, \mathbf{b} = \mathbf{i} + 2\mathbf{j} + \mathbf{k}$

32. $\mathbf{a} = \mathbf{i} + 3\mathbf{j} - 2\mathbf{k}, \mathbf{b} = 2\mathbf{i} - 3\mathbf{k}$

In exercises 33–36, compute the cross product $\mathbf{a} \times \mathbf{b}$.

33. $\mathbf{a} = \langle 1, -2, 1\rangle, \mathbf{b} = \langle 2, 0, 1\rangle$

34. $\mathbf{a} = \langle 1, -2, 0\rangle, \mathbf{b} = \langle 1, 0, -2\rangle$

35. $\mathbf{a} = 2\mathbf{j} + \mathbf{k}, \mathbf{b} = 4\mathbf{i} + 2\mathbf{j} - \mathbf{k}$

36. $\mathbf{a} = \mathbf{i} - 2\mathbf{j} - 3\mathbf{k}, \mathbf{b} = 2\mathbf{i} - \mathbf{j}$

In exercises 37 and 38, find two unit vectors orthogonal to both given vectors.

37. $\mathbf{a} = 2\mathbf{i} + \mathbf{k}, \mathbf{b} = -\mathbf{i} + 2\mathbf{j} - \mathbf{k}$

38. $\mathbf{a} = 3\mathbf{i} + \mathbf{j} - 2\mathbf{k}, \mathbf{b} = 2\mathbf{i} - \mathbf{j}$

39. A force of $\langle 40, -30\rangle$ pounds moves an object in a straight line from $(1, 0)$ to $(60, 22)$. Compute the work done.

40. Use vectors to find the angles in the triangle with vertices $(0, 0)$, $(3, 1)$ and $(1, 4)$.

In exercises 41 and 42, find the distance from the point Q to the given line.

41. $Q = (1, -1, 0)$, line $\begin{cases} x = t + 1 \\ y = 2t - 1 \\ z = 3 \end{cases}$

42. $Q = (0, 1, 0)$, line $\begin{cases} x = 2t - 1 \\ y = 4t \\ z = 3t + 2 \end{cases}$

In exercises 43 and 44, find the indicated area or volume.

43. Area of the parallelogram with adjacent edges formed by $\langle 2, 0, 1\rangle$ and $\langle 0, 1, -3\rangle$

44. Volume of the parallelepiped with three adjacent edges formed by $\langle 1, -1, 2\rangle$, $\langle 0, 0, 4\rangle$ and $\langle 3, 0, 1\rangle$

45. A force of magnitude 50 pounds is applied at the end of a 6-inch-long wrench at an angle of $\frac{\pi}{6}$ to the wrench. Find the magnitude of the torque applied to the bolt.

46. A ball is struck with backspin. Find the direction of the Magnus force and describe the effect on the ball.

In exercises 47–50, find (a) parametric equations and (b) symmetric equations of the line.

47. The line through $(2, -1, -3)$ and $(0, 2, -3)$

48. The line through $(-1, 0, 2)$ and $(-3, 0, -2)$

49. The line through $(2, -1, 1)$ and parallel to $\frac{x-1}{2} = 2y = \frac{z+2}{-3}$

50. The line through $(0, 2, 1)$ and normal to the plane $2x - 3y + z = 4$

In exercises 51 and 52, find the angle between the lines.

51. $\begin{cases} x = 4 + t \\ y = 2 \\ z = 3 + 2t \end{cases}$ and $\begin{cases} x = 4 + 2s \\ y = 2 + 2s \\ z = 3 + 4s \end{cases}$

Review Exercises

52. $\begin{cases} x = 3 + t \\ y = 3 + 3t \\ z = 4 - t \end{cases}$ and $\begin{cases} x = 3 - s \\ y = 3 - 2s \\ z = 4 + 2s \end{cases}$

In exercises 53 and 54, determine whether the lines are parallel, skew or intersect.

53. $\begin{cases} x = 2t \\ y = 3 + t \\ z = -1 + 4t \end{cases}$ and $\begin{cases} x = 4 \\ y = 4 + s \\ z = 3 + s \end{cases}$

54. $\begin{cases} x = 1 - t \\ y = 2t \\ z = 5 - t \end{cases}$ and $\begin{cases} x = 3 + 3s \\ y = 2 \\ z = 1 - 3s \end{cases}$

In exercises 55–58, find an equation of the given plane.

55. The plane containing the point $(-5, 0, 1)$ with normal vector $\langle 4, 1, -2 \rangle$

56. The plane containing the point $(2, -1, 2)$ with normal vector $\langle 3, -1, 0 \rangle$

57. The plane containing the points $(2, 1, 3)$, $(2, -1, 2)$ and $(3, 3, 2)$

58. The plane containing the points $(2, -1, 2)$, $(1, -1, 4)$ and $(3, -1, 2)$

In exercises 59–72, sketch and identify the surface.

59. $9x^2 + y^2 + z = 9$

60. $x^2 + y + z^2 = 1$

61. $y^2 + z^2 = 1$

62. $x^2 + 4y^2 = 4$

63. $x^2 - 2x + y^2 + z^2 = 3$

64. $x^2 + (y + 2)^2 + z^2 = 6$

65. $y = 2$

66. $z = 5$

67. $2x - y + z = 4$

68. $3x + 2y - z = 6$

69. $x^2 - y^2 + 4z^2 = 4$

70. $x^2 - y^2 - z = 1$

71. $x^2 - y^2 - 4z^2 = 4$

72. $x^2 + y^2 - z = 1$

EXPLORATORY EXERCISES

1. Suppose that a piece of pottery is made in the shape of $z = 4 - x^2 - y^2$ for $z \geq 0$. A light source is placed at $(2, 2, 100)$. Draw a sketch showing the pottery and the light source. Based on this picture, which parts of the pottery would be brightly lit and which parts would be poorly lit? This can be quantified, as follows. For several points of your choice on the pottery, find the vector **a** that connects the point to the light source and the normal vector **n** to the tangent plane at that point, and then find the angle between the vectors. For points with relatively large angles, are the points well lit or poorly lit? Develop a rule for using the angle between vectors to determine the lighting level. Find the point that is best lit and the point that is most poorly lit.

2. As we focus on three-dimensional geometry throughout the balance of the book, some projections will be difficult but important to visualize. In this exercise, we contrast the curves C_1 and C_2 defined parametrically by $\begin{cases} x = \cos t \\ y = \cos t \\ z = \sin t \end{cases}$ and $\begin{cases} x = \cos t \\ y = \cos t \\ z = \sqrt{2}\sin t \end{cases}$, respectively. If you have access to three-dimensional graphics, try sketching each curve from a variety of perspectives. Our question will be whether either curve is a circle. For both curves, note that $x = y$. Describe in words and sketch a graph of the plane $x = y$. Next, note that the projection of C_1 back into the yz-plane is a circle ($y = \cos t$, $z = \sin t$). If C_1 is actually a circle in the plane $x = y$, discuss what its projection (shadow) in the yz-plane would look like. Given this, explain whether C_1 is actually a circle or an ellipse. Compare your description of the projection of a circle into the yz-plane to the projection of C_2 into the yz-plane. To make this more quantitative, we can use the general rule that for a two-dimensional region, the area of its projection onto a plane equals the area of the region multiplied by $\cos\theta$, where θ is the angle between the plane in which the region lies and the plane into which it is being projected. Given this, compute the radius of the circle C_2.

Vector-Valued Functions

CHAPTER

11

RoboCup is the international championship of robot soccer. Unlike the remote-controlled destructive robots that you may have seen on television, the robots in this competition are engineered and programmed to respond automatically to the positions of the ball, goal and other players. Once play starts, the robots are completely on their own to analyze the field of play and use teamwork to outmaneuver their opponents and score goals. RoboCup is a challenge for robotics engineers and artificial intelligence researchers. The competitive aspect of RoboCup focuses teams of researchers, while the annual tournament provides invaluable opportunities for feedback and exchange of information.

The competition is divided into several categories, each with its own unique challenges. Overall, one of the greatest difficulties has been providing the robots with adequate vision. In the small size category, "vision" is provided by overhead cameras, with information relayed wirelessly to the robots. With the formidable sight problem removed, the focus is on effective movement of the robots and on providing artificial intelligence for teamwork. The remarkable abilities of these

robots are demonstrated by the sequence of frames shown on the previous page, where a robot on the Cornell Big Red team of 2001 hits a wide-open teammate with a perfect pass that leads to a goal.

There is a considerable amount of mathematics behind this play. To tell whether a teammate is truly open or not, a robot needs to take into account the positions and velocities of each robot, since opponents and teammates could move into the way by the time a pass is executed. In the sequence of photos, notice that all the robots are in motion. Both position and velocity can be described using vectors, but a different vector may be required for each time. In this chapter, we introduce *vector-valued functions,* which assign a vector to each value of the time variable. The calculus introduced in this chapter is essential background knowledge for the programmers of RoboCup.

11.1 VECTOR-VALUED FUNCTIONS

To describe the location of the airplane following the circuitous path indicated in Figure 11.1a, you might consider using a point (x, y, z) in three dimensions. However, it turns out to be more convenient to describe its location at any given time by the endpoint of a vector whose initial point is located at the origin (a **position vector**). (See Figure 11.1b for vectors indicating the location of the plane at a number of times.) Notice that a *function* that gives us a vector in V_3 for each time t would do the job nicely. This is the concept of a vector-valued function, which we define more precisely in Definition 1.1.

DEFINITION 1.1

A **vector-valued function** $\mathbf{r}(t)$ is a mapping from its domain $D \subset \mathbb{R}$ to its range $R \subset V_3$, so that for each t in D, $\mathbf{r}(t) = \mathbf{v}$ for exactly one vector $\mathbf{v} \in V_3$. We can always write a vector-valued function as

$$\mathbf{r}(t) = f(t)\mathbf{i} + g(t)\mathbf{j} + h(t)\mathbf{k}, \tag{1.1}$$

for some scalar functions f, g and h (called the **component functions** of $\mathbf{r}$).

For each t, we regard $\mathbf{r}(t)$ as a position vector. The endpoint of $\mathbf{r}(t)$ then can be viewed as tracing out a curve, as illustrated in Figure 11.1b. Observe that for $\mathbf{r}(t)$ as defined

FIGURE 11.1a
Airplane's flight path

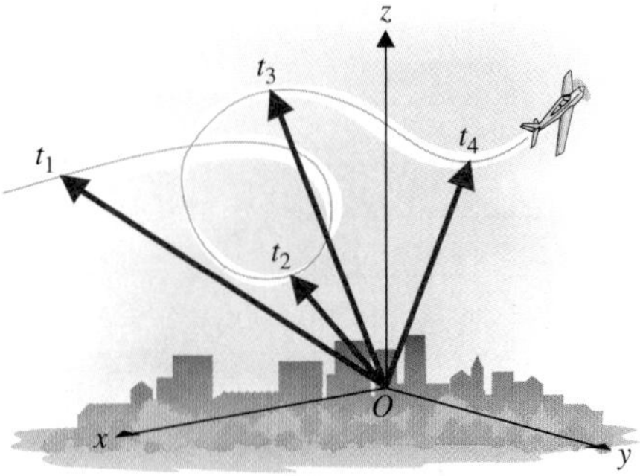

FIGURE 11.1b
Vectors indicating plane's position at several times

in (1.1), this curve is the same as that described by the parametric equations $x = f(t)$, $y = g(t)$ and $z = h(t)$. In three dimensions, such a curve is referred to as a **space curve.**

We can likewise define a vector-valued function $\mathbf{r}(t)$ in V_2 by

$$\mathbf{r}(t) = f(t)\mathbf{i} + g(t)\mathbf{j},$$

for some scalar functions f and g.

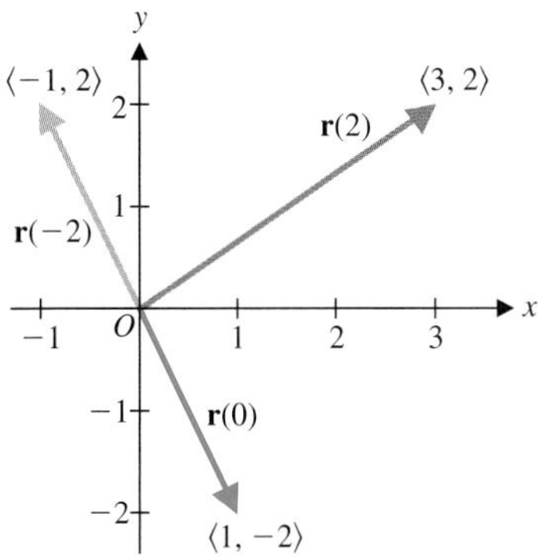

FIGURE 11.2a
Some values of
$\mathbf{r}(t) = (t + 1)\mathbf{i} + (t^2 - 2)\mathbf{j}$

REMARK 1.1

We routinely use the variable t to represent the independent variable for vector-valued functions, since in many applications t represents *time.*

EXAMPLE 1.1 Sketching the Curve Defined by a Vector-Valued Function

Sketch a graph of the curve traced out by the endpoint of the two-dimensional vector-valued function

$$\mathbf{r}(t) = (t + 1)\mathbf{i} + (t^2 - 2)\mathbf{j}.$$

Solution Substituting some values for t, we have $\mathbf{r}(0) = \mathbf{i} - 2\mathbf{j} = \langle 1, -2\rangle$, $\mathbf{r}(2) = 3\mathbf{i} + 2\mathbf{j} = \langle 3, 2\rangle$ and $\mathbf{r}(-2) = \langle -1, 2\rangle$. We plot these in Figure 11.2a. The endpoints of all position vectors $\mathbf{r}(t)$ lie on the curve C, described parametrically by

$$C: x = t + 1, \quad y = t^2 - 2, \quad t \in \mathbb{R}.$$

We can eliminate the parameter by solving for t in terms of x:

$$t = x - 1.$$

The curve is then given by

$$y = t^2 - 2 = (x - 1)^2 - 2.$$

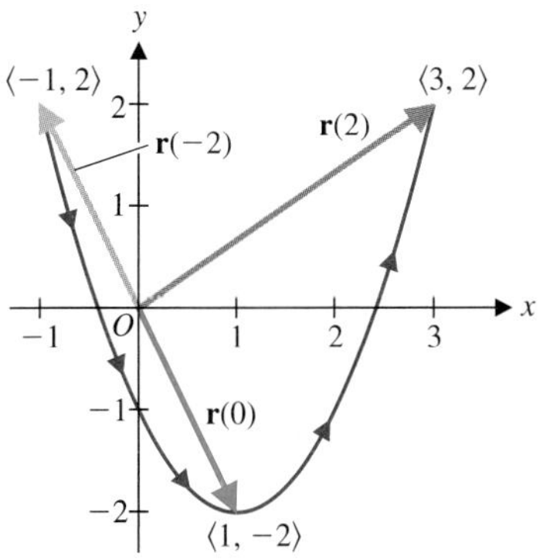

FIGURE 11.2b
Curve defined by
$\mathbf{r}(t) = (t + 1)\mathbf{i} + (t^2 - 2)\mathbf{j}$

Notice that the graph of this is a parabola opening up, with vertex at the point $(1, -2)$, as seen in Figure 11.2b. The small arrows marked on the graph indicate the **orientation,** that is, the direction of increasing values of t. If the curve describes the path of an object, then the orientation indicates the direction in which the object traverses the path. In this case, we can easily determine the orientation from the parametric representation of the curve. Since $x = t + 1$, observe that x increases as t increases. ■

You may recall from your experience with parametric equations in Chapter 9 that eliminating the parameter from the parametric representation of a curve is not always so easy as it was in example 1.1. We illustrate this in example 1.2.

EXAMPLE 1.2 A Vector-Valued Function Defining an Ellipse

Sketch a graph of the curve traced out by the endpoint of the vector-valued function $\mathbf{r}(t) = 4\cos t\,\mathbf{i} - 3\sin t\,\mathbf{j}$, $t \in \mathbb{R}$.

Solution In this case, the curve can be written parametrically as

$$x = 4\cos t, \quad y = -3\sin t, \quad t \in \mathbb{R}.$$

Instead of solving for the parameter t, it often helps to look for some relationship between the variables. Here,

$$\left(\frac{x}{4}\right)^2 + \left(\frac{y}{3}\right)^2 = \cos^2 t + \sin^2 t = 1$$

or

$$\left(\frac{x}{4}\right)^2 + \left(\frac{y}{3}\right)^2 = 1,$$

which is the equation of an ellipse (see Figure 11.3). To determine the orientation of the curve here, you'll need to look carefully at both parametric equations. First, fix a starting place on the curve, for convenience, say (4, 0). This corresponds to $t = 0, \pm 2\pi, \pm 4\pi, \ldots$. As t increases, notice that $\cos t$ (and hence, x) decreases initially, while $\sin t$ increases, so that $y = -3\sin t$ decreases (initially). With both x and y decreasing initially, we get the clockwise orientation indicated in Figure 11.3. ■

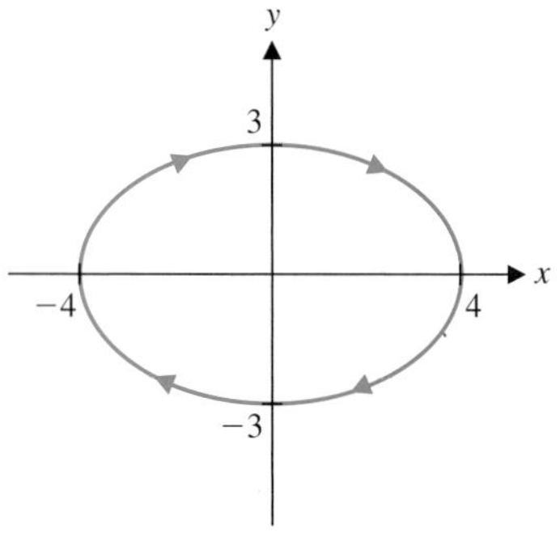

FIGURE 11.3
Curve defined by
$\mathbf{r}(t) = 4\cos t\mathbf{i} - 3\sin t\mathbf{j}$

Just as the endpoint of a vector-valued function in two dimensions traces out a curve, if we were to plot the value of $\mathbf{r}(t) = f(t)\mathbf{i} + g(t)\mathbf{j} + h(t)\mathbf{k}$ for every value of t, the endpoints of the vectors would trace out a curve in three dimensions.

EXAMPLE 1.3 A Vector-Valued Function Defining an Elliptical Helix

Plot the curve traced out by the vector-valued function $\mathbf{r}(t) = \sin t\mathbf{i} - 3\cos t\mathbf{j} + 2t\mathbf{k}$, $t \geq 0$.

Solution The curve is given parametrically by

$$x = \sin t, \quad y = -3\cos t, \quad z = 2t, \quad t \geq 0.$$

While most curves in three dimensions are difficult to recognize, you should notice that there is a relationship between x and y here, namely,

$$x^2 + \left(\frac{y}{3}\right)^2 = \sin^2 t + \cos^2 t = 1. \qquad (1.2)$$

In two dimensions, this is the equation of an ellipse. In three dimensions, since the equation does not involve z, (1.2) is the equation of an elliptic cylinder whose axis is the z-axis. This says that every point on the curve defined by $\mathbf{r}(t)$ lies on this cylinder. From the parametric equations for x and y (in two dimensions), the ellipse is traversed in the counterclockwise direction. This says that the curve will wrap itself around the cylinder (counterclockwise, as you look down the positive z-axis toward the origin), as t increases. Finally, since $z = 2t$, z will increase as t increases and so, the curve will wind its way up the cylinder, as t increases. We show the curve and the elliptical cylinder in Figure 11.4a. We call this curve an **elliptical helix.** In Figure 11.4b, we display a computer-generated graph of the same helix. There, rather than the usual x-, y- and z-axes, we show a framed graph, where the values of x, y and z are indicated on three adjacent edges of a box containing the graph. ■

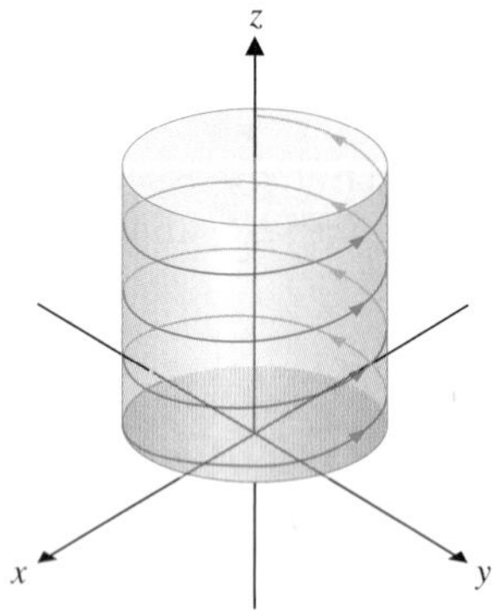

FIGURE 11.4a
Elliptical helix:
$\mathbf{r}(t) = \sin t\mathbf{i} - 3\cos t\mathbf{j} + 2t\mathbf{k}$

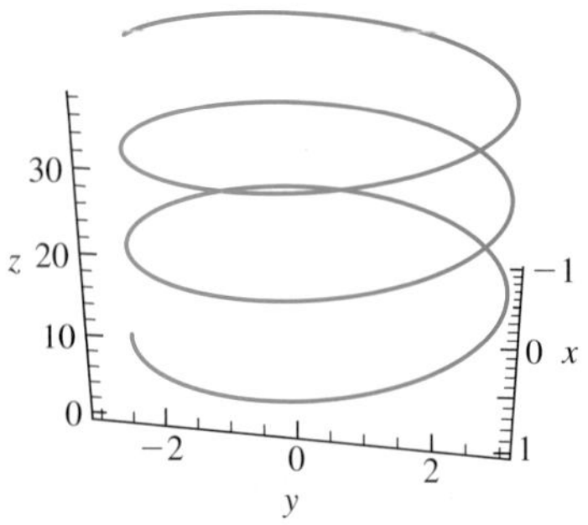

FIGURE 11.4b
Computer sketch:
$\mathbf{r}(t) = \sin t\mathbf{i} - 3\cos t\mathbf{j} + 2t\mathbf{k}$

We can use vector-valued functions as a convenient representation of some very familiar curves, as we see in example 1.4.

EXAMPLE 1.4 A Vector-Valued Function Defining a Line

Plot the curve traced out by the vector-valued function

$$\mathbf{r}(t) = \langle 3 + 2t, 5 - 3t, 2 - 4t \rangle, \quad t \in \mathbb{R}.$$

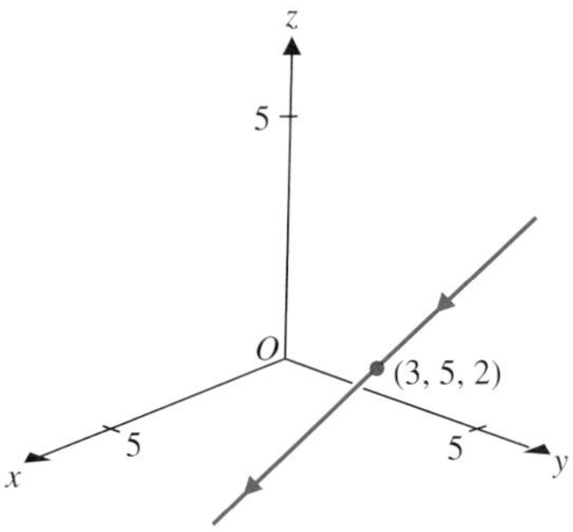

FIGURE 11.5
Straight line:
$\mathbf{r}(t) = \langle 3 + 2t, 5 - 3t, 2 - 4t \rangle$

Solution Notice that the curve is given parametrically by

$$x = 3 + 2t, \quad y = 5 - 3t, \quad z = 2 - 4t, \quad t \in \mathbb{R}.$$

You should recognize these equations as parametric equations for the straight line parallel to the vector $\langle 2, -3, -4 \rangle$ and passing through the point (3, 5, 2), as seen in Figure 11.5. ■

Most three-dimensional graphs are very challenging to sketch by hand. You will probably want to use computer-generated graphics for most sketches. Even so, you will need to be knowledgeable enough to know when to zoom in or out or rotate a graph to uncover a hidden feature. You should be able to draw several basic curves by hand, like those in examples 1.3 and 1.4. More importantly, you should be able to recognize the effects various components have on the graph of a three-dimensional curve. In example 1.5, we walk you through matching four vector-valued functions with their computer-generated graphs.

EXAMPLE 1.5 Matching a Vector-Valued Function to Its Graph

Match each of the vector-valued functions $\mathbf{f}_1(t) = \langle \cos t, \ln t, \sin t \rangle$, $\mathbf{f}_2(t) = \langle t \cos t, t \sin t, t \rangle$, $\mathbf{f}_3(t) = \langle 3 \sin 2t, t, t \rangle$ and $\mathbf{f}_4(t) = \langle 5 \sin^3 t, 5 \cos^3 t, t \rangle$ with the corresponding computer-generated graph.

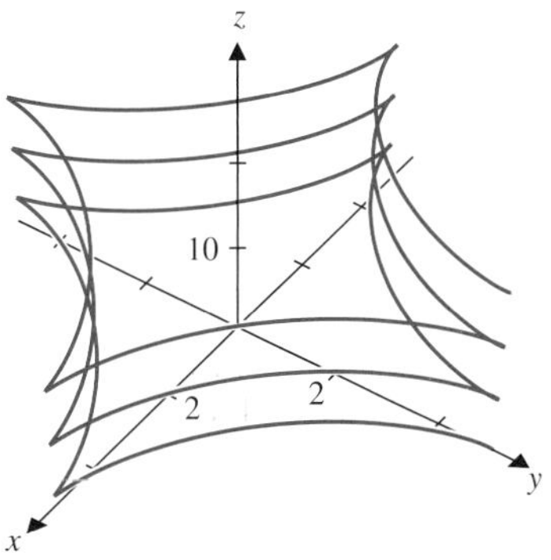

GRAPH A

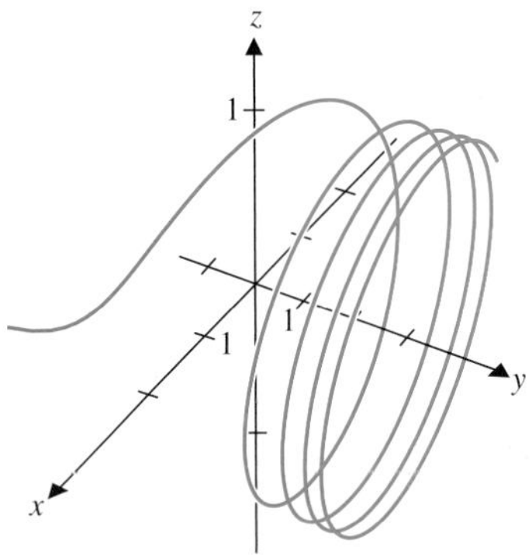

GRAPH B

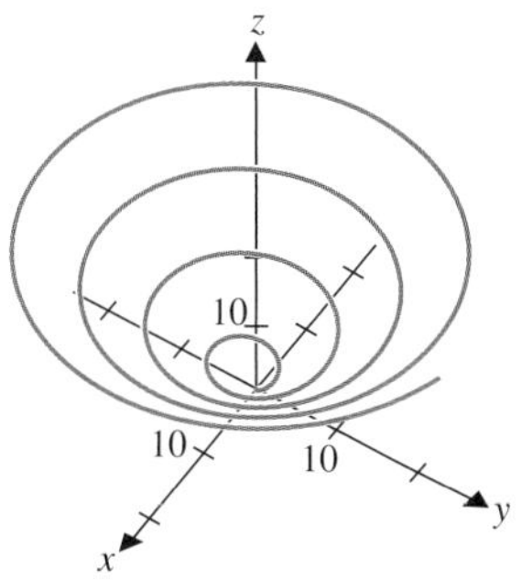

GRAPH C

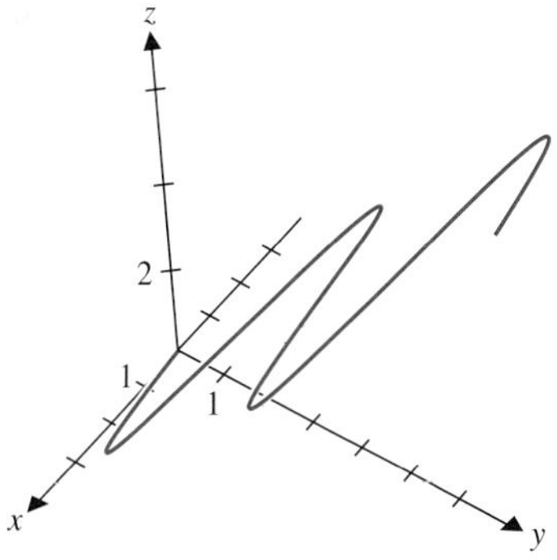

GRAPH D

Solution First, realize that there is no single, correct procedure for solving this problem. Look for familiar functions and match them with familiar graphical properties.

From example 1.3, recall that certain combinations of sines and cosines will produce curves that lie on a cylinder. Notice that for the function $\mathbf{f}_1(t)$, $x = \cos t$ and $z = \sin t$, so that

$$x^2 + z^2 = \cos^2 t + \sin^2 t = 1.$$

This says that every point on the curve lies on the cylinder $x^2 + z^2 = 1$ (the right circular cylinder of radius 1 whose axis is the y-axis). Further, the function $y = \ln t$ tends rapidly to $-\infty$ as $t \to 0^+$ and increases slowly as t increases beyond $t = 1$. Notice that the curve in Graph B appears to lie on a right circular cylinder and that the spirals get closer together as you move to the right (as $y \to \infty$) and move very far apart as you move to the left (as $y \to -\infty$). At first glance, you might expect the curve traced out by $\mathbf{f}_2(t)$ also to lie on a right circular cylinder, but look more closely. Here, we have $x = t \cos t$, $y = t \sin t$ and $z = t$, so that

$$x^2 + y^2 = t^2 \cos^2 t + t^2 \sin^2 t = t^2 = z^2.$$

This says that the curve lies on the surface defined by $x^2 + y^2 = z^2$ (a right circular cone with axis along the z-axis). Notice that only the curve shown in Graph C fits this description. Next, notice that for $\mathbf{f}_3(t)$, the y and z components are identical and so, the curve must lie in the plane $y = z$. Replacing t by y, we have $x = 3 \sin 2t = 3 \sin 2y$, a sine curve lying in the plane $y = z$. Clearly, the curve in Graph D matches this description. Although Graph A is the only curve remaining to match with $\mathbf{f}_4(t)$, notice that if the cosine and sine terms weren't cubed, we'd simply have a helix, as in example 1.3. Since $z = t$, each point on the curve is a point on the cylinder defined parametrically by $x = 5 \sin^3 t$ and $y = 5 \cos^3 t$. You need only look at the graph of the cross section of the cylinder shown in Figure 11.6 (found by graphing the parametric equations $x = 5 \sin^3 t$ and $y = 5 \cos^3 t$ in two dimensions) to decide that Graph A is the obvious choice. ■

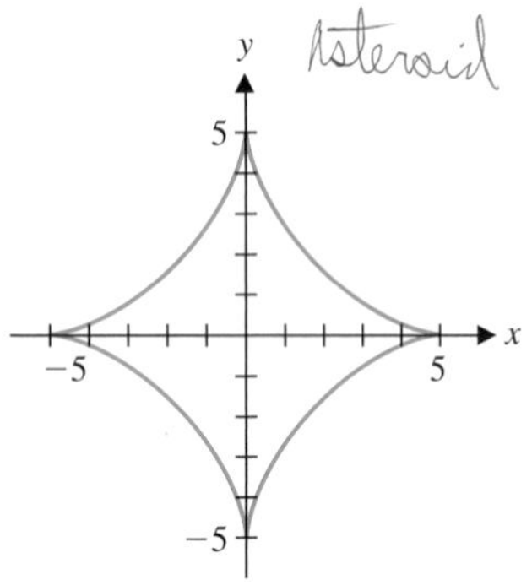

FIGURE 11.6
A cross section of the cylinder $x = 5 \sin^3 t$, $y = 5 \cos^3 t$

○ Arc Length in $\mathbb{R}^3$

A natural question to ask about a curve is, "How long is it?" Recall from section 5.4 that if f and f' are continuous on the interval $[a, b]$, then the arc length of the curve $y = f(x)$ on that interval is given by

$$s = \int_a^b \sqrt{1 + [f'(x)]^2}\, dx.$$

In section 9.3, we extended this to the case of a curve defined parametrically by $x = f(t)$, $y = g(t)$, where f, f', g and g' are all continuous for $t \in [a, b]$. In this case, we showed that if the curve is traversed exactly once as t increases from a to b, then the arc length is given by

$$s = \int_a^b \sqrt{[f'(t)]^2 + [g'(t)]^2}\, dt. \qquad (1.3)$$

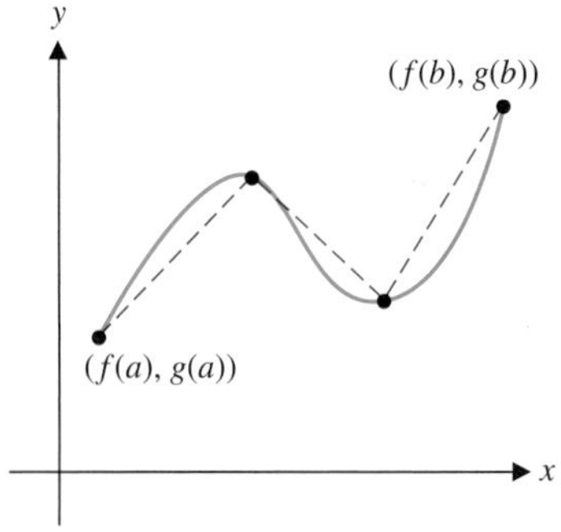

FIGURE 11.7a
Approximate arc length in $\mathbb{R}^2$

In both cases, recall that we developed the arc length formula by first breaking the curve into small pieces (i.e., we *partitioned* the interval $[a, b]$) and then approximating the length with the sum of the lengths of small line segments connecting successive points (see Figure 11.7a). Finally, we made the approximation exact by taking a limit as the number of points in the partition tended to infinity. This says that if the curve C in $\mathbb{R}^2$ is traced out exactly once by the endpoint of the vector-valued function $\mathbf{r}(t) = \langle f(t), g(t) \rangle$ for $t \in [a, b]$, then the arc length is given by (1.3).

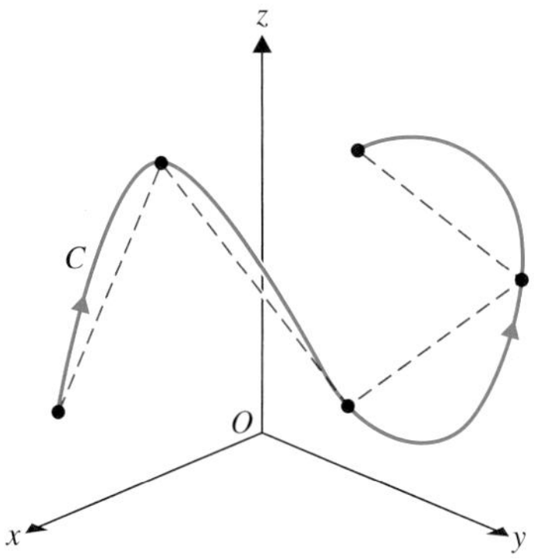

FIGURE 11.7b
Approximate arc length in $\mathbb{R}^3$

The situation in three dimensions is a straightforward extension of the two-dimensional case. Suppose that a curve is traced out by the endpoint of the vector-valued function $\mathbf{r}(t) = \langle f(t), g(t), h(t)\rangle$, where f, f', g, g', h and h' are all continuous for $t \in [a, b]$ and where the curve is traversed exactly once as t increases from a to b. As we have done countless times now, we begin by approximating the quantity of interest, in this case, the arc length. To do this, we partition the interval $[a, b]$ into n subintervals of equal size: $a = t_0 < t_1 < \cdots < t_n = b$, where $t_i - t_{i-1} = \Delta t = \frac{b-a}{n}$, for all $i = 1, 2, \ldots, n$. Next, for each $i = 1, 2, \ldots, n$, we approximate the arc length s_i of that portion of the curve joining the points $(f(t_{i-1}), g(t_{i-1}), h(t_{i-1}))$ and $(f(t_i), g(t_i), h(t_i))$ by the straight-line distance between the points. (See Figure 11.7b for an illustration of the case where $n = 4$.) From the distance formula, we have

$$\begin{aligned} s_i &\approx d\{(f(t_{i-1}), g(t_{i-1}), h(t_{i-1})), (f(t_i), g(t_i), h(t_i))\} \\ &= \sqrt{[f(t_i) - f(t_{i-1})]^2 + [g(t_i) - g(t_{i-1})]^2 + [h(t_i) - h(t_{i-1})]^2}. \end{aligned}$$

Applying the Mean Value Theorem three times (why can we do this?), we get

$$f(t_i) - f(t_{i-1}) = f'(c_i)(t_i - t_{i-1}) = f'(c_i)\,\Delta t,$$
$$g(t_i) - g(t_{i-1}) = g'(d_i)(t_i - t_{i-1}) = g'(d_i)\,\Delta t$$

and

$$h(t_i) - h(t_{i-1}) = h'(e_i)(t_i - t_{i-1}) = h'(e_i)\,\Delta t,$$

for some points c_i, d_i and e_i in the interval (t_{i-1}, t_i). This gives us

$$\begin{aligned} s_i &\approx \sqrt{[f(t_i) - f(t_{i-1})]^2 + [g(t_i) - g(t_{i-1})]^2 + [h(t_i) - h(t_{i-1})]^2} \\ &= \sqrt{[f'(c_i)\,\Delta t]^2 + [g'(d_i)\,\Delta t]^2 + [h'(e_i)\,\Delta t]^2} \\ &= \sqrt{[f'(c_i)]^2 + [g'(d_i)]^2 + [h'(e_i)]^2}\,\Delta t. \end{aligned}$$

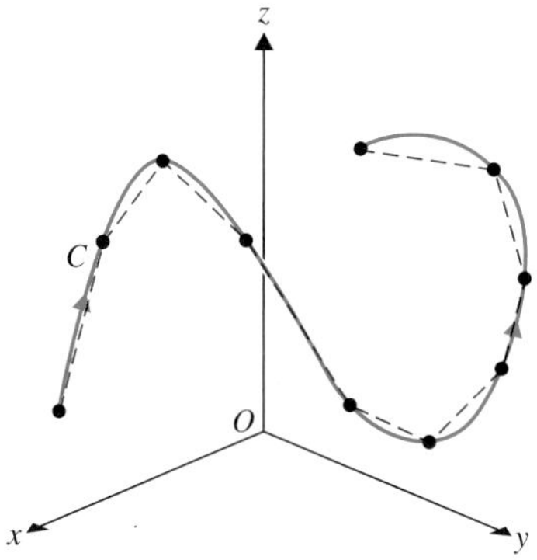

FIGURE 11.7c
Improved arc length approximation

Notice that if Δt is small, then all of c_i, d_i and e_i are very close and we can make the further approximation

$$s_i \approx \sqrt{[f'(c_i)]^2 + [g'(c_i)]^2 + [h'(c_i)]^2}\,\Delta t,$$

for each $i = 1, 2, \ldots, n$. The total arc length is then approximately

$$s \approx \sum_{i=1}^{n} \sqrt{[f'(c_i)]^2 + [g'(c_i)]^2 + [h'(c_i)]^2}\,\Delta t,$$

where the total error in the approximation of arc length tends to 0, as $\Delta t \to 0$. (Carefully consider Figures 11.7b and 11.7c to see why.)

Taking the limit as $n \to \infty$ gives the exact arc length:

$$s = \lim_{n\to\infty} \sum_{i=1}^{n} \sqrt{[f'(c_i)]^2 + [g'(c_i)]^2 + [h'(c_i)]^2}\,\Delta t,$$

provided the limit exists. You should recognize this as the definite integral

Arc length

$$s = \int_a^b \sqrt{[f'(t)]^2 + [g'(t)]^2 + [h'(t)]^2}\,dt. \tag{1.4}$$

Observe that the arc length formula for a curve in $\mathbb{R}^2$ (1.3) is a special case of (1.4). Unfortunately, the integral in (1.4) can only rarely be computed exactly and we must typically be satisfied with a numerical approximation. Example 1.6 illustrates one of the very few arc lengths in $\mathbb{R}^3$ that can be computed exactly.

EXAMPLE 1.6 Computing Arc Length in $\mathbb{R}^3$

Find the arc length of the curve traced out by the endpoint of the vector-valued function $\mathbf{r}(t) = \langle 2t, \ln t, t^2 \rangle$, for $1 \le t \le e$.

Solution First, notice that for $x(t) = 2t$, $y(t) = \ln t$ and $z(t) = t^2$, we have $x'(t) = 2$, $y'(t) = \frac{1}{t}$ and $z'(t) = 2t$, and the curve is traversed exactly once for $1 \le t \le e$. (To see why, observe that $x = 2t$ is an increasing function.) From (1.4), we now have

$$\begin{aligned} s &= \int_1^e \sqrt{2^2 + \left(\frac{1}{t}\right)^2 + (2t)^2}\, dt = \int_1^e \sqrt{4 + \frac{1}{t^2} + 4t^2}\, dt \\ &= \int_1^e \sqrt{\frac{1 + 4t^2 + 4t^4}{t^2}}\, dt = \int_1^e \sqrt{\frac{(1 + 2t^2)^2}{t^2}}\, dt \\ &= \int_1^e \frac{1 + 2t^2}{t}\, dt = \int_1^e \left(\frac{1}{t} + 2t\right) dt \\ &= \left(\ln|t| + 2\frac{t^2}{2}\right)\Bigg|_1^e = (\ln e + e^2) - (\ln 1 + 1) = e^2. \end{aligned}$$

We show a graph of the curve for $1 \le t \le e$ in Figure 11.8. ■

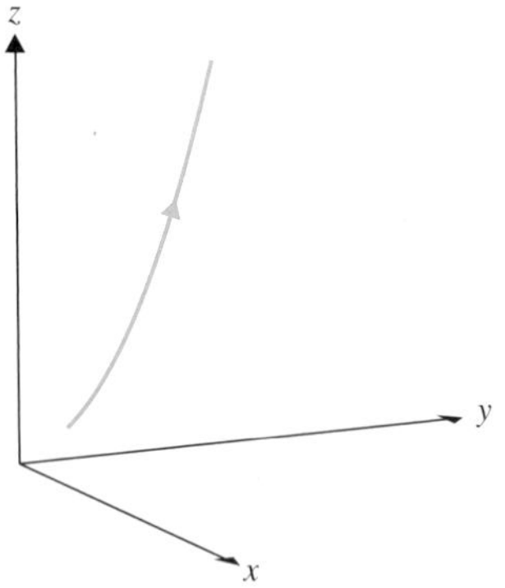

FIGURE 11.8
The curve defined by $\mathbf{r}(t) = \langle 2t, \ln t, t^2 \rangle$

The arc length integral in example 1.7 is typical, in that we need a numerical approximation.

EXAMPLE 1.7 Approximating Arc Length in $\mathbb{R}^3$

Find the arc length of the curve traced out by the endpoint of the vector-valued function $\mathbf{r}(t) = \langle e^{2t}, \sin t, t \rangle$, for $0 \le t \le 2$.

Solution First, note that for $x(t) = e^{2t}$, $y(t) = \sin t$ and $z(t) = t$, we have $x'(t) = 2e^{2t}$, $y'(t) = \cos t$ and $z'(t) = 1$, and that the curve is traversed exactly once for $0 \le t \le 2$. From (1.4), we now have

$$s = \int_0^2 \sqrt{(2e^{2t})^2 + (\cos t)^2 + 1^2}\, dt = \int_0^2 \sqrt{4e^{4t} + \cos^2 t + 1}\, dt.$$

Since you don't know how to evaluate this integral exactly (which is typically the case), you can approximate the integral using Simpson's Rule or the numerical integration routine built into your calculator or computer algebra system, to find that the arc length is approximately $s \approx 53.8$. ■

Often, the curve of interest is determined by the intersection of two surfaces. Parametric equations can give us simple representations of many such curves.

EXAMPLE 1.8 Finding Parametric Equations for an Intersection of Surfaces

Find the arc length of the portion of the curve determined by the intersection of the cone $z = \sqrt{x^2 + y^2}$ and the plane $y + z = 2$ in the first octant.

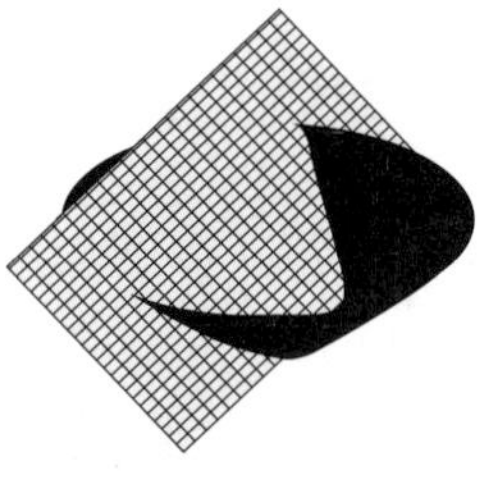

FIGURE 11.9a
Intersection of cone and plane

Solution The cone and plane are shown in Figure 11.9a. From your knowledge of conic sections, note that this curve could be a parabola or an ellipse. Parametric equations for the curve must satisfy both $z = \sqrt{x^2 + y^2}$ and $y + z = 2$. Eliminating z by solving for it in each equation, we get

$$z = \sqrt{x^2 + y^2} = 2 - y.$$

Squaring both sides and gathering terms, we get

$$x^2 + y^2 = (2 - y)^2 = 4 - 4y + y^2$$

or

$$x^2 = 4 - 4y.$$

Solving for y now gives us

$$y = 1 - \frac{x^2}{4},$$

which is clearly the equation of a parabola in two dimensions. To obtain the equation for the three-dimensional parabola, let x be the parameter, which gives us the parametric equations

$$x = t, \quad y = 1 - \frac{t^2}{4} \quad \text{and} \quad z = \sqrt{t^2 + (1 - t^2/4)^2} = 1 + \frac{t^2}{4}.$$

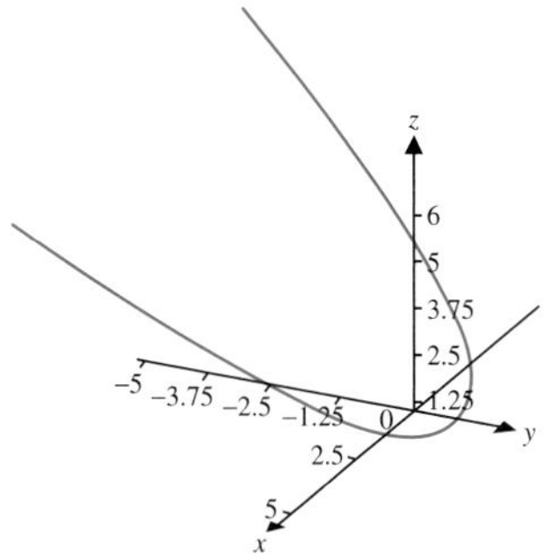

FIGURE 11.9b
Curve of intersection

A graph is shown in Figure 11.9b. The portion of the parabola in the first octant must have $x \geq 0$ (so $t \geq 0$), $y \geq 0$ (so $t^2 \leq 4$) and $z \geq 0$ (always true). This occurs if $0 \leq t \leq 2$. The arc length is then

$$s = \int_0^2 \sqrt{1 + (-t/2)^2 + (t/2)^2}\, dt = \frac{\sqrt{2}}{2} \ln\left(\sqrt{2} + \sqrt{3}\right) + \sqrt{3} \approx 2.54,$$

where we leave the details of the integration to you. ■

BEYOND FORMULAS

If you think that examples 1.1 and 1.2 look very much like parametric equations examples, you're exactly right. The ideas presented there are not new; only the notation and terminology are new. However, the vector notation lets us easily extend these ideas into three dimensions, where the graphs can be more complicated.

EXERCISES 11.1

WRITING EXERCISES

1. Discuss the differences, if any, between the curve traced out by the terminal point of the vector-valued function $\mathbf{r}(t) = \langle f(t), g(t)\rangle$ and the curve defined parametrically by $x = f(t)$, $y = g(t)$.

2. In example 1.3, describe the "shadow" of the helix in the xy-plane (the shadow created by shining a light down from the "top" of the z-axis). Equivalently, if the helix is collapsed down into the xy-plane, describe the resulting curve. Compare this curve to the ellipse defined parametrically by $x = \sin t$, $y = -3\cos t$.

3. Discuss how you would compute the arc length of a curve in four or more dimensions. Specifically, for the curve traced out by the terminal point of the n-dimensional vector-valued function $\mathbf{r}(t) = \langle f_1(t), f_2(t), \ldots, f_n(t)\rangle$ for $n \geq 4$, state the arc

length formula and discuss how it relates to the n-dimensional distance formula.

4. The helix in Figure 11.4a is shown from a standard viewpoint (above the xy-plane, in between the x- and y-axes). Describe what an observer at the point $(0, 0, -1000)$ would see. Also, describe what observers at the points $(1000, 0, 0)$ and $(0, 1000, 0)$ would see.

In exercises 1–4, plot the values of the vector-valued function at the indicated values of t.

1. $\mathbf{r}(t) = \langle 3t, t^2, 2t - 1\rangle, t = 0, t = 1, t = 2$
2. $\mathbf{r}(t) = (4 - t)\mathbf{i} + (1 - t^2)\mathbf{j} + (t^3 - 1)\mathbf{k}, t = -2, t = 0, t = 2$
3. $\mathbf{r}(t) = \langle \cos 3t, 2, \sin 2t - 1\rangle, t = -\frac{\pi}{2}, t = 0, t = \frac{\pi}{2}$
4. $\mathbf{r}(t) = \langle e^{2-t}, 1 - t, 3\rangle, t = -1, t = 0, t = 1$

In exercises 5–18, sketch the curve traced out by the given vector-valued function by hand.

5. $\mathbf{r}(t) = \langle 2\cos t, \sin t - 1\rangle$
6. $\mathbf{r}(t) = \langle \sin t - 2, 4\cos t\rangle$
7. $\mathbf{r}(t) = \langle 2\cos t, 2\sin t, 3\rangle$
8. $\mathbf{r}(t) = \langle \cos 2t, \sin 2t, 1\rangle$
9. $\mathbf{r}(t) = \langle t, t^2 + 1, -1\rangle$
10. $\mathbf{r}(t) = \langle 3, t, t^2 - 1\rangle$
11. $\mathbf{r}(t) = \langle t, 1, 3t^2\rangle$
12. $\mathbf{r}(t) = \langle t + 2, 2t - 1, t + 2\rangle$
13. $\mathbf{r}(t) = \langle 4t - 1, 2t + 1, -6t\rangle$
14. $\mathbf{r}(t) = \langle -2t, 2t, 3 - t\rangle$
15. $\mathbf{r}(t) = \langle 3\cos t, 3\sin t, t\rangle$
16. $\mathbf{r}(t) = \langle 2\cos t, \sin t, 3t\rangle$
17. $\mathbf{r}(t) = \langle 2\cos t, 3\sin t, 2t\rangle$
18. $\mathbf{r}(t) = \langle -1, 2\cos t, 2\sin t\rangle$

In exercises 19–30, use graphing technology to sketch the curve traced out by the given vector-valued function.

19. $\mathbf{r}(t) = \langle 2\cos t + \sin 2t, 2\sin t + \cos 2t\rangle$
20. $\mathbf{r}(t) = \langle 2\cos 3t + \sin 5t, 2\sin 3t + \cos 5t\rangle$
21. $\mathbf{r}(t) = \langle 4\cos 4t - 6\cos t, 4\sin 4t - 6\sin t\rangle$
22. $\mathbf{r}(t) = \langle 8\cos t + 2\cos 7t, 8\sin t + 2\sin 7t\rangle$
23. $\mathbf{r}(t) = \langle t\cos 2t, t\sin 2t, 2t\rangle$
24. $\mathbf{r}(t) = \langle t\cos t, 2t, t\sin t\rangle$
25. $\mathbf{r}(t) = \langle \cos 5t, \sin t, \sin 6t\rangle$
26. $\mathbf{r}(t) = \langle 3\cos 2t, \sin t, \cos 3t\rangle$
27. $\mathbf{r}(t) = \langle t, t, 2t^2 - 1\rangle$
28. $\mathbf{r}(t) = \langle t^3 - t, t^2, 2t - 4\rangle$
29. $\mathbf{r}(t) = \langle \tan t, \sin t^2, \cos t\rangle$
30. $\mathbf{r}(t) = \langle \sin t, -\csc t, \cot t\rangle$

31. In parts a–f, match the vector-valued function with its graph. Give reasons for your choices.
    a. $\mathbf{r}(t) = \langle \cos t^2, t, t\rangle$
    b. $\mathbf{r}(t) = \langle \cos t, \sin t, \sin t^2\rangle$
    c. $\mathbf{r}(t) = \langle \sin 16\sqrt{t}, \cos 16\sqrt{t}, t\rangle$
    d. $\mathbf{r}(t) = \langle \sin t^2, \cos t^2, t\rangle$
    e. $\mathbf{r}(t) = \langle t, t, 6 - 4t^2\rangle$
    f. $\mathbf{r}(t) = \langle t^3 - t, 0.5t^2, 2t - 4\rangle$

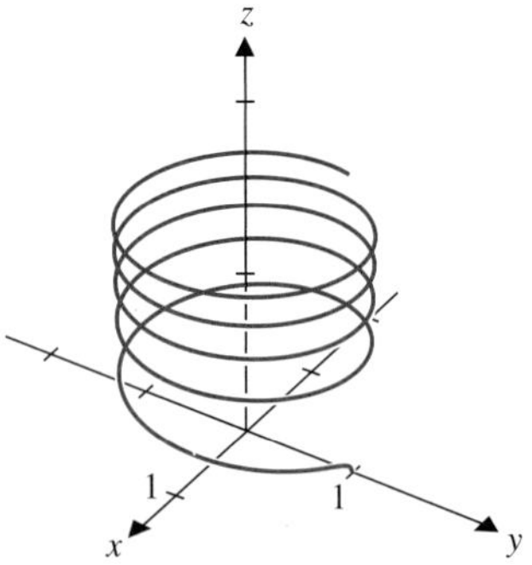

GRAPH A

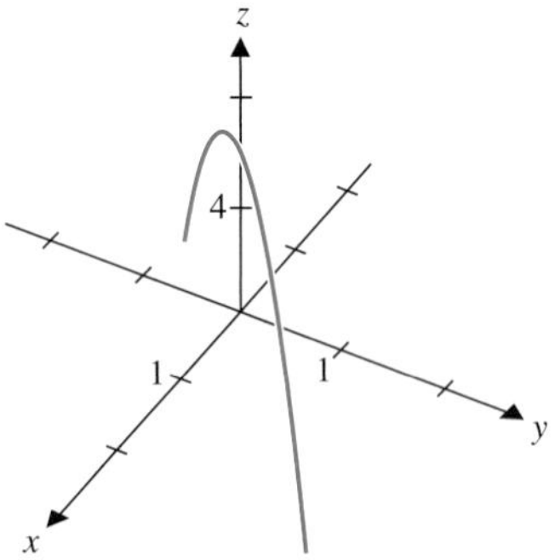

GRAPH B

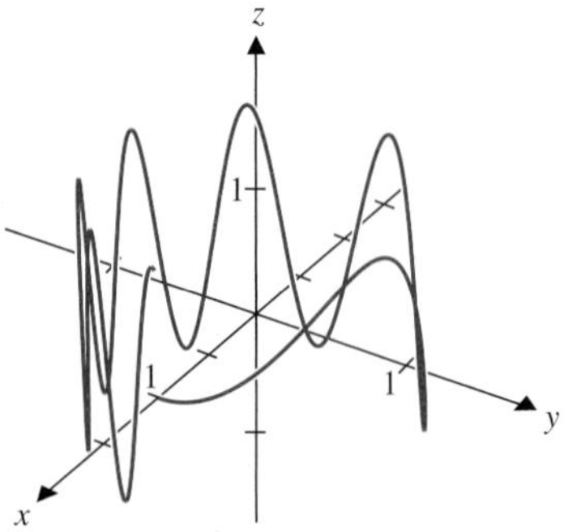

GRAPH C

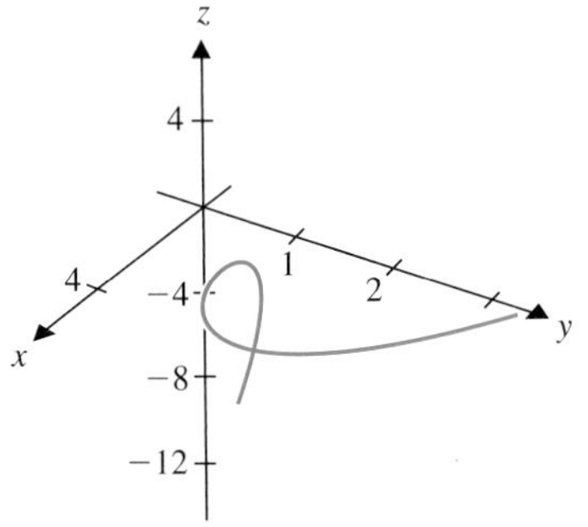

GRAPH D

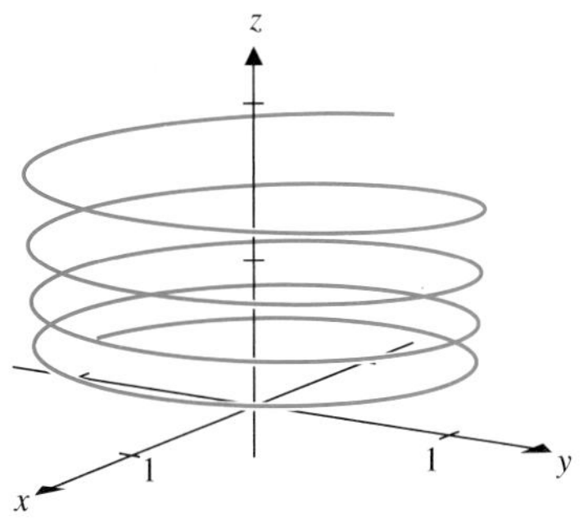

GRAPH E

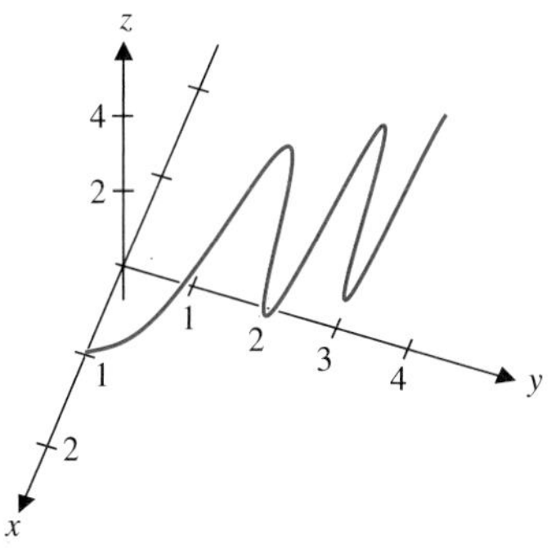

GRAPH F

32. Of the functions in exercise 31, which are periodic? Which are bounded?

In exercises 33–38, use a CAS to sketch the curve and estimate its arc length.

33. $\mathbf{r}(t) = \langle \cos t, \sin t, \cos 2t \rangle, 0 \le t \le 2\pi$

34. $\mathbf{r}(t) = \langle \cos t, \sin t, \sin t + \cos t \rangle, 0 \le t \le 2\pi$

35. $\mathbf{r}(t) = \langle \cos \pi t, \sin \pi t, \cos 16t \rangle, 0 \le t \le 2$

36. $\mathbf{r}(t) = \langle \cos \pi t, \sin \pi t, \cos 16t \rangle, 0 \le t \le 4$

37. $\mathbf{r}(t) = \langle t, t^2 - 1, t^3 \rangle, 0 \le t \le 2$

38. $\mathbf{r}(t) = \langle t^2 + 1, 2t, t^2 - 1 \rangle, 0 \le t \le 2$

39. Show that the curve in exercise 33 lies on the hyperbolic paraboloid $z = x^2 - y^2$. Use a CAS to sketch both the surface and the curve.

40. Show that the curve in exercise 34 lies on the plane $z = x + y$. Use a CAS to sketch both the plane and the curve.

41. Show that the curve $\mathbf{r}(t) = \langle 2t, 4t^2 - 1, 8t^3 \rangle, 0 \le t \le 1$, has the same arc length as the curve in exercise 37.

42. Show that the curve $\mathbf{r}(t) = \langle t + 1, 2\sqrt{t}, t - 1 \rangle, 0 \le t \le 4$, has the same arc length as the curve in exercise 38.

43. Compare the graphs of $\mathbf{r}(t) = \langle t, t^2, t^2 \rangle$, $\mathbf{g}(t) = \langle \cos t, \cos^2 t, \cos^2 t \rangle$ and $\mathbf{h}(t) = \langle \sqrt{t}, t, t \rangle$. Explain the similarities and the differences.

44. Compare the graphs of $\mathbf{r}(t) = \langle 2t - 1, t^2, t \rangle$, $\mathbf{g}(t) = \langle 2 \sin t - 1, \sin^2 t, \sin t \rangle$ and $\mathbf{h}(t) = \langle 2e^t - 1, e^{2t}, e^t \rangle$. Explain the similarities and the differences.

In exercises 45–48, find parametric equations for the indicated curve. If you have access to a graphing utility, graph the surfaces and the resulting curve.

45. The intersection of $z = \sqrt{x^2 + y^2}$ and $z = 2$

46. The intersection of $z = \sqrt{x^2 + y^2}$ and $y + 2z = 2$

47. The intersection of $x^2 + y^2 = 9$ and $y + z = 2$

48. The intersection of $y^2 + z^2 = 9$ and $x = 2$

49. A spiral staircase makes two complete turns as it rises 10 feet between floors. A handrail at the outside of the staircase is located 3 feet from the center pole of the staircase. Use parametric equations for a helix to compute the length of the handrail.

50. Exercise 49 can be worked without calculus. (You might have suspected this since the integral in exercise 49 simplifies dramatically.) Imagine unrolling the staircase so that the handrail is a line segment. Use the formula for the hypotenuse of a right triangle to compute its length.

51. Use a graphing utility to sketch the graph of $\mathbf{r}(t) = \langle \cos t, \cos t, \sin t \rangle$ with $0 \le t \le 2\pi$. Explain why the graph should be the same with $0 \le t \le T$, for any $T \ge 2\pi$. Try several larger domains ($0 \le t \le 2\pi, 0 \le t \le 10\pi, 0 \le t \le 50\pi$, etc.) with your graphing utility. Eventually, the ellipse should start looking thicker and for large enough domains you will see a mess of jagged lines. Explain what has gone wrong with the graphing utility.

52. It may surprise you that the curve in exercise 51 is not a circle. Show that the shadows in the xz-plane and yz-plane are circles. Show that the curve lies in the plane $x = y$. Sketch a graph showing the plane $x = y$ and a circular shadow in the yz-plane. To draw a curve in the plane $x = y$ with the circular shadow, explain why the curve must be wider in the xy-direction than in the z-direction. In other words, the curve is not circular.

53. Show that the arc length of the helix $\langle \cos t, \sin t, t \rangle$, for $0 \le t \le 2\pi$, is $2\pi\sqrt{2}$, equal to the length of the diagonal of a square of side 2π. Show this graphically.

EXPLORATORY EXERCISES

1. In contrast to exercises 51 and 52, the graph of $\mathbf{r}(t) = \langle \cos t, \cos t, \sqrt{2} \sin t \rangle$ *is* a circle. To verify this, start by showing that $\|\mathbf{r}(t)\| = \sqrt{2}$, for all t. Then observe that the curve lies in the plane $x = y$. Explain why this proves that the graph is a (portion of a) circle. A little more insight can be gained by looking at basis vectors. The circle lies in the plane $x = y$, which contains the vector $\mathbf{u} = \frac{1}{\sqrt{2}}\langle 1, 1, 0 \rangle$. The plane $x = y$ also contains the vector $\mathbf{v} = \langle 0, 0, 1 \rangle$. Show that *any* vector $\mathbf{w}$ in the plane $x = y$ can be written as $\mathbf{w} = c_1\mathbf{u} + c_2\mathbf{v}$ for some constants c_1 and c_2. Also, show that $\mathbf{r}(t) = (\sqrt{2}\cos t)\mathbf{u} + (\sqrt{2}\sin t)\mathbf{v}$. Recall that in two dimensions, a circle of radius r centered at the origin can be written parametrically as $(r\cos t)\mathbf{i} + (r \sin t)\mathbf{j}$. In general, suppose that $\mathbf{u}$ and $\mathbf{v}$ are any orthogonal unit vectors. If $\mathbf{r}(t) = (r\cos t)\mathbf{u} + (r\sin t)\mathbf{v}$, show that $\mathbf{r}(t) \cdot \mathbf{r}(t) = r^2$.

2. Referring to exercises 21 and 22, examine the graphs of several vector-valued functions of the form $\mathbf{r}(t) = \langle a\cos ct + b\cos dt, a \sin ct + b \sin dt \rangle$, for constants a, b, c and d. Determine the values of these constants that produce graphs of different types. For example, starting with the graph of $\langle 4\cos 4t - 6\cos t, 4\sin 4t - 6\sin t \rangle$, change $c = 4$ to $c = 3, c = 5, c = 2$, etc. Conjecture a relationship between the number of loops and the difference between c and d. Test this conjecture on other vector-valued functions. Returning to $\langle 4\cos 4t - 6\cos t, 4\sin 4t - 6\sin t \rangle$, change $a = 4$ to other values. Conjecture a relationship between the size of the loops and the value of a.

11.2 THE CALCULUS OF VECTOR-VALUED FUNCTIONS

In this section, we begin to explore the calculus of vector-valued functions, beginning with the notion of limit and progressing to continuity, derivatives and finally, integrals. Take careful note of how our presentation parallels our development of the calculus of scalar functions in Chapters 1, 2 and 4. We follow this same progression again when we examine functions of several variables in Chapter 12. We define everything in this section in terms of vector-valued functions in three dimensions. The definitions can be interpreted for vector-valued functions in two dimensions in the obvious way, by simply dropping the third component everywhere.

For a vector-valued function $\mathbf{r}(t) = \langle f(t), g(t), h(t) \rangle$, when we write

$$\lim_{t \to a} \mathbf{r}(t) = \mathbf{u},$$

we mean that as t gets closer and closer to a, the vector $\mathbf{r}(t)$ is getting closer and closer to the vector $\mathbf{u}$. For $\mathbf{u} = \langle u_1, u_2, u_3 \rangle$, this means that

$$\lim_{t \to a} \mathbf{r}(t) = \lim_{t \to a} \langle f(t), g(t), h(t) \rangle = \mathbf{u} = \langle u_1, u_2, u_3 \rangle.$$

Notice that for this to occur, we must have that $f(t)$ is approaching u_1, $g(t)$ is approaching u_2 and $h(t)$ is approaching u_3. In view of this, we make the following definition.

DEFINITION 2.1

For a vector-valued function $\mathbf{r}(t) = \langle f(t), g(t), h(t) \rangle$, the **limit** of $\mathbf{r}(t)$ as t approaches a is given by

$$\lim_{t \to a} \mathbf{r}(t) = \lim_{t \to a} \langle f(t), g(t), h(t) \rangle = \left\langle \lim_{t \to a} f(t), \lim_{t \to a} g(t), \lim_{t \to a} h(t) \right\rangle, \tag{2.1}$$

provided *all* of the indicated limits exist. If any of the limits indicated on the right-hand side of (2.1) fail to exist, then $\lim_{t \to a} \mathbf{r}(t)$ **does not exist.**

In example 2.1, we see that calculating a limit of a vector-valued function simply consists of calculating three separate limits of scalar functions.

EXAMPLE 2.1 Finding the Limit of a Vector-Valued Function

Find $\lim_{t\to 0}\langle t^2+1, 5\cos t, \sin t\rangle$.

Solution Here, each of the component functions is continuous (for all t) and so, we can calculate their limits simply by substituting the value for t. We have

$$\lim_{t\to 0}\langle t^2+1, 5\cos t, \sin t\rangle = \left\langle \lim_{t\to 0}(t^2+1), \lim_{t\to 0}(5\cos t), \lim_{t\to 0}\sin t\right\rangle$$
$$= \langle 1, 5, 0\rangle. \ \blacksquare$$

EXAMPLE 2.2 A Limit That Does Not Exist

Find $\lim_{t\to 0}\langle e^{2t}+5, t^2+2t-3, 1/t\rangle$.

Solution Notice that the limit of the third component is $\lim_{t\to 0}\frac{1}{t}$, which does not exist. So, even though the limits of the first two components exist, the limit of the vector-valued function does not exist. ■

Recall that for a scalar function f, we say that f is *continuous* at a if and only if

$$\lim_{t\to a} f(t) = f(a).$$

That is, a scalar function is continuous at a point whenever the limit and the value of the function are the same. We define the continuity of vector-valued functions in the same way.

DEFINITION 2.2

The vector-valued function $\mathbf{r}(t) = \langle f(t), g(t), h(t)\rangle$ is **continuous** at $t = a$ whenever

$$\lim_{t\to a}\mathbf{r}(t) = \mathbf{r}(a)$$

(i.e., whenever the limit exists and equals the value of the vector-valued function).

Notice that in terms of the components of $\mathbf{r}$, this says that $\mathbf{r}(t)$ is continuous at $t = a$ whenever

$$\lim_{t\to a}\langle f(t), g(t), h(t)\rangle = \langle f(a), g(a), h(a)\rangle.$$

Further, since

$$\lim_{t\to a}\langle f(t), g(t), h(t)\rangle = \left\langle \lim_{t\to a} f(t), \lim_{t\to a} g(t), \lim_{t\to a} h(t)\right\rangle,$$

it follows that $\mathbf{r}$ is continuous at $t = a$ if and only if

$$\left\langle \lim_{t\to a} f(t), \lim_{t\to a} g(t), \lim_{t\to a} h(t)\right\rangle = \langle f(a), g(a), h(a)\rangle,$$

which occurs if and only if

$$\lim_{t\to a} f(t) = f(a), \quad \lim_{t\to a} g(t) = g(a) \quad \text{and} \quad \lim_{t\to a} h(t) = h(a).$$

Look carefully at what we have just said, and observe that we just proved the following theorem.

THEOREM 2.1

A vector-valued function $\mathbf{r}(t) = \langle f(t), g(t), h(t)\rangle$ is continuous at $t = a$ if and only if *all* of f, g and h are continuous at $t = a$.

Theorem 2.1 says that to determine where a vector-valued function is continuous, you need only check the continuity of each component function (something you already know how to do). We demonstrate this in examples 2.3 and 2.4.

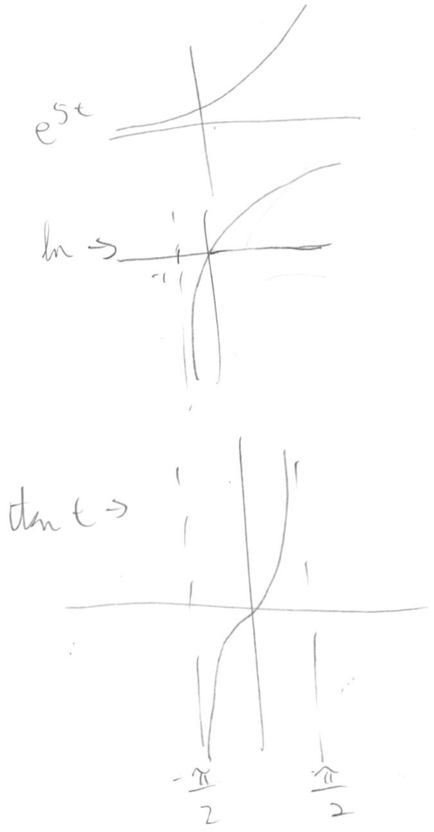

EXAMPLE 2.3 Determining Where a Vector-Valued Function Is Continuous

Determine for what values of t the vector-valued function $\mathbf{r}(t) = \langle e^{5t}, \ln(t+1), \cos t\rangle$ is continuous.

Solution From Theorem 2.1, $\mathbf{r}(t)$ will be continuous wherever *all* its components are continuous. We have: e^{5t} is continuous for all t, $\ln(t+1)$ is continuous for $t > -1$ and $\cos t$ is continuous for all t. So, $\mathbf{r}(t)$ is continuous for $t > -1$. ■

EXAMPLE 2.4 A Vector-Valued Function with Infinitely Many Discontinuities

Determine for what values of t the vector-valued function $\mathbf{r}(t) = \left\langle \tan t, |t+3|, \frac{1}{t-2}\right\rangle$ is continuous.

Solution First, note that $\tan t$ is continuous, except at $t = \dfrac{(2n+1)\pi}{2}$, for $n = 0, \pm 1, \pm 2, \ldots$ (i.e., except at odd multiples of $\frac{\pi}{2}$). The second component $|t+3|$ is continuous for all t (although it's not differentiable at $t = -3$). Finally, the third component $\dfrac{1}{t-2}$ is continuous except at $t = 2$. Since all three components must be continuous in order for $\mathbf{r}(t)$ to be continuous, we have that $\mathbf{r}(t)$ is continuous, except at $t = 2$ and $t = \dfrac{(2n+1)\pi}{2}$, for $n = 0, \pm 1, \pm 2, \ldots$. ■

Recall that in Chapter 2, we defined the derivative of a scalar function f to be

$$f'(t) = \lim_{h \to 0} \frac{f(t+h) - f(t)}{h}.$$

Replacing h by Δt, we can rewrite this as

$$f'(t) = \lim_{\Delta t \to 0} \frac{f(t+\Delta t) - f(t)}{\Delta t}.$$

You may be wondering why we want to change from a perfectly nice variable like h to something more unusual like Δt. The only reason is that we want to use the notation to emphasize that Δt is an *increment* of the variable t. In Chapter 12, we'll be defining partial derivatives of functions of more than one variable, where we'll use this type of notation to make it clear which variable is being incremented.

We now define the derivative of a vector-valued function in the expected way.

DEFINITION 2.3

The **derivative** $\mathbf{r}'(t)$ of the vector-valued function $\mathbf{r}(t)$ is defined by

$$\mathbf{r}'(t) = \lim_{\Delta t \to 0} \frac{\mathbf{r}(t + \Delta t) - \mathbf{r}(t)}{\Delta t}, \tag{2.2}$$

for any values of t for which the limit exists. When the limit exists for $t = a$, we say that $\mathbf{r}$ is **differentiable** at $t = a$.

Fortunately, you will not need to learn any new differentiation rules, as the derivative of a vector-valued function is found directly from the derivatives of the individual components, as we see in Theorem 2.2.

THEOREM 2.2

Let $\mathbf{r}(t) = \langle f(t), g(t), h(t) \rangle$ and suppose that the components f, g and h are all differentiable for some value of t. Then $\mathbf{r}$ is also differentiable at that value of t and its derivative is given by

$$\mathbf{r}'(t) = \langle f'(t), g'(t), h'(t) \rangle. \tag{2.3}$$

PROOF

From the definition of derivative of a vector-valued function (2.2), we have

$$\begin{aligned} \mathbf{r}'(t) &= \lim_{\Delta t \to 0} \frac{\mathbf{r}(t + \Delta t) - \mathbf{r}(t)}{\Delta t} \\ &= \lim_{\Delta t \to 0} \frac{1}{\Delta t} [\langle f(t + \Delta t), g(t + \Delta t), h(t + \Delta t) \rangle - \langle f(t), g(t), h(t) \rangle] \\ &= \lim_{\Delta t \to 0} \frac{1}{\Delta t} \langle f(t + \Delta t) - f(t), g(t + \Delta t) - g(t), h(t + \Delta t) - h(t) \rangle, \end{aligned}$$

from the definition of vector subtraction. Distributing the scalar $\frac{1}{\Delta t}$ into each component and using the definition of limit of a vector-valued function (2.1), we have

$$\begin{aligned} \mathbf{r}'(t) &= \lim_{\Delta t \to 0} \frac{1}{\Delta t} \langle f(t + \Delta t) - f(t), g(t + \Delta t) - g(t), h(t + \Delta t) - h(t) \rangle \\ &= \lim_{\Delta t \to 0} \left\langle \frac{f(t + \Delta t) - f(t)}{\Delta t}, \frac{g(t + \Delta t) - g(t)}{\Delta t}, \frac{h(t + \Delta t) - h(t)}{\Delta t} \right\rangle \\ &= \left\langle \lim_{\Delta t \to 0} \frac{f(t + \Delta t) - f(t)}{\Delta t}, \lim_{\Delta t \to 0} \frac{g(t + \Delta t) - g(t)}{\Delta t}, \lim_{\Delta t \to 0} \frac{h(t + \Delta t) - h(t)}{\Delta t} \right\rangle \\ &= \langle f'(t), g'(t), h'(t) \rangle, \end{aligned}$$

where in the last step we recognized the definition of the derivatives of each of the component functions f, g and h. ■

We illustrate this in example 2.5.

EXAMPLE 2.5   Finding the Derivative of a Vector-Valued Function

Find the derivative of $\mathbf{r}(t) = \langle \sin(t^2), e^{\cos t}, t \ln t \rangle$.

Solution Applying the chain rule to the first two components and the product rule to the third, we have (for $t > 0$):

$$\mathbf{r}'(t) = \left\langle \frac{d}{dt}[\sin(t^2)], \frac{d}{dt}(e^{\cos t}), \frac{d}{dt}(t \ln t) \right\rangle$$

$$= \left\langle \cos(t^2)\frac{d}{dt}(t^2), e^{\cos t}\frac{d}{dt}(\cos t), \frac{d}{dt}(t)\ln t + t\frac{d}{dt}(\ln t) \right\rangle$$

$$= \left\langle \cos(t^2)(2t), e^{\cos t}(-\sin t), (1)\ln t + t\frac{1}{t} \right\rangle$$

$$= \langle 2t\cos(t^2), -\sin t\, e^{\cos t}, \ln t + 1 \rangle. \quad \blacksquare$$

For the most part, to compute derivatives of vector-valued functions, we need only to use the already familiar rules for differentiation of scalar functions. There are several special derivative rules, however, which we state in Theorem 2.3.

THEOREM 2.3

Suppose that $\mathbf{r}(t)$ and $\mathbf{s}(t)$ are differentiable vector-valued functions, $f(t)$ is a differentiable scalar function and c is any scalar constant. Then

(i) $\dfrac{d}{dt}[\mathbf{r}(t) + \mathbf{s}(t)] = \mathbf{r}'(t) + \mathbf{s}'(t)$

(ii) $\dfrac{d}{dt}[c\,\mathbf{r}(t)] = c\,\mathbf{r}'(t)$

(iii) $\dfrac{d}{dt}[f(t)\mathbf{r}(t)] = f'(t)\mathbf{r}(t) + f(t)\mathbf{r}'(t)$

(iv) $\dfrac{d}{dt}[\mathbf{r}(t) \cdot \mathbf{s}(t)] = \mathbf{r}'(t) \cdot \mathbf{s}(t) + \mathbf{r}(t) \cdot \mathbf{s}'(t)$ and

(v) $\dfrac{d}{dt}[\mathbf{r}(t) \times \mathbf{s}(t)] = \mathbf{r}'(t) \times \mathbf{s}(t) + \mathbf{r}(t) \times \mathbf{s}'(t).$

  ↳ not commutative

Notice that parts (iii), (iv) and (v) are the product rules for the various kinds of products we can define. In each of these three cases, it's important to recognize that these follow the same pattern as the usual product rule for the derivative of the product of two scalar functions.

PROOF

(i) For $\mathbf{r}(t) = \langle f_1(t), g_1(t), h_1(t) \rangle$ and $\mathbf{s}(t) = \langle f_2(t), g_2(t), h_2(t) \rangle$, we have from (2.3) and the rules for vector addition that

$$\frac{d}{dt}[\mathbf{r}(t) + \mathbf{s}(t)] = \frac{d}{dt}[\langle f_1(t), g_1(t), h_1(t) \rangle + \langle f_2(t), g_2(t), h_2(t) \rangle]$$

$$= \frac{d}{dt}\langle f_1(t) + f_2(t), g_1(t) + g_2(t), h_1(t) + h_2(t) \rangle$$

$$= \langle f_1'(t) + f_2'(t), g_1'(t) + g_2'(t), h_1'(t) + h_2'(t)\rangle$$
$$= \langle f_1'(t), g_1'(t), h_1'(t)\rangle + \langle f_2'(t), g_2'(t), h_2'(t)\rangle$$
$$= \mathbf{r}'(t) + \mathbf{s}'(t).$$

(iv) From the definition of dot product and the usual product rule for the product of two scalar functions, we have

$$\begin{aligned}\frac{d}{dt}[\mathbf{r}(t)\cdot\mathbf{s}(t)] &= \frac{d}{dt}[\langle f_1(t), g_1(t), h_1(t)\rangle \cdot \langle f_2(t), g_2(t), h_2(t)\rangle]\\ &= \frac{d}{dt}[f_1(t)f_2(t) + g_1(t)g_2(t) + h_1(t)h_2(t)]\\ &= f_1'(t)f_2(t) + f_1(t)f_2'(t) + g_1'(t)g_2(t) + g_1(t)g_2'(t)\\ &\quad + h_1'(t)h_2(t) + h_1(t)h_2'(t)\\ &= [f_1'(t)f_2(t) + g_1'(t)g_2(t) + h_1'(t)h_2(t)]\\ &\quad + [f_1(t)f_2'(t) + g_1(t)g_2'(t) + h_1(t)h_2'(t)]\\ &= \mathbf{r}'(t)\cdot\mathbf{s}(t) + \mathbf{r}(t)\cdot\mathbf{s}'(t).\end{aligned}$$

We leave the proofs of (ii), (iii) and (v) as exercises. ■

We say that the curve traced out by the vector-valued function $\mathbf{r}(t) = \langle f(t), g(t), h(t)\rangle$ on an interval I is **smooth** if $\mathbf{r}'$ is continuous on I and $\mathbf{r}'(t) \neq \mathbf{0}$, except possibly at any endpoints of I. Notice that this says that the curve is smooth provided f', g' and h' are all continuous on I and $f'(t)$, $g'(t)$ and $h'(t)$ are not *all* zero at the same point in I.

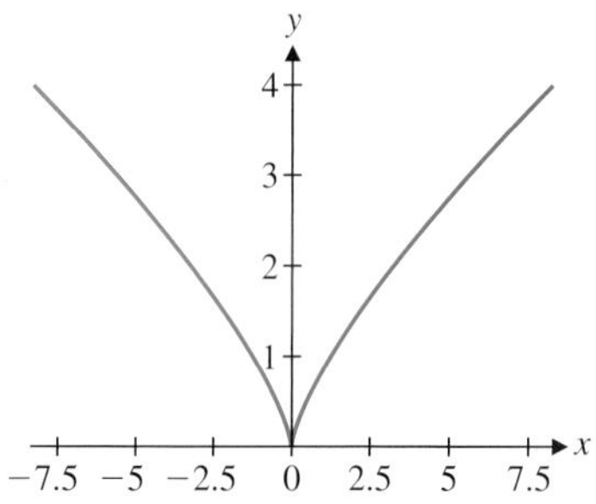

FIGURE 11.10
The curve traced out by $\mathbf{r}(t) = \langle t^3, t^2\rangle$

EXAMPLE 2.6 Determining Where a Curve Is Smooth

Determine where the plane curve traced out by the vector-valued function $\mathbf{r}(t) = \langle t^3, t^2\rangle$ is smooth.

Solution We show a graph of the curve in Figure 11.10.

Here, $\mathbf{r}'(t) = \langle 3t^2, 2t\rangle$ is continuous everywhere and $\mathbf{r}'(t) = \mathbf{0}$ if and only if $t = 0$. This says that the curve is smooth in any interval not including $t = 0$. Referring to Figure 11.10, observe that the curve is smooth except at the cusp located at the origin. ■

We next explore an important graphical interpretation of the derivative of a vector-valued function. First, recall that one interpretation of the derivative of a scalar function is that the value of the derivative at a point gives the slope of the tangent line to the curve at that point. For the case of the vector-valued function $\mathbf{r}(t)$, notice that from (2.2), the derivative of $\mathbf{r}(t)$ at $t = a$ is given by

$$\mathbf{r}'(a) = \lim_{\Delta t\to 0} \frac{\mathbf{r}(a+\Delta t) - \mathbf{r}(a)}{\Delta t}.$$

Again, recall that the endpoint of the vector-valued function $\mathbf{r}(t)$ traces out a curve C in $\mathbb{R}^3$. In Figure 11.11a (on the following page), we show the position vectors $\mathbf{r}(a)$, $\mathbf{r}(a+\Delta t)$ and $\mathbf{r}(a+\Delta t) - \mathbf{r}(a)$, for some fixed $\Delta t > 0$, using our graphical interpretation of vector subtraction, developed in Chapter 10. (How does the picture differ if $\Delta t < 0$?) Notice that for $\Delta t > 0$, the vector $\dfrac{\mathbf{r}(a+\Delta t) - \mathbf{r}(a)}{\Delta t}$ points in the same direction as $\mathbf{r}(a+\Delta t) - \mathbf{r}(a)$.

If we take smaller and smaller values of Δt, $\dfrac{\mathbf{r}(a+\Delta t) - \mathbf{r}(a)}{\Delta t}$ will approach $\mathbf{r}'(a)$. We illustrate this graphically in Figures 11.11b and 11.11c.

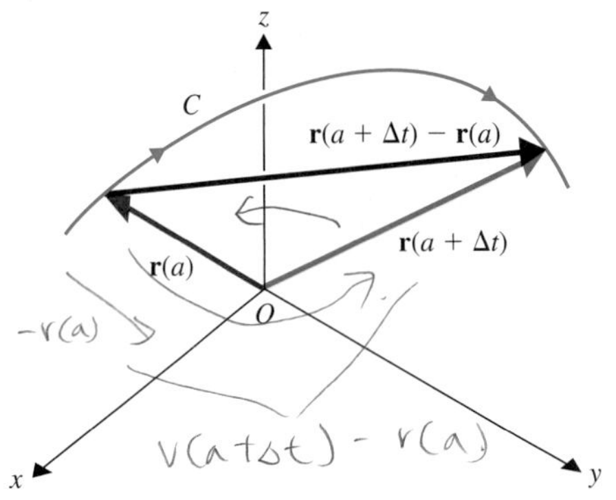

FIGURE 11.11a

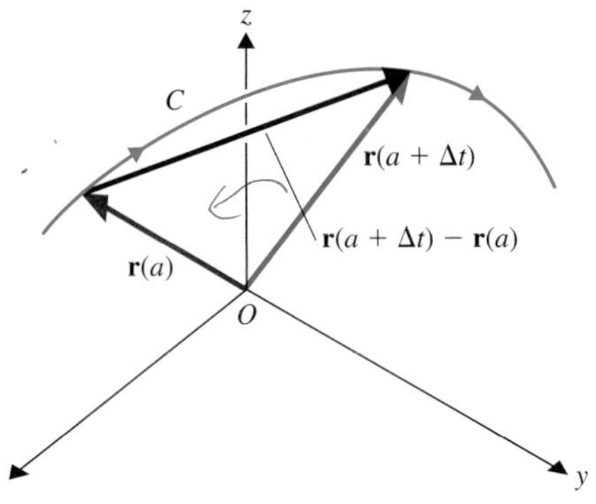

FIGURE 11.11b

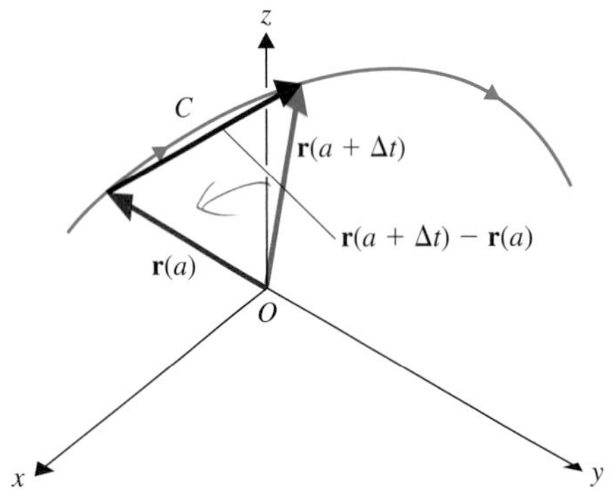

FIGURE 11.11c

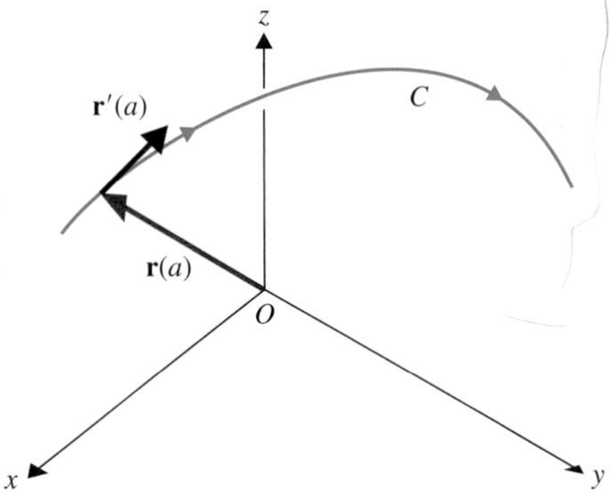

FIGURE 11.11d
The tangent vector $\mathbf{r}'(a)$

As $\Delta t \to 0$, notice that the vector $\dfrac{\mathbf{r}(a+\Delta t)-\mathbf{r}(a)}{\Delta t}$ approaches a vector that is tangent to the curve C at the terminal point of $\mathbf{r}(a)$, as seen in Figure 11.11d. We refer to $\mathbf{r}'(a)$ as a **tangent vector** to the curve C at the point corresponding to $t = a$. Be sure to observe that $\mathbf{r}'(a)$ lies along the tangent line to the curve at $t = a$ and points in the direction of the orientation of C. (Recognize that Figures 11.11a, 11.11b and 11.11c are all drawn so that $\Delta t > 0$. What changes in each of the figures if $\Delta t < 0$?)

We illustrate this notion for a simple curve in $\mathbb{R}^2$ in example 2.7.

EXAMPLE 2.7 Drawing Position and Tangent Vectors

For $\mathbf{r}(t) = \langle -\cos 2t, \sin 2t\rangle$, plot the curve traced out by the endpoint of $\mathbf{r}(t)$ and draw the position vector and tangent vector at $t = \frac{\pi}{4}$.

Solution First, notice that

$$\mathbf{r}'(t) = \langle 2\sin 2t, 2\cos 2t\rangle.$$

Also, the curve traced out by $\mathbf{r}(t)$ is given parametrically by

$$C\colon x = -\cos 2t, \quad y = \sin 2t, \quad t \in \mathbb{R}.$$

Observe that here,

$$x^2 + y^2 = \cos^2 2t + \sin^2 2t = 1,$$

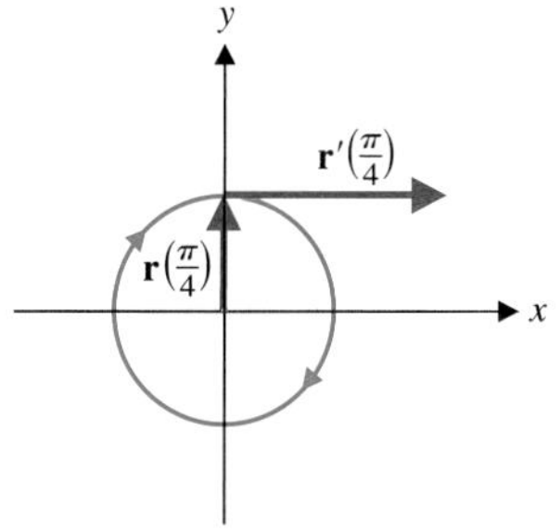

FIGURE 11.12
Position and tangent vectors

so that the curve is the circle of radius 1, centered at the origin. Further, from the parameterization, you can see that the orientation is clockwise. The position and tangent vectors at $t = \frac{\pi}{4}$ are given by

$$\mathbf{r}\left(\frac{\pi}{4}\right) = \left\langle -\cos\frac{\pi}{2}, \sin\frac{\pi}{2}\right\rangle = \langle 0, 1\rangle$$

and

$$\mathbf{r}'\left(\frac{\pi}{4}\right) = \left\langle 2\sin\frac{\pi}{2}, 2\cos\frac{\pi}{2}\right\rangle = \langle 2, 0\rangle,$$

respectively. We show the curve, along with the vectors $\mathbf{r}(\frac{\pi}{4})$ and $\mathbf{r}'(\frac{\pi}{4})$ in Figure 11.12. In particular, you might note that

$$\mathbf{r}\left(\frac{\pi}{4}\right) \cdot \mathbf{r}'\left(\frac{\pi}{4}\right) = 0,$$

so that $\mathbf{r}(\frac{\pi}{4})$ and $\mathbf{r}'(\frac{\pi}{4})$ are orthogonal. In fact, $\mathbf{r}(t)$ and $\mathbf{r}'(t)$ are orthogonal for every t, as follows:

$$\begin{aligned}\mathbf{r}(t) \cdot \mathbf{r}'(t) &= \langle -\cos 2t, \sin 2t\rangle \cdot \langle 2\sin 2t, 2\cos 2t\rangle \\ &= -2\cos 2t \sin 2t + 2\sin 2t \cos 2t = 0. \blacksquare\end{aligned}$$

Were you surprised to find in example 2.7 that the position vector and the tangent vector were orthogonal at every point? As it turns out, this is a special case of a more general result, which we state in Theorem 2.4.

THEOREM 2.4

$\|\mathbf{r}(t)\| =$ constant if and only if $\mathbf{r}(t)$ and $\mathbf{r}'(t)$ are orthogonal, for all t.

PROOF

(i) Suppose that $\|\mathbf{r}(t)\| = c$, for some constant c. Recall that

$$\mathbf{r}(t) \cdot \mathbf{r}(t) = \|\mathbf{r}(t)\|^2 = c^2. \tag{2.4}$$

Differentiating both sides of (2.4), we get

$$\frac{d}{dt}[\mathbf{r}(t) \cdot \mathbf{r}(t)] = \frac{d}{dt}c^2 = 0.$$

From Theorem 2.3 (iv), we now have

$$0 = \frac{d}{dt}[\mathbf{r}(t) \cdot \mathbf{r}(t)] = \mathbf{r}'(t) \cdot \mathbf{r}(t) + \mathbf{r}(t) \cdot \mathbf{r}'(t) = 2\mathbf{r}(t) \cdot \mathbf{r}'(t),$$

so that $\mathbf{r}(t) \cdot \mathbf{r}'(t) = 0$, as desired.

(ii) We leave the proof of the converse as an exercise. ■

Note that in two dimensions, if $\|\mathbf{r}(t)\| = c$ for all t (where c is a constant), then the curve traced out by the position vector $\mathbf{r}(t)$ must lie on the circle of radius c, centered at the origin. Theorem 2.4 then says that the path traced out by $\mathbf{r}(t)$ lies on a circle centered at the origin if and only if the tangent vector is orthogonal to the position vector at every point on the curve. Likewise, in three dimensions, if $\|\mathbf{r}(t)\| = c$ for all t (where c is a constant), the curve traced out by $\mathbf{r}(t)$ lies on the sphere of radius c centered at the origin. In this case, Theorem 2.4 says that the curve traced out by $\mathbf{r}(t)$ lies on a sphere centered at the origin if and only if the tangent vector is orthogonal to the position vector at every point on the curve.

We conclude this section by making a few straightforward definitions. Recall that when we say that the scalar function $F(t)$ is an antiderivative of the scalar function $f(t)$, we mean that F is any function such that $F'(t) = f(t)$. We now extend this notion to vector-valued functions.

DEFINITION 2.4

The vector-valued function $\mathbf{R}(t)$ is an **antiderivative** of the vector-valued function $\mathbf{r}(t)$ whenever $\mathbf{R}'(t) = \mathbf{r}(t)$.

Notice that if $\mathbf{r}(t) = \langle f(t), g(t), h(t)\rangle$ and f, g and h have antiderivatives F, G and H, respectively, then

$$\frac{d}{dt}\langle F(t), G(t), H(t)\rangle = \langle F'(t), G'(t), H'(t)\rangle = \langle f(t), g(t), h(t)\rangle.$$

That is, $\langle F(t), G(t), H(t)\rangle$ is an antiderivative of $\mathbf{r}(t)$. In fact, $\langle F(t) + c_1, G(t) + c_2, H(t) + c_3\rangle$ is also an antiderivative of $\mathbf{r}(t)$, for any choice of constants c_1, c_2 and c_3. This leads us to Definition 2.5.

DEFINITION 2.5

If $\mathbf{R}(t)$ is any antiderivative of $\mathbf{r}(t)$, the **indefinite integral** of $\mathbf{r}(t)$ is defined to be

$$\int \mathbf{r}(t)\,dt = \mathbf{R}(t) + \mathbf{c},$$

where $\mathbf{c}$ is an arbitrary constant vector.

As in the scalar case, $\mathbf{R}(t) + \mathbf{c}$ is the most general antiderivative of $\mathbf{r}(t)$. (Why is that?) Notice that this says that

Indefinite integral of a vector-valued function

$$\int \mathbf{r}(t)\,dt = \int \langle f(t), g(t), h(t)\rangle\,dt = \left\langle \int f(t)\,dt, \int g(t)\,dt, \int h(t)\,dt \right\rangle. \tag{2.5}$$

That is, you integrate a vector-valued function by integrating each of the individual components.

EXAMPLE 2.8 Evaluating the Indefinite Integral of a Vector-Valued Function

Evaluate the indefinite integral $\int \langle t^2 + 2, \sin 2t, 4te^{t^2}\rangle\,dt$.

Solution From (2.5), we have

$$\begin{aligned}\int \langle t^2 + 2, \sin 2t, 4te^{t^2}\rangle\,dt &= \left\langle \int (t^2+2)\,dt, \int \sin 2t\,dt, \int 4te^{t^2}\,dt \right\rangle \\ &= \left\langle \frac{1}{3}t^3 + 2t + c_1, -\frac{1}{2}\cos 2t + c_2, 2e^{t^2} + c_3 \right\rangle \\ &= \left\langle \frac{1}{3}t^3 + 2t, -\frac{1}{2}\cos 2t, 2e^{t^2} \right\rangle + \mathbf{c},\end{aligned}$$

where $\mathbf{c} = \langle c_1, c_2, c_3\rangle$ is an arbitrary constant vector. ■

Similarly, we define the definite integral of a vector-valued function in the obvious way.

DEFINITION 2.6

For the vector-valued function $\mathbf{r}(t) = \langle f(t), g(t), h(t)\rangle$, we define the **definite integral** of $\mathbf{r}(t)$ on the interval $[a, b]$ by

$$\int_a^b \mathbf{r}(t)\,dt = \int_a^b \langle f(t), g(t), h(t)\rangle dt = \left\langle \int_a^b f(t)\,dt, \int_a^b g(t)\,dt, \int_a^b h(t)\,dt \right\rangle. \tag{2.6}$$

This says simply that the definite integral of a vector-valued function $\mathbf{r}(t)$ is the vector whose components are the definite integrals of the corresponding components of $\mathbf{r}(t)$. With this in mind, we now extend the Fundamental Theorem of Calculus to vector-valued functions.

THEOREM 2.5

Suppose that $\mathbf{R}(t)$ is an antiderivative of $\mathbf{r}(t)$ on the interval $[a, b]$. Then,

$$\int_a^b \mathbf{r}(t)\,dt = \mathbf{R}(b) - \mathbf{R}(a).$$

PROOF

The proof is straightforward and we leave this as an exercise. ■

EXAMPLE 2.9 Evaluating the Definite Integral of a Vector-Valued Function

Evaluate $\int_0^1 \langle \sin \pi t, 6t^2 + 4t\rangle\,dt$.

Solution Notice that an antiderivative for the integrand is

$$\left\langle -\frac{1}{\pi}\cos \pi t, \frac{6t^3}{3} + 4\frac{t^2}{2} \right\rangle = \left\langle -\frac{1}{\pi}\cos \pi t, 2t^3 + 2t^2 \right\rangle.$$

From Theorem 2.5, we have that

$$\begin{aligned}\int_0^1 \langle \sin \pi t, 6t^2 + 4t\rangle\,dt &= \left\langle -\frac{1}{\pi}\cos \pi t, 2t^3 + 2t^2 \right\rangle\Big|_0^1 \\ &= \left\langle -\frac{1}{\pi}\cos \pi, 2 + 2 \right\rangle - \left\langle -\frac{1}{\pi}\cos 0, 0 \right\rangle \\ &= \left\langle \frac{1}{\pi} + \frac{1}{\pi}, 4 - 0 \right\rangle = \left\langle \frac{2}{\pi}, 4 \right\rangle.\end{aligned}$$

■

BEYOND FORMULAS

Theorem 2.4 illustrates the importance of good notation. While we could have derived the same result using parametric equations, the vector notation greatly simplifies both the statement and proof of the theorem. The simplicity of the notation allows us to make connections and use our geometric intuition, instead of floundering in a mess of equations. We can visualize the graph of a vector-valued function $\mathbf{r}(t)$ more easily than we can try to keep track of separate equations $x(t)$, $y(t)$ and $z(t)$.

EXERCISES 11.2

WRITING EXERCISES

1. Suppose that $\mathbf{r}(t) = \langle f(t), g(t), h(t)\rangle$, where $\lim_{t\to 0} f(t) = \lim_{t\to 0} g(t) = 0$ and $\lim_{t\to 0} h(t) = \infty$. Describe what is happening graphically as $t \to 0$ and explain why (even though the limits of two of the component functions exist) the limit of $\mathbf{r}(t)$ as $t \to 0$ does not exist.
2. In example 2.3, describe what is happening graphically for $t \le -1$. Explain why we don't say that $\mathbf{r}(t)$ is continuous for $t \le -1$.
3. Suppose that $\mathbf{r}(t)$ is a vector-valued function such that $\mathbf{r}(0) = \langle a, b, c\rangle$ and $\mathbf{r}'(0)$ exists. Imagine zooming in on the curve traced out by $\mathbf{r}(t)$ near the point (a, b, c). Describe what the curve will look like and how it relates to the tangent vector $\mathbf{r}'(0)$.
4. There is a quotient rule corresponding to the product rule in Theorem 2.3, part (iii). State this rule and describe in words how you would prove it. Explain why there isn't a quotient rule corresponding to the product rules in parts (iv) and (v) of Theorem 2.3.

In exercises 1–6, find the limit if it exists.

1. $\lim_{t\to 0}\langle t^2 - 1, e^{2t}, \sin t\rangle$
2. $\lim_{t\to 1}\left\langle t^2, e^{2t}, \sqrt{t^2+2t}\right\rangle$
3. $\lim_{t\to 0}\left\langle \frac{\sin t}{t}, \cos t, \frac{t+1}{t-1}\right\rangle$
4. $\lim_{t\to 1}\left\langle \sqrt{t-1}, t^2+3, \frac{t+1}{t-1}\right\rangle$
5. $\lim_{t\to 0}\left\langle \ln t, \sqrt{t^2+1}, t-3\right\rangle$
6. $\lim_{t\to \pi/2}\langle \cos t, t^2+3, \tan t\rangle$

In exercises 7–12, determine all values of t at which the given vector-valued function is continuous.

7. $\mathbf{r}(t) = \left\langle \frac{t+1}{t-1}, t^2, 2t\right\rangle$
8. $\mathbf{r}(t) = \left\langle \sin t, \cos t, \frac{3}{t}\right\rangle$
9. $\mathbf{r}(t) = \langle \tan t, \sin t^2, \cos t\rangle$
10. $\mathbf{r}(t) = \langle \cos 5t, \tan t, 6\sin t\rangle$
11. $\mathbf{r}(t) = \langle 4\cos t, \sqrt{t}, 4\sin t\rangle$
12. $\mathbf{r}(t) = \langle \sin t, -\csc t, \cot t\rangle$

In exercises 13–18, find the derivative of the given vector-valued function.

13. $\mathbf{r}(t) = \left\langle t^4, \sqrt{t+1}, \frac{3}{t^2}\right\rangle$
14. $\mathbf{r}(t) = \left\langle \frac{t-3}{t+1}, te^{2t}, t^3\right\rangle$
15. $\mathbf{r}(t) = \langle \sin t, \sin t^2, \cos t\rangle$
16. $\mathbf{r}(t) = \langle \cos 5t, \tan t, 6\sin t\rangle$
17. $\mathbf{r}(t) = \langle e^{t^2}, t^2, \sec 2t\rangle$
18. $\mathbf{r}(t) = \left\langle \sqrt{t^2+1}, \cos t, e^{-3t}\right\rangle$

In exercises 19–22, sketch the curve traced out by the endpoint of the given vector-valued function and plot position and tangent vectors at the indicated points.

19. $\mathbf{r}(t) = \langle \cos t, \sin t\rangle, t = 0, t = \frac{\pi}{2}, t = \pi$
20. $\mathbf{r}(t) = \langle t, t^2 - 1\rangle, t = 0, t = 1, t = 2$
21. $\mathbf{r}(t) = \langle \cos t, t, \sin t\rangle, t = 0, t = \frac{\pi}{2}, t = \pi$
22. $\mathbf{r}(t) = \langle t, t, t^2 - 1\rangle, t = 0, t = 1, t = 2$

In exercises 23–32, evaluate the given indefinite or definite integral.

23. $\int \left\langle 3t - 1, \sqrt{t}\right\rangle dt$
24. $\int \left\langle \frac{3}{t^2}, \frac{4}{t}\right\rangle dt$
25. $\int \left\langle \cos 3t, \sin t, e^{4t}\right\rangle dt$
26. $\int \left\langle e^{-3t}, \sin 5t, t^{3/2}\right\rangle dt$
27. $\int \left\langle te^{t^2}, 3t\sin t, \frac{3t}{t^2+1}\right\rangle dt$

28. $\int \langle e^{-3t}, t^2 \cos t^3, t \cos t \rangle \, dt$

29. $\int_0^1 \langle t^2 - 1, 3t \rangle \, dt$

30. $\int_1^4 \langle \sqrt{t}, 5 \rangle \, dt$

31. $\int_0^2 \left\langle \frac{4}{t+1}, e^{t-2}, te^t \right\rangle dt$

32. $\int_0^4 \left\langle 2te^{4t}, t^2 - 1, \frac{4t}{t^2+1} \right\rangle dt$

In exercises 33–36, find t such that $\mathbf{r}(t)$ and $\mathbf{r}'(t)$ are perpendicular.

33. $\mathbf{r}(t) = \langle \cos t, \sin t \rangle$

34. $\mathbf{r}(t) = \langle 2\cos t, \sin t \rangle$

35. $\mathbf{r}(t) = \langle t, t, t^2 - 1 \rangle$

36. $\mathbf{r}(t) = \langle t^2, t, t^2 - 5 \rangle$

37. In each of exercises 33 and 34, show that there are no values of t such that $\mathbf{r}(t)$ and $\mathbf{r}'(t)$ are parallel.

38. In each of exercises 35 and 36, show that there are no values of t such that $\mathbf{r}(t)$ and $\mathbf{r}'(t)$ are parallel.

In exercises 39–42, find all values of t such that $\mathbf{r}'(t)$ is parallel to the xy-plane.

39. $\mathbf{r}(t) = \langle t, t, t^3 - 3 \rangle$

40. $\mathbf{r}(t) = \langle t^2, t, \sin t^2 \rangle$

41. $\mathbf{r}(t) = \langle \cos t, \sin t, \sin 2t \rangle$

42. $\mathbf{r}(t) = \langle \sqrt{t+1}, \cos t, t^4 - 8t^2 \rangle$

43. Prove Theorem 2.3, part (ii).

44. In Theorem 2.3, part (ii), replace the scalar product $c\mathbf{r}(t)$ with the dot product $\mathbf{c} \cdot \mathbf{r}(t)$, for a constant vector $\mathbf{c}$ and prove the results.

45. Prove Theorem 2.3, parts (ii) and (iii).

46. Prove Theorem 2.3, part (v).

47. Label as true or false and explain why. If $\mathbf{u}(t) = \frac{1}{\|\mathbf{r}(t)\|}\mathbf{r}(t)$ and $\mathbf{u}(t) \cdot \mathbf{u}'(t) = 0$ then $\mathbf{r}(t) \cdot \mathbf{r}'(t) = 0$.

48. Label as true or false and explain why. If $\mathbf{r}(t_0) \cdot \mathbf{r}'(t_0) = 0$ for some t_0, then $\|\mathbf{r}(t)\|$ is constant.

49. Prove that if $\mathbf{r}(t)$ and $\mathbf{r}'(t)$ are orthogonal for all t, then $\|\mathbf{r}(t)\| = \text{constant}$ [Theorem 2.4, part (ii)].

50. Prove Theorem 2.5.

51. Define the ellipse C with parametric equations $x = a\cos t$ and $y = b\sin t$, for positive constants a and b. For a fixed value of t, define the points $P = (a\cos t, b\sin t)$, $Q = (a\cos(t+\pi/2), b\sin(t+\pi/2))$ and $Q' = (a\cos(t-\pi/2), b\sin(t-\pi/2))$. Show that the vector QQ' (called the **conjugate diameter**) is parallel to the tangent vector to C at the point P. Sketch a graph and show the relationship between P, Q and Q'.

52. Repeat exercise 51 for the general angle θ, so that the points are $P = (a\cos t, b\sin t)$, $Q = (a\cos(t+\theta), b\sin(t+\theta))$ and $Q' = (a\cos(t-\theta), b\sin(t-\theta))$.

53. Find $\frac{d}{dt}[\mathbf{f}(t) \cdot (\mathbf{g}(t) \times \mathbf{h}(t))]$.

54. Determine whether the following is true or false: $\int_a^b \mathbf{f}(t) \cdot \mathbf{g}(t)\,dt = \int_a^b \mathbf{f}(t)\,dt \cdot \int_a^b \mathbf{g}(t)\,dt$.

EXPLORATORY EXERCISES

1. Find all values of t such that $\mathbf{r}'(t) = \mathbf{0}$ for each function: (a) $\mathbf{r}(t) = \langle t, t^2 - 1 \rangle$, (b) $\mathbf{r}(t) = \langle 2\cos t + \sin 2t, 2\sin t + \cos 2t \rangle$, (c) $\mathbf{r}(t) = \langle 2\cos 3t + \sin 5t, 2\sin 3t + \cos 5t \rangle$, (d) $\mathbf{r}(t) = \langle t^2, t^4 - 1 \rangle$ and (e) $\mathbf{r}(t) = \langle t^3, t^6 - 1 \rangle$. Based on your results, conjecture the graphical significance of having the derivative of a vector-valued function equal the zero vector. If $\mathbf{r}(t)$ is the position function of some object in motion, explain the physical significance of having a zero derivative. Explain your geometric interpretation in light of your physical interpretation.

2. You may recall that a scalar function has either a discontinuity, a "sharp corner" or a cusp at places where the derivative doesn't exist. In this exercise, we look at the analogous *smoothness* of graphs of vector-valued functions. We have said that a curve C is *smooth* if it is traced out by a vector-valued function $\mathbf{r}(t)$, where $\mathbf{r}'(t)$ is continuous and $\mathbf{r}'(t) \neq \mathbf{0}$ for all values of t. Sketch the graph of $\mathbf{r}(t) = \langle t, \sqrt[3]{t^2} \rangle$ and explain why we include the requirement that $\mathbf{r}'(t)$ be continuous. Sketch the graph of $\mathbf{r}(t) = \langle 2\cos t + \sin 2t, 2\sin t + \cos 2t \rangle$ and show that $\mathbf{r}'(0) = \mathbf{0}$. Explain why we include the requirement that $\mathbf{r}'(t)$ be nonzero. Sketch the graph of $\mathbf{r}(t) = \langle 2\cos 3t + \sin 5t, 2\sin 3t + \cos 5t \rangle$ and show that $\mathbf{r}'(t)$ never equals the zero vector. By zooming in on the edges of the graph, show that this curve is accurately described as smooth. Sketch the graphs of $\mathbf{r}(t) = \langle t, t^2 - 1 \rangle$ and $\mathbf{g}(t) = \langle t^2, t^4 - 1 \rangle$ for $t \geq 0$ and observe that they trace out the same curve. Show that $\mathbf{g}'(0) = \mathbf{0}$, but that the curve is smooth at $t = 0$. Explain why this says that the requirement that $\mathbf{r}'(t) \neq \mathbf{0}$ need not hold for *every* $\mathbf{r}(t)$ tracing out the curve. [This requirement needs to hold for only one such $\mathbf{r}(t)$.] Determine which of the following curves are smooth. If the curve is not smooth, identify the graphical characteristic that is "unsmooth": $\mathbf{r}(t) = \langle \cos t, \sin t, t \rangle$, $\mathbf{r}(t) = \langle \cos t, \sin t, \sqrt[3]{t^2} \rangle$, $\mathbf{r}(t) = \langle \tan t, \sin t^2, \cos t \rangle$, $\mathbf{r}(t) = \langle 5\sin^3 t, 5\cos^3 t, t \rangle$ and $\mathbf{r}(t) = \langle \cos t, t^2 e^{-t}, \cos^2 t \rangle$.

11.3 MOTION IN SPACE

We are finally at a point where we have sufficient mathematical machinery to describe the motion of an object in a three-dimensional setting. Problems such as this, dealing with motion, were one of the primary focuses of Newton and many of his contemporaries. Newton used his newly invented calculus to explain all kinds of motion, from the motion of a projectile (such as a ball) hurled through the air, to the motion of the planets. His stunning achievements in this field unlocked mysteries that had eluded the greatest minds for centuries and form the basis of our understanding of mechanics today.

Suppose that an object moves along a curve traced out by the endpoint of the vector-valued function

$$\mathbf{r}(t) = \langle f(t), g(t), h(t) \rangle,$$

where t represents time and where $t \in [a, b]$. We observed in section 11.2 that the value of $\mathbf{r}'(t)$ for any given value of t is a tangent vector pointing in the direction of the orientation of the curve. We can now give another interpretation of this. From (2.3), we have

$$\mathbf{r}'(t) = \langle f'(t), g'(t), h'(t) \rangle$$

and the magnitude of this vector-valued function is

$$\|\mathbf{r}'(t)\| = \sqrt{[f'(t)]^2 + [g'(t)]^2 + [h'(t)]^2}.$$

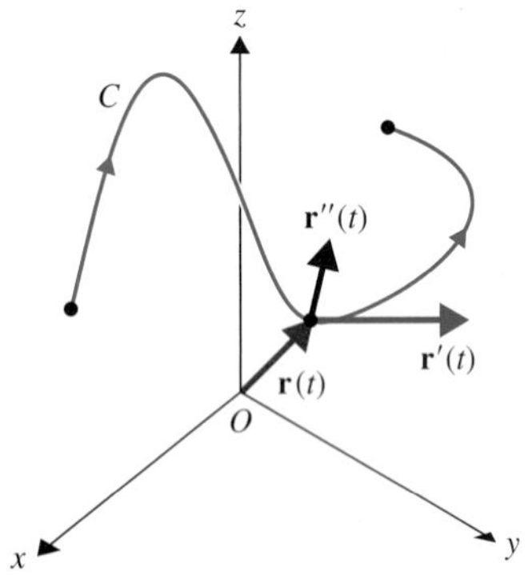

FIGURE 11.13
Position, velocity and acceleration vectors

(Where have you seen this expression before?) Notice that from (1.4), given any number $t_0 \in [a, b]$, the arc length of the portion of the curve from $u = t_0$ up to $u = t$ is given by

$$s(t) = \int_{t_0}^{t} \sqrt{[f'(u)]^2 + [g'(u)]^2 + [h'(u)]^2}\, du. \tag{3.1}$$

Part II of the Fundamental Theorem of Calculus says that if we differentiate both sides of (3.1), we get

$$s'(t) = \sqrt{[f'(t)]^2 + [g'(t)]^2 + [h'(t)]^2} = \|\mathbf{r}'(t)\|.$$

Since $s(t)$ represents arc length, $s'(t)$ gives the instantaneous rate of change of arc length with respect to time, that is, the **speed** of the object as it moves along the curve. So, for any given value of t, $\mathbf{r}'(t)$ is a tangent vector pointing in the direction of the orientation of C (i.e., the direction followed by the object) and whose magnitude gives the speed of the object. So, we call $\mathbf{r}'(t)$ the **velocity** vector, denoted $\mathbf{v}(t)$. Finally, we refer to the derivative of the velocity vector $\mathbf{v}'(t) = \mathbf{r}''(t)$ as the **acceleration** vector, denoted $\mathbf{a}(t)$. When drawing the velocity and acceleration vectors, we locate both of their initial points at the terminal point of $\mathbf{r}(t)$ (i.e., at the point on the curve), as shown in Figure 11.13.

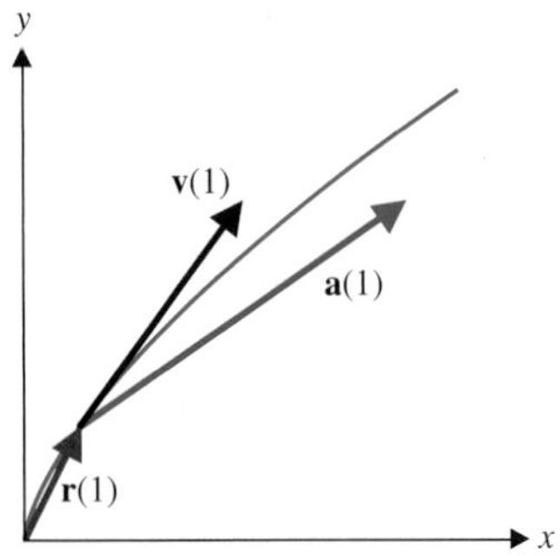

FIGURE 11.14
Position, velocity and acceleration vectors

EXAMPLE 3.1 Finding Velocity and Acceleration Vectors

Find the velocity and acceleration vectors if the position of an object moving in the xy-plane is given by $\mathbf{r}(t) = \langle t^3, 2t^2 \rangle$.

Solution We have

$$\mathbf{v}(t) = \mathbf{r}'(t) = \langle 3t^2, 4t \rangle \quad \text{and} \quad \mathbf{a}(t) = \mathbf{r}''(t) = \langle 6t, 4 \rangle.$$

In particular, this says that at $t = 1$, we have $\mathbf{r}(1) = \langle 1, 2 \rangle$, $\mathbf{v}(1) = \mathbf{r}'(1) = \langle 3, 4 \rangle$ and $\mathbf{a}(1) = \mathbf{r}''(1) = \langle 6, 4 \rangle$. We plot the curve and these vectors in Figure 11.14. ■

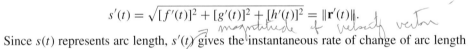

Just as in the case of one-dimensional motion, given the acceleration vector, we can determine the velocity and position vectors, provided we have some additional information.

EXAMPLE 3.2   Finding Velocity and Position from Acceleration

Find the velocity and position of an object at any time t, given that its acceleration is $\mathbf{a}(t) = \langle 6t, 12t + 2, e^t \rangle$, its initial velocity is $\mathbf{v}(0) = \langle 2, 0, 1 \rangle$ and its initial position is $\mathbf{r}(0) = \langle 0, 3, 5 \rangle$.

Solution   Since $\mathbf{a}(t) = \mathbf{v}'(t)$, we integrate once to obtain

$$\mathbf{v}(t) = \int \mathbf{a}(t)\,dt = \int [6t\mathbf{i} + (12t + 2)\mathbf{j} + e^t\mathbf{k}]\,dt$$

$$= 3t^2\mathbf{i} + (6t^2 + 2t)\mathbf{j} + e^t\mathbf{k} + \mathbf{c}_1, \longrightarrow \text{vector } c$$

where $\mathbf{c}_1$ is an arbitrary constant vector. To determine the value of $\mathbf{c}_1$, we use the initial velocity:

$$= \langle 3t^2 + C_1, \ 6t^2 + 2t + C_2, \ e^t + C_3 \rangle$$

$$\langle 2, 0, 1 \rangle = \mathbf{v}(0) = (0)\mathbf{i} + (0)\mathbf{j} + (1)\mathbf{k} + \mathbf{c}_1,$$

so that $\mathbf{c}_1 = \langle 2, 0, 0 \rangle$. This gives us the velocity

$$\mathbf{v}(t) = (3t^2 + 2)\mathbf{i} + (6t^2 + 2t)\mathbf{j} + e^t\mathbf{k}.$$

Since $\mathbf{v}(t) = \mathbf{r}'(t)$, we integrate again, to obtain

$$\mathbf{r}(t) = \int \mathbf{v}(t)\,dt = \int [(3t^2 + 2)\mathbf{i} + (6t^2 + 2t)\mathbf{j} + e^t\mathbf{k}]\,dt$$

$$= (t^3 + 2t)\mathbf{i} + (2t^3 + t^2)\mathbf{j} + e^t\mathbf{k} + \mathbf{c}_2,$$

where $\mathbf{c}_2$ is an arbitrary constant vector. We can use the initial position to determine the value of $\mathbf{c}_2$, as follows:

$$\langle t^3 + 2t + C_1, \ 2t^3 + t^2 + C_2, \ e^t + C_3 \rangle$$

$$\langle 0, 3, 5 \rangle = \mathbf{r}(0) = (0)\mathbf{i} + (0)\mathbf{j} + (1)\mathbf{k} + \mathbf{c}_2,$$

so that $\mathbf{c}_2 = \langle 0, 3, 4 \rangle$. This gives us the position vector

$$\mathbf{r}(t) = (t^3 + 2t)\mathbf{i} + (2t^3 + t^2 + 3)\mathbf{j} + (e^t + 4)\mathbf{k}.$$

We show the curve and indicate sample vectors for $\mathbf{r}(t)$, $\mathbf{v}(t)$ and $\mathbf{a}(t)$ in Figure 11.15.

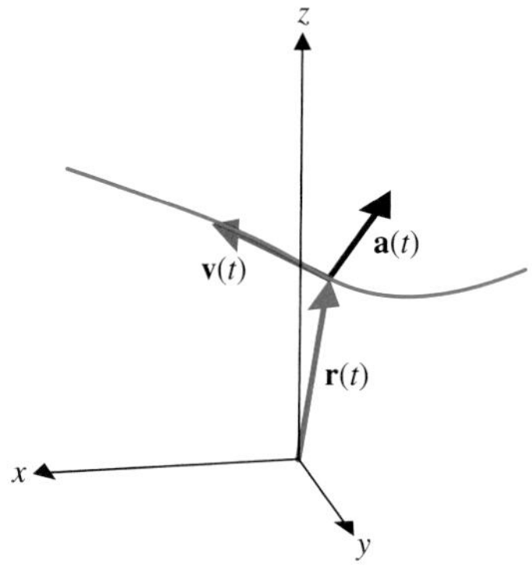

FIGURE 11.15
Position, velocity and acceleration vectors

We have already seen **Newton's second law of motion** several times now. In the case of one-dimensional motion, we had that the net force acting on an object equals the product of the mass and the acceleration ($F = ma$). In the case of motion in two or more dimensions, we have the vector form of Newton's second law:

$$\mathbf{F} = m\mathbf{a}.$$

Here, m is the mass, $\mathbf{a}$ is the acceleration vector and $\mathbf{F}$ is the vector representing the net force acting on the object.

EXAMPLE 3.3   Finding the Force Acting on an Object

Find the force acting on an object moving along a circular path of radius b, with constant angular speed.

Solution   For simplicity, we will take the circular path to lie in the xy-plane and have its center at the origin. Here, by constant **angular speed,** we mean that if θ is the angle

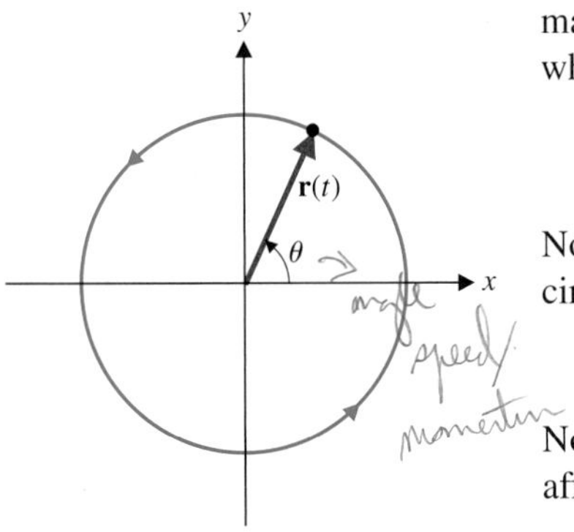

FIGURE 11.16a
Motion along a circle

made by the position vector and the positive x-axis and t is time (see Figure 11.16a, where the indicated orientation is for the case where $\omega > 0$), then we have that

$$\frac{d\theta}{dt} = \omega \text{ (constant)}.$$

Notice that this says that $\theta = \omega t + c$, for some constant c. Further, we can think of the circular path as the curve traced out by the endpoint of the vector-valued function

$$\mathbf{r}(t) = \langle b\cos\theta, b\sin\theta\rangle = \langle b\cos(\omega t + c), b\sin(\omega t + c)\rangle.$$

Notice that the path is the same for every value of c. (Think about what the value of c affects.) For simplicity, we take $\theta = 0$ when $t = 0$, so that $\theta = \omega t$ and

$$\mathbf{r}(t) = \langle b\cos\omega t, b\sin\omega t\rangle.$$

Now that we know the position at any time t, we can differentiate to find the velocity and acceleration. We have

$$\mathbf{v}(t) = \mathbf{r}'(t) = \langle -b\omega\sin\omega t, b\omega\cos\omega t\rangle,$$

so that the speed is $\|\mathbf{v}(t)\| = \omega b$ and

$$\begin{aligned}\mathbf{a}(t) = \mathbf{v}'(t) = \mathbf{r}''(t) &= \langle -b\omega^2\cos\omega t, -b\omega^2\sin\omega t\rangle \\ &= -\omega^2\langle b\cos\omega t, b\sin\omega t\rangle = -\omega^2\mathbf{r}(t).\end{aligned}$$

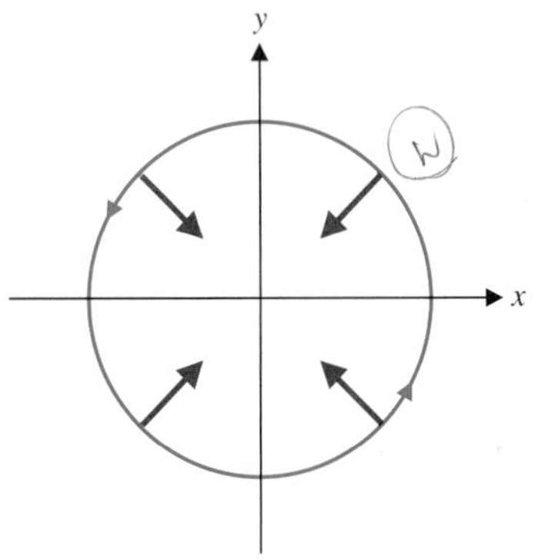

FIGURE 11.16b
Centripetal force

From Newton's second law of motion, we now have

$$\mathbf{F}(t) = m\mathbf{a}(t) = -m\omega^2\mathbf{r}(t).$$

Notice that since $m\omega^2 > 0$, this says that the force acting on the object points in the direction opposite the position vector. That is, at any point on the path, it points in toward the origin (see Figure 11.16b). We call such a force a **centripetal** (center-seeking) force. Finally, observe that on this circular path, $\|\mathbf{r}(t)\| = b$, so that at every point on the path, the force vector has constant magnitude:

$$\|\mathbf{F}(t)\| = \|-m\omega^2\mathbf{r}(t)\| = m\omega^2\|\mathbf{r}(t)\| = m\omega^2 b. \blacksquare$$

Notice that one consequence of the result $\mathbf{F}(t) = -m\omega^2\mathbf{r}(t)$ from example 3.3 is that the magnitude of the force increases as the rotation rate ω increases. You have experienced this if you have been on a roller coaster with tight turns or loops. The faster you are going, the stronger the force that your seat exerts on you. Alternatively, since the speed is $\|\mathbf{v}(t)\| = \omega b$, the tighter the turn (i.e., the smaller b is), the larger ω must be to obtain a given speed. So, on a roller coaster, a tighter turn requires a larger value of ω, which in turn increases the centripetal force.

Just as we did in the one-dimensional case, we can use Newton's second law of motion to determine the position of an object given only a knowledge of the forces acting on it. For instance, an important problem faced by the military is how to aim a projectile (e.g., a missile) so that it will end up hitting its intended target. This problem is harder than it sounds, particularly when the target is an aircraft moving faster than the speed of sound. We present the simplest possible case (where neither the target nor the source of the projectile is moving) in example 3.4.

EXAMPLE 3.4 Analyzing the Motion of a Projectile

A projectile is launched with an initial speed of 140 feet per second from ground level at an angle of $\frac{\pi}{4}$ to the horizontal. Assuming that the only force acting on the object is gravity (i.e., there is no air resistance, etc.), find the maximum altitude, the horizontal range and the speed at impact of the projectile.

Solution Notice that here, the motion is in a single plane (so that we need only consider two dimensions) and the only force acting on the object is the force of gravity, which acts straight downward. Although this is not constant, it is nearly so at altitudes reasonably close to sea level. We will assume that

$$\mathbf{F}(t) = -mg\mathbf{j},$$

where g is the constant acceleration due to gravity, $g \approx 32$ feet/second^2. From Newton's second law of motion, we have

$$-mg\mathbf{j} = \mathbf{F}(t) = m\mathbf{a}(t).$$

We now have

$$\mathbf{v}'(t) = \mathbf{a}(t) = -32\mathbf{j}.$$

Integrating this once gives us

$$\mathbf{v}(t) = \int \mathbf{a}(t)\,dt = -32t\mathbf{j} + \mathbf{c}_1, \tag{3.2}$$

where $\mathbf{c}_1$ is an arbitrary constant vector. If we knew the initial velocity vector $\mathbf{v}(0)$, we could use this to solve for $\mathbf{c}_1$, but we know only the initial speed (i.e., the magnitude of the velocity vector). Referring to Figure 11.17a, notice that you can read off the components of $\mathbf{v}(0)$, using the definitions of the sine and cosine functions:

$$\mathbf{v}(0) = \left\langle 140\cos\frac{\pi}{4}, 140\sin\frac{\pi}{4}\right\rangle = \langle 70\sqrt{2}, 70\sqrt{2}\rangle.$$

From (3.2), we now have

$$\langle 70\sqrt{2}, 70\sqrt{2}\rangle = \mathbf{v}(0) = (-32)(0)\mathbf{j} + \mathbf{c}_1 = \mathbf{c}_1.$$

Substituting this back into (3.2), we have

$$\mathbf{v}(t) = -32t\mathbf{j} + \langle 70\sqrt{2}, 70\sqrt{2}\rangle = \langle 70\sqrt{2}, 70\sqrt{2} - 32t\rangle. \tag{3.3}$$

Integrating (3.3) will give us the position vector

$$\mathbf{r}(t) = \int \mathbf{v}(t)\,dt = \langle 70\sqrt{2}t, 70\sqrt{2}t - 16t^2\rangle + \mathbf{c}_2,$$

where $\mathbf{c}_2$ is an arbitrary constant vector. Since the initial location was not specified, we choose it to be the origin (for simplicity). This gives us

$$\mathbf{0} = \mathbf{r}(0) = \mathbf{c}_2,$$

so that

$$\mathbf{r}(t) = \langle 70\sqrt{2}t, 70\sqrt{2}t - 16t^2\rangle. \tag{3.4}$$

We show a graph of the path of the projectile in Figure 11.17b (on the following page). Now that we have found expressions for the position and velocity vectors for any time, we can answer the physical questions. Notice that the maximum altitude occurs at the instant when the object stops moving up (just before it starts to fall). This says that the vertical ($\mathbf{j}$) component of velocity must be zero. From (3.3), we get

$$0 = 70\sqrt{2} - 32t,$$

so that the time at the maximum altitude is

$$t = \frac{70\sqrt{2}}{32}.$$

117.1875

TODAY IN MATHEMATICS

Evelyn Granville (1924–)
An American mathematician who has made important contributions to the space program and the teaching of mathematics. Growing up poor, she and her sister "accepted education as the means to rise above the limitations that a prejudiced society endeavored to place upon us." She was the first black American woman to be awarded a Ph.D. in mathematics. Upon graduation, she was employed as a computer programmer and in the early 1960s helped NASA write programs to track the paths of vehicles in space for Project Mercury. She then turned to education, her first love. Granville has coauthored an influential textbook on the teaching of mathematics.

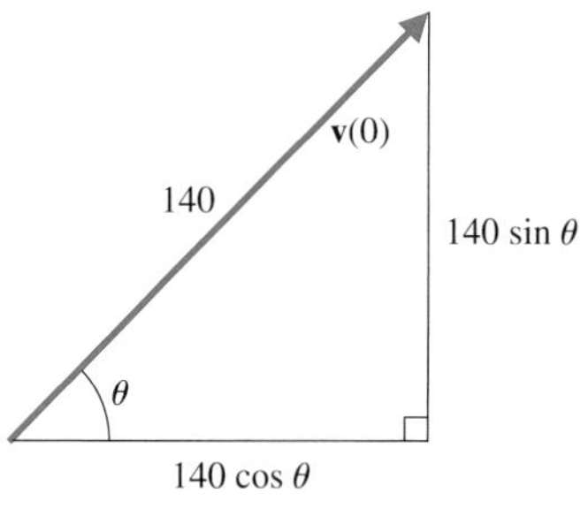

FIGURE 11.17a
Initial velocity vector

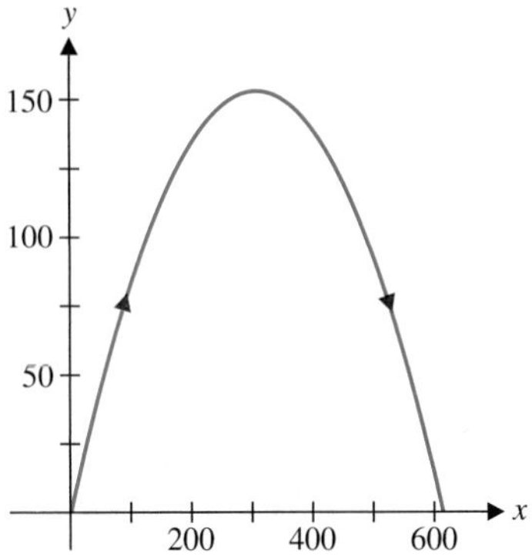

FIGURE 11.17b
Path of a projectile

The maximum altitude is then found from the vertical component of the position vector at this time:

$$\text{Maximum altitude} = 70\sqrt{2}t - 16t^2\Big|_{t=\frac{70\sqrt{2}}{32}} = 70\sqrt{2}\left(\frac{70\sqrt{2}}{32}\right) - 16\left(\frac{70\sqrt{2}}{32}\right)^2$$

$$= \frac{1225}{8} = 153.125 \text{ feet.}$$

To determine the horizontal range, we first need to determine the instant at which the object strikes the ground. Notice that this occurs when the vertical component of the position vector is zero (i.e., when the height above the ground is zero). From (3.4), we see that this occurs when

$$0 = 70\sqrt{2}t - 16t^2 = 2t(35\sqrt{2} - 8t).$$

There are two solutions of this equation: $t = 0$ (the time at which the projectile is launched) and $t = \frac{35\sqrt{2}}{8}$ (the time of impact). The horizontal range is then the horizontal component of position at this time:

$$\text{Range} = 70\sqrt{2}t\Big|_{t=\frac{35\sqrt{2}}{8}} = \left(70\sqrt{2}\right)\left(\frac{35\sqrt{2}}{8}\right) = \frac{1225}{2} = 612.5 \text{ feet.}$$

Finally, the speed at impact is the magnitude of the velocity vector at the time of impact:

$$\left\|\mathbf{v}\left(\frac{35\sqrt{2}}{8}\right)\right\| = \left\|\left\langle 70\sqrt{2}, 70\sqrt{2} - 32\left(\frac{35\sqrt{2}}{8}\right)\right\rangle\right\|$$

$$= \left\|\langle 70\sqrt{2}, -70\sqrt{2}\rangle\right\| = 140 \text{ ft/sec.} \blacksquare$$

You might have noticed in example 3.4 that the speed at impact was the same as the initial speed. Don't expect this to always be the case. Generally, this will be true only for a projectile of constant mass that is fired from ground level and returns to ground level and that is not subject to air resistance or other forces.

Equations of Motion

We now derive the equations of motion for a projectile in a slightly more general setting than that described in example 3.4. Consider a projectile launched from an altitude h above the ground at an angle θ to the horizontal and with initial speed v_0. We can use Newton's second law of motion to determine the position of the projectile at any time t and once we have this, we can answer any questions about the motion.

We again start with Newton's second law and assume that the only force acting on the object is gravity. We have

$$-mg\mathbf{j} = \mathbf{F}(t) = m\mathbf{a}(t).$$

This gives us (as in example 3.4)

$$\mathbf{v}'(t) = \mathbf{a}(t) = -g\mathbf{j}. \tag{3.5}$$

Integrating (3.5) gives us

$$\mathbf{v}(t) = \int \mathbf{a}(t)\,dt = -gt\mathbf{j} + \mathbf{c}_1, \tag{3.6}$$

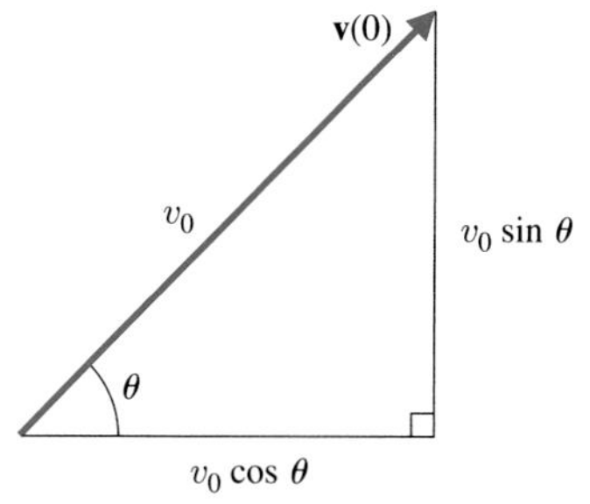

FIGURE 11.18a
Initial velocity

where $\mathbf{c}_1$ is an arbitrary constant vector. In order to solve for $\mathbf{c}_1$, we need the value of $\mathbf{v}(t)$ for some t, but we are given only the initial speed v_0 and the angle at which the projectile is fired. Notice that from the definitions of sine and cosine, we can read off the components of $\mathbf{v}(0)$ from Figure 11.18a. From this and (3.6), we have

$$\langle v_0 \cos\theta, v_0 \sin\theta \rangle = \mathbf{v}(0) = \mathbf{c}_1.$$

This gives us the velocity vector

$$\mathbf{v}(t) = \langle v_0 \cos\theta, v_0 \sin\theta - gt \rangle. \quad (3.7)$$

Since $\mathbf{r}'(t) = \mathbf{v}(t)$, we integrate (3.7) to get the position:

$$\mathbf{r}(t) = \int \mathbf{v}(t)\,dt = \left\langle (v_0 \cos\theta)t, (v_0 \sin\theta)t - \frac{gt^2}{2} \right\rangle + \mathbf{c}_2.$$

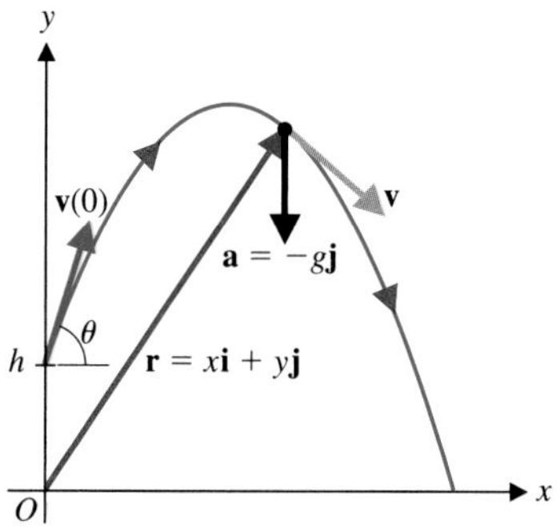

FIGURE 11.18b
Path of the projectile

To solve for $\mathbf{c}_2$, we want to use the initial position $\mathbf{r}(0)$, but we're not given it. We're told only that the projectile starts from an altitude of h feet above the ground. If we select the origin to be the point on the ground directly below the launching point, we have

$$\langle 0, h \rangle = \mathbf{r}(0) = \mathbf{c}_2,$$

so that

$$\mathbf{r}(t) = \left\langle (v_0 \cos\theta)t, (v_0 \sin\theta)t - \frac{gt^2}{2} \right\rangle + \langle 0, h \rangle$$

$$= \left\langle (v_0 \cos\theta)t, h + (v_0 \sin\theta)t - \frac{gt^2}{2} \right\rangle. \quad (3.8)$$

Notice that the path traced out by $\mathbf{r}(t)$ (from $t = 0$ until impact) is a portion of a parabola. (See Figure 11.18b.)

Now that we have derived (3.7) and (3.8), we have all we need to answer any further questions about the motion. For instance, if we need to know the maximum altitude, this occurs at the time at which the vertical component of velocity is zero (i.e., at the time when the projectile stops rising). From (3.7), we solve

$$0 = v_0 \sin\theta - gt,$$

so that the time at which the maximum altitude is reached is given by

Time to reach maximum altitude

$$t_{max} = \frac{v_0 \sin\theta}{g}.$$

The maximum altitude itself is the vertical component of the position vector at this time. From (3.8), we have

Maximum altitude

$$\text{Maximum altitude} = h + (v_0 \sin\theta)t - \frac{gt^2}{2}\bigg|_{t=t_{max}}$$

$$= h + (v_0 \sin\theta)\left(\frac{v_0 \sin\theta}{g}\right) - \frac{g}{2}\left(\frac{v_0 \sin\theta}{g}\right)^2$$

$$= h + \frac{1}{2}\frac{v_0^2 \sin^2\theta}{g}.$$

To find the horizontal range or the speed at impact, we must first find the time of impact. To get this, we set the vertical component of position to zero. From (3.8), we have

$$0 = h + (v_0 \sin\theta)t - \frac{gt^2}{2}.$$

Notice that this is simply a quadratic equation for t. Given v_0, θ and h, we can solve for the time t using the quadratic formula.

In all of the foregoing analysis, we left the constant acceleration due to gravity as g. You will usually use one of the two approximations:

$$g \approx 32 \text{ ft/sec}^2 \quad \text{or} \quad g \approx 9.8 \text{ m/sec}^2.$$

When using any other units, simply adjust the units to feet or meters and the time scale to seconds or make the corresponding adjustments to the value of g.

Just as with bodies moving in a straight line, we can use the calculus to analyze the motion of a body rotating about an axis. It's not hard to see why this is an important problem; just think of a gymnast performing a complicated routine. If the body is considered as a single point moving in three dimensions, the motion can be analyzed as in example 3.4. However, we are also quite interested in the rotational movement of the body. In the case of a gymnast, of course, the twists and turns that are performed are an important consideration. This is an example of rotational motion.

We use a rotational version of Newton's second law to analyze the motion. Torque (denoted by $\boldsymbol{\tau}$) is defined in section 10.4. In the case of an object rotating in two dimensions, the torque has magnitude (denoted by $\tau = \|\boldsymbol{\tau}\|$) given by the product of the force acting in the direction of the motion and the distance from the rotational center. The moment of inertia I of a body is a measure of how much force must be applied to cause the object to start rotating. This is determined by the mass and the distance of the mass from the center of rotation and is examined in some detail in section 13.2. In rotational motion, the primary variable that we track is an angle of displacement, denoted by θ. For a rotating body, the angle measured from some fixed ray changes with time t, so that the angle is a function $\theta(t)$. We define the **angular velocity** to be $\omega(t) = \theta'(t)$ and the **angular acceleration** to be $\alpha(t) = \omega'(t) = \theta''(t)$. The equation of rotational motion is then

$$\tau = I\alpha. \tag{3.9}$$

Notice how closely this resembles Newton's second law, $F = ma$. The calculus used in example 3.5 should look familiar.

EXAMPLE 3.5 The Rotational Motion of a Merry-Go-Round

A stationary merry-go-round of radius 5 feet is started in motion by a push consisting of a force of 10 pounds on the outside edge, tangent to the circular edge of the merry-go-round, for 1 second. The moment of inertia of the merry-go-round is $I = 25$. Find the resulting angular velocity of the merry-go-round.

Solution We first compute the torque of the push. The force is applied 5 feet from the center of rotation, so that the torque has magnitude

$$\tau = (\text{Force})(\text{Distance from axis of rotation}) = (10)(5) = 50 \text{ foot-pounds}.$$

From (3.9), we have

$$50 = 25\alpha,$$

so that $\alpha = 2$. Since the force is applied for one second, this equation holds for $0 \le t \le 1$. Integrating both sides of the equation $\omega' = \alpha$ from $t = 0$ to $t = 1$, we have by the Fundamental Theorem of Calculus that

$$\omega(1) - \omega(0) = \int_0^1 \alpha\, dt = \int_0^1 2\, dt = 2. \tag{3.10}$$

If the merry-go-round is initially stationary, then $\omega(0) = 0$ and (3.10) becomes simply $\omega(1) = 2$ rad/s. ■

Notice that we could draw a more general conclusion from (3.10). Even if the merry-go-round is already in motion, applying a force of 10 pounds tangentially to the edge for 1 second will increase the rotation rate by 2 rad/s.

For rotational motion in three dimensions, the calculations are somewhat more complicated. Recall that we had defined the torque $\boldsymbol{\tau}$ due to a force $\mathbf{F}$ applied at position $\mathbf{r}$ to be

$$\boldsymbol{\tau} = \mathbf{r} \times \mathbf{F}.$$

Example 3.6 relates torque to angular momentum. The **linear momentum** of an object of mass m with velocity $\mathbf{v}$ is given by $\mathbf{p} = m\mathbf{v}$. The **angular momentum** is defined by $\mathbf{L}(t) = \mathbf{r}(t) \times m\mathbf{v}(t)$.

EXAMPLE 3.6 Relating Torque and Angular Momentum

Show that torque equals the derivative of angular momentum.

Solution From the definition of angular momentum and the product rule for the derivative of a cross product [Theorem 2.3 (v)], we have

$$\begin{aligned} \mathbf{L}'(t) &= \frac{d}{dt}[\mathbf{r}(t) \times m\mathbf{v}(t)] \\ &= \mathbf{r}'(t) \times m\mathbf{v}(t) + \mathbf{r}(t) \times m\mathbf{v}'(t) \\ &= \mathbf{v}(t) \times m\mathbf{v}(t) + \mathbf{r}(t) \times m\mathbf{a}(t). \end{aligned}$$

Notice that the first term on the right-hand side is the zero vector, since it is the cross product of parallel vectors. From Newton's second law, we have $\mathbf{F}(t) = m\mathbf{a}(t)$, so we have

$$\mathbf{L}'(t) = \mathbf{r}(t) \times m\mathbf{a}(t) = \mathbf{r} \times \mathbf{F} = \boldsymbol{\tau}. \blacksquare$$

From this result, it is a short step to the principle of **conservation of angular momentum,** which states that, in the absence of torque, angular momentum remains constant. This is left as an exercise.

In example 3.7, we examine a fully three-dimensional projectile motion problem for the first time.

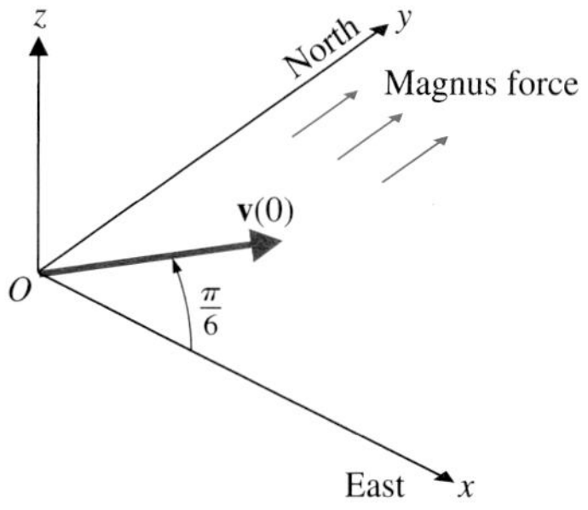

FIGURE 11.19a
The initial velocity and Magnus force vectors

EXAMPLE 3.7 Analyzing the Motion of a Projectile in Three Dimensions

A projectile of mass 1 kg is launched from ground level toward the east at 200 meters/second, at an angle of $\frac{\pi}{6}$ to the horizontal. If the spinning of the projectile applies a steady northerly Magnus force of 2 newtons to the projectile, find the landing location of the projectile and its speed at impact.

Solution Notice that because of the Magnus force, the motion is fully three-dimensional. We orient the x-, y- and z-axes so that the positive y-axis points north, the positive x-axis points east and the positive z-axis points up, as in Figure 11.19a, where we also show the initial velocity vector and vectors indicating the Magnus force. The two forces acting on the projectile are gravity (in the negative z-direction with magnitude $9.8m = 9.8$ newtons) and the Magnus force (in the y-direction with magnitude 2 newtons). Newton's second law is $\mathbf{F} = m\mathbf{a} = \mathbf{a}$. We have

$$\mathbf{a}(t) = \mathbf{v}'(t) = \langle 0, 2, -9.8 \rangle$$

Integrating gives us the velocity function

$$\mathbf{v}(t) = \langle 0, 2t, -9.8t \rangle + \mathbf{c}_1, \tag{3.11}$$

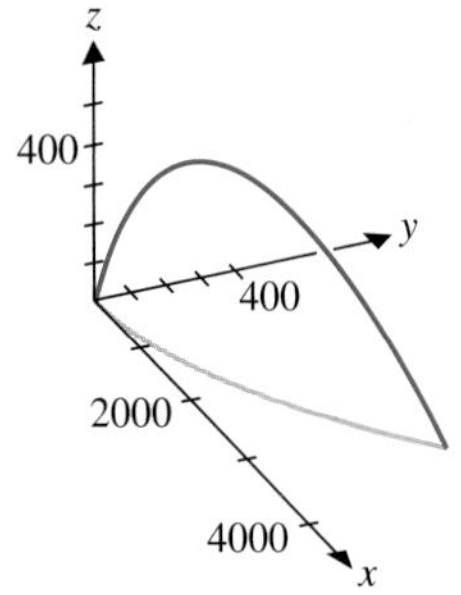

FIGURE 11.19b
Path of the projectile

where $\mathbf{c}_1$ is an arbitrary constant vector. Note that the initial velocity is

$$\mathbf{v}(0) = \left\langle 200\cos\frac{\pi}{6}, 0, 200\sin\frac{\pi}{6}\right\rangle = \langle 100\sqrt{3}, 0, 100\rangle.$$

From (3.11), we now have

$$\langle 100\sqrt{3}, 0, 100\rangle = \mathbf{v}(0) = \mathbf{c}_1,$$

which gives us

$$\mathbf{v}(t) = \langle 100\sqrt{3}, 2t, 100 - 9.8t\rangle.$$

We integrate this to get the position vector:

$$\mathbf{r}(t) = \langle 100\sqrt{3}t, t^2, 100t - 4.9t^2\rangle + \mathbf{c}_2,$$

for a constant vector $\mathbf{c}_2$. Taking the initial position to be the origin, we get

$$\mathbf{0} = \mathbf{r}(0) = \mathbf{c}_2,$$

so that

$$\mathbf{r}(t) = \langle 100\sqrt{3}t, t^2, 100t - 4.9t^2\rangle. \tag{3.12}$$

Note that the projectile strikes the ground when the $\mathbf{k}$ component of position is zero. From (3.12), we have that this occurs when

$$0 = 100t - 4.9t^2 = t(100 - 4.9t).$$

So, the projectile is on the ground when $t = 0$ (time of launch) and when $t = \frac{100}{4.9} \approx 20.4$ seconds (the time of impact). The location of impact is then the endpoint of the vector $\mathbf{r}\left(\frac{100}{4.9}\right) \approx \langle 3534.8, 416.5, 0\rangle$ and the speed at impact is

$$\left\|\mathbf{v}\left(\frac{100}{4.9}\right)\right\| \approx 204 \text{ m/s}.$$

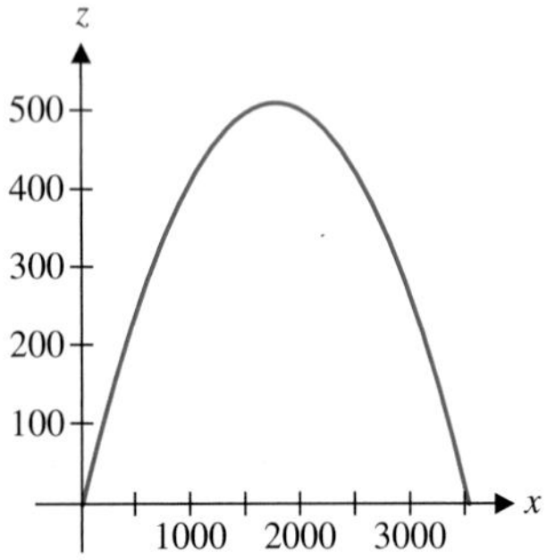

FIGURE 11.19c
Projection of path onto the xz-plane

We show a computer-generated graph of the path of the projectile in Figure 11.19b. In this figure, we also indicate the shadow made by the path of the projectile on the ground. In Figure 11.19c, we show the projection of the projectile's path onto the xz-plane. Observe that this parabola is analogous to the parabola shown in Figure 11.17b. ■

EXERCISES 11.3

WRITING EXERCISES

1. Explain why it makes sense in example 3.4 that the speed at impact equals the initial speed. (Hint: What force would slow the object down?) If the projectile were launched from above ground, discuss how the speed at impact would compare to the initial speed.

2. For an actual projectile, taking into account air resistance, explain why the speed at impact would be less than the initial speed.

3. In this section, we assumed that the acceleration due to gravity is constant. By contrast, air resistance is a function of velocity. (The faster the object goes, the more air resistance there is.) Explain why including air resistance in our Newton's law model of projectile motion would make the mathematics *much* more complicated.

4. In example 3.7, use the x- and y-components of the position function to explain why the projection of the projectile's path onto the xy-plane would be a parabola. The projection onto the xz-plane is also a parabola. Discuss whether or not the path in Figure 11.19b is a parabola. If you were watching the projectile, would the path appear to be parabolic?

In exercises 1–6, find the velocity and acceleration functions for the given position function.

1. $\mathbf{r}(t) = \langle 5\cos 2t, 5\sin 2t\rangle$

2. $\mathbf{r}(t) = \langle 2\cos t + \sin 2t, 2\sin t + \cos 2t\rangle$

3. $\mathbf{r}(t) = \langle 25t, -16t^2 + 15t + 5\rangle$

4. $\mathbf{r}(t) = \langle 25te^{-2t}, -16t^2 + 10t + 20\rangle$

5. $\mathbf{r}(t) = \langle 4te^{-2t}, 2e^{-2t}, -16t^2\rangle$

6. $\mathbf{r}(t) = \langle 3e^{-3t}, \sin 2t, t^3 - 3t\rangle$

In exercises 7–14, find the position function from the given velocity or acceleration function.

7. $\mathbf{v}(t) = \langle 10, -32t + 4\rangle$, $\mathbf{r}(0) = \langle 3, 8\rangle$

8. $\mathbf{v}(t) = \langle 4t, t^2 - 1\rangle$, $\mathbf{r}(0) = \langle 10, -2\rangle$

9. $\mathbf{a}(t) = \langle 0, -32\rangle$, $\mathbf{v}(0) = \langle 5, 0\rangle$, $\mathbf{r}(0) = \langle 0, 16\rangle$

10. $\mathbf{a}(t) = \langle t, \sin t\rangle$, $\mathbf{v}(0) = \langle 2, -6\rangle$, $\mathbf{r}(0) = \langle 10, 4\rangle$

11. $\mathbf{v}(t) = \langle 10, 3e^{-t}, -32t + 4\rangle$, $\mathbf{r}(0) = \langle 0, -6, 20\rangle$

12. $\mathbf{v}(t) = \langle t + 2, t^2, e^{-t/3}\rangle$, $\mathbf{r}(0) = \langle 4, 0, -3\rangle$

13. $\mathbf{a}(t) = \langle t, 0, -16\rangle$, $\mathbf{v}(0) = \langle 12, -4, 0\rangle$, $\mathbf{r}(0) = \langle 5, 0, 2\rangle$

14. $\mathbf{a}(t) = \langle e^{-3t}, t, \sin t\rangle$, $\mathbf{v}(0) = \langle 4, -2, 4\rangle$, $\mathbf{r}(0) = \langle 0, 4, -2\rangle$

In exercises 15–18, find the centripetal force on an object of mass 10 kg with the given position function (in units of meters and seconds).

15. $\mathbf{r}(t) = \langle 4\cos 2t, 4\sin 2t\rangle$

16. $\mathbf{r}(t) = \langle 3\cos 5t, 3\sin 5t\rangle$

17. $\mathbf{r}(t) = \langle 6\cos 4t, 6\sin 4t\rangle$

18. $\mathbf{r}(t) = \langle 2\cos 3t, 2\sin 3t\rangle$

In exercises 19–22, find the force acting on an object of mass 10 kg with the given position function (in units of meters and seconds).

19. $\mathbf{r}(t) = \langle 3\cos 2t, 5\sin 2t\rangle$

20. $\mathbf{r}(t) = \langle 3\cos 4t, 2\sin 5t\rangle$

21. $\mathbf{r}(t) = \langle 3t^2 + t, 3t - 1\rangle$

22. $\mathbf{r}(t) = \langle 20t - 3, -16t^2 + 2t + 30\rangle$

In exercises 23–28, a projectile is fired with initial speed v_0 feet per second from a height of h feet at an angle of θ above the horizontal. Assuming that the only force acting on the object is gravity, find the maximum altitude, horizontal range and speed at impact.

23. $v_0 = 100, h = 0, \theta = \frac{\pi}{3}$

24. $v_0 = 100, h = 0, \theta = \frac{\pi}{6}$

25. $v_0 = 160, h = 10, \theta = \frac{\pi}{4}$

26. $v_0 = 120, h = 10, \theta = \frac{\pi}{3}$

27. $v_0 = 320, h = 10, \theta = \frac{\pi}{4}$

28. $v_0 = 240, h = 10, \theta = \frac{\pi}{3}$

29. Based on your answers to exercises 25 and 27, what effect does doubling the initial speed have on the horizontal range?

30. The angles $\frac{\pi}{3}$ and $\frac{\pi}{6}$ are symmetric about $\frac{\pi}{4}$; that is, $\frac{\pi}{4} - \frac{\pi}{6} = \frac{\pi}{3} - \frac{\pi}{4}$. Based on your answers to exercises 23 and 24, how do horizontal ranges for symmetric angles compare?

31. Beginning with Newton's second law of motion, derive the equations of motion for a projectile fired from altitude h above the ground at an angle θ to the horizontal and with initial speed v_0.

32. For the general projectile of exercise 31, with $h = 0$, (a) show that the horizontal range is $\dfrac{v_0^2 \sin 2\theta}{g}$ and (b) find the angle that produces the maximum horizontal range.

In exercises 33–40, neglect all forces except gravity. In all these situations, the effect of air resistance is actually significant, but your calculations will give a good first approximation.

33. A baseball is hit from a height of 3 feet with initial speed 120 feet per second and at an angle of 30 degrees above the horizontal. Find a vector-valued function describing the position of the ball t seconds after it is hit. To be a home run, the ball must clear a wall that is 385 feet away and 6 feet tall. Determine whether this is a home run.

34. Repeat exercise 33 if the ball is launched with an initial angle of 31 degrees.

35. A baseball pitcher throws a pitch horizontally from a height of 6 feet with an initial speed of 130 feet per second. Find a vector-valued function describing the position of the ball t seconds after release. If home plate is 60 feet away, how high is the ball when it crosses home plate?

36. If a person drops a ball from height 6 feet at the same time the pitcher of exercise 35 releases the ball, how high will the dropped ball be when the pitch crosses home plate?

37. A tennis serve is struck horizontally from a height of 8 feet with initial speed 120 feet per second. For the serve to count (be "in"), it must clear a net that is 39 feet away and 3 feet high and must land before the service line 60 feet away. Find a vector function for the position of the ball and determine whether this serve is in or out.

38. Repeat exercise 37 if the ball is struck with an initial speed of (a) 80 ft/s or (b) 65 ft/s.

39. A football punt is launched at an angle of 50 degrees with an initial speed of 55 mph. Assuming the punt is launched from ground level, compute the "hang time" (the amount of time in the air) for the punt.

40. Compute the extra hang time if the punt in exercise 39 has an initial speed of 60 mph.

41. Find the landing point in exercise 23 if the object has mass 1 slug, is launched due east and there is a northerly Magnus force of 8 pounds.

42. Find the landing point in exercise 24 if the object has mass 1 slug, is launched due east and there is a southerly Magnus force of 4 pounds.

43. Suppose an airplane is acted on by three forces: gravity, wind and engine thrust. Assume that the force vector for gravity is $m\mathbf{g} = m\langle 0, 0, -32\rangle$, the force vector for wind is $\mathbf{w} = \langle 0, 1, 0\rangle$ for $0 \le t \le 1$ and $\mathbf{w} = \langle 0, 2, 0\rangle$ for $t > 1$, and the force vector for engine thrust is $\mathbf{e} = \langle 2t, 0, 24\rangle$. Newton's second law of motion gives us $m\mathbf{a} = m\mathbf{g} + \mathbf{w} + \mathbf{e}$. Assume that $m = 1$ and the initial velocity vector is $\mathbf{v}(0) = \langle 100, 0, 10\rangle$. Show that the velocity vector for $0 \le t \le 1$ is $\mathbf{v}(t) = \langle t^2 + 100, t, 10 - 8t\rangle$. For $t > 1$, integrate the equation $\mathbf{a} = \mathbf{g} + \mathbf{w} + \mathbf{e}$, to get $\mathbf{v}(t) = \langle t^2 + a, 2t + b, -8t + c\rangle$, for constants a, b and c. Explain (on physical grounds) why the function $\mathbf{v}(t)$ should be continuous and find the values of the constants that make it so. Show that $\mathbf{v}(t)$ is not differentiable. Given the nature of the force function, why does this make sense?

44. Find the position function for the airplane in exercise 43.

45. A roller coaster is designed to travel a circular loop of radius 100 feet. If the riders feel weightless at the top of the loop, what is the speed of the roller coaster?

46. A roller coaster travels at variable angular speed $\omega(t)$ and radius $r(t)$ but constant speed $c = \omega(t)r(t)$. For the centripetal force $F(t) = m\omega^2(t)r(t)$, show that $F'(t) = m\omega(t)r(t)\omega'(t)$. Conclude that entering a tight curve with $r'(t) < 0$ but maintaining constant speed, the centripetal force increases.

47. A jet pilot executing a circular turn experiences an acceleration of "5 g's" (that is, $\|\mathbf{a}\| = 5g$). If the jet's speed is 900 km/hr, what is the radius of the turn?

48. For the jet pilot of exercise 47, how many g's would be experienced if the speed were 1800 km/hr?

49. A force of 20 pounds is applied to the outside of a stationary merry-go-round of radius 5 feet for 0.5 second. The moment of inertia is $I = 10$. Find the resultant change in angular velocity of the merry-go-round.

50. A merry-go-round of radius 5 feet and moment of inertia $I = 10$ rotates at 4 rad/s. Find the constant force needed to stop the merry-go-round in 2 seconds.

51. A golfer rotates a club with constant angular acceleration α through an angle of π radians. If the angular velocity increases from 0 to 15 rad/s, find α.

52. For the golf club in exercise 51, find the increase in angular velocity if the club is rotated through an angle of $\frac{3\pi}{2}$ radians with the same angular acceleration. Describe one advantage of a long swing.

53. Softball pitchers such as Jennie Finch often use a double windmill to generate arm speed. At a constant angular acceleration, compare the speeds obtained rotating through an angle of 2π versus rotating through an angle of 4π.

54. As the softball in exercise 53 rotates, its linear speed v is related to the angular velocity ω by $v = r\omega$, where r is the distance of the ball from the center of rotation. The picture shows the pitcher with arm fully extended. Explain why this is a good technique for throwing a fast pitch.

55. Use the result of example 3.6 to prove the **Law of Conservation of Angular Momentum:** if there is zero (net) torque on an object, its angular momentum remains constant.

56. Prove the **Law of Conservation of Linear Momentum:** if there is zero (net) force on an object, its linear momentum remains constant.

57. If acceleration is parallel to position ($\mathbf{a}\|\mathbf{r}$), show that there is no torque. Explain this result in terms of the change in angular momentum. (Hint: If $\mathbf{a}\|\mathbf{r}$, would angular velocity or linear velocity be affected?)

58. If the acceleration $\mathbf{a}$ is constant, show that $\mathbf{L}''' = \mathbf{0}$.

59. Example 3.3 is a model of a satellite orbiting the earth. In this case, the force $\mathbf{F}$ is the gravitational attraction of the earth on the satellite. The magnitude of the force is $\frac{mMG}{b^2}$, where m is the mass of the satellite, M is the mass of the earth and G

is the universal gravitational constant. Using example 3.3, this should be equal to $m\omega^2 b$. For a **geosynchronous orbit,** the frequency ω is such that the satellite completes one orbit in one day. (By orbiting at the same rate as the earth spins, the satellite can remain directly above the same point on the earth.) For a sidereal day of 23 hours, 56 minutes and 4 seconds, find ω. Using $MG \approx 39.87187 \times 10^{13}$ N-m²/kg, find b for a geosynchronous orbit (the units of b will be m).

60. Example 3.3 can also model a jet executing a turn. For a jet traveling at 1000 km/h, find the radius b such that the pilot feels 7 g's of force; that is, the magnitude of the force is 7 mg.

61. We have seen how we can find the trajectory of a projectile given its initial position and initial velocity. For military personnel tracking an incoming missile, the only data available correspond to various points on the trajectory, while the initial position (where the enemy gun is located) is unknown but very important. Assume that a projectile follows a parabolic path (after launch, the only force is gravity). If the projectile passes through points (x_1, y_1, z_1) at time t_1 and (x_2, y_2, z_2) at time t_2, find the initial position $(x_0, y_0, 0)$.

62. Use the result of exercise 61 to identify the initial velocity and launch position of a projectile that passes through (1, 2, 4) at $t = 1$ and (3, 6, 6) at $t = 2$.

63. For a satellite in earth orbit, the speed v in miles per second is related to the height h miles above the surface of the earth by $v = \sqrt{\frac{95{,}600}{4000+h}}$. Suppose a satellite is in orbit 15,000 miles above the surface of the earth. How much does the speed need to decrease to raise the orbit to a height of 20,000 miles?

EXPLORATORY EXERCISES

1. A ball rolls off a table of height 3 feet. Its initial velocity is horizontal with speed v_0. Determine where the ball hits the ground and the velocity vector of the ball at the moment of impact. Find the angle between the horizontal and the impact velocity vector. Next, assume that the next bounce of the ball starts with the ball being launched from the ground with initial conditions determined by the impact velocity. The launch speed equals 0.6 times the impact speed (so the ball won't bounce forever) and the launch angle equals the (positive) angle between the horizontal and the impact velocity vector. Using these conditions, determine where the ball next hits the ground. Continue on to find the third point at which the ball bounces.

2. In many sports such as golf and ski jumping, it is important to determine the range of a projectile on a slope. Suppose that the ground passes through the origin and slopes at an angle of α to the horizontal. Show that an equation of the ground is $y = -(\tan\alpha)x$. An object is launched at height $h = 0$ with initial speed v_0 at an angle of θ from the horizontal. Referring to exercise 31, show that the landing condition is now $y = -(\tan\alpha)x$. Find the x-coordinate of the landing point and show that the range (the distance along the ground) is given by $R = \frac{2}{g}v_0^2 \sec\alpha\cos\theta(\sin\theta + \tan\alpha\cos\theta)$. Use trigonometric identities to rewrite this as $R = \frac{1}{g}v_0^2\sec^2\alpha[\sin\alpha + \sin(\alpha + 2\theta)]$. Use this formula to find the value of θ that maximizes the range. For flat ground ($\alpha = 0$), the optimal angle is 45°. State an easy way of taking the value of α (say, $\alpha = 10°$ or $\alpha = -8°$) and adjusting from 45° to the optimal angle.

11.4 CURVATURE

Imagine that you are designing a new highway. Nearly all roads have curves, to avoid both natural and human-made obstacles. So that cars are able to maintain a reasonable speed on your new road, you should avoid curves that are too sharp. To do this, it would help to have some concept of how sharp a given curve is. In this section, we develop a measure of how much a curve is twisting and turning at any given point. First, realize that any given curve has infinitely many different parameterizations. For instance, the parametric equations $x = t^2$ and $y = t$ describe a parabola that opens to the right. In fact, for any real number $a > 0$, the equations $x = (at)^2$ and $y = at$ describe the same parabola. So, any measure of how sharp a curve is should be independent of the parameterization. The simplest choice of a parameter (for conceptual purposes, but not for computational purposes) is arc length. Further, observe that this is the correct parameter to use, as we measure how sharp a curve is by seeing how much it twists and turns per unit length. (Think about it this way: a turn of 90° over a quarter mile is not particularly sharp in comparison with a turn of 90° over a distance of 30 feet.)

For the curve traced out by the endpoint of the vector-valued function $\mathbf{r}(t) = \langle f(t), g(t), h(t)\rangle$, for $a \le t \le b$, we define the arc length parameter $s(t)$ to be the arc length of that portion of the curve from $u = a$ up to $u = t$. That is, from (1.4),

$$s(t) = \int_a^t \sqrt{[f'(u)]^2 + [g'(u)]^2 + [h'(u)]^2}\, du.$$

Recognizing that $\sqrt{[f'(u)]^2 + [g'(u)]^2 + [h'(u)]^2} = \|\mathbf{r}'(u)\|$, we can write this more simply as

$$s(t) = \int_a^t \|\mathbf{r}'(u)\|\, du. \tag{4.1}$$

Although explicitly finding an arc length parameterization of a curve is not the central thrust of our discussion here, we briefly pause now to construct such a parameterization, for the purpose of illustration.

EXAMPLE 4.1 Parameterizing a Curve in Terms of Arc Length

Find an arc length parameterization of the circle of radius 4 centered at the origin.

Solution Note that one parameterization of this circle is

$$C: x = f(t) = 4\cos t, \quad y = g(t) = 4\sin t, \quad 0 \le t \le 2\pi.$$

In this case, the arc length from $u = 0$ to $u = t$ is given by

$$\begin{aligned} s(t) &= \int_0^t \sqrt{[f'(u)]^2 + [g'(u)]^2}\, du \\ &= \int_0^t \sqrt{[-4\sin u]^2 + [4\cos u]^2}\, du = 4\int_0^t 1\, du = 4t. \end{aligned}$$

That is, $t = s/4$, so that an arc length parameterization for C is

$$C: x = 4\cos\left(\frac{s}{4}\right), \quad y = 4\sin\left(\frac{s}{4}\right), \quad 0 \le s \le 8\pi. \blacksquare$$

Consider the smooth curve C traced out by the endpoint of the vector-valued function $\mathbf{r}(t)$. Recall that for each t, $\mathbf{v}(t) = \mathbf{r}'(t)$ can be thought of as both the velocity vector and a tangent vector, pointing in the direction of motion (i.e., the orientation of C). Notice that

Unit tangent vector

$$\boxed{\mathbf{T}(t) = \frac{\mathbf{r}'(t)}{\|\mathbf{r}'(t)\|}} \tag{4.2}$$

is also a tangent vector, but has length one ($\|\mathbf{T}(t)\| = 1$). We call $\mathbf{T}(t)$ the **unit tangent vector** to the curve C. That is, for each t, $\mathbf{T}(t)$ is a tangent vector of length one pointing in the direction of the orientation of C.

EXAMPLE 4.2 Finding a Unit Tangent Vector

Find the unit tangent vector to the curve determined by $\mathbf{r}(t) = \langle t^2 + 1, t\rangle$.

Solution We have

$$\mathbf{r}'(t) = \langle 2t, 1\rangle,$$

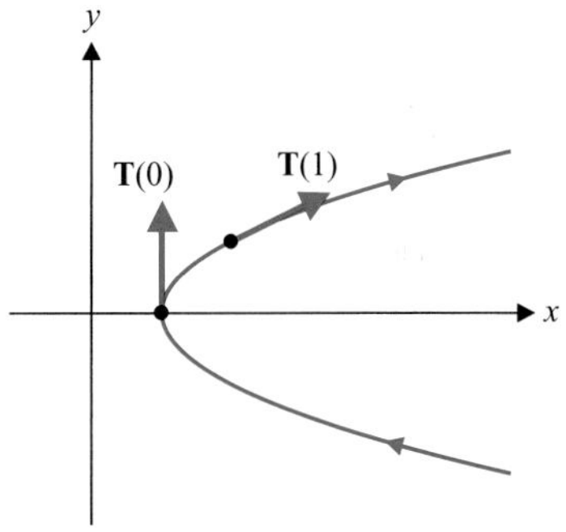

FIGURE 11.20
Unit tangent vectors

so that $$\|\mathbf{r}'(t)\| = \sqrt{(2t)^2 + 1} = \sqrt{4t^2 + 1}.$$

From (4.2), the unit tangent vector is given by

$$\mathbf{T}(t) = \frac{\mathbf{r}'(t)}{\|\mathbf{r}'(t)\|} = \frac{\langle 2t, 1\rangle}{\sqrt{4t^2+1}} = \left\langle \frac{2t}{\sqrt{4t^2+1}}, \frac{1}{\sqrt{4t^2+1}} \right\rangle.$$

In particular, we have $\mathbf{T}(0) = \langle 0, 1\rangle$ and $\mathbf{T}(1) = \left\langle \frac{2}{\sqrt{5}}, \frac{1}{\sqrt{5}} \right\rangle$. We indicate both of these in Figure 11.20. ■

In Figures 11.21a and 11.21b, we show two curves, both connecting the points A and B. Think about driving a car along roads in the shape of these two curves. The curve in Figure 11.21b indicates a much sharper turn than the curve in Figure 11.21a. The question before us is to see how to mathematically describe this degree of "sharpness." You should get an idea of this from Figures 11.21c and 11.21d. These are the same curves as those shown in Figures 11.21a and 11.21b, respectively, but we have drawn in a number of unit tangent vectors at equally spaced points on the curves. Notice that the unit tangent vectors change very slowly along the gentle curve in Figure 11.21c, but twist and turn quite rapidly in the vicinity of the sharp curve in Figure 11.21d. Based on our analysis of Figures 11.21c and 11.21d, notice that the rate of change of the unit tangent vectors with respect to arc length along the curve will give us a measure of sharpness. To this end, we make the following definition.

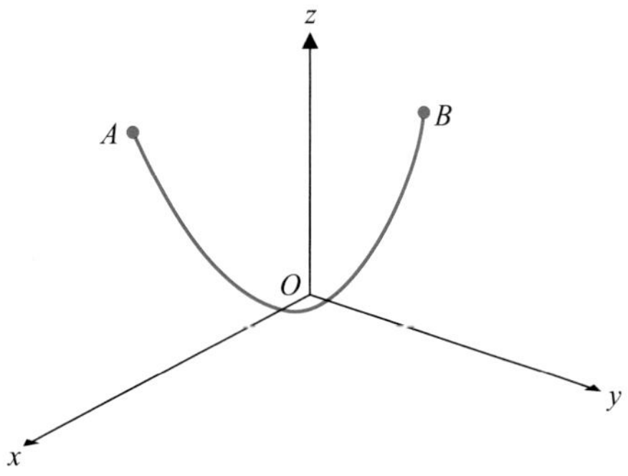

FIGURE 11.21a
Gentle curve

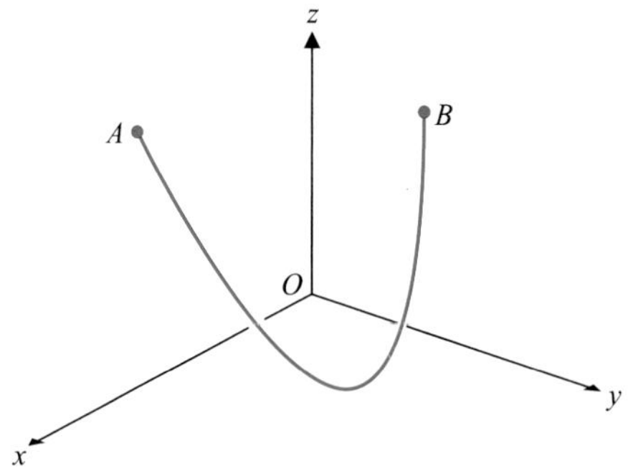

FIGURE 11.21b
Sharp curve

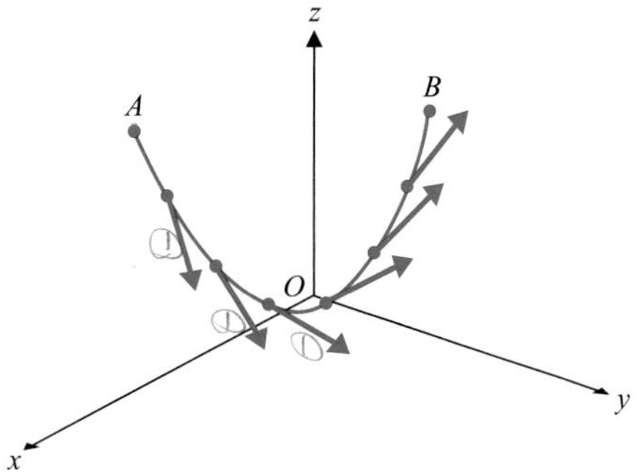

FIGURE 11.21c
Unit tangent vectors

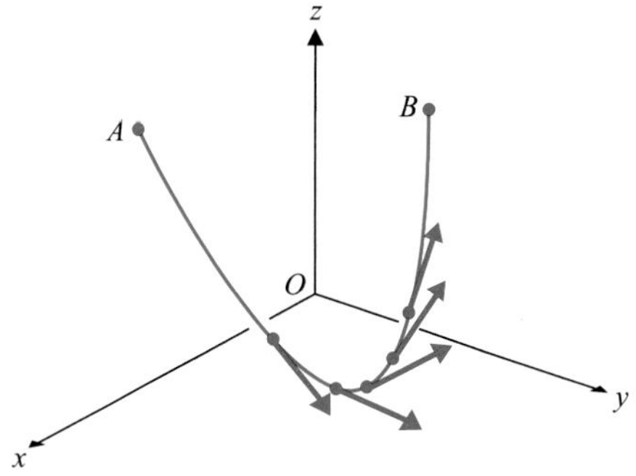

FIGURE 11.21d
Unit tangent vectors

DEFINITION 4.1

The **curvature** κ of a curve is the scalar quantity

$$\kappa = \left\| \frac{d\mathbf{T}}{ds} \right\|. \tag{4.3}$$

Note that, while the definition of curvature makes sense intuitively, it is not a simple matter to compute κ directly from (4.3). To do so, we would need to first find the arc length parameter and the unit tangent vector $\mathbf{T}(t)$, rewrite $\mathbf{T}(t)$ in terms of the arc length parameter s and then differentiate with respect to s. This is not usually done. Instead, observe that by the chain rule,

$$\mathbf{T}'(t) = \frac{d\mathbf{T}}{dt} = \frac{d\mathbf{T}}{ds}\frac{ds}{dt},$$

so that when $\frac{ds}{dt} \neq 0$,

$$\kappa = \left\| \frac{d\mathbf{T}}{ds} \right\| = \frac{\|\mathbf{T}'(t)\|}{\left|\frac{ds}{dt}\right|}. \tag{4.4}$$

Now, from (4.1), we had

$$s(t) = \int_a^t \|\mathbf{r}'(u)\|\, du,$$

so that by part II of the Fundamental Theorem of Calculus,

$$\frac{ds}{dt} = \|\mathbf{r}'(t)\|. \tag{4.5}$$

From (4.4) and (4.5), we now have

Curvature

$$\kappa = \frac{\|\mathbf{T}'(t)\|}{\|\mathbf{r}'(t)\|}, \tag{4.6}$$

where $\|\mathbf{r}'(t)\| \neq 0$, for a smooth curve. Notice that it should be comparatively simple to use (4.6) to compute the curvature. We illustrate this in example 4.3.

EXAMPLE 4.3 Finding the Curvature of a Straight Line

Find the curvature of a straight line.

Solution First, think about what we're asking. Straight lines are, well, straight, so their curvature should be zero at every point. Let's see. Suppose that the line is traced out by the vector-valued function $\mathbf{r}(t) = \langle at + b, ct + d, et + f \rangle$, for some constants a, b, c, d, e and f (where at least one of a, c or e is nonzero). Then,

$$\mathbf{r}'(t) = \langle a, c, e \rangle$$

and so,

$$\|\mathbf{r}'(t)\| = \sqrt{a^2 + c^2 + e^2} = \text{constant} \neq 0.$$

The unit tangent vector is then

$$\mathbf{T}(t) = \frac{\mathbf{r}'(t)}{\|\mathbf{r}'(t)\|} = \frac{\langle a, c, e \rangle}{\sqrt{a^2 + c^2 + e^2}},$$

which is a constant vector. This gives us $\mathbf{T}'(t) = \mathbf{0}$, for all t. From (4.6), we now have

$$\kappa = \frac{\|\mathbf{T}'(t)\|}{\|\mathbf{r}'(t)\|} = \frac{\|\mathbf{0}\|}{\sqrt{a^2 + c^2 + e^2}} = 0,$$

as expected. ■

Well, if a line has zero curvature, can you think of a curve with lots of curvature? The first one to come to mind is likely a circle, which we discuss next.

EXAMPLE 4.4 Finding the Curvature of a Circle

Find the curvature for a circle of radius $a > 0$.

Solution We leave it as an exercise to show that the curvature does not depend on the location of the center of the circle. (Intuitively, it certainly should not.) So, for simplicity, we assume that the circle is centered at the origin. Notice that the circle of radius a centered at the origin is traced out by the vector-valued function $\mathbf{r}(t) = \langle a \cos t, a \sin t \rangle$. Differentiating, we get

$$\mathbf{r}'(t) = \langle -a \sin t, a \cos t \rangle$$

and
$$\|\mathbf{r}'(t)\| = \sqrt{(-a \sin t)^2 + (a \cos t)^2} = a\sqrt{\sin^2 t + \cos^2 t} = a.$$

The unit tangent vector is then given by

$$\mathbf{T}(t) = \frac{\mathbf{r}'(\mathbf{t})}{\|\mathbf{r}'(t)\|} = \frac{\langle -a \sin t, a \cos t \rangle}{a} = \langle -\sin t, \cos t \rangle.$$

Differentiating this gives us

$$\mathbf{T}'(t) = \langle -\cos t, -\sin t \rangle$$

and from (4.6), we have

$$\kappa = \frac{\|\mathbf{T}'(t)\|}{\|\mathbf{r}'(t)\|} = \frac{\|\langle -\cos t, -\sin t \rangle\|}{a} = \frac{\sqrt{(-\cos t)^2 + (-\sin t)^2}}{a} = \frac{1}{a}.$$ ■

Notice that the result of example 4.4 is consistent with your intuition. First, observe that you should be able to drive a car around a circular track while holding the steering wheel in a fixed position. (That is, the curvature should be constant.) Further, the smaller that the radius of a circular track is, the sharper you will need to turn (that is, the larger the curvature). On the other hand, on a circular track of very large radius, it would seem as if you were driving fairly straight (i.e., the curvature will be close to 0).

You probably noticed that computing the curvature of the curves in examples 4.3 and 4.4 was just the slightest bit tedious. We simplify this process somewhat with the result of Theorem 4.1.

THEOREM 4.1

The curvature of the smooth curve traced out by the vector-valued function $\mathbf{r}(t)$ is given by

$$\kappa = \frac{\|\mathbf{r}'(t) \times \mathbf{r}''(t)\|}{\|\mathbf{r}'(t)\|^3}. \tag{4.7}$$

The proof of Theorem 4.1 is rather long and involved and so, we omit it at this time, in the interest of brevity. We return to this in section 11.5, where the proof becomes a simple consequence of another result.

Notice that it is a relatively simple matter to use (4.7) to compute the curvature for nearly any three-dimensional curve.

EXAMPLE 4.5 Finding the Curvature of a Helix

Find the curvature of the helix traced out by $\mathbf{r}(t) = \langle 2\sin t, 2\cos t, 4t\rangle$.

Solution A graph of the helix is indicated in Figure 11.22. We have

$$\mathbf{r}'(t) = \langle 2\cos t, -2\sin t, 4\rangle$$

and

$$\mathbf{r}''(t) = \langle -2\sin t, -2\cos t, 0\rangle.$$

Now,

$$\begin{aligned}\mathbf{r}'(t)\times\mathbf{r}''(t) &= \langle 2\cos t, -2\sin t, 4\rangle \times \langle -2\sin t, -2\cos t, 0\rangle\\ &= \begin{vmatrix} \mathbf{i} & \mathbf{j} & \mathbf{k}\\ 2\cos t & -2\sin t & 4\\ -2\sin t & -2\cos t & 0\end{vmatrix}\\ &= \langle 8\cos t, -8\sin t, -4\cos^2 t - 4\sin^2 t\rangle\\ &= \langle 8\cos t, -8\sin t, -4\rangle.\end{aligned}$$

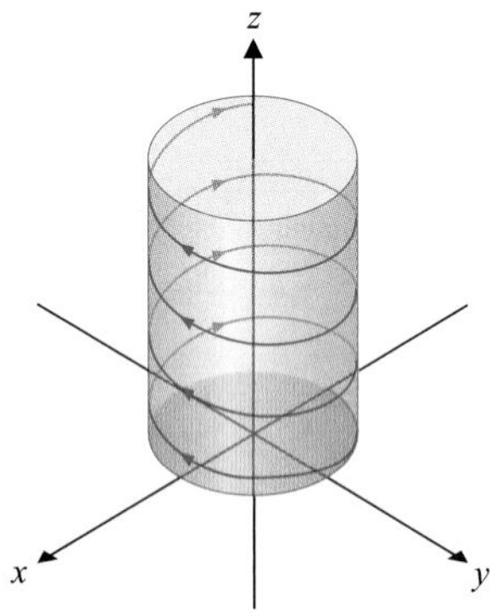

FIGURE 11.22
Circular helix

From (4.7), we get that the curvature is

$$\begin{aligned}\kappa &= \frac{\|\mathbf{r}'(t)\times\mathbf{r}''(t)\|}{\|\mathbf{r}'(t)\|^3}\\ &= \frac{\|\langle 8\cos t, -8\sin t, -4\rangle\|}{\|\langle 2\cos t, -2\sin t, 4\rangle\|^3} = \frac{\sqrt{80}}{\left(\sqrt{20}\right)^3} = \frac{1}{10}.\end{aligned}$$

Note that this says that the helix has a constant curvature, as you should suspect from the graph in Figure 11.22. ■

In the case of a plane curve that is the graph of a function, $y = f(x)$, we can derive a particularly simple formula for the curvature. Notice that such a curve is traced out by the vector-valued function $\mathbf{r}(t) = \langle t, f(t), 0\rangle$, where the third component is 0, since the curve lies completely in the xy-plane. Further, $\mathbf{r}'(t) = \langle 1, f'(t), 0\rangle$ and $\mathbf{r}''(t) = \langle 0, f''(t), 0\rangle$. From (4.7), we have

$$\begin{aligned}\kappa &= \frac{\|\mathbf{r}'(t)\times\mathbf{r}''(t)\|}{\|\mathbf{r}'(t)\|^3} = \frac{\|\langle 1, f'(t), 0\rangle \times \langle 0, f''(t), 0\rangle\|}{\|\langle 1, f'(t), 0\rangle\|^3}\\ &= \frac{|f''(t)|}{\{1+[f'(t)]^2\}^{3/2}},\end{aligned}$$

where we have left the calculation of the cross product as a simple exercise. Since the parameter $t = x$, we can write the curvature as

Curvature for the plane curve $y = f(x)$

$$\kappa = \frac{|f''(x)|}{\{1+[f'(x)]^2\}^{3/2}}. \tag{4.8}$$

EXAMPLE 4.6 Finding the Curvature of a Parabola

Find the curvature of the parabola $y = ax^2 + bx + c$. Also, find the limiting value of the curvature as $x \to \infty$.

Solution Taking $f(x) = ax^2 + bx + c$, we have that $f'(x) = 2ax + b$ and $f''(x) = 2a$. From (4.8), we have that

$$\kappa = \frac{|2a|}{[1 + (2ax + b)^2]^{3/2}}.$$

Taking the limit as $x \to \infty$, we have

$$\lim_{x \to \infty} \kappa = \lim_{x \to \infty} \frac{|2a|}{[1 + (2ax + b)^2]^{3/2}} = 0.$$

In other words, as $x \to \infty$, the parabola straightens out. You've certainly observed this in the graphs of parabolas for some time. Now, we have verified that this is not some sort of optical illusion; it's reality. It is a straightforward exercise to show that the maximum curvature occurs at the vertex of the parabola ($x = -b/2a$). ■

BEYOND FORMULAS

You can think of curvature as being loosely related to concavity, although there are important differences. The precise relationship for curves of the form $y = f(x)$ is given in equation (4.8). Curvature applies to curves in any dimension, whereas concavity applies only to two dimensions. More importantly, curvature measures the amount of curving as you move along the curve, regardless of where the curve goes. Concavity measures curving as you move along the x-axis, not the curve and thus requires that the curve correspond to a function $f(x)$. Given the generality of the curvature measurement, the formulas derived in this section are actually remarkably simple.

EXERCISES 11.4

WRITING EXERCISES

1. Explain what it means for a curve to have zero curvature (a) at a point and (b) on an interval of t-values.

2. Throughout our study of calculus, we have looked at tangent line approximations to curves. Some tangent lines approximate a curve well over a fairly lengthy interval while some stay close to a curve for only very short intervals. If the curvature at $x = a$ is large, would you expect the tangent line at $x = a$ to approximate the curve well over a lengthy interval or a short interval? What if the curvature is small? Explain.

3. Discuss the relationship between curvature and concavity for a function $y = f(x)$.

4. Explain why the curvature $\kappa = \frac{1}{10}$ of the helix in example 4.5 is less than the curvature of the circle $\langle 2\sin t, 2\cos t\rangle$ in two dimensions.

In exercises 1–4, find an arc length parameterization of the given two-dimensional curve.

1. The circle of radius 2 centered at the origin

2. The circle of radius 5 centered at the origin

3. The line segment from the origin to the point (3, 4)

4. The line segment from (1, 2) to the point (5, −2)

In exercises 5–10, find the unit tangent vector to the curve at the indicated points.

5. $\mathbf{r}(t) = \langle 3t, t^2\rangle, t = 0, t = -1, t = 1$

6. $\mathbf{r}(t) = \langle 2t^3, \sqrt{t}\rangle, t = 1, t = 2, t = 3$

7. $\mathbf{r}(t) = \langle 3\cos t, 2\sin t\rangle, t = 0, t = -\frac{\pi}{2}, t = \frac{\pi}{2}$

8. $\mathbf{r}(t) = \langle 4\sin t, 2\cos t\rangle, t = -\pi, t = 0, t = \pi$

9. $\mathbf{r}(t) = \langle 3t, \cos 2t, \sin 2t\rangle, t = 0, t = -\pi, t = \pi$

10. $\mathbf{r}(t) = \langle 4t, 2t, t^2\rangle, t = -1, t = 0, t = 1$

11. Sketch the curve in exercise 7 along with the vectors $\mathbf{r}(0)$, $\mathbf{T}(0)$, $\mathbf{r}\left(\frac{\pi}{2}\right)$ and $\mathbf{T}\left(\frac{\pi}{2}\right)$.

12. Sketch the curve in exercise 8 along with the vectors $\mathbf{r}(0)$, $\mathbf{T}(0)$, $\mathbf{r}\left(\frac{\pi}{2}\right)$ and $\mathbf{T}\left(\frac{\pi}{2}\right)$.

13. Sketch the curve in exercise 9 along with the vectors $\mathbf{r}(0)$, $\mathbf{T}(0)$, $\mathbf{r}(\pi)$ and $\mathbf{T}(\pi)$.

14. Sketch the curve in exercise 10 along with the vectors $\mathbf{r}(0)$, $\mathbf{T}(0)$, $\mathbf{r}(1)$ and $\mathbf{T}(1)$.

In exercises 15–22, find the curvature at the given point.

15. $\mathbf{r}(t) = \langle e^{-2t}, 2t, 4\rangle, t = 0$

16. $\mathbf{r}(t) = \langle 2, \sin \pi t, \ln t\rangle, t = 1$

17. $\mathbf{r}(t) = \langle t, \sin 2t, 3t\rangle, t = 0$

18. $\mathbf{r}(t) = \langle t, t^2 + t - 1, t\rangle, t = 0$

19. $f(x) = 3x^2 - 1, x = 1$

20. $f(x) = x^3 + 2x - 1, x = 2$

21. $f(x) = \sin x, x = \frac{\pi}{2}$

22. $f(x) = e^{-3x}, x = 0$

23. For $f(x) = \sin x$, (see exercise 21), show that the curvature is the same at $x = \frac{\pi}{2}$ and $x = \frac{3\pi}{2}$. Use the graph of $y = \sin x$ to predict whether the curvature would be larger or smaller at $x = \pi$.

24. For $f(x) = e^{-3x}$ (see exercise 22), show that the curvature is larger at $x = 0$ than at $x = 2$. Use the graph of $y = e^{-3x}$ to predict whether the curvature would be larger or smaller at $x = 4$.

In exercises 25–28, sketch the curve and compute the curvature at the indicated points.

25. $\mathbf{r}(t) = \langle 2\cos 2t, 2\sin 2t, 3t\rangle, t = 0, t = \frac{\pi}{2}$

26. $\mathbf{r}(t) = \langle \cos 2t, 2\sin 2t, 4t\rangle, t = 0, t = \frac{\pi}{2}$

27. $\mathbf{r}(t) = \langle t, t, t^2 - 1\rangle, t = 0, t = 2$

28. $\mathbf{r}(t) = \langle 2t - 1, t + 2, t - 3\rangle, t = 0, t = 2$

In exercises 29–32, sketch the curve and find any points of maximum or minimum curvature.

29. $\mathbf{r}(t) = \langle 2\cos t, 3\sin t\rangle$

30. $\mathbf{r}(t) = \langle 4\cos t, 3\sin t\rangle$

31. $y = 4x^2 - 3$

32. $y = \sin x$

In exercises 33–36, graph the curvature function $\kappa(x)$ and find the limit of the curvature as $x \to \infty$.

33. $y = e^{2x}$

34. $y = e^{-2x}$

35. $y = x^3$

36. $y = \sqrt{x}$

37. Explain how the answers to exercises 33–36 relate to the graphs.

38. Find the curvature of the circular helix $\langle a\cos t, a\sin t, bt\rangle$.

39. Label as true or false and explain: at a relative extremum of $y = f(x)$, the curvature is either a minimum or maximum.

40. Label as true or false and explain: at an inflection point of $y = f(x)$, the curvature is zero.

41. Label as true or false and explain: the curvature of the two-dimensional curve $y = f(x)$ is the same as the curvature of the three-dimensional curve $\mathbf{r}(t) = \langle t, f(t), c\rangle$ for any constant c.

42. Label as true or false and explain: the curvature of the two-dimensional curve $y = f(x)$ is the same as the curvature of the three-dimensional curve $\mathbf{r}(t) = \langle t, f(t), t\rangle$.

43. Show that the curvature of the polar curve $r = f(\theta)$ is given by

$$\kappa = \frac{|2[f'(\theta)]^2 - f(\theta)f''(\theta) + [f(\theta)]^2|}{\{[f'(\theta)]^2 + [f(\theta)]^2\}^{3/2}}.$$

44. If $f(0) = 0$, show that the curvature of the polar curve $r = f(\theta)$ at $\theta = 0$ is given by $\kappa = \dfrac{2}{|f'(0)|}$.

In exercises 45–48, use exercises 43 and 44 to find the curvature of the polar curve at the indicated points.

45. $r = \sin 3\theta, \theta = 0, \theta = \frac{\pi}{6}$

46. $r = 3 + 2\cos\theta, \theta = 0, \theta = \frac{\pi}{2}$

47. $r = 3e^{2\theta}, \theta = 0, \theta = 1$

48. $r = 1 - 2\sin\theta, \theta = 0, \theta = \frac{\pi}{2}$

49. Find the curvature of the helix traced out by $\mathbf{r}(t) = \langle 2\sin t, 2\cos t, 0.4t\rangle$ and compare to the result of example 4.5.

50. Find the limit as $n \to 0$ of the curvature of $\mathbf{r}(t) = \langle 2\sin t, 2\cos t, nt\rangle$ for $n > 0$. Explain this result graphically.

51. The cycloid is a curve with parametric equations $x = t - \sin t$, $y = 1 - \cos t$. Show that the curvature of the cycloid equals $\frac{1}{\sqrt{8y}}$, for $y \neq 0$.

52. Find the curvature of $\mathbf{r}(t) = \langle \cosh t, \sinh t, 2\rangle$. Find the limit of the curvature as $t \to \infty$. Use a property of hyperbolas to explain this result.

53. For the logarithmic spiral $r = ae^{b\theta}$, show that the curvature equals $\kappa = \dfrac{e^{-b\theta}}{a\sqrt{1+b^2}}$. Show that as $b \to 0$, the spiral approaches a circle.

EXPLORATORY EXERCISES

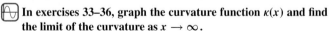

1. In this exercise, we explore an unusual two-dimensional parametric curve sometimes known as the **Cornu spiral.** Define the vector-valued function $\mathbf{r}(t) = \left\langle \int_0^t \cos\left(\frac{\pi u^2}{2}\right) du, \int_0^t \sin\left(\frac{\pi u^2}{2}\right) du \right\rangle$. Use a graphing utility to sketch the graph of $\mathbf{r}(t)$ for $-\pi \le t \le \pi$. Compute the arc length of the curve from $t = 0$ to $t = c$ and compute the curvature at $t = c$. What is the remarkable property that you find?

2. Assume that $f(x)$ has three continuous derivatives. Prove that at a local minimum of $y = f(x)$, $\kappa = f''(x)$ and $\kappa'(x) = f'''(x)$. Prove that at a local maximum of $y = f(x)$, $\kappa = -f''(x)$ and $\kappa'(x) = -f'''(x)$.

11.5 TANGENT AND NORMAL VECTORS

Up to this point, we have used a single frame of reference for all of our work with vectors. That is, we have written all vectors in terms of the standard unit basis vectors **i**, **j** and **k**. However, this is not always the most convenient framework for describing vectors. For instance, such a fixed frame of reference would be particularly inconvenient when investigating the forces acting on an aircraft as it flies across the sky. A much better frame of reference would be one that moves along with the aircraft. As it turns out, such a moving frame of reference sheds light on a wide variety of problems. In this section, we will construct this moving reference frame and see how it immediately provides useful information regarding the forces acting on an object in motion.

Consider an object moving along a smooth curve traced out by the vector-valued function $\mathbf{r}(t) = \langle f(t), g(t), h(t)\rangle$. If we are to define a reference frame that moves with the object, we will need to have (at each point on the curve) three mutually orthogonal unit vectors. One of these should point in the direction of motion (i.e., in the direction of the orientation of the curve). In section 11.4, we defined the unit tangent vector $\mathbf{T}(t)$ by

$$\mathbf{T}(t) = \frac{\mathbf{r}'(t)}{\|\mathbf{r}'(t)\|}.$$

Further, recall from Theorem 2.4 that since $\mathbf{T}(t)$ is a unit vector (and consequently has the constant magnitude of 1), $\mathbf{T}(t)$ must be orthogonal to $\mathbf{T}'(t)$ for each t. This gives us a second unit vector in our moving frame of reference, as follows.

DEFINITION 5.1

The **principal unit normal vector** $\mathbf{N}(t)$ is a unit vector having the same direction as $\mathbf{T}'(t)$ and is defined by

$$\mathbf{N}(t) = \frac{\mathbf{T}'(t)}{\|\mathbf{T}'(t)\|}. \tag{5.1}$$

T(t) is unit vector does not mean T'(t) is a unit vector

You might wonder about the direction in which $\mathbf{N}(t)$ points. Simply saying that it's orthogonal to $\mathbf{T}(t)$ is not quite enough. After all, in three dimensions, there are infinitely many directions that are orthogonal to $\mathbf{T}(t)$. (In two dimensions, there are only two possible directions.) We can clarify this with the following observation.

Recall that from (4.5), we have that $\frac{ds}{dt} = \|\mathbf{r}'(t)\| > 0$. (This followed from the definition of the arc length parameter in (4.1).) In particular, this says that $\left|\frac{ds}{dt}\right| = \frac{ds}{dt}$. From the chain rule, we have

$$\mathbf{T}'(t) = \frac{d\mathbf{T}}{dt} = \frac{d\mathbf{T}}{ds}\frac{ds}{dt}.$$

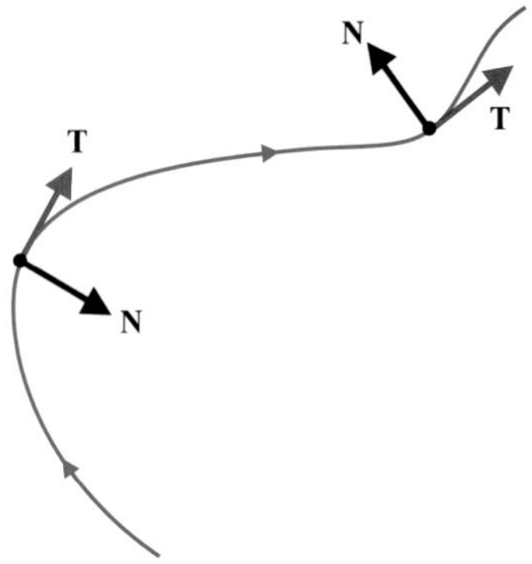

FIGURE 11.23
Principal unit normal vectors

This gives us

$$\mathbf{N}(t) = \frac{\mathbf{T}'(t)}{\|\mathbf{T}'(t)\|} = \frac{\frac{d\mathbf{T}}{ds}\frac{ds}{dt}}{\left\|\frac{d\mathbf{T}}{ds}\right\| \left|\frac{ds}{dt}\right|} = \frac{\frac{d\mathbf{T}}{ds}}{\left\|\frac{d\mathbf{T}}{ds}\right\|}$$

or equivalently,

$$\mathbf{N}(t) = \frac{1}{\kappa}\frac{d\mathbf{T}}{ds}, \tag{5.2}$$

where we have used the definition of curvature in (4.3), $\kappa = \left\|\frac{d\mathbf{T}}{ds}\right\|$.

Although (5.2) is not particularly useful as a formula for computing $\mathbf{N}(t)$ (why not?), we can use it to interpret the meaning of $\mathbf{N}(t)$. Since $\kappa > 0$, in order for (5.2) to make sense, $\mathbf{N}(t)$ will have the same direction as $\frac{d\mathbf{T}}{ds}$. Note that $\frac{d\mathbf{T}}{ds}$ is the instantaneous rate of change of the unit tangent vector with respect to arc length. This says that $\frac{d\mathbf{T}}{ds}$ (and consequently also, $\mathbf{N}$) points in the direction in which $\mathbf{T}$ is turning as arc length increases. That is, $\mathbf{N}(t)$ will always point to the *concave* side of the curve (see Figure 11.23).

EXAMPLE 5.1 Finding Unit Tangent and Principal Unit Normal Vectors

Find the unit tangent and principal unit normal vectors to the curve defined by $\mathbf{r}(t) = \langle t^2, t\rangle$.

Solution Notice that $\mathbf{r}'(t) = \langle 2t, 1\rangle$ and so from (4.2), we have

$$\mathbf{T}(t) = \frac{\mathbf{r}'(t)}{\|\mathbf{r}'(t)\|} = \frac{\langle 2t, 1\rangle}{\|\langle 2t, 1\rangle\|} = \frac{\langle 2t, 1\rangle}{\sqrt{4t^2+1}} = \frac{2t}{\sqrt{4t^2+1}}\mathbf{i} + \frac{1}{\sqrt{4t^2+1}}\mathbf{j}.$$

Using the quotient rule, we have

$$\begin{aligned}\mathbf{T}'(t) &= \frac{2\sqrt{4t^2+1} - 2t\left(\frac{1}{2}\right)(4t^2+1)^{-1/2}(8t)}{4t^2+1}\mathbf{i} - \frac{1}{2}(4t^2+1)^{-3/2}(8t)\mathbf{j} \\ &= 2(4t^2+1)^{-1/2}\frac{(4t^2+1)-4t^2}{4t^2+1}\mathbf{i} - (4t^2+1)^{-3/2}(4t)\mathbf{j} \\ &= 2(4t^2+1)^{-3/2}\langle 1, -2t\rangle.\end{aligned}$$

Further,

$$\begin{aligned}\|\mathbf{T}'(t)\| &= 2(4t^2+1)^{-3/2}\|\langle 1, -2t\rangle\| \\ &= 2(4t^2+1)^{-3/2}\sqrt{1+4t^2} = 2(4t^2+1)^{-1}.\end{aligned}$$

From (5.1), the principal unit normal is then

$$\begin{aligned}\mathbf{N}(t) &= \frac{\mathbf{T}'(t)}{\|\mathbf{T}'(t)\|} = \frac{2(4t^2+1)^{-3/2}\langle 1, -2t\rangle}{2(4t^2+1)^{-1}} \\ &= (4t^2+1)^{-1/2}\langle 1, -2t\rangle.\end{aligned}$$

In particular, for $t = 1$, we get $\mathbf{T}(1) = \left\langle \frac{2}{\sqrt{5}}, \frac{1}{\sqrt{5}}\right\rangle$ and $\mathbf{N}(1) = \left\langle \frac{1}{\sqrt{5}}, -\frac{2}{\sqrt{5}}\right\rangle$. We sketch the curve and these two sample vectors in Figure 11.24. ■

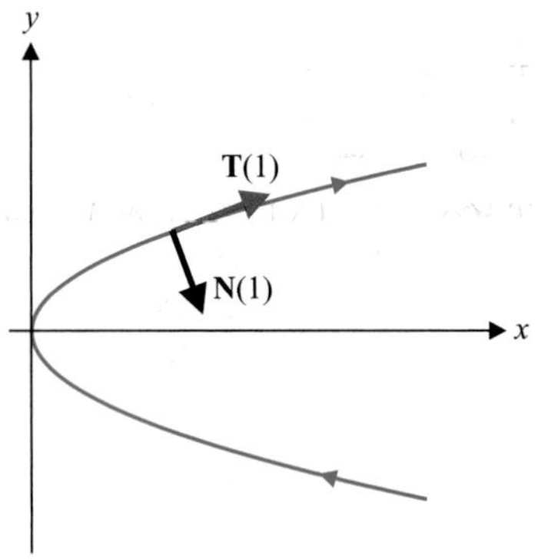

FIGURE 11.24
Unit tangent and principal unit normal vectors

The calculations are similar in three dimensions, as we illustrate in example 5.2.

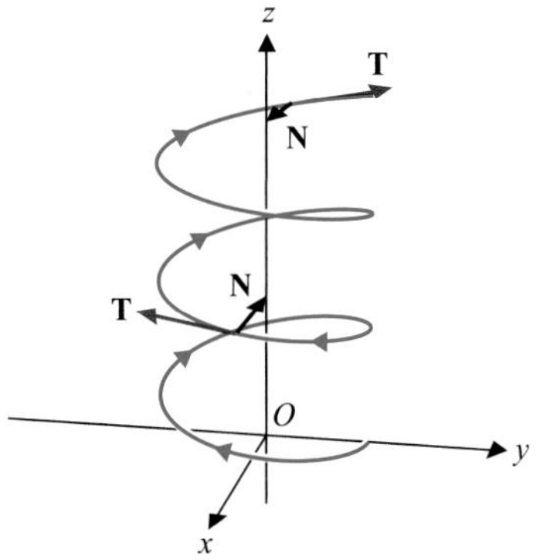

FIGURE 11.25
Unit tangent and principal unit normal vectors

EXAMPLE 5.2 Finding Unit Tangent and Principal Unit Normal Vectors

Find the unit tangent and principal unit normal vectors to the curve determined by $\mathbf{r}(t) = \langle \sin 2t, \cos 2t, t \rangle$.

Solution First, observe that $\mathbf{r}'(t) = \langle 2\cos 2t, -2\sin 2t, 1 \rangle$ and so, we have from (4.2) that

$$\mathbf{T}(t) = \frac{\mathbf{r}'(t)}{\|\mathbf{r}'(t)\|} = \frac{\langle 2\cos 2t, -2\sin 2t, 1\rangle}{\|\langle 2\cos 2t, -2\sin 2t, 1\rangle\|} = \frac{1}{\sqrt{5}}\langle 2\cos 2t, -2\sin 2t, 1\rangle.$$

This gives us
$$\mathbf{T}'(t) = \frac{1}{\sqrt{5}}\langle -4\sin 2t, -4\cos 2t, 0\rangle$$

and so, from (5.1), the principal unit normal is

$$\mathbf{N}(t) = \frac{\mathbf{T}'(t)}{\|\mathbf{T}'(t)\|} = \frac{1}{4}\langle -4\sin 2t, -4\cos 2t, 0\rangle = \langle -\sin 2t, -\cos 2t, 0\rangle.$$

Notice that the curve here is a circular helix and that at each point, $\mathbf{N}(t)$ points straight back toward the z-axis (see Figure 11.25). ■

To get a third unit vector orthogonal to both $\mathbf{T}(t)$ and $\mathbf{N}(t)$, we simply take their cross product.

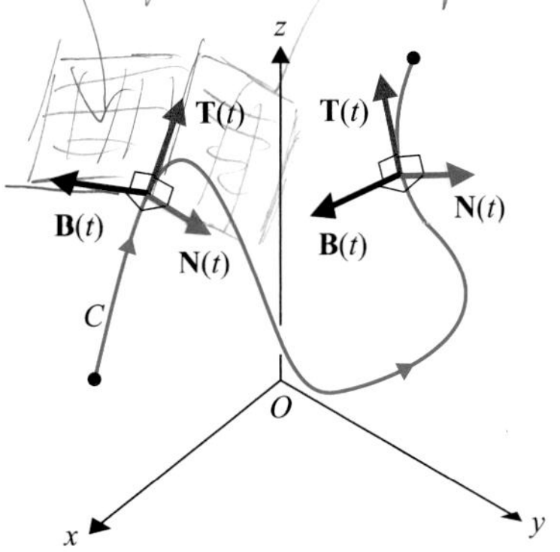

FIGURE 11.26
The **TNB** frame

DEFINITION 5.2

We define the **binormal** vector $\mathbf{B}(t)$ to be

$$\mathbf{B}(t) = \mathbf{T}(t) \times \mathbf{N}(t).$$

Notice that by definition, $\mathbf{B}(t)$ is orthogonal to both $\mathbf{T}(t)$ and $\mathbf{N}(t)$ and by Theorem 4.4 in Chapter 10, its magnitude is given by

$$\|\mathbf{B}(t)\| = \|\mathbf{T}(t) \times \mathbf{N}(t)\| = \|\mathbf{T}(t)\|\,\|\mathbf{N}(t)\| \sin\theta,$$

where θ is the angle between $\mathbf{T}(t)$ and $\mathbf{N}(t)$. However, since $\mathbf{T}(t)$ and $\mathbf{N}(t)$ are both unit vectors, $\|\mathbf{T}(t)\| = \|\mathbf{N}(t)\| = 1$. Further, $\mathbf{T}(t)$ and $\mathbf{N}(t)$ are orthogonal, so that $\sin\theta = 1$ and consequently, $\|\mathbf{B}(t)\| = 1$, too. This triple of three unit vectors $\mathbf{T}(t)$, $\mathbf{N}(t)$ and $\mathbf{B}(t)$ forms a frame of reference, called the **TNB frame** (or the **moving trihedral**), that moves along the curve defined by $\mathbf{r}(t)$ (see Figure 11.26). This has particular importance in a branch of mathematics called *differential geometry* and is used in the navigation of spacecraft.

As you can see, the definition of the binormal vector is certainly straightforward. We illustrate this now for the curve from example 5.2.

EXAMPLE 5.3 Finding the Binormal Vector

Find the binormal vector $\mathbf{B}(t)$ for the curve traced out by $\mathbf{r}(t) = \langle \sin 2t, \cos 2t, t\rangle$.

Solution Recall from example 5.2 that the unit tangent vector is given by $\mathbf{T}(t) = \frac{1}{\sqrt{5}}\langle 2\cos 2t, -2\sin 2t, 1\rangle$ and the principal unit normal vector is given by

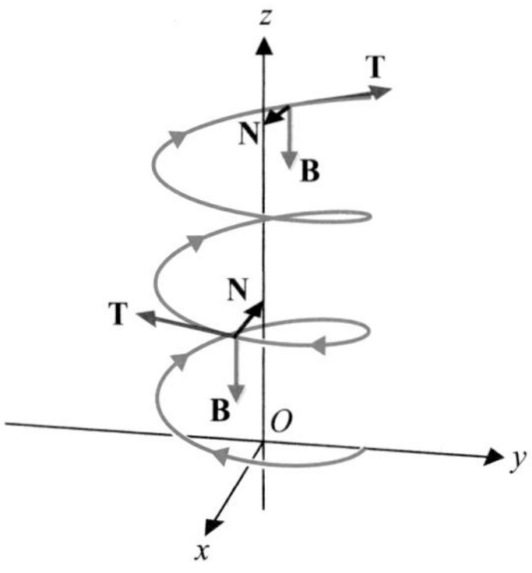

FIGURE 11.27
The **TNB** frame for $\mathbf{r}(t) = \langle \sin 2t, \cos 2t, t \rangle$

$\mathbf{N}(t) = \langle -\sin 2t, -\cos 2t, 0 \rangle$. The binormal vector is then

$$\mathbf{B}(t) = \mathbf{T}(t) \times \mathbf{N}(t) = \frac{1}{\sqrt{5}} \langle 2\cos 2t, -2\sin 2t, 1 \rangle \times \langle -\sin 2t, -\cos 2t, 0 \rangle$$

$$= \frac{1}{\sqrt{5}} \begin{vmatrix} \mathbf{i} & \mathbf{j} & \mathbf{k} \\ 2\cos 2t & -2\sin 2t & 1 \\ -\sin 2t & -\cos 2t & 0 \end{vmatrix}$$

$$= \frac{1}{\sqrt{5}} [\mathbf{i}(\cos 2t) - \mathbf{j}(\sin 2t) + \mathbf{k}(-2\cos^2 2t - 2\sin^2 2t)]$$

$$= \frac{1}{\sqrt{5}} \langle \cos 2t, -\sin 2t, -2 \rangle.$$

We illustrate the **TNB** frame for this curve in Figure 11.27. ■

For each point on a curve, the plane passing through that point and determined by $\mathbf{N}(t)$ and $\mathbf{B}(t)$ is called the **normal plane.** Accordingly, the normal plane to a curve at a given point contains all of the lines that are orthogonal to the tangent vector at that point. For each point on a curve, the plane determined by $\mathbf{T}(t)$ and $\mathbf{N}(t)$ is called the **osculating plane.** For a two-dimensional curve, the osculating plane is simply the xy-plane.

For a given value of t, say $t = t_0$, if the curvature κ of the curve at the point P corresponding to t_0 is nonzero, then the circle of radius $\rho = \frac{1}{\kappa}$ lying completely in the osculating plane and whose center lies a distance of $\frac{1}{\kappa}$ from P along the normal $\mathbf{N}(t)$ is called the **osculating circle** (or the **circle of curvature**). Recall from example 4.4 that the curvature of a circle is the reciprocal of its radius. This says that the osculating circle has the same tangent and curvature at P as the curve. Further, since the normal vector always points to the concave side of the curve, the osculating circle lies on the concave side of the curve. In this sense, then, the osculating circle is the circle that "best fits" the curve at the point P (see Figure 11.28). The radius of the osculating circle is called the **radius of curvature** and the center of the circle is called the **center of curvature.**

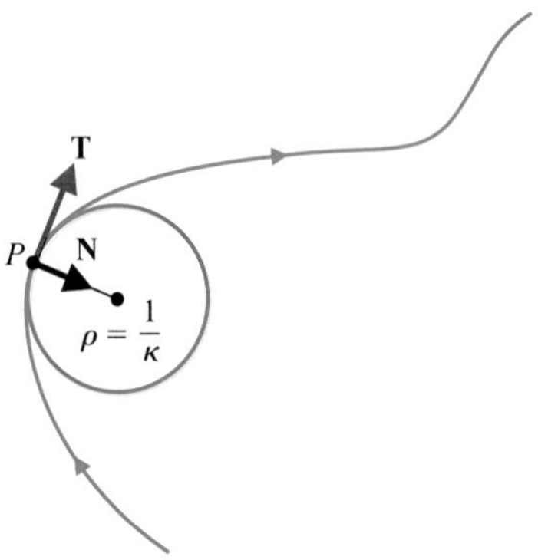

FIGURE 11.28
Osculating circle

EXAMPLE 5.4 Finding the Osculating Circle

Find the osculating circle for the parabola defined by $\mathbf{r}(t) = \langle t^2, t \rangle$ at $t = 0$.

Solution In example 5.1, we found that the unit tangent vector is

$$\mathbf{T}(t) = (4t^2 + 1)^{-1/2} \langle 2t, 1 \rangle,$$
$$\mathbf{T}'(t) = 2(4t^2 + 1)^{-3/2} \langle 1, -2t \rangle$$

and the principal unit normal is

$$\mathbf{N}(t) = (4t^2 + 1)^{-1/2} \langle 1, -2t \rangle.$$

So, from (4.6), the curvature is given by

$$\kappa(t) = \frac{\|\mathbf{T}'(t)\|}{\|\mathbf{r}'(t)\|}$$
$$= \frac{2(4t^2 + 1)^{-3/2}(1 + 4t^2)^{1/2}}{(4t^2 + 1)^{1/2}} = 2(4t^2 + 1)^{-3/2}.$$

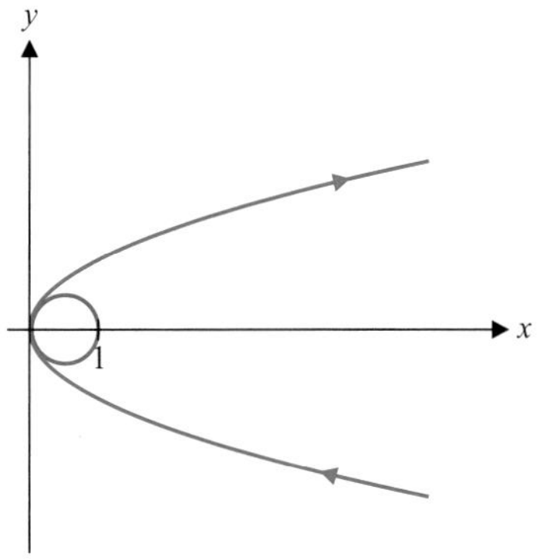

FIGURE 11.29
Osculating circle

We now have $\kappa(0) = 2$, so that the radius of curvature for $t = 0$ is $\rho = \frac{1}{\kappa} = \frac{1}{2}$. Further, $\mathbf{N}(0) = \langle 1, 0 \rangle$ and $\mathbf{r}(0) = \langle 0, 0 \rangle$, so that the center of curvature is located $\rho = \frac{1}{2}$ unit from the origin in the direction of $\mathbf{N}(0)$ (i.e., along the positive x-axis). We draw the curve and the osculating circle in Figure 11.29. ■

○ Tangential and Normal Components of Acceleration

Now that we have defined the unit tangent and principal unit normal vectors, we can make a remarkable observation about the motion of an object. In particular, we'll see how this observation helps to explain the behavior of an automobile as it travels along a curved stretch of road.

Suppose that an object moves along a smooth curve traced out by the vector-valued function $\mathbf{r}(t)$, where t represents time. Recall from the definition of the unit tangent vector that $\mathbf{T}(t) = \frac{\mathbf{r}'(t)}{\|\mathbf{r}'(t)\|}$ and from (4.5), $\|\mathbf{r}'(t)\| = \frac{ds}{dt}$, where s represents arc length. The velocity of the object is then given by

$$\mathbf{v}(t) = \mathbf{r}'(t) = \|\mathbf{r}'(t)\|\mathbf{T}(t) = \frac{ds}{dt}\mathbf{T}(t).$$

Using the product rule [Theorem 2.3 (iii)], the acceleration is given by

$$\mathbf{a}(t) = \mathbf{v}'(t) = \frac{d}{dt}\left(\frac{ds}{dt}\mathbf{T}(t)\right) = \frac{d^2s}{dt^2}\mathbf{T}(t) + \frac{ds}{dt}\mathbf{T}'(t). \tag{5.3}$$

Recall that we had defined the principal unit normal by $\mathbf{N}(t) = \frac{\mathbf{T}'(t)}{\|\mathbf{T}'(t)\|}$, so that

$$\mathbf{T}'(t) = \|\mathbf{T}'(t)\|\mathbf{N}(t). \tag{5.4}$$

Further, by the chain rule,

$$\|\mathbf{T}'(t)\| = \left\|\frac{d\mathbf{T}}{dt}\right\| = \left\|\frac{d\mathbf{T}}{ds}\frac{ds}{dt}\right\|$$
$$= \left|\frac{ds}{dt}\right|\left\|\frac{d\mathbf{T}}{ds}\right\| = \kappa\frac{ds}{dt}, \tag{5.5}$$

where we have also used the definition of the curvature κ given in (4.3) and the fact that $\frac{ds}{dt} > 0$. Putting together (5.4) and (5.5), we now have that

$$\mathbf{T}'(t) = \|\mathbf{T}'(t)\|\,\mathbf{N}(t) = \kappa\frac{ds}{dt}\mathbf{N}(t).$$

Using this together with (5.3), we now get

$$\boxed{\mathbf{a}(t) = \frac{d^2s}{dt^2}\mathbf{T}(t) + \kappa\left(\frac{ds}{dt}\right)^2\mathbf{N}(t).} \tag{5.6}$$

Equation (5.6) provides us with a surprising wealth of insight into the motion of an object. First, notice that since $\mathbf{a}(t)$ is written as a sum of a vector parallel to $\mathbf{T}(t)$ and a vector parallel to $\mathbf{N}(t)$, the acceleration vector always lies in the plane determined by $\mathbf{T}(t)$ and $\mathbf{N}(t)$ (i.e., the osculating plane). In particular, this says that the acceleration is always orthogonal to the binormal $\mathbf{B}(t)$. We call the coefficient of $\mathbf{T}(t)$ in (5.6) the **tangential component of**

TODAY IN MATHEMATICS

Edward Witten (1951–)
An American theoretical physicist who is one of the world's experts in string theory. He earned the Fields Medal in 1990 for his contributions to mathematics. Michael Atiyah, a mathematics colleague, wrote, "Although he is definitely a physicist (as his list of publications clearly shows), his command of mathematics is rivalled by few mathematicians, and his ability to interpret physical ideas in mathematical form is quite unique. Time and again he has surprised the mathematical community by his brilliant application of physical insight leading to new and deep mathematical theorems." In addition, Atiyah wrote, "In his hands, physics is once again providing a rich source of inspiration and insight in mathematics."

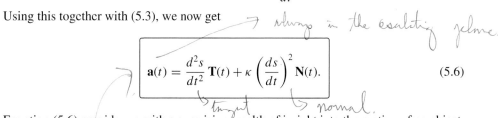

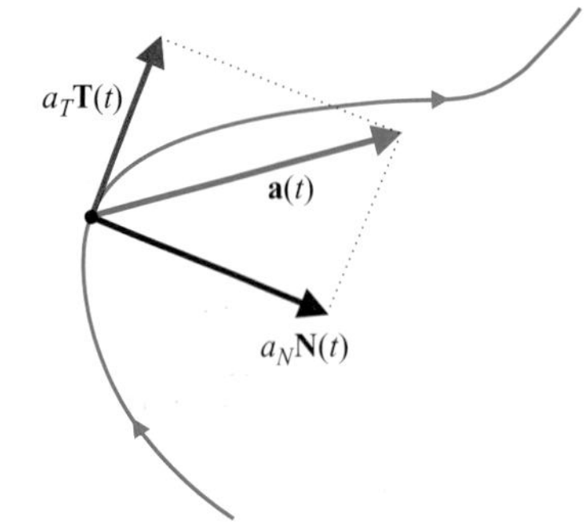

FIGURE 11.30
Tangential and normal components of acceleration

FIGURE 11.31
Driving around a curve

acceleration a_T and the coefficient of $\mathbf{N}(t)$ the **normal component of acceleration** a_N. That is,

$$a_T = \frac{d^2s}{dt^2} \quad \text{and} \quad a_N = \kappa \left(\frac{ds}{dt}\right)^2. \tag{5.7}$$

See Figure 11.30 for a graphical depiction of this decomposition of $\mathbf{a}(t)$ into tangential and normal components.

We now discuss (5.6) in the familiar context of a car driving around a curve in the road (see Figure 11.31). From Newton's second law of motion, the net force acting on the car at any given time t is $\mathbf{F}(t) = m\mathbf{a}(t)$, where m is the mass of the car. From (5.6), we have

$$\mathbf{F}(t) = m\mathbf{a}(t) = m\frac{d^2s}{dt^2}\,\mathbf{T}(t) + m\kappa\left(\frac{ds}{dt}\right)^2 \mathbf{N}(t).$$

Since $\mathbf{T}(t)$ points in the direction of the path of motion, you want the component of the force acting in the direction of $\mathbf{T}(t)$ to be as large as possible compared to the component of the force acting in the direction of the normal $\mathbf{N}(t)$. (If the normal component of the force is too large, it may exceed the normal component of the force of friction between the tires and the highway, causing the car to skid off the road.) Notice that the only way to minimize the force applied in this direction is to make $\left(\frac{ds}{dt}\right)^2$ small, where $\frac{ds}{dt}$ is the speed. So, reducing speed is the only way to reduce the normal component of the force. To have a larger tangential component of the force, you will need to have $\frac{d^2s}{dt^2}$ (the instantaneous rate of change of speed with respect to time) larger. So, to maximize the tangential component of the force, you need to be accelerating while in the curve. You have probably noticed advisory signs on the highway advising you to slow down *before* you enter a sharp curve. Notice that reducing speed (i.e., reducing $\frac{ds}{dt}$) before the curve and then gently accelerating (keeping $\frac{d^2s}{dt^2} > 0$) once you're in the curve keeps the resultant force $\mathbf{F}(t)$ pointing in the general direction you are moving. Alternatively, waiting until you're in the curve to slow down keeps $\frac{d^2s}{dt^2} < 0$, which makes $\frac{d^2s}{dt^2}\mathbf{T}(t)$ point in the *opposite* direction as $\mathbf{T}(t)$. The net force $\mathbf{F}(t)$ will then point away from the direction of motion (see Figure 11.32).

FIGURE 11.32
Net force: $\frac{d^2s}{dt^2} < 0$

EXAMPLE 5.5 Finding Tangential and Normal Components of Acceleration

Find the tangential and normal components of acceleration for an object with position vector $\mathbf{r}(t) = \langle 2\sin t, 2\cos t, 4t\rangle$.

Solution In example 4.5, we found that the curvature of this curve is $\kappa = \frac{1}{10}$. We also have $\mathbf{r}'(t) = \langle 2\cos t, -2\sin t, 4\rangle$, so that

$$\frac{ds}{dt} = \|\mathbf{r}'(t)\| = \sqrt{20}$$

and so, $\dfrac{d^2s}{dt^2} = 0$, for all t. From (5.6), we have that the acceleration is

$$\begin{aligned}\mathbf{a}(t) &= \frac{d^2s}{dt^2}\mathbf{T}(t) + \kappa\left(\frac{ds}{dt}\right)^2\mathbf{N}(t)\\ &= (0)\,\mathbf{T}(t) + \frac{1}{10}\left(\sqrt{20}\right)^2\mathbf{N}(t) = 2\mathbf{N}(t).\end{aligned}$$

So, here we have $a_T = 0$ and $a_N = 2$. ■

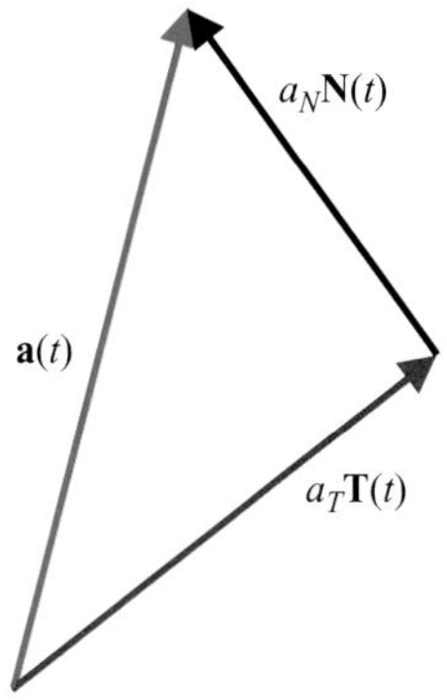

FIGURE 11.33
Components of $\mathbf{a}(t)$

Notice that it's reasonably simple to compute $a_T = \dfrac{d^2s}{dt^2}$. You must only calculate $\dfrac{ds}{dt} = \|\mathbf{r}'(t)\|$ and then differentiate the result. On the other hand, computing a_N is a bit more complicated, since it requires you to first compute the curvature κ. We can simplify the calculation of a_N with the following observation. From (5.6), we have

$$\mathbf{a}(t) = \frac{d^2s}{dt^2}\mathbf{T}(t) + \kappa\left(\frac{ds}{dt}\right)^2\mathbf{N}(t) = a_T\mathbf{T}(t) + a_N\mathbf{N}(t).$$

This says that $\mathbf{a}(t)$ is the vector resulting from adding the **orthogonal** vectors $a_T\mathbf{T}(t)$ and $a_N\mathbf{N}(t)$. (See Figure 11.33, where we have drawn the vectors so that the initial point of $a_N\mathbf{N}(t)$ is located at the terminal point of $a_T\mathbf{T}(t)$.) From the Pythagorean Theorem, we have that

$$\begin{aligned}\|\mathbf{a}(t)\|^2 &= \|a_T\mathbf{T}(t)\|^2 + \|a_N\mathbf{N}(t)\|^2\\ &= a_T^2 + a_N^2,\end{aligned}\tag{5.8}$$

since $\mathbf{T}(t)$ and $\mathbf{N}(t)$ are unit vectors (i.e., $\|\mathbf{T}(t)\| = \|\mathbf{N}(t)\| = 1$). Solving (5.8) for a_N, we get

$$a_N = \sqrt{\|\mathbf{a}(t)\|^2 - a_T^2},\tag{5.9}$$

where we have taken the positive root since $a_N = \kappa\left(\dfrac{ds}{dt}\right)^2 \geq 0$. Once you know $\mathbf{a}(t)$ and a_T, you can use (5.9) to quickly calculate a_N, without first computing the curvature. As an alternative, observe that a_T is the component of $\mathbf{a}(t)$ along the velocity vector $\mathbf{v}(t)$. Further, from (5.7) and (5.9), we can compute a_N and κ. This allows us to compute a_T, a_N and κ without first computing the derivative of the speed.

EXAMPLE 5.6 Finding Tangential and Normal Components of Acceleration

Find the tangential and normal components of acceleration for an object whose path is defined by $\mathbf{r}(t) = \langle t, 2t, t^2\rangle$. In particular, find these components at $t = 1$. Also, find the curvature.

Solution   First, we compute the velocity $\mathbf{v}(t) = \mathbf{r}'(t) = \langle 1, 2, 2t \rangle$ and the acceleration $\mathbf{a}(t) = \langle 0, 0, 2 \rangle$. This gives us

$$\frac{ds}{dt} = \|\mathbf{r}'(t)\| = \|\langle 1, 2, 2t \rangle\| = \sqrt{1^2 + 2^2 + (2t)^2} = \sqrt{5 + 4t^2}.$$

The tangential component of acceleration a_T is the component of $\mathbf{a}(t) = \langle 0, 0, 2 \rangle$ along $\mathbf{v}(t) = \langle 1, 2, 2t \rangle$:

$a_T = comp_{\vec{v}} \vec{a} = \dfrac{\vec{a} \cdot \vec{v}}{\|\vec{v}\|}$ $a_T = \langle 0, 0, 2 \rangle \cdot \dfrac{\langle 1, 2, 2t \rangle}{\sqrt{5 + 4t^2}} = \dfrac{4t}{\sqrt{5 + 4t^2}}.$

From (5.9), we have that the normal component of acceleration is

$$a_N = \sqrt{\|\mathbf{a}(t)\|^2 - a_T^2} = \sqrt{2^2 - \frac{16t^2}{5 + 4t^2}}$$

$$= \sqrt{\frac{4(5 + 4t^2) - 16t^2}{5 + 4t^2}} = \frac{\sqrt{20}}{\sqrt{5 + 4t^2}}.$$

Think about computing a_N from its definition in (5.7) and notice how much simpler it was to use (5.9). Further, at $t = 1$, we have

$$a_T = \frac{4}{3} \quad \text{and} \quad a_N = \frac{\sqrt{20}}{3}.$$

Finally, from (5.7), the curvature is

$$\kappa = \frac{a_N}{\left(\dfrac{ds}{dt}\right)^2} = \frac{\sqrt{20}}{\sqrt{5 + 4t^2}} \frac{1}{\left(\sqrt{5 + 4t^2}\right)^2}$$

$$= \frac{\sqrt{20}}{(5 + 4t^2)^{3/2}}.$$

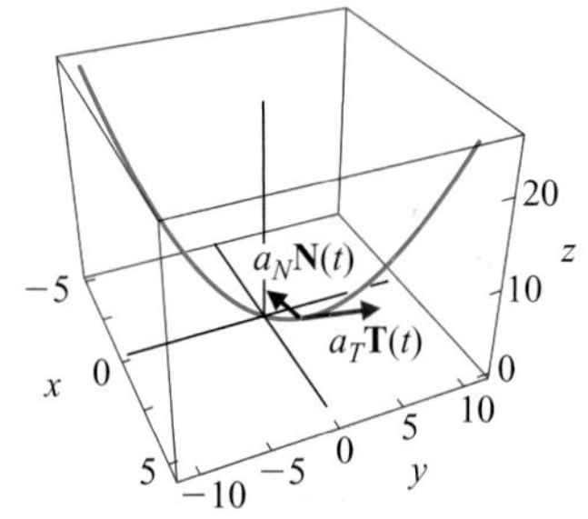

FIGURE 11.34

Tangential and normal components of acceleration at $t = 1$

Notice how easy it was to compute the curvature in this way. In Figure 11.34, we show a plot of the curve traced out by $\mathbf{r}(t)$, along with the tangential and normal components of acceleration at $t = 1$. ∎

Equation (5.6) has a wealth of applications. Among many others, it provides us with a relatively simple proof of Theorem 4.1, which we had deferred until now. You may recall that the result says that the curvature of a path traced out by the vector-valued function $\mathbf{r}(t)$ is given by

$$\kappa = \frac{\|\mathbf{r}'(t) \times \mathbf{r}''(t)\|}{\|\mathbf{r}'(t)\|^3}. \tag{5.10}$$

PROOF

From (5.6), we have

$$\mathbf{a}(t) = \frac{d^2 s}{dt^2} \mathbf{T}(t) + \kappa \left(\frac{ds}{dt}\right)^2 \mathbf{N}(t).$$

Taking the cross product of both sides of this equation with $\mathbf{T}(t)$ gives us

$$\mathbf{T}(t) \times \mathbf{a}(t) = \frac{d^2 s}{dt^2} \mathbf{T}(t) \times \mathbf{T}(t) + \kappa \left(\frac{ds}{dt}\right)^2 \mathbf{T}(t) \times \mathbf{N}(t)$$

$$= \kappa \left(\frac{ds}{dt}\right)^2 \mathbf{T}(t) \times \mathbf{N}(t),$$

since the cross product of any vector with itself is the zero vector. Taking the magnitude of both sides and recognizing that $\mathbf{T}(t) \times \mathbf{N}(t) = \mathbf{B}(t)$, we get

$$\begin{aligned}\|\mathbf{T}(t) \times \mathbf{a}(t)\| &= \kappa \left(\frac{ds}{dt}\right)^2 \|\mathbf{T}(t) \times \mathbf{N}(t)\| \\ &= \kappa \left(\frac{ds}{dt}\right)^2 \|\mathbf{B}(t)\| = \kappa \left(\frac{ds}{dt}\right)^2,\end{aligned}$$

since the binormal vector $\mathbf{B}(t)$ is a unit vector. Recalling that $\mathbf{T}(t) = \dfrac{\mathbf{r}'(t)}{\|\mathbf{r}'(t)\|}$, $\mathbf{a}(t) = \mathbf{r}''(t)$ and $\dfrac{ds}{dt} = \|\mathbf{r}'(t)\|$ gives us

$$\frac{\|\mathbf{r}'(t) \times \mathbf{r}''(t)\|}{\|\mathbf{r}'(t)\|} = \kappa \|\mathbf{r}'(t)\|^2.$$

Solving this for κ leaves us with (5.10), as desired. ■

○ Kepler's Laws

We are now in a position to present one of the most profound discoveries ever made by humankind. For hundreds of years, people believed that the Sun, the other stars and the planets all revolved around the Earth. The year 1543 saw the publication of the astronomer Copernicus' theory that the Earth and other planets, in fact, revolved around the Sun. Sixty years later, based on a very careful analysis of a massive number of astronomical observations, the German astronomer Johannes Kepler formulated three laws that he reasoned must be followed by every planet. We present these now.

KEPLER'S LAWS OF PLANETARY MOTION

1. Each planet follows an elliptical orbit, with the Sun at one focus.
2. The line segment joining the Sun to a planet sweeps out equal areas in equal times.
3. If T is the time required for a given planet to make one orbit of the Sun and if the length of the major axis of its elliptical orbit is $2a$, then $T^2 = ka^3$, for some constant k (i.e., T^2 is proportional to a^3).

Kepler's exhaustive analysis of the data changed our perception of our place in the universe. While Kepler's work was empirical in nature, Newton's approach to the same problem was not. In 1687, in his book *Principia Mathematica*, Newton showed how to use his calculus to derive Kepler's three laws from two of Newton's laws: his second law of motion and his law of universal gravitation. You should not underestimate the significance of this achievement. With this work, Newton shed light on some of the fundamental physical laws that govern our universe.

In order to simplify our analysis, we assume that we are looking at a solar system consisting of one sun and one planet. This is a reasonable assumption, since the gravitational attraction of the sun is far greater than that of any other body (planet, moon, comet, etc.), owing to the sun's far greater mass. (As it turns out, the gravitational attraction of other bodies does have an effect. In fact, it was an observation of the irregularities in the orbit of Uranus that led astronomers to hypothesize the existence of Neptune before it had ever been observed in a telescope.)

We assume that the center of mass of the sun is located at the origin and that the center of mass of the planet is located at the terminal point of the vector-valued function $\mathbf{r}(t)$. The

HISTORICAL NOTES

Johannes Kepler (1571–1630) German astronomer and mathematician whose discoveries revolutionized Western science. Kepler's remarkable mathematical ability and energy produced connections among many areas of research. A study of observations of the moon led to work in optics that included the first fundamentally correct description of the operation of the human eye. Kepler's model of the solar system used an ingenious nesting of the five Platonic solids to describe the orbits of the six known planets. (Kepler invented the term "satellite" to describe the moon, which his system demoted from planetary status.) A study of the wine casks opened at his wedding led Kepler to compute volumes of solids of revolution, inventing techniques that were vital to the subsequent development of calculus.

velocity vector for the planet is then $\mathbf{v}(t) = \mathbf{r}'(t)$, with the acceleration given by $\mathbf{a}(t) = \mathbf{r}''(t)$. From Newton's second law of motion, we have that the net (gravitational) force $\mathbf{F}(t)$ acting on the planet is

$$\mathbf{F}(t) = m\mathbf{a}(t),$$

where m is the mass of the planet. From Newton's law of universal gravitation, we have that if M is the mass of the sun, then the gravitational attraction between the two bodies satisfies

$$\mathbf{F}(t) = -\frac{GmM}{\|\mathbf{r}(t)\|^2}\frac{\mathbf{r}(t)}{\|\mathbf{r}(t)\|},$$

where G is the **universal gravitational constant.**[1] We have written $\mathbf{F}(t)$ in this form so that you can see that at each point, the gravitational attraction acts in the direction **opposite** the position vector $\mathbf{r}(t)$. Further, the gravitational attraction is jointly proportional to the masses of the sun and the planet and inversely proportional to the square of the distance between the sun and the planet. For simplicity, we will let $r = \|\mathbf{r}\|$ and not explicitly indicate the t-variable. Taking $\mathbf{u}(t) = \dfrac{\mathbf{r}(t)}{\|\mathbf{r}(t)\|}$ (a unit vector in the direction of $\mathbf{r}(t)$), we can then write Newton's laws as simply

$$\mathbf{F} = m\mathbf{a} \quad \text{and} \quad \mathbf{F} = -\frac{GmM}{r^2}\mathbf{u}.$$

We begin by demonstrating that the orbit of a planet lies in a plane. Equating the two expressions above for $\mathbf{F}$ and canceling out the common factor of m, we have

$$\mathbf{a} = -\frac{GM}{r^2}\mathbf{u}. \tag{5.11}$$

Notice that this says that the acceleration $\mathbf{a}$ always points in the **opposite** direction from $\mathbf{r}$, so that the force of gravity accelerates the planet toward the sun at all times. Since $\mathbf{a}$ and $\mathbf{r}$ are parallel, we have that

$$\mathbf{r} \times \mathbf{a} = \mathbf{0}. \tag{5.12}$$

Next, from the product rule [Theorem 2.3 (v)], we have

$$\frac{d}{dt}(\mathbf{r} \times \mathbf{v}) = \frac{d\mathbf{r}}{dt} \times \mathbf{v} + \mathbf{r} \times \frac{d\mathbf{v}}{dt}$$
$$= \mathbf{v} \times \mathbf{v} + \mathbf{r} \times \mathbf{a} = \mathbf{0},$$

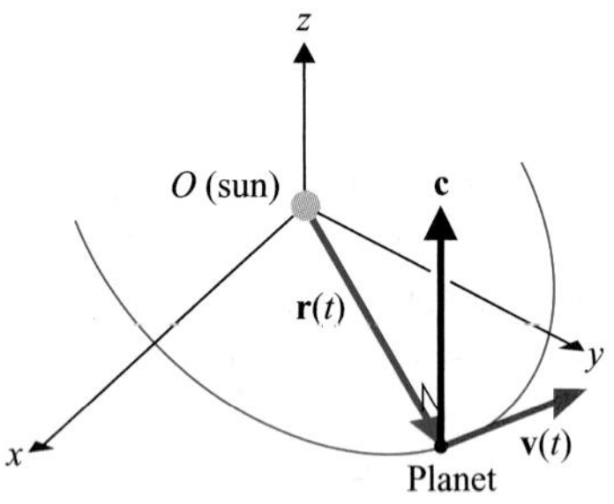

FIGURE 11.35
Position and velocity vectors for planetary motion

in view of (5.12) and since $\mathbf{v} \times \mathbf{v} = \mathbf{0}$. Integrating both sides of this expression gives us

$$\mathbf{r} \times \mathbf{v} = \mathbf{c}, \tag{5.13}$$

for some constant vector $\mathbf{c}$. This says that for each t, $\mathbf{r}(t)$ is orthogonal to the constant vector $\mathbf{c}$. In particular, then, the terminal point of $\mathbf{r}(t)$ (and consequently, the orbit of the planet) lies in the plane orthogonal to the vector $\mathbf{c}$ and containing the origin.

Now that we have established that a planet's orbit lies in a plane, we are in a position to prove Kepler's first law. For the sake of simplicity, we assume that the plane containing the orbit is the xy-plane, so that $\mathbf{c}$ is parallel to the z-axis (see Figure 11.35). Now, observe that since $\mathbf{r} = r\mathbf{u}$, we have by the product rule [Theorem 2.3 (iii)] that

$$\mathbf{v} = \frac{d\mathbf{r}}{dt} = \frac{d}{dt}(r\mathbf{u}) = \frac{dr}{dt}\mathbf{u} + r\frac{d\mathbf{u}}{dt}.$$

[1] If we measure mass in kilograms, force in newtons and distance in meters, G is given approximately by $G \approx 6.672 \times 10^{-11}$ N m^2/kg^2.

Substituting this into (5.13), and replacing $\mathbf{r}$ by $r\,\mathbf{u}$, we have

$$\begin{aligned}\mathbf{c} = \mathbf{r}\times\mathbf{v} = r\mathbf{u}\times\left(\frac{dr}{dt}\mathbf{u} + r\frac{d\mathbf{u}}{dt}\right)\\ = r\frac{dr}{dt}(\mathbf{u}\times\mathbf{u}) + r^2\left(\mathbf{u}\times\frac{d\mathbf{u}}{dt}\right)\\ = r^2\left(\mathbf{u}\times\frac{d\mathbf{u}}{dt}\right),\end{aligned}$$

since $\mathbf{u}\times\mathbf{u} = \mathbf{0}$. Together with (5.11), this gives us

$$\begin{aligned}\mathbf{a}\times\mathbf{c} &= -\frac{GM}{r^2}\mathbf{u}\times r^2\left(\mathbf{u}\times\frac{d\mathbf{u}}{dt}\right)\\ &= -GM\mathbf{u}\times\left(\mathbf{u}\times\frac{d\mathbf{u}}{dt}\right)\\ &= -GM\left[\left(\mathbf{u}\cdot\frac{d\mathbf{u}}{dt}\right)\mathbf{u} - (\mathbf{u}\cdot\mathbf{u})\frac{d\mathbf{u}}{dt}\right],\end{aligned} \tag{5.14}$$

where we have rewritten the vector triple product using Theorem 4.3 (vi) in Chapter 10. There are two other things to note here. First, since $\mathbf{u}$ is a unit vector, $\mathbf{u}\cdot\mathbf{u} = \|\mathbf{u}\|^2 = 1$. Further, from Theorem 2.4, since $\mathbf{u}$ is a vector-valued function of constant magnitude, $\mathbf{u}\cdot\dfrac{d\mathbf{u}}{dt} = 0$. Consequently, (5.14) simplifies to

$$\mathbf{a}\times\mathbf{c} = GM\frac{d\mathbf{u}}{dt} = \frac{d}{dt}(GM\mathbf{u}),$$

since G and M are constants. Observe that using the definition of $\mathbf{a}$, we can write

$$\mathbf{a}\times\mathbf{c} = \frac{d\mathbf{v}}{dt}\times\mathbf{c} = \frac{d}{dt}(\mathbf{v}\times\mathbf{c}),$$

since $\mathbf{c}$ is a constant vector. Equating these last two expressions for $\mathbf{a}\times\mathbf{c}$ gives us

$$\frac{d}{dt}(\mathbf{v}\times\mathbf{c}) = \frac{d}{dt}(GM\mathbf{u}).$$

Integrating both sides gives us

$$\mathbf{v}\times\mathbf{c} = GM\mathbf{u} + \mathbf{b}, \tag{5.15}$$

for some constant vector $\mathbf{b}$. Now, note that $\mathbf{v}\times\mathbf{c}$ must be orthogonal to $\mathbf{c}$ and so, $\mathbf{v}\times\mathbf{c}$ must lie in the xy-plane. (Recall that we had chosen the orientation of the xy-plane so that $\mathbf{c}$ was a vector orthogonal to the plane. This says further that every vector orthogonal to $\mathbf{c}$ must lie in the xy-plane.) From (5.15), since $\mathbf{u}$ and $\mathbf{v}\times\mathbf{c}$ lie in the xy-plane, $\mathbf{b}$ must also lie in the same plane. (Think about why this must be so.) Next, align the x-axis so that the positive x-axis points in the same direction as $\mathbf{b}$ (see Figure 11.36). Also, let θ be the angle from the positive x-axis to $\mathbf{r}(t)$, so that (r, θ) are polar coordinates for the endpoint of the position vector $\mathbf{r}(t)$, as indicated in Figure 11.36.

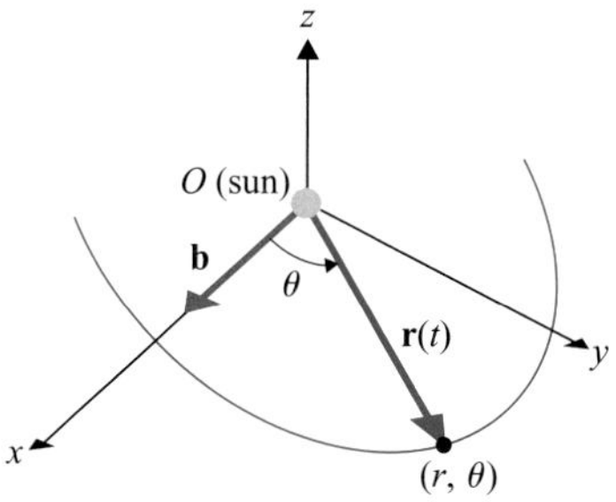

FIGURE 11.36
Polar coordinates for the position of the planet

Next, let $b = \|\mathbf{b}\|$ and $c = \|\mathbf{c}\|$. Then, from (5.13), we have

$$c^2 = \mathbf{c}\cdot\mathbf{c} = (\mathbf{r}\times\mathbf{v})\cdot\mathbf{c} = \mathbf{r}\cdot(\mathbf{v}\times\mathbf{c}),$$

where we have rewritten the scalar triple product using Theorem 4.3 (v) in Chapter 10. Putting this together with (5.15), and writing $\mathbf{r} = r\mathbf{u}$, we get

$$\begin{aligned}c^2 &= \mathbf{r}\cdot(\mathbf{v}\times\mathbf{c}) = r\mathbf{u}\cdot(GM\mathbf{u} + \mathbf{b})\\ &= rGM\mathbf{u}\cdot\mathbf{u} + r\mathbf{u}\cdot\mathbf{b}.\end{aligned} \tag{5.16}$$

Since **u** is a unit vector, $\mathbf{u} \cdot \mathbf{u} = \|\mathbf{u}\|^2 = 1$ and by Theorem 3.2 in Chapter 10,

$$\mathbf{u} \cdot \mathbf{b} = \|\mathbf{u}\| \|\mathbf{b}\| \cos \theta = b \cos \theta,$$

where θ is the angle between **b** and **u** (i.e., the angle between the positive x-axis and **r**). Together with (5.16), this gives us

$$c^2 = rGM + rb \cos \theta = r(GM + b \cos \theta).$$

Solving this for r gives us

$$r = \frac{c^2}{GM + b \cos \theta}.$$

Dividing numerator and denominator by GM reduces this to

$$r = \frac{ed}{1 + e \cos \theta}, \tag{5.17}$$

where $e = \frac{b}{GM}$ and $d = \frac{c^2}{b}$. Recall from Theorem 7.2 in Chapter 9 that (5.17) is a polar equation for a conic section with focus at the origin and eccentricity e. Finally, since the orbit of a planet is a closed curve, this must be the equation of an ellipse, since the other conic sections (parabolas and hyperbolas) are not closed curves. We have now proved that (assuming one sun and one planet and no other celestial bodies), the orbit of a planet is an ellipse with one focus located at the center of mass of the sun.

You may be thinking what a long derivation this was. (We haven't been keeping score, but it was probably one of the longest derivations of anything in this book.) Take a moment, though, to realize the enormity of what we have done. Thanks to the genius of Newton and his second law of motion and his law of universal gravitation, we have in only a few pages used the calculus to settle one of the most profound questions of our existence: How do the mechanics of a solar system work? Through the power of reason and the use of considerable calculus, we have found an answer that is consistent with the observed motion of the planets, first postulated by Kepler. This magnificent achievement came more than 300 years ago and was one of the earliest (and most profound) success stories for the calculus. Since that time, the calculus has proven to be an invaluable tool for countless engineers, physicists, mathematicians and others.

EXERCISES 11.5

WRITING EXERCISES

1. Suppose that you are driving a car, going slightly uphill as the road curves to the left. Describe the directions of the unit tangent, principal unit normal and binormal vectors. What changes if the road curves to the right?

2. If the components of $\mathbf{r}(t)$ are linear functions, explain why you can't compute the principal unit normal vector. Describe graphically why it is impossible to define a single direction for the principal unit normal.

3. Previously in your study of calculus, you have approximated curves with lines and the graphs of other polynomials (Taylor polynomials). Discuss possible circumstances in which the osculating circle would be a better or worse approximation of a curve than the graph of a polynomial.

4. Suppose that you are flying in a fighter jet and an enemy jet is headed straight at you with velocity vector parallel to your principal unit normal vector. Discuss how much danger you are in and what maneuver(s) you might want to make to avoid danger.

In exercises 1–8, find the unit tangent and principal unit normal vectors at the given points.

1. $\mathbf{r}(t) = \langle t, t^2 \rangle$ at $t = 0, t = 1$

2. $\mathbf{r}(t) = \langle t, t^3 \rangle$ at $t = 0, t = 1$

3. $\mathbf{r}(t) = \langle \cos 2t, \sin 2t \rangle$ at $t = 0, t = \frac{\pi}{4}$

4. $\mathbf{r}(t) = \langle 2\cos t, 3\sin t\rangle$ at $t = 0, t = \frac{\pi}{4}$

5. $\mathbf{r}(t) = \langle \cos 2t, t, \sin 2t\rangle$ at $t = 0, t = \frac{\pi}{2}$

6. $\mathbf{r}(t) = \langle \cos t, \sin t, \sin t\rangle$ at $t = 0, t = \frac{\pi}{2}$

7. $\mathbf{r}(t) = \langle t, t^2 - 1, t\rangle$ at $t = 0, t = 1$

8. $\mathbf{r}(t) = \langle t, t, 3\sin 2t\rangle$ at $t = 0, t = -\pi$

In exercises 9–12, find the osculating circle at the given points.

9. $\mathbf{r}(t) = \langle t, t^2\rangle$ at $t = 0$

10. $\mathbf{r}(t) = \langle t, t^3\rangle$ at $t = 0$

11. $\mathbf{r}(t) = \langle \cos 2t, \sin 2t\rangle$ at $t = \frac{\pi}{4}$

12. $\mathbf{r}(t) = \langle 2\cos t, 3\sin t\rangle$ at $t = \frac{\pi}{4}$

In exercises 13–16, find the tangential and normal components of acceleration for the given position functions at the given points.

13. $\mathbf{r}(t) = \langle 8t, 16t - 16t^2\rangle$ at $t = 0, t = 1$

14. $\mathbf{r}(t) = \langle \cos 2t, \sin 2t\rangle$ at $t = 0, t = 2$

15. $\mathbf{r}(t) = \langle \cos 2t, t^2, \sin 2t\rangle$ at $t = 0, t = \frac{\pi}{4}$

16. $\mathbf{r}(t) = \langle 2\cos t, 3\sin t, t^2\rangle$ at $t = 0, t = \frac{\pi}{4}$

17. In exercise 15, determine whether the speed of the object is increasing or decreasing at the given points.

18. In exercise 16, determine whether the speed of the object is increasing or decreasing at the given points.

19. For the circular helix traced out by $\mathbf{r}(t) = \langle a\cos t, a\sin t, bt\rangle$, find the tangential and normal components of acceleration.

20. For the linear path traced out by $\mathbf{r}(t) = \langle a + bt, c + dt, e + ft\rangle$, find the tangential and normal components of acceleration.

In exercises 21–24, find the binormal vector $\mathbf{B}(t) = \mathbf{T}(t) \times \mathbf{N}(t)$ at $t = 0$ and $t = 1$. Also, sketch the curve traced out by $\mathbf{r}(t)$ and the vectors T, N and B at these points.

21. $\mathbf{r}(t) = \langle t, 2t, t^2\rangle$

22. $\mathbf{r}(t) = \langle t, 2t, t^3\rangle$

23. $\mathbf{r}(t) = \langle 4\cos \pi t, 4\sin \pi t, t\rangle$

24. $\mathbf{r}(t) = \langle 3\cos 2\pi t, t, \sin 2\pi t\rangle$

In exercises 25–28, label the statement as true (i.e., always true) or false and explain your answer.

25. $\mathbf{T} \cdot \dfrac{d\mathbf{T}}{ds} = 0$

26. $\mathbf{T} \cdot \mathbf{B} = 0$

27. $\dfrac{d}{ds}(\mathbf{T} \cdot \mathbf{T}) = 0$

28. $\mathbf{T} \cdot (\mathbf{N} \times \mathbf{B}) = 1$

The friction force required to keep a car from skidding on a curve is given by $\mathbf{F}_s(t) = ma_N\mathbf{N}(t)$. In exercises 29–32, find the friction force needed to keep a car of mass $m = 100$ (slugs) from skidding.

29. $\mathbf{r}(t) = \langle 100\cos \pi t, 100\sin \pi t\rangle$

30. $\mathbf{r}(t) = \langle 200\cos \pi t, 200\sin \pi t\rangle$

31. $\mathbf{r}(t) = \langle 100\cos 2\pi t, 100\sin 2\pi t\rangle$

32. $\mathbf{r}(t) = \langle 300\cos 2t, 300\sin 2t\rangle$

33. Based on your answers to exercises 29 and 30, how does the required friction force change when the radius of a turn is doubled?

34. Based on your answers to exercises 29 and 31, how does the required friction force change when the speed of a car on a curve is doubled?

35. Compare the radii of the osculating circles for $y = \cos x$ at $x = 0$ and $x = \frac{\pi}{4}$. Compute the concavity of the curve at these points and use this information to explain why one circle is larger than the other.

36. Compare the osculating circles for $y = \cos x$ at $x = 0$ and $x = \pi$. Compute the concavity of the curve at these points and use this information to help explain why the circles have the same radius.

37. For $y = x^2$, show that each center of curvature lies on the curve traced out by $\mathbf{r}(t) = \langle 4t^3, \frac{1}{2} + 3t^2\rangle$. Graph this curve.

38. For $r = e^{a\theta}$, $a > 0$, show that the radius of curvature is $e^{a\theta}\sqrt{a^2 + 1}$. Show that each center of curvature lies on the curve traced out by $ae^{at}\langle -\sin t, \cos t\rangle$ and graph the curve.

39. In this exercise, we prove Kepler's second law. Denote the (two-dimensional) path of the planet in polar coordinates by $\mathbf{r} = (r\cos\theta)\mathbf{i} + (r\sin\theta)\mathbf{j}$. Show that $\mathbf{r} \times \mathbf{v} = r^2\dfrac{d\theta}{dt}\mathbf{k}$. Conclude that $r^2\dfrac{d\theta}{dt} = \|\mathbf{r} \times \mathbf{v}\|$. Recall that in polar coordinates, the area swept out by the curve $r = r(\theta)$ is given by $A = \displaystyle\int_a^b \frac{1}{2}r^2\,d\theta$ and show that $\dfrac{dA}{dt} = \dfrac{1}{2}r^2\dfrac{d\theta}{dt}$. From $\dfrac{dA}{dt} = \dfrac{1}{2}\|\mathbf{r} \times \mathbf{v}\|$, conclude that equal areas are swept out in equal times.

40. In this exercise, we prove Kepler's third law. Recall that the area of the ellipse $\dfrac{x^2}{a^2} + \dfrac{y^2}{b^2} = 1$ is πab. From exercise 39, the rate at which area is swept out is given by $\dfrac{dA}{dt} = \dfrac{1}{2}\|\mathbf{r} \times \mathbf{v}\|$. Conclude that the period of the orbit is $T = \dfrac{\pi ab}{\frac{1}{2}\|\mathbf{r} \times \mathbf{v}\|}$ and so, $T^2 = \dfrac{4\pi^2a^2b^2}{\|\mathbf{r} \times \mathbf{v}\|^2}$. Use (5.17) to show that the minimum value of r is $r_{\min} = \dfrac{ed}{1 + e}$ and that the maximum value of r is $r_{\max} = \dfrac{ed}{1 - e}$. Explain why $2a = r_{\min} + r_{\max}$ and use this to show that $a = \dfrac{ed}{1 - e^2}$. Given that $1 - e^2 = \dfrac{b^2}{a^2}$, show that $\dfrac{b^2}{a} = ed$. From $e = \dfrac{b}{GM}$ and $d = \dfrac{c^2}{b}$, show that $ed = \dfrac{\|\mathbf{r} \times \mathbf{v}\|^2}{GM}$. It then follows that $\dfrac{b^2}{a} = \dfrac{\|\mathbf{r} \times \mathbf{v}\|^2}{GM}$. Finally, show that $T^2 = ka^3$, where the constant $k = \dfrac{4\pi^2}{GM}$ does not depend on the specific orbit of the planet.

41. (a) Show that $\dfrac{d\mathbf{B}}{ds}$ is orthogonal to $\mathbf{T}$. (b) Show that $\dfrac{d\mathbf{B}}{ds}$ is orthogonal to $\mathbf{B}$.

42. Use the result of exercise 41 to show that $\dfrac{d\mathbf{B}}{ds} = -\tau\mathbf{N}$, for some scalar τ. (τ is called the **torsion,** which measures how much a curve twists.) Also, show that $\tau = -\dfrac{d\mathbf{B}}{ds} \cdot \mathbf{N}$.

43. Show that the torsion for the curve traced out by $\mathbf{r}(t) = \langle f(t), g(t), k\rangle$ is zero for any constant k. (In general, the torsion is zero for any curve that lies in a single plane.)

44. The following three formulas (called the **Frenet-Serret formulas**) are of great significance in the field of differential geometry:

a. $\dfrac{d\mathbf{T}}{ds} = \kappa\mathbf{N}$ [equation (5.2)]

b. $\dfrac{d\mathbf{B}}{ds} = -\tau\mathbf{N}$ (see exercise 42)

c. $\dfrac{d\mathbf{N}}{ds} = -\kappa\mathbf{T} + \tau\mathbf{B}$

Use the fact that $\mathbf{N} = \mathbf{B} \times \mathbf{T}$ and the product rule [Theorem 2.3 (v)] to establish (c).

45. Use the Frenet-Serret formulas (see exercise 44) to establish each of the following formulas:

a. $\mathbf{r}''(t) = s''(t)\mathbf{T} + \kappa[s'(t)]^2\mathbf{N}$

b. $\mathbf{r}'(t) \times \mathbf{r}''(t) = \kappa[s'(t)]^3\mathbf{B}$

c. $\mathbf{r}'''(t) = \{s'''(t) - \kappa^2[s'(t)]^3\}\mathbf{T} + \{3\kappa s'(t)s''(t) + \kappa'(t)[s'(t)]^2\}\mathbf{N} + \kappa\tau[s'(t)]^3\mathbf{B}$

d. $\tau = \dfrac{[\mathbf{r}'(t) \times \mathbf{r}''(t)] \cdot \mathbf{r}'''(t)}{\|\mathbf{r}'(t) \times \mathbf{r}''(t)\|^2}$

46. Show that the torsion for the helix traced out by $\mathbf{r}(t) = \langle a\cos t, a\sin t, bt\rangle$ is given by $\tau = \dfrac{b}{a^2+b^2}$. [Hint: See exercise 45 (d).]

EXPLORATORY EXERCISES

1. In this exercise, we explore some ramifications of the precise form of Newton's law of universal gravitation. Suppose that the gravitational force between objects is $\mathbf{F} = -\dfrac{GMm}{r^n}\mathbf{u}$, for some positive integer $n \geq 1$ (the actual law has $n = 2$). Show that the path of the planet would still be planar and that Kepler's second law still holds. Also, show that the circular orbit $\mathbf{r} = \langle r\cos kt, r\sin kt\rangle$ (where r is a constant) satisfies the equation $\mathbf{F} = m\mathbf{a}$ and hence, is a potential path for the orbit. For this path, find the relationship between the period of the orbit and the radius of the orbit.

2. In this exercise, you will find the locations of three of the five **Lagrange points.** These are equilibrium solutions of the "restricted three-body problem" in which a large body S of mass M_1 is orbited by a smaller body E of mass $M_2 < M_1$. A third object H of very small mass m orbits S such that the relative positions of S, E and H remain constant. Place S at the origin, E at $(1, 0)$ and H at $(x, 0)$ as shown.

Assume that H and E have circular orbits about the center of mass $(c, 0)$. Show that the gravitational force on H is $F = -\dfrac{GM_1m}{x^2} + \dfrac{GM_2m}{(1-x)^2}$. As shown in example 3.3, for circular motion $F = -m\omega^2(x-c)$, where ω is the angular velocity of H. Analyzing the orbit of E, show that $GM_1M_2 = M_2\omega^2(1-c)$. In particular, explain why E has the same angular velocity ω. Combining the three equations, show that $\dfrac{1}{x^2} - \dfrac{k}{(1-x)^2} = \dfrac{x-c}{1-c}$, where $k = \dfrac{M_2}{M_1}$. Given that $c = \dfrac{M_2}{M_1+M_2}$, show that

$$(1+k)x^5 - (3k+2)x^4 + (3k+1)x^3 - x^2 + 2x - 1 = 0.$$

For the Sun-Earth system with $k = 0.000002$, estimate x, the location of the L_1 Lagrange point and the location of NASA's SOHO solar observatory satellite. Then derive and estimate solutions for the L_2 Lagrange point with $x > 1$ and L_3 with $x < 0$.

11.6 PARAMETRIC SURFACES

Throughout this chapter, we have emphasized the connection between vector-valued functions and parametric equations. In this section, we extend the notion of parametric equations to those with two independent parameters. This also means that we will be working with simple cases of functions of two variables, which are developed in more detail in Chapter 12. We will make use of parametric surfaces throughout the remainder of the book.

We have already seen the helix defined by the parametric equations $x = \cos t$, $y = \sin t$ and $z = t$. This curve winds around the cylinder $x^2 + y^2 = 1$. Now, suppose that we wanted to obtain parametric equations that describe the entire cylinder. Given $x = \cos t$ and $y = \sin t$, we have that

$$x^2 + y^2 = \cos^2 t + \sin^2 t = 1.$$

So, any point (x, y, z) with x and y defined in this way must lie on the cylinder. To describe the entire cylinder (i.e., every point on the cylinder), we must allow z to be any real number, not just $z = t$. In other words, z needs to be independent of x and y. Assigning z its own parameter will accomplish this. Using the parameters u and v, we have the parametric equations

$$x = \cos u, \quad y = \sin u \quad \text{and} \quad z = v$$

for the cylinder. In general, parametric equations with two independent parameters correspond to a three-dimensional surface. Examples 6.1 through 6.3 explore some basic but important surfaces.

EXAMPLE 6.1 Graphing a Parametric Surface

Identify and sketch a graph of the surface defined by the parametric equations $x = 2\cos u \sin v$, $y = 2\sin u \sin v$ and $z = 2\cos v$.

Solution Given the cosine and sine terms in both parameters, you should be expecting circular cross sections. Notice that we can eliminate the u parameter, by observing that

$$\begin{aligned} x^2 + y^2 &= (2\cos u \sin v)^2 + (2\sin u \sin v)^2 = 4\cos^2 u \sin^2 v + 4\sin^2 u \sin^2 v \\ &= 4(\cos^2 u + \sin^2 u)\sin^2 v = 4\sin^2 v. \end{aligned}$$

So, for each fixed value of z (which also means a fixed value for v), $x^2 + y^2$ is constant. That is, cross sections of the surface parallel to the xy-plane are circular, with radius $|2\sin v|$. Since $z = 2\cos v$, we also have circular cross sections parallel to either of the other two coordinate planes. Of course, one surface that you've seen that has circular cross sections in all directions is a sphere. To determine that the given parametric equations represent a sphere, observe that

$$\begin{aligned} x^2 + y^2 + z^2 &= 4\cos^2 u \sin^2 v + 4\sin^2 u \sin^2 v + 4\cos^2 v \\ &= 4(\cos^2 u + \sin^2 u)\sin^2 v + 4\cos^2 v \\ &= 4\sin^2 v + 4\cos^2 v = 4. \end{aligned}$$

You should recognize $x^2 + y^2 + z^2 = 4$ as an equation of the sphere centered at the origin with radius 2. A computer-generated sketch is shown in Figure 11.37. ■

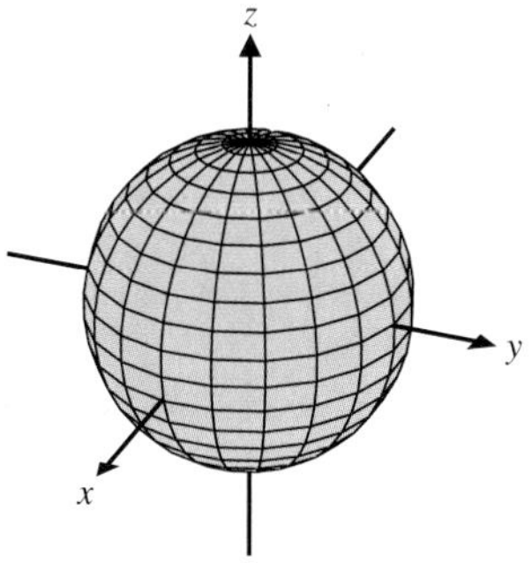

FIGURE 11.37
$x^2 + y^2 + z^2 = 4$

There are several important points to make about example 6.1. First, we did not actually demonstrate that the surface defined by the given parametric equations is the entire sphere. Rather, we showed that points lying on the parametric surface were also on the sphere. In other words, the parametric surface is (at least) part of the sphere. In the exercises, we will supply the missing steps to this puzzle, showing that the parametric equations from example 6.1 do, in fact, describe the entire sphere. In this instance, the equations are a special case of something called **spherical coordinates,** which we will introduce in Chapter 13. An understanding of spherical coordinates makes it simple to find parametric equations for half of a sphere or some other portion of a sphere. Next, as with parametric equations of curves, there are other parametric equations representing the same sphere. In the exercises, we will see that the roles of cosine and sine can be reversed in these equations, with the resulting

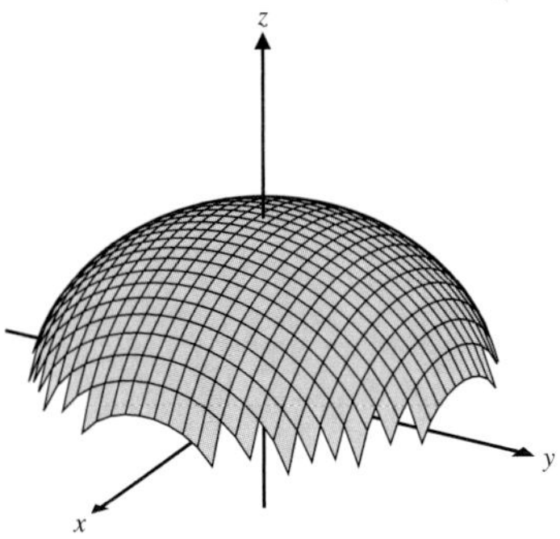

FIGURE 11.38
$z = \sqrt{4 - x^2 - y^2}$

equations still describing a sphere. Finally, to repeat a point made in section 10.6, parametric equations can be used to produce many interesting graphs. Notice the smooth contours and clearly defined circular cross sections in Figure 11.37, compared with the jagged graph in Figure 11.38. Figure 11.38 is a computer-generated graph of $z = \sqrt{4 - x^2 - y^2}$, which should be the top half of the sphere.

For parametric equations of hyperboloids and hyperbolic paraboloids, it is convenient to use the hyperbolic functions $\cosh x$ and $\sinh x$. Recall that

$$\cosh x = \frac{e^x + e^{-x}}{2} \quad \text{and} \quad \sinh x = \frac{e^x - e^{-x}}{2}.$$

A little algebra shows that $\cosh^2 x - \sinh^2 x = 1$, which is an identity needed in example 6.2.

EXAMPLE 6.2 Graphing a Parametric Surface

Sketch the surface defined parametrically by $x = 2\cos u \cosh v$, $y = 2\sin u \cosh v$ and $z = 2\sinh v$, $0 \le u \le 2\pi$ and $-\infty < v < \infty$.

Solution A sketch such as the one we show in Figure 11.39 can be obtained from a computer algebra system.

Notice that this looks like a hyperboloid of one sheet wrapped around the z-axis. To verify that this is correct, observe that

$$\begin{aligned} x^2 + y^2 - z^2 &= 4\cos^2 u \cosh^2 v + 4\sin^2 u \cosh^2 v - 4\sinh^2 v \\ &= 4(\cos^2 u + \sin^2 u)\cosh^2 v - 4\sinh^2 v \\ &= 4\cosh^2 v - 4\sinh^2 v = 4, \end{aligned}$$

where we have used the identities $\cos^2 u + \sin^2 u = 1$ and $\cosh^2 v - \sinh^2 v = 1$. Recall that the graph of $x^2 + y^2 - z^2 = 4$ is indeed a hyperboloid of one sheet. ■

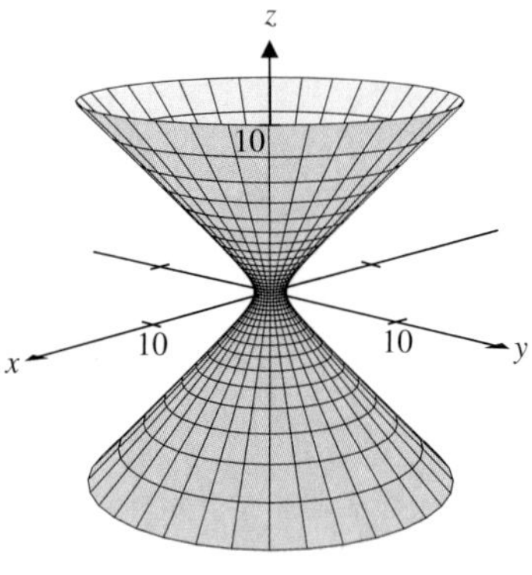

FIGURE 11.39
$x^2 + y^2 - z^2 = 4$

Again, it is instructive to compare the graph in Figure 11.39 with the computer-generated graph of $z = \sqrt{x^2 + y^2 - 4}$ shown in Figure 11.40.

Most of the time when we use parametric representations of surfaces, the task is the opposite of that in example 6.2. That is, given a particular surface, we may need to find a convenient parametric representation of the surface. Example 6.2 gives us an important clue for working example 6.3.

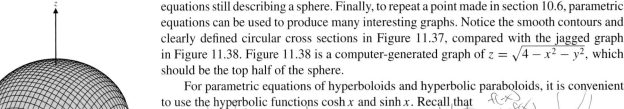

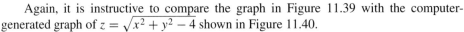

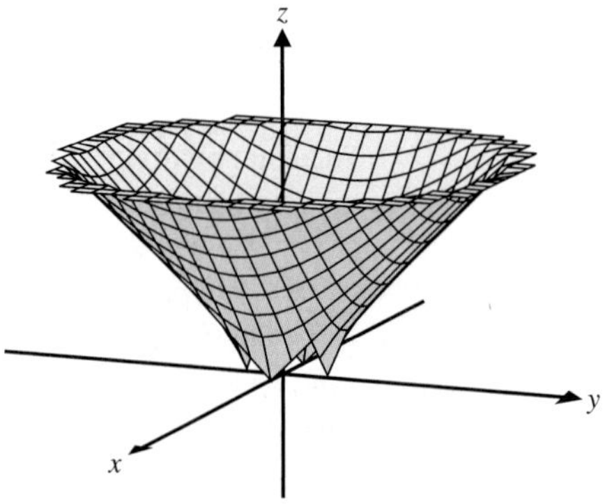

FIGURE 11.40
$z = \sqrt{x^2 + y^2 - 4}$

EXAMPLE 6.3 Finding a Parametric Representation of a Hyperbolic Paraboloid

Find parametric equations for the hyperbolic paraboloid $z = x^2 - y^2$.

Solution It helps to understand the surface with which we are working. Observe that for any value of k, the trace in the plane $z = k$ is a hyperbola. The spread of the hyperbola depends on whether $|k|$ is large or small. To get hyperbolas in x and y, we can start with $x = \cosh u$ and $y = \sinh u$. To enlarge or shrink the hyperbola, we can multiply $\cosh u$ and $\sinh u$ by a constant. We now have $x = v \cosh u$ and $y = v \sinh u$. To get $z = x^2 - y^2$, simply compute

$$x^2 - y^2 = v^2 \cosh^2 u - v^2 \sinh^2 u = v^2(\cosh^2 u - \sinh^2 u) = v^2,$$

since $\cosh^2 u - \sinh^2 u = 1$. This gives us the parametric equations

$$x = v \cosh u, \quad y = v \sinh u \quad \text{and} \quad z = v^2.$$

A graph of the parametric equations is shown in Figure 11.41a. However, notice that this is only the top half of the surface, since $z = v^2 \geq 0$. To get the bottom half of the surface, we set $x = v \sinh u$ and $y = v \cosh u$ so that $z = x^2 - y^2 = -v^2 \leq 0$. Figure 11.41b shows both halves of the surface.

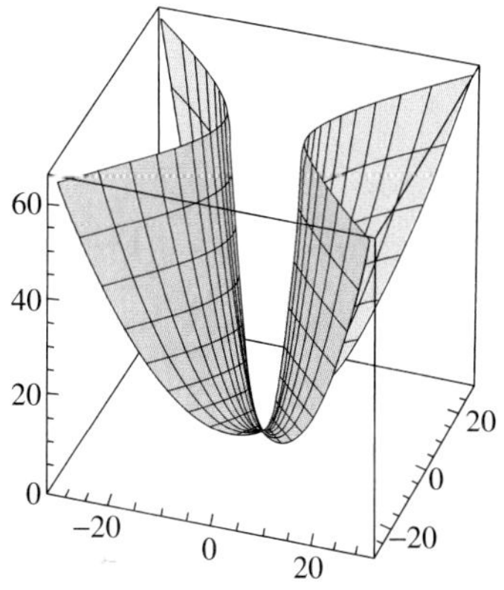

FIGURE 11.41a
Top half of the surface

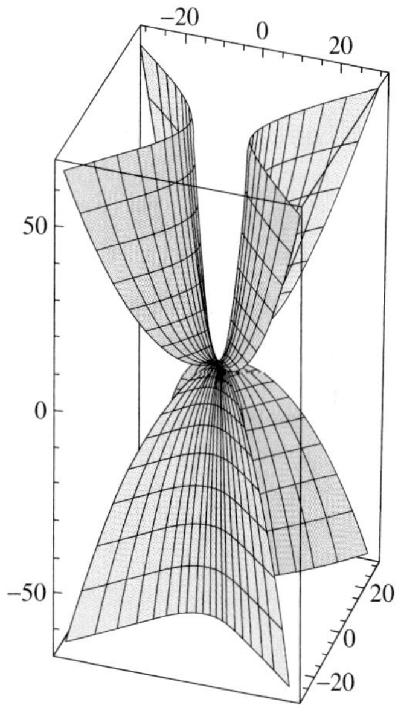

FIGURE 11.41b
$z = x^2 - y^2$

In many cases, the parametric equations that we use are determined by the geometry of the surface. Recall that in two dimensions, certain curves (especially circles) are more easily described in polar coordinates than in rectangular coordinates. We use this fact in example 6.4. Polar coordinates are essentially the parametric equations for circles that we have used over and over again. In particular, the polar coordinates r and θ are related to x and y by

$$x = r \cos\theta, \quad y = r \sin\theta \quad \text{and} \quad r = \sqrt{x^2 + y^2}.$$

So, the equation for the circle $x^2 + y^2 = 4$ can be written in polar coordinates simply as $r = 2$.

EXAMPLE 6.4 Finding Parametric Representations of Surfaces

Find a parametric representation of each surface: (a) the portion of $z = \sqrt{x^2 + y^2}$ inside $x^2 + y^2 = 4$ and (b) the portion of $z = 9 - x^2 - y^2$ above the xy-plane with $y \geq 0$.

Solution For part (a), a graph indicating the cone and the cylinder is shown in Figure 11.42a. Notice that the equations for both surfaces include the term $x^2 + y^2$ and x and y appear only in this combination. This suggests the use of polar coordinates. Taking $x = r\cos\theta$ and $y = r\sin\theta$, the equation of the cone $z = \sqrt{x^2 + y^2}$ becomes $z = r$ and the equation of the cylinder $x^2 + y^2 = 4$ becomes $r = 2$. Since the surface in question is that portion of the cone lying inside the cylinder, every point on the surface lies on the cone. So, every point on the surface satisfies $x = r\cos\theta$, $y = r\sin\theta$ and $z = r$. Observe that the cylinder cuts off the cone, something like a cookie cutter. Instead of all r-values being possible, the cylinder limits us to $r \leq 2$. (Think about why this is so.) A parametric representation for (a) is then

$$x = r\cos\theta, \quad y = r\sin\theta \quad \text{and} \quad z = r, \quad \text{for} \quad 0 \leq r \leq 2 \quad \text{and} \quad 0 \leq \theta \leq 2\pi.$$

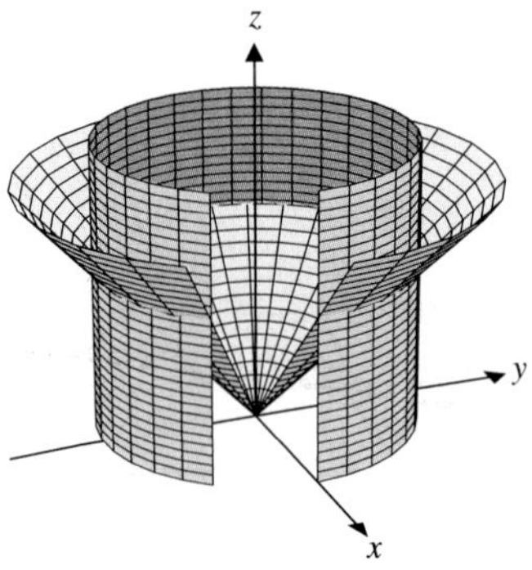

FIGURE 11.42a
Portion of $z = \sqrt{x^2 + y^2}$ inside $x^2 + y^2 = 4$

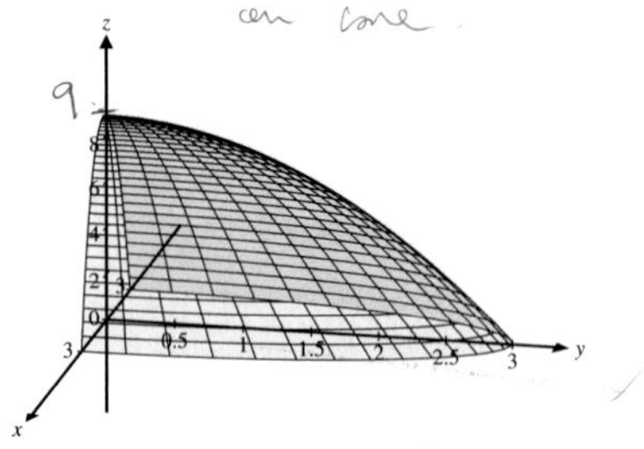

FIGURE 11.42b
Portion of $z = 9 - x^2 - y^2$ above the xy-plane, with $y \geq 0$

For part (b), a graph is shown in Figure 11.42b. Again, the presence of the term $x^2 + y^2$ in the defining equation suggests polar coordinates. Taking $x = r\cos\theta$ and $y = r\sin\theta$, the equation of the paraboloid becomes

$$z = 9 - (x^2 + y^2) = 9 - r^2.$$

To stay above the xy-plane, we need $z > 0$ or $9 - r^2 > 0$ or $|r| < 3$. Choosing positive r-values, we have $0 \leq r < 3$. Then $y \geq 0$, if $\sin\theta \geq 0$. One choice of θ that gives this is $0 \leq \theta \leq \pi$. A parametric representation for the surface is then

$$x = r\cos\theta, \quad y = r\sin\theta \quad \text{and} \quad z = 9 - r^2, \quad \text{for} \quad 0 \leq r < 3 \quad \text{and} \quad 0 \leq \theta \leq \pi.$$

EXERCISES 11.6

WRITING EXERCISES

1. Suppose that a surface has parametric equations $x = f(u)$, $y = g(u)$ and $z = h(v)$. Explain why the surface must be a cylinder. If the range of the function h consists of all real numbers, discuss whether or not the parametric equations $x = f(u)$, $y = g(u)$ and $z = v$ would describe the same surface.

2. In this exercise, we want to understand why parametric equations with two parameters typically graph as surfaces. If $x = f(u, v)$, $y = g(u, v)$ and $z = h(u, v)$, substitute in some constant $v = v_1$ and let C_1 be the curve with parametric equations $x = f(u, v_1)$, $y = g(u, v_1)$ and $z = h(u, v_1)$. Similarly, for a different constant v_2 let C_2 be the curve with parametric equations $x = f(u, v_2)$, $y = g(u, v_2)$ and $z = h(u, v_2)$. Sketch a picture showing what C_1 and C_2 might look like if v_1 and v_2 are close together. Then fill in what you would expect other curves would look like for v-values between v_1 and v_2. Discuss whether you are generating a surface.

3. To show that two parameters do not always produce a surface, sketch the curve with $x = u + v$, $y = u + v$ and $z = u + v$. Discuss the special feature of these equations that enables us to replace the parameters u and v with a single parameter t.

4. The graph of $x = f(t)$ and $y = g(t)$ is typically a curve in two dimensions. The graph of $x = f(u, v)$, $y = g(u, v)$ and $z = h(u, v)$ is typically a surface in three dimensions. Discuss what the graph of $x = f(r, s, t)$, $y = g(r, s, t)$, $z = h(r, s, t)$ and $w = d(r, s, t)$ would look like.

In exercises 1–12, identify and sketch a graph of the parametric surface.

1. $x = u,\ y = v,\ z = u^2 + 2v^2$
2. $x = u,\ y = v,\ z = 4 - u^2 - v^2$
3. $x = u\cos v,\ y = u\sin v,\ z = u^2$
4. $x = u\cos v,\ y = u\sin v,\ z = u$
5. $x = u,\ y = \sin u \cos v,\ z = \sin u \sin v$
6. $x = \cos u \cos v,\ y = u,\ z = \cos u \sin v$
7. $x = 2\sin u \cos v,\ y = 2\sin u \sin v,\ z = 2\cos u$
8. $x = u\cos v,\ y = u\sin v,\ z = v$
9. $x = v\sinh u,\ y = 4v^2,\ z = v\cosh u$
10. $x = \sinh v,\ y = \cos u \cosh v,\ z = \sin u \cosh v$
11. $x = 2\cos u \sinh v,\ y = 2\sin u \sinh v,\ z = 2\cosh v$
12. $x = 2\sinh u,\ y = v,\ z = 2\cosh u$

In exercises 13–22, find a parametric representation of the surface.

13. $z = 3x + 4y$
14. $x^2 + y^2 + z^2 = 4$
15. $x^2 + y^2 - z^2 = 1$
16. $z^2 = x^2 + y^2$
17. The portion of $x^2 + y^2 = 4$ from $z = 0$ to $z = 2$
18. The portion of $y^2 + z^2 = 9$ from $x = -1$ to $x = 1$
19. The portion of $z = 4 - x^2 - y^2$ above the xy-plane
20. The portion of $z = x^2 + y^2$ below $z = 4$
21. $x^2 - y^2 - z^2 = 1$
22. $x = 4y^2 - z^2$
23. Match the parametric equations with the surface.
    a. $x = u\cos v,\ y = u\sin v,\ z = v^2$
    b. $x = v,\ y = u\cos v,\ z = u\sin v$
    c. $x = u,\ y = u\cos v,\ z = u\sin v$

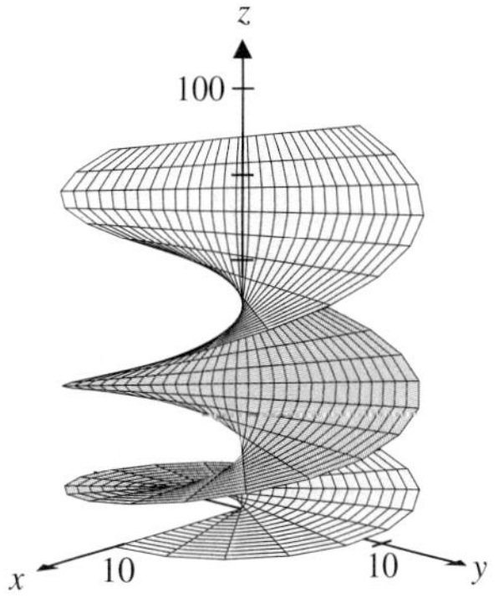

SURFACE A

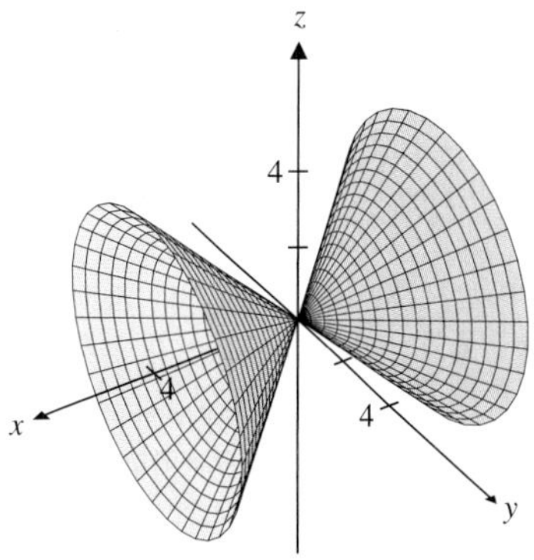

SURFACE B

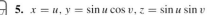

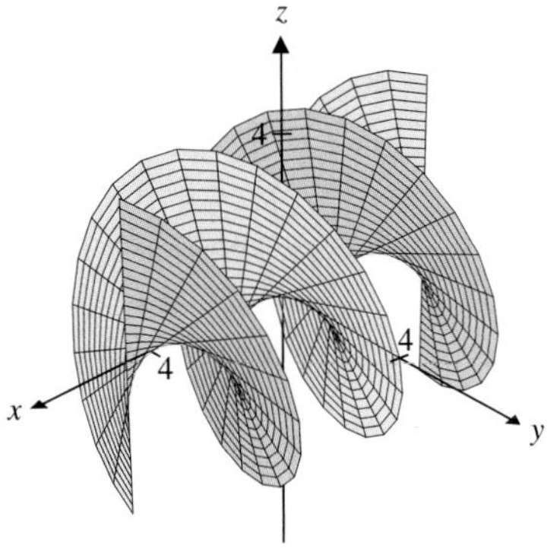

SURFACE C

24. Show that changing the parametric equations of example 6.1 to $x = 2 \sin u \cos v$, $y = 2 \cos u \cos v$ and $z = 2 \sin v$ does not change the fact that $x^2 + y^2 + z^2 = 4$.

25. To show that the surface in example 6.1 is the entire sphere $x^2 + y^2 + z^2 = 4$, start by finding the trace of the sphere in the plane $z = k$ for $-2 \le k \le 2$. If $z = 2 \cos v = k$, determine as fully as possible the value of $2 \sin v$ and then determine the trace in the plane $z = k$ for $x = 2 \cos u \sin v$, $y = 2 \sin u \sin v$ and $z = 2 \cos v$. If the traces are the same, then the surfaces are the same.

26. Investigate whether $x = 2 \sin u \cos v$, $y = 2 \cos u \cos v$ and $z = 2 \sin u$ also are parametric equations for the sphere $x^2 + y^2 + z^2 = 4$. If not, use computer-generated graphs to describe the surface.

Exercises 27–40 relate to spherical coordinates defined by $x = \rho \cos\theta \sin\phi$, $y = \rho \sin\theta \sin\phi$ and $z = \rho \cos\phi$, where $0 \le \rho$, $0 \le \theta \le 2\pi$ and $0 \le \phi \le \pi$.

27. Replace ρ with $\rho = 3$ and determine the surface with parametric equations $x = 3 \cos\theta \sin\phi$, $y = 3 \sin\theta \sin\phi$ and $z = 3 \cos\phi$.

28. Use the results of example 6.1 and exercise 27 to determine the surface defined by $\rho = k$, where k is some positive constant.

29. Replace ϕ with $\phi = \frac{\pi}{4}$ and determine the surface with parametric equations $x = \rho \cos\theta \sin\frac{\pi}{4}$, $y = \rho \sin\theta \sin\frac{\pi}{4}$ and $z = \rho \cos\frac{\pi}{4}$.

30. Replace ϕ with $\phi = \frac{\pi}{6}$ and determine the surface with parametric equations $x = \rho \cos\theta \sin\frac{\pi}{6}$, $y = \rho \sin\theta \sin\frac{\pi}{6}$ and $z = \rho \cos\frac{\pi}{6}$.

31. Replace θ with $\theta = \frac{\pi}{4}$ and determine the surface with parametric equations $x = \rho \cos\frac{\pi}{4} \sin\phi$, $y = \rho \sin\frac{\pi}{4} \sin\phi$ and $z = \rho \cos\phi$.

32. Replace θ with $\theta = \frac{3\pi}{4}$ and determine the surface with parametric equations $x = \rho \cos\frac{3\pi}{4} \sin\phi$, $y = \rho \sin\frac{3\pi}{4} \sin\phi$ and $z = \rho \cos\phi$.

33. Use the results of exercises 29 and 30 to determine the surface $\phi = k$ for some constant k with $0 < k < \pi$.

34. Use the results of exercises 31 and 32 to determine the surface $\theta = k$ for some constant k with $0 < k < 2\pi$.

35. Use the results of exercises 28, 33 and 34 to find parametric equations for the top half-sphere $z = \sqrt{9 - x^2 - y^2}$.

36. Use the results of exercises 28, 33 and 34 to find parametric equations for the right half-sphere $y = \sqrt{9 - x^2 - z^2}$.

37. Use the results of exercises 28, 33 and 34 to find parametric equations for the cone $z = \sqrt{x^2 + y^2}$.

38. Use the results of exercises 28, 33 and 34 to find parametric equations for the cone $z = -\sqrt{x^2 + y^2}$.

39. Find parametric equations for the region that lies above $z = \sqrt{x^2 + y^2}$ and below $x^2 + y^2 + z^2 = 4$.

40. Find parametric equations for the sphere $x^2 + y^2 + (z - 1)^2 = 1$.

Exercises 41–46 relate to parametric equations of a plane.

41. Sketch the plane with parametric equations $x = 2 + u + 2v$, $y = -1 + 2u - v$ and $z = 3 - 3u + 2v$. Show that the points $(2, -1, 3)$, $(3, 1, 0)$ and $(4, -2, 5)$ are on the plane by finding the correct values of u and v. Sketch these points along with the plane and the displacement vectors $\langle 1, 2, -3\rangle$ and $\langle 2, -1, 2\rangle$.

42. Use the results of exercise 41 to find a normal vector to the plane and an equation of the plane in terms of x, y, and z.

43. The parametric equations of exercise 41 can be written as $\mathbf{r} = \langle 2, -1, 3\rangle + u\langle 1, 2, -3\rangle + v\langle 2, -1, 2\rangle$ for $\mathbf{r} = \langle x, y, z\rangle$. For the general equation $\mathbf{r} = \mathbf{r}_0 + u\mathbf{v}_1 + v\mathbf{v}_2$, sketch a plane and show how the vectors $\mathbf{r}_0$, $\mathbf{v}_1$ and $\mathbf{v}_2$ would relate to the plane. In terms of $\mathbf{v}_1$ and $\mathbf{v}_2$, find a normal vector to the plane.

44. Find an equation in x, y and z for the plane defined by $\mathbf{r} = \langle 3, 1, -1\rangle + u\langle 2, -4, 1\rangle + v\langle 2, 0, 1\rangle$.

45. Find parametric equations for the plane through the point $(3, 1, 1)$ and containing the vectors $\langle 2, -1, 3\rangle$ and $\langle 4, 2, 1\rangle$.

46. Find parametric equations for the plane through the point $(0, -1, 2)$ and containing the vectors $\langle -2, 4, 0\rangle$ and $\langle 3, -2, 5\rangle$.

Exercises 47–54 relate to cylindrical coordinates defined by $x = r\cos\theta$, $y = r\sin\theta$ and $z = z$.

47. Sketch the two-dimensional polar graph $r = \cos 2\theta$. Sketch the solid in three dimensions defined by $x = r\cos\theta$, $y = r\sin\theta$ and $z = z$ with $r = \cos 2\theta$ and $0 \le z \le 1$, and compare it to the polar graph. Show that parametric equations for the solid are $x = \cos 2u \cos u$, $y = \cos 2u \sin u$ and $z = v$ with $0 \le u \le 2\pi$ and $0 \le v \le 1$.

48. Sketch the solid defined by $x = (2 - 2\cos u)\cos u$, $y = (2 - 2\cos u)\sin u$ and $z = v$ with $0 \le u \le 2\pi$ and $0 \le v \le 1$. (Hint: Use polar coordinates as in exercise 47.)

49. Find parametric equations for the wedge in the first octant bounded by $y = 0$, $y = x$, $x^2 + y^2 = 4$, $z = 0$ and $z = 1$.

50. Find parametric equations for the portion of the parabola $z = x^2 + y^2$ below $z = 4$, with $y \geq 0$.

51. Sketch the (two-dimensional) graph of $f(t) = e^{-t^2}$ for $t \geq 0$. Sketch the surface $z = e^{-x^2-y^2}$ and compare it to the graph of $f(t)$. Show that parametric equations of the surface are $x = u\cos v$, $y = u\sin v$ and $z = e^{-u^2}$.

52. Sketch the surface defined by $x = u\cos v$, $y = u\sin v$ and $z = ue^{-u^2}$. [Hint: Use the graph of $f(t) = te^{-t^2}$.]

53. Find parametric equations for the surface $z = \sin\sqrt{x^2 + y^2}$.

54. Find parametric equations for the surface $z = \cos(x^2 + y^2)$.

EXPLORATORY EXERCISES

1. If $x = 3\sin u\cos v$, $y = 3\cos u$ and $z = 3\sin u\sin v$, show that $x^2 + y^2 + z^2 = 9$. Explain why this equation doesn't guarantee that the parametric surface defined is the entire sphere, but it does guarantee that all points on the surface are also on the sphere. In this case, the parametric surface is the entire sphere. To verify this in graphical terms, sketch a picture showing geometric interpretations of the "spherical coordinates" u and v. To see what problems can occur, sketch the surface defined by $x = 3\sin\frac{u^2}{u^2+1}\cos v$, $y = 3\cos\frac{u^2}{u^2+1}$ and $z = 3\sin\frac{u^2}{u^2+1}\sin v$. Explain why you do not get the entire sphere. To see a more subtle example of the same problem, sketch the surface $x = \cos u \cosh v$, $y = \sinh v$, $z = \sin u\cosh v$. Use identities to show that $x^2 - y^2 + z^2 = 1$ and identify the surface. Then sketch the surface $x = \cos u\cosh v$, $y = \cos u\sinh v$, $z = \sin u$ and use identities to show that $x^2 - y^2 + z^2 = 1$. Explain why the second surface is not the entire hyperboloid. Explain in words and pictures exactly what the second surface is.

Review Exercises

WRITING EXERCISES

The following list includes terms that are defined and theorems that are stated in this chapter. For each term or theorem, (1) give a precise definition or statement, (2) state in general terms what it means and (3) describe the types of problems with which it is associated.

Vector-valued function	Tangential component	Normal component
		Continuous $\mathbf{F}(x)$
Tangent vector	Arc length	Speed
Angular velocity	Velocity vector	Angular momentum
Arc length parameter	Angular acceleration	Principal unit normal
	Curvature	Osculating circle
Binormal vector	Radius of curvature	Parametric surface

TRUE OR FALSE

State whether each statement is true or false and briefly explain why. If the statement is false, try to "fix it" by modifying the given statement to a new statement that is true.

1. The graph of the vector-valued function $\langle\cos t, \sin t, f(t)\rangle$, for some function f lies on the circle $x^2 + y^2 = 1$.
2. For vector-valued functions, derivatives are found component by component and all of the usual rules (product, quotient, chain) apply.
3. The derivative of a vector-valued function gives the slope of the tangent line.
4. Newton's laws apply only to straight-line motion and not to rotational motion.
5. The greater the osculating circle, the greater the curvature.
6. While driving a car, the vector **T** would be "straight ahead," **N** in the direction you are turning and **B** straight up.
7. If parametric equations for a surface are $x = x(u, v)$, $y = y(u, v)$ and $z = z(u, v)$ with $x^2 + y^2 + z^2 = 1$, the surface is a sphere.

In exercises 1 and 2, sketch the curve and plot the values of the vector-valued function.

1. $\mathbf{r}(t) = \langle t^2, 2 - t^2, 1\rangle$, $t = 0$, $t = 1$, $t = 2$
2. $\mathbf{r}(t) = \langle\sin t, 2\cos t, 3\rangle$, $t = -\pi$, $t = 0$, $t = \pi$

In exercises 3–12, sketch the curve traced out by the given vector-valued function.

3. $\mathbf{r}(t) = \langle 3\cos t + 1, \sin t\rangle$
4. $\mathbf{r}(t) = \langle 2\sin t, \cos t + 2\rangle$
5. $\mathbf{r}(t) = \langle 3\cos t + 2\sin 3t, 3\sin t + 2\cos 3t\rangle$
6. $\mathbf{r}(t) = \langle 3\cos t + \sin 3t, 3\sin t + \cos 3t\rangle$
7. $\mathbf{r}(t) = \langle 2\cos t, 3, 3\sin t\rangle$
8. $\mathbf{r}(t) = \langle 3\cos t, -2, 2\sin t\rangle$
9. $\mathbf{r}(t) = \langle 4\cos 3t + 6\cos t, 6\sin t, 4\sin 3t\rangle$

Review Exercises

10. $\mathbf{r}(t) = \langle \sin \pi t, \sqrt{t^2 + t^3}, \cos \pi t \rangle$

11. $\mathbf{r}(t) = \langle \tan t, 4 \cos t, 4 \sin t \rangle$

12. $\mathbf{r}(t) = \langle \cos 5t, \tan t, 6 \sin t \rangle$

13. In parts (a)–(f), match the vector-valued function with its graph.

a. $\mathbf{r}(t) = \langle \sin t, t, \sin 2t \rangle$ **b.** $\mathbf{r}(t) = \langle t, \sin t, \sin 2t \rangle$

c. $\mathbf{r}(t) = \langle 6 \sin \pi t, t, 6 \cos \pi t \rangle$ **d.** $\mathbf{r}(t) = \langle \sin^5 t, \sin^2 t, \cos t \rangle$

e. $\mathbf{r}(t) = \langle \cos t, 1 - \cos^2 t, \cos t \rangle$

f. $\mathbf{r}(t) = \langle t^2 + 1, t^2 + 2, t - 1 \rangle$

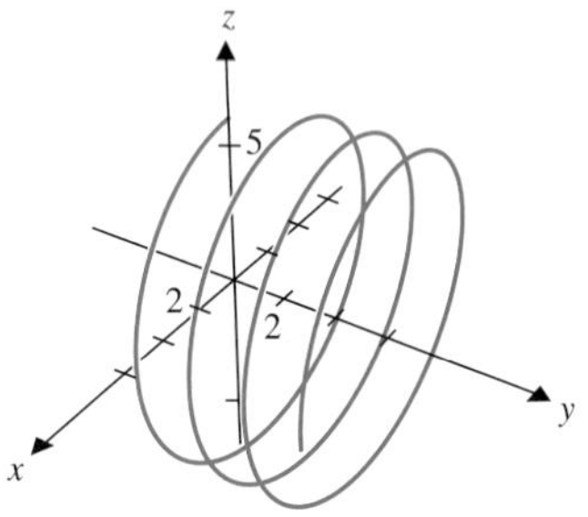

GRAPH A

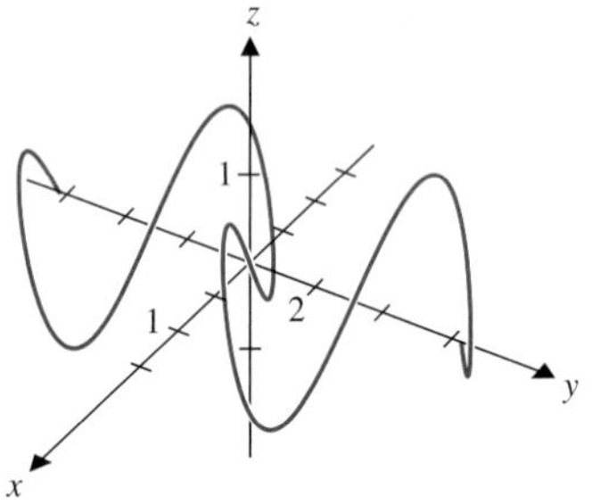

GRAPH B

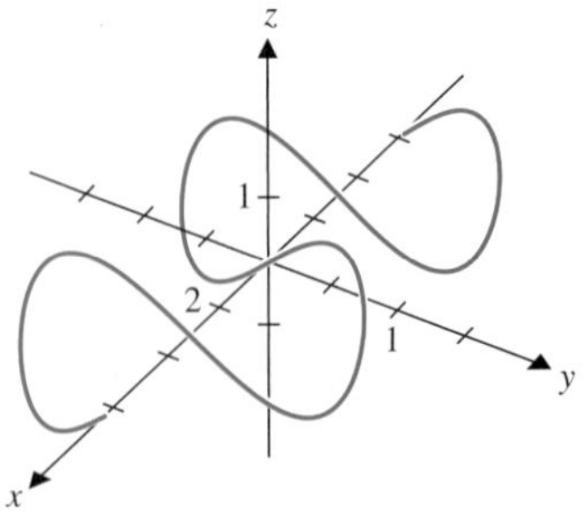

GRAPH C

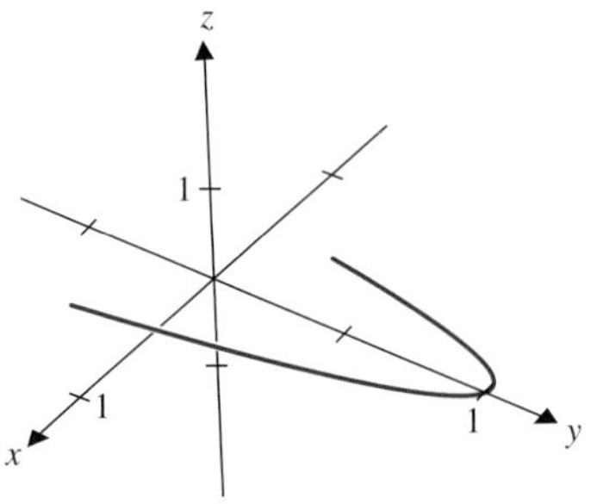

GRAPH D

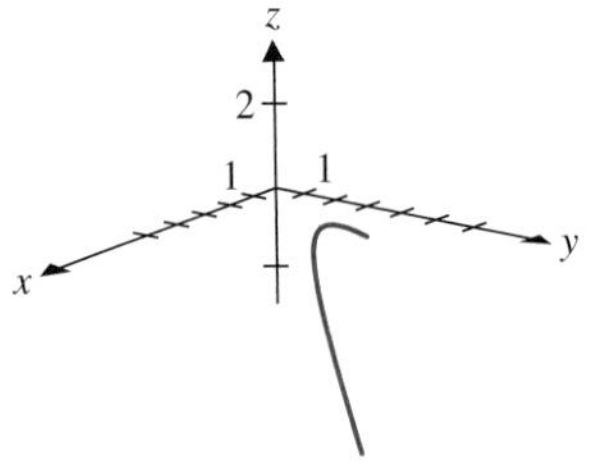

GRAPH E

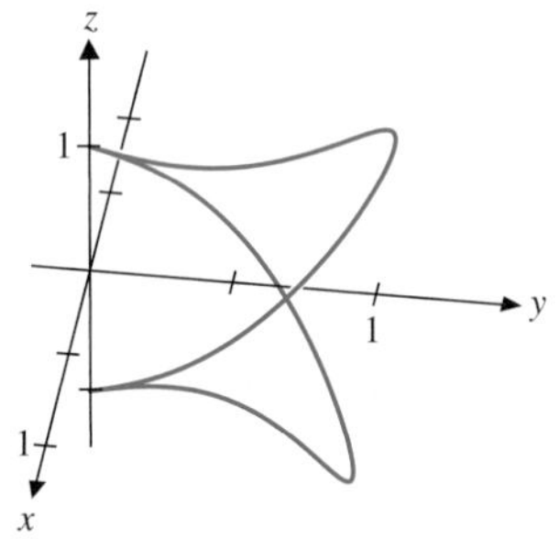

GRAPH F

In exercises 14–16, sketch the curve and find its arc length.

14. $\mathbf{r}(t) = \langle \cos \pi t, \sin \pi t, \cos 4\pi t \rangle, 0 \le t \le 2$

15. $\mathbf{r}(t) = \langle \cos t, \sin t, 6t \rangle, 0 \le t \le 2\pi$

16. $\mathbf{r}(t) = \langle t, 4t - 1, 2 - 6t \rangle, 0 \le t \le 2$

In exercises 17 and 18, find the limit if it exists.

17. $\lim\limits_{t \to 1} \langle t^2 - 1, e^{2t}, \cos \pi t \rangle$

18. $\lim\limits_{t \to 1} \langle e^{-2t}, \csc \pi t, t^3 - 5t \rangle$

Review Exercises

In exercises 19 and 20, determine all values of t at which the given vector-valued function is continuous.

19. $\mathbf{r}(t) = \langle e^{4t}, \ln t^2, 2t \rangle$

20. $\mathbf{r}(t) = \left\langle \sin t, \tan 2t, \dfrac{3}{t^2 - 1} \right\rangle$

In exercises 21 and 22, find the derivative of the given vector-valued function.

21. $\mathbf{r}(t) = \left\langle \sqrt{t^2 + 1}, \sin 4t, \ln 4t \right\rangle$

22. $\mathbf{r}(t) = \langle te^{-2t}, t^3, 5 \rangle$

In exercises 23–26, evaluate the given indefinite or definite integral.

23. $\displaystyle\int \left\langle e^{-4t}, \frac{2}{t^3}, 4t - 1 \right\rangle dt$

24. $\displaystyle\int \left\langle \frac{2t^2}{t^3 + 2}, \sqrt{t + 1} \right\rangle dt$

25. $\displaystyle\int_0^1 \langle \cos \pi t, 4t, 2 \rangle \, dt$

26. $\displaystyle\int_0^2 \langle e^{-3t}, 6t^2 \rangle \, dt$

In exercises 27 and 28, find the velocity and acceleration vectors for the given position vector.

27. $\mathbf{r}(t) = \langle 4\cos 2t, 4\sin 2t, 4t \rangle$

28. $\mathbf{r}(t) = \langle t^2 + 2, 4, t^3 \rangle$

In exercises 29–32, find the position vector from the given velocity or acceleration vector.

29. $\mathbf{v}(t) = \langle 2t + 4, -32t \rangle$, $\mathbf{r}(0) = \langle 2, 1 \rangle$

30. $\mathbf{v}(t) = \langle 4, t^2 - 1 \rangle$, $\mathbf{r}(0) = \langle -4, 2 \rangle$

31. $\mathbf{a}(t) = \langle 0, -32 \rangle$, $\mathbf{v}(0) = \langle 4, 3 \rangle$, $\mathbf{r}(0) = \langle 2, 6 \rangle$

32. $\mathbf{a}(t) = \langle t, e^{2t} \rangle$, $\mathbf{v}(0) = \langle 2, 0 \rangle$, $\mathbf{r}(0) = \langle 4, 0 \rangle$

In exercises 33 and 34, find the force acting on an object of mass 4 with the given position vector.

33. $\mathbf{r}(t) = \langle 12t, 12 - 16t^2 \rangle$

34. $\mathbf{r}(t) = \langle 3\cos 2t, 2\sin 2t \rangle$

In exercises 35 and 36, a projectile is fired with initial speed v_0 feet per second from a height of h feet at an angle of θ above the horizontal. Assuming that the only force acting on the object is gravity, find the maximum altitude, horizontal range and speed at impact.

35. $v_0 = 80$, $h = 0$, $\theta = \frac{\pi}{12}$

36. $v_0 = 80$, $h = 6$, $\theta = \frac{\pi}{4}$

In exercises 37 and 38, find the unit tangent vector to the curve at the indicated points.

37. $\mathbf{r}(t) = \langle e^{-2t}, 2t, 4 \rangle$, $t = 0$, $t = 1$

38. $\mathbf{r}(t) = \langle 2, \sin \pi t^2, \ln t \rangle$, $t = 1$, $t = 2$

In exercises 39–42, find the curvature of the curve at the indicated points.

39. $\mathbf{r}(t) = \langle \cos t, \sin t, \sin t \rangle$, $t = 0$, $t = \frac{\pi}{4}$

40. $\mathbf{r}(t) = \langle 4\cos 2t, 3\sin 2t \rangle$, $t = 0$, $t = \frac{\pi}{4}$

41. $\mathbf{r}(t) = \langle 4, 3t \rangle$, $t = 0$, $t = 1$

42. $\mathbf{r}(t) = \langle t^2, t^3, t^4 \rangle$, $t = 0$, $t = 2$

In exercises 43 and 44, find the unit tangent and principal unit normal vectors at the given points.

43. $\mathbf{r}(t) = \langle \cos t, \sin t, \sin t \rangle$ at $t = 0$

44. $\mathbf{r}(t) = \langle \cos t, \sin t, \sin t \rangle$ at $t = \frac{\pi}{2}$

In exercises 45 and 46, find the tangential and normal components of acceleration at the given points.

45. $\mathbf{r}(t) = \langle 2t, t^2, 2 \rangle$ at $t = 0$, $t = 1$

46. $\mathbf{r}(t) = \langle t^2, 3, 2t \rangle$ at $t = 0$, $t = 2$

In exercises 47 and 48, the friction force required to keep a car from skidding on a curve is given by $\mathbf{F}_s(t) = ma_N\mathbf{N}(t)$. Find the friction force needed to keep a car of mass $m = 120$ (slugs) from skidding for the given position vectors.

47. $\mathbf{r}(t) = \langle 80\cos 6t, 80\sin 6t \rangle$

48. $\mathbf{r}(t) = \langle 80\cos 4t, 80\sin 4t \rangle$

In exercises 49–52, sketch a graph of the parametric surface.

49. $x = 2\sin u$, $y = v^2$, $z = 3\cos u$

50. $x = \cos u \sin v$, $y = \sin u \sin v$, $z = \sin v$

Review Exercises

51. $x = u^2, y = v^2, z = u + 2v$

52. $x = (3 + 2\cos u)\cos v,\ y = (3 + 2\cos u)\sin v,\ z = 2\cos v$

53. Match the parametric equations with the surfaces.

a. $x = u^2, y = u + v, z = v^2$

b. $x = u^2, y = u + v, z = v$

c. $x = u, y = u + v, z = v^2$

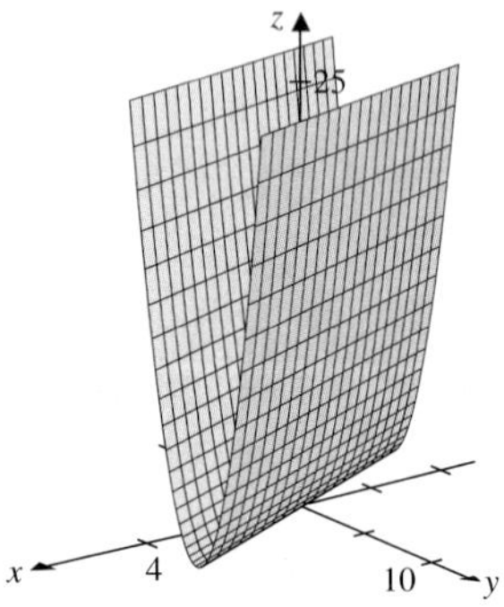

SURFACE A

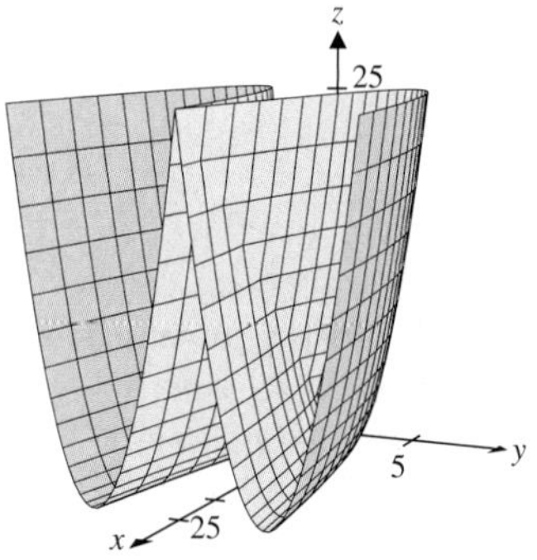

SURFACE B

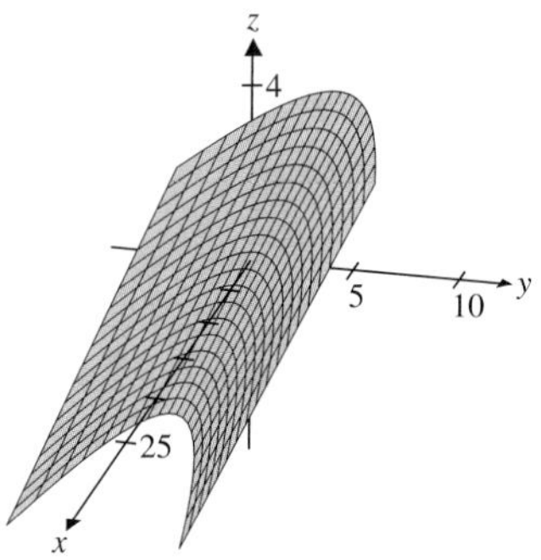

SURFACE C

54. Find a parametric representation of $x^2 + y^2 + z^2 = 9$.

EXPLORATORY EXERCISES

1. In this three-dimensional projectile problem, think of the x-axis as pointing to the right, the y-axis as pointing straight ahead and the z-axis as pointing up. Suppose that a projectile of mass $m = 1$ kg is launched from the ground with an initial velocity of 100 m/s in the yz-plane at an angle of $\frac{\pi}{6}$ above the horizontal. The spinning of the projectile produces a constant Magnus force of $\langle 0.1, 0, 0\rangle$ newtons. Find a vector for the position of the projectile at time $t \geq 0$. Assuming level ground, find the time of flight T for the projectile and find its landing place. Find the curvature for the path of the projectile at time $t \geq 0$. Find the times of minimum and maximum curvature of the path for $0 \leq t \leq T$.

2. A tennis serve is struck at an angle θ below the horizontal from a height of 8 feet and with initial speed 120 feet per second. For the serve to count (be "in"), it must clear a net that is 39 feet away and 3 feet high and must land before the service line 60 feet away. Find the range of angles for which the serve is in.

3. A baseball pitcher throws a pitch at an angle θ below the horizontal from a height of 6 feet with an initial speed of 130 feet per second. Home plate is 60 feet away. For the pitch to be a strike, the ball must cross home plate between 20″ and 42″ above the ground. Find the range of angles for which the pitch will be a strike.

Functions of Several Variables and Partial Differentiation

CHAPTER

12

Few things in baseball are as exciting as a home run. In the summer of 2004, Barry Bonds chased the all-time home run record held by Hank Aaron. Every time Bonds hit the ball, fans watched in anxious anticipation as the ball reached its peak height and then slowly dropped back to the field. Would the ball clear the fence and stay fair for a home run? Through years of experience, players can usually tell exactly where the ball will land. However, since baseballs do not follow simple parabolic paths, most spectators must wait for the ball to land to see whether a given fly ball is a home run.

Think for a minute about the factors that determine how far the ball goes. In our studies of projectile motion, we identified three forces that affect the path: gravity, drag and the Magnus force. If we know the initial velocity (both speed and angle) and initial spin, we can write down a differential equation whose solution closely approximates the flight of the ball. This gives us distance as a function of speed, angle and spin. In this chapter, we introduce some of the basic techniques needed to analyze functions of two or more variables. While many of the ideas are familiar, the details change as we move from one to two or more variables.

You have probably realized that our situation is really far more complicated than outlined here. Air drag depends on environmental factors such as temperature and humidity. Other factors include the type of pitch thrown, the wind velocity and the type of bat used. With all of these factors, we would need a function of ten or more variables! Fortunately, the calculus of functions of ten variables is very similar to the calculus of functions of two or three variables. The theory presented in this chapter is easily extended to as many variables as are needed in a particular application.

After studying the basic calculus for functions of several variables, you should be able to find extrema of relatively simple functions. Perhaps more importantly, you should understand enough about such functions to be able to approximate extrema of more complicated functions. Of course, in real applications, you are rarely given a convenient formula. Even so, the understanding of multivariable calculus that you develop here will help you to make sense of a broad range of complex phenomena.

12.1 FUNCTIONS OF SEVERAL VARIABLES

The first ten chapters of this book focused on functions $f(x)$ whose domain and range were subsets of the real numbers. In Chapter 11, we studied vector-valued functions $\mathbf{F}(t)$ whose domain was a subset of the real numbers, but whose range was a set of vectors in two or more dimensions. In this section, we expand our concept of function to include functions that depend on more than one variable, that is, functions whose *domain* is multidimensional.

A **function of two variables** is a rule that assigns a real number $f(x, y)$ to each ordered pair of real numbers (x, y) in the domain of the function. For a function f defined on the domain $D \subset \mathbb{R}^2$, we sometimes write $f: D \subset \mathbb{R}^2 \to \mathbb{R}$, to indicate that f maps points in two dimensions to real numbers. You may think of such a function as a rule whose input is a pair of real numbers and whose output is a single real number. For instance, $f(x, y) = xy^2$ and $g(x, y) = x^2 - e^y$ are both functions of the two variables x and y.

Likewise, a **function of three variables** is a rule that assigns a real number $f(x, y, z)$ to each ordered triple of real numbers (x, y, z) in the domain $D \subset \mathbb{R}^3$ of the function. We sometimes write $f: D \subset \mathbb{R}^3 \to \mathbb{R}$ to indicate that f maps points in three dimensions to real numbers. For instance, $f(x, y, z) = xy^2 \cos z$ and $g(x, y, z) = 3zx^2 - e^y$ are both functions of the three variables x, y and z.

We can similarly define functions of four (or five or more) variables. Our focus here is on functions of two and three variables, although most of our results can be easily extended to higher dimensions.

Unless specifically stated otherwise, the domain of a function of several variables is taken to be the set of all values of the variables for which the given expression is defined.

multivariable → # of independent variables

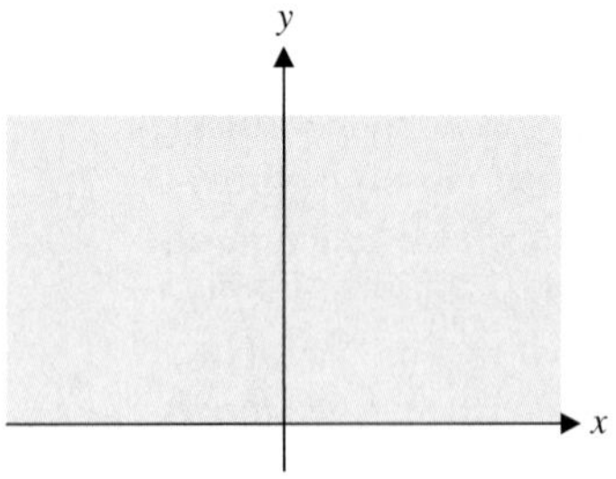

FIGURE 12.1a
The domain of $f(x, y) = x \ln y$

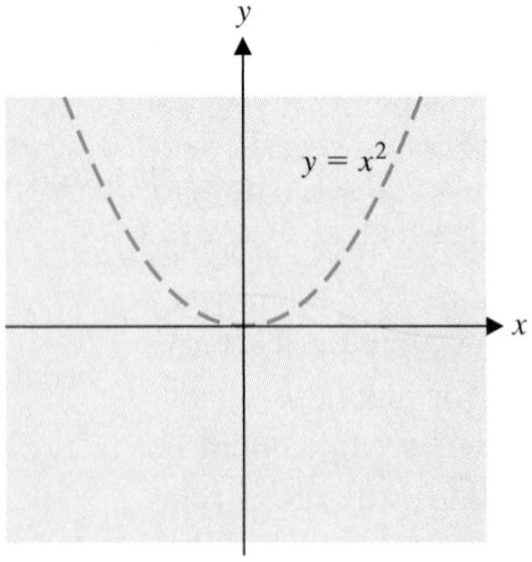

FIGURE 12.1b
The domain of $g(x, y) = \dfrac{2x}{y - x^2}$

EXAMPLE 1.1 Finding the Domain of a Function of Two Variables

Find and sketch the domain for (a) $f(x, y) = x \ln y$ and (b) $g(x, y) = \dfrac{2x}{y - x^2}$.

Solution (a) For $f(x, y) = x \ln y$, recall that $\ln y$ is defined only for $y > 0$. The domain of f is then the set $D = \{(x, y) | y > 0\}$, that is, the half-plane lying above the x-axis (see Figure 12.1a).

(b) For $g(x, y) = \dfrac{2x}{y - x^2}$, note that g is defined unless there is a division by zero, which occurs when $y - x^2 = 0$. The domain of g is then $\{(x, y) | y \neq x^2\}$, which is the entire xy-plane with the parabola $y = x^2$ removed (see Figure 12.1b). ■

EXAMPLE 1.2 Finding the Domain of a Function of Three Variables

Find and describe in graphical terms the domains of (a) $f(x, y, z) = \dfrac{\cos(x + z)}{xy}$ and (b) $g(x, y, z) = \sqrt{9 - x^2 - y^2 - z^2}$.

Solution (a) For $f(x, y, z) = \dfrac{\cos(x+z)}{xy}$, there is a division by zero if $xy = 0$, which occurs if $x = 0$ or $y = 0$. The domain is then $\{(x, y, z) | x \neq 0 \text{ and } y \neq 0\}$, which is all of three-dimensional space, excluding the yz-plane ($x = 0$) and the xz-plane ($y = 0$).

(b) Notice that for $g(x, y, z) = \sqrt{9 - x^2 - y^2 - z^2}$ to be defined, you'll need to have $9 - x^2 - y^2 - z^2 \geq 0$, or $x^2 + y^2 + z^2 \leq 9$. The domain of g is then the sphere of radius 3 centered at the origin and its interior. ■

In many applications, you won't have a formula representing a function of interest. Rather, you may know values of the function at only a relatively small number of points, as in example 1.3.

EXAMPLE 1.3 A Function Defined by a Table of Data

A computer simulation of the flight of a baseball provided the data displayed in the accompanying table for the range in feet of a ball hit with initial velocity v ft/s and backspin rate of ω rpm. Each ball is struck at an angle of 30° above the horizontal.

v \ ω	0	1000	2000	3000	4000
150	294	312	333	350	367
160	314	334	354	373	391
170	335	356	375	395	414
180	355	376	397	417	436

Range of baseball in feet

Thinking of the range as a function $R(v, \omega)$, find $R(180, 0)$, $R(160, 0)$, $R(160, 4000)$ and $R(160, 2000)$. Discuss the results in baseball terms.

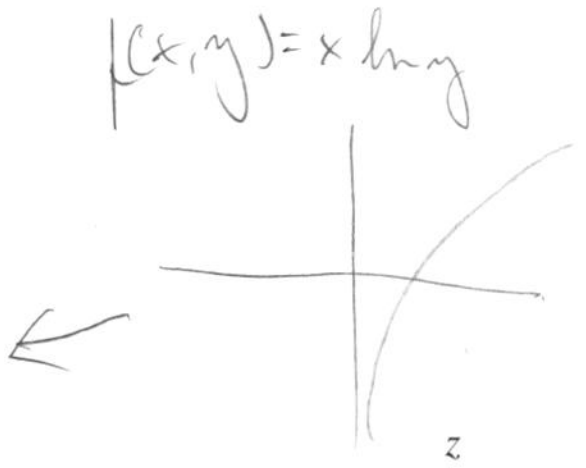

Solution The function values are found by looking in the row with the given value of v and the column with the given value of ω. Thus, $R(180, 0) = 355$, $R(160, 0) = 314$, $R(160, 4000) = 391$ and $R(160, 2000) = 354$. This says that a ball with no backspin and initial velocity 180 ft/s flies 41 ft farther than one with initial velocity 160 ft/s (no surprise there). However, observe that if a 160 ft/s ball also has backspin of 4000 rpm, it actually flies 36 ft farther than the 180 ft/s ball with no backspin. (The backspin gives the ball a lift force that keeps it in the air longer.) The combination of 160 ft/s and 2000 rpm produces almost exactly the same distance as 180 ft/s with no spin. (Watts and Bahill estimate that hitting the ball $\frac{1}{4}''$ below center produces 2000 rpm.) Thus, both initial velocity and spin have significant effects on the distance the ball flies. ■

The **graph** of the function $f(x, y)$ is the graph of the equation $z = f(x, y)$. This is not new, as you have already graphed a number of quadric surfaces that represent functions of two variables.

EXAMPLE 1.4 Graphing Functions of Two Variables

Graph (a) $f(x, y) = x^2 + y^2$ and (b) $g(x, y) = \sqrt{4 - x^2 + y^2}$.

Solution (a) For $f(x, y) = x^2 + y^2$, you may recognize the surface $z = x^2 + y^2$ as a circular paraboloid. Notice that the traces in the planes $z = k > 0$ are circles, while the traces in the planes $x = k$ and $y = k$ are parabolas. A graph is shown in Figure 12.2a.

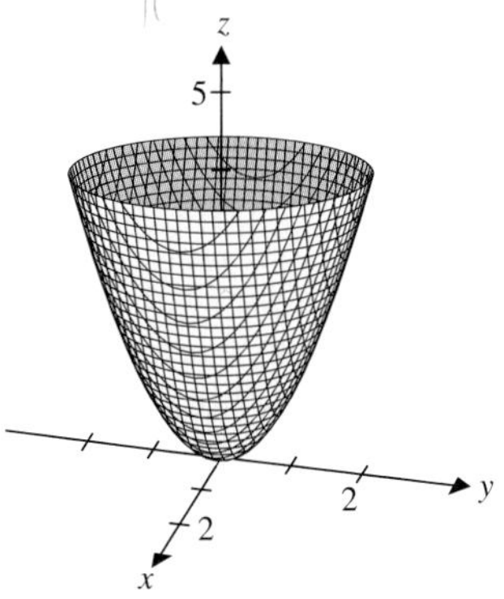

FIGURE 12.2a
$z = x^2 + y^2$

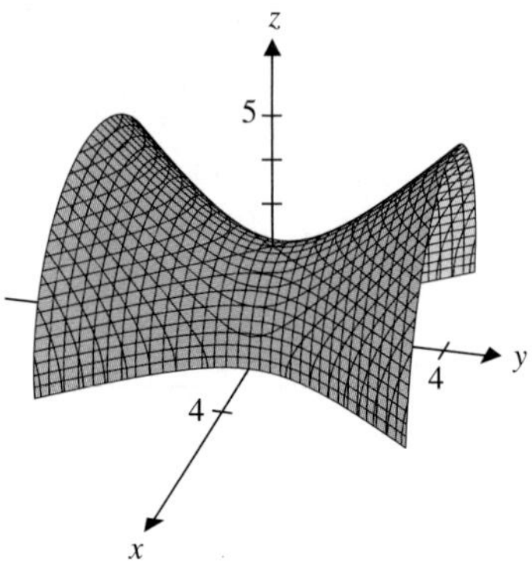

FIGURE 12.2b
$z = \sqrt{4 - x^2 + y^2}$

(b) For $g(x, y) = \sqrt{4 - x^2 + y^2}$, note that the surface $z = \sqrt{4 - x^2 + y^2}$ is the top half of the surface $z^2 = 4 - x^2 + y^2$ or $x^2 - y^2 + z^2 = 4$. Here, observe that the traces in the planes $x = k$ and $z = k$ are hyperbolas, while the traces in the planes $y = k$ are circles. This gives us a hyperboloid of one sheet, wrapped around the y-axis. The graph of $z = g(x, y)$ is the top half of the hyperboloid, as shown in Figure 12.2b. ■

Recall from your earlier experience drawing surfaces in three dimensions that an analysis of traces is helpful in sketching many graphs.

EXAMPLE 1.5 Graphing Functions in Three Dimensions

Graph (a) $f(x, y) = \sin x \cos y$ and (b) $g(x, y) = e^{-x^2}(y^2 + 1)$.

Solution (a) For $f(x, y) = \sin x \cos y$, notice that the traces in the planes $y = k$ are the sine curves $z = \sin x \cos k$, while its traces in the planes $x = k$ are the cosine curves $z = \sin k \cos y$. The traces in the planes $z = k$ are the curves $k = \sin x \cos y$. These are a bit more unusual, as seen in Figure 12.3a (which is computer-generated) for $k = 0.5$. The surface should look like a sine wave in all directions, as shown in the computer-generated plot in Figure 12.3b.

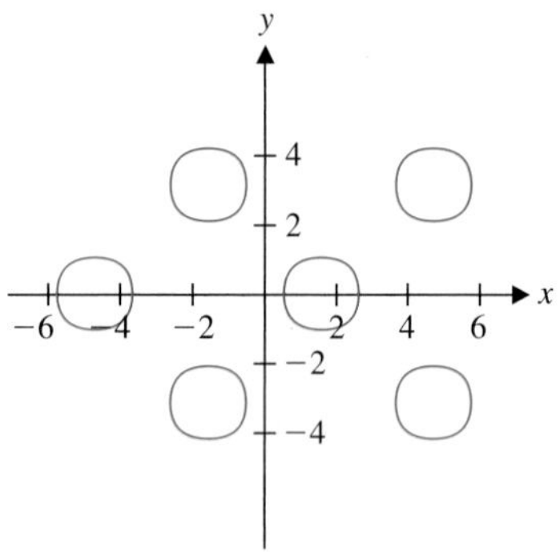

FIGURE 12.3a
The traces of the surface $z = \sin x \cos y$ in the plane $z = 0.5$

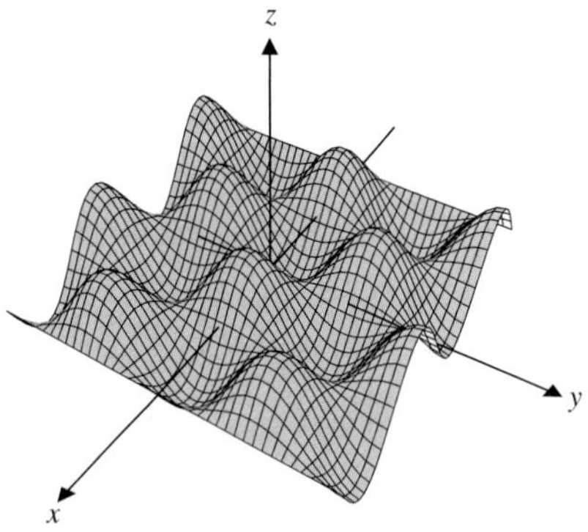

FIGURE 12.3b
$z = \sin x \cos y$

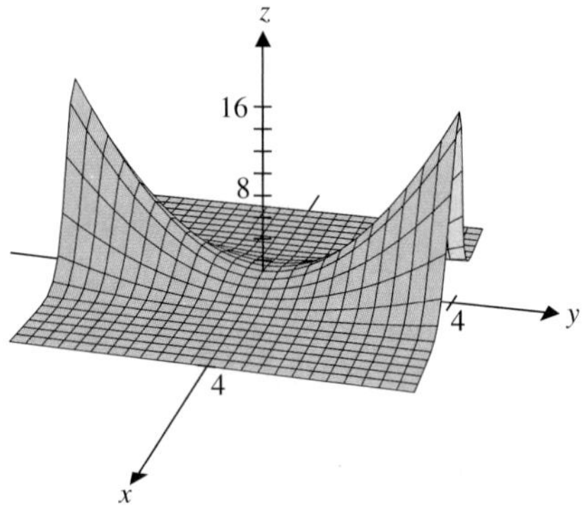

FIGURE 12.3c
$z = e^{-x^2}(y^2 + 1)$

look along x axis   look along y axis

(b) For $g(x, y) = e^{-x^2}(y^2 + 1)$, observe that the traces of the surface in the planes $x = k$ are parabolic, while the traces in the planes $y = k$ are proportional to $z = e^{-x^2}$, which are bell-shaped curves. The traces in the planes $z = k$ are not particularly helpful here. A sketch of the surface is shown in Figure 12.3c. ■

Graphing functions of more than one variable is not a simple business. For most functions of two variables, you must take hints from the expressions and try to piece together the clues to identify the surface. Your knowledge of functions of one variable is critical here.

EXAMPLE 1.6 Matching a Function of Two Variables to Its Graph

Match the functions $f_1(x, y) = \cos(x^2 + y^2)$, $f_2(x, y) = \cos(e^x + e^y)$, $f_3(x, y) = \ln(x^2 + y^2)$ and $f_4(x, y) = e^{-xy}$ to the surfaces shown in Figures 12.4a–12.4d.

NOTES

You may be tempted to use computer-generated graphs throughout this chapter. However, we must emphasize that our goal is an *understanding* of three-dimensional graphs, which can best be obtained by sketching many graphs by hand. Doing this will help you know whether a computer-generated graph is accurate or misleading. Even when you use a graphing utility to produce a three-dimensional graph, we urge you to think through the traces, as we did in example 1.5.

Solution There are two properties of $f_1(x, y)$ that you should immediately notice. First, since the cosine of any angle lies between -1 and 1, $z = f_1(x, y)$ must always lie between -1 and 1. Second, the expression $x^2 + y^2$ is significant. Given any value of r, and any point (x, y) on the circle $x^2 + y^2 = r^2$, the height of the surface at the point (x, y) is a constant, given by $z = f_1(x, y) = \cos(r^2)$. Look for a surface that is bounded (this rules out Figure 12.4a) and has circular cross sections parallel to the xy-plane (ruling out Figures 12.4b and 12.4d). That leaves Figure 12.4c for the graph of $z = f_1(x, y)$.

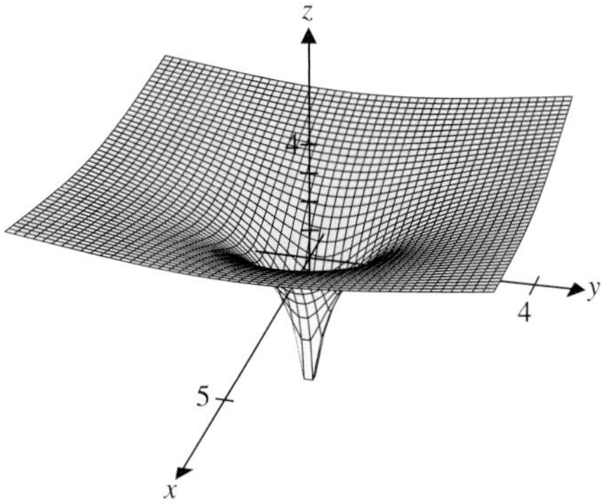

FIGURE 12.4a

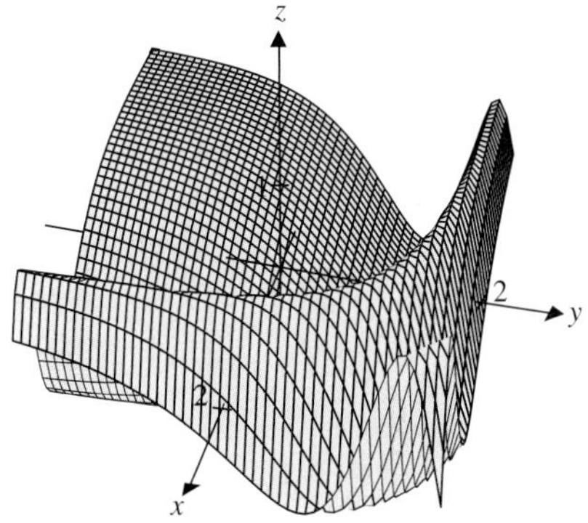

FIGURE 12.4b

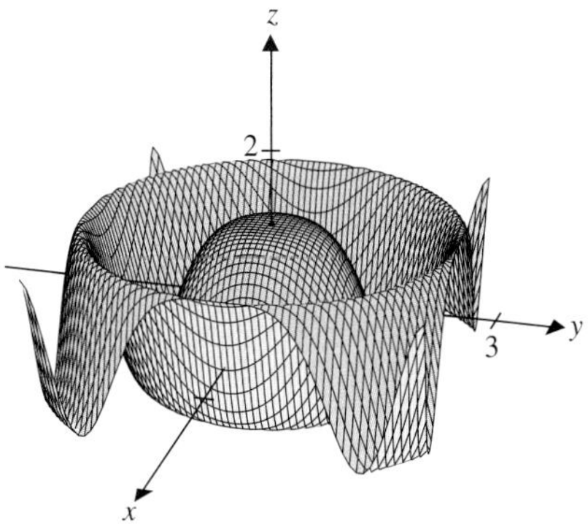

FIGURE 12.4c

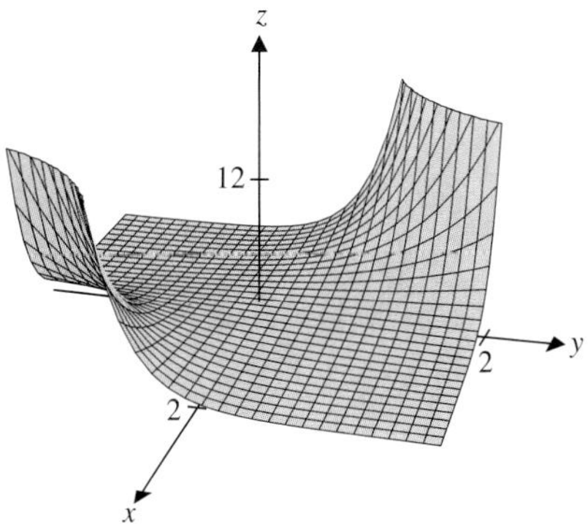

FIGURE 12.4d

You should notice that $y = f_3(x, y)$ also has circular cross sections parallel to the xy-plane, again because of the expression $x^2 + y^2$. (Think of polar coordinates.) Another important property of $f_3(x, y)$ for you to recognize is that the logarithm tends to $-\infty$ as its argument (in this case, $x^2 + y^2$) approaches 0. This appears to be what is indicated in Figure 12.4a, with the surface dropping sharply toward the center of the sketch. So, $z = f_3(x, y)$ corresponds to Figure 12.4a.

The remaining two functions involve exponentials. The most important distinction between them is that $f_2(x, y)$ lies between -1 and 1, due to the cosine term. This suggests that the graph of $f_2(x, y)$ is given in Figure 12.4b. To avoid jumping to a decision prematurely (after all, the domains used to produce these figures are all slightly different and could be misleading), make sure that the properties of $f_4(x, y)$ correspond to Figure 12.4d. Note that $e^{-xy} \to 0$ as $xy \to \infty$ and $e^{-xy} \to \infty$ as $xy \to -\infty$. As you move away from the origin in regions where x and y have the same sign, the surface should approach the xy-plane ($z = 0$). In regions where x and y have opposite signs, the surface should rise sharply. Notice that this behavior is exactly what you are seeing in Figure 12.4d. ■

REMARK 1.1

The analysis we used in example 1.6 may seem a bit slow, but we urge you to practice this on your own. The more you think (carefully) about how the properties of functions correspond to the structures of surfaces in three dimensions, the easier this chapter will be.

As with any use of technology, the creation of informative three-dimensional graphs can require a significant amount of knowledge and trial-and-error exploration. Even when you have an idea of what a graph should look like (and most often you won't), you may need to change the viewing window several times before you can clearly see a particular feature. The wireframe graph in Figure 12.5a is a poor representation of $f(x, y) = x^2 + y^2$. Notice that this graph shows numerous traces in the planes $x = c$ and $y = c$ for $-5 \le c \le 5$. However, no traces are drawn in planes parallel to the xy-plane, so you get no sense that the figure has circular cross sections. One way to improve this is to limit the range of z-values to $0 \le z \le 20$, as in Figure 12.5b. Observe that cutting off the graph here (i.e., not displaying all values of z for the displayed values of x and y) reveals the circular cross section at $z = 20$. An even better plot is obtained by using the parametric representation $x = u \cos v$, $y = u \sin v$, $z = u^2$, as shown in Figure 12.5c.

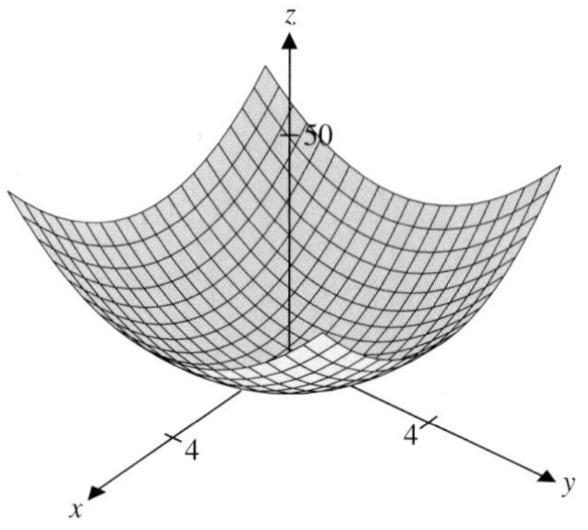

FIGURE 12.5a
$z = x^2 + y^2$

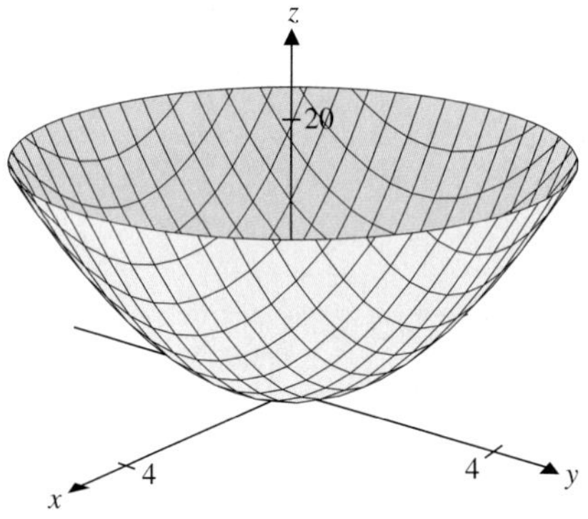

FIGURE 12.5b
$z = x^2 + y^2$

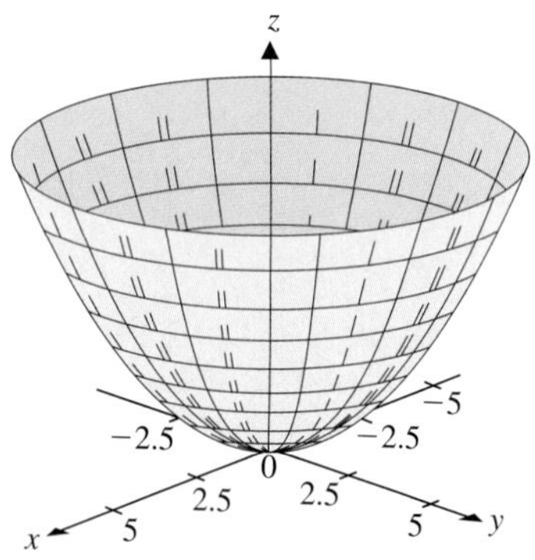

FIGURE 12.5c
$z = x^2 + y^2$

An important feature of three-dimensional graphs that is not present in two-dimensional graphs is the **viewpoint** from which the graph is drawn. In Figures 12.5a and 12.5b, we are looking at the paraboloid from a viewpoint that is above the xy-plane and between the positive x- and y-axes. This is the default viewpoint for many graphing utilities and is very similar to the way we have drawn graphs by hand. Figure 12.3c shows the default viewpoint of $f(x, y) = e^{-x^2}(y^2 + 1)$. In Figure 12.6a, we switch the viewpoint to the positive y-axis, from which we can see the bell-shaped profile of the graph. This viewpoint shows us several traces with $y = c$, so that we see a number of curves of the form $z = ke^{-x^2}$. In Figure 12.6b, the viewpoint is the positive x-axis, so that we see parabolic traces of the form $z = k(y^2 + 1)$. Figure 12.6c shows the view from high above the x-axis.

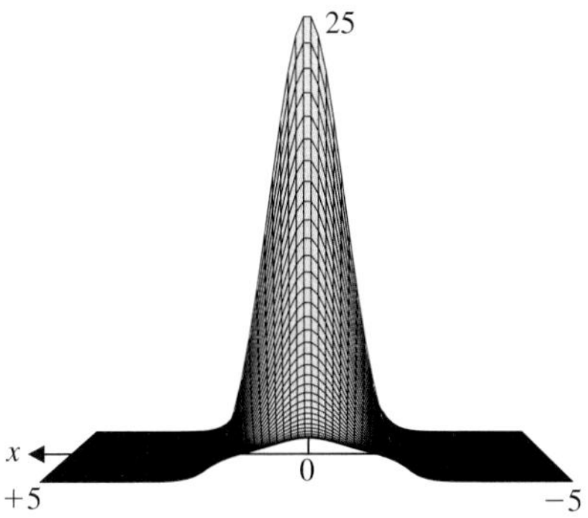

FIGURE 12.6a
$z = e^{-x^2}(y^2 + 1)$

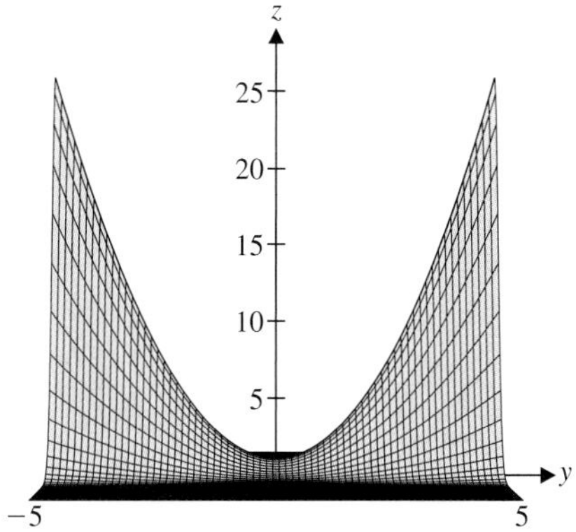

FIGURE 12.6b
$z = e^{-x^2}(y^2 + 1)$

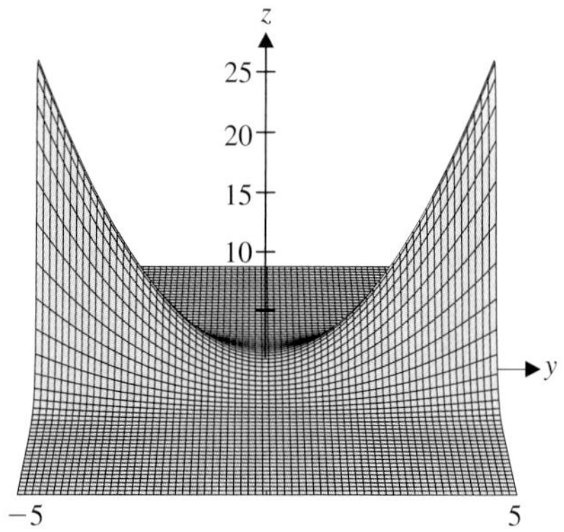

FIGURE 12.6c
$z = e^{-x^2}(y^2 + 1)$

Many graphing utilities offer alternatives to wireframe graphs. One deficiency of wireframe graphs is the lack of traces parallel to the xy-plane. This is not a problem in Figures 12.6a to 12.6c, where traces in the planes $z = c$ are too complicated to be helpful. However, in Figures 12.5a and 12.5b, the circular cross sections provide valuable information about the structure of the graph. To see such traces, many graphing utilities provide a "contour mode" or "parametric surface" option. These are shown in Figures 12.7a and 12.7b for $f(x, y) = x^2 + y^2$ and are explored further in the exercises.

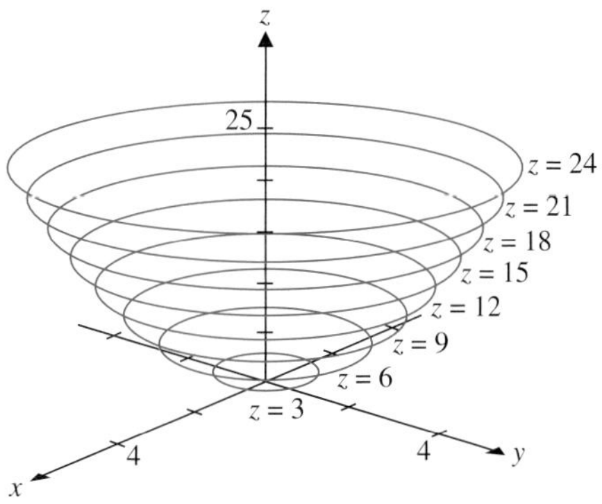

FIGURE 12.7a
$z = x^2 + y^2$ (contour mode)

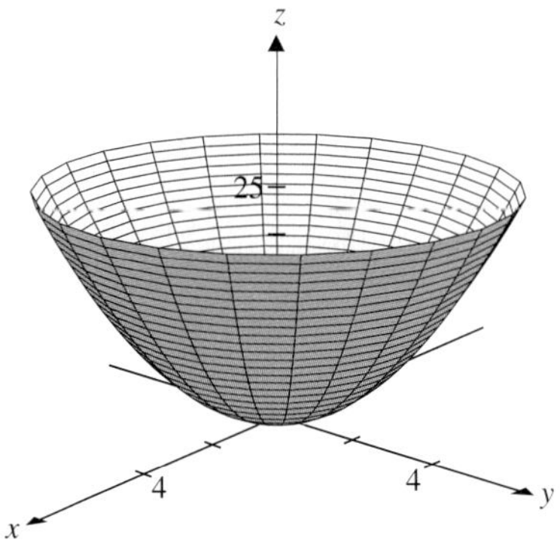

FIGURE 12.7b
$z = x^2 + y^2$ (parametric plot)

Two other types of graphs, the **contour plot** and the **density plot,** provide the same information condensed into a two-dimensional picture. Recall that for two of the surfaces in example 1.6, it was important to recognize that the surface had circular cross sections, since x and y appeared only in the combination $x^2 + y^2$. The contour plot and the density plot will aid in identifying features such as this.

A **level curve** of the function $f(x, y)$ is the (two-dimensional) graph of the equation $f(x, y) = c$, for some constant c. (So, the level curve $f(x, y) = c$ is a two-dimensional graph of the trace of the surface $z = f(x, y)$ in the plane $z = c$.) A **contour plot** of $f(x, y)$ is a graph of numerous level curves $f(x, y) = c$, for representative values of c.

EXAMPLE 1.7 Sketching Contour Plots

Sketch contour plots for (a) $f(x, y) = -x^2 + y$ and (b) $g(x, y) = x^2 + y^2$.

Solution (a) First, note that the level curves of $f(x, y)$ are defined by $-x^2 + y = c$, where c is a constant. Solving for y, you can identify the level curves as the parabolas $y = x^2 + c$. A contour plot with $c = -4, -2, 0, 2$ and 4 is shown in Figure 12.8a.

(b) The level curves for $g(x, y)$ are the circles $x^2 + y^2 = c$. In this case, note that there are level curves *only* for $c \geq 0$. A contour plot with $c = 1, 4, 7$ and 10 is shown in Figure 12.8b. ■

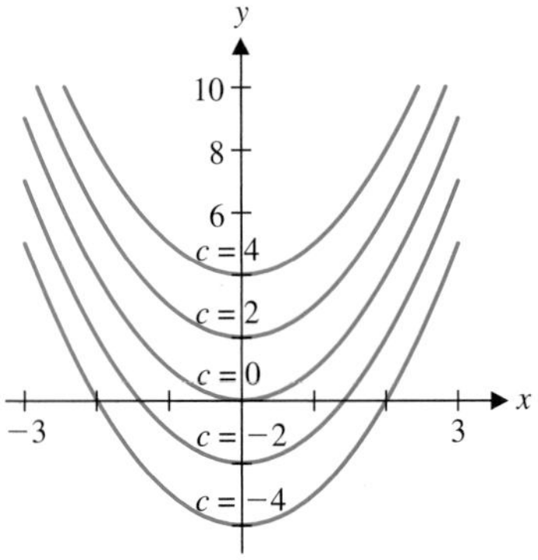

FIGURE 12.8a
Contour plot of $f(x, y) = -x^2 + y$

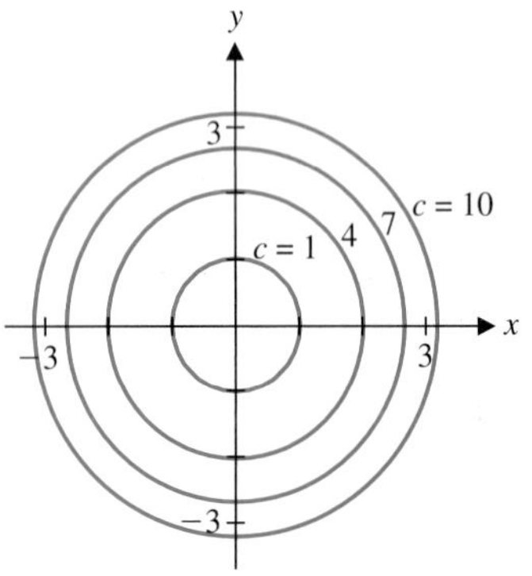

FIGURE 12.8b
Contour plot of $g(x, y) = x^2 + y^2$

Note that in example 1.7, we used values for c that were equally spaced. There is no requirement that you do so, but it can help you to get a sense for how the level curves would "stack up" to produce the three-dimensional graph. We show a more extensive contour plot for $g(x, y) = x^2 + y^2$ in Figure 12.9a. In Figure 12.9b, we show a plot of the surface, with a number of traces drawn (in planes parallel to the xy-plane). Notice that the projections of these traces onto the xy-plane correspond to the contour plot in Figure 12.9a.

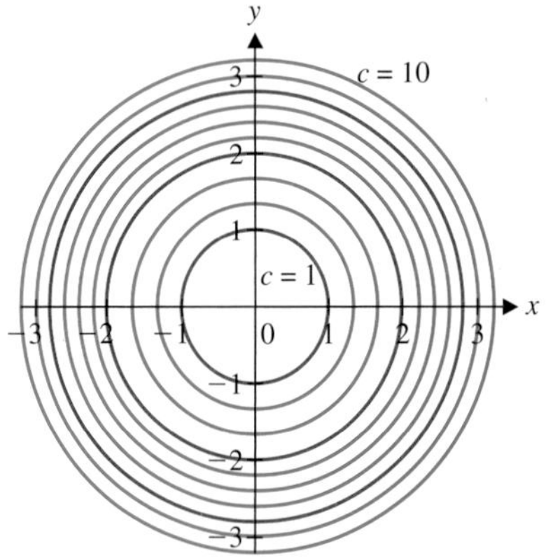

FIGURE 12.9a
Contour plot of $g(x, y) = x^2 + y^2$

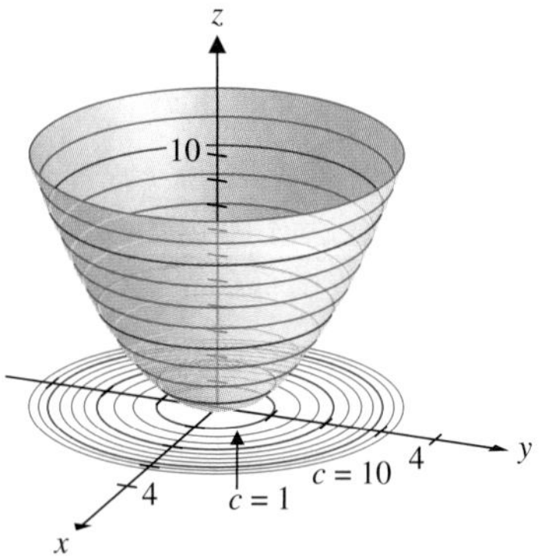

FIGURE 12.9b
$z = x^2 + y^2$

Look carefully at Figure 12.9a and observe that the contour plot indicates that the increase in the radii of the circles is not constant as c increases.

As you might expect, for more complicated functions, the process of matching contour plots with surfaces becomes more challenging.

EXAMPLE 1.8 Matching Surfaces to Contour Plots

Match the surfaces of example 1.6 to the contour plots shown in Figures 12.10a–12.10d.

Solution In Figures 12.4a and 12.4c, the level curves are circular, so these surfaces correspond to the contour plots in Figures 12.10a and 12.10b, but, which is which? The principal feature of the surface in Figure 12.4a is the behavior near the z-axis. Because of the rapid change in the function near the z-axis, there will be a large number of level curves near the origin. (Think about this.) By contrast, the oscillations in Figure 12.4c would produce level curves that alternately get closer together and farther apart. We can conclude that Figure 12.4a matches with Figure 12.10a, while Figure 12.4c matches with Figure 12.10b. Now, consider the two remaining surfaces and level curves. Imagine intersecting the surface in Figure 12.4d with the plane $z = 4$. You

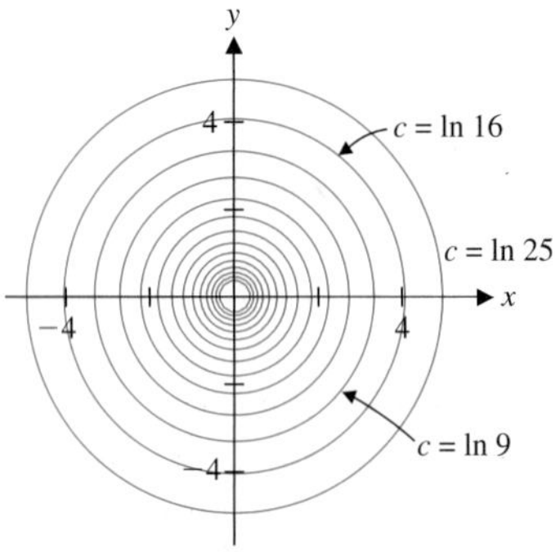

FIGURE 12.10a

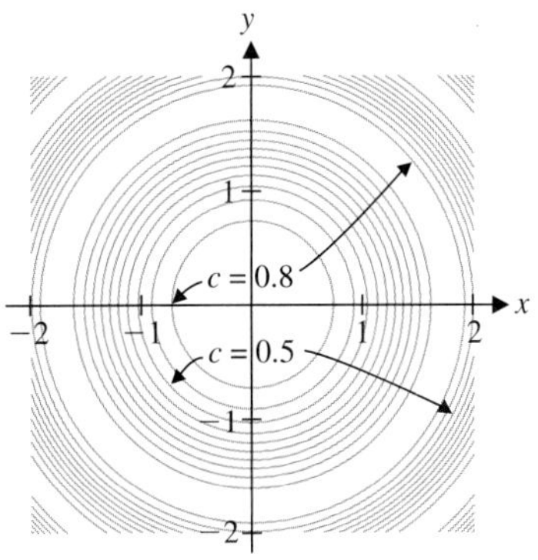

FIGURE 12.10b

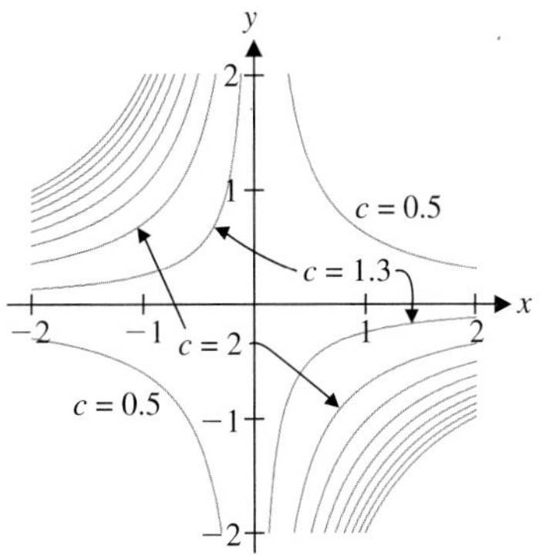

FIGURE 12.10c

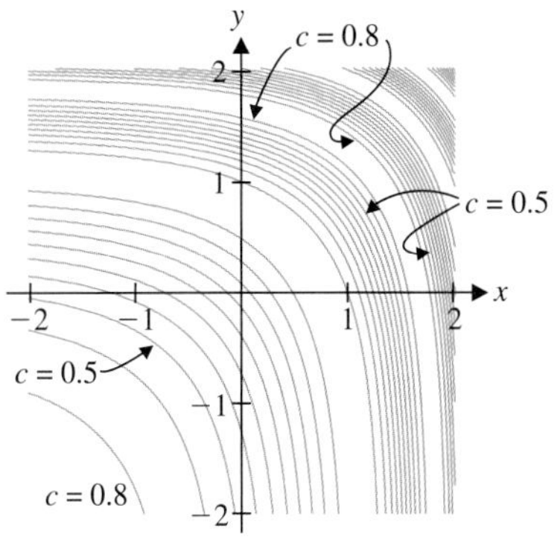

FIGURE 12.10d

would get two separate curves that open in opposite directions (to the lower left and upper right of Figure 12.4d). These correspond to the hyperbolas seen in Figure 12.10c. The final match of Figure 12.10d to Figure 12.4b is more difficult to see, but notice how the curves of Figure 12.10d correspond to the curve of the peaks in Figure 12.4b. (To see this, you will need to adjust for the y-axis pointing up in Figure 12.10d and to the right in Figure 12.4b.) As an additional means of distinguishing the last two graphs, notice that Figure 12.4d is very flat near the origin. This corresponds to the lack of level curves near the origin in Figure 12.10c. By contrast, Figure 12.4b shows oscillation near the origin and there are several level curves near the origin in Figure 12.10d. ■

REMARK 1.2

If the level curves in a contour plot are plotted for equally spaced values of z, observe that a tightly packed region of the contour plot will correspond to a region of rapid change in the function. Alternatively, blank space in the contour plot corresponds to a region of slow change in the function. For this reason, we typically draw contour plots using equally spaced values of z.

A **density plot** is closely related to a contour plot, in that they are both two-dimensional representations of a surface in three dimensions. For a density plot, each pixel is shaded according to the size of the function value at a point representing the pixel, with different colors and shades indicating different function values. In a density plot, notice that level curves can be seen as curves formed by a specific shade.

EXAMPLE 1.9 Matching Functions and Density Plots

Match the density plots in Figures 12.11a–12.11c with the functions

$$f_1(x, y) = \frac{1}{y^2 - x^2}, f_2(x, y) = \frac{2x}{y - x^2} \text{ and } f_3(x, y) = \cos(x^2 + y^2).$$

Solution As we did with contour plots, we start with the most obvious properties of the functions and try to identify the corresponding properties in the density plots. Both $f_1(x, y)$ and $f_2(x, y)$ have gaps in their domains due to divisions by zero. Near the discontinuities, you should expect large function values. Notice that Figure 12.11b shows a lighter color band in the shape of a hyperbola (like $y^2 - x^2 = c$ for a small

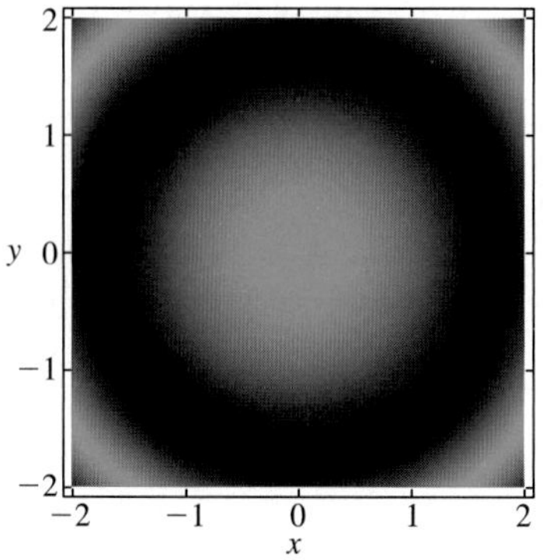

FIGURE 12.11a

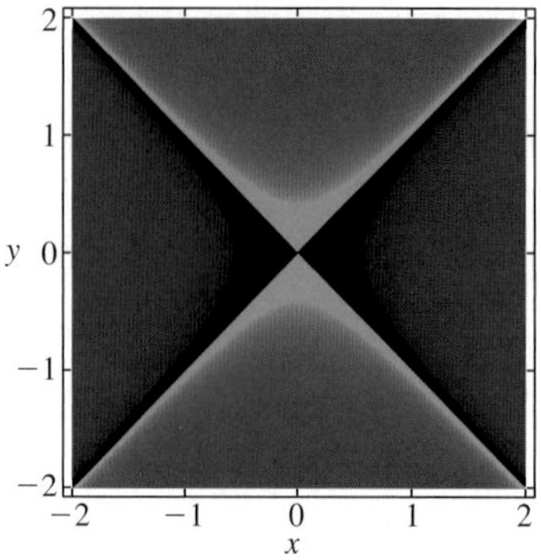

FIGURE 12.11b

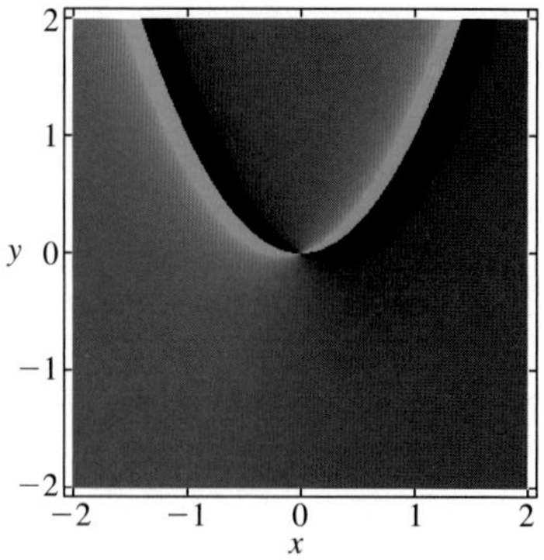

FIGURE 12.11c

number c) and Figure 12.11c shows a lighter color band in the shape of a parabola (like $y - x^2 = 0$). This tells you that the density plot for $f_1(x, y)$ is Figure 12.11b and the density plot for $f_2(x, y)$ is Figure 12.11c. That leaves Figure 12.11a for $f_3(x, y)$. You should be able to see the circular bands in the density plot arising from the $x^2 + y^2$ term in $f_3(x, y)$. ■

There are many examples of contour plots and density plots that you see every day. Weather maps often show level curves of atmospheric pressure (see Figure 12.12a). In this setting, the level curves are called **isobars** (that is, curves along which the barometric pressure is constant). Other weather maps represent temperature or degree of wetness with color coding (see Figure 12.12b), which are essentially density plots.

Scientists also use density plots while studying other climatic phenomena. For instance, in Figures 12.12c and 12.12d, we show two density plots indicating sea-surface height (which correlates with ocean heat content) indicating changes in the El Niño phenomenon over a period of several weeks.

We close this section by briefly looking at the graphs of functions of three variables, $f(x, y, z)$. We won't actually graph any such functions, since a true graph would require

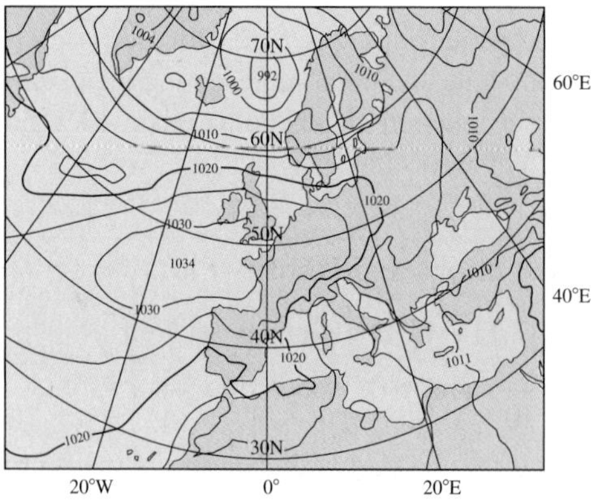

FIGURE 12.12a
Weather map showing barometric pressure

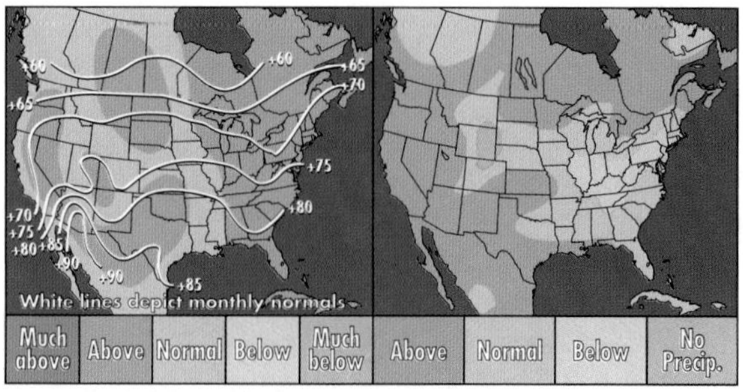

FIGURE 12.12b
Weather maps showing bands of temperature and precipitation

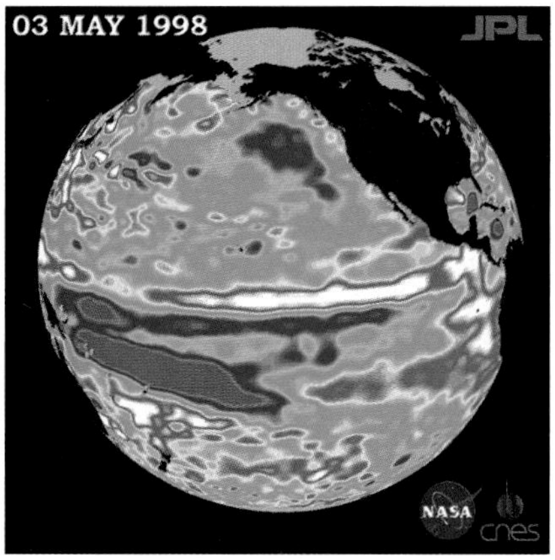

FIGURE 12.12c
Ocean heat content

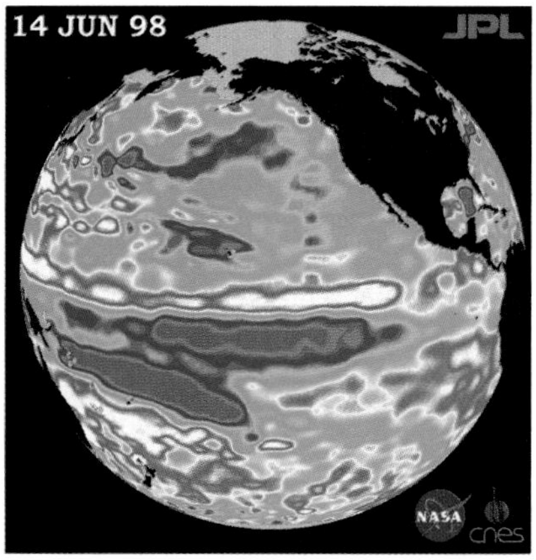

FIGURE 12.12d
Ocean heat content

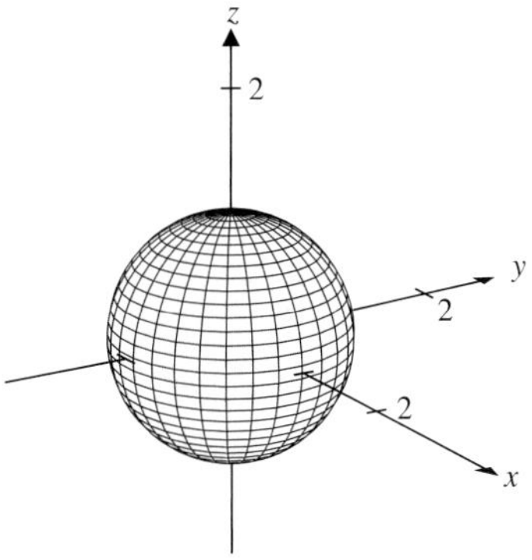

FIGURE 12.13a
$x^2 + y^2 + z^2 = 1$

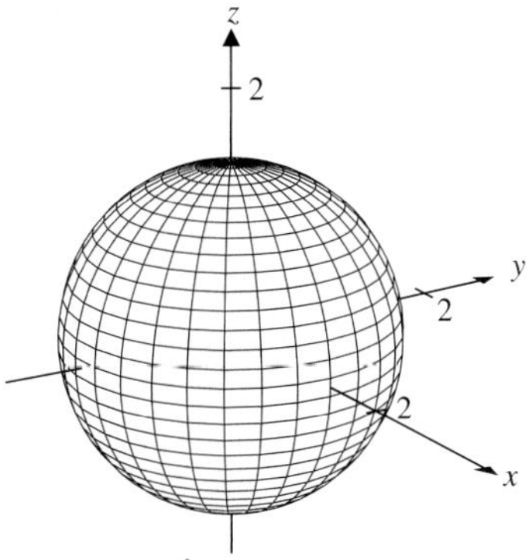

FIGURE 12.13b
$x^2 + y^2 + z^2 = 2$

four dimensions (three independent variables plus one dependent variable). We can, however, gain important information from looking at graphs of the **level surfaces** of a function f. These are the graphs of the equation $f(x, y, z) = c$, for different choices of the constant c. Much as level curves do for functions of two variables, level surfaces can help you identify symmetries and regions of rapid or slow change in a function of three variables.

EXAMPLE 1.10 Sketching Level Surfaces

Sketch several level surfaces of $f(x, y, z) = x^2 + y^2 + z^2$.

Solution The level surfaces are described by the equation $x^2 + y^2 + z^2 = c$. Of course, these are spheres of radius $\sqrt{c}$ for $c > 0$. Surfaces with $c = 1$ and $c = 2$ are shown in Figures 12.13a and 12.13b, respectively. ■

Notice that the function in example 1.10 measures the square of the distance from the origin. If you didn't recognize this at first, carefully plotting a few of the level surfaces would clearly show you the symmetry and gradual increase of the function.

BEYOND FORMULAS

Our main way of thinking about surfaces in three dimensions is to analyze two-dimensional cross sections and build them up into a three-dimensional image. This allows us to use our experience with equations and graphs in two dimensions to determine properties of the graphs. Contour plots and density plots do essentially the same thing, with the one restriction that the cross sections represented are in parallel planes (for example, all parallel to the xy-plane). These two-dimensional plots do not show the distortions that result from trying to represent a three-dimensional object on two-dimensional paper. Thus, we can often draw better conclusions from a contour plot than from a three-dimensional graph.

EXERCISES 12.1

WRITING EXERCISES

1. In example 1.4, we sketched a paraboloid and the top half of a hyperboloid as examples of graphs of functions of two variables. Explain why neither a full hyperboloid nor an ellipsoid would be the graph of a function of two variables. Develop a "vertical line test" for determining whether a given surface is the graph of a function of two variables.

2. In example 1.4, we used traces to help sketch the surface, but in example 1.5 the traces were less helpful. Discuss the differences in the functions involved and how you can tell whether or not traces will be helpful.

3. In examples 1.7 and 1.8, we discussed how to identify a contour plot given the formula for a function. In this exercise, you will discuss the inverse problem. That is, given a contour plot, what can be said about the function? For example, explain why a contour plot without labels (identifying the value of z) could correspond to more than one function. If the contour plot shows a set of concentric circles around a point, explain why you would expect that point to be the location of a local extremum. Explain why, without labels, you could not distinguish a local maximum from a local minimum.

4. For this exercise, imagine a contour plot that shows level curves for equally spaced z-values (e.g., $z = 0$, $z = 2$ and $z = 4$). Near point A, the level curves are very close together, but near point B, there are no level curves showing at all. Discuss the behavior of the function near points A and B, especially commenting on whether the function is changing rapidly or slowly.

In exercises 1–6, describe and sketch the domain of the function.

1. $f(x, y) = \dfrac{1}{x+y}$

2. $f(x, y) = \dfrac{3xy}{y - x^2}$

3. $f(x, y) = \ln(2 + x + y)$

4. $f(x, y) = \sqrt{1 - x^2 - y^2}$

5. $f(x, y, z) = \dfrac{2xz}{\sqrt{4 - x^2 - y^2 - z^2}}$

6. $f(x, y, z) = \dfrac{e^{yz}}{z - x^2 - y^2}$

In exercises 7–10, describe the range of the function.

7. $f(x, y) = \sqrt{2 + x - y}$

8. $f(x, y) = \cos(x^2 + y^2)$

9. $f(x, y) = x^2 + y^2 - 1$

10. $f(x, y) = e^{x-y}$

In exercises 11–14, compute the indicated function values.

11. $f(x, y) = x^2 + y$; $f(1, 2)$, $f(0, 3)$

12. $f(x, y, z) = \dfrac{x+y}{z}$; $f(1, 2, 3)$, $f(5, -4, 3)$

13. $f(w, x, y, z) = \cos w - \dfrac{2xz}{y+z}$; $f(0, 1, 2, 3)$, $f(\pi, 2, 0, -1)$

14. $f(x_1, x_2, x_3, x_4, x_5) = \dfrac{x_1 + x_2 + x_3}{x_4^2 + x_5^2}$; $f(1, -1, 2, 3, 4)$, $f(5, -4, 3, -2, 1)$

In exercises 15 and 16, use the table in example 1.3.

15. Find (a) $R(150, 1000)$, (b) $R(150, 2000)$ and (c) $R(150, 3000)$. (d) Based on your answers, how much extra distance is gained from an additional 1000 rpm of backspin?

16. Find (a) $R(150, 2000)$, (b) $R(160, 2000)$ and (c) $R(170, 2000)$. (d) Based on your answers, how much extra distance is gained from an additional 10 ft/s of initial velocity?

In exercises 17–20, sketch the indicated traces and graph $z = f(x, y)$.

17. $f(x, y) = x^2 + y^2$; $z = 1, z = 4, z = 9, x = 0$

18. $f(x, y) = x^2 - y^2$; $z = 0, z = 1, y = 0, y = 2$

19. $f(x, y) = \sqrt{x^2 + y^2}$; $z = 1, z = 2, z = 3, y = 0$

20. $f(x, y) = x - 2y$; $z = 0, z = 1, x = 0, y = 0$

In exercises 21–30, use a graphing utility to sketch graphs of $z = f(x, y)$ from two different viewpoints, showing different features of the graphs.

21. $f(x, y) = x^2 + y^3$

22. $f(x, y) = x^2 + y^4$

23. $f(x, y) = x^2 + y^2 - x^4$

24. $f(x, y) = \dfrac{x^2}{x^2 + y^2 + 1}$

25. $f(x, y) = \cos\sqrt{x^2 + y^2}$

26. $f(x, y) = \sin^2 x + \cos^2 y$

27. $f(x, y) = xye^{-x^2-y^2}$

28. $f(x, y) = ye^x$

29. $f(x, y) = \ln(x^2 + y^2 - 1)$

30. $f(x, y) = 2x \sin xy \ln y$

31. The **topographical map** seen at the top of the following page shows level curves for the height of a hill. For each point indicated, identify the height and sketch a short arrow indicating the direction from that point that corresponds to "straight up" the hill; that is, show the direction of the largest rate of increase in height.

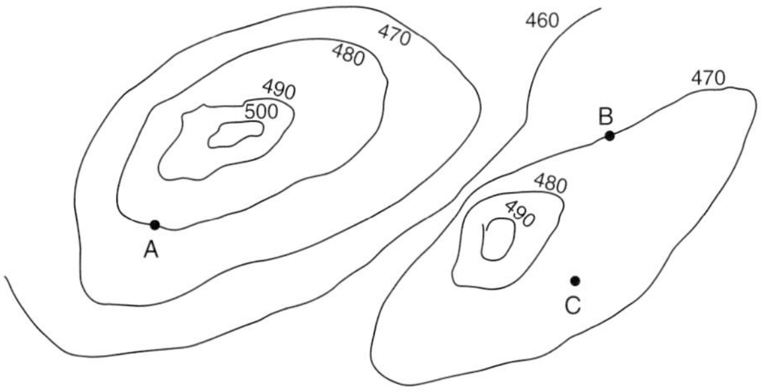

For exercise 31

32. For the topographical map from exercise 31, there are two peaks shown. Identify the locations of the peaks and use the labels to approximate the height of each peak.

33. **Doppler radar** is used by meteorologists to track storms. The radar can measure the position and velocity of water and dust particles with enough accuracy to identify characteristics of the storm. The image on the left shows reflectivity, the amount of microwave energy reflected. The red represents the highest levels of reflectivity. In this case, a tornado is on-screen and the particles with the highest energy are debris from the tornado. From this image, conjecture the center of the tornado and the direction of movement of the tornado.

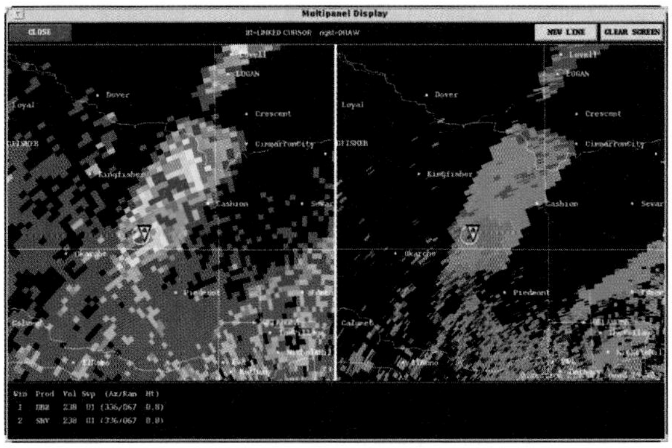

34. Referring to exercise 33, the image on the right shows the same tornado as the image on the left. In this image, velocity is color-coded with positive velocity in green and negative velocity in red (positive and negative mean toward the radar location and away from the radar, respectively). The area where red and green spiral together is considered a **tornado signature.** Explain why this set of velocity measurements indicates a tornado.

35. The **heat index** is a combination of temperature and humidity that measures how effectively the human body is able to dissipate heat; in other words, the heat index is a measure of how hot it feels. The more humidity there is, the harder it is for the body to evaporate moisture and cool off, so the hotter you feel. The table shows the heat index for selected temperatures and humidities in shade with a light breeze. For the function $H(t, h)$, find $H(80, 20)$, $H(80, 40)$ and $H(80, 60)$. At 80°, approximately how many degrees does an extra 20% humidity add to the heat index?

	20%	40%	60%	80%
70°	65.1	66.9	68.8	70.7
80°	77.4	80.4	82.8	85.9
90°	86.5	92.3	100.5	112.0
100°	98.8	111.2	129.5	154.0

36. Use the preceding heat index table to find $H(90, 20)$, $H(90, 40)$ and $H(90, 60)$. At 90°, approximately how many degrees does an extra 20% humidity add to the heat index? This answer is larger than the answer to exercise 35. Discuss what this means in terms of the danger of high humidity.

In exercises 37–42, sketch a contour plot.

37. $f(x, y) = x^2 + 4y^2$ **38.** $f(x, y) = \cos\sqrt{x^2 + y^2}$

39. $f(x, y) = y - 4x^2$ **40.** $f(x, y) = y^3 - 2x$

41. $f(x, y) = e^{y - x^3}$ **42.** $f(x, y) = ye^x$

In exercises 43–46, use a CAS to sketch a contour plot.

43. $f(x, y) = xye^{-x^2-y^2}$ **44.** $f(x, y) = x^3 - 3xy + y^2$

45. $f(x, y) = \sin x \sin y$ **46.** $f(x, y) = \sin(y - x^2)$

47. In parts a–d, match the surfaces to the contour plots.

a.

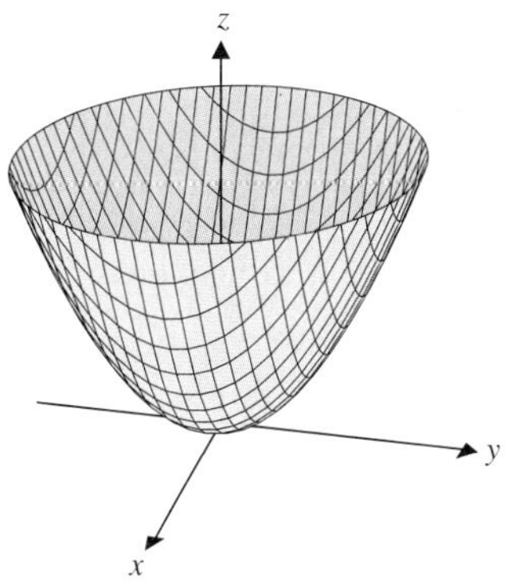

b.

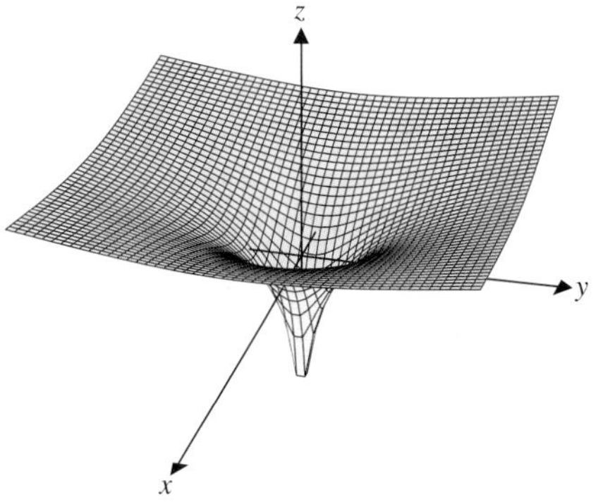

c.

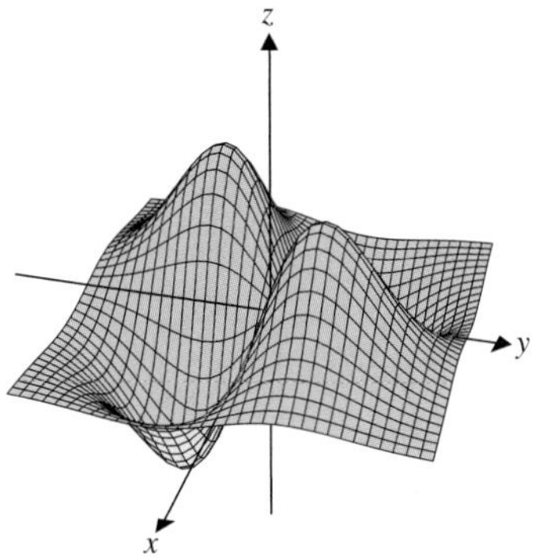

d.

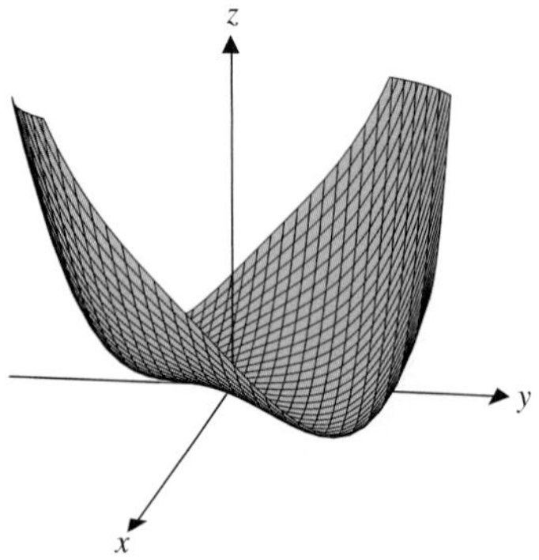

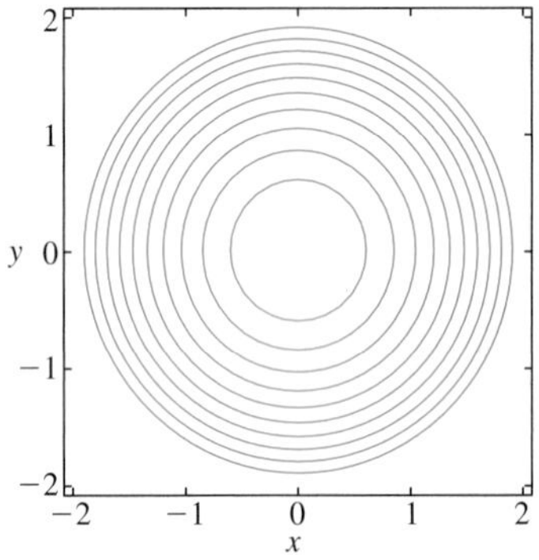

CONTOUR A

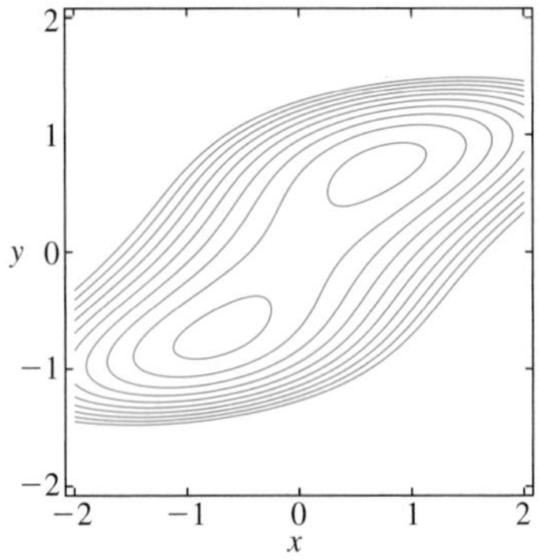

CONTOUR B

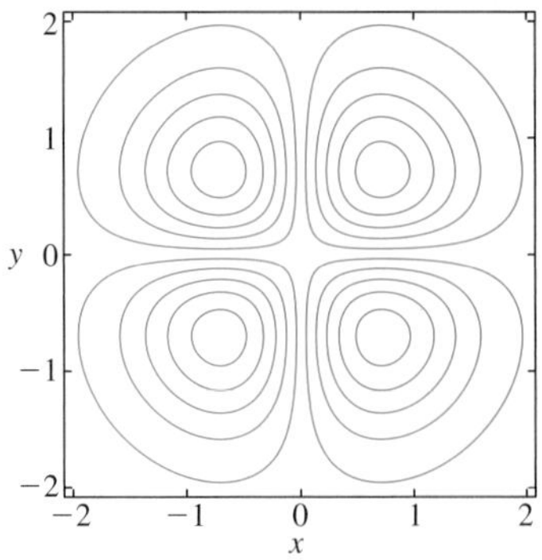

CONTOUR C

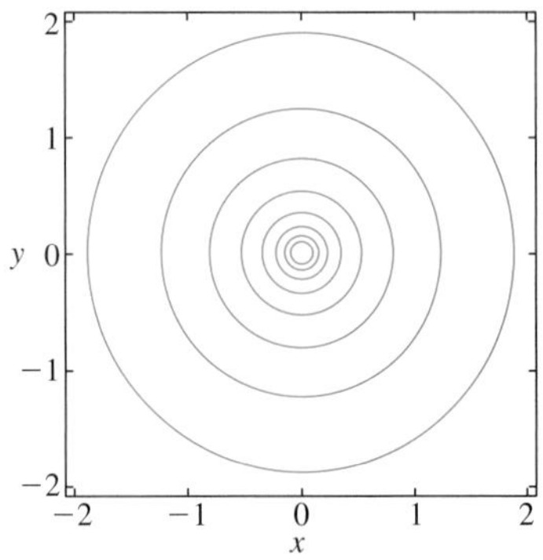

CONTOUR D

48. In parts a–d, match the density plots to the contour plots of exercise 47.

a.

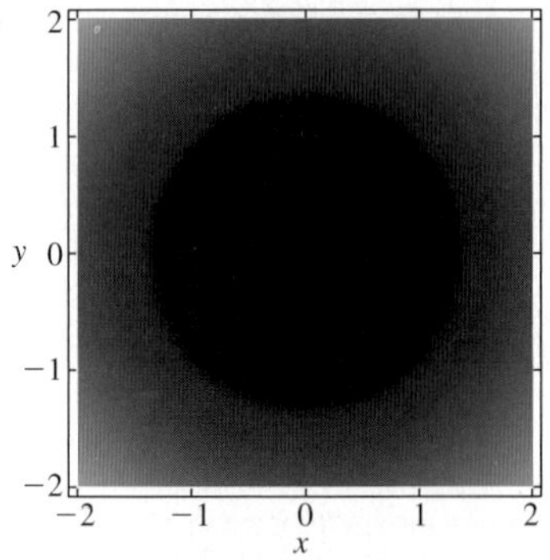

b.

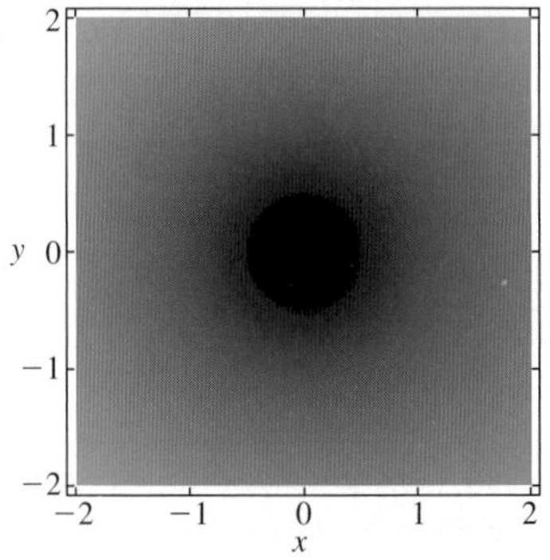

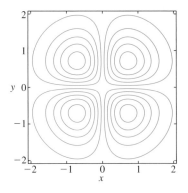

c.

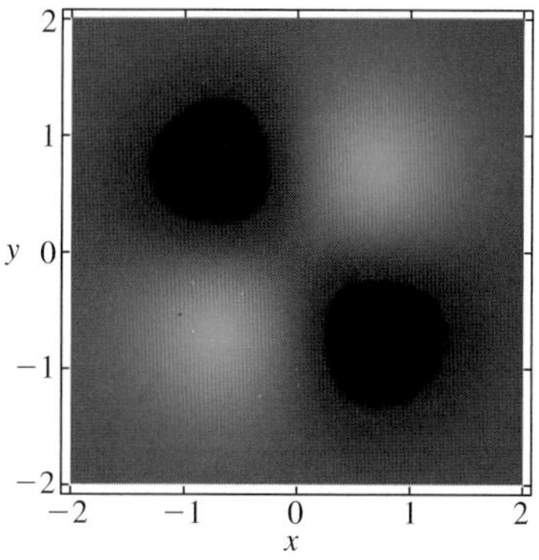

d.

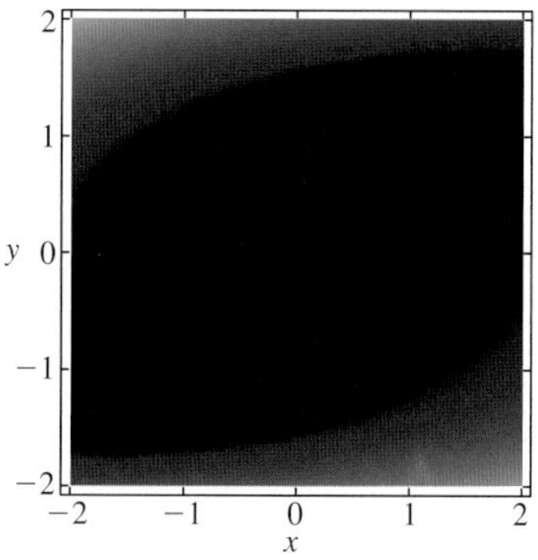

49. In parts a–f, match the functions to the surfaces.

a. $f(x, y) = x^2 + 3x^7$

b. $f(x, y) = x^2 - y^3$

c. $f(x, y) = \cos^2 x + y^2$

d. $f(x, y) = \cos(x^2 + y^2)$

e. $f(x, y) = \sin(x^2 + y^2)$

f. $f(x, y) = e^{-x^2-y^2}$

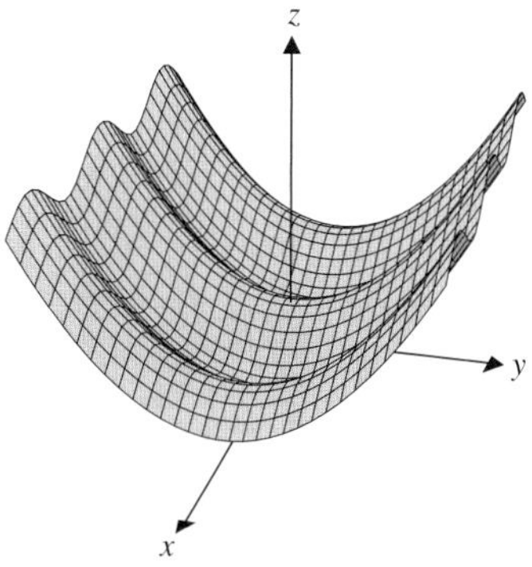

SURFACE A

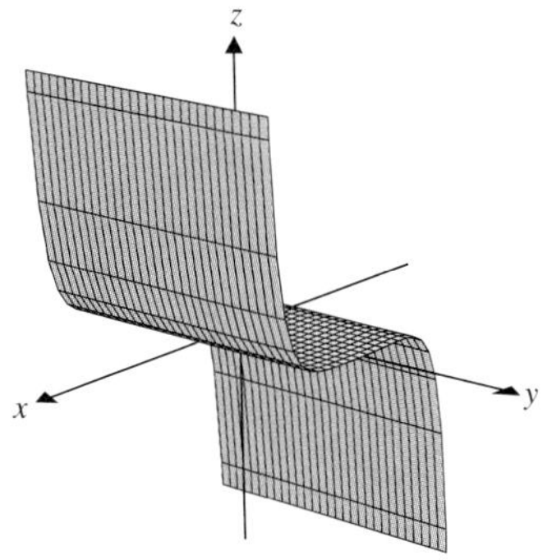

SURFACE B

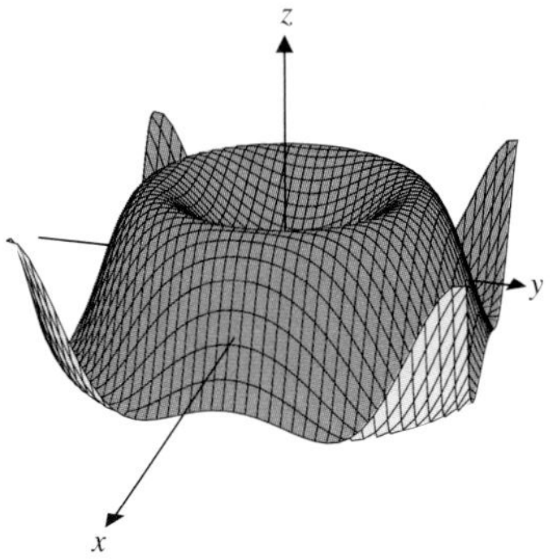

SURFACE C

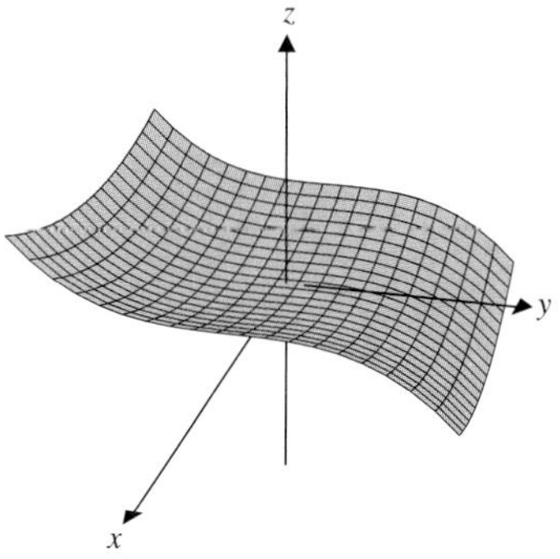

SURFACE D

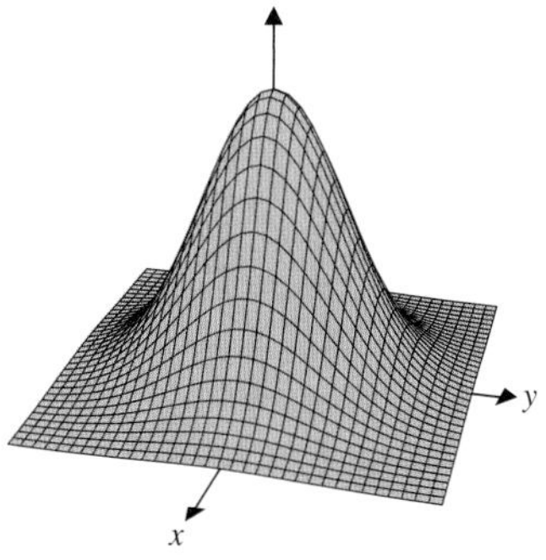

SURFACE E

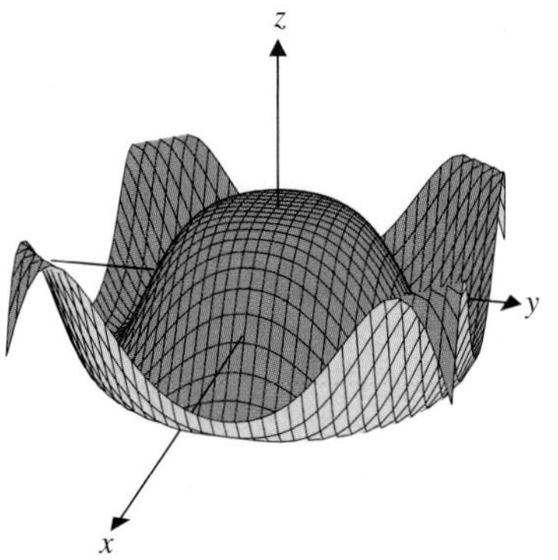

SURFACE F

In exercises 50–52, sketch several level surfaces of the given function.

50. $f(x, y, z) = x^2 - y^2 + z^2$

51. $f(x, y, z) = x^2 + y^2 - z$

52. $f(x, y, z) = z - \sqrt{x^2 + y^2}$

53. The graph of $f(x, y) = x^2 - y^2$ is shown from two different viewpoints. Identify which is viewed from (a) the positive x-axis and (b) the positive y-axis.

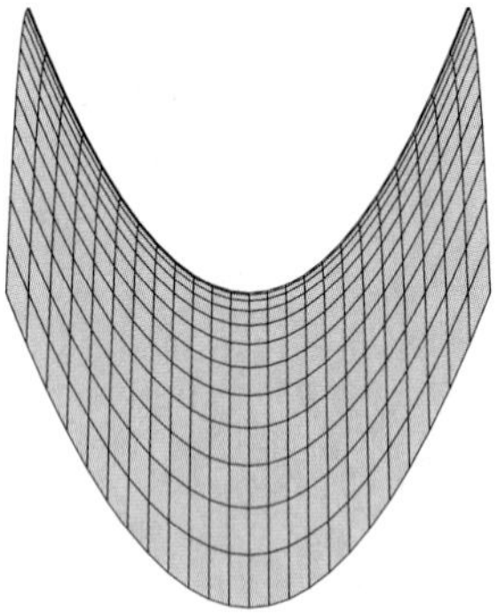

VIEW A

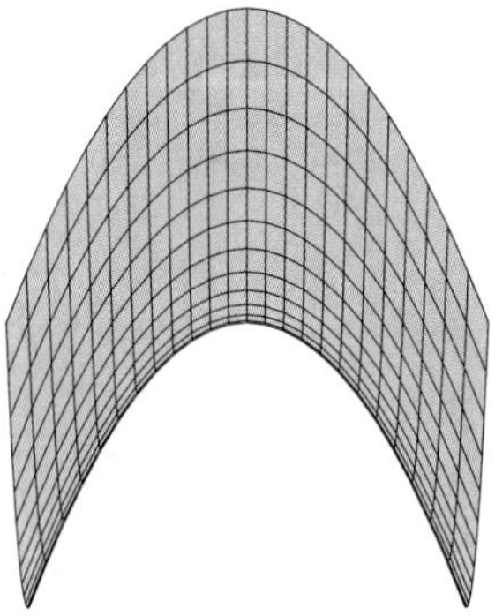

VIEW B

54. The graph of $f(x, y) = x^2y^2 - y^4 + x^3$ is shown from two different viewpoints. Identify which is viewed from (a) the positive x-axis and (b) the positive y-axis.

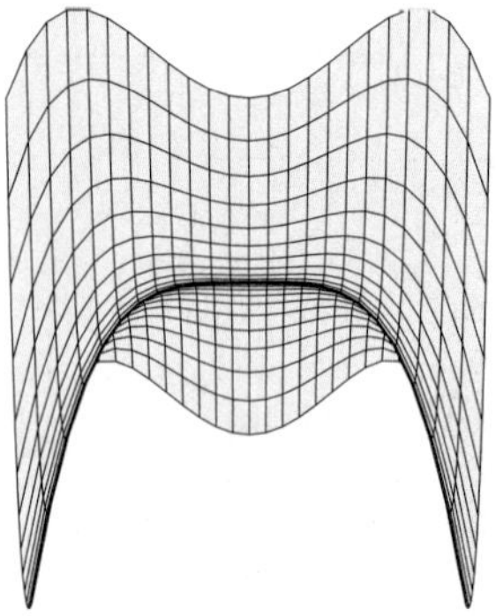

VIEW A

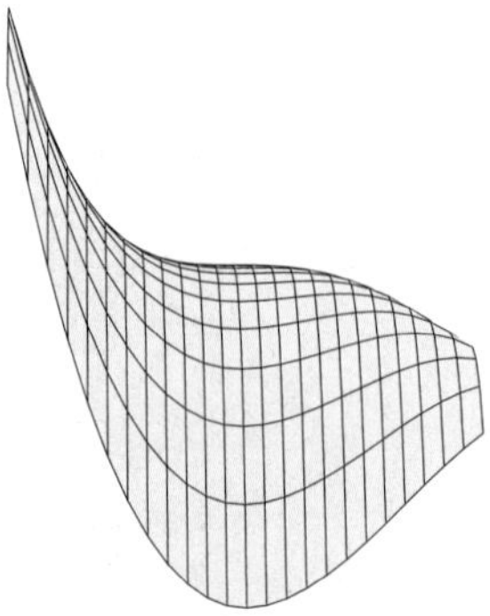

VIEW B

55. For the graphs in exercises 53 and 54, most software that produces wireframe graphs will show the view from the z-axis as a square grid. Explain why this is an accurate (although not very helpful) representation.

56. Suppose that you are shining a flashlight down at a surface from the positive z-axis. Explain why the result will be similar to a density plot.

57. Describe in words the graph of $z = \sin(x + y)$. In which direction is the "wave" traveling? Explain why the wireframe graph as viewed from the point (100, 100, 0) appears to be rectangular.

58. Find a viewpoint from which a wireframe graph of $z = \sin(x + y)$ shows only a single sine wave.

59. Find a viewpoint from which a wireframe graph of $z = (y - \sqrt{3}x)^2$ shows only a single parabola.

60. Find all viewpoints from which a wireframe graph of $z = e^{-x^2-y^2}$ shows a bell-shaped curve.

61. Suppose that the accompanying contour plot represents the population density in a city at a particular time in the evening. If there is a large rock concert that evening, locate the stadium. Speculate on what might account for other circular level curves and the linear level curves.

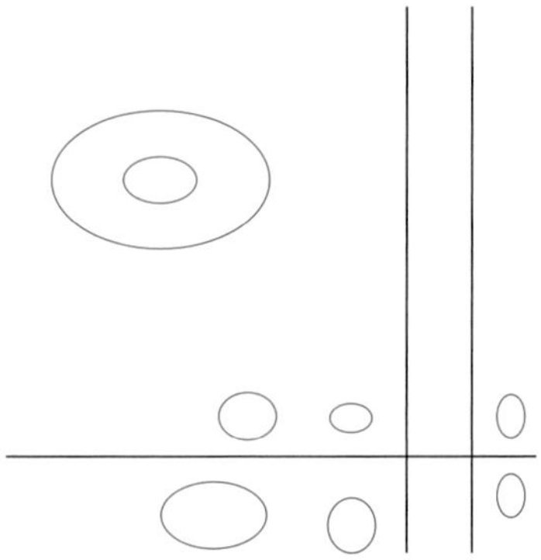

62. Suppose that the accompanying contour plot represents the temperature in a room. If it is winter, identify likely positions for a heating vent and a window. Speculate on what the circular level curves might represent.

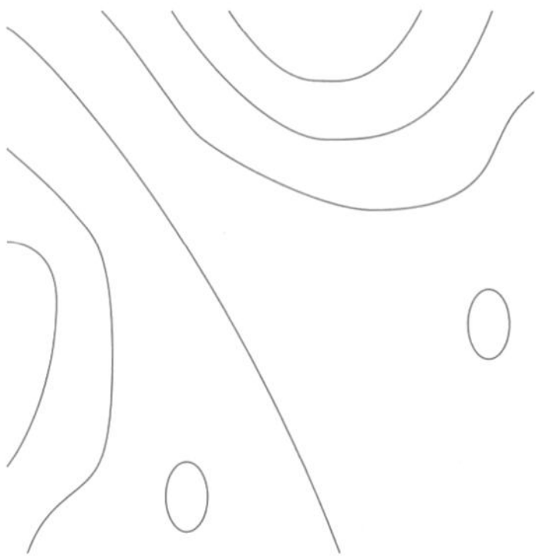

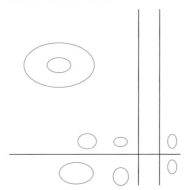

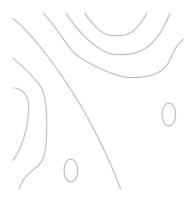

63. Suppose that the accompanying contour plot represents the coefficient of restitution (the "bounciness") at various locations on a tennis racket. Locate the point of maximum power for the racket, and explain why you know it's *maximum* power and not minimum power. Racket manufacturers sometimes call one of the level curves the "sweet spot" of the racket. Explain why this is reasonable.

64. Suppose that the accompanying contour plot represents the elevation on a golf putting green. Assume that the elevation increases as you move up the contour plot. If the hole is at point H, describe what putts from points A, B and C would be like.

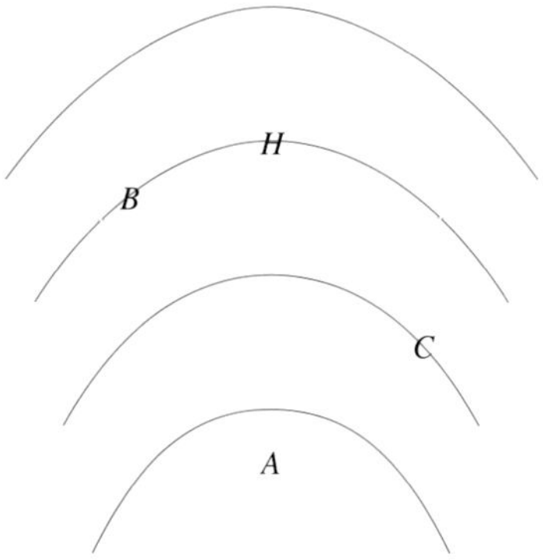

65. A well-known college uses the following formula to predict the grade average of prospective students:

$$\text{PGA} = 0.708(\text{HS}) + 0.0018(\text{SATV}) + 0.001(\text{SATM}) - 1.13$$

Here, PGA is the predicted grade average, HS is the student's high school grade average (in core academic courses, on a four-point scale), SATV is the student's SAT verbal score and SATM is the student's SAT math score. Use your scores to compute your own predicted grade average. Determine whether it is possible to have a predicted average of 4.0, or a negative predicted grade average. In this formula, the predicted grade average is a function of three variables. State which variable you think is the most important and explain why you think so.

66. In *The Hidden Game of Football,* Carroll, Palmer and Thorn give the following formula for the probability p that the team with the ball will win the game:

$$\ln\left(\frac{p}{1-p}\right) = 0.6s + 0.084\frac{s}{\sqrt{t/60}} - 0.0073(y - 74).$$

Here, s is the current score differential (+ if you're winning, – if you're losing), t is the number of minutes remaining and y is the number of yards to the goal line. For the function $p(s, t, y)$, compute $p(2, 10, 40)$, $p(3, 10, 40)$, $p(3, 10, 80)$ and $p(3, 20, 40)$, and interpret the differences in football terms.

67. Suppose that you drive x mph for d miles and then y mph for d miles. Show that your average speed S is given by $S(x, y) = \dfrac{2xy}{x+y}$ mph. On a 40-mile trip, if you average 30 mph for the first 20 miles, how fast must you go to average 40 mph for the entire trip? How fast must you go to average 60 mph for the entire trip?

68. The **price-to-earnings ratio** of a stock is defined by $R = \frac{P}{E}$, where P is the price per share of the stock and E is the earnings. The yield of the stock is defined by $Y = \frac{d}{P}$, where d is the dividends per share. Find the yield as a function of R, d and E.

69. If your graphing utility can draw three-dimensional parametric graphs, compare the wireframe graph of $z = x^2 + y^2$ with the parametric graph of $x(r, t) = r\cos t$, $y(r, t) = r\sin t$ and $z(r, t) = r^2$. (Change parameter letters from r and t to whichever letters your utility uses.)

70. If your graphing utility can draw three-dimensional parametric graphs, compare the wireframe graph of $z = \ln(x^2 + y^2)$ with the parametric graph of $x(r, t) = r\cos t$, $y(r, t) = r\sin t$ and $z(r, t) = \ln(r^2)$.

71. If your graphing utility can draw three-dimensional parametric graphs, find parametric equations for $z = \cos(x^2 + y^2)$ and compare the wireframe and parametric graphs.

72. If your graphing utility can draw three-dimensional parametric graphs, compare the wireframe graphs of $z = \pm\sqrt{1 - x^2 - y^2}$ with the parametric graph of $x(u, v) = \cos u \sin v$, $y(u, v) = \sin u \sin v$ and $z(u, v) = \cos v$.

EXPLORATORY EXERCISES

1. Graphically explore the results of the transformations $g_1(x, y) = f(x, y) + c$, $g_2(x, y) = f(x, y + c)$ and $g_3(x, y) = f(x + c, y)$. [Hint: Take a specific function like $f(x, y) = x^2 + y^2$ and look at the graphs of the transformed functions $x^2 + y^2 + 2$, $x^2 + (y + 2)^2$ and $(x + 2)^2 + y^2$.] Determine what changes occur when the constant is added. Test your hypothesis for other constants (be sure to try negative constants, too). Then, explore the transformations $g_4(x, y) = cf(x, y)$ and $g_5(x, y) = f(c_1x, c_2y)$.

2. One common use of functions of two or more variables is in image processing. For instance, to digitize a black-and-white photograph, you can superimpose a rectangular grid and label each subrectangle with a number representing the brightness of that portion of the photograph. The grid defines the x- and y-values and the brightness numbers are the function values. Briefly describe how this function differs from other functions in this section. (Hint: How many x- and y-values are there?) Near the soccer jersey in the photograph shown, describe how the brightness function behaves. To "sharpen" the picture by increasing the contrast, should you transform the function values to make them closer together or farther apart?

PHOTO WITH GRID

B & W PHOTO

7	8	8	9
5	2	5	6
6	7	7	6
6	6	4	6

DIGITIZED PHOTO

12.2 LIMITS AND CONTINUITY

At the beginning of our study of the calculus and again when we introduced vector-valued functions, we have followed the same progression of topics, beginning with graphs of functions, then limits, continuity, derivatives and integrals. We continue this progression now by extending the concept of limit to functions of two (and then three) variables. As you will see, the increase in dimension causes some interesting complications.

First, recall that for a function of a single variable, if we write $\lim_{x \to a} f(x) = L$, we mean that as x gets closer and closer to a, $f(x)$ gets closer and closer to the number L. Here, for functions of several variables, the idea is very similar. When we write

$$\lim_{(x,y)\to(a,b)} f(x, y) = L,$$

we mean that as (x, y) gets closer and closer to (a, b), $f(x, y)$ is getting closer and closer to the number L. In this case, (x, y) may approach (a, b) along any of the infinitely many different paths passing through (a, b).

For instance, $\lim_{(x,y)\to(2,3)} (xy - 2)$ asks us to identify what happens to the function $xy - 2$ as x approaches 2 and y approaches 3. Clearly, $xy - 2$ approaches $2(3) - 2 = 4$ and we write

$$\lim_{(x,y)\to(2,3)} (xy - 2) = 4.$$

Similarly, you can reason that

$$\lim_{(x,y)\to(-1,\pi)} (\sin xy - x^2 y) = \sin(-\pi) - \pi = -\pi.$$

In other words, for many (nice) functions, we can compute limits simply by substituting into the function.

However, as with functions of a single variable, the limits in which we're most interested cannot be computed by simply substituting values for x and y. For instance, for

$$\lim_{(x,y)\to(1,0)} \frac{y}{x+y-1},$$

substituting in $x = 1$ and $y = 0$ gives the indeterminate form $\frac{0}{0}$. To evaluate this limit, we must investigate further.

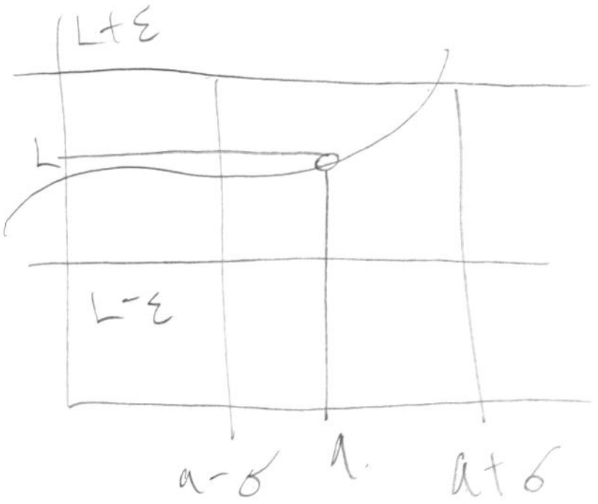

You may recall from our discussion in section 1.6 that for a function f of a single variable defined on an open interval containing a (but not necessarily at a), we say that $\lim_{x\to a} f(x) = L$ if given any number $\varepsilon > 0$, there is another number $\delta > 0$ such that $|f(x) - L| < \varepsilon$ whenever $0 < |x - a| < \delta$. In other words, no matter how close you wish to make $f(x)$ to L (we represent this distance by ε), you can make it that close, just by making x sufficiently close to a (i.e., within a distance δ of a).

The definition of the limit of a function of two variables is completely analogous to the definition for a function of a single variable. We say that $\lim_{(x,y)\to(a,b)} f(x, y) = L$, if we can make $f(x, y)$ as close as desired to L by making the point (x, y) sufficiently close to (a, b). We make this more precise in Definition 2.1.

DEFINITION 2.1 (Formal Definition of Limit)

Let f be defined on the interior of a circle centered at the point (a, b), except possibly at (a, b) itself. We say that $\lim_{(x,y)\to(a,b)} f(x, y) = L$ if for every $\varepsilon > 0$ there exists a $\delta > 0$ such that $|f(x, y) - L| < \varepsilon$ whenever $0 < \sqrt{(x-a)^2 + (y-b)^2} < \delta$.

We illustrate the definition in Figure 12.14.

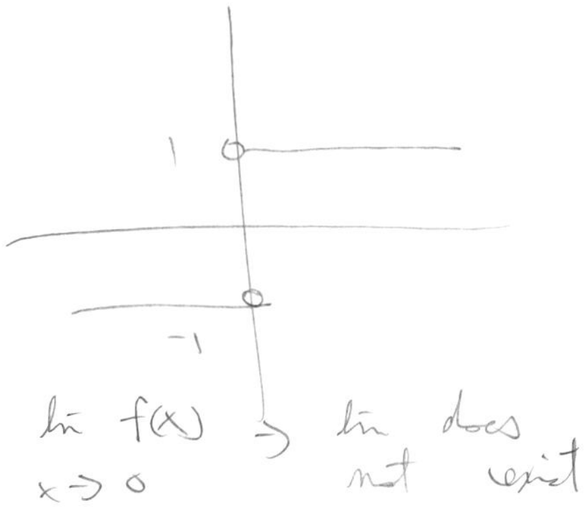

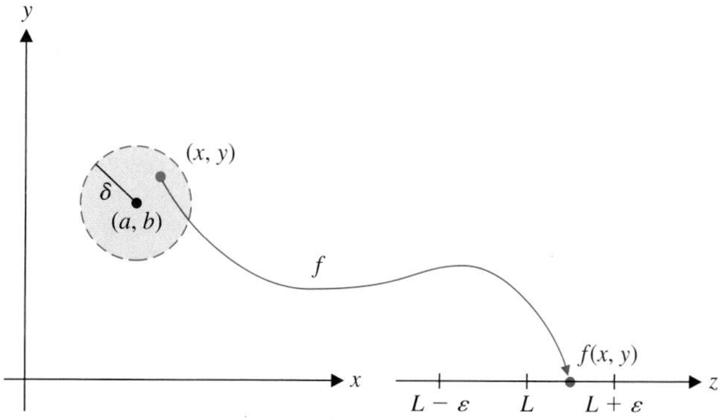

FIGURE 12.14
The definition of limit

Notice that the definition says that given any desired degree of closeness $\varepsilon > 0$, you must be able to find another number $\delta > 0$, so that *all* points lying within a distance δ of (a, b) are mapped by f to points within distance ε of L on the real line.

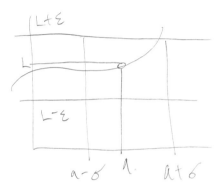

EXAMPLE 2.1 Using the Definition of Limit

Verify that $\lim_{(x,y)\to(a,b)} x = a$ and $\lim_{(x,y)\to(a,b)} y = b$.

Solution Certainly, both of these limits are intuitively quite clear. We can use Definition 2.1 to verify them, however. Given any number $\varepsilon > 0$, we must find another number $\delta > 0$ so that $|x - a| < \varepsilon$ whenever $0 < \sqrt{(x-a)^2 + (y-b)^2} < \delta$. Notice that

$$\sqrt{(x-a)^2 + (y-b)^2} \geq \sqrt{(x-a)^2} = |x - a|,$$

and so, taking $\delta = \varepsilon$, we have that

$$|x - a| = \sqrt{(x-a)^2} \leq \sqrt{(x-a)^2 + (y-b)^2} < \varepsilon,$$

whenever $0 < \sqrt{(x-a)^2 + (y-b)^2} < \delta$. Likewise, we can show that $\lim_{(x,y)\to(a,b)} y = b$. ■

With this definition of limit, we can prove the usual results for limits of sums, products and quotients. That is, if $f(x, y)$ and $g(x, y)$ both have limits as (x, y) approaches (a, b), we have

$$\lim_{(x,y)\to(a,b)} [f(x, y) \pm g(x, y)] = \lim_{(x,y)\to(a,b)} f(x, y) \pm \lim_{(x,y)\to(a,b)} g(x, y)$$

(i.e., the limit of a sum or difference is the sum or difference of the limits),

$$\lim_{(x,y)\to(a,b)} [f(x, y)g(x, y)] = \left[\lim_{(x,y)\to(a,b)} f(x, y)\right]\left[\lim_{(x,y)\to(a,b)} g(x, y)\right]$$

(i.e., the limit of a product is the product of the limits) and

$$\lim_{(x,y)\to(a,b)} \frac{f(x, y)}{g(x, y)} = \frac{\lim_{(x,y)\to(a,b)} f(x, y)}{\lim_{(x,y)\to(a,b)} g(x, y)}$$

(i.e., the limit of a quotient is the quotient of the limits), *provided* $\lim_{(x,y)\to(a,b)} g(x, y) \neq 0$.

A **polynomial** in the two variables x and y is any sum of terms of the form $cx^n y^m$, where c is a constant and n and m are nonnegative integers. Using the preceding results and example 2.1, we can show that the limit of any polynomial always exists and is found simply by substitution.

EXAMPLE 2.2 Finding a Simple Limit

Evaluate $\lim_{(x,y)\to(2,1)} \frac{2x^2y + 3xy}{5xy^2 + 3y}$.

Solution First, note that this is the limit of a rational function (i.e., the quotient of two polynomials). Since the limit in the denominator is

$$\lim_{(x,y)\to(2,1)} (5xy^2 + 3y) = 10 + 3 = 13 \neq 0,$$

we have

$$\lim_{(x,y)\to(2,1)} \frac{2x^2y + 3xy}{5xy^2 + 3y} = \frac{\lim_{(x,y)\to(2,1)} (2x^2y + 3xy)}{\lim_{(x,y)\to(2,1)} (5xy^2 + 3y)} = \frac{14}{13}.$$ ■

Think about the implications of Definition 2.1 (even if you are a little unsure of the role of ε and δ). If there is *any* way to approach the point (a, b) without the function values approaching the value L (e.g., by virtue of the function values blowing up, oscillating or by approaching some other value), then the limit will not equal L. For the limit to equal L, the function has to approach L along *every* possible path. This gives us a simple method for determining that a limit does not exist.

REMARK 2.1

If $f(x, y)$ approaches L_1 as (x, y) approaches (a, b) along a path P_1 and $f(x, y)$ approaches $L_2 \neq L_1$ as (x, y) approaches (a, b) along a path P_2, then $\lim\limits_{(x,y)\to(a,b)} f(x, y)$ *does not exist.*

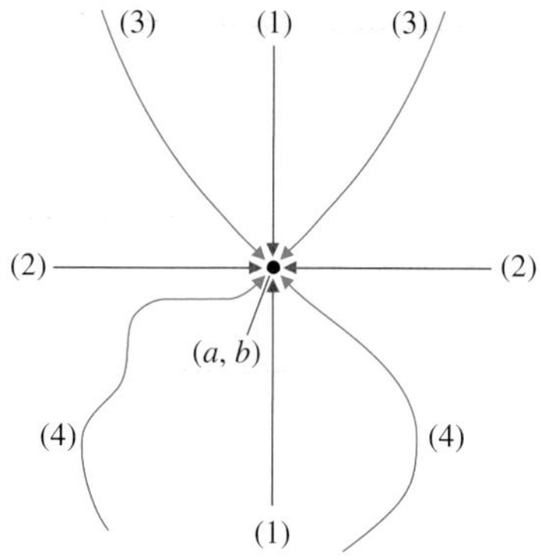

FIGURE 12.15
Various paths to (a, b)

Unlike the case for functions of a single variable where we must consider left- and right-hand limits, in two dimensions, instead of just two paths approaching a given point, there are infinitely many (and you obviously can't check each one individually). In practice, when you suspect that a limit does not exist, you should check the limit along the simplest paths first. We will use the following guidelines.

REMARK 2.2

The simplest paths to try are (1) $x = a, y \to b$ (vertical lines); (2) $y = b, x \to a$ (horizontal lines); (3) $y = g(x), x \to a$ [where $b = g(a)$] and (4) $x = g(y)$, $y \to b$ [where $a = g(b)$]. Several of these paths are illustrated in Figure 12.15.

EXAMPLE 2.3 A Limit That Does Not Exist

Evaluate $\lim\limits_{(x,y)\to(1,0)} \dfrac{y}{x + y - 1}$.

Solution First, we consider the vertical line path along the line $x = 1$ and compute the limit as y approaches 0. If $(x, y) \to (1, 0)$ along the line $x = 1$, we have

$$\lim_{(1,y)\to(1,0)} \frac{y}{1 + y - 1} = \lim_{y\to0} 1 = 1.$$

We next consider the path along the horizontal line $y = 0$ and compute the limit as x approaches 1. Here, we have

$$\lim_{(x,0)\to(1,0)} \frac{0}{x + 0 - 1} = \lim_{x\to0} 0 = 0.$$

Since the function approaches two different values along two different paths to the point $(1, 0)$, the limit does not exist. ∎

Many of our examples and exercises have (x, y) approaching $(0, 0)$. In this case, notice that another simple path passing through $(0, 0)$ is the line $y = x$.

EXAMPLE 2.4 A Limit That Is the Same Along Two Paths but Does Not Exist

Evaluate $\lim\limits_{(x,y)\to(0,0)} \dfrac{xy}{x^2 + y^2}$.

Solution First, we consider the limit along the path $x = 0$. We have

$$\lim_{(0,y)\to(0,0)} \frac{0}{0 + y^2} = \lim_{y\to 0} 0 = 0.$$

Similarly, for the path $y = 0$, we have

$$\lim_{(x,0)\to(0,0)} \frac{0}{x^2 + 0} = \lim_{x\to 0} 0 = 0.$$

Be careful; just because the limits along the first two paths you try are the same does *not* mean that the limit exists. For a limit to exist, the limit must be the same along *all* paths through (0, 0) (not just along two). Here, we may simply need to look at more paths. Notice that for the path $y = x$, we have

$$\lim_{(x,x)\to(0,0)} \frac{x(x)}{x^2 + x^2} = \lim_{x\to 0} \frac{x^2}{2x^2} = \frac{1}{2}.$$

Since the limit along this path doesn't match the limit along the first two paths, the limit does not exist. ■

As you've seen in examples 2.3 and 2.4, substitutions for particular paths often result in the function reducing to a constant. When choosing paths, you should look for substitutions that will simplify the function dramatically.

EXAMPLE 2.5 A Limit Problem Requiring a More Complicated Choice of Path

Evaluate $\displaystyle\lim_{(x,y)\to(0,0)} \frac{xy^2}{x^2 + y^4}$.

Solution First, we consider the path $x = 0$ and get

$$\lim_{(0,y)\to(0,0)} \frac{0}{0 + y^4} = \lim_{y\to 0} 0 = 0.$$

Similarly, following the path $y = 0$, we get

$$\lim_{(x,0)\to(0,0)} \frac{0}{x^2 + 0} = \lim_{x\to 0} 0 = 0.$$

Since the limits along the first two paths are the same, we try another path. As in example 2.4, we next try the line $y = x$. As it turns out, this limit is

$$\lim_{(x,x)\to(0,0)} \frac{x^3}{x^2 + x^4} = \lim_{x\to 0} \frac{x}{1 + x^2} = 0,$$

also. In exercise 51, you will show that the limit along *every* straight line through the origin is 0. However, we still cannot conclude that the limit is 0. For this to happen, the limit along *all* paths (not just along all straight-line paths) must be 0. At this point, there are two possibilities: either the limit exists (and equals 0) or the limit does not exist, in which case, we must discover a path through (0, 0) along which the limit is not 0. Notice that along the path $x = y^2$, the terms x^2 and y^4 will be equal. We then have

$$\lim_{(y^2,y)\to(0,0)} \frac{y^2(y^2)}{(y^2)^2 + y^4} = \lim_{y\to 0} \frac{y^4}{2y^4} = \frac{1}{2}.$$

Since this limit does not agree with the limits along the earlier paths, the original limit does not exist. ■

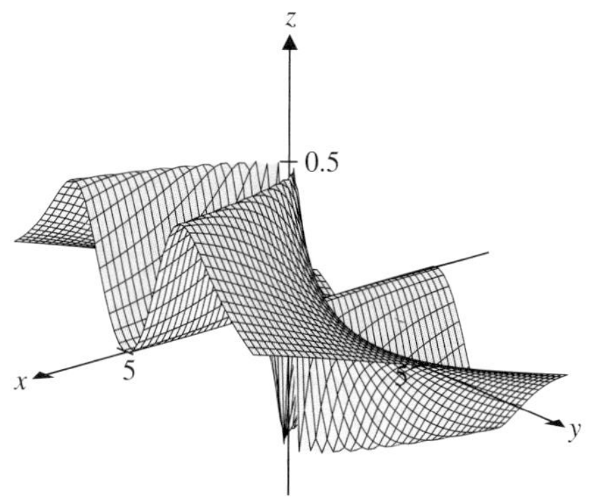

FIGURE 12.16a

$z = \dfrac{xy^2}{x^2 + y^4}$, for $-5 \le x \le 5$, $-5 \le y \le 5$

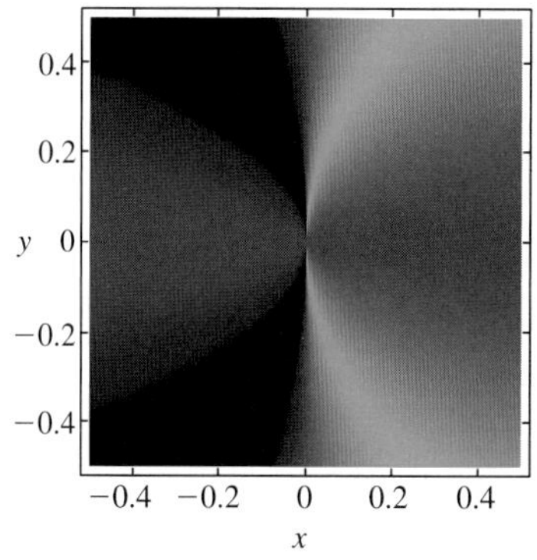

FIGURE 12.16b

Density plot of $f(x, y) = \dfrac{xy^2}{x^2 + y^4}$

Before discussing how to show that a limit *does* exist, we pause to explore example 2.5 graphically. First, try to imagine what the graph of $f(x, y) = \dfrac{xy^2}{x^2 + y^4}$ might look like. The function is defined except at the origin, it approaches 0 along the x-axis, y-axis and along any line $y = kx$ through the origin. Yet, $f(x, y)$ approaches $\frac{1}{2}$ along the parabola $x = y^2$. A standard sketch of the surface $z = f(x, y)$ with $-5 \le x \le 5$ and $-5 \le y \le 5$ is helpful, but you need to know what you're looking for. You can see part of the ridge at $z = 0.5$, as well as a trough at $z = -0.5$ corresponding to $x = -y^2$, in Figure 12.16a. A density plot clearly shows the parabola of large function values in light blue and a parabola of small function values in black (see Figure 12.16b). Near the origin, the surface has a ridge at $x = y^2$, $z = \frac{1}{2}$, dropping off quickly to a smooth surface that approaches the origin. The ridge is in two pieces ($y > 0$ and $y < 0$) separated by the origin.

The procedure we followed in examples 2.3, 2.4 and 2.5 was used to show that a limit does *not* exist. What if a limit does exist? Of course, you'll never be able to establish that a limit exists by computing limits along specific paths. There are infinitely many paths through any given point and you'll never be able to exhaust all of the possibilities. However, after following a number of paths and getting the same limit along each of them, you should begin to suspect that the limit just might exist. One tool you can use is the following generalization of the Squeeze Theorem presented in section 1.3.

THEOREM 2.1

Suppose that $|f(x, y) - L| \le g(x, y)$ for all (x, y) in the interior of some circle centered at (a, b), except possibly at (a, b). If $\lim_{(x,y)\to(a,b)} g(x, y) = 0$, then $\lim_{(x,y)\to(a,b)} f(x, y) = L$.

PROOF

For any given $\varepsilon > 0$, we know from the definition of $\lim_{(x,y)\to(a,b)} g(x, y) = 0$, that there is a number $\delta > 0$ such that $0 < \sqrt{(x-a)^2 + (y-b)^2} < \delta$ guarantees that $|g(x, y) - 0| < \varepsilon$.

For any such points (x, y), we have

$$|f(x, y) - L| \le g(x, y) < \varepsilon.$$

It now follows from the definition of limit that $\lim_{(x,y)\to(a,b)} f(x, y) = L$. ■

In other words, the theorem simply states that if $|f(x, y) - L|$ is trapped between 0 (the absolute value is never negative) and a function (g) that approaches 0, then $|f(x, y) - L|$ must also have a limit of 0.

To use Theorem 2.1, you start with a conjecture for the limit L (obtained for instance, by calculating the limit along several simple paths). Then, look for a simpler function that is larger than $|f(x, y) - L|$ and that tends to zero as (x, y) approaches (a, b).

EXAMPLE 2.6 Proving That a Limit Exists

Evaluate $\lim_{(x,y)\to(0,0)} \dfrac{x^2 y}{x^2 + y^2}$.

Solution As we did in earlier examples, we start by looking at the limit along several paths through (0, 0). Along the path $x = 0$, we have

$$\lim_{(0,y)\to(0,0)} \frac{0}{0 + y^2} = 0.$$

Similarly, along the path $y = 0$, we have

$$\lim_{(x,0)\to(0,0)} \frac{0}{x^2 + 0} = 0.$$

Further, along the path $y = x$, we have

$$\lim_{(x,x)\to(0,0)} \frac{x^3}{x^2 + x^2} = \lim_{x\to 0} \frac{x}{2} = 0.$$

We know that if the limit exists, it must equal 0. Our last calculation gives an important clue that the limit does exist. After simplifying the expression, there remained an extra power of x in the numerator forcing the limit to 0. To show that the limit equals 0, consider

$$|f(x, y) - L| = |f(x, y) - 0| = \left|\frac{x^2 y}{x^2 + y^2}\right|.$$

Notice that without the y^2 term in the denominator, we could cancel the x^2 terms. Since $x^2 + y^2 \ge x^2$, we have that for $x \ne 0$

$$|f(x, y) - L| = \left|\frac{x^2 y}{x^2 + y^2}\right| \le \left|\frac{x^2 y}{x^2}\right| = |y|.$$

Since $\lim_{(x,y)\to(0,0)} |y| = 0$, Theorem 2.1 gives us $\lim_{(x,y)\to(0,0)} \dfrac{x^2 y}{x^2 + y^2} = 0$, also. ■

When (x, y) approaches a point other than (0, 0), the idea is the same as in example 2.6, but the algebra may get messier, as we see in example 2.7.

EXAMPLE 2.7 Finding a Limit of a Function of Two Variables

Evaluate $\displaystyle\lim_{(x,y)\to(1,0)} \frac{(x-1)^2 \ln x}{(x-1)^2+y^2}$.

Solution Along the path $x = 1$, we have

$$\lim_{(1,y)\to(1,0)} \frac{0}{y^2} = 0.$$

Along the path $y = 0$, we have

$$\lim_{(x,0)\to(1,0)} \frac{(x-1)^2 \ln x}{(x-1)^2} = \lim_{x\to 1} \ln x = 0.$$

A third path through $(1, 0)$ is the line $y = x - 1$ (note that in this case, we must have $y \to 0$ as $x \to 1$). We have

$$\lim_{(x,x-1)\to(1,0)} \frac{(x-1)^2 \ln x}{(x-1)^2+(x-1)^2} = \lim_{x\to 1} \frac{(x-1)^2 \ln x}{2(x-1)^2} = \lim_{x\to 1} \frac{\ln x}{2} = 0.$$

At this point, you should begin to suspect that the limit just might be 0. You never know, though, until you find another path along which the limit is different or until you prove that the limit actually is 0. To show this, we consider

$$|f(x,y) - L| = \left| \frac{(x-1)^2 \ln x}{(x-1)^2+y^2} \right|.$$

Notice that if the y^2 term were not present in the denominator, then we could cancel the $(x-1)^2$ terms. We have

$$|f(x,y) - L| = \left| \frac{(x-1)^2 \ln x}{(x-1)^2+y^2} \right| \le \left| \frac{(x-1)^2 \ln x}{(x-1)^2} \right| = |\ln x|$$

Since $\displaystyle\lim_{(x,y)\to(1,0)} |\ln x| = 0$, it follows from Theorem 2.1 that

$$\lim_{(x,y)\to(1,0)} \frac{(x-1)^2 \ln x}{(x-1)^2+y^2} = 0, \text{ also.} \quad ■$$

As with functions of one variable and (more recently) vector-valued functions, the concept of continuity is closely connected to limits. Recall that in these cases, a function (or vector-valued function) is continuous at a point whenever the limit and the value of the function are the same. This same characterization applies to continuous functions of several variables, as we see in Definition 2.2.

DEFINITION 2.2

Suppose that $f(x, y)$ is defined in the interior of a circle centered at the point (a, b). We say that f is **continuous** at (a, b) if $\displaystyle\lim_{(x,y)\to(a,b)} f(x,y) = f(a,b)$.

If $f(x, y)$ is not continuous at (a, b), then we call (a, b) a **discontinuity** of f.

This definition is completely analogous to our previous definitions of continuity for the cases of functions of one variable and vector-valued functions. The graphical interpretation is similar, although three-dimensional graphs can be more complicated. Still, the idea is

* Continuity is always at a point

FIGURE 12.17a
Open disk

FIGURE 12.17b
Closed disk

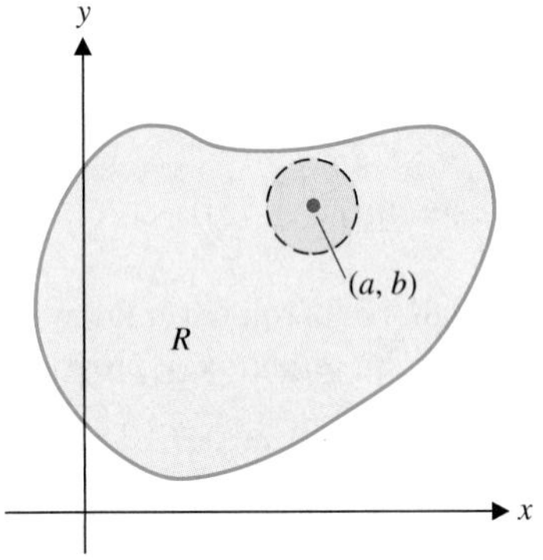

FIGURE 12.18a
Interior point

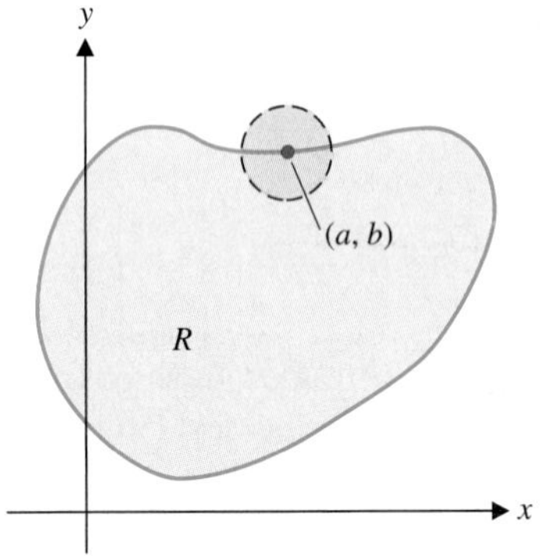

FIGURE 12.18b
Boundary point

that for a continuous function $f(x, y)$, if (x, y) changes slightly, then $f(x, y)$ also changes slightly.

Before we define the concept of continuity on a region $R \subset \mathbb{R}^2$, we first need to define open and closed regions in two dimensions. We refer to the interior of a circle (i.e., the set of all points inside but not on the circle) as an **open disk** (see Figure 12.17a). A **closed disk** consists of the circle and its interior (see Figure 12.17b). These are the two-dimensional analogs of open and closed intervals, respectively, of the real line. For a given two-dimensional region R, a point (a, b) in R is called an **interior point** of R if there is an open disk centered at (a, b) that lies *completely* inside of R (see Figure 12.18a). A point (a, b) in R is called a **boundary point** of R if *every* open disk centered at (a, b) contains points in R *and* points outside R (see Figure 12.18b). A set R is **closed** if it contains *all* of its boundary points. Alternatively, R is **open** if it contains *none* of its boundary points. Note that these are analogous to closed and open intervals of the real line: closed intervals include all (both) of their boundary points (endpoints), while open intervals include none (neither) of their boundary points.

If the domain of a function contains any of its boundary points, we will need to modify our definition of continuity slightly, to ensure that the limit is calculated over paths that lie inside the domain only. (Recall that this is essentially what we did to define continuity of a function of a single variable on a closed interval.) If (a, b) is a boundary point of the domain D of a function f, we say that f is continuous at (a, b) if

$$\lim_{\substack{(x,y)\to(a,b)\\(x,y)\in D}} f(x, y) = f(a, b).$$

This notation indicates that the limit is taken only along paths lying completely inside D. Note that this limit requires a slight modification of Definition 2.1, as follows.

We say that

$$\lim_{\substack{(x,y)\to(a,b)\\(x,y)\in D}} f(x, y) = L$$

if for every $\varepsilon > 0$ there exists a $\delta > 0$ such that $|f(x) - L| < \varepsilon$ whenever $(x, y) \in D$ and $0 < \sqrt{(x-a)^2 + (y-b)^2} < \delta$.

We say that a function $f(x, y)$ is **continuous on a region** R if it is continuous at each point in R.

Notice that because we define continuity in terms of limits, we immediately have the following results, which follow directly from the corresponding results for limits. If $f(x, y)$

and $g(x, y)$ are continuous at (a, b), then $f + g$, $f - g$ and $f \cdot g$ are all continuous at (a, b). Further, f/g is continuous at (a, b), if, in addition, $g(a, b) \neq 0$. We leave the proofs of these statements as exercises.

In many cases, determining where a function is continuous involves identifying where the function isn't defined and using our continuity results for functions of a single variable.

EXAMPLE 2.8 Determining Where a Function of Two Variables Is Continuous

Find all points where the given function is continuous: (a) $f(x, y) = \dfrac{x}{x^2 - y}$ and

(b) $g(x, y) = \begin{cases} \dfrac{x^4}{x(x^2 + y^2)}, & \text{if } (x, y) \neq (0, 0) \\ 0, & \text{if } (x, y) = (0, 0) \end{cases}.$

Solution For (a), notice that $f(x, y)$ is a quotient of two polynomials (i.e., a rational function) and so, it is continuous at any point where we don't divide by 0. Since division by zero occurs only when $y = x^2$, we have that f is continuous at all points (x, y) with $y \neq x^2$. For (b), the function g is also a quotient of polynomials, except at the origin. Notice that there is a division by 0 whenever $x = 0$. We must consider the point (0, 0) separately, however, since the function is not defined by the rational expression there. We can verify that $\lim_{(x,y)\to(0,0)} g(x, y) = 0$ using the following string of inequalities. Notice that for $(x, y) \neq (0, 0)$,

$$|g(x, y)| = \left|\frac{x^4}{x(x^2 + y^2)}\right| \leq \left|\frac{x^4}{x(x^2)}\right| = |x|$$

and $|x| \to 0$ as $(x, y) \to (0, 0)$. By Theorem 2.1, we have that

$$\lim_{(x,y)\to(0,0)} g(x, y) = 0 = g(0, 0),$$

so that g is continuous at (0, 0). Putting this all together, we get that g is continuous at the origin and also at all points (x, y) with $x \neq 0$. ■

Theorem 2.2 shows that we can use all of our established continuity results for functions of a single variable when considering functions of several variables.

THEOREM 2.2

Suppose that $f(x, y)$ is continuous at (a, b) and $g(x)$ is continuous at the point $f(a, b)$. Then

$$h(x, y) = (g \circ f)(x, y) = g(f(x, y))$$

is continuous at (a, b).

SKETCH OF THE PROOF

We leave the proof as an exercise, but it goes something like this. Notice that if (x, y) is close to (a, b), then by the continuity of f at (a, b), $f(x, y)$ will be close to $f(a, b)$. By the continuity of g at the point $f(a, b)$, it follows that $g(f(x, y))$ will be close to $g(f(a, b))$, so that $g \circ f$ is also continuous at (a, b). ■

EXAMPLE 2.9 Determining Where a Composition of Functions Is Continuous

Determine where $f(x, y) = e^{x^2 y}$ is continuous.

Solution Notice that $f(x, y) = g(h(x, y))$, where $g(t) = e^t$ and $h(x, y) = x^2 y$. Since g is continuous for all values of t and h is a polynomial in x and y (and hence continuous for all x and y), it follows from Theorem 2.2 that f is continuous for all x and y. ■

REMARK 2.3

All of the foregoing analysis is extended to functions of three (or more) variables in the obvious fashion.

DEFINITION 2.3

Let the function $f(x, y, z)$ be defined on the interior of a sphere, centered at the point (a, b, c), except possibly at (a, b, c) itself. We say that $\lim_{(x,y,z)\to(a,b,c)} f(x, y, z) = L$ if for every $\varepsilon > 0$ there exists a $\delta > 0$ such that $|f(x, y, z) - L| < \varepsilon$ whenever

$$0 < \sqrt{(x-a)^2 + (y-b)^2 + (z-c)^2} < \delta.$$

Observe that, as with limits of functions of two variables, Definition 2.3 says that in order to have $\lim_{(x,y,z)\to(a,b,c)} f(x, y, z) = L$, we must have that $f(x, y, z)$ approaches L along every possible path through the point (a, b, c). Just as with functions of two variables, notice that if a function of three variables approaches different limits along two particular paths, then the limit does not exist.

EXAMPLE 2.10 A Limit in Three Dimensions That Does Not Exist

Evaluate $\lim_{(x,y,z)\to(0,0,0)} \frac{x^2 + y^2 - z^2}{x^2 + y^2 + z^2}$.

Solution First, we consider the path $x = y = 0$ (the z-axis). There, we have

$$\lim_{(0,0,z)\to(0,0,0)} \frac{0^2 + 0^2 - z^2}{0^2 + 0^2 + z^2} = \lim_{z\to 0} \frac{-z^2}{z^2} = -1.$$

Along the path $x = z = 0$ (the y-axis), we have

$$\lim_{(0,y,0)\to(0,0,0)} \frac{0^2 + y^2 - 0^2}{0^2 + y^2 + 0^2} = \lim_{y\to 0} \frac{y^2}{y^2} = 1.$$

Since the limits along these two specific paths do not agree, the limit does not exist. ■

We extend the definition of continuity to functions of three variables in the obvious way, as follows.

DEFINITION 2.4

Suppose that $f(x, y, z)$ is defined in the interior of a sphere centered at (a, b, c). We say that f is **continuous** at (a, b, c) if $\lim_{(x,y,z)\to(a,b,c)} f(x, y, z) = f(a, b, c)$.

If $f(x, y, z)$ is not continuous at (a, b, c), then we call (a, b, c) a **discontinuity** of f.

As you can see, limits and continuity for functions of three variables work essentially the same as they do for functions of two variables. You will examine these in more detail in the exercises.

EXAMPLE 2.11 Continuity for a Function of Three Variables

Find all points where $f(x, y, z) = \ln(9 - x^2 - y^2 - z^2)$ is continuous.

Solution Notice that $f(x, y, z)$ is defined only for $9 - x^2 - y^2 - z^2 > 0$. On this domain, f is a composition of continuous functions, which is also continuous. So, f is continuous for $9 > x^2 + y^2 + z^2$, which you should recognize as the interior of the sphere of radius 3 centered at $(0, 0, 0)$. ■

BEYOND FORMULAS

The examples in this section illustrate an important principle of logic. In this case, a limit exists if you get the same limiting value *for all possible paths* through the point. To disprove such a "for all" statement, you need only to find one specific counterexample. Finding two paths with different limits *proves* that the limit does not exist. However, to prove a "for all" statement, you must demonstrate that a general (arbitrary) example produces the desired result. This is typically a more elaborate task than finding a counterexample. To see what we mean, compare examples 2.3 and 2.6. To understand the different methods for proving that a limit does or does not exist, you need the more basic understanding of the logic of a universal "for all" statement.

EXERCISES 12.2

WRITING EXERCISES

1. Choosing between the paths $y = x$ and $x = y^2$, explain why $y = x$ is a better choice in example 2.4 but $x = y^2$ is a better choice in example 2.5.
2. In terms of Definition 2.1, explain why the limit in example 2.5 does not exist. That is, explain why making (x, y) close to $(0, 0)$ doesn't guarantee that $f(x, y)$ is close to 0.
3. A friend claims that a limit equals 0, but you found that it does not exist. Looking over your friend's work, you see that the path with $x = 0$ and the path with $y = 0$ both produce a limit of 0. No other work is shown. Explain to your friend why other paths must be checked.
4. Explain why the path $y = x$ is not a valid path for the limit in example 2.7.

In exercises 1–6, compute the indicated limit.

1. $\lim_{(x,y)\to(1,3)} \frac{x^2 y}{4x^2 - y}$
2. $\lim_{(x,y)\to(2,-1)} \frac{x + y}{x^2 - 2xy}$
3. $\lim_{(x,y)\to(\pi,1)} \frac{\cos xy}{y^2 + 1}$
4. $\lim_{(x,y)\to(-3,0)} \frac{e^{xy}}{x^2 + y^2}$
5. $\lim_{(x,y,z)\to(1,0,2)} \frac{4xz}{y^2 + z^2}$
6. $\lim_{(x,y,z)\to(1,1,2)} \frac{e^{x+y-z}}{x - z}$

In exercises 7–22, show that the indicated limit does not exist.

7. $\lim_{(x,y)\to(0,0)} \frac{3x^2}{x^2+y^2}$

8. $\lim_{(x,y)\to(0,0)} \frac{2y^2}{2x^2-y^2}$

9. $\lim_{(x,y)\to(0,0)} \frac{4xy}{3y^2-x^2}$

10. $\lim_{(x,y)\to(0,0)} \frac{2xy}{x^2+2y^2}$

11. $\lim_{(x,y)\to(0,0)} \frac{2x^2y}{x^4+y^2}$

12. $\lim_{(x,y)\to(0,0)} \frac{3x^3\sqrt{y}}{x^4+y^2}$

13. $\lim_{(x,y)\to(0,0)} \frac{\sqrt[3]{x}y^2}{x+y^3}$

14. $\lim_{(x,y)\to(0,0)} \frac{2xy^3}{x^2+8y^6}$

15. $\lim_{(x,y)\to(0,0)} \frac{y\sin x}{x^2+y^2}$

16. $\lim_{(x,y)\to(0,0)} \frac{x(\cos y-1)}{x^3+y^3}$

17. $\lim_{(x,y)\to(1,2)} \frac{xy-2x-y+2}{x^2-2x+y^2-4y+5}$

18. $\lim_{(x,y)\to(2,0)} \frac{2y^2}{(x-2)^2+y^2}$

19. $\lim_{(x,y,z)\to(0,0,0)} \frac{3x^2}{x^2+y^2+z^2}$

20. $\lim_{(x,y,z)\to(0,0,0)} \frac{x^2+y^2+z^2}{x^2-y^2+z^2}$

21. $\lim_{(x,y,z)\to(0,0,0)} \frac{xyz}{x^3+y^3+z^3}$

22. $\lim_{(x,y,z)\to(0,0,0)} \frac{x^2yz}{x^4+y^4+z^4}$

In exercises 23–30, show that the indicated limit exists.

23. $\lim_{(x,y)\to(0,0)} \frac{xy^2}{x^2+y^2}$

24. $\lim_{(x,y)\to(0,0)} \frac{x^2y}{x^2+y^2}$

25. $\lim_{(x,y)\to(0,0)} \frac{2x^2\sin y}{2x^2+y^2}$

26. $\lim_{(x,y)\to(0,0)} \frac{x^3y+x^2y^3}{x^2+y^2}$

27. $\lim_{(x,y)\to(0,0)} \frac{x^3+4x^2+2y^2}{2x^2+y^2}$

28. $\lim_{(x,y)\to(0,0)} \frac{x^2y-x^2-y^2}{x^2+y^2}$

29. $\lim_{(x,y,z)\to(0,0,0)} \frac{3x^3}{x^2+y^2+z^2}$

30. $\lim_{(x,y,z)\to(0,0,0)} \frac{x^2y^2z^2}{x^2+y^2+z^2}$

In exercises 31–34, use graphs and density plots to explain why the limit in the indicated exercise does not exist.

31. Exercise 7

32. Exercise 8

33. Exercise 9

34. Exercise 10

In exercises 35–44, determine all points at which the given function is continuous.

35. $f(x,y)=\sqrt{9-x^2-y^2}$

36. $f(x,y,z)=\frac{x^3}{y}+\sin z$

37. $f(x,y)=\ln(3-x^2+y)$

38. $f(x,y)=\tan(x+y)$

39. $f(x,y,z)=\sqrt{x^2+y^2+z^2-4}$

40. $f(x,y,z)=\sqrt{z-x^2-y^2}$

41. $f(x,y)=\begin{cases}(y-2)\cos\left(\frac{1}{x^2}\right), & \text{if } x\neq 0\\ 0, & \text{if } x=0\end{cases}$

42. $f(x,y)=\begin{cases}\dfrac{\sin\sqrt{1-x^2-y^2}}{\sqrt{1-x^2-y^2}}, & \text{if } x^2+y^2<1\\ 1, & \text{if } x^2+y^2=1\end{cases}$

43. $f(x,y)=\begin{cases}\dfrac{x^2-y^2}{x-y}, & \text{if } x\neq y\\ 2x, & \text{if } x=y\end{cases}$

44. $f(x,y)=\begin{cases}\cos\left(\dfrac{1}{x^2+y^2}\right), & \text{if } (x,y)\neq(0,0)\\ 1, & \text{if } (x,y)=(0,0)\end{cases}$

In exercises 45 and 46, estimate the indicated limit numerically.

45. $\lim_{(x,y)\to(0,0)} \frac{1-\cos xy}{x^2y^2+x^2y^3}$

46. $\lim_{(x,y)\to(0,0)} \frac{3\sin xy^2}{x^2y^2+xy^2}$

In exercises 47–50, label the statement as true or false and explain.

47. If $\lim_{(x,y)\to(a,b)} f(x,y)=L$, then $\lim_{x\to a} f(x,b)=L$.

48. If $\lim_{x\to a} f(x,b)=L$, then $\lim_{(x,y)\to(a,b)} f(x,y)=L$.

49. If $\lim_{x\to a} f(x,b)=\lim_{y\to b} f(a,y)=L$, then $\lim_{(x,y)\to(a,b)} f(x,y)=L$.

50. If $\lim_{(x,y)\to(0,0)} f(x,y)=0$, then $\lim_{(x,y)\to(0,0)} f(cx,y)=0$ for any constant c.

51. In example 2.5, show that for any k, if the limit is evaluated along the line $y=kx$, $\lim_{(x,y)\to(0,0)} \frac{xy^2}{x^2+y^4}=0$.

52. Repeat exercise 51 for the limit in exercise 14.

53. Show that the function $f(x,y)=\begin{cases}\dfrac{xy^2}{x^2+y^4}, & \text{if } (x,y)\neq(0,0)\\ 0, & \text{if } (x,y)=(0,0)\end{cases}$ is not continuous at (0, 0). Notice that this function is closely related to that of example 2.5.

54. Show that the function in exercise 53 "acts" continuous at the origin along any straight line through the origin, in the sense that for any such line l with the limit restricted to points (x,y) on l, $\lim_{(x,y)\to(0,0)} f(x,y)=f(0,0)$.

In exercises 55–58, use polar coordinates to find the indicated limit, if it exists. Note that $(x,y)\to(0,0)$ is equivalent to $r\to 0$.

55. $\lim_{(x,y)\to(0,0)} \frac{\sqrt{x^2+y^2}}{\sin\sqrt{x^2+y^2}}$

56. $\lim_{(x,y)\to(0,0)} \frac{e^{x^2+y^2}-1}{x^2+y^2}$

57. $\lim_{(x,y)\to(0,0)} \frac{xy^2}{x^2+y^2}$

58. $\lim_{(x,y)\to(0,0)} \frac{x^2y}{x^2+y^2}$

EXPLORATORY EXERCISES

1. In this exercise, you will explore how the patterns of contour plots relate to the existence of limits. Start by showing that $\lim_{(x,y)\to(0,0)} \frac{x^2}{x^2+y^2}$ doesn't exist and $\lim_{(x,y)\to(0,0)} \frac{x^2y}{x^2+y^2} = 0$. Then sketch several contour plots for each function while zooming in on the point (0, 0). For a function whose limit exists as (x, y) approaches (a, b), what should be happening to the range of function values as you zoom in on the point (a, b)? Describe the appearance of each contour plot for $\frac{x^2y}{x^2+y^2}$ near (0, 0). By contrast, what should be happening to the range of function values as you zoom in on a point at which the limit doesn't exist? Explain how this appears in the contour plots for $\frac{x^2}{x^2+y^2}$. Use contour plots to conjecture whether or not the following limits exist: $\lim_{(x,y)\to(0,0)} \frac{xy}{x^2+y}$ and $\lim_{(x,y)\to(0,0)} \frac{x \sin y}{x^2+y^2}$.

2. Find a function $g(y)$ such that $f(x, y)$ is continuous for

$$f(x, y) = \begin{cases} (1+xy)^{1/x}, & \text{if } x \neq 0 \\ g(y), & \text{if } x = 0 \end{cases}.$$

12.3 PARTIAL DERIVATIVES

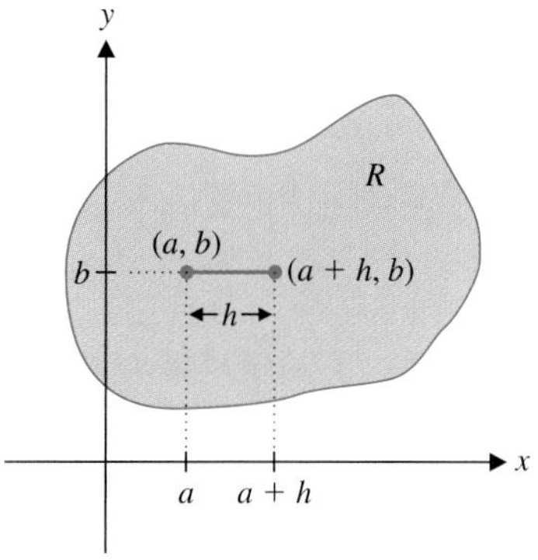

FIGURE 12.19
Average temperature on a horizontal line segment

In this section, we generalize the notion of derivative to functions of more than one variable. First, recall that for a function f of a single variable, we define the derivative function as

$$f'(x) = \lim_{h\to 0} \frac{f(x+h) - f(x)}{h},$$

for any values of x for which the limit exists. At any particular value $x = a$, we interpret $f'(a)$ as the instantaneous rate of change of the function with respect to x at that point.

Consider a flat metal plate in the shape of the region $R \subset \mathbb{R}^2$. Suppose that the temperature at any point $(x, y) \in R$ is given by $f(x, y)$. If you move along the horizontal line segment from (a, b) to $(a + h, b)$, what is the average rate of change of the temperature with respect to the horizontal distance x (see Figure 12.19)? Notice that on this line segment, y is a constant ($y = b$). So, the average rate of change on this line segment is given by

$$\frac{f(a+h, b) - f(a, b)}{h}.$$

To get the instantaneous rate of change of f in the x-direction at the point (a, b), we take the limit as $h \to 0$:

$$\lim_{h\to 0} \frac{f(a+h, b) - f(a, b)}{h}.$$

You should recognize this limit as a derivative. Since f is a function of two variables and we have held the one variable fixed ($y = b$), we call this the **partial derivative of f with respect to x** at the point (a, b), denoted

$$\frac{\partial f}{\partial x}(a, b) = \lim_{h\to 0} \frac{f(a+h, b) - f(a, b)}{h}.$$

This says that $\frac{\partial f}{\partial x}(a, b)$ gives the instantaneous rate of change of f with respect to x (i.e., in the x-direction) at the point (a, b). Graphically, observe that in defining $\frac{\partial f}{\partial x}(a, b)$, we are looking only at points in the plane $y = b$. The intersection of $z = f(x, y)$ and $y = b$ is a curve, as shown in Figures 12.20a and 12.20b (on the following page). The partial derivative $\frac{\partial f}{\partial x}(a, b)$ then gives the slope of the tangent line to this curve at $x = a$, as indicated in Figure 12.20b.

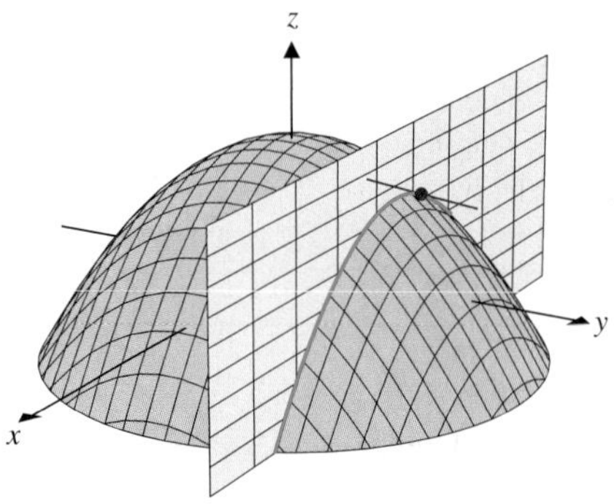

FIGURE 12.20a
Intersection of the surface $z = f(x, y)$ with the plane $y = b$

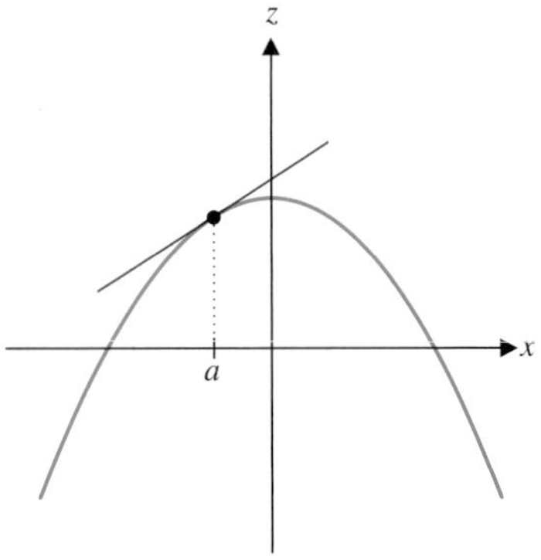

FIGURE 12.20b
The curve $z = f(x, b)$

Alternatively, if we move along a vertical line segment from (a, b) to $(a, b + h)$ (see Figure 12.21), the average rate of change of f along this segment is given by

$$\frac{f(a, b+h) - f(a, b)}{h}.$$

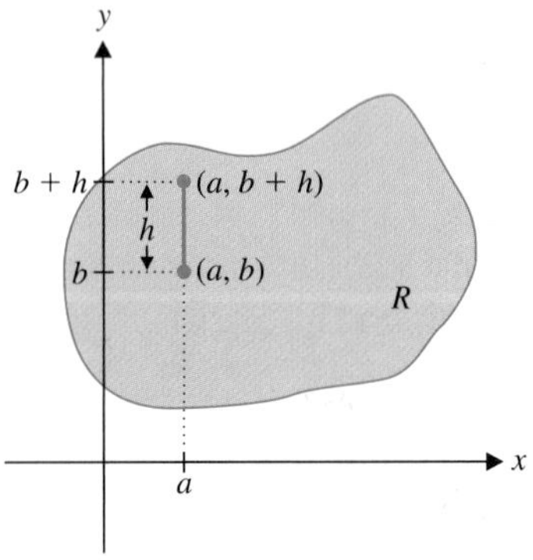

FIGURE 12.21
Average temperature on a vertical line segment

The instantaneous rate of change of f in the y-direction at the point (a, b) is then given by

$$\lim_{h \to 0} \frac{f(a, b+h) - f(a, b)}{h},$$

which you should again recognize as a derivative. In this case, however, we have held the value of x fixed ($x = a$) and refer to this as the **partial derivative of f with respect to y** at the point (a, b), denoted

$$\frac{\partial f}{\partial y}(a, b) = \lim_{h \to 0} \frac{f(a, b+h) - f(a, b)}{h}.$$

Graphically, observe that in defining $\frac{\partial f}{\partial y}(a, b)$, we are looking only at points in the plane $x = a$. The intersection of $z = f(x, y)$ and $x = a$ is a curve, as shown in Figures 12.22a and 12.22b. In this case, notice that the partial derivative $\frac{\partial f}{\partial y}(a, b)$ gives the slope of the tangent line to the curve at $y = b$, as shown in Figure 12.22b.

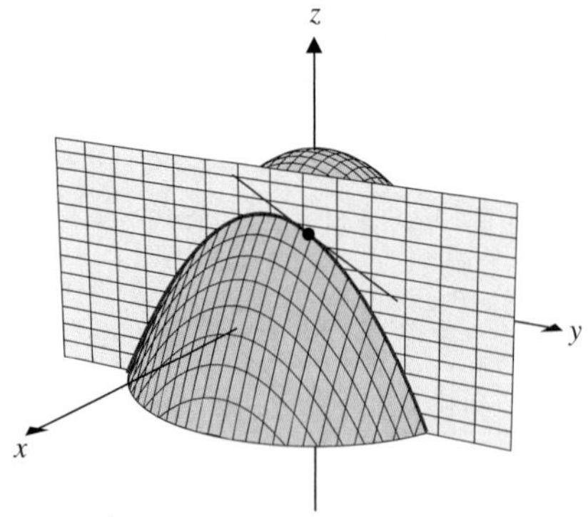

FIGURE 12.22a
The intersection of the surface $z = f(x, y)$ with the plane $x = a$

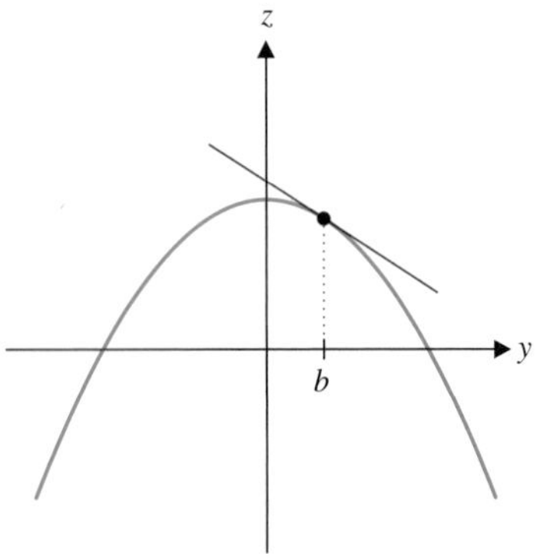

FIGURE 12.22b
The curve $z = f(a, y)$

More generally, we define the partial derivative functions as follows.

DEFINITION 3.1

The **partial derivative of $f(x, y)$ with respect to x,** written $\frac{\partial f}{\partial x}$, is defined by

$$\frac{\partial f}{\partial x}(x, y) = \lim_{h \to 0} \frac{f(x+h, y) - f(x, y)}{h},$$

for any values of x and y for which the limit exists.

The **partial derivative of $f(x, y)$ with respect to y,** written $\frac{\partial f}{\partial y}$, is defined by

$$\frac{\partial f}{\partial y}(x, y) = \lim_{h \to 0} \frac{f(x, y+h) - f(x, y)}{h},$$

for any values of x and y for which the limit exists.

Since we are now dealing with functions of several variables, we can no longer use the prime notation for denoting partial derivatives. [Which partial derivative would $f'(x, y)$ denote?] We introduce several convenient types of notation here. For $z = f(x, y)$, we write

$$\frac{\partial f}{\partial x}(x, y) = f_x(x, y) = \frac{\partial z}{\partial x}(x, y) = \frac{\partial}{\partial x}[f(x, y)].$$

The expression $\frac{\partial}{\partial x}$ is a **partial differential operator.** It tells you to take the partial derivative (with respect to x) of whatever expression follows it. Similarly, we have

$$\frac{\partial f}{\partial y}(x, y) = f_y(x, y) = \frac{\partial z}{\partial y}(x, y) = \frac{\partial}{\partial y}[f(x, y)].$$

Look carefully at how we defined these derivatives and you'll see that we can compute partial derivatives using all of our usual rules for computing ordinary derivatives. Notice that in the definition of $\frac{\partial f}{\partial x}$, the value of y is held constant, say at $y = b$. If we define $g(x) = f(x, b)$, then

$$\frac{\partial f}{\partial x}(x, b) = \lim_{h \to 0} \frac{f(x+h, b) - f(x, b)}{h} = \lim_{h \to 0} \frac{g(x+h) - g(x)}{h} = g'(x).$$

That is, to compute the partial derivative $\frac{\partial f}{\partial x}$, you simply take an ordinary derivative with respect to x, while treating y as a constant. Similarly, you can compute $\frac{\partial f}{\partial y}$ by taking an ordinary derivative with respect to y, while treating x as a constant.

EXAMPLE 3.1 Computing Partial Derivatives

For $f(x, y) = 3x^2 + x^3y + 4y^2$, compute $\frac{\partial f}{\partial x}(x, y)$, $\frac{\partial f}{\partial y}(x, y)$, $f_x(1, 0)$ and $f_y(2, -1)$.

Solution Compute $\frac{\partial f}{\partial x}$ by treating y as a constant. We have

$$\frac{\partial f}{\partial x} = \frac{\partial}{\partial x}(3x^2 + x^3y + 4y^2) = 6x + (3x^2)y + 0 = 6x + 3x^2y.$$

The partial derivative of $4y^2$ with respect to x is 0, since $4y^2$ is treated as if it were a constant when differentiating with respect to x. Next, we compute $\dfrac{\partial f}{\partial y}$ by treating x as a constant. We have

$$\frac{\partial f}{\partial y} = \frac{\partial}{\partial y}(3x^2 + x^3 y + 4y^2) = 0 + x^3(1) + 8y = x^3 + 8y.$$

Substituting values for x and y, we get

$$f_x(1, 0) = \frac{\partial f}{\partial x}(1, 0) = 6 + 0 = 6$$

and

$$f_y(2, -1) = \frac{\partial f}{\partial y}(2, -1) = 8 - 8 = 0. \blacksquare$$

Since we are holding one of the variables fixed when we compute a partial derivative, we have the product rules:

$$\frac{\partial}{\partial x}(uv) = \frac{\partial u}{\partial x}v + u\frac{\partial v}{\partial x}$$

and

$$\frac{\partial}{\partial y}(uv) = \frac{\partial u}{\partial y}v + u\frac{\partial v}{\partial y}$$

and the quotient rule:

$$\frac{\partial}{\partial x}\left(\frac{u}{v}\right) = \frac{\dfrac{\partial u}{\partial x}v - u\dfrac{\partial v}{\partial x}}{v^2},$$

with a corresponding quotient rule holding for $\dfrac{\partial}{\partial y}\left(\dfrac{u}{v}\right)$.

EXAMPLE 3.2   Computing Partial Derivatives

For $f(x, y) = e^{xy} + \dfrac{x}{y}$, compute $\dfrac{\partial f}{\partial x}$ and $\dfrac{\partial f}{\partial y}$.

Solution Recall that if $g(x) = e^{4x} + \dfrac{x}{4}$, then $g'(x) = 4e^{4x} + \dfrac{1}{4}$, from the chain rule. Replacing the 4's with y and treating them as we would any other constant, we have

$$\frac{\partial f}{\partial x} = \frac{\partial}{\partial x}\left(e^{xy} + \frac{x}{y}\right) = ye^{xy} + \frac{1}{y}.$$

For the y-partial derivative, recall that if $h(y) = \dfrac{4}{y}$, then $h'(y) = -\dfrac{4}{y^2}$. Replacing the 4 with x and treating it as you would any other constant, we have

$$\frac{\partial f}{\partial y} = \frac{\partial}{\partial y}\left(e^{xy} + \frac{x}{y}\right) = xe^{xy} - \frac{x}{y^2}. \blacksquare$$

We interpret partial derivatives as rates of change, in the same way as we interpret ordinary derivatives of functions of a single variable.

EXAMPLE 3.3 An Application of Partial Derivatives to Thermodynamics

For a real gas, van der Waals' equation states that

$$\left(P + \frac{n^2 a}{V^2}\right)(V - nb) = nRT.$$

Here, P is the pressure of the gas, V is the volume of the gas, T is the temperature (in degrees Kelvin), n is the number of moles of gas, R is the universal gas constant and a and b are constants. Compute and interpret $\frac{\partial P}{\partial V}$ and $\frac{\partial T}{\partial P}$.

Solution We first solve for P to get

$$P = \frac{nRT}{V - nb} - \frac{n^2 a}{V^2}$$

and compute

$$\frac{\partial P}{\partial V} = \frac{\partial}{\partial V}\left(\frac{nRT}{V - nb} - \frac{n^2 a}{V^2}\right) = -\frac{nRT}{(V - nb)^2} + 2\frac{n^2 a}{V^3}.$$

Notice that this gives the rate of change of pressure relative to a change in volume (with temperature held constant). Next, solving van der Waals' equation for T, we get

$$T = \frac{1}{nR}\left(P + \frac{n^2 a}{V^2}\right)(V - nb)$$

and compute

$$\frac{\partial T}{\partial P} = \frac{\partial}{\partial P}\left[\frac{1}{nR}\left(P + \frac{n^2 a}{V^2}\right)(V - nb)\right] = \frac{1}{nR}(V - nb).$$

This gives the rate of change of temperature relative to a change in pressure (with volume held constant). In exercise 22, you will have an opportunity to discover an interesting fact about these partial derivatives. ■

TODAY IN MATHEMATICS

Shing-Tung Yau (1949–)
A Chinese-born mathematician who earned a Fields Medal for his contributions to algebraic geometry and partial differential equations. A strong supporter of mathematics education in China, he established the Institute of Mathematical Science in Hong Kong, the Morningside Center of Mathematics of the Chinese Academy of Sciences and the Center of Mathematical Sciences at Zhejiang University. His work has been described by colleagues as "extremely deep and powerful" and as showing "enormous technical power and insight. He has cracked problems on which progress has been stopped for years."

Notice that the partial derivatives found in the preceding examples are themselves functions of two variables. We have seen that second- and higher-order derivatives of functions of a single variable provide much significant information. Not surprisingly, **higher-order partial derivatives** are also very important in applications.

For functions of two variables, there are four different second-order partial derivatives. The partial derivative with respect to x of $\frac{\partial f}{\partial x}$ is $\frac{\partial}{\partial x}\left(\frac{\partial f}{\partial x}\right)$, usually abbreviated as $\frac{\partial^2 f}{\partial x^2}$ or f_{xx}. Similarly, taking two successive partial derivatives with respect to y gives us $\frac{\partial}{\partial y}\left(\frac{\partial f}{\partial y}\right) = \frac{\partial^2 f}{\partial y^2} = f_{yy}$. For **mixed second-order partial derivatives,** one derivative is taken with respect to each variable. If the first partial derivative is taken with respect to x, we have $\frac{\partial}{\partial y}\left(\frac{\partial f}{\partial x}\right)$, abbreviated as $\frac{\partial^2 f}{\partial y \partial x}$, or $(f_x)_y = f_{xy}$. If the first partial derivative is taken with respect to y, we have $\frac{\partial}{\partial x}\left(\frac{\partial f}{\partial y}\right)$, abbreviated as $\frac{\partial^2 f}{\partial x \partial y}$, or $(f_y)_x = f_{yx}$.

EXAMPLE 3.4 Computing Second-Order Partial Derivatives

Find all second-order partial derivatives of $f(x, y) = x^2y - y^3 + \ln x$.

Solution We start by computing the first-order partial derivatives: $\frac{\partial f}{\partial x} = 2xy + \frac{1}{x}$ and $\frac{\partial f}{\partial y} = x^2 - 3y^2$. We then have

$$\frac{\partial^2 f}{\partial x^2} = \frac{\partial}{\partial x}\left(\frac{\partial f}{\partial x}\right) = \frac{\partial}{\partial x}\left(2xy + \frac{1}{x}\right) = 2y - \frac{1}{x^2},$$

$$\frac{\partial^2 f}{\partial y \partial x} = \frac{\partial}{\partial y}\left(\frac{\partial f}{\partial x}\right) = \frac{\partial}{\partial y}\left(2xy + \frac{1}{x}\right) = 2x,$$

$$\frac{\partial^2 f}{\partial x \partial y} = \frac{\partial}{\partial x}\left(\frac{\partial f}{\partial y}\right) = \frac{\partial}{\partial x}(x^2 - 3y^2) = 2x$$

and finally,

$$\frac{\partial^2 f}{\partial y^2} = \frac{\partial}{\partial y}\left(\frac{\partial f}{\partial y}\right) = \frac{\partial}{\partial y}(x^2 - 3y^2) = -6y.$$

Notice in example 3.4 that $\frac{\partial^2 f}{\partial y \partial x} = \frac{\partial^2 f}{\partial x \partial y}$. It turns out that this is true for most, but *not all*, of the functions that you will encounter. (See exercise 61 for a counterexample.) The proof of the following result can be found in most texts on advanced calculus.

THEOREM 3.1

If $f_{xy}(x, y)$ and $f_{yx}(x, y)$ are continuous on an open set containing (a, b), then $f_{xy}(a, b) = f_{yx}(a, b)$.

We can, of course, compute third-, fourth- or even higher-order partial derivatives. Theorem 3.1 can be extended to show that as long as the partial derivatives are all continuous in an open set, the order of differentiation doesn't matter. With higher-order partial derivatives, notations such as $\frac{\partial^3 f}{\partial x \partial y \partial x}$ become quite awkward and so, we usually use f_{xyx} instead.

EXAMPLE 3.5 Computing Higher-Order Partial Derivatives

For $f(x, y) = \cos(xy) - x^3 + y^4$, compute f_{xyy} and f_{xyyy}.

Solution We have

$$f_x = \frac{\partial}{\partial x}\left[\cos(xy) - x^3 + y^4\right] = -y\sin(xy) - 3x^2.$$

Differentiating f_x with respect to y gives us

$$f_{xy} = \frac{\partial}{\partial y}[-y\sin(xy) - 3x^2] = -\sin(xy) - xy\cos(xy)$$

and

$$f_{xyy} = \frac{\partial}{\partial y}[-\sin(xy) - xy\cos(xy)] = -2x\cos(xy) + x^2y\sin(xy).$$

Finally, we have

$$f_{xyyy} = \frac{\partial}{\partial y}[-2x\cos(xy) + x^2 y\sin(xy)]$$
$$= 2x^2\sin(xy) + x^2\sin(xy) + x^3 y\cos(xy) = 3x^2\sin(xy) + x^3 y\cos(xy). \quad\blacksquare$$

Thus far, we have worked with partial derivatives of functions of two variables. The extensions to functions of three or more variables are completely analogous to what we have discussed here. In example 3.6, you can see that the calculations proceed just as you would expect.

EXAMPLE 3.6 Partial Derivatives of Functions of Three Variables

For $f(x, y, z) = \sqrt{xy^3 z} + 4x^2 y$, defined for $x, y, z \geq 0$, compute f_x, f_{xy} and f_{xyz}.

Solution To keep x, y and z as separate as possible, we first rewrite f as

$$f(x, y, z) = x^{1/2} y^{3/2} z^{1/2} + 4x^2 y.$$

To compute the partial derivative with respect to x, we treat y and z as constants and obtain

$$f_x = \frac{\partial}{\partial x}(x^{1/2} y^{3/2} z^{1/2} + 4x^2 y) = \left(\frac{1}{2} x^{-1/2}\right) y^{3/2} z^{1/2} + 8xy.$$

Next, treating x and z as constants, we get

$$f_{xy} = \frac{\partial}{\partial y}\left(\frac{1}{2} x^{-1/2} y^{3/2} z^{1/2} + 8xy\right) = \left(\frac{1}{2} x^{-1/2}\right)\left(\frac{3}{2} y^{1/2}\right) z^{1/2} + 8x.$$

Finally, treating x and y as constants, we get

$$f_{xyz} = \frac{\partial}{\partial z}\left[\left(\frac{1}{2} x^{-1/2}\right)\left(\frac{3}{2} y^{1/2}\right) z^{1/2} + 8x\right] = \left(\frac{1}{2} x^{-1/2}\right)\left(\frac{3}{2} y^{1/2}\right)\left(\frac{1}{2} z^{-1/2}\right).$$

Notice that this derivative is defined for $x, z > 0$ and $y \geq 0$. Further, you can show that all first-, second- and third-order partial derivatives are continuous for $x, y, z > 0$, so that the order in which we take the partial derivatives is irrelevant in this case. $\blacksquare$

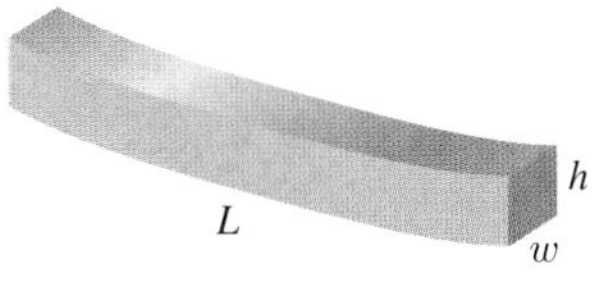

FIGURE 12.23
A horizontal beam

EXAMPLE 3.7 An Application of Partial Derivatives to a Sagging Beam

The sag in a beam of length L, width w and height h (see Figure 12.23) is given by $S(L, w, h) = c\dfrac{L^4}{wh^3}$ for some constant c. Show that $\dfrac{\partial S}{\partial L} = \dfrac{4}{L} S$, $\dfrac{\partial S}{\partial w} = -\dfrac{1}{w} S$ and $\dfrac{\partial S}{\partial h} = -\dfrac{3}{h} S$. Use this result to determine which variable has the greatest proportional effect on the sag.

Solution We start by computing

$$\frac{\partial S}{\partial L} = \frac{\partial}{\partial L}\left(c\frac{L^4}{wh^3}\right) = c\frac{4L^3}{wh^3}.$$

To rewrite this in terms of S, multiply top and bottom by L to get

$$\frac{\partial S}{\partial L} = c\frac{4L^3}{wh^3} = c\frac{4L^4}{wh^3 L} = \frac{4}{L} c\frac{L^4}{wh^3} = \frac{4}{L} S.$$

The other calculations are similar and are left as exercises. To interpret the results, suppose that a small change ΔL in length produces a small change ΔS in the sag. We now have that $\frac{\Delta S}{\Delta L} \approx \frac{\partial S}{\partial L} = \frac{4}{L}S$. Rearranging the terms, we have

$$\frac{\Delta S}{S} \approx 4\frac{\Delta L}{L}.$$

That is, the proportional change in S is approximately four times the proportional change in L. Similarly, we have that in absolute value, the proportional change in S is approximately the proportional change in w and three times the proportional change in h. Proportionally then, a change in the length has the greatest effect on the amount of sag. In this sense, length is the most *important* of the three dimensions. ■

In many applications, no formula for the function is available and we can only estimate the value of the partial derivatives from a small collection of data points.

EXAMPLE 3.8 Estimating Partial Derivatives from a Table of Data

A computer simulation of the flight of a baseball provided the data displayed in the table for the range $f(v, \omega)$ in feet of a ball hit with initial velocity v ft/s and backspin rate of ω rpm. Each ball is struck at an angle of 30° above the horizontal.

v \ ω	0	1000	2000	3000	4000
150	294	312	333	350	367
160	314	334	354	373	391
170	335	356	375	395	414
180	355	376	397	417	436

Use the data to estimate $\frac{\partial f}{\partial v}(160, 2000)$ and $\frac{\partial f}{\partial \omega}(160, 2000)$. Interpret both quantities in baseball terms.

Solution From the definition of partial derivative, we know that

$$\frac{\partial f}{\partial v}(160, 2000) = \lim_{h\to 0} \frac{f(160+h, 2000) - f(160, 2000)}{h},$$

so we can approximate the value of the partial derivative by computing the difference quotient $\frac{f(160+h, 2000) - f(160, 2000)}{h}$ for as small a value of h as possible. Since data points are provided for $v = 150$, we can compute the difference quotient for $h = -10$, to get

$$\frac{\partial f}{\partial v}(160, 2000) \approx \frac{f(150, 2000) - f(160, 2000)}{150 - 160} = \frac{333 - 354}{150 - 160} = 2.1.$$

We can also use the data point for $v = 170$, to get

$$\frac{\partial f}{\partial v}(160, 2000) \approx \frac{f(170, 2000) - f(160, 2000)}{170 - 160} = \frac{375 - 354}{170 - 160} = 2.1.$$

Since both estimates equal 2.1, we make the estimate $\frac{\partial f}{\partial v}(160, 2000) \approx 2.1$. The data point $f(160, 2000) = 354$ tells us that a ball struck with initial velocity 160 ft/s and backspin 2000 rpm will fly 354 feet. The partial derivative tells us that increasing the initial velocity by 1 ft/s will add approximately 2.1 feet to the distance.

Similarly, to estimate $\frac{\partial f}{\partial \omega}(160, 2000)$, we note that the closest data values to $\omega = 2000$ are $\omega = 1000$ and $\omega = 3000$. We get

$$\frac{\partial f}{\partial \omega}(160, 2000) \approx \frac{f(160,1000) - f(160,2000)}{1000 - 2000} = \frac{334 - 354}{1000 - 2000} = 0.02$$

and

$$\frac{\partial f}{\partial \omega}(160, 2000) \approx \frac{f(160,3000) - f(160,2000)}{3000 - 2000} = \frac{373 - 354}{3000 - 2000} = 0.019.$$

Reasonable estimates for $\frac{\partial f}{\partial \omega}(160, 2000)$ are then 0.02, 0.019 or 0.0195 (the average of the two calculations). Using 0.02 as our approximation, we can interpret this to mean that an increase in backspin of 1 rpm will add approximately 0.02 ft to the distance. A simpler way to interpret this is to say that an increase of 100 rpm will add approximately 2 ft to the distance. ■

BEYOND FORMULAS

When you think about partial derivatives, it helps to use the Rule of Three, which suggests that mathematical topics should be explored from symbolic, graphical and numerical viewpoints, where appropriate. Symbolically, you have all of the usual derivative formulas at your disposal. Graphically, you can view the value of a partial derivative at a particular point as the slope of the tangent line to a cross section of the surface $z = f(x, y)$. Numerically, you can approximate the value of a partial derivative at a point using a difference quotient, as in example 3.8.

HW #15 Pg 957 - 958 #1-16 odd, 17, 19.

EXERCISES 12.3

WRITING EXERCISES

1. Suppose that the function $f(x, y)$ is a sum of terms where each term contains x or y but not both. Explain why $f_{xy} = 0$.

2. In Definition 3.1, explain how to remember which partial derivative involves the term $f(x + h, y)$ and which involves the term $f(x, y + h)$.

3. In section 2.8, we computed derivatives implicitly, by using the chain rule and differentiating both sides of an equation with respect to x. In the process of doing so, we made calculations such as $(x^2y^2)' = 2xy^2 + 2x^2yy'$. Explain why this derivative is computed differently than the partial derivatives of this section.

4. For $f(x, y, z) = x^3e^{4x \sin y} + y^2 \sin xy + 4xyz$, you could compute f_{xyz} in a variety of orders. Discuss how many different orders are possible and which order(s) would be the easiest.

In exercises 1–8, find all first-order partial derivatives.

1. $f(x, y) = x^3 - 4xy^2 + y^4$
2. $f(x, y) = x^2y^3 - 3x$
3. $f(x, y) = x^2 \sin xy - 3y^3$
4. $f(x, y) = 3e^{x^2y} - \sqrt{x - 1}$
5. $f(x, y) = 4e^{x/y} - \frac{y}{x}$
6. $f(x, y) = \frac{x - 3}{y} + x^2 \tan y$
7. $f(x, y, z) = 3x \sin y + 4x^3y^2z$
8. $f(x, y, z) = \frac{2}{\sqrt{x^2 + y^2 + z^2}}$

In exercises 9–16, find the indicated partial derivatives.

9. $f(x, y) = x^3 - 4xy^2 + 3y;\ \frac{\partial^2 f}{\partial x^2}, \frac{\partial^2 f}{\partial y^2}, \frac{\partial^2 f}{\partial y \partial x}$

10. $f(x, y) = x^2 y - 4x + 3\sin y;\ \frac{\partial^2 f}{\partial x^2}, \frac{\partial^2 f}{\partial y^2}, \frac{\partial^2 f}{\partial y \partial x}$

11. $f(x, y) = x^4 - 3x^2y^3 + 5y;\ f_{xx}, f_{xy}, f_{xyy}$

12. $f(x, y) = e^{4x} - \sin y^2 - \sqrt{xy};\ f_{xx}, f_{xy}, f_{yyx}$

13. $f(x, y, z) = x^3y^2 - \sin yz;\ f_{xx}, f_{yz}, f_{xyz}$

14. $f(x, y, z) = e^{2xy} - \frac{z^2}{y} + xz\sin y;\ f_{xx}, f_{yy}, f_{yyzz}$

15. $f(w, x, y, z) = w^2xy - e^{wz};\ f_{ww}, f_{wxy}, f_{wwxyz}$

16. $f(w, x, y, z) = \sqrt{wyz} - x^3\sin w;\ f_{xx}, f_{yy}, f_{wxyz}$

In exercises 17–20, (a) sketch the graph of $z = f(x, y)$ and (b) on this graph, highlight the appropriate two-dimensional trace and interpret the partial derivative as a slope.

17. $f(x, y) = 4 - x^2 - y^2,\ \frac{\partial f}{\partial x}(1, 1)$

18. $f(x, y) = \sqrt{x^2 + y^2},\ \frac{\partial f}{\partial x}(1, 0)$

19. $f(x, y) = 4 - x^2 - y^2,\ \frac{\partial f}{\partial y}(2, 0)$

20. $f(x, y) = \sqrt{x^2 + y^2},\ \frac{\partial f}{\partial y}(0, 2)$

21. Compute and interpret $\frac{\partial V}{\partial T}$ for van der Waals' equation (see example 3.3).

22. For van der Waals' equation, show that $\frac{\partial T}{\partial P}\frac{\partial P}{\partial V}\frac{\partial V}{\partial T} = -1$. If you misunderstood the chain rule, why might you expect this product to equal 1?

23. For the specific case of van der Waals' equation given by $\left(P + \frac{14}{V^2}\right)(V - 0.004) = 12T$, use the partial derivative $\frac{\partial P}{\partial T}$ to estimate the change in pressure due to an increase of one degree.

24. For the specific case of van der Waals' equation given by $\left(P + \frac{14}{V^2}\right)(V - 0.004) = 12T$, use the partial derivative $\frac{\partial T}{\partial V}$ to estimate the change in temperature due to an increase in volume of one unit.

25. In example 3.7, show that $\frac{\partial S}{\partial w} = -\frac{1}{w}S$.

26. In example 3.7, show that $\frac{\partial S}{\partial h} = -\frac{3}{h}S$.

27. If the sag in the beam of example 3.7 were given by $S(L, w, h) = c\frac{L^3}{wh^4}$, determine which variable would have the greatest proportional effect.

28. Based on example 3.7 and your result in exercise 27, state a simple rule for determining which variable has the greatest proportional effect.

In exercises 29–32, find all points at which $\frac{\partial f}{\partial x} = \frac{\partial f}{\partial y} = 0$ and interpret the significance of the points graphically.

29. $f(x, y) = x^2 + y^2$

30. $f(x, y) = x^2 + y^2 - x^4$

31. $f(x, y) = \sin x \sin y$

32. $f(x, y) = e^{-x^2-y^2}$

In exercises 33–36, use the contour plot to estimate $\frac{\partial f}{\partial x}$ and $\frac{\partial f}{\partial y}$ at the origin.

33.

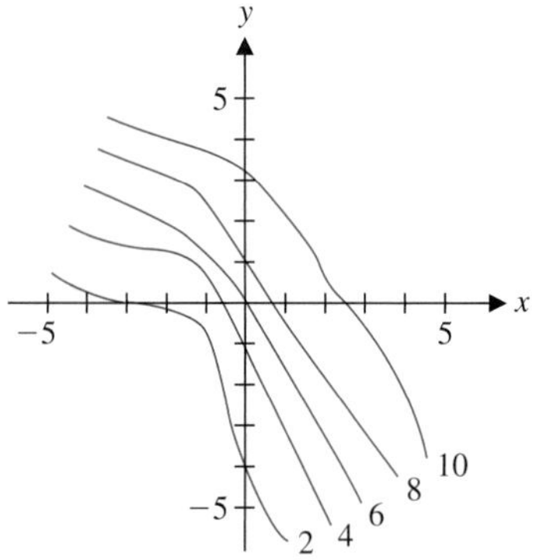

34.

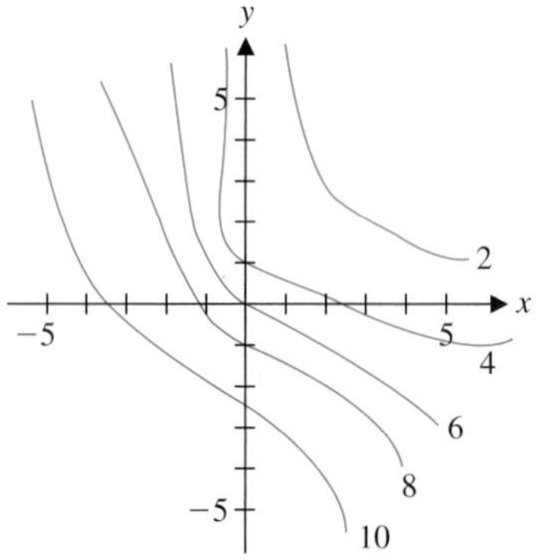

35.

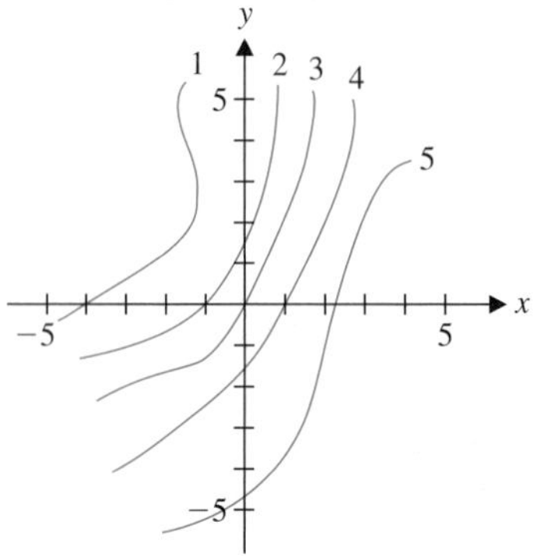

36.

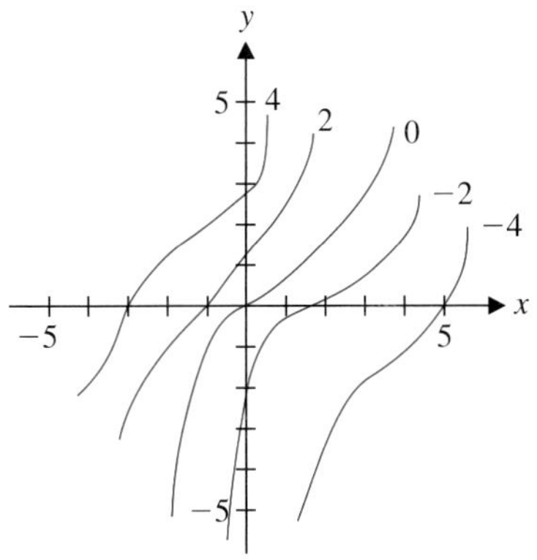

37. The table shows wind chill (how cold it "feels" outside) as a function of temperature (degrees Fahrenheit) and wind speed (mph). We can think of this as a function $C(t, s)$. Estimate the partial derivatives $\frac{\partial C}{\partial t}(10, 10)$ and $\frac{\partial C}{\partial s}(10, 10)$. Interpret each partial derivative and explain why it is surprising that $\frac{\partial C}{\partial t}(10, 10) \neq 1$.

Speed \ *Temp*	**30**	**20**	**10**	**0**	**−10**
0	30	20	10	0	−10
5	27	16	6	−5	−15
10	16	4	−9	−24	−33
15	9	−5	−18	−32	−45
20	4	−10	−25	−39	−53
25	0	−15	−29	−44	−59
30	−2	−18	−33	−48	−63

38. Rework exercise 37 using the point (10, 20). Explain the significance of the inequality $\left|\frac{\partial C}{\partial s}(10, 10)\right| > \left|\frac{\partial C}{\partial s}(10, 20)\right|$.

39. Using the baseball data in example 3.8, estimate and interpret $\frac{\partial f}{\partial v}(170, 3000)$ and $\frac{\partial f}{\partial \omega}(170, 3000)$.

40. According to the data in example 3.8, a baseball with initial velocity 170 ft/s and backspin 3000 rpm flies 395 ft. Suppose that the ball must go 400 ft to clear the fence for a home run. Based on your answers to exercise 39, how much extra backspin is needed for a home run?

41. Carefully write down a definition for the three first-order partial derivatives of a function of three variables $f(x, y, z)$.

42. Determine how many second-order partial derivatives there are of $f(x, y, z)$. Assuming a result analogous to Theorem 3.1, how many of these second-order partial derivatives are actually different?

43. Show that the functions $f_n(x, t) = \sin n\pi x \cos n\pi ct$ satisfy the **wave equation** $c^2 \frac{\partial^2 f}{\partial x^2} = \frac{\partial^2 f}{\partial t^2}$, for any positive integer n and any constant c.

44. Show that if $f(x)$ is a function with a continuous second derivative, then $f(x - ct)$ and $f(x + ct)$ are both solutions of the wave equation of exercise 43. If x represents position and t represents time, explain why c can be interpreted as the velocity of the wave.

45. The value of an investment of \$1000 invested at a constant 10% rate for 5 years is $V = 1000\left[\frac{1 + 0.1(1 - T)}{1 + I}\right]^5$, where T is the tax rate and I is the inflation rate. Compute $\frac{\partial V}{\partial I}$ and $\frac{\partial V}{\partial T}$, and discuss whether the tax rate or the inflation rate has a greater influence on the value of the investment.

46. The value of an investment of \$1000 invested at a rate r for 5 years with a tax rate of 28% is $V = 1000\left[\frac{1 + 0.72r}{1 + I}\right]^5$, where I is the inflation rate. Compute $\frac{\partial V}{\partial r}$ and $\frac{\partial V}{\partial I}$, and discuss whether the investment rate or the inflation rate has a greater influence on the value of the investment.

47. Suppose that the position of a guitar string of length L varies according to $p(x, t) = \sin x \cos t$, where x represents the distance along the string, $0 \le x \le L$, and t represents time. Compute and interpret $\frac{\partial p}{\partial x}$ and $\frac{\partial p}{\partial t}$.

48. Suppose that the concentration of some pollutant in a river as a function of position x and time t is given by $p(x, t) = p_0(x - ct)e^{-\mu t}$ for constants p_0, c and μ. Show that $\frac{\partial p}{\partial t} = -c\frac{\partial p}{\partial x} - \mu p$. Interpret both $\frac{\partial p}{\partial t}$ and $\frac{\partial p}{\partial x}$, and explain how this equation relates the change in pollution at a specific location to the current of the river and the rate at which the pollutant decays.

49. In a chemical reaction, the temperature T, entropy S, Gibbs free energy G and enthalpy H are related by $G = H - TS$. Show that $\frac{\partial(G/T)}{\partial T} = -\frac{H}{T^2}$.

50. For the chemical reaction of exercise 49, show that $\frac{\partial(G/T)}{\partial(1/T)} = H$. Chemists measure the enthalpy of a reaction by measuring this rate of change.

51. Suppose that three resistors are in parallel in an electrical circuit. If the resistances are R_1, R_2 and R_3 ohms, respectively, then the net resistance in the circuit equals $R = \frac{R_1 R_2 R_3}{R_1 R_2 + R_1 R_3 + R_2 R_3}$. Compute and interpret the partial derivative $\frac{\partial R}{\partial R_1}$. Given this partial derivative, explain how to quickly write down the partial derivatives $\frac{\partial R}{\partial R_2}$ and $\frac{\partial R}{\partial R_3}$.

52. The ideal gas law relating pressure, temperature and volume is $P = \frac{cT}{V}$, for some constant c. Show that $T\frac{\partial P}{\partial T}\frac{\partial V}{\partial T} = c$.

53. A process called **tag-and-recapture** is used to estimate populations of animals in the wild. First, some number T of the animals are captured, tagged and released into the wild. Later, a number S of the animals are captured, of which t are observed to be tagged. The estimate of the total population is then $P(T, S, t) = \frac{TS}{t}$. Compute $P(100, 60, 15)$; the proportion of tagged animals in the recapture is $\frac{15}{60} = \frac{1}{4}$. Based on your estimate of the total population, what proportion of the total population has been tagged? Now compute $\frac{\partial P}{\partial t}(100, 60, 15)$ and use it to estimate how much your population estimate would change if one more recaptured animal were tagged.

54. Let $T(x, y)$ be the temperature at longitude x and latitude y in the United States. In general, explain why you would expect to have $\frac{\partial T}{\partial y} < 0$. If a cold front is moving from east to west, would you expect $\frac{\partial T}{\partial x}$ to be positive or negative?

55. Suppose that L hours of labor and K dollars of investment by a company result in a productivity of $P = L^{0.75}K^{0.25}$. Compute the marginal productivity of labor, defined by $\frac{\partial P}{\partial L}$ and the marginal productivity of capital, defined by $\frac{\partial P}{\partial K}$.

56. For the production function in exercise 55, show that $\frac{\partial^2 P}{\partial L^2} < 0$ and $\frac{\partial^2 P}{\partial K^2} < 0$. Interpret this in terms of diminishing returns on investments in labor and capital. Show that $\frac{\partial^2 P}{\partial L \partial K} > 0$ and interpret it in economic terms.

57. Suppose that the demand for flour is given by $D_1 = 300 + \frac{10}{p_1+4} - 5p_2$ and the demand for bread is given by $D_2 = 250 + \frac{6}{p_2+2} - 6p_1$, where p_1 is the price of a pound of flour and p_2 is the price of a loaf of bread. Show that $\frac{\partial D_1}{\partial p_2}$ and $\frac{\partial D_2}{\partial p_1}$ are both negative. This is the definition of **complementary commodities.** Interpret the partial derivatives and explain why the word *complementary* is appropriate.

58. Suppose that $D_1(p_1, p_2)$ and $D_2(p_1, p_2)$ are demand functions for commodities with prices p_1 and p_2, respectively. If $\frac{\partial D_1}{\partial p_2}$ and $\frac{\partial D_2}{\partial p_1}$ are both positive, explain why the commodities are called **substitute commodities.**

59. Suppose that the output of a factory is given by $P = 20K^{1/3}L^{1/2}$, where K is the capital investment in thousands of dollars and L is the labor force in thousands of workers. If $K = 125$ and $L = 900$, use a partial derivative to estimate the effect of adding a thousand workers.

60. Suppose that the output of a factory is given by $P = 80K^{1/4}L^{3/4}$, where K is the capital investment in thousands of dollars and L is the labor force in thousands of workers. If $K = 256$ and $L = 10,000$, use a partial derivative to estimate the effect of increasing capital by one thousand dollars.

61. For the function

$$f(x, y) = \begin{cases} \frac{xy(x^2 - y^2)}{x^2 + y^2}, & \text{if } (x, y) \neq (0, 0) \\ 0, & \text{if } (x, y) = (0, 0) \end{cases}$$

use the limit definitions of partial derivatives to show that $f_{xy}(0, 0) = -1$ but $f_{yx}(0, 0) = 1$. Determine which assumption in Theorem 3.1 is not true.

62. For $f(x, y) = \begin{cases} \frac{xy^2}{x^2 + y^4}, & \text{if } (x, y) \neq (0, 0) \\ 0, & \text{if } (x, y) = (0, 0) \end{cases}$, show that $\frac{\partial f}{\partial x}(0, 0) = \frac{\partial f}{\partial y}(0, 0) = 0$. [Note that we have previously shown that this function is not continuous at $(0, 0)$.]

63. Sometimes the order of differentiation makes a practical difference. For $f(x, y) = \frac{1}{x}\sin(xy^2)$, show that $\frac{\partial^2 f}{\partial x \partial y} = \frac{\partial^2 f}{\partial y \partial x}$ but that the ease of calculations is not the same.

64. For a rectangle of length L and perimeter P, show that the area is given by $A = \frac{1}{2}LP - L^2$. Compute $\frac{\partial A}{\partial L}$. A simpler formula for area is $A = LW$, where W is the width of the rectangle. Compute $\frac{\partial A}{\partial L}$ and show that your answer is not equivalent to the previous derivative. Explain the difference by noting that in one case the width is held constant while L changes, whereas in the other case the perimeter is held constant while L changes.

65. Suppose that $f(x, y)$ is a function with continuous second-order partial derivatives. Consider the curve obtained by intersecting the surface $z = f(x, y)$ with the plane $y = y_0$. Explain how the slope of this curve at the point $x = x_0$ relates to $\frac{\partial f}{\partial x}(x_0, y_0)$. Relate the concavity of this curve at the point $x = x_0$ to $\frac{\partial^2 f}{\partial x^2}(x_0, y_0)$.

66. As in exercise 65, develop a graphical interpretation of $\frac{\partial^2 f}{\partial y^2}(x_0, y_0)$.

EXPLORATORY EXERCISES

1. In exercises 65 and 66, you interpreted the second-order partial derivatives f_{xx} and f_{yy} in terms of concavity. In this exercise, you will develop a geometric interpretation of the mixed partial derivative f_{xy}. (More information can be found in the article "What is f_{xy}?" by Brian McCartin in the March 1998 issue of the journal *PRIMUS*.) Start by using Taylor's Theorem (see section 8.7) to show that

$$\lim_{k\to 0}\lim_{h\to 0}\frac{f(x,y)-f(x+h,y)-f(x,y+k)+f(x+h,y+k)}{hk}=f_{xy}(x,y).$$

[Hint: Treating y as a constant, you have $f(x+h,y)=f(x,y)+hf_x(x,y)+h^2g(x,y)$, for some function $g(x,y)$. Similarly, expand the other terms in the numerator.] Therefore, for small h and k, $f_{xy}(x,y)\approx\dfrac{f_0-f_1-f_2+f_3}{hk}$, where $f_0=f(x,y)$, $f_1=f(x+h,y)$, $f_2=f(x,y+k)$ and $f_3=f(x+h,y+k)$. The four points $P_0=(x,y,f_0)$, $P_1=(x+h,y,f_1)$, $P_2=(x,y+k,f_2)$ and $P_3=(x+h,y+k,f_3)$ determine a parallelepiped, as shown in the figure below.

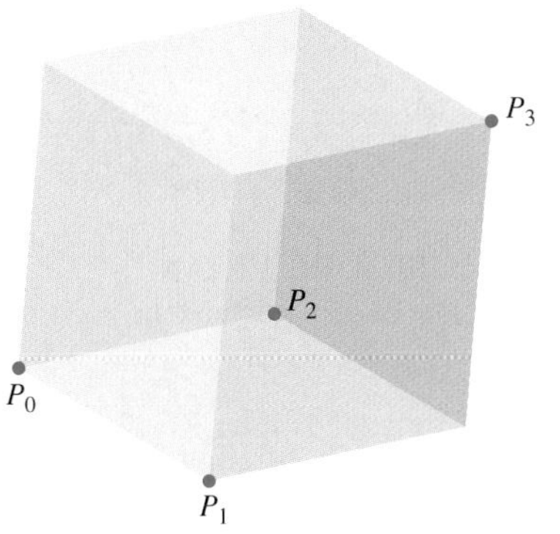

Recalling that the volume of a parallelepiped formed by vectors **a**, **b** and **c** is given by $|\mathbf{a}\cdot(\mathbf{b}\times\mathbf{c})|$, show that the volume of this box equals $|(f_0-f_1-f_2+f_3)hk|$. That is, the volume is approximately equal to $|f_{xy}(x,y)|(hk)^2$. Conclude that the larger $|f_{xy}(x,y)|$ is, the greater the volume of the box and hence, the farther the point P_3 is from the plane determined by the points P_0, P_1 and P_2. To see what this means graphically, start with the function $f(x,y)=x^2+y^2$ at the point $(1, 1, 2)$. With $h=k=0.1$, show that the points $(1, 1, 2)$, $(1.1, 1, 2.21)$, $(1, 1.1, 2.21)$ and $(1.1, 1.1, 2.42)$ all lie in the same plane. The derivative $f_{xy}(1,1)=0$ indicates that at the point $(1.1, 1.1, 2.42)$, the graph does not curve away from the plane of the points $(1, 1, 2)$, $(1.1, 1, 2.21)$ and $(1, 1.1, 2.21)$. Contrast this to the behavior of the function $f(x,y)=x^2+xy$ at the point $(1, 1, 2)$. This says that f_{xy} measures the amount of curving of the surface as you sequentially change x and y by small amounts.

2. A ball, such as a baseball, flying through the air encounters air resistance in the form of **air drag.** The magnitude of the drag force is typically the product of a number (called the drag coefficient) and the square of the velocity. The drag coefficient is not actually a constant. The figure (reprinted from *Keep Your Eye on the Ball* by Watts and Bahill) shows experimental data for the drag coefficient as a function of the roughness of the ball (measured by ε/D, where ε is the size of the bumps on the ball and D is the diameter of the ball) and the Reynolds number (Re, which is proportional to velocity). We'll call the drag coefficient f, rename $u=\varepsilon/D$ and $v=\text{Re}$ and consider $f(u,v)$. Use the graph to estimate $\dfrac{\partial f}{\partial u}(0.005, 1.5\times 10^5)$ and $\dfrac{\partial f}{\partial v}(0.005, 1.5\times 10^5)$ and interpret each partial derivative. All golf balls have "dimples" that make the surface of the golf ball rougher. Explain why a golf ball with dimples, traveling at a velocity corresponding to a Reynolds number of about 0.9×10^5, will fly much farther than a ball with no dimples.

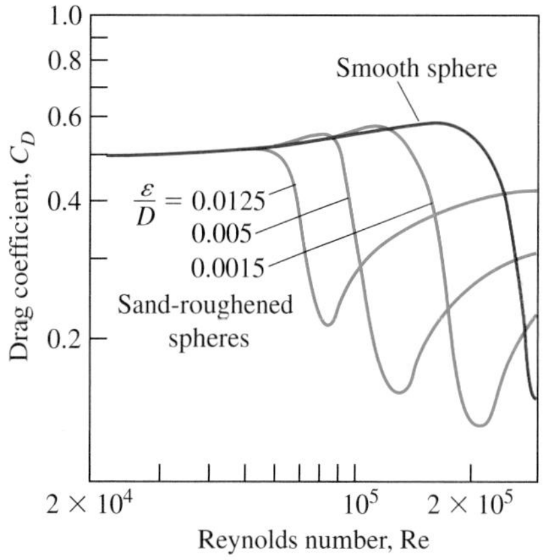

3. For a function $g(x,y)$, define $F(x)=\int_a^b g(x,y)\,dy$. In this exercise, you will explore the question of whether or not $F'(x)=\int_a^b \frac{\partial g}{\partial x}(x,y)\,dy$. (a) Show that this is true for $g(x,y)=e^{xy}$. (b) Show that it is true for $g(x,y)=h(x)k(y)$ if k is continuous and h is differentiable. (c) Show that it is true for $g(x,y)=\frac{1}{x}e^{xy}$ on the interval $[0, 2]$. (d) Find numerically that it is not true for $g(x,y)=\frac{1}{y}e^{xy}$. (e) Conjecture conditions on the function $g(x,y)$ for which the statement is true. (f) A mathematician would say that the underlying issue in this problem is the interchangeability of limits and integrals. Explain how limits are involved.

12.4 TANGENT PLANES AND LINEAR APPROXIMATIONS

Recall that the tangent line to the curve $y = f(x)$ at $x = a$ stays close to the curve near the point of tangency. This enables us to use the tangent line to approximate values of the function close to the point of tangency (see Figure 12.24a). The equation of the tangent line is given by

$$y = f(a) + f'(a)(x - a)\,. \tag{4.1}$$

In section 3.1, we called this the *linear approximation* to $f(x)$ at $x = a$.

In much the same way, we can approximate the value of a function of two variables near a given point using the tangent *plane* to the surface at that point. For instance, the graph of $z = 6 - x^2 - y^2$ and its tangent plane at the point $(1, 2, 1)$ are shown in Figure 12.24b. Notice that near the point $(1, 2, 1)$, the surface and the tangent plane are very close together.

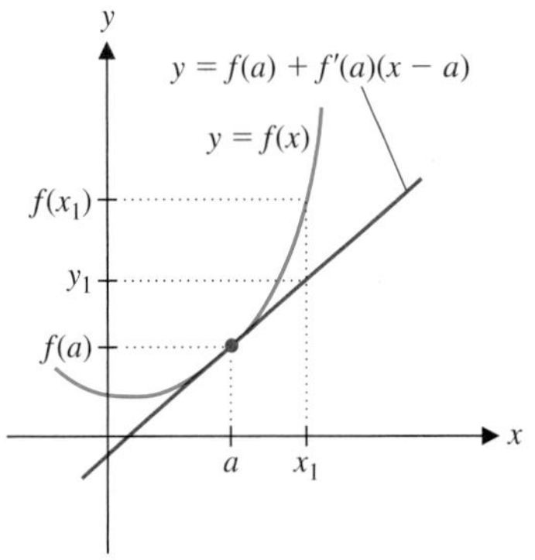

FIGURE 12.24a
Linear approximation

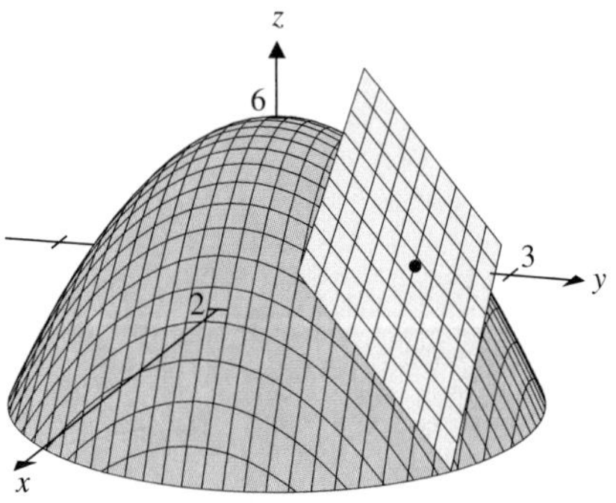

FIGURE 12.24b
$z = 6 - x^2 - y^2$ and the tangent plane at $(1, 2, 1)$

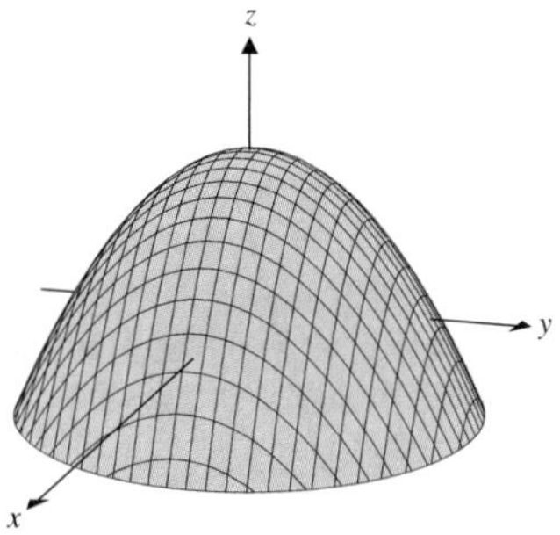

FIGURE 12.25a
$z = 6 - x^2 - y^2$, with $-3 \le x \le 3$ and $-3 \le y \le 3$

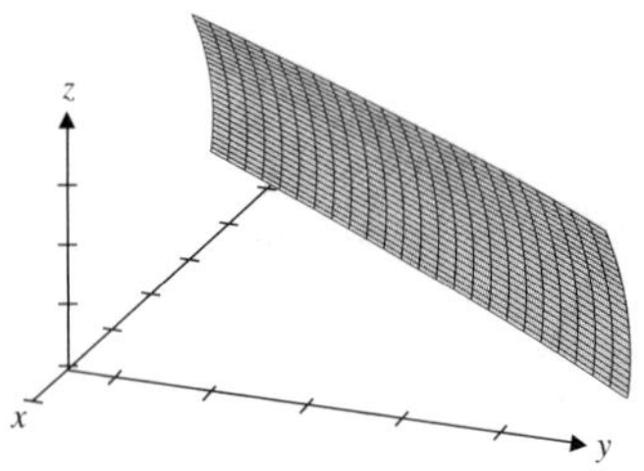

FIGURE 12.25b
$z = 6 - x^2 - y^2$, with $0.9 \le x \le 1.1$ and $1.9 \le y \le 2.1$

Refer to Figures 12.25a and 12.25b to visualize the process. Starting from a standard graphing window (Figure 12.25a shows $z = 6 - x^2 - y^2$ with $-3 \le x \le 3$ and $-3 \le y \le 3$), zoom in on the point $(1, 2, 1)$, as in Figure 12.25b (showing $z = 6 - x^2 - y^2$ with $0.9 \le x \le 1.1$ and $1.9 \le y \le 2.1$). The surface in Figure 12.25b looks like a plane, since we have zoomed in sufficiently far that the surface and its tangent plane are difficult to distinguish visually. This suggests that for points (x, y) close to the point of tangency, we can use the corresponding z-value on the tangent plane as an approximation to the value of the function at that point. More generally, we begin by looking for an equation of the tangent plane to $z = f(x, y)$ at the point $(a, b, f(a, b))$, where f_x and f_y are continuous at (a, b). For this, we'll need a point in the plane and a vector normal to the plane. One point lying in the tangent plane is, of course, the point of tangency $(a, b, f(a, b))$. To find a normal vector, we will find two vectors lying in the plane and then take their cross product to find a vector orthogonal to both (and thus, orthogonal to the plane).

Imagine intersecting the surface $z = f(x, y)$ with the plane $y = b$, as shown in Figure 12.26a. As we observed in section 12.3, the result is a curve in the plane $y = b$ whose slope at $x = a$ is given by $f_x(a, b)$. Along the tangent line at $x = a$, a change of 1 unit in x corresponds to a change of $f_x(a, b)$ in z. Since we're looking at a curve that lies in the plane $y = b$, the value of y doesn't change at all along the curve. A vector with the same direction as the tangent line is then $\langle 1, 0, f_x(a, b)\rangle$. This vector must then be parallel to

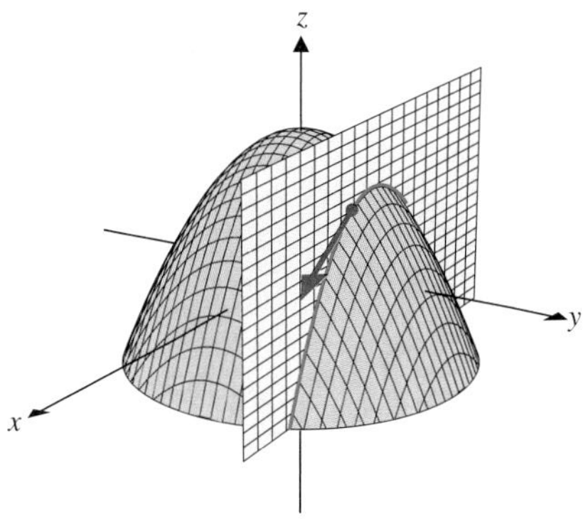

FIGURE 12.26a
The intersection of the surface $z = f(x, y)$ with the plane $y = b$

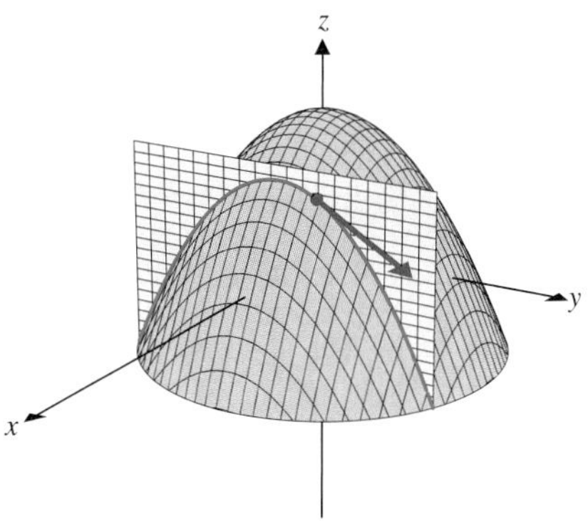

FIGURE 12.26b
The intersection of the surface $z = f(x, y)$ with the plane $x = a$

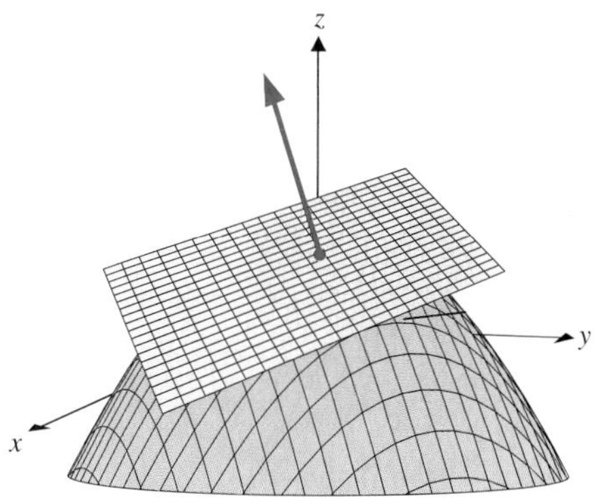

FIGURE 12.26c
Tangent plane and normal vector

the tangent plane. (Think about this some.) Similarly, intersecting the surface $z = f(x, y)$ with the plane $x = a$, as shown in Figure 12.26b, we get a curve lying in the plane $x = a$, whose slope at $y = b$ is given by $f_y(a, b)$. A vector with the same direction as the tangent line at $y = b$ is then $\langle 0, 1, f_y(a, b)\rangle$.

We have now found two vectors that are parallel to the tangent plane: $\langle 1, 0, f_x(a, b)\rangle$ and $\langle 0, 1, f_y(a, b)\rangle$. A vector normal to the plane is then given by the cross product:

$$\langle 0, 1, f_y(a, b)\rangle \times \langle 1, 0, f_x(a, b)\rangle = \langle f_x(a, b), f_y(a, b), -1\rangle.$$

We indicate the tangent plane and normal vector at a point in Figure 12.26c. We have the following result.

REMARK 4.1

Notice the similarity between the equation of the tangent plane given in (4.2) and the equation of the tangent line to $y = f(x)$ given in (4.1).

THEOREM 4.1

Suppose that $f(x, y)$ has continuous first partial derivatives at (a, b). A normal vector to the tangent plane to $z = f(x, y)$ at (a, b) is then $\langle f_x(a, b), f_y(a, b), -1\rangle$. Further, an equation of the tangent plane is given by

$$z - f(a, b) = f_x(a, b)(x - a) + f_y(a, b)(y - b)$$

or

$$z = f(a, b) + f_x(a, b)(x - a) + f_y(a, b)(y - b). \tag{4.2}$$

Observe that since we now know a normal vector to the tangent plane, the line orthogonal to the tangent plane and passing through the point $(a, b, f(a, b))$ is given by

$$x = a + f_x(a, b)t, \quad y = b + f_y(a, b)t, \quad z = f(a, b) - t. \tag{4.3}$$

This line is called the **normal line** to the surface at the point $(a, b, f(a, b))$.

It's now a simple matter to use Theorem 4.1 to construct the equations of a tangent plane and normal line to nearly any surface, as we illustrate in examples 4.1 and 4.2.

EXAMPLE 4.1 Finding Equations of the Tangent Plane and the Normal Line

Find equations of the tangent plane and the normal line to $z = 6 - x^2 - y^2$ at the point $(1, 2, 1)$.

a b f(a,b)

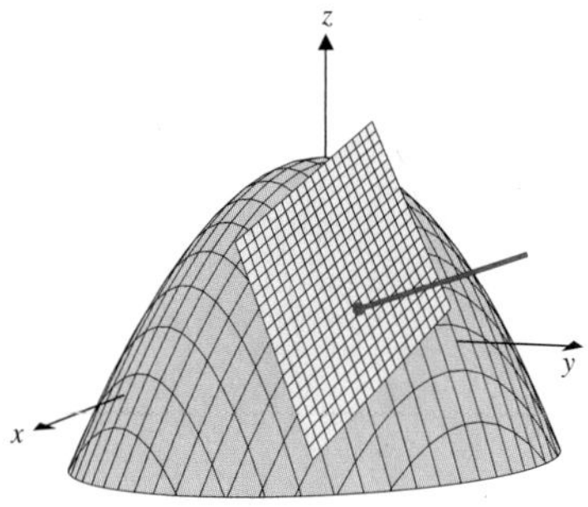

FIGURE 12.27
Surface, tangent plane and normal line at the point (1, 2, 1)

Solution For $f(x, y) = 6 - x^2 - y^2$, we have $f_x = -2x$ and $f_y = -2y$. This gives us $f_x(1, 2) = -2$ and $f_y(1, 2) = -4$. A normal vector is then $\langle -2, -4, -1 \rangle$ and from (4.2), an equation of the tangent plane is

$$z = 1 - 2(x - 1) - 4(y - 2).$$

From (4.3), equations of the normal line are

$$x = 1 - 2t, \quad y = 2 - 4t, \quad z = 1 - t.$$

A sketch of the surface, the tangent plane and the normal line is shown in Figure 12.27. ■

EXAMPLE 4.2 Finding Equations of the Tangent Plane and the Normal Line

Find equations of the tangent plane and the normal line to $z = x^3 + y^3 + \frac{x^2}{y}$ at (2, 1, 13).

Solution Here, $f_x = 3x^2 + \frac{2x}{y}$ and $f_y = 3y^2 - \frac{x^2}{y^2}$, so that $f_x(2, 1) = 12 + 4 = 16$ and $f_y(2, 1) = 3 - 4 = -1$. A normal vector is then $\langle 16, -1, -1 \rangle$ and from (4.2), an equation of the tangent plane is

$$z = 13 + 16(x - 2) - (y - 1).$$

From (4.3), equations of the normal line are

$$x = 2 + 16t, \quad y = 1 - t, \quad z = 13 - t.$$

A sketch of the surface, the tangent plane and the normal line is shown in Figure 12.28.

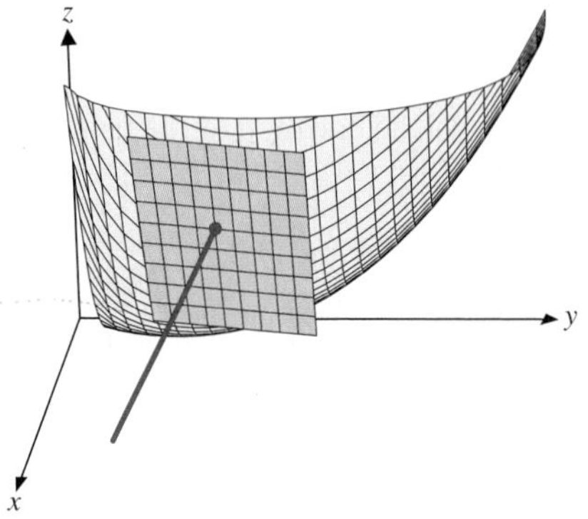

FIGURE 12.28
Surface, tangent plane and normal line at the point (2, 1, 13)

■

In Figures 12.27 and 12.28, the tangent plane appears to stay close to the surface near the point of tangency. This says that the z-values on the tangent plane should be close to the corresponding z-values on the surface, which are given by the function values $f(x, y)$, at least for (x, y) close to the point of tangency. Further, the simple form of the equation for the tangent plane makes it ideal for approximating the value of complicated functions.

We define the **linear approximation** $L(x, y)$ of $f(x, y)$ at the point (a, b) to be the function defining the z-values on the tangent plane, namely,

$$L(x, y) = f(a, b) + f_x(a, b)(x - a) + f_y(a, b)(y - b), \tag{4.4}$$

from (4.2). We illustrate this with example 4.3.

EXAMPLE 4.3 Finding a Linear Approximation

Compute the linear approximation of $f(x, y) = 2x + e^{x^2-y}$ at (0, 0). Compare the linear approximation to the actual function values for (a) $x = 0$ and y near 0; (b) $y = 0$ and x near 0; (c) $y = x$, with both x and y near 0 and (d) $y = 2x$, with both x and y near 0.

Solution Here, $f_x = 2 + 2xe^{x^2-y}$ and $f_y = -e^{x^2-y}$, so that $f_x(0, 0) = 2$ and $f_y(0, 0) = -1$. Also, $f(0, 0) = 1$. From (4.4), the linear approximation is then given by

$$L(x, y) = 1 + 2(x - 0) - (y - 0) = 1 + 2x - y.$$

The following table compares values of $L(x, y)$ and $f(x, y)$ for a number of points of the form $(0, y)$, $(x, 0)$, (x, x) and $(x, 2x)$.

(x, y)	$f(x, y)$	$L(x, y)$
(0, 0.1)	0.905	0.9
(0, 0.01)	0.99005	0.99
(0, −0.1)	1.105	1.1
(0, −0.01)	1.01005	1.01
(0.1, 0)	1.21005	1.2
(0.01, 0)	1.02010	1.02
(−0.1, 0)	0.81005	0.8
(−0.01, 0)	0.98010	0.98

(x, y)	$f(x, y)$	$L(x, y)$
(0.1, 0.1)	1.11393	1.1
(0.01, 0.01)	1.01015	1.01
(−0.1, −0.1)	0.91628	0.9
(−0.01, −0.01)	0.99015	0.99
(0.1, 0.2)	1.02696	1.0
(0.01, 0.02)	1.00030	1.0
(−0.1, −0.2)	1.03368	1.0
(−0.01, −0.02)	1.00030	1.0

Notice that the closer a given point is to the point of tangency, the more accurate the linear approximation tends to be at that point. This is typical of this type of approximation. We will explore this further in the exercises. ■

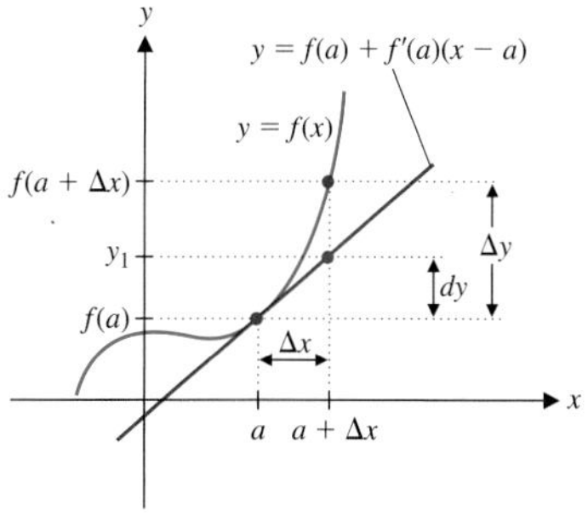

FIGURE 12.29
Increments and differentials for a function of one variable

Increments and Differentials

Now that we have examined linear approximations from a graphical perspective, we will examine them in a symbolic fashion. First, we remind you of the notation and some alternative language that we used in section 3.1 for functions of a single variable. We defined the *increment* Δy of the function $f(x)$ at $x = a$ to be

$$\Delta y = f(a + \Delta x) - f(a).$$

Referring to Figure 12.29, notice that for Δx small,

$$\Delta y \approx dy = f'(a)\,\Delta x,$$

where we referred to dy as the *differential* of y. Further, observe that if f is differentiable at $x = a$ and $\varepsilon = \dfrac{\Delta y - dy}{\Delta x}$, then we have

$$\begin{aligned}\varepsilon &= \frac{\Delta y - dy}{\Delta x} = \frac{f(a+\Delta x) - f(a) - f'(a)\,\Delta x}{\Delta x} \\ &= \frac{f(a+\Delta x) - f(a)}{\Delta x} - f'(a) \to 0,\end{aligned}$$

as $\Delta x \to 0$. (You'll need to recognize the definition of derivative here!) Finally, solving for Δy in terms of ε, we have

$$\Delta y = dy + \varepsilon\,\Delta x,$$

where $\varepsilon \to 0$, as $\Delta x \to 0$. We can make a similar observation for functions of several variables, as follows.

For $z = f(x, y)$, we define the **increment** of f at (a, b) to be

$$\Delta z = f(a + \Delta x, b + \Delta y) - f(a, b).$$

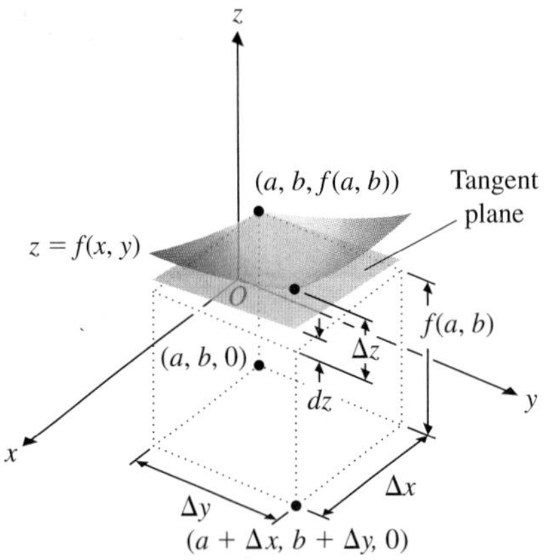

FIGURE 12.30
Linear approximation

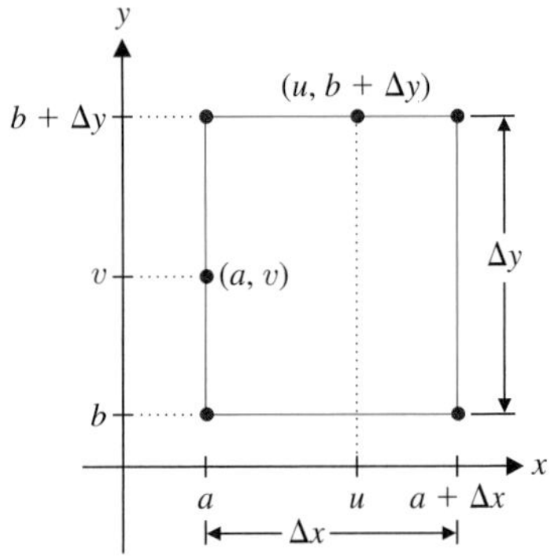

FIGURE 12.31
Intermediate points from the Mean Value Theorem

That is, Δz is the change in z that occurs when a is incremented by Δx and b is incremented by Δy, as illustrated in Figure 12.30. Notice that as long as f is continuous in some open region containing (a, b) and f has first partial derivatives on that region, we can write

$$\begin{aligned}\Delta z &= f(a + \Delta x, b + \Delta y) - f(a, b) \\ &= [f(a + \Delta x, b + \Delta y) - f(a, b + \Delta y)] + [f(a, b + \Delta y) - f(a, b)]\end{aligned}$$

Adding and subtracting $f(a, b + \Delta y)$.

$$= f_x(u, b + \Delta y)[(a + \Delta x) - a] + f_y(a, v)[(b + \Delta y) - b]$$

Applying the Mean Value Theorem to both terms.

$$= f_x(u, b + \Delta y)\,\Delta x + f_y(a, v)\,\Delta y,$$

by the Mean Value Theorem. Here, u is some value between a and $a + \Delta x$, and v is some value between b and $b + \Delta y$ (see Figure 12.31). This gives us

$$\begin{aligned}\Delta z &= f_x(u, b + \Delta y)\,\Delta x + f_y(a, v)\,\Delta y \\ &= \{f_x(a, b) + [f_x(u, b + \Delta y) - f_x(a, b)]\}\,\Delta x \\ &\quad + \{f_y(a, b) + [f_y(a, v) - f_y(a, b)]\}\,\Delta y,\end{aligned}$$

which we rewrite as

$$\Delta z = f_x(a,b)\,\Delta x + f_y(a,b)\,\Delta y + \varepsilon_1 \Delta x + \varepsilon_2\,\Delta y,$$

where $\quad \varepsilon_1 = f_x(u, b+\Delta y) - f_x(a,b) \quad$ and $\quad \varepsilon_2 = f_y(a,v) - f_y(a,b).$

Finally, observe that if f_x and f_y are both continuous in some open region containing (a,b), then ε_1 and ε_2 will both tend to 0, as $(\Delta x, \Delta y) \to (0,0)$. In fact, you should recognize that since $\varepsilon_1, \varepsilon_2 \to 0$, as $(\Delta x, \Delta y) \to (0,0)$, the products $\varepsilon_1\,\Delta x$ and $\varepsilon_2\,\Delta y$ both tend to 0 even faster than do $\varepsilon_1, \varepsilon_2, \Delta x$ or Δy individually. (Think about this!)

We have now established the following result.

THEOREM 4.2

Suppose that $z = f(x,y)$ is defined on the rectangular region $R = \{(x,y) | x_0 < x < x_1, y_0 < y < y_1\}$ and f_x and f_y are defined on R and are continuous at $(a,b) \in R$. Then for $(a+\Delta x, b+\Delta y) \in R$,

$$\Delta z = f_x(a,b)\,\Delta x + f_y(a,b)\,\Delta y + \varepsilon_1\,\Delta x + \varepsilon_2\,\Delta y, \tag{4.5}$$

where ε_1 and ε_2 are functions of Δx and Δy that both tend to zero, as $(\Delta x, \Delta y) \to (0,0)$.

For some very simple functions, we can compute Δz by hand, as illustrated in example 4.4.

EXAMPLE 4.4 Computing the Increment Δz

For $z = f(x,y) = x^2 - 5xy$, find Δz and write it in the form indicated in Theorem 4.2.

Solution We have

$$\begin{aligned}
\Delta z &= f(x+\Delta x, y+\Delta y) - f(x,y) \\
&= [(x+\Delta x)^2 - 5(x+\Delta x)(y+\Delta y)] - (x^2 - 5xy) \\
&= x^2 + 2x\,\Delta x + (\Delta x)^2 - 5(xy + x\,\Delta y + y\,\Delta x + \Delta x\,\Delta y) - x^2 + 5xy \\
&= \underbrace{(2x - 5y)}_{f_x}\,\Delta x + \underbrace{(-5x)}_{f_y}\,\Delta y + \underbrace{(\Delta x)}_{\varepsilon_1}\,\Delta x + \underbrace{(-5\Delta x)}_{\varepsilon_2}\,\Delta y \\
&= f_x(x,y)\,\Delta x + f_y(x,y)\,\Delta y + \varepsilon_1\,\Delta x + \varepsilon_2\,\Delta y,
\end{aligned}$$

where $\varepsilon_1 = \Delta x$ and $\varepsilon_2 = -5\Delta x$ both tend to zero, as $(\Delta x, \Delta y) \to (0,0)$, as indicated in Theorem 4.2. You should observe here that by grouping the terms differently, we would get different choices for ε_1 and ε_2. ■

Look closely at the first two terms in the expansion of the increment Δz given in (4.5). If we take $\Delta x = x - a$ and $\Delta y = y - b$, then they correspond to the linear approximation of $f(x,y)$. In this context, we give this a special name. If we increment x by the amount $dx = \Delta x$ and increment y by $dy = \Delta y$, then we define the **differential** of z to be

$$\boxed{dz = f_x(x,y)\,dx + f_y(x,y)\,dy.}$$

This is sometimes referred to as a **total differential.** Notice that for dx and dy small, we have from (4.5) that

$$\Delta z \approx dz.$$

You should recognize that this is the same approximation as the linear approximation developed in the beginning of this section. In this case, though, we have developed this from an analytical perspective, rather than the geometrical one used in the beginning of the section.

In Definition 4.1, we give a special name to functions that can be approximated linearly in the above fashion.

DEFINITION 4.1

Let $z = f(x, y)$. We say that f is **differentiable** at (a, b) if we can write

$$\Delta z = f_x(a, b)\,\Delta x + f_y(a, b)\,\Delta y + \varepsilon_1\,\Delta x + \varepsilon_2\,\Delta y,$$

where ε_1 and ε_2 are both functions of Δx and Δy and $\varepsilon_1, \varepsilon_2 \to 0$, as $(\Delta x, \Delta y) \to (0, 0)$. We say that f is differentiable on a region $R \subset \mathbb{R}^2$ whenever f is differentiable at every point in R.

Although this definition of a differentiable function may not appear to be an obvious generalization of the corresponding definition for a function of a single variable, in fact, it is. We explore this in the exercises.

Note that from Theorem 4.2, if f_x and f_y are defined on some open rectangle R containing the point (a, b) and if f_x and f_y are continuous at (a, b), then f will be differentiable at (a, b). Just as with functions of a single variable, it can be shown that if f is differentiable at a point (a, b), then it is also continuous at (a, b). Further, owing to Theorem 4.2, if a function is differentiable at a point, then the linear approximation (differential) at that point provides a good approximation to the function near that point. Be very careful of what this does *not* say, however. If a function has partial derivatives at a point, it need *not* be differentiable or even continuous at that point. (In exercises 31 and 32, you will see examples of a function with partial derivatives defined everywhere, but that is not differentiable at a point.)

The idea of a linear approximation extends easily to three or more dimensions. We lose the graphical interpretation of a tangent plane approximating a surface, but the definition should make sense.

DEFINITION 4.2

The **linear approximation** to $f(x, y, z)$ at the point (a, b, c) is given by

$$L(x, y, z) = f(a, b, c) + f_x(a, b, c)(x - a) + f_y(a, b, c)(y - b) + f_z(a, b, c)(z - c).$$

We can write the linear approximation in the context of increments and differentials, as follows. If we increment x by Δx, y by Δy and z by Δz, then the increment of $w = f(x, y, z)$

is given by

$$\Delta w = f(x + \Delta x, y + \Delta y, z + \Delta z) - f(x, y, z)$$
$$\approx dw = f_x(x, y, z)\,\Delta x + f_y(x, y, z)\,\Delta y + f_z(x, y, z)\,\Delta z.$$

A good way to interpret (and remember!) the linear approximation is that each partial derivative represents the change in the function relative to the change in that variable. The linear approximation starts with the function value at the known point and adds in the approximate changes corresponding to each of the independent variables.

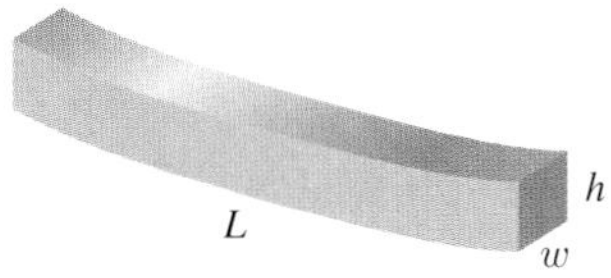

FIGURE 12.32
A typical beam

EXAMPLE 4.5 Approximating the Sag in a Beam

Suppose that the sag in a beam of length L, width w and height h is given by $S(L, w, h) = 0.0004\dfrac{L^4}{wh^3}$, with all lengths measured in inches. We illustrate the beam in Figure 12.32. A beam is supposed to measure $L = 36$, $w = 2$ and $h = 6$ with a corresponding sag of 1.5552 inches. Due to weathering and other factors, the manufacturer only guarantees measurements with error tolerances $L = 36 \pm 1$, $w = 2 \pm 0.4$ and $h = 6 \pm 0.8$. Use a linear approximation to estimate the possible range of sags in the beam.

Solution We first compute $\dfrac{\partial S}{\partial L} = 0.0016\dfrac{L^3}{wh^3}$, $\dfrac{\partial S}{\partial w} = -0.0004\dfrac{L^4}{w^2h^3}$ and $\dfrac{\partial S}{\partial h} = -0.0012\dfrac{L^4}{wh^4}$. At the point (36, 2, 6), we then have $\dfrac{\partial S}{\partial L}(36, 2, 6) = 0.1728$, $\dfrac{\partial S}{\partial w}(36, 2, 6) = -0.7776$ and $\dfrac{\partial S}{\partial h}(36, 2, 6) = -0.7776$. From Definition 4.2, the linear approximation of the sag is then given by

$$S \approx 1.5552 + 0.1728(L - 36) - 0.7776(w - 2) - 0.7776(h - 6).$$

Notice that we could have written this in differential form using Definition 4.1.

From the stated tolerances, $L - 36$ must be between -1 and 1, $w - 2$ must be between -0.4 and 0.4 and $h - 6$ must be between -0.8 and 0.8. Notice that the maximum sag then occurs with $L - 36 = 1$, $w - 2 = -0.4$ and $h - 6 = -0.8$. The linear approximation predicts that

$$S - 1.5552 \approx 0.1728 + 0.31104 + 0.62208 = 1.10592.$$

Similarly, the minimum sag occurs with $L - 36 = -1$, $w - 2 = 0.4$ and $h - 6 = 0.8$. The linear approximation predicts that

$$S - 1.5552 \approx -0.1728 - 0.31104 - 0.62208 = -1.10592.$$

Based on the linear approximation, the sag is 1.5552 ± 1.10592, or between 0.44928 and 2.66112. As you can see, in this case, the uncertainty in the sag is substantial. ■

In many real-world situations, we do not have a formula for the quantity we are interested in computing. Even so, given sufficient information, we can still use linear approximations to estimate the desired quantity.

EXAMPLE 4.6 Estimating the Gauge of a Sheet of Metal

Manufacturing plants create rolls of metal of a desired gauge (thickness) by feeding the metal through very large rollers. The thickness of the resulting metal depends on the gap between the working rollers, the speed at which the rollers turn and the temperature of the metal. Suppose that for a certain metal, a gauge of 4 mm is produced by a gap of 4 mm, a speed of 10 m/s and a temperature of 900°. Experiments show that an increase in speed of 0.2 m/s increases the gauge by 0.06 mm and an increase in temperature of 10° decreases the gauge by 0.04 mm. Use a linear approximation to estimate the gauge at 10.1 m/s and 880°.

Solution With no change in gap, we assume that the gauge is a function $g(s, t)$ of the speed s and the temperature t. Based on our data, $\frac{\partial g}{\partial s} \approx \frac{0.06}{0.2} = 0.3$ and $\frac{\partial g}{\partial t} \approx \frac{-0.04}{10} = -0.004$. From Definition 4.2, the linear approximation of $g(s, t)$ is given by

$$g(s, t) \approx 4 + 0.3(s - 10) - 0.004(t - 900).$$

With $s = 10.1$ and $t = 880$, we get the estimate

$$g(10.1, 880) \approx 4 + 0.3(0.1) - 0.004(-20) = 4.11. \blacksquare$$

BEYOND FORMULAS

You should think of linear approximations more in terms of example 4.6 than example 4.5. That is, linear approximations are most commonly used when there is no known formula for the function *f*. You can then read equation (4.4) or Definition 4.2 as a recipe that tells you which ingredients (i.e., function values and derivatives) you need to approximate a function value. The visual image behind this formula, shown in Figure 12.30, gives you information about how good your approximation is.

EXERCISES 12.4

HW pg 970-971
#1, 3, 5, 9, 27, 29, 31

WRITING EXERCISES

1. Describe which graphical properties of the surface $z = f(x, y)$ would cause the linear approximation of $f(x, y)$ at (a, b) to be particularly accurate or inaccurate.
2. Temperature varies with longitude (x), latitude (y) and altitude (z). Speculate whether or not the temperature function would be differentiable and what significance the answer would have for weather prediction.
3. Imagine a surface $z = f(x, y)$ with a ridge of discontinuities along the line $y = x$. Explain in graphical terms why $f(x, y)$ would not be differentiable at (0, 0) or any other point on the line $y = x$.
4. The function in exercise 3 might have first partial derivatives $f_x(0, 0)$ and $f_y(0, 0)$. Explain why the slopes along $x = 0$ and $y = 0$ could have limits as x and y approach 0. If *differentiable* is intended to describe functions with smooth graphs, explain why differentiability is not defined in terms of the existence of partial derivatives.

In exercises 1–6, find equations of the tangent plane and normal line to the surface at the given point.

1. $z = x^2 + y^2 - 1$ at (a) (2, 1, 4) and (b) (0, 2, 3)
2. $z = e^{-x^2-y^2}$ at (a) (0, 0, 1) and (b) $(1, 1, e^{-2})$
3. $z = \sin x \cos y$ at (a) $(0, \pi, 0)$ and (b) $(\frac{\pi}{2}, \pi, -1)$
4. $z = x^3 - 2xy$ at (a) (−2, 3, 4) and (b) (1, −1, 3)
5. $z = \sqrt{x^2 + y^2}$ at (a) (−3, 4, 5) and (b) (8, −6, 10)
6. $z = \frac{4x}{y}$ at (a) (1, 2, 2) and (b) (−1, 4, −1)

In exercises 7–12, compute the linear approximation of the function at the given point.

7. $f(x, y) = \sqrt{x^2 + y^2}$ at (a) (3, 0) and (b) (0, −3)
8. $f(x, y) = \sin x \cos y$ at (a) $(0, \pi)$ and (b) $(\frac{\pi}{2}, \pi)$
9. $f(x, y) = xe^{xy^2} + 3y^2$ at (a) (0, 1) and (b) (2, 0)

10. $f(x, y, z) = xe^{yz} - \sqrt{x - y^2}$ at (a) $(4, 1, 0)$ and (b) $(1, 0, 2)$

11. $f(w, x, y, z) = w^2xy - e^{wyz}$ at (a) $(-2, 3, 1, 0)$ and (b) $(0, 1, -1, 2)$

12. $f(w, x, y, z) = \cos xyz - w^3x^2$ at (a) $(2, -1, 4, 0)$ and (b) $(2, 1, 0, 1)$

In exercises 13–16, compare the linear approximation from the indicated exercise to the exact function value at the given points.

13. Exercise 7 part (a) at $(3, -0.1)$, $(3.1, 0)$, $(3.1, -0.1)$

14. Exercise 7 part (b) at $(0.1, -3)$, $(0, -3.1)$, $(0.1, -3.1)$

15. Exercise 8 part (a) at $(0, 3)$, $(0.1, \pi)$, $(0.1, 3)$

16. Exercise 9 part (b) at $(2.1, 0)$, $(2, 0.2)$, $(1, -1)$

17. Use a linear approximation to estimate the range of sags in the beam of example 4.5 if the error tolerances are $L = 36 \pm 0.5$, $w = 2 \pm 0.2$ and $h = 6 \pm 0.5$.

18. Use a linear approximation to estimate the range of sags in the beam of example 4.5 if the error tolerances are $L = 32 \pm 0.4$, $w = 2 \pm 0.3$ and $h = 8 \pm 0.4$.

19. Use a linear approximation to estimate the gauge of the metal in example 4.6 at 9.9 m/s and 930°.

20. Use a linear approximation to estimate the gauge of the metal in example 4.6 at 10.2 m/s and 910°.

21. Suppose that for a metal similar to that of example 4.6, an increase in speed of 0.3 m/s increases the gauge by 0.03 mm and an increase in temperature of 20° decreases the gauge by 0.02 mm. Use a linear approximation to estimate the gauge at 10.2 m/s and 890°.

22. Suppose that for the metal in example 4.6, a decrease of 0.05 mm in the gap between the working rolls decreases the gauge by 0.04 mm. Use a linear approximation in three variables to estimate the gauge at 10.15 m/s, 905° and a gap of 3.98 mm.

In exercises 23–26, find the increment Δz and write it in the form given in Theorem 4.2.

23. $f(x, y) = 2xy + y^2$

24. $f(x, y) = (x + y)^2$

25. $f(x, y) = x^2 + y^2$

26. $f(x, y) = x^3 - 3xy$

27. Determine whether or not $f(x, y) = x^2 + 3xy$ is differentiable.

28. Determine whether or not $f(x, y) = xy^2$ is differentiable.

In exercises 29 and 30, find the total differential of $f(x, y)$.

29. $f(x, y) = ye^x + \sin x$

30. $f(x, y) = \sqrt{x + y}$

In exercises 31 and 32, show that the partial derivatives $f_x(0, 0)$ and $f_y(0, 0)$ both exist, but the function $f(x, y)$ is not differentiable at $(0, 0)$.

31. $$f(x, y) = \begin{cases} \dfrac{2xy}{x^2 + y^2}, & \text{if } (x, y) \neq (0, 0) \\ 0, & \text{if } (x, y) = (0, 0) \end{cases}$$

32. $$f(x, y) = \begin{cases} \dfrac{xy^2}{x^2 + y^2}, & \text{if } (x, y) \neq (0, 0) \\ 0, & \text{if } (x, y) = (0, 0) \end{cases}$$

33. In this exercise, we visualize the linear approximation of example 4.3. Start with a contour plot of $f(x, y) = 2x + e^{x^2 - y}$ with $-1 \leq x \leq 1$ and $-1 \leq y \leq 1$. Then zoom in on the point $(0, 0)$ of the contour plot until the level curves appear straight and equally spaced. (Level curves for z-values between 0.9 and 1.1 with a graphing window of $-0.1 \leq x \leq 0.1$ and $-0.1 \leq y \leq 0.1$ should work.) You will need the z-values for the level curves. Notice that to move from the $z = 1$ level curve to the $z = 1.05$ level curve you move 0.025 unit to the right. Then $\dfrac{\partial f}{\partial x}(0, 0) \approx \dfrac{\Delta z}{\Delta x} = \dfrac{0.05}{0.025} = 2$. Verify graphically that $\dfrac{\partial f}{\partial y}(0, 0) \approx -1$. Explain how to use the contour plot to reproduce the linear approximation $1 + 2x - y$.

34. Use the graphical method of exercise 33 to find the linear approximation of $f(x, y) = \sin(x^2 + 2xy)$ at the point $(1, 3)$.

In exercises 35–38, use the given contour plot to estimate the linear approximation of $f(x, y)$ at $(0, 0)$.

35.

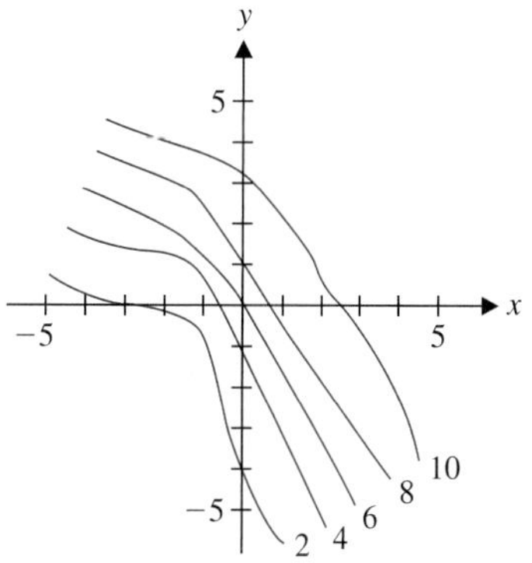

36.

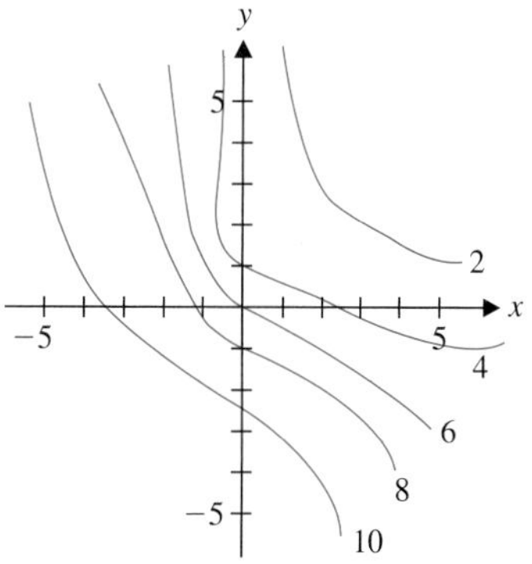

37.

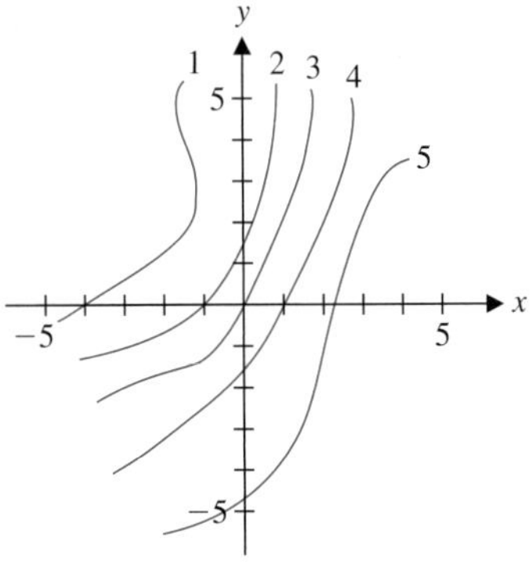

38.

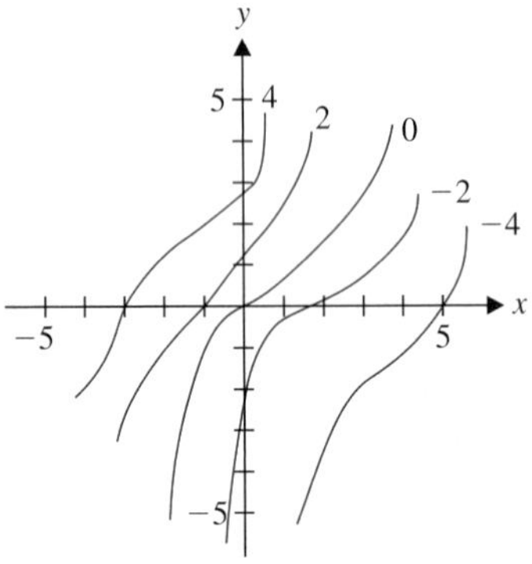

39. The table here gives wind chill (how cold it "feels" outside) as a function of temperature (degrees Fahrenheit) and wind speed (mph). We can think of this as a function $w(t, s)$. Estimate the partial derivatives $\frac{\partial w}{\partial t}(10, 10)$ and $\frac{\partial w}{\partial s}(10, 10)$ and the linear approximation of $w(t, s)$ at $(10, 10)$. Use the linear approximation to estimate the wind chill at $(12, 13)$.

Speed \ *Temp*	30	20	10	0	−10
0	30	20	10	0	−10
5	27	16	6	−5	−15
10	16	4	−9	−24	−33
15	9	−5	−18	−32	−45
20	4	−10	−25	−39	−53
25	0	−15	−29	−44	−59
30	−2	−18	−33	−48	−63

40. Estimate the linear approximation of wind chill at $(10, 15)$ and use it to estimate the wind chill at $(12, 13)$. Explain any differences between this answer and that of exercise 39.

41. In exercise 33, we specified that you zoom in on the contour plot until the level curves appear linear *and* equally spaced. To see why the second condition is necessary, sketch a contour plot of $f(x, y) = e^{x-y}$ with $-1 \le x \le 1$ and $-1 \le y \le 1$. Use this plot to estimate $\frac{\partial f}{\partial x}(0, 0)$ and $\frac{\partial f}{\partial y}(0, 0)$ and compare to the exact values. Zoom in until the level curves are equally spaced and estimate again. Explain why this estimate is much better.

42. Show that $\left\langle 0, 1, \frac{\partial f}{\partial y}(a, b)\right\rangle \times \left\langle 1, 0, \frac{\partial f}{\partial x}(a, b)\right\rangle = \left\langle \frac{\partial f}{\partial x}(a, b), \frac{\partial f}{\partial y}(a, b), -1\right\rangle$.

43. Show that $f(x, y) = \begin{cases} \dfrac{x^2 y}{x^2 + y^2}, & \text{if } (x, y) \ne (0, 0) \\ 0, & \text{if } (x, y) = (0, 0) \end{cases}$ is continuous but not differentiable at $(0, 0)$.

44. Let S be a surface defined parametrically by $\mathbf{r}(u, v) = \langle x(u, v), y(u, v), z(u, v)\rangle$. Define $\mathbf{r}_u(u, v) = \left\langle \frac{\partial x}{\partial u}(u, v), \frac{\partial y}{\partial u}(u, v), \frac{\partial z}{\partial u}(u, v)\right\rangle$ and $\mathbf{r}_v(u, v) = \left\langle \frac{\partial x}{\partial v}(u, v), \frac{\partial y}{\partial v}(u, v), \frac{\partial z}{\partial v}(u, v)\right\rangle$. Show that $\mathbf{r}_u \times \mathbf{r}_v$ is a normal vector to the tangent plane at the point $(x(u, v), y(u, v), z(u, v))$.

In exercises 45–48, use the result of exercise 44 to find an equation of the tangent plane to the parametric surface at the indicated point.

45. S is defined by $x = 2u$, $y = v$ and $z = 4uv$; at $u = 1$ and $v = 2$.

46. S is defined by $x = 2u^2$, $y = uv$ and $z = 4uv^2$; at $u = -1$ and $v = 1$.

47. S is the cylinder $x^2 + y^2 = 1$ with $0 \le z \le 2$; at $(1, 0, 1)$.

48. S is the cylinder $y^2 = 2x$ with $0 \le z \le 2$; at $(2, 2, 1)$.

EXPLORATORY EXERCISES

For exercises 1 and 2, we need to use the notation of matrix algebra. First, we define the 2 × 2 matrix A to be a two-dimensional array of real numbers, $A = \begin{bmatrix} a & b \\ c & d \end{bmatrix}$, we define a column vector x to be a one-dimensional array of real numbers, $\mathbf{x} = \begin{bmatrix} x_1 \\ x_2 \end{bmatrix}$ and define the product of a matrix and column vector to be the (column) vector $\begin{bmatrix} a & b \\ c & d \end{bmatrix}\begin{bmatrix} x_1 \\ x_2 \end{bmatrix} = \begin{bmatrix} ax_1 + bx_2 \\ cx_1 + dx_2 \end{bmatrix}$. You can learn much more about matrix algebra by looking at one of the many textbooks on the subject, including those bearing the title *Linear Algebra*.

1. We can extend the linear approximation of this section to quadratic approximations. Define the **Hessian matrix** $H = \begin{bmatrix} f_{xx} & f_{xy} \\ f_{yx} & f_{yy} \end{bmatrix}$, the **gradient vector** $\nabla f(x_0, y_0) = \langle f_x(x_0, y_0), f_y(x_0, y_0)\rangle$, the column vector $\mathbf{x} = \begin{bmatrix} x \\ y \end{bmatrix}$, the vector $\mathbf{x}_0 = \begin{bmatrix} x_0 \\ y_0 \end{bmatrix}$ and the transpose vector

$\mathbf{x}^T = [x\ y]$. The **quadratic approximation** of $f(x, y)$ at the point (x_0, y_0) is defined by

$$Q(x, y) = f(x_0, y_0) + \nabla f(x_0, y_0) \cdot (\mathbf{x} - \mathbf{x}_0) + \frac{1}{2}(\mathbf{x} - \mathbf{x}_0)^T H(x_0, y_0)(\mathbf{x} - \mathbf{x}_0).$$

Find the quadratic approximation of $f(x, y) = 2x + e^{x^2-y}$ and compute $Q(x, y)$ for the points in the table of example 4.3.

2. An important application of linear approximations of functions $f(x)$ is Newton's method for finding solutions of equations of the form $f(x) = 0$. In this exercise, we extend Newton's method to functions of several variables. To be specific, suppose that $f_1(x, y)$ and $f_2(x, y)$ are functions of two variables with continuous partial derivatives. To solve the equations $f_1(x, y) = 0$ and $f_2(x, y) = 0$ simultaneously, start with a guess $x = x_0$ and $y = y_0$. The idea is to replace $f_1(x, y)$ and $f_2(x, y)$ with their linear approximations $L_1(x, y)$ and $L_2(x, y)$ and solve the (simpler) equations $L_1(x, y) = 0$ and $L_2(x, y) = 0$ simultaneously. Write out the linear approximations and show that we want

$$\frac{\partial f_1}{\partial x}(x_0, y_0)(x - x_0) + \frac{\partial f_1}{\partial y}(x_0, y_0)(y - y_0) = -f_1(x_0, y_0)$$
$$\frac{\partial f_2}{\partial x}(x_0, y_0)(x - x_0) + \frac{\partial f_2}{\partial y}(x_0, y_0)(y - y_0) = -f_2(x_0, y_0).$$

Recall that there are several ways (substitution and elimination are popular) to solve two linear equations in two unknowns. The simplest way symbolically is to use matrices. If we define the **Jacobian matrix**

$$J(\mathbf{x}_0) = \begin{bmatrix} \frac{\partial f_1}{\partial x}(\mathbf{x}_0) & \frac{\partial f_1}{\partial y}(\mathbf{x}_0) \\ \frac{\partial f_2}{\partial x}(\mathbf{x}_0) & \frac{\partial f_2}{\partial y}(\mathbf{x}_0) \end{bmatrix}$$

where $\mathbf{x}_0$ represents the point (x_0, y_0), the preceding equations can be written as $J(\mathbf{x}_0)(\mathbf{x} - \mathbf{x}_0) = -\mathbf{f}(\mathbf{x}_0)$, which has solution $\mathbf{x} - \mathbf{x}_0 = -J^{-1}(\mathbf{x}_0)\mathbf{f}(\mathbf{x}_0)$ or $\mathbf{x} = \mathbf{x}_0 - J^{-1}(\mathbf{x}_0)\mathbf{f}(\mathbf{x}_0)$. Here, the matrix $J^{-1}(\mathbf{x}_0)$ is called the **inverse** of the matrix $J(\mathbf{x}_0)$ and $\mathbf{f}(\mathbf{x}_0) = \begin{bmatrix} f_1(\mathbf{x}_0) \\ f_2(\mathbf{x}_0) \end{bmatrix}$. The inverse A^{-1} of a matrix A (when it is defined) is the matrix for which $\mathbf{a} = A\mathbf{b}$ if and only if $\mathbf{b} = A^{-1}\mathbf{a}$, for all column vectors $\mathbf{a}$ and $\mathbf{b}$. In general, Newton's method is defined by the iteration

$$\mathbf{x}_{n+1} = \mathbf{x}_n - J^{-1}(\mathbf{x}_n)\mathbf{f}(\mathbf{x}_n).$$

Use Newton's method with an initial guess of $\mathbf{x}_0 = (-1, 0.5)$ to approximate a solution of the equations $x^2 - 2y = 0$ and $x^2 y - \sin y = 0$.

12.5 THE CHAIN RULE

You already are quite familiar with the chain rule for functions of a single variable. For instance, to differentiate the function $e^{\sin(x^2)}$, we have

$$\frac{d}{dx}[e^{\sin(x^2)}] = e^{\sin(x^2)} \underbrace{\frac{d}{dx}[\sin(x^2)]}_{\text{the derivative of the } inside}$$
$$= e^{\sin(x^2)} \cos(x^2) \underbrace{\frac{d}{dx}(x^2)}_{\text{the derivative of the } inside}$$
$$= e^{\sin(x^2)} \cos(x^2)(2x).$$

The general form of the chain rule says that for differentiable functions f and g,

$$\frac{d}{dx}[f(g(x))] = f'(g(x)) \underbrace{g'(x)}_{\text{the derivative of the } inside}.$$

We now extend the chain rule to functions of several variables. This takes several slightly different forms, depending on the number of independent variables, but each is a variation of the already familiar chain rule for functions of a single variable.

For a differentiable function $f(x, y)$, where x and y are both in turn, differentiable functions of a single variable t, to find the derivative of $f(x, y)$ with respect to t, we first write $g(t) = f(x(t), y(t))$. Then, from the definition of (an ordinary) derivative, we have

$$\frac{d}{dt}[f(x(t), y(t))] = g'(t) = \lim_{\Delta t \to 0} \frac{g(t + \Delta t) - g(t)}{\Delta t}$$
$$= \lim_{\Delta t \to 0} \frac{f(x(t + \Delta t), y(t + \Delta t)) - f(x(t), y(t))}{\Delta t}.$$

For simplicity, we write $\Delta x = x(t + \Delta t) - x(t)$, $\Delta y = y(t + \Delta t) - y(t)$ and $\Delta z = f(x(t + \Delta t), y(t + \Delta t)) - f(x(t), y(t))$. This gives us

$$\frac{d}{dt}[f(x(t), y(t))] = \lim_{\Delta t \to 0} \frac{\Delta z}{\Delta t}.$$

Since f is a differentiable function of x and y, we have (from the definition of differentiability) that

$$\Delta z = \frac{\partial f}{\partial x}\Delta x + \frac{\partial f}{\partial y}\Delta y + \varepsilon_1 \Delta x + \varepsilon_2 \Delta y,$$

where ε_1 and ε_2 both tend to 0, as $(\Delta x, \Delta y) \to (0, 0)$. Dividing through by Δt gives us

$$\frac{\Delta z}{\Delta t} = \frac{\partial f}{\partial x}\frac{\Delta x}{\Delta t} + \frac{\partial f}{\partial y}\frac{\Delta y}{\Delta t} + \varepsilon_1 \frac{\Delta x}{\Delta t} + \varepsilon_2 \frac{\Delta y}{\Delta t}.$$

Taking the limit as $\Delta t \to 0$ now gives us

$$\frac{d}{dt}[f(x(t), y(t))] = \lim_{\Delta t \to 0} \frac{\Delta z}{\Delta t}$$
$$= \frac{\partial f}{\partial x} \lim_{\Delta t \to 0} \frac{\Delta x}{\Delta t} + \frac{\partial f}{\partial y} \lim_{\Delta t \to 0} \frac{\Delta y}{\Delta t}$$
$$+ \lim_{\Delta t \to 0} \varepsilon_1 \lim_{\Delta t \to 0} \frac{\Delta x}{\Delta t} + \lim_{\Delta t \to 0} \varepsilon_2 \lim_{\Delta t \to 0} \frac{\Delta y}{\Delta t}. \tag{5.1}$$

Notice that $$\lim_{\Delta t \to 0} \frac{\Delta x}{\Delta t} = \lim_{\Delta t \to 0} \frac{x(t + \Delta t) - x(t)}{\Delta t} = \frac{dx}{dt}$$

and $$\lim_{\Delta t \to 0} \frac{\Delta y}{\Delta t} = \lim_{\Delta t \to 0} \frac{y(t + \Delta t) - y(t)}{\Delta t} = \frac{dy}{dt}.$$

Further, notice that since $x(t)$ and $y(t)$ are differentiable, they are also continuous and so,

$$\lim_{\Delta t \to 0} \Delta x = \lim_{\Delta t \to 0} [x(t + \Delta t) - x(t)] = 0.$$

Likewise, $\lim\limits_{\Delta t \to 0} \Delta y = 0$, also. Consequently, since $(\Delta x, \Delta y) \to (0, 0)$, as $\Delta t \to 0$, we have

$$\lim_{\Delta t \to 0} \varepsilon_1 = \lim_{\Delta t \to 0} \varepsilon_2 = 0.$$

From (5.1), we now have

$$\frac{d}{dt}[f(x(t), y(t))] = \frac{\partial f}{\partial x} \lim_{\Delta t \to 0} \frac{\Delta x}{\Delta t} + \frac{\partial f}{\partial y} \lim_{\Delta t \to 0} \frac{\Delta y}{\Delta t}$$
$$+ \lim_{\Delta t \to 0} \varepsilon_1 \lim_{\Delta t \to 0} \frac{\Delta x}{\Delta t} + \lim_{\Delta t \to 0} \varepsilon_2 \lim_{\Delta t \to 0} \frac{\Delta y}{\Delta t}$$
$$= \frac{\partial f}{\partial x}\frac{dx}{dt} + \frac{\partial f}{\partial y}\frac{dy}{dt}.$$

We summarize the chain rule for the derivative of $f(x(t), y(t))$ in Theorem 5.1.

THEOREM 5.1 (Chain Rule)

If $z = f(x(t), y(t))$, where $x(t)$ and $y(t)$ are differentiable and $f(x, y)$ is a differentiable function of x and y, then

$$\frac{dz}{dt} = \frac{d}{dt}[f(x(t), y(t))] = \frac{\partial f}{\partial x}(x(t), y(t))\frac{dx}{dt} + \frac{\partial f}{\partial y}(x(t), y(t))\frac{dy}{dt}.$$

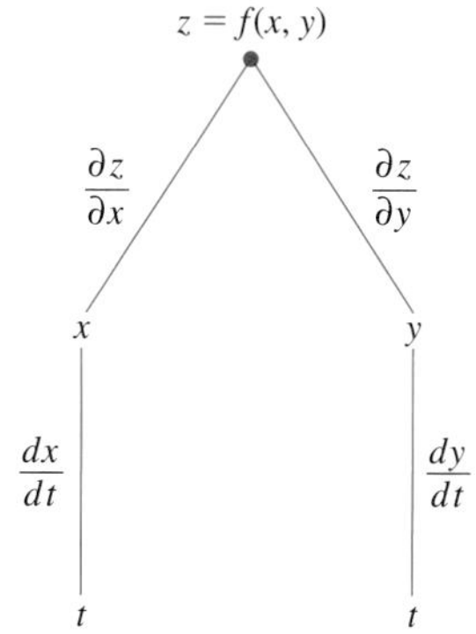

As a convenient device for remembering the chain rule, we sometimes use a **tree diagram** like the one shown in the margin. Notice that if $z = f(x, y)$ and x and y are both functions of the variable t, then t is the independent variable. We consider x and y to be **intermediate variables,** since they both depend on t. In the tree diagram, we list the dependent variable z at the top, followed by each of the intermediate variables x and y, with the independent variable t at the bottom level, with each of the variables connected by a path. Next to each of the paths, we indicate the corresponding derivative $\left(\text{i.e., between } z \text{ and } x\text{, we indicate } \frac{\partial z}{\partial x}\right)$. The chain rule then gives $\frac{dz}{dt}$ as the sum of all of the products of the derivatives along each path to t. That is,

$$\frac{dz}{dt} = \frac{\partial z}{\partial x}\frac{dx}{dt} + \frac{\partial z}{\partial y}\frac{dy}{dt}.$$

This device is especially useful for functions of several variables that are in turn functions of several other variables, as we will see shortly.

We illustrate the use of this new chain rule in example 5.1.

EXAMPLE 5.1 Using the Chain Rule

For $z = f(x, y) = x^2e^y$, $x(t) = t^2 - 1$ and $y(t) = \sin t$, find the derivative of $g(t) = f(x(t), y(t))$.

Solution We first compute the derivatives $\frac{\partial z}{\partial x} = 2xe^y$, $\frac{\partial z}{\partial y} = x^2e^y$, $x'(t) = 2t$ and $y'(t) = \cos t$. The chain rule (Theorem 5.1) then gives us

$$\begin{aligned} g'(t) &= \frac{\partial z}{\partial x}\frac{dx}{dt} + \frac{\partial z}{\partial y}\frac{dy}{dt} = 2xe^y(2t) + x^2e^y\cos t \\ &= 2(t^2 - 1)e^{\sin t}(2t) + (t^2 - 1)^2e^{\sin t}\cos t. \quad \blacksquare \end{aligned}$$

In example 5.1, notice that you could have first substituted for x and y and then computed the derivative of $g(t) = (t^2 - 1)^2e^{\sin t}$, using the usual rules of differentiation. In example 5.2, you don't have any alternative but to use the chain rule.

EXAMPLE 5.2 A Case Where the Chain Rule Is Needed

Suppose the production of a firm is modeled by the **Cobb-Douglas production function** $P(k, l) = 20k^{1/4}l^{3/4}$, where k measures capital (in millions of dollars) and l measures the labor force (in thousands of workers). Suppose that when $l = 2$ and $k = 6$, the labor force is decreasing at the rate of 20 workers per year and capital is growing at the rate of \$400,000 per year. Determine the rate of change of production.

Solution Suppose that $g(t) = P(k(t), l(t))$. From the chain rule, we have

$$g'(t) = \frac{\partial P}{\partial k}k'(t) + \frac{\partial P}{\partial l}l'(t).$$

Notice that $\frac{\partial P}{\partial k} = 5k^{-3/4}l^{3/4}$ and $\frac{\partial P}{\partial l} = 15k^{1/4}l^{-1/4}$. With $l = 2$ and $k = 6$, this gives us $\frac{\partial P}{\partial k}(6, 2) \approx 2.1935$ and $\frac{\partial P}{\partial l}(6, 2) \approx 19.7411$. Since k is measured in millions of dollars and l is measured in thousands of workers, we have $k'(t) = 0.4$ and $l'(t) = -0.02$. From the chain rule, we now have

$$\begin{aligned} g'(t) &= \frac{\partial P}{\partial k}k'(t) + \frac{\partial P}{\partial l}l'(t) \\ &\approx 2.1935(0.4) + 19.7411(-0.02) = 0.48258. \end{aligned}$$

This indicates that the production is increasing at the rate of approximately one-half unit per year. ■

We can easily extend Theorem 5.1 to the case of a function $f(x, y)$, where x and y are both functions of the two independent variables s and t, $x = x(s, t)$ and $y = y(s, t)$. Notice that if we differentiate with respect to s, we treat t as a constant. Applying Theorem 5.1 (while holding t fixed), we have

$$\frac{\partial}{\partial s}[f(x, y)] = \frac{\partial f}{\partial x}\frac{\partial x}{\partial s} + \frac{\partial f}{\partial y}\frac{\partial y}{\partial s}.$$

Similarly, we can find a chain rule for $\frac{\partial}{\partial t}[f(x, y)]$. This gives us the following more general form of the chain rule.

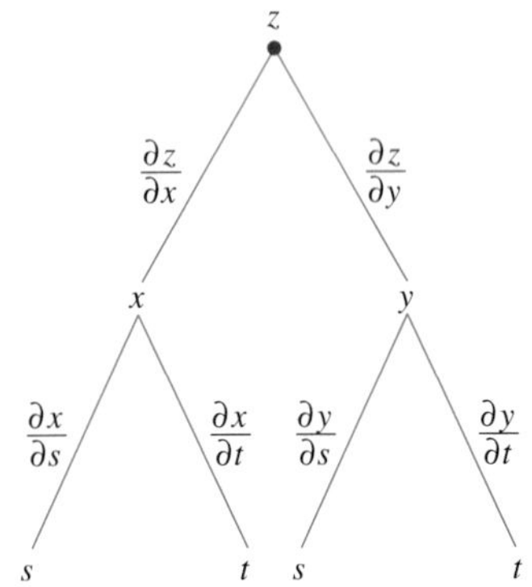

THEOREM 5.2 (Chain Rule)

Suppose that $z = f(x, y)$, where f is a differentiable function of x and y and where $x = x(s, t)$ and $y = y(s, t)$ both have first-order partial derivatives. Then we have the chain rules:

$$\frac{\partial z}{\partial s} = \frac{\partial z}{\partial x}\frac{\partial x}{\partial s} + \frac{\partial z}{\partial y}\frac{\partial y}{\partial s}$$

and

$$\frac{\partial z}{\partial t} = \frac{\partial z}{\partial x}\frac{\partial x}{\partial t} + \frac{\partial z}{\partial y}\frac{\partial y}{\partial t}.$$

The tree diagram shown in the margin serves as a convenient reminder of the chain rules indicated in Theorem 5.2, again by summing the products of the indicated partial derivatives along each path from z to s or t, respectively.

The chain rule is easily extended to functions of three or more variables. You will explore this in the exercises.

EXAMPLE 5.3 Using the Chain Rule

Suppose that $f(x, y) = e^{xy}$, $x(u, v) = 3u \sin v$ and $y(u, v) = 4v^2u$. For $g(u, v) = f(x(u, v), y(u, v))$, find the partial derivatives $\frac{\partial g}{\partial u}$ and $\frac{\partial g}{\partial v}$.

Solution We first compute the partial derivatives $\frac{\partial f}{\partial x} = ye^{xy}$, $\frac{\partial f}{\partial y} = xe^{xy}$, $\frac{\partial x}{\partial u} = 3 \sin v$ and $\frac{\partial y}{\partial u} = 4v^2$. The chain rule (Theorem 5.2) gives us

$$\frac{\partial g}{\partial u} = \frac{\partial f}{\partial x}\frac{\partial x}{\partial u} + \frac{\partial f}{\partial y}\frac{\partial y}{\partial u} = ye^{xy}(3\sin v) + xe^{xy}(4v^2).$$

Substituting for x and y, we get

$$\frac{\partial g}{\partial u} = 4v^2ue^{12u^2v^2\sin v}(3\sin v) + 3u\sin ve^{12u^2v^2\sin v}(4v^2).$$

For the partial derivative of g with respect to v, we compute $\frac{\partial x}{\partial v} = 3u\cos v$ and $\frac{\partial y}{\partial v} = 8vu$. Here, the chain rule gives us

$$\frac{\partial g}{\partial v} = ye^{xy}(3u\cos v) + xe^{xy}(8vu).$$

Substituting for x and y, we have

$$\frac{\partial g}{\partial v} = 4v^2ue^{12u^2v^2\sin v}(3u\cos v) + 3u\sin ve^{12u^2v^2\sin v}(8vu).$$ ■

Once again, it is often simpler to first substitute in the expressions for x and y. We leave it as an exercise to show that you get the same derivatives either way. On the other hand, there are plenty of times where the general forms of the chain rule seen in Theorems 5.1 and 5.2 are indispensable. You will see some of these in the exercises, while we present several important uses next.

EXAMPLE 5.4 Converting from Rectangular to Polar Coordinates

For a differentiable function $f(x, y)$ with continuous partial derivatives, $x = r\cos\theta$ and $y = r\sin\theta$, show that $f_r = f_x\cos\theta + f_y\sin\theta$ and
$f_{rr} = f_{xx}\cos^2\theta + 2f_{xy}\cos\theta\sin\theta + f_{yy}\sin^2\theta$.

Solution First, notice that $\frac{\partial x}{\partial r} = \cos\theta$ and $\frac{\partial y}{\partial r} = \sin\theta$. From Theorem 5.2, we now have

$$f_r = \frac{\partial f}{\partial r} = \frac{\partial f}{\partial x}\frac{\partial x}{\partial r} + \frac{\partial f}{\partial y}\frac{\partial y}{\partial r} = f_x\cos\theta + f_y\sin\theta.$$

Be very careful when computing the second partial derivative. Using the expression we have already found for f_r and Theorem 5.2, we have

$$\begin{aligned} f_{rr} &= \frac{\partial(f_r)}{\partial r} = \frac{\partial}{\partial r}(f_x\cos\theta + f_y\sin\theta) = \frac{\partial}{\partial r}(f_x)\cos\theta + \frac{\partial}{\partial r}(f_y)\sin\theta \\ &= \left[\frac{\partial}{\partial x}(f_x)\frac{\partial x}{\partial r} + \frac{\partial}{\partial y}(f_x)\frac{\partial y}{\partial r}\right]\cos\theta + \left[\frac{\partial}{\partial x}(f_y)\frac{\partial x}{\partial r} + \frac{\partial}{\partial y}(f_y)\frac{\partial y}{\partial r}\right]\sin\theta \\ &= (f_{xx}\cos\theta + f_{xy}\sin\theta)\cos\theta + (f_{yx}\cos\theta + f_{yy}\sin\theta)\sin\theta \\ &= f_{xx}\cos^2\theta + 2f_{xy}\cos\theta\sin\theta + f_{yy}\sin^2\theta, \end{aligned}$$

as desired. ■

In the exercises, you will use the chain rule to compute other partial derivatives in polar coordinates. One important exercise is to show that we can write (for $r \neq 0$)

$$f_{xx} + f_{yy} = f_{rr} + \frac{1}{r}f_r + \frac{1}{r^2}f_{\theta\theta}.$$

This particular combination of second partial derivatives, $f_{xx} + f_{yy}$, is called the **Laplacian** of f and appears frequently in equations describing heat conduction and wave propagation, among others.

A slightly different use of a change of variables is demonstrated in example 5.5. An important strategy in solving some equations is to first rewrite and solve them in the most general form possible. One convenient approach to this is to convert to **dimensionless variables.** As the name implies, these are typically combinations of variables such that all the units cancel. One example would be for an object with (one-dimensional) velocity v ft/s and initial velocity $v(0) = v_0$ ft/s. The variable $V = \frac{v}{v_0}$ is dimensionless because the units of V are ft/s divided by ft/s, leaving no units. Often, a change to dimensionless variables will simplify an equation.

EXAMPLE 5.5 Dimensionless Variables

An object moves in two dimensions according to the equations of motion: $x''(t) = 0$, $y''(t) = -g$, with initial velocity $x'(0) = v_0 \cos\theta$ and $y'(0) = v_0 \sin\theta$ and initial position $x(0) = y(0) = 0$. Rewrite the equations and initial conditions in terms of the variables $X = \frac{g}{v_0^2}x$, $Y = \frac{g}{v_0^2}y$ and $T = \frac{g}{v_0}t$. Show that the variables X, Y and T are dimensionless, assuming that x and y are given in feet and t in seconds.

Solution To transform the equations, we first need to rewrite the derivatives $x'' = \frac{d^2x}{dt^2}$ and $y'' = \frac{d^2y}{dt^2}$ in terms of X, Y and T. From the chain rule, we have

$$\frac{dx}{dt} = \frac{dx}{dT}\frac{dT}{dt} = \frac{d\left(v_0^2X/g\right)}{dT}\frac{d(gt/v_0)}{dt} = \frac{v_0^2}{g}\frac{dX}{dT}\frac{g}{v_0} = v_0\frac{dX}{dT}.$$

Again, we must be careful computing the second derivative. We have

$$\begin{aligned}\frac{d^2x}{dt^2} &= \frac{d}{dt}\left(\frac{dx}{dt}\right) = \frac{d}{dt}\left(v_0\frac{dX}{dT}\right) = \frac{d}{dT}\left(v_0\frac{dX}{dT}\right)\frac{dT}{dt} \\ &= v_0\frac{d^2X}{dT^2}\frac{g}{v_0} = g\frac{d^2X}{dT^2}.\end{aligned}$$

You should verify that similar calculations give $\frac{dy}{dt} = v_0\frac{dY}{dT}$ and $\frac{d^2y}{dt^2} = g\frac{d^2Y}{dT^2}$. The differential equation $x''(t) = 0$ then becomes $g\frac{d^2X}{dT^2} = 0$ or simply, $\frac{d^2X}{dT^2} = 0$. Similarly, the differential equation $y''(t) = -g$ becomes $g\frac{d^2Y}{dT^2} = -g$ or $\frac{d^2Y}{dT^2} = -1$. Further, the initial condition $x'(0) = v_0\cos\theta$ becomes $v_0\frac{dX}{dT}(0) = v_0\cos\theta$ or $\frac{dX}{dT}(0) = \cos\theta$ and the initial condition $y'(0) = v_0\sin\theta$ becomes $v_0\frac{dY}{dT}(0) = v_0\sin\theta$ or $\frac{dY}{dT}(0) = \sin\theta$. The initial value problem is now

$$\frac{d^2X}{dT^2} = 0,\ \frac{d^2Y}{dT^2} = -1,\ \frac{dX}{dT}(0) = \cos\theta,\ \frac{dY}{dT}(0) = \sin\theta,\ X(0) = 0,\ Y(0) = 0.$$

Notice that the only parameter left in the entire set of equations is θ, which is measured in radians (which is considered unitless). So, we would not need to know which unit system is being used to solve this initial value problem. Finally, to show that the variables are indeed dimensionless, we look at the units. In the English system, the initial speed v_0 has units ft/s and g has units ft/s^2 (the same as acceleration). Then $X = \frac{g}{v_0^2}x$ has units

$$\frac{\text{ft/s}^2}{(\text{ft/s})^2}(\text{ft}) = \frac{\text{ft}^2/\text{s}^2}{\text{ft}^2/\text{s}^2} = 1.$$

Similarly, Y has no units. Finally $T = \frac{g}{v_0}t$ has units

$$\frac{\text{ft/s}^2}{\text{ft/s}}(\text{s}) = \frac{\text{ft/s}}{\text{ft/s}} = 1. \blacksquare$$

Implicit Differentiation

Suppose that the equation $F(x, y) = 0$ defines y implicitly as a function of x, say $y = f(x)$. In section 2.8, we saw how to calculate $\frac{dy}{dx}$ in such a case. We can use the chain rule for functions of several variables to obtain an alternative method for calculating this. Moreover, this will provide us with new insights into when this can be done and, more important yet, this will generalize to functions of several variables defined implicitly by an equation.

We let $z = F(x, y)$, where $x = t$ and $y = f(t)$. From Theorem 5.1, we have

$$\frac{dz}{dt} = F_x\frac{dx}{dt} + F_y\frac{dy}{dt}.$$

But, since $z = F(x, y) = 0$, we have $\frac{dz}{dt} = 0$, too. Further, since $x = t$, we have $\frac{dx}{dt} = 1$ and $\frac{dy}{dt} = \frac{dy}{dx}$. This leaves us with

$$0 = F_x + F_y\frac{dy}{dx}.$$

Notice that we can solve this for $\frac{dy}{dx}$, provided $F_y \neq 0$. In this case, we have

$$\frac{dy}{dx} = -\frac{F_x}{F_y}.$$

Recognize that we already know how to calculate $\frac{dy}{dx}$ implicitly, so this doesn't appear to give us anything new. However, it turns out that the **Implicit Function Theorem** (proved in a course in advanced calculus) says that if F_x and F_y are continuous on an open disk containing the point (a, b) where $F(a, b) = 0$ and $F_y(a, b) \neq 0$, then the equation $F(x, y) = 0$ implicitly defines y as a function of x nearby the point (a, b).

More significantly, we can extend this notion to functions of several variables defined implicitly, as follows. Suppose that the equation $F(x, y, z) = 0$ implicitly defines a function $z = f(x, y)$, where f is differentiable. Then, we can find the partial derivatives f_x and f_y using the chain rule, as follows. We first let $w = F(x, y, z)$. From the chain rule, we have

$$\frac{\partial w}{\partial x} = F_x\frac{\partial x}{\partial x} + F_y\frac{\partial y}{\partial x} + F_z\frac{\partial z}{\partial x}.$$

Notice that since $w = F(x, y, z) = 0$, $\frac{\partial w}{\partial x} = 0$. Also, $\frac{\partial x}{\partial x} = 1$ and $\frac{\partial y}{\partial x} = 0$, since x and y are independent variables. This gives us

$$0 = F_x + F_z \frac{\partial z}{\partial x}.$$

We can solve this for $\frac{\partial z}{\partial x}$, as long as $F_z \neq 0$, to obtain

$$\frac{\partial z}{\partial x} = -\frac{F_x}{F_z}. \tag{5.2}$$

Likewise, differentiating w with respect to y leads us to

$$\frac{\partial z}{\partial y} = -\frac{F_y}{F_z}, \tag{5.3}$$

again, as long as $F_z \neq 0$. Much as in the two-variable case, the Implicit Function Theorem for functions of three variables says that if F_x, F_y and F_z are continuous inside a sphere containing the point (a, b, c) where $F(a, b, c) = 0$ and $F_z(a, b, c) \neq 0$, then the equation $F(x, y, z) = 0$ implicitly defines z as a function of x and y nearby the point (a, b, c).

EXAMPLE 5.6 Finding Partial Derivatives Implicitly

Find $\frac{\partial z}{\partial x}$ and $\frac{\partial z}{\partial y}$, given that $F(x, y, z) = xy^2 + z^3 + \sin(xyz) = 0$.

Solution First, note that using the usual chain rule, we have

$$F_x = y^2 + yz\cos(xyz),$$

$$F_y = 2xy + xz\cos(xyz)$$

and

$$F_z = 3z^2 + xy\cos(xyz).$$

From (5.2), we now have

$$\frac{\partial z}{\partial x} = -\frac{F_x}{F_z} = -\frac{y^2 + yz\cos(xyz)}{3z^2 + xy\cos(xyz)}.$$

Likewise, from (5.3), we have

$$\frac{\partial z}{\partial y} = -\frac{F_y}{F_z} = -\frac{2xy + xz\cos(xyz)}{3z^2 + xy\cos(xyz)}.$$ ■

Notice that, much like implicit differentiation with two variables, implicit differentiation with three variables yields expressions for the derivatives that depend on all three variables.

BEYOND FORMULAS

There are many more examples of the chain rule than the two written out in Theorems 5.1 and 5.2. We ask you to write out other forms of the chain rule in the exercises. All of these variations would be impossible to memorize, but all you need to do to reproduce whichever rule you need is construct the appropriate tree diagram and remember the general format, "derivative of the outside times derivative of the inside."

EXERCISES 12.5

HW 17 25-27 recommended

981 → #3, 5, 7, 9, 17, 19, 21

WRITING EXERCISES

1. In example 5.1, we mentioned that direct substitution followed by differentiation was an option (see exercises 1 and 2 below) and is often preferable. Discuss the advantages and disadvantages of direct substitution versus the method of example 5.1.

2. In example 5.6, we treated z as a function of x and y. Explain how to modify our results from the Implicit Function Theorem for treating x as a function of y and z.

---

1. Repeat example 5.1 by first substituting $x = t^2 - 1$ and $y = \sin t$ and then computing $g'(t)$.

2. Repeat example 5.3 by first substituting $x = 3u \sin v$ and $y = 4v^2 u$ and then computing $\frac{\partial g}{\partial u}$ and $\frac{\partial g}{\partial v}$.

In exercises 3–6, use the chain rule to find the indicated derivative(s).

3. $g'(t)$, where $g(t) = f(x(t), y(t))$, $f(x, y) = x^2 y - \sin y$, $x(t) = \sqrt{t^2 + 1}$, $y(t) = e^t$

4. $g'(t)$, where $g(t) = f(x(t), y(t))$, $f(x, y) = \sqrt{x^2 + y^2}$, $x(t) = \sin t$, $y(t) = t^2 + 2$

5. $\frac{\partial g}{\partial u}$ and $\frac{\partial g}{\partial v}$, where $g(u, v) = f(x(u, v), y(u, v))$, $f(x, y) = 4x^2 y^3$, $x(u, v) = u^3 - v \sin u$, $y(u, v) = 4u^2$

6. $\frac{\partial g}{\partial u}$ and $\frac{\partial g}{\partial v}$, where $g(u, v) = f(x(u, v), y(u, v))$, $f(x, y) = xy^3 - 4x^2$, $x(u, v) = e^{u^2}$, $y(u, v) = \sqrt{v^2 + 1} \sin u$

In exercises 7–10, state the chain rule for the general composite function.

7. $g(t) = f(x(t), y(t), z(t))$

8. $g(u, v) = f(x(u, v), y(u, v), z(u, v))$

9. $g(u, v, w) = f(x(u, v, w), y(u, v, w))$

10. $g(u, v, w) = f(x(u, v, w), y(u, v, w), z(u, v, w))$

11. In example 5.2, suppose that $l = 4$ and $k = 6$, the labor force is decreasing at the rate of 60 workers per year and capital is growing at the rate of \$100,000 per year. Determine the rate of change of production.

12. In example 5.2, suppose that $l = 3$ and $k = 4$, the labor force is increasing at the rate of 80 workers per year and capital is decreasing at the rate of \$200,000 per year. Determine the rate of change of production.

13. Suppose the production of a firm is modeled by $P(k, l) = 16k^{1/3} l^{2/3}$, with k and l defined as in example 5.2. Suppose that $l = 3$ and $k = 4$, the labor force is increasing at the rate of 80 workers per year and capital is decreasing at the rate of \$200,000 per year. Determine the rate of change of production.

14. Suppose the production of a firm is modeled by $P(k, l) = 16k^{1/3} l^{2/3}$, with k and l defined as in example 5.2. Suppose that $l = 2$ and $k = 5$, the labor force is increasing at the rate of 40 workers per year and capital is decreasing at the rate of \$100,000 per year. Determine the rate of change of production.

15. For a business product, income is the product of the quantity sold and the price, which we can write as $I = qp$. If the quantity sold increases at a rate of 5% and the price increases at a rate of 3%, show that income increases at a rate of 8%.

16. Assume that $I = qp$ as in exercise 15. If the quantity sold decreases at a rate of 3% and price increases at a rate of 5%, determine the rate of increase or decrease in income.

In exercises 17–20, use the chain rule twice to find the indicated derivative.

17. $g(t) = f(x(t)), y(t))$, find $g''(t)$

18. $g(t) = f(x(t), y(t), z(t))$, find $g''(t)$

19. $g(u, v) = f(x(u, v), y(u, v))$, find $\frac{\partial^2 g}{\partial u^2}$

20. $g(u, v) = f(x(u, v), y(u, v))$, find $\frac{\partial^2 g}{\partial u \partial v}$

In exercises 21–24, use implicit differentiation to find $\frac{\partial z}{\partial x}$ and $\frac{\partial z}{\partial y}$.

21. $3x^2 z + 2z^3 - 3yz = 0$

22. $xyz - 4y^2 z^2 + \cos xy = 0$

23. $3e^{xyz} - 4xz^2 + x \cos y = 2$

24. $3yz^2 - e^{4x} \cos 4z - 3y^2 = 4$

25. For a differentiable function $f(x, y)$ with continuous partial derivatives, $x = r \cos\theta$ and $y = r \sin\theta$, show that $f_\theta = -f_x r \sin\theta + f_y r \cos\theta$.

26. For a differentiable function $f(x, y)$ with continuous partial derivatives, $x = r \cos\theta$ and $y = r \sin\theta$, show that $f_{\theta\theta} = f_{xx} r^2 \sin^2\theta - 2f_{xy} r^2 \cos\theta \sin\theta + f_{yy} r^2 \cos^2\theta - f_x r \cos\theta - f_y r \sin\theta$.

27. For a differentiable function $f(x, y)$ with continuous partial derivatives, $x = r \cos\theta$ and $y = r \sin\theta$, use the results of exercises 25 and 26 and example 5.4 to show that $f_{xx} + f_{yy} = f_{rr} + \frac{1}{r} f_r + \frac{1}{r^2} f_{\theta\theta}$. This expression is called the **Laplacian** of f.

28. Given that $r = \sqrt{x^2 + y^2}$, show that $\frac{\partial r}{\partial x} = \frac{x}{\sqrt{x^2+y^2}} = \frac{x}{r} = \cos\theta$. Starting from $r = \frac{x}{\cos\theta}$, does it follow that $\frac{\partial r}{\partial x} = \frac{1}{\cos\theta}$? Explain why it's not possible for both calculations to be correct. Find all mistakes.

29. The **heat equation** for the temperature $u(x, t)$ of a thin rod of length L is $\alpha^2 u_{xx} = u_t$, $0 < x < L$, for some constant α^2, called the **thermal diffusivity.** Make the change of variables $X = \frac{x}{L}$ and $T = \frac{\alpha^2}{L^2}t$ to simplify the equation. Show that X and T are dimensionless, given that the units of α^2 are ft²/s.

30. The **wave equation** for the displacement $u(x, t)$ of a vibrating string of length L is $a^2 u_{xx} = u_{tt}$, $0 < x < L$, for some constant a^2. Make the change of variables $X = \frac{x}{L}$ and $T = \frac{a}{L}t$ to simplify the equation. Assuming that X and T are dimensionless, find the dimensions of a^2.

Exercises 31–40 relate to Taylor series for functions of two or more variables.

31. Suppose that $f(x, y)$ is a function with all partial derivatives continuous. For constants u_1 and u_2, define $g(h) = f(x + hu_1, y + hu_2)$. We will construct the Taylor series for $g(h)$ about $h = 0$. First, show that $g(0) = f(x, y)$. Then show that $g'(0) = f_x(x, y)u_1 + f_y(x, y)u_2$. Next, show that $g''(0) = f_{xx}u_1^2 + 2f_{xy}u_1u_2 + f_{yy}u_2^2$, where the functions f_{xx}, f_{xy} and f_{yy} are all evaluated at (x, y). Evaluate $g'''(0)$ and $g^{(4)}(0)$, and briefly describe the pattern of terms that emerges.

32. Use the result of exercise 31 with $hu_1 = \Delta x$ and $hu_2 = \Delta y$ to show that

$$f(x + \Delta x, y + \Delta y) = f(x, y) + f_x(x, y)\Delta x + f_y(x, y)\Delta y + \tfrac{1}{2}[f_{xx}(x, y)\Delta x^2 + 2f_{xy}(x, y)\Delta x\Delta y + f_{yy}(x, y)\Delta y^2] + \tfrac{1}{3!}[f_{xxx}(x, y)\Delta x^3 + 3f_{xxy}(x, y)\Delta x^2\Delta y + 3f_{xyy}(x, y)\Delta x\Delta y^2 + f_{yyy}(x, y)\Delta y^3] + \cdots,$$

which is the form of Taylor series for functions of two variables about the center (x, y).

33. Use the result of exercise 32 to write out the third-order Taylor polynomial for $f(x, y) = \sin x \cos y$ about $(0, 0)$.

34. Compare your answer in exercise 33 to a term-by-term multiplication of the Maclaurin series (Taylor series with center 0) for $\sin x$ and $\cos y$. Write out the fourth-order and fifth-order terms for this product.

35. Write out the third-order polynomial for $f(x, y) = \sin xy$ about $(0, 0)$.

36. Compare your answer in exercise 35 to the Maclaurin series for $\sin u$ with the substitution $u = xy$.

37. Write out the third-order polynomial for $f(x, y) = e^{2x+y}$ about $(0, 0)$.

38. Compare your answer in exercise 37 to the Maclaurin series for e^u with the substitution $u = 2x + y$.

39. The Environmental Protection Agency uses the 55/45 rule for combining a car's highway gas mileage rating h and its city gas mileage rating c into a single rating R for fuel efficiency using the formula $R = \frac{1}{0.55/c + 0.45/h}$. Find the first-order Taylor series (terms for Δc and Δh but not Δc^2) for $R(c, h)$ about $(1, 1)$.

40. Given the answer to exercise 39, explain why it's surprising that the EPA would use the complicated formula it does. To see why, consider a car with $h = 40$ and graphically compare the actual rating R to the Taylor approximation for $0 \le c \le 40$. If c is approximately the same as h, is there much difference in the graphs? As c approaches 0, how do the graphs compare? The EPA wants to convey useful information to the public. If a car got 40 mpg on the highway and 5 mpg in the city, would you want the overall rating to be (relatively) high or low?

41. The pressure, temperature, volume and enthalpy of a gas are all interrelated. Enthalpy is determined by pressure and temperature, so $E = f(P, T)$, for some function f. Pressure is determined by temperature and volume, so $P = g(T, V)$, for some function g. Show that $E = h(T, V)$ where h is a composition of f and g. Chemists write $\frac{\partial f}{\partial T}$ as $\left(\frac{\partial E}{\partial T}\right)_P$ to show that P is being held constant. Similarly, $\left(\frac{\partial E}{\partial T}\right)_V$ would refer to $\frac{\partial h}{\partial T}$. Using this convention, show that $\left(\frac{\partial E}{\partial T}\right)_V = \left(\frac{\partial E}{\partial T}\right)_P + \left(\frac{\partial E}{\partial P}\right)_T \left(\frac{\partial P}{\partial T}\right)_V$.

42. An economist analyzing the relationship among capital expenditure, labor and production in an industry might start with production $p(x, y)$ as a function of capital x and labor y. An additional assumption is that if labor and capital are doubled, the production should double. This translates to $p(2x, 2y) = 2p(x, y)$. This can be generalized to the relationship $p(kx, ky) = kp(x, y)$, for any positive constant k. Differentiate both sides of this equation with respect to k and show that $p(x, y) = xp_x(x, y) + yp_y(x, y)$. This would be stated by the economist as, "The total production equals the sum of the costs of capital and labor paid at their level of marginal product." Match each term in the quote with the corresponding term in the equation.

43. A baseball player who has h hits in b at bats has a batting average of $a = \frac{h}{b}$. For example, 100 hits in 400 at bats would be an average of 0.250. It is traditional to carry three decimal places and to describe this average as being "250 points." To use the chain rule to estimate the change in batting average after a player gets a hit, assume that h and b are functions of time and that getting a hit means $h' = b' = 1$. Show that $a' = \frac{b-h}{b^2}$.

Early in a season, a typical batter might have 50 hits in 200 at bats. Show that getting a hit will increase batting average by about 4 points. Find the approximate increase in batting average later in the season for a player with 100 hits in 400 at bats. In general, if b and h are both doubled, how does a' change?

44. For the baseball players of exercise 43, approximate the number of points that the batting average will decrease by making an out.

45. Find the general form for the derivative of $g(t) = u(t)^{v(t)}$ for differentiable functions u and v. (Hint: Start with $f(u, v) = u^v$.) Apply the result to find the derivative of $(2t + 1)^{3t^2}$.

EXPLORATORY EXERCISES

1. We have previously done calculations of the amount of work done by some force. Recall that if a scalar force $F(x)$ is applied as x increases from $x = a$ to $x = b$, then the work done equals $W = \int_a^b F(x)\,dx$. If the position x is a differentiable function of time, then we can write $W = \int_0^T F(x(t))x'(t)\,dt$, where $x(0) = a$ and $x(T) = b$. **Power** is defined as the time derivative of work. Work is sometimes measured in foot-pounds, so power could be measured in foot-pounds per second (ft-lb/s). One horsepower is equal to 550 ft-lb/s. Show that if force and velocity are constant, then power is the product of force and velocity, Determine how many pounds of force are required to maintain 400 hp at 80 mph. For a variable force and velocity, use the chain rule to compute power.

2. Engineers and physicists (and therefore mathematicians) spend countless hours studying the properties of forced oscillators. Two physical situations that are well modeled by the same mathematical equations are a spring oscillating due to some force and a simple electrical circuit with a voltage source. A general solution of a forced oscillator can have the form $u(t) = g(t) - \int_0^t g(u)e^{-(t-u)/2}[\cos\frac{\sqrt{3}}{2}(t-u) + \frac{2}{3}\sin\frac{\sqrt{3}}{2}(t-u)]\,du$. If $g(0) = 1$ and $g'(0) = 2$, compute $u(0)$ and $u'(0)$.

12.6 THE GRADIENT AND DIRECTIONAL DERIVATIVES

Suppose that you are hiking in rugged terrain. You can think of your altitude at the point given by longitude x and latitude y as defining a function $f(x, y)$. Although you are unlikely to have a handy formula for this function, you can learn more about this function than you might expect. If you face due east (in the direction of the positive x-axis), the slope of the terrain is given by the partial derivative $\frac{\partial f}{\partial x}(x, y)$. Similarly, facing due north, the slope of the terrain is given by $\frac{\partial f}{\partial y}(x, y)$. However, in terms of $f(x, y)$, how would you compute the slope in some other direction, say north-by-northwest? In this section, we develop the notion of *directional derivative,* which will answer this question.

Suppose that we want to find the instantaneous rate of change of $f(x, y)$ at the point $P(a, b)$ and in the direction given by the *unit* vector $\mathbf{u} = \langle u_1, u_2 \rangle$. Let $Q(x, y)$ be any point on the line through $P(a, b)$ in the direction of $\mathbf{u}$. Notice that the vector $\overrightarrow{PQ}$ is then parallel to $\mathbf{u}$. Since two vectors are parallel if and only if one is a scalar multiple of the other, we have that $\overrightarrow{PQ} = h\mathbf{u}$, for some scalar h, so that

$$\overrightarrow{PQ} = \langle x - a, y - b \rangle = h\mathbf{u} = h\langle u_1, u_2 \rangle = \langle hu_1, hu_2 \rangle.$$

It then follows that $x - a = hu_1$ and $y - b = hu_2$, so that

$$x = a + hu_1 \quad \text{and} \quad y = b + hu_2.$$

The point Q is then described by $(a + hu_1, b + hu_2)$, as indicated in Figure 12.33 (on the following page). Notice that the average rate of change of $z = f(x, y)$ along the line from P to Q is then

$$\frac{f(a + hu_1, b + hu_2) - f(a, b)}{h}.$$

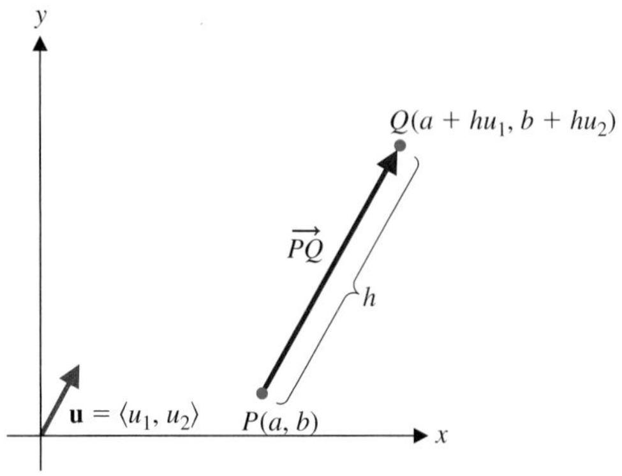

FIGURE 12.33
The vector $\overrightarrow{PQ}$

The instantaneous rate of change of $f(x, y)$ at the point $P(a, b)$ and in the direction of the unit vector $\mathbf{u}$ is then found by taking the limit as $h \to 0$. We give this limit a special name in Definition 6.1.

↳ instantaneous change

DEFINITION 6.1

The **directional derivative of** $f(x, y)$ at the point (a, b) and in the direction of the unit vector $\mathbf{u} = \langle u_1, u_2 \rangle$ is given by

$$D_{\mathbf{u}} f(a, b) = \lim_{h \to 0} \frac{f(a + hu_1, b + hu_2) - f(a, b)}{h},$$

provided the limit exists.

Notice that this limit resembles the definition of partial derivative, except that in this case, both variables may change. Further, you should observe that the directional derivative in the direction of the positive x-axis (i.e., in the direction of the unit vector $\mathbf{u} = \langle 1, 0 \rangle$) is

$$D_{\mathbf{u}} f(a, b) = \lim_{h \to 0} \frac{f(a + h, b) - f(a, b)}{h},$$

which you should recognize as the partial derivative $\frac{\partial f}{\partial x}$. Likewise, the directional derivative in the direction of the positive y-axis (i.e., in the direction of the unit vector $\mathbf{u} = \langle 0, 1 \rangle$) is $\frac{\partial f}{\partial y}$. It turns out that any directional derivative can be calculated simply, in terms of the first partial derivatives, as we see in Theorm 6.1.

THEOREM 6.1 $f: \mathbb{R}^2 \to \mathbb{R}$.

Suppose that f is differentiable at (a, b) and $\mathbf{u} = \langle u_1, u_2 \rangle$ is any unit vector. Then, we can write

$$D_{\mathbf{u}} f(a, b) = f_x(a, b) u_1 + f_y(a, b) u_2.$$

PROOF

Let $g(h) = f(a + hu_1, b + hu_2)$. Then, $g(0) = f(a, b)$ and so, from Definition 6.1, we have

$$D_{\mathbf{u}} f(a, b) = \lim_{h \to 0} \frac{f(a + hu_1, b + hu_2) - f(a, b)}{h} = \lim_{h \to 0} \frac{g(h) - g(0)}{h} = g'(0).$$

If we define $x = a + hu_1$ and $y = b + hu_2$, we have $g(h) = f(x, y)$. From the chain rule (Theorem 5.1), we have

$$g'(h) = \frac{\partial f}{\partial x}\frac{dx}{dh} + \frac{\partial f}{\partial y}\frac{dy}{dh} = \frac{\partial f}{\partial x}u_1 + \frac{\partial f}{\partial y}u_2.$$

Finally, taking $h = 0$ gives us

$$D_{\mathbf{u}} f(a, b) = g'(0) = \frac{\partial f}{\partial x}(a, b)u_1 + \frac{\partial f}{\partial y}(a, b)u_2,$$

as desired. ■

EXAMPLE 6.1 Computing Directional Derivatives

For $f(x, y) = x^2 y - 4y^3$, compute $D_{\mathbf{u}} f(2, 1)$ for the directions (a) $\mathbf{u} = \left\langle \frac{\sqrt{3}}{2}, \frac{1}{2} \right\rangle$ and (b) $\mathbf{u}$ in the direction from (2, 1) to (4, 0).

Solution Regardless of the direction, we first need to compute the first partial derivatives $\frac{\partial f}{\partial x} = 2xy$ and $\frac{\partial f}{\partial y} = x^2 - 12y^2$. Then, $f_x(2, 1) = 4$ and $f_y(2, 1) = -8$. For (a), the unit vector is given as $\mathbf{u} = \left\langle \frac{\sqrt{3}}{2}, \frac{1}{2} \right\rangle$ and so, from Theorem 6.1 we have

$$D_{\mathbf{u}} f(2, 1) = f_x(2, 1)u_1 + f_y(2, 1)u_2 = 4\frac{\sqrt{3}}{2} - 8\left(\frac{1}{2}\right) = 2\sqrt{3} - 4 \approx -0.5.$$

Notice that this says that the function is decreasing in this direction.

For (b), we must first find the unit vector $\mathbf{u}$ in the indicated direction. Observe that the vector from (2, 1) to (4, 0) corresponds to the position vector $\langle 2, -1 \rangle$ and so, the unit vector in that direction is $\mathbf{u} = \left\langle \frac{2}{\sqrt{5}}, -\frac{1}{\sqrt{5}} \right\rangle$. We then have from Theorem 6.1 that

$$D_{\mathbf{u}} f(2, 1) = f_x(2, 1)u_1 + f_y(2, 1)u_2 = 4\frac{2}{\sqrt{5}} - 8\left(-\frac{1}{\sqrt{5}}\right) = \frac{16}{\sqrt{5}}.$$

So, the function is increasing rapidly in this direction. ■

For convenience, we define the **gradient** of a function to be the vector-valued function whose components are the first-order partial derivatives of f, as specified in Definition 6.2. We denote the gradient of a function f by **grad** f or ∇f (read "del f").

DEFINITION 6.2

The **gradient** of $f(x, y)$ is the vector-valued function

$$\nabla f(x, y) = \left\langle \frac{\partial f}{\partial x}, \frac{\partial f}{\partial y} \right\rangle = \frac{\partial f}{\partial x}\mathbf{i} + \frac{\partial f}{\partial y}\mathbf{j},$$

provided both partial derivatives exist.

Using the gradient, we can write a directional derivative as the dot product of the gradient and the unit vector in the direction of interest, as follows. For any unit vector $\mathbf{u} = \langle u_1, u_2 \rangle$,

$$\begin{aligned} D_{\mathbf{u}} f(x, y) &= f_x(x, y)u_1 + f_y(x, y)u_2 \\ &= \langle f_x(x, y), f_y(x, y) \rangle \cdot \langle u_1, u_2 \rangle \\ &= \nabla f(x, y) \cdot \mathbf{u}. \end{aligned}$$

We state this result in Theorem 6.2.

THEOREM 6.2

If f is a differentiable function of x and y and $\mathbf{u}$ is any unit vector, then

$$D_{\mathbf{u}} f(x, y) = \nabla f(x, y) \cdot \mathbf{u}.$$

Writing directional derivatives as a dot product has many important consequences, one of which we see in example 6.2.

EXAMPLE 6.2 Finding Directional Derivatives

For $f(x, y) = x^2 + y^2$, find $D_{\mathbf{u}} f(1, -1)$ for (a) $\mathbf{u}$ in the direction of $\mathbf{v} = \langle -3, 4 \rangle$ and (b) $\mathbf{u}$ in the direction of $\mathbf{v} = \langle 3, -4 \rangle$.

Solution First, note that

$$\nabla f = \left\langle \frac{\partial f}{\partial x}, \frac{\partial f}{\partial y} \right\rangle = \langle 2x, 2y \rangle.$$

At the point $(1, -1)$, we have $\nabla f(1, -1) = \langle 2, -2 \rangle$. For (a), a unit vector in the same direction as $\mathbf{v}$ is $\mathbf{u} = \left\langle -\frac{3}{5}, \frac{4}{5} \right\rangle$. The directional derivative of f in this direction at the point $(1, -1)$ is then

$$D_{\mathbf{u}} f(1, -1) = \langle 2, -2 \rangle \cdot \left\langle -\frac{3}{5}, \frac{4}{5} \right\rangle = \frac{-6 - 8}{5} = -\frac{14}{5}.$$

For (b), the unit vector is $\mathbf{u} = \left\langle \frac{3}{5}, -\frac{4}{5} \right\rangle$ and so, the directional derivative of f in this direction at $(1, -1)$ is

$$D_{\mathbf{u}} f(1, -1) = \langle 2, -2 \rangle \cdot \left\langle \frac{3}{5}, -\frac{4}{5} \right\rangle = \frac{6 + 8}{5} = \frac{14}{5}.$$ ■

A graphical interpretation of the directional derivatives in example 6.2 is given in Figure 12.34a. Suppose we intersect the surface $z = f(x, y)$ with a plane passing through the point $(1, -1, 2)$, which is perpendicular to the xy-plane and parallel to the vector $\mathbf{u}$ (see Figure 12.34a). Notice that the intersection is a curve in two dimensions. Sketch this curve on a new set of coordinate axes, chosen so that the new origin corresponds to the point $(1, -1, 2)$, the new vertical axis is in the z-direction and the new positive horizontal axis points in the direction of the vector $\mathbf{u}$. In Figure 12.34b, we show the case for $\mathbf{u} = \left\langle -\frac{3}{5}, \frac{4}{5} \right\rangle$ and in Figure 12.34c, we show the case for $\mathbf{u} = \left\langle \frac{3}{5}, -\frac{4}{5} \right\rangle$. In each case, the directional derivative gives the slope of the curve at the origin (in the new coordinate system). Notice that the direction vectors in example 6.2 parts (a) and (b) differ only by sign and the resulting curves in Figures 12.34b and 12.34c are exact mirror images of each other.

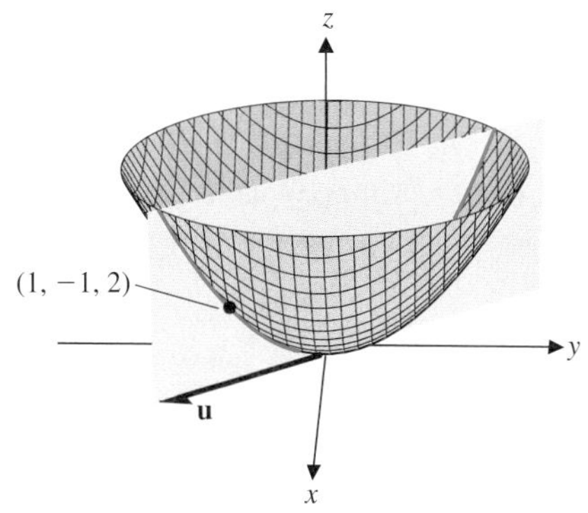

FIGURE 12.34a
Intersection of surface with plane

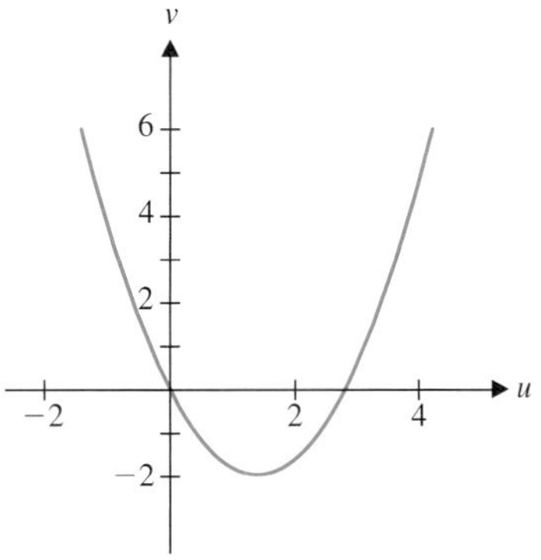

FIGURE 12.34b
$\mathbf{u} = \langle -\frac{3}{5}, \frac{4}{5} \rangle$

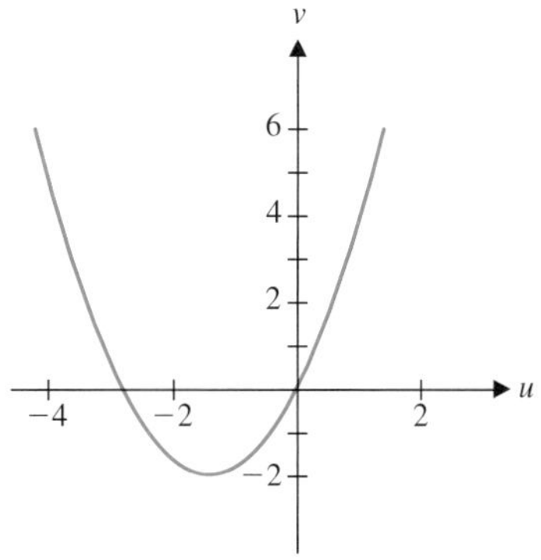

FIGURE 12.34c
$\mathbf{u} = \langle \frac{3}{5}, -\frac{4}{5} \rangle$

We can use a contour plot to estimate the value of a directional derivative, as we illustrate in example 6.3.

EXAMPLE 6.3 Directional Derivatives and Level Curves

Use a contour plot of $z = x^2 + y^2$ to estimate $D_{\mathbf{u}} f(1, -1)$ for $\mathbf{u} = \langle -\frac{3}{5}, \frac{4}{5} \rangle$.

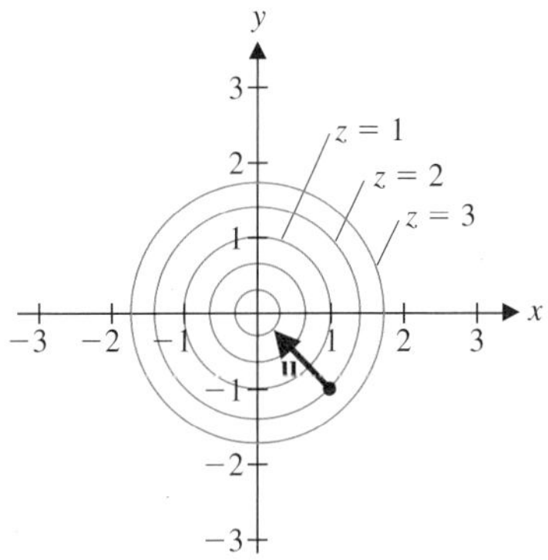

FIGURE 12.35
Contour plot of $z = x^2 + y^2$

Solution A contour plot of $z = x^2 + y^2$ is shown in Figure 12.35 with the direction vector $\mathbf{u} = \langle -\frac{3}{5}, \frac{4}{5} \rangle$ sketched in with its initial point located at the point $(1, -1)$. The level curves shown correspond to $z = 0.2, 0.5, 1, 2$ and 3. From the graph, you can approximate the directional derivative by estimating $\frac{\Delta z}{\Delta u}$, where Δu is the distance traveled along the unit vector $\mathbf{u}$. For the unit vector shown, $\Delta u = 1$. Further, the vector appears to extend from the $z = 2$ level curve to the $z = 0.2$ level curve. In this case, $\Delta z = 0.2 - 2 = -1.8$ and our estimate of the directional derivative is $\frac{\Delta z}{\Delta u} = -1.8$. Compared to the actual directional derivative of $-\frac{14}{5} = -2.8$ (found in example 6.2), this is not very accurate. A better estimate could be obtained with a smaller Δu. For example, to get from the $z = 2$ level curve to the $z = 1$ level curve, it appears that we travel along about 40% of the unit vector. Then $\frac{\Delta z}{\Delta u} \approx \frac{1-2}{0.4} = -2.5$. You could continue this process by drawing more level curves, corresponding to values of z closer to $z = 2$. ■

Keep in mind that a directional derivative gives the rate of change of a function in a given direction. So, it's reasonable to ask in what direction a given function has its maximum or minimum rate of increase. First, recall from Theorem 3.2 in Chapter 10 that for any two vectors $\mathbf{a}$ and $\mathbf{b}$, we have $\mathbf{a} \cdot \mathbf{b} = \|\mathbf{a}\| \|\mathbf{b}\| \cos\theta$, where θ is the angle between the vectors $\mathbf{a}$ and $\mathbf{b}$. Applying this to the form of the directional derivative given in Theorem 6.2, we have

$$\begin{aligned} D_{\mathbf{u}} f(a, b) &= \nabla f(a, b) \cdot \mathbf{u} \\ &= \|\nabla f(a, b)\| \|\mathbf{u}\| \cos\theta = \|\nabla f(a, b)\| \cos\theta, \end{aligned}$$

where θ is the angle between the gradient vector at (a, b) and the direction vector $\mathbf{u}$.

Notice now that $\|\nabla f(a, b)\| \cos\theta$ has its maximum value when $\theta = 0$, so that $\cos\theta = 1$. The directional derivative is then $\|\nabla f(a, b)\|$. Further, observe that the angle $\theta = 0$ when $\nabla f(a, b)$ and $\mathbf{u}$ are in the *same* direction, so that $\mathbf{u} = \frac{\nabla f(a, b)}{\|\nabla f(a, b)\|}$. Similarly, the minimum value of the directional derivative occurs when $\theta = \pi$, so that $\cos\theta = -1$. In this case, $\nabla f(a, b)$ and $\mathbf{u}$ have *opposite* directions, so that $\mathbf{u} = -\dfrac{\nabla f(a, b)}{\|\nabla f(a, b)\|}$. Finally, observe that when $\theta = \frac{\pi}{2}$, $\mathbf{u}$ is perpendicular to $\nabla f(a, b)$ and the directional derivative in this direction is zero. Since the level curves are curves in the xy-plane on which f is constant, notice that a zero directional derivative at a point indicates that $\mathbf{u}$ is tangent to a level curve. We summarize these observations in Theorem 6.3.

THEOREM 6.3

Suppose that f is a differentiable function of x and y at the point (a, b). Then

(i)   the maximum rate of change of f at (a, b) is $\|\nabla f(a, b)\|$, occuring in the direction of the gradient;
(ii) the minimum rate of change of f at (a, b) is $-\|\nabla f(a, b)\|$, occuring in the direction opposite the gradient;
(iii) the rate of change of f at (a, b) is 0 in the directions orthogonal to $\nabla f(a, b)$ and
(iv) the gradient $\nabla f(a, b)$ is orthogonal to the level curve $f(x, y) = c$ at the point (a, b), where $c = f(a, b)$.

In using Theorem 6.3, remember that the directional derivative corresponds to the rate of change of the function $f(x, y)$ in the given direction.

EXAMPLE 6.4   Finding Maximum and Minimum Rates of Change

Find the maximum and minimum rates of change of the function $f(x, y) = x^2 + y^2$ at the point $(1, 3)$.

Solution   We first compute the gradient $\nabla f = \langle 2x, 2y \rangle$ and evaluate it at the point $(1, 3)$: $\nabla f(1, 3) = \langle 2, 6 \rangle$. From Theorem 6.3, the maximum rate of change of f at $(1, 3)$ is $\|\nabla f(1, 3)\| = \|\langle 2, 6 \rangle\| = \sqrt{40}$ and occurs in the direction of

$$\mathbf{u} = \frac{\nabla f(1, 3)}{\|\nabla f(1, 3)\|} = \frac{\langle 2, 6 \rangle}{\sqrt{40}}.$$

Similarly, the minimum rate of change of f at $(1, 3)$ is $-\|\nabla f(1, 3)\| = -\|\langle 2, 6 \rangle\| = -\sqrt{40}$, which occurs in the direction of

$$\mathbf{u} = -\frac{\nabla f(1, 3)}{\|\nabla f(1, 3)\|} = \frac{-\langle 2, 6 \rangle}{\sqrt{40}}. \qquad \blacksquare$$

Notice that the direction of maximum increase in example 6.4 points away from the origin, since the displacement vector from $(0, 0)$ to $(1, 3)$ is parallel to $\mathbf{u} = \langle 2, 6 \rangle / \sqrt{40}$. This should make sense given the familiar shape of the paraboloid. The contour plot of $f(x, y)$ shown in Figure 12.36 indicates that the gradient is perpendicular to the level curves. We expand on this idea in example 6.5.

EXAMPLE 6.5   Finding the Direction of Steepest Ascent

The contour plot of $f(x, y) = 3x - x^3 - 3xy^2$ shown in Figure 12.37 indicates several level curves near a relative maximum at $(1, 0)$. Find the direction of maximum increase from the point $A(0.6, -0.7)$ and sketch in the path of steepest ascent.

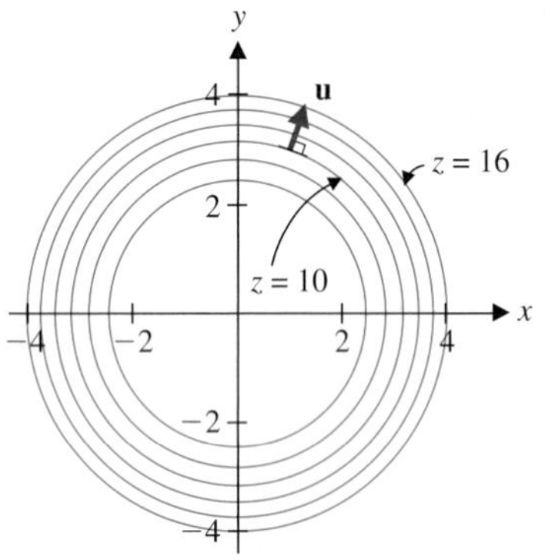

FIGURE 12.36
Contour plot of $z = x^2 + y^2$

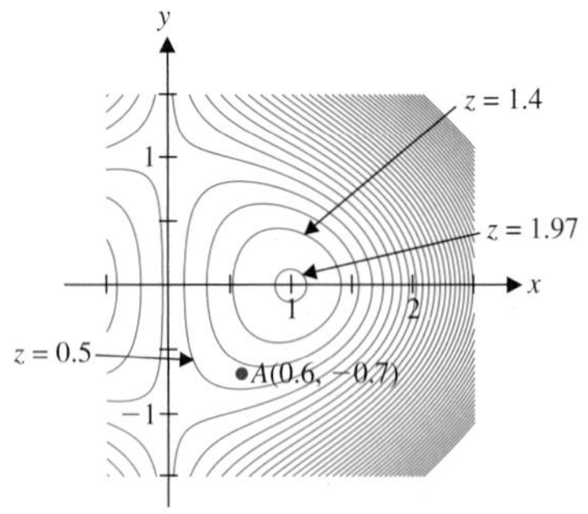

FIGURE 12.37
Contour plot of
$z = 3x - x^3 - 3xy^2$

level curve: z values are the same at a cut, its level everywhere.

Solution From Theorem 6.3, the direction of maximum increase at $(0.6, -0.7)$ is given by the gradient $\nabla f(0.6, -0.7)$. We have $\nabla f = \langle 3 - 3x^2 - 3y^2, -6xy \rangle$ and so, $\nabla f(0.6, -0.7) = \langle 0.45, 2.52 \rangle$. The unit vector in this direction is then $\mathbf{u} = \langle 0.176, 0.984 \rangle$. A vector in this direction (not drawn to scale) at the point $(0.6, -0.7)$ is shown in Figure 12.38a. Notice that this vector *does not* point toward the maximum at $(1, 0)$. (By analogy, on a mountain, the steepest path from a given point will not always point toward the actual peak.) The **path of steepest ascent** is a curve that remains perpendicular to each level curve through which it passes. Notice that at the tip of the vector drawn in Figure 12.38a, the vector is no longer perpendicular to the level curve. Finding an equation for the path of steepest ascent is challenging. In Figure 12.38b, we sketch in a plausible path of steepest ascent. ■

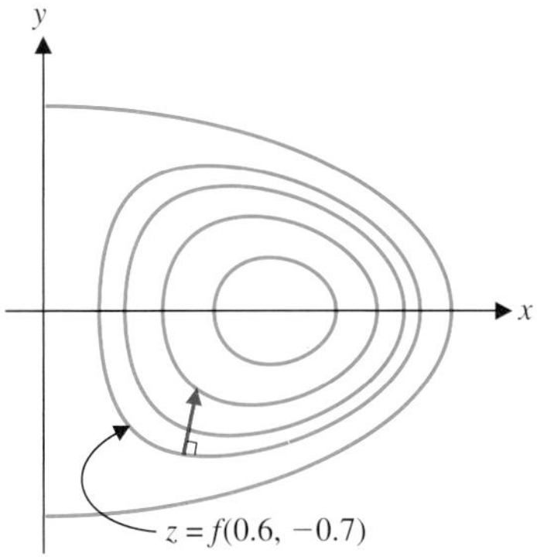

FIGURE 12.38a
Direction of steepest ascent at $(0.6, -0.7)$

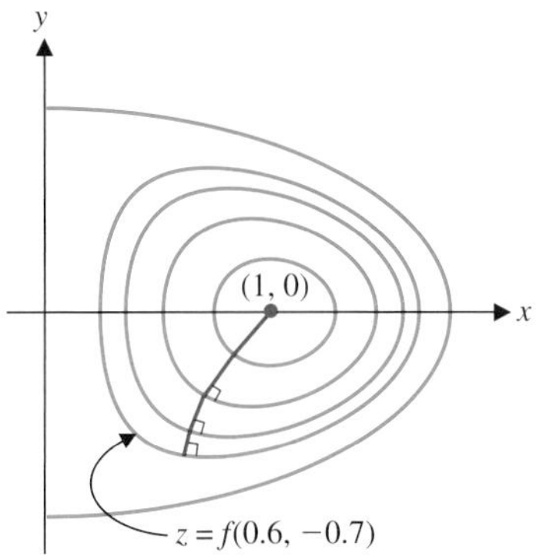

FIGURE 12.38b
Path of steepest ascent

Most of the results of this section extend easily to functions of any number of variables.

DEFINITION 6.3

The **directional derivative** of $f(x, y, z)$ at the point (a, b, c) and in the direction of the unit vector $\mathbf{u} = \langle u_1, u_2, u_3 \rangle$ is given by

$$D_{\mathbf{u}} f(a, b, c) = \lim_{h \to 0} \frac{f(a + hu_1, b + hu_2, c + hu_3) - f(a, b, c)}{h},$$

provided the limit exists.

The **gradient** of $f(x, y, z)$ is the vector-valued function

$$\nabla f(x, y, z) = \left\langle \frac{\partial f}{\partial x}, \frac{\partial f}{\partial y}, \frac{\partial f}{\partial z} \right\rangle = \frac{\partial f}{\partial x}\mathbf{i} + \frac{\partial f}{\partial y}\mathbf{j} + \frac{\partial f}{\partial z}\mathbf{k},$$

provided all the partial derivatives are defined.

As was the case for functions of two variables, the gradient gives us a simple representation of directional derivatives in three dimensions.

THEOREM 6.4

If f is a differentiable function of x, y and z and $\mathbf{u}$ is any unit vector, then

$$D_{\mathbf{u}} f(x, y, z) = \nabla f(x, y, z) \cdot \mathbf{u}. \tag{6.1}$$

As in two dimensions, we have that

$$\begin{aligned} D_{\mathbf{u}} f(x, y, z) &= \nabla f(x, y, z) \cdot \mathbf{u} = \|\nabla f(x, y, z)\| \|\mathbf{u}\| \cos\theta \\ &= \|\nabla f(x, y, z)\| \cos\theta, \end{aligned}$$

where θ is the angle between the vectors $\nabla f(x, y, z)$ and $\mathbf{u}$. For precisely the same reasons as in two dimensions, it follows that the direction of maximum increase at any given point is given by the gradient at that point.

EXAMPLE 6.6 Finding the Direction of Maximum Increase

If the temperature at point (x, y, z) is given by $T(x, y, z) = 85 + (1 - z/100)e^{-(x^2+y^2)}$, find the direction from the point $(2, 0, 99)$ in which the temperature increases most rapidly.

Solution We first compute the gradient

$$\nabla f = \left\langle \frac{\partial f}{\partial x}, \frac{\partial f}{\partial y}, \frac{\partial f}{\partial z} \right\rangle$$

$$= \left\langle -2x\left(1 - \frac{z}{100}\right)e^{-(x^2+y^2)}, -2y\left(1 - \frac{z}{100}\right)e^{-(x^2+y^2)}, -\left(\frac{1}{100}\right)e^{-(x^2+y^2)} \right\rangle$$

and $\nabla f(2, 0, 99) = \langle -\frac{1}{25}e^{-4}, 0, -\frac{1}{100}e^{-4} \rangle$. To find a unit vector in this direction, you can simplify the algebra by canceling the common factor of e^{-4} (think about why this makes sense) and multiplying by 100. A unit vector in the direction of $\langle -4, 0, -1 \rangle$ and also in the direction of $\nabla f(2, 0, 99)$, is then $\dfrac{\langle -4, 0, -1 \rangle}{\sqrt{17}}$. ■

Recall that for any constant k, the equation $f(x, y, z) = k$ defines a level surface of the function $f(x, y, z)$. Now, suppose that $\mathbf{u}$ is any unit vector lying in the tangent plane to the level surface $f(x, y, z) = k$ at a point (a, b, c) on the level surface. Then, it follows that the rate of change of f in the direction of $\mathbf{u}$ at (a, b, c) [given by the directional derivative $D_{\mathbf{u}} f(a, b, c)$] is zero, since f is constant on a level surface. From (6.1), we now have that

$$0 = D_{\mathbf{u}} f(a, b, c) = \nabla f(a, b, c) \cdot \mathbf{u}.$$

This occurs only when the vectors $\nabla f(a, b, c)$ and $\mathbf{u}$ are orthogonal. Since $\mathbf{u}$ was taken to be any vector lying in the tangent plane, we now have that $\nabla f(a, b, c)$ is orthogonal to every vector lying in the tangent plane at the point (a, b, c). Observe that this says that $\nabla f(a, b, c)$ is a normal vector to the tangent plane to the surface $f(x, y, z) = k$ at the point (a, b, c). This proves Theorem 6.5.

THEOREM 6.5

Suppose that $f(x, y, z)$ has continuous partial derivatives at the point (a, b, c) and $\nabla f(a, b, c) \neq \mathbf{0}$. Then, $\nabla f(a, b, c)$ is a normal vector to the tangent plane to the surface $f(x, y, z) = k$, at the point (a, b, c). Further, the equation of the tangent plane is

$$0 = f_x(a, b, c)(x - a) + f_y(a, b, c)(y - b) + f_z(a, b, c)(z - c).$$

We refer to the line through (a, b, c) in the direction of $\nabla f(a, b, c)$ as the **normal line** to the surface at the point (a, b, c). Observe that this has parametric equations

$$x = a + f_x(a, b, c)t, \quad y = b + f_y(a, b, c)t, \quad z = c + f_z(a, b, c)t.$$

In example 6.7, we illustrate the use of the gradient at a point to find the tangent plane and normal line to a surface at that point.

EXAMPLE 6.7 Using a Gradient to Find a Tangent Plane and Normal Line to a Surface

Find equations of the tangent plane and the normal line to $x^3y - y^2 + z^2 = 7$ at the point $(1, 2, 3)$.

Solution If we interpret the surface as a level surface of the function $f(x, y, z) = x^3y - y^2 + z^2$, a normal vector to the tangent plane at the point $(1, 2, 3)$ is given by $\nabla f(1, 2, 3)$. We have $\nabla f = \langle 3x^2y, x^3 - 2y, 2z \rangle$ and

$\nabla f(1, 2, 3) = \langle 6, -3, 6 \rangle$. Given the normal vector $\langle 6, -3, 6 \rangle$ and point $(1, 2, 3)$, an equation of the tangent plane is

$$6(x - 1) - 3(y - 2) + 6(z - 3) = 0.$$

The normal line has parametric equations

$$x = 1 + 6t, \quad y = 2 - 3t, \quad z = 3 + 6t. \blacksquare$$

Recall that in section 12.4, we found that a normal vector to the tangent plane to the surface $z = f(x, y)$ at the point $(a, b, f(a, b))$ is given by $\left\langle \frac{\partial f}{\partial x}(a, b), \frac{\partial f}{\partial y}(a, b), -1 \right\rangle$. Note that this is simply a special case of the gradient formula of Theorem 6.5, as follows. First, observe that we can rewrite the equation $z = f(x, y)$ as $f(x, y) - z = 0$. We can then think of this surface as a level surface of the function $g(x, y, z) = f(x, y) - z$, which at the point $(a, b, f(a, b))$ has normal vector

$$\nabla g(a, b, f(a, b)) = \left\langle \frac{\partial f}{\partial x}(a, b), \frac{\partial f}{\partial y}(a, b), -1 \right\rangle.$$

Just as it is important to constantly think of ordinary derivatives as slopes of tangent lines and as instantaneous rates of change, it is crucial to keep in mind at all times the interpretations of gradients. *Always* think of gradients as vector-valued functions whose values specify the direction of maximum increase of a function and whose values provide normal vectors (to the level curves in two dimensions and to the level surfaces in three dimensions).

EXAMPLE 6.8 Using a Gradient to Find a Tangent Plane to a Surface

Find an equation of the tangent plane to $z = \sin(x + y)$ at the point $(\pi, \pi, 0)$.

Solution We rewrite the equation of the surface as $g(x, y, z) = \sin(x + y) - z = 0$ and compute $\nabla g(x, y, z) = \langle \cos(x + y), \cos(x + y), -1 \rangle$. At the point $(\pi, \pi, 0)$, the normal vector to the surface is given by $\nabla g(\pi, \pi, 0) = \langle 1, 1, -1 \rangle$. An equation of the tangent plane is then

$$(x - \pi) + (y - \pi) - z = 0. \blacksquare$$

BEYOND FORMULAS

The term *gradient* shows up in a large number of applications. Standard usage of the term is very close to our development in this section. By its use in directional derivatives, the gradient gives all the information you need to determine the change in a quantity as you move in some direction from your current position. Because the gradient gives the direction of maximum increase, any process that depends on maximizing or minimizing some quantity may be described with the gradient. When you see *gradient* in an application, think of these properties.

EXERCISES 12.6

HW 18 Pg 992
1, 5, 9, 13, 17, 21, 25, 29, 33

WRITING EXERCISES

1. Pick an area outside your classroom that has a small hill. Starting at the bottom of the hill, describe how to follow the gradient path to the top. In particular, describe how to determine the direction in which the gradient points at a given point on the hill. In general, should you be looking ahead or down at the ground? Should individual blades of grass count? What should you do if you encounter a wall?

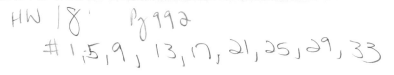

2. Discuss whether the gradient path described in exercise 1 is guaranteed to get you to the top of the hill. Discuss whether the gradient path is the shortest path, the quickest path or the easiest path.

3. Use the sketch in Figure 12.34a to explain why the curves in Figures 12.34b and 12.34c are different.

4. Suppose the function $f(x, y)$ represents the altitude at various points on a ski slope. Explain in physical terms why the direction of maximum increase is 180° opposite the direction of maximum decrease, with the direction of zero change halfway in between. If $f(x, y)$ represents altitude on a rugged mountain instead of a ski slope, explain why the results (which are still true) are harder to visualize.

In exercises 1–4, find the gradient of the given function.

1. $f(x, y) = x^2 + 4xy^2 - y^5$
2. $f(x, y) = x^3 e^{3y} - y^4$
3. $f(x, y) = xe^{xy^2} + \cos y^2$
4. $f(x, y) = e^{3y/x} - x^2y^3$

In exercises 5–10, find the gradient of the given function at the indicated point.

5. $f(x, y) = 2e^{4x/y} - 2x$, $(2, -1)$
6. $f(x, y) = \sin 3xy + y^2$, $(\pi, 1)$
7. $f(x, y, z) = 3x^2y - z\cos x$, $(0, 2, -1)$
8. $f(x, y, z) = z^2e^{2x-y} - 4xz^2$, $(1, 2, 2)$
9. $f(w, x, y, z) = w^2\cos x + 3ye^{xz}$, $(2, \pi, 1, 4)$
10. $f(x_1, x_2, x_3, x_4, x_5) = \sin\left(\dfrac{x_1}{x_2}\right) - 3x_3^2x_4x_5 - 2\sqrt{x_1x_3}$, $(2, 1, 2, -1, 4)$

In exercises 11–26, compute the directional derivative of f at the given point in the direction of the indicated vector.

11. $f(x, y) = x^2y + 4y^2$, $(2, 1)$, $\mathbf{u} = \left\langle \frac{1}{2}, \frac{\sqrt{3}}{2} \right\rangle$
12. $f(x, y) = x^3y - 4y^2$, $(2, -1)$, $\mathbf{u} = \left\langle \frac{1}{\sqrt{2}}, \frac{1}{\sqrt{2}} \right\rangle$
13. $f(x, y) = \sqrt{x^2 + y^2}$, $(3, -4)$, $\mathbf{u}$ in the direction of $\langle 3, -2\rangle$
14. $f(x, y) = e^{4x^2-y}$, $(1, 4)$, $\mathbf{u}$ in the direction of $\langle -2, -1\rangle$
15. $f(x, y) = \cos(2x - y)$, $(\pi, 0)$, $\mathbf{u}$ in the direction from $(\pi, 0)$ to $(2\pi, \pi)$
16. $f(x, y) = x^2\sin 4y$, $(-2, \frac{\pi}{8})$, $\mathbf{u}$ in the direction from $(-2, \frac{\pi}{8})$ to $(0, 0)$
17. $f(x, y) = x^2 - 2xy + y^2$, $(-2, -1)$, $\mathbf{u}$ in the direction from $(-2, -1)$ to $(2, -3)$
18. $f(x, y) = y^2 + 2ye^{4x}$, $(0, -2)$, $\mathbf{u}$ in the direction from $(0, -2)$ to $(-4, 4)$
19. $f(x, y, z) = x^3yz^2 - 4xy$, $(1, -1, 2)$, $\mathbf{u}$ in the direction of $\langle 2, 0, -1\rangle$
20. $f(x, y, z) = \sqrt{x^2 + y^2 + z^2}$, $(1, -4, 8)$, $\mathbf{u}$ in the direction of $\langle 1, 1, -2\rangle$
21. $f(x, y, z) = e^{xy+z}$, $(1, -1, 1)$, $\mathbf{u}$ in the direction of $\langle 4, -2, 3\rangle$
22. $f(x, y, z) = \cos xy + z$, $(0, -2, 4)$, $\mathbf{u}$ in the direction of $\langle 0, 3, -4\rangle$
23. $f(w, x, y, z) = w^2\sqrt{x^2 + 1} + 3ze^{xz}$, $(2, 0, 1, 0)$, $\mathbf{u}$ in the direction of $\langle 1, 3, 4, -2\rangle$
24. $f(w, x, y, z) = \cos(w^2xy) + 3z - \tan 2z$, $(2, -1, 1, 0)$, $\mathbf{u}$ in the direction of $\langle -2, 0, 1, 4\rangle$
25. $f(x_1, x_2, x_3, x_4, x_5) = \dfrac{x_1^2}{x_2} - \sin^{-1}2x_3 + 3\sqrt{x_4x_5}$, $(2, 1, 0, 1, 4)$, $\mathbf{u}$ in the direction of $\langle 1, 0, -2, 4, -2\rangle$
26. $f(x_1, x_2, x_3, x_4, x_5) = 3x_1x_2^3x_3 - e^{4x_3} + \ln\sqrt{x_4x_5}$, $(-1, 2, 0, 4, 1)$, $\mathbf{u}$ in the direction of $\langle 2, -1, 0, 1, -2\rangle$

In exercises 27–36, find the directions of maximum and minimum change of f at the given point, and the values of the maximum and minimum rates of change.

27. $f(x, y) = x^2 - y^3$, $(2, 1)$
28. $f(x, y) = x^2 - y^3$, $(-1, -2)$
29. $f(x, y) = y^2e^{4x}$, $(0, -2)$
30. $f(x, y) = y^2e^{4x}$, $(3, -1)$
31. $f(x, y) = x\cos 3y$, $(2, 0)$
32. $f(x, y) = x\cos 3y$, $(-2, \pi)$
33. $f(x, y) = \sqrt{2x^2 - y}$, $(3, 2)$
34. $f(x, y) = \sqrt{x^2 + y^2}$, $(3, -4)$
35. $f(x, y, z) = 4x^2yz^3$, $(1, 2, 1)$
36. $f(x, y, z) = \sqrt{x^2 + y^2 + z^2}$, $(1, 2, -2)$

37. In exercises 34 and 36, compare the gradient direction to the position vector from the origin to the given point. Explain in terms of the graph of f why this relationship should hold.

38. Suppose that $g(x)$ is a differentiable function and $f(x, y) = g(x^2 + y^2)$. Show that $\nabla f(a, b)$ is parallel to $\langle a, b\rangle$. Explain this in graphical terms.

39. Graph $z = \sin(x + y)$. Compute $\nabla \sin(x + y)$ and explain why the gradient gives you the direction that the sine wave travels. In which direction would the sine wave travel for $z = \sin(2x - y)$?

40. Show that the vector $\langle 100, -100\rangle$ is perpendicular to $\nabla \sin(x + y)$. Explain why the directional derivative of $\sin(x + y)$ in the direction of $\langle 100, -100\rangle$ must be zero.

Sketch a wireframe graph of $z = \sin(x + y)$ from the viewpoint $(100, -100, 0)$. Explain why you only see one trace. Find a viewpoint from which $z = \sin(2x - y)$ only shows one trace.

In exercises 41–44, find equations of the tangent plane and normal line to the surface at the given point.

41. $z = x^2 + y^3$ at $(1, -1, 0)$

42. $z = \sqrt{x^2 + y^2}$ at $(3, -4, 5)$

43. $x^2 + y^2 + z^2 = 6$ at $(-1, 2, 1)$

44. $z^2 = x^2 - y^2$ at $(5, -3, -4)$

In exercises 45 and 46, find all points at which the tangent plane to the surface is parallel to the xy-plane. Discuss the graphical significance of each point.

45. $z = 2x^2 - 4xy + y^4$

46. $z = \sin x \cos y$

In exercises 47–50, sketch in the path of steepest ascent from the indicated point.

47.

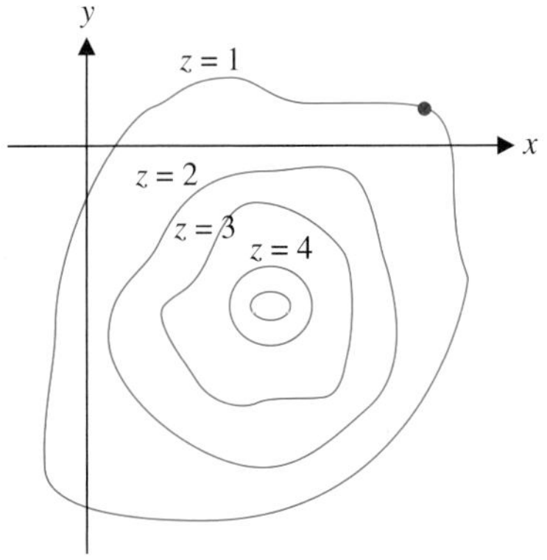

48.

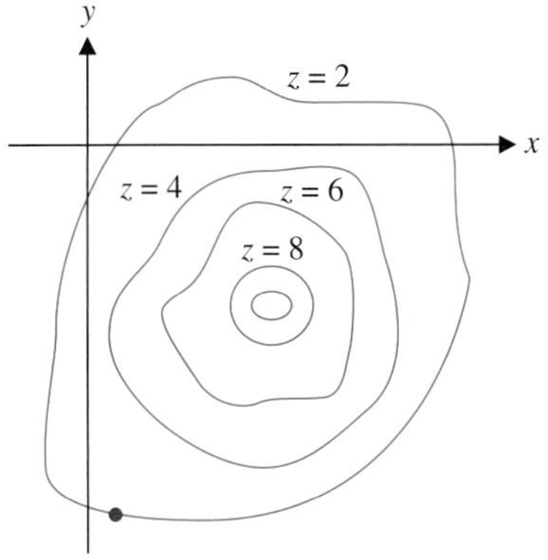

49.

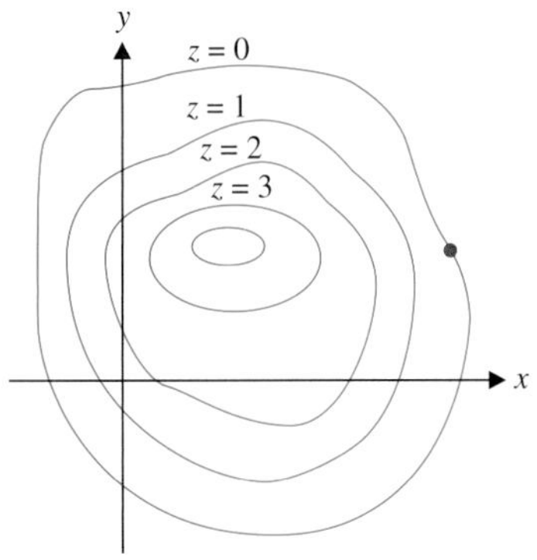

50.

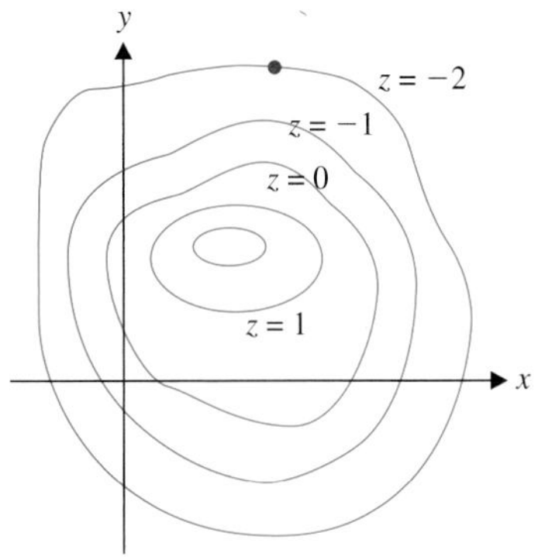

In exercises 51 and 52, use the contour plot to estimate $\nabla f(0, 0)$.

51.

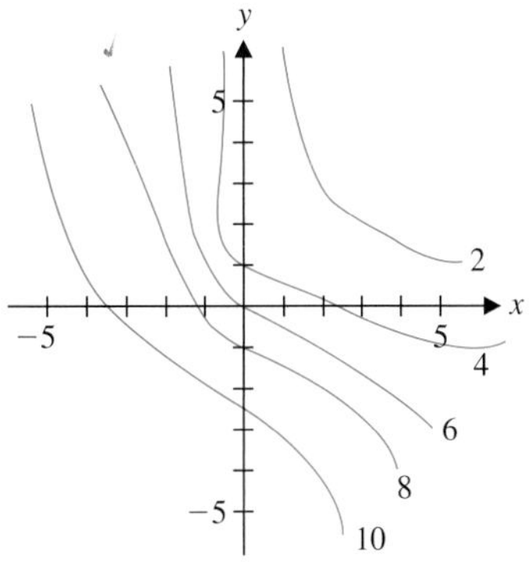

52.

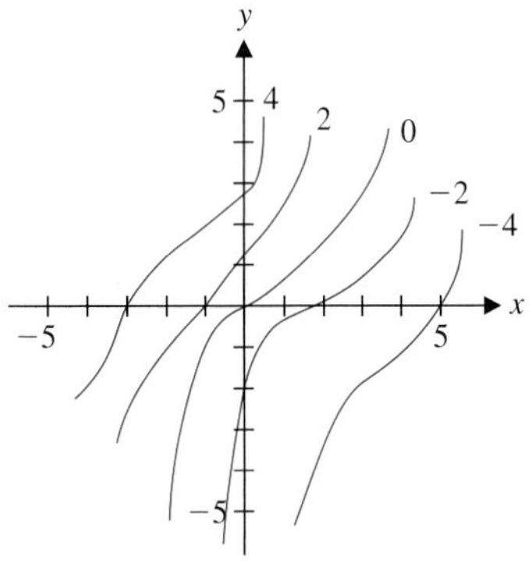

In exercises 53 and 54, use the table to estimate $\nabla f(0, 0)$.

53.

$y \backslash x$	−0.2	−0.1	0	0.1	0.2
−0.4	2.1	2.5	2.8	3.1	3.4
−0.2	1.9	2.2	2.4	2.6	2.9
0	1.6	1.8	2.0	2.2	2.5
0.2	1.3	1.4	1.6	1.8	2.1
0.4	1.1	1.2	1.1	1.4	1.7

54.

$y \backslash x$	−0.4	−0.2	0	0.2	0.4
−0.6	2.4	2.1	1.8	1.3	1.0
−0.3	2.6	2.2	1.9	1.5	1.2
0	2.7	2.4	2.0	1.6	1.3
0.3	2.9	2.5	2.1	1.7	1.5
0.6	3.1	2.7	2.3	1.9	1.7

55. At a certain point on a mountain, a surveyor sights due east and measures a 10° drop-off, then sights due north and measures a 6° rise. Find the direction of steepest ascent and compute the degree rise in that direction.

56. At a certain point on a mountain, a surveyor sights due west and measures a 4° rise, then sights due north and measures a 3° rise. Find the direction of steepest ascent and compute the degree rise in that direction.

57. Suppose that the elevation on a hill is given by $f(x, y) = 200 - y^2 - 4x^2$. From the site at (1, 2), in which direction will the rain run off?

58. For the hill of exercise 57, if a level road is to be built at elevation 100, find the shape of the road.

59. Suppose the temperature at each point (x, y, z) on a surface S is given by the function $T(x, y, z)$. Physics tells us heat flows from hot to cold and that the greater the temperature difference, the greater the flow. Explain why these facts would lead you to conclude that the maximum heat flow occurs in the direction $-\nabla T$ and, by Fourier's Law of Heat Flow, that the maximum heat flow is proportional to $\|\nabla T\|$.

60. If the temperature at the point (x, y, z) is given by $T(x, y, z) = 80 + 5e^{-z}(x^{-2} + y^{-1})$, find the direction from the point $(1, 4, 8)$ in which the temperature decreases most rapidly.

61. Suppose that a spacecraft is slightly off course. The function f is an error function that measures how far off course the spacecraft is as a function of its position $\langle x, y, z\rangle$ and velocity $\langle v_x, v_y, v_z\rangle$. That is, f is a function of six variables. Writing $f(x, y, z, v_x, v_y, v_z)$, if the gradient of f at a particular time is $\nabla f = \langle 0, 2, 0, -3, 0, 0\rangle$, identify the change in position and change in velocity needed to correct the error.

62. Suppose that a person has money invested in five stocks. Let x_i be the number of shares held in stock i and let $f(x_1, x_2, x_3, x_4, x_5)$ equal the total value of the stocks. If $\nabla f = \langle 2, -1, 6, 0, -2\rangle$, indicate which stocks should be sold and which should be bought, and indicate the relative amounts of each sale or buy.

63. Sharks find their prey through a keen sense of smell and an ability to detect small electrical impluses. If $f(x, y, z)$ indicates the electrical charge in the water at position (x, y, z) and a shark senses that $\nabla f = \langle 12, -20, 5\rangle$, in which direction should the shark swim to find its prey?

64. The speed S of a tennis serve depends on the speed v of the tennis racket, the tension t of the strings of the racket, the liveliness e of the ball and the angle θ at which the racket is held. Writing $S(v, t, e, \theta)$, if $\nabla S = \langle 12, -2, 3, -3\rangle$, discuss the relative contributions of each factor. That is, for each variable, if the variable is increased, does the ball speed increase or decrease?

65. Label each as true or false and explain why. (a) $\nabla(f + g) = \nabla f + \nabla g$, (b) $\nabla(fg) = (\nabla f)g + f(\nabla g)$

66. Show that for $f(x, y) = \begin{cases} \frac{x^2 y}{x^6 + 2y^2}, & \text{if } (x, y) \neq (0, 0) \\ 0, & \text{if } (x, y) = (0, 0) \end{cases}$ and any $\mathbf{u}$, the directional derivative $D_{\mathbf{u}} f(0, 0)$ exists, but f is not continuous at (0, 0).

67. In example 4.6 of this chapter, we looked at a manufacturing process. Suppose that a gauge of 4 mm results from a gap of 4 mm, a speed of 10 m/s and a temperature of 900°. Further, suppose that an increase in gap of 0.05 mm increases the gauge by 0.04 mm, an increase in speed of 0.2 m/s increases the gauge by 0.06 mm and an increase in temperature of 10° decreases the gauge by 0.04 mm. Thinking of gauge as a function of gap, speed and temperature, find the direction of maximum increase of gauge.

68. The **Laplacian** of a function $f(x, y)$ is defined by $\nabla^2 f(x, y) = f_{xx}(x, y) + f_{yy}(x, y)$. Compute $\nabla^2 f(x, y)$ for $f(x, y) = x^3 - 2xy + y^2$.

EXPLORATORY EXERCISES

1. The horizontal range of a baseball that has been hit depends on its launch angle and the rate of backspin on the ball. The accompanying figure (reprinted from *Keep Your Eye on the Ball* by Watts and Bahill) shows level curves for the range as a function of angle and spin rate for an initial speed of 110 mph. Watts and Bahill suggest using the dashed line to find the best launch angle for a given spin rate. For example, start at $\omega = 2000$, move horizontally to the dashed line and then vertically down to $\theta = 30$. For a spin rate of 2000 rpm, the greatest range is achieved with a launch angle of 30°. To understand why, note that the dashed line intersects level curves at points where the level curves have horizontal tangents. Start at a point where the dashed line intersects a level curve and explain why you can conclude from the graph that changing the angle would decrease the range. Therefore, the dashed line indicates optimal angles. As ω increases, does the optimal angle increase or decrease? Explain in physical terms why this makes sense. Explain why you know that the dashed line does not follow a gradient path and explain what a gradient path would represent.

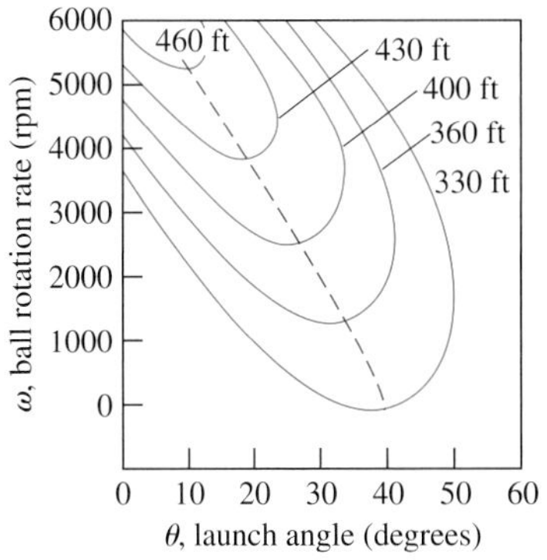

2. With the computer revolution of the 1990s came a new need to generate realistic-looking graphics. In this exercise, we look at one of the basic principles of three-dimensional graphics. We have often used wireframe graphs such as Figure A to visualize surfaces in three dimensions. Certainly, the graphic in Figure A is crude, but even this sketch is quite informative to us, as we can clearly see a local maximum. By having the computer plot more points, as in Figure B, we can smooth out some of the rough edges. Still, there is something missing, isn't there?

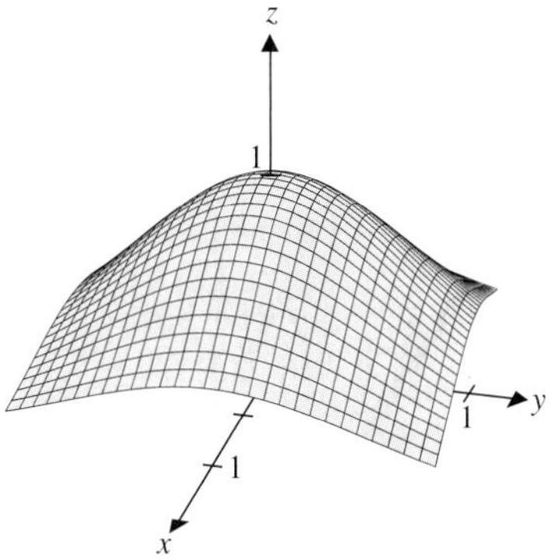

FIGURE A

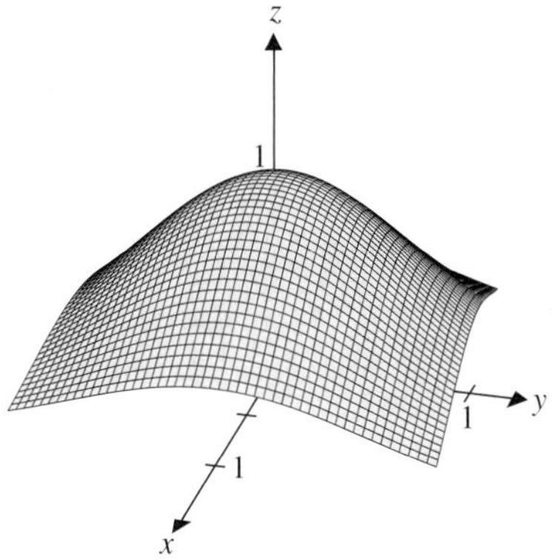

FIGURE B

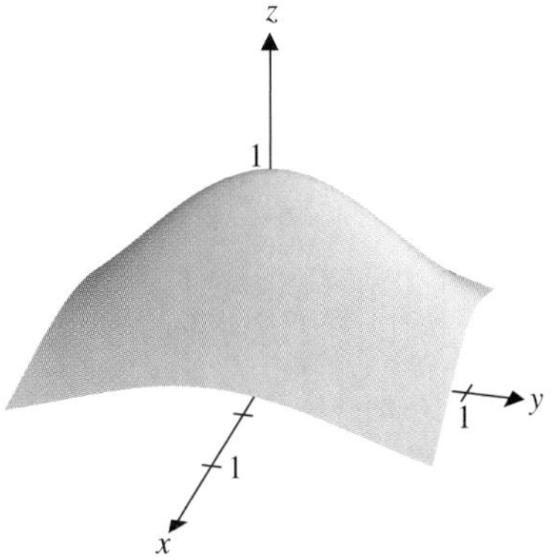

FIGURE C

Almost everything we see in real life is shaded by a light source from above. This shading gives us very important clues about the three-dimensional structure of the surface. In Figure C, we have simply added some shading to Figure B. There is more work to be done in smoothing out Figure C, but for now we want to understand how the shading works. In particular, we'll discuss a basic type of shading called **Lambert shading.** The idea is to shade a portion of the picture based on the size of the angle between the normal to the surface and the line to the light source. The larger the angle, the darker the portion of the picture should be. Explain why this works. For the surface $z = e^{-x^2-y^2}$ (shown in Figures A–C with $-1 \le x \le 1$ and $-1 \le y \le 1$) and a light source at $(0, 0, 100)$, compute the angle at the points $(0, 0, 1)$, $(0, 1, e^{-1})$ and $(1, 0, e^{-1})$. Show that all points with $x^2 + y^2 = 1$ have the same angle and explain why, in terms of the symmetry of the surface. If the position of the light source is changed, will these points remain equally well lit? Based on Figure C, try to determine where the light source is located.

12.7 EXTREMA OF FUNCTIONS OF SEVERAL VARIABLES

You have seen optimization problems reappear in a number of places, since we first introduced the idea in section 3.7. In this section, we introduce the mathematical basis for optimizing functions of several variables.

Carefully examine the surface $z = xe^{-x^2/2-y^3/3+y}$, shown in Figure 12.39a for $-2 \le x \le 4$ and $-1 \le y \le 4$. From the graph, notice that you can identify both a peak and a valley. We can zoom in to get a better view of the peak. (See Figure 12.39b for $0.9 \le x \le 1.1$ and $0.9 \le y \le 1.1$.) Referring to Figure 12.39a, we can zoom in to get a better view of the valley. (See Figure 12.39c for $-1.1 \le x \le -0.9$ and $0.9 \le y \le 1.1$.) Such points are referred to as local extrema, which we define as follows.

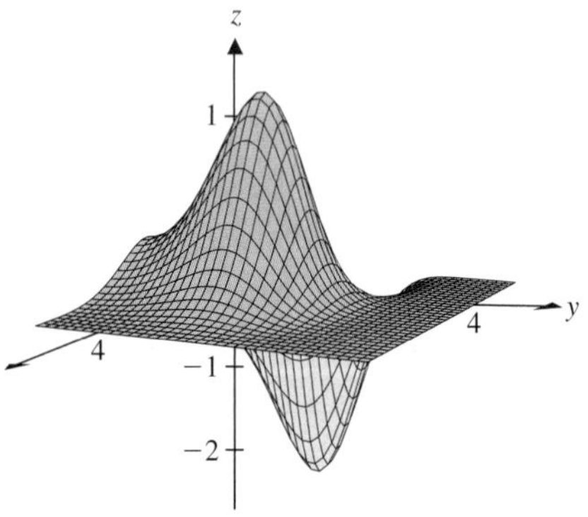

FIGURE 12.39a
$z = xe^{-x^2/2-y^3/3+y}$

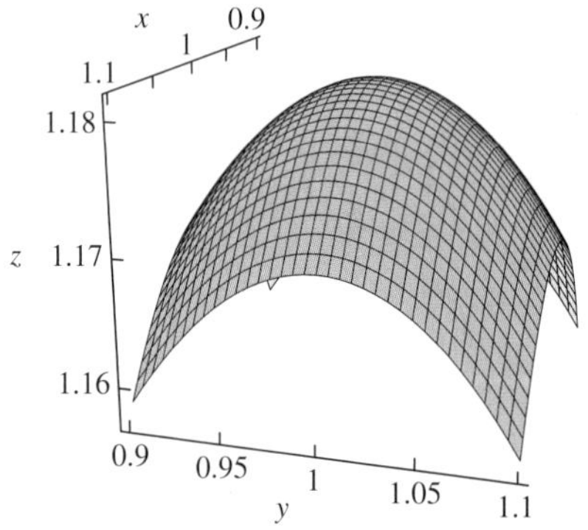

FIGURE 12.39b
Local maximum

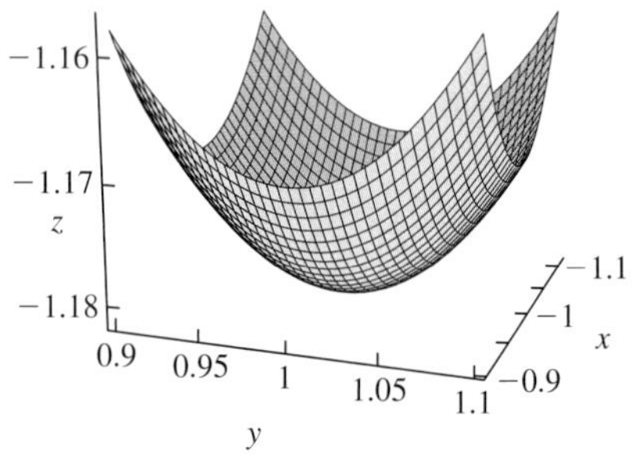

FIGURE 12.39c
Local minimum

> **DEFINITION 7.1**
>
> We call $f(a, b)$ a **local maximum** of f if there is an open disk R centered at (a, b), for which $f(a, b) \ge f(x, y)$ for all $(x, y) \in R$. Similarly, $f(a, b)$ is called a **local minimum** of f if there is an open disk R centered at (a, b), for which $f(a, b) \le f(x, y)$ for all $(x, y) \in R$. In either case, $f(a, b)$ is called a **local extremum** of f.

Note the similarity between Definition 7.1 and the definition of local extrema given in section 3.3. The idea here is the same as it was in Chapter 3. That is, if $f(a, b) \ge f(x, y)$ for all (x, y) "near" (a, b), we call $f(a, b)$ a local maximum.

Look carefully at Figures 12.39b and 12.39c; it appears that at both local extrema, the tangent plane is horizontal. Think about this for a moment and convince yourself that if the tangent plane was tilted, the function would be increasing in one direction and decreasing in another direction, which can't happen at a local extremum (maximum or minimum). Much as with functions of one variable, it turns out that local extrema can occur only where the first (partial) derivatives are zero or do not exist.

DEFINITION 7.2

The point (a, b) is a **critical point** of the function $f(x, y)$ if (a, b) is in the domain of f and either $\frac{\partial f}{\partial x}(a, b) = \frac{\partial f}{\partial y}(a, b) = 0$ or one or both of $\frac{\partial f}{\partial x}$ and $\frac{\partial f}{\partial y}$ do not exist at (a, b).

Recall that for a function $f(x)$ of a single variable, if f has a local extremum at $x = a$, then a must be a critical number of f [i.e., $f'(a) = 0$ or $f'(a)$ is undefined]. Similarly, if $f(a, b)$ is a local extremum (local maximum or local minimum), then (a, b) must be a critical point of f. Be careful, though; although local extrema can occur only at critical points, every critical point need *not* correspond to a local extremum. For this reason, we refer to critical points as *candidates* for local extrema.

THEOREM 7.1

If $f(x, y)$ has a local extremum at (a, b), then (a, b) must be a critical point of f.

PROOF

Suppose that $f(x, y)$ has a local extremum at (a, b). Holding y constant at $y = b$, notice that the function $g(x) = f(x, b)$ has a local extremum at $x = a$. By Fermat's Theorem (Theorem 3.2 in Chapter 3), either $g'(a) = 0$ or $g'(a)$ doesn't exist. Note that $g'(a) = \frac{\partial f}{\partial x}(a, b)$. Likewise, holding x constant at $x = a$, observe that the function $h(y) = f(a, y)$ has a local extremum at $y = b$. It follows that $h'(b) = 0$ or $h'(b)$ doesn't exist. Note that $h'(b) = \frac{\partial f}{\partial y}(a, b)$. Combining these two observations, we have that each of $\frac{\partial f}{\partial x}(a, b)$ and $\frac{\partial f}{\partial y}(a, b)$ equals 0 or doesn't exist. We can then conclude that (a, b) must be a critical point of f. ■

When looking for local extrema, you must first find all critical points, since local extrema can occur only at critical points. Then, analyze each critical point to determine whether it is the location of a local maximum, local minimum or neither. We now return to the function $f(x, y) = xe^{-x^2/2-y^3/3+y}$ discussed in the introduction to the section.

EXAMPLE 7.1 Finding Local Extrema Graphically

Find all critical points of $f(x, y) = xe^{-x^2/2-y^3/3+y}$ and analyze each critical point graphically.

Solution First, we compute the first partial derivatives:

$$\frac{\partial f}{\partial x} = e^{-x^2/2-y^3/3+y} + x(-x)e^{-x^2/2-y^3/3+y} = (1 - x^2)e^{-x^2/2-y^3/3+y}$$

and

$$\frac{\partial f}{\partial y} = x(-y^2 + 1)e^{-x^2/2-y^3/3+y}.$$

Since exponentials are always positive, we have $\frac{\partial f}{\partial x} = 0$ if and only if $1 - x^2 = 0$, that is, when $x = \pm 1$. We have $\frac{\partial f}{\partial y} = 0$ if and only if $x(-y^2 + 1) = 0$, that is, when $x = 0$ or $y = \pm 1$. Notice that both partial derivatives exist for all (x, y) and so, the only critical points are solutions of $\frac{\partial f}{\partial x} = \frac{\partial f}{\partial y} = 0$. For this to occur, we need $x = \pm 1$ and either $x = 0$ or $y = \pm 1$. However, if $x = 0$, then $\frac{\partial f}{\partial x} \neq 0$, so there are no critical points with $x = 0$. This leaves all combinations of $x = \pm 1$ and $y = \pm 1$ as critical points: $(1, 1)$, $(-1, 1)$, $(1, -1)$ and $(-1, -1)$. Keep in mind that the critical points are only candidates for local extrema; we must look further to determine whether they correspond to extrema. We zoom in on each critical point in turn, to graphically identify any local extrema. We have already seen (see Figures 12.39b and 12.39c) that $f(x, y)$ has a local maximum at $(1, 1)$ and a local minimum at $(-1, 1)$. Figures 12.40a and 12.40b show $z = f(x, y)$ zoomed in on $(1, -1)$ and $(-1, -1)$, respectively. In Figure 12.40a, notice that in the plane $x = 1$ (extending left to right), the point at $(1, -1)$ is a local minimum. However, in the plane $y = -1$ (extending back to front), the point at $(1, -1)$ is a local maximum. This point is therefore not a local extremum. We refer to such a point as a *saddle point*. (It looks like a saddle, doesn't it?) Similarly, in Figure 12.40b, notice that in the plane $x = -1$ (extending left to right), the point at $(-1, -1)$ is a local maximum. However, in the plane $y = -1$ (extending back to front), the point at $(-1, -1)$ is a local minimum. Again, at $(-1, -1)$ we have a saddle point.

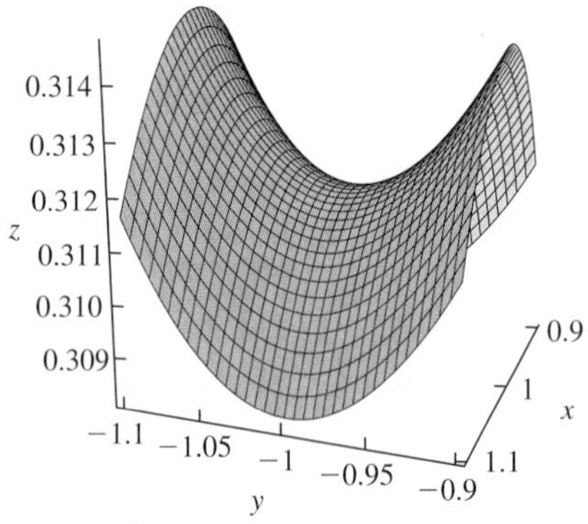

FIGURE 12.40a
Saddle point at $(1, -1)$

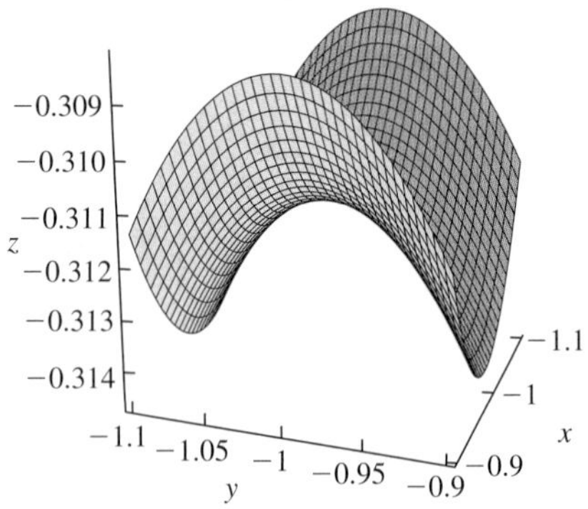

FIGURE 12.40b
Saddle point at $(-1, -1)$

We now pause to carefully define saddle points.

DEFINITION 7.3

The point $P(a, b, f(a, b))$ is a **saddle point** of $z = f(x, y)$ if (a, b) is a critical point of f and if every open disk centered at (a, b) contains points (x, y) in the domain of f for which $f(x, y) < f(a, b)$ and points (x, y) in the domain of f for which $f(x, y) > f(a, b)$.

To further explore example 7.1 graphically, we show a contour plot of $f(x, y) = xe^{-x^2/2-y^3/3+y}$ in Figure 12.41. Notice that near the local maximum at $(1, 1)$

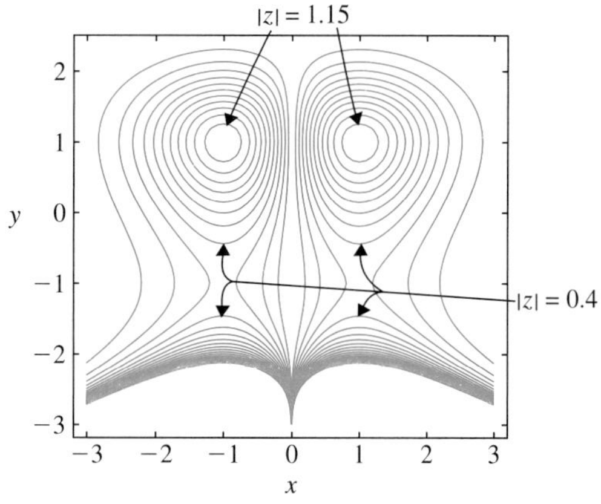

FIGURE 12.41
Contour plot of $f(x, y) = xe^{-x^2/2-y^3/3+y}$

and the local minimum at $(-1, 1)$ the level curves resemble concentric circles. This corresponds to the paraboloid-like shape of the surface near these points (see Figures 12.39b and 12.39c). Concentric ovals are characteristic of local extrema. Notice that, without the level curves labeled, there is no way to tell from the contour plot which is the maximum and which is the minimum. Saddle points are typically characterized by the hyperbolic-looking curves seen near $(-1, -1)$ and $(1, -1)$.

Of course, we can't rely on interpreting three-dimensional graphs for finding local extrema. Recall that for functions of a single variable, we developed two tests (the first derivative test and the second derivative test) for determining when a given critical number corresponds to a local maximum or a local minimum or neither. The following result, which we prove at the end of the section, is surprisingly simple and is a generalization of the second derivative test for functions of a single variable.

THEOREM 7.2 (Second Derivatives Test)

Suppose that $f(x, y)$ has continuous second-order partial derivatives in some open disk containing the point (a, b) and that $f_x(a, b) = f_y(a, b) = 0$. Define the **discriminant** D for the point (a, b) by

$$D(a, b) = f_{xx}(a, b)f_{yy}(a, b) - [f_{xy}(a, b)]^2.$$

(i) If $D(a, b) > 0$ and $f_{xx}(a, b) > 0$, then f has a local minimum at (a, b).
(ii) If $D(a, b) > 0$ and $f_{xx}(a, b) < 0$, then f has a local maximum at (a, b).
(iii) If $D(a, b) < 0$, then f has a saddle point at (a, b).
(iv) If $D(a, b) = 0$, then no conclusion can be drawn.

It's important to make some sense of this result (in other words, to understand it and not just memorize it). Note that to have $D(a, b) > 0$, we must have *both* $f_{xx}(a, b) > 0$ and $f_{yy}(a, b) > 0$ *or* both $f_{xx}(a, b) < 0$ and $f_{yy}(a, b) < 0$. In the first case, notice that the surface $z = f(x, y)$ will be concave up in the plane $y = b$ and concave up in the plane $x = a$. In this case, the surface will look like an upward-opening paraboloid near the point (a, b). Consequently, f has a local minimum at (a, b). In the second case, both $f_{xx}(a, b) < 0$ and $f_{yy}(a, b) < 0$. This says that the surface $z = f(x, y)$ will be concave down in the plane

$y = b$ and concave down in the plane $x = a$. So, in this case, the surface looks like a downward-opening paraboloid near the point (a, b) and hence, f has a local maximum at (a, b). Observe that one way to get $D(a, b) < 0$ is for $f_{xx}(a, b)$ and $f_{yy}(a, b)$ to have opposite signs (one positive and one negative). To have opposite concavities in the planes $x = a$ and $y = b$ means that there is a saddle point at (a, b), as in Figures 12.40a and 12.40b. We note that having $f_{xx}(a, b) > 0$ and $f_{yy}(a, b) > 0$, without having $D(a, b) > 0$ does not say that $f(a, b)$ is a local minimum. We explore this in the exercises.

EXAMPLE 7.2 Using the Discriminant to Find Local Extrema

Locate and classify all critical points for $f(x, y) = 2x^2 - y^3 - 2xy$.

Solution We first compute the first partial derivatives: $f_x = 4x - 2y$ and $f_y = -3y^2 - 2x$. Since both f_x and f_y are defined for all (x, y), the critical points are solutions of the two equations:

$$f_x = 4x - 2y = 0$$

and

$$f_y = -3y^2 - 2x = 0.$$

Point	(0, 0)	$\left(-\frac{1}{6}, -\frac{1}{3}\right)$
$f_{xx} = 4$	4	4
$f_{yy} = -6y$	0	2
$f_{xy} = -2$	−2	−2
$D(a, b)$	−4	4

Solving the first equation for y, we get $y = 2x$. Substituting this into the second equation, we have

$$0 = -3(4x^2) - 2x = -12x^2 - 2x$$
$$= -2x(6x + 1),$$

so that $x = 0$ or $x = -\frac{1}{6}$. The corresponding y-values are $y = 0$ and $y = -\frac{1}{3}$. The only two critical points are then $(0, 0)$ and $\left(-\frac{1}{6}, -\frac{1}{3}\right)$. To classify these points, we first compute the second partial derivatives: $f_{xx} = 4$, $f_{yy} = -6y$ and $f_{xy} = -2$, and then test the discriminant. We have

$$D(0, 0) = (4)(0) - (-2)^2 = -4 < 0$$

and

$$D\left(-\tfrac{1}{6}, -\tfrac{1}{3}\right) = (4)(2) - (-2)^2 = 4 > 0.$$

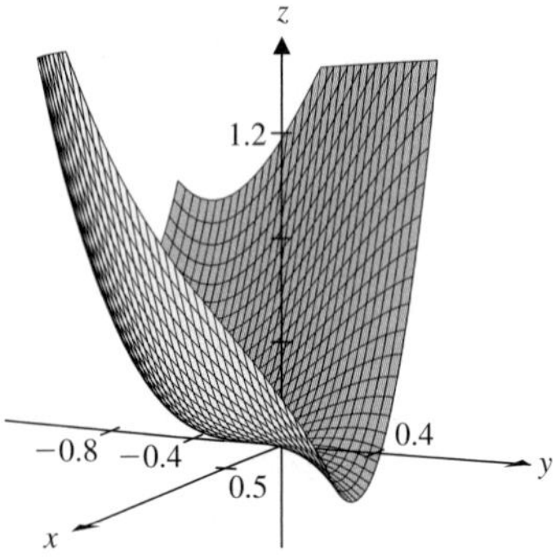

FIGURE 12.42
$z = 2x^2 - y^3 - 2xy$

From Theorem 7.2, we conclude that there is a saddle point of f at $(0, 0)$, since $D(0, 0) < 0$. Further, there is a local minimum at $\left(-\frac{1}{6}, -\frac{1}{3}\right)$ since $D\left(-\frac{1}{6}, -\frac{1}{3}\right) > 0$ and $f_{xx}\left(-\frac{1}{6}, -\frac{1}{3}\right) > 0$. The surface is shown in Figure 12.42. ■

As we see in example 7.3, the Second Derivatives Test does not always help us to classify a critical point.

EXAMPLE 7.3 Classifying Critical Points

Locate and classify all critical points for $f(x, y) = x^3 - 2y^2 - 2y^4 + 3x^2y$.

Solution Here, we have $f_x = 3x^2 + 6xy$ and $f_y = -4y - 8y^3 + 3x^2$. Since both f_x and f_y exist for all (x, y), the critical points are solutions of the two equations:

$$f_x = 3x^2 + 6xy = 0$$

and

$$f_y = -4y - 8y^3 + 3x^2 = 0.$$

From the first equation, we have

$$0 = 3x^2 + 6xy = 3x(x + 2y),$$

so that at a critical point, $x = 0$ or $x = -2y$. Substituting $x = 0$ into the second equation, we have

$$0 = -4y - 8y^3 = -4y(1 + 2y^2).$$

The only (real) solution of this equation is $y = 0$. This says that for $x = 0$, we have only one critical point: (0, 0). Substituting $x = -2y$ into the second equation, we have

$$0 = -4y - 8y^3 + 3(4y^2) = -4y(1 + 2y^2 - 3y) = -4y(2y - 1)(y - 1).$$

The solutions of this equation are $y = 0$, $y = \frac{1}{2}$ and $y = 1$, with corresponding critical points (0, 0), $\left(-1, \frac{1}{2}\right)$ and $(-2, 1)$. To classify the critical points, we compute the second partial derivatives, $f_{xx} = 6x + 6y$, $f_{yy} = -4 - 24y^2$ and $f_{xy} = 6x$, and evaluate the discriminant at each critical point. We have

$$D(0, 0) = (0)(-4) - (0)^2 = 0,$$

$$D\left(-1, \frac{1}{2}\right) = (-3)(-10) - (-6)^2 = -6 < 0$$

and

$$D(-2, 1) = (-6)(-28) - (-12)^2 = 24 > 0.$$

From Theorem 7.2, we conclude that f has a saddle point at $\left(-1, \frac{1}{2}\right)$, since $D\left(-1, \frac{1}{2}\right) < 0$. Further, f has a local maximum at $(-2, 1)$ since $D(-2, 1) > 0$ and $f_{xx}(-2, 1) < 0$. Unfortunately, Theorem 7.2 gives us no information about the critical point (0, 0), since $D(0, 0) = 0$. However, notice that in the plane $y = 0$ we have $f(x, y) = x^3$. In two dimensions, the curve $z = x^3$ has an inflection point at $x = 0$. This shows that there is no local extremum at (0, 0). The surface near (0, 0) is shown in Figure 12.43a. The surface near $(-2, 1)$ and $\left(-1, \frac{1}{2}\right)$ is shown in Figures 12.43b and 12.43c, respectively. Since the graphs are not especially clear, it's good that we have done the analysis!

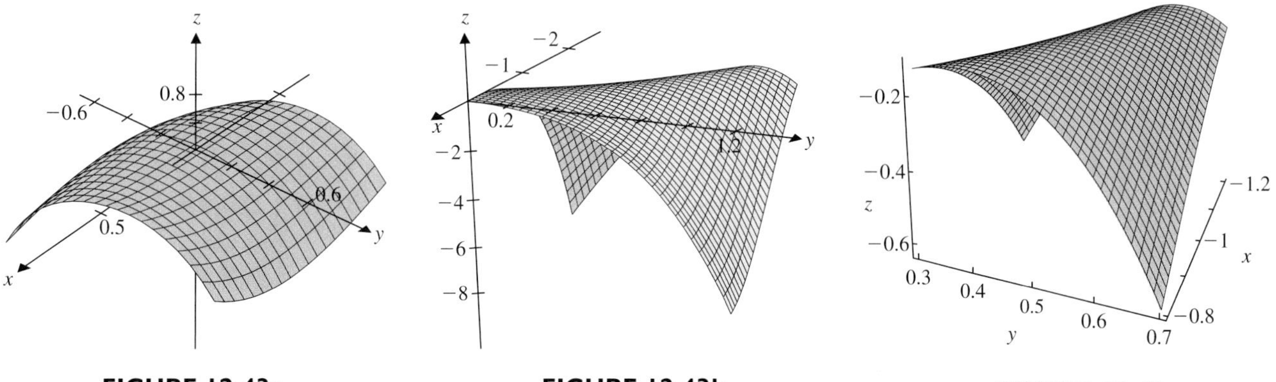

FIGURE 12.43a
$z = f(x, y)$ near (0, 0)

FIGURE 12.43b
$z = f(x, y)$ near $(-2, 1)$

FIGURE 12.43c
$z = f(x, y)$ near $\left(-1, \frac{1}{2}\right)$

One commonly used application of the theory of local extrema is the statistical technique of **least squares.** This technique (or, more accurately, this criterion) is essential to many commonly used curve-fitting and data analysis procedures. The following example illustrates the use of least squares in **linear regression.**

EXAMPLE 7.4 Linear Regression

Population data from the U.S. census are shown in the following table.

Year	*Population*
1960	179,323,175
1970	203,302,031
1980	226,542,203
1990	248,709,873

x	y
0	179
1	203
2	227
3	249

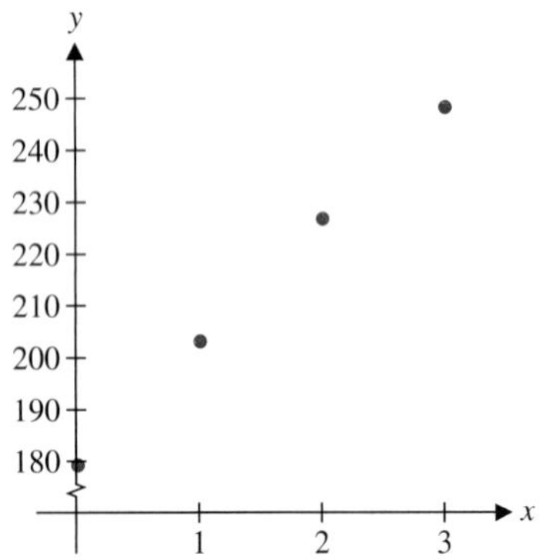

FIGURE 12.44
U.S. population since 1960 (in millions)

Find the straight line that "best" fits the data.

Solution To make the data more manageable, we first transform the raw data into variables x (the number of decades since 1960) and y (population, in millions of people, rounded off to the nearest whole number). We display the transformed data in the table in the margin. A plot of x and y is shown in Figure 12.44. From the plot, it would appear that the population data are nearly linear. Our goal is to find the line that "best" fits the data. (This is called the **regression line.**) The criterion for "best" fit is the least-squares criterion, as defined below. We take the equation of the line to be $y = ax + b$, with constants a and b to be determined. For a value of x represented in the data, the error (or **residual**) is given by the difference between the actual y-value and the predicted value $ax + b$. The least-squares criterion is to choose a and b to minimize the sum of the squares of all the residuals. (In a sense, this minimizes the total error.) For the given data, the residuals are shown in the following table.

x	$ax + b$	y	*Residual*
0	b	179	$b - 179$
1	$a + b$	203	$a + b - 203$
2	$2a + b$	227	$2a + b - 227$
3	$3a + b$	249	$3a + b - 249$

The sum of the squares of the residuals is then given by the function

$$f(a, b) = (b - 179)^2 + (a + b - 203)^2 + (2a + b - 227)^2 + (3a + b - 249)^2.$$

From Theorem 7.1, we must have $\frac{\partial f}{\partial a} = \frac{\partial f}{\partial b} = 0$ at the minimum point, since f_a and f_b are defined everywhere. We have

$$0 = \frac{\partial f}{\partial a} = 2(a + b - 203) + 4(2a + b - 227) + 6(3a + b - 249)$$

and

$$0 = \frac{\partial f}{\partial b} = 2(b - 179) + 2(a + b - 203) + 2(2a + b - 227) + 2(3a + b - 249).$$

After multiplying out all terms, we have

$$28a + 12b = 2808$$

and

$$12a + 8b = 1716.$$

The second equation reduces to $3a + 2b = 429$, so that $a = 143 - \frac{2}{3}b$. Substituting this into the first equation, we have

$$28\left(143 - \frac{2}{3}b\right) + 12b = 2808,$$

or

$$4004 - 2808 = \left(\frac{56}{3} - 12\right)b.$$

This gives us $b = \frac{897}{5} = 179.4$, so that

$$a = 143 - \frac{2}{3}\left(\frac{897}{5}\right) = \frac{117}{5} = 23.4.$$

The regression line with these coefficients is

$$y = 23.4x + 179.4.$$

Realize that all we have determined so far is that (a, b) is a critical point, a candidate for a local extremum. To verify that our choice of a and b gives the *minimum* function value, note that the surface $z = f(x, y)$ is a paraboloid opening toward the positive z-axis (see Figure 12.45) and the only critical point of an upward-curving paraboloid is an absolute minimum. Alternatively, you can show that $D(a, b) = 80 > 0$ and $f_{aa} > 0$. A plot of the regression line $y = 23.4x + 179.4$ with the data points is shown in Figure 12.46. Look carefully and notice that the line matches the data quite well. This also gives us confidence that we have found the minimum sum of the squared residuals. ■

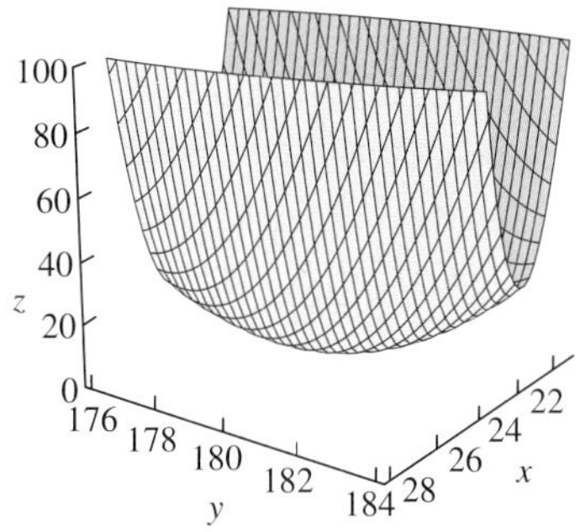

FIGURE 12.45
$z = f(x, y)$

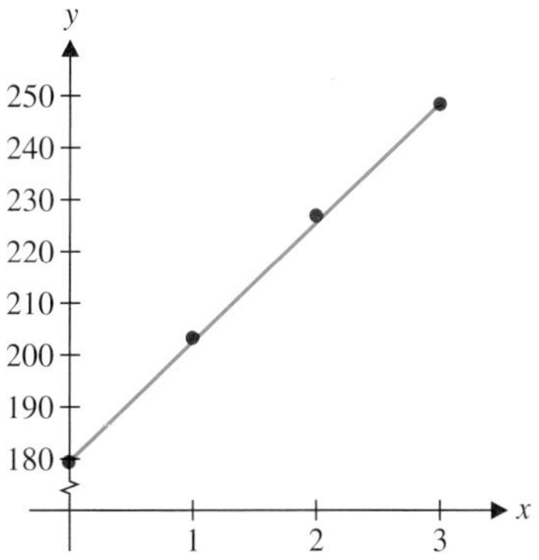

FIGURE 12.46
The regression line

As you will see in the exercises, finding critical points of even simple functions of several variables can be challenging. For the complicated functions that often arise in applications, finding critical points by hand can be nearly impossible. Because of this, numerical procedures for estimating maxima and minima are essential. We briefly introduce one such method here.

Given a function $f(x, y)$, make your best guess (x_0, y_0) of the location of a local maximum (or minimum). We call this your **initial guess** and want to use this to obtain a more precise estimate of the location of the maximum (or minimum). How might we do that? Well, recall that the direction of maximum increase of the function from the point (x_0, y_0) is given by the gradient $\nabla f(x_0, y_0)$. So, starting at (x_0, y_0), if we move in the direction of $\nabla f(x_0, y_0)$, f should be increasing, but how far should we go in this direction? One strategy (the method of **steepest ascent**) is to continue moving in the direction of the gradient until the function stops increasing. We call this stopping point (x_1, y_1). Starting anew from (x_1, y_1), we repeat the process, by computing a new gradient $\nabla f(x_1, y_1)$ and following it until $f(x, y)$ stops increasing, at some point (x_2, y_2). We then continue this process until the change in function values from $f(x_n, y_n)$ to $f(x_{n+1}, y_{n+1})$ is insignificant. Likewise, to find a local minimum, follow the path of **steepest descent,** by moving in the direction opposite the gradient, $-\nabla f(x_0, y_0)$ (the direction of maximum decrease of the function). We illustrate the steepest ascent algorithm in example 7.5.

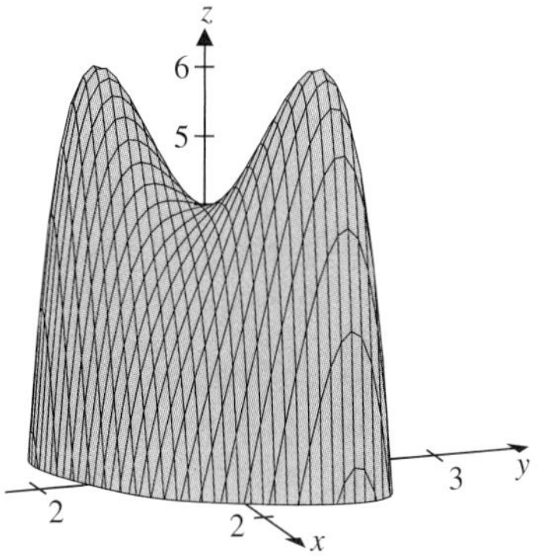

FIGURE 12.47
$z = 4xy - x^4 - y^4 + 4$

EXAMPLE 7.5 Method of Steepest Ascent

Use the steepest ascent algorithm to estimate the maximum of $f(x, y) = 4xy - x^4 - y^4 + 4$ in the first octant.

Solution A sketch of the surface is shown in Figure 12.47. We will estimate the maximum on the right by starting with an initial guess of $(2, 3)$, where $f(2, 3) = -69$.

(Note that this is obviously not the maximum, but it will suffice as a crude initial guess.) From this point, we want to follow the path of steepest ascent and move in the direction of $\nabla f(2, 3)$. We have

$$\nabla f(x, y) = \langle 4y - 4x^3, 4x - 4y^3 \rangle$$

and so, $\nabla f(2, 3) = \langle -20, -100 \rangle$. Note that every point lying on the line through (2, 3) in the direction of $\langle -20, -100 \rangle$ will have the form $(2 - 20h, 3 - 100h)$, for some value of $h > 0$. (Think about this!) Our goal is to move in this direction until $f(x, y)$ stops increasing. Notice that this puts us at a critical point for function values on the line of points $(2 - 20h, 3 - 100h)$. Since the function values along this line are given by $g(h) = f(2 - 20h, 3 - 100h)$, we find the smallest positive h such that $g'(h) = 0$. From the chain rule, we have

$$\begin{aligned} g'(h) &= -20\frac{\partial f}{\partial x}(2 - 20h, 3 - 100h) - 100\frac{\partial f}{\partial y}(2 - 20h, 3 - 100h) \\ &= -20[4(3 - 100h) - 4(2 - 20h)^3] - 100[4(2 - 20h) - 4(3 - 100h)^3]. \end{aligned}$$

Solving the equation $g'(h) = 0$ (we did it numerically), we get $h \approx 0.02$. This moves us to the point $(x_1, y_1) = (2 - 20h, 3 - 100h) = (1.6, 1)$, with function value $f(x_1, y_1) = 2.8464$. A contour plot of $f(x, y)$ with this first step is shown in Figure 12.48a. Notice that since $f(x_1, y_1) > f(x_0, y_0)$, we have found an improved approximation of the local maximum. To improve this further, we repeat the process starting with the new point. In this case, we have $\nabla f(1.6, 1) = \langle -12.384, 2.4 \rangle$ and we look for a critical point for the new function $g(h) = f(1.6 - 12.384h, 1 + 2.4h)$, for $h > 0$. Again, from the chain rule, we have

$$g'(h) = -12.384\frac{\partial f}{\partial x}(1.6 - 12.384h, 1 + 2.4h) + 2.4\frac{\partial f}{\partial y}(1.6 - 12.384h, 1 + 2.4h).$$

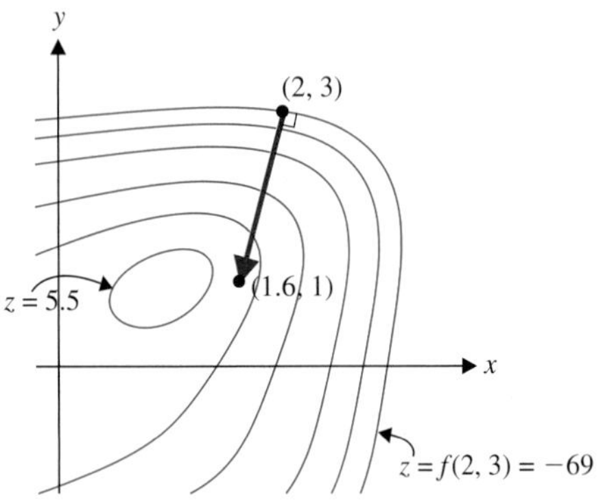

FIGURE 12.48a
First step of steepest ascent

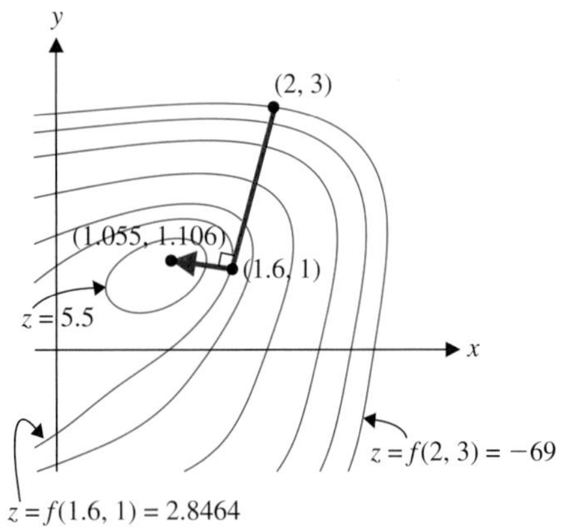

FIGURE 12.48b
Second step of steepest ascent

Solving $g'(h) = 0$ numerically gives us $h \approx 0.044$. This moves us to the point $(x_2, y_2) = (1.6 - 12.384h, 1 + 2.4h) = (1.055, 1.106)$, with function value $f(x_2, y_2) = 5.932$. Notice that we have again improved our approximation of the local maximum. A contour plot of $f(x, y)$ with the first two steps is shown in Figure 12.48b. From the contour plot, it appears that we are now very near a local maximum. In practice, you continue this process until you are no longer improving the approximation significantly. (This is easily implemented on a computer.) In the accompanying table, we show the first seven steps of steepest ascent. We leave it as an exercise to show that the local maximum is actually at $(1, 1)$ with function value $f(1, 1) = 6$.

n	x_n	y_n	$f(x_n, y_n)$
0	2	3	−69
1	1.6	1	2.846
2	1.055	1.106	5.932
3	1.0315	1.0035	5.994
4	1.0049	1.0094	5.9995
5	1.0029	1.0003	5.99995
6	1.0005	1.0009	5.999995
7	1.0003	1.0003	5.9999993

We define absolute extrema in a similar fashion to local extrema.

DEFINITION 7.4

We call $f(a, b)$ the **absolute maximum** of f on the region R if $f(a, b) \geq f(x, y)$ for all $(x, y) \in R$. Similarly, $f(a, b)$ is called the **absolute minimum** of f on R if $f(a, b) \leq f(x, y)$ for all $(x, y) \in R$. In either case, $f(a, b)$ is called an **absolute extremum** of f.

Recall that for a function f of a single variable, we observed that whenever f is continuous on the closed interval $[a, b]$, it will assume a maximum and minimum value on $[a, b]$. Further, we proved that absolute extrema must occur at either critical numbers of f or at the endpoints of the interval $[a, b]$. The situation for absolute extrema of functions of two variables is very similar. First, we need some terminology. We say that a region $R \subset \mathbb{R}^2$ is **bounded** if there is a disk that completely contains R. We now have the following result (whose proof can be found in more advanced texts).

THEOREM 7.3 (Extreme Value Theorem)

Suppose that $f(x, y)$ is continuous on the closed and bounded region $R \subset \mathbb{R}^2$. Then f has both an absolute maximum and an absolute minimum on R. Further, an absolute extremum may only occur at a critical point in R or at a point on the boundary of R.

Note that if $f(a, b)$ is an absolute extremum of f in R and (a, b) is in the interior of R, then (a, b) is also a local extremum of f, in which case, (a, b) must be a critical point. This says that all of the absolute extrema of a function f in a region R occur either at critical points (and we already know how to find these) or on the boundary of the region. Observe that this also provides us with a method for locating absolute extrema of continuous functions on closed and bounded regions. That is, we find the extrema on the boundary and compare these against the local extrema. We examine this in example 7.6, where the basic steps are as follows:

- Find all critical points of f in the region R.
- Find the maximum and minimum values of f on the boundary of R.
- Compare the values of f at the critical points with the maximum and minimum values of f on the boundary of R.

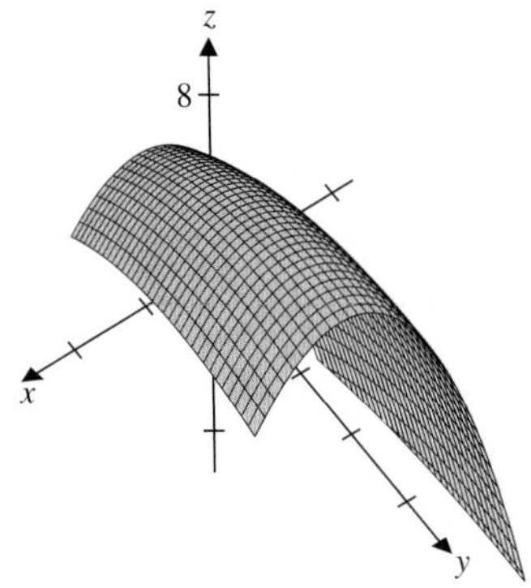

FIGURE 12.49a
The surface
$z = 5 + 4x - 2x^2 + 3y - y^2$

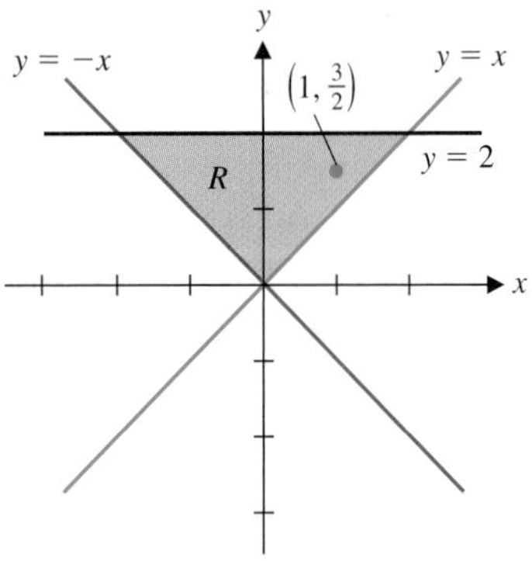

FIGURE 12.49b
The region R

EXAMPLE 7.6 Finding Absolute Extrema

Find the absolute extrema of $f(x, y) = 5 + 4x - 2x^2 + 3y - y^2$ on the region R bounded by the lines $y = 2$, $y = x$ and $y = -x$.

Solution We show a sketch of the surface in Figure 12.49a and a sketch of the region R in Figure 12.49b. From the sketch of the surface, notice that the absolute minimum appears to occur on the line $x = -2$ and the absolute maximum occurs somewhere near the line $x = 1$. Since an extremum can occur only at a critical point or at a point on the boundary of R, we first check to see whether there are any interior critical points. We have $f_x = 4 - 4x = 0$ for $x = 1$ and $f_y = 3 - 2y = 0$ for $y = \frac{3}{2}$. So, there is only one critical point $\left(1, \frac{3}{2}\right)$ and it is located in the interior of R. Next, we look for the maximum and minimum values of f on the boundary of R. In this case, the boundary consists of three separate pieces: the portion of the line $y = 2$ for $-2 \le x \le 2$, the portion of the line $y = x$ for $0 \le x \le 2$ and the portion of the line $y = -x$ for $-2 \le x \le 0$. We look for the maximum value of f on each of these separately. On the portion of the line $y = 2$ for $-2 \le x \le 2$, we have

$$f(x, y) = f(x, 2) = 5 + 4x - 2x^2 + 6 - 4 = 7 + 4x - 2x^2 = g(x).$$

To find the maximum and minimum values of f on this portion of the boundary, we need only find the maximum and minimum values of g on the interval $[-2, 2]$. We have $g'(x) = 4 - 4x = 0$ only for $x = 1$. Comparing the value of g at the endpoints and at the only critical number in the interval, we have: $g(-2) = -9$, $g(2) = 7$ and $g(1) = 9$. So, the maximum value of f on this portion of the boundary is 9 and the minimum value is -9.

On the portion of the line $y = x$ for $0 \le x \le 2$, we have

$$f(x, y) = f(x, x) = 5 + 7x - 3x^2 = h(x).$$

We have $h'(x) = 7 - 6x = 0$, only for $x = \frac{7}{6}$, which is in the interval. Comparing the values of h at the endpoints and the critical number, we have: $h(0) = 5$, $h(2) = 7$ and $h\left(\frac{7}{6}\right) \approx 9.08$. So, the maximum value of f on this portion of the boundary is approximately 9.08 and its minimum value is 5.

On the portion of the line $y = -x$ for $-2 \le x \le 0$, we have

$$f(x, y) = f(x, -x) = 5 + x - 3x^2 = k(x).$$

We have $k'(x) = 1 - 6x = 0$, only for $x = \frac{1}{6}$, which is *not* in the interval $[-2, 0]$ under consideration. Comparing the values of k at the endpoints, we have $k(-2) = -9$ and $k(0) = 5$, so that the maximum value of f on this portion of the boundary is 5 and its minimum value is -9.

Finally, we compute the value of f at the lone critical point in the interior of R: $f\left(1, \frac{3}{2}\right) = \frac{37}{4} = 9.25$. The largest of all these values we have computed is the absolute maximum in R and the smallest is the absolute minimum. So, the absolute maximum is $f\left(1, \frac{3}{2}\right) = 9.25$ and the absolute minimum is $f(-2, 2) = -9$. Note that these are also consistent with what we observed in Figure 12.49a. ■

We close this section with a proof of the Second Derivatives Test (Theorem 7.2). To keep the notation to a minimum, we will assume that the critical point to be tested is (0, 0). The proof can be extended to any critical point by a change of variables.

Proof of the Second Derivatives Test

Suppose that (0, 0) is a critical point of $f(x, y)$ with $f_x(0, 0) = f_y(0, 0) = 0$. We will look at the change in $f(x, y)$ from (0, 0) in the direction of the unit vector $\mathbf{u} = \frac{\langle k, 1\rangle}{\sqrt{k^2+1}}$, for some constant k. (Note that $\mathbf{u}$ can point in any direction *except* the direction of $\mathbf{i}$.) In this direction, notice that $x = ky$. If we define $g(x) = f(kx, x)$, then by the chain rule, we have

$$g'(x) = kf_x(kx, x) + f_y(kx, x) \tag{7.1}$$

and $$g''(x) = k^2 f_{xx}(kx, x) + kf_{xy}(kx, x) + kf_{yx}(kx, x) + f_{yy}(kx, x).$$

At $x = 0$, this gives us

$$g''(0) = k^2 f_{xx}(0, 0) + 2kf_{xy}(0, 0) + f_{yy}(0, 0), \tag{7.2}$$

where we have $f_{xy} = f_{yx}$, since f was assumed to have continuous second partial derivatives. Since $f_x(0, 0) = f_y(0, 0) = 0$, we have from (7.1) that

$$g'(0) = kf_x(0, 0) + f_y(0, 0) = 0.$$

Using the second derivative test for functions of a single variable, the sign of $g''(0)$ can tell us whether there is a local maximum or a local minimum of g at $x = 0$. Observe that using (7.2), we can write $g''(0)$ as

$$g''(0) = ak^2 + 2bk + c = p(k),$$

where a, b and c are the constants $a = f_{xx}(0, 0)$, $b = f_{xy}(0, 0)$ and $c = f_{yy}(0, 0)$. Of course, the graph of $p(k)$ is a parabola. Recall that for any parabola, if $a > 0$, then $p(k)$ has a minimum at $k = -\frac{b}{a}$, given by $p\left(-\frac{b}{a}\right) = -\frac{b^2}{a} + c$. (Hint: Complete the square.) In case (i) of the theorem, we assume that the discriminant satisfies

$$0 < D(0, 0) = f_{xx}(0, 0)f_{yy}(0, 0) - [f_{xy}(0, 0)]^2 = ac - b^2,$$

so that $-\frac{b^2}{a} + c > 0$. In this case,

$$p(k) \geq p\left(-\frac{b}{a}\right) = -\frac{b^2}{a} + c > 0.$$

We have shown that, in case (i), when $D(0, 0) > 0$ and $f_{xx}(0, 0) > 0$, $g''(0) = p(k) > 0$ for all k. So, g has a local minimum at 0 and consequently, in all directions, the point at (0, 0) is a local minimum of f. For case (ii), where $D(0, 0) > 0$ and $f_{xx}(0, 0) < 0$, we consider $p(k)$ with $a < 0$. In a similar fashion, we can show that here, $p(k) \leq -\frac{b^2}{a} + c < 0$. Given that we have $g''(0) = p(k) < 0$ for all k, we conclude that the point at (0, 0) is a local maximum of f. For case (iii), where the discriminant $D(0, 0) < 0$, the parabola $p(k)$ will assume both positive and negative values. For some values of k, we have $g''(0) > 0$ and the point (0, 0) is a local minimum along the path $x = ky$, while for other values of k, we have $g''(0) < 0$ and the point (0, 0) is a local maximum along the path $x = ky$. Taken together, this says that the point at (0, 0) must be a saddle point of f. To complete the proof, we must only consider the case where $\mathbf{u} = \mathbf{i}$. In this case, the preceding proof is easily revised to show the same results and we leave the details as an exercise.

BEYOND FORMULAS

You can think about local extrema for functions of n variables in the same way for any value of n. Critical points are points where either all of the first partial derivatives are zero or where one is undefined. These provide the candidates for local extrema, in the sense that a local extremum may occur only at a critical point. However, further testing is needed to determine whether a function has a local maximum, a local minimum or neither at a given critical point.

EXERCISES 12.7

WRITING EXERCISES

1. If $f(x, y)$ has a local minimum at (a, b), explain why the point $(a, b, f(a, b))$ is a local minimum in the intersection of $z = f(x, y)$ with any vertical plane passing through the point. Explain why the condition $f_x(a, b) = f_y(a, b) = 0$ guarantees that (a, b) is a critical point in any such plane.

2. Suppose that $f_x(a, b) \neq 0$. Explain why the tangent plane to $z = f(x, y)$ at (a, b) must be "tilted", so that there is not a local extremum at (a, b).

3. Suppose that $f_x(a, b) = f_y(a, b) = 0$ and $f_{xx}(a, b) f_{yy}(a, b) < 0$. Explain why there must be a saddle point at (a, b).

4. Explain why the center of a set of concentric circles in a contour plot will often represent a local extremum.

In exercises 1–8, locate all critical points and classify them using Theorem 7.2.

1. $f(x, y) = e^{-x^2}(y^2 + 1)$
2. $f(x, y) = \cos^2 x + y^2$
3. $f(x, y) = x^3 - 3xy + y^3$
4. $f(x, y) = 4xy - x^4 - y^4 + 4$
5. $f(x, y) = y^2 + x^2y + x^2 - 2y$
6. $f(x, y) = 2x^2 + y^3 - x^2y - 3y$
7. $f(x, y) = e^{-x^2-y^2}$
8. $f(x, y) = x \sin y$

In exercises 9–14, locate all critical points and analyze each graphically. If you have a CAS, use Theorem 7.2 to classify each point.

9. $f(x, y) = x^2 - \dfrac{4xy}{y^2 + 1}$
10. $f(x, y) = \dfrac{x + y}{x^2 + y^2 + 1}$
11. $f(x, y) = xe^{-x^2-y^2}$
12. $f(x, y) = x^2e^{-x^2-y^2}$
13. $f(x, y) = xye^{-x^2-y^2}$
14. $f(x, y) = xye^{-x^2-y^4}$

In exercises 15–18, numerically approximate all critical points. Classify each point graphically or with Theorem 7.2.

15. $f(x, y) = xy^2 - x^2 - y + \frac{1}{16}x^4$
16. $f(x, y) = 2y(x + 2) - x^2 + y^4 - 9y^2$
17. $f(x, y) = (x^2 - y^3)e^{-x^2-y^2}$
18. $f(x, y) = (x^2 - 3x)e^{-x^2-y^2}$

19. Show that for data $(x_1, y_1), (x_2, y_2), \ldots, (x_n, y_n)$, the least-squares equations become

$$\left(\sum_{k=1}^{n} x_k\right) a + \left(\sum_{k=1}^{n} 1\right) b = \sum_{k=1}^{n} y_k$$

$$\left(\sum_{k=1}^{n} x_k^2\right) a + \left(\sum_{k=1}^{n} x_k\right) b = \sum_{k=1}^{n} x_k y_k$$

20. Solve the equations in exercise 19 for a and b.

In exercises 21–27, use least squares as in example 7.4 to find a linear model of the data.

21. A famous mental calculation prodigy named Jacques Inaudi was timed in 1894 at various mental arithmetic problems. His times are shown below. (Think about what your times might be!) The data are taken from *The Number Sense* by Stanislas Dehaene. Treat the number of operations as the independent variable (x) and time as the dependent variable (y).

Number of operations	1	4	9	16
Time (sec)	0.6	2.0	6.4	21
Example	$3 \cdot 7$	$63 \cdot 58$	$638 \cdot 823$	$7286 \cdot 5397$

22. Repeat exercise 21 with the following data point added: 36 operations in 240 seconds (an example is $729{,}856 \cdot 297{,}143$). How much effect does this last point have on the linear model?

23. The Dow Jones Industrial averages for several days starting in June 1998 are shown. Use your linear model to predict the average on day 12. Linear models of similar data can be found in information supplied by financial consulting firms, typically with the warning to not use the linear model for forecasting. Explain why this warning is appropriate.

Date (number of days)	0	2	4	6	8
Dow Jones average	8910	8800	9040	9040	9050

24. The following data show the average price of a gallon of regular gasoline in California. Use the linear model to predict the price in 1990 and 1995. The actual prices were \$1.09 and \$1.23. Explain why your forecasts were not accurate.

Year	1970	1975	1980	1985
Price	\$0.34	\$0.59	\$1.23	\$1.11

25. The following data show the height and weight of a small number of people. Use the linear model to predict the weight of a 6′8″ person and a 5′0″ person. Comment on how accurate you think the model is.

Height (inches)	68	70	70	71
Weight (pounds)	160	172	184	180

26. The following data show the age and income for a small number of people. Use the linear model to predict the income of a 45-year-old and of an 80-year-old. Comment on how accurate you think the model is.

Age (years)	24	32	40	56
Income (\$)	30,000	34,000	52,000	82,000

27. The accompanying data show the average number of points professional football teams score when starting different distances from the opponents' goal line. (For more information, see Hal Stern's "A Statistician Reads the Sports Pages" in *Chance,* Summer 1998. The number of points is determined by the next score, so that if the opponent scores next, the number of points is negative.) Use the linear model to predict the average number of points starting (a) 60 yards from the goal line and (b) 40 yards from the goal line.

Yards from goal	15	35	55	75	95
Average points	4.57	3.17	1.54	0.24	−1.25

28. In *The Hidden Game of Pro Football,* authors Carroll, Palmer and Thorn claim that the data presented in exercise 27 support the conclusion that when a team loses a fumble they lose an average of 4 points *regardless of where they are on the field.* That is, a fumble at the 50-yard line costs the same number of points as a fumble at the opponents' 10-yard line. Use your result from exercise 27 to verify this claim.

In exercises 29–32, calculate the first two steps of the steepest ascent algorithm from the given starting point.

29. $f(x, y) = 2xy - 2x^2 + y^3$, $(0, -1)$

30. $f(x, y) = 3xy - x^3 - y^2$, $(1, 1)$

31. $f(x, y) = x - x^2y^4 + y^2$, $(1, 1)$

32. $f(x, y) = xy^2 - x^2 - y$, $(1, 0)$

33. Calculate one step of the steepest ascent algorithm for $f(x, y) = 2xy - 2x^2 + y^3$, starting at $(0, 0)$. Explain in graphical terms what goes wrong.

34. Define a **steepest descent algorithm** for finding local minima.

In exercises 35–38, find the absolute extrema of the function on the region.

35. $f(x, y) = x^2 + 3y - 3xy$, region bounded by $y = x$, $y = 0$ and $x = 2$

36. $f(x, y) = x^2 + y^2 - 4xy$, region bounded by $y = x$, $y = -3$ and $x = 3$

37. $f(x, y) = x^2 + y^2$, region bounded by $(x - 1)^2 + y^2 = 4$

38. $f(x, y) = x^2 + y^2 - 2x - 4y$, region bounded by $y = x$, $y = 3$ and $x = 0$

39. A box is to be constructed out of 96 square feet of material. Find the dimensions x, y and z that maximize the volume of the box.

40. If the bottom of the box in exercise 39 must be reinforced by doubling up the material (essentially, there are two bottoms of the box), find the dimensions that maximize the volume of the box.

41. Heron's formula gives the area of a triangle with sides of lengths a, b and c as $A = \sqrt{s(s-a)(s-b)(s-c)}$, where $s = \frac{1}{2}(a + b + c)$. For a given perimeter, find the triangle of maximum area.

42. Find the maximum of $x^2 + y^2$ on the square with $-1 \le x \le 1$ and $-1 \le y \le 1$. Use your result to explain why a computer graph of $z = x^2 + y^2$ with the graphing window $-1 \le x \le 1$ and $-1 \le y \le 1$ does not show a circular cross section at the top.

43. Find all critical points of $f(x, y) = x^2y^2$ and show that Theorem 7.2 fails to identify any of them. Use the form of the function to determine what each critical point represents. Repeat for $f(x, y) = x^{2/3}y^2$.

44. Complete the square to identify all local extrema of
(a) $f(x, y) = x^2 + 2x + y^2 - 4y + 1$,
(b) $f(x, y) = x^4 - 6x^2 + y^4 + 2y^2 - 1$.

45. In exercise 3, there is a saddle point at (0, 0). This means that there is (at least) one trace of $z = x^3 - 3xy + y^3$ with a local minimum at (0, 0) and (at least) one trace with a local maximum at (0, 0). To analyze traces in the planes $y = kx$ (for some constant k), substitute $y = kx$ and show that $z = (1 + k^3)x^3 - 3kx^2$. Show that $f(x) = (1 + k^3)x^3 - 3kx^2$ has a local minimum at $x = 0$ if $k < 0$ and a local maximum at $x = 0$ if $k > 0$. (Hint: Use the Second Derivative Test from section 3.5.)

46. In exercise 4, there is a saddle point at (0, 0). As in exercise 45, find traces such that there is a local maximum at (0, 0) and traces such that there is a local minimum at (0, 0).

47. In example 7.3, (0, 0) is a critical point but is not classified by Theorem 7.2. Use the technique of exercise 45 to analyze this saddle point.

48. Repeat exercise 47 for $f(x, y) = x^2 - 3xy^2 + 4x^3y$.

49. For $f(x, y, z) = xz - x + y^3 - 3y$, show that (0, 1, 1) is a critical point. To classify this critical point, show that $f(0+\Delta x, 1+\Delta y, 1+\Delta z) = \Delta x\Delta z + 3\Delta y^2 + \Delta y^3 + f(0, 1, 1)$. Setting $\Delta y = 0$ and $\Delta x\Delta z > 0$, conclude that $f(0, 1, 1)$ is not a local maximum. Setting $\Delta y = 0$ and $\Delta x\Delta z < 0$, conclude that $f(0, 1, 1)$ is not a local minimum.

50. Repeat exercise 49 for the point (0, −1, 1).

In exercises 51–54, label the statement as true or false and explain why.

51. If $f(x, y)$ has a local maximum at (a, b), then $\frac{\partial f}{\partial x}(a, b) = \frac{\partial f}{\partial y}(a, b) = 0$.

52. If $\frac{\partial f}{\partial x}(a, b) = \frac{\partial f}{\partial y}(a, b) = 0$, then $f(x, y)$ has a local maximum at (a, b).

53. In between any two local maxima of $f(x, y)$ there must be at least one local minimum of $f(x, y)$.

54. If $f(x, y)$ has exactly two critical points, they can't both be local maxima.

55. In the contour plot, the locations of four local extrema and nine saddle points are visible. Identify these critical points.

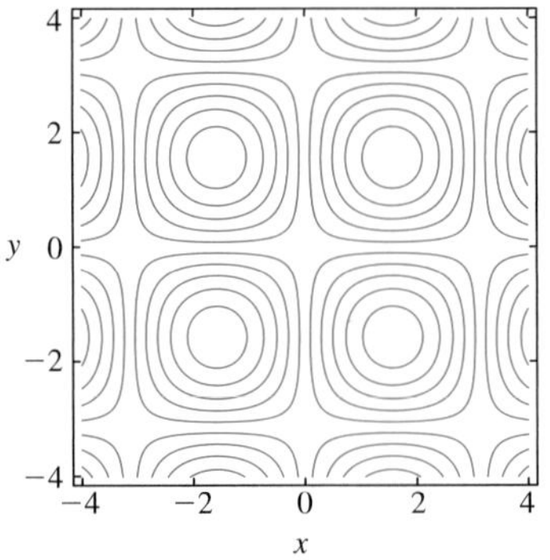

56. In the contour plot, the locations of one local extremum and one saddle point are visible. Identify each critical point.

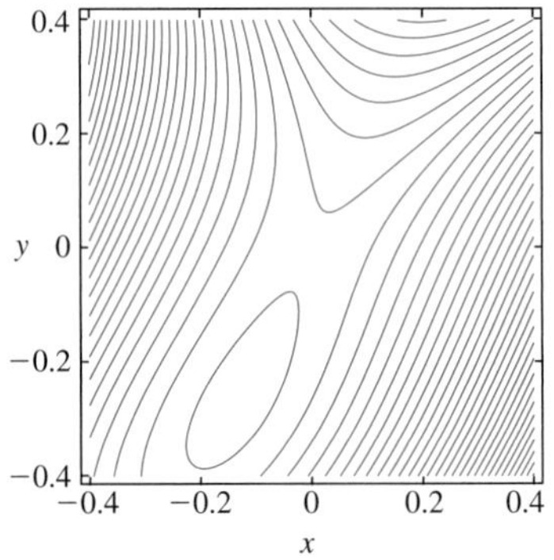

In exercises 57–60, use the contour plot to conjecture the locations of all local extrema and saddle points.

57.

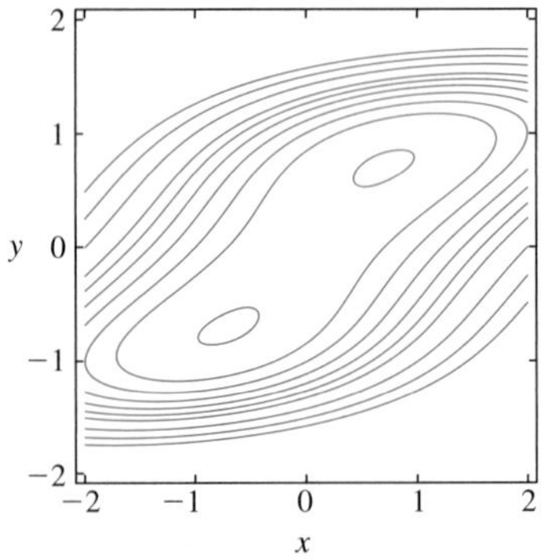

58.

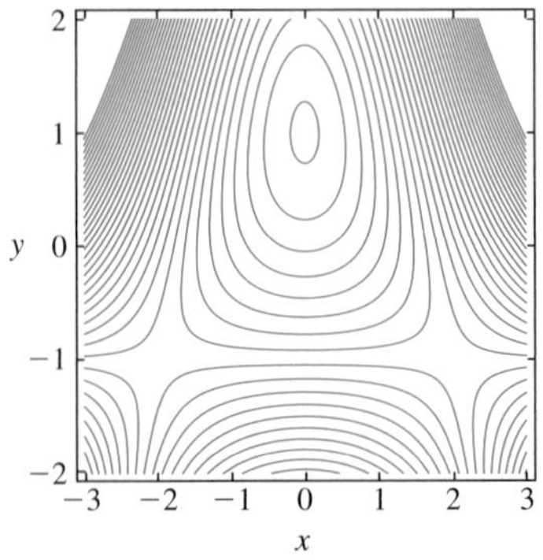

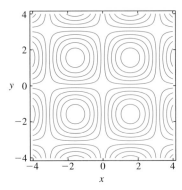

59.

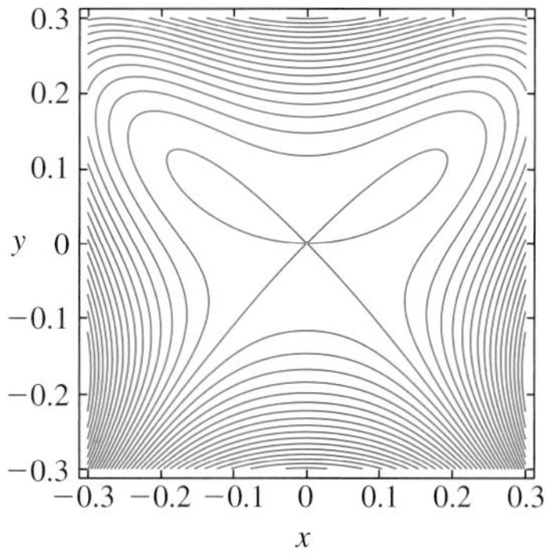

60.

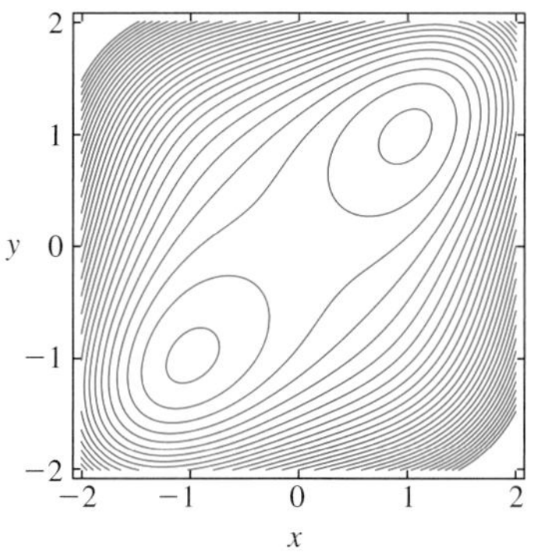

61. Construct the function $d(x, y)$ giving the distance from a point (x, y, z) on the paraboloid $z = 4 - x^2 - y^2$ to the point $(3, -2, 1)$. Then determine the point that minimizes $d(x, y)$.

62. Use the method of exercise 61 to find the closest point on the cone $z = \sqrt{x^2 + y^2}$ to the point $(2, -3, 0)$.

63. Use the method of exercise 61 to find the closest point on the sphere $x^2 + y^2 + z^2 = 9$ to the point $(2, 1, -3)$.

64. Use the method of exercise 61 to find the closest point on the plane $3x - 4y + 3z = 12$ to the origin.

65. Show that the function $f(x, y) = 5xe^y - x^5 - e^{5y}$ has exactly one critical point, which is a local maximum but not an absolute maximum.

66. Show that the function $f(x, y) = 2x^4 + e^{4y} - 4x^2e^y$ has exactly two critical points, both of which are local minima.

67. Prove that the situation of exercise 66 (two local minima without a local maximum) can never occur for differentiable functions of one variable.

68. The **Hardy-Weinberg law** of genetics describes the relationship between the proportions of different genes in populations. Suppose that a certain gene has three types (e.g., blood types of A, B and O). If the three types have proportions p, q and r, respectively, in the population, then the Hardy-Weinberg law states that the proportion of people who carry two different types of genes equals $f(p, q, r) = 2pq + 2pr + 2qr$. Explain why $p + q + r = 1$ and then show that the maximum value of $f(p, q, r)$ is $\frac{2}{3}$.

EXPLORATORY EXERCISES

1. In example 7.4, we found the "best" linear fit to population data using the least-squares criterion. Use the least-squares criterion to find the best quadratic fit to the data. That is, for functions of the form $ax^2 + bx + c$, find the values of the constants a, b and c that minimize the sum of the squares of the residuals. For the given data, show that the sum of the squares of the residuals for the quadratic model is less than for the linear model. Explain why this has to be true mathematically. In spite of this, explain why the linear model might be preferable to the quadratic model. (Hint: Use both models to predict 100 years into the future and backtrack 100 years into the past.)

2. Use the least-squares criterion to find the best exponential fit ($y = ae^{bx}$) to the data of example 7.4. Compare the actual residuals of this model to the actual residuals of the linear model. Explain why there is some theoretical justification for using an exponential model, and then discuss the advantages and disadvantages of the exponential and linear models.

3. A practical flaw with the method of steepest ascent presented in example 7.5 is that the equation $g'(h) = 0$ may be difficult to solve. An alternative is to use Newton's method to approximate a solution. A method commonly used in practice is to approximate h using one iteration of Newton's method with initial guess $h = 0$. We derive the resulting formula here. Recall that $g(h) = f(x_k + ah, y_k + bh)$, where (x_k, y_k) is the current point and $\langle a, b\rangle = \nabla f(x_k, y_k)$. Newton's method applied to $g'(h) = 0$ with $h_0 = 0$ is given by $h_1 = -\frac{g'(0)}{g''(0)}$. Show that $g'(0) = af_x(x_k, y_k) + bf_y(x_k, y_k) = a^2 + b^2 = \nabla f(x_k, y_k) \cdot \nabla f(x_k, y_k)$. Also, show that $g''(0) = a^2 f_{xx}(x_k, y_k) + 2abf_{xy}(x_k, y_k) + b^2 f_{yy}(x_k, y_k) = \nabla f(x_k, y_k) \cdot H(x_k, y_k)\nabla f(x_k, y_k)$, where the Hessian matrix is defined by $H = \begin{bmatrix} f_{xx} & f_{xy} \\ f_{yx} & f_{yy} \end{bmatrix}$. Putting this together with the work in example 7.5, the method of steepest ascent becomes $\mathbf{v}_{k+1} = \mathbf{v}_k - \frac{\nabla f(\mathbf{v}_k) \cdot \nabla f(\mathbf{v}_k)}{\nabla f(\mathbf{v}_k) \cdot H(\mathbf{v}_k)\nabla f(\mathbf{v}_k)} \nabla f(\mathbf{v}_k)$, where $\mathbf{v}_k = \begin{bmatrix} x_k \\ y_k \end{bmatrix}$.

12.8 CONSTRAINED OPTIMIZATION AND LAGRANGE MULTIPLIERS

The local extrema we found in section 12.7 form only one piece of the optimization puzzle. In many applications, the goal is not to identify theoretical maximum or minimum values, but to achieve the absolute best possible product given a large set of constraints such as limited resources or technology. For example, an automotive engineer's objective might be to minimize wind drag in the design of a car. However, new designs are severely limited by customer demands of luxury and attractiveness and particularly by manufacturing constraints such as cost. In this section, we develop a technique for finding the maximum or minimum of a function, given one or more constraints on the function's domain.

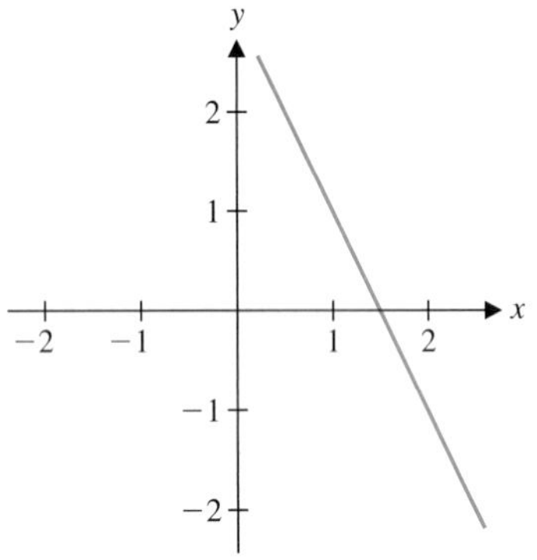

FIGURE 12.50a
$y = 3 - 2x$

We first consider the two-dimensional geometric problem of finding the point on the line $y = 3 - 2x$ that is closest to the origin. A graph of the line is shown in Figure 12.50a. Notice that the set of points that are 1 unit from the origin form the circle $x^2 + y^2 = 1$. In Figure 12.50b, you can see that the line $y = 3 - 2x$ lies entirely outside this circle. This tells us that every point on the line $y = 3 - 2x$ lies more than 1 unit from the origin. Looking at the circle $x^2 + y^2 = 4$ in Figure 12.50c, you can clearly see that there are infinitely many points on the line that are less than 2 units from the origin. If we shrink the circle in Figure 12.50c (or enlarge the circle in Figure 12.50b), it will eventually reach a size at which the line is tangent to the circle (see Figure 12.50d). The point of tangency is the closest point on the line to the origin, since all other points on the line are outside the circle and hence, are farther away from the origin.

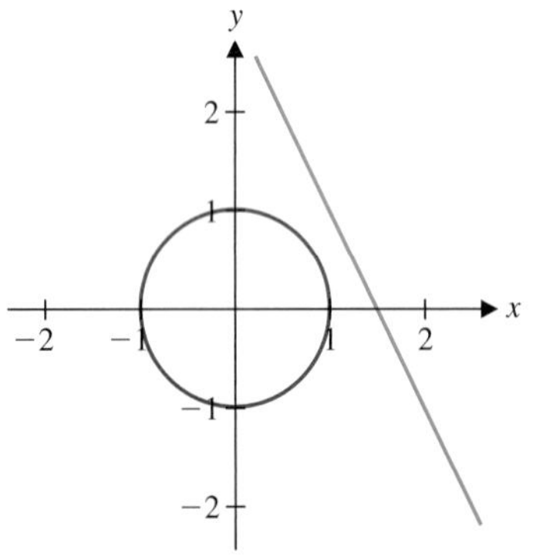

FIGURE 12.50b
$y = 3 - 2x$ and the circle of radius 1 centered at (0, 0)

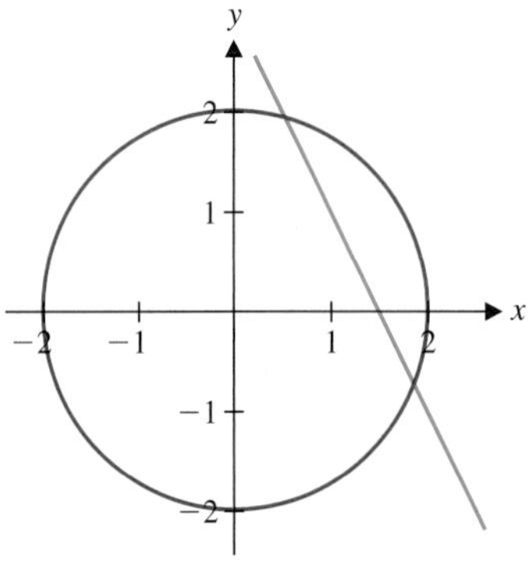

FIGURE 12.50c
$y = 3 - 2x$ and the circle of radius 2 centered at (0, 0)

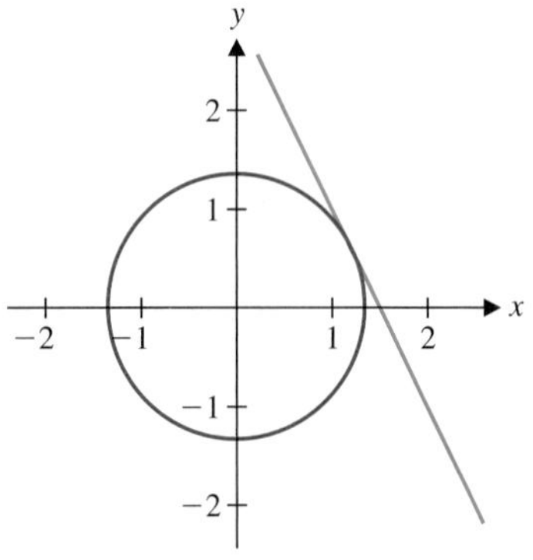

FIGURE 12.50d
$y = 3 - 2x$ and a circle tangent to the line

Translating the preceding geometric argument into the language of calculus, we want to minimize the distance from the point (x, y) to the origin, given by $\sqrt{x^2 + y^2}$. Before we continue, observe that the distance is minimized at exactly the same point at which the square of the distance is minimized. Minimizing the square of the distance, given by $x^2 + y^2$, avoids the mess created by the square root in the distance formula. So, instead, we minimize $f(x, y) = x^2 + y^2$, subject to the constraint that the point lie on the line (i.e., that $y = 3 - 2x$) or $g(x, y) = 2x + y - 3 = 0$. We have already argued that at the closest point, the line and circle are tangent. Since the gradient vector for a given function is orthogonal to its level curves at any given point, for a level curve of f to be tangent to the constraint curve $g(x, y) = 0$, the gradients of f and g must be parallel. That is, at the closest point (x, y) on the line to the origin, we must have $\nabla f(x, y) = \lambda \nabla g(x, y)$, for some constant λ. We illustrate this in example 8.1.

EXAMPLE 8.1 Finding a Minimum Distance

Use the relationship $\nabla f(x, y) = \lambda \nabla g(x, y)$ and the constraint $y = 3 - 2x$ to find the point on the line $y = 3 - 2x$ that is closest to the origin.

Solution For $f(x, y) = x^2 + y^2$, we have $\nabla f(x, y) = \langle 2x, 2y \rangle$ and for $g(x, y) = 2x + y - 3$, we have $\nabla g(x, y) = \langle 2, 1 \rangle$. The vector equation $\nabla f(x, y) = \lambda \nabla g(x, y)$ becomes

$$\langle 2x, 2y \rangle = \lambda \langle 2, 1 \rangle,$$

from which it follows that

$$2x = 2\lambda \quad \text{and} \quad 2y = \lambda.$$

The second equation gives us $\lambda = 2y$. The first equation then gives us $x = \lambda = 2y$. Substituting $x = 2y$ into the constraint equation $y = 3 - 2x$, we have $y = 3 - 2(2y)$, or $5y = 3$. The solution is $y = \frac{3}{5}$, giving us $x = 2y = \frac{6}{5}$. The closest point is then $(\frac{6}{5}, \frac{3}{5})$. Look carefully at Figure 12.50d and recognize that this is consistent with our graphical solution. Also, note that the line described parametrically by $x = \lambda$, $y = \frac{\lambda}{2}$ is the line through the origin and perpendicular to $y = 3 - 2x$. ■

The technique illustrated in example 8.1 can be applied to a wide variety of constrained optimization problems. We will now develop this method, referred to as the **method of Lagrange multipliers.**

Suppose that we want to find maximum or minimum values of the function $f(x, y, z)$, subject to the constraint that $g(x, y, z) = 0$. We assume that both f and g have continuous first partial derivatives. Now, suppose that f has an extremum at (x_0, y_0, z_0) lying on the level surface S defined by $g(x, y, z) = 0$. Let C be any curve lying on the level surface and passing through the point (x_0, y_0, z_0). Assume that C is traced out by the terminal point of the vector-valued function $\mathbf{r}(t) = \langle x(t), y(t), z(t) \rangle$ and that $\mathbf{r}(t_0) = \langle x_0, y_0, z_0 \rangle$. Define a function of the single variable t by

$$h(t) = f(x(t), y(t), z(t)).$$

Notice that if $f(x, y, z)$ has an extremum at (x_0, y_0, z_0), then $h(t)$ must have an extremum at t_0 and so, $h'(t_0) = 0$. From the chain rule, we get that

$$\begin{aligned} 0 = h'(t_0) &= f_x(x_0, y_0, z_0)x'(t_0) + f_y(x_0, y_0, z_0)y'(t_0) + f_z(x_0, y_0, z_0)z'(t_0) \\ &= \langle f_x(x_0, y_0, z_0), f_y(x_0, y_0, z_0), f_z(x_0, y_0, z_0) \rangle \cdot \langle x'(t_0), y'(t_0), z'(t_0) \rangle \\ &= \nabla f(x_0, y_0, z_0) \cdot \mathbf{r}'(t_0). \end{aligned}$$

That is, if $f(x_0, y_0, z_0)$ is an extremum, the gradient of f at (x_0, y_0, z_0) is orthogonal to the tangent vector $\mathbf{r}'(t_0)$. Since C was an arbitrary curve lying on the level surface S, it follows that $\nabla f(x_0, y_0, z_0)$ must be orthogonal to every curve lying on the level surface S and so, too is orthogonal to S. Recall from Theorem 6.5 that ∇g is also orthogonal to the level surface $g(x, y, z) = 0$, so that $\nabla f(x_0, y_0, z_0)$ and $\nabla g(x_0, y_0, z_0)$ must be parallel. This proves the following result.

HISTORICAL NOTES

Joseph-Louis Lagrange (1736–1813) Mathematician who developed many fundamental techniques in the calculus of variations, including the method that bears his name. Lagrange was largely self-taught, but quickly attracted the attention of the great mathematician Leonhard Euler. At age 19, Lagrange was appointed Professor of Mathematics at the Royal Artillery School in his native Turin. Over a long and outstanding career, Lagrange made contributions to probability, differential equations and fluid mechanics, for which he introduced what is now known as the Lagrangian function.

THEOREM 8.1

Suppose that $f(x, y, z)$ and $g(x, y, z)$ are functions with continuous first partial derivatives and $\nabla g(x, y, z) \neq \mathbf{0}$ on the surface $g(x, y, z) = 0$. Suppose that either

(i) the minimum value of $f(x, y, z)$ subject to the constraint $g(x, y, z) = 0$ occurs at (x_0, y_0, z_0); or
(ii) the maximum value of $f(x, y, z)$ subject to the constraint $g(x, y, z) = 0$ occurs at (x_0, y_0, z_0).

Then $\nabla f(x_0, y_0, z_0) = \lambda \nabla g(x_0, y_0, z_0)$, for some constant λ (called a **Lagrange multiplier**).

Note that Theorem 8.1 says that if $f(x, y, z)$ has an extremum at a point (x_0, y_0, z_0) on the surface $g(x, y, z) = 0$, we will have for $(x, y, z) = (x_0, y_0, z_0)$,

$$f_x(x, y, z) = \lambda g_x(x, y, z),$$
$$f_y(x, y, z) = \lambda g_y(x, y, z),$$
$$f_z(x, y, z) = \lambda g_z(x, y, z)$$

and

$$g(x, y, z) = 0.$$

Finding such extrema then boils down to solving these four equations for the four unknowns x, y, z and λ. (Actually, we need only find the values of x, y and z.)

It's important to recognize that this method only produces *candidates* for extrema. Along with finding a solution(s) to the above four equations, you need to verify (graphically as we did in example 8.1 or by some other means) that the solution you found in fact represents the desired optimal point.

Notice that the Lagrange multiplier method we have just developed can also be applied to functions of two variables, by ignoring the third variable in Theorem 8.1. That is, if $f(x, y)$ and $g(x, y)$ have continuous first partial derivatives and $f(x_0, y_0)$ is an extremum of f, subject to the constraint $g(x, y) = 0$, then we must have

$$\nabla f(x_0, y_0) = \lambda \nabla g(x_0, y_0),$$

for some constant λ. Graphically, this says that if $f(x_0, y_0)$ is an extremum, the level curve of f passing through (x_0, y_0) is tangent to the constraint curve $g(x, y) = 0$ at (x_0, y_0). We illustrate this in Figure 12.51. In this case, we end up with the three equations

$$f_x(x, y) = \lambda g_x(x, y), \quad f_y(x, y) = \lambda g_y(x, y) \quad \text{and} \quad g(x, y) = 0,$$

for the three unknowns x, y and λ. We illustrate this in example 8.2.

EXAMPLE 8.2 Finding the Optimal Thrust of a Rocket

A rocket is launched with a constant thrust corresponding to an acceleration of u ft/s^2. Ignoring air resistance, the rocket's height after t seconds is given by $f(t, u) = \frac{1}{2}(u - 32)t^2$ feet. Fuel usage for t seconds is proportional to $u^2 t$ and the limited fuel capacity of the rocket satisfies the equation $u^2 t = 10{,}000$. Find the value of u that maximizes the height that the rocket reaches when the fuel runs out.

Solution From Theorem 8.1, we look for solutions of $\nabla f(t, u) = \lambda \nabla g(t, u)$, where $g(t, u) = u^2 t - 10{,}000 = 0$ is the constraint equation. We have $\nabla f(t, u) = \left\langle (u - 32)t, \frac{1}{2}t^2 \right\rangle$

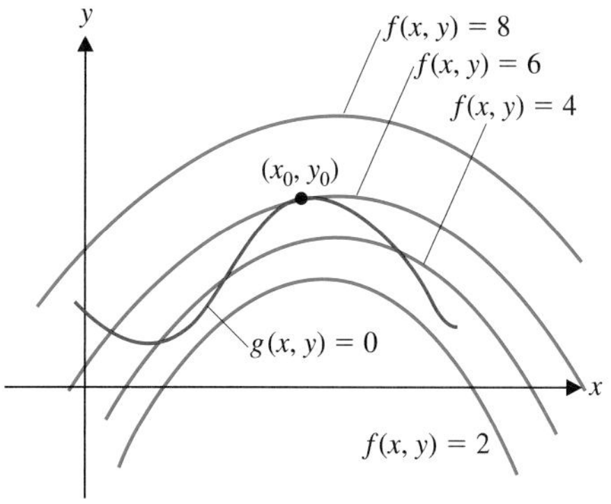

FIGURE 12.51
Level curve tangent to constraint curve at an extremum

and $\nabla g(t, u) = \langle u^2, 2ut \rangle$. From Theorem 8.1, we must have

$$\left\langle (u - 32)t, \frac{1}{2}t^2 \right\rangle = \lambda \langle u^2, 2ut \rangle,$$

for some constant λ. It follows that

$$(u - 32)t = \lambda u^2 \quad \text{and} \quad \frac{1}{2}t^2 = \lambda 2ut.$$

Solving both equations for λ, we have

$$\lambda = \frac{(u - 32)t}{u^2} = \frac{\frac{1}{2}t^2}{2ut}.$$

This gives us

$$2u(u - 32)t^2 = \frac{1}{2}t^2u^2.$$

Solutions include $t = 0$ and $u = 0$, but neither of these satisfies the constraint $u^2t = 10{,}000$. Canceling the factors of t^2 and u, we have $4(u - 32) = u$. The solution to this is $u = \frac{128}{3}$. With this value of u, the engines can burn for

$$t = \frac{10{,}000}{u^2} = \frac{10{,}000}{(128/3)^2} \approx 5.5 \text{ seconds},$$

with the rocket reaching a height of $z = \frac{1}{2}(\frac{128}{3} - 32)(5.5)^2 \approx 161$ feet. ■

We left example 8.2 unfinished. (Can you tell what's missing?) It is very difficult to argue that our solution actually represents a *maximum* height. (Could it be a saddle point?) What we do know is that by Theorem 8.1, *if* there is a maximum, we found it. Returning to a discussion of the physical problem, it should be completely reasonable that with a limited amount of fuel, there is a maximum attainable altitude and so, we did indeed find the maximum altitude.

Theorem 8.1 provides another major piece of the optimization puzzle. We next optimize a function subject to an inequality constraint of the form $g(x, y) \le c$. To understand our technique, recall how we solved for absolute extrema of functions of several variables on a closed and bounded region in section 12.7. We found critical points in the interior of the region and compared the values of the function at the critical points with the maximum and minimum function values on the boundary of the region. To find the extrema of $f(x, y)$ subject to a constraint of the form $g(x, y) \le c$, we first find the critical points of $f(x, y)$ that satisfy the constraint, then find the extrema of the function on the boundary $g(x, y) = c$

(the constraint curve) and finally, compare the function values. We illustrate this in example 8.3.

EXAMPLE 8.3 Optimization with an Inequality Constraint

Suppose that the temperature of a metal plate is given by $T(x, y) = x^2 + 2x + y^2$, for points (x, y) on the elliptical plate defined by $x^2 + 4y^2 \le 24$. Find the maximum and minimum temperatures on the plate.

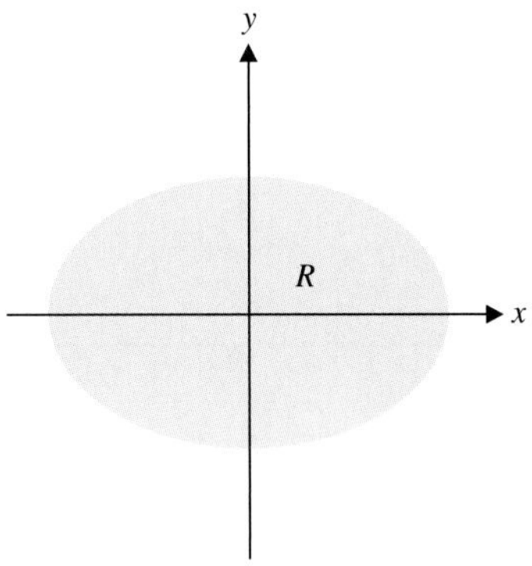

FIGURE 12.52
A metal plate

Solution The plate corresponds to the shaded region R shown in Figure 12.52. We first look for critical points of $T(x, y)$ inside the region R. We have $\nabla T(x, y) = \langle 2x + 2, 2y \rangle = \langle 0, 0 \rangle$ if $(x, y) = (-1, 0)$, which is in R. At this point, $T(-1, 0) = -1$. We next look for the extrema of $T(x, y)$ on the ellipse $x^2 + 4y^2 = 24$. We first rewrite the constraint equation as $g(x, y) = x^2 + 4y^2 - 24 = 0$. From Theorem 8.1, any extrema on the ellipse will satisfy the Lagrange multiplier equation: $\nabla T(x, y) = \lambda \nabla g(x, y)$ or

$$\langle 2x + 2, 2y \rangle = \lambda \langle 2x, 8y \rangle = \langle 2\lambda x, 8\lambda y \rangle.$$

This occurs when $$2x + 2 = 2\lambda x \quad \text{and} \quad 2y = 8\lambda y.$$

Notice that the second equation holds when $y = 0$ or $\lambda = \frac{1}{4}$. If $y = 0$, the constraint $x^2 + 4y^2 = 24$ gives $x = \pm\sqrt{24}$. If $\lambda = \frac{1}{4}$, the first equation becomes $2x + 2 = \frac{1}{2}x$ so that $x = -\frac{4}{3}$. The constraint $x^2 + 4y^2 = 24$ now gives $y = \pm\frac{\sqrt{50}}{3}$. Finally, we compare the function values at all of these points (the one interior critical point and the candidates for boundary extrema):

$$T(-1, 0) = -1,$$
$$T(\sqrt{24}, 0) = 24 + 2\sqrt{24} \approx 33.8,$$
$$T(-\sqrt{24}, 0) = 24 - 2\sqrt{24} \approx 14.2,$$
$$T\left(-\frac{4}{3}, \frac{\sqrt{50}}{3}\right) = \frac{14}{3} \approx 4.7$$

and $$T\left(-\frac{4}{3}, -\frac{\sqrt{50}}{3}\right) = \frac{14}{3} \approx 4.7.$$

From this list, it's easy to identify the minimum value of -1 at the point $(-1, 0)$ and the maximum value of $24 + 2\sqrt{24}$ at the point $(\sqrt{24}, 0)$. ■

In example 8.4, we illustrate the use of Lagrange multipliers for functions of three variables. In the course of doing so, we develop an interpretation of the Lagrange multiplier λ.

EXAMPLE 8.4 Finding an Optimal Level of Production

For a business that produces three products, suppose that when producing x, y and z thousand units of the products, the profit of the company (in thousands of dollars) can be modeled by $P(x, y, z) = 4x + 8y + 6z$. Manufacturing constraints force $x^2 + 4y^2 + 2z^2 \le 800$. Find the maximum profit for the company. Rework the problem with the constraint $x^2 + 4y^2 + 2z^2 \le 801$ and use the result to interpret the meaning of λ.

Solution We start with $\nabla P(x, y, z) = \langle 4, 8, 6\rangle$ and note that there are no critical points. This says that the extrema must lie on the boundary of the constraint region. That is, they must satisfy the constraint equation $g(x, y, z) = x^2 + 4y^2 + 2z^2 - 800 = 0$. From Theorem 8.1, the Lagrange multiplier equation is $\nabla P(x, y, z) = \lambda \nabla g(x, y, z)$ or

$$\langle 4, 8, 6\rangle = \lambda\langle 2x, 8y, 4z\rangle = \langle 2\lambda x, 8\lambda y, 4\lambda z\rangle.$$

This occurs when $4 = 2\lambda x, \quad 8 = 8\lambda y \quad \text{and} \quad 6 = 4\lambda z.$

From the first equation, we get $x = \frac{2}{\lambda}$. The second equation gives us $y = \frac{1}{\lambda}$ and the third equation gives us $z = \frac{3}{2\lambda}$. From the constraint equation $x^2 + 4y^2 + 2z^2 = 800$, we now have

$$800 = \left(\frac{2}{\lambda}\right)^2 + 4\left(\frac{1}{\lambda}\right)^2 + 2\left(\frac{3}{2\lambda}\right)^2 = \frac{25}{2\lambda^2},$$

so that $\lambda^2 = \frac{25}{1600}$ and $\lambda = \frac{1}{8}.$

(Why did we choose the positive sign for λ? Hint: Think about what x, y and z represent. Since $x > 0$, we must have $\lambda = \frac{2}{x} > 0$.) The only candidate for an extremum is then

$$x = \frac{2}{\lambda} = 16, \quad y = \frac{1}{\lambda} = 8 \quad \text{and} \quad z = \frac{3}{2\lambda} = 12,$$

and the corresponding profit is

$$P(16, 8, 9) = 4(16) + 8(8) + 6(12) = 200.$$

Observe that this is the maximum profit. Notice that if the constant on the right-hand side of the constraint equation is changed to 801, the first difference occurs in solving for λ, where we now get

$$801 = \frac{25}{2\lambda^2},$$

so that $\lambda \approx 0.12492$, $x = \frac{2}{\lambda} \approx 16.009997$, $y = \frac{1}{\lambda} \approx 8.004998$ and $z = \frac{3}{2\lambda} \approx 12.007498$. In this case, the maximum profit is

$$P\left(\frac{2}{\lambda}, \frac{1}{\lambda}, \frac{3}{2\lambda}\right) \approx 200.12496.$$

It is interesting to observe that the increase in profit is

$$P\left(\frac{2}{\lambda}, \frac{1}{\lambda}, \frac{3}{2\lambda}\right) - P(16, 8, 9) \approx 200.12496 - 200 = 0.12496 \approx \lambda.$$

As you might suspect from this observation, the Lagrange multiplier λ actually gives you the instantaneous rate of change of the profit with respect to a change in the production constraint. ■

We close this section by considering the case of finding the minimum or maximum value of a differentiable function $f(x, y, z)$ subject to two constraints $g(x, y, z) = 0$ and $h(x, y, z) = 0$, where g and h are also differentiable. Notice that for both constraints to be satisfied at a point (x, y, z), the point must lie on both surfaces defined by the constraints. Consequently, in order for there to be a solution, we must assume that the two surfaces intersect. We further assume that ∇g and ∇h are nonzero and are not parallel, so that the two surfaces intersect in a curve C and are not tangent to one another. As we have already

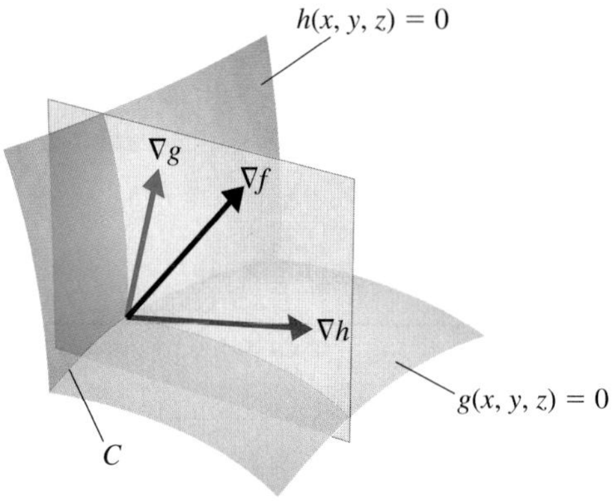

FIGURE 12.53
Constraint surfaces and the plane determined by the normal vectors ∇g and ∇h

seen, if f has an extremum at a point (x_0, y_0, z_0) on a curve C, then $\nabla f(x_0, y_0, z_0)$ must be normal to the curve. Notice that since C lies on both constraint surfaces, $\nabla g(x_0, y_0, z_0)$ and $\nabla h(x_0, y_0, z_0)$ are both orthogonal to C at (x_0, y_0, z_0). This says that $\nabla f(x_0, y_0, z_0)$ must lie in the plane determined by $\nabla g(x_0, y_0, z_0)$ and $\nabla h(x_0, y_0, z_0)$ (see Figure 12.53). That is, for $(x, y, z) = (x_0, y_0, z_0)$ and some constants λ and μ (Lagrange multipliers),

$$\nabla f(x, y, z) = \lambda \nabla g(x, y, z) + \mu \nabla h(x, y, z).$$

The method of Lagrange multipliers for the case of two constraints then consists of finding the point (x, y, z) and the Lagrange multipliers λ and μ (for a total of five unknowns) satisfying the five equations defined by:

$$f_x(x, y, z) = \lambda g_x(x, y, z) + \mu h_x(x, y, z),$$
$$f_y(x, y, z) = \lambda g_y(x, y, z) + \mu h_y(x, y, z),$$
$$f_z(x, y, z) = \lambda g_z(x, y, z) + \mu h_z(x, y, z),$$
$$g(x, y, z) = 0$$

and

$$h(x, y, z) = 0.$$

We illustrate the use of Lagrange multipliers for the case of two constraints in example 8.5.

EXAMPLE 8.5 Optimization with Two Constraints

The plane $x + y + z = 12$ intersects the paraboloid $z = x^2 + y^2$ in an ellipse. Find the point on the ellipse that is closest to the origin.

Solution We illustrate the intersection of the plane with the paraboloid in Figure 12.54. Observe that minimizing the distance to the origin is equivalent to minimizing $f(x, y, z) = x^2 + y^2 + z^2$ [the *square* of the distance from the point (x, y, z) to the origin]. Further, the constraints may be written as $g(x, y, z) = x + y + z - 12 = 0$ and $h(x, y, z) = x^2 + y^2 - z = 0$. At any extremum, we must have that

$$\nabla f(x, y, z) = \lambda \nabla g(x, y, z) + \mu \nabla h(x, y, z)$$

or

$$\langle 2x, 2y, 2z \rangle = \lambda \langle 1, 1, 1 \rangle + \mu \langle 2x, 2y, -1 \rangle.$$

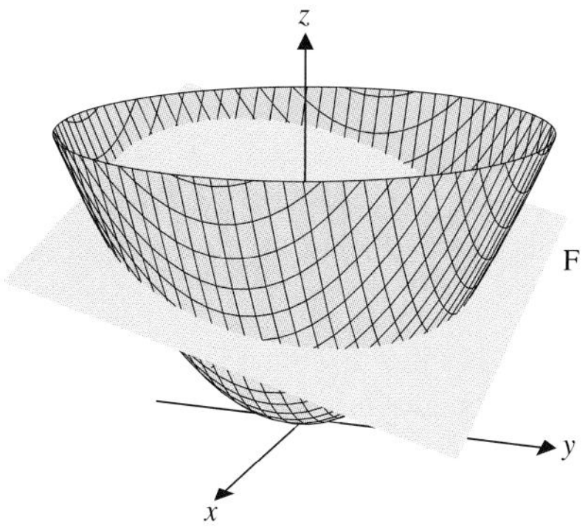

FIGURE 12.54
Intersection of a paraboloid and a plane

Together with the constraint equations, we now have the system of equations

$$2x = \lambda + 2\mu x, \quad (8.1)$$
$$2y = \lambda + 2\mu y, \quad (8.2)$$
$$2z = \lambda - \mu, \quad (8.3)$$
$$x + y + z - 12 = 0 \quad (8.4)$$

and

$$x^2 + y^2 - z = 0. \quad (8.5)$$

From (8.1), we have $\lambda = 2x(1 - \mu)$,

while from (8.2), we have $\lambda = 2y(1 - \mu)$.

Setting these two expressions for λ equal gives us

$$2x(1 - \mu) = 2y(1 - \mu),$$

from which it follows that either $\mu = 1$ (in which case $\lambda = 0$) or $x = y$. However, if $\mu = 1$ and $\lambda = 0$, we have from (8.3) that $z = -\frac{1}{2}$, which contradicts (8.5). Consequently, the only possibility is to have $x = y$, from which it follows from (8.5) that $z = 2x^2$. Substituting this into (8.4) gives us

$$\begin{aligned} 0 &= x + y + z - 12 = x + x + 2x^2 - 12 \\ &= 2x^2 + 2x - 12 = 2(x^2 + x - 6) = 2(x + 3)(x - 2), \end{aligned}$$

so that $x = -3$ or $x = 2$. Since $y = x$ and $z = 2x^2$, we have that (2, 2, 8) and (−3, −3, 18) are the only candidates for extrema. Finally, since

$$f(2, 2, 8) = 72 \quad \text{and} \quad f(-3, -3, 18) = 342,$$

the closest point on the intersection of the two surfaces to the origin is (2, 2, 8). By the same reasoning, observe that the farthest point on the intersection of the two surfaces from the origin is (−3, −3, 18). Notice that these are also consistent with what you can see in Figure 12.54. ■

The method of Lagrange multipliers can be extended in a straightforward fashion to the case of minimizing or maximizing a function of any number of variables subject to any number of constraints.

EXERCISES 12.8

WRITING EXERCISES

1. Explain why the point of tangency in Figure 12.50d must be the closest point to the origin.
2. Explain why in example 8.1 you know that the critical point found corresponds to the minimum distance and not the maximum distance or a saddle point.
3. In example 8.2, explain in physical terms why there would be a value of u that would maximize the rocket's height. In particular, explain why a larger value of u wouldn't *always* produce a larger height.
4. In example 8.4, we showed that the Lagrange multiplier λ corresponds to the rate of change of profit with respect to a change in production level. Explain how knowledge of this value (positive, negative, small, large) would be useful to a plant manager.

In exercises 1–8, use Lagrange multipliers to find the closest point on the given curve to the indicated point.

1. $y = 3x - 4$, origin
2. $y = 2x + 1$, origin
3. $y = 3 - 2x$, $(4, 0)$
4. $y = x - 2$, $(0, 2)$
5. $y = x^2$, $(3, 0)$
6. $y = x^2$, $(0, 2)$
7. $y = x^2$, $\left(2, \frac{1}{2}\right)$
8. $y = x^2 - 1$, $(1, 2)$

In exercises 9–16, use Lagrange multipliers to find the maximum and minimum of the function $f(x, y)$ subject to the constraint $g(x, y) = c$.

9. $f(x, y) = 4xy$ subject to $x^2 + y^2 = 8$
10. $f(x, y) = 4xy$ subject to $4x^2 + y^2 = 8$
11. $f(x, y) = 4x^2y$ subject to $x^2 + y^2 = 3$
12. $f(x, y) = 2x^3y$ subject to $x^2 + y^2 = 4$
13. $f(x, y) = xe^y$ subject to $x^2 + y^2 = 2$
14. $f(x, y) = e^{2x+y}$ subject to $x^2 + y^2 = 5$
15. $f(x, y) = x^2e^y$ subject to $x^2 + y^2 = 3$
16. $f(x, y) = x^2y^2$ subject to $x^2 + 4y^2 = 24$

In exercises 17–20, find the maximum and minimum of the function $f(x, y)$ subject to the constraint $g(x, y) \le c$.

17. $f(x, y) = 4xy$ subject to $x^2 + y^2 \le 8$
18. $f(x, y) = 4xy$ subject to $4x^2 + y^2 \le 8$
19. $f(x, y) = 4x^2y$ subject to $x^2 + y^2 \le 3$
20. $f(x, y) = 2x^3y$ subject to $x^2 + y^2 \le 4$
21. Rework example 8.2 with extra fuel, so that $u^2t = 11{,}000$.
22. In exercise 21, compute λ. Comparing solutions to example 8.2 and exercise 21, compute the change in z divided by the change in u^2t.
23. Solve example 8.2 by substituting $t = 10{,}000/u^2$ into the height equation. Be sure to show that your solution represents a *maximum* height.
24. In example 8.2, the general constraint is $u^2t = k$ and the resulting maximum height is $h(k)$. Use the technique of exercise 23 and the results of example 8.2 to show that $\lambda = h'(k)$.
25. Suppose that the business in example 8.4 has profit function $P(x, y, z) = 3x + 6y + 6z$ and manufacturing constraint $2x^2 + y^2 + 4z^2 \le 8800$. Maximize the profits.
26. Suppose that the business in example 8.4 has profit function $P(x, y, z) = 3xz + 6y$ and manufacturing constraint $x^2 + 2y^2 + z^2 \le 6$. Maximize the profits.
27. In exercise 25, show that the Lagrange multiplier gives the rate of change of the profit relative to a change in the production constraint.
28. Use the value of λ (do not solve any equations) to determine the amount of profit if the constraint in exercise 26 is changed to $x^2 + 2y^2 + z^2 \le 7$.
29. Minimize $2x + 2y$ subject to the constraint $xy = c$ for some constant $c > 0$ and conclude that for a given area, the rectangle with smallest perimeter is the square.
30. As in exercise 29, find the rectangular box of a given volume that has the minimum surface area.
31. Maximize $y - x$ subject to the constraint $x^2 + y^2 = 1$.
32. Maximize e^{x+y} subject to the constraint $x^2 + y^2 = 2$.
33. In the picture, a sailboat is sailing into a crosswind. The wind is blowing out of the north; the sail is at an angle α to the east of due north and at an angle β north of the hull of the boat. The hull, in turn, is at an angle θ to the north of due east. Explain why $\alpha + \beta + \theta = \frac{\pi}{2}$. If the wind is blowing with speed w, then the northward component of the wind's force on the boat is given by $w \sin\alpha \sin\beta \sin\theta$. If this component is positive, the boat can travel "against the wind." Taking $w = 1$ for convenience, maximize $\sin\alpha \sin\beta \sin\theta$ subject to the constraint $\alpha + \beta + \theta = \frac{\pi}{2}$.

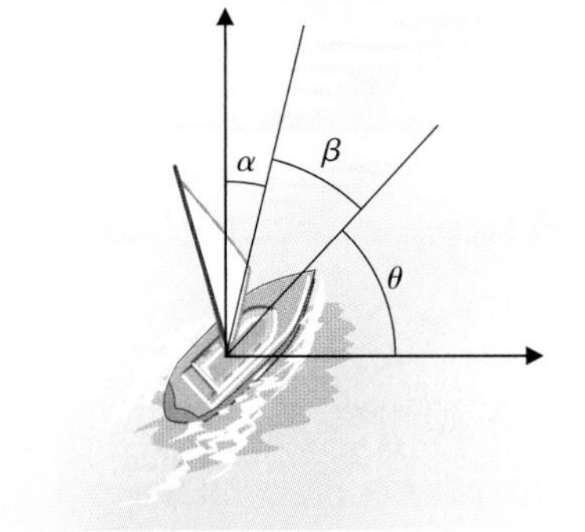

34. Suppose a music company sells two types of speakers. The profit for selling x speakers of style A and y speakers of style B is modeled by $f(x, y) = x^3 + y^3 - 5xy$. The company can't manufacture more than k speakers total in a given month for some constant $k > 5$. Show that the maximum profit is $\dfrac{k^2(k-5)}{4}$ and show that $\lambda = \dfrac{df}{dk}$.

35. Consider the problem of finding extreme values of xy^2 subject to $x + y = 0$. Show that the Lagrange multiplier method identifies $(0, 0)$ as a critical point. Show that this point is neither a local minimum nor a local maximum.

36. Make the substitution $y = -x$ in the function $f(x, y) = xy^2$. Show that $x = 0$ is a critical point and determine what type point is at $x = 0$. Explain why the Lagrange multiplier method fails in exercise 35.

37. The production of a company is given by the Cobb-Douglas function $P = 200L^{2/3}K^{1/3}$. Cost constraints on the business force $2L + 5K \le 150$. Find the values of the labor L and capital K to maximize production.

38. Maximize the profit $P = 4x + 5y$ of a business given the production possibilities constraint curve $2x^2 + 5y^2 \le 32{,}500$.

In exercises 39 and 40, you will illustrate the least-cost rule.

39. Minimize the cost function $C = 25L + 100K$, given the production constraint $P = 60L^{2/3}K^{1/3} = 1920$.

40. In exercise 39, show that the minimum cost occurs when the ratio of marginal productivity of labor, $\dfrac{\partial P}{\partial L}$, to the marginal productivity of capital, $\dfrac{\partial P}{\partial K}$, equals the ratio of the price of labor, $\dfrac{\partial C}{\partial L}$, to the price of capital, $\dfrac{\partial C}{\partial K}$.

41. A person has \$300 to spend on entertainment. Assume that CDs cost \$10 apiece, DVDs cost \$15 apiece and the person's utility function is $10c^{0.4}d^{0.6}$ for buying c CDs and d DVDs. Find c and d to maximize the utility function.

42. To generalize exercise 41, suppose that on a fixed budget of \$$k$ you buy x units of product A purchased at \$$a$ apiece and y units of product B purchased at \$$b$ apiece. For the utility function $x^p y^q$ with $p + q = 1$ and $0 < p < 1$, show that the utility function is maximized with $x = \dfrac{kp}{a}$ and $y = \dfrac{kq}{b}$.

Exercises 43–46 involve optimization with two constraints.

43. Minimize $f(x, y, z) = x^2 + y^2 + z^2$, subject to the constraints $x + 2y + 3z = 6$ and $y + z = 0$.

44. Interpret the function $f(x, y, z)$ of exercise 43 in terms of the distance from a point (x, y, z) to the origin. Sketch the two planes given in exercise 43. Interpret exercise 43 as finding the closest point on a line to the origin.

45. Maximize $f(x, y, z) = xyz$, subject to the constraints $x + y + z = 4$ and $x + y - z = 0$.

46. Maximize $f(x, y, z) = 3x + y + 2z$, subject to the constraints $y^2 + z^2 = 1$ and $x + y - z = 1$.

47. Find the points on the intersection of $x^2 + y^2 = 1$ and $x^2 + z^2 = 1$ that are (a) closest to and (b) farthest from the origin.

48. Find the point on the intersection of $x + 2y + z = 2$ and $y = x$ that is closest to the origin.

49. Use Lagrange multipliers to explore the problem of finding the closest point on $y = x^n$ to the point $(0, 1)$, for some positive integer n. Show that $(0, 0)$ is always a solution to the Lagrange multiplier equation. Show that $(0, 0)$ is the location of a local maximum for $n = 2$, but a local minimum for $n > 2$. As $n \to \infty$, show that the difference between the absolute minimum and the local minimum at $(0, 0)$ goes to 0.

50. Repeat example 8.4 with constraints $x \ge 0$, $y \ge 0$ and $z \ge 0$. Note that you can find the maximum on the boundary $x = 0$ by maximizing $8y + 6z$ subject to $4y^2 + 2z^2 \le 800$.

51. Estimate the closest point on the paraboloid $z = x^2 + y^2$ to the point $(1, 0, 0)$.

52. Estimate the closest point on the hyperboloid $x^2 + y^2 - z^2 = 1$ to the point $(0, 2, 0)$.

EXPLORATORY EXERCISES

1. (This exercise was suggested by Adel Faridani of Oregon State University.) The maximum height found in example 8.2 is actually the height at the time the fuel runs out. A different problem is to find the maximum total height, including the extra height gained after the fuel runs out. In this exercise, we find the value of u that maximizes the total height. First, find the velocity and height when the fuel runs out. (Hint: This should be a function of u only.) Then find the total height of a rocket with that initial height and initial velocity, again assuming that gravity is the only force. Find u to maximize this function. Compare this u-value with that found in example 8.2. Explain in physical terms why this one is larger.

2. Find the maximum value of $f(x, y, z) = xyz$ subject to $x + y + z = 1$ with $x > 0$, $y > 0$ and $z > 0$. Conclude that $\sqrt[3]{xyz} \le \frac{x+y+z}{3}$. The expression on the left is called the **geometric mean** of x, y and z while the expression on the right is the more familiar **arithmetic mean.** Generalize this result in two ways. First, show that the geometric mean of any three positive numbers does not exceed the arithmetic mean. Then, show that the geometric mean of any number of positive numbers does not exceed the arithmetic mean.

Review Exercises

WRITING EXERCISES

The following list includes terms that are defined and theorems that are stated in this chapter. For each term or theorem, (1) give a precise definition or statement, (2) state in general terms what it means and (3) describe the types of problems with which it is associated.

Level curve	Limit of $f(x, y)$	Continuous
Level surface	Tangent plane	Normal line
Partial derivative	Differential	Differentiable
Linear approximation	Laplacian	Implicit differentiation
Chain rule	Gradient	Local extremum
Directional derivative	Saddle point	Second Derivatives Test
Critical point	Extreme Value Theorem	Lagrange multiplier
Linear regression	Contour plot	Density plot

TRUE OR FALSE

State whether each statement is true or false and briefly explain why. If the statement is false, try to "fix it" by modifying the given statement to a new statement that is true.

1. Quadric surfaces are examples of graphs of functions of two variables.

2. Level curves are traces in planes $z = c$ of the surface $z = f(x, y)$.

3. If a function is continuous on every line through (a, b), then it is continuous at (a, b).

4. $\frac{\partial f}{\partial x}(a, b)$ equals the slope of the tangent line to $z = f(x, y)$ at (a, b).

5. For the partial derivative $\frac{\partial^2 f}{\partial x \partial y}$, the order of partial derivatives does not matter.

6. The normal vector to the tangent plane is given by the partial derivatives.

7. A linear approximation is an equation for a tangent plane.

8. The gradient vector is perpendicular to all level curves.

9. If $D_{\mathbf{u}} f(a, b) < 0$, then $f(a + u_1, b + u_2) < f(a, b)$.

10. A normal vector to the tangent plane to $z = f(x, y)$ is $\nabla f(x, y)$.

11. If $\frac{\partial f}{\partial x}(a, b) > 0$ and $\frac{\partial f}{\partial y}(a, b) < 0$, then there is a saddle point at (a, b).

12. The maximum of f on a region R occurs either at a critical point or on the boundary of R.

13. Solving the equation $\nabla f = \lambda \nabla g$ gives the maximum of f subject to $g = 0$.

In exercises 1–10, sketch the graph of $z = f(x, y)$.

1. $f(x, y) = x^2 - y^2$

2. $f(x, y) = \sqrt{x^2 + y^2}$

3. $f(x, y) = 2 - x^2 - y^2$

4. $f(x, y) = \sqrt{2 - x^2 - y^2}$

5. $f(x, y) = \frac{3}{x^2} + \frac{2}{y^2}$

6. $f(x, y) = \frac{x^5}{y}$

7. $f(x, y) = \sin(x^2 y)$

8. $f(x, y) = \sin(y - x^2)$

9. $f(x, y) = 3xe^y - x^3 - e^{3y}$

10. $f(x, y) = 4x^2 e^y - 2x^4 - e^{4y}$

11. In parts a–f, match the functions to the surfaces.

a. $f(x, y) = \sin xy$

b. $f(x, y) = \sin(x/y)$

c. $f(x, y) = \sin \sqrt{x^2 + y^2}$

d. $f(x, y) = x \sin y$

e. $f(x, y) = \frac{4}{2x^2 + 3y^2 - 1}$

f. $f(x, y) = \frac{4}{2x^2 + 3y^2 + 1}$

Review Exercises

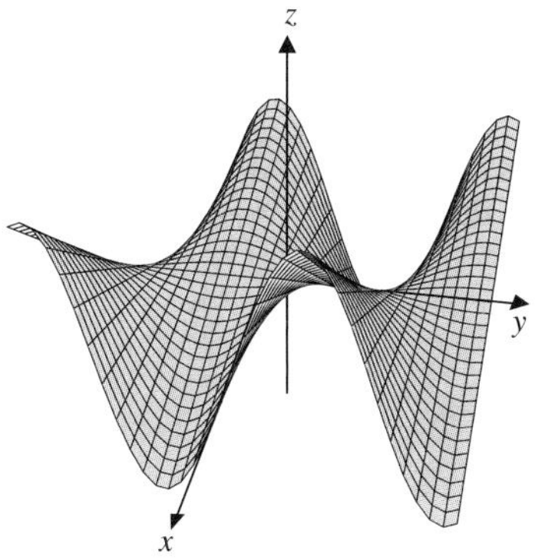

SURFACE A

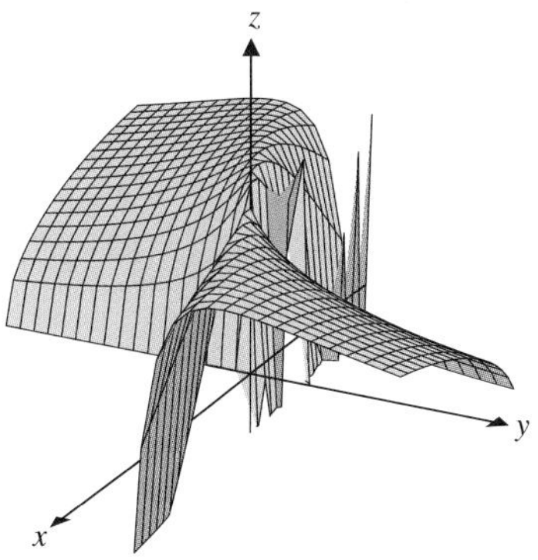

SURFACE B

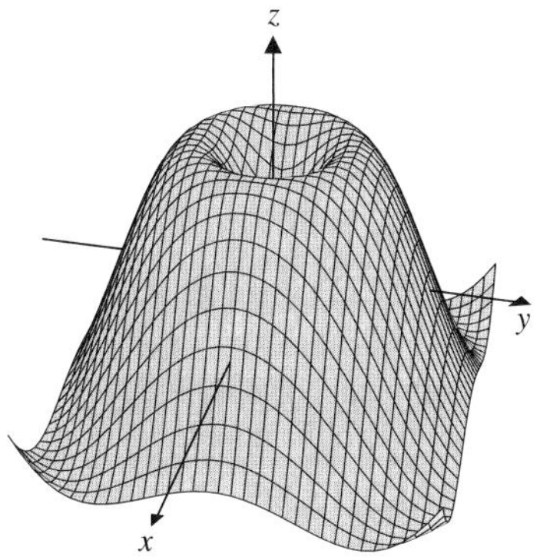

SURFACE C

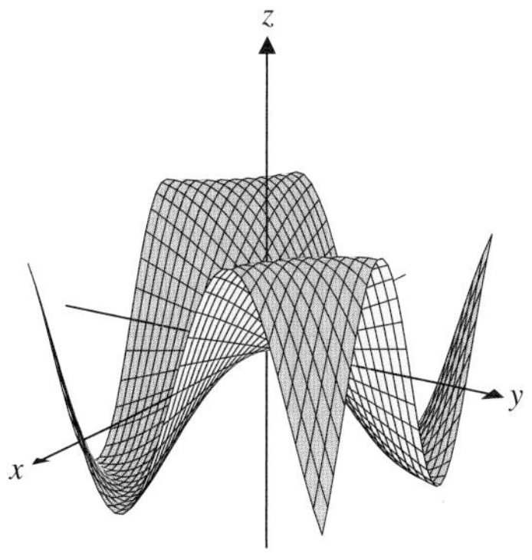

SURFACE D

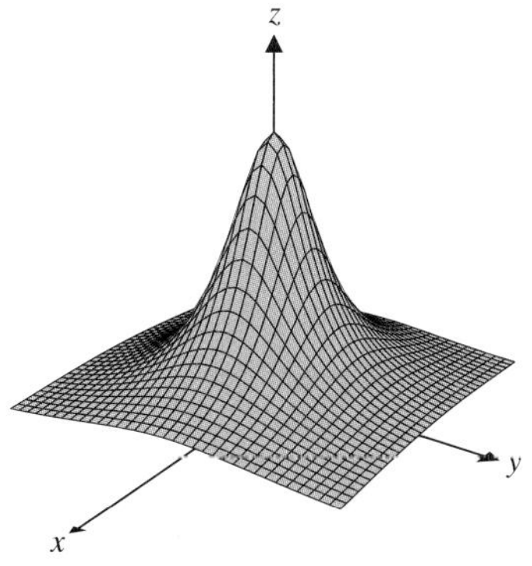

SURFACE E

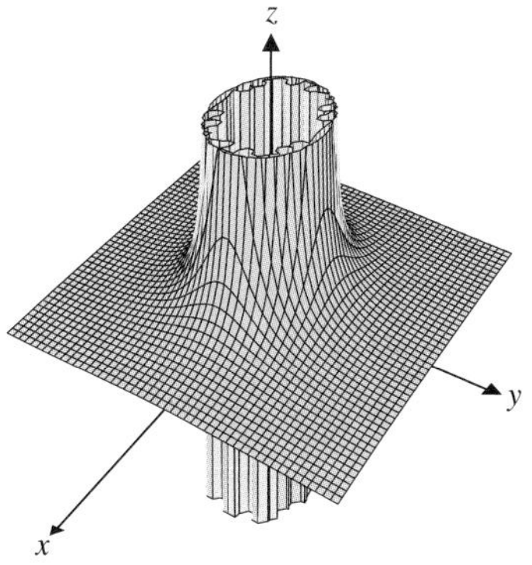

SURFACE F

Review Exercises

12. In parts a–d, match the surfaces to the contour plots.

a.

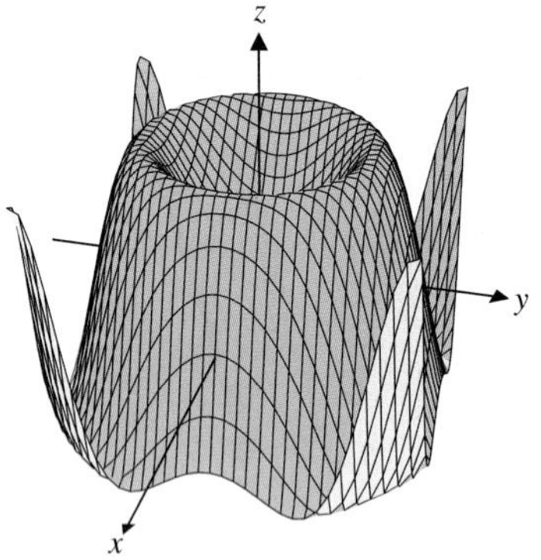

b.

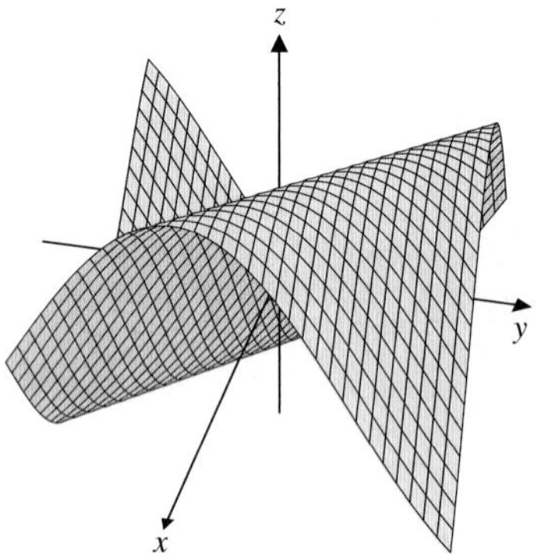

c.

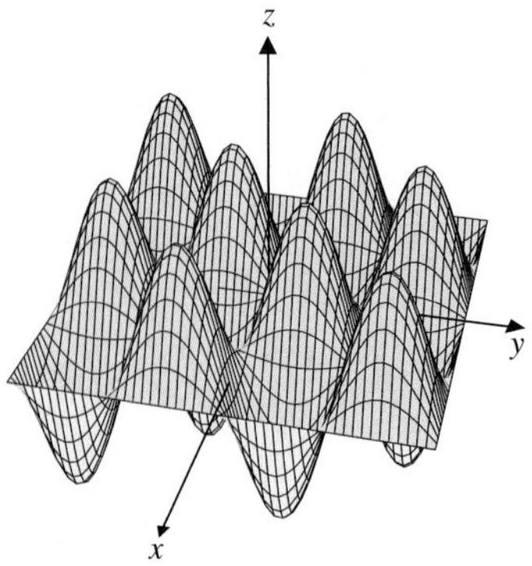

d.

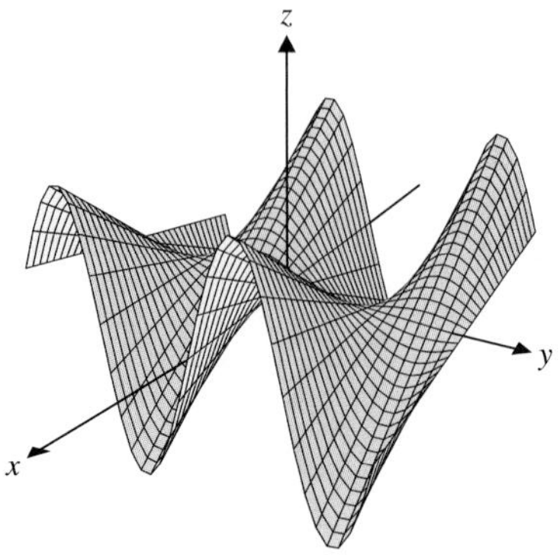

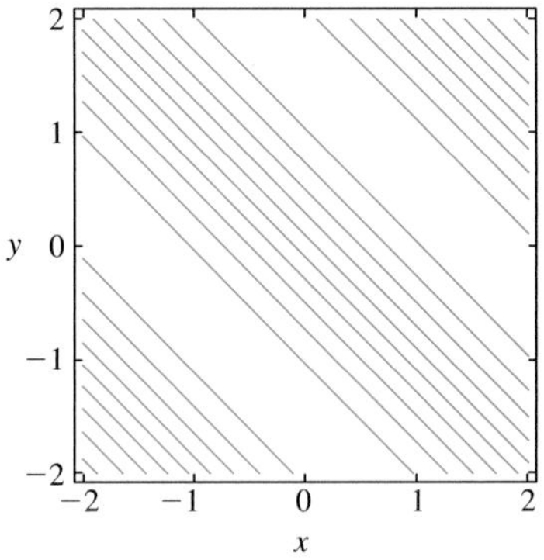

CONTOUR A

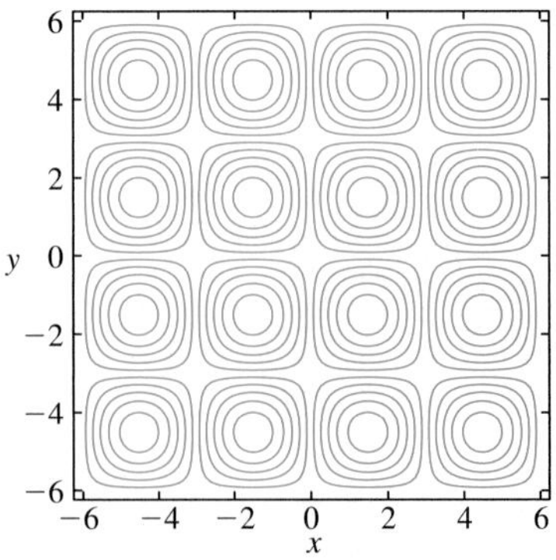

CONTOUR B

Review Exercises

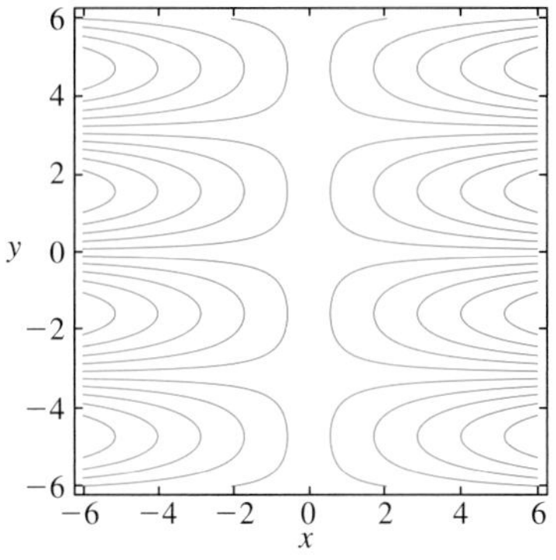

CONTOUR C

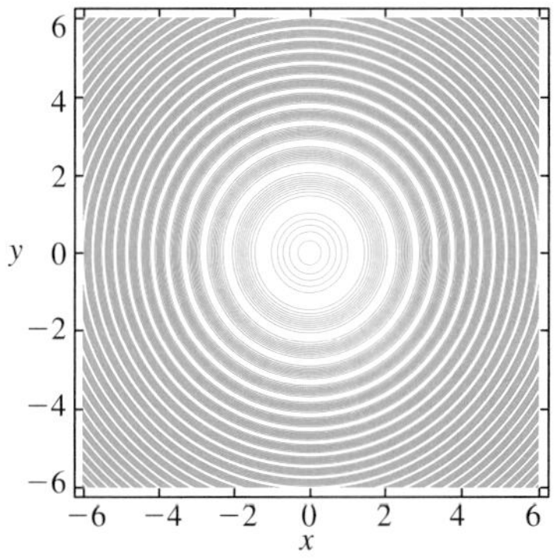

CONTOUR D

13. In parts a–d, match the density plots to the contour plots of exercise 12.

a.

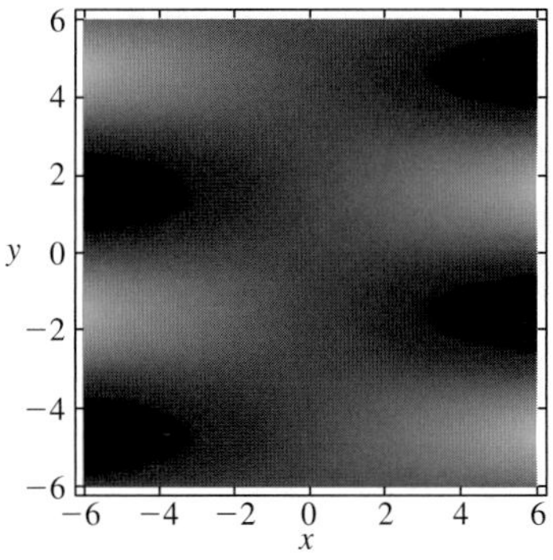

b.

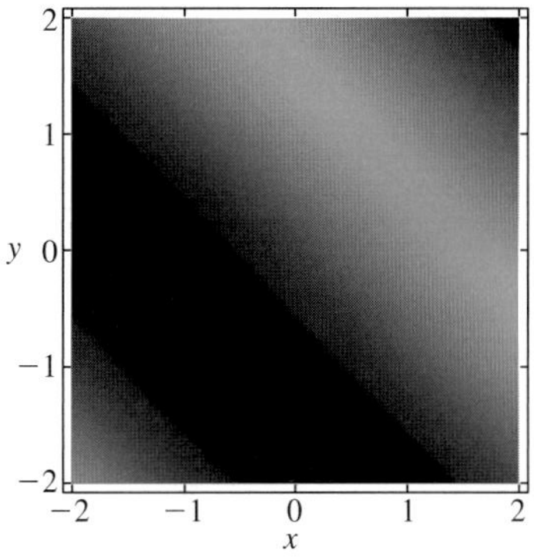

c.

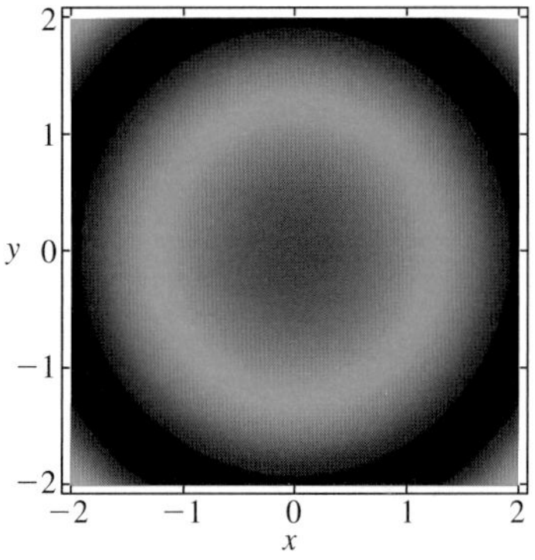

d.

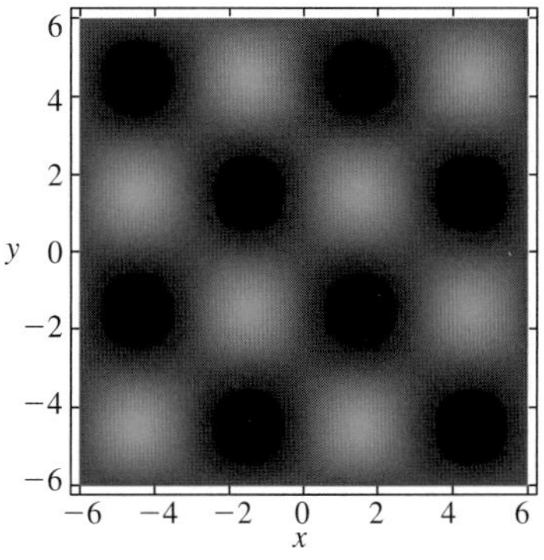

14. Compute the indicated limit.

a. $\lim_{(x,y)\to(0,2)} \dfrac{3x}{y^2+1}$

b. $\lim_{(x,y)\to(1,\pi)} \dfrac{xy-1}{\cos xy}$

Review Exercises

In exercises 15–18, show that the indicated limit does not exist.

15. $\lim_{(x,y)\to(0,0)} \frac{3x^2y}{x^4+y^2}$

16. $\lim_{(x,y)\to(0,0)} \frac{2xy^{3/2}}{x^2+y^3}$

17. $\lim_{(x,y)\to(0,0)} \frac{x^2+y^2}{x^2+xy+y^2}$

18. $\lim_{(x,y)\to(0,0)} \frac{x^2}{x^2+xy+y^2}$

In exercises 19 and 20, show that the indicated limit exists.

19. $\lim_{(x,y)\to(0,0)} \frac{x^3+xy^2}{x^2+y^2}$

20. $\lim_{(x,y)\to(0,0)} \frac{3y^2\ln(x+1)}{x^2+3y^2}$

In exercises 21 and 22, find the region on which the function is continuous.

21. $f(x,y) = 3x^2e^{4y} - \frac{3y}{x}$

22. $f(x,y) = \sqrt{4-4x^2-y^2}$

In exercises 23–26, find both first-order partial derivatives.

23. $f(x,y) = \frac{4x}{y} + xe^{xy}$

24. $f(x,y) = xe^{xy} + 3y^2$

25. $f(x,y) = 3x^2y\cos y - \sqrt{x}$

26. $f(x,y) = \sqrt{x^3y} + 3x - 5$

27. Show that the function $f(x,y) = e^x \sin y$ satisfies **Laplace's equation** $\frac{\partial^2 f}{\partial x^2} + \frac{\partial^2 f}{\partial y^2} = 0$.

28. Show that the function $f(x,y) = e^x \cos y$ satisfies Laplace's equation. (See exercise 27.)

In exercises 29 and 30, use the chart to estimate the partial derivatives.

29. $\frac{\partial f}{\partial x}(0,0)$ and $\frac{\partial f}{\partial y}(0,0)$

30. $\frac{\partial f}{\partial x}(10,0)$ and $\frac{\partial f}{\partial y}(10,0)$

y \ x	−20	−10	0	10	20
−20	2.4	2.1	0.8	0.5	1.0
−10	2.6	2.2	1.4	1.0	1.2
0	2.7	2.4	2.0	1.6	1.2
10	2.9	2.5	2.6	2.2	1.8
20	3.1	2.7	3.0	2.9	2.7

In exercises 31–34, compute the linear approximation of the function at the given point.

31. $f(x,y) = 3y\sqrt{x^2+5}$ at $(-2, 5)$

32. $f(x,y) = \frac{x+2}{4y-2}$ at $(2, 3)$

33. $f(x,y) = \tan(x+2y)$ at $(\pi, \frac{\pi}{2})$

34. $f(x,y) = \ln(x^2+3y)$ at $(4, 2)$

In exercises 35 and 36, find the indicated derivatives.

35. $f(x,y) = 2x^4y + 3x^2y^2$; f_{xx}, f_{yy}, f_{xy}

36. $f(x,y) = x^2e^{3y} - \sin y$; f_{xx}, f_{yy}, f_{yyx}

In exercises 37–40, find an equation of the tangent plane.

37. $z = x^2y + 2x - y^2$ at $(1, -1, 0)$

38. $z = \sqrt{x^2+y^2}$ at $(3, -4, 5)$

39. $x^2 + 2xy + y^2 + z^2 = 5$ at $(0, 2, 1)$

40. $x^2z - y^2x + 3y - z = -4$ at $(1, -1, 2)$

In exercises 41 and 42, use the chain rule to find the indicated derivative(s).

41. $g'(t)$ where $g(t) = f(x(t), y(t))$, $f(x,y) = x^2y + y^2$, $x(t) = e^{4t}$ and $y(t) = \sin t$

42. $\frac{\partial g}{\partial u}$ and $\frac{\partial g}{\partial v}$ where $g(u,v) = f(x(u,v), y(u,v))$, $f(x,y) = 4x^2 - y$, $x(u,v) = u^3v + \sin u$ and $y(u,v) = 4v^2$

In exercises 43 and 44, state the chain rule for the general composite function.

43. $g(t) = f(x(t), y(t), z(t), w(t))$

44. $g(u,v) = f(x(u,v), y(u,v))$

In exercises 45 and 46, use implicit differentiation to find $\frac{\partial z}{\partial x}$ and $\frac{\partial z}{\partial y}$.

45. $x^2 + 2xy + y^2 + z^2 = 1$

46. $x^2z - y^2x + 3y - z = -4$

In exercises 47 and 48, find the gradient of the given function at the indicated point.

47. $f(x,y) = 3x\sin 4y - \sqrt{xy}$, (π, π)

48. $f(x,y,z) = 4xz^2 - 3\cos x + 4y^2$, $(0, 1, -1)$

In exercises 49–52, compute the directional derivative of f at the given point in the direction of the indicated vector.

49. $f(x,y) = x^3y - 4y^2$, $(-2, 3)$, $\mathbf{u} = \langle \frac{3}{5}, \frac{4}{5} \rangle$

50. $f(x,y) = x^2 + xy^2$, $(2, 1)$, $\mathbf{u}$ in the direction of $\langle 3, -2 \rangle$

Review Exercises

51. $f(x, y) = e^{3xy} - y^2$, $(0, -1)$, **u** in the direction from $(2, 3)$ to $(3, 1)$

52. $f(x, y) = \sqrt{x^2 + xy^2}$, $(2, 1)$, **u** in the direction of $\langle 1, -2 \rangle$

In exercises 53–56, find the directions of maximum and minimum change of f at the given point, and the values of the maximum and minimum rates of change.

53. $f(x, y) = x^3y - 4y^2$, $(-2, 3)$

54. $f(x, y) = x^2 + xy^2$, $(2, 1)$

55. $f(x, y) = \sqrt{x^4 + y^4}$, $(2, 0)$

56. $f(x, y) = x^2 + xy^2$, $(1, 2)$

57. Suppose that the elevation on a hill is given by $f(x, y) = 100 - 4x^2 - 2y$. From the site at $(2, 1)$, in which direction will the rain run off?

58. If the temperature at the point (x, y, z) is given by $T(x, y, z) = 70 + 5e^{-z^2}(4x + 3y^{-1})$, find the direction from the point $(1, 2, 1)$ in which the temperature decreases most rapidly.

In exercises 59–62, find all critical points and use Theorem 7.2 (if applicable) to classify them.

59. $f(x, y) = 2x^4 - xy^2 + 2y^2$

60. $f(x, y) = 2x^4 + y^3 - x^2y$

61. $f(x, y) = 4xy - x^3 - 2y^2$

62. $f(x, y) = 3xy - x^3y + y^2 - y$

63. The following data show the height and weight of a small number of people. Use the linear model to predict the weight of a 6′2″ person and a 5′0″ person. Comment on how accurate you think the model is.

Height (inches)	64	66	70	71
Weight (pounds)	140	156	184	190

64. The following data show the age and income for a small number of people. Use the linear model to predict the income of a 20-year-old and of a 60-year-old. Comment on how accurate you think the model is.

Age (years)	28	32	40	56
Income ($)	36,000	34,000	88,000	104,000

In exercises 65 and 66, find the absolute extrema of the function on the given region.

65. $f(x, y) = 2x^4 - xy^2 + 2y^2$, $0 \le x \le 4$, $0 \le y \le 2$

66. $f(x, y) = 2x^4 + y^3 - x^2y$, region bounded by $y = 0$, $y = x$ and $x = 2$

In exercises 67–70, use Lagrange multipliers to find the maximum and minimum of the function $f(x, y)$, subject to the constraint $g(x, y) = c$.

67. $f(x, y) = x + 2y$, subject to $x^2 + y^2 = 5$

68. $f(x, y) = 2x^2y$, subject to $x^2 + y^2 = 4$

69. $f(x, y) = xy$, subject to $x^2 + y^2 = 1$

70. $f(x, y) = x^2 + 2y^2 - 2x$, subject to $x^2 + y^2 = 1$

In exercises 71 and 72, use Lagrange multipliers to find the closest point on the given curve to the indicated point.

71. $y = x^3$, $(4, 0)$

72. $y = x^3$, $(2, 1)$

EXPLORATORY EXERCISES

1. The graph (from the excellent book *Tennis Science for Tennis Players* by Howard Brody) shows the vertical angular acceptance of a tennis serve as a function of velocity and the height at which the ball is hit. With vertical angular acceptance, Brody is measuring a margin of error. For example, if serves with angles ranging from 5° to 8° will land in the service box (for a given height and velocity), the vertical angular acceptance is 3°. For a given height, does the angular acceptance increase or decrease as velocity increases? Explain why this is reasonable. For a given velocity, does the angular acceptance increase or decrease as height increases? Explain why this is reasonable.

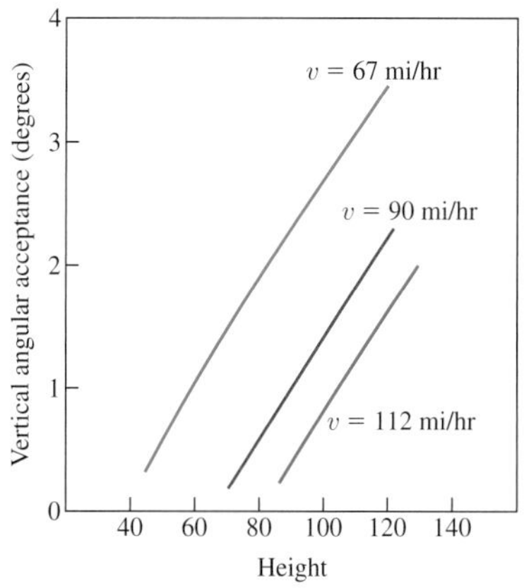

Review Exercises

2. The graphic in exercise 1 is somewhat like a contour plot. Assuming that angular acceptance is the dependent variable, explain what is different about this plot from the contour plots drawn in this chapter. Which type of plot do you think is easier to read? The accompanying plot shows angular acceptance in terms of a number of variables. Identify the independent variables and compare this plot to the level surfaces drawn in this chapter.

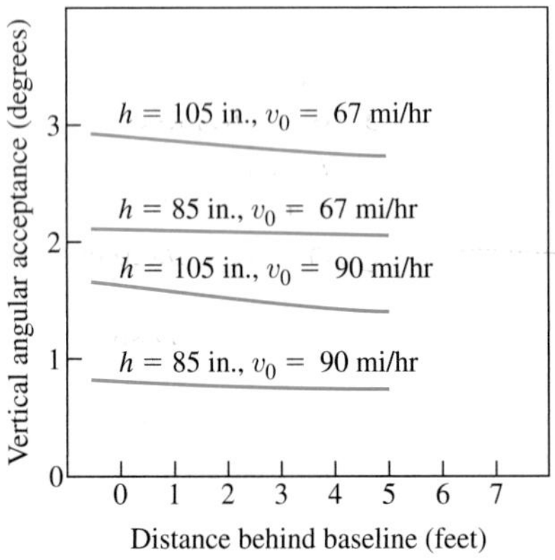

3. The horizontal range of a golf ball or baseball depends on the launch angle and the rate of backspin on the ball. The accompanying figure, reprinted from *Keep Your Eye on the Ball* by Watts and Bahill, shows level curves for this relationship for an initial velocity of 110 mph, although the dependent variable (range) is graphed vertically and the level curves represent constant values of one of the independent variables. Estimate the partial derivatives of range at 30° and 1910 rpm and use them to find a linear approximation of range. Predict the range at 25° and 2500 rpm, and also at 40° and 4000 rpm. Discuss the accuracy of each prediction.

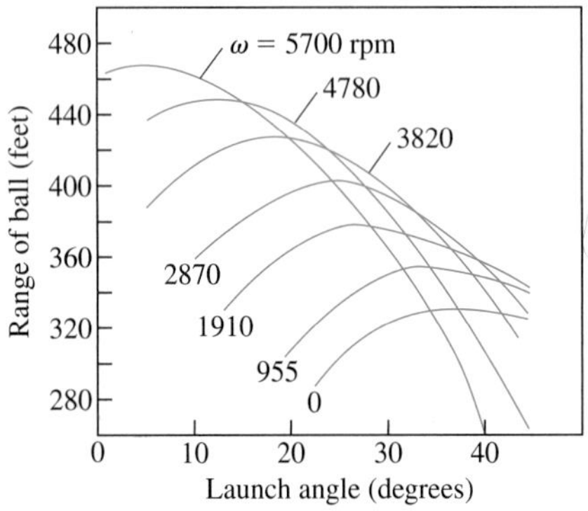

Multiple Integrals

CHAPTER

13

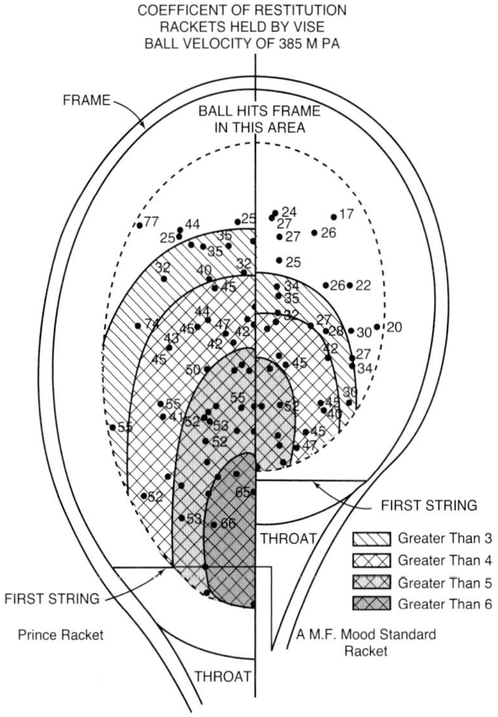

The design of modern sports equipment has become a sophisticated engineering enterprise. Many innovations can be traced back to a brilliant engineer but mediocre athlete named Howard Head. As an aircraft engineer in the 1940s, Head became frustrated learning to ski on the wooden skis of the day. Following years of experimentation, Head revolutionized the ski industry by introducing metal skis designed using principles borrowed from aircraft design.

By 1970, Head had retired from the Head Ski Company as a wealthy ski mogul. He quickly became frustrated by his slow progress learning to play tennis, a sport then played exclusively with wooden rackets. Head again focused on his equipment, reasoning that a larger racket would twist less and therefore be easier to control. However, years of experimentation showed that large wooden rackets either broke easily or were too heavy to swing.

Given that Head's metal skis were successful largely because they reduced the twisting of the skis in turns, it is not surprising that his experimentation turned to oversized metal tennis rackets. The rackets that Head eventually marketed as Prince rackets revolutionized tennis racket design. As the accompanying diagram shows, the sweet spot of the oversized racket is considerably larger than the sweet spot of the smaller wooden racket.

In this chapter, we introduce double and triple integrals, which are needed to compute the mass, moment of inertia and other important properties of three-dimensional solids. The moment of inertia is a measure of the resistance of an object to rotation. As shown in the exercises in section 13.2, compared to smaller rackets, the larger Head rackets have a larger moment of inertia and thus, twist less on off-center shots. Engineers use similar calculations as they test new materials for strength and weight for the next generation of sports equipment.

13.1 DOUBLE INTEGRALS

Before we introduce the idea of a double integral for a function of two variables, we first briefly remind you of the definition of definite integral for a function of a single variable and then generalize the definition slightly. Recall that we defined the definite integral while looking for the area A under the graph of a

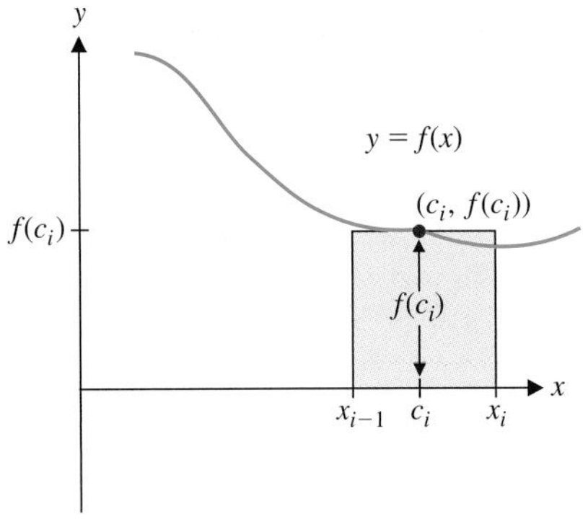

FIGURE 13.1a
Approximating the area on the subinterval $[x_{i-1}, x_i]$

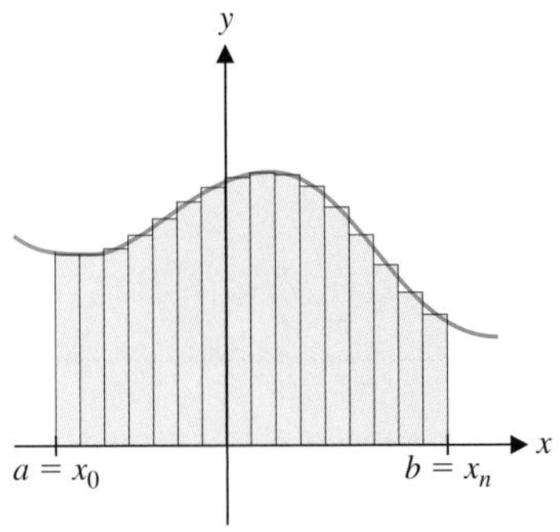

FIGURE 13.1b
Area under the curve

continuous function f defined on an interval $[a, b]$, where $f(x) \geq 0$ on $[a, b]$. We did this by *partitioning* the interval $[a, b]$ into n subintervals $[x_{i-1}, x_i]$, for $i = 1, 2, \ldots, n$, of equal width $\Delta x = \dfrac{b-a}{n}$, where

doesn't have to be equal but equal is better.

$$a = x_0 < x_1 < \cdots < x_n = b.$$

On each subinterval $[x_{i-1}, x_i]$, for $i = 1, 2, \ldots, n$, we approximated the area under the curve by the area of the rectangle of height $f(c_i)$, for some point $c_i \in [x_{i-1}, x_i]$, as indicated in Figure 13.1a. Adding together the areas of all of these rectangles, we obtain an approximation of the area, as indicated in Figure 13.1b:

$$A \approx \sum_{i=1}^{n} f(c_i)\Delta x.$$

Finally, taking the limit as $n \to \infty$ (which also means that $\Delta x \to 0$), we get the exact area (assuming that the limit exists and is the same for all choices of the evaluation points c_i):

$$A = \lim_{n\to\infty} \sum_{i=1}^{n} f(c_i)\Delta x.$$

We defined the definite integral as this limit:

$$\int_a^b f(x)\,dx = \lim_{n\to\infty} \sum_{i=1}^{n} f(c_i)\,\Delta x. \tag{1.1}$$

We generalize this by allowing partitions that are **irregular** (that is, where not all subintervals have the same width). We need this kind of generalization, among other reasons, for more sophisticated numerical methods for approximating definite integrals. This generalization is also needed for theoretical purposes; this is pursued in a more advanced course. We proceed essentially as above, except that we allow different subintervals to have different widths and define the width of the ith subinterval $[x_{i-1}, x_i]$ to be $\Delta x_i = x_i - x_{i-1}$ (see Figure 13.2 for the case where $n = 7$).

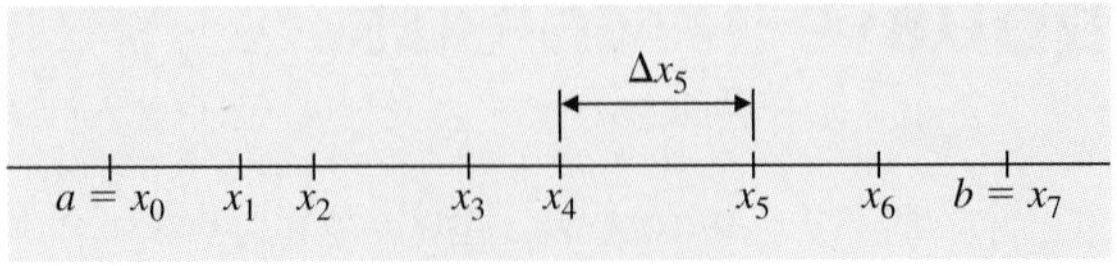

FIGURE 13.2
Irregular partition of $[a, b]$

An approximation of the area is then (essentially, as before)

$$A \approx \sum_{i=1}^{n} f(c_i)\,\Delta x_i,$$

for any choice of the evaluation points $c_i \in [x_{i-1}, x_i]$, for $i = 1, 2, \ldots, n$. To get the exact area, we need to let $n \to \infty$, but since the partition is irregular, this alone will not guarantee that all of the Δx_i's will approach zero. We take a little extra care, by defining $\|P\|$ (the **norm of the partition**) to be the *largest* of all the Δx_i's. We then arrive at the following more general definition of definite integral.

DEFINITION 1.1

For any function f defined on the interval $[a, b]$, the **definite integral** of f on $[a, b]$ is

$$\int_a^b f(x)\,dx = \lim_{\|P\|\to 0} \sum_{i=1}^{n} f(c_i)\,\Delta x_i,$$

provided the limit exists and is the same for all choices of the evaluation points $c_i \in [x_{i-1}, x_i]$, for $i = 1, 2, \ldots, n$. In this case, we say that f is **integrable** on $[a, b]$.

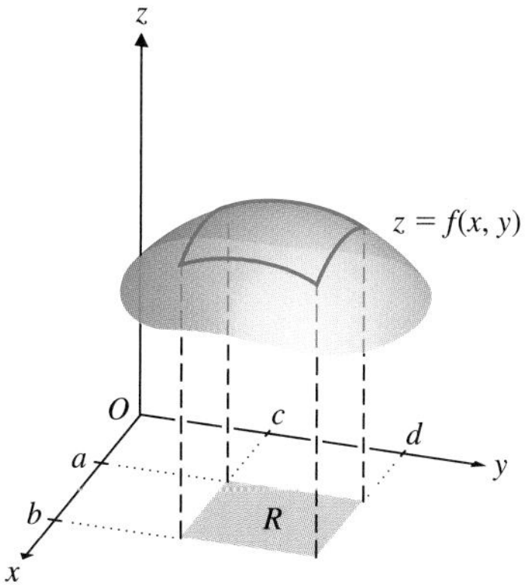

FIGURE 13.3
Volume under the surface $z = f(x, y)$

Here, by saying that the limit in Definition 1.1 equals some value L, we mean that we can make $\sum_{i=1}^{n} f(c_i)\,\Delta x_i$ as close as needed to L, just by making $\|P\|$ sufficiently small. How close must the sum get to L? We must be able to make the sum within any specified distance $\varepsilon > 0$ of L. More precisely, given any $\varepsilon > 0$, there must be a $\delta > 0$ (depending on the choice of ε), such that

$$\left|\sum_{i=1}^{n} f(c_i)\,\Delta x_i - L\right| < \varepsilon,$$

for *every* partition P with $\|P\| < \delta$. Notice that this is only a very slight generalization of our original notion of definite integral. All we have done is to allow the partitions to be irregular and then defined $\|P\|$ to ensure that $\Delta x_i \to 0$, for every i.

While you would likely never use Definition 1.1 to compute an area, your computer or calculator software probably does use irregular partitions to estimate integrals. Definition 1.1 will help us see how to generalize the notion of integral to functions of several variables.

Double Integrals over a Rectangle

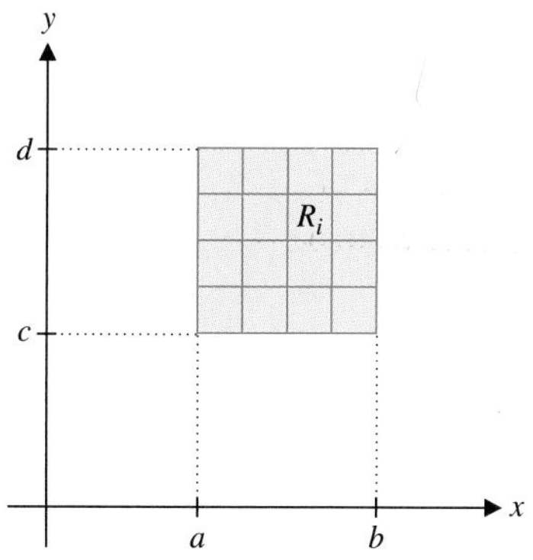

FIGURE 13.4a
Partition of R

We developed the definite integral as a natural by-product of our method for finding area under a curve in the xy-plane. Likewise, we are guided in our development of the double integral by a corresponding problem. For a function $f(x, y)$, where f is continuous and $f(x, y) \ge 0$ for all $a \le x \le b$ and $c \le y \le d$, we wish to find the *volume* of the solid lying below the surface $z = f(x, y)$ and above the rectangle $R = \{(x, y) | a \le x \le b$ and $c \le y \le d\}$ in the xy-plane (see Figure 13.3).

We proceed essentially as we did to find the area under a curve. First, we partition the rectangle R by laying down a grid on top of R consisting of n smaller rectangles (see Figure 13.4a). (Note: The rectangles in the grid need not be all of the same size.) Call the smaller rectangles $R_1, R_2, \ldots, R_n$. (The order in which you number them is irrelevent.) For each rectangle R_i $(i = 1, 2, \ldots, n)$ in the partition, we want to find an approximation to the volume V_i lying beneath the surface $z = f(x, y)$ and above the rectangle R_i. The sum of these approximate volumes is then an approximation to the total volume. Above each rectangle R_i in the partition, construct a rectangular box whose height is $f(u_i, v_i)$, for some point $(u_i, v_i) \in R_i$ (see Figure 13.4b). Notice that the volume V_i beneath the surface and above R_i is approximated by the volume of the box:

$$V_i \approx \text{Height} \times \text{Area of base} = f(u_i, v_i)\,\Delta A_i,$$

where ΔA_i denotes the area of the rectangle R_i.

The total volume is then approximately

$$V \approx \sum_{i=1}^{n} f(u_i, v_i)\,\Delta A_i. \tag{1.2}$$

As in our development of the definite integral in Chapter 4, we call the sum in (1.2) a **Riemann sum.** We illustrate the approximation of the volume under a surface by a Riemann sum in Figures 13.4c and 13.4d. Notice that the larger number of rectangles used in Figure 13.4d appears to give a better approximation of the volume.

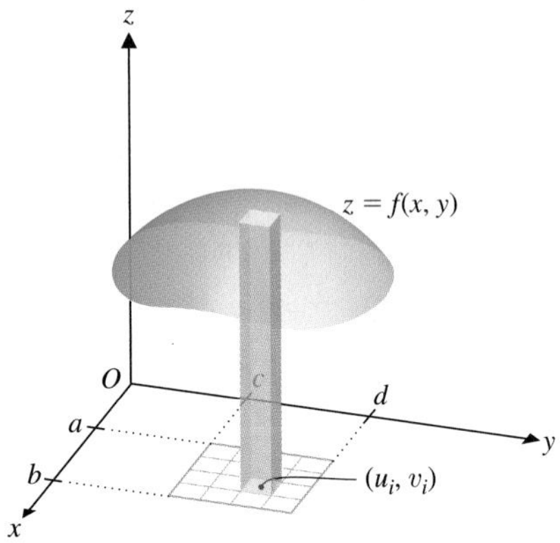

FIGURE 13.4b
Approximating the volume above R_i by a rectangular box

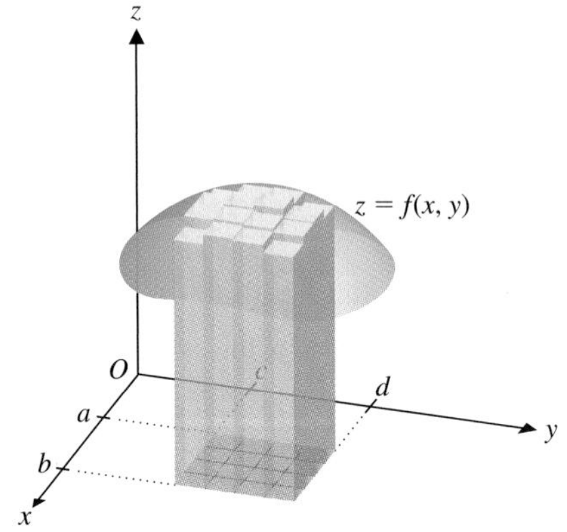

FIGURE 13.4c
Approximate volume

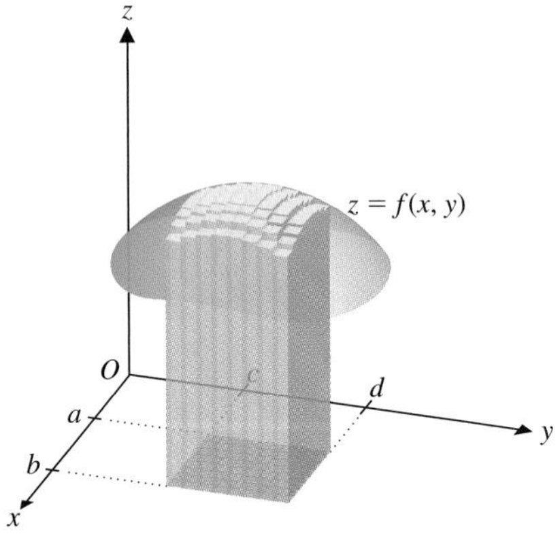

FIGURE 13.4d
Approximate volume

EXAMPLE 1.1 Approximating the Volume Lying Beneath a Surface

Approximate the volume lying beneath the surface $z = x^2 \sin \dfrac{\pi y}{6}$ and above the rectangle $R = \{(x, y) | 0 \le x \le 6, 0 \le y \le 6\}$.

Solution First, note that f is continuous and $f(x, y) = x^2 \sin \dfrac{\pi y}{6} \ge 0$ on R (see Figure 13.5a). Next, a simple partition of R is a partition into four squares of equal size, as indicated in Figure 13.5b. We choose the evaluation points (u_i, v_i) to be the centers of each of the four squares, that is, $\left(\frac{3}{2}, \frac{3}{2}\right)$, $\left(\frac{9}{2}, \frac{3}{2}\right)$, $\left(\frac{3}{2}, \frac{9}{2}\right)$ and $\left(\frac{9}{2}, \frac{9}{2}\right)$.

Since the four squares are the same size, we have $\Delta A_i = 9$, for each i. For $f(x, y) = x^2 \sin \dfrac{\pi y}{6}$, we have from (1.2) that

$$\begin{aligned} V &\approx \sum_{i=1}^{4} f(u_i, v_i)\,\Delta A_i \\ &= f\left(\frac{3}{2}, \frac{3}{2}\right)(9) + f\left(\frac{9}{2}, \frac{3}{2}\right)(9) + f\left(\frac{3}{2}, \frac{9}{2}\right)(9) + f\left(\frac{9}{2}, \frac{9}{2}\right)(9) \\ &= 9\left[\left(\frac{3}{2}\right)^2 \sin\left(\frac{\pi}{4}\right) + \left(\frac{9}{2}\right)^2 \sin\left(\frac{\pi}{4}\right) + \left(\frac{3}{2}\right)^2 \sin\left(\frac{3\pi}{4}\right) + \left(\frac{9}{2}\right)^2 \sin\left(\frac{3\pi}{4}\right)\right] \\ &= \frac{405}{2}\sqrt{2} \approx 286.38. \end{aligned}$$

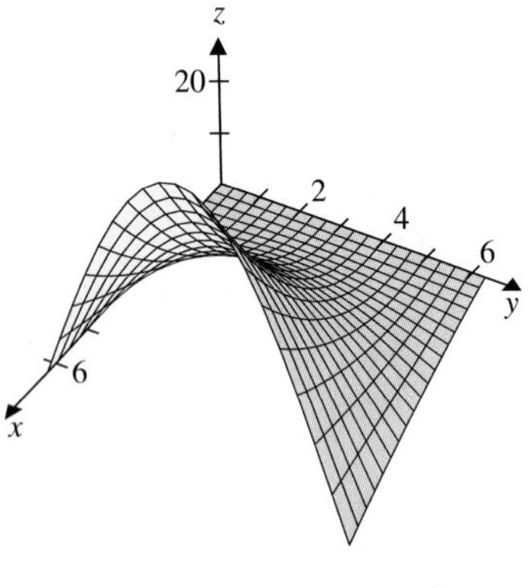

FIGURE 13.5a
$z = x^2 \sin \dfrac{\pi y}{6}$

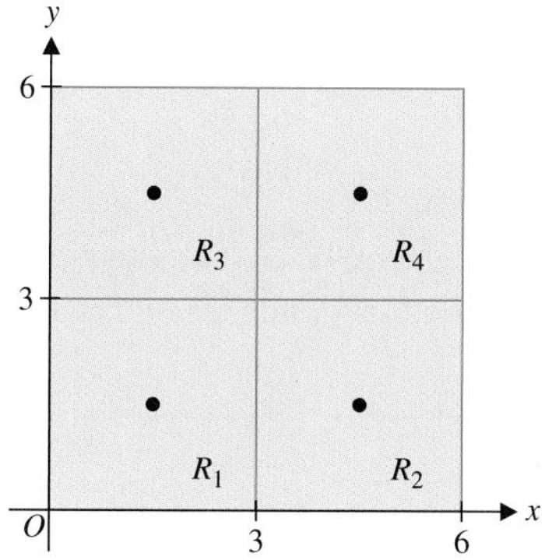

FIGURE 13.5b
Partition of R into four equal squares

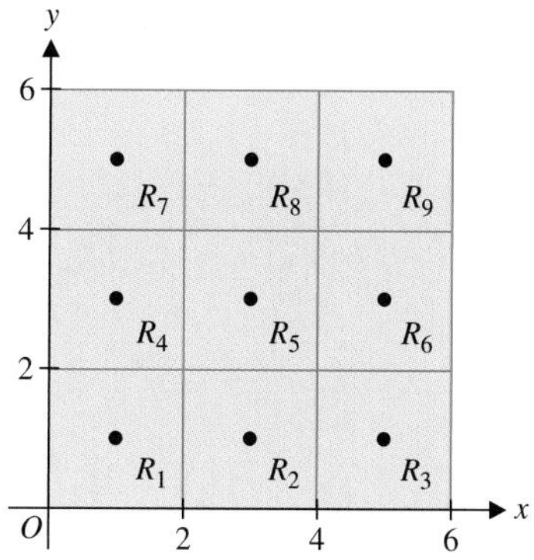

FIGURE 13.5c
Partition of R into nine equal squares

We can improve on this approximation by increasing the number of rectangles in the partition. For instance, if we partition R into nine squares of equal size (see Figure 13.5c) and again use the center of each square as the evaluation point, we have $\Delta A_i = 4$ for each i and

$$\begin{aligned} V &\approx \sum_{i=1}^{9} f(u_i, v_i)\,\Delta A_i \\ &= 4[f(1,1) + f(3,1) + f(5,1) + f(1,3) + f(3,3) + f(5,3) \\ &\quad + f(1,5) + f(3,5) + f(5,5)] \\ &= 4\left[1^2 \sin\left(\frac{\pi}{6}\right) + 3^2 \sin\left(\frac{\pi}{6}\right) + 5^2 \sin\left(\frac{\pi}{6}\right) + 1^2 \sin\left(\frac{3\pi}{6}\right) + 3^2 \sin\left(\frac{3\pi}{6}\right)\right. \\ &\quad \left. + 5^2 \sin\left(\frac{3\pi}{6}\right) + 1^2 \sin\left(\frac{5\pi}{6}\right) + 3^2 \sin\left(\frac{5\pi}{6}\right) + 5^2 \sin\left(\frac{5\pi}{6}\right)\right] \\ &= 280. \end{aligned}$$

No. of Squares in Partition	*Approximate Volume*
4	286.38
9	280.00
36	276.25
144	275.33
400	275.13
900	275.07

Continuing in this fashion to divide R into more and more squares of equal size and using the center of each square as the evaluation point, we construct continually better and better approximations of the volume (see the table in the margin). From the table, it appears that a reasonable approximation to the volume is slightly less than 275.07. In fact, the exact volume is $\frac{864}{\pi} \approx 275.02$. (We'll show you how to find this shortly.) ■

NOTES

The choice of the center of each square as the evaluation point, as used in example 1.1, corresponds to the Midpoint rule for approximating the value of a definite integral for a function of a single variable (discussed in section 4.7). This choice of evaluation points generally produces a reasonably good approximation.

Now, how can we turn (1.2) into an exact formula for volume? Note that it takes more than simply letting $n \to \infty$. We need to have *all* of the rectangles in the partition shrink to zero area. A convenient way of doing this is to define the **norm of the partition** $\|P\|$ to be the largest diagonal of any rectangle in the partition. Note that if $\|P\| \to 0$, then *all* of the rectangles must shrink to zero area. We can now make the volume approximation (1.2) exact:

$$V = \lim_{\|P\| \to 0} \sum_{i=1}^{n} f(u_i, v_i)\,\Delta A_i,$$

assuming the limit exists and is the same for every choice of the evaluation points. Here, by saying that this limit equals V, we mean that we can make $\sum_{i=1}^{n} f(u_i, v_i)\,\Delta A_i$ as close as needed to V, just by making $\|P\|$ sufficiently small. More precisely, this says that given any

$\varepsilon > 0$, there is a $\delta > 0$ (depending on the choice of ε), such that

$$\left| \sum_{i=1}^{n} f(u_i, v_i)\, \Delta A_i - V \right| < \varepsilon,$$

for every partition P with $\|P\| < \delta$. More generally, we have the following definition, which applies even when the function takes on negative values.

DEFINITION 1.2

For any function $f(x, y)$ defined on the rectangle $R = \{(x, y) | a \le x \le b \text{ and } c \le y \le d\}$, we define the **double integral** of f over R by

$$\iint_R f(x, y)\, dA = \lim_{\|P\| \to 0} \sum_{i=1}^{n} f(u_i, v_i)\, \Delta A_i,$$

provided the limit exists and is the same for every choice of the evaluation points (u_i, v_i) in R_i, for $i = 1, 2, \ldots, n$. When this happens, we say that f is **integrable** over R.

REMARK 1.1

It can be shown that if f is continuous on R, then it is also integrable over R. The proof can be found in more advanced texts.

There's one small problem with this new double integral. Just as when we first defined the definite integral of a function of one variable, we don't yet know how to compute it! For complicated regions R, this is a little bit tricky, but for a rectangle, it's a snap, as we see in the following.

We first consider the special case where $f(x, y) \ge 0$ on the rectangle $R = \{(x, y) | a \le x \le b \text{ and } c \le y \le d\}$. Notice that here, $\iint_R f(x, y)\, dA$ represents the volume lying beneath the surface $z = f(x, y)$ and above the region R. Recall that we already know how to compute this volume, from our work in section 5.2. We can do this by slicing the solid with planes parallel to the yz-plane, as indicated in Figure 13.6a. If we denote the area of the cross section of the solid for a given value of x by $A(x)$, then we have from equation (2.1) in section 5.2 that the volume is given by

$$V = \int_a^b A(x)\, dx.$$

Now, note that for each *fixed* value of x, the area of the cross section is simply the area under the curve $z = f(x, y)$ for $c \le y \le d$, which is given by the integral

$$A(x) = \int_c^d f(x, y)\, dy.$$

This integration is called a **partial integration** with respect to y, since x is held fixed and $f(x, y)$ is integrated with respect to y. This leaves us with

$$V = \int_a^b A(x)\, dx = \int_a^b \left[\int_c^d f(x, y)\, dy \right] dx. \tag{1.3}$$

Likewise, if we instead slice the solid with planes parallel to the xz-plane, as indicated in Figure 13.6b, we get that the volume is given by

$$V = \int_c^d A(y)\, dy = \int_c^d \left[\int_a^b f(x, y)\, dx \right] dy. \tag{1.4}$$

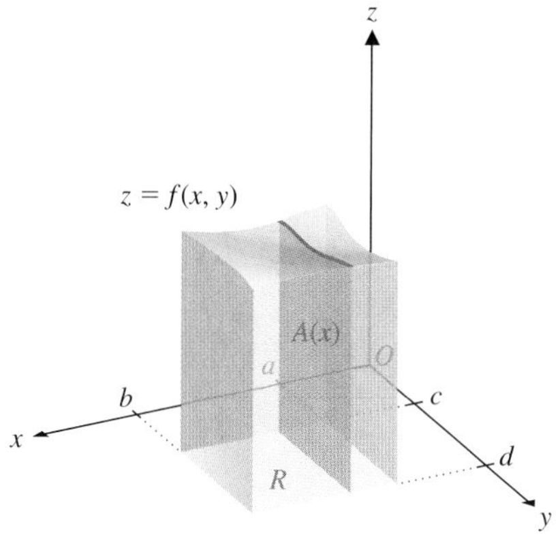

FIGURE 13.6a
Slicing the solid parallel to the yz-plane

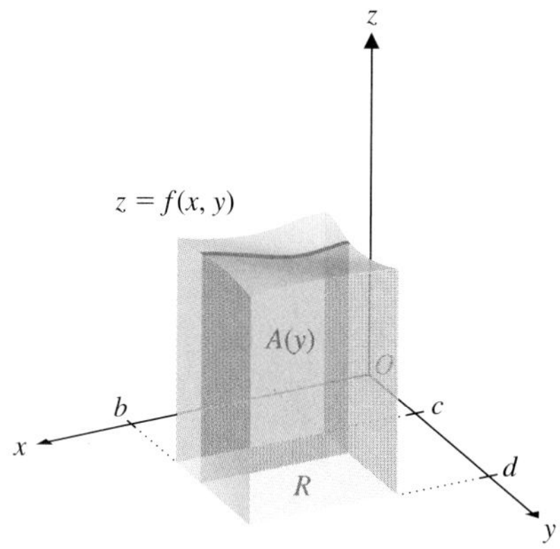

FIGURE 13.6b
Slicing the solid parallel to the xz-plane

The integrals in (1.3) and (1.4) are called **iterated integrals.** Note that each of these indicates a partial integration with respect to the inner variable (i.e., you first integrate with respect to the inner variable, treating the outer variable as a constant), to be followed by an integration with respect to the outer variable.

For simplicity, we ordinarily write the iterated integrals without the brackets:

$$\int_a^b \left[\int_c^d f(x, y)\,dy\right] dx = \int_a^b \int_c^d f(x, y)\,dy\,dx$$

and

$$\int_c^d \left[\int_a^b f(x, y)\,dx\right] dy = \int_c^d \int_a^b f(x, y)\,dx\,dy.$$

As indicated, these integrals are evaluated inside out, using the methods of integration we've already established for functions of a single variable. This now establishes the following result for the special case where $f(x, y) \geq 0$. The proof of the result for the general case is rather lengthy and we omit it.

THEOREM 1.1 (Fubini's Theorem)

Suppose that f is integrable over the rectangle $R = \{(x, y) | a \leq x \leq b \text{ and } c \leq y \leq d\}$. Then we can write the double integral of f over R as either of the iterated integrals:

$$\iint_R f(x, y)\,dA = \int_a^b \int_c^d f(x, y)\,dy\,dx = \int_c^d \int_a^b f(x, y)\,dx\,dy. \tag{1.5}$$

Fubini's Theorem simply tells you that you can always rewrite a double integral over a rectangle as either one of a pair of iterated integrals. We illustrate this in example 1.2.

EXAMPLE 1.2 Double Integral over a Rectangle

If $R = \{(x, y) | 0 \leq x \leq 2 \text{ and } 1 \leq y \leq 4\}$, evaluate $\iint_R (6x^2 + 4xy^3)\,dA$.

HISTORICAL NOTES

Guido Fubini (1879–1943)
Italian mathematician who made wide-ranging contributions to mathematics, physics and engineering. Fubini's early work was in differential geometry, but he quickly diversified his research to include analysis, the calculus of variations, group theory, non-Euclidean geometry and mathematical physics. Mathematics was the family business, as his father was a mathematics teacher and his sons became engineers. Fubini moved to the United States in 1939 to escape the persecution of Jews in Italy. He was working on an engineering textbook inspired by his sons' work when he died.

Solution From (1.5), we have

$$\iint_R (6x^2 + 4xy^3)\,dA = \int_1^4 \int_0^2 (6x^2 + 4xy^3)\,dx\,dy$$

$$= \int_1^4 \left[\int_0^2 (6x^2 + 4xy^3)\,dx\right] dy$$

$$= \int_1^4 \left(6\frac{x^3}{3} + 4\frac{x^2}{2}y^3\right)\Bigg|_{x=0}^{x=2} dy$$

$$= \int_1^4 (16 + 8y^3)\,dy$$

$$= \left(16y + 8\frac{y^4}{4}\right)\Bigg|_1^4$$

$$= [16(4) + 2(4)^4] - [16(1) + 2(1)^4] = 558.$$

Note that we evaluated the first integral above by integrating with respect to x, while treating y as a constant. We leave it as an exercise to show that you get the same value by integrating first with respect to y, that is, that

$$\iint_R (6x^2 + 4xy^3)\,dA = \int_0^2 \int_1^4 (6x^2 + 4xy^3)\,dy\,dx = 558,$$

also. ■

○ Double Integrals over General Regions

So, what if we wanted to extend the notion of double integral to a bounded, nonrectangular region like the one shown in Figure 13.7a? (Recall that a region is bounded if it fits inside a circle of some finite radius.) We begin, as we did for the case of rectangular regions, by looking for the volume lying beneath the surface $z = f(x, y)$ and lying above the region R, where $f(x, y) \geq 0$ and f is continuous on R. First, notice that the grid we used initially to partition a rectangular region must be modified, since such a rectangular grid won't "fit" a nonrectangular region, as shown in Figure 13.7b.

We resolve this problem by considering only those rectangular subregions that lie *completely* inside the region R (see Figure 13.7c, where we have labeled these rectangles).

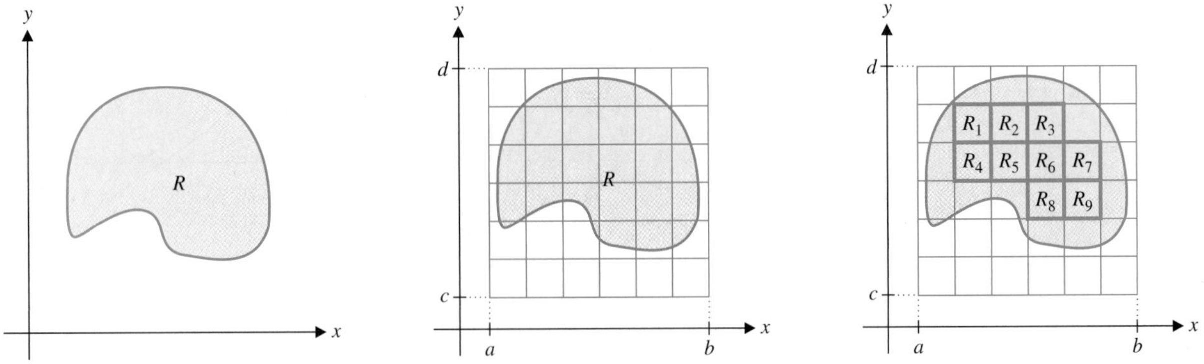

FIGURE 13.7a
Nonrectangular region

FIGURE 13.7b
Grid for a general region

FIGURE 13.7c
Inner partition

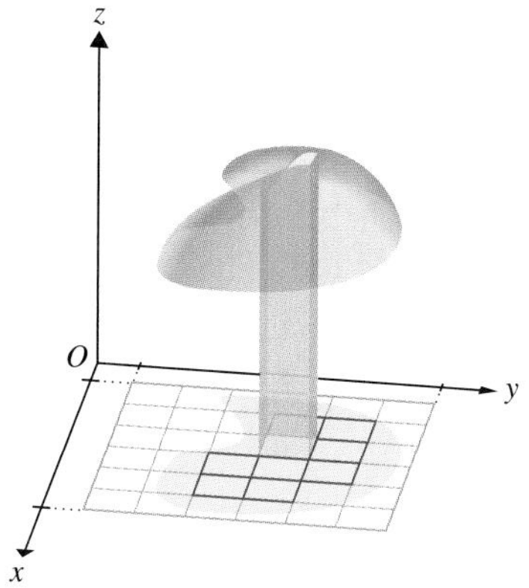

FIGURE 13.7d
Sample volume box

We call the collection of these rectangles an **inner partition** of R. For instance, in the inner partition indicated in Figure 13.7c, there are nine subregions.

From this point on, we proceed essentially as we did for the case of a rectangular region. That is, on each rectangular subregion R_i $(i = 1, 2, \ldots, n)$ in an inner partition, we construct a rectangular box of height $f(u_i, v_i)$, for some point $(u_i, v_i) \in R_i$ (see Figure 13.7d for a sample box). The volume V_i beneath the surface and above R_i is then approximately

$$V_i \approx \text{Height} \times \text{Area of base} = f(u_i, v_i)\,\Delta A_i,$$

where we again denote the area of R_i by ΔA_i. The total volume V lying beneath the surface and above the region R is then approximately

$$V \approx \sum_{i=1}^{n} f(u_i, v_i)\,\Delta A_i. \tag{1.6}$$

We define the norm of the inner partition $\|P\|$ to be the length of the largest diagonal of any of the rectangles $R_1, R_2, \ldots, R_n$. Notice that as we make $\|P\|$ smaller and smaller, the inner partition fills in R nicely (see Figure 13.8a) and the approximate volume given by (1.6) should get closer and closer to the actual volume. (See Figure 13.8b.) We then have

$$V = \lim_{\|P\| \to 0} \sum_{i=1}^{n} f(u_i, v_i)\,\Delta A_i,$$

assuming the limit exists and is the same for every choice of the evaluation points.

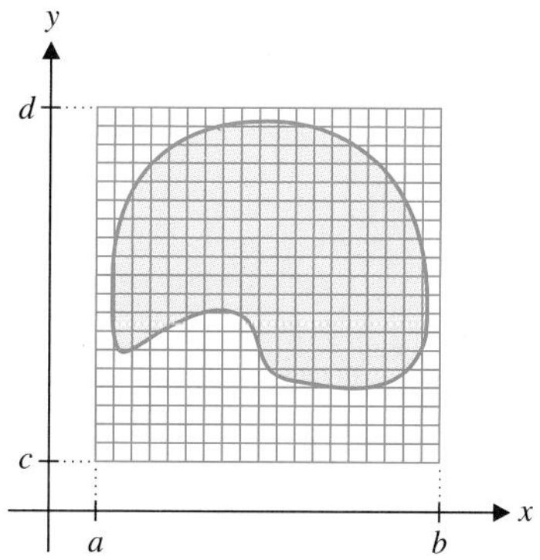

FIGURE 13.8a
Refined grid

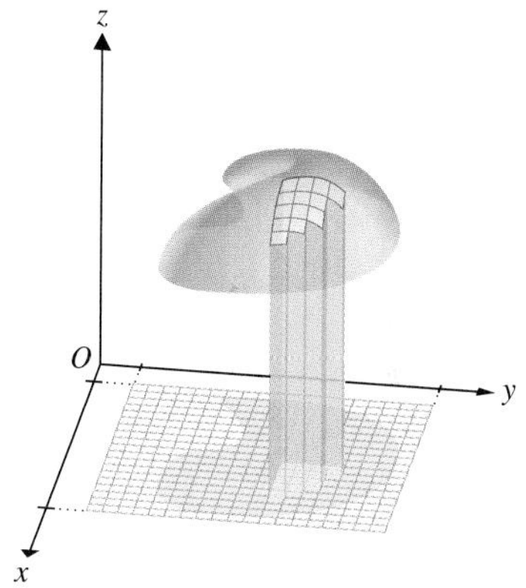

FIGURE 13.8b
Approximate volume

More generally, we have Definition 1.3.

DEFINITION 1.3

For any function $f(x, y)$ defined on a bounded region $R \subset \mathbb{R}^2$, we define the **double integral** of f over R by

$$\iint_R f(x, y)\,dA = \lim_{\|P\| \to 0} \sum_{i=1}^{n} f(u_i, v_i)\,\Delta A_i, \tag{1.7}$$

provided the limit exists and is the same for every choice of the evaluation points (u_i, v_i) in R_i, for $i = 1, 2, \ldots, n$. In this case, we say that f is **integrable** over R.

REMARK 1.2

Once again, it can be shown that if f is continuous on R, then it is integrable over R, although the proof is beyond the level of this course.

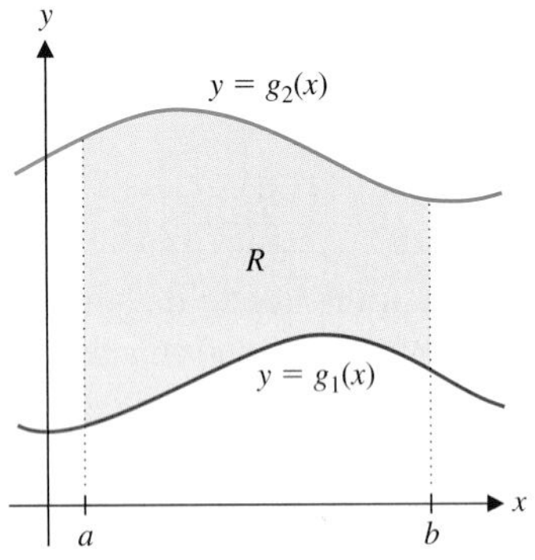

FIGURE 13.9a
The region R

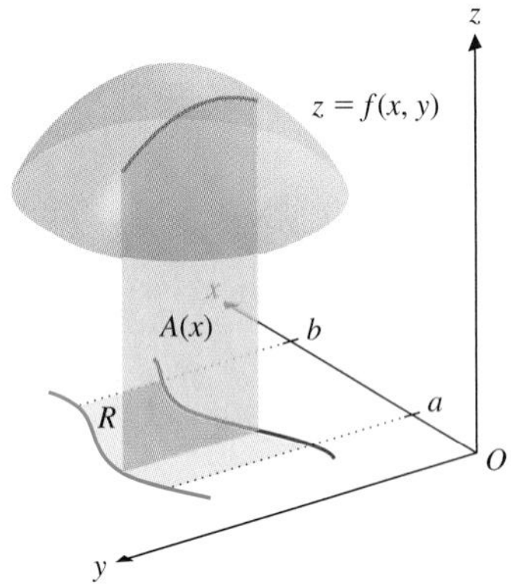

FIGURE 13.9b
Volume by slicing

CAUTION

Be sure to draw a reasonably good sketch of the region R before you try to write down the iterated integrals. Without doing this, you may be lucky enough (or clever enough) to get the first few exercises to work out, but you will be ultimately doomed to failure. It is *essential* that you have a clear picture of the region in order to set up the integrals correctly.

The question remains as to how we can calculate a double integral over a nonrectangular region. The answer is a bit more complicated than it was for the case of a rectangular region and depends on the exact form of R.

We first consider the case where the region R lies between the vertical lines $x = a$ and $x = b$, with $a < b$, has a top defined by the curve $y = g_2(x)$ and a bottom defined by $y = g_1(x)$, where $g_1(x) \le g_2(x)$ for all x in (a, b). That is, R has the form

$$R = \{(x, y) | a \le x \le b \text{ and } g_1(x) \le y \le g_2(x)\}.$$

See Figure 13.9a for a typical region of this form lying in the first quadrant of the xy-plane. Think about this for the special case where $f(x, y) \ge 0$ on R. Here, the double integral of f over R gives the volume lying beneath the surface $z = f(x, y)$ and above the region R in the xy-plane. We can find this volume by the method of slicing, just as we did for the case of a double integral over a rectangular region.

From Figure 13.9b, observe that for each *fixed* $x \in [a, b]$, the area of the slice lying above the line segment indicated and below the surface $z = f(x, y)$ is given by

$$A(x) = \int_{g_1(x)}^{g_2(x)} f(x, y)\, dy.$$

The volume of the solid is then given by equation (2.1) in section 5.2 to be

$$V = \int_a^b A(x)\, dx = \int_a^b \int_{g_1(x)}^{g_2(x)} f(x, y)\, dy\, dx.$$

Recognizing the volume as $V = \iint_R f(x, y)\, dA$ proves the following theorem, for the special case where $f(x, y) \ge 0$ on R.

THEOREM 1.2

Suppose that $f(x, y)$ is continuous on the region R defined by $R = \{(x, y) | a \le x \le b \text{ and } g_1(x) \le y \le g_2(x)\}$, for continuous functions g_1 and g_2, where $g_1(x) \le g_2(x)$, for all x in $[a, b]$. Then,

$$\iint_R f(x, y)\, dA = \int_a^b \int_{g_1(x)}^{g_2(x)} f(x, y)\, dy\, dx.$$

Although the general proof of Theorem 1.2 is beyond the level of this text, the derivation given above for the special case where $f(x, y) \ge 0$ should help to make some sense of why it is true.

Notice that once again, we have managed to write a double integral as an iterated integral. This allows us to use all of our techniques of integration for functions of a single variable to help evaluate double integrals.

We illustrate the process of writing a double integral as an iterated integral in example 1.3.

EXAMPLE 1.3 Evaluating a Double Integral

Let R be the region bounded by the graphs of $y = x$, $y = 0$ and $x = 4$. Evaluate

$$\iint_R (4e^{x^2} - 5 \sin y)\, dA.$$

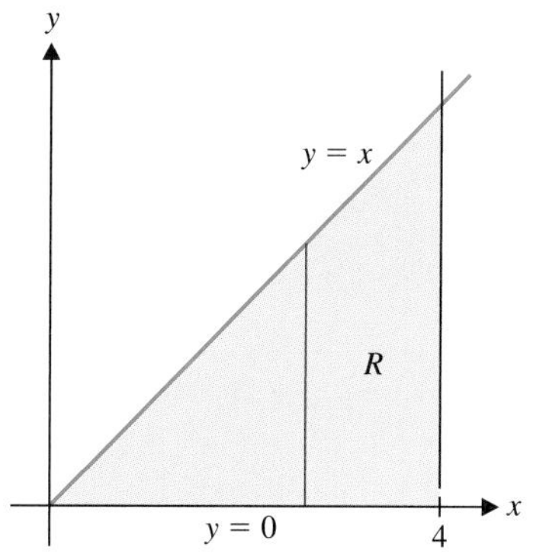

FIGURE 13.10
The region R

Solution First, we draw a graph of the region R in Figure 13.10. To help with determining the limits of integration, we have drawn a line segment illustrating that for each fixed value of x on the interval [0, 4], the y-values range from 0 up to x. From Theorem 1.2, we have

$$\begin{aligned}\iint_R (4e^{x^2} - 5\sin y)\,dA &= \int_0^4 \int_0^x (4e^{x^2} - 5\sin y)\,dy\,dx \qquad (1.8)\\ &= \int_0^4 (4ye^{x^2} + 5\cos y)\Big|_{y=0}^{y=x} dx\\ &= \int_0^4 [(4xe^{x^2} + 5\cos x) - (0 + 5\cos 0)]\,dx\\ &= \int_0^4 (4xe^{x^2} + 5\cos x - 5)\,dx\\ &= (2e^{x^2} + 5\sin x - 5x)\Big|_0^4\\ &= 2e^{16} + 5\sin 4 - 22 \approx 1.78 \times 10^7.\end{aligned}$$

Keep in mind that the inner integration above (with respect to y) is a partial integration with respect to y, so that we hold x fixed.

Be *very* careful here; there are plenty of traps to fall into. The most common error is to simply look for the minimum and maximum values of x and y and mistakenly write

$$\iint_R f(x, y)\,dA = \int_0^4 \int_0^4 f(x, y)\,dy\,dx. \quad \text{This is incorrect!}$$

Compare this last iterated integral to the correct expression in (1.8). Notice that instead of integrating over the region R shown in Figure 13.10, it corresponds to integration over the rectangle $0 \le x \le 4$, $0 \le y \le 4$. (This is close, but no cigar!) ■

As with any other integral, iterated integrals often cannot be evaluated symbolically (even with a very good computer algebra system). In such cases, we must rely on approximate methods. If you can, evaluate the inner integral symbolically and then use a numerical method (e.g., Simpson's Rule) to approximate the outer integral.

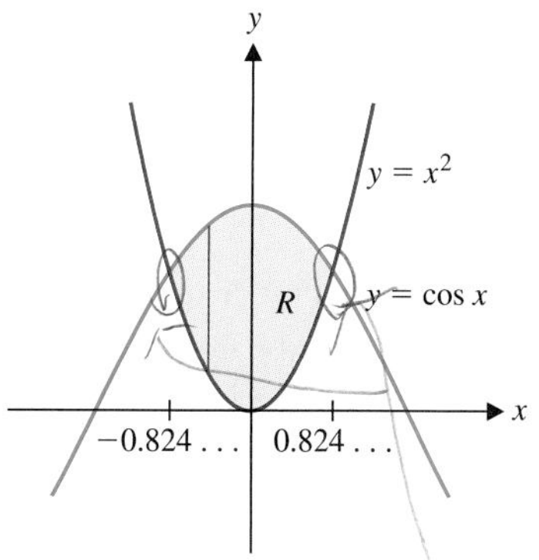

FIGURE 13.11
The region R

EXAMPLE 1.4 Approximate Limits of Integration

Evaluate $\iint_R (x^2 + 6y)\,dA$, where R is the region bounded by the graphs of $y = \cos x$ and $y = x^2$.

Solution We show a graph of the region R in Figure 13.11. Notice that the inner limits of integration are easy to see from the figure; for each fixed x, y ranges from x^2 up to $\cos x$. However, the outer limits of integration are not quite so clear. To find these, we must find the intersections of the two curves by solving the equation $\cos x = x^2$. We can't solve this exactly, but using a numerical procedure (e.g., Newton's method or one built into your calculator or computer algebra system), we get approximate intersections of $x \approx \pm 0.82413$. From Theorem 1.2, we now have

$$\iint_R (x^2 + 6y)\, dA \approx \int_{-0.82413}^{0.82413} \int_{x^2}^{\cos x} (x^2 + 6y)\, dy\, dx$$
$$= \int_{-0.82413}^{0.82413} \left(x^2 y + 6\frac{y^2}{2} \right) \Bigg|_{y=x^2}^{y=\cos x} dx$$
$$= \int_{-0.82413}^{0.82413} [(x^2 \cos x + 3\cos^2 x) - (x^4 + 3x^4)]\, dx$$
$$\approx 3.659765588,$$

where we have evaluated the last integral approximately, even though it could be done exactly, using integration by parts and a trigonometric identity. ■

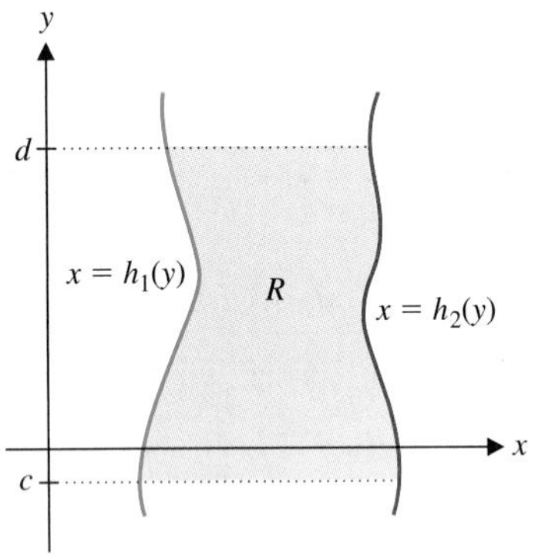

FIGURE 13.12
Typical region

Not all double integrals can be computed using the technique of examples 1.3 and 1.4. Often, it is necessary (or at least convenient) to think of the geometry of the region R in a different way.

Suppose that the region R has the form

$$R = \{(x, y) | c \le y \le d \text{ and } h_1(y) \le x \le h_2(y)\}.$$

See Figure 13.12 for a typical region of this form. Then, much as in Theorem 1.2, we can write double integrals as iterated integrals, as in Theorem 1.3.

THEOREM 1.3

Suppose that $f(x, y)$ is continuous on the region R defined by $R = \{(x, y) | c \le y \le d \text{ and } h_1(y) \le x \le h_2(y)\}$, for continuous functions h_1 and h_2, where $h_1(y) \le h_2(y)$, for all y in $[c, d]$. Then,

$$\iint_R f(x, y)\, dA = \int_c^d \int_{h_1(y)}^{h_2(y)} f(x, y)\, dx\, dy.$$

The general proof of this theorem is beyond the level of this course, although the reasonableness of this result should be apparent from Theorem 1.2 and the analysis preceding that theorem, for the special case where $f(x, y) \ge 0$ on R.

TODAY IN MATHEMATICS

Mary Ellen Rudin (1924–)
An American mathematician who published more than 70 research papers while supervising Ph.D. students, raising four children and earning the love and respect of students and colleagues. As a child, she and her friends played games that were "very elaborate and purely in the imagination. I think actually that that is something that contributes to making a mathematician—having time to think and being in the habit of imagining all sorts of complicated things." She says, "I'm very geometric in my thinking. I'm not really interested in numbers." She describes her teaching style as, "I bubble and I get students enthusiastic."

EXAMPLE 1.5 Integrating First with Respect to x

Write $\iint_R f(x, y)\, dA$ as an iterated integral, where R is the region bounded by the graphs of $x = y^2$ and $x = 2 - y$.

Solution First, we sketch a graph of the region (see Figure 13.13a). Notice that integrating first with respect to y is not a very good choice, since the upper boundary of the region is $y = \sqrt{x}$ for $0 \le x \le 1$ and $y = 2 - x$ for $1 \le x \le 4$. A more reasonable choice is to use Theorem 1.3 and integrate first with respect to x. In Figure 13.13b, we have included a horizontal line segment indicating the inner limits of integration: for each fixed y, x runs from $x = y^2$ over to $x = 2 - y$. The value of y then runs between the values at the intersections of the two curves. To find these, we solve $y^2 = 2 - y$ or

$$0 = y^2 + y - 2 = (y + 2)(y - 1),$$

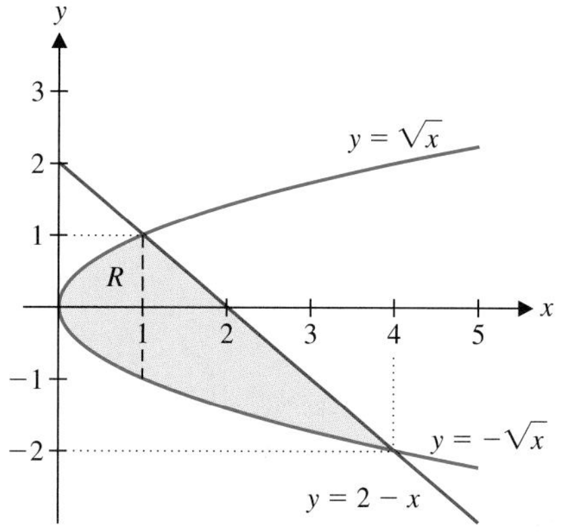

FIGURE 13.13a
The region R

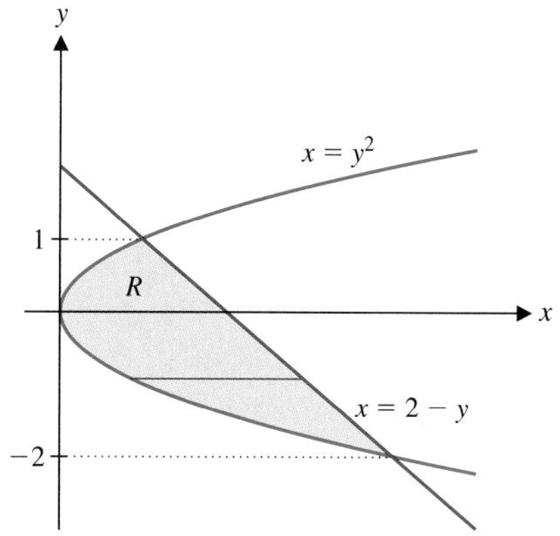

FIGURE 13.13b
The region R

so that the intersections are at $y = -2$ and $y = 1$. From Theorem 1.3, we now have

$$\iint_R f(x, y)\,dA = \int_{-2}^{1}\int_{y^2}^{2-y} f(x, y)\,dx\,dy.$$

You will often have to choose which variable to integrate with respect to first. Sometimes, you make your choice on the basis of the region. Often, a double integral can be set up either way but is much easier to calculate one way than the other. This is the case in example 1.6.

EXAMPLE 1.6 Evaluating a Double Integral

Let R be the region bounded by the graphs of $y = \sqrt{x}$, $x = 0$ and $y = 3$. Evaluate $\iint_R (2xy^2 + 2y\cos x)\,dA$.

Solution We show a graph of the region in Figure 13.14. From Theorem 1.3, we have

$$\begin{aligned}\iint_R (2xy^2 + 2y\cos x)\,dA &= \int_0^3\int_0^{y^2} (2xy^2 + 2y\cos x)\,dx\,dy \\ &= \int_0^3 (x^2y^2 + 2y\sin x)\Big|_{x=0}^{x=y^2} dy \\ &= \int_0^3 [(y^6 + 2y\sin y^2) - (0 + 2y\sin 0)]\,dy \\ &= \int_0^3 (y^6 + 2y\sin y^2)\,dy \\ &= \left(\frac{y^7}{7} - \cos y^2\right)\Big|_0^3 \\ &= \frac{3^7}{7} - \cos 9 + \cos 0 \approx 314.3.\end{aligned}$$

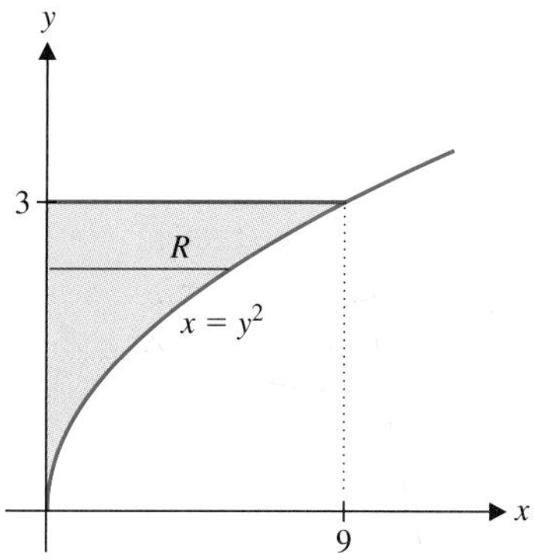

FIGURE 13.14
The region R

Alternatively, integrating with respect to y first, we get

$$\iint_R (2xy^2 + 2y\cos x)\,dA = \int_0^9 \int_{\sqrt{x}}^3 (2xy^2 + 2y\cos x)\,dy\,dx$$
$$= \int_0^9 \left(2x\frac{y^3}{3} + y^2\cos x\right)\Bigg|_{y=\sqrt{x}}^{y=3} dx$$
$$= \int_0^9 \left[\frac{2}{3}x(27 - x^{3/2}) + (3^2 - x)\cos x\right] dx,$$

which leaves you with an integration by parts to carry out. We leave the details as an exercise. Which way do you think is easier? ■

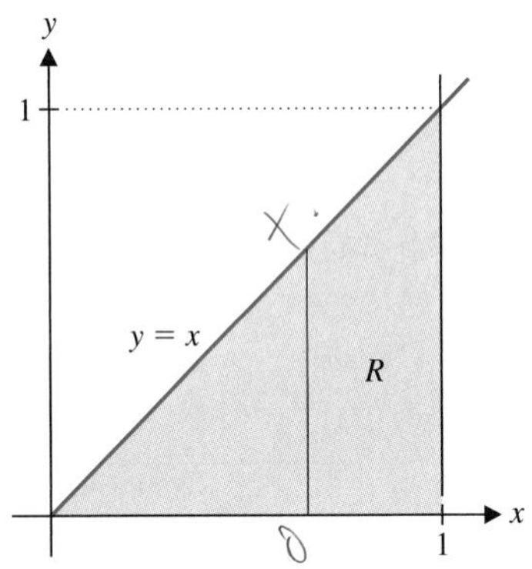

FIGURE 13.15
The region R

In example 1.6, we saw that changing the order of integration may make a given double integral easier to compute. As we see in example 1.7, sometimes you will *need* to change the order of integration in order to evaluate a double integral.

EXAMPLE 1.7 A Case Where We Must Switch the Order of Integration

Evaluate the iterated integral $\int_0^1 \int_y^1 e^{x^2}\,dx\,dy$.

Solution First, note that we cannot evaluate the integral the way it is presently written, as we don't know an antiderivative for e^{x^2}. On the other hand, if we switch the order of integration, the integral becomes quite simple, as follows. First, recognize that for each fixed y on the interval [0, 1], x ranges from y over to 1, giving us the triangular region of integration shown in Figure 13.15. If we switch the order of integration, notice that for each fixed x in the interval [0, 1], y ranges from 0 up to x and we get the double iterated integral:

$$\int_0^1 \int_y^1 e^{x^2}\,dx\,dy = \int_0^1 \int_0^x e^{x^2}\,dy\,dx$$
$$= \int_0^1 e^{x^2} y\Big|_{y=0}^{y=x} dx$$
$$= \int_0^1 e^{x^2} x\,dx.$$

Notice that we can evaluate this last integral with the substitution $u = x^2$, since $du = 2x\,dx$ and the first integration has conveniently provided us with the needed factor of x. We have

$$\int_0^1 \int_y^1 e^{x^2}\,dx\,dy = \frac{1}{2}\int_0^1 \underbrace{e^{x^2}}_{e^u} \underbrace{(2x)\,dx}_{du}$$
$$= \frac{1}{2}e^{x^2}\Big|_{x=0}^{x=1} = \frac{1}{2}(e^1 - 1).$$ ■

CAUTION

Carefully study the steps we used to change the order of integration in example 1.7. Notice that we did not simply swap the two integrals, nor did we just switch y's to x's on the inside limits. When you change the order of integration, it is extremely important that you sketch the region over which you are integrating, as in Figure 13.15. This allows you to see the orientation of the different parts of the boundary of the region. Failing to do this is the single most common error made by students at this point. This is a skill you need to practice, as you will use it throughout the rest of the course. (Sketching a picture takes only a few moments and will help you to avoid many fatal errors. So, do this routinely!)

We complete the section by stating several simple properties of double integrals.

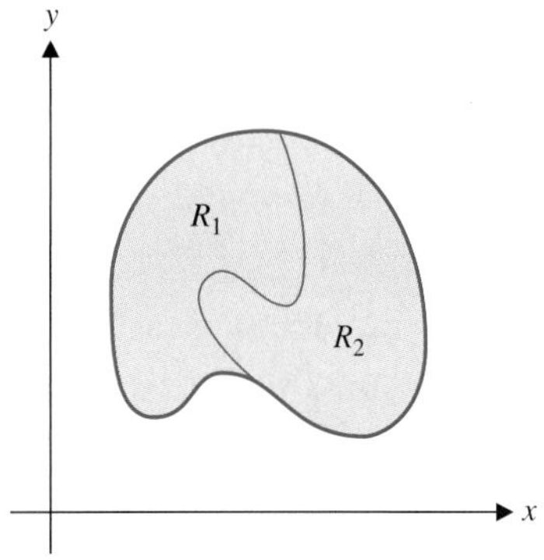

FIGURE 13.16
$R = R_1 \cup R_2$

THEOREM 1.4

Let $f(x, y)$ and $g(x, y)$ be integrable over the region $R \subset \mathbb{R}^2$ and let c be any constant. Then, the following hold:

(i) $\displaystyle\iint_R cf(x, y)\, dA = c \iint_R f(x, y)\, dA,$

(ii) $\displaystyle\iint_R [f(x, y) + g(x, y)]\, dA = \iint_R f(x, y)\, dA + \iint_R g(x, y)\, dA$ and

(iii) if $R = R_1 \cup R_2$, where R_1 and R_2 are nonoverlapping regions (see Figure 13.16), then

$$\iint_R f(x, y)\, dA = \iint_{R_1} f(x, y)\, dA + \iint_{R_2} f(x, y)\, dA.$$

Each of these follows directly from the definition of double integral in (1.7) and the proofs are left as exercises.

BEYOND FORMULAS

You should think of double integrals in terms of the Rule of Three: symbolic, graphical and numerical interpretations. Symbolically, you compute double integrals as iterated integrals, where the greatest challenge is correctly setting up the limits of integration. Graphically, the volume calculation that motivates Definition 1.2 is analogous to the area interpretation of single integrals. Numerically, double integrals can be approximated by Riemann sums. From your experience with single integrals and partial derivatives in Chapter 12, what percentage of double integrals do you expect to be able to evaluate symbolically?

EXERCISES 13.1

HW 1g 1043-1044
7, 9, 15, 19, 25, 27, 31, 37, 39, 41, 45

WRITING EXERCISES

1. If $f(x, y) \geq 0$ on a region R, then $\iint_R f(x, y)\, dA$ gives the volume of the solid above the region R in the xy-plane and below the surface $z = f(x, y)$. If $f(x, y) \geq 0$ on a region R_1 and $f(x, y) \leq 0$ on a region R_2, discuss the geometric meaning of $\iint_{R_2} f(x, y)\, dA$ and $\iint_R f(x, y)\, dA$, where $R = R_1 \cup R_2$.
2. The definition of $\iint_R f(x, y)\, dA$ requires that the norm of the partition $\|P\|$ approaches 0. Explain why it is not enough to simply require that the number of rectangles n in the partition approaches ∞.
3. When computing areas between curves in section 5.1, we discussed strategies for deciding whether to integrate with respect to x or y. Compare these strategies to those given in this section for deciding which variable to use as the inside variable of a double integral.
4. Suppose you (or your software) are using Riemann sums to approximate a particularly difficult double integral $\iint_R f(x, y)\, dA$. Further, suppose that $R = R_1 \cup R_2$ and the function $f(x, y)$ is nearly constant on R_1 but oscillates wildly on R_2, where R_1 and R_2 are nonoverlapping regions. Explain why you would need more rectangles in R_2 than R_1 to get equally accurate approximations. Thus, irregular partitions can be used to improve the efficiency of numerical integration routines.

In exercises 1–4, compute the Riemann sum for the given function and region, a partition with n equal-sized rectangles and the given evaluation rule.

1. $f(x, y) = x + 2y^2$, $0 \leq x \leq 2$, $-1 \leq y \leq 1$, $n = 4$, evaluate at midpoint
2. $f(x, y) = 4x^2 + y$, $1 \leq x \leq 5$, $0 \leq y \leq 2$, $n = 4$, evaluate at midpoint
3. $f(x, y) = x + 2y^2$, $0 \leq x \leq 2$, $-1 \leq y \leq 1$, $n = 16$, evaluate at midpoint

4. $f(x, y) = 4x^2 + y, 1 \le x \le 5, 0 \le y \le 2, n = 16,$ evaluate at midpoint

In exercises 5 and 6, compute the Riemann sum for the given function, the irregular partition shown and midpoint evaluation.

5. $f(x, y) = 3x - y$

6. $f(x, y) = 2x + y$

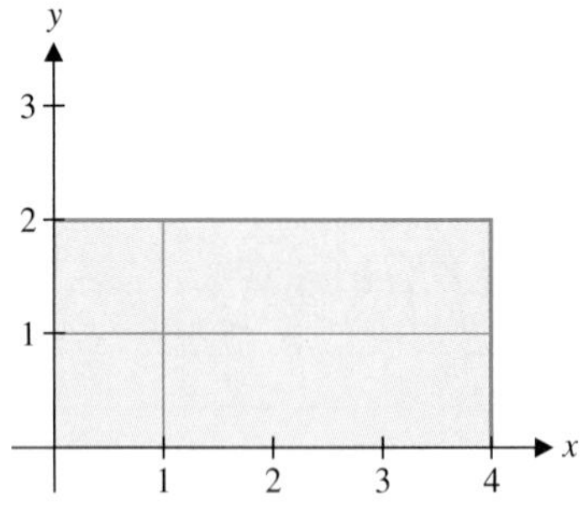

In exercises 7–10, evaluate the double integral.

7. $\iint_R (x^2 - 2y)\, dA$, where $R = \{0 \le x \le 2, -1 \le y \le 1\}$

8. $\iint_R 4xe^{2y} dA$, where $R = \{2 \le x \le 4, 0 \le y \le 1\}$

9. $\iint_R (1 - ye^{xy})\, dA$, where $R = \{0 \le x \le 2, 0 \le y \le 3\}$

10. $\iint_R (3x - 4x\sqrt{xy})\, dA$, where $R = \{0 \le x \le 4, 0 \le y \le 9\}$

In exercises 11–14, sketch the solid whose volume is given by the iterated integral.

11. $\int_{-1}^{1} \int_{0}^{1} (6 - 2x - 3y)\, dy\, dx$

12. $\int_{0}^{2} \int_{-1}^{1} (2 + x + 2y)\, dy\, dx$

13. $\int_{0}^{2} \int_{0}^{3} (x^2 + y^2)\, dy\, dx$

14. $\int_{-1}^{1} \int_{-1}^{1} (4 - x^2 - y^2)\, dy\, dx$

In exercises 15–22, evaluate the iterated integral.

15. $\int_{0}^{1} \int_{0}^{2x} (x + 2y)\, dy\, dx$

16. $\int_{0}^{2} \int_{0}^{x^2} (x + 3)\, dy\, dx$

17. $\int_{0}^{1} \int_{0}^{2y} (4x\sqrt{y} + y)\, dx\, dy$

18. $\int_{0}^{\pi} \int_{0}^{2} y \sin(xy)\, dx\, dy$

19. $\int_{0}^{2} \int_{0}^{2y} e^{y^2} dx\, dy$

20. $\int_{1}^{2} \int_{0}^{2/x} e^{xy} dy\, dx$

21. $\int_{1}^{4} \int_{0}^{1/x} \cos(xy)\, dy\, dx$

22. $\int_{0}^{1} \int_{0}^{y^2} \frac{3}{4 + y^3} dx\, dy$

23. Show that $\int_{0}^{1} \int_{0}^{2x} x^2\, dy\, dx \ne \int_{0}^{2} \int_{0}^{y/2} x^2 dx\, dy$.

24. Sketch the solids whose volumes are given in exercise 23 and explain why the volumes are not equal.

In exercises 25–32, find an integral equal to the volume of the solid bounded by the given surfaces and evaluate the integral.

25. $z = x^2 + y^2, z = 0, y = 1, y = 4, x = 0, x = 3$

26. $z = 3x^2 + 2y, z = 0, y = 0, y = 1, x = 1, x = 3$

27. $z = x^2 + y^2, z = 0, y = x^2, y = 1$

28. $z = 3x^2 + 2y, z = 0, y = 1 - x^2, y = 0$

29. $z = 6 - x - y, z = 0, x = 4 - y^2, x = 0$

30. $z = 4 - 2y, z = 0, x = y^4, x = 1$

31. $z = y^2, z = 0, y = 0, y = x, x = 2$

32. $z = x^2, z = 0, y = x, y = 4, x = 0$

In exercises 33–36, approximate the double integral.

33. $\iint_R (2x - y)\, dA$, where R is bounded by $y = \sin x$ and $y = 1 - x^2$

34. $\iint_R (2x - y)\, dA$, where R is bounded by $y = e^x$ and $y = 2 - x^2$

35. $\iint_R e^{x^2} dA$, where R is bounded by $y = x^2$ and $y = 1$

36. $\iint_R \sqrt{y^2 + 1}\, dA$, where R is bounded by $x = 4 - y^2$ and $x = 0$

In exercises 37–42, change the order of integration.

37. $\int_{0}^{1} \int_{0}^{2x} f(x, y)\, dy\, dx$

38. $\int_{0}^{1} \int_{2x}^{2} f(x, y)\, dy\, dx$

39. $\int_{0}^{2} \int_{2y}^{4} f(x, y)\, dx\, dy$

40. $\int_{0}^{1} \int_{0}^{2y} f(x, y)\, dx\, dy$

41. $\int_{0}^{\ln 4} \int_{e^x}^{4} f(x, y)\, dy\, dx$

42. $\int_{1}^{2} \int_{0}^{\ln y} f(x, y)\, dx\, dy$

In exercises 43–46, evaluate the iterated integral by first changing the order of integration.

43. $\int_{0}^{2} \int_{x}^{2} 2e^{y^2} dy\, dx$

44. $\int_{0}^{1} \int_{\sqrt{x}}^{1} \frac{3}{4 + y^3} dy\, dx$

45. $\int_{0}^{1} \int_{y}^{1} 3xe^{x^3} dx\, dy$

46. $\int_{0}^{1} \int_{\sqrt{y}}^{1} \cos x^3 dx\, dy$

47. Determine whether your CAS can evaluate the integrals $\int_{x}^{2} 2e^{y^2} dy$ and $\int_{0}^{2} \int_{x}^{2} 2e^{y^2} dy\, dx$.

48. Explain why a CAS would have trouble evaluating the first integral in exercise 47. Based on your result in exercise 47, can your CAS switch orders of integration to evaluate a double integral?

In exercises 49–52, sketch the solid whose volume is described by the given iterated integral.

49. $\int_{0}^{3} \int_{0}^{6-2x} (6 - 2x - y)\, dy\, dx$

50. $\int_0^4 \int_0^{4-x} (4 - x - y)\, dy\, dx$

51. $\int_{-2}^2 \int_{-\sqrt{4-x^2}}^{\sqrt{4-x^2}} (4 - x^2 - y^2)\, dy\, dx$

52. $\int_0^1 \int_0^{\sqrt{1-x^2}} (x^2 + y^2)\, dy\, dx$

53. Explain why $\int_0^1 \int_0^{2x} f(x, y)\, dy\, dx$ is not generally equal to $\int_0^1 \int_0^{2y} f(x, y)\, dx\, dy$.

54. Give an example of a function for which the integrals in exercise 53 are equal. As generally as possible, describe what property such a function must have.

55. Compute the iterated integral by sketching a graph and using a basic geometric formula:

$$\int_{-1}^1 \int_{-\sqrt{1-x^2}}^{\sqrt{1-x^2}} \sqrt{1 - x^2 - y^2}\, dy\, dx.$$

56. Prove Theorem 1.4.

57. Prove that $\int_a^b \int_c^d f(x)g(y)\, dy\, dx = \left(\int_a^b f(x)\, dx\right)\left(\int_c^d g(y)\, dy\right)$ for continuous functions f and g.

58. Use the result of exercise 57 to quickly evaluate $\int_0^{2\pi} \int_{15}^{38} e^{-4y^2} \sin x\, dy\, dx$.

59. For the table of function values here, use upper-left corner evaluations to estimate $\int_0^1 \int_0^1 f(x, y)\, dy\, dx$.

y \ x	0.0	0.25	0.5	0.75	1.0
0.0	2.2	2.0	1.7	1.4	1.0
0.25	2.3	2.1	1.8	1.6	1.1
0.5	2.5	2.3	2.0	1.8	1.4
0.75	2.8	2.6	2.3	2.2	1.8
1.0	3.2	3.0	2.8	2.7	2.5

60. Repeat exercise 59 with lower-right corner evaluations.

61. For the table of function values in exercise 59, use upper-left corner evaluations to estimate $\int_0^1 \int_0^{0.5} f(x, y)\, dy\, dx$.

62. Repeat exercise 61 with lower-right corner evaluations.

63. For the function in exercise 59, use an inner partition and lower-right corner evaluations to estimate $\int_0^1 \int_0^{\sqrt{1-x^2}} f(x, y)\, dy\, dx$.

64. Use the function in exercise 59, an inner partition and upper-right corner evaluations to estimate $\int_0^1 \int_0^{1-y} f(x, y)\, dx\, dy$.

65. Use the average of the function values at all four corners to approximate the integral in exercise 59.

66. Use the average of the function values at all four corners to approximate the integral in exercise 61.

67. Use the contour plot to determine which is the best estimate of $\int_{-1}^1 \int_0^1 f(x, y)\, dy\, dx$: (a) 1, (b) 2 or (c) 4.

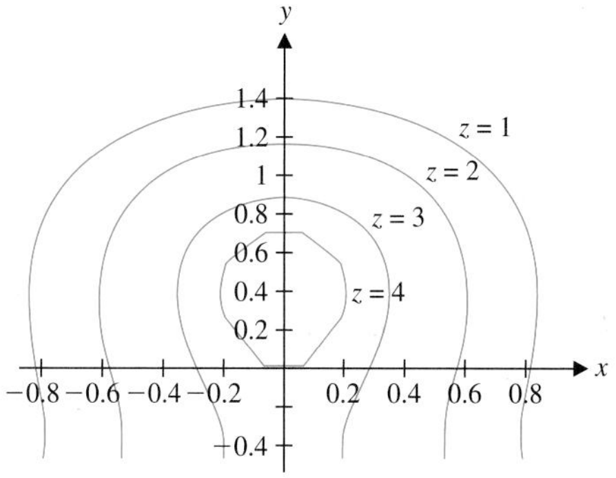

68. Use the contour plot to determine which is the best estimate of $\int_0^1 \int_0^{1-x} f(x, y)\, dy\, dx$: (a) 1, (b) 2 or (c) 4.

69. From the Fundamental Theorem of Calculus, we have $\int_a^b f'(x)\, dx = f(b) - f(a)$. Find the corresponding rule for evaluating the double integral $\int_c^d \int_a^b f_{xy}(x, y)\, dx\, dy$. Use this rule to evaluate $\int_0^1 \int_0^1 24xy^2 dx\, dy$, with $f(x, y) = 3x + 4x^2y^3 + y^2$.

70. Determine whether the rule from exercise 69 holds for double integrals over nonrectangular regions. Test it on $\int_0^1 \int_0^x 24xy^2 dy\, dx$.

71. Evaluate $\int_0^2 [\tan^{-1}(4 - x) - \tan^{-1} x]\, dx$ by rewriting it as a double integral and switching the order of integration.

72. Evaluate $\int_0^{1/2} [\sin^{-1}(1 - x) - \sin^{-1} x]\, dx$ by rewriting it as a double integral and switching the order of integration.

73. Evaluate $\int_0^2 \int_0^{2y} f(x, y)\, dx\, dy$ for $f(x, y) = \min\{2x, y\}$.

74. Evaluate $\int_0^2 \int_0^{2x} f(x, y)\, dy\, dx$ for $f(x, y) = \min\{y, x^2\}$.

EXPLORATORY EXERCISES

1. Set up a double integral for the volume of the solid bounded by the graphs of $z = 4 - x^2 - y^2$ and $z = x^2 + y^2$. Note that you actually have two tasks. First, the general rule for finding the volume between two surfaces is analogous to the general rule for finding the area between two curves. The greater challenge here is to find the limits of integration.

2. As mentioned in the text, numerical methods for approximating double integrals can be troublesome. The **Monte Carlo method** makes clever use of probability theory to approximate $\iint_R f(x, y)\, dA$ for a bounded region R. Suppose, for example, that R is contained within the rectangle $0 \le x \le 1, 0 \le y \le 1$. Generate two random numbers a and b from the uniform distribution on [0, 1]; this means that every number between 0 and 1 is in some sense equally likely. Determine whether or not the point (a, b) is in the region R and then repeat the process a large number of times. If, for example, 64 out of 100 points generated were within R, explain why a reasonable estimate of the area of R is 0.64 times the area of the rectangle $0 \le x \le 1, 0 \le y \le 1$. For each point (a, b) that is within R, compute $f(a, b)$. If the average of all of these function values is 13.6, explain why a reasonable estimate of $\iint_R f(x, y)\, dA$ is $(0.64)(13.6) = 8.704$. Use the Monte Carlo method to estimate $\int_1^2 \int_{\ln x}^{\sqrt{x}} \sin(xy)\, dy\, dx$. (Hint: Show that y is between $\ln 1 = 0$ and $\sqrt{2} < 2$.)

3. Improper double integrals can be treated much like improper single integrals. Evaluate $\int_0^\infty \int_0^\infty e^{-2x-3y} dx\, dy$ by first evaluating the inside integral as $\lim_{R\to\infty} \int_0^R e^{-2x-3y} dx$. To explore whether the integral is well defined, evaluate the integral as $\lim_{R\to\infty} \left(\int_0^R \int_0^R e^{-2x-3y} dx\, dy\right)$ and $\lim_{R\to\infty} \left(\int_0^{2R} \int_0^{R^2} e^{-2x-3y} dx\, dy\right)$. Then evaluate $\iint_R e^{-x^2-y} dA$, where R is the portion of the xy-plane with $0 \le x \le y$.

13.2 AREA, VOLUME AND CENTER OF MASS

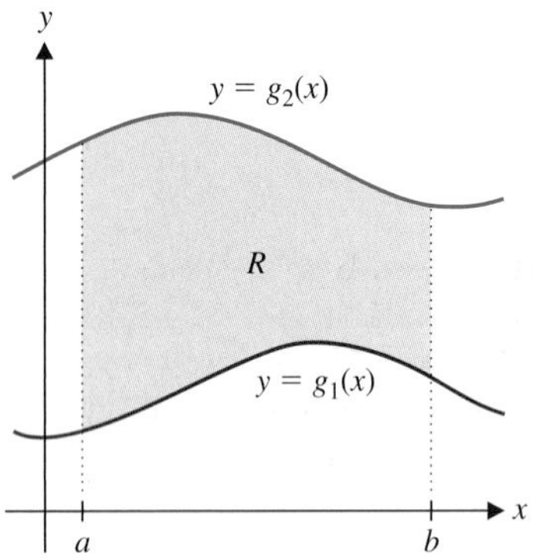

FIGURE 13.17
The region R

To use double integrals to solve problems, it's very important that you recognize what each component of the integral represents. For this reason, we pause briefly to set up a double iterated integral as a double sum. Consider the case of a continuous function $f(x, y) \ge 0$ on some region $R \subset \mathbb{R}^2$. If R has the form

$$R = \{(x, y) | a \le x \le b \text{ and } g_1(x) \le y \le g_2(x)\},$$

as indicated in Figure 13.17, then we have from our work in section 13.1 that the volume V lying beneath the surface $z = f(x, y)$ and above the region R is given by

$$V = \int_a^b A(x)\, dx = \int_a^b \int_{g_1(x)}^{g_2(x)} f(x, y)\, dy\, dx. \tag{2.1}$$

Here, for each fixed x, $A(x)$ is the area of the cross section of the solid corresponding to that particular value of x. Our aim is to write the volume integral in (2.1) in a slightly different way from our derivation in section 13.1. First, notice that by the definition of definite integral, we have that

$$\int_a^b A(x)\, dx = \lim_{\|P_1\|\to 0} \sum_{i=1}^n A(c_i)\, \Delta x_i, \tag{2.2}$$

where P_1 represents a partition of the interval $[a, b]$, c_i is some point in the ith subinterval $[x_{i-1}, x_i]$ and $\Delta x_i = x_i - x_{i-1}$ (the width of the ith subinterval). For each fixed $x \in [a, b]$, since $A(x)$ is the area of the cross section, we have that

$$A(x) = \int_{g_1(x)}^{g_2(x)} f(x, y)\, dy = \lim_{\|P_2\|\to 0} \sum_{j=1}^m f(x, v_j)\Delta y_j, \tag{2.3}$$

where P_2 represents a partition of the interval $[g_1(x), g_2(x)]$, v_j is some point in the jth subinterval $[y_{j-1}, y_j]$ of the partition P_2 and $\Delta y_j = y_j - y_{j-1}$ (the width of the jth

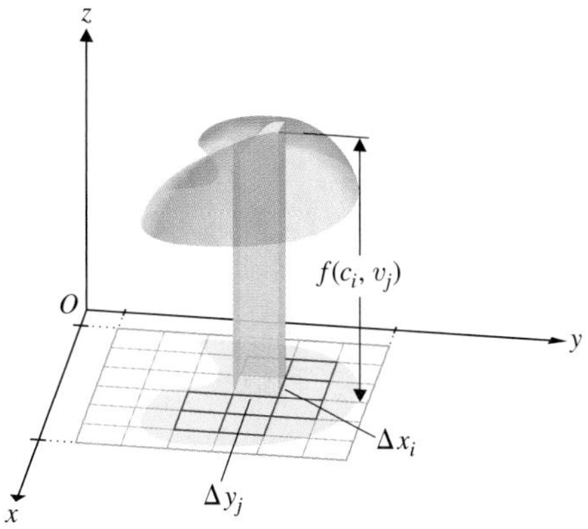

FIGURE 13.18
Volume of a typical box

subinterval). Putting (2.1), (2.2), and (2.3) together, we get

$$\begin{aligned} V &= \lim_{\|P_1\|\to 0} \sum_{i=1}^{n} A(c_i)\Delta x_i \\ &= \lim_{\|P_1\|\to 0} \sum_{i=1}^{n} \left[\lim_{\|P_2\|\to 0} \sum_{j=1}^{m} f(c_i, v_j)\Delta y_j \right] \Delta x_i \\ &= \lim_{\|P_1\|\to 0} \lim_{\|P_2\|\to 0} \sum_{i=1}^{n} \sum_{j=1}^{m} f(c_i, v_j)\Delta y_j \Delta x_i. \end{aligned} \tag{2.4}$$

The double summation in (2.4) is called a **double Riemann sum.** Notice that each term corresponds to the volume of a box of length Δx_i, width Δy_j and height $f(c_i, v_j)$. (See Figure 13.18.) Observe that by superimposing the two partitions, we have produced an inner partition of the region R. If we represent this inner partition of R by P and the norm of the partition P by $\|P\|$, the length of the longest diagonal of any rectangle in the partition, we can write (2.4) with only one limit, as

$$V = \lim_{\|P\|\to 0} \sum_{i=1}^{n} \sum_{j=1}^{m} f(c_i, v_j)\Delta y_j \Delta x_i. \tag{2.5}$$

When you write down an iterated integral representing volume, you can use (2.5) to help identify each of the components as follows:

$$\begin{aligned} V &= \lim_{\|P\|\to 0} \sum_{i=1}^{n} \sum_{j=1}^{m} \underbrace{f(c_i, v_j)}_{\text{height}} \underbrace{\Delta y_j}_{\text{width}} \underbrace{\Delta x_i}_{\text{length}} \\ &= \int_a^b \int_{g_1(x)}^{g_2(x)} \underbrace{f(x, y)}_{\text{height}} \underbrace{dy}_{\text{width}} \underbrace{dx}_{\text{length}}. \end{aligned} \tag{2.6}$$

You should make at least a mental picture of the components of the integral in (2.6), keeping in mind the corresponding components of the Riemann sum. We leave it as an exercise to show that for a region of the form

$$R = \{(x, y) | c \le y \le d \text{ and } h_1(y) \le x \le h_2(y)\},$$

we get a corresponding interpretation of the iterated integral:

$$V = \lim_{\|P\|\to 0} \sum_{j=1}^{m} \sum_{i=1}^{n} \underbrace{f(c_i, v_j)}_{\text{height}} \underbrace{\Delta x_i}_{\text{length}} \underbrace{\Delta y_j}_{\text{width}}$$

$$= \int_c^d \int_{h_1(y)}^{h_2(y)} \underbrace{f(x, y)}_{\text{height}} \underbrace{dx}_{\text{length}} \underbrace{dy}_{\text{width}}. \tag{2.7}$$

Observe that for any bounded region $R \subset \mathbb{R}^2$, $\iint_R 1\, dA$, which we sometimes write simply as $\iint_R dA$, gives the volume under the surface $z = 1$ and above the region R in the xy-plane. Since all of the cross sections parallel to the xy-plane are the same, the solid is a cylinder and so, its volume is the product of its height (1) and its cross-sectional area. That is,

$$\iint_R dA = (1)\,(\text{Area of } R) = \text{Area of } R. \tag{2.8}$$

So, we now have the option of using a double integral to find the area of a plane region.

EXAMPLE 2.1 Using a Double Integral to Find Area

Find the area of the plane region bounded by the graphs of $x = y^2$, $y - x = 3$, $y = -3$ and $y = 2$ (see Figure 13.19).

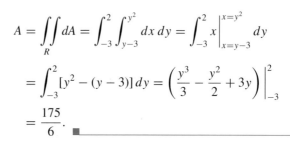

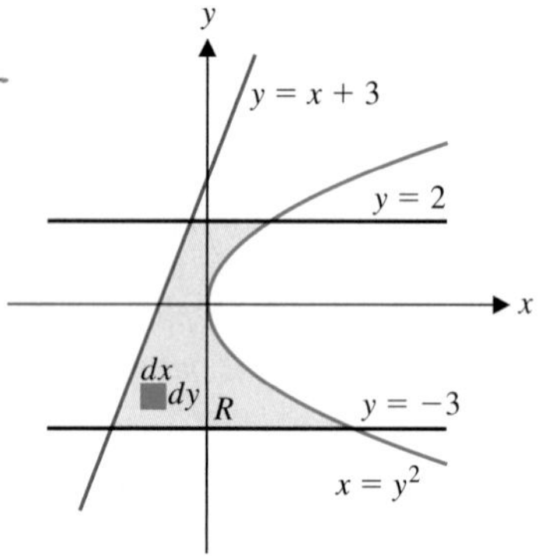

FIGURE 13.19
The region R

Solution Note that we have indicated in the figure a small rectangle with sides dx and dy, respectively. This helps to indicate the limits for the iterated integral. From (2.8), we have

$$A = \iint_R dA = \int_{-3}^{2} \int_{y-3}^{y^2} dx\, dy = \int_{-3}^{2} x \Big|_{x=y-3}^{x=y^2} dy$$

$$= \int_{-3}^{2} [y^2 - (y - 3)]\, dy = \left(\frac{y^3}{3} - \frac{y^2}{2} + 3y \right) \Big|_{-3}^{2}$$

$$= \frac{175}{6}. \quad ■$$

Think about example 2.1 a little further. Recall that we had worked similar problems in section 5.1 using single integrals. In fact, you might have set up the desired area directly as

$$A = \int_{-3}^{2} [y^2 - (y - 3)]\, dy,$$

exactly as you see in the second line of work above. While we will sometimes use double integrals to more easily solve familiar problems, double integrals will allow us to solve many new problems as well.

We have already developed formulas for calculating the volume of a solid lying below a surface of the form $z = f(x, y)$ and above a region R (of several different forms), lying in the xy-plane. So, what's the problem, then? As you will see in examples 2.2–2.4, the challenge in setting up the iterated integrals comes in seeing the region R that the solid lies above and then determining the limits of integration for the iterated integrals.

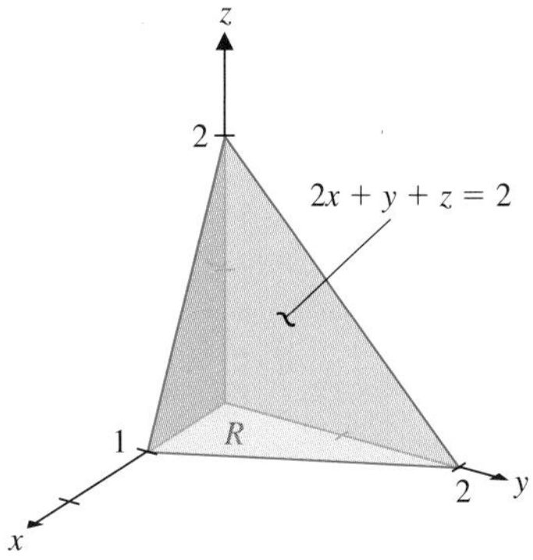

FIGURE 13.20a
Tetrahedron

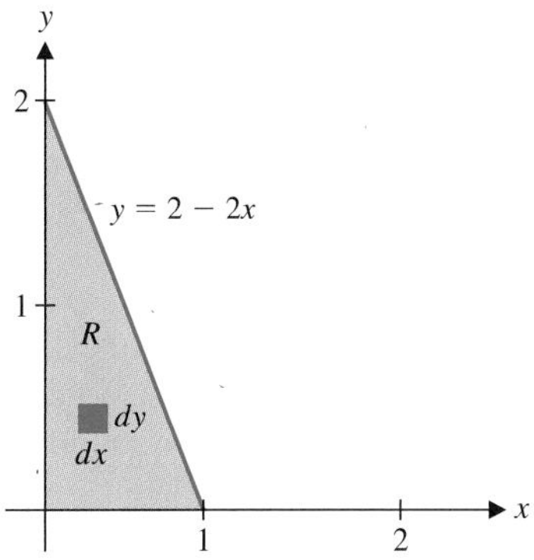

FIGURE 13.20b
The region R

EXAMPLE 2.2 Using a Double Integral to Find Volume

Find the volume of the tetrahedron bounded by the plane $2x + y + z = 2$ and the three coordinate planes.

Solution First, we need to draw a sketch of the solid. Since the plane $2x + y + z = 2$ intersects the coordinate axes at the points (1, 0, 0), (0, 2, 0) and (0, 0, 2), a sketch is easy to draw. Simply connect the three points of intersection with the coordinate axes and you'll get the graph of the tetrahedron (a four-sided object with all triangular sides) seen in Figure 13.20a. In order to use our volume formula, though, we'll first need to visualize the tetrahedron as a solid lying below a surface of the form $z = f(x, y)$ and lying above some region R in the xy-plane. Notice that the solid lies below the plane $z = 2 - 2x - y$ and above the triangular region R in the xy-plane, as indicated in Figure 13.20a. Although we're not simply handed R, you can see that R is the triangular region bounded by the x- and y-axes and the trace of the plane $2x + y + z = 2$ in the xy-plane. The trace is found by simply setting $z = 0$: $2x + y = 2$ (see Figure 13.20b). From (2.6), the volume is then

$$\frac{2-0}{0-1} = -2 \quad y = -2x+2$$

$$V = \int_0^1 \int_0^{2-2x} \underbrace{(2 - 2x - y)}_{\text{height}} \underbrace{dy}_{\text{width}} \underbrace{dx}_{\text{length}}$$

$$= \int_0^1 \left(2y - 2xy - \frac{y^2}{2}\right)\Bigg|_{y=0}^{y=2-2x} dx$$

$$= \int_0^1 \left[2(2 - 2x) - 2x(2 - 2x) - \frac{(2 - 2x)^2}{2}\right] dx$$

$$= \frac{2}{3},$$

where we leave the routine details of the final calculation to you. ■

We cannot emphasize enough the need to draw reasonable sketches of the solid and particularly of the base of the solid in the xy-plane. You may be lucky enough to guess the limits of integration for a few of these problems, but don't be deceived: you need to draw good sketches and look carefully to determine the limits of integration correctly.

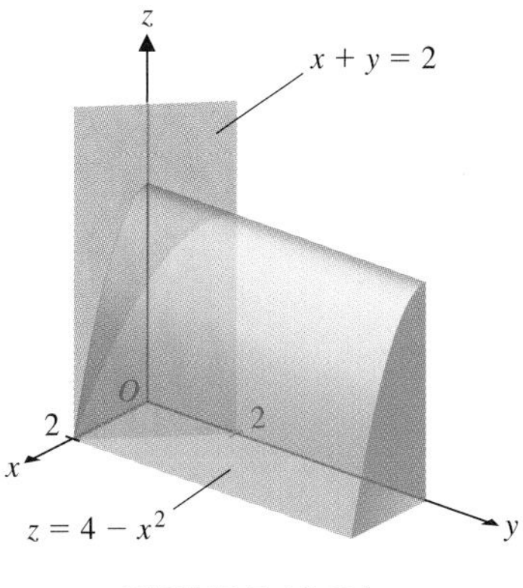

FIGURE 13.21a
Solid in the first octant

EXAMPLE 2.3 Finding the Volume of a Solid

Find the volume of the solid lying in the first octant and bounded by the graphs of $z = 4 - x^2$, $x + y = 2$, $x = 0$, $y = 0$ and $z = 0$.

Solution First, draw a sketch of the solid. You should note that $z = 4 - x^2$ is a cylinder (since there's no y term), $x + y = 2$ is a plane and $x = 0$, $y = 0$ and $z = 0$ are the coordinate planes. (See Figure 13.21a.) Notice that the solid lies below the surface $z = 4 - x^2$ and above the triangular region R in the xy-plane formed by the x- and y-axes and the trace of the plane $x + y = 2$ in the xy-plane (i.e., the line

$y=0,\ z=0,\ x=0$

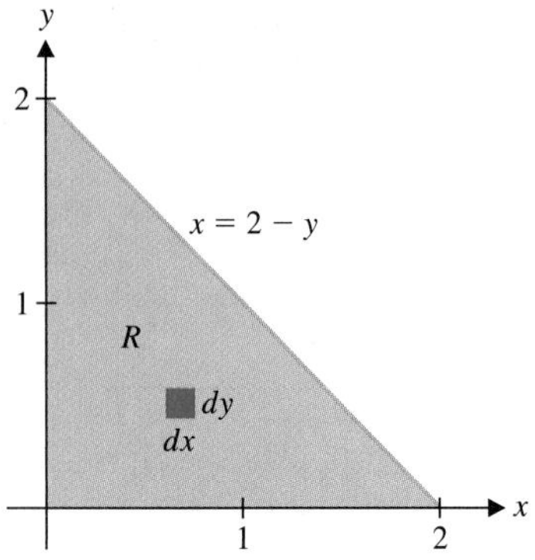

FIGURE 13.21b
The region R

$x + y = 2$). This is shown in Figure 13.21b. Although we could integrate with respect to either x or y first, we integrate with respect to x first. From (2.7), we have

$$V = \int_0^2 \int_0^{2-y} \underbrace{(4 - x^2)}_{\text{height}} \underbrace{dx}_{\text{length}} \underbrace{dy}_{\text{width}}$$

$$= \int_0^2 \left(4x - \frac{x^3}{3} \right) \Bigg|_{x=0}^{x=2-y} dy$$

$$= \int_0^2 \left[4(2 - y) - \frac{(2 - y)^3}{3} \right] dy$$

$$= \frac{20}{3}. \quad \blacksquare$$

EXAMPLE 2.4   Finding the Volume of a Solid Bounded
                    Above the *xy*-Plane

Find the volume of the solid bounded by the graphs of $z = 2$, $z = x^2 + 1$, $y = 0$ and $x + y = 2$.

Solution   First, observe that the graph of $z = x^2 + 1$ is a parabolic cylinder with axis parallel to the y-axis. It intersects the plane $z = 2$ where $x^2 + 1 = 2$ or $x = \pm 1$. This forms a long trough, which is cut off by the planes $y = 0$ (the xz-plane) and $x + y = 2$. A sketch of the solid is shown in Figure 13.22a. The solid lies below $z = 2$ and above the cylinder $z = x^2 + 1$. You can view the integrand $f(x, y)$ in (2.6) as the height of the solid above the point (x, y). Drawing a vertical line from the xy-plane through the solid in Figure 13.22a shows that the height of the solid is the difference between 2 and $x^2 + 1$, so that $f(x, y) = 2 - (x^2 + 1) = 1 - x^2$. In Figure 13.22a, notice that the solid lies above the region R in the xy-plane bounded by $y = 0$, $x + y = 2$, $x = -1$ and $x = 1$. (See Figure 13.22b.)

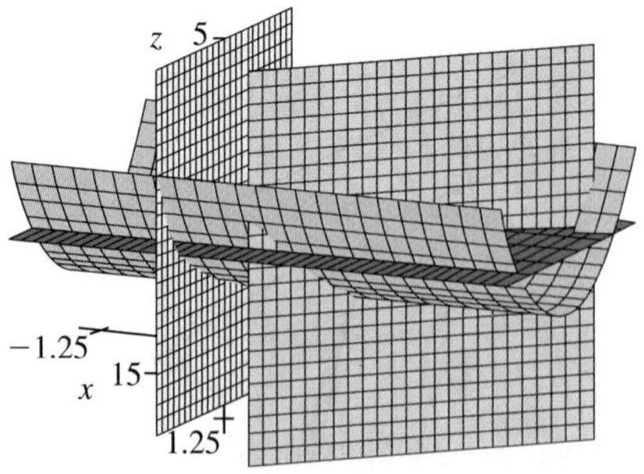

FIGURE 13.22a
The solid

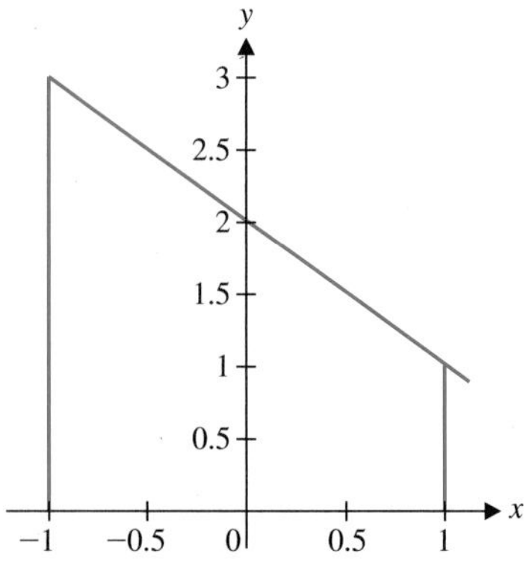

FIGURE 13.22b
The region R

NOTES

Notice in example 2.4 that the limits of integration come from the two defining surfaces for y (that is, $y = 0$ and $y = 2 - x$) and the x-values for the intersection of the other two defining surfaces $z = 2$ and $z = x^2 + 1$. The defining surfaces and intersections are the sources of the limits of integration, but don't just guess which one to put where: use a graph of the surface to see how to arrange these elements.

It's easy to see from Figure 13.22b that we should integrate with respect to y first. For each fixed x in the interval $[-1, 1]$, y runs from 0 to $2 - x$. The volume is then

$$
\begin{aligned}
V &= \int_{-1}^{1} \int_{0}^{2-x} (1 - x^2) \, dy \, dx \\
&= \int_{-1}^{1} (1 - x^2) y \Big|_{y=0}^{y=2-x} \, dx \\
&= \int_{-1}^{1} (1 - x^2)(2 - x) \, dx \\
&= \frac{8}{3}. \quad \blacksquare
\end{aligned}
$$

Double integrals are used to calculate numerous quantities of interest in applications. We present one application in example 2.5, while others can be found in the exercises.

EXAMPLE 2.5 Estimating Population

Suppose that $f(x, y) = 20{,}000 y e^{-x^2 - y^2}$ models the population density (population per square mile) of a species of small animals, with x and y measured in miles. Estimate the population in the triangular-shaped habitat with vertices $(1, 1)$, $(2, 1)$ and $(1, 0)$.

Solution The population in any region R is estimated by

$$
\iint\limits_{R} f(x, y) \, dA = \iint\limits_{R} 20{,}000 y e^{-x^2 - y^2} \, dA.
$$

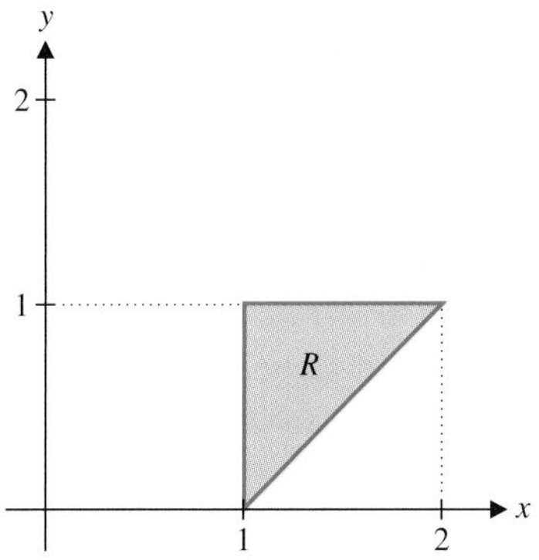

FIGURE 13.23
Habitat region

[As a quick check on the reasonableness of this formula, note that $f(x, y)$ is measured in units of population per square mile and the area increment dA carries units of square miles, so that the product $f(x, y) \, dA$ carries the desired units of population.] Notice that the integrand is $20{,}000 y e^{-x^2 - y^2} = 20{,}000 e^{-x^2} y e^{-y^2}$, which suggests that we should integrate with respect to y first. As always, we first sketch a graph of the region R (shown in Figure 13.23). Notice that the line through the points $(1, 0)$ and $(2, 1)$ has the equation $y = x - 1$, so that R extends from $y = x - 1$ up to $y = 1$, as x increases from 1 to 2. We now have

$$
\iint\limits_{R} f(x, y) \, dA = \int_{1}^{2} \int_{x-1}^{1} 20{,}000 e^{-x^2} y e^{-y^2} \, dy \, dx
$$

$$
= \int_{1}^{2} 10{,}000 e^{-x^2} [e^{-(x-1)^2} - e^{-1}] \, dx
$$

$$
\approx 698,
$$

where we approximated the last integral numerically. $\blacksquare$

○ Moments and Center of Mass

We close this section by briefly discussing a physical application of double integrals. Consider a thin, flat plate (a **lamina**) in the shape of the region $R \subset \mathbb{R}^2$ whose density (mass per unit area) varies throughout the plate (i.e., some areas of the plate are more dense than others). From an engineering standpoint, it's often important to determine where you could place a support to balance the plate. We call this point the **center of mass** of the lamina. We'll first need to find the total mass of the plate. For a real plate, we'd simply place it on a scale, but for our theoretical plate, we'll need to be more clever. Suppose the lamina has the shape of the region R shown in Figure 13.24a and has mass density (mass per unit area) given by the function $\rho(x, y)$. Construct an inner partition of R, as in Figure 13.24b. Notice that if the norm of the partition $\|P\|$ is small, then the density will be nearly constant on each rectangle of the inner partition. So, for each $i = 1, 2, \ldots, n$, pick some point $(u_i, v_i) \in R_i$. Then, the mass m_i of the portion of the lamina corresponding to the rectangle R_i is given approximately by

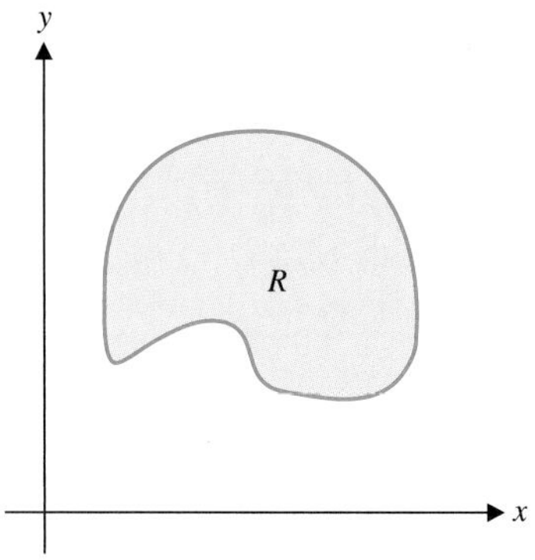

FIGURE 13.24a
Lamina

$$m_i \approx \underbrace{\rho(u_i, v_i)}_{\text{mass/unit area}} \underbrace{\Delta A_i}_{\text{area}},$$

where ΔA_i denotes the area of R_i. The total mass m of the lamina is then given approximately by

$$m \approx \sum_{i=1}^{n} \rho(u_i, v_i)\,\Delta A_i.$$

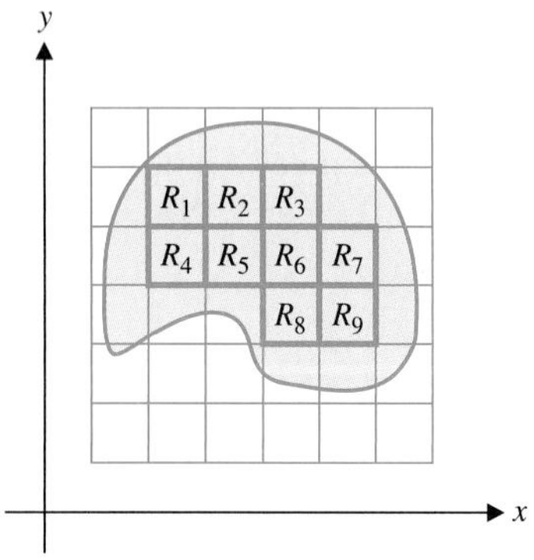

FIGURE 13.24b
Inner partition of R

Notice that if $\|P\|$ is small, then this should be a reasonable approximation of the total mass.

To get the mass exactly, we take the limit as $\|P\|$ tends to zero, which you should recognize as a double integral:

$$m = \lim_{\|P\|\to 0} \sum_{i=1}^{n} \rho(u_i, v_i)\,\Delta A_i = \iint_R \rho(x, y)\,dA. \tag{2.9}$$

Notice that if you want to balance a lamina like the one shown in Figure 13.24a, you'll need to balance it both from left to right and from top to bottom. In the language of our previous discussion of center of mass in section 5.6, we'll need to find the first moments: both left to right (we call this the **moment with respect to the *y*-axis**) and top to bottom (the **moment with respect to the *x*-axis**). First, we approximate the moment M_y with respect to the y-axis. Assuming that the mass in the ith rectangle of the partition is concentrated at the point (u_i, v_i), we have

$$M_y \approx \sum_{i=1}^{n} u_i \rho(u_i, v_i)$$

(i.e., the sum of the products of the masses and their directed distances from the y-axis). Taking the limit as $\|P\|$ tends to zero, we get

$$M_y = \lim_{\|P\|\to 0} \sum_{i=1}^{n} u_i \rho(u_i, v_i) = \iint_R x\rho(x, y)\,dA. \tag{2.10}$$

Similarly, looking at the sum of the products of the masses and their directed distances from the x-axis, we get the moment M_x with respect to the x-axis,

$$M_x = \lim_{\|P\|\to 0} \sum_{i=1}^{n} v_i \rho(u_i, v_i) = \iint_R y\rho(x, y)\,dA. \tag{2.11}$$

The center of mass is the point $(\bar{x}, \bar{y})$ defined by

$$\bar{x} = \frac{M_y}{m} \quad \text{and} \quad \bar{y} = \frac{M_x}{m}. \tag{2.12}$$

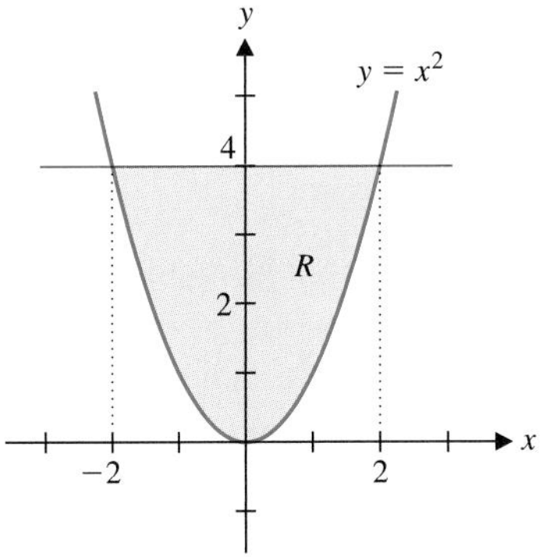

FIGURE 13.25
Lamina

EXAMPLE 2.6 Finding the Center of Mass of a Lamina

Find the center of mass of the lamina in the shape of the region bounded by the graphs of $y = x^2$ and $y = 4$, having mass density given by $\rho(x, y) = 1 + 2y + 6x^2$.

Solution We sketch the region in Figure 13.25. From (2.9), we have that the total mass of the lamina is given by

$$\begin{aligned} m &= \iint_R \rho(x, y)\,dA = \int_{-2}^{2}\int_{x^2}^{4} (1 + 2y + 6x^2)\,dy\,dx \\ &= \int_{-2}^{2} \left(y + 2\frac{y^2}{2} + 6x^2 y \right)\Bigg|_{y=x^2}^{y=4} dx \\ &= \int_{-2}^{2} [(4 + 16 + 24x^2) - (x^2 + x^4 + 6x^4)]\,dx \\ &= \frac{1696}{15} \approx 113.1. \end{aligned}$$

We compute the moment M_y from (2.10):

$$\begin{aligned} M_y &= \iint_R x\rho(x, y)\,dA = \int_{-2}^{2}\int_{x^2}^{4} x(1 + 2y + 6x^2)\,dy\,dx \\ &= \int_{-2}^{2}\int_{x^2}^{4} (x + 2xy + 6x^3)\,dy\,dx \\ &= \int_{-2}^{2} (xy + xy^2 + 6x^3 y)\Big|_{y=x^2}^{y=4} dx \\ &= \int_{-2}^{2} [(4x + 16x + 24x^3) - (x^3 + x^5 + 6x^5)]\,dx = 0. \end{aligned}$$

Note that from (2.12), this says that the x-coordinate of the center of mass is $\bar{x} = \frac{M_y}{m} = \frac{0}{113.1} = 0$. This should not surprise you since both the region *and* the mass density are symmetric with respect to the y-axis. [Notice that $\rho(-x, y) = \rho(x, y)$.] Next, from (2.11), we have

$$\begin{aligned} M_x &= \iint_R y\rho(x, y)\,dA = \int_{-2}^{2}\int_{x^2}^{4} y(1 + 2y + 6x^2)\,dy\,dx \\ &= \int_{-2}^{2}\int_{x^2}^{4} (y + 2y^2 + 6x^2 y)\,dy\,dx \\ &= \int_{-2}^{2} \left(\frac{y^2}{2} + 2\frac{y^3}{3} + 6x^2\frac{y^2}{2} \right)\Bigg|_{y=x^2}^{y=4} dx \\ &= \int_{-2}^{2} \left[\left(8 + \frac{128}{3} + 48x^2 \right) - \left(\frac{x^4}{2} + \frac{2}{3}x^6 + 3x^6 \right) \right] dx \\ &= \frac{11{,}136}{35} \approx 318.2 \end{aligned}$$

and so, from (2.12) we have $\bar{y} = \dfrac{M_x}{m} \approx \dfrac{318.2}{113.1} \approx 2.8$. The center of mass is then located at approximately

$$(\bar{x}, \bar{y}) \approx (0, 2.8). \blacksquare$$

In example 2.6, we computed the first moments M_y and M_x to find the balance point (center of mass) of the lamina in Figure 13.25. Further physical properties of this lamina can be determined using the **second moments** I_y and I_x. Much as we defined the first moments in equations (2.10) and (2.11), the second moment about the y-axis (often called the **moment of inertia about the y-axis**) of a lamina in the shape of the region R, with density function $\rho(x, y)$ is defined by

$$I_y = \iint_R x^2 \rho(x, y)\, dA.$$

Similarly, the second moment about the x-axis (also called the **moment of inertia about the x-axis**) of a lamina in the shape of the region R, with density function $\rho(x, y)$ is defined by

$$I_x = \iint_R y^2 \rho(x, y)\, dA.$$

Physics tells us that the larger I_y is, the more difficult it is to rotate the lamina about the y-axis. Similarly, the larger I_x is, the more difficult it is to rotate the lamina about the x-axis. We explore this briefly in example 2.7.

EXAMPLE 2.7 Finding the Moments of Inertia of a Lamina

Find the moments of inertia I_y and I_x for the lamina in example 2.6.

Solution The region R is the same as in example 2.6 (see Figure 13.25), so that the limits of integration are the same. We have

$$\begin{aligned} I_y &= \int_{-2}^{2} \int_{x^2}^{4} x^2(1 + 2y + 6x^2)\, dy\, dx \\ &= \int_{-2}^{2} (20x^2 + 23x^4 - 7x^6)\, dx \\ &= \frac{2176}{15} \approx 145.07 \end{aligned}$$

and

$$\begin{aligned} I_x &= \int_{-2}^{2} \int_{x^2}^{4} y^2(1 + 2y + 6x^2)\, dy\, dx \\ &= \int_{-2}^{2} \left(\frac{448}{3} + 128x^2 - \frac{1}{3}x^6 - \frac{5}{2}x^8 \right) dx \\ &= \frac{61{,}952}{63} \approx 983.37. \end{aligned}$$

A comparison of the two moments of inertia shows that it is much more difficult to rotate the lamina of Figure 13.25 about the x-axis than about the y-axis. Examine the figure and the density function to be sure that this makes sense to you. $\blacksquare$

EXERCISES 13.2

HW Pg 1055 #1-18 odd.

WRITING EXERCISES

1. The double Riemann sum in (2.5) disguises the fact that the order of integration is important. Explain how the order of integration affects the details of the double Riemann sum.

2. Many double integrals can be set up in two steps: first identify the function $f(x, y)$, then identify the two-dimensional region R and set up the limits of integration. Explain how these two steps are separated in examples 2.2, 2.3 and 2.4.

3. The sketches in examples 2.2, 2.3 and 2.4 are essential, but somewhat difficult to draw. Explain each sketch, including which surface should be drawn first, second and so on. Also, when a previously drawn surface is cut in half by a plane, explain how to identify which half of the cut surface to keep.

4. The moment M_y is the moment about the y-axis, but is used to find the x-coordinate of the center of mass. Explain why it is M_y and not M_x that is used to compute the x-coordinate of the center of mass.

In exercises 1–6, use a double integral to compute the area of the region bounded by the curves.

1. $y = x^2, y = 8 - x^2$
2. $y = x^2, y = x + 2$
3. $y = 2x, y = 3 - x, y = 0$
4. $y = 3x, y = 5 - 2x, y = 0$
5. $y = x^2, x = y^2$
6. $y = x^3, y = x^2$

In exercises 7–18, compute the volume of the solid bounded by the given surfaces.

7. $2x + 3y + z = 6$ and the three coordinate planes
8. $x + 2y - 3z = 6$ and the three coordinate planes
9. $z = 4 - x^2 - y^2$ and $z = 0$, with $-1 \le x \le 1$ and $-1 \le y \le 1$
10. $z = x^2 + y^2, z = 0, x = 0, x = 1, y = 0, y = 1$
11. $z = 1 - y, z = 0, y = 0, x = 1, x = 2$
12. $z = 2 + x, z = 0, x = 0, y = 0, y = 1$
13. $z = 1 - y^2, x + y = 1$ and the three coordinate planes (first octant)
14. $z = 1 - x^2 - y^2, x + y = 1$ and the three coordinate planes
15. $z = x^2 + y^2 + 3, z = 1, y = x^2, y = 4$
16. $z = x^2 + y^2 + 1, z = -1, y = x^2, y = 2x + 3$
17. $z = x + 2, z = y - 2, x = y^2 - 2, x = y$
18. $z = 2x + y + 1, z = -2x, x = y^2, x = 1$

In exercises 19–22, set up a double integral for the volume bounded by the given surfaces and estimate it numerically.

19. $z = \sqrt{x^2 + y^2}, y = 4 - x^2$, first octant
20. $z = \sqrt{4 - x^2 - y^2}$, inside $x^2 + y^2 = 1$, first octant
21. $z = e^{xy}, x + 2y = 4$ and the three coordinate planes
22. $z = e^{x^2+y^2}, z = 0$ and $x^2 + y^2 = 4$

In exercises 23–28, find the mass and center of mass of the lamina with the given density.

23. Lamina bounded by $y = x^3$ and $y = x^2$, $\rho(x, y) = 4$
24. Lamina bounded by $y = x^4$ and $y = x^2$, $\rho(x, y) = 4$
25. Lamina bounded by $x = y^2$ and $x = 1$, $\rho(x, y) = y^2 + x + 1$
26. Lamina bounded by $x = y^2$ and $x = 4$, $\rho(x, y) = y + 3$
27. Lamina bounded by $y = x^2$ $(x > 0)$, $y = 4$ and $x = 0$, $\rho(x, y) =$ distance from y-axis
28. Lamina bounded by $y = x^2 - 4$ and $y = 5$, $\rho(x, y) =$ square of the distance from the y-axis
29. The laminae of exercises 25 and 26 are both symmetric about the x-axis. Explain why it is not true in both exercises that the center of mass is located on the x-axis.
30. Suppose that a lamina is symmetric about the x-axis. State a condition on the density function $\rho(x, y)$ that guarantees that the center of mass is located on the x-axis.
31. Suppose that a lamina is symmetric about the y-axis. State a condition on the density function $\rho(x, y)$ that guarantees that the center of mass is located on the y-axis.
32. Give an example of a lamina that *is* symmetric about the y-axis but that *does not* have its center of mass on the y-axis.
33. Suppose that $f(x, y) = 15{,}000xe^{-x^2-y^2}$ is the population density of a species of small animals. Estimate the population in the triangular region with vertices (1, 1), (2, 1) and (1, 0).
34. Suppose that $f(x, y) = 15{,}000xe^{-x^2-y^2}$ is the population density of a species of small animals. Estimate the population in the region bounded by $y = x^2$, $y = 0$ and $x = 1$.

35. Suppose that $f(x, t) = 20e^{-t/6}$ is the yearly rate of change of the price per barrel of oil. If x is the number of billions of barrels and t is the number of years since 2000, compute and interpret $\int_0^{10} \int_0^4 f(x, t)\, dt\, dx$.

36. Repeat exercise 35 for $f(x, t) = \begin{cases} 20e^{-t/6}, & \text{if } 0 \le x \le 4 \\ 14e^{-t/6}, & \text{if } x > 4 \end{cases}$.

37. Find the mass and moments of inertia I_y and I_x for a lamina in the shape of the region bounded by $y = x^2$ and $y = 4$ with density $\rho(x, y) = 1$.

38. Find the mass and moments of inertia I_y and I_x for a lamina in the shape of the region bounded by $y = \frac{1}{4}x^2$ and $y = 1$ with density $\rho(x, y) = 4$. Comparing your answer with exercise 37, you should have found the same mass but different moments of inertia. Use the shapes of the regions to explain why this makes sense.

39. Figure skaters can control their rate of spin ω by varying their body positions, utilizing the principle of **conservation of angular momentum.** This states that in the absence of outside forces, the quantity $I_y\omega$ remains constant. Thus, reducing I_y by a factor of 2 will increase spin rate by a factor of 2. Compare the spin rates of the following two crude models of a figure skater, the first with arms extended (use $\rho = 1$) and the second with arms raised and legs crossed (use $\rho = 2$).

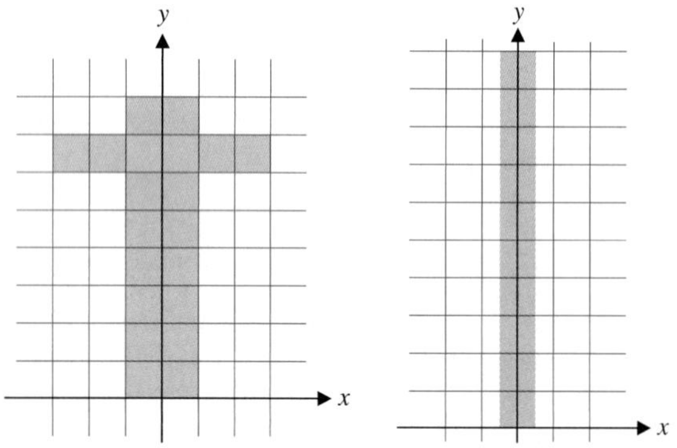

40. Lamina A is in the shape of the rectangle $-1 \le x \le 1$ and $-5 \le y \le 5$, with density $\rho(x, y) = 1$. It models a diver in the "layout" position. Lamina B is in the shape of the rectangle $-1 \le x \le 1$ and $-2 \le y \le 2$ with density $\rho(x, y) = 2.5$. It models a diver in the "tuck" position. Find the moment of inertia I_x for each lamina, and explain why divers use the tuck position to do multiple rotation dives.

41. Estimate the moment of inertia about the y-axis of the two ellipses R_1 bounded by $x^2 + 4y^2 = 16$ and R_2 bounded by $x^2 + 4y^2 = 36$. Assuming a constant density of $\rho = 1$, R_1 and R_2 can be thought of as models of two tennis racket heads. The rackets have the same shape, but the second racket is much bigger than the first (the difference in size is about the same as the difference between rackets of the 1960s and rackets of the 1990s).

42. For the tennis rackets in exercise 41, a rotation about the y-axis would correspond to the racket twisting in your hand, which is undesirable. Compare the tendency of each racket to twist. As related in Blandig and Monteleone's *What Makes a Boomerang Come Back,* the larger moment of inertia is what motivated a sore-elbowed Howard Head to construct large-headed tennis rackets in the 1970s.

In exercises 43–50, define the average value of $f(x, y)$ on a region R of area a by $\dfrac{1}{a}\iint\limits_R f(x, y)\, dA$.

43. Compute the average value of $f(x, y) = y$ on the region bounded by $y = x^2$ and $y = 4$.

44. Compute the average value of $f(x, y) = y^2$ on the region bounded by $y = x^2$ and $y = 4$.

45. In exercise 43, compare the average value of $f(x, y)$ to the y-coordinate of the center of mass of a lamina with the same shape and constant density.

46. In exercise 44, R extends from $y = 0$ to $y = 4$. Explain why the average value of $f(x, y)$ corresponds to a y-value larger than 2.

47. Compute the average value of $f(x, y) = \sqrt{x^2 + y^2}$ on the region bounded by $y = x^2 - 4$ and $y = 3x$.

48. Interpret the geometric meaning of the average value in exercise 47. (Hint: What does $\sqrt{x^2 + y^2}$ represent geometrically?)

49. Suppose the temperature at the point (x, y) in a region R is given by $T(x, y) = 50 + \cos(2x + y)$, where R is bounded by $y = x^2$ and $y = 8 - x^2$. Estimate the average temperature in R.

50. Suppose the elevation at the point (x, y) in a region R is given by $h(x, y) = 2300 + 50 \sin x \cos y$, where R is bounded by $y = x^2$ and $y = 2x$. Estimate the average elevation in R.

51. Suppose that the function $f(x, y)$ gives the rainfall per unit area at the point (x, y) in a region R. State in words what (a) $\iint\limits_R f(x, y)\, dA$ and (b) $\dfrac{\iint\limits_R f(x, y)\, dA}{\iint\limits_R 1\, dA}$ represent.

52. Suppose that the function $p(x, y)$ gives the population density at the point (x, y) in a region R. State in words what (a) $\iint\limits_R p(x, y)\, dA$ and (b) $\dfrac{\iint\limits_R p(x, y)\, dA}{\iint\limits_R 1\, dA}$ represent.

53. A triangular lamina has vertices $(0, 0)$, $(0, 1)$ and $(c, 0)$ for some positive constant c. Assuming constant mass density, show that the y-coordinate of the center of mass of the lamina is independent of the constant c.

54. Find the x-coordinate of the center of mass of the lamina of exercise 53 as a function of c.

55. Let T be the tetrahedron with vertices $(0, 0, 0)$, $(a, 0, 0)$, $(0, b, 0)$ and $(0, 0, c)$. Let B be the rectangular box with the same vertices plus $(a, b, 0)$, $(a, 0, c)$, $(0, b, c)$, and (a, b, c). Show that the volume of T is $\frac{1}{6}$ the volume of B.

56. Explain how to slice the box B of exercise 55 to get the tetrahedron T. Identify the percentage of volume that is sliced off each time.

In exercises 57–64, use the following definition of joint pdf (probability density function): a function $f(x, y)$ is a joint pdf on the region S if $f(x, y) \geq 0$ for all (x, y) in S and $\iint_S f(x, y)\, dA = 1$. Then for any region $R \subset S$, the probability that (x, y) is in R is given by $\iint_R f(x, y)\, dA$.

57. Show that $f(x, y) = e^{-x}e^{-y}$ is a joint pdf in the first quadrant $x \geq 0$, $y \geq 0$. (Hint: You will need to evaluate an improper double integral as iterated improper integrals.)

58. Show that $f(x, y) = 0.3x + 0.4y$ is a joint pdf on the rectangle $0 \leq x \leq 2$, $0 \leq y \leq 1$.

59. Find a constant c such that $f(x, y) = c(x + 2y)$ is a joint pdf on the triangle with vertices $(0, 0)$, $(2, 0)$ and $(2, 6)$.

60. Find a constant c such that $f(x, y) = c(x^2 + y)$ is a joint pdf on the region bounded by $y = x^2$ and $y = 4$.

61. Suppose that $f(x, y)$ is a joint pdf on the region bounded by $y = x^2$, $y = 0$ and $x = 2$. Set up a double integral for the probability that $y < x$.

62. Suppose that $f(x, y)$ is a joint pdf on the region bounded by $y = x^2$, $y = 0$ and $x = 2$. Set up a double integral for the probability that $y < 2$.

63. A point is selected at random from the region bounded by $y = 4 - x^2 (x > 0)$, $x = 0$ and $y = 0$. This means that the joint pdf for the point is constant, $f(x, y) = c$. Find the value of c. Then compute the probability that $y > x$ for the randomly chosen point.

64. A point is selected at random from the region bounded by $y = 4 - x^2 (x > 0)$, $x = 0$ and $y = 0$. Compute the probability that $y > 2$.

65. When solving projectile motion problems, we track the motion of an object's center of mass. For a high jumper, the athlete's entire body must clear the bar. Amazingly, a high jumper can accomplish this without raising his or her center of mass above the bar. To see how, suppose the athelete's body is bent into a shape modeled by the region between $y = \sqrt{9 - x^2}$ and $y = \sqrt{8 - x^2}$ with the bar at the point $(0, 2)$. Assuming constant mass density, show that the center of mass is below the bar, but the body does not touch the bar.

66. Show that $V_1 = V_2$, where V_1 is the volume under $z = 4 - x^2 - y^2$ and above the xy-plane and V_2 is the volume between $z = x^2 + y^2$ and $z = 4$. Illustrate this with a graph.

EXPLORATORY EXERCISES

1. A function $f(x, y)$ is a **joint probability density function** on a region R if $f(x, y) \geq 0$ for all (x, y) in R and $\iint_R f(x, y)dA = 1$. Suppose that a person playing darts is aiming at the bull's-eye but is not very accurate. Suppose that the bull's-eye is centered at the origin and the dartboard is the region R bounded by $x^2 + y^2 = 64$ (units are inches), and the joint density function for the resulting position of the dart is $f(x, y) = ce^{-x^2-y^2}$, for some constant c. Estimate the value of the constant c such that $f(x, y)$ is a joint density function on R. For a region U contained within R, the probability that the dart lands in U is given by $\iint_U f(x, y)\, dA$.

Estimate the probability that the dart hits inside the bull's-eye circle $x^2 + y^2 = \frac{1}{4}$. Estimate the probability that the dart accidentally lands in the "triple 20" band bounded by $x^2 + y^2 = 16$, $x^2 + y^2 = 14$, $y = 6.3x$ and $y = -6.3x$. Explain why all of the regions in this exercise would be easily described in polar coordinates. (Then start reading the next section!)

2. In this exercise, we explore an important issue in rocket design. We will work with the crude model shown, where the main tower of the rocket is 1 unit by 8 units and each triangular fin has height 1 and width w. First, find the y-coordinate y_1 of the center of mass, assuming a constant density $\rho(x, y) = 1$. Second, find the y-coordinate y_2 of the center of mass assuming the following density structure: the top half of the main tower has density $\rho = 1$, the bottom half of the main tower has density $\rho = 2$ and the fins have density $\rho = \frac{1}{4}$. Find the smallest value of w such that $y_1 < y_2$. In this case, if the rocket tilts slightly, air drag will push the rocket back in line. This stability criterion explains why model rockets have large, lightweight fins.

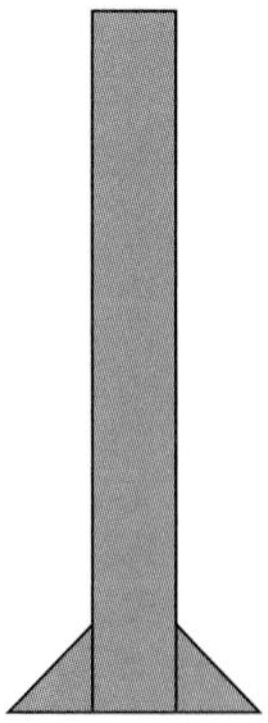

13.3 DOUBLE INTEGRALS IN POLAR COORDINATES

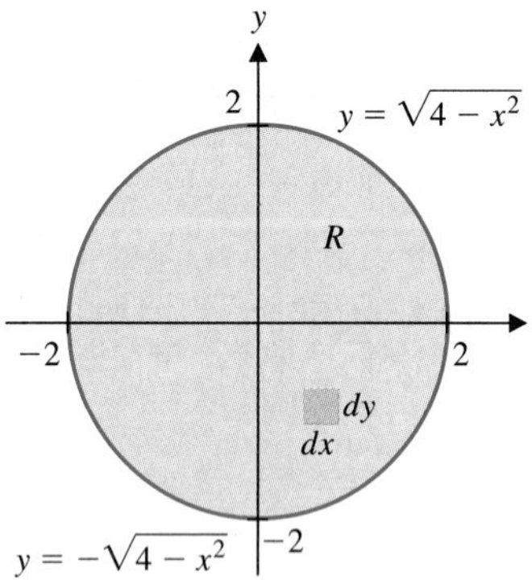

FIGURE 13.26
A circular region

Polar coordinates prove to be particularly useful for dealing with certain double integrals. This happens for several reasons. Most importantly, if the region over which you are integrating is in some way circular, polar coordinates may be exactly what you need for dealing with an otherwise intractable integration problem. For instance, you might need to evaluate

$$\iint_R (x^2 + y^2 + 3)\, dA.$$

This certainly looks simple enough, until we tell you that R is the circle of radius 2, centered at the origin, as shown in Figure 13.26. We write the top half of the circle as the graph of $y = \sqrt{4 - x^2}$ and the bottom half as $y = -\sqrt{4 - x^2}$. The double integral in question now becomes

$$\begin{aligned}\iint_R (x^2 + y^2 + 3)\, dA &= \int_{-2}^{2} \int_{-\sqrt{4-x^2}}^{\sqrt{4-x^2}} (x^2 + y^2 + 3)\, dy\, dx \\ &= \int_{-2}^{2} \left(x^2 y + \frac{y^3}{3} + 3y \right) \Big|_{y=-\sqrt{4-x^2}}^{y=\sqrt{4-x^2}} dx \\ &= 2 \int_{-2}^{2} \left[(x^2 + 3)\sqrt{4 - x^2} + \frac{1}{3}(4 - x^2)^{3/2} \right] dx. \qquad (3.1)\end{aligned}$$

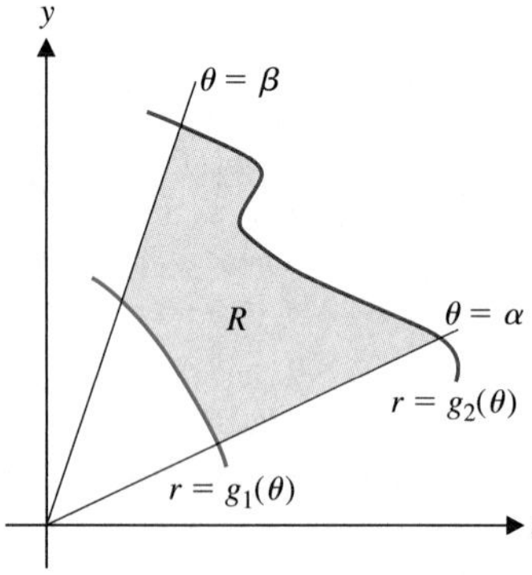

FIGURE 13.27a
Polar region R

We probably don't need to convince you that the integral in (3.1) is most unpleasant. On the other hand, as we'll see shortly, this double integral is simple when it's written in polar coordinates. We consider several types of polar regions.

Suppose the region R can be written in the form

$$R = \{(r, \theta) | \alpha \le \theta \le \beta \text{ and } g_1(\theta) \le r \le g_2(\theta)\},$$

where $0 \le g_1(\theta) \le g_2(\theta)$, for all θ in $[\alpha, \beta]$, as pictured in Figure 13.27a. As our first step, we partition R, but rather than use a rectangular grid, as we have done with rectangular

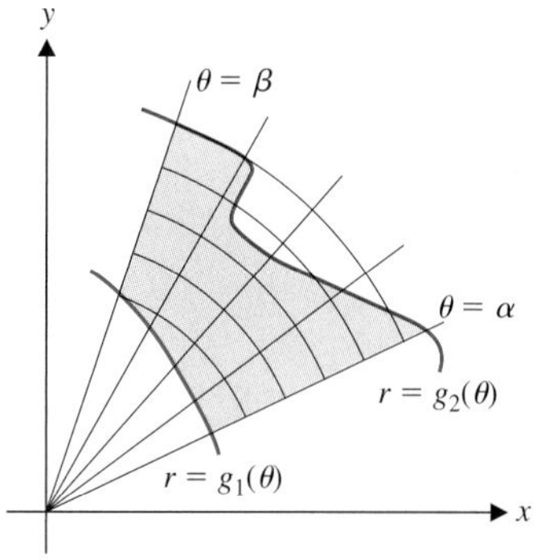

FIGURE 13.27b
Partition of R

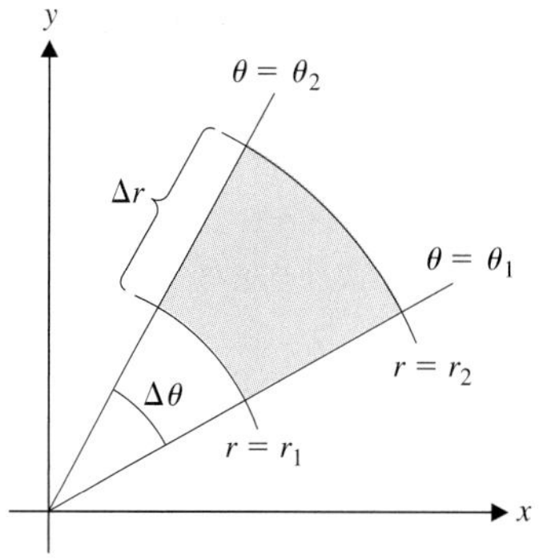

FIGURE 13.27c
Elementary polar region

coordinates, we use a partition consisting of a number of concentric circular arcs (of the form $r =$ constant) and rays (of the form $\theta =$ constant). We indicate such a partition of the region R in Figure 13.27b.

Notice that rather than consisting of rectangles, the "grid" in this case is made up of **elementary polar regions,** each bounded by two circular arcs and two rays (as shown in Figure 13.27c). In an **inner partition,** we include only those elementary polar regions that lie completely inside R.

We pause now briefly to calculate the area ΔA of the elementary polar region indicated in Figure 13.27c. Let $\bar{r} = \frac{1}{2}(r_1 + r_2)$ be the average radius of the two concentric circular arcs $r = r_1$ and $r = r_2$. Recall that the area of a circular sector is given by $A = \frac{1}{2}\theta r^2$, where $r =$ radius and θ is the central angle of the sector. Consequently, we have that

$$\begin{aligned}
\Delta A &= \text{Area of outer sector} - \text{Area of inner sector} \\
&= \frac{1}{2}\Delta\theta r_2^2 - \frac{1}{2}\Delta\theta r_1^2 \\
&= \frac{1}{2}\left(r_2^2 - r_1^2\right)\Delta\theta \\
&= \frac{1}{2}(r_2 + r_1)(r_2 - r_1)\Delta\theta \\
&= \bar{r}\,\Delta r\,\Delta\theta. \qquad (3.2)
\end{aligned}$$

As a familiar starting point, we first consider the problem of finding the volume lying beneath a surface $z = f(r, \theta)$, where f is continuous and $f(r, \theta) \geq 0$ on R. Using (3.2), we find that the volume V_i lying beneath the surface $z = f(r, \theta)$ and above the ith elementary polar region in the partition is then approximately the volume of the cylinder:

$$V_i \approx \underbrace{f(r_i, \theta_i)}_{\text{height}} \quad \underbrace{\Delta A_i}_{\text{area of base}} = f(r_i, \theta_i)\, r_i\, \Delta r_i\, \Delta\theta_i,$$

where (r_i, θ_i) is a point in R_i and r_i is the average radius in R_i. We get an approximation to the total volume V by summing over all the regions in the inner partition:

$$V \approx \sum_{i=1}^{n} f(r_i, \theta_i)\, r_i\, \Delta r_i\, \Delta\theta_i.$$

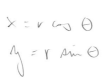

As we have done a number of times now, we obtain the exact volume by taking the limit as the norm of the partition $\|P\|$ tends to zero and recognizing the iterated integral:

$$V = \lim_{\|P\|\to 0} \sum_{i=1}^{n} f(r_i, \theta_i)\, r_i\, \Delta r_i\, \Delta\theta_i$$
$$= \int_\alpha^\beta \int_{g_1(\theta)}^{g_2(\theta)} f(r,\theta)\, r\, dr\, d\theta.$$

In this case, $\|P\|$ is the longest diagonal of any elementary polar region in the inner partition. More generally, we have the result in Theorem 3.1, which holds regardless of whether or not $f(r,\theta) \ge 0$ on R.

NOTES

Theorem 3.1 says that to write a double integral in polar coordinates, we write $x = r\cos\theta$, $y = r\sin\theta$, find the limits of integration for r and θ and replace dA by $r\,dr\,d\theta$. Be certain not to omit the factor of r in $dA = r\,dr\,d\theta$; this is a very common error.

THEOREM 3.1 (Fubini's Theorem)

Suppose that $f(r,\theta)$ is continuous on the region $R = \{(r,\theta) \mid \alpha \le \theta \le \beta$ and $g_1(\theta) \le r \le g_2(\theta)\}$, where $0 \le g_1(\theta) \le g_2(\theta)$ for all θ in $[\alpha, \beta]$. Then,

$$\iint_R f(r,\theta)\, dA = \int_\alpha^\beta \int_{g_1(\theta)}^{g_2(\theta)} f(r,\theta)\, r\, dr\, d\theta. \quad (3.3)$$

The proof of this result is beyond the level of this text. However, the result should seem reasonable from our development for the case where $f(r,\theta) \ge 0$.

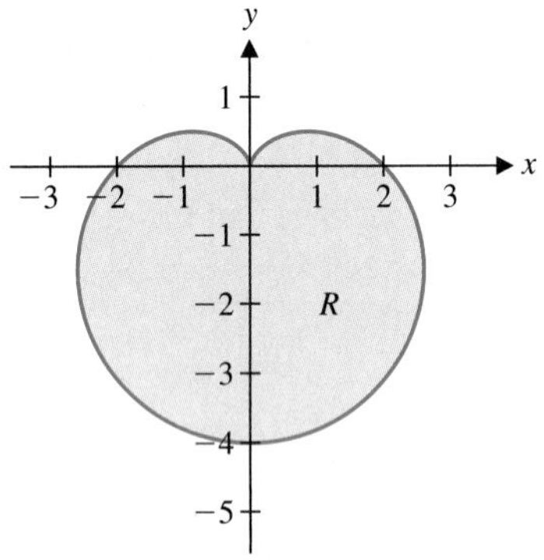

FIGURE 13.28
$r = 2 - 2\sin\theta$

EXAMPLE 3.1 Computing Area in Polar Coordinates

Find the area inside the curve defined by $r = 2 - 2\sin\theta$.

Solution First, we sketch a graph of the region in Figure 13.28. For each fixed θ, r ranges from 0 (corresponding to the origin) to $2 - 2\sin\theta$ (corresponding to the cardioid). To go all the way around the cardioid, exactly once, θ ranges from 0 to 2π. From (3.3), we then have

$$A = \iint_R \underbrace{dA}_{r\,dr\,d\theta} = \int_0^{2\pi} \int_0^{2-2\sin\theta} r\, dr\, d\theta$$
$$= \int_0^{2\pi} \frac{r^2}{2}\Bigg|_{r=0}^{r=2-2\sin\theta} d\theta$$
$$= \frac{1}{2}\int_0^{2\pi} [(2-2\sin\theta)^2 - 0]\, d\theta = 6\pi,$$

where we have left the details of the final calculation as an exercise. ■

We now return to our introductory example and show how the introduction of polar coordinates can dramatically simplify certain double integrals in rectangular coordinates.

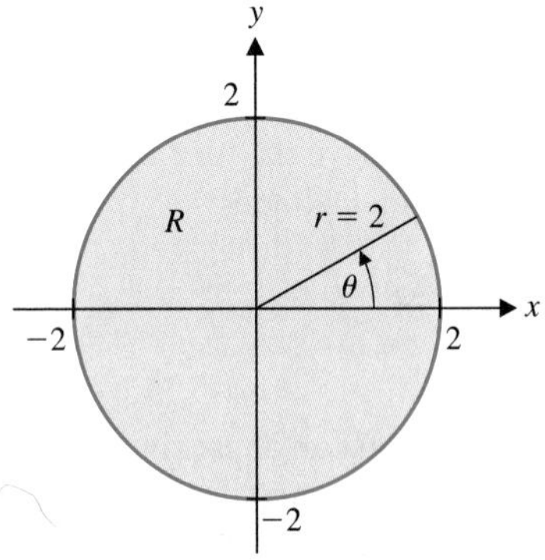

FIGURE 13.29
The region R

EXAMPLE 3.2 Evaluating a Double Integral in Polar Coordinates

Evaluate $\iint_R (x^2 + y^2 + 3)\, dA$, where R is the circle of radius 2 centered at the origin.

Solution First, recall from this section's introduction that in rectangular coordinates as in (3.1), this integral is extremely messy. From the region of integration shown in Figure 13.29, it's easy to see that for each fixed θ, r ranges from 0 (corresponding to the origin) to 2 (corresponding to a point on the circle). Then, in order to go around the circle exactly once, θ ranges from 0 to 2π. Finally, notice that the integrand contains the quantity $x^2 + y^2$, which you should recognize as r^2 in polar coordinates. From (3.3), we now have

$$\iint_R \underbrace{(x^2+y^2+3)}_{r^2+3}\,\underbrace{dA}_{r\,dr\,d\theta} = \int_0^{2\pi}\int_0^2 (r^2+3)\,r\,dr\,d\theta$$

$$= \int_0^{2\pi}\int_0^2 (r^3+3r)\,dr\,d\theta$$

$$= \int_0^{2\pi} \left(\frac{r^4}{4} + 3\frac{r^2}{2}\right)\Bigg|_{r=0}^{r=2} d\theta$$

$$= \int_0^{2\pi} \left[\left(\frac{2^4}{4} + 3\frac{2^2}{2}\right) - 0\right] d\theta$$

$$= 10\int_0^{2\pi} d\theta = 20\pi.$$

Notice how simple this iterated integral was, as compared to the corresponding integral in rectangular coordinates in (3.1). ■

NOTES

For double integrals of the form $\int_a^b \int_c^d f(r)dr\,d\theta$, note that the inner integral does not depend on θ. As a result, we can rewrite the double integral as

$$\left(\int_a^b 1\,d\theta\right)\left(\int_c^d f(r)\,dr\right) = (b-a)\int_c^d f(r)\,dr.$$

When dealing with double integrals, you should always consider whether the region over which you're integrating is in some way circular. If it is a circle or some portion of a circle, consider using polar coordinates.

EXAMPLE 3.3 Finding Volume Using Polar Coordinates

Find the volume inside the paraboloid $z = 9 - x^2 - y^2$, outside the cylinder $x^2 + y^2 = 4$ and above the xy-plane.

Solution Notice that the paraboloid has its vertex at the point (0, 0, 9) and the axis of the cylinder is the z-axis. (See Figure 13.30a.) You should observe that the solid lies below the paraboloid and above the region in the xy-plane lying between the traces of the cylinder and the paraboloid in the xy-plane, that is, between the circles of radius 2 and 3, both centered at the origin. So, for each fixed $\theta \in [0, 2\pi]$, r ranges from 2 to 3. We call such a region a **circular annulus** (see Figure 13.30b). From (3.3), we have

$$V = \iint_R \underbrace{(9-x^2-y^2)}_{9-r^2}\,\underbrace{dA}_{r\,dr\,d\theta} = \int_0^{2\pi}\int_2^3 (9-r^2)\,r\,dr\,d\theta$$

$$= \int_0^{2\pi}\int_2^3 (9r - r^3)\,dr\,d\theta = 2\pi\int_2^3 (9r-r^3)\,dr$$

$$= 2\pi\left(9\frac{r^2}{2} - \frac{r^4}{4}\right)\Bigg|_{r=2}^{r=3} = \frac{25}{2}\pi. \quad ■$$

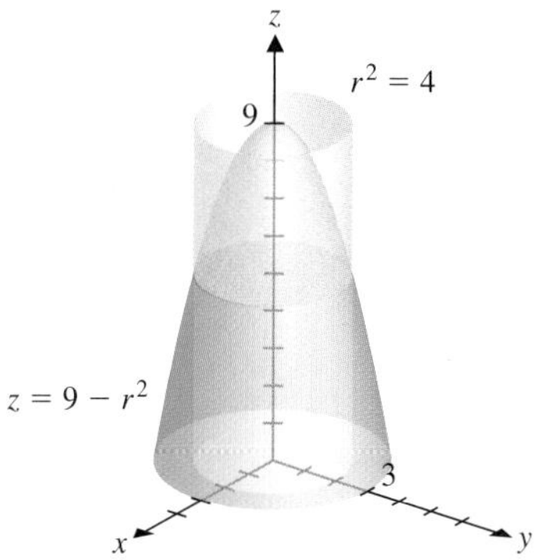

FIGURE 13.30a
Volume outside the cylinder and inside the paraboloid

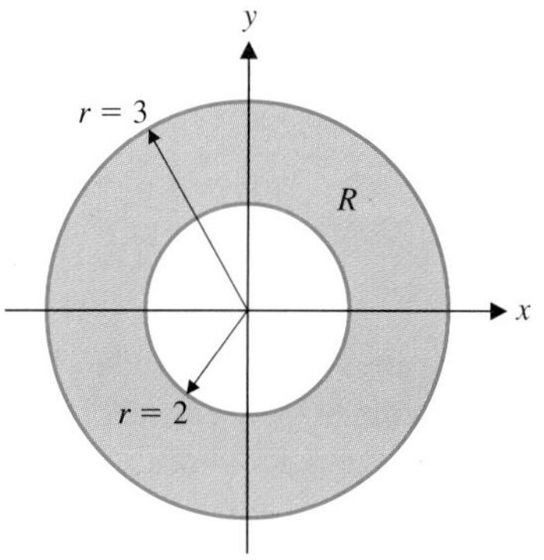

FIGURE 13.30b
Circular annulus

There are actually two things that you should look for when you are considering using polar coordinates for a double integral. The first is most obvious: Is the geometry of the region circular? The other is: Does the integral contain the expression $x^2 + y^2$ (particularly inside of other functions such as square roots, exponentials, etc.)? Since $r^2 = x^2 + y^2$, changing to polar coordinates will often simplify terms of this form.

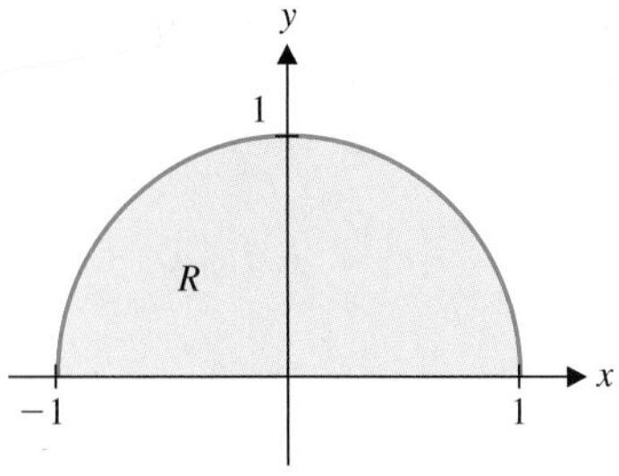

FIGURE 13.31
The region R

EXAMPLE 3.4 Changing a Double Integral to Polar Coordinates

Evaluate the iterated integral $\int_{-1}^{1}\int_{0}^{\sqrt{1-x^2}} x^2(x^2+y^2)^2 dy\,dx$.

Solution First, you should recognize that evaluating this integral in rectangular coordinates is nearly hopeless. (Try it and see why!) On the other hand, it does have a term of the form $x^2 + y^2$, which we discussed above. Even more significantly, the region over which you're integrating turns out to be a semicircle, as follows. Reading the inside limits of integration first, observe that for each fixed x between -1 and 1, y ranges from $y = 0$ up to $y = \sqrt{1-x^2}$ (the top half of the circle of radius 1 centered at the origin). We sketch the region in Figure 13.31. From (3.3), we have

$$\int_{-1}^{1}\int_{0}^{\sqrt{1-x^2}} x^2(x^2+y^2)^2 dy\,dx = \iint_R \underbrace{x^2}_{r^2\cos^2\theta}\ \underbrace{(x^2+y^2)^2}_{(r^2)^2}\ \underbrace{dA}_{r\,dr\,d\theta}$$

Since $x = r\cos\theta$.

$$= \int_0^{\pi}\int_0^{1} r^7\cos^2\theta\, dr\, d\theta$$

$$= \int_0^{\pi} \frac{r^8}{8}\Big|_{r=0}^{r=1} \cos^2\theta\, d\theta$$

$$= \frac{1}{8}\int_0^{\pi}\frac{1}{2}(1+\cos 2\theta)\, d\theta$$

Since $\cos^2\theta = \frac{1}{2}(1+\cos 2\theta)$.

$$= \frac{1}{16}\left(\theta + \frac{1}{2}\sin 2\theta\right)\Big|_0^{\pi} = \frac{\pi}{16}.$$ ■

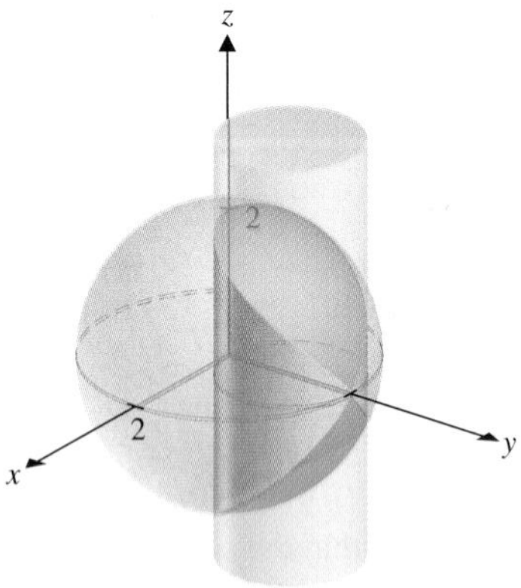

FIGURE 13.32a
Volume inside the sphere and inside the cylinder

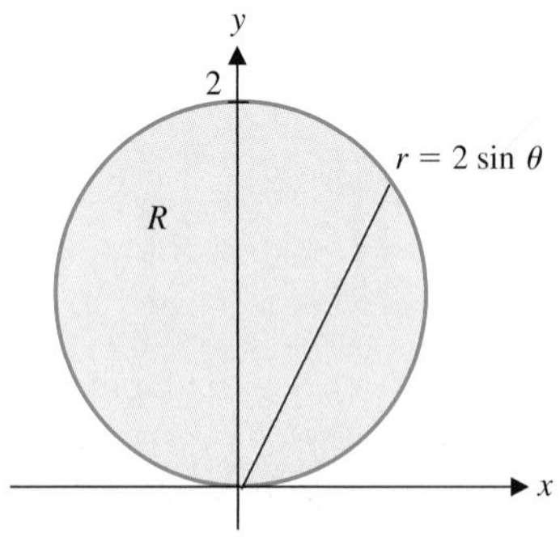

FIGURE 13.32b
The region R

EXAMPLE 3.5 Finding Volume Using Polar Coordinates

Find the volume cut out of the sphere $x^2 + y^2 + z^2 = 4$ by the cylinder $x^2 + y^2 = 2y$.

Solution We show a sketch of the solid in Figure 13.32a. (If you complete the square on the equation of the cylinder, you'll see that it is a circular cylinder of radius 1, whose axis is the line: $x = 0$, $y = 1$, $z = t$.) Notice that equal portions of the volume lie above and below the circle of radius 1 centered at (0, 1), indicated in Figure 13.32b. So, we compute the volume lying below the top hemisphere $z = \sqrt{4-x^2-y^2}$ and above the region R indicated in Figure 13.32b and double it. We have

$$V = 2\iint_R \sqrt{4-x^2-y^2}\, dA.$$

Since R is a circle and the integrand includes a term of the form x^2+y^2, we introduce polar coordinates. Since $y = r\sin\theta$, the circle $x^2 + y^2 = 2y$ becomes $r^2 = 2r\sin\theta$ or $r = 2\sin\theta$. This gives us

$$V = 2\int_0^{\pi}\int_0^{2\sin\theta} \sqrt{4-r^2}\, r\, dr\, d\theta,$$

since the entire circle $r = 2\sin\theta$ is traced out for $0 \le \theta \le \pi$ and since for each fixed $\theta \in [0, \pi]$, r ranges from $r = 0$ to $r = 2\sin\theta$. Notice further that by symmetry, we get

$$\begin{aligned} V &= 4\int_0^{\pi/2}\int_0^{2\sin\theta} \sqrt{4-r^2}\, r\, dr\, d\theta \\ &= -2\int_0^{\pi/2} \left[\frac{2}{3}(4-r^2)^{3/2}\right]_{r=0}^{r=2\sin\theta} d\theta \\ &= -\frac{4}{3}\int_0^{\pi/2} \left[(4-4\sin^2\theta)^{3/2} - 4^{3/2}\right] d\theta \\ &= -\frac{32}{3}\int_0^{\pi/2} [(\cos^2\theta)^{3/2} - 1]\, d\theta \\ &= -\frac{32}{3}\int_0^{\pi/2} (\cos^3\theta - 1)\, d\theta \\ &= -\frac{64}{9} + \frac{16}{3}\pi \approx 9.644. \end{aligned}$$

There are several things to observe here. First, our use of symmetry was crucial. By restricting the integral to the interval $[0, \frac{\pi}{2}]$, we could write $(\cos^2\theta)^{3/2} = \cos^3\theta$, which is *not* true on the entire interval $[0, \pi]$. (Why not?) Second, if you think that this integral was messy, consider what it looks like in rectangular coordinates. (It's not pretty!) ■

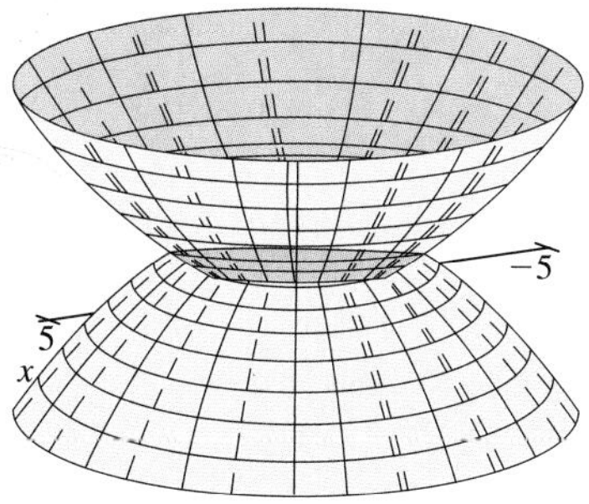

FIGURE 13.33a
Intersecting paraboloids

EXAMPLE 3.6 Finding the Volume Between Two Paraboloids

Find the volume of the solid bounded by $z = 8 - x^2 - y^2$ and $z = x^2 + y^2$.

Solution Observe that the surface $z = 8 - x^2 - y^2$ is a paraboloid with vertex at $z = 8$ and opening downward, while $z = x^2 + y^2$ is a paraboloid with vertex at the origin and opening upward. The solid is shown in Figure 13.33a. At a given point (x, y), the height of the solid is given by

$$(8 - x^2 - y^2) - (x^2 + y^2) = 8 - 2x^2 - 2y^2.$$

We now have
$$V = \iint_R (8 - 2x^2 - 2y^2)\, dA,$$

where the region of integration R is the shadow of the solid in the xy-plane. The solid is widest at the intersection of the two paraboloids, which occurs where $8 - x^2 - y^2 = x^2 + y^2$ or $x^2 + y^2 = 4$. The region of integration R is then the disk shown in Figure 13.33b and is most easily described in polar coordinates. The integrand becomes $8 - 2x^2 - 2y^2 = 8 - 2r^2$ and we have

$$\begin{aligned} V &= \int_0^{2\pi}\int_0^2 (8 - 2r^2) r\, dr\, d\theta \\ &= 16\pi. \end{aligned}$$ ■

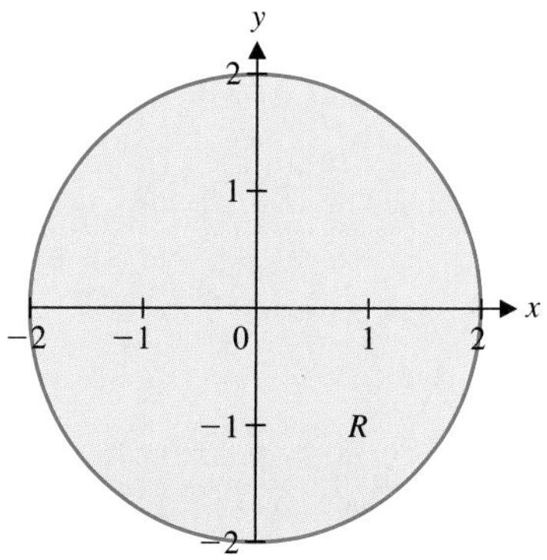

FIGURE 13.33b
The region R

Finally, we observe that we can also evaluate double integrals in polar coordinates by integrating first with respect to θ. Although such integrals are uncommon (given the way in which we change variables from rectangular to polar coordinates), we provide this for the sake of completeness.

Suppose the region R can be written in the form

$$R = \{(r, \theta) | 0 \le a \le r \le b \text{ and } h_1(r) \le \theta \le h_2(r)\},$$

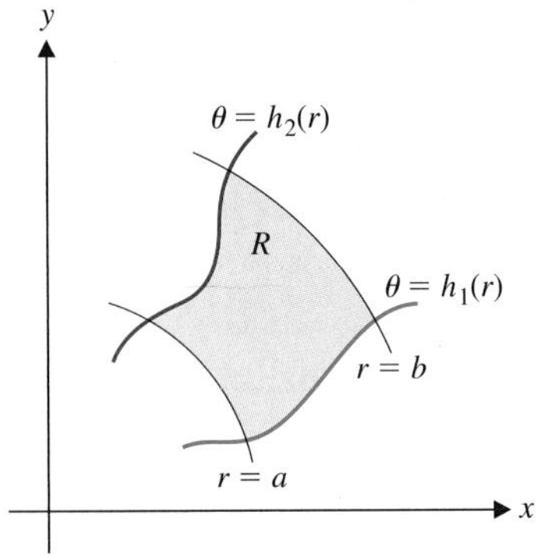

FIGURE 13.34
The region R

where $h_1(r) \le h_2(r)$, for all r in $[a, b]$, as pictured in Figure 13.34. Then, it can be shown that if $f(r, \theta)$ is continuous on R, we have

$$\iint_R f(r,\theta)\,dA = \int_a^b \int_{h_1(r)}^{h_2(r)} f(r,\theta)\,r\,d\theta\,dr. \tag{3.4}$$

BEYOND FORMULAS

This section may change the way you think of polar coordinates. While they allow us to describe a variety of unusual curves (roses, cardioids and so on) in a convenient form, polar coordinates are an essential computational tool for double integrals. In section 13.6, they serve the same role in triple integrals. In general, polar coordinates are useful in applications where some form of radial symmetry is present. Can you describe any situations in engineering, physics or chemistry where a structure or force has radial symmetry?

HW Pg 1064-1065
#1, 5, 9, 13, 21, 27

EXERCISES 13.3

WRITING EXERCISES

1. Thinking of $dy\,dx$ as representing the area dA of a small rectangle, explain in geometric terms why

$$\iint_R f(x,y)\,dA \ne \iint_R f(r\cos\theta, r\sin\theta)\,dr\,d\theta.$$

2. In all of the examples in this section, we integrated with respect to r first. It is perfectly legitimate to integrate with respect to θ first. Explain why it is unlikely that it will ever be necessary to do so. [Hint: If θ is on the inside, you need functions of the form $\theta(r)$ for the limits of integration.]

3. Given a double integral in rectangular coordinates as in example 3.2 or 3.4, identify at least two indications that the integral would be easier to evaluate in polar coordinates.

4. In section 9.5, we derived a formula $A = \int_a^b \frac{1}{2}[f(\theta)]^2 d\theta$ for the area bounded by the polar curve $r = f(\theta)$ and rays $\theta = a$ and $\theta = b$. Discuss how this formula relates to the formula used in example 3.1. Discuss which formula is easier to remember and which formula is more generally useful.

In exercises 1–6, find the area of the region bounded by the given curves.

1. $r = 3 + 2\sin\theta$
2. $r = 2 - 2\cos\theta$
3. one leaf of $r = \sin 3\theta$
4. $r = 3\cos\theta$
5. inside $r = 2\sin 3\theta$, outside $r = 1$, first quadrant
6. inside $r = 1$ and outside $r = 2 - 2\cos\theta$

In exercises 7–12, use polar coordinates to evaluate the double integral.

7. $\iint_R \sqrt{x^2+y^2}\,dA$, where R is the disk $x^2 + y^2 \le 9$
8. $\iint_R \sqrt{x^2+y^2+1}\,dA$, where R is the disk $x^2 + y^2 \le 16$
9. $\iint_R e^{-x^2-y^2}\,dA$, where R is the disk $x^2 + y^2 \le 4$
10. $\iint_R e^{-\sqrt{x^2+y^2}}\,dA$, where R is the disk $x^2 + y^2 \le 1$
11. $\iint_R y\,dA$, where R is bounded by $r = 2 - \cos\theta$
12. $\iint_R x\,dA$, where R is bounded by $r = 1 - \sin\theta$

In exercises 13–16, use the most appropriate coordinate system to evaluate the double integral.

13. $\iint_R (x^2+y^2)\,dA$, where R is bounded by $x^2 + y^2 = 9$
14. $\iint_R 2xy\,dA$, where R is bounded by $y = 4 - x^2$ and $y = 0$
15. $\iint_R (x^2+y^2)\,dA$, where R is bounded by $y = x$, $y = 0$ and $x = 2$
16. $\iint_R \cos\sqrt{x^2+y^2}\,dA$, where R is bounded by $x^2 + y^2 = 9$

In exercises 17–26, use an appropriate coordinate system to compute the volume of the indicated solid.

17. Below $z = x^2 + y^2$, above $z = 0$, inside $x^2 + y^2 = 9$
18. Below $z = x^2 + y^2 - 4$, above $z = 0$, inside $x^2 + y^2 = 9$

19. Below $z = \sqrt{x^2 + y^2}$, above $z = 0$, inside $x^2 + y^2 = 4$

20. Below $z = \sqrt{x^2 + y^2}$, above $z = 0$, inside $x^2 + (y - 1)^2 = 1$

21. Below $z = \sqrt{4 - x^2 - y^2}$, above $z = 1$, inside $x^2 + y^2 = \frac{1}{4}$

22. Below $z = 8 - x^2 - y^2$, above $z = 3x^2 + 3y^2$

23. Below $z = 6 - x - y$, in the first octant

24. Below $z = 4 - x^2 - y^2$, between $y = x$, $y = 0$ and $x = 1$

25. Below $z = 4 - x^2 - y^2$, above $z = x^2 + y^2$, between $y = 0$ and $y = x$, in the first octant

26. Above $z = \sqrt{x^2 + y^2}$, below $z = 4$, above the xy-plane, between $y = x$ and $y = 2x$, in the first octant

In exercises 27–32, evaluate the iterated integral by converting to polar coordinates.

27. $\displaystyle\int_{-2}^{2}\int_{-\sqrt{4-x^2}}^{\sqrt{4-x^2}} \sqrt{x^2 + y^2}\, dy\, dx$

28. $\displaystyle\int_{-2}^{2}\int_{0}^{\sqrt{4-x^2}} \sin(x^2 + y^2)\, dy\, dx$

29. $\displaystyle\int_{0}^{2}\int_{-\sqrt{4-x^2}}^{\sqrt{4-x^2}} e^{-x^2-y^2}\, dy\, dx$

30. $\displaystyle\int_{0}^{2}\int_{-\sqrt{4-x^2}}^{0} y\, dy\, dx$

31. $\displaystyle\int_{0}^{2}\int_{x}^{\sqrt{8-x^2}} (x^2 + y^2)^{3/2}\, dy\, dx$

32. $\displaystyle\int_{0}^{1}\int_{y}^{\sqrt{2y-y^2}} x\, dx\, dy$

In exercises 33–36, compute the probability that a dart lands in the region R, assuming that the probability is given by $\iint_R \frac{1}{\pi} e^{-x^2-y^2} dA$.

33. A double bull's-eye, R is the region inside $r = \frac{1}{4}$ (inch)

34. A single bull's-eye, R bounded by $r = \frac{1}{4}$ and $r = \frac{1}{2}$

35. A triple-20, R bounded by $r = 3\frac{3}{4}$, $r = 4$, $\theta = \frac{9\pi}{20}$ and $\theta = \frac{11\pi}{20}$

36. A double-20, R bounded by $r = 6\frac{1}{4}$, $r = 6\frac{1}{2}$, $\theta = \frac{9\pi}{20}$ and $\theta = \frac{11\pi}{20}$

37. Find the area of the triple-20 region described in exercise 35.

38. Find the area of the double-20 region described in exercise 36.

39. Find the center of mass of a lamina in the shape of $x^2 + (y - 1)^2 = 1$, with density $\rho(x, y) = 1/\sqrt{x^2 + y^2}$.

40. Find the center of mass of a lamina in the shape of $r = 2 - 2\cos\theta$, with density $\rho(x, y) = x^2 + y^2$.

41. Suppose that $f(x, y) = 20{,}000\, e^{-x^2-y^2}$ is the population density of a species of small animals. Estimate the population in the region bounded by $x^2 + y^2 = 1$.

42. Suppose that $f(x, y) = 15{,}000\, e^{-x^2-y^2}$ is the population density of a species of small animals. Estimate the population in the region bounded by $(x - 1)^2 + y^2 = 1$.

43. Find the moment of inertia I_y of the circular lamina bounded by $x^2 + y^2 = R^2$, with density $\rho(x, y) = 1$. If the radius doubles, by what factor does the moment of inertia increase?

44. Repeat exercise 43 for the density function $\rho(x, y) = \sqrt{x^2 + y^2}$.

45. Use a double integral to derive the formula for the volume of a sphere of radius a.

46. Use a double integral to derive the formula for the volume of a right circular cone of height h and base radius a. (Hint: Show that the desired volume equals the volume under $z = h$ and above $z = \frac{h}{a}\sqrt{x^2 + y^2}$.)

47. Find the volume cut out of the sphere $x^2 + y^2 + z^2 = 9$ by the cylinder $x^2 + y^2 = 2x$.

48. Find the volume of the wedge sliced out of the sphere $x^2 + y^2 + z^2 = 4$ by the planes $y = x$ and $y = 2x$. (Keep the portion with $x \geq 0$.)

49. Set up a double integral for the volume of the piece sliced off of the top of $x^2 + y^2 + z^2 = 4$ by the plane $y + z = 2$.

50. Set up a double integral for the volume of the portion of the region below $x + 2y + 3z = 6$ and above $z = 0$ cut out by the cylinder $x^2 + 4y^2 = 4$.

51. Show that the volume under the cone $z = k - r$ and above the xy-plane (where $k > 0$) grows as a cubic function of k. Show that the volume under the paraboloid $z = k - r^2$ and above the xy-plane (where $k > 0$) grows as a quadratic function of k. Explain why this volume increases less rapidly than that of the cone.

52. Show that the volume under the surface $z = k - r^n$ and above the xy-plane (where $k > 0$) approaches a linear function of k as $n \to \infty$. Explain why this makes sense.

53. Evaluate $\displaystyle\iint_R \frac{2}{1 + x^2 + y^2} dA$ where R is outside $r = 1$ and inside $r = 2\sin\theta$.

54. Evaluate $\displaystyle\iint_R \frac{\ln(x^2 + y^2)}{x^2 + y^2} dA$ where R is bounded by $r = 1$ and $r = 2$.

EXPLORATORY EXERCISES

1. Suppose that the following data give the density of a lamina at different locations. Estimate the mass of the lamina.

r \ θ	0	$\frac{\pi}{2}$	π	$\frac{3\pi}{2}$	2π
$\frac{1}{2}$	1.0	1.4	1.4	1.2	1.0
1	0.8	1.2	1.0	1.0	0.8
$\frac{3}{2}$	1.0	1.3	1.4	1.3	1.2
2	1.2	1.6	1.6	1.4	1.2

2. One of the most important integrals in probability theory is $\int_{-\infty}^{\infty} e^{-x^2}\,dx$. Since there is no antiderivative of e^{-x^2} among the elementary functions, we can't evaluate this integral directly. A clever use of polar coordinates is needed. Start by giving the integral a name,

$$\int_{-\infty}^{\infty} e^{-x^2}\,dx = I.$$

Now, assuming that all the integrals converge, argue that $\int_{-\infty}^{\infty} e^{-y^2}\,dy = I$ and

$$\int_{-\infty}^{\infty} e^{-x^2}\,dx \int_{-\infty}^{\infty} e^{-y^2}\,dy = \int_{-\infty}^{\infty}\int_{-\infty}^{\infty} e^{-x^2-y^2}\,dy\,dx = I^2.$$

Convert the iterated integral to polar coordinates and evaluate it. The desired integral I is simply the square root of the iterated integral. Explain why the same trick can't be used to evaluate $\int_{-1}^{1} e^{-x^2}\,dx$.

13.4 SURFACE AREA

Recall that in section 5.4, we devised a method of finding the surface area for a surface of revolution. In this section, we consider how to find surface area in a more general setting. Suppose that $f(x, y) \geq 0$ and f has continuous first partial derivatives in some region R in the xy-plane. We would like to find a way to calculate the surface area of that portion of the surface $z = f(x, y)$ lying above R. As we have done innumerable times now, we begin by forming an inner partition of R, consisting of the rectangles $R_1, R_2, \ldots, R_n$. Our strategy is to approximate the surface area lying above each R_i, for $i = 1, 2, \ldots, n$ and then sum the individual approximations to obtain an approximation of the total surface area. We proceed as follows.

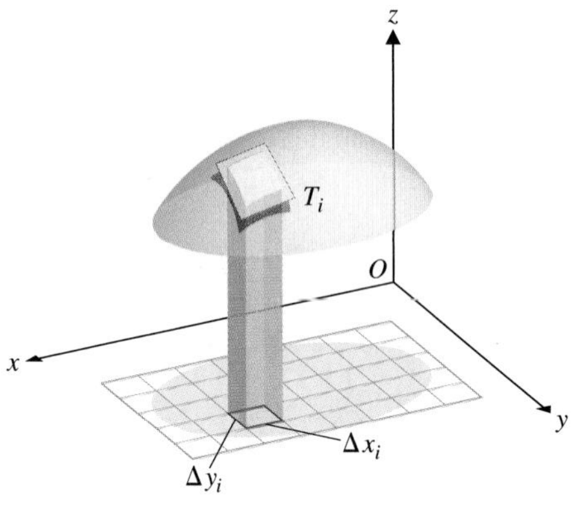

FIGURE 13.35a
Surface area

For each $i = 1, 2, \ldots, n$, let $(x_i, y_i, 0)$ represent the corner of R_i closest to the origin and construct the tangent plane to the surface $z = f(x, y)$ at the point $(x_i, y_i, f(x_i, y_i))$. Since the tangent plane stays close to the surface near the point of tangency, the area ΔT_i of that portion of the tangent plane that lies above R_i is an approximation to the surface area above R_i (see Figure 13.35a). Notice, too that the portion of the tangent plane lying above R_i is a parallelogram, T_i, whose area ΔT_i you should be able to easily compute. Adding together these approximations, we get that the total surface area S is approximately

$$S \approx \sum_{i=1}^{n} \Delta T_i.$$

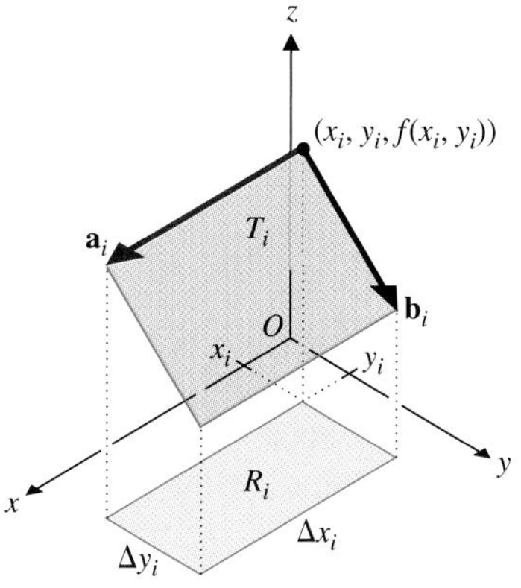

FIGURE 13.35b
Portion of the tangent plane above R_i

Also note that as the norm of the partition $\|P\|$ tends to zero, the approximations should approach the exact surface area and so we have

$$S = \lim_{\|P\| \to 0} \sum_{i=1}^{n} \Delta T_i, \tag{4.1}$$

assuming the limit exists. The only remaining question is how to find the values of ΔT_i, for $i = 1, 2, \ldots, n$. Let the dimensions of R_i be Δx_i and Δy_i, and let the vectors $\mathbf{a}_i$ and $\mathbf{b}_i$ form two adjacent sides of the parallelogram T_i, as indicated in Figure 13.35b. Recall from our discussion of tangent planes in section 12.4 that the tangent plane is given by

$$z - f(x_i, y_i) = f_x(x_i, y_i)(x - x_i) + f_y(x_i, y_i)(y - y_i). \tag{4.2}$$

Look carefully at Figure 13.35b; the vector $\mathbf{a}_i$ has its initial point at $(x_i, y_i, f(x_i, y_i))$. Its terminal point is the point on the tangent plane corresponding to $x = x_i + \Delta x_i$ and $y = y_i$. From (4.2), we get that the z-coordinate of the terminal point satisfies

$$\begin{aligned} z - f(x_i, y_i) &= f_x(x_i, y_i)(x_i + \Delta x_i - x_i) + f_y(x_i, y_i)(y_i - y_i) \\ &= f_x(x_i, y_i)\Delta x_i. \end{aligned}$$

This says that the vector $\mathbf{a}_i$ is given by

$$\mathbf{a}_i = \langle \Delta x_i, 0, f_x(x_i, y_i)\,\Delta x_i \rangle.$$

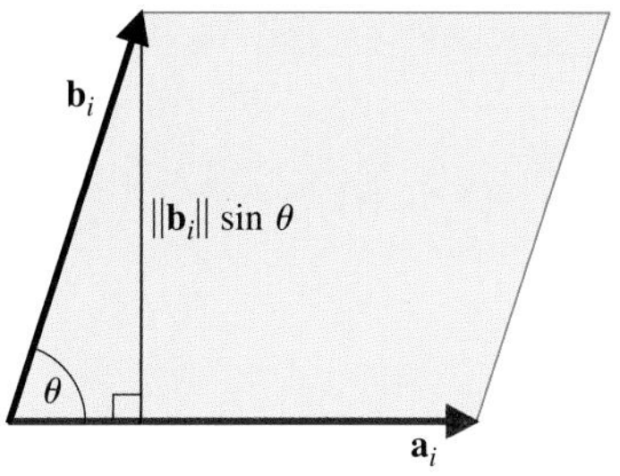

FIGURE 13.36
The parallelogram T_i

Likewise, $\mathbf{b}_i$ has its initial point at $(x_i, y_i, f(x_i, y_i))$, but has its terminal point at the point on the tangent plane corresponding to $x = x_i$ and $y = y_i + \Delta y_i$. Again, using (4.2), we get that the z-coordinate of this point is given by

$$\begin{aligned} z - f(x_i, y_i) &= f_x(x_i, y_i)\,(x_i - x_i) + f_y(x_i, y_i)\,(y_i + \Delta y_i - y_i) \\ &= f_y(x_i, y_i)\,\Delta y_i. \end{aligned}$$

This says that $\mathbf{b}_i$ is given by

$$\mathbf{b}_i = \langle 0, \Delta y_i, f_y(x_i, y_i)\Delta y_i \rangle.$$

Notice that ΔT_i is the area of the parallelogram shown in Figure 13.36, which you should recognize as

$$\Delta T_i = \|\mathbf{a}_i\|\|\mathbf{b}_i\| \sin\theta = \|\mathbf{a}_i \times \mathbf{b}_i\|,$$

where θ indicates the angle between $\mathbf{a}_i$ and $\mathbf{b}_i$. We have

$$\begin{aligned} \mathbf{a}_i \times \mathbf{b}_i &= \begin{vmatrix} \mathbf{i} & \mathbf{j} & \mathbf{k} \\ \Delta x_i & 0 & f_x(x_i, y_i)\,\Delta x_i \\ 0 & \Delta y_i & f_y(x_i, y_i)\,\Delta y_i \end{vmatrix} \\ &= -f_x(x_i, y_i)\,\Delta x_i\,\Delta y_i\mathbf{i} - f_y(x_i, y_i)\,\Delta x_i\,\Delta y_i\mathbf{j} + \Delta x_i\,\Delta y_i\mathbf{k}. \end{aligned}$$

This gives us

$$\Delta T_i = \|\mathbf{a}_i \times \mathbf{b}_i\| = \sqrt{[f_x(x_i, y_i)]^2 + [f_y(x_i, y_i)]^2 + 1}\ \underbrace{\Delta x_i\,\Delta y_i},$$

where $\Delta A_i = \Delta x_i\,\Delta y_i$ is the area of the rectangle R_i. From (4.1), we now have that the total surface area is given by

$$\begin{aligned} S &= \lim_{\|P\| \to 0} \sum_{i=1}^{n} \Delta T_i \\ &= \lim_{\|P\| \to 0} \sum_{i=1}^{n} \sqrt{[f_x(x_i, y_i)]^2 + [f_y(x_i, y_i)]^2 + 1}\,\Delta A_i. \end{aligned}$$

You should recognize this limit as the double integral

Surface area

$$S = \iint_R \sqrt{[f_x(x, y)]^2 + [f_y(x, y)]^2 + 1}\, dA. \tag{4.3}$$

There are several things to note here. First, you can easily show that the surface area formula (4.3) also holds for the case where $f(x, y) \le 0$ on R. Second, you should note the similarity to the arc length formula derived in section 5.4. Further, recall that $\mathbf{n} = \langle f_x(x, y), f_y(x, y), -1\rangle$ is a normal vector for the tangent plane to the surface $z = f(x, y)$ at (x, y). With this in mind, recognize that you can think of the integrand in (4.3) as $\|\mathbf{n}\|$, an idea we'll develop more fully in Chapter 14.

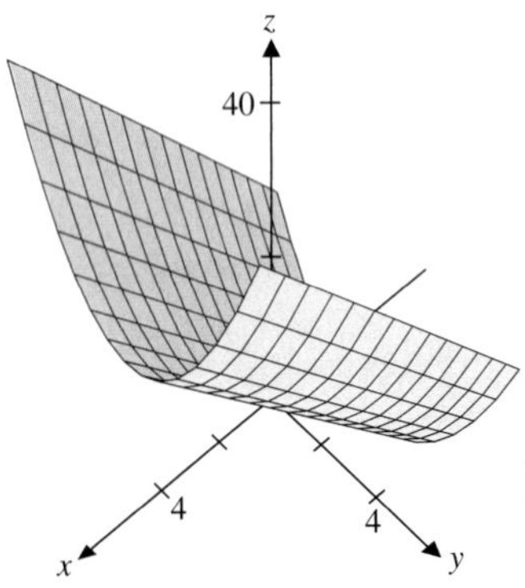

FIGURE 13.37a
The surface $z = y^2 + 4x$

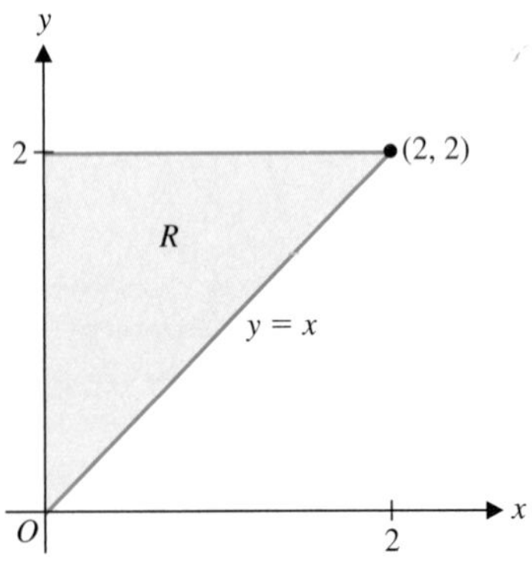

FIGURE 13.37b
The region R

EXAMPLE 4.1 Calculating Surface Area

Find the surface area of that portion of the surface $z = y^2 + 4x$ lying above the triangular region R in the xy-plane with vertices at (0, 0), (0, 2) and (2, 2).

Solution We show a computer-generated sketch of the surface in Figure 13.37a and the region R in Figure 13.37b. If we take $f(x, y) = y^2 + 4x$, then we have $f_x(x, y) = 4$ and $f_y(x, y) = 2y$. From (4.3), we now have

$$S = \iint_R \sqrt{[f_x(x, y)]^2 + [f_y(x, y)]^2 + 1}\, dA$$
$$= \iint_R \sqrt{4^2 + 4y^2 + 1}\, dA.$$

Looking carefully at Figure 13.37b, you can read off the limits of integration, to obtain

$$S = \int_0^2 \int_0^y \sqrt{4y^2 + 17}\, dx\, dy = \int_0^2 \sqrt{4y^2 + 17}\, x \Big|_{x=0}^{x=y} dy$$
$$= \int_0^2 y\sqrt{4y^2 + 17}\, dy = \frac{1}{8}(4y^2 + 17)^{3/2}\left(\frac{2}{3}\right)\Big|_0^2$$
$$= \frac{1}{12}\left[[4(2^2) + 17]^{3/2} - [4(0)^2 + 17]^{3/2}\right] \approx 9.956.$$

■

Computing surface area requires more than simply substituting into formula (4.3). You will also need to carefully determine the region over which you're integrating and the best coordinate system to use, as in example 4.2.

EXAMPLE 4.2 Finding Surface Area Using Polar Coordinates

Find the surface area of that portion of the paraboloid $z = 1 + x^2 + y^2$ that lies below the plane $z = 5$.

Solution First, note that we have not given you the region of integration; you'll need to determine that from a careful analysis of the graph (see Figure 13.38a). Next, observe that the plane $z = 5$ intersects the paraboloid in a circle of radius 2, parallel to the xy-plane and centered at the point (0, 0, 5). (Simply plug $z = 5$ into the equation of the

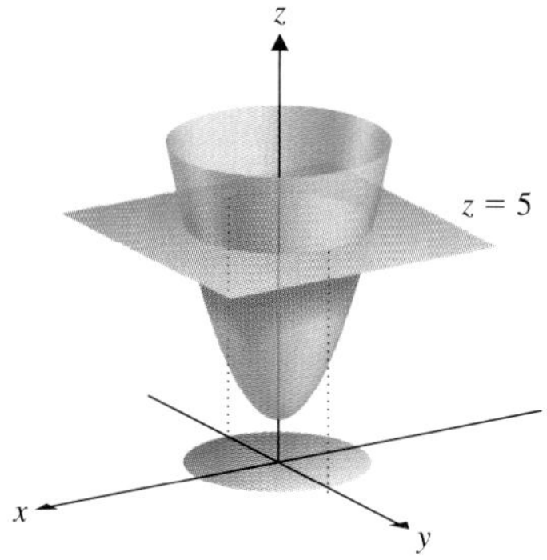

FIGURE 13.38a
Intersection of the paraboloid with the plane $z = 5$

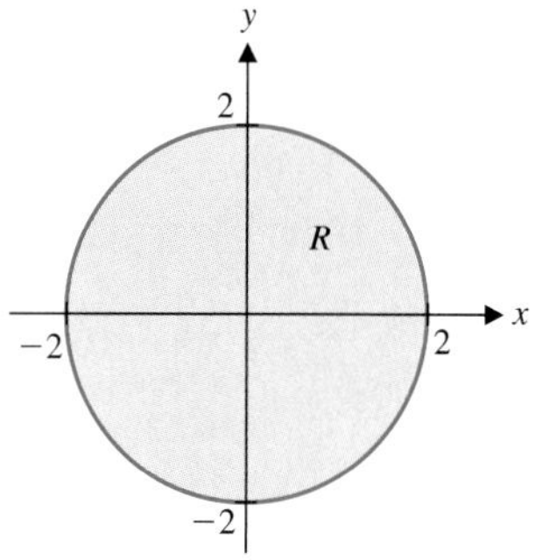

FIGURE 13.38b
The region R

paraboloid to see why.) So, the surface area *below* the plane $z = 5$ lies *above* the circle in the xy-plane of radius 2, centered at the origin. We show the region of integration R in Figure 13.38b. Taking $f(x, y) = 1 + x^2 + y^2$, we have $f_x(x, y) = 2x$ and $f_y(x, y) = 2y$, so that from (4.3), we have

$$\begin{aligned} S &= \iint_R \sqrt{[f_x(x, y)]^2 + [f_y(x, y)]^2 + 1}\, dA \\ &= \iint_R \sqrt{4x^2 + 4y^2 + 1}\, dA. \end{aligned}$$

Note that since the region of integration is circular and the integrand contains the term $x^2 + y^2$, polar coordinates are indicated. We have

$$\begin{aligned} S &= \iint_R \underbrace{\sqrt{4(x^2 + y^2) + 1}}_{\sqrt{4r^2+1}}\ \underbrace{dA}_{r\,dr\,d\theta} \\ &= \int_0^{2\pi} \int_0^2 \sqrt{4r^2 + 1}\, r\, dr\, d\theta \\ &= \frac{1}{8}\int_0^{2\pi} \left(\frac{2}{3}\right)(4r^2 + 1)^{3/2}\Big|_{r=0}^{r=2} d\theta \\ &= \frac{1}{12}\int_0^{2\pi} (17^{3/2} - 1^{3/2})\, d\theta \\ &= \frac{2\pi}{12}(17^{3/2} - 1) \approx 36.18. \end{aligned}$$

We must point out that (just as with arc length) most surface area integrals cannot be computed exactly. Most of the time, you must rely on numerical approximations of the integrals. Although your computer algebra system no doubt can approximate even iterated integrals numerically, you should try to evaluate at least one of the iterated integrals and then approximate the second integral numerically (e.g., using Simpson's Rule). This is the situation in example 4.3.

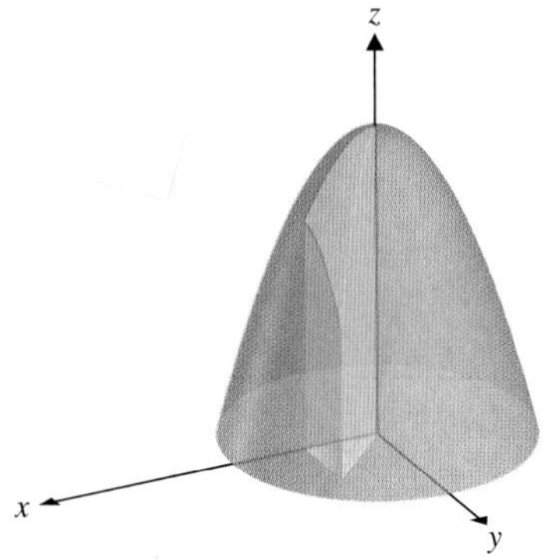

FIGURE 13.39a
$z = 4 - x^2 - y^2$

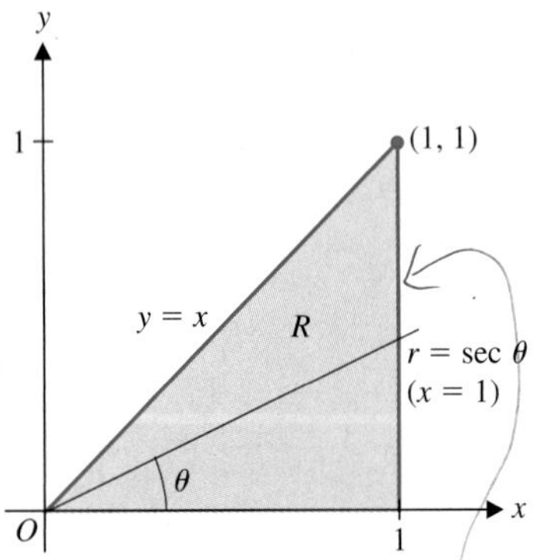

FIGURE 13.39b
The region R

EXAMPLE 4.3 Surface Area That Must Be Approximated Numerically

Find the surface area of that portion of the paraboloid $z = 4 - x^2 - y^2$ that lies above the triangular region R in the xy-plane with vertices at the points (0, 0), (1, 1) and (1, 0).

Solution We sketch the paraboloid and the region R in Figure 13.39a. Taking $f(x, y) = 4 - x^2 - y^2$, we get $f_x(x, y) = -2x$ and $f_y(x, y) = -2y$. From (4.3), we have

$$S = \iint_R \sqrt{[f_x(x, y)]^2 + [f_y(x, y)]^2 + 1}\, dA$$
$$= \iint_R \sqrt{4x^2 + 4y^2 + 1}\, dA.$$

Note that you have little hope of evaluating this double integral in rectangular coordinates. (Think about this!) Even though the region of integration is not circular, we'll try polar coordinates, since the integrand contains the term $x^2 + y^2$. We indicate the region R in Figure 13.39b. The difficulty here is in describing the region R in terms of polar coordinates. Look carefully at Figure 13.39b and notice that for each fixed angle θ, the radius r varies from 0 out to a point on the line $x = 1$. Since in polar coordinates $x = r\cos\theta$, the line $x = 1$ corresponds to $r\cos\theta = 1$ or $r = \sec\theta$, in polar coordinates. Further, θ varies from $\theta = 0$ (the x-axis) to $\theta = \frac{\pi}{4}$ (the line $y = x$). The surface area integral now becomes

$$S = \iint_R \underbrace{\sqrt{4x^2 + 4y^2 + 1}}_{\sqrt{4r^2+1}}\ \underbrace{dA}_{r\,dr\,d\theta}$$
$$= \int_0^{\pi/4} \int_0^{\sec\theta} \sqrt{4r^2 + 1}\, r\, dr\, d\theta$$
$$= \frac{1}{8}\int_0^{\pi/4} \left(\frac{2}{3}\right)(4r^2 + 1)^{3/2}\Big|_{r=0}^{r=\sec\theta}\, d\theta$$
$$= \frac{1}{12}\int_0^{\pi/4} [(4\sec^2\theta + 1)^{3/2} - 1]\, d\theta$$
$$\approx 0.93078,$$

where we have approximated the value of the final integral, since no exact means of integration was available. You can arrive at this approximation using Simpson's Rule or using your computer algebra system. ■

BEYOND FORMULAS

The surface area calculations in this section are important in their own right. Builders often need to know the surface area of the structure they are designing. However, for our purposes the ideas in this section will assume more importance when we introduce surface integrals in section 14.6. For surface integrals, surface area is a basic component used in the setup of the integral. This is similar to how the arc length formula is incorporated in the formula for the surface area of a surface of revolution in section 5.4.

EXERCISES 13.4

HW Pg 1070 #1 – 12 odd.

WRITING EXERCISES

1. Starting at equation (4.1), there are several ways to estimate ΔT_i. Explain why it is important that we were able to find an approximation of the form $f(x_i, y_i)\Delta x_i \Delta y_i$.

2. In example 4.3, we evaluated the inner integral before estimating the remaining integral numerically. Discuss the number of calculations that would be necessary to use a rule such as Simpson's Rule to estimate an iterated integral. Explain why we thought it important to evaluate the inner integral first.

In exercises 1–12, find the surface area of the indicated surface.

1. The portion of $z = x^2 + 2y$ between $y = x$, $y = 0$ and $x = 4$.

2. The portion of $z = 4y + 3x^2$ between $y = 2x$, $y = 0$ and $x = 2$.

3. The portion of $z = 4 - x^2 - y^2$ above the xy-plane.

4. The portion of $z = x^2 + y^2$ below $z = 4$.

5. The portion of $z = \sqrt{x^2 + y^2}$ below $z = 2$.

6. The portion of $z = \sqrt{x^2 + y^2}$ between $y = x^2$ and $y = 4$.

7. The portion of $x + 3y + z = 6$ in the first octant.

8. The portion of $2x + y + z = 8$ in the first octant.

9. The portion of $x - y - 2z = 4$ with $x \geq 0$, $y \leq 0$ and $z \leq 0$.

10. The portion of $2x + y - 4z = 4$ with $x \geq 0$, $y \geq 0$ and $z \leq 0$.

11. The portion of $z = \sqrt{4 - x^2 - y^2}$ above $z = 0$.

12. The portion of $z = \sin x + \cos y$ with $0 \leq x \leq 2\pi$ and $0 \leq y \leq 2\pi$.

In exercises 13–20, numerically estimate the surface area.

13. The portion of $z = e^{x^2+y^2}$ inside of $x^2 + y^2 = 4$.

14. The portion of $z = e^{-x^2-y^2}$ inside of $x^2 + y^2 = 1$.

15. The portion of $z = x^2 + y^2$ between $z = 5$ and $z = 7$.

16. The portion of $z = x^2 + y^2$ inside $r = 2 - 2\cos\theta$.

17. The portion of $z = y^2$ below $z = 4$ and between $x = -2$ and $x = 2$.

18. The portion of $z = 4 - x^2$ above $z = 0$ and between $y = 0$ and $y = 4$.

19. The portion of $z = \sin x \cos y$ with $0 \leq x \leq \pi$ and $0 \leq y \leq \pi$.

20. The portion of $z = \sqrt{x^2 + y^2 - 4}$ below $z = 1$.

21. In exercises 5 and 6, determine the surface area of the cone as a function of the area A of the base R of the solid and the height of the cone.

22. Use your solution to exercise 21 to quickly find the surface area of the portion of $z = \sqrt{x^2 + y^2}$ above the rectangle $0 \leq x \leq 2$, $1 \leq y \leq 4$.

23. In exercises 9 and 10, determine the surface area of the portion of the plane indicated as a function of the area A of the base R of the solid and the angle θ between the given plane and the xy-plane.

24. Use your solution to exercise 23 to quickly find the surface area of the portion of $z = 1 + y$ above the rectangle $-1 \leq x \leq 3$, $0 \leq y \leq 2$.

25. Generalizing exercises 17 and 18, determine the surface area of the portion of the cylinder indicated as a function of the arc length L of the base (two-dimensional) curve of the cylinder and the height h of the surface in the third dimension.

26. Use your solution to exercise 25 to quickly find the surface area of the portion of the cylinder with triangular cross sections parallel to the triangle with vertices $(1, 0, 0)$, $(0, 1, 0)$ and the origin and lying between the planes $z = 0$ and $z = 4$.

27. In example 4.2, find the value of k such that the plane $z = k$ slices off half of the surface area. Before working the problem, explain why $k = 3$ (halfway between $z = 1$ and $z = 5$) won't work.

28. Find the value of k such that the indicated surface area equals that of example 4.2: the surface area of that portion of the paraboloid $z = x^2 + y^2$ that lies below the plane $z = k$.

Exercises 29–32 involve parametic surfaces.

29. Let S be a surface defined by parametric equations $\mathbf{r}(u, v) = \langle x(u, v), y(u, v), z(u, v)\rangle$, for $a \leq u \leq b$ and $c \leq v \leq d$. Show that the surface area of S is given by $\int_c^d \int_a^b \|\mathbf{r}_u \times \mathbf{r}_v\| \, du \, dv$, where $\mathbf{r}_u(u, v) = \left\langle \frac{\partial x}{\partial u}(u, v), \frac{\partial y}{\partial u}(u, v), \frac{\partial z}{\partial u}(u, v)\right\rangle$ and $\mathbf{r}_v(u, v) = \left\langle \frac{\partial x}{\partial v}(u, v), \frac{\partial y}{\partial v}(u, v), \frac{\partial z}{\partial v}(u, v)\right\rangle$.

30. Use the formula from exercise 29 to find the surface area of the portion of the hyperboloid defined by parametric equations $x = 2\cos u \cosh v$, $y = 2\sin u \cosh v$, $z = 2\sinh v$ for $0 \leq u \leq 2\pi$ and $-1 \leq v \leq 1$. (Hint: Set up the double integral and approximate it numerically.)

31. Use the formula from exercise 29 to find the surface area of the surface defined by $x = u$, $y = v\cos u$, $z = v\sin u$ for $0 \leq u \leq 2\pi$ and $0 \leq v \leq 1$.

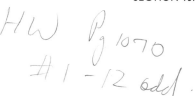

32. Use the formula from exercise 29 to find the surface area of the surface defined by $x = u$, $y = v + 2$, $z = 2uv$ for $0 \le u \le 2$ and $0 \le v \le 1$.

EXPLORATORY EXERCISES

1. An old joke tells of the theoretical mathematician hired to improve dairy production who starts his report with the assumption, "Consider a spherical cow." In this exercise, we will approximate an animal's body with ellipsoids. Estimate the volume and surface area of the ellipsoids $16x^2 + y^2 + 4z^2 = 16$ and $16x^2 + y^2 + 4z^2 = 36$. Note that the second ellipsoid retains the proportions of the first ellipsoid, but the length of each dimension is multiplied by $\frac{3}{2}$. Show that the volume increases by a much greater proportion than does the surface area. In general, volume increases as the cube of length (in this case, $\left(\frac{3}{2}\right)^3 = 3.375$) and surface area increases as the square of length (in this case, $\left(\frac{3}{2}\right)^2 = 2.25$). This has implications for the sizes of animals, since volume tends to be proportional to weight and surface area tends to be proportional to strength. Explain why a cow increased in size proportionally by a factor of $\frac{3}{2}$ might collapse under its weight.

2. For a surface $z = f(x, y)$, recall that a normal vector to the tangent plane at $(a, b, f(a, b))$ is $\langle f_x(a, b), f_y(a, b), -1\rangle$. Show that the surface area formula can be rewritten as

$$\text{Surface area} = \iint_R \frac{\|\mathbf{n}\|}{|\mathbf{n}\cdot\mathbf{k}|}\, dA,$$

where $\mathbf{n}$ is the unit normal vector to the surface. Use this formula to set up a double integral for the surface area of the top half of the sphere $x^2 + y^2 + z^2 = 4$ and compare this to the work required to set up the same integral in exercise 17. (Hint: Use the gradient to compute the normal vector and substitute $z = \sqrt{4 - x^2 - y^2}$ to write the integral in terms of x and y.) For a surface such as $y = 4 - x^2 - z^2$, it is convenient to think of y as the dependent variable and double integrate with respect to x and z. Write out the surface area formula in terms of the normal vector for this orientation and use it to compute the surface area of the portion of $y = 4 - x^2 - z^2$ inside $x^2 + z^2 = 1$ and to the right of the xz-plane.

13.5 TRIPLE INTEGRALS

We developed the definite integral of a function of one variable $f(x)$ initially to compute the area under the curve $y = f(x)$. Similarly, we first devised the double integral of a function of two variables $f(x, y)$ to compute the volume lying beneath the surface $z = f(x, y)$. We have no comparable geometric motivation for defining the triple integral of a function of three variables $f(x, y, z)$, since the graph of $u = f(x, y, z)$ is a **hypersurface** in four dimensions. (We can't even visualize a graph in four dimensions.) Despite this lack of immediate geometric significance, integrals of functions of three variables have many very significant applications to studying the three-dimensional world in which we live. We'll consider two of these applications (finding the mass and center of mass of a solid) at the end of this section.

We pattern our development of the triple integral of a function of three variables after our development of the double integral of a function of two variables. We first consider the relatively simple case of a function $f(x, y, z)$ defined on a rectangular box Q in three-dimensional space defined by

$$Q = \{(x, y, z) | a \le x \le b, c \le y \le d \text{ and } r \le z \le s\}.$$

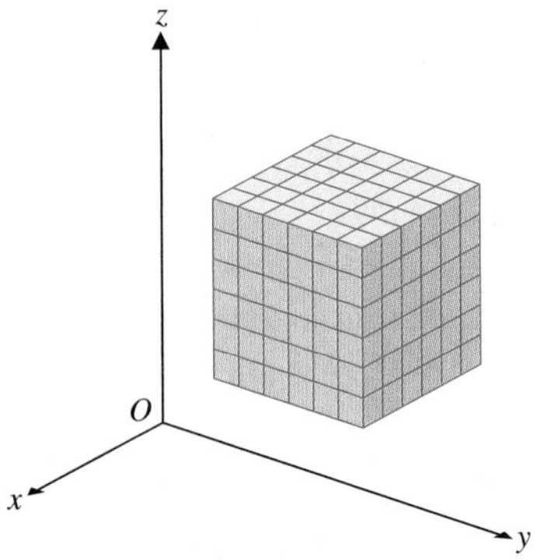

FIGURE 13.40a
Partition of the box Q

We begin by partitioning the region Q by slicing it by planes parallel to the xy-plane, planes parallel to the xz-plane and planes parallel to the yz-plane. Notice that this divides Q into a number of smaller boxes (see Figure 13.40a). Number the smaller boxes in any order: $Q_1, Q_2, \ldots, Q_n$. For each box Q_i $(i = 1, 2, \ldots, n)$, call the x, y and z dimensions of the box Δx_i, Δy_i and Δz_i, respectively (see Figure 13.40b). The volume of the box Q_i is then $\Delta V_i = \Delta x_i \Delta y_i \Delta z_i$. As we did in both one and two dimensions, we pick any point

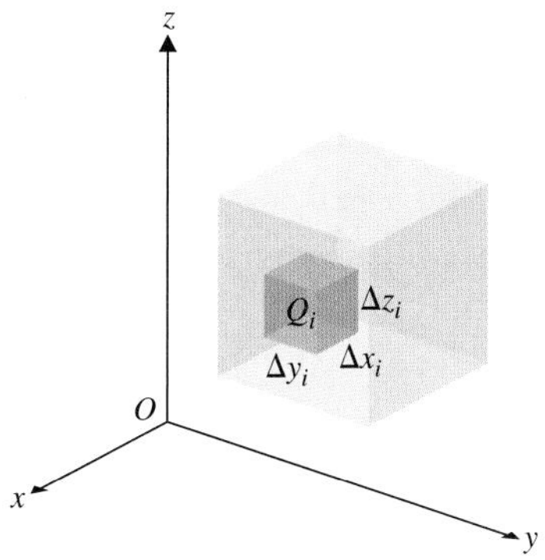

FIGURE 13.40b
Typical box Q_i

REMARK 5.1

It can be shown that as long as f is continuous over Q, f will be integrable over Q.

(u_i, v_i, w_i) in the box Q_i and form the Riemann sum

$$\sum_{i=1}^{n} f(u_i, v_i, w_i)\Delta V_i.$$

In this three-dimensional case, we define the norm of the partition $\|P\|$ to be the longest diagonal of any of the boxes Q_i, $i = 1, 2, \ldots, n$. We can now define the triple integral of $f(x, y, z)$ over Q.

DEFINITION 5.1

For any function $f(x, y, z)$ defined on the rectangular box Q, we define the **triple integral** of f over Q by

$$\iiint_Q f(x, y, z)\, dV = \lim_{\|P\|\to 0} \sum_{i=1}^{n} f(u_i, v_i, w_i)\, \Delta V_i, \tag{5.1}$$

provided the limit exists and is the same for every choice of evaluation points (u_i, v_i, w_i) in Q_i, for $i = 1, 2, \ldots, n$. When this happens, we say that f is **integrable** over Q.

Now that we have defined a triple integral, how can we calculate the value of one? The answer should prove to be no surprise. Just as a double integral can be written as two iterated integrals, a triple integral turns out to be equivalent to *three* iterated integrals.

THEOREM 5.1 (Fubini's Theorem)

Suppose that $f(x, y, z)$ is continuous on the box Q defined by $Q = \{(x, y, z) | a \le x \le b, c \le y \le d \text{ and } r \le z \le s\}$. Then, we can write the triple integral over Q as a triple iterated integral:

$$\iiint_Q f(x, y, z)\, dV = \int_r^s \int_c^d \int_a^b f(x, y, z)\, dx\, dy\, dz. \tag{5.2}$$

As was the case for double integrals, the three iterated integrals in (5.2) are evaluated from the inside out, using partial integrations. That is, in the innermost integral, we hold y and z fixed and integrate with respect to x and in the second integration, we hold z fixed and integrate with respect to y. Notice also that in this simple case (where Q is a rectangular box) the order of the integrations in (5.2) is irrelevant, so that we might just as easily write the triple integral as

$$\iiint_Q f(x, y, z)\, dV = \int_a^b \int_c^d \int_r^s f(x, y, z)\, dz\, dy\, dx,$$

or in any of the four remaining orders.

EXAMPLE 5.1 Triple Integral Over a Rectangular Box

Evaluate the triple integral $\iiint_Q 2xe^y \sin z\, dV$, where Q is the rectangle defined by

$$Q = \{(x, y, z) \,|\, 1 \le x \le 2, 0 \le y \le 1 \text{ and } 0 \le z \le \pi\}.$$

Solution From (5.2), we have

$$\iiint_Q 2xe^y \sin z\, dV = \int_0^\pi \int_0^1 \int_1^2 2xe^y \sin z\, dx\, dy\, dz$$
$$= \int_0^\pi \int_0^1 e^y \sin z \frac{2x^2}{2}\Big|_{x=1}^{x=2} dy\, dz$$
$$= 3\int_0^\pi \sin z\, e^y \Big|_{y=0}^{y=1} dz$$
$$= 3(e^1 - 1)(-\cos z)\Big|_{z=0}^{z=\pi}$$
$$= 3(e-1)(-\cos\pi + \cos 0)$$
$$= 6(e-1).$$

You should pick one of the other five possible orders of integration and show that you get the same result. ■

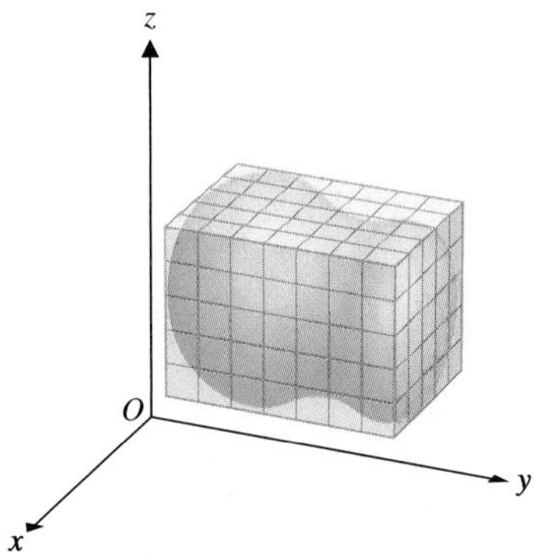

FIGURE 13.41a
Partition of a solid

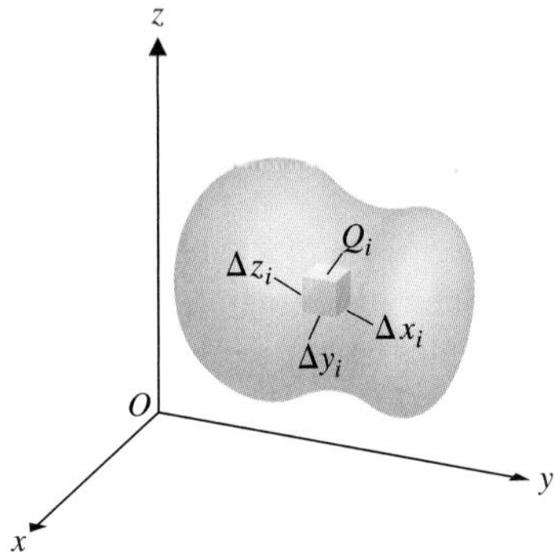

FIGURE 13.41b
Typical rectangle in inner partition of solid

As we did for double integrals, we can define triple integrals for more general regions in three dimensions by using an inner partition of the region. For any bounded solid Q in three dimensions, we partition Q by slicing it with planes parallel to the three coordinate planes. As in the case where Q was a box, these planes form a number of boxes (see Figures 13.41a and 13.41b). In this case, we consider only those boxes $Q_1, Q_2, \ldots, Q_n$ that lie *entirely* in Q and call this an **inner partition** of the solid Q. For each $i = 1, 2, \ldots, n$, we pick any point $(u_i, v_i, w_i) \in Q_i$ and form the Riemann sum

$$\sum_{i=1}^{n} f(u_i, v_i, w_i)\,\Delta V_i,$$

where $\Delta V_i = \Delta x_i\, \Delta y_i\, \Delta z_i$ represents the volume of Q_i. We can then define a triple integral over a general region Q as the limit of Riemann sums, as follows.

DEFINITION 5.2

For a function $f(x, y, z)$ defined on the (bounded) solid Q, we define the triple integral of $f(x, y, z)$ over Q by

$$\iiint_Q f(x, y, z)\, dV = \lim_{\|P\|\to 0} \sum_{i=1}^{n} f(u_i, v_i, w_i)\,\Delta V_i, \tag{5.3}$$

provided the limit exists and is the same for every choice of the evaluation points (u_i, v_i, w_i) in Q_i, for $i = 1, 2, \ldots, n$. When this happens, we say that f is **integrable** over Q.

Observe that (5.3) is identical to (5.1), except that in (5.3), we are summing over an inner partition of Q.

The (very) big remaining question is how to evaluate triple integrals over more general regions. The fact that there are six different orders of integration possible in a triple iterated integral makes it difficult to write down a single result that will allow us to evaluate all triple integrals. So, rather than write down an exhaustive list, we'll indicate the general idea by

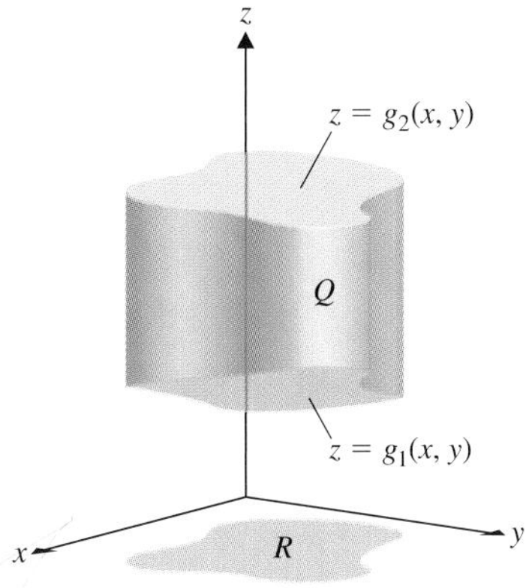

FIGURE 13.42
Solid with defined top and bottom surfaces

looking at several specific cases. For instance, if the region Q can be written in the form

$$Q = \{(x, y, z) \mid (x, y) \in R \text{ and } g_1(x, y) \le z \le g_2(x, y)\},$$

where R is some region in the xy-plane and where $g_1(x, y) \le g_2(x, y)$ for all (x, y) in R (see Figure 13.42), then it can be shown that

$$\iiint_Q f(x, y, z)\, dV = \iint_R \int_{g_1(x,y)}^{g_2(x,y)} f(x, y, z)\, dz\, dA. \tag{5.4}$$

As we have seen before, the innermost integration in (5.4) is a partial integration, where we hold x and y fixed and integrate with respect to z, and the outer double integral is evaluated using the methods we have already developed in sections 13.1 and 13.3.

EXAMPLE 5.2 Triple Integral Over a Tetrahedron

Evaluate $\iiint_Q 6xy\, dV$, where Q is the tetrahedron bounded by the planes $x = 0$, $y = 0$, $z = 0$ and $2x + y + z = 4$ (see Figure 13.43a).

Solution Notice that each point in the solid lies above the triangular region R in the xy-plane indicated in Figures 13.43a and 13.43b. You can think of R as forming the *base* of the solid. Notice that for each fixed point $(x, y) \in R$, z ranges from $z = 0$ up to $z = 4 - 2x - y$. It helps to draw a vertical line from the base and through the top surface of the solid, as we have indicated in Figure 13.43a. The line first enters the solid on the xy-plane ($z = 0$) and exits the solid on the plane $z = 4 - 2x - y$. This tells you that the innermost limits of integration (given that the first integration is with respect to z) are $z = 0$ and $z = 4 - 2x - y$. From (5.4), we now have

$$\iiint_Q 6xy\, dV = \iint_R \int_0^{4-2x-y} 6xy\, dz\, dA.$$

This leaves us with setting up the double integral over the triangular region shown in Figure 13.43b. Notice that for each fixed $x \in [0, 2]$, y ranges from 0 up to $y = 4 - 2x$.

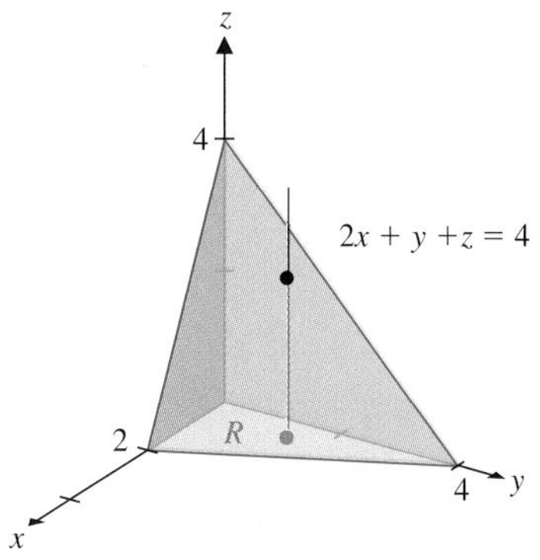

FIGURE 13.43a
Tetrahedron

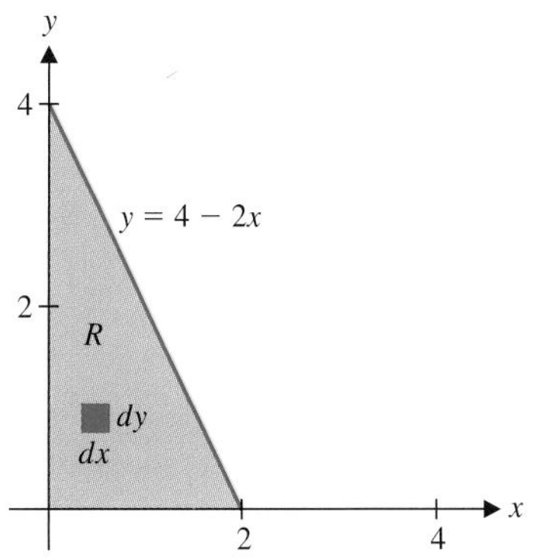

FIGURE 13.43b
The base of the solid in the xy-plane

NOTES

Observe that in example 5.2, the boundary of R consists of $x = 0$ and $y = 0$ (corresponding to the defining surfaces of Q that *do not* involve z) and $y = 4 - 2x$ (corresponding to the intersection of the two defining surfaces of Q that *do* involve z). The limits of integration for the outer two integrals can typically be found in this fashion.

We now have

$$\begin{aligned}\iiint_Q 6xy\,dV &= \iint_R \int_0^{4-2x-y} 6xy\,dz\,dA \\ &= \int_0^2 \int_0^{4-2x} \int_0^{4-2x-y} 6xy\,dz\,dy\,dx \\ &= \int_0^2 \int_0^{4-2x} (6xyz)\Big|_{z=0}^{z=4-2x-y} dy\,dx \\ &= \int_0^2 \int_0^{4-2x} 6xy(4-2x-y)\,dy\,dx \\ &= \int_0^2 6\left(4x\frac{y^2}{2} - 2x^2\frac{y^2}{2} - x\frac{y^3}{3}\right)\Big|_{y=0}^{y=4-2x} dx \\ &= \int_0^2 \left[12x(4-2x)^2 - 6x^2(4-2x)^2 - 2x(4-2x)^3\right] dx \\ &= \frac{64}{5},\end{aligned}$$

where we leave the details of the last integration to you. ■

The greatest challenge in setting up a triple integral is to get the limits of integration correct. You can improve your chances of doing this by taking the time to draw a good sketch of the solid and identifying either the base of the solid in one of the coordinate planes (as we did in example 5.2) or top and bottom boundaries of the solid when both lie above or below the same region R in one of the coordinate planes. In particular, if the solid extends from $z = f(x, y)$ to $z = g(x, y)$ for each (x, y) in some two-dimensional region R, then z is a good choice for the innermost variable of integration. This may seem like a lot to keep in mind, but we'll illustrate these ideas generously in the examples that follow and in the exercises. Be sure that you don't rely on making guesses. Guessing may get you through the first several exercises, but will not work in general.

Once you have identified a base or a top and bottom surface of a solid, draw a line from a representative point in the base (or bottom surface) through the top surface of the solid, as we did in Figure 13.43a, indicating the limits for the innermost integral. To illustrate this, we take several different views of example 5.2.

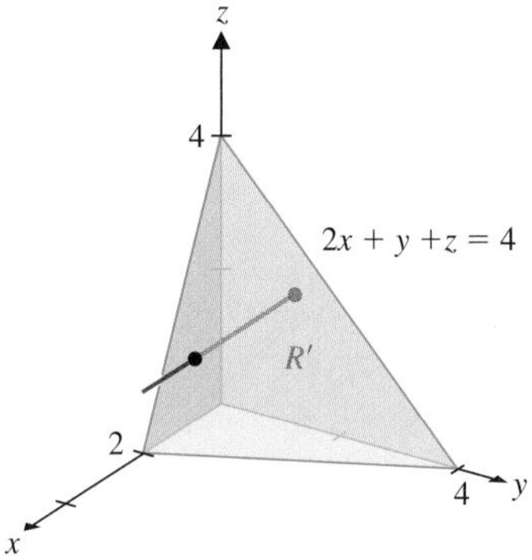

FIGURE 13.44a
Tetrahedron viewed with base in the yz-plane

EXAMPLE 5.3 A Triple Integral Where the First Integration Is with Respect to x

Evaluate $\iiint_Q 6xy\,dV$, where Q is the tetrahedron bounded by the planes $x = 0$, $y = 0$, $z = 0$ and $2x + y + z = 4$, as in example 5.2, but this time, integrate first with respect to x.

Solution You might object that our only evaluation result for triple integrals (5.4) is for integration with respect to z first. While this is true, you need to realize that x, y and z are simply variables that we represent by letters of the alphabet. Who cares which letter is which? Notice that we can think of the tetrahedron as a solid with its base in the triangular region R' of the yz-plane, as indicated in Figure 13.44a. In this case, we draw a line orthogonal to the yz-plane, which enters the solid in the yz-plane ($x = 0$) and exits in the plane $x = \frac{1}{2}(4 - y - z)$. Adapting (5.4) to this situation

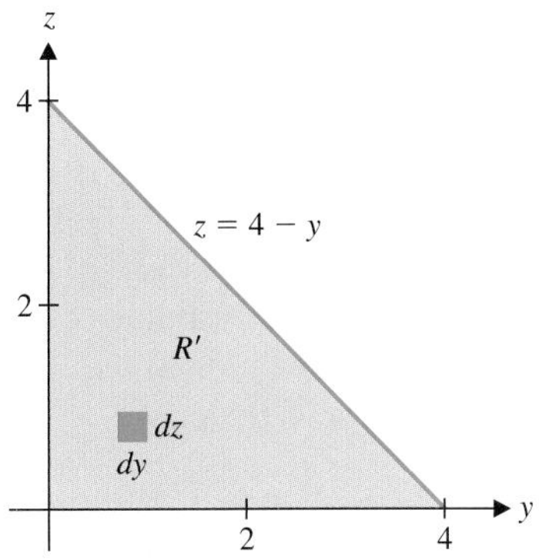

FIGURE 13.44b
The region R'

(i.e., interchanging the roles of x and z), we have

$$\iiint_Q 6xy\,dV = \iint_{R'} \int_0^{\frac{1}{2}(4-y-z)} 6xy\,dx\,dA$$
$$= \iint_{R'} \left(6\frac{x^2}{2}y\right)\Big|_{x=0}^{x=\frac{1}{2}(4-y-z)} dA$$
$$= \iint_{R'} 3\,\frac{(4-y-z)^2}{4}\,y\,dA.$$

To evaluate the remaining double integral, we look at the region R' in the yz-plane, as shown in Figure 13.44b. We now have

$$\iiint_Q 6xy\,dV = \frac{3}{4}\int_0^4 \int_0^{4-y} (4-y-z)^2 y\,dz\,dy = \frac{64}{5},$$

where we have left the routine details for you to verify. Finally, we leave it to you to show that we can also write this triple integral as a triple iterated integral where we integrate with respect to y first, as in

$$\iiint_Q 6xy\,dV = \int_0^2 \int_0^{4-2x} \int_0^{4-2x-z} 6xy\,dy\,dz\,dx.$$

NOTES

Notice in example 5.3 that the boundary of R' consists of $y = 0$ and $z = 0$ (the defining surfaces of Q that *do not* involve x) and $z = 4 - y$ (the intersection of the two defining surfaces that *do* involve x). These are the typical sources of surfaces for the limits of integration in the outer two integrals.

We want to emphasize again that the challenge here is to get the correct limits of integration. While you can always use a computer algebra system to evaluate the integrals (at least approximately), no computer algebra system will set up the limits of integration for you! Keep in mind that the innermost limits of integration correspond to two three-dimensional surfaces. (You can think of these as the top and the bottom of the solid, if you orient yourself properly.) These limits can involve either or both (or neither) of the two outer variables of integration. The limits of integration for the middle integral represent two curves in one of the coordinate planes and can involve only the outermost variable of integration. Realize, too, that once you integrate with respect to a given variable, that variable is eliminated from subsequent integrations (since you've evaluated the result of the integration between two specific values of that variable). Keep these ideas in mind as you work through the examples and exercises and make sure you work lots of problems. Triple integrals can look intimidating at first and *the only way to become proficient with these is to work plenty of problems*! Multiple integrals form the basis of much of the remainder of the book, so don't skimp on your effort now.

EXAMPLE 5.4 Evaluating a Triple Integral by Changing the Order of Integration

Evaluate $\displaystyle\int_0^4 \int_x^4 \int_0^y \frac{6}{1+48z-z^3}\,dz\,dy\,dx.$

Solution First, notice that evaluating the innermost integral requires a partial fractions decomposition, which produces three natural logarithm terms. The second integration is not pretty. We can significantly simplify the integral by changing the order of integration, but we must first identify the surfaces that bound the solid over which we

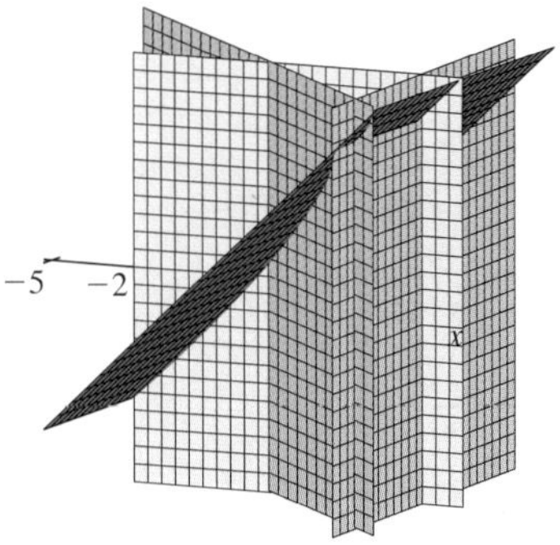

FIGURE 13.45a

are integrating. Starting with the inside limits, observe that the slanted plane $z = y$ forms the top of the solid and $z = 0$ forms the bottom.

The middle limits of integration indicate that the solid is also bounded by the planes $y = x$ and $y = 4$. The outer limits of $x = 0$ and $x = 4$ indicate that the solid is also bounded by the plane $x = 0$. (Here, $x = 4$ corresponds to the intersection of $y = x$ and $y = 4$.) A sketch of the solid is shown in Figure 13.45a. Notice that since y is involved in three different boundary planes, it is a poor choice for the inner variable of integration. To integrate with respect to x first, notice that a ray in the direction of the positive x-axis enters the solid through the plane $x = 0$ and exits through the plane $x = y$. We now have

$$\int_0^4 \int_x^4 \int_0^y \frac{6}{1 + 48z - z^3}\, dz\, dy\, dx = \iint_R \int_0^y \frac{6}{1 + 48z - z^3}\, dx\, dA,$$

where R is the triangle bounded by $z = y$, $z = 0$ and $y = 4$. (See Figure 13.45b.) In R, y extends from $y = z$ to $y = 4$, as z ranges from $z = 0$ to $z = 4$. The integral then becomes

$$\begin{aligned}\int_0^4 \int_x^4 \int_0^y \frac{6}{1 + 48z - z^3}\, dz\, dy\, dx &= \int_0^4 \int_z^4 \int_0^y \frac{6}{1 + 48z - z^3}\, dx\, dy\, dz \\ &= \int_0^4 \int_z^4 \frac{6}{1 + 48z - z^3}\, y\, dy\, dz \\ &= \int_0^4 \frac{6}{1 + 48z - z^3} \frac{y^2}{2}\bigg|_{y=z}^{y=4} dz \\ &= \int_0^4 \frac{48 - 3z^2}{1 + 48z - z^3}\, dz \\ &= \ln\left|1 + 48z - z^3\right|\Big|_{z=0}^{z=4} \\ &= \ln 129. \blacksquare\end{aligned}$$

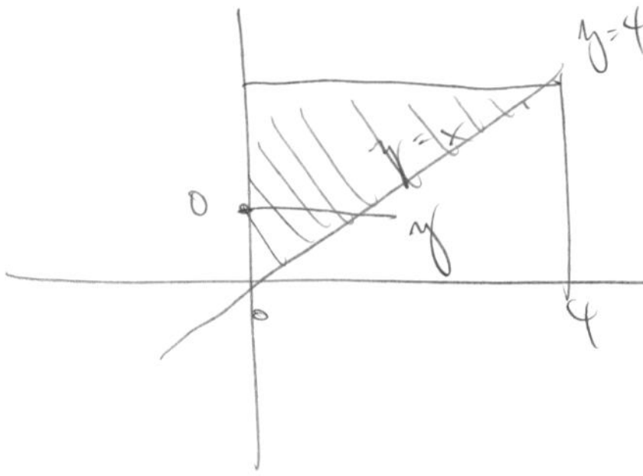

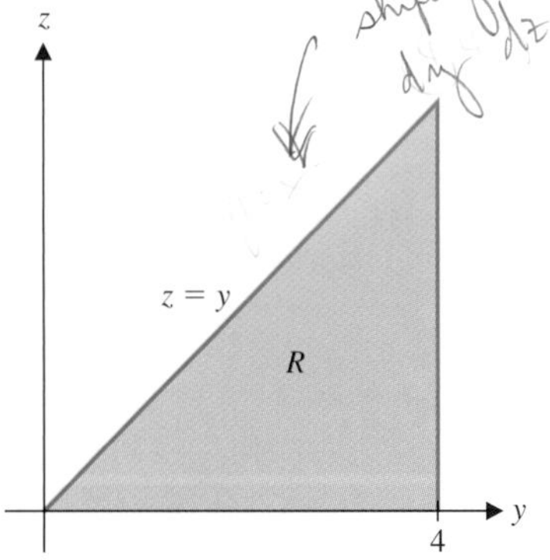

FIGURE 13.45b
The region R

NOTES

Try the original triple integral in example 5.4 on your CAS. Many integration packages are unable to evaluate this triple integral exactly. However, most packages will correctly return ln 129 if you ask them to evaluate the integral with the order of integration reversed. Technology does not replace an understanding of calculus techniques.

As you can see from example 5.4, there are clear advantages to considering alternative approaches for calculating triple integrals. So, take an extra moment to look at a sketch of a solid and consider your alternatives before jumping into the problem (i.e., look before you leap).

Recall that for double integrals, we had found that $\iint_R dA$ gives the area of the region R. Similarly, observe that if $f(x, y, z) = 1$ for all $(x, y, z) \in Q$, then from (5.3), we have

$$\iiint_Q 1\, dV = \lim_{\|P\| \to 0} \sum_{i=1}^{n} \Delta V_i = V, \qquad (5.5)$$

where V is the volume of the solid Q.

EXAMPLE 5.5 Using a Triple Integral to Find Volume

Find the volume of the solid bounded by the graphs of $z = 4 - y^2$, $x + z = 4$, $x = 0$ and $z = 0$.

Solution We show a sketch of the solid in Figure 13.46a. First, observe that we can consider the base of the solid to be the region R formed by the projection of the solid onto the yz-plane ($x = 0$). Notice that this is the region bounded by the parabola $z = 4 - y^2$ and the y-axis (see Figure 13.46b). Then, for each fixed y and z, the

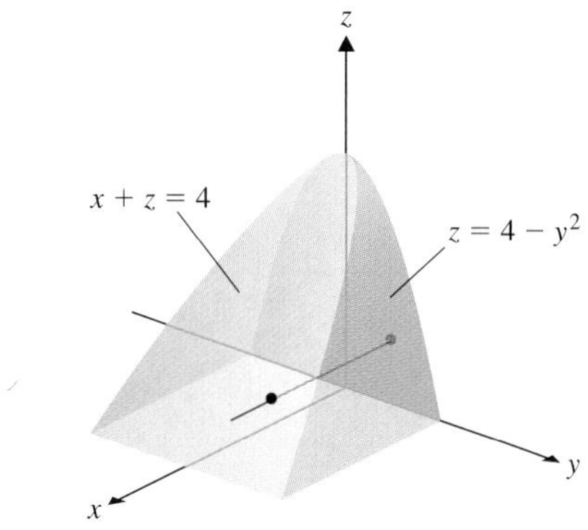

FIGURE 13.46a
The solid Q

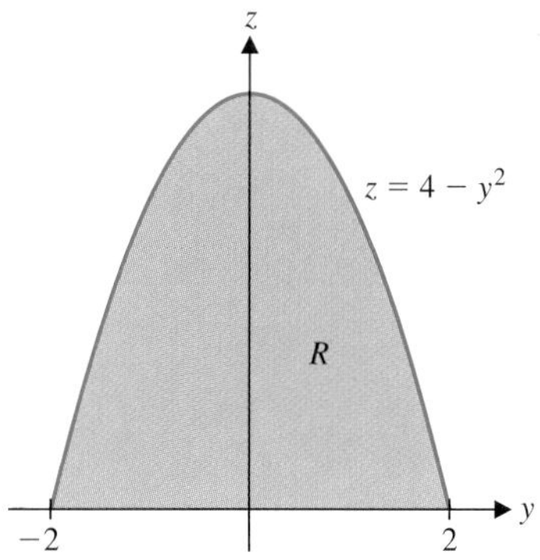

FIGURE 13.46b
Base R of the solid

> **NOTES**
>
> To integrate with respect to z first, you must identify surfaces forming the top and bottom of the solid. To integrate with respect to y first, you must identify surfaces forming (from the standard viewpoint) the right and left sides of the solid. To integrate with respect to x first, you must identify surfaces forming the front and back of the solid. Often, the easiest pair of surfaces to identify will indicate the best choice of variable for the innermost integration.

corresponding values of x range from 0 to $4 - z$. The volume of the solid is then given by

$$\begin{aligned} V &= \iiint_Q dV = \iint_R \int_0^{4-z} dx\, dA \\ &= \int_{-2}^{2} \int_0^{4-y^2} \int_0^{4-z} dx\, dz\, dy \\ &= \int_{-2}^{2} \int_0^{4-y^2} (4-z)\, dz\, dy \\ &= \int_{-2}^{2} \left(4z - \frac{z^2}{2} \right) \Bigg|_{z=0}^{z=4-y^2} dy \\ &= \int_{-2}^{2} \left[4(4-y^2) - \frac{1}{2}(4-y^2)^2 \right] dy \\ &= \frac{128}{5}, \end{aligned}$$

where we have left the details of the last integration to you. ■

○ Mass and Center of Mass

In section 13.2, we discussed finding the mass and center of mass of a lamina (a thin, flat plate). We now pause briefly to extend these results to three dimensions. Suppose that a solid Q has mass density given by $\rho(x, y, z)$ (in units of mass per unit volume). To find the total mass of a solid, we proceed (as we did for laminas) by constructing an inner partition of the solid: $Q_1, Q_2, \ldots, Q_n$. Realize that if each box Q_i is small (see Figure 13.47 on the following page), then the density should be nearly constant on Q_i and so, it is reasonable to approximate the mass m_i of Q_i by

$$m_i \approx \underbrace{\rho(u_i, v_i, w_i)}_{\text{mass/unit volume}} \underbrace{\Delta V_i}_{\text{volume}},$$

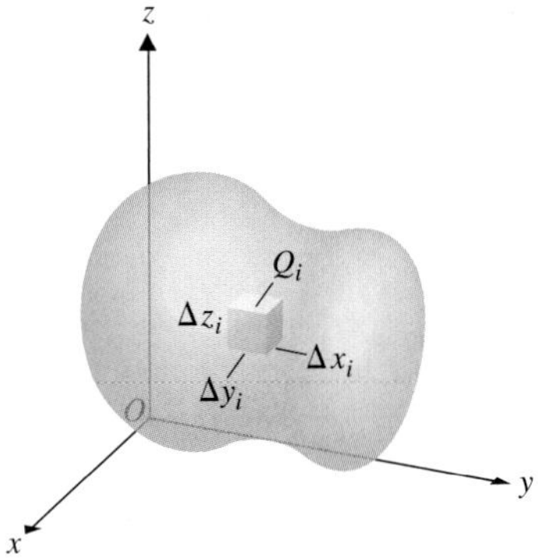

FIGURE 13.47
One box Q_i of the inner partition of Q

for any point $(u_i, v_i, w_i) \in Q_i$, where ΔV_i is the volume of Q_i. The total mass m of Q is then given approximately by

$$m \approx \sum_{i=1}^{n} \rho(u_i, v_i, w_i)\,\Delta V_i.$$

Letting the norm of the partition $\|P\|$ approach zero, we get the exact mass, which we recognize as a triple integral:

$$m = \lim_{\|P\|\to 0} \sum_{i=1}^{n} \rho\,(u_i, v_i, w_i)\,\Delta V_i = \iiint_Q \rho\,(x, y, z)\,dV. \tag{5.6}$$

Now, recall that the center of mass of a lamina was the point at which the lamina will balance. For an object in three dimensions, you can think of this as balancing it left to right (i.e., along the y-axis), front to back (i.e., along the x-axis) and top to bottom (i.e., along the z-axis). To do this, we need to find first moments with respect to each of the three coordinate planes. We define these moments as

$$M_{yz} = \iiint_Q x\rho(x, y, z)\,dV, \quad M_{xz} = \iiint_Q y\rho(x, y, z)\,dV \tag{5.7}$$

and

$$M_{xy} = \iiint_Q z\rho(x, y, z)\,dV, \tag{5.8}$$

the **first moments** with respect to the yz-plane, the xz-plane and the xy-plane, respectively. The **center of mass** is then given by the point $(\bar{x}, \bar{y}, \bar{z})$, where

$$\bar{x} = \frac{M_{yz}}{m}, \quad \bar{y} = \frac{M_{xz}}{m}, \quad \bar{z} = \frac{M_{xy}}{m}. \tag{5.9}$$

Notice that these are straightforward generalizations of the corresponding formulas for the center of mass of a lamina.

EXAMPLE 5.6 Center of Mass of a Solid

Find the center of mass of the solid of constant mass density ρ bounded by the graphs of the right circular cone $z = \sqrt{x^2 + y^2}$ and the plane $z = 4$ (see Figure 13.48a).

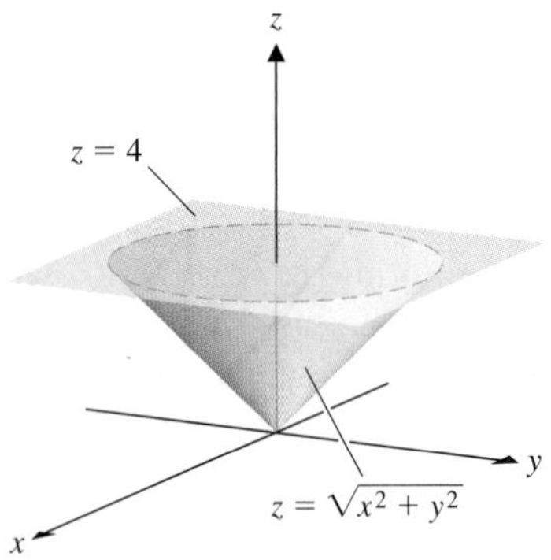

FIGURE 13.48a
The solid Q

Solution Notice that the projection R of the solid onto the xy-plane is the disk of radius 4 centered at the origin (see Figure 13.48b). Further, for each $(x, y) \in R$, z ranges from the cone $(z = \sqrt{x^2 + y^2})$ up to the plane $z = 4$. From (5.6), the total mass of the solid is given by

$$m = \iiint_Q \rho\,(x, y, z)\,dV = \rho \iint_R \int_{\sqrt{x^2+y^2}}^{4} dz\,dA$$

$$= \rho \iint_R \left(4 - \sqrt{x^2 + y^2}\right) dA,$$

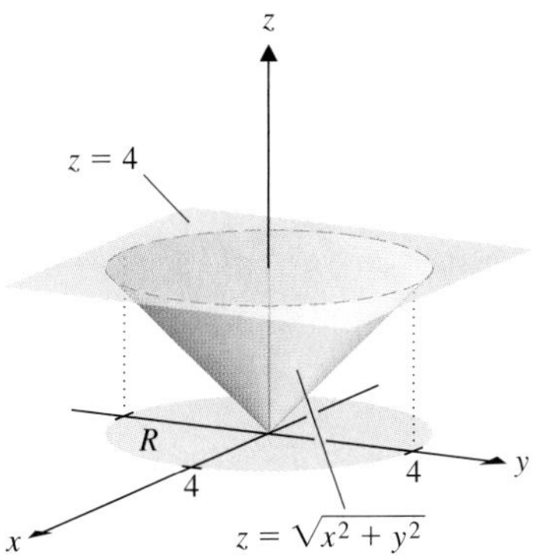

FIGURE 13.48b
Projection of the solid onto the xy-plane

where R is the disk of radius 4 in the xy-plane, centered at the origin, as indicated in Figure 13.48b. Since the region R is circular and since the integrand contains a term of the form $\sqrt{x^2+y^2}$, we use polar coordinates for the remaining double integral. We have

$$\begin{aligned} m &= \rho \iint_R \Big(4 - \underbrace{\sqrt{x^2+y^2}}_{r}\Big) \underbrace{dA}_{r\,dr\,d\theta} \\ &= \rho \int_0^{2\pi} \int_0^4 (4 - r)\, r\, dr\, d\theta \\ &= \rho \int_0^{2\pi} \left(4\frac{r^2}{2} - \frac{r^3}{3}\right)\Bigg|_{r=0}^{r=4} d\theta \\ &= \rho \left(32 - \frac{4^3}{3}\right)(2\pi) = \frac{64}{3}\pi\rho. \end{aligned}$$

From (5.8), we get that the moment with respect to the xy-plane is

$$\begin{aligned} M_{xy} &= \iiint_Q z\, \rho(x, y, z)\, dV = \rho \iint_R \int_{\sqrt{x^2+y^2}}^4 z\, dz\, dA \\ &= \rho \iint_R \frac{z^2}{2}\Bigg|_{\sqrt{x^2+y^2}}^4 dA \\ &= \frac{\rho}{2} \iint_R [16 - (x^2 + y^2)]\, dA. \end{aligned}$$

For the same reasons as when we computed the mass, we change to polar coordinates in the double integral to get

$$\begin{aligned} M_{xy} &= \frac{\rho}{2} \iint_R \Big[16 - \underbrace{(x^2+y^2)}_{r^2}\Big] \underbrace{dA}_{r\,dr\,d\theta} \\ &= \frac{\rho}{2} \int_0^{2\pi} \int_0^4 (16 - r^2)\, r\, dr\, d\theta \\ &= \frac{\rho}{2} \int_0^{2\pi} \left(16\frac{r^2}{2} - \frac{r^4}{4}\right)\Bigg|_{r=0}^{r=4} d\theta \\ &= 32\rho(2\pi) = 64\pi\rho. \end{aligned}$$

Notice that the solid is symmetric with respect to both the xz-plane and the yz-plane and so, the moments with respect to both of those planes are zero, since the density is constant. (Why does constant density matter?) That is, $M_{xz} = M_{yz} = 0$. From (5.9), the center of mass is then given by

$$(\bar{x}, \bar{y}, \bar{z}) = \left(\frac{M_{yz}}{m}, \frac{M_{xz}}{m}, \frac{M_{xy}}{m}\right) = \left(0, 0, \frac{64\pi\rho}{64\pi\rho/3}\right) = (0, 0, 3).$$ ■

HW Pg 1082

#1, 5, 9, 17

EXERCISES 13.5

WRITING EXERCISES

1. Discuss the importance of having a reasonably accurate sketch to help determine the limits (and order) of integration. Identify which features of a sketch are essential and which are not. Discuss whether it's important for your sketch to distinguish between two surfaces like $z = 4 - x^2 - y^2$ and $z = \sqrt{4 - x^2 - y^2}$.

2. In example 5.2, explain why all six orders of integration are equally simple. Given this choice, most people prefer to integrate in the order of example 5.2 ($dz\,dy\,dx$). Discuss the visual advantages of this order.

3. In example 5.4, identify any clues in the problem statement that might indicate that y should be the innermost variable of integration. In example 5.5, identify any clues that might indicate that z should *not* be the innermost variable of integration. (Hint: With how many surfaces is each variable associated?)

4. In example 5.6, we used polar coordinates in x and y. Explain why this is permissible and when it is likely to be convenient to do so.

In exercises 1–14, evaluate the triple integral $\iiint\limits_Q f(x, y, z)\,dV$.

1. $f(x, y, z) = 2x + y - z$, $Q = \{(x, y, z)\,|\,0 \le x \le 2, -2 \le y \le 2, 0 \le z \le 2\}$

2. $f(x, y, z) = 2x^2 + y^3$, $Q = \{(x, y, z)\,|\,0 \le x \le 3, -2 \le y \le 1, 1 \le z \le 2\}$

3. $f(x, y, z) = \sqrt{y} - 3z^2$, $Q = \{(x, y, z)\,|\,2 \le x \le 3, 0 \le y \le 1, -1 \le z \le 1\}$

4. $f(x, y, z) = 2xy - 3xz^2$, $Q = \{(x, y, z)\,|\,0 \le x \le 2, -1 \le y \le 1, 0 \le z \le 2\}$

5. $f(x, y, z) = 4yz$, Q is the tetrahedron bounded by $x + 2y + z = 2$ and the coordinate planes

6. $f(x, y, z) = 3x - 2y$, Q is the tetrahedron bounded by $4x + y + 3z = 12$ and the coordinate planes

7. $f(x, y, z) = 3y^2 - 2z$, Q is the tetrahedron bounded by $3x + 2y - z = 6$ and the coordinate planes

8. $f(x, y, z) = 6xz^2$, Q is the tetrahedron bounded by $-2x + y + z = 4$ and the coordinate planes

9. $f(x, y, z) = 2xy$, Q is bounded by $z = 1 - x^2 - y^2$ and $z = 0$

10. $f(x, y, z) = x - y$, Q is bounded by $z = x^2 + y^2$ and $z = 4$

11. $f(x, y, z) = 2yz$, Q is bounded by $z + x = 2$, $z - x = 2$, $z = 1$, $y = -2$ and $y = 2$

12. $f(x, y, z) = x^3 y$, Q is bounded by $z = 1 - y^2$, $z = 0$, $x = -1$ and $x = 1$

13. $f(x, y, z) = 15$, Q is bounded by $2x + y + z = 4$, $z = 0$, $x = 1 - y^2$ and $x = 0$

14. $f(x, y, z) = 2x + y$, Q is bounded by $z = 6 - x - y$, $z = 0$, $y = 2 - x$, $y = 0$ and $x = 0$

15. Sketch the region Q in exercise 9 and explain why the triple integral equals 0. Would the integral equal 0 for $f(x, y, z) = 2x^2 y$? For $f(x, y, z) = 2x^2 y^2$?

16. Show that $\iiint\limits_Q (z - x)\,dV = 0$, where Q is bounded by $z = 6 - x - y$ and the coordinate planes. Explain geometrically why this is correct.

In exercises 17–28, compute the volume of the solid bounded by the given surfaces.

17. $z = x^2$, $z = 1$, $y = 0$ and $y = 2$

18. $z = 1 - y^2$, $z = 0$, $x = 2$ and $x = 4$

19. $z = 1 - y^2$, $z = 0$, $z = 4 - 2x$ and $x = 4$

20. $z = x^2$, $z = x + 2$, $y + z = 5$ and $y = -1$

21. $y = 4 - x^2$, $z = 0$ and $z - y = 6$

22. $x = y^2$, $x = 4$, $x + z = 6$ and $x + z = 8$

23. $y = 3 - x$, $y = 0$, $z = x^2$ and $z = 1$

24. $x = y^2$, $x = 4$, $z = 2 + x$ and $z = 0$

25. $z = 1 + x$, $z = 1 - x$, $z = 1 + y$, $z = 1 - y$ and $z = 0$ (a pyramid)

26. $z = 5 - y^2$, $z = 6 - x$, $z = 6 + x$ and $z = 1$

27. $z = 4 - x^2 - y^2$ and the xy-plane

28. $z = 6 - x - y$, $x^2 + y^2 = 1$ and $z = -1$

In exercises 29–32, find the mass and center of mass of the solid with density $\rho(x, y, z)$ and the given shape.

29. $\rho(x, y, z) = 4$, solid bounded by $z = x^2 + y^2$ and $z = 4$

30. $\rho(x, y, z) = 2 + x$, solid bounded by $z = x^2 + y^2$ and $z = 4$

31. $\rho(x, y, z) = 10 + x$, tetrahedron bounded by $x + 3y + z = 6$ and the coordinate planes

32. $\rho(x, y, z) = 1 + x$, tetrahedron bound by $2x + y + 4z = 4$ and the coordinate planes

33. Explain why the x-coordinate of the center of mass in exercise 29 is zero, but the x-coordinate in exercise 30 is not zero.

34. In exercise 29, if $\rho(x, y, z) = 2 + x^2$, is the x-coordinate of the center of mass zero? Explain.

35. In exercise 5, evaluate the integral in three different ways, using each variable as the innermost variable once.

36. In exercise 6, evaluate the integral in three different ways, using each variable as the innermost variable once.

In exercises 37–42, sketch the solid whose volume is given and rewrite the iterated integral using a different innermost variable.

37. $\int_0^2 \int_0^{4-2y} \int_0^{4-2y-z} dx\, dz\, dy$

38. $\int_0^1 \int_0^{2-2y} \int_0^{2-x-2y} dz\, dx\, dy$

39. $\int_0^1 \int_0^{\sqrt{1-x^2}} \int_0^{\sqrt{1-x^2-y^2}} dz\, dy\, dx$

40. $\int_0^1 \int_0^{1-x^2} \int_0^{2-x} dy\, dz\, dx$

41. $\int_0^2 \int_0^{\sqrt{4-z^2}} \int_{x^2+z^2}^{4} dy\, dx\, dz$

42. $\int_0^2 \int_0^{\sqrt{4-z^2}} \int_{\sqrt{y^2+z^2}}^{2} dx\, dy\, dz$

43. Suppose that the density of an airborne pollutant in a room is given by $f(x, y, z) = xyze^{-x^2-2y^2-4z^2}$ grams per cubic foot for $0 \le x \le 12$, $0 \le y \le 12$ and $0 \le z \le 8$. Find the total amount of pollutant in the room. Divide by the volume of the room to get the average density of pollutant in the room.

44. If the danger level for the pollutant in exercise 43 is 1 gram per 1000 cubic feet, show that the room on the whole is below the danger level, but there is a portion of the room that is well above the danger level.

Exercises 45–48 involve probability.

45. A function $f(x, y, z)$ is a pdf on the three-dimensional region Q if $f(x, y, z) \ge 0$ for all (x, y, z) in Q and $\iiint_Q f(x, y, z)\, dV = 1$. Find c such that $f(x, y, z) = c$ is a pdf on the tetrahedron bounded by $x + 2y + z = 2$ and the coordinate planes.

46. If a point is chosen at random from the tetrahedron in exercise 45, find the probability that $z < 1$.

47. Find the value of k such that the probability that $z < k$ in exercise 45 equals $\frac{1}{2}$.

48. Compare your answer to exercise 47 to the z-coordinate of the center of mass of the tetrahedron Q with constant density.

49. Write $\int_a^b \int_c^d \int_r^s f(x)g(y)h(z)\, dz\, dy\, dx$ as a product of three single integrals. In general, can any triple integral with integrand $f(x)g(y)h(z)$ be factored as the product of three single integrals?

50. Compute $\iiint_Q f(x, y, z)\, dV$, where Q is the tetrahedron bounded by $2x + y + 3z = 6$ and the coordinate planes, and $f(x, y, z) = \max\{x, y, z\}$.

51. Let T be the tetrahedron in the first octant with vertices $(0, 0, 0)$, $(a, 0, 0)$, $(0, b, 0)$ and $(0, 0, c)$, for positive constants a, b and c. Let C be the parallelepiped in the first octant with the same vertices. Show that the volume of T is one-sixth the volume of C.

EXPLORATORY EXERCISES

1. In this exercise, you will examine several models of baseball bats. Sketch the region extending from $y = 0$ to $y = 32$ with distance from the y-axis given by $r = \frac{1}{2} + \frac{3}{128}y$. This should look vaguely like a baseball bat, with 32 representing the 32-inch length of a typical bat. Assume a constant **weight density** of $\rho = 0.39$ ounce per cubic inch. Compute the weight of the bat and the center of mass of the bat. (Hint: Compute the y-coordinate and argue that the x- and z-coordinates are zero.) Sketch each of the following regions, explain what the name means and compute the mass and center of mass. (a) **Long bat:** same as the original except y extends from $y = 0$ to $y = 34$. (b) **Choked up:** y goes from -2 to 30 with $r = \frac{35}{64} + \frac{3}{128}y$. (c) **Corked bat:** same as the original with the cylinder $26 \le y \le 32$ and $0 \le r \le \frac{1}{4}$ removed. (d) **Aluminum bat:** same as the original with the section from $r = 0$ to $r = \frac{3}{8} + \frac{3}{128}y$, $0 \le y \le 32$ removed and density $\rho = 1.56$. Explain why it makes sense that the choked-up bat has the center of mass 2 inches to the left of the original bat. Part of the "folklore" of baseball is that batters with aluminum bats can hit "inside" pitches better than batters with traditional wood bats. If "inside" means smaller values of y and the center of mass represents the "sweet spot" of the bat (the best place to hit the ball), discuss whether your calculations support baseball's folk wisdom.

2. In this exercise, we continue with the baseball bats of exercise 1. This time, we want to compute the moment of inertia $\iiint_Q y^2 \rho\, dV$ for each of the bats. The smaller the moment of inertia is, the easier it is to swing the bat. Use your calculations to answer the following questions. How much harder is it to swing a slightly longer bat? How much easier is it to swing a bat that has been choked up 2 inches? Does corking really make a noticeable difference in the ease with which a bat can be swung? How much easier is it to swing a hollow aluminum bat, even if it weighs the same as a regular bat?

13.6 CYLINDRICAL COORDINATES

In example 5.6 of section 13.5, we found it convenient to introduce polar coordinates in order to evaluate the outer double integral in a triple integral problem. Sometimes, this is more than a mere convenience, as we see in example 6.1.

EXAMPLE 6.1 A Triple Integral Requiring Polar Coordinates

Evaluate $\iiint_Q e^{x^2+y^2}\,dV$, where Q is the solid bounded by the cylinder $x^2+y^2=9$, the xy-plane and the plane $z=5$.

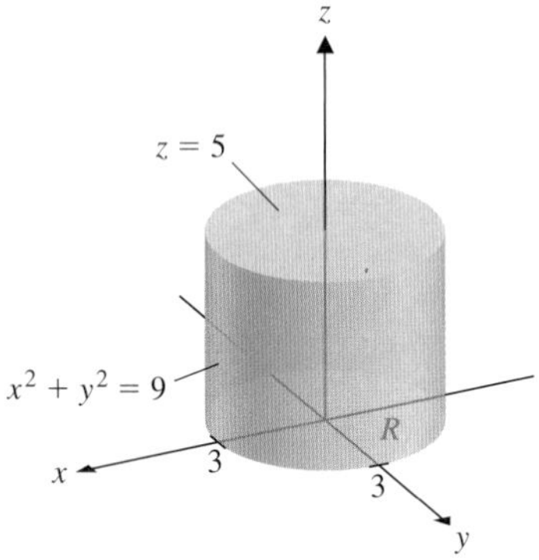

FIGURE 13.49a
The solid Q

Solution We show a sketch of the solid in Figure 13.49a. This might seem simple enough; certainly the solid is not particularly complicated. Unfortunately, the integral is rather troublesome. Notice that the base of the solid is the circle R of radius 3 centered at the origin and lying in the xy-plane. Further, for each (x, y) in R, z ranges from 0 up to 5. So, we have

$$\iiint_Q e^{x^2+y^2}\,dV = \iint_R \int_0^5 e^{x^2+y^2}\,dz\,dA$$
$$= 5\iint_R e^{x^2+y^2}\,dA.$$

The challenge lies in evaluating the remaining double integral. From Figure 13.49b, observe that for each fixed $x \in [-3, 3]$, y ranges from $-\sqrt{9-x^2}$ (the bottom semicircle) up to $\sqrt{9-x^2}$ (the top semicircle). We now have

$$\iiint_Q e^{x^2+y^2}\,dV = 5\iint_R e^{x^2+y^2}\,dA = 5\int_{-3}^{3}\int_{-\sqrt{9-x^2}}^{\sqrt{9-x^2}} e^{x^2+y^2}\,dy\,dx.$$

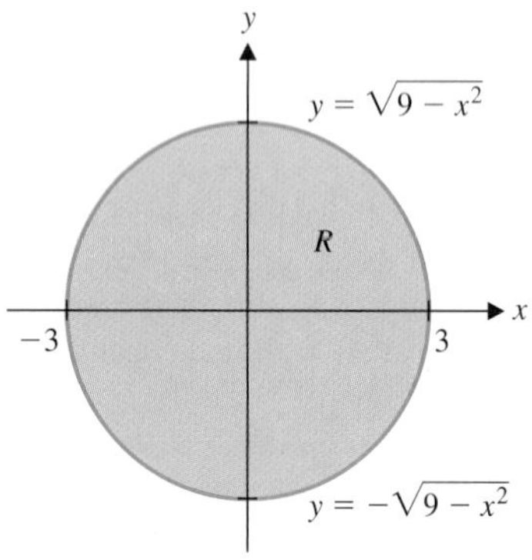

FIGURE 13.49b
The region R

Without polar coordinates, we're at a dead end, since we don't know an antiderivative for $e^{x^2+y^2}$. Even the authors' computer algebra system has difficulty with this, giving a nearly indecipherable answer in terms of an integral of the *error* function, which you have likely never seen before. Even so, our computer algebra system could not handle the second integration, except approximately. On the other hand, if we introduce polar coordinates: $x = r\cos\theta$ and $y = r\sin\theta$, we get that for each $\theta \in [0, 2\pi]$, r ranges from 0 up to 3. We now have an integral requiring only a simple substitution:

$$\iiint_Q e^{x^2+y^2}\,dV = 5\iint_R \underbrace{e^{x^2+y^2}}_{e^{r^2}}\,\underbrace{dA}_{r\,dr\,d\theta}$$
$$= 5\int_0^{2\pi}\int_0^3 e^{r^2}\,r\,dr\,d\theta$$
$$= \frac{5}{2}\int_0^{2\pi} e^{r^2}\Big|_{r=0}^{r=3}\,d\theta$$
$$= 5\pi(e^9 - 1)$$
$$\approx 1.27 \times 10^5,$$

which is a much more acceptable answer. ■

The process of replacing two of the variables in a three-dimensional coordinate system by polar coordinates, as we illustrated in example 6.1, is so common that we give this coordinate system a name: cylindrical coordinates.

To be precise, we specify a point $P(x, y, z) \in \mathbb{R}^3$ by identifying polar coordinates for the point $(x, y) \in \mathbb{R}^2$: $x = r\cos\theta$ and $y = r\sin\theta$, where $r^2 = x^2 + y^2$ and θ is the angle made by the line segment connecting the origin and the point $(x, y, 0)$ with the positive x-axis, as indicated in Figure 13.50. Then, $\tan\theta = \frac{y}{x}$. We refer to (r, θ, z) as **cylindrical coordinates** for the point P.

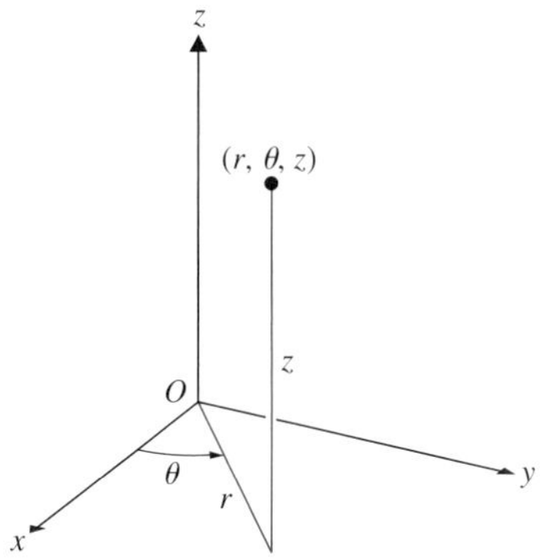

FIGURE 13.50
Cylindrical coordinates

EXAMPLE 6.2 Equation of a Cylinder in Cylindrical Coordinates

Write the equation for the cylinder $x^2 + y^2 = 16$ (see Figure 13.51) in cylindrical coordinates.

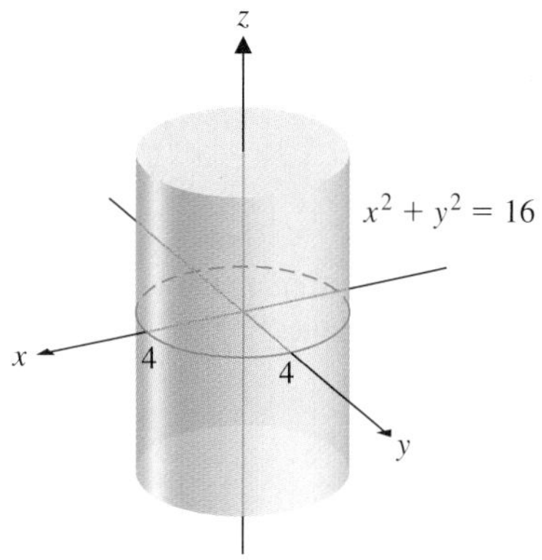

FIGURE 13.51
The cylinder $r = 4$

Solution In cylindrical coordinates $r^2 = x^2 + y^2$, so the cylinder becomes $r^2 = 16$ or $r = \pm 4$. But note that since θ is not specified, the equation $r = 4$ describes the same cylinder. ■

EXAMPLE 6.3 Equation of a Cone in Cylindrical Coordinates

Write the equation for the cone $z^2 = x^2 + y^2$ (see Figure 13.52) in cylindrical coordinates.

Solution Since $x^2 + y^2 = r^2$, the cone becomes $z^2 = r^2$ or $z = \pm r$. In cases where we need r to be positive, we write separate equations for the top cone $z = r$ and the bottom cone $z = -r$. ■

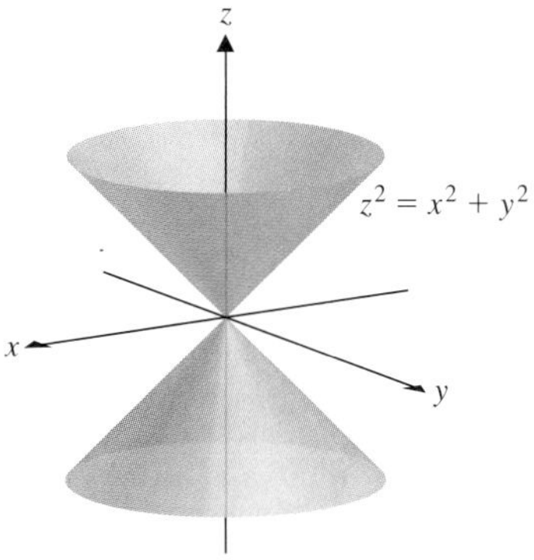

FIGURE 13.52
The cone $z = r$

As we did on a case-by-case basis in several examples in section 13.5 and in example 6.1, we can use cylindrical coordinates to simplify certain triple integrals. For instance, suppose that we can write the solid Q as

$$Q = \{(r, \theta, z) \mid (r, \theta) \in R \text{ and } k_1(r, \theta) \le z \le k_2(r, \theta)\},$$

where $k_1(r, \theta) \le k_2(r, \theta)$, for all (r, θ) in the region R of the xy-plane defined by

$$R = \{(r, \theta) | \alpha \le \theta \le \beta \text{ and } g_1(\theta) \le r \le g_2(\theta)\},$$

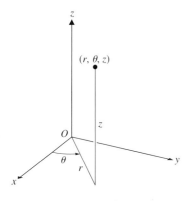

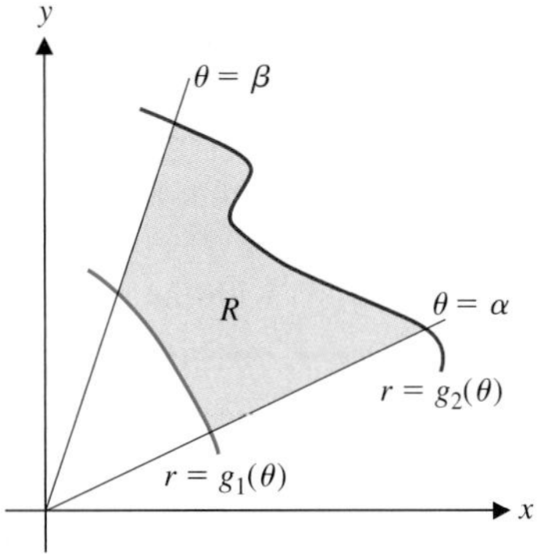

FIGURE 13.53
The region R

where $0 \leq g_1(\theta) \leq g_2(\theta)$, for all θ in $[\alpha, \beta]$, as shown in Figure 13.53. Then, notice that from (5.4), we can write

$$\iiint_Q f(r, \theta, z)\, dV = \iint_R \left[\int_{k_1(r,\theta)}^{k_2(r,\theta)} f(r, \theta, z)\, dz\right] dA.$$

Since the outer double integral is a double integral in polar coordinates, we already know how to write it as an iterated integral. We have

$$\iiint_Q f(r, \theta, z)\, dV = \iint_R \left[\int_{k_1(r,\theta)}^{k_2(r,\theta)} f(r, \theta, z)\, dz\right] \underbrace{dA}_{r\,dr\,d\theta}$$

$$= \int_\alpha^\beta \int_{g_1(\theta)}^{g_2(\theta)} \left[\int_{k_1(r,\theta)}^{k_2(r,\theta)} f(r, \theta, z)\, dz\right] r\, dr\, d\theta.$$

This gives us an evaluation formula for triple integrals in cylindrical coordinates:

$$\boxed{\iiint_Q f(r, \theta, z)\, dV = \int_\alpha^\beta \int_{g_1(\theta)}^{g_2(\theta)} \int_{k_1(r,\theta)}^{k_2(r,\theta)} f(r, \theta, z)\, r\, dz\, dr\, d\theta.} \tag{6.1}$$

In setting up triple integrals in cylindrical coordinates, it often helps to visualize the volume element $dV = r\, dz\, dr\, d\theta$. (See Figure 13.54.)

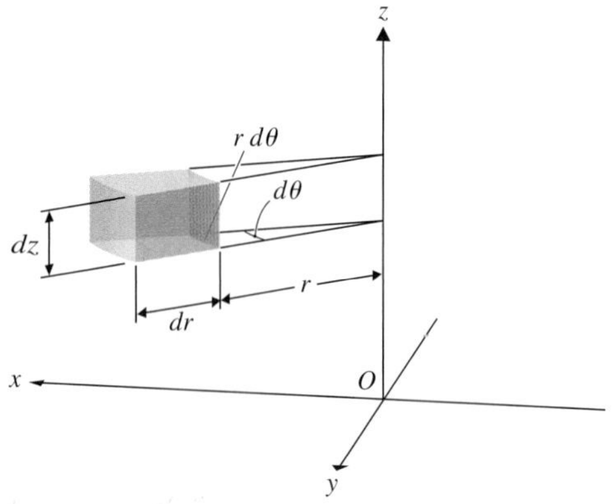

FIGURE 13.54
Volume element for cylindrical coordinates

EXAMPLE 6.4 A Triple Integral in Cylindrical Coordinates

Write $\iiint_Q f(r, \theta, z)\, dV$ as a triple iterated integral in cylindrical coordinates if

$$Q = \left\{(x, y, z) \mid \sqrt{x^2 + y^2} \leq z \leq \sqrt{18 - x^2 - y^2}\right\}.$$

Solution The first task in setting up any iterated multiple integral is to draw a sketch of the region over which you are integrating. Here, $z = \sqrt{x^2 + y^2}$ is the top half of a right circular cone, with vertex at the origin and axis lying along the z-axis, and $z = \sqrt{18 - x^2 - y^2}$ is the top hemisphere of radius $\sqrt{18}$ centered at the origin. So, we are looking for the set of all points lying above the cone and below the hemisphere. (See Figure 13.55a.) Recognize that in cylindrical coordinates, the cone is written $z = r$ and the hemisphere becomes $z = \sqrt{18 - r^2}$, since $x^2 + y^2 = r^2$. This says that for each r and θ, z ranges from r up to $\sqrt{18 - r^2}$. Notice that the cone and the hemisphere intersect when

$$\sqrt{18 - r^2} = r$$

or

$$18 - r^2 = r^2,$$

so that

$$18 = 2r^2 \quad \text{or} \quad r = 3.$$

That is, the two surfaces intersect in a circle of radius 3 lying in the plane $z = 3$. The projection of the solid down onto the xy-plane is then the circle of radius 3 centered at the origin (see Figure 13.55b) and we have

$$\iiint_Q f(r, \theta, z)\, dV = \int_0^{2\pi} \int_0^3 \int_r^{\sqrt{18-r^2}} f(r, \theta, z)\, r\, dz\, dr\, d\theta.$$

$x^2 + y^2 = 18 - x^2 - y^2$
$2x^2 + 2y^2 = 18$
$x^2 + y^2 = 9.$

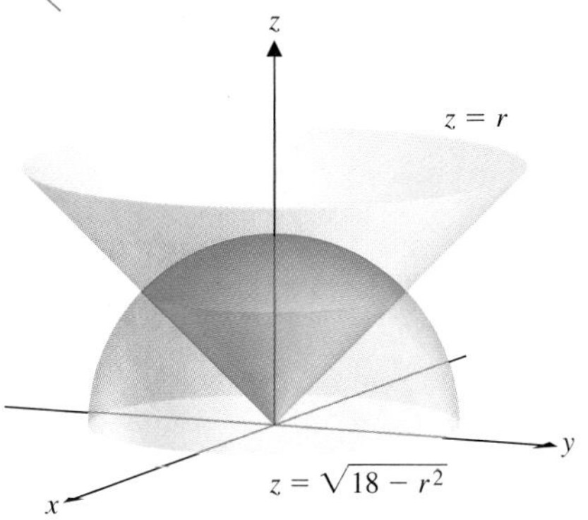

FIGURE 13.55a
The solid Q

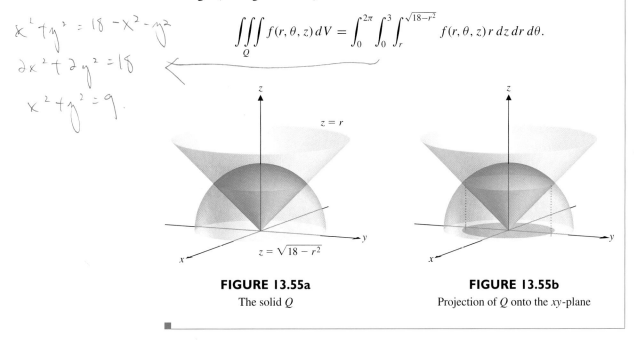

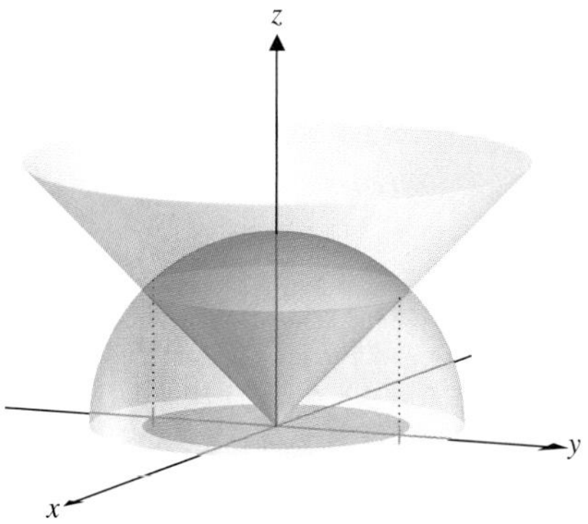

FIGURE 13.55b
Projection of Q onto the xy-plane

Very often, a triple integral in rectangular coordinates would be simpler in cylindrical coordinates. You must then recognize how to write the solid in cylindrical coordinates, as well as how to rewrite the integral.

EXAMPLE 6.5 Changing from Rectangular to Cylindrical Coordinates

Evaluate the triple iterated integral $\displaystyle\int_{-1}^{1} \int_{-\sqrt{1-x^2}}^{\sqrt{1-x^2}} \int_{x^2+y^2}^{2-x^2-y^2} (x^2 + y^2)^{3/2}\, dz\, dy\, dx.$

Solution As written, the integral is virtually impossible to evaluate exactly. (Even our computer algebra system had trouble with it.) Notice that the integrand involves $x^2 + y^2$, which is simply r^2 in cylindrical coordinates. You should also try to visualize the region over which you are integrating. First, from the innermost limits of integration, notice that $z = 2 - x^2 - y^2$ is a paraboloid opening downward, with vertex at the point $(0, 0, 2)$ and $z = x^2 + y^2$ is a paraboloid opening upward with vertex at the origin. So, the solid is some portion of the solid bounded by the two paraboloids. The two paraboloids intersect when

$$2 - x^2 - y^2 = x^2 + y^2$$

or

$$1 = x^2 + y^2.$$

So, the intersection forms a circle of radius 1 lying in the plane $z = 1$ and centered at the point $(0, 0, 1)$. Looking at the outer two integrals, note that for each $x \in [-1, 1]$, y ranges from $-\sqrt{1 - x^2}$ (the bottom semicircle of radius 1 centered at the origin) to $\sqrt{1 - x^2}$ (the top semicircle of radius 1 centered at the origin). Since this corresponds to the projection of the circle of intersection onto the xy-plane, the triple integral is over the entire solid below the one paraboloid and above the other. (See Figure 13.56.)

TODAY IN MATHEMATICS

Enrico Bombieri (1940–)
An Italian mathematician who proved what is now known as the Bombieri-Vinogradov Mean Value Theorem on the distribution of prime numbers. Bombieri is known as a solver of "deep" problems, which are fundamental questions that baffle the world's best mathematicians. This is a risky enterprise, because many such problems are unsolvable. One of Bombieri's special talents is a seemingly instinctual sense of which problems are both solvable and interesting. To see why this is a true talent, think about how hard it would be to write double and triple integrals that are at the same time interesting, challenging and solvable (antiderivatives found and evaluated) and then imagine performing an equivalent task for problems that nobody has ever solved.

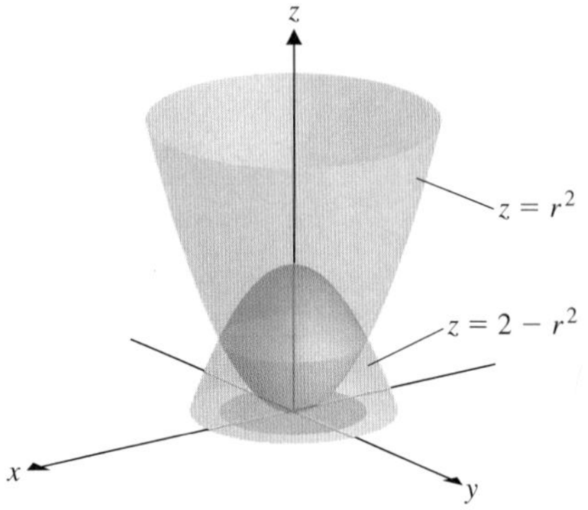

FIGURE 13.56
The solid Q

In cylindrical coordinates, the top paraboloid becomes $z = 2 - x^2 - y^2 = 2 - r^2$ and the bottom paraboloid becomes $z = x^2 + y^2 = r^2$. So, for each fixed value of r and θ, z varies from r^2 up to $2 - r^2$. Further, since the projection of the solid onto the xy-plane is the circle of radius 1 centered at the origin, r varies from 0 to 1 and θ varies from 0 to 2π. We can now write the triple integral in cylindrical coordinates as

$$\begin{aligned}
\int_{-1}^{1}\int_{-\sqrt{1-x^2}}^{\sqrt{1-x^2}}\int_{x^2+y^2}^{2-x^2-y^2} (x^2+y^2)^{3/2}\,dz\,dy\,dx &= \int_0^{2\pi}\int_0^1\int_{r^2}^{2-r^2} (r^2)^{3/2}\,r\,dz\,dr\,d\theta \\
&= \int_0^{2\pi}\int_0^1\int_{r^2}^{2-r^2} r^4\,dz\,dr\,d\theta \\
&= \int_0^{2\pi}\int_0^1 r^4(2-2r^2)\,dr\,d\theta \\
&= 2\int_0^{2\pi}\left(\frac{r^5}{5} - \frac{r^7}{7}\right)\Bigg|_{r=0}^{r=1} d\theta = \frac{8\pi}{35}.
\end{aligned}$$

Evaluating the triple integral in cylindrical coordinates was easy, compared to evaluating the original integral directly. ■

When converting an iterated integral from rectangular to cylindrical coordinates, it's important to carefully visualize the solid over which you are integrating. While we have so far defined cylindrical coordinates by replacing x and y by their polar coordinate representations, we can do this with any two of the three variables, as we see in example 6.6.

EXAMPLE 6.6 Using a Triple Integral to Find Volume

Use a triple integral to find the volume of the solid Q bounded by the graph of $y = 4 - x^2 - z^2$ and the xz-plane.

Solution Notice that the graph of $y = 4 - x^2 - z^2$ is a paraboloid with vertex at $(0, 4, 0)$, whose axis is the y-axis and that opens toward the negative y-axis. We show the solid in Figure 13.57a. Without thinking too much about it, we might consider integration with respect to z first. In this case, the projection of the solid onto the xy-plane is the parabola formed by the intersection of the paraboloid with the xy-plane (see Figure 13.57b). Notice from Figure 13.57b that for each fixed x and y, the line through the point $(x, y, 0)$ and perpendicular to the xy-plane enters the solid on the bottom half of the paraboloid ($z = -\sqrt{4 - x^2 - y}$) and exits the solid on the top surface of the paraboloid ($z = \sqrt{4 - x^2 - y}$). This gives you the innermost limits of integration. We get the rest from looking at the projection of the paraboloid onto the xy-plane. (See Figure 13.57c.)

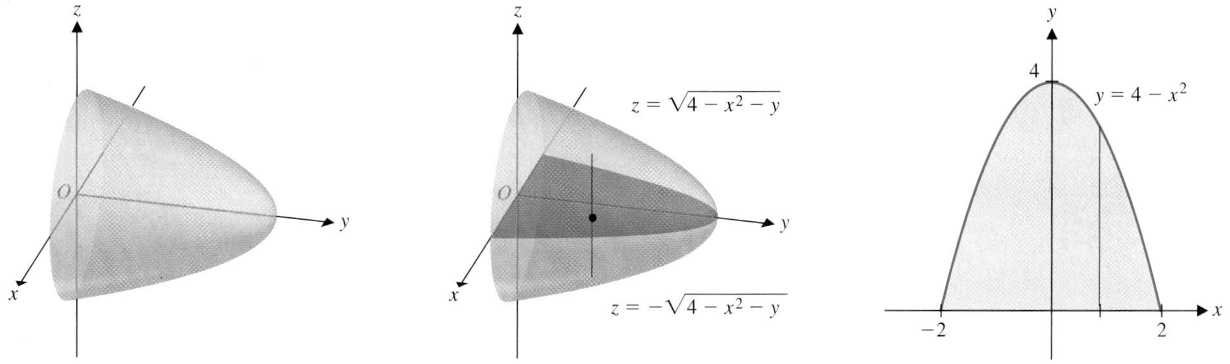

FIGURE 13.57a
The solid Q

FIGURE 13.57b
Solid, showing projection onto xy-plane

FIGURE 13.57c
Projection onto the xy-plane

Reading the outer limits of integration from Figure 13.57c and using (5.5), we get

$$V = \iiint_Q dV = \int_{-2}^{2}\int_{0}^{4-x^2}\int_{-\sqrt{4-x^2-y}}^{\sqrt{4-x^2-y}} dz\,dy\,dx$$

$$= \int_{-2}^{2}\int_{0}^{4-x^2} z\Big|_{z=-\sqrt{4-x^2-y}}^{z=\sqrt{4-x^2-y}} dy\,dx$$

$$= \int_{-2}^{2}\int_{0}^{4-x^2} 2\sqrt{4-x^2-y}\,dy\,dx$$

$$= \int_{-2}^{2} (-2)\left(\frac{2}{3}\right)(4-x^2-y)^{3/2}\Big|_{y=0}^{y=4-x^2} dx$$

$$= \frac{4}{3}\int_{-2}^{2}(4-x^2)^{3/2}dx = 8\pi.$$

Notice that the last integration here is challenging. (We used a CAS to carry it out.) Alternatively, if you look at Figure 13.57a and turn your head to the side, you should see a paraboloid with a circular base in the xz-plane. This suggests that you might want

NOTES

Example 6.6 suggests that polar coordinates may be used with any two of x, y and z, to produce a cylindrical coordinate representation of a solid.

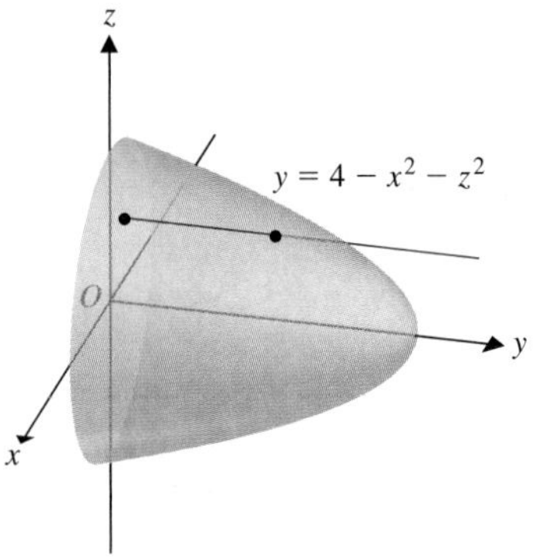

FIGURE 13.57d
Paraboloid with base in the xz-plane

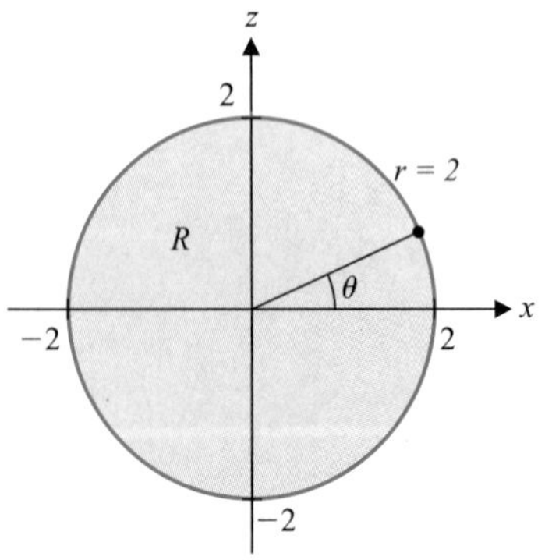

FIGURE 13.57e
Base of the solid

to integrate first with respect to y. Referring to Figure 13.57d, observe that for each point in the base of the solid in the xz-plane, y ranges from 0 to $4 - x^2 - z^2$. Notice that the base in this case is formed by the intersection of the paraboloid with the xz-plane $(y = 0)$: $0 = 4 - x^2 - z^2$ or $x^2 + z^2 = 4$ (i.e., the circle of radius 2 centered at the origin; see Figure 13.57e).

We can now write the volume as

$$V = \iiint_Q dV = \iint_R \int_0^{4-x^2-z^2} dy\, dA$$
$$= \iint_R (4 - x^2 - z^2)\, dA,$$

where R is the disk indicated in Figure 13.57e. Since the region R is a circle and the integrand contains the combination of variables $x^2 + z^2$, we define polar coordinates $x = r\cos\theta$ and $z = r\sin\theta$. From Figure 13.57e, notice that for each fixed angle $\theta \in [0, 2\pi]$, r runs from 0 to 2. This gives us

$$V = \iint_R \underbrace{(4 - x^2 - z^2)}_{4-r^2} \underbrace{dA}_{r\,dr\,d\theta}$$
$$= \int_0^{2\pi}\int_0^2 (4 - r^2)\, r\, dr\, d\theta$$
$$= -\frac{1}{2}\int_0^{2\pi} \frac{(4-r^2)^2}{2}\bigg|_{r=0}^{r=2} d\theta$$
$$= 4\int_0^{2\pi} d\theta = 8\pi.$$

Notice that in some sense, viewing the solid as having its base in the xz-plane is more natural than our first approach to the problem and the integrations are much simpler. ■

H.W Pg 1091 #21, 23, 25, 27, 33, 37

EXERCISES 13.6

WRITING EXERCISES

1. Using the examples in this section as a guide, make a short list of figures that are easily described in cylindrical coordinates.
2. The three-dimensional solid bounded by $x = a, x = b, y = c, y = d, z = e$ and $z = f$ is a rectangular box. Given this, speculate on how **cylindrical coordinates** got its name. Specifically, identify what type of cylinder is the basic figure of cylindrical coordinates.
3. In example 6.4, explain why the outer double integration limits are determined by the intersection of the cone and the hemisphere (and not, for example, by the trace of the hemisphere in the xy-plane).
4. Carefully examine the rectangular and cylindrical limits of integration in example 6.5. Note that in both integrals, z is the innermost variable of integration. In this case, explain why the innermost limits of integration for the triple integral in cylindrical coordinates can be obtained by substituting polar coordinates into the innermost limits of integration of the triple integral in rectangular coordinates. Explain why this would not be the case if the order of integration had been changed.

In exercises 1–8, write the given equation in cylindrical coordinates.

1. $x^2 + y^2 = 16$
2. $x^2 + y^2 = 1$
3. $(x-2)^2 + y^2 = 4$
4. $x^2 + (y-3)^2 = 9$
5. $z = x^2 + y^2$
6. $z = \sqrt{x^2 + y^2}$
7. $y = 2x$
8. $z = e^{-x^2-y^2}$

In exercises 9–20, set up the triple integral $\iiint_Q f(x,y,z)\,dV$ in cylindrical coordinates.

9. Q is the region above $z=\sqrt{x^2+y^2}$ and below $z=\sqrt{8-x^2-y^2}$.

10. Q is the region above $z=-\sqrt{x^2+y^2}$, below $z=0$ and inside $x^2+y^2=4$.

11. Q is the region above the xy-plane and below $z=9-x^2-y^2$.

12. Q is the region above the xy-plane and below $z=4-x^2-y^2$ in the first octant.

13. Q is the region above $z=x^2+y^2-1$, below $z=8$ and between $x^2+y^2=3$ and $x^2+y^2=8$.

14. Q is the region above $z=x^2+y^2-4$ and below $z=-x^2-y^2$.

15. Q is the region bounded by $y=4-x^2-z^2$ and $y=0$.

16. Q is the region bounded by $y=\sqrt{x^2+z^2}$ and $y=9$.

17. Q is the region bounded by $x=y^2+z^2$ and $x=2-y^2-z^2$.

18. Q is the region bounded by $x=\sqrt{y^2+z^2}$ and $x=4$.

19. Q is the frustum of a cone bounded by $z=2$, $z=3$ and $z=\sqrt{x^2+y^2}$.

20. Q is the region bounded by $z=1$ and $z=3$ and under $z=4-x^2-y^2$.

In exercises 21–32, set up and evaluate the indicated triple integral in the appropriate coordinate system.

21. $\iiint_Q e^{x^2+y^2}\,dV$, where Q is the region inside $x^2+y^2=4$ and between $z=1$ and $z=2$.

22. $\iiint_Q ze^{\sqrt{x^2+y^2}}\,dV$, where Q is the region inside $x^2+y^2=4$, outside $x^2+y^2=1$ and between $z=0$ and $z=3$.

23. $\iiint_Q (x+z)\,dV$, where Q is the region below $x+2y+3z=6$ in the first octant.

24. $\iiint_Q (y+2)\,dV$, where Q is the region below $x+z=4$ in the first octant between $y=1$ and $y=2$.

25. $\iiint_Q z\,dV$, where Q is the region between $z=\sqrt{x^2+y^2}$ and $z=\sqrt{4-x^2-y^2}$.

26. $\iiint_Q \sqrt{x^2+y^2}\,dV$, where Q is the region between $z=\sqrt{x^2+y^2}$ and $z=0$ and inside $x^2+y^2=4$.

27. $\iiint_Q (x+y)\,dV$, where Q is the tetrahedron bounded by $x+2y+z=4$ and the coordinate planes.

28. $\iiint_Q (2x-y)\,dV$, where Q is the tetrahedron bounded by $3x+y+2z=6$ and the coordinate planes.

29. $\iiint_Q e^z\,dV$, where Q is the region above $z=-\sqrt{4-x^2-y^2}$, below the xy-plane and outside $x^2+y^2=3$.

30. $\iiint_Q \sqrt{x^2+y^2}e^z\,dV$, where Q is the region inside $x^2+y^2=1$ and between $z=(x^2+y^2)^{3/2}$ and $z=0$.

31. $\iiint_Q 2x\,dV$, where Q is the region between $z=\sqrt{x^2+y^2}$ and $z=0$ and inside $x^2+(y-1)^2=1$.

32. $\iiint_Q y\,dV$, where Q is the region between $z=x^2+y^2$ and $z=0$ and inside $(x-2)^2+y^2=4$.

In exercises 33–38, evaluate the iterated integral after changing coordinate systems.

33. $\displaystyle\int_{-1}^{1}\int_{-\sqrt{1-x^2}}^{\sqrt{1-x^2}}\int_0^{\sqrt{x^2+y^2}} 3z^2\,dz\,dy\,dx$

34. $\displaystyle\int_0^{1}\int_{-\sqrt{1-x^2}}^{\sqrt{1-x^2}}\int_0^{2-x^2-y^2} \sqrt{x^2+y^2}\,dz\,dy\,dx$

35. $\displaystyle\int_0^{2}\int_{-\sqrt{4-y^2}}^{\sqrt{4-y^2}}\int_{\sqrt{x^2+y^2}}^{\sqrt{8-x^2-y^2}} 2\,dz\,dx\,dy$

36. $\displaystyle\int_0^{1}\int_0^{\sqrt{1-x^2}}\int_{1-x^2-y^2}^{4} \sqrt{x^2+y^2}\,dz\,dy\,dx$

37. $\displaystyle\int_{-3}^{3}\int_{-\sqrt{9-x^2}}^{0}\int_0^{x^2+z^2} (x^2+z^2)\,dy\,dz\,dx$

38. $\displaystyle\int_{-2}^{0}\int_{-\sqrt{4-z^2}}^{\sqrt{4-z^2}}\int_{y^2+z^2}^{4} (y^2+z^2)^{3/2}\,dx\,dy\,dz$

In exercises 39–46, sketch graphs of the cylindrical equations.

39. $z=r$ **40.** $z=r^2$ **41.** $z=4-r^2$

42. $z=\sqrt{4-r^2}$ **43.** $r=2\sec\theta$ **44.** $r=2\sin\theta$

45. $\theta=\pi/4$ **46.** $r=4$

In exercises 47–50, find the mass and center of mass of the solid with the given density and bounded by the graphs of the indicated equations.

47. $\rho(x,y,z)=\sqrt{x^2+y^2}$, bounded by $z=\sqrt{x^2+y^2}$ and $z=4$.

48. $\rho(x,y,z)=e^{-x^2-y^2}$, bounded by $z=\sqrt{4-x^2-y^2}$ and the xy-plane.

49. $\rho(x,y,z)=4$, between $z=x^2+y^2$ and $z=4$ and inside $x^2+(y-1)^2=1$.

50. $\rho(x,y,z)=\sqrt{x^2+z^2}$, bounded by $y=\sqrt{x^2+z^2}$ and $y=\sqrt{8-x^2-z^2}$.

Exercises 51–60, relate to unit basis vectors in cylindrical coordinates.

51. For the position vector $\mathbf{r}=\langle x,y,0\rangle=\langle r\cos\theta, r\sin\theta, 0\rangle$ in cylindrical coordinates, compute the unit vector $\hat{\mathbf{r}}=\dfrac{\mathbf{r}}{r}$, where $r=\|\mathbf{r}\|\neq 0$.

52. Referring to exercise 51, for the unit vector $\hat{\boldsymbol{\theta}} = \langle -\sin\theta, \cos\theta, 0\rangle$, show that $\hat{\mathbf{r}}$, $\hat{\boldsymbol{\theta}}$ and $\mathbf{k}$ are mutually orthogonal.

53. The unit vectors $\hat{\mathbf{r}}$ and $\hat{\boldsymbol{\theta}}$ in exercises 51 and 52 are not constant vectors. This changes many of our calculations and interpretations. For an object in motion (that is, where r, θ and z are functions of time), compute the derivatives of $\hat{\mathbf{r}}$ and $\hat{\boldsymbol{\theta}}$ in terms of each other.

54. For the vector $\mathbf{v}$ from $(0, 0, 0)$ to $(2, 2, 0)$, show that $\mathbf{v} = r\hat{\mathbf{r}}$.

55. For the vector $\mathbf{v}$ from $(1, 1, 0)$ to $(3, 3, 0)$, find a constant c such that $\mathbf{v} = c\hat{\mathbf{r}}$. Compare to exercise 54.

56. For the vector $\mathbf{v}$ from $(-1, -1, 0)$ to $(1, 1, 0)$, find a constant c such that $\mathbf{v} = c\hat{\mathbf{r}}$. Compare to exercise 55.

57. For the vector $\mathbf{v}$ from $(1, -1, 0)$ to $(1, 1, 0)$, find a constant c such that $\mathbf{v} = c\int_{-\pi/4}^{\pi/4} \hat{\boldsymbol{\theta}} d\theta$.

58. For the vector $\mathbf{v}$ from $(-1, -1, 0)$ to $(1, 1, 0)$, write $\mathbf{v}$ in the form $c\int_a^b \hat{\boldsymbol{\theta}}\, d\theta$. Compare to exercise 56.

59. For the point $(-1, -1, 0)$, sketch the vectors $\hat{\mathbf{r}}$ and $\hat{\boldsymbol{\theta}}$. Illustrate graphically how the vector $\mathbf{v}$ from $(-1, -1, 0)$ to $(1, 1, 0)$ can be represented both in terms of $\hat{\mathbf{r}}$ and in terms of $\hat{\boldsymbol{\theta}}$.

60. For the vector $\mathbf{v}$ from $(-1, -1, 0)$ to $(1, \sqrt{3}, 1)$, find constants a, b, θ_1, θ_2 and c such that $\mathbf{v} = a\hat{\mathbf{r}} + b\int_{\theta_1}^{\theta_2} \hat{\boldsymbol{\theta}}\, d\theta + c\mathbf{k}$.

EXPLORATORY EXERCISES

1. Many computer graphing packages will sketch graphs in cylindrical coordinates, with one option being to have r as a function of z and θ. In some cases, the graphs are very familiar. Sketch the following and solve for z to write the equation in the notation of this section: (a) $r = \sqrt{z}$, (b) $r = z^2$, (c) $r = \ln z$, (d) $r = \sqrt{4 - z^2}$, (e) $r = z^2 \cos\theta$. By leaving z out altogether, some old polar curves get an interesting three-dimensional extension: (f) $r = \sin^2\theta, 0 \le z \le 4$, (g) $r = 2 - 2\cos\theta$, $0 \le z \le 4$. Many graphs are simply new. Explore the following graphs and others of your own creation: (h) $r = \cos\theta - \ln z$, (i) $r = z^2 \ln(\theta + 1)$, (j) $r = ze^{\theta/8}$, (k) $r = \theta e^{-z}$.

2. In this exercise, you will explore a class of surfaces known as **Plücker's conoids.** In parametric equations, the conoid with n folds is given by $x = r\cos\theta$, $y = r\sin\theta$ and $z = \sin(n\theta)$. Use a CAS to sketch the conoid with 2 folds. Show that on this surface $z = \frac{2xy}{x^2+y^2}$. In vector notation, the parametric equations can be written as $\langle 0, 0, \sin(n\theta)\rangle + \langle r\cos\theta, r\sin\theta, 0\rangle$, with the interpretation that the conoid is generated by moving a line around and perpendicular to the circle $\langle \cos\theta, \sin\theta, 0\rangle$. For $n = 2$, sketch a parametric graph with $1 \le r \le 2$ and $0 \le \theta \le 2\pi$ and compare the surface to a Möbius strip. Explain how the line segment moving around the circle rotates according to the function $\sin 2\theta$. Sketch similar graphs for $n = 3$, $n = 4$ and $n = 5$, and explain why n is referred to as the number of folds.

13.7 SPHERICAL COORDINATES

We introduce here another common coordinate system that is frequently more convenient than either rectangular or cylindrical coordinates. In particular, some triple integrals that cannot be calculated exactly in either rectangular or cylindrical coordinates can be dealt with easily in spherical coordinates.

We can specify a point P with rectangular coordinates (x, y, z) by the corresponding **spherical coordinates** (ρ, ϕ, θ). Here, ρ is defined to be the distance from the origin,

$$\rho = \sqrt{x^2 + y^2 + z^2}. \tag{7.1}$$

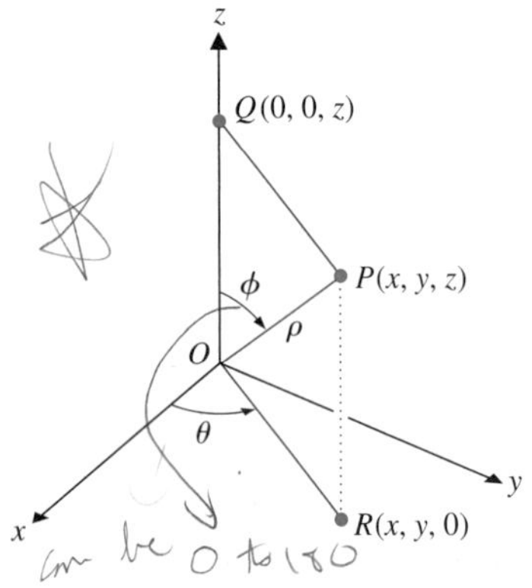

FIGURE 13.58
Spherical coordinates

Note that specifying the distance a point lies away from the origin specifies a sphere on which the point must lie (i.e., the equation $\rho = \rho_0 > 0$ represents the sphere of radius ρ_0 centered at the origin). To name a specific point on the sphere, we further specify two angles, ϕ and θ, as indicated in Figure 13.58. Notice that ϕ is the angle from the positive z-axis to the vector $\overrightarrow{OP}$ and θ is the angle from the positive x-axis to the vector $\overrightarrow{OR}$, where R is the point lying in the xy-plane with rectangular coordinates $(x, y, 0)$ (i.e., R is the projection of P onto the xy-plane). You should observe from this description that

$$\rho \ge 0 \quad \text{and} \quad 0 \le \phi \le \pi.$$

If you look closely at Figure 13.58, you can see how to relate rectangular and spherical coordinates. Notice that

$$x = \|\overrightarrow{OR}\| \cos\theta = \|\overrightarrow{QP}\| \cos\theta.$$

Looking at the triangle OQP, we find that $\|\overrightarrow{QP}\| = \rho\sin\phi$, so that

$$x = \rho\sin\phi\cos\theta. \tag{7.2}$$

Similarly, we have
$$y = \|\overrightarrow{OR}\| \sin\theta = \rho\sin\phi\sin\theta. \tag{7.3}$$

Finally, focusing again on triangle OQP, we have

$$z = \rho\cos\phi. \tag{7.4}$$

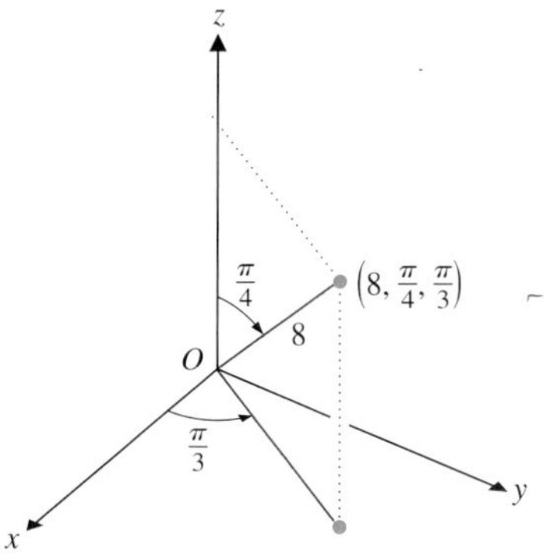

FIGURE 13.59
The point $(8, \frac{\pi}{4}, \frac{\pi}{3})$

EXAMPLE 7.1 Converting from Spherical to Rectangular Coordinates

Find rectangular coordinates for the point described by $(8, \pi/4, \pi/3)$ in spherical coordinates.

Solution We show a sketch of the point in Figure 13.59. From (7.2), (7.3) and (7.4), we have

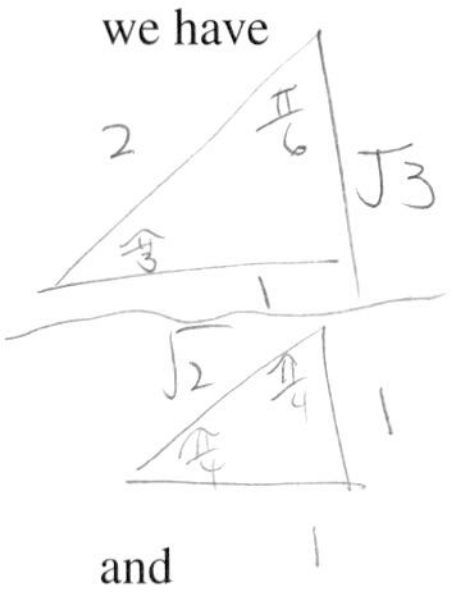

$$x = 8\sin\frac{\pi}{4}\cos\frac{\pi}{3} = 8\left(\frac{\sqrt{2}}{2}\right)\left(\frac{1}{2}\right) = 2\sqrt{2},$$

$$y = 8\sin\frac{\pi}{4}\sin\frac{\pi}{3} = 8\left(\frac{\sqrt{2}}{2}\right)\left(\frac{\sqrt{3}}{2}\right) = 2\sqrt{6}$$

and

$$z = 8\cos\frac{\pi}{4} = 8\left(\frac{\sqrt{2}}{2}\right) = 4\sqrt{2}.$$ ■

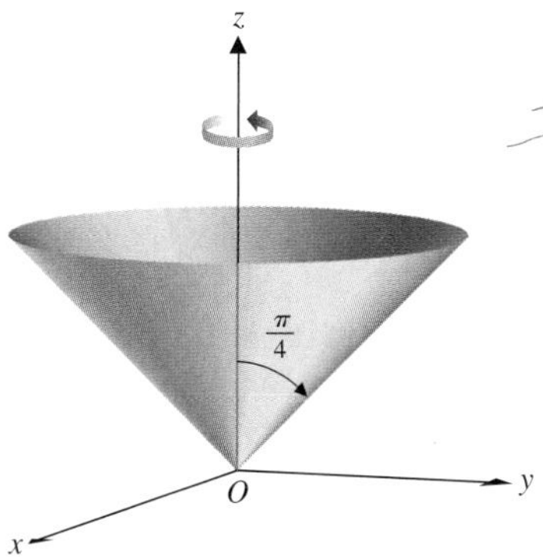

FIGURE 13.60a
Top half-cone $\phi = \frac{\pi}{4}$

It's often very helpful (especially when dealing with triple integrals) to represent common surfaces in spherical coordinates.

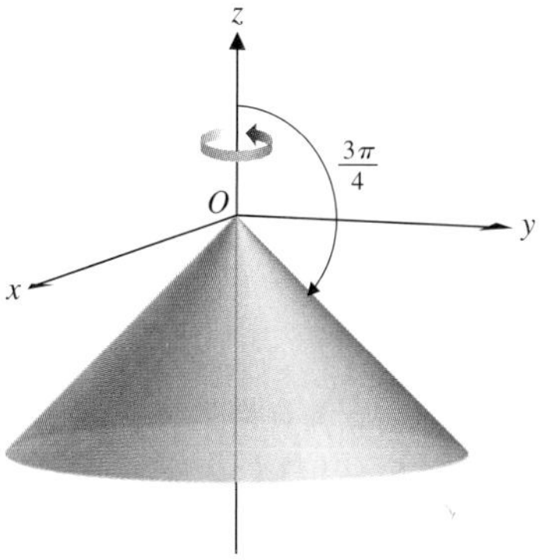

FIGURE 13.60b
Bottom half-cone $\phi = \frac{3\pi}{4}$

EXAMPLE 7.2 Equation of a Cone in Spherical Coordinates

Rewrite the equation of the cone $z^2 = x^2 + y^2$ in spherical coordinates.

Solution Using (7.2), (7.3) and (7.4), the equation of the cone becomes

$$\begin{aligned}\rho^2\cos^2\phi &= \rho^2\sin^2\phi\cos^2\theta + \rho^2\sin^2\phi\sin^2\theta \\ &= \rho^2\sin^2\phi(\cos^2\theta + \sin^2\theta) \\ &= \rho^2\sin^2\phi. \qquad \text{Since } \cos^2\theta + \sin^2\theta = 1.\end{aligned}$$

Notice that in order to have $\rho^2\cos^2\phi = \rho^2\sin^2\phi$, we must either have $\rho = 0$ (which corresponds to the origin) or $\cos^2\phi = \sin^2\phi$. For the latter to occur, we must have $\phi = \frac{\pi}{4}$ or $\phi = \frac{3\pi}{4}$. (Recall that $0 \le \phi \le \pi$.) Observe that taking $\phi = \frac{\pi}{4}$ (and allowing ρ and θ to be anything) describes the top half of the cone, as shown in Figure 13.60a. You can think of this as taking a single ray (say in the yz-plane) with $\phi = \frac{\pi}{4}$ and revolving this around the z-axis. (This is the effect of letting θ run from 0 to 2π.) Similarly, $\phi = \frac{3\pi}{4}$ describes the bottom half cone, as seen in Figure 13.60b. ■

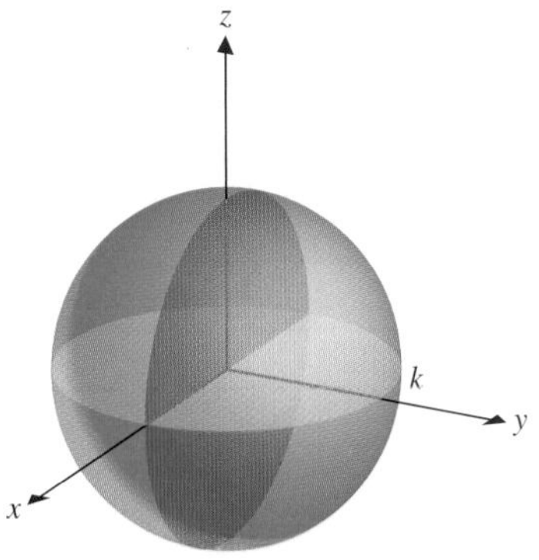

FIGURE 13.61a
The sphere $\rho = k$

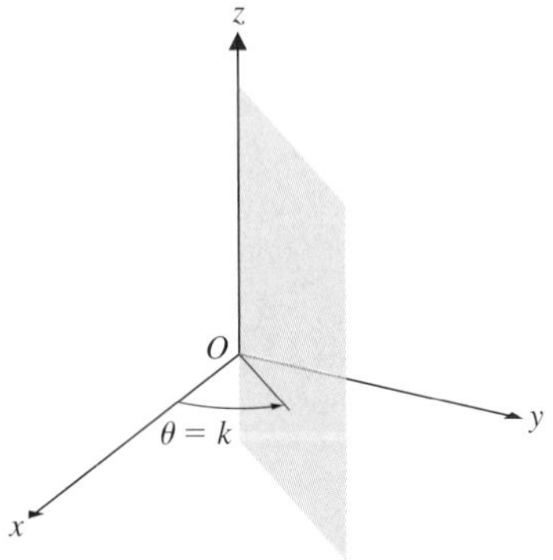

FIGURE 13.61b
The half-plane $\theta = k$

Notice that, in general, the equation $\rho = k$ (for any constant $k > 0$) represents the sphere of radius k, centered at the origin. (See Figure 13.61a.) The equation $\theta = k$ (for any constant k) represents a vertical half-plane, with its edge along the z-axis. (See Figure 13.61b.) Further, the equation $\phi = k$ (for any constant k) represents the top half of a cone if $0 < k < \frac{\pi}{2}$ (see Figure 13.62a) and represents the bottom half of a cone if $\frac{\pi}{2} < k < \pi$ (see Figure 13.62b). Finally, note that $\phi = \frac{\pi}{2}$ describes the xy-plane. Can you think of what the equations $\phi = 0$ and $\phi = \pi$ represent?

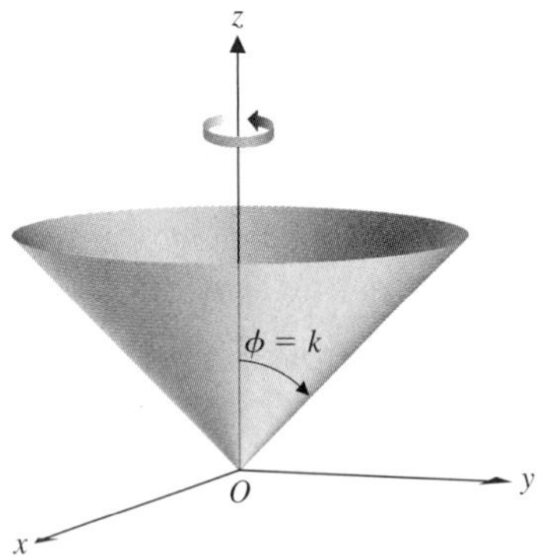

FIGURE 13.62a
Top half-cone $\phi = k$, where $0 < k < \frac{\pi}{2}$

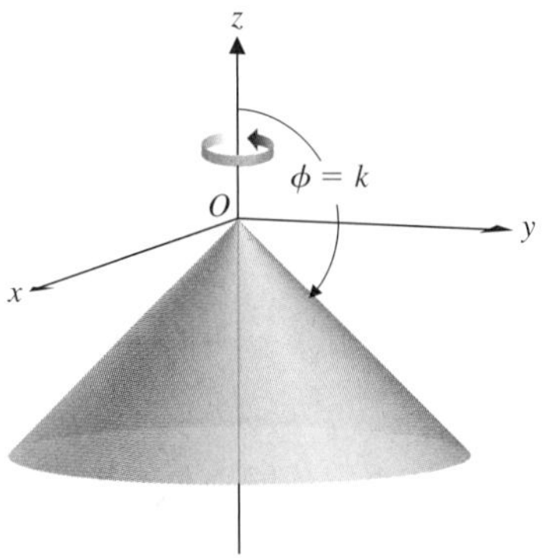

FIGURE 13.62b
Bottom half-cone $\phi = k$, where $\frac{\pi}{2} < k < \pi$

○ Triple Integrals in Spherical Coordinates

Just as polar coordinates are indispensable in calculating double integrals over circular regions, especially when the integrand involves the particular combination of variables $x^2 + y^2$, spherical coordinates are an indispensable aid in dealing with triple integrals over spherical regions, particularly with those where the integrand involves the combination $x^2 + y^2 + z^2$. Integrals of this type are encountered frequently in applications. For instance, consider the triple integral

$$\iiint_Q \cos(x^2 + y^2 + z^2)^{3/2}\, dV,$$

where Q is the **unit ball:** $x^2 + y^2 + z^2 \le 1$. No matter which order you choose for the integrations, you will arrive at a triple iterated integral that looks like

$$\int_{-1}^{1} \int_{-\sqrt{1-x^2}}^{\sqrt{1-x^2}} \int_{-\sqrt{1-x^2-y^2}}^{\sqrt{1-x^2-y^2}} \cos(x^2 + y^2 + z^2)^{3/2}\, dz\, dy\, dx.$$

In rectangular coordinates (or cylindrical coordinates, for that matter), you have little hope of calculating this integral exactly. In spherical coordinates, however, this integral is a snap. First, we need to see how to write triple integrals in spherical coordinates.

For the integral $\iiint_Q f(\rho, \phi, \theta)\, dV$, we begin, as we have many times before, by constructing an inner partition of the solid Q. But, rather than slicing up Q using planes parallel to the three coordinate planes, we divide Q by slicing it with spheres of the form $\rho = \rho_k$, half-planes of the form $\theta = \theta_k$ and half-cones of the form $\phi = \phi_k$. Notice that instead of subdividing Q into a number of rectangular boxes, this divides Q into a number of

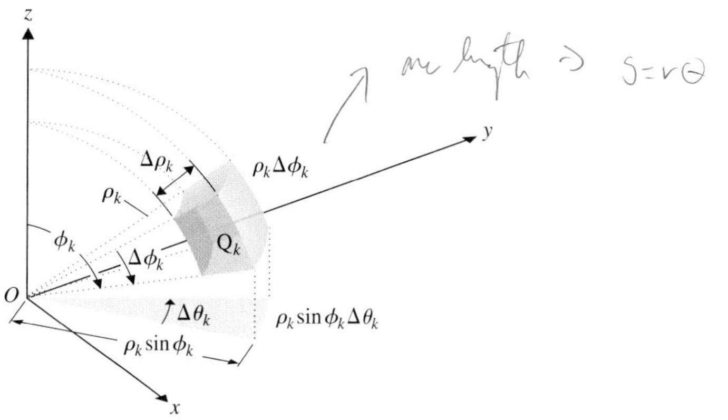

FIGURE 13.63
The spherical wedge Q_k

spherical wedges of the form:

$$Q_k = \{(\rho, \phi, \theta) | \rho_{k-1} \le \rho \le \rho_k, \phi_{k-1} \le \phi \le \phi_k, \theta_{k-1} \le \theta \le \theta_k\},$$

as depicted in Figure 13.63. Here, we have $\Delta\rho_k = \rho_k - \rho_{k-1}$, $\Delta\phi_k = \phi_k - \phi_{k-1}$ and $\Delta\theta_k = \theta_k - \theta_{k-1}$. Notice that Q_k is nearly a rectangular box and so, its volume ΔV_k is approximately the same as that of a rectangular box with the same dimensions:

$$\begin{aligned} \Delta V_k &\approx \Delta\rho_k(\rho_k\, \Delta\phi_k)\,(\rho_k \sin\phi_k\, \Delta\theta_k) \\ &= \rho_k^2 \sin\phi_k\, \Delta\rho_k \Delta\phi_k \Delta\theta_k. \end{aligned}$$

We consider only those wedges that lie completely inside Q, to form an inner partition $Q_1, Q_2, \ldots, Q_n$ of the solid Q. Summing over the inner partition and letting the norm of the partition $\|P\|$ (here, the longest diagonal of any of the wedges in the inner partition) approach zero, we get

$$\begin{aligned} \iiint_Q f(\rho, \phi, \theta)\, dV &= \lim_{\|P\|\to 0} \sum_{k=1}^{n} f(\rho_k, \phi_k, \theta_k)\Delta V_k \\ &= \lim_{\|P\|\to 0} \sum_{k=1}^{n} f(\rho_k, \phi_k, \theta_k)\rho_k^2 \sin\phi_k\, \Delta\rho_k\, \Delta\phi_k\, \Delta\theta_k \\ &= \iiint_Q f(\rho, \phi, \theta)\rho^2 \sin\phi\, d\rho\, d\phi\, d\theta, \end{aligned} \tag{7.5}$$

where the limits of integration for each of the three iterated integrals are found in much the same way as we have done for other coordinate systems. From (7.5), notice that the volume element in spherical coordinates is given by

$$dV = \rho^2 \sin\phi\, d\rho\, d\phi\, d\theta.$$

We can now return to our introductory example.

EXAMPLE 7.3 A Triple Integral in Spherical Coordinates

Evaluate the triple integral $\iiint_Q \cos(x^2 + y^2 + z^2)^{3/2}\, dV$, where Q is the unit ball: $x^2 + y^2 + z^2 \le 1$.

Solution Notice that since Q is the unit ball, ρ (the radial distance from the origin) ranges from 0 to 1. Further, the angle ϕ ranges from 0 to π (where $\phi = 0$ starts us at the top of the sphere, $\phi \in [0, \pi/2]$ corresponds to the top hemisphere and $\phi \in [\pi/2, \pi]$ corresponds to the bottom hemisphere). Finally (to get all the way around the sphere), the angle θ ranges from 0 to 2π. From (7.5), we have that since $x^2 + y^2 + z^2 = \rho^2$,

$$\iiint_Q \cos(\underbrace{x^2 + y^2 + z^2}_{\rho^2})^{3/2} \underbrace{dV}_{\rho^2 \sin\phi\, d\rho\, d\phi\, d\theta}$$

$$= \int_0^{2\pi}\int_0^{\pi}\int_0^1 \cos(\rho^2)^{3/2}\rho^2 \sin\phi\, d\rho\, d\phi\, d\theta$$

$$= \frac{1}{3}\int_0^{2\pi}\int_0^{\pi}\int_0^1 \cos(\rho^3)\,(3\rho^2)\sin\phi\, d\rho\, d\phi\, d\theta$$

$$= \frac{1}{3}\int_0^{2\pi}\int_0^{\pi} \sin(\rho^3)\Big|_{\rho=0}^{\rho=1} \sin\phi\, d\phi\, d\theta$$

$$= \frac{\sin 1}{3}\int_0^{2\pi}\int_0^{\pi} \sin\phi\, d\phi\, d\theta$$

$$= -\frac{\sin 1}{3}\int_0^{2\pi} \cos\phi\Big|_{\phi=0}^{\phi=\pi} d\theta$$

$$= -\frac{\sin 1}{3}\int_0^{2\pi} (\cos\pi - \cos 0)\, d\theta$$

$$= \frac{2}{3}(\sin 1)(2\pi) \approx 3.525. \quad \blacksquare$$

NOTES

Since the integrand $\rho^2\cos(\rho^3)\sin\phi$ is in the factored form $f(\rho)g(\phi)h(\theta)$ and the limits of integration are all constant (which is frequently the case for integrals in spherical coordinates), the triple integral may be rewritten as:

$$\left(\int_0^1 \rho^2\cos(\rho^3)\,d\rho\right)\left(\int_0^{\pi} \sin\phi\, d\phi\right)\left(\int_0^{2\pi} 1\, d\theta\right).$$

Generally, spherical coordinates are useful in triple integrals when the solid over which you are integrating is in some way spherical and particularly when the integrand contains the term $x^2 + y^2 + z^2$. In example 7.4, we use spherical coordinates to simplify the calculation of a volume.

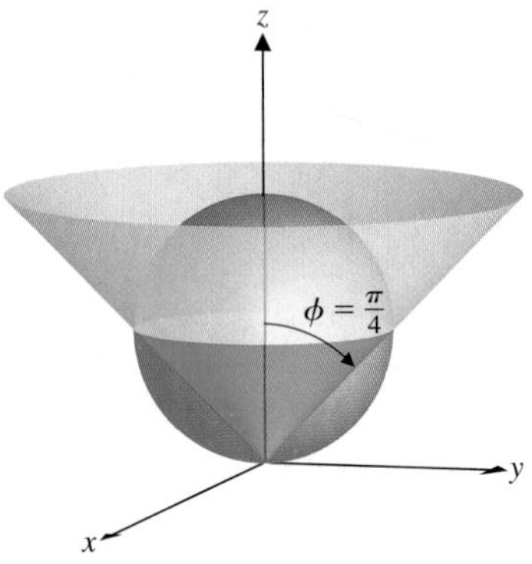

FIGURE 13.64
The cone $\phi = \frac{\pi}{4}$ and the sphere $\rho = 2\cos\phi$

EXAMPLE 7.4 Finding a Volume Using Spherical Coordinates

Find the volume lying inside the sphere $x^2 + y^2 + z^2 = 2z$ and inside the cone $z^2 = x^2 + y^2$.

Solution Notice that by completing the square in the equation of the sphere, we get

$$x^2 + y^2 + (z^2 - 2z + 1) = 1$$

or

$$x^2 + y^2 + (z - 1)^2 = 1,$$

the sphere of radius 1, centered at the point (0, 0, 1). Notice, too, that since the sphere sits completely above the xy-plane, only the top half of the cone, $z = \sqrt{x^2 + y^2}$ intersects the sphere. See Figure 13.64 for a sketch of the solid. You might try to find the volume using rectangular coordinates (but, don't spend much time on it). Because of the spherical geometry, we consider the problem in spherical coordinates. (Keep in mind that cones have a very simple representation in spherical coordinates.) From (7.1), (7.4) and the original equation of the sphere, we get

$$\underbrace{x^2 + y^2 + z^2}_{\rho^2} = 2\underbrace{z}_{\rho\cos\phi}$$

or

$$\rho^2 = 2\rho\cos\phi.$$

This equation is satisfied when $\rho = 0$ (corresponding to the origin) or when $\rho = 2\cos\phi$ (the equation of the sphere in spherical coordinates). For the top half of the cone, we have $z = \sqrt{x^2 + y^2}$, or in spherical coordinates $\phi = \frac{\pi}{4}$, as discussed in example 7.2.

Referring again to Figure 13.64, notice that to stay inside the cone and inside the sphere, we have that for each fixed ϕ and θ, ρ can range from 0 up to $2\cos\phi$. For each fixed θ, to stay inside the cone, ϕ must range from 0 to $\frac{\pi}{4}$. Finally, to get all the way around the solid, θ ranges from 0 to 2π. The volume of the solid is then given by

$$
\begin{aligned}
V &= \iiint_Q \underbrace{dV}_{\rho^2 \sin\phi\, d\rho\, d\phi\, d\theta} \\
&= \int_0^{2\pi} \int_0^{\pi/4} \int_0^{2\cos\phi} \rho^2 \sin\phi\, d\rho\, d\phi\, d\theta \\
&= \int_0^{2\pi} \int_0^{\pi/4} \left.\frac{1}{3}\rho^3\right|_{\rho=0}^{\rho=2\cos\phi} \sin\phi\, d\phi\, d\theta \\
&= \frac{8}{3} \int_0^{2\pi} \int_0^{\pi/4} \cos^3\phi \sin\phi\, d\phi\, d\theta \\
&= -\frac{8}{3} \int_0^{2\pi} \left.\frac{\cos^4\phi}{4}\right|_{\phi=0}^{\phi=\pi/4} d\theta = -\frac{2}{3} \int_0^{2\pi} \left(\cos^4\frac{\pi}{4} - 1\right) d\theta \\
&= -\frac{4\pi}{3}\left(\cos^4\frac{\pi}{4} - 1\right) = -\frac{4\pi}{3}\left(\frac{1}{4} - 1\right) = \pi. \quad \blacksquare
\end{aligned}
$$

EXAMPLE 7.5   Changing an Integral from Rectangular to Spherical Coordinates

Evaluate the triple iterated integral $\displaystyle\int_{-2}^{2} \int_0^{\sqrt{4-x^2}} \int_0^{\sqrt{4-x^2-y^2}} (x^2 + y^2 + z^2)\, dz\, dy\, dx$.

Solution   Although the integrand is simply a polynomial, the limits of integration make the second and third integrations very messy. Notice that the integrand contains the combination of variables $x^2 + y^2 + z^2$, which equals ρ^2 in spherical coordinates. Further, the solid over which we are integrating is a portion of a sphere, as follows. Notice that for each x in the interval $[-2, 2]$ indicated by the outermost limits of integration, y varies from 0 (corresponding to the x-axis) to $y = \sqrt{4-x^2}$ (the top semicircle of radius 2 centered at the origin). Finally, z varies from 0 (corresponding to the xy-plane) up to $z = \sqrt{4-x^2-y^2}$ (the top hemisphere of radius 2 centered at the origin). The solid Q over which we are integrating is then the half of the hemisphere that lies above the first and second quadrants of the xy-plane, as illustrated in Figure 13.65. In spherical coordinates, this portion of the sphere is obtained if we let ρ range from 0 up to 2, ϕ range from 0 up to $\frac{\pi}{2}$ and θ range from 0 to π. The integral then becomes

$$
\begin{aligned}
&\int_{-2}^{2} \int_0^{\sqrt{4-x^2}} \int_0^{\sqrt{4-x^2-y^2}} (x^2 + y^2 + z^2)\, dz\, dy\, dx \\
&= \iiint_Q \underbrace{(x^2 + y^2 + z^2)}_{\rho^2}\ \underbrace{dV}_{\rho^2 \sin\phi\, d\rho\, d\phi\, d\theta} \\
&= \int_0^{\pi} \int_0^{\pi/2} \int_0^{2} \rho^2(\rho^2 \sin\phi)\, d\rho\, d\phi\, d\theta = \frac{32}{5}\pi,
\end{aligned}
$$

where we leave the details of this relatively simple integration to you. $\blacksquare$

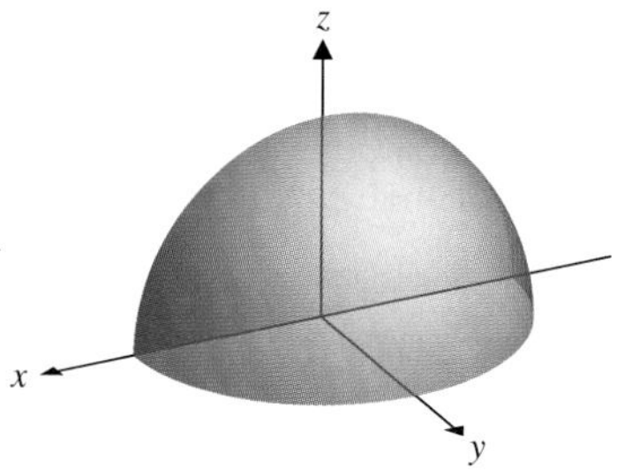

FIGURE 13.65
The solid Q

BEYOND FORMULAS

Spherical coordinates are used in a variety of applications, usually to take advantage of certain symmetries present in the structure or force being analyzed. In particular, if the value of a function $f(x, y, z)$ depends only on the distance of the point (x, y, z) from the origin, then spherical coordinates can be convenient to use. This is analogous to the use of polar coordinates in two dimensions to take advantage of radial symmetry. In what way is $r = c$ in two dimensions analogous to $\rho = c$ in three dimensions?

HW #25
Pg 1098 - 1099
#33, 37, 39, 43, 45, 53

EXERCISES 13.7

WRITING EXERCISES

1. Discuss the relationship between the spherical coordinates angles ϕ and θ and the longitude and latitude angles on a map of the earth. Satellites in geosynchronous orbit remain at a constant distance above a fixed point on the earth. Discuss how spherical coordinates could be used to represent the position of the satellite.

2. Explain why any point in $\mathbb{R}^3$ can be represented in spherical coordinates with $\rho \geq 0, 0 \leq \theta \leq 2\pi$ and $0 \leq \phi \leq \pi$. In particular, explain why it is not necessary to allow $\rho < 0$ or $\pi < \phi \leq 2\pi$. Discuss whether the ranges $\rho \geq 0, 0 \leq \theta \leq \pi$ and $0 \leq \phi \leq 2\pi$ would suffice to describe all points.

3. Explain why in spherical coordinates the equation $\theta = k$ represents a half-plane (see Figure 13.61b) and not a whole plane.

4. Using the examples in this section as a guide, make a short list of surfaces that are simple to describe in spherical coordinates.

In exercises 1–8, convert the spherical point (ρ, ϕ, θ) into rectangular coordinates.

1. $(4, 0, \pi)$
2. $(4, \frac{\pi}{2}, \pi)$
3. $(4, \frac{\pi}{2}, 0)$
4. $(4, \pi, \frac{\pi}{2})$
5. $(2, \frac{\pi}{4}, 0)$
6. $(2, \frac{\pi}{4}, \frac{2\pi}{3})$
7. $(\sqrt{2}, \frac{\pi}{6}, \frac{\pi}{3})$
8. $(\sqrt{2}, \frac{\pi}{6}, \frac{2\pi}{3})$

In exercises 9–16, convert the equation into spherical coordinates.

9. $x^2 + y^2 + z^2 = 9$
10. $x^2 + y^2 + z^2 = 6$
11. $y = x$
12. $z = 0$
13. $z = 2$
14. $x^2 + y^2 + (z - 1)^2 = 1$
15. $z = \sqrt{3(x^2 + y^2)}$
16. $z = -\sqrt{x^2 + y^2}$

In exercises 17–24, sketch the graph of the spherical equation and give a corresponding *xy*-equation.

17. $\rho = 2$
18. $\rho = 4$
19. $\phi = \frac{\pi}{4}$
20. $\phi = \frac{\pi}{2}$
21. $\theta = 0$
22. $\theta = \frac{\pi}{4}$
23. $\phi = \frac{\pi}{3}$
24. $\theta = \frac{\pi}{10}$

In exercises 25–32, sketch the region defined by the given ranges.

25. $0 \leq \rho \leq 4, 0 \leq \phi \leq \frac{\pi}{4}, 0 \leq \theta \leq \pi$
26. $0 \leq \rho \leq 4, 0 \leq \phi \leq \frac{\pi}{2}, 0 \leq \theta \leq 2\pi$
27. $0 \leq \rho \leq 2, 0 \leq \phi \leq \frac{\pi}{2}, 0 \leq \theta \leq \pi$
28. $0 \leq \rho \leq 2, 0 \leq \phi \leq \frac{\pi}{4}, \pi \leq \theta \leq 2\pi$
29. $0 \leq \rho \leq 3, 0 \leq \phi \leq \pi, 0 \leq \theta \leq \pi$
30. $0 \leq \rho \leq 3, 0 \leq \phi \leq \frac{3\pi}{4}, 0 \leq \theta \leq 2\pi$
31. $2 \leq \rho \leq 3, 0 \leq \phi \leq \pi, 0 \leq \theta \leq 2\pi$
32. $2 \leq \rho \leq 3, 0 \leq \phi \leq \frac{\pi}{2}, 0 \leq \theta \leq 2\pi$

In exercises 33–42, set up and evaluate the indicated triple integral in an appropriate coordinate system.

33. $\iiint_Q e^{(x^2+y^2+z^2)^{3/2}} dV$, where Q is bounded by the hemisphere $z = \sqrt{4 - x^2 - y^2}$ and the *xy*-plane.

34. $\iiint_Q \sqrt{x^2 + y^2 + z^2}\, dV$, where Q is bounded by the hemisphere $z = -\sqrt{9 - x^2 - y^2}$ and the *xy*-plane.

35. $\iiint_Q (x^2 + y^2 + z^2)^{5/2} dV$, where Q is inside $x^2 + y^2 + z^2 = 2$ and outside $x^2 + y^2 = 1$.

36. $\iiint_Q e^{\sqrt{x^2+y^2+z^2}} dV$, where Q is bounded by $y = \sqrt{4 - x^2 - z^2}$ and $y = 0$.

37. $\iiint_Q (x^2 + y^2 + z^2)\, dV$, where Q is the cube with $0 \leq x \leq 1$, $1 \leq y \leq 2$ and $3 \leq z \leq 4$.

38. $\iiint_Q (x + y + z)\, dV$, where Q is the tetrahedron bounded by $x + 2y + z = 4$ and the coordinate planes.

39. $\iiint\limits_Q (x^2+y^2)\,dV$, where Q is bounded by $z = 4 - x^2 - y^2$ and the xy-plane.

40. $\iiint\limits_Q e^{x^2+y^2}\,dV$, where Q is bounded by $x^2+y^2=4$, $z=0$ and $z=2$.

41. $\iiint\limits_Q \sqrt{x^2+y^2+z^2}\,dV$, where Q is bounded by $z=\sqrt{x^2+y^2}$ and $z=\sqrt{2-x^2-y^2}$.

42. $\iiint\limits_Q (x^2+y^2+z^2)^{3/2}\,dV$, where Q is the region below $z=-\sqrt{x^2+y^2}$ and inside $z=-\sqrt{4-x^2-y^2}$.

In exercises 43–52, use an appropriate coordinate system to find the volume of the given solid.

43. The region below $x^2+y^2+z^2=4z$ and above $z=\sqrt{x^2+y^2}$

44. The region above $z=\sqrt{x^2+y^2}$ and below $x^2+y^2+z^2=4$

45. The region inside $z=\sqrt{2x^2+2y^2}$ and between $z=2$ and $z=4$

46. The region bounded by $z=4x^2+4y^2$, $z=0$, $x^2+y^2=1$ and $x^2+y^2=2$

47. The region under $z=\sqrt{x^2+y^2}$ and above the square $-1\le x\le 1, -1\le y\le 1$

48. The region bounded by $x+2y+z=4$ and the coordinate planes

49. The region below $x^2+y^2+z^2=4$, above $z=\sqrt{x^2+y^2}$ in the first octant

50. The region below $x^2+y^2+z^2=4$, above $z=\sqrt{x^2+y^2}$, between $y=x$ and $x=0$ with $y\ge 0$

51. The region below $z=\sqrt{x^2+y^2}$, above the xy-plane and inside $x^2+y^2=4$

52. The region between $z=4-x^2-y^2$ and the xy-plane

In exercises 53–56, evaluate the iterated integral by changing coordinate systems.

53. $$\int_0^1\int_{-\sqrt{1-x^2}}^{\sqrt{1-x^2}}\int_{-\sqrt{1-x^2-y^2}}^{\sqrt{1-x^2-y^2}}\sqrt{x^2+y^2+z^2}\,dz\,dy\,dx$$

54. $$\int_{-1}^1\int_{-\sqrt{1-x^2}}^{\sqrt{1-x^2}}\int_{1-\sqrt{1-x^2-y^2}}^{1+\sqrt{1-x^2-y^2}}(x^2+y^2+z^2)^{3/2}\,dz\,dy\,dx$$

55. $$\int_{-2}^2\int_0^{\sqrt{4-x^2}}\int_{\sqrt{x^2+y^2}}^{\sqrt{8-x^2-y^2}}(x^2+y^2+z^2)^{3/2}\,dz\,dy\,dx$$

56. $$\int_0^4\int_0^{\sqrt{16-x^2}}\int_{\sqrt{x^2+y^2}}^{4}\sqrt{x^2+y^2+z^2}\,dz\,dy\,dx$$

57. Find the center of mass of the solid with constant density and bounded by $z=\sqrt{x^2+y^2}$ and $z=\sqrt{4-x^2-y^2}$.

58. Find the center of mass of the solid with constant density in the first quadrant and bounded by $z=\sqrt{x^2+y^2}$ and $z=\sqrt{4-x^2-y^2}$.

Exercises 59–64 relate to unit basis vectors in spherical coordinates.

59. For the position vector $\mathbf{r}=\langle x,y,z\rangle=\langle\rho\cos\theta\sin\phi, \rho\sin\theta\sin\phi, \rho\cos\phi\rangle$ in spherical coordinates, compute the unit vector $\hat{\rho}=\dfrac{\mathbf{r}}{r}$, where $r=\|\mathbf{r}\|\ne 0$.

60. For the unit vectors $\hat{\theta}=\langle-\sin\theta,\cos\theta,0\rangle$ and $\hat{\phi}=\langle\cos\theta\cos\phi,\sin\theta\cos\phi,-\sin\phi\rangle$, show that $\hat{\rho}$, $\hat{\theta}$ and $\hat{\phi}$ are mutually orthogonal.

61. For the vector $\mathbf{v}$ from $(1,1,\sqrt{2})$ to $(2,2,2\sqrt{2})$, find a constant c such that $\mathbf{v}=c\hat{\rho}$.

62. For the vector $\mathbf{v}$ from $(1,1,\sqrt{2})$ to $(-1,1,\sqrt{2})$, find a constant c such that $\mathbf{v}=c\int_{\pi/4}^{3\pi/4}\hat{\theta}\,d\theta$. Compare to exercise 61.

63. For the vector $\mathbf{v}$ from $(1,1,\sqrt{2})$ to $(\sqrt{2},\sqrt{2},0)$, find a constant c such that $\mathbf{v}=c\int_{\pi/4}^{\pi/2}\hat{\phi}\,d\phi$.

64. From the point $(1,1,\sqrt{2})$, sketch the vectors $\hat{\rho}$, $\hat{\theta}$ and $\hat{\phi}$. Illustrate graphically the vectors $\mathbf{v}$ in exercises 61–63.

65. Sketch the cardioid $r=2-2\sin\theta$ in the xy-plane. Define $\tilde{r}$ and $\tilde{\theta}$ to be polar coordinates in the yz-plane (that is, $y=\tilde{r}\cos\tilde{\theta}$ and $z=\tilde{r}\sin\tilde{\theta}$). Show that $\tilde{\theta}=\frac{\pi}{2}-\phi$ and $\cos\phi=\sin\tilde{\theta}$. Use this information to graph $\rho=2-2\cos\phi$.

66. As in exercise 65, relate the graph of $\rho=\cos^2 3\phi$ to the two-dimensional graph of $r=\sin^2 3\theta$ and use this information to sketch the three-dimensional surface.

67. As in exercise 65, sketch the surface $\rho=\sin^2\phi$ by relating it to a two-dimensional polar curve.

68. As in exercise 65, sketch the surface $\rho=1+\sin 3\phi$ by relating it to a two-dimensional polar curve.

EXPLORATORY EXERCISES

1. If you have a graphing utility that will graph surfaces of the form $\rho=f(\phi,\theta)$, graph $\rho=2\phi$ and $\rho=\left(\phi-\frac{\pi}{2}\right)^2$. Discuss the symmetry that results from the variable θ not appearing in the equation. Discuss the changes in ρ as you move down from $\phi=0$ to $\phi=\pi$. Using what you have learned, try graphing the following by hand and then compare your sketches to those of your graphing utility: (a) $\rho=e^{-\phi}$, (b) $\rho=e^{\phi}$, (c) $\rho=\sin^2\phi$ and (d) $\rho=\sin^2\left(\phi-\frac{\pi}{2}\right)$.

2. Use a graphing utility to graph $\rho=5\cos\theta$ and $\rho=\sqrt{\cos\theta}$. Discuss the symmetry that results from the variable ϕ not appearing in the equation. Discuss the changes in ρ as you move around from $\theta=0$ to $\theta=2\pi$. Using what you have learned, try graphing the following by hand and then compare your sketches to those of your graphing utility: (a) $\rho = \sin^2\theta$, (b) $\rho = \sin^2\left(\theta - \frac{\pi}{2}\right)$, (c) $\rho = e^{\theta}$ and (d) $\rho = e^{-\theta}$.

3. Use a graphing utility to graph each of the following. Adjust the graphing window as needed to get a good idea of what the graph looks like. (a) $\rho = \sin(\phi + \theta)$, (b) $\rho = \phi\sin\theta$, (c) $\rho = \sin^2\theta\cos\phi$, (d) $\rho = 4\cos^2\theta + 2\sin\phi - 3\sin\theta$ and (e) $\rho = 4\cos\theta\sin 5\phi + 3\cos^2\phi$. There are innumerable interesting and unusual graphs in spherical coordinates. Experiment and find your own!

13.8 CHANGE OF VARIABLES IN MULTIPLE INTEGRALS

One of our most basic tools for evaluating a definite integral is substitution. For instance, to evaluate the integral $\int_0^2 2xe^{x^2+3}dx$, you would make the substitution $u = x^2 + 3$. While you can probably do this in your head, let's consider the details one more time. Here, $du = 2x\,dx$ and don't forget: when you change variables in a definite integral, you must also change the limits of integration to suit the new variable. In this case, when $x = 0$, we have $u = 0^2 + 3 = 3$ and when $x = 2$, $u = 2^2 + 3 = 7$. This leaves us with

$$\int_0^2 2xe^{x^2+3}dx = \int_0^2 \underbrace{e^{x^2+3}}_{e^u}\underbrace{(2x)\,dx}_{du}$$
$$= \int_3^7 e^u du = e^u\Big|_3^7 = e^7 - e^3.$$

The primary reason for making the preceding change of variable was to simplify the integrand, so that it was easier to find an antiderivative. Notice too that we not only transformed the integrand, but we also changed the interval over which we were integrating.

You should recognize that we have already implemented changes of variables in multiple integrals in the very special cases of polar coordinates (for double integrals) and cylindrical and spherical coordinates (for triple integrals). There were several reasons for doing this. In the case of double integrals in rectangular coordinates, if the integrand contains the term $x^2 + y^2$ or if the region over which you are integrating is in some way circular, then polar coordinates may be indicated. For instance, consider the iterated integral

$$\int_0^3\int_0^{\sqrt{9-x^2}} \cos(x^2 + y^2)\,dy\,dx.$$

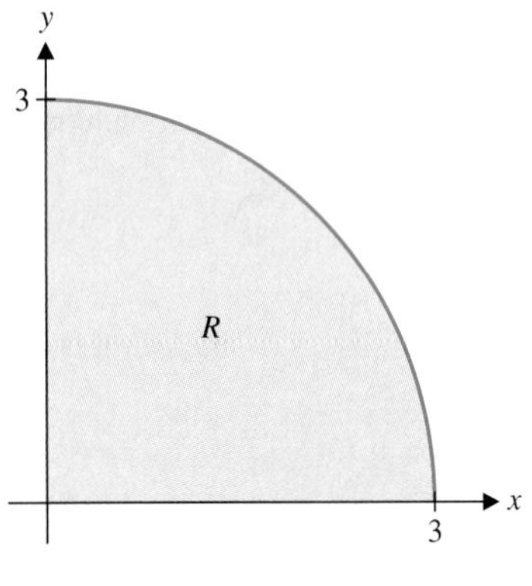

FIGURE 13.66a
The region of integration in the xy-plane

There is really no way to evaluate this integral as it is written in rectangular coordinates. (Try it!) However, recognizing that the region of integration R is the portion of the circle of radius 3 centered at the origin that lies in the first quadrant (see Figure 13.66a), and since the integrand includes the term $x^2 + y^2$, it's a good bet that polar coordinates will help. In fact, we have

$$\int_0^3\int_0^{\sqrt{9-x^2}} \cos(x^2 + y^2)\,dy\,dx = \iint_R \cos\underbrace{(x^2 + y^2)}_{r^2}\underbrace{dA}_{r\,dr\,d\theta}$$
$$= \int_0^{\pi/2}\int_0^3 \cos(r^2)\,r\,dr\,d\theta,$$

which is now an easy integral to evaluate. Notice that there are two things that happened here. First, we simplified the integrand (into one with a known antiderivative) and second, we transformed the region over which we integrated, as follows. In the xy-plane, we integrated over the circular sector indicated in Figure 13.66a. In the $r\theta$-plane, we integrated over

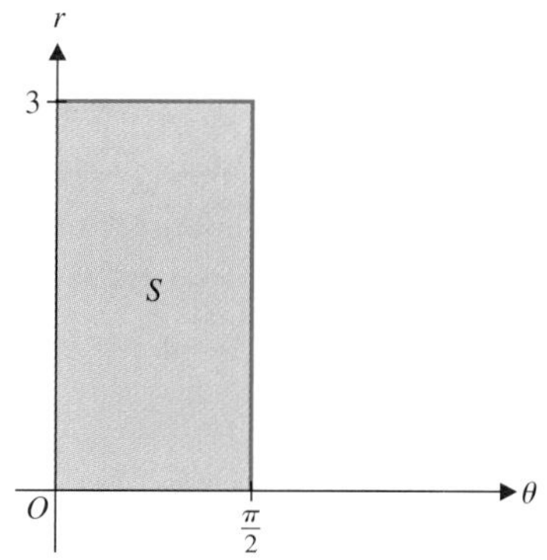

FIGURE 13.66b
The region of integration in the $r\theta$-plane

a (simpler) region: the rectangle S defined by $S = \{(r, \theta) | 0 \le r \le 3 \text{ and } 0 \le \theta \le \frac{\pi}{2}\}$, as indicated in Figure 13.66b.

Recall that we changed to cylindrical or spherical coordinates in triple integrals for similar reasons. In each case, we ended up simplifying the integrand and transforming the region of integration. More generally, how do we change variables in a multiple integral? Before we answer this question, we must first explore the concept of transformation in several variables.

A **transformation** T from the uv-plane to the xy-plane is a function that maps points in the uv-plane to points in the xy-plane, so that

$$T(u, v) = (x, y),$$

where

$$x = g(u, v) \quad \text{and} \quad y = h(u, v),$$

for some functions g and h. We consider changes of variables in double integrals as defined by a transformation T from a region S in the uv-plane onto a region R in the xy-plane (see Figure 13.67). We refer to R as the **image** of S under the transformation T. We say that T is **one-to-one** on S if for every point (x, y) in R there is exactly one point (u, v) in S such that $T(u, v) = (x, y)$. Notice that this says that (at least in principle), we can solve for u and v in terms of x and y. Further, we consider only transformations for which g and h have continuous first partial derivatives in the region S.

The primary reason for introducing a change of variables in a multiple integral is to simplify the calculation of the integral. This is accomplished by simplifying the integrand, the region over which you are integrating or both. Before exploring the effect of a transformation on a multiple integral, we examine several examples of how a transformation can simplify a region in two dimensions.

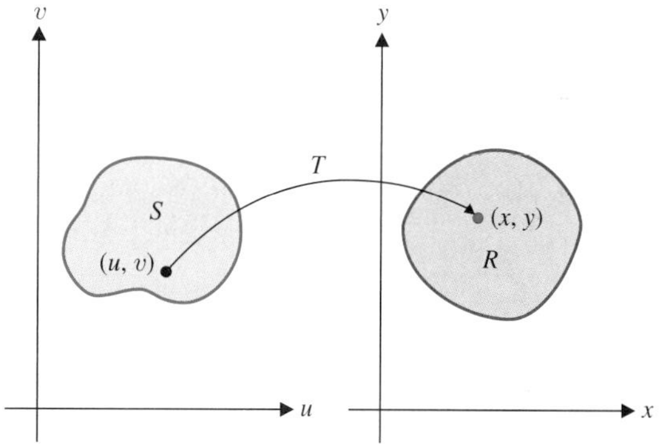

FIGURE 13.67
The transformation T mapping S onto R

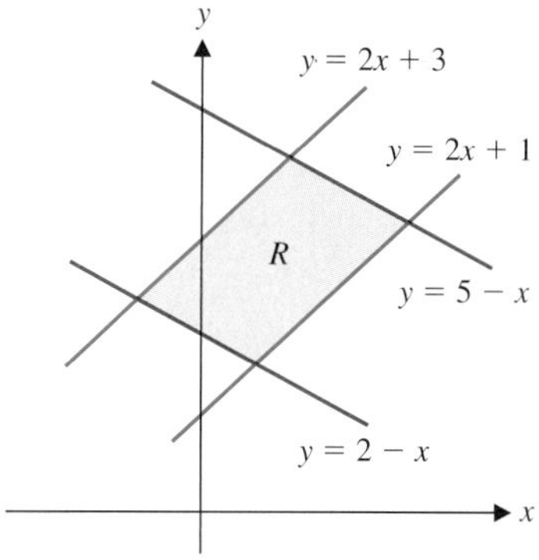

FIGURE 13.68a
The region R in the xy-plane

EXAMPLE 8.1 The Transformation of a Simple Region

Let R be the region bounded by the straight lines $y = 2x + 3$, $y = 2x + 1$, $y = 5 - x$ and $y = 2 - x$. Find a transformation T mapping a region S in the uv-plane onto R, where S is a rectangular region, with sides parallel to the u- and v-axes.

Solution First, notice that the region R is a parallelogram in the xy-plane. (See Figure 13.68a.) We can rewrite the equations for the lines forming the boundaries of R as $y - 2x = 3$, $y - 2x = 1$, $y + x = 5$ and $y + x = 2$. This suggests the

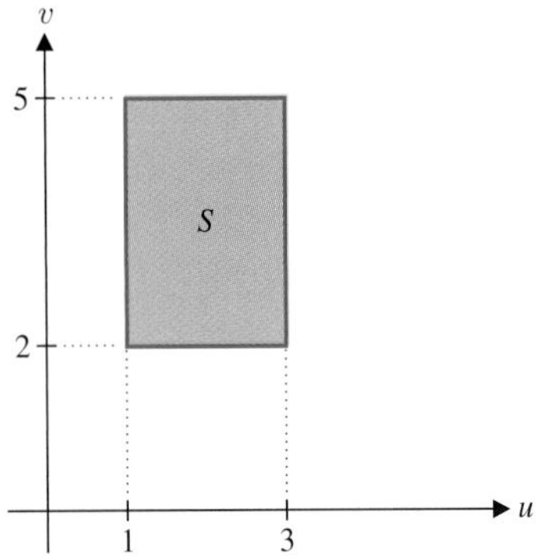

FIGURE 13.68b
The region S in the uv-plane

change of variables

$$u = y - 2x \quad \text{and} \quad v = y + x. \tag{8.1}$$

Observe that the lines forming the boundaries of R then correspond to the lines $u = 3$, $u = 1$, $v = 5$ and $v = 2$, respectively, forming the boundaries of the corresponding region S in the uv-plane. (See Figure 13.68b.) Solving equations (8.1) for x and y, we have the transformation T defined by

$$x = \frac{1}{3}(v - u) \quad \text{and} \quad y = \frac{1}{3}(2v + u).$$

Note that the transformation maps the four corners of the rectangle S to the vertices of the parallelogram R, as follows:

$$T(1, 2) = \left(\frac{1}{3}(2 - 1), \frac{1}{3}[2(2) + 1]\right) = \left(\frac{1}{3}, \frac{5}{3}\right),$$

$$T(3, 2) = \left(\frac{1}{3}(2 - 3), \frac{1}{3}[2(2) + 3]\right) = \left(-\frac{1}{3}, \frac{7}{3}\right),$$

$$T(1, 5) = \left(\frac{1}{3}(5 - 1), \frac{1}{3}[2(5) + 1]\right) = \left(\frac{4}{3}, \frac{11}{3}\right)$$

and

$$T(3, 5) = \left(\frac{1}{3}(5 - 3), \frac{1}{3}[2(5) + 3]\right) = \left(\frac{2}{3}, \frac{13}{3}\right).$$

We leave it as an exercise to verify that the above four points are indeed the vertices of the parallelogram R. (To do this, simply solve the system of equations for the points of intersection.) ■

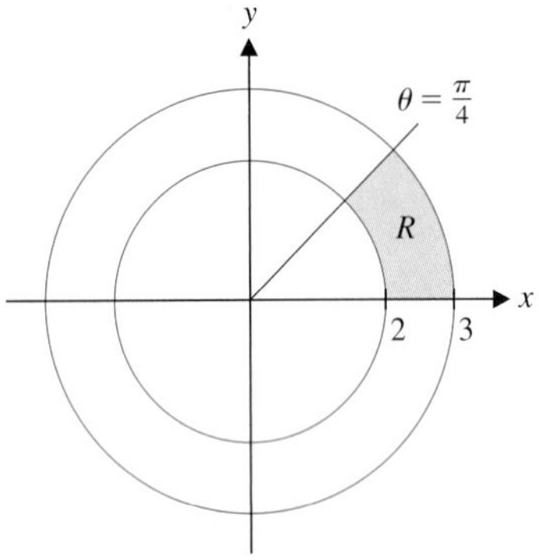

FIGURE 13.69a
The region R in the xy-plane

In example 8.2, we see how polar coordinates can be used to transform a rectangle in the $r\theta$-plane into a sector of a circular annulus.

EXAMPLE 8.2 A Transformation Involving Polar Coordinates

Let R be the region inside the circle $x^2 + y^2 = 9$ and outside the circle $x^2 + y^2 = 4$ and lying in the first quadrant between the lines $y = 0$ and $y = x$. Find a transformation T from a rectangular region S in the $r\theta$-plane to the region R.

Solution First, we picture the region R (a sector of a circular annulus) in Figure 13.69a. The obvious transformation is accomplished with polar coordinates. We let $x = r\cos\theta$ and $y = r\sin\theta$, so that $x^2 + y^2 = r^2$. The inner and outer circles forming a portion of the boundary of R then correspond to $r = 2$ and $r = 3$, respectively. Further, the line $y = x$ corresponds to the line $\theta = \frac{\pi}{4}$ and the line $y = 0$ corresponds to the line $\theta = 0$. We show the region S in Figure 13.69b. ■

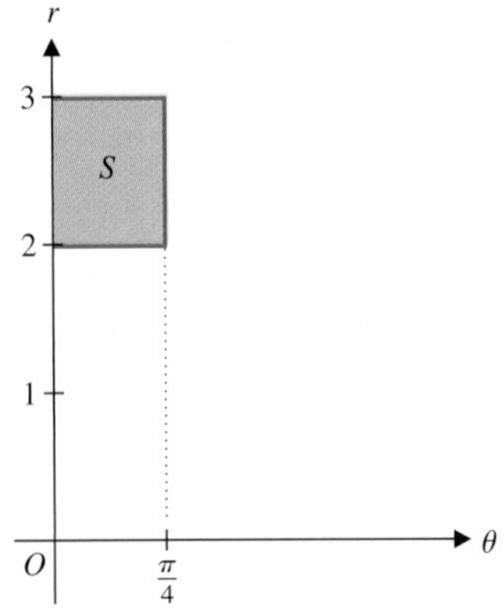

FIGURE 13.69b
The region S in the $r\theta$-plane

Now that we have introduced transformations, we consider our primary goal for this section: to determine how a change of variables in a multiple integral will affect the integral. We consider the double integral

$$\iint_R f(x, y)\, dA,$$

where f is continuous on R. Further, we assume that R is the image of a region S in the uv-plane under the one-to-one transformation T. Recall that we originally constructed the double integral by forming an inner partition of R and taking a limit of the corresponding Riemann sums. We now consider an inner partition of the region S in the uv-plane, consisting of the n rectangles $S_1, S_2, \ldots, S_n$, as depicted in Figure 13.70a. We denote the lower left

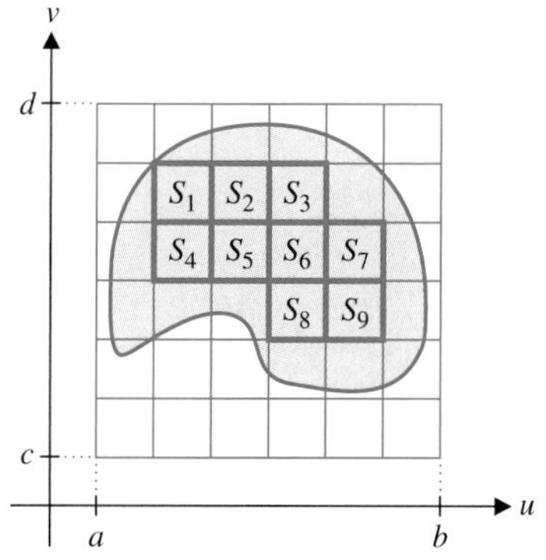

FIGURE 13.70a
An inner partition of the region S in the uv-plane

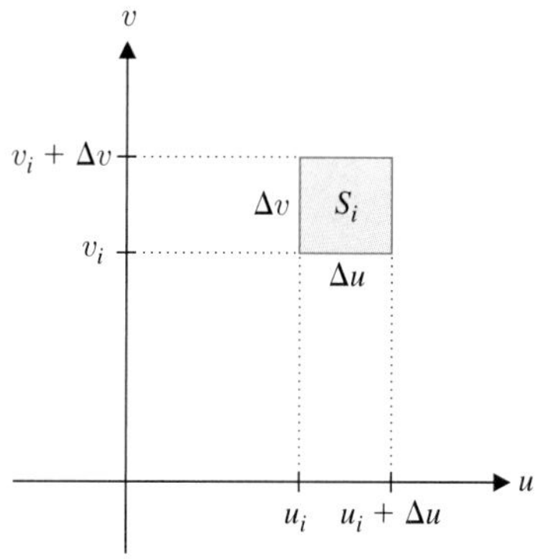

FIGURE 13.70b
The rectangle S_i

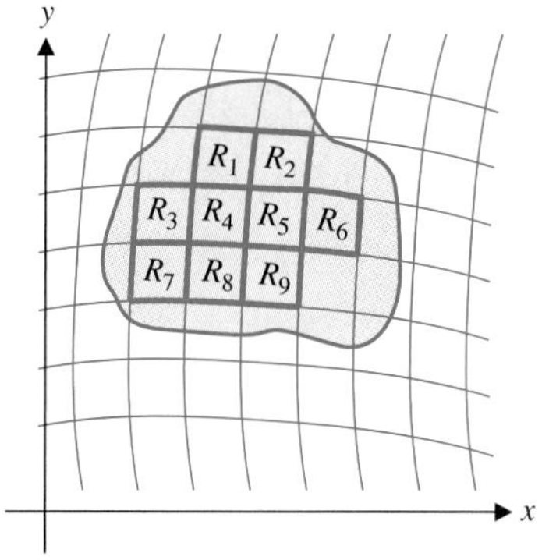

FIGURE 13.71
Curvilinear inner partition of the region R in the xy-plane

corner of each rectangle S_i by $(u_i, v_i)(i = 1, 2, \ldots, n)$ and take all of the rectangles to have the same dimensions Δu by Δv, as indicated in Figure 13.70b. Let $R_1, R_2, \ldots, R_n$ be the images of $S_1, S_2, \ldots, S_n$, respectively, under the transformation T and let the points $(x_1, y_1), (x_2, y_2), \ldots, (x_n, y_n)$ be the images of $(u_1, v_1), (u_2, v_2), \ldots, (u_n, v_n)$, respectively. Notice that $R_1, R_2, \ldots, R_n$ will then form an inner partition of the region R in the xy-plane (although it will not generally consist of rectangles), as indicated in Figure 13.71. In particular, the image of the rectangle S_i under T is the curvilinear region R_i. From our development of the double integral, we know that

$$\iint_R f(x, y)\, dA \approx \sum_{i=1}^{n} f(x_i, y_i)\, \Delta A_i, \tag{8.2}$$

where ΔA_i is the area of R_i, for $i = 1, 2, \ldots, n$. The only problem with this approximation is that we don't know how to find ΔA_i, since the regions R_i are not generally rectangles. We can, however, find a reasonable approximation, as follows.

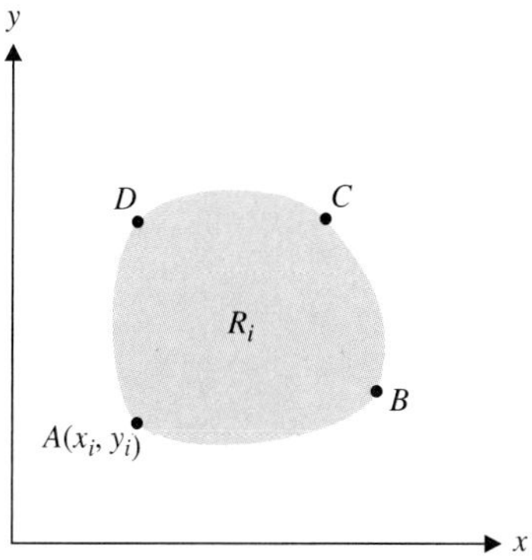

FIGURE 13.72a
The region R_i

Notice that T maps the four corners of S_i: $(u_i, v_i), (u_i + \Delta u, v_i), (u_i + \Delta u, v_i + \Delta v)$ and $(u_i, v_i + \Delta v)$ to four points denoted A, B, C and D, respectively, on the boundary of R_i, as indicated below:

$$(u_i, v_i) \xrightarrow{T} A(g(u_i, v_i), h(u_i, v_i)) = A(x_i, y_i),$$

$$(u_i + \Delta u, v_i) \xrightarrow{T} B(g(u_i + \Delta u, v_i), h(u_i + \Delta u, v_i)),$$

$$(u_i + \Delta u, v_i + \Delta v) \xrightarrow{T} C(g(u_i + \Delta u, v_i + \Delta v), h(u_i + \Delta u, v_i + \Delta v))$$

and

$$(u_i, v_i + \Delta v) \xrightarrow{T} D(g(u_i, v_i + \Delta v), h(u_i, v_i + \Delta v)).$$

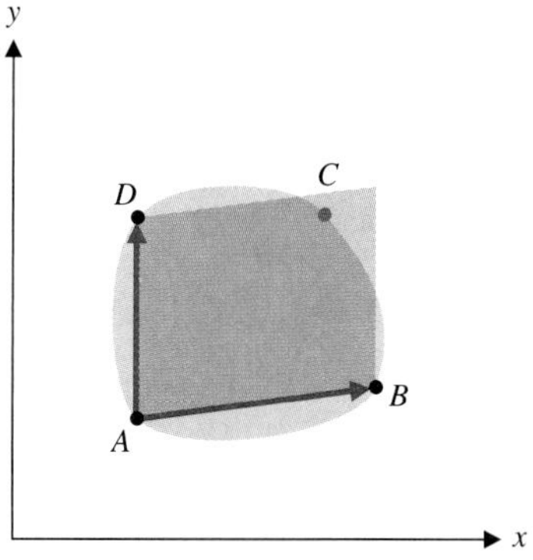

FIGURE 13.72b
The parallelogram determined by the vectors $\overrightarrow{AB}$ and $\overrightarrow{AD}$

We indicate these four points and a typical curvilinear region R_i in Figure 13.72a. Notice that as long as Δu and Δv are small, we can approximate the area of R_i by the area of the parallelogram determined by the vectors $\overrightarrow{AB}$ and $\overrightarrow{AD}$, as indicated in Figure 13.72b. If we consider $\overrightarrow{AB}$ and $\overrightarrow{AD}$ as three-dimensional vectors (with zero **k** components), recall from our discussion in section 10.4 that the area of the parallelogram is simply $\|\overrightarrow{AB} \times \overrightarrow{AD}\|$. We will take this as an approximation of the area ΔA_i. First, notice that

$$\overrightarrow{AB} = \langle g(u_i + \Delta u, v_i) - g(u_i, v_i), h(u_i + \Delta u, v_i) - h(u_i, v_i)\rangle \tag{8.3}$$

and

$$\overrightarrow{AD} = \langle g(u_i, v_i + \Delta v) - g(u_i, v_i), h(u_i, v_i + \Delta v) - h(u_i, v_i)\rangle. \tag{8.4}$$

From the definition of partial derivative, we have

$$g_u(u_i, v_i) = \lim_{\Delta u \to 0} \frac{g(u_i + \Delta u, v_i) - g(u_i, v_i)}{\Delta u}.$$

This tells us that for Δu small,

$$g(u_i + \Delta u, v_i) - g(u_i, v_i) \approx g_u(u_i, v_i)\,\Delta u.$$

Likewise, we have

$$h(u_i + \Delta u, v_i) - h(u_i, v_i) \approx h_u(u_i, v_i)\,\Delta u.$$

Similarly, for Δv small, we have

$$g(u_i, v_i + \Delta v) - g(u_i, v_i) \approx g_v(u_i, v_i)\,\Delta v$$

and

$$h(u_i, v_i + \Delta v) - h(u_i, v_i) \approx h_v(u_i, v_i)\,\Delta v.$$

Together with (8.3) and (8.4), these give us

$$\overrightarrow{AB} \approx \langle g_u(u_i, v_i)\,\Delta u, h_u(u_i, v_i)\,\Delta u\rangle = \Delta u \langle g_u(u_i, v_i), h_u(u_i, v_i)\rangle$$

and

$$\overrightarrow{AD} \approx \langle g_v(u_i, v_i)\,\Delta v, h_v(u_i, v_i)\,\Delta v\rangle = \Delta v \langle g_v(u_i, v_i), h_v(u_i, v_i)\rangle.$$

An approximation of the area of R_i is then given by

$$\Delta A_i \approx \|\overrightarrow{AB} \times \overrightarrow{AD}\|, \tag{8.5}$$

where

$$\overrightarrow{AB} \times \overrightarrow{AD} \approx \begin{vmatrix} \mathbf{i} & \mathbf{j} & \mathbf{k} \\ \Delta u g_u(u_i, v_i) & \Delta u h_u(u_i, v_i) & 0 \\ \Delta v g_v(u_i, v_i) & \Delta v h_v(u_i, v_i) & 0 \end{vmatrix}$$

$$= \begin{vmatrix} g_u(u_i, v_i) & h_u(u_i, v_i) \\ g_v(u_i, v_i) & h_v(u_i, v_i) \end{vmatrix} \Delta u\,\Delta v\,\mathbf{k}. \tag{8.6}$$

HISTORICAL NOTES

Carl Gustav Jacobi (1804–1851) German mathematician who made important advances in the functional determinant named for him. As a child, Jacobi was academically advanced beyond what the German educational system could offer and did much independent research. He obtained a university teaching post after converting from Judaism to Christianity. He was considered the world's top researcher on elliptic functions and made contributions to number theory and partial differential equations. An inspiring teacher, Jacobi and his students revived German mathematics.

For simplicity, we write the determinant as

$$\begin{vmatrix} g_u(u_i, v_i) & h_u(u_i, v_i) \\ g_v(u_i, v_i) & h_v(u_i, v_i) \end{vmatrix} = \begin{vmatrix} g_u(u_i, v_i) & g_v(u_i, v_i) \\ h_u(u_i, v_i) & h_v(u_i, v_i) \end{vmatrix} = \begin{vmatrix} \frac{\partial x}{\partial u} & \frac{\partial x}{\partial v} \\ \frac{\partial y}{\partial u} & \frac{\partial y}{\partial v} \end{vmatrix} (u_i, v_i).$$

We give this determinant a name and introduce some new notation in Definition 8.1.

DEFINITION 8.1

The determinant $\begin{vmatrix} \frac{\partial x}{\partial u} & \frac{\partial x}{\partial v} \\ \frac{\partial y}{\partial u} & \frac{\partial y}{\partial v} \end{vmatrix}$ is referred to as the **Jacobian** of the transformation T and is written using the notation $\dfrac{\partial(x, y)}{\partial(u, v)}$.

From (8.5) and (8.6), we now have (since $\mathbf{k}$ is a unit vector) that

$$\Delta A_i \approx \|\overrightarrow{AB} \times \overrightarrow{AD}\| = \left|\frac{\partial(x, y)}{\partial(u, v)}\right| \Delta u\,\Delta v,$$

where the determinant is evaluated at the point (u_i, v_i). From (8.2), we now have

$$\iint_R f(x,y)\,dA \approx \sum_{i=1}^n f(x_i,y_i)\,\Delta A_i \approx \sum_{i=1}^n f(x_i,y_i)\left|\frac{\partial(x,y)}{\partial(u,v)}\right| \Delta u\,\Delta v$$

$$= \sum_{i=1}^n f(g(u_i,v_i),h(u_i,v_i))\left|\frac{\partial(x,y)}{\partial(u,v)}\right| \Delta u\,\Delta v.$$

You should recognize this last expression as a Riemann sum for the double integral

$$\iint_S f(g(u,v),h(u,v))\left|\frac{\partial(x,y)}{\partial(u,v)}\right| du\,dv.$$

The preceding analysis is a sketch of the more extensive proof of Theorem 8.1.

THEOREM 8.1 (Change of Variables in Double Integrals)

Suppose that the region S in the uv-plane is mapped onto the region R in the xy-plane by the one-to-one transformation T defined by $x = g(u,v)$ and $y = h(u,v)$, where g and h have continuous first partial derivatives on S. If f is continuous on R and the Jacobian $\frac{\partial(x,y)}{\partial(u,v)}$ is nonzero on S, then

$$\iint_R f(x,y)\,dA = \iint_S f(g(u,v),\,h(u,v))\left|\frac{\partial(x,y)}{\partial(u,v)}\right| du\,dv.$$

We first observe that the change of variables to polar coordinates in a double integral is just a special case of Theorem 8.1.

EXAMPLE 8.3 Changing Variables to Polar Coordinates

Use Theorem 8.1 to derive the evaluation formula for polar coordinates ($r > 0$):

$$\iint_R f(x,y)\,dA = \iint_S f(r\cos\theta, r\sin\theta)\,r\,dr\,d\theta.$$

Solution First, recognize that a change of variables to polar coordinates consists of the transformation from the $r\theta$-plane to the xy-plane, defined by $x = r\cos\theta$ and $y = r\sin\theta$. This gives us the Jacobian

$$\frac{\partial(x,y)}{\partial(r,\theta)} = \begin{vmatrix} \frac{\partial x}{\partial r} & \frac{\partial x}{\partial \theta} \\ \frac{\partial y}{\partial r} & \frac{\partial y}{\partial \theta} \end{vmatrix} = \begin{vmatrix} \cos\theta & -r\sin\theta \\ \sin\theta & r\cos\theta \end{vmatrix} = r\cos^2\theta + r\sin^2\theta = r.$$

By Theorem 8.1, we now have the familiar formula

$$\iint_R f(x,y)\,dA = \iint_S f(r\cos\theta, r\sin\theta)\left|\frac{\partial(x,y)}{\partial(r,\theta)}\right| dr\,d\theta$$

$$= \iint_S f(r\cos\theta, r\sin\theta)\,r\,dr\,d\theta.$$

■

In example 8.4, we show how a change of variables can be used to simplify the region of integration (thereby also simplifying the integral).

EXAMPLE 8.4 Changing Variables to Transform a Region

Evaluate the integral $\iint_R (x^2 + 2xy)\,dA$, where R is the region bounded by the lines $y = 2x + 3$, $y = 2x + 1$, $y = 5 - x$ and $y = 2 - x$.

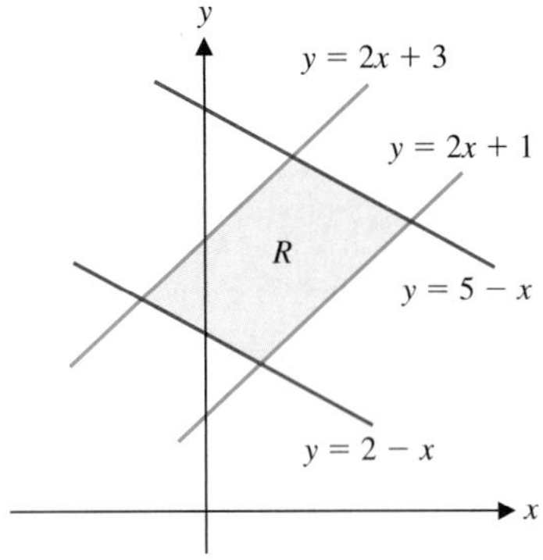

FIGURE 13.73
The region R

Solution The difficulty in evaluating this integral is that the region of integration (see Figure 13.73) requires us to break the integral into three pieces. (Think about this some!) An alternative is to find a change of variables corresponding to a transformation from a rectangle in the uv-plane to R in the xy-plane. Recall that we did just this in example 8.1. There, we had found that the change of variables

$$x = \frac{1}{3}(v - u) \quad \text{and} \quad y = \frac{1}{3}(2v + u)$$

maps the rectangle $S = \{(u, v) | 1 \le u \le 3 \text{ and } 2 \le v \le 5\}$ to R. Notice that the Jacobian of this transformation is

$$\frac{\partial(x, y)}{\partial(u, v)} = \begin{vmatrix} \frac{\partial x}{\partial u} & \frac{\partial x}{\partial v} \\ \frac{\partial y}{\partial u} & \frac{\partial y}{\partial v} \end{vmatrix} = \begin{vmatrix} -\frac{1}{3} & \frac{1}{3} \\ \frac{1}{3} & \frac{2}{3} \end{vmatrix} = -\frac{1}{3}.$$

By Theorem 8.1, we now have:

$$\begin{aligned} \iint_R (x^2 + 2xy)\,dA &= \iint_S \left[\frac{1}{9}(v - u)^2 + \frac{2}{9}(v - u)(2v + u)\right] \left|\frac{\partial(x, y)}{\partial(u, v)}\right| du\,dv \\ &= \frac{1}{27}\int_2^5 \int_1^3 [(v - u)^2 + 2(2v^2 - uv - u^2)]\,du\,dv \\ &= \frac{196}{27}, \end{aligned}$$

where we leave the calculation of the final (routine) iterated integral to you. ■

Recall that for single definite integrals, we often must introduce a change of variable in order to find an antiderivative for the integrand. This is also the case in double integrals, as we see in example 8.5.

EXAMPLE 8.5 A Change of Variables Required to Find an Antiderivative

Evaluate the double integral $\iint_R \frac{e^{x-y}}{x + y}\,dA$, where R is the rectangle bounded by the lines $y = x$, $y = x + 5$, $y = 2 - x$ and $y = 4 - x$.

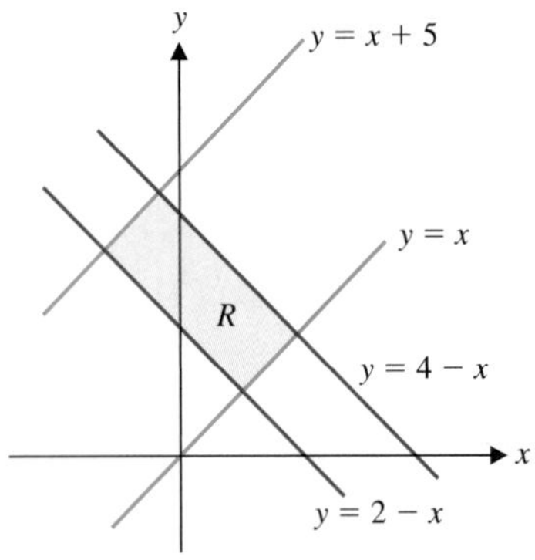

FIGURE 13.74a
The region R

Solution First, notice that although the region over which you are to integrate is simply a rectangle in the xy-plane, its sides are not parallel to the x- and y-axes. (See Figure 13.74a.) This is the least of your problems right now, though. If you look carefully at the integrand, you'll recognize that you do not know an antiderivative for this integrand (no matter which variable you integrate with respect to first). A straightforward change of variables is to let $u = x - y$ and $v = x + y$. Solving these equations for x and y gives us

$$x = \frac{1}{2}(u + v) \quad \text{and} \quad y = \frac{1}{2}(v - u). \tag{8.7}$$

The Jacobian of this transformation is then

$$\frac{\partial(x, y)}{\partial(u, v)} = \begin{vmatrix} \frac{\partial x}{\partial u} & \frac{\partial x}{\partial v} \\ \frac{\partial y}{\partial u} & \frac{\partial y}{\partial v} \end{vmatrix} = \begin{vmatrix} \frac{1}{2} & \frac{1}{2} \\ -\frac{1}{2} & \frac{1}{2} \end{vmatrix} = \frac{1}{2}.$$

The next issue is to find the region S in the uv-plane that is mapped onto the region R in the xy-plane by this transformation. Remember that the boundary curves of the region S are mapped to the boundary curves of R. From (8.7), we have that $y = x$ corresponds to

$$\frac{1}{2}(v - u) = \frac{1}{2}(u + v) \quad \text{or} \quad u = 0.$$

Likewise, $y = x + 5$ corresponds to

$$\frac{1}{2}(v - u) = \frac{1}{2}(u + v) + 5 \quad \text{or} \quad u = -5,$$

$y = 2 - x$ corresponds to

$$\frac{1}{2}(v - u) = 2 - \frac{1}{2}(u + v) \quad \text{or} \quad v = 2$$

and $y = 4 - x$ corresponds to

$$\frac{1}{2}(v - u) = 4 - \frac{1}{2}(u + v) \quad \text{or} \quad v = 4.$$

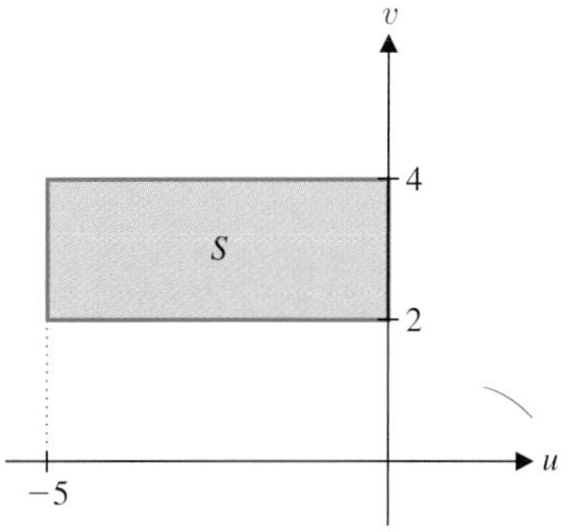

FIGURE 13.74b
The region S

This says that the region S in the uv-plane corresponding to the region R in the xy-plane is the rectangle

$$S = \{(u, v) | -5 \le u \le 0 \text{ and } 2 \le v \le 4\},$$

as indicated in Figure 13.74b. You can now easily read off the limits of integration in the uv-plane. By Theorem 8.1, we have

$$\begin{aligned} \iint_R \frac{e^{x-y}}{x+y}\, dA &= \iint_S \frac{e^u}{v} \left| \frac{\partial(x, y)}{\partial(u, v)} \right| du\, dv = \frac{1}{2} \int_2^4 \int_{-5}^0 \frac{e^u}{v}\, du\, dv \\ &= \frac{1}{2} \int_2^4 \frac{1}{v} e^u \Big|_{u=-5}^{u=0} dv = \frac{1}{2}(e^0 - e^{-5}) \int_2^4 \frac{1}{v}\, dv \\ &= \frac{1}{2}(1 - e^{-5}) \ln|v| \Big|_{v=2}^{v=4} = \frac{1}{2}(1 - e^{-5})(\ln 4 - \ln 2) \\ &\approx 0.34424. \blacksquare \end{aligned}$$

Much as we have now done in two dimensions, we can develop a change of variables formula for triple integrals. The proof of Theorem 8.2 can be found in most texts on advanced calculus. We first define the Jacobian of a transformation in three dimensions.

For a transformation T from a region S of uvw-space onto a region R in xyz-space, defined by $x = g(u, v, w)$, $y = h(u, v, w)$ and $z = l(u, v, w)$, the **Jacobian** of the transformation is the determinant $\frac{\partial(x, y, z)}{\partial(u, v, w)}$ defined by

$$\frac{\partial(x, y, z)}{\partial(u, v, w)} = \begin{vmatrix} \frac{\partial x}{\partial u} & \frac{\partial x}{\partial v} & \frac{\partial x}{\partial w} \\ \frac{\partial y}{\partial u} & \frac{\partial y}{\partial v} & \frac{\partial y}{\partial w} \\ \frac{\partial z}{\partial u} & \frac{\partial z}{\partial v} & \frac{\partial z}{\partial w} \end{vmatrix}.$$

Theorem 8.2 presents a result for triple integrals that corresponds to Theorem 8.1.

THEOREM 8.2 (Change of Variables in Triple Integrals)

Suppose that the region S in uvw-space is mapped onto the region R in xyz-space by the one-to-one transformation T defined by $x = g(u, v, w)$, $y = h(u, v, w)$ and $z = l(u, v, w)$, where g, h and l have continuous first partial derivatives in S. If f is continuous in R and the Jacobian $\frac{\partial(x, y, z)}{\partial(u, v, w)}$ is nonzero in S, then

$$\iiint_R f(x, y, z)\, dV = \iiint_S f(g(u,v,w), h(u,v,w), l(u,v,w)) \left| \frac{\partial(x, y, z)}{\partial(u, v, w)} \right| du\, dv\, dw.$$

We introduce a change of variables in a triple integral for precisely the same reasons as we do in double integrals: in order to simplify the integrand or the region of integration or both. In example 8.6, we use Theorem 8.2 to derive the change of variables formula for the conversion from rectangular to spherical coordinates and see that this is simply a special case of the general change of variables process given in Theorem 8.2.

EXAMPLE 8.6 Deriving the Evaluation Formula for Spherical Coordinates

Use Theorem 8.2 to derive the evaluation formula for triple integrals in spherical coordinates:

$$\iiint_R f(x, y, z)\, dV = \iiint_S f(\rho \sin\phi \cos\theta, \rho \sin\phi \sin\theta, \rho \cos\phi)\rho^2 \sin\phi\, d\rho\, d\phi\, d\theta.$$

Solution Suppose that the region R in xyz-space is the image of the region S in $\rho\phi\theta$-space under the transformation T defined by the change to spherical coordinates. Recall that we have

$$x = \rho \sin\phi \cos\theta, \quad y = \rho \sin\phi \sin\theta \quad \text{and} \quad z = \rho \cos\phi.$$

The Jacobian of this transformation is then

$$\frac{\partial(x, y, z)}{\partial(\rho, \phi, \theta)} = \begin{vmatrix} \frac{\partial x}{\partial \rho} & \frac{\partial x}{\partial \phi} & \frac{\partial x}{\partial \theta} \\ \frac{\partial y}{\partial \rho} & \frac{\partial y}{\partial \phi} & \frac{\partial y}{\partial \theta} \\ \frac{\partial z}{\partial \rho} & \frac{\partial z}{\partial \phi} & \frac{\partial z}{\partial \theta} \end{vmatrix} = \begin{vmatrix} \sin\phi \cos\theta & \rho \cos\phi \cos\theta & -\rho \sin\phi \sin\theta \\ \sin\phi \sin\theta & \rho \cos\phi \sin\theta & \rho \sin\phi \cos\theta \\ \cos\phi & -\rho \sin\phi & 0 \end{vmatrix}.$$

For the sake of convenience, we expand this determinant along to the third row, rather than the first row. This gives us

$$\begin{aligned}\frac{\partial(x,y,z)}{\partial(\rho,\phi,\theta)} &= \cos\phi\begin{vmatrix}\rho\cos\phi\cos\theta & -\rho\sin\phi\sin\theta\\ \rho\cos\phi\sin\theta & \rho\sin\phi\cos\theta\end{vmatrix}\\ &\quad+\rho\sin\phi\begin{vmatrix}\sin\phi\cos\theta & -\rho\sin\phi\sin\theta\\ \sin\phi\sin\theta & \rho\sin\phi\cos\theta\end{vmatrix}\\ &= \cos\phi\,(\rho^2\sin\phi\cos\phi\cos^2\theta+\rho^2\sin\phi\cos\phi\sin^2\theta)\\ &\quad+\rho\sin\phi\,(\rho\sin^2\phi\cos^2\theta+\rho\sin^2\phi\sin^2\theta)\\ &= \rho^2\sin\phi\cos^2\phi+\rho^2\sin^3\phi=\rho^2\sin\phi\,(\cos^2\phi+\sin^2\phi)\\ &= \rho^2\sin\phi.\end{aligned}$$

From Theorem 8.2, we now have that

$$\begin{aligned}\iiint_R f(x,y,z)\,dV &= \iiint_S f(\rho\sin\phi\cos\theta,\rho\sin\phi\sin\theta,\rho\cos\phi)\left|\frac{\partial(x,y,z)}{\partial(\rho,\phi,\theta)}\right|d\rho\,d\phi\,d\theta\\ &= \iiint_S f(\rho\sin\phi\cos\theta,\rho\sin\phi\sin\theta,\rho\cos\phi)\rho^2\sin\phi\,d\rho\,d\phi\,d\theta,\end{aligned}$$

where we have used the fact that $0\le\phi\le\pi$ to write $|\sin\phi|=\sin\phi$. Notice that this is the same evaluation formula that we developed in section 13.7. ■

EXERCISES 13.8

WRITING EXERCISES

1. Explain what is meant by a "rectangular region" in the uv-plane. In particular, explain what is rectangular about the polar region $1\le r\le 2$ and $0\le\theta\le\pi$.
2. The order of variables in the Jacobian is not important in the sense that $\left|\frac{\partial(x,y)}{\partial(v,u)}\right|=\left|\frac{\partial(x,y)}{\partial(u,v)}\right|$ but the order is very important in the sense that $\left|\frac{\partial(x,y)}{\partial(u,v)}\right|\ne\left|\frac{\partial(u,v)}{\partial(x,y)}\right|$. Give a geometric explanation of why $\left|\frac{\partial(x,y)}{\partial(u,v)}\right|\left|\frac{\partial(u,v)}{\partial(x,y)}\right|=1$.

In exercises 1–12, find a transformation from a rectangular region S in the uv-plane to the region R.

1. R is bounded by $y=4x+2$, $y=4x+5$, $y=3-2x$ and $y=1-2x$
2. R is bounded by $y=2x-1$, $y=2x+5$, $y=1-3x$ and $y=-1-3x$
3. R is bounded by $y=1-3x$, $y=3-3x$, $y=x-1$ and $y=x-3$
4. R is bounded by $y=2x-1$, $y=2x+1$, $y=3$ and $y=1$
5. R is inside $x^2+y^2=4$, outside $x^2+y^2=1$ and in the first quadrant
6. R is inside $x^2+y^2=4$, outside $x^2+y^2=1$ and in the first quadrant between $y=x$ and $x=0$
7. R is inside $x^2+y^2=9$, outside $x^2+y^2=4$ and between $y=x$ and $y=-x$ with $y\ge 0$
8. R is inside $x^2+y^2=9$ with $x\ge 0$
9. R is bounded by $y=x^2$, $y=x^2+2$, $y=4-x^2$ and $y=2-x^2$ with $x\ge 0$
10. R is bounded by $y=x^2$, $y=x^2+2$, $y=3-x^2$ and $y=2-x^2$ with $x\le 0$
11. R is bounded by $y=e^x$, $y=e^x+1$, $y=3-e^x$ and $y=5-e^x$
12. R is bounded by $y=2x^2+1$, $y=2x^2+3$, $y=2-x^2$ and $y=4-x^2$ with $x\ge 0$

In exercises 13–22, evaluate the double integral.

13. $\iint_R (y-4x)\,dA$, where R is given in exercise 1.
14. $\iint_R (y+3x)\,dA$, where R is given in exercise 2.

15. $\iint_R (y+3x)^2\,dA$, where R is given in exercise 3.

16. $\iint_R e^{y-x}\,dA$, where R is given in exercise 4.

17. $\iint_R x\,dA$, where R is given in exercise 5.

18. $\iint_R e^{y-e^x}\,dA$, where R is given in exercise 11.

19. $\displaystyle\iint_R \frac{e^{y-4x}}{y+2x}\,dA$, where R is given in exercise 1.

20. $\displaystyle\iint_R \frac{e^{y+3x}}{y-2x}\,dA$, where R is given in exercise 2.

21. $\iint_R (x+y)\,dA$, where R is given in exercise 1.

22. $\iint_R (x+2y)\,dA$, where R is given in exercise 2.

In exercises 23–26, find the Jacobian of the given transformation.

23. $x = ue^v,\ y = ue^{-v}$

24. $x = 2uv,\ y = 3u - v$

25. $x = u/v,\ y = v^2$

26. $x = 4u + v^2,\ y = 2uv$

In exercises 27 and 28, find a transformation from a (three-dimensional) rectangular region *S* in *uvw*-space to the solid *Q*.

27. Q is bounded by $x+y+z=1$, $x+y+z=2$, $x+2y=0$, $x+2y=1$, $y+z=2$ and $y+z=4$.

28. Q is bounded by $x+z=1$, $x+z=2$, $2y+3z=0$, $2y+3z=1$, $y+2z=2$ and $y+2z=4$.

In exercises 29 and 30, find the volume of the given solid.

29. Q in exercise 27

30. Q in exercise 28

31. In Theorem 8.1, we required that the Jacobian be nonzero. To see why this is necessary, consider a transformation where $x = u - v$ and $y = 2v - 2u$. Show that the Jacobian is zero. Then try solving for u and v.

32. Compute the Jacobian for the spherical-like transformation $x = \rho\sin\phi$, $y = \rho\cos\phi\cos\theta$ and $z = \rho\cos\phi\sin\theta$.

33. The integral $\displaystyle\int_0^1\int_0^1 \frac{1}{1-(xy)^2}\,dx\,dy$ arises in the study of the Riemann-zeta function. Use the transformation $x = \dfrac{\sin u}{\cos v}$ and $y = \dfrac{\sin v}{\cos u}$ to write this integral in the form $\displaystyle\int_0^{\pi/2}\int_0^{\pi/2-v} f(u,v)\,du\,dv$ and then evaluate the integral.

34. Show that the transformation $x = \dfrac{\sin u}{\cos v}$ and $y = \dfrac{\sin v}{\cos u}$ in exercise 33 transforms the square $0 \le x \le 1$, $0 \le y \le 1$ into the triangle $0 \le u \le \frac{\pi}{2} - v$, $0 \le v \le \frac{\pi}{2}$. (Hint: Transform each side of the square separately.)

EXPLORATORY EXERCISES

1. Transformations are involved in many important applications of mathematics. The **direct linear transformation** discussed in this exercise was used by Titleist golf researchers Gobush, Pelletier and Days to study the motion of golf balls (see *Science and Golf II,* 1996). Bright dots are drawn onto golf balls. The dots are tracked by a pair of cameras as the ball is hit. The challenge is to use this information to reconstruct the exact position of the ball at various times, allowing the researchers to estimate the speed, spin rate and launch angle of the ball. In the direct linear transformation model developed by Abdel-Aziz and Karara, a dot at actual position (x, y, z) will appear at pixel (u_1, v_1) of camera 1's digitized image where

$$u_1 = \frac{c_{11}x + c_{21}y + c_{31}z + c_{41}}{d_{11}x + d_{21}y + d_{31}z + 1} \quad \text{and}$$

$$v_1 = \frac{c_{51}x + c_{61}y + c_{71}z + c_{81}}{d_{11}x + d_{21}y + d_{31}z + 1},$$

for constants $c_{11}, c_{21}, \ldots, c_{81}$ and d_{11}, d_{21} and d_{31}. Similarly, camera 2 "sees" this dot at pixel (u_2, v_2) where

$$u_2 = \frac{c_{12}x + c_{22}y + c_{32}z + c_{42}}{d_{12}x + d_{22}y + d_{32}z + 1} \quad \text{and}$$

$$v_2 = \frac{c_{52}x + c_{62}y + c_{72}z + c_{82}}{d_{12}x + d_{22}y + d_{32}z + 1},$$

for a different set of constants $c_{12}, c_{22}, \ldots, c_{82}$ and d_{12}, d_{22} and d_{32}. The constants are determined by taking a series of measurements of motionless balls to calibrate the model. Given that the model for each camera consists of eleven constants, explain why in theory, six different measurements would more than suffice to determine the constants. In reality, more measurements are taken and a least-squares criterion is used to find the best fit of the model to the data. Suppose that this procedure gives us the model

$$u_1 = \frac{2x+y+z+1}{x+y+2z+1}, \quad v_1 = \frac{3x+z}{x+y+2z+1},$$

$$u_2 = \frac{x+z+6}{2x+3z+1}, \quad v_2 = \frac{4x+y+3}{2x+3z+1}.$$

If the screen coordinates of a dot are $(u_1, v_1) = (0, -3)$ and $(u_2, v_2) = (5, 0)$, solve for the actual position (x, y, z) of the dot. Actually, a dot would not show up as a single pixel, but as a somewhat blurred image over several pixels. The dot is officially located at the pixel nearest the center of mass of the pixels involved. Suppose that a dot's image activates the following pixels: (34, 42), (35, 42), (32, 41), (33, 41), (34, 41), (35, 41), (36, 41), (34, 40), (35, 40), (36, 40) and (36, 39). Find the center of mass of these pixels and round off to determine the "location" of the dot.

Review Exercises

WRITING EXERCISES

The following list includes terms that are defined and theorems that are stated in this chapter. For each term or theorem, (1) give a precise definition or statement, (2) state in general terms what it means and (3) describe the types of problems with which it is associated.

Irregular partition	Definite integral	Double integral
Fubini's Theorem	Double Riemann sum	Volume
Center of mass	First moment	Moment of inertia
Surface area	Triple integral	Mass
Cylindrical coordinates	Spherical coordinates	Rectangular coordinates
Jacobian	Transformation	

TRUE OR FALSE

State whether each statement is true or false and briefly explain why. If the statement is false, try to "fix it" by modifying the given statement to a new statement that is true.

1. When using a double integral to compute volume, the choice of integration variables and order is determined by the geometry of the region.
2. $\iint_R f(x, y)\, dA$ gives the volume between $z = f(x, y)$ and the xy-plane.
3. When using a double integral to compute area, the choice of integration variables and order is determined by the geometry of the region.
4. A line through the center of mass of a region divides the region into subregions of equal area.
5. If R is bounded by a circle, $\iint_R f(x, y)\, dA$ should be computed using polar coordinates.
6. The surface area of a region is approximately equal to the area of the projection of the region into the xy-plane.
7. A triple integral in rectangular coordinates has three possible orders of integration.
8. The choice of coordinate systems for a triple integral is determined by the function being integrated.
9. If a region or a function involves $x^2 + y^2$, you should use cylindrical coordinates.
10. For a triple integral in spherical coordinates, the order of integration does not matter.
11. Transforming a double integral in xy-coordinates to one in uv-coordinates, you need formulas for u and v in terms of x and y.

In exercises 1 and 2, compute the Riemann sum for the given function and region, a partition with n equal-sized rectangles and the given evaluation rule.

1. $f(x, y) = 5x - 2y$, $1 \le x \le 3$, $0 \le y \le 1$, $n = 4$, evaluate at midpoint
2. $f(x, y) = 4x^2 + y$, $0 \le x \le 1$, $1 \le y \le 3$, $n = 4$, evaluate at midpoint

In exercises 3–10, evaluate the double integral.

3. $\iint_R (4x + 9x^2y^2)\, dA$, where $R = \{(x, y) \mid 0 \le x \le 3, 1 \le y \le 2\}$
4. $\iint_R 2e^{4x+2y}\, dA$, where $R = \{(x, y) \mid 0 \le x \le 1, 0 \le y \le 1\}$
5. $\iint_R e^{-x^2-y^2}\, dA$, where $R = \{(x, y) \mid 1 \le x^2 + y^2 \le 4\}$
6. $\iint_R 2xy\, dA$, where R is bounded by $y = x$, $y = 2 - x$ and $y = 0$
7. $\displaystyle\int_{-1}^{1} \int_{x^2}^{2x} (2xy - 1)\, dy\, dx$
8. $\displaystyle\int_{0}^{1} \int_{2x}^{2} (3y^2x + 4)\, dy\, dx$

Review Exercises

9. $\iint_R xy\,dA$, where R is bounded by $r = 2\cos\theta$

10. $\iint_R \sin(x^2 + y^2)\,dA$, where R is bounded by $x^2 + y^2 = 4$

In exercises 11 and 12, approximate the double integral.

11. $\iint_R 4xy\,dA$, where R is bounded by $y = x^2 - 4$ and $y = \ln x$

12. $\iint_R 6x^2y\,dA$, where R is bounded by $y = \cos x$ and $y = x^2 - 1$

In exercises 13–24, compute the volume of the solid.

13. Bounded by $z = 1 - x^2$, $z = 0$, $y = 0$ and $y = 1$

14. Bounded by $z = 4 - x^2 - y^2$, $z = 0$, $x = 0$, $x + y = 1$ and $y = 0$

15. Between $z = x^2 + y^2$ and $z = 8 - x^2 - y^2$

16. Under $z = e^{\sqrt{x^2+y^2}}$ and inside $x^2 + y^2 = 4$

17. Bounded by $x + 2y + z = 8$ and the coordinate planes

18. Bounded by $x + 5y + 7z = 1$ and the coordinate planes

19. Bounded by $z = \sqrt{x^2 + y^2}$ and $z = 4$

20. Bounded by $x = \sqrt{y^2 + z^2}$ and $x = 2$

21. Between $z = \sqrt{x^2 + y^2}$ and $x^2 + y^2 + z^2 = 4$

22. Inside $x^2 + y^2 + z^2 = 4z$ and below $z = 1$

23. Under $z = 6 - x^2 - y^2$, above $z = 0$ and inside $x^2 + y^2 = 1$

24. Under $z = x$ and inside $r = \cos\theta$

In exercises 25 and 26, change the order of integration.

25. $\displaystyle\int_0^2 \int_0^{x^2} f(x, y)\,dy\,dx$

26. $\displaystyle\int_0^2 \int_{x^2}^4 f(x, y)\,dy\,dx$

In exercises 27 and 28, convert to polar coordinates and evaluate the integral.

27. $\displaystyle\int_0^2 \int_{-\sqrt{4-x^2}}^{\sqrt{4-x^2}} 2x\,dy\,dx$

28. $\displaystyle\int_0^2 \int_0^{\sqrt{4-x^2}} 2\sqrt{x^2 + y^2}\,dy\,dx$

In exercises 29–32, find the mass and center of mass.

29. The lamina bounded by $y = 2x$, $y = x$ and $x = 2$, $\rho(x, y) = 2x$

30. The lamina bounded by $y = x$, $y = 4 - x$ and $y = 0$, $\rho(x, y) = 2y$

31. The solid bounded by $z = 1 - x^2$, $z = 0$, $y = 0$, $y + z = 2$, $\rho(x, y, z) = 2$

32. The solid bounded by $x = \sqrt{y^2 + z^2}$, $x = 2$, $\rho(x, y, z) = 3x$

In exercises 33 and 34, use a double integral to find the area.

33. Bounded by $y = x^2$, $y = 2 - x$ and $y = 0$

34. One leaf of $r = \sin 4\theta$

In exercises 35 and 36, find the average value of the function on the indicated region.

35. $f(x, y) = x^2$, region bounded by $y = 2x$, $y = x$ and $x = 1$

36. $f(x, y) = \sqrt{x^2 + y^2}$, region bounded by $x^2 + y^2 = 1$, $x = 0$, $y = 0$

In exercises 37–42, evaluate or estimate the surface area.

37. The portion of $z = 2x + 4y$ between $y = x$, $y = 2$ and $x = 0$

38. The portion of $z = x^2 + 6y$ between $y = x^2$ and $y = 4$

39. The portion of $z = xy$ inside $x^2 + y^2 = 8$, in the first octant

40. The portion of $z = \sin(x^2 + y^2)$ inside $x^2 + y^2 = \pi$

41. The portion of $z = \sqrt{x^2 + y^2}$ below $z = 4$

42. The portion of $x + 2y + 3z = 6$ in the first octant

In exercises 43–50, set up the triple integral $\iiint_Q f(x, y, z)\,dV$ in an appropriate coordinate system. If $f(x, y, z)$ is given, evaluate the integral.

43. $f(x, y, z) = z(x + y)$, $Q = \{(x, y, z) | 0 \le x \le 2, -1 \le y \le 1, -1 \le z \le 1\}$

44. $f(x, y, z) = 2xye^{yz}$, $Q = \{(x, y, z) | 0 \le x \le 2, 0 \le y \le 1, 0 \le z \le 1\}$

45. $f(x, y, z) = \sqrt{x^2 + y^2 + z^2}$, Q is above $z = \sqrt{x^2 + y^2}$ and below $x^2 + y^2 + z^2 = 4$.

46. $f(x, y, z) = 3x$, Q is the region below $z = \sqrt{x^2 + y^2}$, above $z = 0$ and inside $x^2 + y^2 = 4$.

Review Exercises

47. Q is bounded by $x + y + z = 6, z = 0, y = x, y = 2$ and $x = 0$.

48. Q is the region below $z = \sqrt{4 - x^2 - y^2}$, above $z = 0$ and inside $x^2 + y^2 = 1$.

49. Q is the region below $z = \sqrt{4 - x^2 - y^2}$ and above $z = 0$.

50. Q is the region below $z = 6 - x - y$, above $z = 0$ and inside $x^2 + y^2 = 8$.

In exercises 51–54, evaluate the integral after changing coordinate systems.

51. $\int_0^1 \int_x^{\sqrt{2-x^2}} \int_0^{\sqrt{x^2+y^2}} e^z \, dz \, dy \, dx$

52. $\int_0^{\sqrt{2}} \int_y^{\sqrt{4-y^2}} \int_0^2 4z \, dz \, dx \, dy$

53. $\int_{-1}^1 \int_0^{\sqrt{1-x^2}} \int_{\sqrt{x^2+y^2}}^{\sqrt{2-x^2-y^2}} \sqrt{x^2 + y^2 + z^2} \, dz \, dy \, dx$

54. $\int_{-2}^2 \int_0^{\sqrt{4-y^2}} \int_0^{\sqrt{4-x^2-y^2}} dz \, dx \, dy$

In exercises 55–60, write the given equation in (a) cylindrical and (b) spherical coordinates.

55. $y = 3$ **56.** $x^2 + y^2 = 9$

57. $x^2 + y^2 + z^2 = 4$ **58.** $y = x$

59. $z = \sqrt{x^2 + y^2}$ **60.** $z = 4$

In exercises 61–66, sketch the graph.

61. $r = 4$ **62.** $\rho = 4$ **63.** $\theta = \frac{\pi}{4}$

64. $\phi = \frac{\pi}{4}$ **65.** $r = 2\cos\theta$ **66.** $\rho = 2\sec\phi$

In exercises 67 and 68, find a transformation from a rectangular region S in the uv-plane to the region R.

67. R bounded by $y = 2x - 1, y = 2x + 1, y = 2 - 2x$ and $y = 4 - 2x$

68. R inside $x^2 + y^2 = 9$, outside $x^2 + y^2 = 4$ and in the second quadrant

In exercises 69 and 70, evaluate the double integral.

69. $\iint_R e^{y-2x} dA$, where R is given in exercise 67

70. $\iint_R (y + 2x)^3 \, dA$, where R is given in exercise 67

In exercises 71 and 72, find the Jacobian of the given transformation.

71. $x = u^2 v, y = 4u + v^2$

72. $x = 4u - 5v, y = 2u + 3v$

EXPLORATORY EXERCISES

1. Let S be a sphere of radius R centered at the origin. Different types of symmetry produce different simplifications in integration. If $f(x, y, z) = -f(-x, y, z)$, show that $\iiint_S f dV = 0$. If $f(x, y, z) = -f(x, -y, z)$, show that $\iiint_S f dV = 0$. If $f(x, y, z) = -f(-x, -y, z)$, show that $\iiint_S f dV = 0$. If $f(x, y, z) = -f(-x, -y, -z)$, what, if anything, can be said about $\iiint_S f \, dV$? Next, suppose that $f(a, b, c) = f(x, y, z)$ whenever $a^2 + b^2 + c^2 = x^2 + y^2 + z^2$. Show that $\iiint_S f(x, y, z) \, dV = 4\pi \int_0^R \rho^2 g(\rho) \, d\rho$ for some function g. (State the relationship between g and f.) If f can be written in cylindrical coordinates as $f(r, \theta, z) = g(r)$, for some continous function g, simplify $\iiint_S f dV$ as much as possible.

2. Let S be the solid bounded above by $z = \sqrt{R^2 - x^2 - y^2}$ and below by $z = \sqrt{x^2 + y^2}$. Set up and simplify as much as possible $\iiint_S f dV$ in each of the following cases: (a) $f(x, y, z) = -f(-x, y, z)$; (b) $f(x, y, z) = -f(x, -y, z)$; (c) $f(x, y, z) = -f(-x, -y, z)$; (d) $f(x, y, z) = -f(-x, -y, -z)$; (e) $f(r, \theta, z) = g(r)$, for some function g and (f) $f(\rho, \phi, \theta) = g(\rho)$, for some function g.

CHAPTER 14

Vector Calculus

The Volkswagen Beetle was one of the most beloved and recognizable cars of the 1950s, 1960s and 1970s. So, Volkswagen's decision to release a redesigned Beetle in 1998 created quite a stir in the automotive world. The new Beetle resembles the classic Beetle, but has been modernized to improve gas mileage, safety, handling and overall performance. The calculus that we introduce in this chapter will provide you with some of the basic tools necessary for designing and analyzing automobiles, aircraft and other types of complex machinery.

Think about how you might redesign an automobile to improve its aerodynamic performance. Engineers have identified many important principles of aerodynamics, but the design of a complicated structure like a car still has an element of trial and error. Before high-speed computers were available, engineers built small-scale or full-scale models of new designs and tested them in a wind tunnel. Unfortunately, such models don't always provide adequate information and can be prohibitively expensive to build, particularly if you have 20 or 30 new ideas you'd like to try.

The old Beetle

The new Beetle

With modern computers, wind tunnel tests can be accurately simulated by sophisticated programs. Mathematical models give engineers the ability to thoroughly test anything from minor modifications to radical changes.

The calculus that goes into a computer simulation of a wind tunnel is beyond what you've seen so far. Such simulations must keep track of the air velocity at each point on and around a car. A function assigning a vector (e.g., a velocity vector) to each point in space is called a vector field, which we introduce in section 14.1. To determine where vortices and turbulence occur in a fluid flow, you must compute line integrals, which are discussed in sections 14.2 and 14.3. The curl and divergence, introduced in section 14.5, allow you to analyze the rotational and linear properties of a fluid flow. Other properties of three-dimensional objects, such as mass and moments of inertia for a thin shell (such as a dome of a building), require the evaluation of surface integrals, which we develop in section 14.6. The relationships among line integrals, surface integrals, double integrals and triple integrals are explored in the remaining sections of the chapter.

In the case of the redesigned Volkswagen Beetle, computer simulations resulted in numerous improvements over the original. One measure of a vehicle's aerodynamic efficiency is its drag coefficient. Without getting into the technicalities, the lower its drag coefficient is, the less the velocity of the car is reduced by air resistance. The original Beetle has a drag coefficient of 0.46 (as reported by Robertson and Crowe in *Engineering Fluid Mechanics*). By comparison, a low-slung (and quite aerodynamic) 1985 Chevrolet Corvette has a drag coefficient of 0.34. Volkswagen's specification sheet for the new Beetle lists a drag coefficient of 0.38, representing a considerable reduction in air drag from the original Beetle. Through careful mathematical analysis, Volkswagen improved the performance of the Beetle while retaining the distinctive shape of the original car.

14.1 VECTOR FIELDS

To analyze the flight characteristics of an airplane, engineers use wind tunnel tests to provide information about the flow of air over the wings and around the fuselage. As you can imagine, to model such a test mathematically, we need to be able to describe the velocity of the air at various points throughout the tunnel. So, we need to define a function that assigns a vector to each point in space. Such a function would have both a multidimensional domain (like the functions of Chapters 12 and 13) and a multidimensional range (like the vector-valued functions introduced in Chapter 11). We call such a function a *vector field.* Although vector fields in higher dimensions can be very useful, we will focus here on vector fields in two and three dimensions.

DEFINITION 1.1

A **vector field** in the plane is a function $\mathbf{F}(x, y)$ mapping points in $\mathbb{R}^2$ into the set of two-dimensional vectors V_2. We write

$$\mathbf{F}(x, y) = \langle f_1(x, y), f_2(x, y)\rangle = f_1(x, y)\mathbf{i} + f_2(x, y)\mathbf{j},$$

for scalar functions $f_1(x, y)$ and $f_2(x, y)$. In space, a **vector field** is a function $\mathbf{F}(x, y, z)$ mapping points in $\mathbb{R}^3$ into the set of three-dimensional vectors V_3. In this case, we write

$$\begin{aligned}\mathbf{F}(x, y, z) &= \langle f_1(x, y, z), f_2(x, y, z), f_3(x, y, z)\rangle \\ &= f_1(x, y, z)\mathbf{i} + f_2(x, y, z)\mathbf{j} + f_3(x, y, z)\mathbf{k},\end{aligned}$$

for scalar functions $f_1(x, y, z)$, $f_2(x, y, z)$ and $f_3(x, y, z)$.

To describe a two-dimensional vector field graphically, we draw a collection of the vectors $\mathbf{F}(x, y)$ for various points (x, y) in the domain, in each case drawing the vector so that its initial point is located at (x, y). We illustrate this in example 1.1.

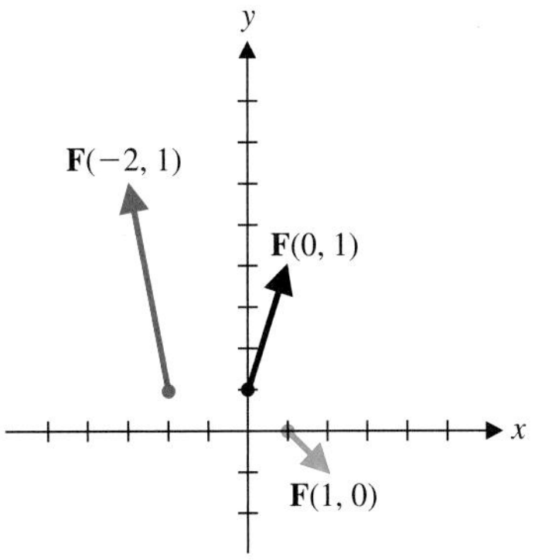

FIGURE 14.1
Values of $\mathbf{F}(x, y)$

EXAMPLE 1.1 Plotting a Vector Field

For the vector field $\mathbf{F}(x, y) = \langle x + y, 3y - x\rangle$, evaluate (a) $\mathbf{F}(1, 0)$, (b) $\mathbf{F}(0, 1)$ and (c) $\mathbf{F}(-2, 1)$. Plot each vector $\mathbf{F}(x, y)$ using the point (x, y) as the initial point.

Solution (a) Taking $x = 1$ and $y = 0$, we have $\mathbf{F}(1, 0) = \langle 1 + 0, 0 - 1\rangle = \langle 1, -1\rangle$. In Figure 14.1, we have plotted the vector $\langle 1, -1\rangle$ with its initial point located at the point $(1, 0)$, so that its terminal point is located at $(2, -1)$.

(b) Taking $x = 0$ and $y = 1$, we have $\mathbf{F}(0, 1) = \langle 0 + 1, 3 - 0\rangle = \langle 1, 3\rangle$. In Figure 14.1, we have also indicated the vector $\langle 1, 3\rangle$, taking the point $(0, 1)$ as its initial point, so that its terminal point is located at $(1, 4)$.

(c) With $x = -2$ and $y = 1$, we have $\mathbf{F}(-2, 1) = \langle -2 + 1, 3 + 2\rangle = \langle -1, 5\rangle$. In Figure 14.1, the vector $\langle -1, 5\rangle$ is plotted by placing its initial point at $(-2, 1)$ and its terminal point at $(-3, 6)$. ■

Graphing vector fields poses something of a problem. Notice that the graph of a two-dimensional vector field would be *four*-dimensional (i.e., two independent variables plus two dimensions for the vectors). Likewise, the graph of a three-dimensional vector field would be *six*-dimensional. Despite this, we can visualize many of the important properties of a vector field by plotting a number of values of the vector field as we had started to do in Figure 14.1. In general, by the **graph of the vector field** $\mathbf{F}(x, y)$, we mean a two-dimensional graph with vectors $\mathbf{F}(x, y)$, plotted with their initial point located at (x, y), for a variety of points (x, y). Many graphing calculators and computer algebra systems have commands to graph vector fields. Notice in example 1.2 that the vectors in the computer-generated graphs are not drawn to the correct length. Instead, some software packages automatically shrink or stretch all of the vectors proportionally to a size that avoids cluttering up the overall graph.

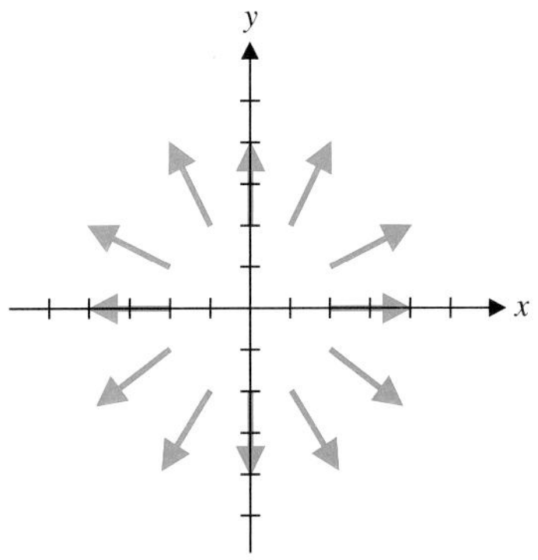

FIGURE 14.2a
$\mathbf{F}(x, y) = \langle x, y\rangle$

EXAMPLE 1.2 Graphing Vector Fields

Graph the vector fields $\mathbf{F}(x, y) = \langle x, y\rangle$, $\mathbf{G}(x, y) = \dfrac{\langle x, y\rangle}{\sqrt{x^2 + y^2}}$ and $\mathbf{H}(x, y) = \langle y, -x\rangle$ and identify any patterns.

Solution First choose a variety of points (x, y), evaluate the vector field at these points and plot the vectors using (x, y) as the initial point. Notice that in the following table we have chosen points on the axes and in each of the four quadrants.

(x, y)	$\langle x, y\rangle$		(x, y)	$\langle x, y\rangle$
$(2, 0)$	$\langle 2, 0\rangle$		$(-2, 1)$	$\langle -2, 1\rangle$
$(1, 2)$	$\langle 1, 2\rangle$		$(-2, 0)$	$\langle -2, 0\rangle$
$(2, 1)$	$\langle 2, 1\rangle$		$(-1, -2)$	$\langle -1, -2\rangle$
$(0, 2)$	$\langle 0, 2\rangle$		$(0, -2)$	$\langle 0, -2\rangle$
$(-1, 2)$	$\langle -1, 2\rangle$		$(1, -2)$	$\langle 1, -2\rangle$
$(-2, -1)$	$\langle -2, -1\rangle$		$(2, -1)$	$\langle 2, -1\rangle$

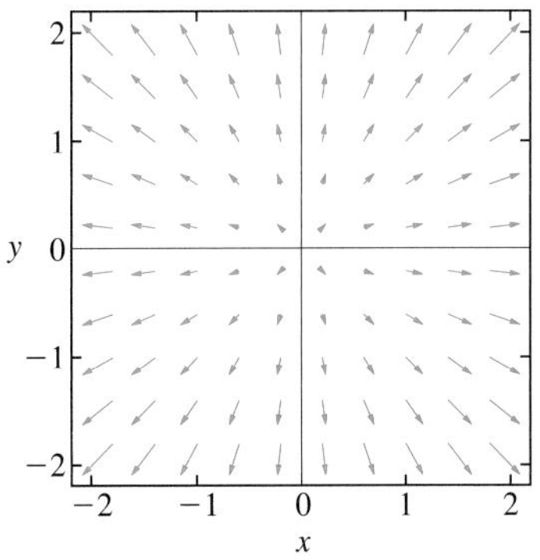

FIGURE 14.2b
$\mathbf{F}(x, y) = \langle x, y\rangle$

The vectors indicated in the table are plotted in Figure 14.2a. A computer-generated plot of the vector field is shown in Figure 14.2b. Notice that the vectors drawn here have not been drawn to scale, in order to improve the readability of the graph.

In both plots, notice that the vectors all point away from the origin and increase in length as the initial points get farther from the origin. In fact, the initial point (x, y) lies

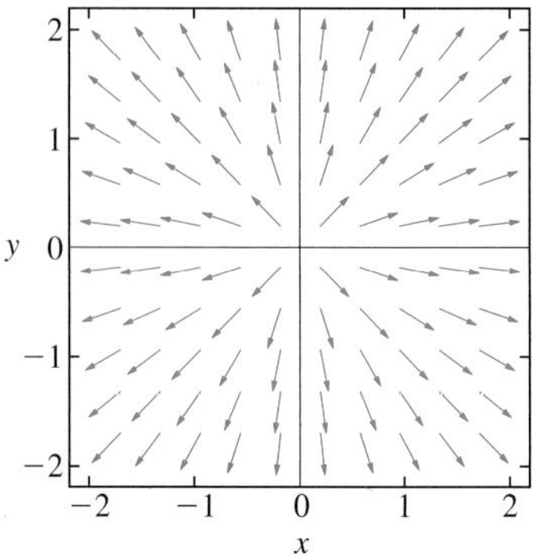

FIGURE 14.2c
$\mathbf{G}(x, y) = \dfrac{\langle x, y\rangle}{\sqrt{x^2 + y^2}}$

a distance $\sqrt{x^2 + y^2}$ from the origin and the vector $\langle x, y\rangle$ has length $\sqrt{x^2 + y^2}$. So, the length of each vector corresponds to the distance from its initial point to the origin. This gives us an important clue about the graph of $\mathbf{G}(x, y)$. Although the formula may look messy, notice that $\mathbf{G}(x, y)$ is the same as $\mathbf{F}(x, y)$ except for the division by $\sqrt{x^2 + y^2}$, which is the magnitude of the vector $\langle x, y\rangle$. Recall that dividing a vector by its magnitude yields a unit vector in the same direction. Thus, for each (x, y), $\mathbf{G}(x, y)$ has the same direction as $\mathbf{F}(x, y)$, but is a unit vector. A computer-generated plot of $\mathbf{G}(x, y)$ is shown in Figure 14.2c.

We compute some sample vectors for $\mathbf{H}(x, y)$ in the following table and plot these in Figure 14.3a.

(x, y)	$\langle y, -x\rangle$		(x, y)	$\langle y, -x\rangle$
$(2, 0)$	$\langle 0, -2\rangle$		$(-2, 1)$	$\langle 1, 2\rangle$
$(1, 2)$	$\langle 2, -1\rangle$		$(-2, 0)$	$\langle 0, 2\rangle$
$(2, 1)$	$\langle 1, -2\rangle$		$(-1, -2)$	$\langle -2, 1\rangle$
$(0, 2)$	$\langle 2, 0\rangle$		$(0, -2)$	$\langle -2, 0\rangle$
$(-1, 2)$	$\langle 2, 1\rangle$		$(1, -2)$	$\langle -2, -1\rangle$
$(-2, -1)$	$\langle -1, 2\rangle$		$(2, -1)$	$\langle -1, -2\rangle$

A computer-generated plot of $\mathbf{H}(x, y)$ is shown in Figure 14.3b. If you think of $\mathbf{H}(x, y)$ as representing the velocity field for a fluid in motion, the vectors suggest a circular rotation of the fluid. Recall that tangent lines to a circle are perpendicular to radius lines. The radius vector from the origin to the point (x, y) is $\langle x, y\rangle$, which is perpendicular to the vector $\langle y, -x\rangle$, since $\langle x, y\rangle \cdot \langle y, -x\rangle = 0$. Also, notice that the vectors are not of constant size. As for $\mathbf{F}(x, y)$, the length of the vector $\langle y, -x\rangle$ is $\sqrt{x^2 + y^2}$, which is the distance from the origin to the initial point (x, y).

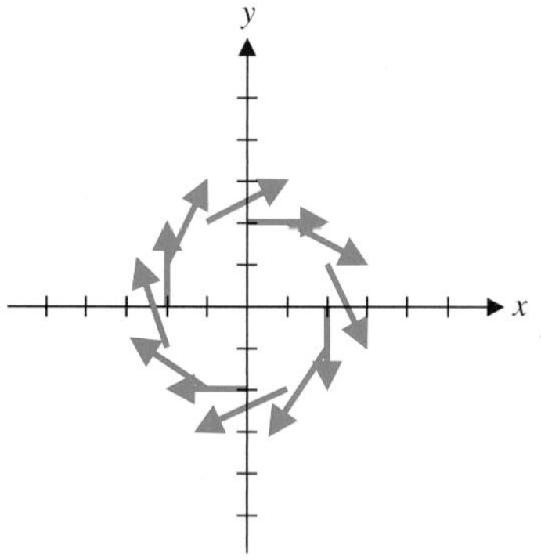

FIGURE 14.3a
$\mathbf{H}(x, y) = \langle y, -x\rangle$

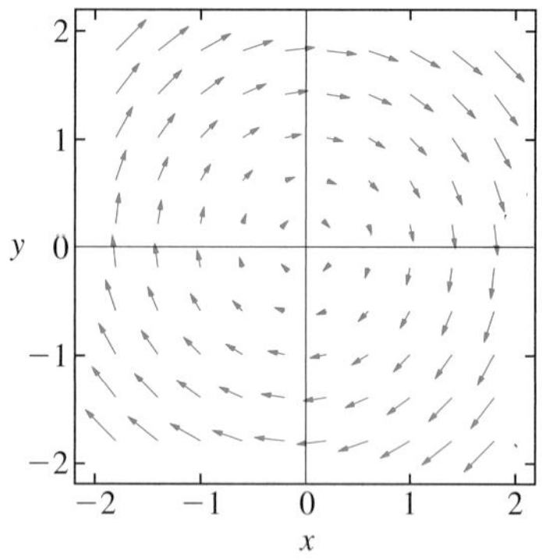

FIGURE 14.3b
$\mathbf{H}(x, y) = \langle y, -x\rangle$

Although the ideas in example 1.2 are very important, most vector fields are too complicated to effectively draw by hand. Example 1.3 illustrates how to relate the component functions of a vector field to its graph.

EXAMPLE 1.3 Matching Vector Fields to Graphs

Match the vector fields $\mathbf{F}(x, y) = \langle y^2, x - 1\rangle$, $\mathbf{G}(x, y) = \langle y + 1, e^{x/6}\rangle$ and $\mathbf{H}(x, y) = \langle y^3, x^2 - 1\rangle$ to the graphs shown.

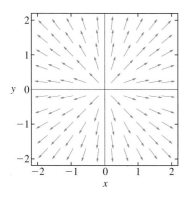

Solution While there is no general procedure for matching vector fields to their graphs, you should look for special features of the components of the vector fields and try to locate these in the graphs. For instance, the first component of $\mathbf{F}(x, y)$ is $y^2 \geq 0$, so the vectors $\mathbf{F}(x, y)$ will never point to the left. Graphs A and C both have vectors with negative first components (in the fourth quadrant), so Graph B must be the graph of $\mathbf{F}(x, y)$. The vectors in Graph B also have small vertical components near $x = 1$, where the second component of $\mathbf{F}(x, y)$ equals zero. Similarly, the second component of $\mathbf{G}(x, y)$ is $e^{x/6} > 0$, so the vectors $\mathbf{G}(x, y)$ will always point upward. Graph A is the only one of these graphs with this property. Further, the vectors in Graph A are almost vertical near $y = -1$, where the first component of $\mathbf{G}(x, y)$ equals zero. That leaves Graph C for $\mathbf{H}(x, y)$, but let's check to be sure this is reasonable. Observe that the first component of $\mathbf{H}(x, y)$ is y^3, which is negative for $y < 0$ and positive for $y > 0$. The vectors then point to the left for $y < 0$ and to the right for $y > 0$, as seen in Graph C. Finally, the vectors in Graph C have small vertical components near $x = 1$ and $x = -1$, where the second component of $\mathbf{H}(x, y)$ equals zero.

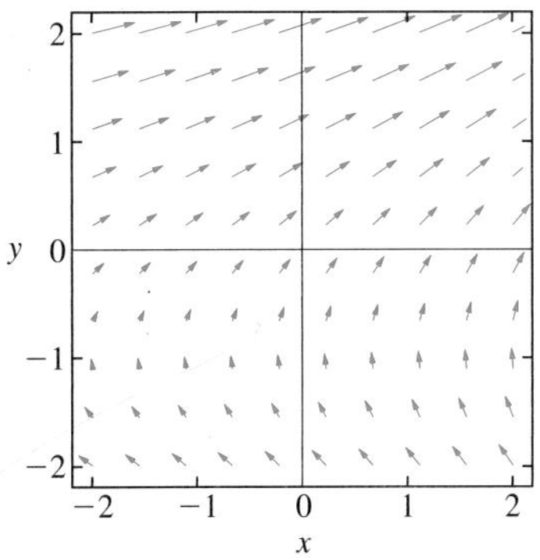

GRAPH A

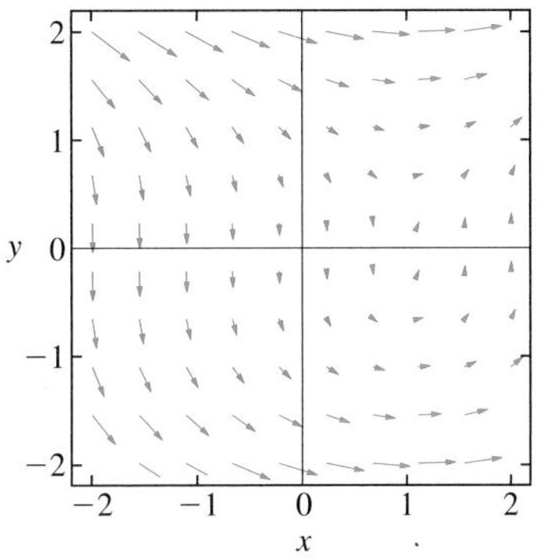

GRAPH B

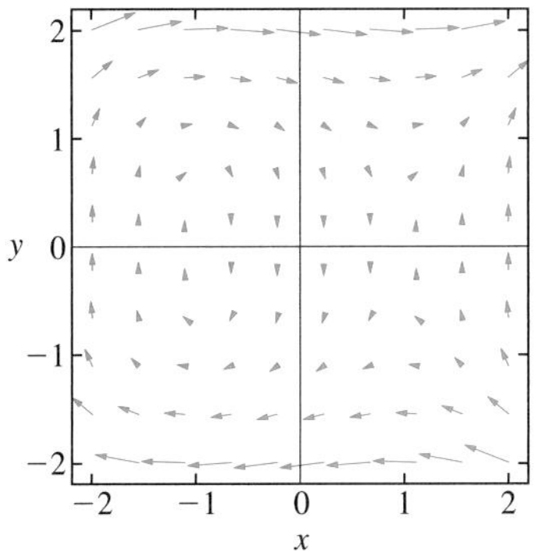

GRAPH C

As you might imagine, vector fields in space are typically more difficult to sketch than vector fields in the plane, but the idea is the same. That is, pick a variety of representative points and plot the vector $\mathbf{F}(x, y, z)$ with its initial point located at (x, y, z). Unfortunately, the difficulties associated with representing three-dimensional vectors on two-dimensional paper reduce the usefulness of these graphs.

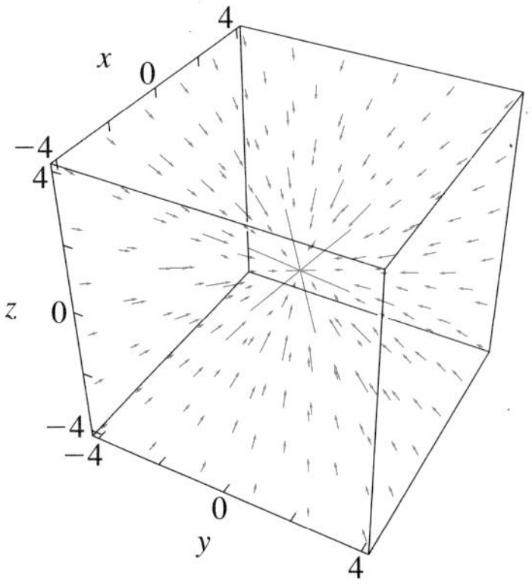

FIGURE 14.4
Gravitational force field

EXAMPLE 1.4 Graphing a Vector Field in Space

Use a CAS to graph the vector field $\mathbf{F}(x, y, z) = \dfrac{\langle -x, -y, -z\rangle}{(x^2 + y^2 + z^2)^{3/2}}$.

Solution In Figure 14.4, we show a computer-generated plot of the vector field $\mathbf{F}(x, y, z)$.

Notice that the vectors all point toward the origin, getting larger near the origin (where the field is undefined). You should get the sense of an attraction to the origin that gets stronger the closer you get. In fact, you might have recognized that $\mathbf{F}(x, y, z)$ describes the gravitational force field for an object located at the origin or the electrical field for a charge located at the origin. ■

If the vector field graphed in Figure 14.4 represents a force field, then the graph indicates that an object acted on by this force field will be drawn toward the origin. However, this does

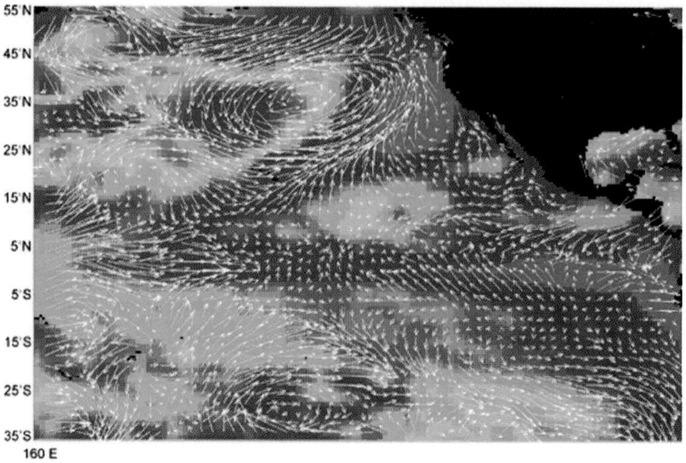

FIGURE 14.5
Velocity field for Pacific Ocean currents (March 1998)

not mean that a given object must move in a straight path toward the origin. For instance, an object with initial position (2, 0, 0) and initial velocity $\langle 0, 2, 0\rangle$ will spiral in toward the origin. To more accurately sketch the *path* followed by an object, we need additional information. Notice that in many cases, we can think of velocity as not explicitly depending on time, but instead depending on location. For instance, imagine watching a mountain stream with waterfalls and whirlpools that don't change (significantly) over time. In this case, the motion of a leaf dropped into the stream would depend on *where* you drop the leaf, rather than *when* you drop the leaf. This says that the velocity of the stream is a function of location. That is, the velocity of any particle located at the point (x, y) in the stream can be described by a vector field $\mathbf{F}(x, y) = \langle f_1(x, y), f_2(x, y)\rangle$, called the **velocity field.** The path of any given particle in the flow starting at the point (x_0, y_0) is then the curve traced out by $\langle x(t), y(t)\rangle$, where $x(t)$ and $y(t)$ are the solutions of the differential equations $x'(t) = f_1(x(t), y(t))$ and $y'(t) = f_2(x(t), y(t))$, with initial conditions $x(t_0) = x_0$ and $y(t_0) = y_0$. In these cases, we can use the velocity field to construct **flow lines,** which indicate the path followed by a particle starting at a given point in the flow.

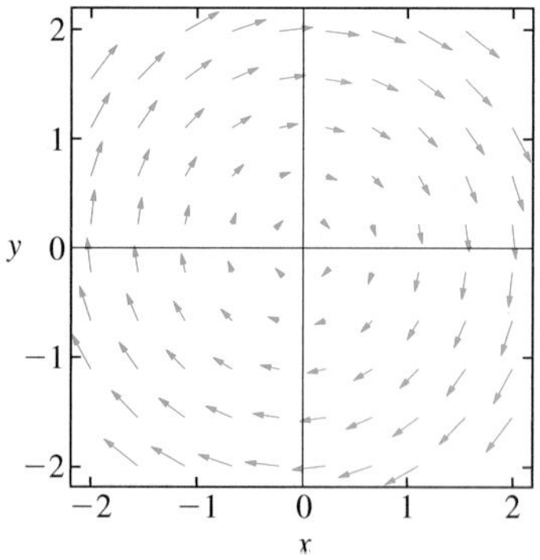

FIGURE 14.6a
$\langle y, -x\rangle$

In practice, one way to visualize the velocity field for a given process is to plot a number of velocity vectors at a single instant in time. Figure 14.5 shows the velocity field of Pacific Ocean currents in March 1998. The picture is color-coded for temperature, with a band of water swinging up from South America to the Pacific northwest representing "El Niño." The velocity field provides information about how the warmer and cooler areas of ocean water are likely to change. Since El Niño is associated with significant climate changes, an understanding of its movement is critically important.

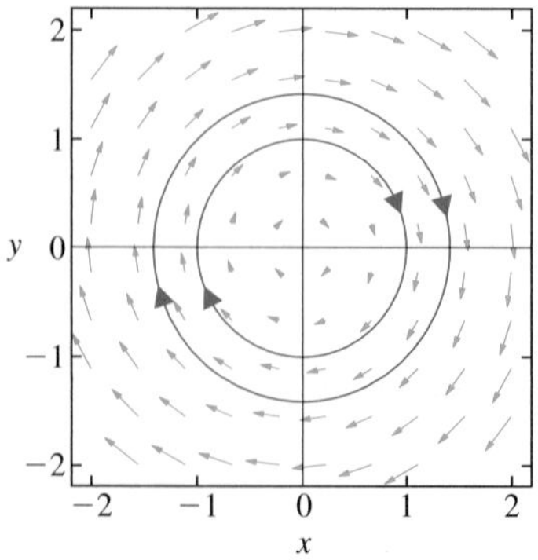

FIGURE 14.6b
Flow lines: $\langle y, -x\rangle$

EXAMPLE 1.5 Graphing Vector Fields and Flow Lines

Graph the vector fields $\langle y, -x\rangle$ and $\langle 2, 1 + 2xy\rangle$ and for each, sketch-in approximate flow lines through the points (0, 1), (0, −1) and (1, 1).

Solution We have previously graphed the vector field $\langle y, -x\rangle$ in example 1.2 and show a computer-generated graph of the vector field in Figure 14.6a. Notice that the plotted vectors nearly join together as concentric circles. In Figure 14.6b, we have superimposed circular paths that stay tangent to the velocity field and pass through the

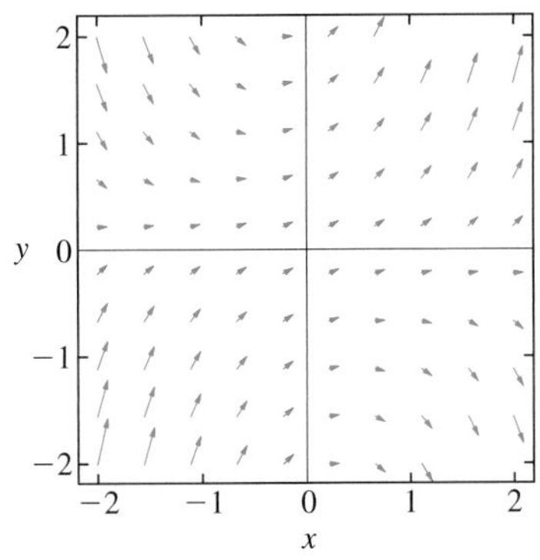

FIGURE 14.7a
$\langle 2, 1 + 2xy\rangle$

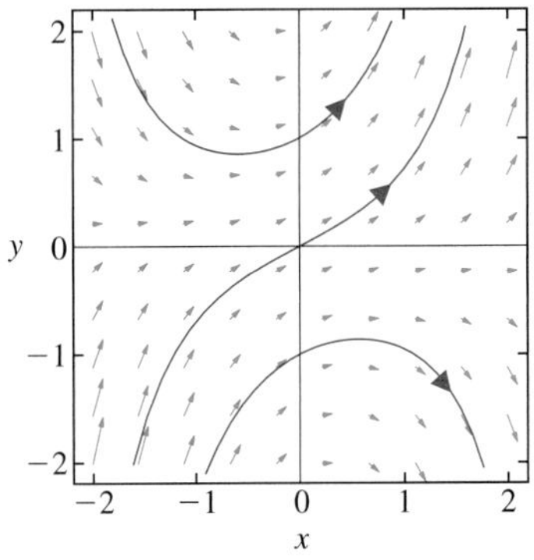

FIGURE 14.7b
Flow lines: $\langle 2, 1 + 2xy\rangle$

points $(0, 1)$, $(0, -1)$ and $(1, 1)$. (Notice that the first two of these paths are the same.) It isn't difficult to verify that the flow lines are indeed circles, as follows. Observe that a circle of radius a centered at the origin with a clockwise orientation (as indicated) can be described by the endpoint of the vector-valued function $\mathbf{r}(t) = \langle a \sin t, a \cos t\rangle$. The velocity vector $\mathbf{r}'(t) = \langle a \cos t, -a \sin t\rangle$ gives a tangent vector to the curve for each t. If we eliminate the parameter, the velocity field for the position vector $\mathbf{r} = \langle x, y\rangle$ is given by $\mathbf{T} = \langle y, -x\rangle$, which is the vector field we are presently plotting.

We show a computer-generated graph of the vector field $\langle 2, 1 + 2xy\rangle$ in Figure 14.7a, which suggests some parabolic-like paths. In Figure 14.7b, we sketch two of these paths through the points $(0, 1)$ and $(0, -1)$. However, the vectors in Figure 14.7a also indicate some paths that look more like cubics, such as the path through $(0, 0)$ sketched in Figure 14.7b. In this case, though, it's more difficult to determine equations for the flow lines. As it turns out, these are neither parabolic nor cubic. We'll explore this further in the exercises. ■

A good sketch of a vector field allows us to visualize at least some of the flow lines. However, even a great sketch can't replace the information available from an exact equation for the flow lines. We can solve for an equation of a flow line by noting that if $\mathbf{F}(x, y) = \langle f_1(x, y), f_2(x, y)\rangle$ is a velocity field and $\langle x(t), y(t)\rangle$ is the position function, then $x'(t) = f_1(x, y)$ and $y'(t) = f_2(x, y)$. By the chain rule, we have

$$\frac{dy}{dx} = \frac{dy/dt}{dx/dt} = \frac{y'(t)}{x'(t)} = \frac{f_2(x, y)}{f_1(x, y)}. \tag{1.1}$$

Equation (1.1) is a first-order differential equation for the unknown function $y(x)$. We refer you to section 7.2, where we developed a technique for solving one group of differential equations, called *separable equations*. In section 7.3, we presented a method (Euler's method) for approximating the solution of any first-order differential equation passing through a given point.

EXAMPLE 1.6 Using a Differential Equation to Construct Flow Lines

Construct the flow lines for the vector field $\langle y, -x\rangle$.

Solution From (1.1), the flow lines are solutions of the differential equation

$$\frac{dy}{dx} = -\frac{x}{y}.$$

From our discussion in section 7.2, this differential equation is separable and can be solved as follows. We first rewrite the equation as

$$y\frac{dy}{dx} = -x.$$

Integrating both sides with respect to x gives us

$$\int y\frac{dy}{dx}\,dx = -\int x\,dx,$$

so that

$$\frac{y^2}{2} = -\frac{x^2}{2} + k.$$

Multiplying both sides by 2 and replacing the constant $2k$ by c, we have

$$y^2 = -x^2 + c$$

or

$$x^2 + y^2 = c.$$

That is, for any choice of the constant $c > 0$, the solution corresponds to a circle centered at the origin. The vector field and the flow lines are then exactly as plotted in Figures 14.6a and 14.6b. ■

In example 1.7, we illustrate the use of Euler's method for constructing an approximate flow line.

EXAMPLE 1.7 Using Euler's Method to Approximate Flow Lines

Use Euler's method with $h = 0.05$ to approximate the flow line for the vector field $\langle 2, 1 + 2xy \rangle$ passing through the point $(0, 1)$, for $0 \le x \le 1$.

Solution Recall that for the differential equation $y' = f(x, y)$ and for any given value of h, Euler's method produces a sequence of approximate values of the solution function $y = y(x)$ corresponding to the points $x_i = x_0 + ih$, for $i = 1, 2, \ldots$. Specifically, starting from an initial point (x_0, y_0), where $y_0 = y(x_0)$, we construct the approximate values $y_i \approx y(x_i)$, where the y_i's are determined iteratively from the equation

$$y_{i+1} = y_i + hf(x_i, y_i), \quad i = 0, 1, 2, \ldots.$$

Since the flow line must pass through the point $(0, 1)$, we start with $x_0 = 0$ and $y_0 = 1$. Further, here we have the differential equation

$$\frac{dy}{dx} = \frac{1 + 2xy}{2} = \frac{1}{2} + xy = f(x, y).$$

In this case (unlike example 1.6), the differential equation is not separable and you do not know how to solve it exactly. For Euler's method, we then have

$$y_{i+1} = y_i + hf(x_i, y_i) = y_i + 0.05\left(\frac{1}{2} + x_i y_i\right),$$

with $x_0 = 0$, $y_0 = 1$. For the first two steps, we have

$$y_1 = y_0 + 0.05\left(\frac{1}{2} + x_0 y_0\right) = 1 + 0.05(0.5) = 1.025,$$

$x_1 = 0.05$,

$$y_2 = y_1 + 0.05\left(\frac{1}{2} + x_1 y_1\right) = 1.025 + 0.05(0.5 + 0.05125) = 1.0525625$$

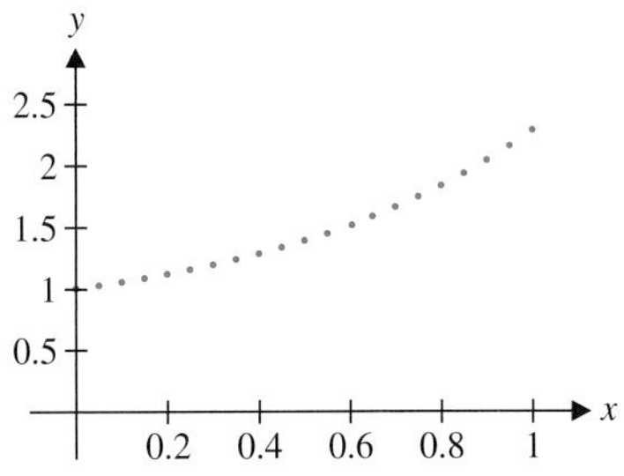

FIGURE 14.8
Approximate flow line through (0, 1)

and $x_2 = 0.1$. Continuing in this fashion, we get the sequence of approximate values indicated in the following table.

x_i	y_i		x_i	y_i		x_i	y_i
0	1		0.35	1.2344		0.70	1.6577
0.05	1.025		0.40	1.2810		0.75	1.7407
0.10	1.0526		0.45	1.3316		0.80	1.8310
0.15	1.0828		0.50	1.3866		0.85	1.9293
0.20	1.1159		0.55	1.4462		0.90	2.0363
0.25	1.1521		0.60	1.5110		0.95	2.1529
0.30	1.1915		0.65	1.5813		1.00	2.2801

A plot of these points is shown in Figure 14.8. Compare this path to the top curve (also through the point (0, 1)) shown in Figure 14.7b. ■

An important type of vector field with which we already have some experience is the *gradient field,* where the vector field is the gradient of some scalar function. Because of the importance of gradient fields, there are a number of terms associated with them. In Definition 1.2, we do not specify the number of independent variables, since the terms can be applied to functions of two, three or more variables.

DEFINITION 1.2

For any scalar function *f*, the vector field $\mathbf{F} = \nabla f$ is called the **gradient field** for the function *f*. We call *f* a **potential function** for **F**. Whenever $\mathbf{F} = \nabla f$, for some scalar function *f*, we refer to **F** as a **conservative vector field.**

It's worth noting that if you read about conservative vector fields and potentials in some applied areas (such as physics and engineering), you will sometimes see the function $-f$ referred to as the potential function. This is a minor difference in terminology, only. In this text (as is traditional in mathematics), we will consistently refer to *f* as the potential function. Rest assured that everything we say here about conservative vector fields is also true in these applications areas. The only slight difference may be that we call *f* the potential function, while others may refer to $-f$ as the potential function.

Finding the gradient field corresponding to a given scalar function is a simple matter.

EXAMPLE 1.8 Finding Gradient Fields

Find the gradient fields corresponding to the functions (a) $f(x, y) = x^2y - e^y$ and (b) $g(x, y, z) = \dfrac{1}{x^2 + y^2 + z^2}$, and use a CAS to sketch the fields.

Solution (a) We first compute the partial derivatives $\dfrac{\partial f}{\partial x} = 2xy$ and $\dfrac{\partial f}{\partial y} = x^2 - e^y$, so that

$$\nabla f(x, y) = \left\langle \frac{\partial f}{\partial x}, \frac{\partial f}{\partial y} \right\rangle = \langle 2xy, x^2 - e^y \rangle.$$

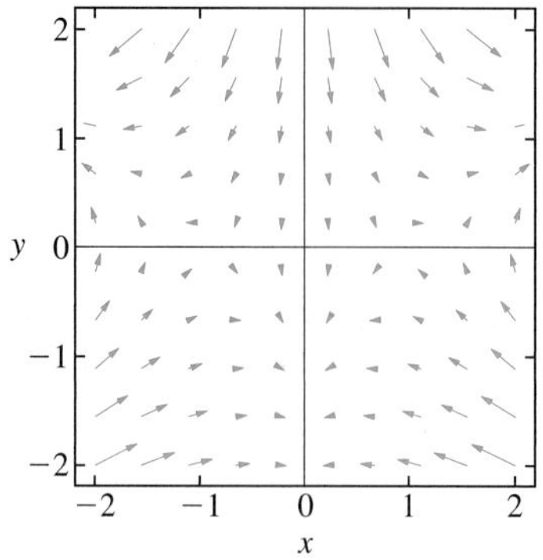

FIGURE 14.9a
$\nabla(x^2y - e^y)$

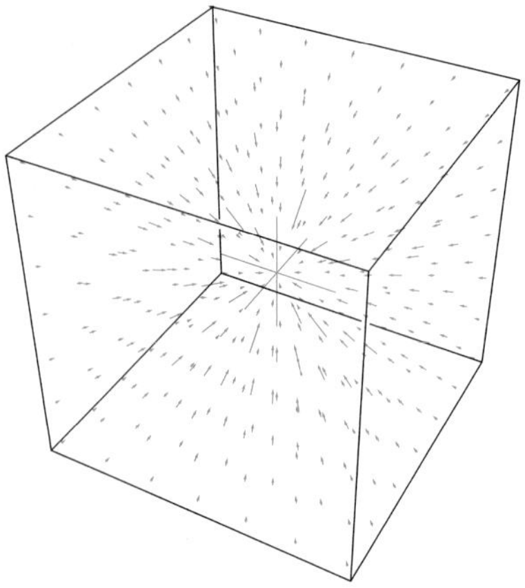

FIGURE 14.9b
$\nabla\left(\frac{1}{x^2+y^2+z^2}\right)$

A computer-generated graph of $\nabla f(x, y)$ is shown in Figure 14.9a.

(b) For $g(x, y, z) = (x^2 + y^2 + z^2)^{-1}$, we have

$$\frac{\partial g}{\partial x} = -(x^2 + y^2 + z^2)^{-2}(2x) = -\frac{2x}{x^2 + y^2 + z^2},$$

and by symmetry (think about this!), we conclude that $\frac{\partial g}{\partial y} = -\frac{2y}{x^2 + y^2 + z^2}$ and $\frac{\partial g}{\partial z} = -\frac{2z}{x^2 + y^2 + z^2}$. This gives us

$$\nabla g(x, y, z) = \left\langle \frac{\partial g}{\partial x}, \frac{\partial g}{\partial y}, \frac{\partial g}{\partial z} \right\rangle = -\frac{2\langle x, y, z\rangle}{x^2 + y^2 + z^2}.$$

A computer-generated graph of $\nabla g(x, y, z)$ is shown in Figure 14.9b. ■

You will discover that many calculations involving vector fields simplify dramatically if the vector field is a gradient field (i.e., if the vector field is conservative). To take full advantage of these simplifications, you will need to be able to construct a potential function that generates a given conservative field. The technique introduced in example 1.9 will work for most of the examples in this chapter.

EXAMPLE 1.9 Finding Potential Functions

Determine whether each of the following vector fields is conservative. If it is, find a corresponding potential function $f(x, y)$: (a) $\mathbf{F}(x, y) = \langle 2xy - 3, x^2 + \cos y\rangle$ and (b) $\mathbf{G}(x, y) = \langle 3x^2y^2 - 2y, x^2y - 2x\rangle$.

Solution The idea here is to try to construct a potential function. In the process of trying to do so, we may instead recognize that there is no potential function for the given vector field. For (a), if $f(x, y)$ is a potential function for $\mathbf{F}(x, y)$, we have that

$$\nabla f(x, y) = \mathbf{F}(x, y) = \langle 2xy - 3, x^2 + \cos y\rangle,$$

so that $$\frac{\partial f}{\partial x} = 2xy - 3 \quad \text{and} \quad \frac{\partial f}{\partial y} = x^2 + \cos y. \tag{1.2}$$

Integrating the first of these two equations with respect to x and treating y as a constant, we get

$$f(x, y) = \int (2xy - 3)\, dx = x^2 y - 3x + g(y). \tag{1.3}$$

Here, we have added an arbitrary function of y alone, $g(y)$, rather than a *constant* of integration, because any function of y is treated as a constant when integrating with respect to x. Differentiating the expression for $f(x, y)$ with respect to y gives us

$$\frac{\partial f}{\partial y}(x, y) = x^2 + g'(y) = x^2 + \cos y,$$

from (1.2). This gives us $g'(y) = \cos y$, so that

$$g(y) = \int \cos y\, dy = \sin y + c.$$

From (1.3), we now have

$$f(x, y) = x^2 y - 3x + \sin y + c,$$

where c is an arbitrary constant. Since we have been able to construct a potential function, the vector field $\mathbf{F}(x, y)$ is conservative.

(b) Again, we assume that there is a potential function g for $\mathbf{G}(x, y)$ and try to construct it. In this case, we have

$$\nabla g(x, y) = \mathbf{G}(x, y) = \langle 3x^2y^2 - 2y, x^2y - 2x \rangle,$$

so that $$\frac{\partial g}{\partial x} = 3x^2y^2 - 2y \quad \text{and} \quad \frac{\partial g}{\partial y} = x^2y - 2x. \tag{1.4}$$

Integrating the first equation in (1.4) with respect to x, we have

$$g(x, y) = \int (3x^2y^2 - 2y)\, dx = x^3y^2 - 2yx + h(y),$$

where h is an arbitrary function of y. Differentiating this with respect to y, we have

$$\frac{\partial f}{\partial y}(x, y) = 2x^3y - 2x + h'(y) = x^2y - 2x,$$

from (1.4). Solving for $h'(y)$, we get

$$h'(y) = x^2y - 2x - 2x^3y + 2x = x^2y - 2x^3y,$$

which is impossible, since $h(y)$ is a function of y alone. We then conclude that there is no potential function for $\mathbf{G}(x, y)$ and so, the vector field is not conservative. ■

REMARK 1.1

To find a potential function, you can either integrate $\frac{\partial f}{\partial x}$ with respect to x or integrate $\frac{\partial f}{\partial y}$ with respect to y. Before choosing which one to integrate, think about which integral will be easier to compute. In section 14.3, we introduce a simple method for determining whether or not a vector field is conservative.

Coulomb's law states that the electrostatic force on a charge q_0 due to a charge q is given by $\mathbf{F} = \frac{qq_0}{r^2}\mathbf{u}$, where r is the distance (in cm) between the charges and $\mathbf{u}$ is a unit vector from q to q_0. The unit of charge is esu and $\mathbf{F}$ is measured in dynes. The **electrostatic**

field E is defined as the force per unit charge, so that

$$\mathbf{E} = \frac{\mathbf{F}}{q_0} = \frac{q}{r^2}\mathbf{u}.$$

In example 1.10, we compute the electrostatic field for an electric dipole.

EXAMPLE 1.10 Electrostatic Field of a Dipole

Find the electrostatic field due to a charge of +1 esu at (1, 0) and a charge of −1 esu at (−1, 0).

Solution The distance r from (1, 0) to an arbitrary point (x, y) is $\sqrt{(x-1)^2 + y^2}$ and a unit vector from (1, 0) to (x, y) is $\frac{1}{r}\langle x-1, y\rangle$. The contribution to **E** from (1, 0) is then $\frac{1}{r^2}\frac{\langle x-1, y\rangle}{r} = \frac{\langle x-1, y\rangle}{[(x-1)^2 + y^2]^{3/2}}$. Similarly, the contribution to **E** from the negative charge at (−1, 0) is $\frac{-1}{r^2}\frac{\langle x+1, y\rangle}{r} = \frac{-\langle x+1, y\rangle}{[(x+1)^2 + y^2]^{3/2}}$. Adding the two terms, we get the electrostatic field

$$\mathbf{E} = \frac{\langle x-1, y\rangle}{[(x-1)^2 + y^2]^{3/2}} - \frac{\langle x+1, y\rangle}{[(x+1)^2 + y^2]^{3/2}}.$$

A computer-generated graph of this vector field is shown in Figures 14.10a to 14.10c.

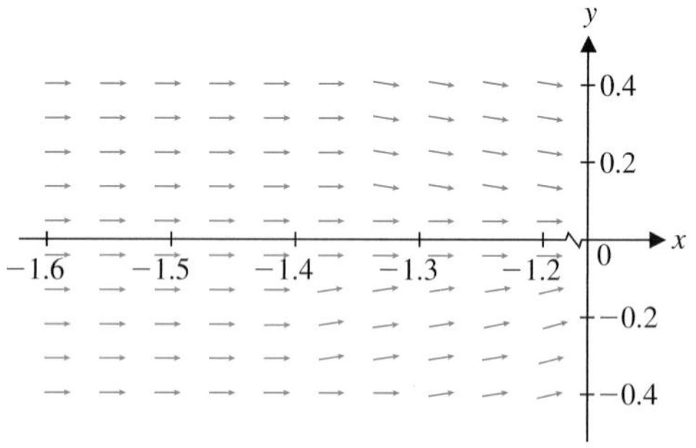

FIGURE 14.10a

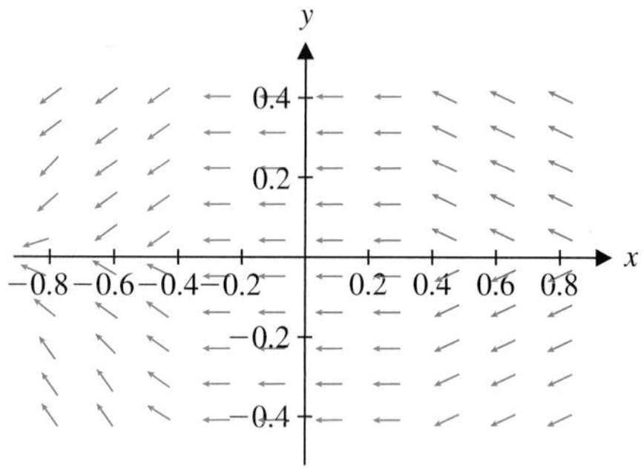

FIGURE 14.10b

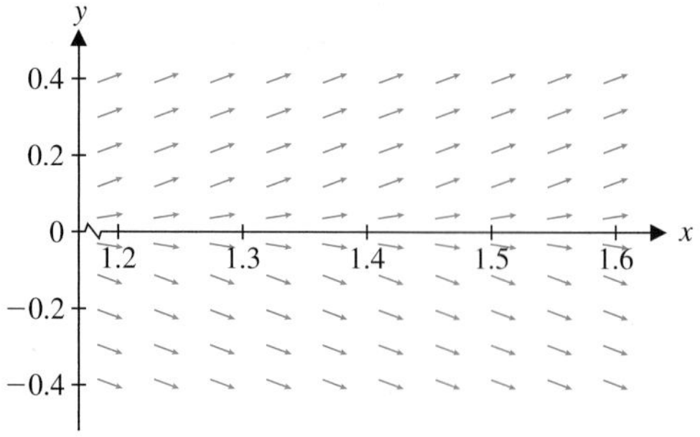

FIGURE 14.10c

Notice that the field lines point away from the positive charge at (1, 0) and toward the negative charge at (−1, 0). ■

HW #26 p1127 #5, 13, 19, 23-34 odd

EXERCISES 14.1

WRITING EXERCISES

1. Compare hand-drawn sketches of the vector fields $\langle y, -x\rangle$ and $\langle 10y, -10x\rangle$. In particular, describe which graph is easier to interpret. Computer-generated graphs of these vector fields are identical when the software "scales" the vector field by dividing out the 10. It may seem odd that computers don't draw accurate graphs, but explain why the software programmers chose to scale the vector fields.

2. The gravitational force field is an example of an "inverse square law." That is, the magnitude of the gravitational force is inversely proportional to the square of the distance from the origin. Explain why the $\frac{3}{2}$ exponent in the denominator of example 1.4 is correct for an inverse square law.

3. Explain why each vector in a vector field graph is tangent to a flow line. Explain why this means that a flow line can be visualized by joining together a large number of small (scaled) vector field vectors.

4. In example 1.9(b), explain why the presence of the x's in the expression for $g'(y)$ proves that there is no potential function.

In exercises 1–10, sketch several vectors in the vector field by hand and verify your sketch with a CAS.

1. $\mathbf{F}(x, y) = \langle -y, x\rangle$

2. $\mathbf{F}(x, y) = \dfrac{\langle -y, x\rangle}{\sqrt{x^2+y^2}}$

3. $\mathbf{F}(x, y) = \langle 0, x^2\rangle$

4. $\mathbf{F}(x, y) = \langle 2x, 0\rangle$

5. $\mathbf{F}(x, y) = 2y\mathbf{i} + \mathbf{j}$

6. $\mathbf{F}(x, y) = -\mathbf{i} + y^2\mathbf{j}$

7. $\mathbf{F}(x, y, z) = \langle 0, z, 1\rangle$

8. $\mathbf{F}(x, y, z) = \langle 2, 0, 0\rangle$

9. $\mathbf{F}(x, y, z) = \dfrac{\langle x, y, z\rangle}{\sqrt{x^2+y^2+z^2}}$

10. $\mathbf{F}(x, y, z) = \dfrac{\langle x, y, z\rangle}{x^2+y^2+z^2}$

11. Match the vector fields with their graphs.

$$\mathbf{F}_1(x, y) = \frac{\langle x, y\rangle}{\sqrt{x^2+y^2}}, \quad \mathbf{F}_2(x, y) = \langle x, y\rangle,$$

$$\mathbf{F}_3(x, y) = \langle e^y, x\rangle, \quad \mathbf{F}_4(x, y) = \langle e^y, y\rangle$$

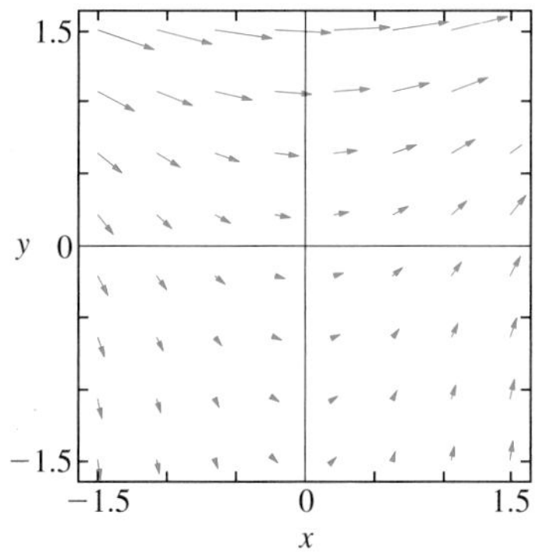

GRAPH A

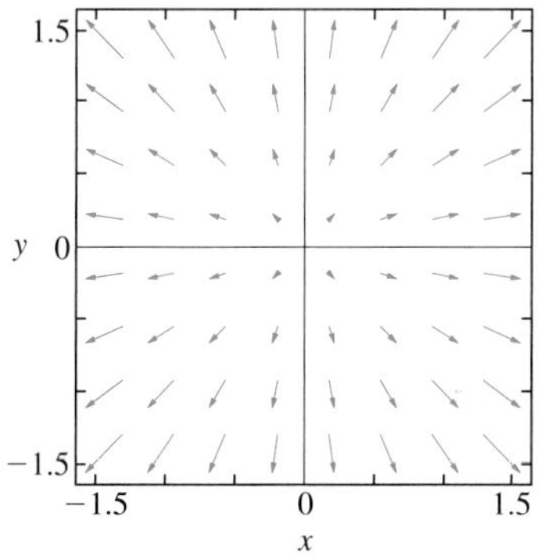

GRAPH B

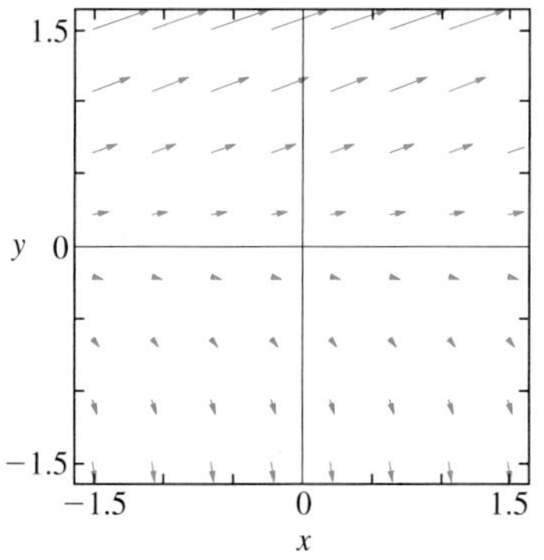

GRAPH C

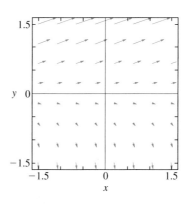

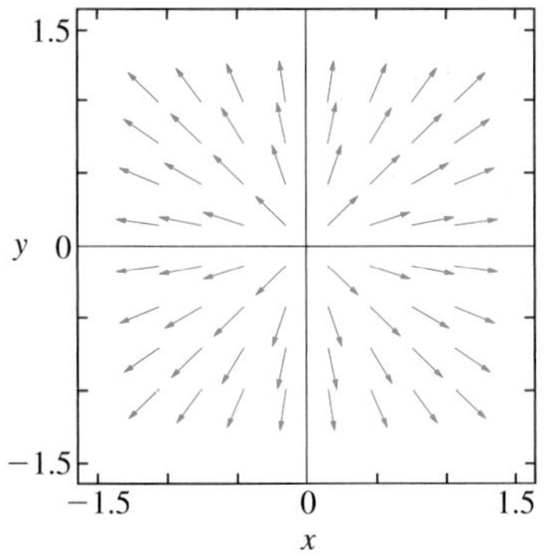

GRAPH D

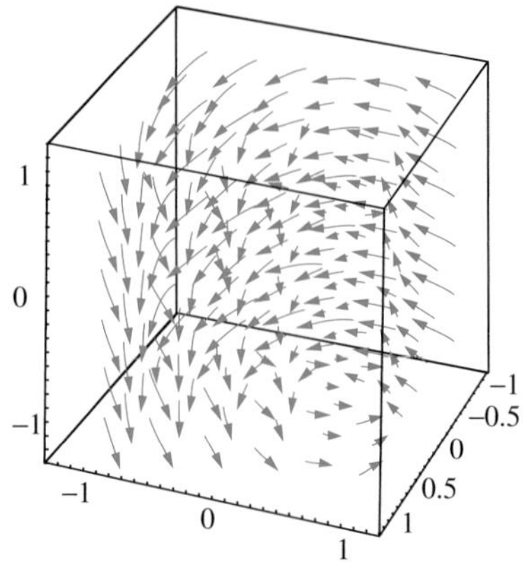

GRAPH C

12. Match the vector fields with their graphs. $\mathbf{F}_1(x, y, z) = \langle 1, x, y \rangle$, $\mathbf{F}_2(x, y, z) = \langle 1, 1, y \rangle$, $\mathbf{F}_3(x, y, z) = \dfrac{\langle y, -x, 0 \rangle}{2 - z}$, $\mathbf{F}_4(x, y, z) = \dfrac{\langle z, 0, -x \rangle}{2 - y}$.

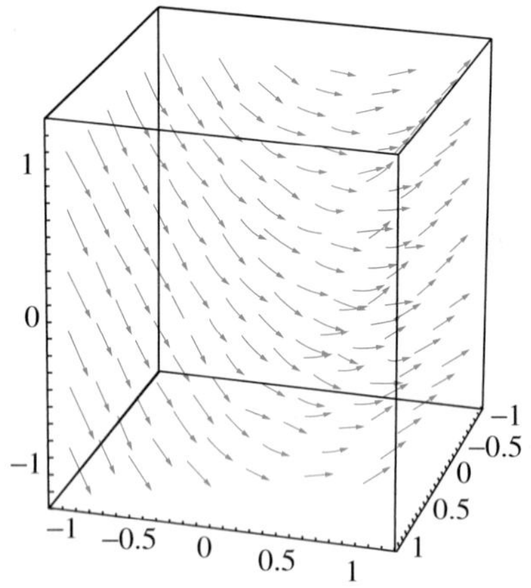

GRAPH A

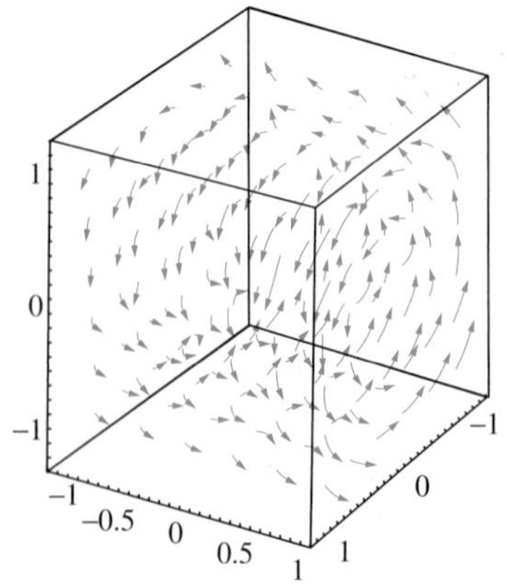

GRAPH B

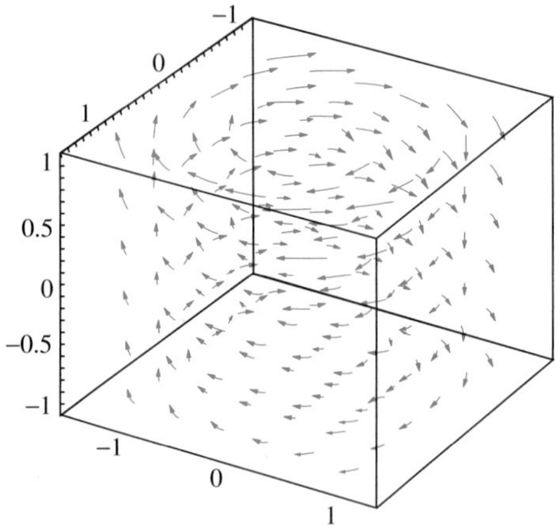

GRAPH D

In exercises 13–22, find the gradient field corresponding to f. Use a CAS to graph it.

13. $f(x, y) = x^2 + y^2$

14. $f(x, y) = x^2 - y^2$

15. $f(x, y) = \sqrt{x^2 + y^2}$

16. $f(x, y) = \sin(x^2 + y^2)$

17. $f(x, y) = xe^{-y}$

18. $f(x, y) = y \sin x$

19. $f(x, y, z) = \sqrt{x^2 + y^2 + z^2}$

20. $f(x, y, z) = xyz$

21. $f(x, y, z) = x^2 y + yz$

22. $f(x, y, z) = (x - y)^2 + z$

In exercises 23–34, determine whether or not the vector field is conservative. If it is, find a potential function.

23. $\langle y, x \rangle$

24. $\langle 2, y \rangle$

25. $\langle y, -x \rangle$

26. $\langle y, 1 \rangle$

27. $(x - 2xy)\mathbf{i} + (y^2 - x^2)\mathbf{j}$

28. $(x^2 - y)\mathbf{i} + (x - y)\mathbf{j}$

29. $\langle y \sin xy, x \sin xy \rangle$

30. $\langle y \cos x, \sin x - y \rangle$

31. $\langle 4x - z, 3y + z, y - x \rangle$

32. $\langle z^2 + 2xy, x^2 - z, 2xz - 1 \rangle$

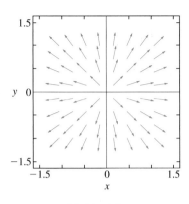

33. $\langle y^2z^2 - 1, 2xyz^2, 4z^3\rangle$

34. $\langle z^2 + 2xy, x^2 + 1, 2xz - 3\rangle$

In exercises 35–42, find equations for the flow lines.

35. $\langle 2, \cos x\rangle$

36. $\langle x^2, 2\rangle$

37. $\langle 2y, 3x^2\rangle$

38. $\langle \frac{1}{y}, 2x\rangle$

39. $y\mathbf{i} + xe^y\mathbf{j}$

40. $e^{-x}\mathbf{i} + 2x\mathbf{j}$

41. $\langle y, y^2 + 1\rangle$

42. $\langle 2, y^2 + 1\rangle$

43. Suppose that $f(x)$, $g(y)$ and $h(z)$ are continuous functions. Show that $\langle f(x), g(y), h(z)\rangle$ is conservative, by finding a potential function.

44. Show that $\langle k_1, k_2\rangle$ is conservative, for constants k_1 and k_2.

In exercises 45–52, use the notation $\mathbf{r} = \langle x, y\rangle$ and $r = \|\mathbf{r}\| = \sqrt{x^2 + y^2}$.

45. Show that $\nabla(r) = \frac{\mathbf{r}}{r}$.

46. Show that $\nabla(r^2) = 2\mathbf{r}$.

47. Find $\nabla(r^3)$.

48. Use exercises 45–47 to conjecture the value of $\nabla(r^n)$, for any positive integer n. Prove that your answer is correct.

49. Show that $\frac{\langle 1, 1\rangle}{r}$ is *not* conservative.

50. Show that $\frac{\langle -y, x\rangle}{r^2}$ is conservative on the domain $y > 0$ by finding a potential function. Show that the potential function can be thought of as the polar angle θ.

51. The current in a wire produces a magnetic field $\mathbf{B} = \frac{k\langle -y, x\rangle}{r^2}$. Draw a sketch showing a wire and its magnetic field.

52. Show that $\frac{\mathbf{r}}{r^n} = \frac{\langle x, y\rangle}{(x^2 + y^2)^{n/2}}$ is conservative, for any integer n.

53. A two-dimensional force acts radially away from the origin with magnitude 3. Write the force as a vector field.

54. A two-dimensional force acts radially toward the origin with magnitude equal to the square of the distance from the origin. Write the force as a vector field.

55. A three-dimensional force acts radially toward the origin with magnitude equal to the square of the distance from the origin. Write the force as a vector field.

56. A three-dimensional force acts radially away from the z-axis (parallel to the xy-plane) with magnitude equal to the cube of the distance from the z-axis. Write the force as a vector field.

57. Derive the electrostatic field for positive charges q at $(-1, 0)$ and $(1, 0)$ and negative charge $-q$ at $(0, 0)$.

58. The figure shows the magnetic field of the earth. Compare this to the electrostatic field of a dipole shown in example 1.10. In what way is a bar magnet similar to an electric dipole?

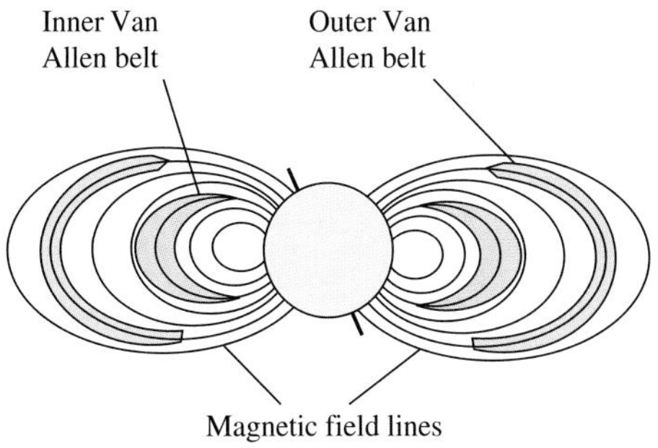

59. If $T(x, y, z)$ gives the temperature at position (x, y, z) in space, the velocity field for heat flow is given by $\mathbf{F} = -k\nabla T$ for a constant $k > 0$. This is known as **Fourier's law.** Use this vector field to determine whether heat flows from hot to cold or vice versa. Would anything change if the law were $\mathbf{F} = k\nabla T$?

60. An isotherm is a curve on a map indicating areas of constant temperature. Given Fourier's law (exercise 59), determine the angle between the velocity field for heat flow and an isotherm.

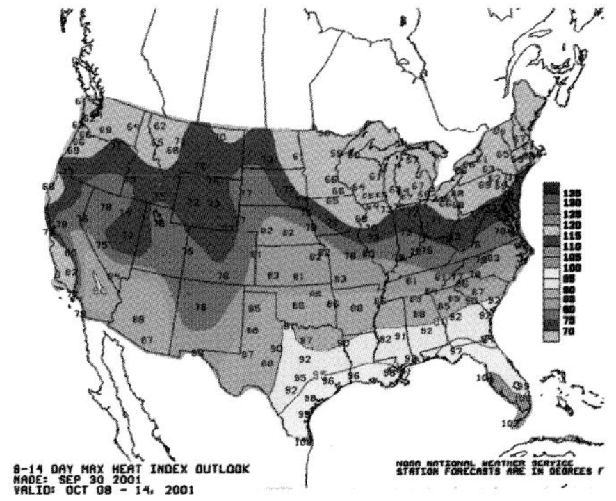

EXPLORATORY EXERCISES

1. Show that the vector field $\mathbf{F}(x, y) = \langle y, x\rangle$ has potential function $f(x, y) = xy$. The curves $f(x, y) = c$ for constants c are called **equipotential curves.** Sketch equipotential curves for several constants (positive and negative). Find the flow lines for this vector field and show that the flow lines and equipotential curves intersect at right angles. This situation is common. To further develop these relationships, show that the potential function and the flow function $g(x, y) = \frac{1}{2}(y^2 - x^2)$ are both solutions of **Laplace's equation** $\nabla^2 u = 0$ where $\nabla^2 u = u_{xx} + u_{yy}$.

2. In example 1.5, we graphed the flow lines for the vector field $\langle 2, 1 + 2xy\rangle$ and mentioned that finding equations for the flow lines was beyond what's been presented in the text. We develop a method for finding the flow lines here by solving **linear ordinary differential equations.** We will illustrate this for an easier vector field, $\langle x, 2x - y\rangle$. First, note that if $x'(t) = x$ and $y'(t) = 2x - y$, then

$$\frac{dy}{dx} = \frac{2x - y}{x} = 2 - \frac{y}{x}.$$

The flow lines will be the graphs of functions $y(x)$ such that $y'(x) = 2 - \frac{y}{x}$, or $y' + \frac{1}{x}y = 2$. The left-hand side of the equation should look a little like a product rule. Our main goal is to multiply by a term called an **integrating factor,** to make the left-hand side exactly a product rule derivative. It turns out that for the equation $y' + f(x)y = g(x)$, an integrating factor is $e^{\int f(x)\,dx}$. In the present case, for $x > 0$, we have $e^{\int 1/x dx} = e^{\ln x} = x$. (We have chosen the integration constant to be 0 to keep the integrating factor simple.) Multiply both sides of the equation by x and show that $xy' + y = 2x$. Show that $xy' + y = (xy)'$. From $(xy)' = 2x$, integrate to get $xy = x^2 + c$, or $y = x + \frac{c}{x}$. To find a flow line passing through the point $(1, 2)$, show that $c = 1$ and thus, $y = x + \frac{1}{x}$. To find a flow line passing through the point $(1, 1)$, show that $c = 0$ and thus, $y = x$. Sketch the vector field and highlight the curves $y = x + \frac{1}{x}$ and $y = x$.

14.2 LINE INTEGRALS

In section 5.6, we used integration to find the mass of a thin rod with variable mass density. There, we had observed that if the rod extends from $x = a$ to $x = b$ and has mass density function $\rho(x)$, then the mass of the rod is given by $\int_a^b \rho(x)\,dx$. This definition is fine for objects that are essentially one-dimensional, but what if we wanted to find the mass of a helical spring (see Figure 14.11)? Calculus is remarkable in that the same technique can solve a wide variety of problems. As you should expect by now, we will derive a solution by first approximating the curve with line segments and then taking a limit.

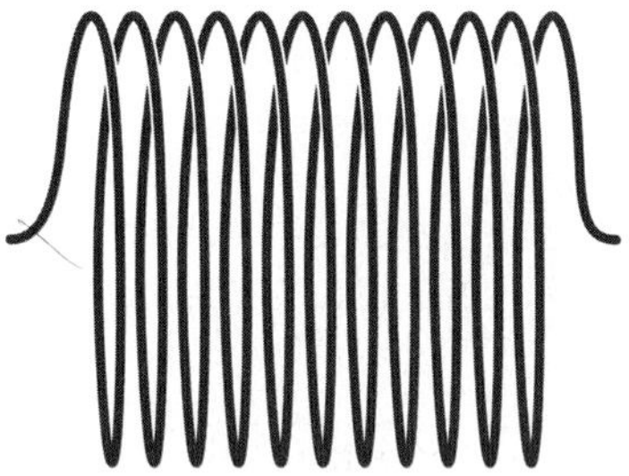

FIGURE 14.11
A helical spring

In this three-dimensional setting, the density function has the form $\rho(x, y, z)$ (where ρ is measured in units of mass per unit length). We assume that the object is in the shape of a curve C in three dimensions with endpoints (a, b, c) and (d, e, f). Further, we assume that the curve is **oriented,** which means that there is a direction to the curve. For example, the curve C might start at (a, b, c) and end at (d, e, f). We first partition the curve into n pieces with endpoints $(a, b, c) = (x_0, y_0, z_0), (x_1, y_1, z_1), (x_2, y_2, z_2), \ldots, (x_n, y_n, z_n) = (d, e, f)$, as indicated in Figure 14.12. We will use the shorthand P_i to denote the point (x_i, y_i, z_i), and C_i for the section of the curve C extending from P_{i-1} to P_i, for each $i = 1, 2, \ldots, n$. Our initial objective is to approximate the mass of the portion of the object along C_i. Note that if the segment C_i is small enough, we can consider the density to be constant on C_i. In this case, the mass of this segment would simply be the product of the density and the length of C_i. For some point (x_i^*, y_i^*, z_i^*) on C_i, we approximate the density on C_i by $\rho(x_i^*, y_i^*, z_i^*)$. The mass of the section C_i is then approximately

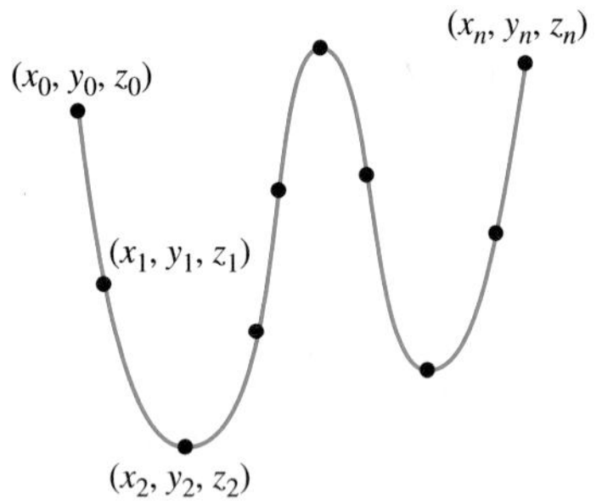

FIGURE 14.12
Partitioned curve

$$\rho(x_i^*, y_i^*, z_i^*)\Delta s_i,$$

where Δs_i represents the arc length of C_i. The mass m of the entire object is then approximately the sum of the masses of the n segments,

$$m \approx \sum_{i=1}^{n} \rho(x_i^*, y_i^*, z_i^*)\Delta s_i.$$

You should expect that this approximation will improve as we divide the curve into more and more segments that are shorter and shorter in length. Finally, taking the norm of the partition $\|P\|$ to be the maximum of the arc lengths $\Delta s_i (i = 1, 2, \ldots, n)$, we have

$$m = \lim_{\|P\|\to 0} \sum_{i=1}^{n} \rho(x_i^*, y_i^*, z_i^*)\Delta s_i, \tag{2.1}$$

provided the limit exists and is the same for every choice of the evaluation points (x_i^*, y_i^*, z_i^*) $(i = 1, 2, \ldots, n)$.

You might recognize that (2.1) looks like the limit of a Riemann sum (an integral). As it turns out, this limit arises naturally in numerous applications. We pause now to give this limit a name and identify some useful properties.

DEFINITION 2.1

The **line integral of $f(x, y, z)$ with respect to arc length** along the oriented curve C in three-dimensional space, written $\int_C f(x, y, z)\, ds$, is defined by

$$\int_C f(x, y, z)\, ds = \lim_{\|P\| \to 0} \sum_{i=1}^{n} f(x_1^*, y_i^*, z_i^*) \Delta s_i,$$

provided the limit exists and is the same for all choices of evaluation points.

We define line integrals of functions $f(x, y)$ of two variables along an oriented curve C in the xy-plane in a similar way. Often, the curve C is specified by parametric equations or you can use your skills developed in Chapters 9 and 11 to construct parametric equations for the curve. In these cases, Theorem 2.1 allows us to evaluate the line integral as a definite integral of a function of one variable.

THEOREM 2.1 (Evaluation Theorem)

Suppose that $f(x, y, z)$ is continuous in a region D containing the curve C and that C is described parametrically by $(x(t), y(t), z(t))$, for $a \le t \le b$, where $x(t)$, $y(t)$ and $z(t)$ have continuous first derivatives. Then,

$$\int_C f(x, y, z)\, ds = \int_a^b f(x(t), y(t), z(t)) \sqrt{[x'(t)]^2 + [y'(t)]^2 + [z'(t)]^2}\, dt.$$

Suppose that $f(x, y)$ is continuous in a region D containing the curve C and that C is described parametrically by $(x(t), y(t))$, for $a \le t \le b$, where $x(t)$ and $y(t)$ have continuous first derivatives. Then

$$\int_C f(x, y)\, ds = \int_a^b f(x(t), y(t)) \sqrt{[x'(t)]^2 + [y'(t)]^2}\, dt.$$

PROOF

We prove the result for the case of a curve in two dimensions and leave the three-dimensional case as an exercise. From Definition 2.1 (adjusted for the two-dimensional case), we have

$$\int_C f(x, y)\, ds = \lim_{\|P\| \to 0} \sum_{i=1}^{n} f(x_i^*, y_i^*) \Delta s_i, \tag{2.2}$$

where Δs_i represents the arc length of the section of the curve C between (x_{i-1}, y_{i-1}) and (x_i, y_i). Choose $t_0, t_1, \ldots, t_n$ so that $x(t_i) = x_i$ and $y(t_i) = y_i$, for $i = 0, 1, \ldots, n$. We approximate the arc length of such a small section of the curve by the straight-line distance:

$$\Delta s_i \approx \sqrt{(x_i - x_{i-1})^2 + (y_i - y_{i-1})^2}.$$

Further, since $x(t)$ and $y(t)$ have continuous first derivatives, we have by the Mean Value Theorem (as in the derivation of the arc length formula in section 9.3), that

$$\Delta s_i \approx \sqrt{(x_i - x_{i-1})^2 + (y_i - y_{i-1})^2} \approx \sqrt{[x'(t_i^*)]^2 + [y'(t_i^*)]^2}\,\Delta t_i,$$

for some $t_i^* \in (t_{i-1}, t_i)$. Together with (2.2), this gives us

$$\int_C f(x, y)\,ds = \lim_{\|P\|\to 0} \sum_{i=1}^{n} f(x(t_i^*), y(t_i^*))\sqrt{[x'(t_i^*)]^2 + [y'(t_i^*)]^2}\,\Delta t_i$$
$$= \int_a^b f(x(t), y(t))\sqrt{[x'(t)]^2 + [y'(t)]^2}\,dt. \quad \blacksquare$$

A curve C is called **smooth** if it satisfies the hypotheses of Theorem 2.1 and one additional condition. Specifically, C is smooth if it can be described parametrically by $x = x(t)$, $y = y(t)$ and $z = z(t)$, for $a \le t \le b$, where $x(t)$, $y(t)$ and $z(t)$ all have continuous first derivatives and $[x'(t)]^2 + [y'(t)]^2 + [z'(t)]^2 \ne 0$ on the interval $[a, b]$. Similarly, a plane curve is smooth if it can be parameterized by $x = x(t)$ and $y = y(t)$, for $a \le t \le b$, where $x(t)$ and $y(t)$ have continuous first derivatives and $[x'(t)]^2 + [y'(t)]^2 \ne 0$ on the interval $[a, b]$.

Notice that for curves in space, Theorem 2.1 says essentially that the arc length element ds can be replaced by

$$ds = \sqrt{[x'(t)]^2 + [y'(t)]^2 + [z'(t)]^2}\,dt. \tag{2.3}$$

The term $\sqrt{[x'(t)]^2 + [y'(t)]^2 + [z'(t)]^2}$ in the integral should be very familiar, having been present in our integral representations of both arc length and surface area. Likewise, for curves in the plane, the arc length element is

$$ds = \sqrt{[x'(t)]^2 + [y'(t)]^2}\,dt. \tag{2.4}$$

EXAMPLE 2.1 Finding the Mass of a Helical Spring

Find the mass of a spring in the shape of the helix defined parametrically by $x = 2\cos t$, $y = t$, $z = 2\sin t$, for $0 \le t \le 6\pi$, with density $\rho(x, y, z) = 2y$.

Solution A graph of the helix is shown in Figure 14.13. The density is

$$\rho(x, y, z) = 2y = 2t,$$

and from (2.3), the arc length element ds is given by

$$ds = \sqrt{[x'(t)]^2 + [y'(t)]^2 + [z'(t)]^2}\,dt = \sqrt{(-2\sin t)^2 + (1)^2 + (2\cos t)^2}\,dt = \sqrt{5}\,dt,$$

where we have used the identity $4\sin^2 t + 4\cos^2 t = 4$. By Theorem 2.1, we have

$$\text{mass} = \int_C \rho(x, y, z)\,ds = \int_0^{6\pi} \underbrace{2t}_{\rho(x,y,z)}\ \underbrace{\sqrt{5}\,dt}_{ds} = 2\sqrt{5}\int_0^{6\pi} t\,dt$$
$$= 2\sqrt{5}\,\frac{(6\pi)^2}{2} = 36\pi^2\sqrt{5}. \quad \blacksquare$$

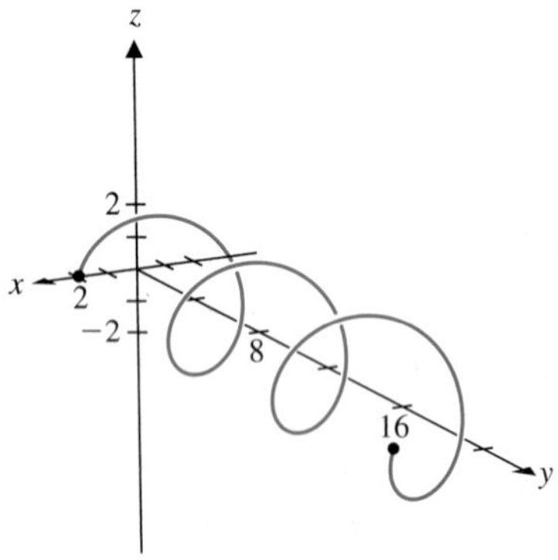

FIGURE 14.13
The helix $x = 2\cos t$, $y = t$, $z = 2\sin t$, $0 \le t \le 6\pi$

We should point out that example 2.1 is unusual in at least one respect: we were able to compute the integral exactly. Most line integrals of the type defined in Definition 2.1 are too complicated to evaluate exactly and will need to be approximated with some numerical method, as in example 2.2.

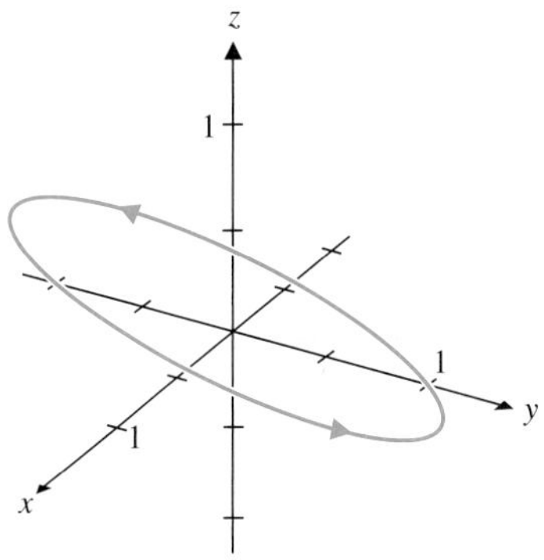

FIGURE 14.14
The curve $x = \cos t$, $y = \sin t$, $z = \cos t$ with $0 \le t \le 2\pi$

EXAMPLE 2.2 Evaluating a Line Integral with Respect to Arc Length

Evaluate the line integral $\int_C (2x^2 - 3yz)\,ds$, where C is the curve defined parametrically by $x = \cos t$, $y = \sin t$, $z = \cos t$ with $0 \le t \le 2\pi$.

oriented curve since they are parametric

Solution A graph of C is shown in Figure 14.14. The integrand is

$$f(x, y, z) = 2x^2 - 3yz = 2\cos^2 t - 3\sin t \cos t.$$

From (2.3), the arc length element is given by

$$\begin{aligned} ds &= \sqrt{[x'(t)]^2 + [y'(t)]^2 + [z'(t)]^2}\,dt \\ &= \sqrt{(-\sin t)^2 + (\cos t)^2 + (-\sin t)^2}\,dt = \sqrt{1 + \sin^2 t}\,dt, \end{aligned}$$

where we have used the identity $\sin^2 t + \cos^2 t = 1$. By Theorem 2.1, we now have

$$\int_C (2x^2 - 3yz)\,ds = \int_0^{2\pi} \underbrace{(2\cos^2 t - 3\sin t \cos t)}_{2x^2 - 3yz} \underbrace{\sqrt{1 + \sin^2 t}\,dt}_{ds} \approx 6.9922,$$

where we approximated the last integral numerically. ■

In example 2.3, you must find parameteric equations for the (two-dimensional) curve before evaluating the line integral. Also, you will discover an important fact about the orientation of the curve C.

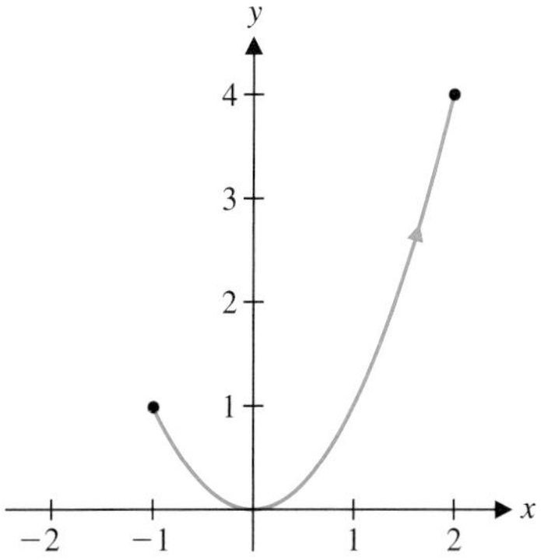

FIGURE 14.15a
$y = x^2$ from $(-1, 1)$ to $(2, 4)$

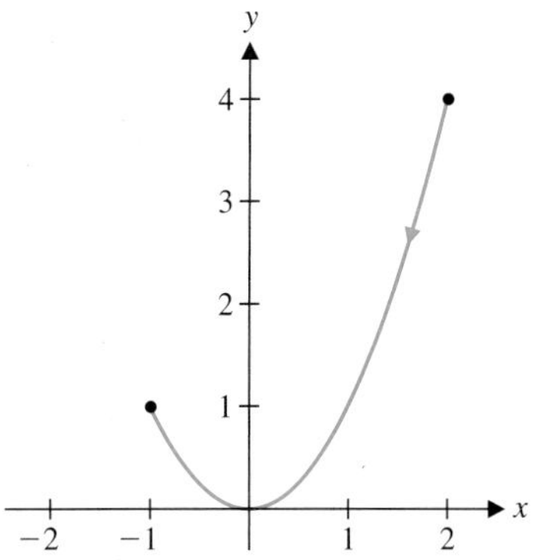

FIGURE 14.15b
$y = x^2$ from $(2, 4)$ to $(-1, 1)$

EXAMPLE 2.3 Evaluating a Line Integral with Respect to Arc Length

Evaluate the line integral $\int_C 2x^2 y\,ds$, where C is (a) the portion of the parabola $y = x^2$ from $(-1, 1)$ to $(2, 4)$ and (b) the portion of the parabola $y = x^2$ from $(2, 4)$ to $(-1, 1)$.

Solution (a) A sketch of the curve is shown in Figure 14.15a. Taking $x = t$ as the parameter (since the curve is already written explicitly in terms of x), we can write parametric equations for the curve as $x = t$ and $y = t^2$, for $-1 \le t \le 2$. Using this, the integrand becomes $2x^2 y = 2t^2 t^2 = 2t^4$ and from (2.4), the arc length element is

$$ds = \sqrt{[x'(t)]^2 + [y'(t)]^2}\,dt = \sqrt{1 + 4t^2}\,dt.$$

The integral is now written as

$$\int_C 2x^2 y\,ds = \int_{-1}^{2} \underbrace{2t^4}_{2x^2 y} \underbrace{\sqrt{1 + 4t^2}\,dt}_{ds} \approx 45.391,$$

where we have again evaluated the integral numerically (although in this case a good CAS can give you an exact answer).

(b) The curve is the same as in part (a), except that the orientation is backward (see Figure 14.15b). In this case, we represent the curve with the parametric equations $x = -t$ and $y = t^2$, for $-2 \le t \le 1$. Observe that everything else in the integral remains the same and we have

$$\int_C 2x^2 y\,ds = \int_{-2}^{1} 2t^4 \sqrt{1 + 4t^2}\,dt \approx 45.391,$$

as before. ■

– infinite parametric for curves

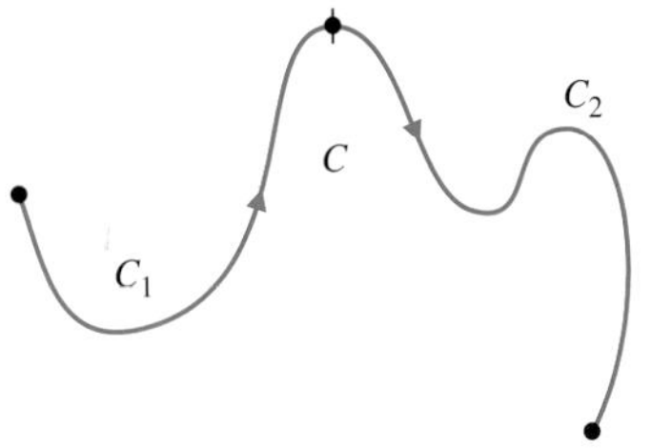

FIGURE 14.16
$C = C_1 \cup C_2$

Notice that in example 2.3, the line integral was the same no matter which orientation we took for the curve. It turns out that this is true in general for all line integrals defined by Definition 2.1 (i.e., line integrals with respect to arc length).

Notice that you can use Theorem 2.1 to rewrite a line integral only when the curve C is smooth. Fortunately, although many curves of interest are not smooth, we can extend the result of Theorem 2.1 to the case where C is a union of a finite number of smooth curves:

$$C = C_1 \cup C_2 \cup \cdots \cup C_n,$$ *where the derivative is continuous everywhere*

where each of $C_1, C_2, \ldots, C_n$ is smooth and where the terminal point of C_i is the same as the initial point of C_{i+1}, for $i = 1, 2, \ldots, n - 1$. We call such a curve C **piecewise-smooth.** Notice that if C_1 and C_2 are oriented curves and the endpoint of C_1 is the same as the initial point of C_2, then the curve $C_1 \cup C_2$ is an oriented curve with the same initial point as C_1 and the same endpoint as C_2. (See Figure 14.16.) The results in Theorem 2.2 should not seem surprising. Here, for an oriented curve C in two or three dimensions, the curve $-C$ denotes the same curve as C, but with the opposite orientation.

THEOREM 2.2

Suppose that $f(x, y, z)$ is a continuous function in some region D containing the oriented curve C. Then, if C is piecewise-smooth, with $C = C_1 \cup C_2 \cup \cdots \cup C_n$, where $C_1, C_2, \ldots, C_n$ are all smooth and where the terminal point of C_i is the same as the initial point of C_{i+1}, for $i = 1, 2, \ldots, n - 1$, we have

(i)

$$\int_{-C} f(x, y, z)\, ds = \int_C f(x, y, z)\, ds$$

and (ii)

piece-wise smooth

$$\int_C f(x, y, z)\, ds = \int_{C_1} f(x, y, z)\, ds + \int_{C_2} f(x, y, z)\, ds + \cdots + \int_{C_n} f(x, y, z)\, ds.$$

We leave the proof of the theorem as an exercise. Notice that the corresponding result will also be true in two dimensions. We use part (ii) of Theorem 2.2 in example 2.4.

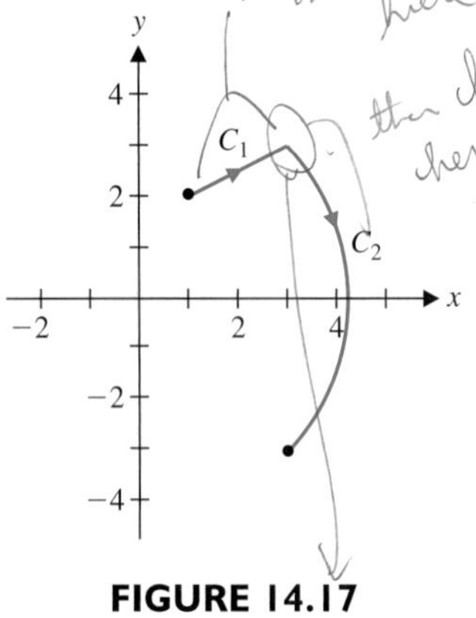

FIGURE 14.17
Piecewise-smooth curve

derivative is constant here
longer than here

EXAMPLE 2.4   Evaluating a Line Integral over a Piecewise-Smooth Curve

Evaluate the line integral $\int_C (3x - y)\, ds$, where C is the line segment from $(1, 2)$ to $(3, 3)$, followed by the portion of the circle $x^2 + y^2 = 18$ traversed from $(3, 3)$ clockwise around to $(3, -3)$.

Solution   A graph of the curve is shown in Figure 14.17. Notice that we'll need to evaluate the line integral separately over the line segment C_1 and the quarter-circle C_2. Further, although C is not smooth, it is piecewise-smooth, since C_1 and C_2 are both smooth. We can write C_1 parametrically as $x = 1 + (3 - 1)t = 1 + 2t$ and $y = 2 + (3 - 2)t = 2 + t$, for $0 \le t \le 1$. Also, on C_1, the integrand is given by

$$3x - y = 3(1 + 2t) - (2 + t) = 1 + 5t$$

and from (2.4), the arc length element is

$$ds = \sqrt{(2)^2 + (1)^2}\, dt = \sqrt{5}\, dt.$$

Putting this together, we have

$$\int_{C_1} f(x, y)\, ds = \int_0^1 \underbrace{(1 + 5t)}_{f(x,\, y)}\, \underbrace{\sqrt{5}\, dt}_{ds} = \frac{7}{2}\sqrt{5}. \tag{2.5}$$

Next, for C_2, the usual parametric equations for a circle of radius r oriented counterclockwise are $x(t) = r\cos t$ and $y(t) = r\sin t$. In the present case, the radius is $\sqrt{18}$ and the curve is oriented clockwise, which means that $y(t)$ has the opposite sign from the usual orientation. So, parametric equations for C_2 are $x(t) = \sqrt{18}\cos t$ and $y(t) = -\sqrt{18}\sin t$. Notice, too, that the initial point $(3, 3)$ corresponds to the angle $-\frac{\pi}{4}$ and the endpoint $(3, -3)$ corresponds to the angle $\frac{\pi}{4}$. Finally, on C_2, the integrand is given by

$$3x - y = 3\sqrt{18}\cos t + \sqrt{18}\sin t$$

and the arc length element is

$$ds = \sqrt{\left(-\sqrt{18}\sin t\right)^2 + \left(-\sqrt{18}\cos t\right)^2}\, dt = \sqrt{18}\, dt,$$

where we have again used the fact that $\sin^2 t + \cos^2 t = 1$. This gives us

$$\int_{C_2} f(x, y)\, ds = \int_{-\pi/4}^{\pi/4} \underbrace{\left(3\sqrt{18}\cos t + \sqrt{18}\sin t\right)}_{f(x,\, y)}\, \underbrace{\sqrt{18}\, dt}_{ds} = 54\sqrt{2}. \tag{2.6}$$

Combining the integrals over the two curves, we have from (2.5) and (2.6) that

$$\int_C f(x, y)\, ds = \int_{C_1} f(x, y)\, ds + \int_{C_2} f(x, y)\, ds = \frac{7}{2}\sqrt{5} + 54\sqrt{2}.\ \blacksquare$$

So far, we have discussed how to calculate line integrals for curves described parametrically and we have given one application of line integrals (calculation of mass). In the exercises, we discuss further applications. We now develop a geometric interpretation of the line integral. Whereas $\int_a^b f(x)\, dx$ corresponds to a limit of sums of the heights of the function $f(x)$ above or below the x-axis for an interval $[a, b]$ of the x-axis, the line integral $\int_C f(x, y)\, ds$ corresponds to a limit of the sums of the heights of the function $f(x, y)$ above or below the xy-plane for a curve C lying in the xy-plane. We depict this in Figures 14.18a and 14.18b. Recall that for $f(x) \geq 0$, $\int_a^b f(x)\, dx$ measures the area under the curve $y = f(x)$

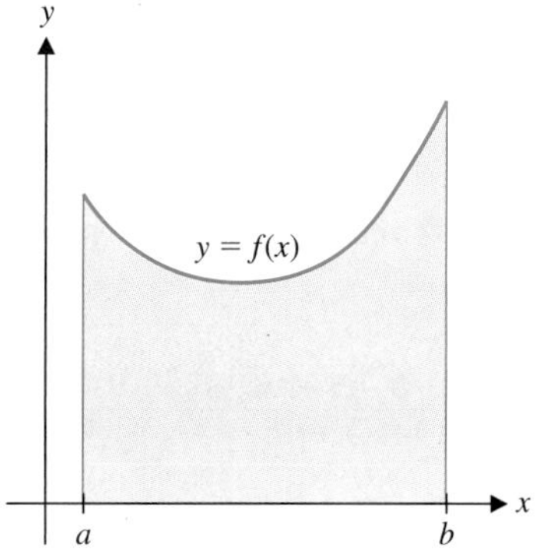

FIGURE 14.18a
$\int_a^b f(x)\, dx$

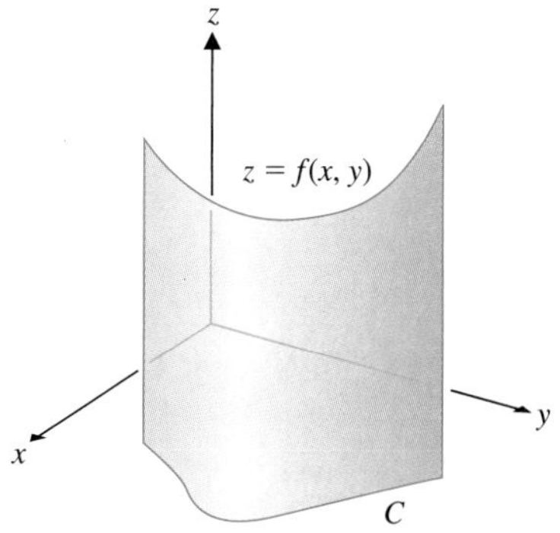

FIGURE 14.18b
$\int_C f(x, y)\, ds$

on the interval $[a, b]$, shaded in Figure 14.18a. Likewise, if $f(x, y) \geq 0$, $\int_C f(x, y)\, ds$ measures the surface area of the shaded surface indicated in Figure 14.18b. In general, $\int_a^b f(x)\, dx$ measures signed area [positive if $f(x) > 0$ and negative if $f(x) < 0$] and the line integral $\int_C f(x, y)\, ds$ measures the (signed) surface area of the surface formed by vertical segments from the xy-plane to the graph of $z = f(x, y)$.

Theorem 2.3, whose proof we leave as a straightforward exercise, gives some geometric significance to line integrals in both two and three dimensions.

THEOREM 2.3

For any piecewise-smooth curve C (in two or three dimensions), $\int_C 1\, ds$ gives the arc length of the curve C.

We have now extended the geometry and certain applications of definite integrals to line integrals of the form $\int_C f(x, y)\, ds$. However, not all of the properties of definite integrals can be extended to line integrals as we have defined them thus far. As we will see, a careful consideration of the calculation of work will force us to define several alternative versions of the line integral.

Recall that if a constant force f is exerted to move an object a distance d in a straight line, the work done is given by $W = f \cdot d$. In section 5.6, we extended this to a variable force $f(x)$ applied to an object as it moves in a straight line from $x = a$ to $x = b$, where the work done by the force is given by

$$W = \int_a^b f(x)\, dx.$$

We now extend this idea to find the work done as an object moves along a curve in three dimensions. Here, force vectors are given by the values of vector fields (force fields) and we want to compute the work done on an object by a force field $\mathbf{F}(x, y, z)$, as the object moves along a curve C. Unfortunately, our present notion of line integral (the line integral with respect to arc length) does not help in this case. As we did for finding mass, we need to start from scratch and so, partition the curve C into n segments $C_1, C_2, \ldots, C_n$. Notice that on each segment C_i $(i = 1, 2, \ldots, n)$, if the segment is small and $\mathbf{F}$ is continuous, then $\mathbf{F}$ will be nearly constant on C_i and so, we can approximate $\mathbf{F}$ by its value at some point (x_i^*, y_i^*, z_i^*) on C_i. The work done along C_i (call it W_i) is then approximately the same as the product of the component of the force $\mathbf{F}(x_i^*, y_i^*, z_i^*)$ in the direction of the unit tangent vector $\mathbf{T}(x, y, z)$ to C at (x_i^*, y_i^*, z_i^*) and the distance traveled. That is,

$$W_i \approx \mathbf{F}(x_i^*, y_i^*, z_i^*) \cdot \mathbf{T}(x_i^*, y_i^*, z_i^*)\, \Delta s_i,$$

where Δs_i is the arc length of the segment C_i. Now, if C_i can be represented parametrically by $x = x(t)$, $y = y(t)$ and $z = z(t)$, for $a \leq t \leq b$, and assuming that C_i is smooth, we have

$$\begin{aligned} W_i &\approx \mathbf{F}(x_i^*, y_i^*, z_i^*) \cdot \mathbf{T}(x_i^*, y_i^*, z_i^*)\, \Delta s_i \\ &= \frac{\mathbf{F}(x_i^*, y_i^*, z_i^*) \cdot \langle x'(t_i^*), y'(t_i^*), z'(t_i^*)\rangle}{\sqrt{[x'(t_i^*)]^2 + [y'(t_i^*)]^2 + [z'(t_i^*)]^2}} \sqrt{[x'(t_i^*)]^2 + [y'(t_i^*)]^2 + [z'(t_i^*)]^2}\, \Delta t \\ &= \mathbf{F}(x_i^*, y_i^*, z_i^*) \cdot \langle x'(t_i^*), y'(t_i^*), z'(t_i^*)\rangle\, \Delta t, \end{aligned}$$

where $(x_i^*, y_i^*, z_i^*) = (x(t_i^*), y(t_i^*), z(t_i^*))$. Next, if

$$\mathbf{F}(x, y, z) = \langle F_1(x, y, z), F_2(x, y, z), F_3(x, y, z)\rangle,$$

we have

$$W_i \approx \langle F_1(x_i^*, y_i^*, z_i^*), F_2(x_i^*, y_i^*, z_i^*), F_3(x_i^*, y_i^*, z_i^*)\rangle \cdot \langle x'(t_i^*), y'(t_i^*), z'(t_i^*)\rangle \,\Delta t.$$

Adding together the approximations of the work done over the various segments of C, we have that the total work done is approximately

$$W \approx \sum_{i=1}^{n} \langle F_1(x_i^*, y_i^*, z_i^*), F_2(x_i^*, y_i^*, z_i^*), F_3(x_i^*, y_i^*, z_i^*)\rangle \cdot \langle x'(t_i^*), y'(t_i^*), z'(t_i^*)\rangle \,\Delta t.$$

Finally, taking the limit as the norm of the partition of C approaches zero, we arrive at

$$\begin{aligned} W &= \lim_{\|P\|\to 0} \sum_{i=1}^{n} \mathbf{F}(x_i^*, y_i^*, z_i^*) \cdot \langle x'(t_i^*), y'(t_i^*), z'(t_i^*)\rangle \,\Delta t \\ &= \lim_{\|P\|\to 0} \sum_{i=1}^{n} [F_1(x_i^*, y_i^*, z_i^*)x'(t_i^*)\Delta t + F_2(x_i^*, y_i^*, z_i^*)y'(t_i^*)\,\Delta t \\ &\qquad + F_3(x_i^*, y_i^*, z_i^*)z'(t_i^*)\,\Delta t] \\ &= \int_a^b F_1(x(t), y(t), z(t))x'(t)\,dt + \int_a^b F_2(x(t), y(t), z(t))y'(t)\,dt \\ &\quad + \int_a^b F_3(x(t), y(t), z(t))z'(t)\,dt. \end{aligned} \tag{2.7}$$

We now define line integrals corresponding to each of the three integrals in (2.7).

In Definition 2.2, the notation is the same as in Definition 2.1, with the added terms $\Delta x_i = x_i - x_{i-1}$, $\Delta y_i = y_i - y_{i-1}$ and $\Delta z_i = z_i - z_{i-1}$.

DEFINITION 2.2

The **line integral of $f(x, y, z)$ with respect to x** along the oriented curve C in three-dimensional space is written as $\int_C f(x, y, z)\,dx$ and is defined by

$$\int_C f(x, y, z)\,dx = \lim_{\|P\|\to 0} \sum_{i=1}^{n} f(x_i^*, y_i^*, z_i^*)\,\Delta x_i,$$

provided the limit exists and is the same for all choices of evaluation points.

Likewise, we define the **line integral of $f(x, y, z)$ with respect to y** along C by

$$\int_C f(x, y, z)\,dy = \lim_{\|P\|\to 0} \sum_{i=1}^{n} f(x_i^*, y_i^*, z_i^*)\,\Delta y_i$$

and the **line integral of $f(x, y, z)$ with respect to z** along C by

$$\int_C f(x, y, z)\,dz = \lim_{\|P\|\to 0} \sum_{i=1}^{n} f(x_i^*, y_i^*, z_i^*)\,\Delta z_i.$$

In each case, the line integral is defined whenever the corresponding limit exists and is the same for all choices of evaluation points.

If we have a parametric representation of the curve C, then we can rewrite each line integral as a definite integral. The proof of Theorem 2.4 is very similar to that of Theorem 2.1 and we leave it as an exercise.

THEOREM 2.4 (Evaluation Theorem)

Suppose that $f(x, y, z)$ is continuous in a region D containing the curve C and that C is described parametrically by $x = x(t)$, $y = y(t)$ and $z = z(t)$, where t ranges from $t = a$ to $t = b$ and $x(t)$, $y(t)$ and $z(t)$ have continuous first derivatives. Then,

$$\int_C f(x, y, z)\,dx = \int_a^b f(x(t), y(t), z(t))\,x'(t)\,dt,$$

$$\int_C f(x, y, z)\,dy = \int_a^b f(x(t), y(t), z(t))\,y'(t)\,dt \quad \text{and}$$

$$\int_C f(x, y, z)\,dz = \int_a^b f(x(t), y(t), z(t))\,z'(t)\,dt.$$

Before returning to the calculation of work, we will examine some simpler examples. Recall that the line integral along a given curve with respect to arc length will not change if we traverse the curve in the opposite direction. On the other hand, as we see in example 2.5, line integrals with respect to x, y or z change sign when the orientation of the curve changes.

EXAMPLE 2.5 Calculating a Line Integral in Space

Compute the line integral $\int_C (4xz + 2y)\,dx$, where C is the line segment (a) from (2, 1, 0) to (4, 0, 2) and (b) from (4, 0, 2) to (2, 1, 0).

Solution First, parametric equations for C for part (a) are

$$x = 2 + (4 - 2)t = 2 + 2t,$$
$$y = 1 + (0 - 1)t = 1 - t \quad \text{and}$$
$$z = 0 + (2 - 0)t = 2t,$$

for $0 \le t \le 1$. The integrand is then

$$4xz + 2y = 4(2 + 2t)(2t) + 2(1 - t) = 16t^2 + 14t + 2$$

and the element dx is given by

$$dx = x'(t)\,dt = 2\,dt.$$

From the Evaluation Theorem, the line integral is now given by

$$\int_C (4xz + 2y)\,dx = \int_0^1 \underbrace{(16t^2 + 14t + 2)}_{4xz+2y}\,\underbrace{(2)\,dt}_{dx} = \frac{86}{3}.$$

For part (b), you can use the fact that the line segment connects the same two points as in part (a), but in the opposite direction. The same parametric equations will then work, with the single change that t will run from $t = 1$ to $t = 0$. This gives us

$$\int_C (4xz + 2y)\,dx = \int_1^0 (16t^2 + 14t + 2)(2)\,dt = -\frac{86}{3},$$

where you should recall that reversing the order of integration changes the sign of the integral. ■

Theorem 2.5 corresponds to Theorem 2.2 for line integrals with respect to arc length, but pay special attention to the minus sign in part (i). We state the result for line integrals with respect to x, with corresponding results holding true for line integrals with respect to y or z, as well. We leave the proof as an exercise.

THEOREM 2.5

Suppose that $f(x, y, z)$ is a continuous function in some region D containing the oriented curve C. Then, the following hold.

(i) If C is piecewise-smooth, then

change of orientation of curve →

$$\int_{-C} f(x, y, z)\,dx = -\int_{C} f(x, y, z)\,dx.$$

(ii) If $C = C_1 \cup C_2 \cup \cdots \cup C_n$, where $C_1, C_2, \ldots, C_n$ are all smooth and the terminal point of C_i is the same as the initial point of C_{i+1}, for $i = 1, 2, \ldots, n-1$, then

$$\int_{C} f(x, y, z)\,dx = \int_{C_1} f(x, y, z)\,dx + \int_{C_2} f(x, y, z)\,dx + \cdots + \int_{C_n} f(x, y, z)\,dx.$$

NOTES

As a convenience, we will usually write

$$\int_{C} f(x, y, z)\,dx + \int_{C} g(x, y, z)\,dy$$
$$+ \int_{C} h(x, y, z)\,dz$$
$$= \int_{C} f(x, y, z)\,dx + g(x, y, z)\,dy$$
$$+ h(x, y, z)\,dz.$$

Line integrals with respect to x, y and z can be very simple when the curve C consists of line segments parallel to the coordinate axes, as we see in example 2.6.

EXAMPLE 2.6   Calculating a Line Integral in Space

Compute $\int_{C} 4x\,dy + 2y\,dz$, where C consists of the line segment from $(0, 1, 0)$ to $(0, 1, 1)$, followed by the line segment from $(0, 1, 1)$ to $(2, 1, 1)$ and followed by the line segment from $(2, 1, 1)$ to $(2, 4, 1)$.

Solution   We show a sketch of the curves in Figure 14.19. Parametric equations for the first segment C_1 are $x = 0$, $y = 1$ and $z = t$ with $0 \le t \le 1$. On this segment, we have $dy = 0\,dt$ and $dz = 1\,dt$. On the second segment C_2, parametric equations are $x = 2t$, $y = 1$ and $z = 1$ with $0 \le t \le 1$. On this segment, we have $dy = dz = 0\,dt$. On the third segment, parametric equations are $x = 2$, $y = 3t + 1$ and $z = 1$ with $0 \le t \le 1$. On this segment, we have $dy = 3\,dt$ and $dz = 0\,dt$. Putting this all together, we have

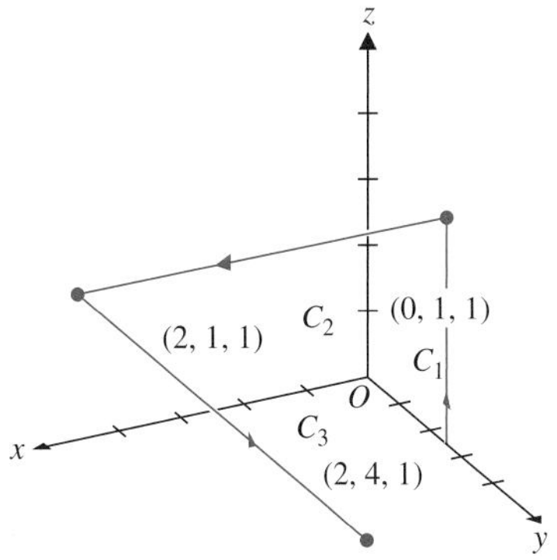

FIGURE 14.19
The path C

$$\int_{C} 4x\,dy + 2y\,dz = \int_{C_1} 4x\,dy + 2y\,dz + \int_{C_2} 4x\,dy + 2y\,dz + \int_{C_3} 4x\,dy + 2y\,dz$$

$$= \int_{0}^{1} [\underbrace{4(0)}_{4x}\ \underbrace{(0)}_{y'(t)} + \underbrace{2(1)}_{2y}\ \underbrace{(1)}_{z'(t)}]\,dt + \int_{0}^{1} [\underbrace{4(2t)}_{4x}\ \underbrace{(0)}_{y'(t)} + \underbrace{2(1)}_{2y}\ \underbrace{(0)}_{z'(t)}]\,dt$$

$$+ \int_{0}^{1} [\underbrace{4(2)}_{4x}\ \underbrace{(3)}_{y'(t)} + \underbrace{2(3t + 1)}_{2y}\ \underbrace{(0)}_{z'(t)}]\,dt$$

$$= \int_{0}^{1} 26\,dt = 26. \quad\blacksquare$$

Notice that a line integral will be zero if the integrand simplifies to 0 or if the variable of integration is constant along the curve. For instance, if z is constant on some curve, then the change in z (given by dz) will be 0 on that curve.

Recall that our motivation for introducing line integrals with respect to the three coordinate variables was to compute the work done by a force field while moving an object along a curve. From (2.7), the work performed by the force field $\mathbf{F}(x, y, z) = \langle F_1(x, y, z), F_2(x, y, z), F_3(x, y, z)\rangle$ along the curve defined parametrically by $x = x(t)$, $y = y(t)$, $z = z(t)$, for $a \le t \le b$, is given by

$$W = \int_a^b F_1(x(t), y(t), z(t))x'(t)\,dt + \int_a^b F_2(x(t), y(t), z(t))y'(t)\,dt + \int_a^b F_3(x(t), y(t), z(t))z'(t)\,dt.$$

You should now recognize that we can rewrite each of the three terms in this expression for work using Theorem 2.4, to obtain

$$W = \int_C F_1(x, y, z)\,dx + \int_C F_2(x, y, z)\,dy + \int_C F_3(x, y, z)\,dz.$$

We now introduce some notation to write such a combination of line integrals in a simpler form.

Suppose that a vector field $\mathbf{F}(x, y, z) = \langle F_1(x, y, z), F_2(x, y, z), F_3(x, y, z)\rangle$. Since we use $\mathbf{r} = x\mathbf{i} + y\mathbf{j} + z\mathbf{k}$, we define

$$d\mathbf{r} = dx\mathbf{i} + dy\mathbf{j} + dz\mathbf{k} \quad \text{or} \quad d\mathbf{r} = \langle dx, dy, dz\rangle.$$

We now define the line integral

$$\begin{aligned}\int_C \mathbf{F}(x, y, z)\cdot d\mathbf{r} &= \int_C F_1(x, y, z)\,dx + F_2(x, y, z)\,dy + F_3(x, y, z)\,dz \\ &= \int_C F_1(x, y, z)\,dx + \int_C F_2(x, y, z)\,dy + \int_C F_3(x, y, z)\,dz.\end{aligned}$$

In the case where $\mathbf{F}(x, y, z)$ is a force field, the work done by $\mathbf{F}$ in moving a particle along the curve C can be written simply as

$$\boxed{W = \int_C \mathbf{F}(x, y, z)\cdot d\mathbf{r}.} \tag{2.8}$$

Notice how the different parts of $\int_C \mathbf{F}(x, y, z)\cdot d\mathbf{r}$ correspond to our knowledge of work. The only way in which the x-component of force affects the work done is when the object moves in the x-direction (i.e., when $dx \neq 0$). Similarly, the y-component of force contributes to the work only when $dy \neq 0$ and the z-component of force contributes to the work only when $dz \neq 0$.

EXAMPLE 2.7 Computing Work

Compute the work done by the force field $\mathbf{F}(x, y, z) = \langle 4y, 2xz, 3y\rangle$ acting on an object as it moves along the helix defined parametrically by $x = 2\cos t$, $y = 2\sin t$ and $z = 3t$, from the point $(2, 0, 0)$ to the point $(-2, 0, 3\pi)$.

Solution From (2.8), the work is given by

$$W = \int_C \mathbf{F}(x, y, z)\cdot d\mathbf{r} = \int_C 4y\,dx + 2xz\,dy + 3y\,dz.$$

We have already provided parametric equations for the curve, but not the range of t-values. Notice that from $z = 3t$, you can determine that $(2, 0, 0)$ corresponds to $t = 0$

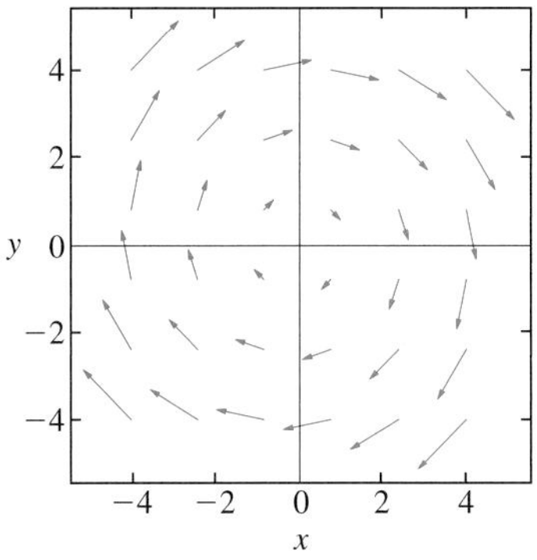

FIGURE 14.20
$\mathbf{F}(x, y) = \langle y, -x \rangle$

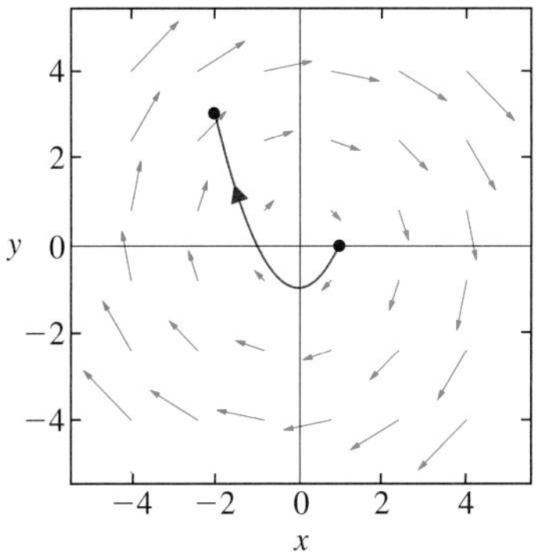

FIGURE 14.21
$\mathbf{F}(x, y) = \langle y, -x \rangle$ and
$x = t, y = t^2 - 1, -2 \le t \le 1$

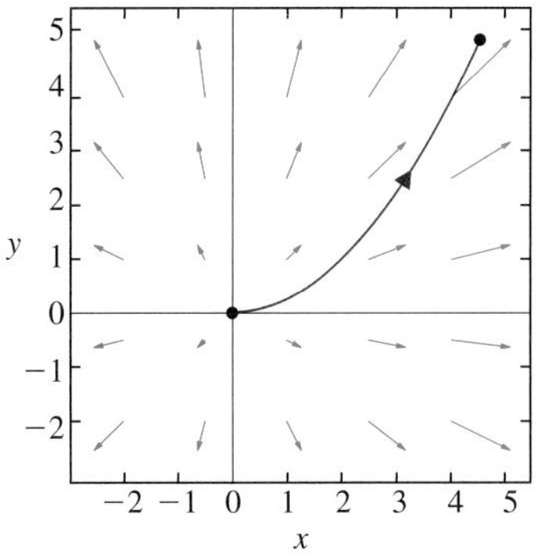

FIGURE A

and $(-2, 0, 3\pi)$ corresponds to $t = \pi$. Substituting in for x, y, z and $dx = -2\sin t\,dt$, $dy = 2\cos t\,dt$ and $dz = 3\,dt$, we have

$$W = \int_C 4y\,dx + 2xz\,dy + 3y\,dz$$
$$= \int_0^{\pi} [\underbrace{4(2\sin t)}_{4y}\underbrace{(-2\sin t)}_{x'(t)} + \underbrace{2(2\cos t)(3t)}_{2xz}\underbrace{(2\cos t)}_{y'(t)} + \underbrace{3(2\sin t)}_{3y}\underbrace{(3)}_{z'(t)}]\,dt$$
$$= \int_0^{\pi} (-16\sin^2 t + 24t\cos^2 t + 18\sin t)\,dt = 36 - 8\pi + 6\pi^2,$$

where we used a computer algebra system to evaluate the final integral. ■

We compute the work performed by a two-dimensional vector field in the same way as we did in three dimensions, as we illustrate in example 2.8.

EXAMPLE 2.8 Computing Work

Compute the work done by the force field $\mathbf{F}(x, y) = \langle y, -x \rangle$ acting on an object as it moves along the parabola $y = x^2 - 1$ from $(1, 0)$ to $(-2, 3)$.

Solution From (2.8), the work is given by

$$W = \int_C \mathbf{F}(x, y) \cdot d\mathbf{r} = \int_C y\,dx - x\,dy.$$

Here, we use $x = t$ and $y = t^2 - 1$ as parametric equations for the curve, with t ranging from $t = 1$ to $t = -2$. In this case, $dx = 1\,dt$ and $dy = 2t\,dt$ and the work is

$$W = \int_C y\,dx - x\,dy = \int_1^{-2} [(t^2 - 1)(1) - (t)(2t)]\,dt = \int_1^{-2} (-t^2 - 1)\,dt = 6.$$

■

A careful look at example 2.8 graphically will show us an important geometric interpretation of the work line integral. A computer-generated graph of the vector field $\mathbf{F}(x, y) = \langle y, -x \rangle$ is shown in Figure 14.20. Thinking of $\mathbf{F}(x, y)$ as describing the velocity field for a fluid in motion, notice that the fluid is rotating clockwise. In Figure 14.21, we superimpose the curve $x = t$, $y = t^2 - 1$, $-2 \le t \le 1$, onto the vector field $\mathbf{F}(x, y)$. Notice that an object moving along the curve from $(1, 0)$ to $(-2, 3)$ is generally moving in the same direction as that indicated by the vectors in the vector field. If $\mathbf{F}(x, y)$ represents a force field, then the force pushes an object moving along C, adding energy to it and therefore doing positive work. If the curve were oriented in the opposite direction, the force would oppose the motion of the object, thereby doing negative work.

EXAMPLE 2.9 Determining the Sign of a Line Integral Graphically

In each of the following graphs, an oriented curve is superimposed onto a vector field $\mathbf{F}(x, y)$. Determine whether $\int_C \mathbf{F}(x, y) \cdot d\mathbf{r}$ is positive or negative.

Solution In Figure A, the curve is oriented in the same direction as the vectors, so the force is making a positive contribution to the object's motion. The work done by the force is then positive. In Figure B, the curve is oriented in the opposite direction as the vectors, so that the force is making a negative contribution to the object's motion. The work done by the force is then negative. In Figure C, the force field vectors are purely horizontal. Since both the curve and the force vectors point to the right, the work is positive. Finally, in Figure D, the force field is the same as in Figure C, but the curve is

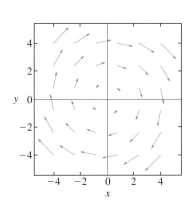

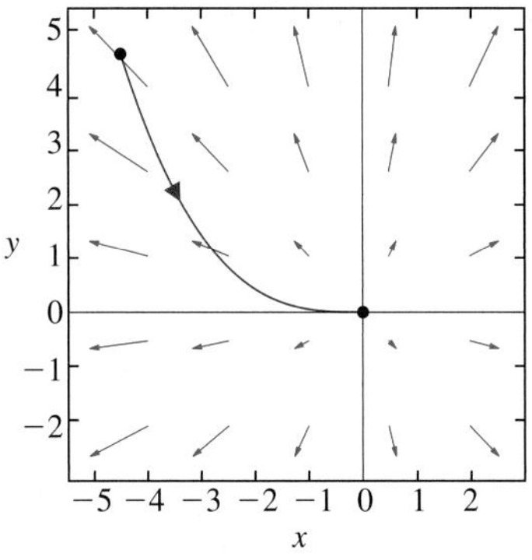

FIGURE B

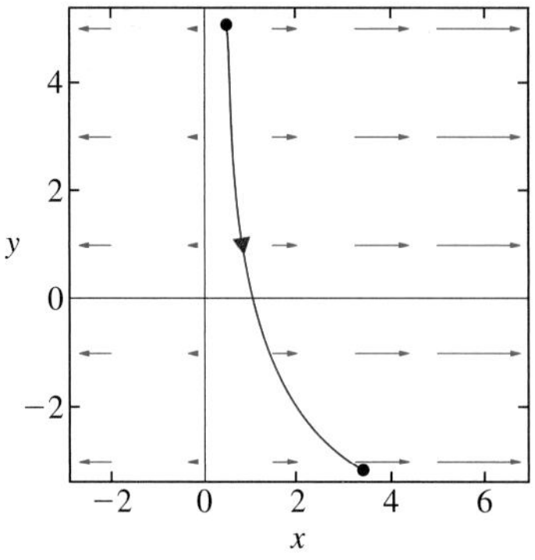

FIGURE C

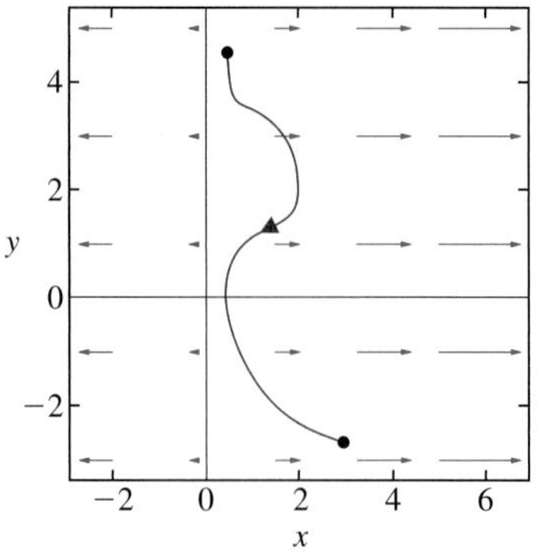

FIGURE D

more complicated. Since the force vectors are horizontal and do not depend on y, the work done as the object moves to the right is exactly canceled when the object doubles back to the left. Comparing the initial and terminal points, the object has made a net movement to the right (the same direction as the vector field), so that the work done is positive. ■

BEYOND FORMULAS

Several different line integrals are defined in this section. To keep them straight, always think of dx as an increment (small change) in x, dy as an increment in y and so on. In particular, ds represents an increment in arc length (distance) along the curve, which is why the arc length formula plays a role in the Evaluation Theorem. In applications, the dx, dy and dz line integrals are useful when the quantity being measured (such as a force) can be broken down into separate x, y and z components. By contrast, the ds line integral is applied to the measurement of a quantity as we move along the curve in three dimensions. What is an example of this situation?

EXERCISES 14.2

HW pg 1142-1143
#1, 7, 25, 29

WRITING EXERCISES

1. It is important to understand why $\int_C f\,ds = \int_{-C} f\,ds$. Think of f as being a density function and the line integral as giving the mass of an object. Explain why the integrals must be equal.

2. For example 2.3, part (a), a different set of parametric equations is $x = -t$ and $y = t^2$, with t running from $t = 1$ to $t = -2$. In light of the Evaluation Theorem, explain why we couldn't use these parametric equations.

3. Explain in words why Theorem 2.5(i) is true. In particular, explain in terms of approximating sums why the integrals in Theorem 2.5(i) have opposite signs but the integrals in Theorem 2.2(i) are the same.

4. In example 2.9, we noted that the force vectors in Figure D are horizontal and independent of y. Explain why this allows us to ignore the vertical component of the curve. Also, explain why the work would be the same for *any* curve with the same initial and terminal points.

In exercises 1–24, evaluate the line integral.

1. $\int_C 2x\,ds$, where C is the line segment from (1, 2) to (3, 5)
2. $\int_C (x - y)\,ds$, where C is the line segment from (1, 0) to (3, 1)
3. $\int_C (3x + y)\,ds$, where C is the line segment from (5, 2) to (1, 1)
4. $\int_C 2xy\,ds$, where C is the line segment from (1, 2) to (−1, 0)
5. $\int_C 2x\,dx$, where C is the line segment from (0, 2) to (2, 6)
6. $\int_C 3y^2\,dy$, where C is the line segment from (2, 0) to (1, 3)

7. $\int_C 3x\,ds$, where C is the quarter-circle $x^2 + y^2 = 4$ from $(2, 0)$ to $(0, 2)$

8. $\int_C (3x - y)\,ds$, where C is the quarter-circle $x^2 + y^2 = 9$ from $(0, 3)$ to $(3, 0)$

9. $\int_C 2x\,dx$, where C is the quarter-circle $x^2 + y^2 = 4$ from $(2, 0)$ to $(0, 2)$

10. $\int_C 3y^2 dy$, where C is the quarter-circle $x^2 + y^2 = 4$ from $(0, 2)$ to $(-2, 0)$

11. $\int_C 3y\,dx$, where C is the half-ellipse $x^2 + 4y^2 = 4$ from $(0, 1)$ to $(0, -1)$ with $x \geq 0$

12. $\int_C x^2\,dy$, where C is the ellipse $4x^2 + y^2 = 4$ oriented counterclockwise

13. $\int_C 3y\,ds$, where C is the portion of $y = x^2$ from $(0, 0)$ to $(2, 4)$

14. $\int_C 2x\,ds$, where C is the portion of $y = x^2$ from $(-2, 4)$ to $(2, 4)$

15. $\int_C 2x\,dx$, where C is the portion of $y = x^2$ from $(2, 4)$ to $(0, 0)$

16. $\int_C 3y^2 dy$, where C is the portion of $y = x^2$ from $(2, 4)$ to $(0, 0)$

17. $\int_C 3y\,dx$, where C is the portion of $x = y^2$ from $(1, 1)$ to $(4, 2)$

18. $\int_C (x + y)\,dy$, where C is the portion of $x = y^2$ from $(1, 1)$ to $(1, -1)$

19. $\int_C 3x\,ds$, where C is the line segment from $(0, 0)$ to $(1, 0)$, followed by the quarter-circle to $(0, 1)$

20. $\int_C 2y\,ds$, where C is the portion of $y = x^2$ from $(0, 0)$ to $(2, 4)$, followed by the line segment to $(3, 0)$

21. $\int_C 4z\,ds$, where C is the line segment from $(1, 0, 1)$ to $(2, -2, 2)$

22. $\int_C xz\,ds$, where C is the line segment from $(2, 1, 0)$ to $(2, 0, 2)$

23. $\int_C 4(x - z)z\,dx$, where C is the portion of $y = x^2$ in the plane $z = 2$ from $(1, 1, 2)$ to $(2, 4, 2)$

24. $\int_C z\,ds$, where C is the intersection of $x^2 + y^2 = 4$ and $z = 0$ (oriented clockwise as viewed from above)

In exercises 25–36, compute the work done by the force field F along the curve C.

25. $\mathbf{F}(x, y) = \langle 2x, 2y \rangle$, C is the line segment from $(3, 1)$ to $(5, 4)$

26. $\mathbf{F}(x, y) = \langle 2y, -2x \rangle$, C is the line segment from $(4, 2)$ to $(0, 4)$

27. $\mathbf{F}(x, y) = \langle 2x, 2y \rangle$, C is the quarter-circle from $(4, 0)$ to $(0, 4)$

28. $\mathbf{F}(x, y) = \langle 2y, -2x \rangle$, C is the upper half-circle from $(-3, 0)$ to $(3, 0)$

$\int_C 2\,dx + x\,dy$

29. $\mathbf{F}(x, y) = \langle 2, x \rangle$, C is the portion of $y = x^2$ from $(0, 0)$ to $(1, 1)$

30. $\mathbf{F}(x, y) = \langle 0, xy \rangle$, C is the portion of $y = x^3$ from $(0, 0)$ to $(1, 1)$

31. $\mathbf{F}(x, y) = \langle 3x, 2 \rangle$, C is the line segment from $(0, 0)$ to $(0, 1)$, followed by the line segment to $(4, 1)$

32. $\mathbf{F}(x, y) = \langle y, x \rangle$, C is the square from $(0, 0)$ to $(1, 0)$ to $(1, 1)$ to $(0, 1)$ to $(0, 0)$

33. $\mathbf{F}(x, y, z) = \langle y, 0, z \rangle$, C is the triangle from $(0, 0, 0)$ to $(2, 1, 2)$ to $(2, 1, 0)$ to $(0, 0, 0)$

34. $\mathbf{F}(x, y, z) = \langle z, y, 0 \rangle$, C is the line segment from $(1, 0, 2)$ to $(2, 4, 2)$

35. $\mathbf{F}(x, y, z) = \langle xy, 3z, 1 \rangle$, C is the helix $x = \cos t$, $y = \sin t$, $z = 2t$ from $(1, 0, 0)$ to $(0, 1, \pi)$

36. $\mathbf{F}(x, y, z) = \langle z, 0, 3x^2 \rangle$, C is the quarter-ellipse $x = 2\cos t$, $y = 3\sin t$, $z = 1$ from $(2, 0, 1)$ to $(0, 3, 1)$

In exercises 37–42, use the graph to determine whether the work done is positive, negative or zero.

37.

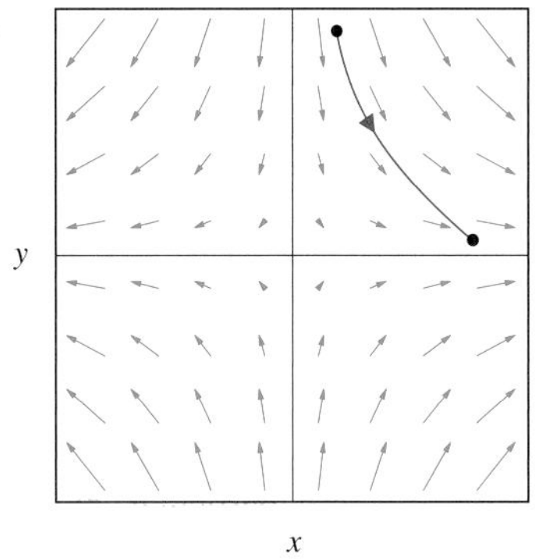

38.

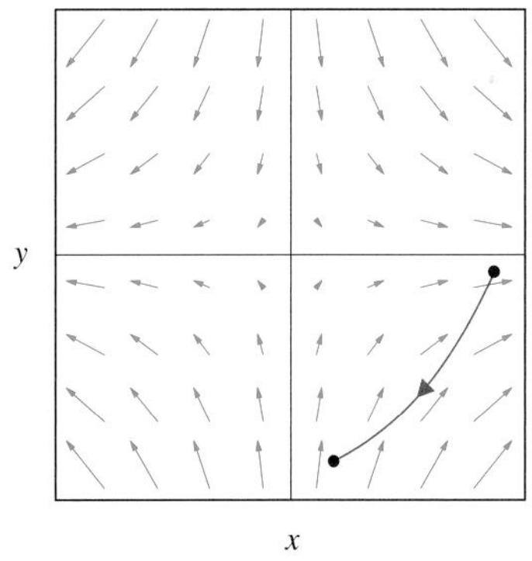

39.

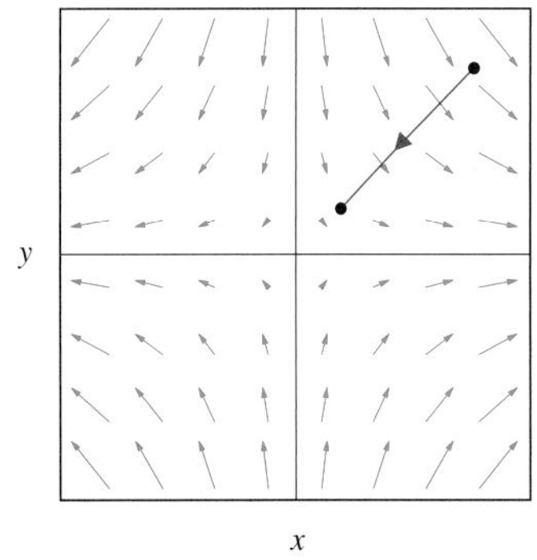

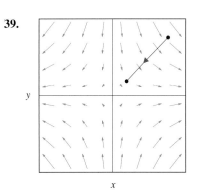

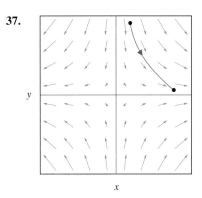

40.

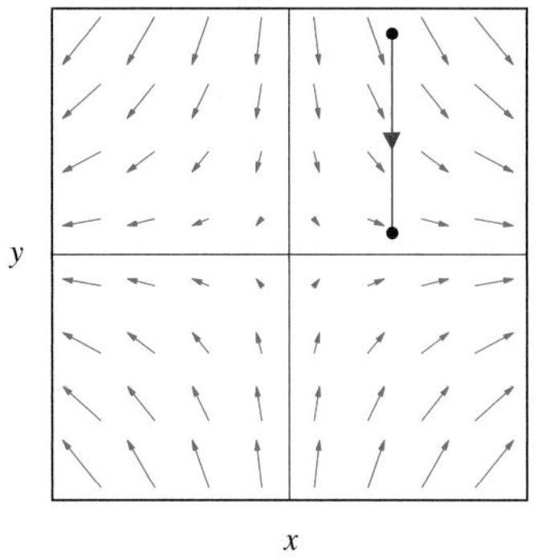

41.

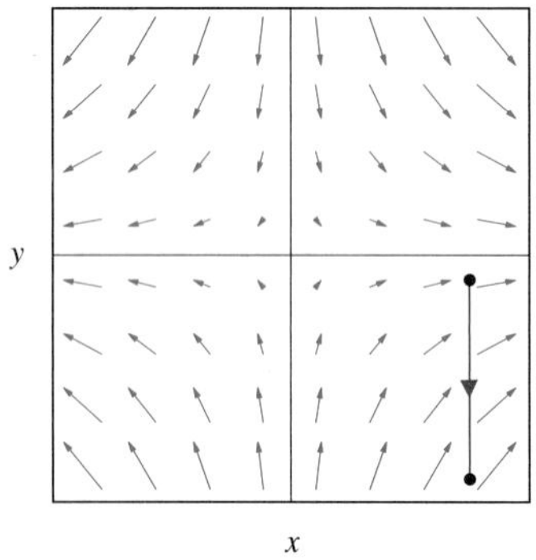

42.

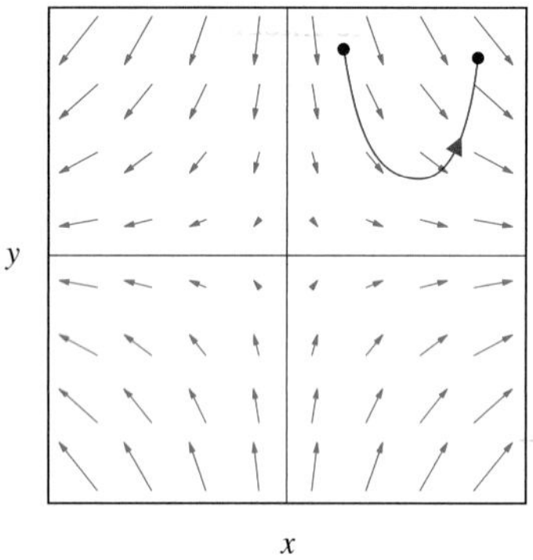

In exercises 43–52, use the formulas $m = \int_C \rho\,ds$, $\bar{x} = \frac{1}{m}\int_C x\rho\,ds$, $\bar{y} = \frac{1}{m}\int_C y\rho\,ds$, $I = \int_C w^2\rho\,ds$.

43. Compute the mass m of a rod with density $\rho(x, y) = x$ in the shape of $y = x^2, 0 \le x \le 3$.

44. Compute the mass m of a rod with density $\rho(x, y) = y$ in the shape of $y = 4 - x^2, 0 \le x \le 2$.

45. Compute the center of mass $(\bar{x}, \bar{y})$ of the rod of exercise 43.

46. Compute the center of mass $(\bar{x}, \bar{y})$ of the rod of exercise 44.

47. Compute the moment of inertia I for rotating the rod of exercise 43 about the y-axis. Here, w is the distance from the point (x, y) to the y-axis.

48. Compute the moment of inertia I for rotating the rod of exercise 44 about the x-axis. Here, w is the distance from the point (x, y) to the x-axis.

49. Compute the moment of inertia I for rotating the rod of exercise 43 about the line $y = 9$. Here, w is the distance from the point (x, y) to $y = 9$.

50. Compute the moment of inertia I for rotating the rod of exercise 44 about the line $x = 2$. Here, w is the distance from the point (x, y) to $x = 2$.

51. Compute the mass m of the helical spring $x = \cos 2t$, $y = \sin 2t$, $z = t$, $0 \le t \le \pi$, with density $\rho = z^2$.

52. Compute the mass m of the ellipse $x = 4\cos t$, $y = 4\sin t$, $z = 4\cos t$, $0 \le t \le 2\pi$, with density $\rho = 4$.

53. Show that the center of mass in exercises 45 and 46 is not located at a point on the rod. Explain why this means that our previous interpretation of center of mass as a balance point is no longer valid. Instead, the center of mass is the point about which the object rotates when a torque is applied.

54. Suppose a torque is applied to the rod in exercise 43 such that the rod rotates but has no other motion. Find parametric equations for the position of the part of the rod that starts at the point (1, 1).

In exercises 55–60, find the surface area extending from the given curve in the xy-plane to the given surface.

55. Above the quarter-circle of radius 2 centered at the origin from (2, 0, 0) to (0, 2, 0) up to the surface $z = x^2 + y^2$

56. Above the portion of $y = x^2$ from (0, 0, 0) to (2, 4, 0) up to the surface $z = x^2 + y^2$

57. Above the line segment from (2, 0, 0) to (−2, 0, 0) up to the surface $z = 4 - x^2 - y^2$

58. Above the line segment from (1, 1, 0) to (−1, 1, 0) up to the surface $z = \sqrt{x^2 + y^2}$

59. Above the unit square $x \in [0, 1]$, $y \in [0, 1]$ up to the plane $z = 4 - x - y$

60. Above the ellipse $x^2 + 4y^2 = 4$ up to the plane $z = 4 - x$

In exercises 61 and 62, estimate the line integrals (a) $\int_C f\,ds$, (b) $\int_C f\,dx$ and (c) $\int_C f\,dy$.

61.

(x, y)	(0, 0)	(1, 0)	(1, 1)	(1.5, 1.5)
$f(x, y)$	2	3	3.6	4.4

(x, y)	(2, 2)	(3, 2)	(4, 1)
$f(x, y)$	5	4	4

62.

(x, y)	(0, 0)	(1, −1)	(2, 0)	(3, 1)
$f(x, y)$	1	0	−1.2	0.4

(x, y)	(4, 0)	(3, −1)	(2, −2)
$f(x, y)$	1.5	2.4	2

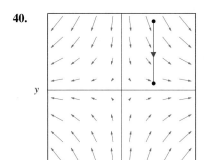

63. Prove Theorem 2.1 in the case of a curve in three dimensions.

64. Prove Theorem 2.2.

65. Prove Theorem 2.3.

66. Prove Theorem 2.4.

67. Prove Theorem 2.5.

68. If C has parametric equations $x = x(t)$, $y = y(t)$, $z = z(t)$, $a \le t \le b$, for differentiable functions x, y and z, show that $\int_C \mathbf{F} \cdot \mathbf{T}\, ds = \int_a^b [F_1(x, y, z)x'(t) + F_2(x, y, z)y'(t) + F_3(x, y, z)z'(t)]\, dt$, which is the work line integral $\int_C \mathbf{F} \cdot d\mathbf{r}$.

69. If the two-dimensional vector $\mathbf{n}$ is normal (perpendicular to the tangent) to the curve C at each point and $\mathbf{F}(x, y) = \langle F_1(x, y), F_2(x, y)\rangle$, show that $\int_C \mathbf{F} \cdot \mathbf{n}\, ds = \int_C F_1\, dy - F_2\, dx$.

70. If $T(x, y)$ is the temperature function, the line integral $\int_C(-k\nabla T) \cdot \mathbf{n}\, ds$ gives the rate of heat loss across C. For $T(x, y) = 60e^{y/50}$ and C the rectangle with sides $x = -20$, $x = 20$, $y = -5$ and $y = 5$, compute the rate of heat loss. Explain in terms of the temperature function why the integral is 0 along two sides of C.

EXPLORATORY EXERCISES

1. Look carefully at the solutions to exercises 5–6, 9–10 and 15–16. Compare the solutions to integrals of the form $\int_a^b 2x\, dx$ and $\int_c^d 3y^2 dy$. Formulate a rule for evaluating line integrals of the form $\int_C f(x)\, dx$ and $\int_C g(y)\, dy$. If the curve C is a closed curve (e.g., a square or a circle), evaluate the line integrals $\int_C f(x)\, dx$ and $\int_C g(y)\, dy$.

14.3 INDEPENDENCE OF PATH AND CONSERVATIVE VECTOR FIELDS

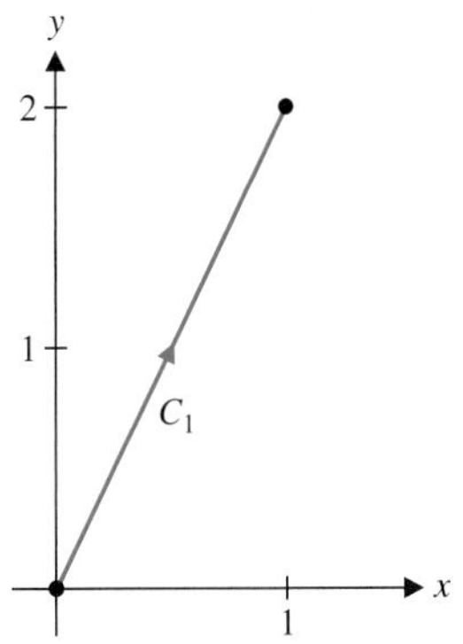

FIGURE 14.22a
The path C_1

As you've seen, there are a lot of steps needed to evaluate a line integral. First, you must parameterize the curve, rewrite the line integral as a definite integral and then evaluate the resulting definite integral. While this process is unavoidable for many line integrals, we will now consider a group of line integrals that are the same along every curve connecting the given endpoints. We'll determine the circumstances under which this occurs and see that when this does happen, there is a simple way to evaluate the integral.

We begin with a simple observation. Consider the line integral $\int_{C_1} \mathbf{F} \cdot d\mathbf{r}$, where $\mathbf{F}(x, y) = \langle 2x, 3y^2\rangle$ and C_1 is the straight line segment joining the two points (0, 0) and (1, 2). (See Figure 14.22a.) To parameterize the curve, we take $x = t$ and $y = 2t$, for $0 \le t \le 1$. We then have

$$\begin{aligned}\int_{C_1} \mathbf{F} \cdot d\mathbf{r} &= \int_{C_1} \langle 2x, 3y^2\rangle \cdot \langle dx, dy\rangle \\ &= \int_{C_1} 2x\, dx + 3y^2\, dy \\ &= \int_0^1 [2t + 12t^2(2)]\, dt = 9,\end{aligned}$$

where we have left the details of the final (routine) calculation to you. For the same vector field $\mathbf{F}(x, y)$, consider now $\int_{C_2} \mathbf{F} \cdot d\mathbf{r}$, where C_2 is made up of the horizontal line segment from (0, 0) to (1, 0) followed by the vertical line segment from (1, 0) to (1, 2). (See Figure 14.22b.) In this case, we have

$$\begin{aligned}\int_{C_2} \mathbf{F} \cdot d\mathbf{r} &= \int_{C_2} \langle 2x, 3y^2\rangle \cdot \langle dx, dy\rangle \\ &= \int_0^1 2x\, dx + \int_0^2 3y^2\, dy = 9,\end{aligned}$$

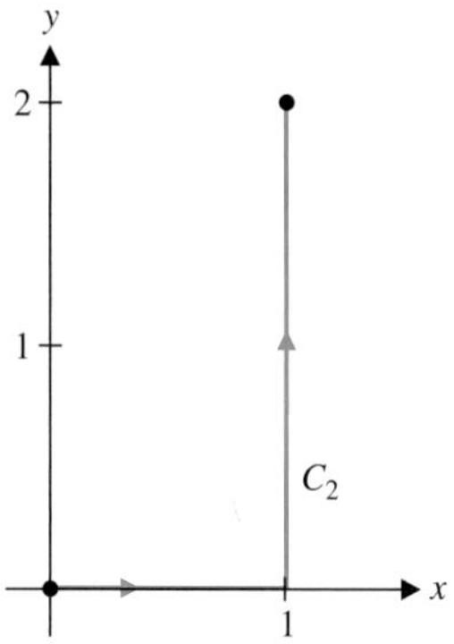

FIGURE 14.22b
The path C_2

where we have again left the final details to you. Look carefully at these two line integrals. Although the integrands are the same and the endpoints of the two curves are the same, the curves followed are quite different. You should try computing this line integral over several additional curves from (0, 0) to (1, 2). You will find that each line integral has the same value: 9. This integral is an example of one that is the same along any curve from (0, 0) to (1, 2).

Let C be any piecewise-smooth curve, traced out by the endpoint of the vector-valued function $\mathbf{r}(t)$, for $a \le t \le b$. In this context, we usually refer to a curve connecting two given points as a **path.** We say that the line integral $\int_C \mathbf{F} \cdot d\mathbf{r}$ is **independent of path** in the domain D if the integral is the same for every path contained in D that has the same beginning and ending points. Before we see when this happens, we need a definition.

> **DEFINITION 3.1**
>
> A region $D \subset \mathbb{R}^n$ (for $n \ge 2$) is called **connected** if every pair of points in D can be connected by a piecewise-smooth curve lying entirely in D.

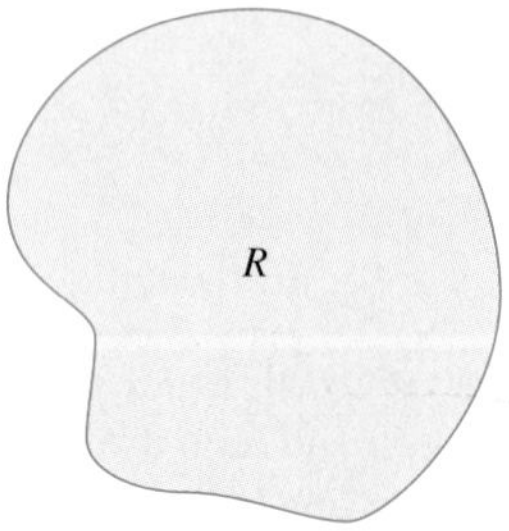

FIGURE 14.23a
Connected region

In Figure 14.23a, we show a region in $\mathbb{R}^2$ that is connected and in Figure 14.23b, we indicate a region that is not connected. We are now in a position to prove a result concerning integrals that are independent of path. While we state and prove the result for line integrals in the plane, the result is valid in any number of dimensions.

> **THEOREM 3.1**
>
> Suppose that the vector field $\mathbf{F}(x, y) = \langle M(x, y), N(x, y)\rangle$ is continuous on the open, connected region $D \subset \mathbb{R}^2$. Then, the line integral $\int_C \mathbf{F}(x, y) \cdot d\mathbf{r}$ is independent of path in D if and only if $\mathbf{F}$ is conservative on D.

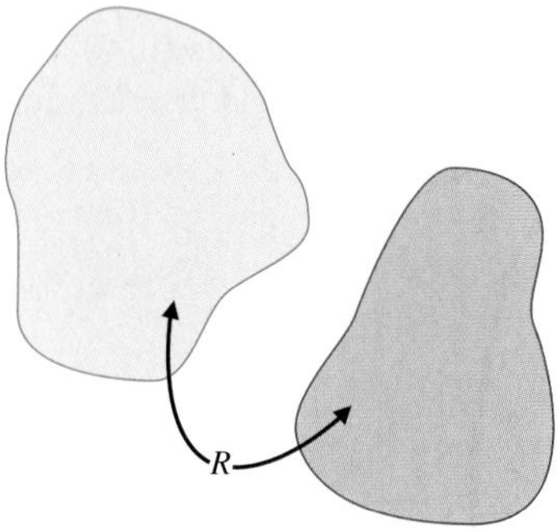

FIGURE 14.23b
Not connected

PROOF

Recall that a vector field $\mathbf{F}$ is conservative whenever $\mathbf{F} = \nabla f$, for some scalar function f (called a potential function for $\mathbf{F}$). You should recognize that there are several things to prove here.

First, suppose that $\mathbf{F}$ is conservative, with $\mathbf{F}(x, y) = \nabla f(x, y)$. Then

$$\mathbf{F}(x, y) = \langle M(x, y), N(x, y)\rangle = \nabla f(x, y) = \langle f_x(x, y), f_y(x, y)\rangle$$

and so, we must have

$$M(x, y) = f_x(x, y) \quad \text{and} \quad N(x, y) = f_y(x, y).$$

Let $A(x_1, y_1)$ and $B(x_2, y_2)$ be any two points in D and let C be any smooth path from A to B, lying in D and defined parametrically by $C: x = g(t), y = h(t)$, where $t_1 \le t \le t_2$. (You can extend this proof to any piecewise-smooth path in the obvious way.) Then, we have

$$\begin{aligned}\int_C \mathbf{F}(x, y) \cdot d\mathbf{r} &= \int_C M(x, y)\,dx + N(x, y)\,dy \\ &= \int_C f_x(x, y)\,dx + f_y(x, y)\,dy \\ &= \int_{t_1}^{t_2} [f_x(g(t), h(t))g'(t) + f_y(g(t), h(t))h'(t)]\,dt. \qquad (3.1)\end{aligned}$$

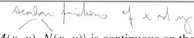

Notice that since f_x and f_y were assumed to be continuous, we have by the chain rule that

$$\frac{d}{dt}[f(g(t), h(t))] = f_x(g(t), h(t))g'(t) + f_y(g(t), h(t))h'(t),$$

which is the integrand in (3.1). By the Fundamental Theorem of Calculus, we now have

$$\begin{aligned}\int_C \mathbf{F}(x, y)\cdot d\mathbf{r} &= \int_{t_1}^{t_2} [f_x(g(t), h(t))g'(t) + f_y(g(t), h(t))h'(t)]\,dt \\ &= \int_{t_1}^{t_2} \frac{d}{dt}[f(g(t), h(t))]\,dt \\ &= f(g(t_2), h(t_2)) - f(g(t_1), h(t_1)) \\ &= f(x_2, y_2) - f(x_1, y_1).\end{aligned}$$

In particular, this says that the value of the integral depends only on the value of the potential function at the two endpoints of the curve and not on the particular path followed. That is, the line integral is independent of path, as desired.

Next, we need to prove that if the integral is independent of path, then the vector field must be conservative. So, suppose that $\int_C \mathbf{F}(x, y)\cdot d\mathbf{r}$ is independent of path in D. For any points (u, v) and $(x_0, y_0) \in D$, define the function

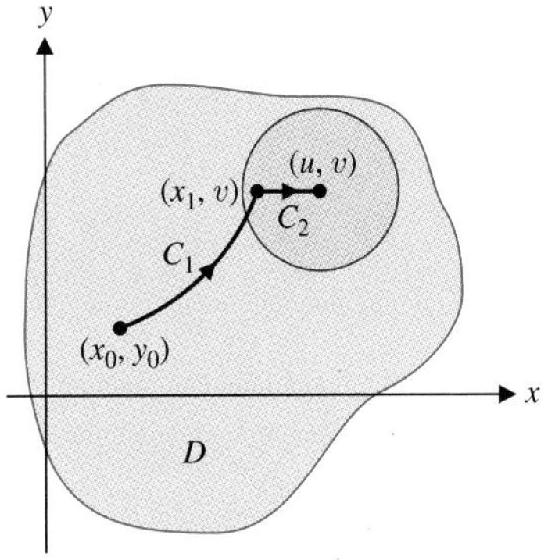

FIGURE 14.24
First path

$$f(u, v) = \int_{(x_0, y_0)}^{(u,v)} \mathbf{F}(x, y)\cdot d\mathbf{r}.$$

(We are using the variables u and v, since the variables x and y inside the integral are dummy variables and cannot be used both inside and outside the line integral.) Notice that since the line integral is independent of path in D, we need not specify a path over which to integrate; it's the same over every path in D connecting these points. (Since D is connected, there is always a path lying in D that connects the points.) Further, since D is open, there is a disk centered at (u, v) and lying completely inside D. Pick any point (x_1, v) in the disk with $x_1 < u$ and let C_1 be any path from (x_0, y_0) to (x_1, v) lying in D. So, in particular, if we integrate over the path consisting of C_1 followed by the horizontal path C_2 indicated in Figure 14.24, we must have

$$f(u, v) = \int_{(x_0, y_0)}^{(x_1,v)} \mathbf{F}(x, y)\cdot d\mathbf{r} + \int_{(x_1, v)}^{(u,v)} \mathbf{F}(x, y)\cdot d\mathbf{r}. \tag{3.2}$$

Observe that the first integral in (3.2) is independent of u. So, taking the partial derivative of both sides of (3.2) with respect to u, we get

$$\begin{aligned}f_u(u, v) &= \frac{\partial}{\partial u}\int_{(x_0, y_0)}^{(x_1,v)} \mathbf{F}(x, y)\cdot d\mathbf{r} + \frac{\partial}{\partial u}\int_{(x_1, v)}^{(u,v)} \mathbf{F}(x, y)\cdot d\mathbf{r} \\ &= 0 + \frac{\partial}{\partial u}\int_{(x_1, v)}^{(u,v)} \mathbf{F}(x, y)\cdot d\mathbf{r} \\ &= \frac{\partial}{\partial u}\int_{(x_1, v)}^{(u,v)} M(x, y)\,dx + N(x, y)\,dy.\end{aligned}$$

Notice that on the second portion of the indicated path, y is a constant and so, $dy = 0$. This gives us

$$f_u(u, v) = \frac{\partial}{\partial u}\int_{(x_1, v)}^{(u,v)} M(x, y)\,dx + N(x, y)\,dy = \frac{\partial}{\partial u}\int_{(x_1, v)}^{(u,v)} M(x, y)\,dx.$$

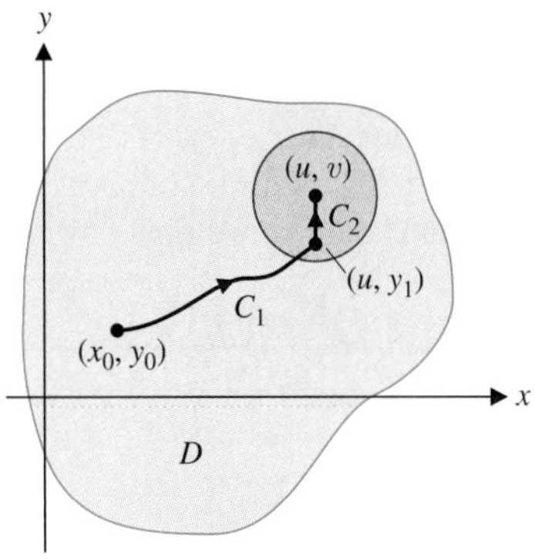

FIGURE 14.25
Second path

Finally, from the second form of the Fundamental Theorem of Calculus, we have

$$f_u(u,v) = \frac{\partial}{\partial u}\int_{(x_1,v)}^{(u,v)} M(x,y)\,dx = M(u,v). \tag{3.3}$$

Similarly, pick any point (u, y_1) in the disk centered at (u, v) with $y_1 < v$ and let C_1 be any path from (x_0, y_0) to (u, y_1) lying in D. Then, integrating over the path consisting of C_1 followed by the vertical path C_2 indicated in Figure 14.25, we find that

$$f(u,v) = \int_{(x_0,y_0)}^{(u,y_1)} \mathbf{F}(x,y)\cdot d\mathbf{r} + \int_{(u,y_1)}^{(u,v)} \mathbf{F}(x,y)\cdot d\mathbf{r}. \tag{3.4}$$

In this case, the first integral is independent of v. So, differentiating both sides of (3.4) with respect to v, we have

$$\begin{aligned} f_v(u,v) &= \frac{\partial}{\partial v}\int_{(x_0,y_0)}^{(u,y_1)} \mathbf{F}(x,y)\cdot d\mathbf{r} + \frac{\partial}{\partial v}\int_{(u,y_1)}^{(u,v)} \mathbf{F}(x,y)\cdot d\mathbf{r} \\ &= 0 + \frac{\partial}{\partial v}\int_{(u,y_1)}^{(u,v)} \mathbf{F}(x,y)\cdot d\mathbf{r} \\ &= \frac{\partial}{\partial v}\int_{(u,y_1)}^{(u,v)} M(x,y)\,dx + N(x,y)\,dy \\ &= \frac{\partial}{\partial v}\int_{(u,y_1)}^{(u,v)} N(x,y)\,dy = N(u,v), \end{aligned} \tag{3.5}$$

by the second form of the Fundamental Theorem of Calculus, where we have used the fact that on the second part of the indicated path, x is a constant, so that $dx = 0$. Replacing u and v by x and y, respectively, in (3.3) and (3.5) establishes that

$$\mathbf{F}(x,y) = \langle M(x,y), N(x,y)\rangle = \langle f_x(x,y), f_y(x,y)\rangle = \nabla f(x,y),$$

so that $\mathbf{F}$ is conservative in D. ■

Notice that in the course of the first part of the proof of Theorem 3.1, we also proved the following result, which corresponds to the Fundamental Theorem of Calculus for definite integrals.

THEOREM 3.2 (Fundamental Theorem for Line Integrals)

Suppose that $\mathbf{F}(x,y) = \langle M(x,y), N(x,y)\rangle$ is continuous in the open, connected region $D \subset \mathbb{R}^2$ and that C is any piecewise-smooth curve lying in D, with initial point (x_1, y_1) and terminal point (x_2, y_2). Then, if $\mathbf{F}$ is conservative on D, with $\mathbf{F}(x,y) = \nabla f(x,y)$, we have

$$\int_C \mathbf{F}(x,y)\cdot d\mathbf{r} = f(x,y)\Big|_{(x_1,y_1)}^{(x_2,y_2)} = f(x_2,y_2) - f(x_1,y_1).$$

You should quickly recognize the advantages presented by Theorem 3.2. For a conservative vector field, you don't need to parameterize the path to compute a line integral; you need only find a potential function and then simply evaluate the potential function between the endpoints of the curve. We illustrate this in example 3.1.

EXAMPLE 3.1 A Line Integral That Is Independent of Path

Show that for $\mathbf{F}(x, y) = \langle 2xy - 3, x^2 + 4y^3 + 5 \rangle$, the line integral $\int_C \mathbf{F}(x, y) \cdot d\mathbf{r}$ is independent of path. Then, evaluate the line integral for any curve C with initial point at $(-1, 2)$ and terminal point at $(2, 3)$.

Solution From Theorem 3.1, the line integral is independent of path if and only if the vector field $\mathbf{F}(x, y)$ is conservative. So, we look for a potential function for $\mathbf{F}$, that is, a function $f(x, y)$ for which

$$\mathbf{F}(x, y) = \langle 2xy - 3, x^2 + 4y^3 + 5 \rangle = \nabla f(x, y) = \langle f_x(x, y), f_y(x, y) \rangle.$$

Of course, this occurs when

$$f_x = 2xy - 3 \quad \text{and} \quad f_y = x^2 + 4y^3 + 5. \tag{3.6}$$

Integrating the first of these two equations with respect to x (note that we might just as easily integrate the second one with respect to y), we get

$$f(x, y) = \int (2xy - 3)\, dx = x^2 y - 3x + g(y), \tag{3.7}$$

where $g(y)$ is some arbitrary function of y alone. (Recall that we get an arbitrary *function* of y instead of a *constant* of integration, since we are integrating a function of x and y with respect to x.) Differentiating with respect to y, we get

$$f_y(x, y) = x^2 + g'(y).$$

Notice that from (3.6), we already have an expression for f_y. Setting these two expressions equal, we get

$$x^2 + g'(y) = x^2 + 4y^3 + 5$$

and subtracting x^2 from both sides, we get

$$g'(y) = 4y^3 + 5.$$

Finally, integrating this last expression with respect to y gives us

$$g(y) = y^4 + 5y + c.$$

We now have from (3.7) that

$$f(x, y) = x^2 y - 3x + y^4 + 5y + c$$

is a potential function for $\mathbf{F}(x, y)$, for any constant c. Now that we have found a potential function, we have by Theorem 3.2 that for any path from $(-1, 2)$ to $(2, 3)$,

$$\int_C \mathbf{F}(x, y) \cdot d\mathbf{r} = f(x, y)\Big|_{(-1,2)}^{(2,3)}$$
$$= [2^2(3) - 3(2) + 3^4 + 5(3) + c] - [2 + 3 + 2^4 + 5(2) + c]$$
$$= 71. \blacksquare$$

Notice that when we evaluated the line integral in example 3.1, the constant c in the expression for the potential function dropped out. For this reason, we usually take the constant to be zero when we write down a potential function.

We consider a curve C to be **closed** if its two endpoints are the same. That is, for a plane curve C defined parametrically by

$$C = \{(x, y) | x = g(t), y = h(t), a \leq t \leq b\},$$

C is closed if $(g(a), h(a)) = (g(b), h(b))$. Theorem 3.3 provides us with an important connection between conservative vector fields and line integrals along closed curves.

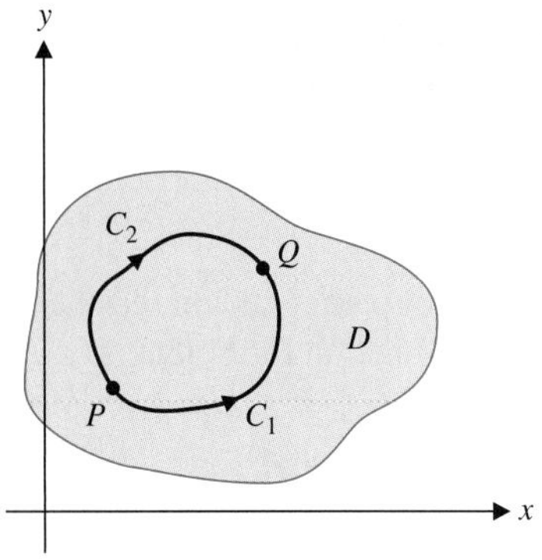

FIGURE 14.26a
Curves C_1 and C_2

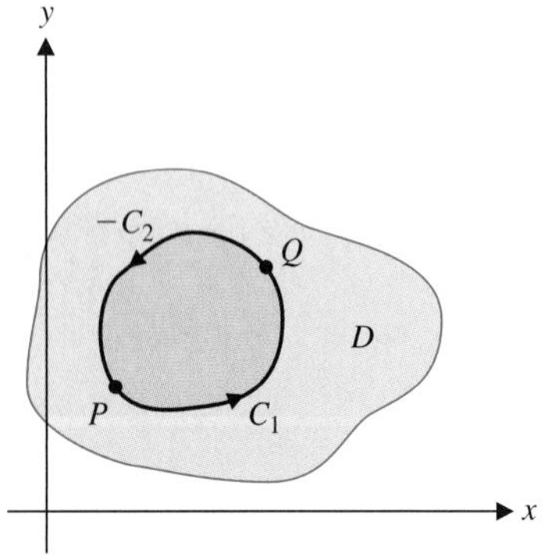

FIGURE 14.26b
The closed curve formed by $C_1 \cup (-C_2)$

> **THEOREM 3.3**
>
> Suppose that $\mathbf{F}(x, y)$ is continuous in the open, connected region $D \subset \mathbb{R}^2$. Then $\mathbf{F}$ is conservative on D if and only if $\int_C \mathbf{F}(x, y) \cdot d\mathbf{r} = 0$ for every piecewise-smooth closed curve C lying in D.

PROOF

Suppose that $\int_C \mathbf{F}(x, y) \cdot d\mathbf{r} = 0$ for every piecewise-smooth closed curve C lying in D. Take any two points P and Q lying in D and let C_1 and C_2 be any two piecewise-smooth closed curves from P to Q that lie in D, as indicated in Figure 14.26a. (Note that since D is connected, there always exist such curves.) Then, the curve C consisting of C_1 followed by $-C_2$ is a piecewise-smooth closed curve lying in D, as indicated in Figure 14.26b. It now follows that

$$0 = \int_C \mathbf{F}(x, y) \cdot d\mathbf{r} = \int_{C_1} \mathbf{F}(x, y) \cdot d\mathbf{r} + \int_{-C_2} \mathbf{F}(x, y) \cdot d\mathbf{r}$$

$$= \int_{C_1} \mathbf{F}(x, y) \cdot d\mathbf{r} - \int_{C_2} \mathbf{F}(x, y) \cdot d\mathbf{r}, \quad \text{From Theorem 2.5}$$

so that

$$\int_{C_1} \mathbf{F}(x, y) \cdot d\mathbf{r} = \int_{C_2} \mathbf{F}(x, y) \cdot d\mathbf{r}.$$

Since C_1 and C_2 were any two curves from P to Q, we have that $\int_C \mathbf{F}(x, y) \cdot d\mathbf{r}$ is independent of path and so, $\mathbf{F}$ is conservative by Theorem 3.1. The second half of the theorem (that $\mathbf{F}$ conservative implies $\int_C \mathbf{F}(x, y) \cdot d\mathbf{r} = 0$ for every piecewise-smooth closed curve C lying in D) is a simple consequence of Theorem 3.2 and is left as an exercise. ■

You have already seen that line integrals are not always independent of path. Said differently, not all vector fields are conservative. In view of this, it would be helpful to have a simple way of deciding whether or not a line integral is independent of path before going through the process of trying to construct a potential function.

Note that by Theorem 3.1, if $\mathbf{F}(x, y) = \langle M(x, y), N(x, y) \rangle$ is continuous on the open, connected region D and the line integral $\int_C \mathbf{F}(x, y) \cdot d\mathbf{r}$ is independent of path, then $\mathbf{F}$ must be conservative. That is, there is a function $f(x, y)$ for which $\mathbf{F}(x, y) = \nabla f(x, y)$, so that

$$M(x, y) = f_x(x, y) \quad \text{and} \quad N(x, y) = f_y(x, y).$$

Differentiating the first equation with respect to y and the second equation with respect to x, we have

$$M_y(x, y) = f_{xy}(x, y) \quad \text{and} \quad N_x(x, y) = f_{yx}(x, y).$$

Notice now that if M_y and N_x are continuous in D, then the mixed second partial derivatives $f_{xy}(x, y)$ and $f_{yx}(x, y)$ must be equal in D, by Theorem 3.1 in Chapter 12. We must then have that

$$M_y(x, y) = N_x(x, y),$$

for all (x, y) in D. As it turns out, if we further assume that D is **simply-connected** (that is, that every closed curve in D encloses only points in D), then the converse of this result is

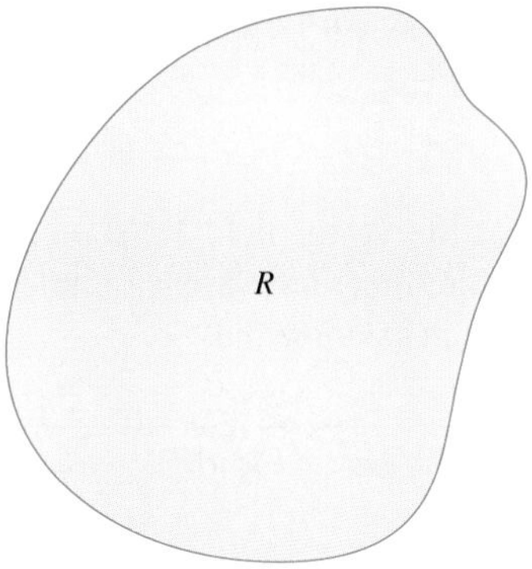

FIGURE 14.27a
Simply-connected

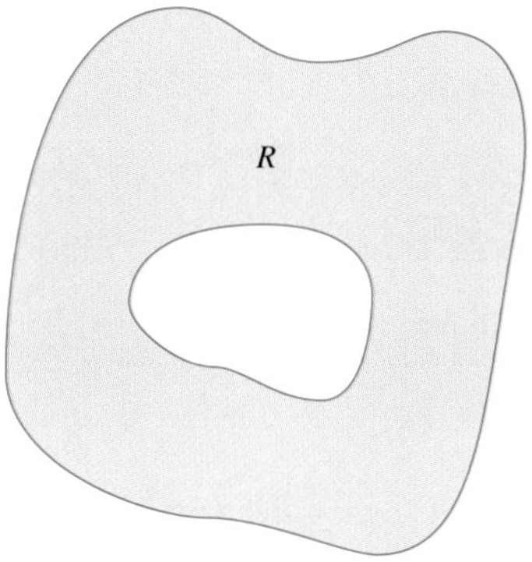

FIGURE 14.27b
Not simply-connected

also true [i.e., $\int_C \mathbf{F}(x, y) \cdot d\mathbf{r}$ is independent of path whenever $M_y = N_x$ in D]. We illustrate a simply-connected region in Figure 14.27a and a region that is not simply-connected in Figure 14.27b. You can think about simply-connected regions as regions that have no holes. We can now state the following result.

THEOREM 3.4

Suppose that $M(x, y)$ and $N(x, y)$ have continuous first partial derivatives on a simply-connected region D. Then, $\int_C M(x, y)\,dx + N(x, y)\,dy$ is independent of path in D if and only if $M_y(x, y) = N_x(x, y)$ for all (x, y) in D.

We have already proved that independence of path implies that $M_y(x, y) = N_x(x, y)$ for all (x, y) in D. We postpone the proof of the second half of the theorem until our presentation of Green's Theorem in section 14.4.

EXAMPLE 3.2 Testing a Line Integral for Independence of Path

Determine whether or not the line integral $\int_C (e^{2x} + x \sin y)\,dx + (x^2 \cos y)\,dy$ is independent of path.

Solution In this case, we have

$$M_y = \frac{\partial}{\partial y}(e^{2x} + x \sin y) = x \cos y$$

and

$$N_x = \frac{\partial}{\partial x}(x^2 \cos y) = 2x \cos y,$$

so that $M_y \neq N_x$. By Theorem 3.4, the line integral is thus not independent of path. ■

CONSERVATIVE VECTOR FIELDS

Before moving on to three-dimensional vector fields, we pause to summarize the results we have developed for two-dimensional vector fields $\mathbf{F}(x, y) = \langle M(x, y), N(x, y) \rangle$, where we assume that $M(x, y)$ and $N(x, y)$ have continuous first partial derivatives on an open, simply-connected region $D \subset \mathbb{R}^2$. In this case, the following five statements are equivalent, meaning that for a given vector field, either all five statements are true or all five statements are false.

1. $\mathbf{F}(x, y)$ is conservative in D.
2. $\mathbf{F}(x, y)$ is a gradient field in D (*i.e.*, $\mathbf{F}(x, y) = \nabla f(x, y)$, for some potential function f, for all $(x, y) \in D$).
3. $\int_C \mathbf{F} \cdot d\mathbf{r}$ is independent of path in D.
4. $\int_C \mathbf{F} \cdot d\mathbf{r} = 0$ for every piecewise-smooth closed curve C lying in D.
5. $M_y(x, y) = N_x(x, y)$, for all $(x, y) \in D$.

All we have said about independence of path and conservative vector fields can be extended to higher dimensions, although the test for when a line integral is independent of path becomes slightly more complicated. For a three-dimensional vector field $\mathbf{F}(x, y, z)$, we say that **F** is **conservative** in a region D whenever there is a scalar function $f(x, y, z)$ for which

$$\mathbf{F}(x, y, z) = \nabla f(x, y, z).$$

As in two dimensions, f is called a **potential function** for the vector field **F**. You can construct a potential function for a conservative vector field in three dimensions in much the same way as you did in two dimensions. We illustrate this in example 3.3.

EXAMPLE 3.3 Showing That a Three-Dimensional Vector Field Is Conservative

Show that the vector field $\mathbf{F}(x, y, z) = \langle 4xe^z, \cos y, 2x^2e^z\rangle$ is conservative, by finding a potential function f.

Solution We need to find a potential function $f(x, y, z)$ for which

$$\begin{aligned}\mathbf{F}(x, y, z) &= \langle 4xe^z, \cos y, 2x^2e^z\rangle = \nabla f(x, y, z)\\ &= \langle f_x(x, y, z), f_y(x, y, z), f_z(x, y, z)\rangle.\end{aligned}$$

This will occur if and only if

$$f_x = 4xe^z, \quad f_y = \cos y \quad \text{and} \quad f_z = 2x^2e^z. \tag{3.8}$$

Integrating the first of these equations with respect to x, we have

$$f(x, y, z) = \int 4xe^z\,dx = 2x^2e^z + g(y, z),$$

where $g(y, z)$ is an arbitrary function of y and z alone. Note that since y and z are treated as constants when integrating or differentiating with respect to x, we add an arbitrary function of y and z (instead of an arbitrary constant) after a partial integration with respect to x. Differentiating this expression with respect to y, we have

$$f_y(x, y, z) = g_y(y, z) = \cos y,$$

from the second equation in (3.8). Integrating $g_y(y, z)$ with respect to y now gives us

$$g(y, z) = \int \cos y\,dy = \sin y + h(z),$$

where $h(z)$ is an arbitrary function of z alone. Notice that here, we got an arbitrary function of z alone, since we were integrating $g(y, z)$ (a function of y and z alone) with respect to y. This now gives us

$$f(x, y, z) = 2x^2e^z + g(y, z) = 2x^2e^z + \sin y + h(z).$$

Differentiating this last equation with respect to z yields

$$f_z(x, y, z) = 2x^2e^z + h'(z) = 2x^2e^z,$$

from the third equation in (3.8). This gives us that $h'(z) = 0$, so that $h(z)$ is a constant. (We'll choose it to be 0.) We now have that a potential function for $\mathbf{F}(x, y, z)$ is

$$f(x, y, z) = 2x^2e^z + \sin y$$

and so, **F** is conservative. ■

We summarize the main results for line integrals for three-dimensional vector fields in Theorem 3.5.

THEOREM 3.5

Suppose that the vector field $\mathbf{F}(x, y, z)$ is continuous on the open, connected region $D \subset \mathbb{R}^3$. Then, the line integral $\int_C \mathbf{F}(x, y, z) \cdot d\mathbf{r}$ is independent of path in D if and only if the vector field $\mathbf{F}$ is conservative in D, that is, $\mathbf{F}(x, y, z) = \nabla f(x, y, z)$, for all (x, y, z) in D, for some scalar function f (a potential function for $\mathbf{F}$). Further, for any piecewise-smooth curve C lying in D, with initial point (x_1, y_1, z_1) and terminal point (x_2, y_2, z_2), we have

$$\int_C \mathbf{F}(x, y, z) \cdot d\mathbf{r} = f(x, y, z)\Big|_{(x_1,y_1,z_1)}^{(x_2,y_2,z_2)} = f(x_2, y_2, z_2) - f(x_1, y_1, z_1).$$

vector valued function

HW pg 1153 #1, 5, 13

EXERCISES 14.3

WRITING EXERCISES

1. You have seen two different methods of determining whether a line integral is independent of path: one in example 3.1 and the other in example 3.2. If you have reason to believe that a line integral will be independent of path, explain which method you would prefer to use.

2. In the situation of exercise 1, if you doubt that a line integral is independent of path, explain which method you would prefer to use. If you have no evidence as to whether the line integral is or isn't independent of path, explain which method you would prefer to use.

3. In section 14.1, we introduced conservative vector fields and stated that some calculations simplified when the vector field is conservative. Discuss one important example of this.

4. Our definition of independence of path applies only to line integrals of the form $\int_C \mathbf{F} \cdot d\mathbf{r}$. Explain why an arc length line integral $\int_C f\, ds$ would not be independent of path (unless $f = 0$).

In exercises 1–12, determine whether F is conservative. If it is, find a potential function f.

1. $\mathbf{F}(x, y) = \langle 2xy - 1, x^2 \rangle$
2. $\mathbf{F}(x, y) = \langle 3x^2y^2, 2x^3y - y \rangle$
3. $\mathbf{F}(x, y) = \left\langle \frac{1}{y} - 2x, y - \frac{x}{y^2} \right\rangle$
4. $\mathbf{F}(x, y) = \langle \sin y - x, x \cos y \rangle$
5. $\mathbf{F}(x, y) = \langle e^{xy} - 1, xe^{xy} \rangle$
6. $\mathbf{F}(x, y) = \langle e^y - 2x, xe^y - x^2y \rangle$
7. $\mathbf{F}(x, y) = \langle ye^{xy}, xe^{xy} + \cos y \rangle$
8. $\mathbf{F}(x, y) = \langle y \cos xy - 2xy, x \cos xy - x^2 \rangle$
9. $\mathbf{F}(x, y, z) = \langle z^2 + 2xy, x^2 + 1, 2xz - 3 \rangle$
10. $\mathbf{F}(x, y, z) = \langle y^2 - x, 2xy + \sin z, y \cos z \rangle$
11. $\mathbf{F}(x, y, z) = \langle y^2z^2 + x, y + 2xyz^2, 2xy^2z \rangle$
12. $\mathbf{F}(x, y, z) = \langle 2xe^{yz} - 1, x^2 + e^{yz}, x^2ye^{yz} \rangle$

In exercises 13–18, show that the line integral is independent of path and use a potential function to evaluate the integral.

13. $\int_C 2xy\, dx + (x^2 - 1)\, dy$, where C runs from $(1, 0)$ to $(3, 1)$
14. $\int_C 3x^2y^2 dx + (2x^3y - 4)\, dy$, where C runs from $(1, 2)$ to $(-1, 1)$
15. $\int_C ye^{xy} dx + (xe^{xy} - 2y)\, dy$, where C runs from $(1, 0)$ to $(0, 4)$
16. $\int_C (2xe^{x^2} - 2y)\, dx + (2y - 2x)\, dy$, where C runs from $(1, 2)$ to $(-1, 1)$
17. $\int_C (z^2 + 2xy)\, dx + x^2 dy + 2xz\, dz$, where C runs from $(2, 1, 3)$ to $(4, -1, 0)$
18. $\int_C (2x \cos z - x^2)\, dx + (z - 2y)\, dy + (y - x^2 \sin z)\, dz$, where C runs from $(3, -2, 0)$ to $(1, 0, \pi)$

In exercises 19–30, evaluate $\int_C \mathbf{F} \cdot d\mathbf{r}$.

19. $\mathbf{F}(x, y) = \langle x^2 + 1, y^3 - 3y + 2 \rangle$, C is the top half-circle from $(-4, 0)$ to $(4, 0)$
20. $\mathbf{F}(x, y) = \langle xe^{x^2} - 2, \sin y \rangle$, C is the portion of the parabola $y = x^2$ from $(-2, 4)$ to $(2, 4)$
21. $\mathbf{F}(x, y, z) = \langle x^2, y^2, z^2 \rangle$, C is the top half-circle from $(1, 4, -3)$ to $(1, 4, 3)$
22. $\mathbf{F}(x, y, z) = \langle \cos x, \sqrt{y} + 1, 4z^3 \rangle$, C is the quarter-circle from $(2, 0, 3)$ to $(2, 3, 0)$
23. $\mathbf{F}(x, y, z) = \dfrac{\langle x, y, z \rangle}{\sqrt{x^2 + y^2 + z^2}}$, C runs from $(1, 3, 2)$ to $(2, 1, 5)$
24. $\mathbf{F}(x, y, z) = \dfrac{\langle x, y, z \rangle}{x^2 + y^2 + z^2}$, C runs from $(2, 0, 0)$ to $(0, 1, -1)$

25. $\mathbf{F}(x, y) = \langle 3x^2y + 1, 3xy^2 \rangle$, C is the bottom half-circle from $(1, 0)$ to $(-1, 0)$

26. $\mathbf{F}(x, y) = \langle 4xy - 2x, 2x^2 - x \rangle$, C is the portion of the parabola $y = x^2$ from $(-2, 4)$ to $(2, 4)$

27. $\mathbf{F}(x, y) = \langle y^2 e^{xy^2} - y, 2xye^{xy^2} - x - 1 \rangle$, C is the line segment from $(2, 3)$ to $(3, 0)$

28. $\mathbf{F}(x, y) = \langle 2ye^{2x} + y^3, e^{2x} + 3xy^2 \rangle$, C is the line segment from $(4, 3)$ to $(1, -3)$

29. $\mathbf{F}(x, y) = \left\langle \frac{1}{y} - e^{2x}, 2y - \frac{x}{y^2} \right\rangle$, C is the circle $(x - 5)^2 + (y + 6)^2 = 16$, oriented counterclockwise

30. $\mathbf{F}(x, y) = \langle 3y - \sqrt{y/x}, 3x - \sqrt{x/y} \rangle$, C is the ellipse $4(x - 4)^2 + 9(y - 4)^2 = 36$, oriented counterclockwise

In exercises 31–36, use the graph to determine whether or not the vector field is conservative.

31.

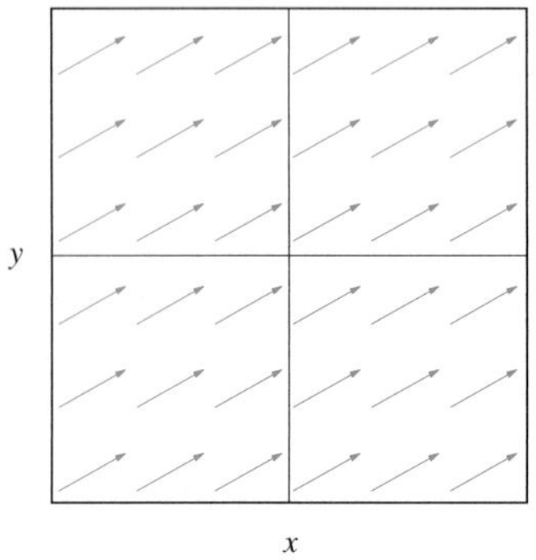

32.

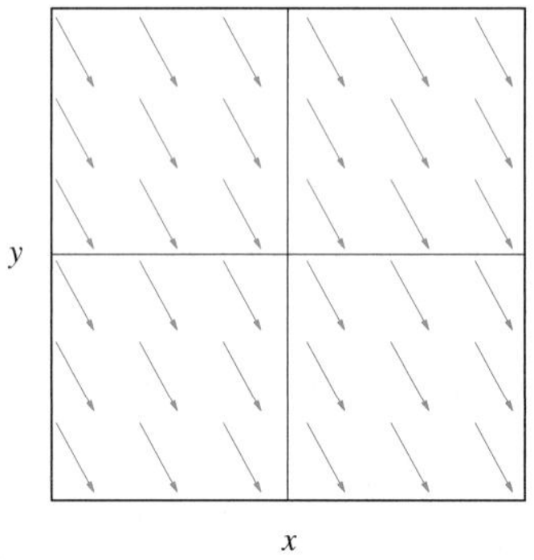

33.

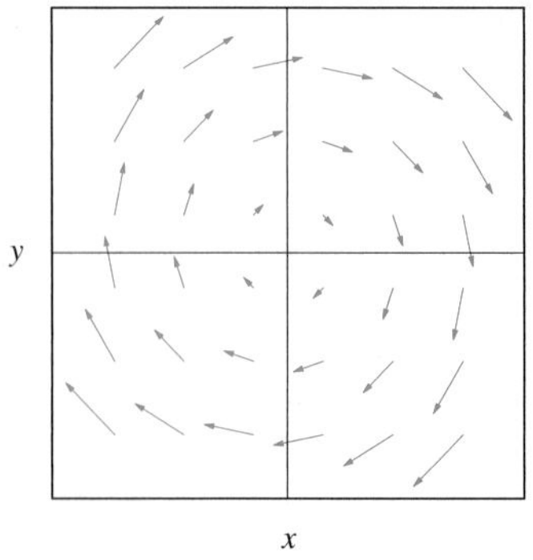

34.

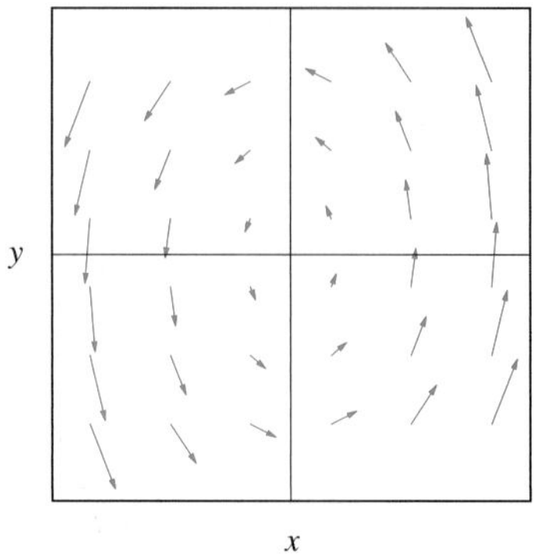

35.

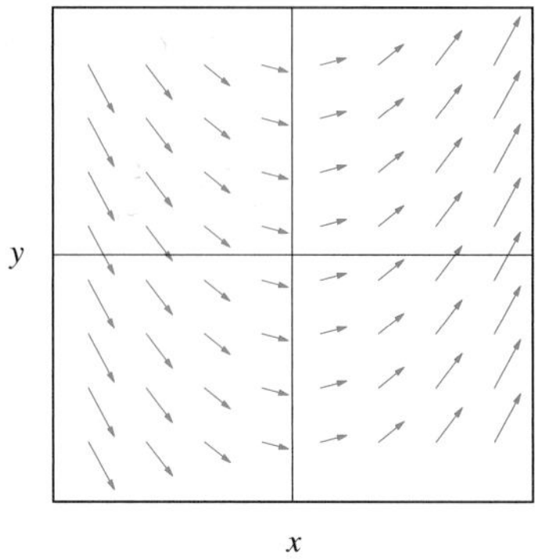

36.

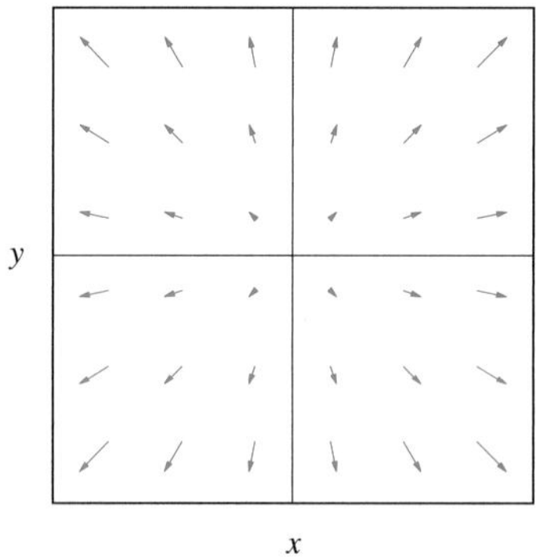

In exercises 37–40, show that the line integral is not independent of path by finding two paths that give different values of the integral.

37. $\int_C y\,dx - x\,dy$, where C goes from $(-2, 0)$ to $(2, 0)$

38. $\int_C 2\,dx + x\,dy$, where C goes from $(1, 4)$ to $(2, -2)$

39. $\int_C y\,dx - 3\,dy$, where C goes from $(-2, 2)$ to $(0, 0)$

40. $\int_C y^2\,dx + x^2\,dy$, where C goes from $(0, 0)$ to $(1, 1)$

In exercises 41–44, label each statement as True or False and briefly explain.

41. If $\mathbf{F}$ is conservative, then $\int_C \mathbf{F} \cdot d\mathbf{r} = 0$.

42. If $\int_C \mathbf{F} \cdot d\mathbf{r}$ is independent of path, then $\mathbf{F}$ is conservative.

43. If $\mathbf{F}$ is conservative, then $\int_C \mathbf{F} \cdot d\mathbf{r} = 0$ for any closed curve C.

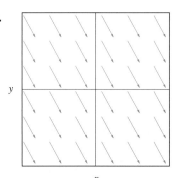

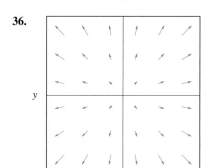

44. If $\mathbf{F}$ is conservative, then $\int_C \mathbf{F} \cdot d\mathbf{r}$ is independent of path.

45. Let $\mathbf{F}(x, y) = \dfrac{1}{x^2 + y^2}\langle -y, x \rangle$. Find a potential function f for $\mathbf{F}$ and carefully note any restrictions on the domain of f. Let C be the unit circle and show that $\int_C \mathbf{F} \cdot d\mathbf{r} = 2\pi$. Explain why the Fundamental Theorem for Line Integrals does not apply to this calculation. Quickly explain how to compute $\int_C \mathbf{F} \cdot d\mathbf{r}$ over the circle $(x - 2)^2 + (y - 3)^2 = 1$.

46. Finish the proof of Theorem 3.3 by showing that if $\mathbf{F}$ is conservative in an open, connected region $D \subset \mathbb{R}^2$, then $\int_C \mathbf{F} \cdot d\mathbf{r} = 0$ for all piecewise-smooth closed curves C lying in D.

47. Determine whether or not each region is simply-connected. (a) $\{(x, y) : x^2 + y^2 < 2\}$ (b) $\{(x, y) : 1 < x^2 + y^2 < 2\}$

48. Determine whether or not each region is simply-connected. (a) $\{(x, y) : 1 < x < 2\}$ (b) $\{(x, y) : 1 < x^2 < 2\}$

49. The Coulomb force for a unit charge at the origin and charge q at point $P_1 = (x_1, y_1, z_1)$ is $\mathbf{F} = \dfrac{kq}{r^2}\hat{\mathbf{r}}$, where $r = \sqrt{x^2 + y^2 + z^2}$ and $\hat{\mathbf{r}} = \dfrac{\langle x, y, z \rangle}{r}$. Show that the work done by $\mathbf{F}$ to move the charge q from P_1 to $P_2 = (x_2, y_2, z_2)$ is equal to $\dfrac{kq}{r_1} - \dfrac{kq}{r_2}$, where $r_1 = \sqrt{x_1^2 + y_1^2 + z_1^2}$ and $r_2 = \sqrt{x_2^2 + y_2^2 + z_2^2}$.

50. Interpret the result of exercise 49 in the case where (a) P_1 is closer to the origin than P_2. (Is the work positive or negative? Why does this make sense physically?) (b) P_2 is closer to the origin than P_1 and (c) P_1 and P_2 are the same distance from the origin.

51. The work done to increase the temperature of a gas from T_1 to T_2 and increase its pressure from P_1 to P_2 is given by $\displaystyle\int_C \left(\frac{RT}{P}dP - R\,dT\right)$. Here, R is a constant, T is temperature, P is pressure and C is the path of (P, T) values as the changes occur. Compare the work done along the following two paths. (a) C_1 consists of the line segment from (P_1, T_1) to (P_1, T_2), followed by the line segment to (P_2, T_2); (b) C_2 consists of the line segment from (P_1, T_1) to (P_2, T_1), followed by the line segment to (P_2, T_2).

52. Based on your answers in exercise 51, is the force field involved in changing the temperature and pressure of the gas conservative?

53. A vector field $\mathbf{F}$ satisfies $\mathbf{F} = \nabla\phi$ (where ϕ is continuous) at every point except P, where it is undefined. Suppose that C_1 is a small closed curve enclosing P, C_2 is a large closed curve enclosing C_1 and C_3 is a closed curve that does not enclose P. (See the figure.) Given that $\int_{C_1} \mathbf{F} \cdot d\mathbf{r} = 0$, explain why $\int_{C_2} \mathbf{F} \cdot d\mathbf{r} = \int_{C_3} \mathbf{F} \cdot d\mathbf{r} = 0$.

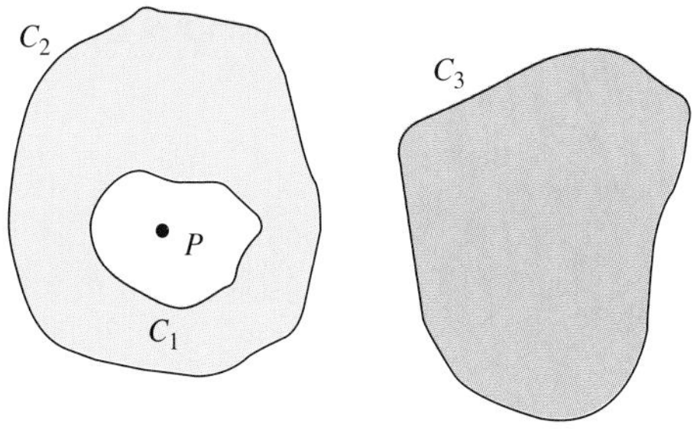

54. The circulation of a fluid with velocity field $\mathbf{v}$ around the closed path C is defined by $\Gamma = \int_C \mathbf{v} \cdot d\mathbf{r}$. For inviscid flow, $\dfrac{d}{dt}\Gamma = \displaystyle\int_C \mathbf{v} \cdot d\mathbf{v}$. Show that in this case $\dfrac{d}{dt}\Gamma = 0$. This is known as **Kelvin's Circulation Theorem** and explains why small whirlpools in a stream stay coherent and move for periods of time.

EXPLORATORY EXERCISES

1. For closed curves, we can take advantage of portions of a line integral that will equal zero. For example, if C is a closed curve, explain why you can simplify $\int_C (x + y^2)\,dx + (y^2 + x)\,dy$ to $\int_C y^2 dx + x\,dy$. In general, explain why the $f(x)$ and $g(y)$ terms can be dropped in the line integral $\int_C (f(x) + y^2)\,dx + (x + g(y))\,dy$. Describe which other terms can be dropped in the line integral over a closed curve. Use the example

$$\int_C (x^3 + y^2 + x^2y^2 + \cos y)\,dx + (y^2 + 2xy - x\sin y + x^3y)\,dy$$

to help organize your thinking.

2. In this exercise, we explore a basic principle of physics called **conservation of energy.** Start with the work integral $\int_C \mathbf{F} \cdot d\mathbf{r}$, where the position function $\mathbf{r}(t)$ is a continuously differentiable function of time. Substitute Newton's second law: $\mathbf{F} = m\dfrac{d\mathbf{v}}{dt}$ and $d\mathbf{r} = \mathbf{r}'(t)\,dt = \mathbf{v}\,dt$ and show that $\int_C \mathbf{F} \cdot d\mathbf{r} = \Delta K$. Here, K is **kinetic energy** defined by $K = \frac{1}{2}m\|\mathbf{v}\|^2$ and ΔK is the change of kinetic energy from the initial point of C to the terminal point of C. Next, assume that $\mathbf{F}$ is conservative with $\mathbf{F} = -\nabla f$, where the function f represents **potential energy.** Show that $\int_C \mathbf{F} \cdot d\mathbf{r} = -\Delta f$ where Δf equals the change in potential energy from the initial point of C to the terminal point of C. Conclude that under these hypotheses (conservative force, continuous acceleration) the net change in energy $\Delta K + \Delta f$ equals 0. Therefore, $K + f$ is constant.

14.4 GREEN'S THEOREM

HISTORICAL NOTES

George Green (1793–1841) English mathematician who discovered Green's Theorem. Green was self-taught, receiving only two years of schooling before going to work in his father's bakery at age 9. He continued to work in and eventually took over the family mill while teaching himself mathematics. In 1828, he published an essay in which he gave potential functions their name and applied them to the study of electricity and magnetism. This little-read essay introduced Green's Theorem and the so-called Green's functions used in the study of partial differential equations. Green was admitted to Cambridge University at age 40 and published several papers before his early death from illness. The significance of his original essay remained unknown until shortly after his death.

In this section, we develop a connection between certain line integrals around a closed curve in the plane and double integrals over the region enclosed by the curve. At first glance, you might think this a strange and abstract connection, one that only a mathematician could care about. Actually, the reverse is true; Green's Theorem is a significant result with far-reaching implications. It is of fundamental importance in the analysis of fluid flows and in the theories of electricity and magnetism.

Before stating the main result, we briefly define some terminology. Recall that for a plane curve C defined parametrically by

$$C = \{(x, y)|x = g(t), y = h(t), a \leq t \leq b\},$$

C is closed if its two endpoints are the same, i.e., $(g(a), h(a)) = (g(b), h(b))$. A curve C is **simple** if it does not intersect itself, except at the endpoints. We illustrate a simple closed curve in Figure 14.28a and a closed curve that is not simple in Figure 14.28b.

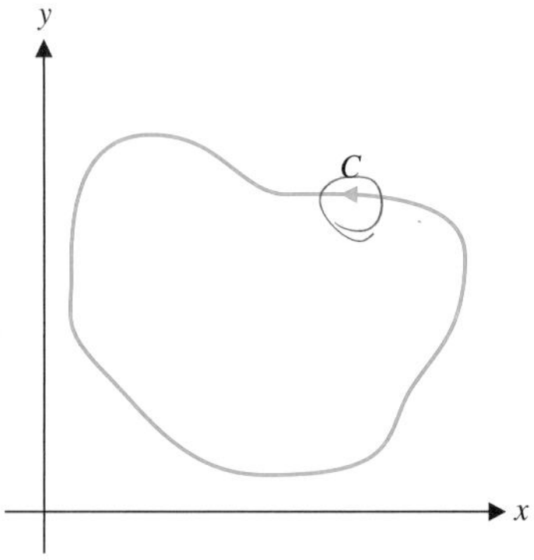

FIGURE 14.28a
Simple closed curve

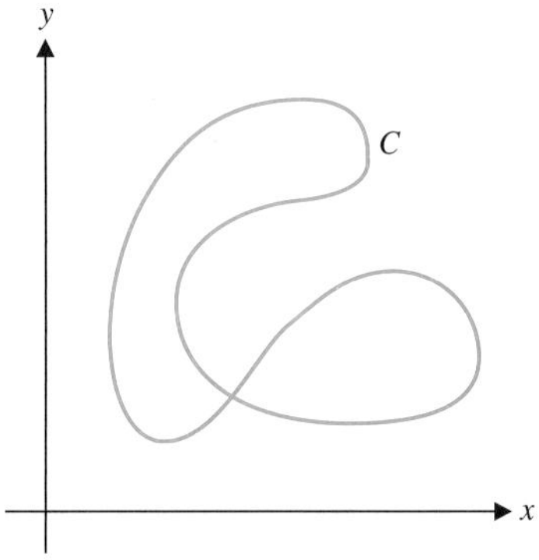

FIGURE 14.28b
Closed, but not simple curve

We say that a simple closed curve C has **positive orientation** if the region R enclosed by C stays to the left of C, as the curve is traversed; a curve has **negative orientation** if the region R stays to the right of C. In Figures 14.29a and 14.29b, we illustrate a simple closed curve with positive orientation and one with negative orientation, respectively.

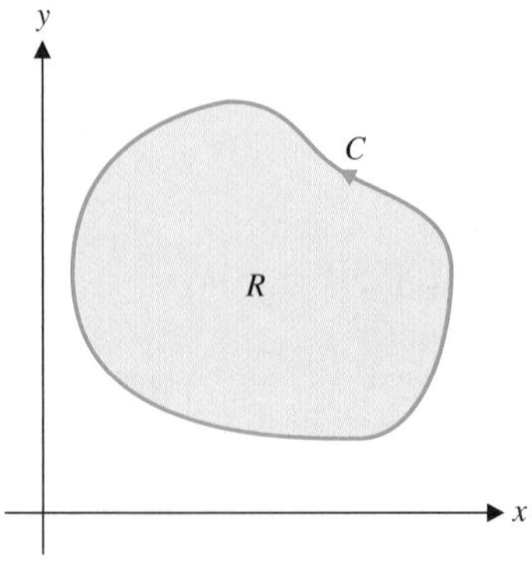

FIGURE 14.29a
Positive orientation

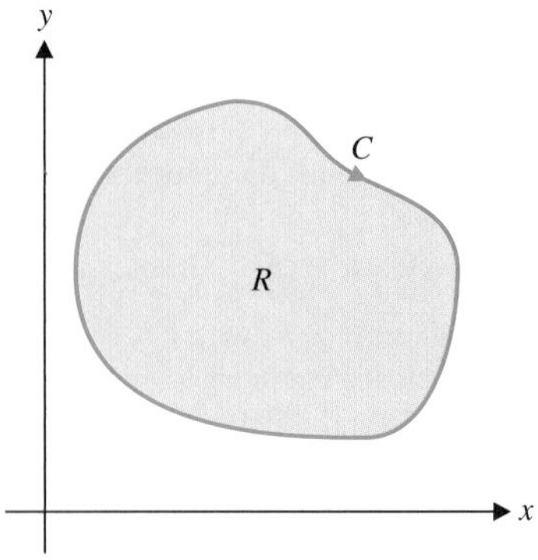

FIGURE 14.29b
Negative orientation

We use the notation

$$\oint_C \mathbf{F}(x, y) \cdot d\mathbf{r}$$

to denote a line integral along a simple closed curve C oriented in the positive direction.

We can now state the main result of the section.

THEOREM 4.1 (Green's Theorem)

Let C be a piecewise-smooth, simple closed curve in the plane with positive orientation and let R be the region enclosed by C, together with C. Suppose that $M(x, y)$ and $N(x, y)$ are continuous and have continuous first partial derivatives in some open region D, with $R \subset D$. Then,

$$\oint_C M(x, y)\,dx + N(x, y)\,dy = \iint_R \left(\frac{\partial N}{\partial x} - \frac{\partial M}{\partial y} \right) dA.$$

You can find a general proof of Green's Theorem in a more advanced text. We prove it here only for a special case.

PROOF

Here, we assume that the region R can be written in the form

$$R = \{(x, y) | a \le x \le b \text{ and } g_1(x) \le y \le g_2(x)\},$$

where $g_1(x) \le g_2(x)$, for all x in $[a, b]$, $g_1(a) = g_2(a)$ and $g_1(b) = g_2(b)$, as illustrated in Figure 14.30a. Notice that we can divide C into the two pieces indicated in Figure 14.30a:

$$C = C_1 \cup C_2,$$

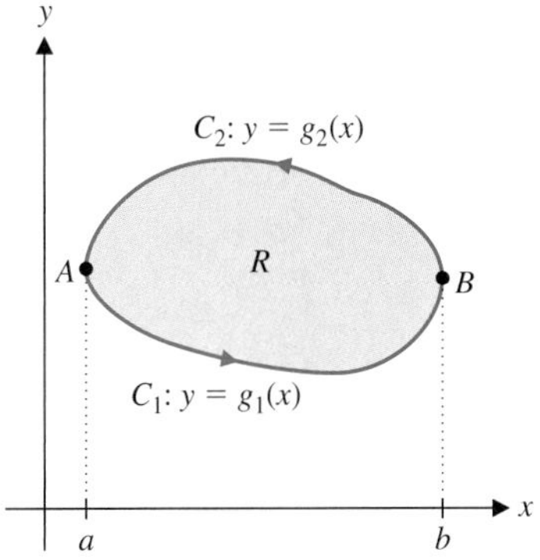

FIGURE 14.30a
The region R

where C_1 is the bottom portion of the curve, defined by

$$C_1 = \{(x, y) | a \le x \le b, y = g_1(x)\}$$

and C_2 is the top portion of the curve, defined by

$$C_2 = \{(x, y) | a \le x \le b, y = g_2(x)\},$$

where the orientation is as indicated in the figure. From the Evaluation Theorem for line integrals (Theorem 2.4), we then have

$$\begin{aligned} \oint_C M(x, y)\,dx &= \int_{C_1} M(x, y)\,dx + \int_{C_2} M(x, y)\,dx \\ &= \int_a^b M(x, g_1(x))\,dx - \int_a^b M(x, g_2(x))\,dx \\ &= \int_a^b [M(x, g_1(x)) - M(x, g_2(x))]\,dx, \end{aligned} \tag{4.1}$$

where the minus sign in front of the second integral comes from our traversing C_2 "backward" (i.e., from right to left). On the other hand, notice that we can write

$$\iint_R \frac{\partial M}{\partial y}\,dA = \int_a^b \int_{g_1(x)}^{g_2(x)} \frac{\partial M}{\partial y}\,dy\,dx$$

$$= \int_a^b M(x,y)\Big|_{y=g_1(x)}^{y=g_2(x)}\,dx \qquad \text{By the Fundamental Theorem of Calculus}$$

$$= \int_a^b [M(x,g_2(x)) - M(x,g_1(x))]\,dx.$$

Together with (4.1), this gives us

$$\oint_C M(x,y)\,dx = -\iint_R \frac{\partial M}{\partial y}\,dA. \tag{4.2}$$

We now assume that we can also write the region R in the form

$$R = \{(x,y)|c \le y \le d \text{ and } h_1(y) \le x \le h_2(y)\},$$

where $h_1(y) \le h_2(y)$ for all y in $[c,d]$, $h_1(c) = h_2(c)$ and $h_1(d) = h_2(d)$. Here, we write $C = C_3 \cup C_4$, as illustrated in Figure 14.30b. In this case, notice that we can write

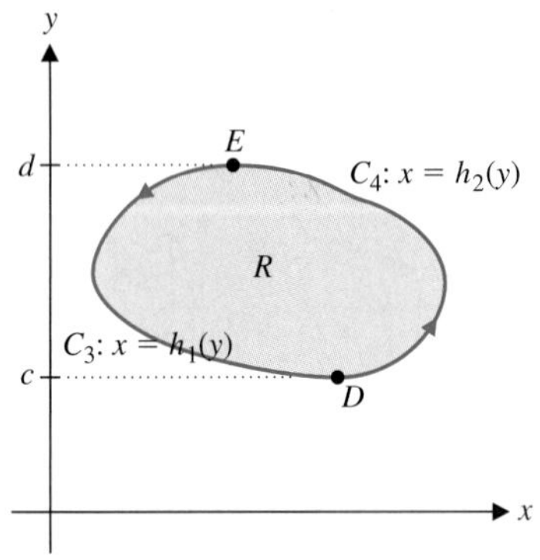

FIGURE 14.30b
The region R

$$\oint_C N(x,y)\,dy = \int_{C_3} N(x,y)\,dy + \int_{C_4} N(x,y)\,dy$$

$$= -\int_c^d N(h_1(y),y)\,dy + \int_c^d N(h_2(y),y)\,dy$$

$$= \int_c^d [N(h_2(y),y) - N(h_1(y),y)]\,dy, \tag{4.3}$$

where the minus sign in front of the first integral accounts for our traversing C_3 "backward" (in this case, from top to bottom). Further, notice that

$$\iint_R \frac{\partial N}{\partial x}\,dA = \int_c^d \int_{h_1(y)}^{h_2(y)} \frac{\partial N}{\partial x}\,dx\,dy$$

$$= \int_c^d [N(h_2(y),y) - N(h_1(y),y)]\,dy.$$

Together with (4.3), this gives us

$$\oint_C N(x,y)\,dy = \iint_R \frac{\partial N}{\partial x}\,dA. \tag{4.4}$$

Adding together (4.2) and (4.4), we have

$$\oint_C M(x,y)\,dx + N(x,y)\,dy = \iint_R \left(\frac{\partial N}{\partial x} - \frac{\partial M}{\partial y}\right) dA,$$

as desired. ■

Although the significance of Green's Theorem lies in the connection it provides between line integrals and double integrals in more theoretical settings, we illustrate the result in example 4.1 by using it to simplify the calculation of a line integral.

EXAMPLE 4.1 Using Green's Theorem

Use Green's Theorem to rewrite and evaluate $\oint_C (x^2 + y^3)\,dx + 3xy^2 dy$, where C consists of the portion of $y = x^2$ from (2, 4) to (0, 0), followed by the line segments from (0, 0) to (2, 0) and from (2, 0) to (2, 4).

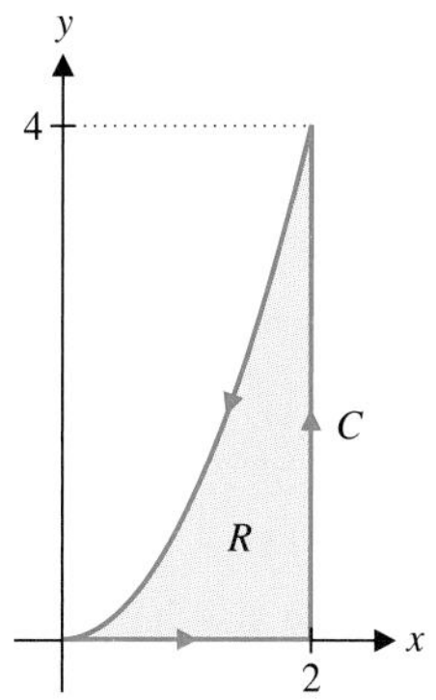

FIGURE 14.31
The region R

Solution We indicate the curve C and the enclosed region R in Figure 14.31. Notice that C is a piecewise-smooth, simple closed curve with positive orientation. Further, for $M(x, y) = x^2 + y^3$ and $N(x, y) = 3xy^2$, M and N are continuous and have continuous first partial derivatives everywhere. Green's Theorem then says that

$$\oint_C (x^2 + y^3)\,dx + 3xy^2 dy = \iint_R \left(\frac{\partial N}{\partial x} - \frac{\partial M}{\partial y}\right) dA$$
$$= \iint_R (3y^2 - 3y^2)\,dA = 0.$$

Notice that in example 4.1, since the integrand of the double integral was zero, evaluating the double integral was far easier than evaluating the line integral directly. There is another simple way of thinking of the line integral in example 4.1. Notice that you can write this as $\oint_C \mathbf{F}(x, y) \cdot d\mathbf{r}$, where $\mathbf{F}(x, y) = \langle x^2 + y^3, 3xy^2\rangle$. Notice further that $\mathbf{F}$ is conservative [with potential function $f(x, y) = \frac{1}{3}x^3 + xy^3$] and so, by Theorem 3.3 in section 14.3, the line integral of $\mathbf{F}$ over any piecewise-smooth, closed curve must be zero.

EXAMPLE 4.2 Evaluating a Challenging Line Integral with Green's Theorem

Evaluate the line integral $\oint_C (7y - e^{\sin x})\,dx + [15x - \sin(y^3 + 8y)]\,dy$, where C is the circle of radius 3 centered at the point $(5, -7)$, as shown in Figure 14.32.

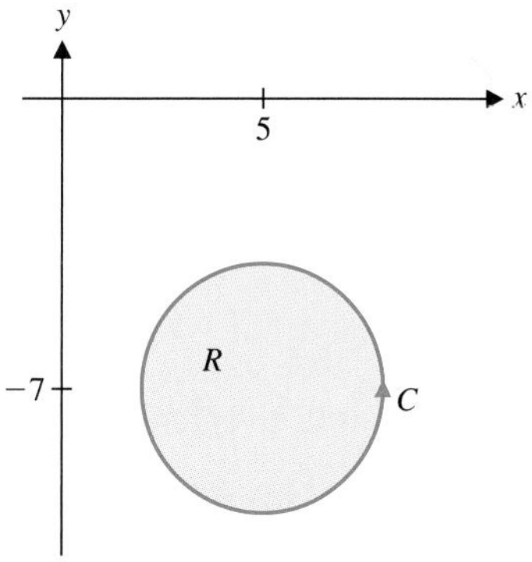

FIGURE 14.32
The region R

Solution First, notice that it will be virtually impossible to evaluate the line integral directly. (Think about this some, but don't spend too much time on it!) However, taking $M(x, y) = 7y - e^{\sin x}$ and $N(x, y) = 15x - \sin(y^3 + 8y)$, notice that M and N are continuous and have continuous first partial derivatives everywhere. So, we may apply Green's Theorem, which gives us

$$\oint_C (7y - e^{\sin x})\,dx + [15x - \sin(y^3 + 8y)]\,dy = \iint_R \left(\frac{\partial N}{\partial x} - \frac{\partial M}{\partial y}\right) dA$$
$$= \iint_R (15 - 7)\,dA$$
$$= 8\iint_R dA = 72\pi,$$

where $\iint_R dA$ is simply the area inside the region R, $\iint_R dA = \pi(3)^2 = 9\pi$.

If you look at example 4.2 critically, you might suspect that the integrand was chosen carefully so that the line integral was impossible to evaluate directly, but so that the integrand of the double integral was trivial. That's true: we did cook up the problem simply to illustrate the power of Green's Theorem. More significantly, Green's Theorem provides us with a wealth of interesting observations. One of these is as follows. Suppose that C is a piecewise-smooth, simple closed curve enclosing the region R. Then, taking $M(x, y) = 0$ and $N(x, y) = x$, we have

$$\oint_C x\,dy = \iint_R \left(\frac{\partial N}{\partial x} - \frac{\partial M}{\partial y}\right) dA = \iint_R dA,$$

which is simply the area of the region R. Alternatively, notice that if we take $M(x, y) = -y$ and $N(x, y) = 0$, we have

$$\oint_C -y\,dx = \iint_R \left(\frac{\partial N}{\partial x} - \frac{\partial M}{\partial y}\right) dA = \iint_R dA,$$

which is again the area of R. Putting these last two results together, we also have

$$\iint_R dA = \frac{1}{2}\oint_C x\,dy - y\,dx. \tag{4.5}$$

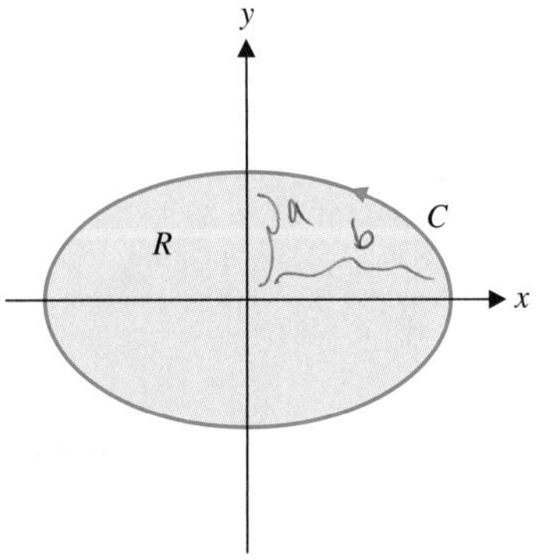

FIGURE 14.33
Elliptical region R

EXAMPLE 4.3 Using Green's Theorem to Find Area

Find the area enclosed by the ellipse $\frac{x^2}{a^2} + \frac{y^2}{b^2} = 1$.

Solution First, observe that the ellipse corresponds to the simple closed curve C defined parametrically by

$$C = \{(x, y) | x = a\cos t,\ y = b\sin t,\ 0 \le t \le 2\pi\},$$

where $a, b > 0$. You should also observe that C is smooth and positively oriented. (See Figure 14.33.) From (4.5), we have that the area A of the ellipse is given by

$$A = \frac{1}{2}\oint_C x\,dy - y\,dx = \frac{1}{2}\int_0^{2\pi} [(a\cos t)(b\cos t) - (b\sin t)(-a\sin t)]\,dt$$
$$= \frac{1}{2}\int_0^{2\pi} (ab\cos^2 t + ab\sin^2 t)\,dt = \pi ab. \ ■$$

-> think of circle -> πr^2

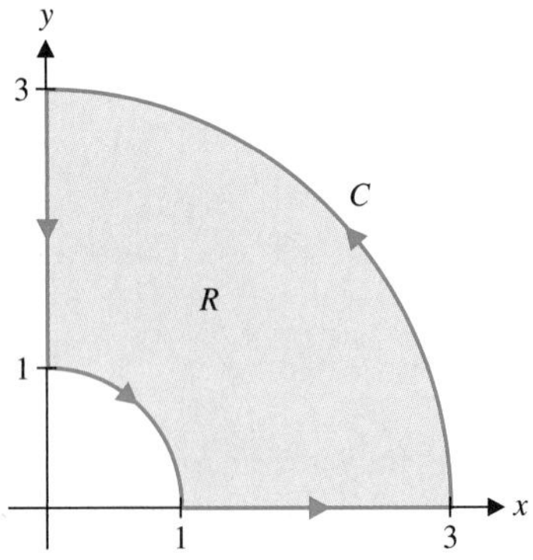

FIGURE 14.34
The region R

EXAMPLE 4.4 Using Green's Theorem to Evaluate a Line Integral

Evaluate the line integral $\oint_C (e^x + 6xy)\,dx + (8x^2 + \sin y^2)\,dy$, where C is the positively-oriented boundary of the region bounded by the circles of radii 1 and 3, centered at the origin and lying in the first quadrant, as indicated in Figure 14.34.

Solution Notice that since C consists of four distinct pieces, evaluating the line integral directly by parameterizing the curve is probably not a good choice. On the other hand, since C is a piecewise-smooth, simple closed curve, we have by Green's Theorem that

$$\oint_C (e^x + 6xy)\,dx + (8x^2 + \sin y^2)\,dy = \iint_R \left[\frac{\partial}{\partial x}(8x^2 + \sin y^2) - \frac{\partial}{\partial y}(e^x + 6xy)\right] dA$$
$$= \iint_R (16x - 6x)\,dA = \iint_R 10x\,dA,$$

where R is the region between the two circles and lying in the first quadrant. Notice that this is easy to compute using polar coordinates, as follows:

$$\oint_C (e^x + 6xy)\,dx + (8x^2 + \sin y^2)\,dy = \iint_R 10 \underbrace{x}_{r\cos\theta}\ \underbrace{dA}_{r\,dr\,d\theta}$$

$$= \int_0^{\pi/2}\int_1^3 (10r\cos\theta)\,r\,dr\,d\theta$$

$$= \int_0^{\pi/2} \cos\theta \frac{10r^3}{3}\Big|_{r=1}^{r=3}\,d\theta$$

$$= \frac{10}{3}(3^3 - 1^3)\sin\theta\Big|_0^{\pi/2}$$

$$= \frac{260}{3}. \ \blacksquare$$

You should notice that in example 4.4, Green's Theorem is not a mere convenience; rather, it is a virtual necessity. Evaluating the line integral directly would prove to be a very significant challenge. (Go ahead and try it to see what we mean.)

For simplicity, we often will use the notation ∂R to refer to the boundary of the region R, oriented in the positive direction. Using this notation, the conclusion of Green's Theorem is written as

$$\oint_{\partial R} M(x, y)\,dx + N(x, y)\,dy = \iint_R \left(\frac{\partial N}{\partial x} - \frac{\partial M}{\partial y}\right)dA.$$

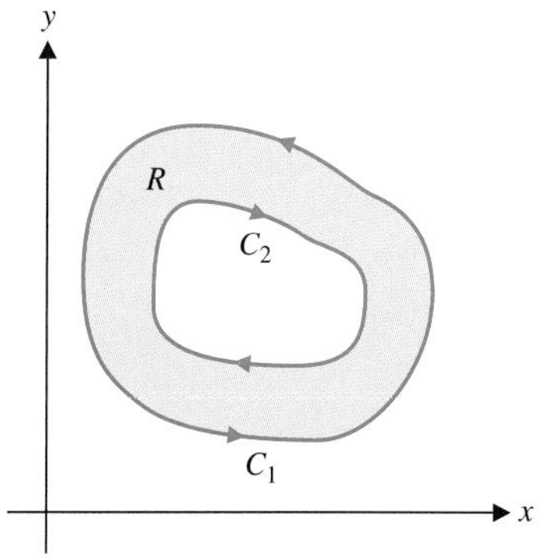

FIGURE 14.35a
Region with a hole

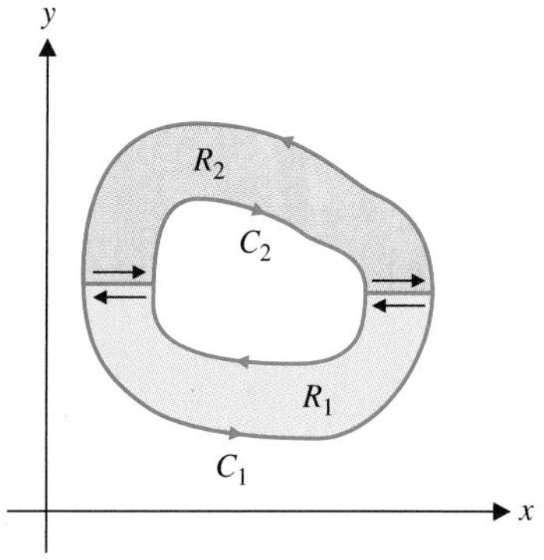

FIGURE 14.35b
$R = R_1 \cup R_2$

We can extend Green's Theorem to the case where a region is not simply-connected (i.e., where the region has one or more holes). We must emphasize that when dealing with such regions, the integration is taken over the *entire* boundary of the region (not just the outermost portion of the boundary!) and that the boundary curve is traversed in the positive direction, always keeping the region to the left. For instance, for the region R illustrated in Figure 14.35a with a single hole, notice that the boundary of R, ∂R, consists of two separate curves, C_1 and C_2, where C_2 is traversed clockwise, in order to keep the orientation positive on all of the boundary. Since the region is not simply-connected, we may not apply Green's Theorem directly. Rather, we first make two horizontal slits in the region, as indicated in Figure 14.35b, dividing R into the two simply-connected regions R_1 and R_2. Notice that we can then apply Green's Theorem in each of R_1 and R_2 separately. Adding the double integrals over R_1 and R_2 gives us the double integral over all of R. We have

$$\iint_R \left(\frac{\partial N}{\partial x} - \frac{\partial M}{\partial y}\right)dA = \iint_{R_1}\left(\frac{\partial N}{\partial x} - \frac{\partial M}{\partial y}\right)dA + \iint_{R_2}\left(\frac{\partial N}{\partial x} - \frac{\partial M}{\partial y}\right)dA$$

$$= \oint_{\partial R_1} M(x, y)\,dx + N(x, y)\,dy$$

$$+ \oint_{\partial R_2} M(x, y)\,dx + N(x, y)\,dy.$$

Further, since the line integrals over the common portions of ∂R_1 and ∂R_2 (i.e., the slits) are traversed in the opposite direction (one way on ∂R_1 and the other on ∂R_2), the line integrals over these portions will cancel out, leaving only the line integrals over C_1 and C_2.

This gives us

$$\iint_R \left(\frac{\partial N}{\partial x} - \frac{\partial M}{\partial y}\right) dA = \oint_{\partial R_1} M(x, y)\, dx + N(x, y)\, dy + \oint_{\partial R_2} M(x, y)\, dx + N(x, y)\, dy$$

$$= \oint_{C_1} M(x, y)\, dx + N(x, y)\, dy + \oint_{C_2} M(x, y)\, dx + N(x, y)\, dy$$

$$= \oint_C M(x, y)\, dx + N(x, y)\, dy.$$

This says that Green's Theorem also holds for regions with a single hole. Of course, we can repeat the preceding argument to extend Green's Theorem to regions with any *finite* number of holes.

EXAMPLE 4.5 An Application of Green's Theorem

For $\mathbf{F}(x, y) = \frac{1}{x^2 + y^2}\langle -y, x\rangle$, show that $\oint_C \mathbf{F}(x, y) \cdot d\mathbf{r} = 2\pi$, for every simple closed curve C enclosing the origin.

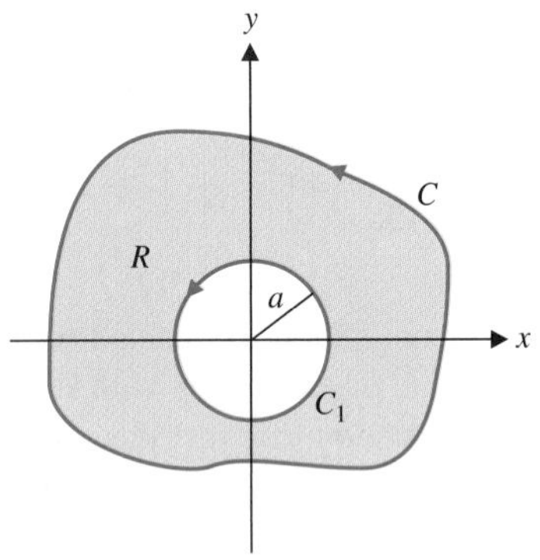

FIGURE 14.36
The region R

Solution Let C be any simple closed curve enclosing the origin and let C_1 be the circle of radius $a > 0$, centered at the origin (and positively oriented), where a is taken to be sufficiently small so that C_1 is completely enclosed by C, as illustrated in Figure 14.36. Further, let R be the region bounded between the curves C and C_1 (and including the curves themselves). Applying our extended version of Green's Theorem in R, we have

$$\oint_C \mathbf{F}(x, y) \cdot d\mathbf{r} - \oint_{C_1} \mathbf{F}(x, y) \cdot d\mathbf{r}$$

$$= \oint_{\partial R} \mathbf{F}(x, y) \cdot d\mathbf{r}$$

$$= \iint_R \left(\frac{\partial N}{\partial x} - \frac{\partial M}{\partial y}\right) dA$$

$$= \iint_R \left[\frac{(1)(x^2 + y^2) - x(2x)}{(x^2 + y^2)^2} - \frac{(-1)(x^2 + y^2) + y(2y)}{(x^2 + y^2)^2}\right] dA$$

$$= \iint_R 0\, dA = 0.$$

This gives us

$$\oint_C \mathbf{F}(x, y) \cdot d\mathbf{r} = \oint_{C_1} \mathbf{F}(x, y) \cdot d\mathbf{r}.$$

Now, we chose C_1 to be a circle because we can easily parameterize a circle and then evaluate the line integral around C_1 explicitly. Notice that C_1 can be expressed parametrically by $x = a\cos t$, $y = a\sin t$, for $0 \le t \le 2\pi$. Noting that on C_1, $x^2 + y^2 = a^2$, this leaves us with an integral that we can easily evaluate, as follows:

$$\oint_C \mathbf{F}(x, y) \cdot d\mathbf{r} = \oint_{C_1} \mathbf{F}(x, y) \cdot d\mathbf{r} = \oint_{C_1} \frac{1}{a^2} \langle -y, x \rangle \cdot d\mathbf{r}$$
$$= \frac{1}{a^2} \oint_{C_1} -y\,dx + x\,dy$$
$$= \frac{1}{a^2} \int_0^{2\pi} (-a \sin t)(-a \sin t) + (a \cos t)(a \cos t)\,dt$$
$$= \int_0^{2\pi} dt = 2\pi. \quad \blacksquare$$

Notice that without Green's Theorem, proving a result such as that developed in example 4.5 would be elusive.

Now that we have Green's Theorem, we are in a position to prove the second half of Theorem 3.4. For convenience, we restate the theorem here.

THEOREM 4.2

Suppose that $M(x, y)$ and $N(x, y)$ have continuous first partial derivatives on a simply-connected region D. Then, $\int_C M(x, y)\,dx + N(x, y)\,dy$ is independent of path if and only if $M_y(x, y) = N_x(x, y)$ for all (x, y) in D.

PROOF

Recall that in section 14.3, we proved the first part of the theorem: that if $\int_C M(x, y)\,dx + N(x, y)\,dy$ is independent of path, then it follows that $M_y(x, y) = N_x(x, y)$ for all (x, y) in D. We now prove that if $M_y(x, y) = N_x(x, y)$ for all (x, y) in D, then it follows that the line integral is independent of path. Let S be any piecewise-smooth closed curve lying in D. If S is simple and positively oriented, then since D is simply-connected, the region R enclosed by S is completely contained in D, so that $M_y(x, y) = N_x(x, y)$ for all (x, y) in R. From Green's Theorem, we now have that

$$\oint_S M(x, y)\,dx + N(x, y)\,dy = \iint_R \left(\frac{\partial N}{\partial x} - \frac{\partial M}{\partial y} \right) dA = 0.$$

That is, for every piecewise-smooth, simple closed curve S lying in D, we have

$$\oint_S M(x, y)\,dx + N(x, y)\,dy = 0. \tag{4.6}$$

If S is not simple, then it intersects itself one or more times, creating two or more loops, each one of which is a simple closed curve. Since the line integral of $M(x, y)\,dx + N(x, y)\,dy$ over each of these is zero by (4. 6), it also follows that $\int_S M(x, y)\,dx + N(x, y)\,dy = 0$. It now follows from Theorem 3.3 that $\mathbf{F}(x, y) = \langle M(x, y), N(x, y) \rangle$ must be conservative in D. Finally, it follows from Theorem 3.1 that $\int_C M(x, y)\,dx + N(x, y)\,dy$ is independent of path. ■

BEYOND FORMULAS

Green's Theorem is the first of three theorems in this chapter that relate different types of integrals. The alternatives given in these results can be helpful both computationally and theoretically. Example 4.2 shows how we can evaluate a difficult line integral by evaluating the equivalent (and simpler) double integral. Perhaps surprisingly, the use of Green's Theorem in example 4.5 is probably more important in applications. The theoretical result can be applied to any relevant problem and results like example 4.5 can sometimes provide important insight into general processes.

HW #29
Pg 1164 ##1,5

EXERCISES 14.4

WRITING EXERCISES

1. Given a line integral to evaluate, briefly describe the circumstances under which you should think about using Green's Theorem to replace the line integral with a double integral. Comment on the properties of the curve C and the functions involved.

2. In example 4.1, Green's Theorem allowed us to quickly show that the line integral equals 0. Following the example, we noted that this was the line integral for a conservative force field. Discuss which method (Green's Theorem, conservative field) you would recommend trying first to determine whether a line integral equals 0.

3. Equation (4.5) shows how to compute area as a line integral. Using example 4.3 as a guide, explain why we wrote the area as $\frac{1}{2}\oint_C x\,dy - y\,dx$ instead of $\oint_C x\,dy$ or $\oint_C -y\,dx$.

4. Suppose that you drive a car to a variety of places for shopping and then return home. If your path formed a simple closed curve, explain how you could use (4.5) to estimate the area enclosed by your path. (Hint: If $\langle x, y\rangle$ represents position, what does $\langle x', y'\rangle$ represent?)

In exercises 1–4, evaluate the indicated line integral (a) directly and (b) using Green's Theorem.

1. $\oint_C (x^2 - y)\,dx + y^2\,dy$, where C is the circle $x^2 + y^2 = 1$ oriented counterclockwise

2. $\oint_C (y^2 + x)\,dx + (3x + 2xy)\,dy$, where C is the circle $x^2 + y^2 = 4$ oriented counterclockwise

3. $\oint_C x^2\,dx - x^3\,dy$, where C is the square from $(0, 0)$ to $(0, 2)$ to $(2, 2)$ to $(2, 0)$ to $(0, 0)$

4. $\oint_C (y^2 - 2x)\,dx + x^2\,dy$, where C is the square from $(0, 0)$ to $(1, 0)$ to $(1, 1)$ to $(0, 1)$ to $(0, 0)$

In exercises 5–20, use Green's Theorem to evaluate the indicated line integral.

5. $\oint_C xe^{2x}\,dx - 3x^2y\,dy$, where C is the rectangle from $(0, 0)$ to $(3, 0)$ to $(3, 2)$ to $(0, 2)$ to $(0, 0)$

6. $\oint_C ye^{2x}\,dx + x^2y^2\,dy$, where C is the rectangle from $(-2, 0)$ to $(3, 0)$ to $(3, 2)$ to $(-2, 2)$ to $(-2, 0)$

7. $\oint_C \left(\frac{x}{x^2+1} - y\right)dx + (3x - 4\tan y/2)\,dy$, where C is the portion of $y = x^2$ from $(-1, 1)$ to $(1, 1)$, followed by the portion of $y = 2 - x^2$ from $(1, 1)$ to $(-1, 1)$

8. $\int_C (xy - e^{2x})\,dx + (2x^2 - 4y^2)\,dy$, where C is formed by $y = x^2$ and $y = 8 - x^2$ oriented clockwise

9. $\oint_C (\tan x - y^3)\,dx + (x^3 - \sin y)\,dy$, where C is the circle $x^2 + y^2 = 2$

10. $\int_C \left(\sqrt{x^2+1} - x^2y\right)dx + (xy^2 - y^{5/3})\,dy$, where C is the circle $x^2 + y^2 = 4$ oriented clockwise

11. $\oint_C \mathbf{F}\cdot d\mathbf{r}$, where $\mathbf{F} = \langle x^3 - y, x + y^3\rangle$ and C is formed by $y = x^2$ and $y = x$

12. $\oint_C \mathbf{F}\cdot d\mathbf{r}$, where $\mathbf{F} = \langle y^2 + 3x^2y, xy + x^3\rangle$ and C is formed by $y = x^2$ and $y = 2x$

13. $\oint_C \mathbf{F}\cdot d\mathbf{r}$, where $\mathbf{F} = \langle e^{x^2} - y, e^{2x} + y\rangle$ and C is formed by $y = 1 - x^2$ and $y = 0$

14. $\oint_C \mathbf{F}\cdot d\mathbf{r}$, where $\mathbf{F} = \langle xe^{xy} + y, ye^{xy} + 2x\rangle$ and C is formed by $y = x^2$ and $y = 4$

15. $\oint_C [y^3 - \ln(x + 1)]\,dx + \left(\sqrt{y^2+1} + 3x\right)dy$, where C is formed by $x = y^2$ and $x = 4$

16. $\oint_C (y \sec^2 x - 2)\,dx + (\tan x - 4y^2)\,dy$, where C is formed by $x = 1 - y^2$ and $x = 0$

17. $\oint_C x^2\,dx + 2x\,dy + (z-2)\,dz$, where C is the triangle from $(0, 0, 2)$ to $(2, 0, 2)$ to $(2, 2, 2)$ to $(0, 0, 2)$

18. $\oint_C 4y\,dx + y^3\,dy + z^4\,dz$, where C is $x^2 + y^2 = 4$ in the plane $z = 0$

19. $\oint_C \mathbf{F} \cdot d\mathbf{r}$, where $\mathbf{F} = \langle x^3 - y^4, e^{x^2+z^2}, x^2 - 16y^2z^2 \rangle$ and C is $x^2 + z^2 = 1$ in the plane $y = 0$

20. $\oint_C \mathbf{F} \cdot d\mathbf{r}$, where $\mathbf{F} = \langle x^3 - y^2 z, \sqrt{x^2 + z^2}, 4xy - z^4 \rangle$ and C is formed by $z = 1 - x^2$ and $z = 0$ in the plane $y = 2$

In exercises 21–26, use a line integral to compute the area of the given region.

21. The ellipse $4x^2 + y^2 = 16$

22. The ellipse $4x^2 + y^2 = 4$

23. The region bounded by $x^{2/3} + y^{2/3} = 1$. (Hint: Let $x = \cos^3 t$ and $y = \sin^3 t$)

24. The region bounded by $x^{2/5} + y^{2/5} = 1$

25. The region bounded by $y = x^2$ and $y = 4$

26. The region bounded by $y = x^2$ and $y = 2x$

27. Use Green's Theorem to show that the center of mass of the region bounded by the positive curve C with constant density is given by $\bar{x} = \frac{1}{2A}\oint_C x^2\,dy$ and $\bar{y} = -\frac{1}{2A}\oint_C y^2\,dx$, where A is the area of the region.

28. Use the result of exercise 27 to find the center of mass of the region in exercise 26, assuming constant density.

29. Use the result of exercise 27 to find the center of mass of the region bounded by the curve traced out by $\langle t^3 - t, 1 - t^2 \rangle$, for $-1 \le t \le 1$, assuming constant density.

30. Use the result of exercise 27 to find the center of mass of the region bounded by the curve traced out by $\langle t^2 - t, t^3 - t \rangle$, for $0 \le t \le 1$, assuming constant density.

31. Use Green's Theorem to prove the change of variables formula

$$\iint_R dA = \iint_S \left| \frac{\partial(x, y)}{\partial(u, v)} \right| du\,dv,$$

where $x = x(u, v)$ and $y = y(u, v)$ are functions with continuous partial derivatives.

32. For $\mathbf{F} = \dfrac{1}{x^2 + y^2}\langle -y, x \rangle$ and C any circle of radius $r > 0$ not containing the origin, show that $\oint_C \mathbf{F} \cdot d\mathbf{r} = 0$.

In exercises 33–36, use the technique of example 4.5 to evaluate the line integral.

33. $\oint_C \mathbf{F} \cdot d\mathbf{r}$, where $\mathbf{F} = \left\langle \dfrac{x}{x^2 + y^2}, \dfrac{y}{x^2 + y^2} \right\rangle$ and C is any positively oriented simple closed curve containing the origin

34. $\oint_C \mathbf{F} \cdot d\mathbf{r}$, where $\mathbf{F} = \left\langle \dfrac{y^2 - x^2}{(x^2 + y^2)^2}, \dfrac{-2xy}{(x^2 + y^2)^2} \right\rangle$ and C is any positively oriented simple closed curve containing the origin

35. $\oint_C \mathbf{F} \cdot d\mathbf{r}$, where $\mathbf{F} = \left\langle \dfrac{x^3}{x^4 + y^4}, \dfrac{y^3}{x^4 + y^4} \right\rangle$ and C is any positively oriented simple closed curve containing the origin

36. $\oint_C \mathbf{F} \cdot d\mathbf{r}$, where $\mathbf{F} = \left\langle \dfrac{y^2 x}{x^4 + y^4}, \dfrac{-x^2 y}{x^4 + y^4} \right\rangle$ and C is any positively oriented simple closed curve containing the origin

37. Where is $\mathbf{F}(x, y) = \left\langle \dfrac{2x}{x^2 + y^2}, \dfrac{2y}{x^2 + y^2} \right\rangle$ defined? Show that $M_y = N_x$ everywhere the partial derivatives are defined. If C is a simple closed curve enclosing the origin, does Green's Theorem guarantee that $\oint_C \mathbf{F} \cdot d\mathbf{r} = 0$? Explain.

38. For the vector field of exercise 37, show that $\oint_C \mathbf{F} \cdot d\mathbf{r}$ is the same for all closed curves enclosing the origin.

39. If $\mathbf{F}(x, y) = \left\langle \dfrac{2x}{x^2 + y^2}, \dfrac{2y}{x^2 + y^2} \right\rangle$ and C is a simple closed curve in the fourth quadrant, does Green's Theorem guarantee that $\oint_C \mathbf{F} \cdot d\mathbf{r} = 0$? Explain.

EXPLORATORY EXERCISES

1. Evaluate $\oint_C \mathbf{F} \cdot d\mathbf{r}$, where

$$\mathbf{F} = \left\langle \frac{-y}{(x^2 + y^2)^2}, \frac{x}{(x^2 + y^2)^2} \right\rangle$$

and C is the circle $x^2 + y^2 = a^2$. Use the result and Green's Theorem to show that $\displaystyle\iint_R \frac{-2}{(x^2 + y^2)^2}\,dA$ diverges, where R is the disk $x^2 + y^2 \le 1$.

14.5 CURL AND DIVERGENCE

We have seen how Green's Theorem relates the line integral of a function over the boundary of a plane region R to the double integral of a related function over the region R. In some cases, the line integral is easier to evaluate, while in other cases, the double integral is easier. More significantly, Green's Theorem provides us with a connection between physical quantities measured on the boundary of a plane region with related quantities in the interior of the region. The goal of the rest of the chapter is to extend Green's Theorem to results that relate triple integrals, double integrals and line integrals. The first step is to understand the vector operations of curl and divergence introduced in this section.

Both the curl and divergence are generalizations of the notion of derivative that are applied to vector fields. Both directly measure important physical quantities related to a vector field $\mathbf{F}(x, y, z)$.

DEFINITION 5.1

The **curl** of the vector field $\mathbf{F}(x, y, z) = \langle F_1(x, y, z), F_2(x, y, z), F_3(x, y, z)\rangle$ is the vector field

$$\text{curl } \mathbf{F} = \left(\frac{\partial F_3}{\partial y} - \frac{\partial F_2}{\partial z}\right)\mathbf{i} + \left(\frac{\partial F_1}{\partial z} - \frac{\partial F_3}{\partial x}\right)\mathbf{j} + \left(\frac{\partial F_2}{\partial x} - \frac{\partial F_1}{\partial y}\right)\mathbf{k},$$

defined at all points at which all the indicated partial derivatives exist.

An easy way to remember curl **F** is to use cross product notation, as follows. Notice that using a determinant, we can write

$$\nabla \times \mathbf{F} = \begin{vmatrix} \mathbf{i} & \mathbf{j} & \mathbf{k} \\ \frac{\partial}{\partial x} & \frac{\partial}{\partial y} & \frac{\partial}{\partial z} \\ F_1 & F_2 & F_3 \end{vmatrix}$$

$$= \left(\frac{\partial F_3}{\partial y} - \frac{\partial F_2}{\partial z}\right)\mathbf{i} - \left(\frac{\partial F_3}{\partial x} - \frac{\partial F_1}{\partial z}\right)\mathbf{j} + \left(\frac{\partial F_2}{\partial x} - \frac{\partial F_1}{\partial y}\right)\mathbf{k}$$

$$= \left\langle \frac{\partial F_3}{\partial y} - \frac{\partial F_2}{\partial z}, \frac{\partial F_1}{\partial z} - \frac{\partial F_3}{\partial x}, \frac{\partial F_2}{\partial x} - \frac{\partial F_1}{\partial y} \right\rangle = \text{curl } \mathbf{F}, \tag{5.1}$$

whenever all of the indicated partial derivatives are defined.

EXAMPLE 5.1 Computing the Curl of a Vector Field

Compute curl **F** for (a) $\mathbf{F}(x, y, z) = \langle x^2y, 3x - yz, z^3\rangle$ and (b) $\mathbf{F}(x, y, z) = \langle x^3 - y, y^5, e^z\rangle$.

Solution Using the cross product notation in (5.1), we have that for (a):

$$\text{curl } \mathbf{F} = \nabla \times \mathbf{F} = \begin{vmatrix} \mathbf{i} & \mathbf{j} & \mathbf{k} \\ \frac{\partial}{\partial x} & \frac{\partial}{\partial y} & \frac{\partial}{\partial z} \\ x^2y & 3x - yz & z^3 \end{vmatrix}$$

$$= \left(\frac{\partial(z^3)}{\partial y} - \frac{\partial(3x - yz)}{\partial z}\right)\mathbf{i} - \left(\frac{\partial(z^3)}{\partial x} - \frac{\partial(x^2y)}{\partial z}\right)\mathbf{j}$$

$$+ \left(\frac{\partial(3x - yz)}{\partial x} - \frac{\partial(x^2y)}{\partial y}\right)\mathbf{k}$$

$$= (0 + y)\mathbf{i} - (0 - 0)\mathbf{j} + (3 - x^2)\mathbf{k} = \langle y, 0, 3 - x^2\rangle.$$

Similarly, for part (b), we have

$$\text{curl }\mathbf{F} = \nabla \times \mathbf{F} = \begin{vmatrix} \mathbf{i} & \mathbf{j} & \mathbf{k} \\ \frac{\partial}{\partial x} & \frac{\partial}{\partial y} & \frac{\partial}{\partial z} \\ x^3 - y & y^5 & e^z \end{vmatrix}$$

$$= \left(\frac{\partial(e^z)}{\partial y} - \frac{\partial(y^5)}{\partial z}\right)\mathbf{i} - \left(\frac{\partial(e^z)}{\partial x} - \frac{\partial(x^3-y)}{\partial z}\right)\mathbf{j} + \left(\frac{\partial(y^5)}{\partial x} - \frac{\partial(x^3-y)}{\partial y}\right)\mathbf{k}$$

$$= (0-0)\mathbf{i} - (0-0)\mathbf{j} + (0+1)\mathbf{k} = \langle 0, 0, 1\rangle. \blacksquare$$

Notice that in part (b) of example 5.1, the only term that contributes to the curl is the term $-y$ in the **i**-component of $\mathbf{F}(x, y, z)$. This illustrates an important property of the curl. Terms in the **i**-component of the vector field involving only x will not contribute to the curl, nor will terms in the **j**-component involving only y nor terms in the **k**-component involving only z. You can use these observations to simplify some calculations of the curl. For instance, notice that

$$\begin{aligned} \text{curl}\langle x^3, \sin^2 y, \sqrt{z^2+1} + x^2\rangle &= \text{curl}\langle 0, 0, x^2\rangle \\ &= \nabla \times \langle 0, 0, x^2\rangle = \langle 0, -2x, 0\rangle. \end{aligned}$$

The simplification discussed above gives an important hint about what the curl measures, since the variables must get "mixed up" to produce a nonzero curl. Example 5.2 provides a clue as to the meaning of the curl of a vector field.

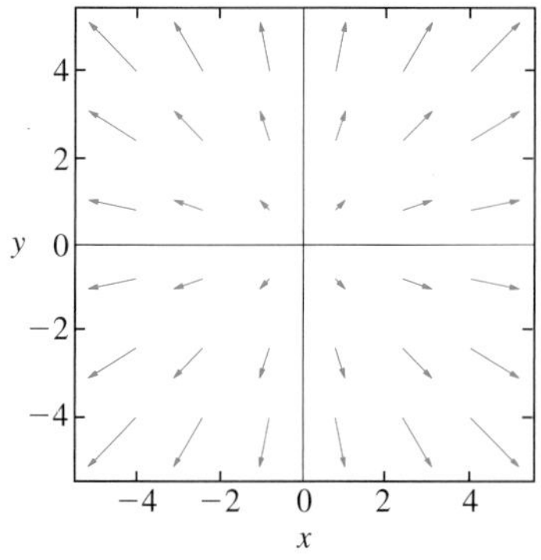

FIGURE 14.37a
Graph of $\langle x, y, 0\rangle$

EXAMPLE 5.2 Interpreting the Curl of a Vector Field

Compute the curl of (a) $\mathbf{F}(x, y, z) = x\mathbf{i} + y\mathbf{j}$ and (b) $\mathbf{G}(x, y, z) = y\mathbf{i} - x\mathbf{j}$, and interpret each graphically.

Solution For (a), we have

$$\nabla \times \mathbf{F} = \begin{vmatrix} \mathbf{i} & \mathbf{j} & \mathbf{k} \\ \frac{\partial}{\partial x} & \frac{\partial}{\partial y} & \frac{\partial}{\partial z} \\ x & y & 0 \end{vmatrix} = \langle 0-0, -(0-0), 0-0\rangle = \langle 0, 0, 0\rangle.$$

For (b), we have

$$\nabla \times \mathbf{G} = \begin{vmatrix} \mathbf{i} & \mathbf{j} & \mathbf{k} \\ \frac{\partial}{\partial x} & \frac{\partial}{\partial y} & \frac{\partial}{\partial z} \\ y & -x & 0 \end{vmatrix} = \langle 0-0, -(0-0), -1-1\rangle = \langle 0, 0, -2\rangle.$$

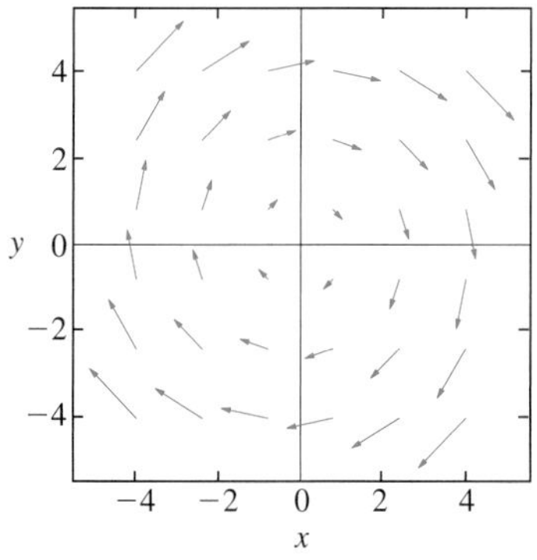

FIGURE 14.37b
Graph of $\langle y, -x, 0\rangle$

Graphs of the vector fields **F** and **G** in two dimensions are shown in Figures 14.37a and 14.37b, respectively. It is helpful to think of each of these vector fields as the velocity field for a fluid in motion across the xy-plane. In this case, the vectors indicated in the graph of the velocity field indicate the direction of flow of the fluid. For the vector field $\langle x, y, 0\rangle$, observe that the fluid flows directly away from the origin, so that the fluid has no rotation and in (a) we found that curl $\mathbf{F} = \mathbf{0}$. By contrast, the vector field $\langle y, -x, 0\rangle$ indicates a clockwise rotation of the fluid, while in (b) we computed a nonzero curl. In particular, notice that if you curl the fingers of your right hand so that your fingertips point in the direction of the flow, your thumb will point into the page, in the direction of $-\mathbf{k}$, which has the same direction as

$$\text{curl}\langle y, -x, 0\rangle = \nabla \times \langle y, -x, 0\rangle = -2\mathbf{k}. \blacksquare$$

As we will see through our discussion of Stokes' Theorem in section 14.8, $\nabla \times \mathbf{F}(x, y, z)$ provides a measure of the tendency of the fluid flow to rotate about an axis parallel to $\nabla \times \mathbf{F}(x, y, z)$. If $\nabla \times \mathbf{F} = \mathbf{0}$, we say that the vector field is **irrotational** at that point. (That is, the fluid does not tend to rotate near the point.)

We noted earlier that there is no contribution to the curl of a vector field $\mathbf{F}(x, y, z)$ from terms in the **i**-component of **F** that involve only x, nor from terms in the **j**-component of **F** involving only y nor terms in the **k**-component of **F** involving only z. By contrast, these terms make important contributions to the **divergence** of a vector field, the other major vector operation introduced in this section.

DEFINITION 5.2

The **divergence** of the vector field $\mathbf{F}(x, y, z) = \langle F_1(x, y, z), F_2(x, y, z), F_3(x, y, z)\rangle$ is the scalar function

$$\text{div}\,\mathbf{F}(x, y, z) = \frac{\partial F_1}{\partial x} + \frac{\partial F_2}{\partial y} + \frac{\partial F_3}{\partial z},$$

defined at all points at which all the indicated partial derivatives exist.

NOTES

Take care to note that, while the curl of a vector field is another vector field, the divergence of a vector field is a scalar function.

While we wrote the curl using cross product notation, note that we can write the divergence of a vector field using dot product notation, as follows:

$$\nabla \cdot \mathbf{F} = \left\langle \frac{\partial}{\partial x}, \frac{\partial}{\partial y}, \frac{\partial}{\partial z} \right\rangle \cdot \langle F_1, F_2, F_3\rangle = \frac{\partial F_1}{\partial x} + \frac{\partial F_2}{\partial y} + \frac{\partial F_3}{\partial z} = \text{div}\,\mathbf{F}(x, y, z). \quad (5.2)$$

EXAMPLE 5.3 Computing the Divergence of a Vector Field

Compute div **F** for (a) $\mathbf{F}(x, y, z) = \langle x^2y, 3x - yz, z^3\rangle$ and (b) $\mathbf{F}(x, y, z) = \langle x^3 - y, z^5, e^y\rangle$.

Solution For (a), we have from (5.2) that

$$\text{div}\,\mathbf{F} = \nabla \cdot \mathbf{F} = \frac{\partial(x^2y)}{\partial x} + \frac{\partial(3x - yz)}{\partial y} + \frac{\partial(z^3)}{\partial z} = 2xy - z + 3z^2.$$

For (b), we have from (5.2) that

$$\text{div}\,\mathbf{F} = \nabla \cdot \mathbf{F} = \frac{\partial(x^3 - y)}{\partial x} + \frac{\partial(z^5)}{\partial y} + \frac{\partial(e^y)}{\partial z} = 3x^2 + 0 + 0 = 3x^2. \blacksquare$$

Notice that in part (b) of example 5.3 the only term contributing to the divergence is the x^3 term in the **i**-component of **F**. Further, observe that in general, the divergence of $\mathbf{F}(x, y, z)$ is not affected by terms in the **i**-component of **F** that do not involve x, terms in the **j**-component of **F** that do not involve y or terms in the **k**-component of **F** that do not involve z. Returning to the two-dimensional vector fields of example 5.2, we can develop a graphical interpretation of the divergence.

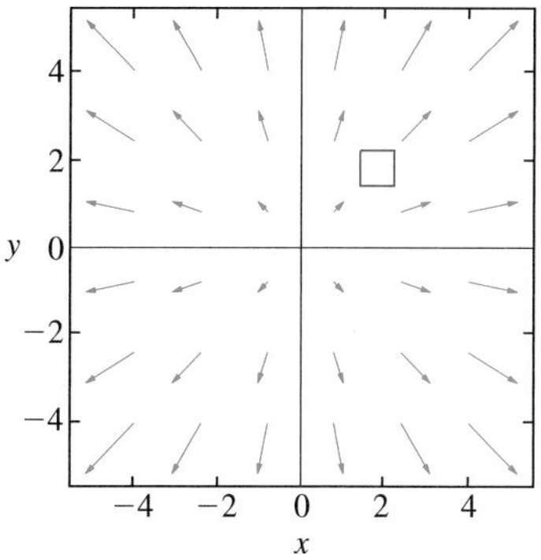

FIGURE 14.38a
Graph of $\langle x, y\rangle$

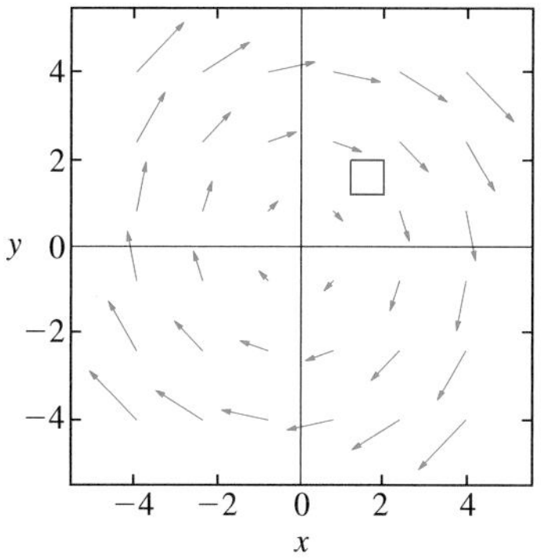

FIGURE 14.38b
Graph of $\langle y, -x\rangle$

EXAMPLE 5.4 Interpreting the Divergence of a Vector Field

Compute the divergence of (a) $\mathbf{F}(x, y) = x\mathbf{i} + y\mathbf{j}$ and (b) $\mathbf{F}(x, y) = y\mathbf{i} - x\mathbf{j}$ and interpret each graphically.

Solution For (a), we have $\nabla \cdot \mathbf{F} = \frac{\partial(x)}{\partial x} + \frac{\partial(y)}{\partial y} = 2$. For (b), we have $\nabla \cdot \mathbf{F} = \frac{\partial(y)}{\partial x} + \frac{\partial(-x)}{\partial y} = 0$. Graphs of the vector fields in (a) and (b) are shown in Figures 14.38a and 14.38b, respectively. Notice the boxes that we have superimposed on the graph of each vector field. If $\mathbf{F}(x, y)$ represents the velocity field of a fluid in motion, try to use the graphs to estimate the net flow of fluid into or out of each box. For $\langle y, -x\rangle$, the fluid is rotating in circular paths, so that the velocity of any particle on a given circle centered at the origin is a constant. This suggests that the flow into the box should equal the flow out of the box and the net flow is 0, which you'll notice is also the value of the divergence of this velocity field. By contrast, for the vector field $\langle x, y\rangle$, notice that the arrows coming into the box are shorter than the arrows exiting the box. This says that the net flow *out of* the box is positive (i.e., there is more fluid exiting the box than entering the box). Notice that in this case, the divergence is positive. ■

We'll show in section 14.7 (using the Divergence Theorem) that the divergence of a vector field at a point (x, y, z) corresponds to the net flow of fluid per unit volume out of a small box centered at (x, y, z). If $\nabla \cdot \mathbf{F}(x, y, z) > 0$, more fluid exits the box than enters (as illustrated in Figure 14.38a) and we call the point (x, y, z) a **source.** If $\nabla \cdot \mathbf{F}(x, y, z) < 0$, more fluid enters the box than exits and we call the point (x, y, z) a **sink.** If $\nabla \cdot \mathbf{F}(x, y, z) = 0$, throughout some region D, then we say that the vector field $\mathbf{F}$ is **source-free** or **incompressible.**

We have now used the "del" operator ∇ for three different derivative-like operations. The gradient of a scalar function f is the vector field ∇f, the curl of a vector field $\mathbf{F}$ is the vector field $\nabla \times \mathbf{F}$ and the divergence of a vector field $\mathbf{F}$ is the scalar function $\nabla \cdot \mathbf{F}$. Pay special attention to the different roles of scalar and vector functions in these operations. An analysis of the possible combinations of these operations will give us further insight into the properties of vector fields.

EXAMPLE 5.5 Vector Fields and Scalar Functions Involving the Gradient

If $f(x, y, z)$ is a scalar function and $\mathbf{F}(x, y, z)$ is a vector field, determine whether each operation is a scalar function, a vector field or undefined: (a) $\nabla \times (\nabla f)$, (b) $\nabla \times (\nabla \cdot \mathbf{F})$, (c) $\nabla \cdot (\nabla f)$.

Solution Examine each of these expressions one step at a time, working from the inside out. In (a), ∇f is a vector field, so the curl of ∇f is defined and gives a vector field. In (b), $\nabla \cdot \mathbf{F}$ is a scalar function, so the curl of $\nabla \cdot \mathbf{F}$ is undefined. In (c), ∇f is a vector field, so the divergence of ∇f is defined and gives a scalar function. ■

We can say more about the two operations defined in example 5.5 parts (a) and (c). If f has continuous second-order partial derivatives, then $\nabla f = \langle f_x, f_y, f_z\rangle$ and the divergence of the gradient is the scalar function

$$\nabla \cdot (\nabla f) = \left\langle \frac{\partial}{\partial x}, \frac{\partial}{\partial y}, \frac{\partial}{\partial z} \right\rangle \cdot \langle f_x, f_y, f_z\rangle = f_{xx} + f_{yy} + f_{zz}.$$

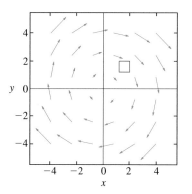

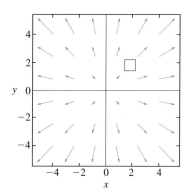

This combination of second partial derivatives arises in many important applications in physics and engineering. We call $\nabla \cdot (\nabla f)$ the **Laplacian** of f and typically use the shorthand notation

$$\nabla \cdot (\nabla f) = \nabla^2 f = f_{xx} + f_{yy} + f_{zz}$$

or $\Delta f = \nabla^2 f$.

Using the same notation, the curl of the gradient of a scalar function f is

$$\nabla \times (\nabla f) = \begin{vmatrix} \mathbf{i} & \mathbf{j} & \mathbf{k} \\ \frac{\partial}{\partial x} & \frac{\partial}{\partial y} & \frac{\partial}{\partial z} \\ f_x & f_y & f_z \end{vmatrix} = \langle f_{zy} - f_{yz}, f_{xz} - f_{zx}, f_{yx} - f_{xy} \rangle = \langle 0, 0, 0 \rangle,$$

assuming the mixed partial derivatives are equal. (We've seen that this occurs whenever all of the second-order partial derivatives are continuous in some open region.) Recall that if $\mathbf{F} = \nabla f$, then we call $\mathbf{F}$ a conservative field. The result $\nabla \times (\nabla f) = \mathbf{0}$ proves Theorem 5.1, which gives us a simple way for determining when a given three-dimensional vector field is not conservative.

THEOREM 5.1

Suppose that $\mathbf{F}(x, y, z) = \langle F_1(x, y, z), F_2(x, y, z), F_3(x, y, z) \rangle$ is a vector field whose components F_1, F_2 and F_3 have continuous first-order partial derivatives throughout an open region $D \subset \mathbb{R}^3$. If $\mathbf{F}$ is conservative, then $\nabla \times \mathbf{F} = \mathbf{0}$.

We can use Theorem 5.1 to determine that a given vector field is not conservative, as we illustrate in example 5.6.

EXAMPLE 5.6 Determining When a Vector Field Is Conservative

Use Theorem 5.1 to determine whether the following vector fields are conservative: (a) $\mathbf{F} = \langle \cos x - z, y^2, xz \rangle$ and (b) $\mathbf{F} = \langle 2xz, 3z^2, x^2 + 6yz \rangle$.

Solution For (a), we have

$$\nabla \times \mathbf{F} = \begin{vmatrix} \mathbf{i} & \mathbf{j} & \mathbf{k} \\ \frac{\partial}{\partial x} & \frac{\partial}{\partial y} & \frac{\partial}{\partial z} \\ \cos x - z & y^2 & xz \end{vmatrix} = \langle 0 - 0, -1 - z, 0 - 0 \rangle \neq \mathbf{0}$$

and so, by Theorem 5.1, $\mathbf{F}$ is not conservative.

For (b), we have

$$\nabla \times \mathbf{F} = \begin{vmatrix} \mathbf{i} & \mathbf{j} & \mathbf{k} \\ \frac{\partial}{\partial x} & \frac{\partial}{\partial y} & \frac{\partial}{\partial z} \\ 2xz & 3z^2 & x^2 + 6yz \end{vmatrix} = \langle 6z - 6z, 2x - 2x, 0 - 0 \rangle = \mathbf{0}.$$

Notice that in this case, Theorem 5.1 does not tell us whether or not **F** is conservative. However, you might notice that

$$\mathbf{F}(x, y, z) = \langle 2xz, 3z^2, x^2 + 6yz \rangle = \nabla(x^2z + 3yz^2).$$

Since we have found a potential function for **F**, we now see that it is indeed a conservative field. ■

Given example 5.6, you might be wondering whether or not the converse of Theorem 5.1 is true. That is, if $\nabla \times \mathbf{F} = \mathbf{0}$, must it follow that **F** is conservative? The answer to this is, "NO". We had an important clue to this in example 4.5. There, we saw that for the two-dimensional vector field $\mathbf{F}(x, y) = \frac{1}{x^2 + y^2}\langle -y, x \rangle$, $\oint_C \mathbf{F}(x, y) \cdot d\mathbf{r} = 2\pi$, for every simple closed curve C enclosing the origin. We follow up on this idea in example 5.7.

EXAMPLE 5.7 An Irrotational Vector Field That Is Not Conservative

For $\mathbf{F}(x, y, z) = \frac{1}{x^2 + y^2}\langle -y, x, 0 \rangle$, show that $\nabla \times \mathbf{F} = \mathbf{0}$ throughout the domain of **F**, but that **F** is not conservative.

Solution First, notice that

$$\nabla \times \mathbf{F} = \begin{vmatrix} \mathbf{i} & \mathbf{j} & \mathbf{k} \\ \frac{\partial}{\partial x} & \frac{\partial}{\partial y} & \frac{\partial}{\partial z} \\ \frac{-y}{x^2 + y^2} & \frac{x}{x^2 + y^2} & 0 \end{vmatrix}$$

$$= \mathbf{i} \begin{vmatrix} \frac{\partial}{\partial y} & \frac{\partial}{\partial z} \\ \frac{x}{x^2 + y^2} & 0 \end{vmatrix} - \mathbf{j} \begin{vmatrix} \frac{\partial}{\partial x} & \frac{\partial}{\partial z} \\ \frac{-y}{x^2 + y^2} & 0 \end{vmatrix} + \mathbf{k} \begin{vmatrix} \frac{\partial}{\partial x} & \frac{\partial}{\partial y} \\ \frac{-y}{x^2 + y^2} & \frac{x}{x^2 + y^2} \end{vmatrix}$$

$$= \mathbf{k}\left[\frac{\partial}{\partial x}\left(\frac{x}{x^2 + y^2}\right) + \frac{\partial}{\partial y}\left(\frac{y}{x^2 + y^2}\right)\right]$$

$$= \mathbf{k}\left[\frac{(x^2 + y^2) - 2x^2}{(x^2 + y^2)^2} + \frac{(x^2 + y^2) - 2y^2}{(x^2 + y^2)^2}\right] = \mathbf{0},$$

so that **F** is irrotational at every point at which it's defined (i.e., everywhere but on the line $x = y = 0$, that is, the z-axis). However, in example 4.5, we already showed that $\oint_C \mathbf{F}(x, y, z) \cdot d\mathbf{r} = 2\pi$, for every simple closed curve C lying in the xy-plane and enclosing the origin. Given this, it follows from Theorem 3.3 that **F** cannot be conservative, since if it were, we would need to have $\oint_C \mathbf{F}(x, y, z) \cdot d\mathbf{r} = 0$ for every piecewise-smooth closed curve C lying in the domain of **F**. ■

Note that in example 5.7, the vector field in question had a singularity (i.e., a point where one or more of the components of the vector field blow up to ∞) at every point on the z-axis. Even though the curves we considered did not pass through any of these singularities, they in some sense "enclosed" the z-axis. This is enough to make the converse of Theorem 5.1 false. As it turns out, the converse is true if we add some additional hypotheses. Specifically, we can say the following.

THEOREM 5.2

Suppose that $\mathbf{F}(x, y, z) = \langle F_1(x, y, z), F_2(x, y, z), F_3(x, y, z)\rangle$ is a vector field whose components F_1, F_2 and F_3 have continuous first partial derivatives throughout all of $\mathbb{R}^3$. Then, $\mathbf{F}$ is conservative if and only if $\nabla \times \mathbf{F} = \mathbf{0}$.

Notice that half of this theorem is already known from Theorem 5.1. Also, notice that we required that the components of $\mathbf{F}$ have continuous first partial derivatives throughout *all* of $\mathbb{R}^3$ (a requirement that was not satisfied by the vector field in example 5.7). The other half of the theorem requires the additional sophistication of Stokes' Theorem and we will prove a more general version of this in section 14.8.

CONSERVATIVE VECTOR FIELDS

We can now summarize a number of equivalent properties for three-dimensional vector fields. Suppose that $\mathbf{F}(x, y, z) = \langle F_1(x, y, z), F_2(x, y, z), F_3(x, y, z)\rangle$ is a vector field whose components F_1, F_2 and F_3 have continuous first partial derivatives throughout all of $\mathbb{R}^3$. Then the following are equivalent:

1. $\mathbf{F}(x, y, z)$ is conservative.
2. $\int_C \mathbf{F} \cdot d\mathbf{r}$ is independent of path.
3. $\int_C \mathbf{F} \cdot d\mathbf{r} = 0$ for every piecewise-smooth closed curve C.
4. $\nabla \times \mathbf{F} = \mathbf{0}$.
5. $\mathbf{F}(x, y, z)$ is a gradient field ($\mathbf{F} = \nabla f$ for some potential function f).

We close this section by rewriting Green's Theorem in terms of the curl and divergence.

First, suppose that $\mathbf{F}(x, y) = \langle M(x, y), N(x, y), 0\rangle$ is a vector field, for some functions $M(x, y)$ and $N(x, y)$. Suppose that R is a region in the xy-plane whose boundary curve C is piecewise-smooth, positively oriented, simple and closed and that M and N are continuous and have continuous first partial derivatives in some open region D, where $R \subset D$. Then, from Green's Theorem, we have

$$\iint_R \left(\frac{\partial N}{\partial x} - \frac{\partial M}{\partial y}\right) dA = \oint_C M\,dx + N\,dy.$$

Notice that the integrand of the double integral, $\frac{\partial N}{\partial x} - \frac{\partial M}{\partial y}$, is the $\mathbf{k}$ component of $\nabla \times \mathbf{F}$. Further, since $dz = 0$ on any curve lying in the xy-plane, we have

$$\oint_C M\,dx + N\,dy = \oint_C \mathbf{F} \cdot d\mathbf{r}.$$

Thus, we can write Green's Theorem in the form

$$\oint_C \mathbf{F} \cdot d\mathbf{r} = \iint_R (\nabla \times \mathbf{F}) \cdot \mathbf{k}\,dA.$$

We generalize this to Stokes' Theorem in section 14.8.

To take Green's Theorem in yet another direction, suppose that $\mathbf{F}$ and R are as just defined and suppose that C is traced out by the endpoint of the vector-valued function

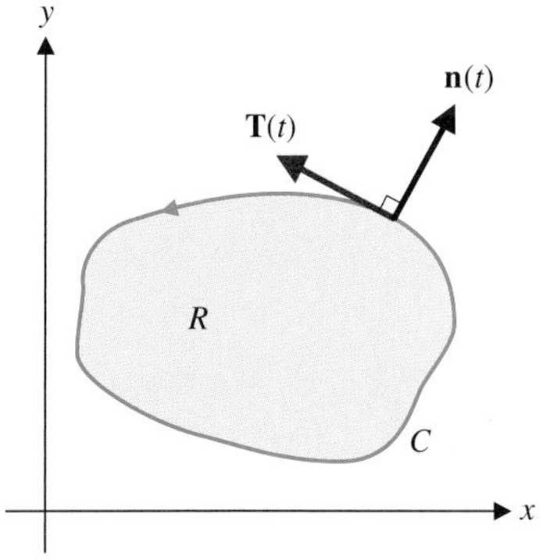

FIGURE 14.39
Unit tangent and exterior unit normal vectors to R

$\mathbf{r}(t) = \langle x(t), y(t)\rangle$, for $a \le t \le b$, where $x(t)$ and $y(t)$ have continuous first derivatives for $a \le t \le b$. Recall that the unit tangent vector to the curve is given by

$$\mathbf{T}(t) = \left\langle \frac{x'(t)}{\|\mathbf{r}'(t)\|}, \frac{y'(t)}{\|\mathbf{r}'(t)\|} \right\rangle.$$

It's then easy to verify that the exterior unit normal vector to C at any point (i.e., the unit normal vector that points out of R) is given by

$$\mathbf{n}(t) = \left\langle \frac{y'(t)}{\|\mathbf{r}'(t)\|}, \frac{-x'(t)}{\|\mathbf{r}'(t)\|} \right\rangle.$$

(See Figure 14.39.) Now, from Theorem 2.1, we have

$$\begin{aligned}\oint_C \mathbf{F}\cdot\mathbf{n}\,ds &= \int_a^b (\mathbf{F}\cdot\mathbf{n})(t)\|\mathbf{r}'(t)\|\,dt \\ &= \int_a^b \left[\frac{M(x(t), y(t))y'(t)}{\|\mathbf{r}'(t)\|} - \frac{N(x(t), y(t))x'(t)}{\|\mathbf{r}'(t)\|}\right]\|\mathbf{r}'(t)\|\,dt \\ &= \int_a^b [M(x(t), y(t))y'(t)\,dt - N(x(t), y(t))x'(t)\,dt] \\ &= \oint_C M(x, y)\,dy - N(x, y)\,dx \\ &= \iint_R \left(\frac{\partial M}{\partial x} + \frac{\partial N}{\partial y}\right) dA,\end{aligned}$$

from Green's Theorem. Finally, recognize that the integrand of the double integral is the divergence of $\mathbf{F}$ and this gives us another vector form of Green's Theorem:

$$\oint_C \mathbf{F}\cdot\mathbf{n}\,ds = \iint_R \nabla\cdot\mathbf{F}(x, y)\,dA. \tag{5.3}$$

This form of Green's Theorem is generalized to the Divergence Theorem in section 14.7.

EXERCISES 14.5

5, 9, 13, 15, 19

WRITING EXERCISES

1. Suppose that $\nabla\times\mathbf{F} = \langle 2, 0, 0\rangle$. Describe what the graph of the vector field $\mathbf{F}$ looks like. Explain how the graph of the vector field $\mathbf{G}$ with $\nabla\times\mathbf{G} = \langle 20, 0, 0\rangle$ compares.
2. If $\nabla\cdot\mathbf{F} > 0$ at a point P and $\mathbf{F}$ is the velocity field of a fluid, explain why the word **source** is a good choice for what's happening at P. Explain why **sink** is a good word if $\nabla\cdot\mathbf{F} < 0$.
3. You now have two ways of determining whether or not a vector field is conservative: try to find the potential or see whether the curl equals $\mathbf{0}$. If you have reason to believe that the vector field is conservative, explain which test you prefer.
4. In the text, we discussed geometrical interpretations of the divergence and curl. Discuss the extent to which the divergence and curl are analogous to tangential and normal components of acceleration.

In exercises 1–12, find the curl and divergence of the given vector field.

1. $x^2\mathbf{i} - 3xy\mathbf{j}$
2. $y^2\mathbf{i} + 4x^2y\mathbf{j}$
3. $2xz\mathbf{i} - 3y\mathbf{k}$
4. $x^2\mathbf{i} - 3xy\mathbf{j} + x\mathbf{k}$
5. $\langle xy, yz, x^2\rangle$
6. $\langle xe^z, yz^2, x + y\rangle$
7. $\langle x^2, y - z, xe^y\rangle$
8. $\langle y, x^2y, 3z + y\rangle$
9. $\langle 3yz, x^2, x\cos y\rangle$
10. $\langle y^2, x^2e^z, \cos xy\rangle$
11. $\langle 2xz, y + z^2, zy^2\rangle$
12. $\langle xy^2, 3y^2z^2, 2x - zy^3\rangle$

In exercises 13–26, determine whether the given vector field is conservative and/or incompressible.

13. $\langle 2x, 2yz^2, 2y^2z\rangle$
14. $\langle 2xy, x^2 - 3y^2z^2, 1 - 2zy^3\rangle$
15. $\langle 3yz, x^2, x\cos y\rangle$
16. $\langle y^2, x^2e^z, \cos xy\rangle$

17. $\langle \sin z, z^2 e^{yz^2}, x\cos z + 2yze^{yz^2} \rangle$

18. $\langle 2xy\cos z, x^2\cos z - 3y^2 z, -x^2 y \sin z - y^3 \rangle$

19. $\left\langle z^2 - 3ye^{3x}, z^2 - e^{3x}, 2z\sqrt{xy} \right\rangle$

20. $\langle 2xz, 3y, x^2 - y \rangle$

21. $\langle xy^2, 3xz, 4 - zy^2 \rangle$

22. $\langle x, y, 1 - 3z \rangle$

23. $\langle 4x, 3y^3, e^z \rangle$

24. $\langle \sin x, 2y^2, \sqrt{z} \rangle$

25. $\langle -2xy, z^2\cos yz^2 - x^2, 2yz\cos yz^2 \rangle$

26. $\langle e^y, xe^y + z^2, 2yz - 1 \rangle$

27. Label each expression as a scalar quantity, a vector quantity or undefined, if f is a scalar function and $\mathbf{F}$ is a vector field.

a. $\nabla \cdot (\nabla f)$ **b.** $\nabla \times (\nabla \cdot \mathbf{F})$ **c.** $\nabla(\nabla \times \mathbf{F})$
d. $\nabla(\nabla \cdot \mathbf{F})$ **e.** $\nabla \times (\nabla f)$

28. Label each expression as a scalar quantity, a vector quantity or undefined, if f is a scalar function and $\mathbf{F}$ is a vector field.

a. $\nabla(\nabla f)$ **b.** $\nabla \cdot (\nabla \cdot \mathbf{F})$ **c.** $\nabla \cdot (\nabla \times \mathbf{F})$
d. $\nabla \times (\nabla \mathbf{F})$ **e.** $\nabla \times (\nabla \times (\nabla \times \mathbf{F}))$

29. If $\mathbf{r} = \langle x, y, z \rangle$, prove that $\nabla \times \mathbf{r} = \mathbf{0}$ and $\nabla \cdot \mathbf{r} = 3$.

30. If $\mathbf{r} = \langle x, y, z \rangle$ and $r = \|\mathbf{r}\|$, prove that $\nabla \cdot (r\mathbf{r}) = 4r$.

In exercises 31–36, conjecture whether the divergence at point *P* is positive, negative or zero.

31.
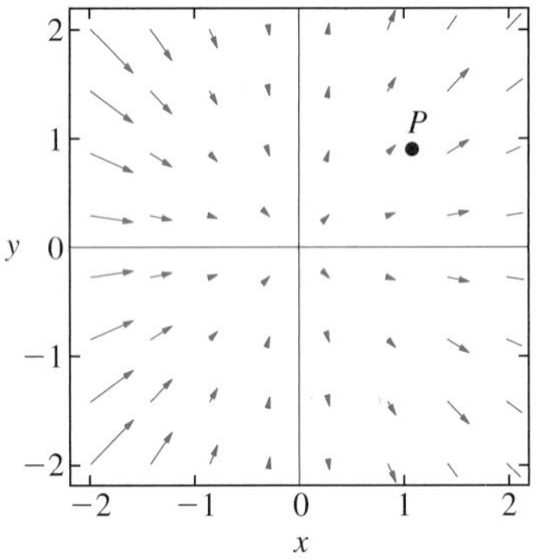

32.
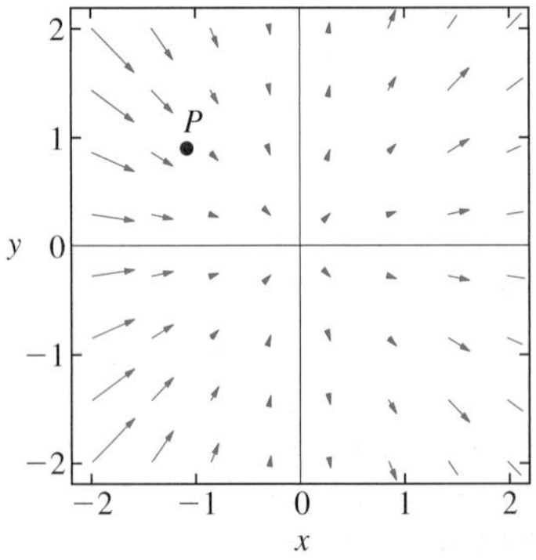

33.
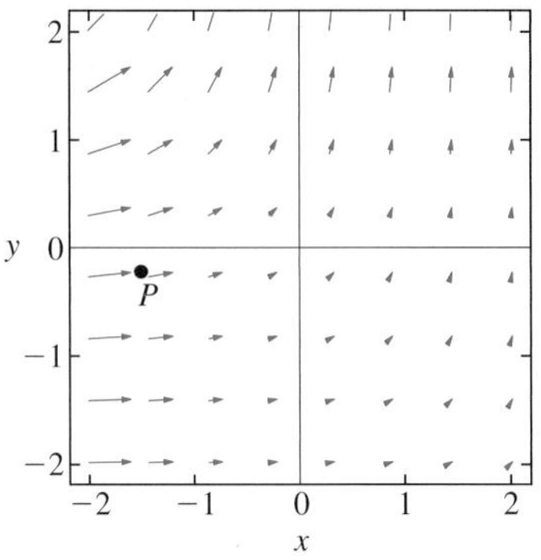

34.
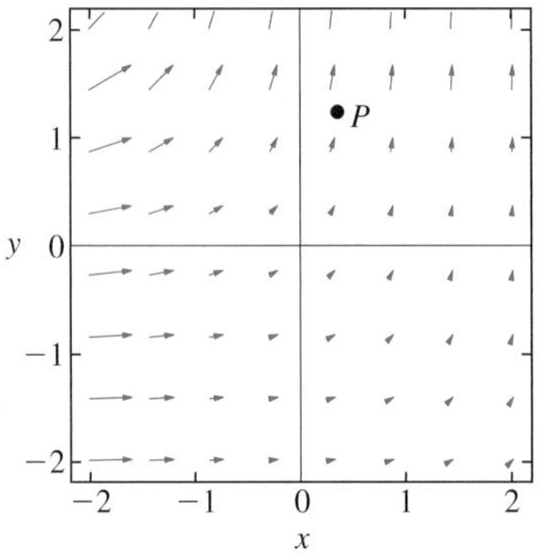

35.
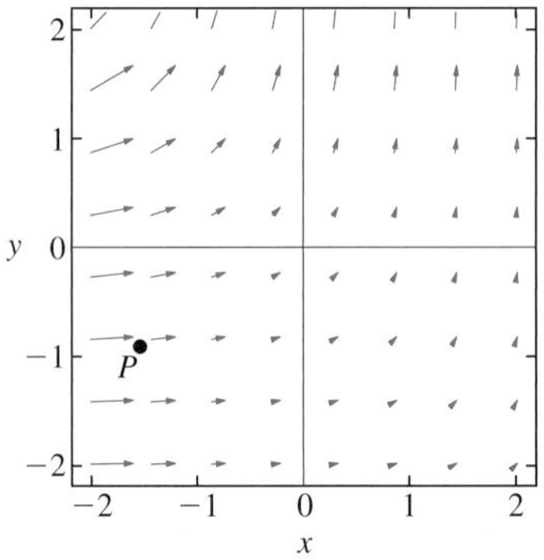

36.
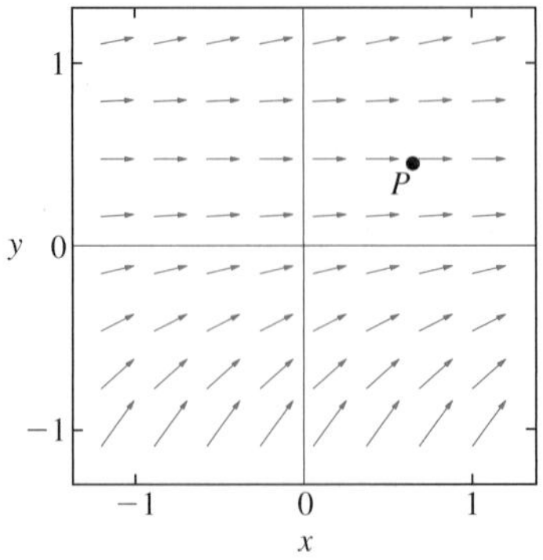

37. If **F** and **G** are vector fields, prove that

$$\nabla \cdot (\mathbf{F} \times \mathbf{G}) = \mathbf{G} \cdot (\nabla \times \mathbf{F}) - \mathbf{F} \cdot (\nabla \times \mathbf{G}).$$

38. If **F** is a vector field, prove that $\nabla \cdot (\nabla \times \mathbf{F}) = 0$.

39. If **F** is a vector field, prove that

$$\nabla \times (\nabla \times \mathbf{F}) = \nabla(\nabla \cdot \mathbf{F}) - \nabla^2 \mathbf{F}.$$

40. If **A** is a constant vector and $\mathbf{r} = \langle x, y, z \rangle$, prove that

$$\nabla \times (\mathbf{A} \times \mathbf{r}) = 2\mathbf{A}.$$

41. If the **j**-component, $\frac{\partial F_1}{\partial z} - \frac{\partial F_3}{\partial x}$, of the curl of **F** is positive, show that there is a closed curve C such that $\int_C \mathbf{F} \cdot d\mathbf{r} \neq 0$.

42. If the **k**-component, $\frac{\partial F_2}{\partial x} - \frac{\partial F_1}{\partial y}$, of the curl of **F** is positive, show that there is a closed curve C such that $\int_C \mathbf{F} \cdot d\mathbf{r} \neq 0$.

43. Prove **Green's first identity:** For $C = \partial R$,

$$\iint_R f \nabla^2 g \, dA = \int_C f(\nabla g) \cdot \mathbf{n} \, ds - \iint_R (\nabla f \cdot \nabla g) \, dA.$$

[Hint: Use the vector form of Green's Theorem in (5.3) applied to $\mathbf{F} = f \nabla g$.]

44. Prove **Green's second identity:** For $C = \partial R$,

$$\iint_R (f \nabla^2 g - g \nabla^2 f) \, dA = \int_C (f \nabla g - g \nabla f) \cdot \mathbf{n} \, ds.$$

(Hint: Use Green's first identity from exercise 43.)

45. For a vector field $\mathbf{F}(x, y) = \langle F_1(x, y), F_2(x, y) \rangle$ and closed curve C with normal vector **n** (that is, **n** is perpendicular to the tangent vector to C at each point), show that $\oint_C \mathbf{F} \cdot \mathbf{n} \, ds = \iint_R \nabla \cdot \mathbf{F} \, dA = \oint_C F_1 \, dy - F_2 \, dx$.

46. If $T(x, y, t)$ is the temperature function at position (x, y) at time t, heat flows across a curve C at a rate given by $\oint_C (-k \nabla T) \cdot \mathbf{n} \, ds$, for some constant k. At steady-state, this rate is zero and the temperature function can be written as $T(x, y)$. In this case, use Green's Theorem to show that $\nabla^2 T = 0$.

47. If f is a scalar function and **F** a vector field, show that

$$\nabla \cdot (f\mathbf{F}) = \nabla f \cdot \mathbf{F} + f(\nabla \cdot \mathbf{F}).$$

48. If f is a scalar function and **F** a vector field, show that

$$\nabla \times (f\mathbf{F}) = \nabla f \times \mathbf{F} + f(\nabla \times \mathbf{F}).$$

49. If $\nabla \cdot \mathbf{F} = 0$, we say that **F** is **solenoidal.** If $\nabla^2 f = 0$, show that ∇f is both solenoidal and irrotational.

50. If **F** and **G** are irrotational, prove that $\mathbf{F} \times \mathbf{G}$ is solenoidal. (Refer to exercise 49.)

51. If f is a scalar function, $\mathbf{r} = \langle x, y \rangle$ and $r = \|\mathbf{r}\|$, show that

$$\nabla f(r) = f'(r) \frac{\mathbf{r}}{r}.$$

52. If f is a scalar function, $\mathbf{r} = \langle x, y \rangle$ and $r = \|\mathbf{r}\|$, show that

$$\nabla^2 f(r) = f''(r) + \frac{1}{r} f'(r).$$

53. Compute the Laplacian Δf for $f(x, y, z) = \sqrt{x^2 + y^2 + z^2}$.

54. Compute the Laplacian Δf for $f(x, y, z) = \dfrac{1}{x^2 + y^2 + z^2}$.

55. Suppose that $\mathbf{F}(x, y) = \langle x^2, y^2 - 4x \rangle$ represents the velocity field of a fluid in motion. For a small box centered at (x, y), determine whether the flow into the box is greater than, less than or equal to the flow out of the box. (a) $(x, y) = (0, 0)$ and (b) $(x, y) = (1, 0)$.

56. Repeat exercise 55 for (a) $(x, y) = (1, 1)$ and (b) $(x, y) = (0, -1)$.

57. Give an example of a vector field **F** such that $\nabla \cdot \mathbf{F}$ is a positive function of y only.

58. Give an example of a vector field **F** such that $\nabla \times \mathbf{F}$ is a function of x only.

59. Gauss' law states that $\nabla \cdot \mathbf{E} = \frac{\rho}{\epsilon_0}$. Here, **E** is an electrostatic field, ρ is the charge density and ϵ_0 is the permittivity. If **E** has a potential function $-\phi$, derive **Poisson's equation** $\nabla^2 \phi = -\frac{\rho}{\epsilon_0}$.

60. For two-dimensional fluid flow, if $\mathbf{v} = \langle v_x(x, y), v_y(x, y) \rangle$ is the velocity field, then **v** has a **stream function** g if $\frac{\partial g}{\partial x} = -v_y$ and $\frac{\partial g}{\partial y} = v_x$. Show that if **v** has a stream function and the components v_x and v_y have continuous partial derivatives, then $\nabla \cdot \mathbf{v} = 0$.

61. For $\mathbf{v} = \langle 2xy, -y^2 + x \rangle$, show that $\nabla \cdot \mathbf{v} = 0$ and find a stream function g.

62. For $\mathbf{v} = \langle xe^{xy} - 1, 2 - ye^{xy} \rangle$, show that $\nabla \cdot \mathbf{v} = 0$ and find a stream function g.

63. Sketch the function $f(x) = \dfrac{1}{1 + x^2}$ and use it to sketch the vector field $\mathbf{F} = \left\langle 0, \dfrac{1}{1 + x^2}, 0 \right\rangle$. If this represents the velocity field of a fluid and a paddle wheel is placed in the fluid at various points near the origin, explain why the paddle wheel would start spinning. Compute $\nabla \times \mathbf{F}$ and label the fluid flow as rotational or irrotational. How does this compare to the motion of the paddle wheel?

64. Sketch the vector field $\mathbf{F} = \left\langle \dfrac{1}{1 + x^2}, 0, 0 \right\rangle$. If this represents the velocity field of a fluid and a paddle wheel is placed in the fluid at various points near the origin, explain why the paddle wheel would not start spinning. Compute $\nabla \times \mathbf{F}$ and label the fluid flow as rotational or irrotational. How does this compare to the motion of the paddle wheel?

65. Show that if $\mathbf{G} = \nabla \times \mathbf{H}$, for some vector field $\mathbf{H}$ with continuous partial derivatives, then $\nabla \cdot \mathbf{G} = 0$.

66. Show the converse of exercise 65; that is, if $\nabla \cdot \mathbf{G} = 0$, then $\mathbf{G} = \nabla \times \mathbf{H}$ for some vector field $\mathbf{H}$. $\left[\text{Hint: Let } \mathbf{H}(x, y, z) = \left\langle 0, \int_0^x G_3(u, y, z)\,du, -\int_0^x G_2(u, y, z)\,du \right\rangle.\right]$

EXPLORATORY EXERCISES

1. In some calculus and engineering books, you will find the vector identity

$$\nabla \times (\mathbf{F} \times \mathbf{G}) = (\mathbf{G} \cdot \nabla)\mathbf{F} - \mathbf{G}(\nabla \cdot \mathbf{F}) - (\mathbf{F} \cdot \nabla)\mathbf{G} + \mathbf{F}(\nabla \cdot \mathbf{G}).$$

Which two of the four terms on the right-hand side look like they should be undefined? Write out the left-hand side as completely as possible, group it into four terms, identify the two familiar terms on the right-hand side and then define the unusual terms on the right-hand side. (Hint: The notation makes sense as a generalization of the definitions in this section.)

2. Prove the vector formula

$$\nabla \times (\nabla \times \mathbf{F}) = \nabla(\nabla \cdot \mathbf{F}) - \nabla^2 \mathbf{F}.$$

As in exercise 1, a major part of the problem is to decipher an unfamiliar notation.

3. Maxwell's laws relate an electric field $\mathbf{E}(t)$ to a magnetic field $\mathbf{H}(t)$. In a region with no charges and no current, the laws state that $\nabla \cdot \mathbf{E} = 0$, $\nabla \cdot \mathbf{H} = 0$, $\nabla \times \mathbf{E} = -\mu \mathbf{H}_t$ and $\nabla \times \mathbf{H} = \mu \mathbf{E}_t$. From these laws, prove that

$$\nabla \times (\nabla \times \mathbf{E}) = -\mu^2 \mathbf{E}_{tt}$$

and

$$\nabla \times (\nabla \times \mathbf{H}) = -\mu^2 \mathbf{H}_{tt}.$$

14.6 SURFACE INTEGRALS

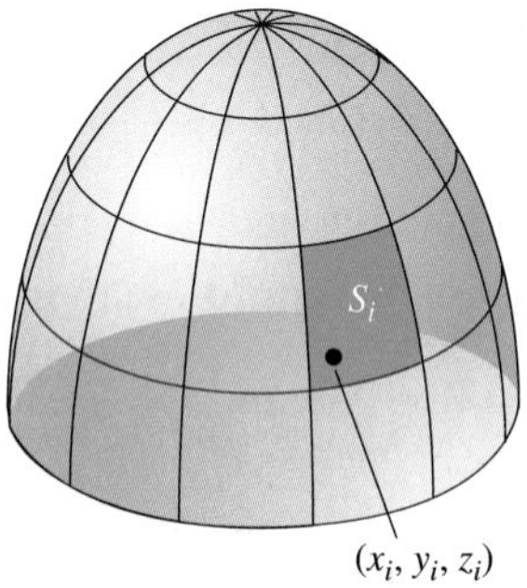

FIGURE 14.40
Partition of a surface

Whether it is the ceiling of the Sistine Chapel, the dome of a college library or the massive roof of the Toronto SkyDome, domes are impressive architectural structures, in part because of their lack of visible support. This feature of domes worries architects, who must be certain that the weight is properly supported. A critical part of an architect's calculation is the mass of the dome.

How would you compute the mass of a dome? You have already seen how to use double integrals to compute the mass of a two-dimensional lamina and triple integrals to find the mass of a three-dimensional solid. However, a dome is a three-dimensional structure more like a thin shell (a surface) than a solid. We hope you're one step ahead of us on this one: if you don't know how to find the mass of a dome exactly, you can try to approximate its mass by slicing it into a number of small sections and estimating the mass of each section. In Figure 14.40, we show a curved surface that has been divided into a number of pieces. If the pieces are small enough, notice that the density of each piece will be approximately constant.

So, first subdivide (partition) the surface into n smaller pieces, $S_1, S_2, \ldots, S_n$. Next, let $\rho(x, y, z)$ be the density function (measured in units of mass per unit area). Further, for each $i = 1, 2, \ldots, n$, let (x_i, y_i, z_i) be a point on the section S_i and let ΔS_i be the surface area of S_i. The mass of the section S_i is then given approximately by $\rho(x_i, y_i, z_i)\Delta S_i$. The total mass m of the surface is given approximately by the sum of these approximate masses,

$$m \approx \sum_{i=1}^{n} \rho(x_i, y_i, z_i)\,\Delta S_i.$$

You should expect that the exact mass is given by the limit of these sums as the size of the pieces gets smaller and smaller. We define the **diameter** of a section S_i to be the maximum distance between any two points on S_i and the norm of the partition $\|P\|$ as the maximum of the diameters of the S_i's. Then we have that

$$m = \lim_{\|P\| \to 0} \sum_{i=1}^{n} \rho\,(x_i, y_i, z_i)\,\Delta S_i.$$

This limit is an example of a new type of integral, the **surface integral,** which is the focus of this section.

DEFINITION 6.1

The **surface integral** of a function $g(x, y, z)$ over a surface $S \subset \mathbb{R}^3$, written $\iint\limits_S g(x, y, z)\,dS$, is given by

$$\iint\limits_S g(x, y, z)\,dS = \lim_{\|P\|\to 0} \sum_{i=1}^{n} g(x_i, y_i, z_i)\,\Delta S_i,$$

provided the limit exists and is the same for all choices of the evaluation points (x_i, y_i, z_i).

Notice how our development of the surface integral parallels our development of the line integral. Whereas the line integral extended a single integral over an interval to an integral over a curve in three dimensions, the surface integral extends a double integral over a two-dimensional region to an integral over a two-dimensional surface in three dimensions. In both cases, we are "curving" our domain into three dimensions.

Now that we have defined the surface integral, how can we calculate one? The basic idea is to rewrite a surface integral as a double integral and then evaluate the double integral using existing techniques. To convert a given surface integral into a double integral, you will have two main tasks:

1. Write the integrand $g(x, y, z)$ as a function of two variables.
2. Write the surface area element dS in terms of the area element dA.

We will develop a general rule for step (2) before considering specific examples.

Consider a surface such as the one pictured in Figure 14.40. For the sake of simplicity, we assume that the surface is the graph of the equation $z = f(x, y)$, where f has continuous first partial derivatives in some region R in the xy-plane. Notice that for an inner partition $R_1, R_2, \ldots, R_n$ of R, if we take the point $(x_i, y_i, 0)$ as the point in R_i closest to the origin, then the portion of the surface S_i lying above R_i will differ very little from the portion T_i of the tangent plane to the surface at $(x_i, y_i, f(x_i, y_i))$ lying above R_i. More to the point, the surface area of S_i will be approximately the same as the area of the parallelogram T_i. In Figure 14.41, we have indicated the portion T_i of the tangent plane lying above R_i.

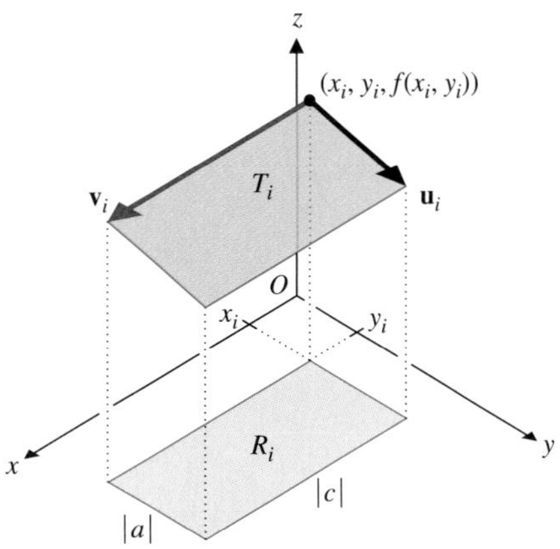

FIGURE 14.41
Portion of the tangent plane lying above R_i

Let the vectors $\mathbf{u}_i = \langle 0, a, b\rangle$ and $\mathbf{v}_i = \langle c, 0, d\rangle$ form two adjacent sides of the parallelogram T_i, as indicated in Figure 14.41. Notice that since $\mathbf{u}_i$ and $\mathbf{v}_i$ lie in the tangent plane, $\mathbf{n}_i = \mathbf{u}_i \times \mathbf{v}_i = \langle ad, bc, -ac\rangle$ is a normal vector to the tangent plane. We saw in section 10.4 that the area of the parallelogram can be written as

$$\Delta S_i = \|\mathbf{u}_i \times \mathbf{v}_i\| = \|\mathbf{n}_i\|.$$

We further observe that the area of R_i is given by $\Delta A_i = |ac|$ and $\mathbf{n}_i \cdot \mathbf{k} = -ac$, so that $|\mathbf{n}_i \cdot \mathbf{k}| = |ac|$. We can now write

$$\Delta S_i = \frac{|ac|\ \|\mathbf{n}_i\|}{|ac|} = \frac{\|\mathbf{n}_i\|}{|\mathbf{n}_i \cdot \mathbf{k}|}\Delta A_i,$$

since $ac \neq 0$. The corresponding expression relating the surface area element dS and the area element dA is then

$$\boxed{dS = \frac{\|\mathbf{n}\|}{|\mathbf{n} \cdot \mathbf{k}|}\, dA.}$$

In the exercises, we will ask you to derive similar formulas for the cases where the surface S is written as a function of x and z or as a function of y and z.

We will consider two main cases of surface integrals. In the first, the surface is defined by a function $z = f(x, y)$. In the second, the surface is defined by parametric equations $x = x(u, v)$, $y = y(u, v)$ and $z = z(u, v)$. In each case, your primary task will be to determine a normal vector to use in the general conversion formula for dS.

If S is the surface $z = f(x, y)$, recall from our discussion in section 12.4 that a normal vector to S is given by $\mathbf{n} = \langle f_x, f_y, -1\rangle$. This is a convenient normal vector for our purposes, since $|\mathbf{n} \cdot \mathbf{k}| = 1$. With $\|\mathbf{n}\| = \sqrt{(f_x)^2 + (f_y)^2 + 1}$, we have the following result.

THEOREM 6.1 (Evaluation Theorem)

If the surface S is given by $z = f(x, y)$ for (x, y) in the region $R \subset \mathbb{R}^2$, where f has continuous first partial derivatives, then

$$\iint_S g(x, y, z)\, dS = \iint_R g(x, y, f(x, y))\sqrt{(f_x)^2 + (f_y)^2 + 1}\, dA.$$

PROOF

From the definition of surface integral in Definition 6.1, we have

$$\begin{aligned}
\iint_S g(x, y, z)\, dS &= \lim_{\|P\|\to 0} \sum_{i=1}^{n} g(x_i, y_i, z_i)\,\Delta S_i \\
&= \lim_{\|P\|\to 0} \sum_{i=1}^{n} g(x_i, y_i, z_i)\frac{\|\mathbf{n}_i\|}{|\mathbf{n}_i \cdot \mathbf{k}|}\,\Delta A_i \\
&= \lim_{\|P\|\to 0} \sum_{i=1}^{n} g(x_i, y_i, f(x_i, y_i))\sqrt{(f_x)^2 + (f_y)^2 + 1}\,\Big|_{(x_i, y_i)} \Delta A_i \\
&= \iint_R g(x, y, f(x, y))\sqrt{(f_x)^2 + (f_y)^2 + 1}\, dA,
\end{aligned}$$

as desired. ■

Theorem 6.1 says that we can evaluate a surface integral by evaluating a related double integral. To convert the surface integral into a double integral, substitute $z = f(x, y)$ in the function $g(x, y, z)$ and replace the surface area element dS with $\|\mathbf{n}\|\, dA$, which for the surface $z = f(x, y)$ is given by

$$dS = \|\mathbf{n}\|\, dA = \sqrt{(f_x)^2 + (f_y)^2 + 1}\, dA. \tag{6.1}$$

EXAMPLE 6.1 Evaluating a Surface Integral

Evaluate $\iint_S 3z\, dS$, where the surface S is the portion of the plane $2x + y + z = 2$ lying in the first octant.

Solution On S, we have $z = 2 - 2x - y$, so we must evaluate $\iint_S 3(2 - 2x - y)\, dS$. Note that a normal vector to the plane $2x + y + z = 2$ is $\mathbf{n} = \langle 2, 1, 1\rangle$, so that in this case, the element of surface area is given by (6.1) to be

$$dS = \|\mathbf{n}\|\, dA = \sqrt{6}\, dA.$$

From Theorem 6.1, we then have

$$\iint_S 3(2 - 2x - y)\, dS = \iint_R 3(2 - 2x - y)\sqrt{6}\, dA,$$

where R is the projection of the surface onto the xy-plane. A graph of the surface S is shown in Figure 14.42a. In this case, notice that R is the triangle indicated in Figure 14.42b. The triangle is bounded by $x = 0$, $y = 0$ and the line $2x + y = 2$ (the intersection of the plane $2x + y + z = 2$ with the plane $z = 0$). If we integrate with respect to y first, the inside integration limits are $y = 0$ and $y = 2 - 2x$, with x ranging from 0 to 1. This gives us

$$\begin{aligned}\iint_S 3(2 - 2x - y)\, dS &= \iint_R 3(2 - 2x - y)\sqrt{6}\, dA \\ &= \int_0^1 \int_0^{2-2x} 3\sqrt{6}(2 - 2x - y)\, dy\, dx \\ &= 2\sqrt{6},\end{aligned}$$

where we leave the routine details of the integration as an exercise. ■

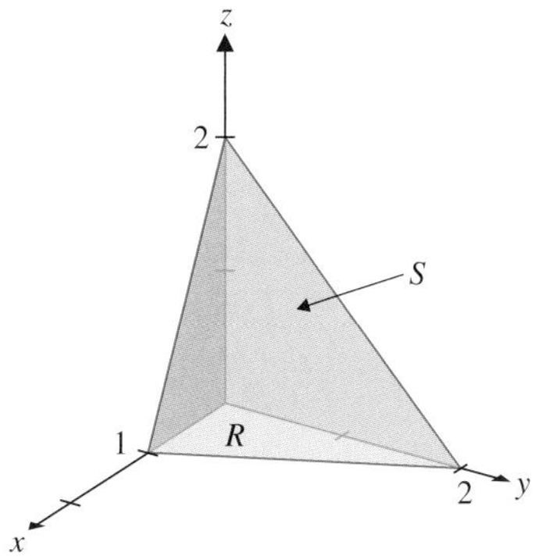

FIGURE 14.42a
$z = 2 - 2x - y$

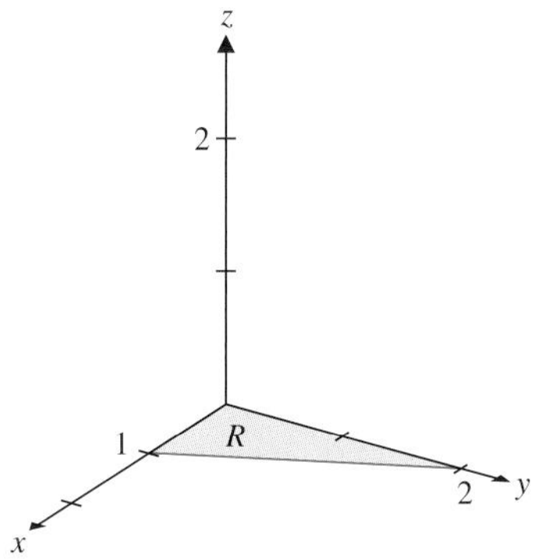

FIGURE 14.42b
The projection R of the surface S onto the xy-plane

In example 6.2, we will need to rewrite the double integral using polar coordinates.

EXAMPLE 6.2 Evaluating a Surface Integral Using Polar Coordinates

Evaluate $\iint_S z\, dS$, where the surface S is the portion of the paraboloid $z = 4 - x^2 - y^2$ lying above the xy-plane.

Solution Substituting $z = 4 - x^2 - y^2$, we have

$$\iint_S z\, dS = \iint_S (4 - x^2 - y^2)\, dS.$$

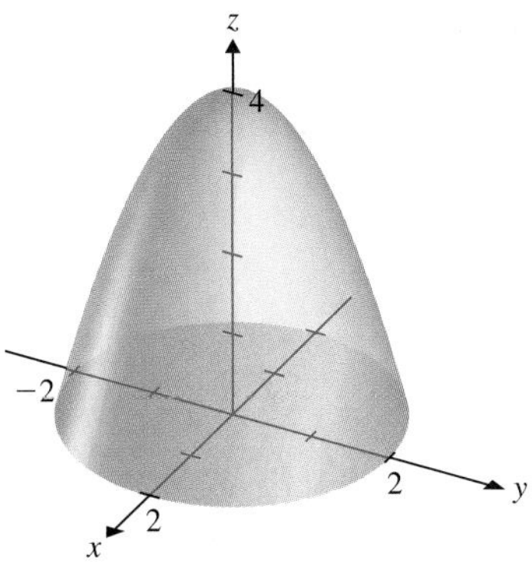

FIGURE 14.43
$z = 4 - x^2 - y^2$

In this case, a normal vector to the surface $z = 4 - x^2 - y^2$ is $\mathbf{n} = \langle -2x, -2y, -1\rangle$, so that

$$dS = \|\mathbf{n}\|\, dA = \sqrt{4x^2 + 4y^2 + 1}\, dA.$$

This gives us

$$\iint_S (4 - x^2 - y^2)\, dS = \iint_R (4 - x^2 - y^2)\sqrt{4x^2 + 4y^2 + 1}\, dA.$$

Here, the region R is enclosed by the intersection of the paraboloid with the xy-plane, which is the circle $x^2 + y^2 = 4$ (see Figure 14.43). With a circular region of integration and the term $x^2 + y^2$ appearing (twice!) in the integrand, you had better be thinking about polar coordinates. We have $4 - x^2 - y^2 = 4 - r^2$, $\sqrt{4x^2 + 4y^2 + 1} = \sqrt{4r^2 + 1}$ and $dA = r\, dr\, d\theta$. For the circle $x^2 + y^2 = 4$, r ranges from 0 to 2 and θ ranges from 0 to 2π. Then, we have

$$\begin{aligned} \iint_S (4 - x^2 - y^2)\, dS &= \iint_R (4 - x^2 - y^2)\sqrt{4x^2 + 4y^2 + 1}\, dA \\ &= \int_0^{2\pi}\int_0^2 (4 - r^2)\sqrt{4r^2 + 1}\, r\, dr\, d\theta \\ &= \frac{289}{60}\pi\sqrt{17} - \frac{41}{60}\pi, \end{aligned}$$

where we leave the details of the final integration to you. ■

○ Parametric Representation of Surfaces

In the remainder of this section, we study parametric representations of surface integrals. Before applying parametric equations to the computation of surface integrals, we need a better understanding of surfaces that have been defined parametrically. You have already seen parametric surfaces in section 11.6. For instance, you can describe the cone $z = \sqrt{x^2 + y^2}$ in cylindrical coordinates by $z = r$, $0 \le \theta \le 2\pi$, which is a parametric representation with parameters r and θ. Similarly, the equation $\rho = 4$, $0 \le \theta \le 2\pi$ and $0 \le \phi \le \pi$, is a parametric representation of the sphere $x^2 + y^2 + z^2 = 16$, with parameters θ and ϕ. It will be helpful to review these graphs as well as to look at some new ones.

Given a particular surface, we may need to find a convenient parametric representation of the surface. The general form for parametric equations representing a surface in three dimensions is $x = x(u, v)$, $y = y(u, v)$ and $z = z(u, v)$ for $u_1 \le u \le u_2$ and $v_1 \le v \le v_2$. The parameters u and v can correspond to familiar coordinates (x and y, or r and θ, for instance), or less familiar expressions. Keep in mind that to fully describe a surface, you will need to define two parameters.

EXAMPLE 6.3 Finding Parametric Representations of a Surface

Find a simple parametric representation for (a) the portion of the cone $z = \sqrt{x^2 + y^2}$ inside the cylinder $x^2 + y^2 = 4$ and (b) the portion of the sphere $x^2 + y^2 + z^2 = 16$ inside of the cone $z = \sqrt{x^2 + y^2}$.

Solution It is important to realize that both parts (a) and (b) have numerous solutions. (In fact, every surface can be represented parametrically in an infinite number of ways.) The solutions we show here are among the simplest and most useful, but they are *not* the only reasonable solutions. In (a), the repeated appearance of the term $x^2 + y^2$

suggests that cylindrical coordinates (r, θ, z) might be convenient. A sketch of the surface is shown in Figure 14.44a. Notice that the cone $z = \sqrt{x^2 + y^2}$ becomes $z = r$ in cylindrical coordinates. Recall that in cylindrical coordinates, $x = r\cos\theta$ and $y = r\sin\theta$. Notice that the parameters r and θ have ranges determined by the cylinder $x^2 + y^2 = 4$, so that $0 \le r \le 2$ and $0 \le \theta \le 2\pi$. Parametric equations for the cone are then $x = r\cos\theta$, $y = r\sin\theta$ and $z = r$ with $0 \le r \le 2$ and $0 \le \theta \le 2\pi$.

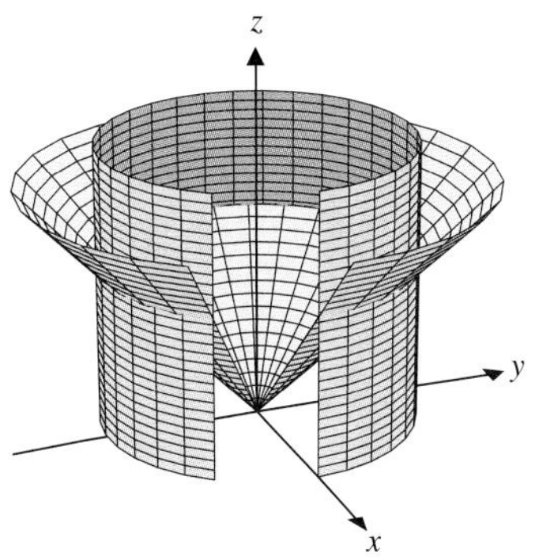

FIGURE 14.44a
The cone $z = \sqrt{x^2 + y^2}$ and the cylinder $x^2 + y^2 = 4$

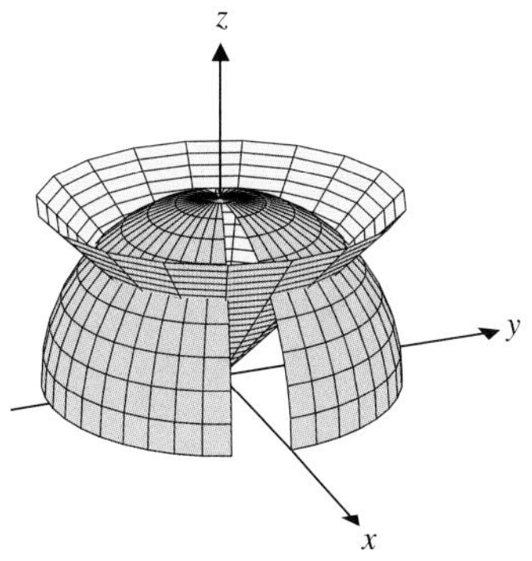

FIGURE 14.44b
The portion of the sphere inside the cone

The surface in part (b) is a portion of a sphere, which suggests (what else?) spherical coordinates: $x = \rho\sin\phi\cos\theta$, $y = \rho\sin\phi\sin\theta$ and $z = \rho\cos\phi$, where $\rho^2 = x^2 + y^2 + z^2$. The equation of the sphere $x^2 + y^2 + z^2 = 16$ is then $\rho = 4$. Using this, a parametric representation of the sphere is $x = 4\sin\phi\cos\theta$, $y = 4\sin\phi\sin\theta$ and $z = 4\cos\phi$, where $0 \le \theta \le 2\pi$ and $0 \le \phi \le \pi$. To find the portion of the sphere inside the cone, observe that the cone can be described in spherical coordinates as $\phi = \frac{\pi}{4}$. Referring to Figure 14.44b, note that the portion of the sphere inside the cone is then described by $x = 4\sin\phi\cos\theta$, $y = 4\sin\phi\sin\theta$ and $z = 4\cos\phi$, where $0 \le \theta \le 2\pi$ and $0 \le \phi \le \frac{\pi}{4}$. ■

Suppose that we have a parametric representation for the surface S: $x = x(u, v)$, $y = y(u, v)$ and $z = z(u, v)$, defined on the rectangle $R = \{(u, v) | a \le u \le b \text{ and } c \le v \le d\}$ in the uv-plane. It is often convenient to use parametric equations to evaluate the surface integral $\iint_S f(x, y, z)\, dS$. Of course, to do this, we must substitute for x, y and z to rewrite the integrand in terms of the parameters u and v, as

$$g(u, v) = f(x(u, v), y(u, v), z(u, v)).$$

We must also write the surface area element dS in terms of the area element dA for the uv-plane. Unfortunately, we can't use (6.1) here, since this holds only for the case of a surface written in the form $z = f(x, y)$. Instead, we'll need to back up just a bit.

First, notice that the position vector for points on the surface S is $\mathbf{r}(u, v) = \langle x(u, v), y(u, v), z(u, v)\rangle$. We define the vectors $\mathbf{r}_u$ and $\mathbf{r}_v$ (the subscripts denote partial derivatives) by

$$\mathbf{r}_u(u, v) = \langle x_u(u, v), y_u(u, v), z_u(u, v)\rangle$$

and

$$\mathbf{r}_v(u, v) = \langle x_v(u, v), y_v(u, v), z_v(u, v)\rangle.$$

Notice that for any fixed (u, v), both of the vectors $\mathbf{r}_u(u, v)$ and $\mathbf{r}_v(u, v)$ lie in the tangent plane to S at the point $(x(u, v), y(u, v), z(u, v))$. So, unless these two vectors are parallel,

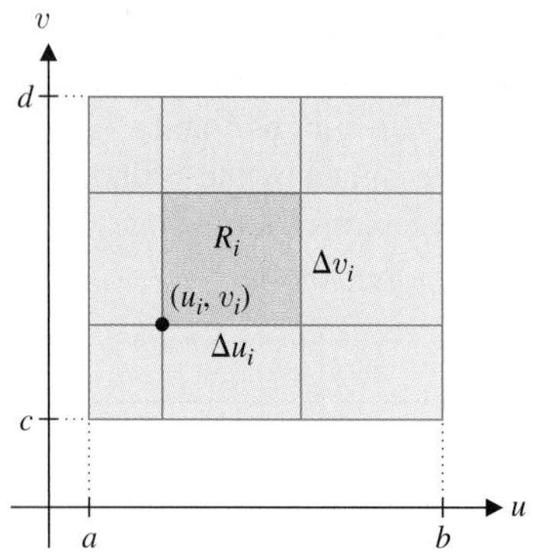

FIGURE 14.45a
Partition of parameter domain (uv-plane)

$\mathbf{n} = \mathbf{r}_u \times \mathbf{r}_v$ is a normal vector to the surface at the point $(x(u, v), y(u, v), z(u, v))$. We say that the surface S is **smooth** if $\mathbf{r}_u$ and $\mathbf{r}_v$ are continuous and $\mathbf{r}_u \times \mathbf{r}_v \neq \mathbf{0}$, for all $(u, v) \in R$. (This says that the surface will not have any corners.) We say that S is **piecewise-smooth** if we can write $S = S_1 \cup S_2 \cup \cdots \cup S_n$, for some smooth surfaces $S_1, S_2, \ldots, S_n$.

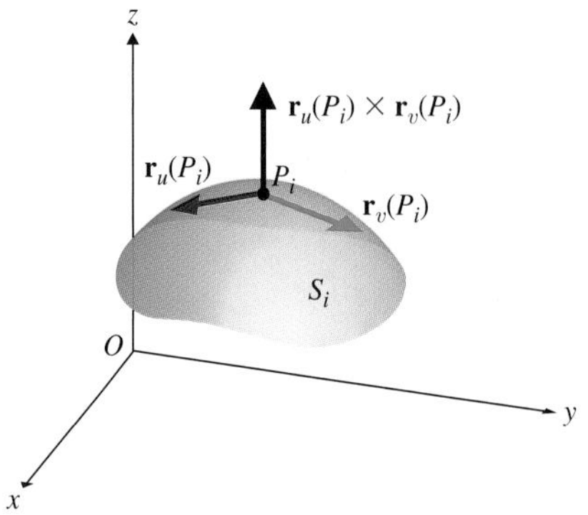

FIGURE 14.45b
Curvilinear region S_i

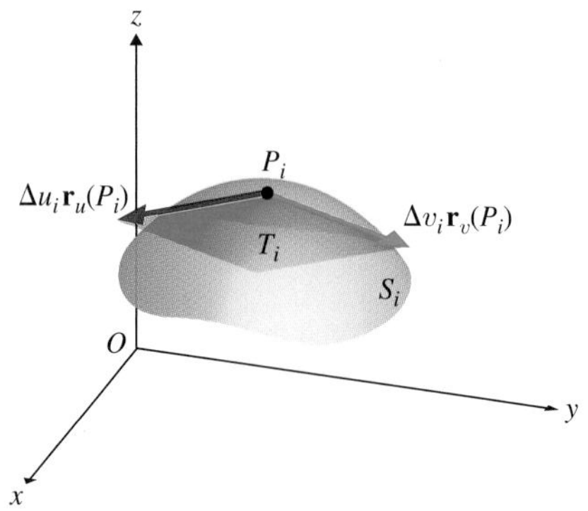

FIGURE 14.45c
The parallelogram T_i

As we have done many times now, we partition the rectangle R in the uv-plane. For each rectangle R_i in the partition, let (u_i, v_i) be the closest point in R_i to the origin, as indicated in Figure 14.45a. Notice that each of the sides of R_i gets mapped to a curve in xyz-space, so that R_i gets mapped to a curvilinear region S_i in xyz-space, as indicated in Figure 14.45b. Observe that if we locate their initial points at the point $P_i(x(u_i, v_i), y(u_i, v_i), z(u_i, v_i))$, the vectors $\mathbf{r}_u(u_i, v_i)$ and $\mathbf{r}_v(u_i, v_i)$ lie tangent to two adjacent curved sides of S_i. So, we can approximate the area ΔS_i of S_i by the area of the parallelogram T_i whose sides are formed by the vectors $\Delta u_i \mathbf{r}_u(u_i, v_i)$ and $\Delta v_i \mathbf{r}_v(u_i, v_i)$. (See Figure 14.45c.) As we know, the area of the parallelogram is given by the magnitude of the cross product

$$\begin{aligned}\|\Delta u_i \mathbf{r}_u(u_i, v_i) \times \Delta v_i \mathbf{r}_v(u_i, v_i)\| &= \|\mathbf{r}_u(u_i, v_i) \times \mathbf{r}_v(u_i, v_i)\| \, \Delta u_i \Delta v_i \\ &= \|\mathbf{r}_u(u_i, v_i) \times \mathbf{r}_v(u_i, v_i)\| \, \Delta A_i,\end{aligned}$$

where ΔA_i is the area of the rectangle R_i. We then have that

$$\Delta S_i \approx \|\mathbf{r}_u(u_i, v_i) \times \mathbf{r}_v(u_i, v_i)\| \Delta A_i$$

and it follows that the element of surface area can be written as

$$dS = \|\mathbf{r}_u \times \mathbf{r}_v\| \, dA. \tag{6.2}$$

Notice that this corresponds closely to (6.1), as $\mathbf{r}_u \times \mathbf{r}_v$ is a normal vector to S. Finally, we developed (6.2) in the comparatively simple case where the parameter domain R (that is, the domain in the uv-plane) was a rectangle. If the parameter domain is not a rectangle, you should recognize that we can do the same thing by constructing an inner partition of the region. We can now evaluate surface integrals using parametric equations, as in example 6.4.

EXAMPLE 6.4 Evaluating a Surface Integral Using Spherical Coordinates

Evaluate $\iint_S (3x^2 + 3y^2 + 3z^2) \, dS$, where S is the sphere $x^2 + y^2 + z^2 = 4$.

Solution Since the surface is a sphere and the integrand contains the term $x^2 + y^2 + z^2$, spherical coordinates are indicated. Notice that the sphere is described by $\rho = 2$ and, on the surface of the sphere, the integrand becomes $3(x^2 + y^2 + z^2) = 12$.

Further, we can describe the sphere $\rho = 2$ with the parametric equations $x = 2\sin\phi\cos\theta$, $y = 2\sin\phi\sin\theta$ and $z = 2\cos\phi$, for $0 \le \theta \le 2\pi$ and $0 \le \phi \le \pi$. This says that the sphere is traced out by the endpoint of the vector-valued function

$$\mathbf{r}(\phi, \theta) = \langle 2\sin\phi\cos\theta, 2\sin\phi\sin\theta, 2\cos\phi\rangle.$$

We then have the partial derivatives

$$\mathbf{r}_\theta = \langle -2\sin\phi\sin\theta, 2\sin\phi\cos\theta, 0\rangle$$

and

$$\mathbf{r}_\phi = \langle 2\cos\phi\cos\theta, 2\cos\phi\sin\theta, -2\sin\phi\rangle.$$

We leave it as an exercise to show that a normal vector to the surface is given by

$$\mathbf{n} = \mathbf{r}_\theta \times \mathbf{r}_\phi = \langle -4\sin^2\phi\cos\theta, -4\sin^2\phi\sin\theta, -4\sin\phi\cos\phi\rangle,$$

so that $\|\mathbf{n}\| = 4|\sin\phi|$. Equation (6.2) now gives us $dS = 4|\sin\phi|\,dA$, so that

$$\iint_S (3x^2 + 3y^2 + 3z^2)\,dS = \iint_R (12)(4)\,|\sin\phi|\,dA$$
$$= \int_0^{2\pi}\int_0^{\pi} 48\sin\phi\,d\phi\,d\theta$$
$$= 192\pi,$$

where we replaced $|\sin\phi|$ by $\sin\phi$ by using the fact that for $0 \le \phi \le \pi$, $\sin\phi \ge 0$. ■

NOTES

For the sphere $x^2+y^2+z^2 = R^2$, the surface area element in spherical coordinates is $dS = R^2\sin\phi\,d\phi\,d\theta$.

If you did the calculation of dS in example 6.4, you may not think that parametric equations lead to simple solutions. (That's why we didn't show all of the details!) However, recall that for changing a triple integral from rectangular to spherical coordinates, you replace $dx\,dy\,dz$ by $\rho^2\sin\phi\,d\rho\,d\phi\,d\theta$. In example 6.4, we have $\rho^2 = 4$ and $dS = 4\sin\phi\,dA$. Looks familiar now, doesn't it? This shortcut is valuable, since surface integrals over spheres are reasonably common.

So, when you evaluate a surface integral, what have you computed? We close the section with three examples. The first is familiar: observe that the surface integral of the function $f(x, y, z) = 1$ over the surface S is simply the surface area of S. That is,

$$\iint_S 1\,dS = \text{Surface area of } S.$$

The proof of this follows directly from the definition of the surface integral and is left as an exercise.

EXAMPLE 6.5 Using a Surface Integral to Compute Surface Area

Compute the surface area of the portion of the hyperboloid $x^2 + y^2 - z^2 = 4$ between $z = 0$ and $z = 2$.

Solution We need to evaluate $\iint_S 1\,dS$. Notice that we can write the hyperboloid parametrically as $x = 2\cos u\cosh v$, $y = 2\sin u\cosh v$ and $z = 2\sinh v$. (You can derive parametric equations in the following way. To get a circular cross section of radius 2 in the xy-plane, start with $x = 2\cos u$ and $y = 2\sin u$. To get a hyperbola in the xz- or yz-plane, multiply x and y by $\cosh v$ and set $z = \sinh v$.) We have $0 \le u \le 2\pi$ to get the circular cross sections and $0 \le v \le \sinh^{-1} 1$ (≈ 0.88). The hyperboloid is traced out by the endpoint of the vector-valued function

$$\mathbf{r}(u, v) = \langle 2\cos u\cosh v, 2\sin u\cosh v, 2\sinh v\rangle,$$

so that $\mathbf{r}_u = \langle -2\sin u\cosh v, 2\cos u\cosh v, 0\rangle$

and $\mathbf{r}_v = \langle 2\cos u\sinh v, 2\sin u\sinh v, 2\cosh v\rangle$.

This gives us the normal vector

$$\mathbf{n} = \mathbf{r}_u \times \mathbf{r}_v = \langle 4\cos u\cosh^2 v, 4\sin u\cosh^2 v, -4\cosh v\sinh v\rangle,$$

where $\|\mathbf{n}\| = 4\cosh v\sqrt{\cosh^2 v + \sinh^2 v}$. We now have

$$\begin{aligned}\iint_S 1\,dS &= \iint_R 4\cosh v\sqrt{\cosh^2 v + \sinh^2 v}\,dA \\ &= \int_0^{\sinh^{-1} 1}\int_0^{2\pi} 4\cosh v\sqrt{\cosh^2 v + \sinh^2 v}\,du\,dv \\ &\approx 31.95,\end{aligned}$$

where we evaluated the final integral numerically. ■

HISTORICAL NOTES

August Ferdinand Möbius (1790–1868) German astronomer and mathematician who gave one of the earliest descriptions of the one-sided surface that bears his name. Möbius' doctoral thesis was in astronomy, but his astronomy teachers included the great mathematician Carl Friedrich Gauss. Möbius did research in both fields. His mathematics publications, primarily in geometry and topology, were exceedingly clear and well developed.

Our next example of a surface integral requires some preliminary discussion. First, we say that a surface S is **orientable** (or **two-sided**) if it is possible to define a unit normal vector $\mathbf{n}$ at each point (x, y, z) not on the boundary of the surface and if $\mathbf{n}$ is a continuous function of (x, y, z). In this case, S has two identifiable sides (a top and a bottom or an inside and an outside). Once we choose a consistent direction for all normal vectors to point, we call the surface **oriented.** For instance, a sphere is a two-sided surface; the two sides of the surface are the inside and the outside. Notice that you can't get from the inside to the outside without passing through the sphere. The **positive orientation** for the sphere (or for any other *closed* surface) is to choose outward normal vectors (normal vectors pointing away from the interior).

All of the surfaces we have seen so far in this course are two-sided, but it's not difficult to construct a one-sided surface. Perhaps the most famous example of a one-sided surface is the **Möbius strip,** named after the German mathematician A. F. Möbius. You can easily construct a Möbius strip by taking a long rectangular strip of paper, giving it a half-twist and then taping the short edges together, as illustrated in Figures 14.46a through 14.46c. Notice that if you started painting the strip, you would eventually return to your starting point, having painted both "sides" of the strip, but without having crossed any edges. This says that the Möbius strip has no inside and no outside and is therefore not orientable.

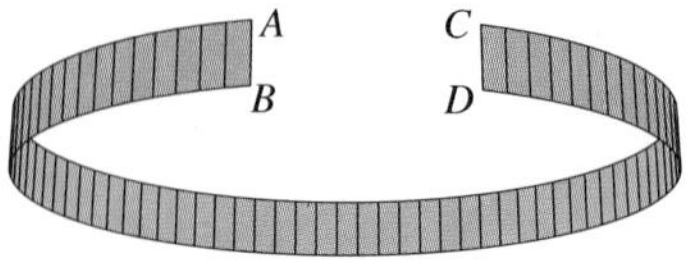

FIGURE 14.46a
A long, thin strip

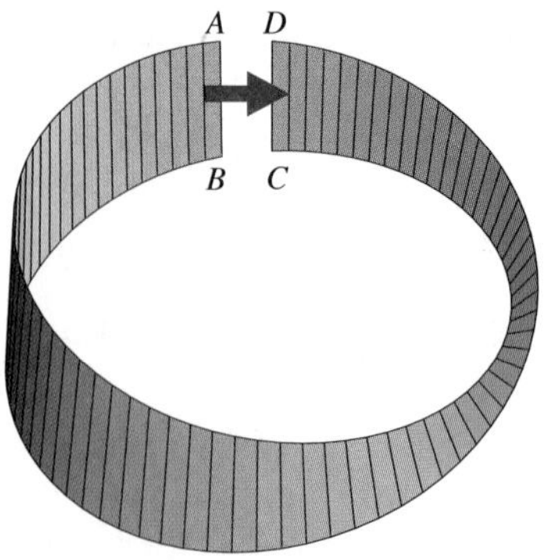

FIGURE 14.46b
Make one half-twist

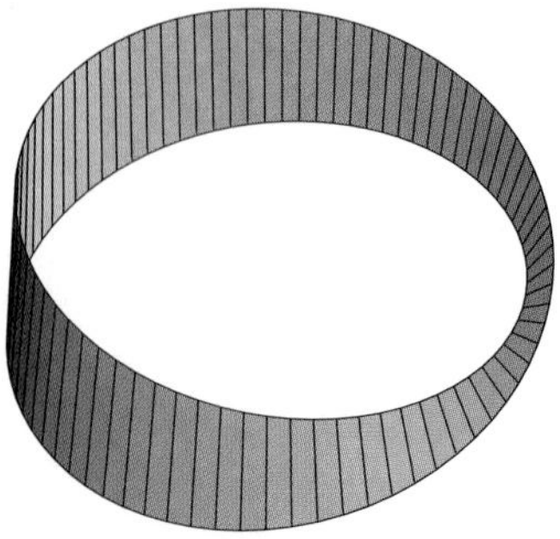

FIGURE 14.46c
A Möbius strip

One reason we need to be able to orient a surface is to compute the **flux** of a vector field. It's easiest to visualize the flux for a vector field representing the velocity field for a fluid in motion. In this context, the flux measures the net flow rate of the fluid across the surface in the direction of the specified normal vectors. (Notice that for this to make sense, the surface must have two identifiable sides. That is, the surface must be orientable.) The orientation of the surface lets us distinguish one direction from the other. In general, we have the following definition.

DEFINITION 6.2

Let $\mathbf{F}(x, y, z)$ be a continuous vector field defined on an oriented surface S with unit normal vector $\mathbf{n}$. The **surface integral of F over** S (or the **flux of F over** S) is given by $\iint_S \mathbf{F} \cdot \mathbf{n}\, dS$.

Think carefully about the role of the unit normal vector in Definition 6.2. Notice that since $\mathbf{n}$ is a unit vector, the integrand $\mathbf{F} \cdot \mathbf{n}$ gives (at any given point on S) the component of $\mathbf{F}$ in the direction of $\mathbf{n}$. So, if $\mathbf{F}$ represents the velocity field for a fluid in motion, $\mathbf{F} \cdot \mathbf{n}$ corresponds to the component of the velocity that moves the fluid across the surface (from one side to the other). Also, note that $\mathbf{F} \cdot \mathbf{n}$ can be positive or negative, depending on which normal vector we have chosen. (Keep in mind that at each point on a surface, there are two unit normal vectors, one pointing toward each side of the surface.) You should recognize that this is why we need to have an oriented surface.

EXAMPLE 6.6 Computing the Flux of a Vector Field

Compute the flux of the vector field $\mathbf{F}(x, y, z) = \langle x, y, 0\rangle$ over the portion of the paraboloid $z = x^2 + y^2$ below $z = 4$ (oriented with upward-pointing normal vectors).

Solution First, observe that at any given point, the normal vectors for the paraboloid $z = x^2 + y^2$ are $\pm\langle 2x, 2y, -1\rangle$. For the normal vector to point upward, we need a positive z-component. In this case,

$$\mathbf{m} = -\langle 2x, 2y, -1\rangle = \langle -2x, -2y, 1\rangle$$

is such a normal vector. A unit vector pointing in the same direction as $\mathbf{m}$ is then

$$\mathbf{n} = \frac{\langle -2x, -2y, 1\rangle}{\sqrt{4x^2 + 4y^2 + 1}}.$$

Before computing $\mathbf{F} \cdot \mathbf{n}$, we use the normal vector $\mathbf{m}$ to write the surface area increment dS in terms of dA. From (6.1), we have

$$dS = \|\mathbf{m}\|\, dA = \sqrt{4x^2 + 4y^2 + 1}\, dA.$$

Putting this all together gives us

$$\begin{aligned}\iint_S \mathbf{F} \cdot \mathbf{n}\, dS &= \iint_R \langle x, y, 0\rangle \cdot \frac{\langle -2x, -2y, 1\rangle}{\sqrt{4x^2 + 4y^2 + 1}} \sqrt{4x^2 + 4y^2 + 1}\, dA \\ &= \iint_R \langle x, y, 0\rangle \cdot \langle -2x, -2y, 1\rangle\, dA \\ &= \iint_R (-2x^2 - 2y^2)\, dA,\end{aligned}$$

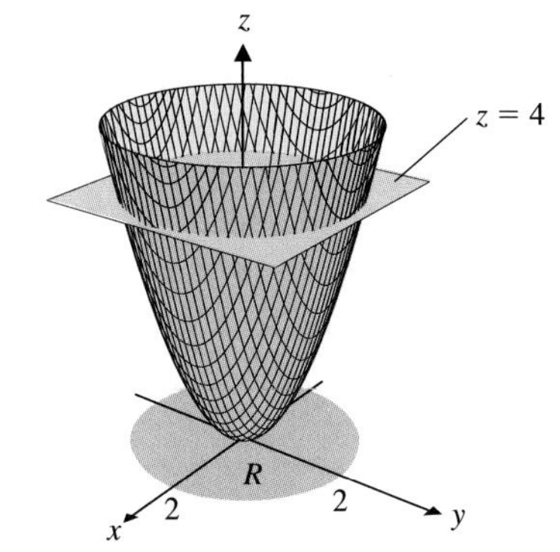

FIGURE 14.47
$z = x^2 + y^2$

where the region R is the projection of the portion of the paraboloid under consideration onto the xy-plane. Note how the square roots arising from the calculation of $\|\mathbf{n}\|$ and dS canceled out one another. Look at the graph in Figure 14.47 and recognize that this projection is bounded by the circle $x^2 + y^2 = 4$. You should quickly realize that the double integral should be set up in polar coordinates. We have

$$\iint_S \mathbf{F} \cdot \mathbf{n}\, dS = \iint_R (-2x^2 - 2y^2)\, dA$$

$$= \int_0^{2\pi} \int_0^2 (-2r^2) r\, dr\, d\theta = -16\pi. \quad \blacksquare$$

Flux integrals enable engineers and physicists to compute the flow of a variety of quantities in three dimensions. If $\mathbf{F}$ represents the velocity field of a fluid, then the flux gives the net amount of fluid crossing the surface S. In example 6.7, we compute heat flow across a surface. If $T(x, y, z)$ gives the temperature at (x, y, z), then the net heat flow across the surface S is the flux of $\mathbf{F} = -k\nabla T$, where the constant k is called the **heat conductivity** of the material.

EXAMPLE 6.7   Computing the Heat Flow Out of a Sphere

For $T(x, y, z) = 30 - \frac{1}{18}z^2$ and $k = 2$, compute the heat flow out of the region bounded by $x^2 + y^2 + z^2 = 9$.

Solution   We compute the flux of $-2\nabla\left(30 - \frac{1}{18}z^2\right) = -2\langle 0, 0, -\frac{1}{9}z\rangle = \langle 0, 0, \frac{2}{9}z\rangle$. Since we want the flow out of the sphere, we need to find an outward unit normal to the sphere. An outward normal is $\nabla(x^2 + y^2 + z^2) = \langle 2x, 2y, 2z\rangle$, so we take $\mathbf{n} = \dfrac{\langle 2x, 2y, 2z\rangle}{\sqrt{4x^2 + 4y^2 + 4z^2}}$. On the sphere, we have $4x^2 + 4y^2 + 4z^2 = 36$, so that the denominator simplifies to $\sqrt{36} = 6$. This gives us $\mathbf{n} = \frac{1}{3}\langle x, y, z\rangle$ and $\mathbf{F} \cdot \mathbf{n} = \frac{2}{27}z^2$. Since the surface is a sphere, we will use spherical coordinates for the surface integral. On the sphere, $z = \rho\cos\phi = 3\cos\phi$, so that $\mathbf{F} \cdot \mathbf{n} = \frac{2}{27}(3\cos\phi)^2 = \frac{2}{3}\cos^2\phi$. Further, $dS = \rho^2\sin\phi\, dA = 9\sin\phi\, dA$. The flux is then

$$\iint_S \mathbf{F} \cdot \mathbf{n}\, ds = \iint_R \left(\frac{2}{3}\cos^2\phi\right) 9\sin\phi\, dA = \int_0^{2\pi} \int_0^\pi 6\cos^2\phi\,\sin\phi\, d\phi\, d\theta = 8\pi.$$

Since the flux is positive, the heat is flowing out of the sphere. Finally, observe that since the temperature decreases as $|z|$ increases, this result should make sense. ■

BEYOND FORMULAS

Surface integrals complete the set of integrals introduced in this book. (There are many more types of integrals used in more advanced mathematics courses and applications areas.) Although the different types of integrals sometimes require different methods of evaluation, the underlying concepts are the same. In each case, the integral is a limit of approximating sums and in some cases can be evaluated directly with some form of an antiderivative. What is being summed in a surface integral? What is being summed in a flux integral?

EXERCISES 14.6

WRITING EXERCISES

1. For definition 6.1, we defined the partition of a surface and took the limit as the norm of the partition tends to 0. Explain why it would not be sufficient to have the number of segments in the partition tend to ∞. (Hint: The pieces of the partition don't have to be the same size.)
2. In example 6.2, you could alternatively start with cylindrical coordinates and use a parametric representation as we did in example 6.4. Discuss which method you think would be simpler.
3. Explain in words why $\iint_S 1\, dS$ equals the surface area of S. (Hint: Although you are supposed to explain in words, you will need to refer to Riemann sums.)
4. For example 6.6, sketch a graph showing the surface S and several normal vectors to the surface. Also, show several vectors in the graph of the vector field $\mathbf{F}$. Explain why the flux is negative.

In exercises 1–8, find a parametric representation of the surface.

1. $z = 3x + 4y$
2. $x^2 + y^2 + z^2 = 4$
3. $x^2 + y^2 - z^2 = 1$
4. $x^2 - y^2 + z^2 = 4$
5. The portion of $x^2 + y^2 = 4$ from $z = 0$ to $z = 2$
6. The portion of $y^2 + z^2 = 9$ from $x = -1$ to $x = 1$
7. The portion of $z = 4 - x^2 - y^2$ above the xy-plane
8. The portion of $z = x^2 + y^2$ below $z = 4$

In exercises 9–16, sketch a graph of the parametric surface.

9. $x = u,\ y = v,\ z = u^2 + 2v^2$
10. $x = u,\ y = v,\ z = 4 - u^2 - v^2$
11. $x = u\cos v,\ y = u\sin v,\ z = u^2$
12. $x = u\cos v,\ y = u\sin v,\ z = u$
13. $x = 2\sin u\cos v,\ y = 2\sin u\sin v,\ z = 2\cos u$
14. $x = 2\cos v,\ y = 2\sin v,\ z = u$
15. $x = u,\ y = \sin u\cos v,\ z = \sin u\sin v$
16. $x = \cos u\cos v,\ y = u,\ z = \cos u\sin v$
17. Match the parametric equations with the surface.
    a. $x = u\cos v,\ y = u\sin v,\ z = v^2$
    b. $x = v,\ y = u\cos v,\ z = u\sin v$
    c. $x = u,\ y = u\cos v,\ z = u\sin v$

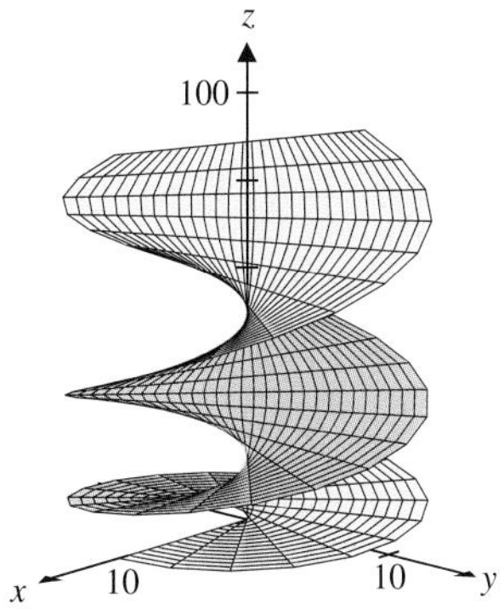

SURFACE A

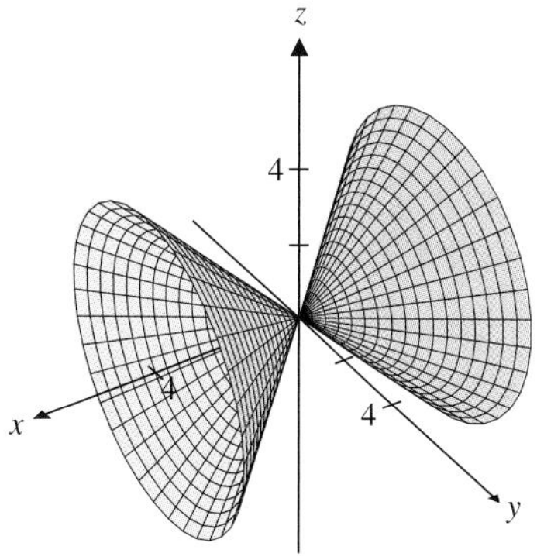

SURFACE B

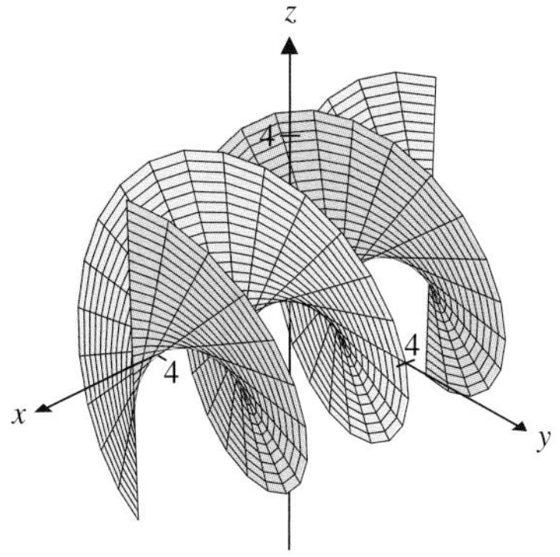

SURFACE C

18. In example 6.4, show that

$$\mathbf{r}_\theta \times \mathbf{r}_\phi = \langle -4\sin^2\phi\cos\theta,\ -4\sin^2\phi\sin\theta,\ -4\sin\phi\cos\phi\rangle$$

and then show that $\|\mathbf{n}\| = 4|\sin\phi|$.

In exercises 19–26, find the surface area of the given surface.

19. The portion of the cone $z = \sqrt{x^2 + y^2}$ below the plane $z = 4$

20. The portion of the paraboloid $z = x^2 + y^2$ below the plane $z = 4$

21. The portion of the plane $3x + y + 2z = 6$ inside the cylinder $x^2 + y^2 = 4$

22. The portion of the plane $x + 2y + z = 4$ above the region bounded by $y = x^2$ and $y = 1$

23. The portion of the cone $z = \sqrt{x^2 + y^2}$ above the triangle with vertices $(0, 0)$, $(1, 0)$ and $(1, 1)$

24. The portion of the paraboloid $z = x^2 + y^2$ inside the cylinder $x^2 + y^2 = 4$

25. The portion of the hemisphere $z = \sqrt{4 - x^2 - y^2}$ above the plane $z = 1$

26. The portion of $y = 4 - x^2$ with $y \geq 0$ and between $z = 0$ and $z = 2$

In exercises 27–36, set up a double integral and evaluate the surface integral $\iint_S g(x, y, z)\, dS$.

27. $\iint_S xz\, dS$, S is the portion of the plane $z = 2x + 3y$ above the rectangle $1 \leq x \leq 2, 1 \leq y \leq 3$

28. $\iint_S (z - y^2)\, dS$, S is the portion of the paraboloid $z = x^2 + y^2$ below $z = 4$

29. $\iint_S (x^2 + y^2 + z^2)^{3/2}\, dS$, S is the lower hemisphere $z = -\sqrt{9 - x^2 - y^2}$

30. $\iint_S \sqrt{x^2 + y^2 + z^2}\, dS$, S is the sphere $x^2 + y^2 + z^2 = 9$

31. $\iint_S (x^2 + y^2 - z)\, dS$, S is the portion of the paraboloid $z = 4 - x^2 - y^2$ between $z = 1$ and $z = 2$

32. $\iint_S z\, dS$, S is the hemisphere $z = -\sqrt{9 - x^2 - y^2}$

33. $\iint_S z^2\, dS$, S is the portion of the cone $z^2 = x^2 + y^2$ between $z = -4$ and $z = 4$

34. $\iint_S z^2\, dS$, S is the portion of the cone $z = \sqrt{x^2 + y^2}$ above the rectangle $0 \leq x \leq 2, -1 \leq y \leq 2$

35. $\iint_S x\, dS$, S is the portion of $x^2 + y^2 - z^2 = 1$ between $z = 0$ and $z = 1$

36. $\iint_S \sqrt{x^2 + y^2 + z^2}\, dS$, S is the portion of $x = -\sqrt{4 - y^2 - z^2}$ between $y = 0$ and $y = x$

In exercises 37–48, evaluate the flux integral $\iint_S \mathbf{F} \cdot \mathbf{n}\, dS$.

37. $\mathbf{F} = \langle x, y, z \rangle$, S is the portion of $z = 4 - x^2 - y^2$ above the xy-plane (**n** upward)

38. $\mathbf{F} = \langle y, -x, 1 \rangle$, S is the portion of $z = x^2 + y^2$ below $z = 4$ (**n** downward)

39. $\mathbf{F} = \langle y, -x, z \rangle$, S is the portion of $z = \sqrt{x^2 + y^2}$ below $z = 3$ (**n** downward)

40. $\mathbf{F} = \langle 0, 1, y \rangle$, S is the portion of $z = -\sqrt{x^2 + y^2}$ inside $x^2 + y^2 = 4$ (**n** upward)

41. $\mathbf{F} = \langle xy, y^2, z \rangle$, S is the boundary of the unit cube with $0 \leq x \leq 1, 0 \leq y \leq 1, 0 \leq z \leq 1$ (**n** outward)

42. $\mathbf{F} = \langle y, z, 0 \rangle$, S is the boundary of the box with $0 \leq x \leq 2$, $0 \leq y \leq 3, 0 \leq z \leq 1$ (**n** outward)

43. $\mathbf{F} = \langle 1, 0, z \rangle$, S is the boundary of the region bounded above by $z = 4 - x^2 - y^2$ and below by $z = 1$ (**n** outward)

44. $\mathbf{F} = \langle x, y, z \rangle$, S is the boundary of the region between $z = 0$ and $z = -\sqrt{4 - x^2 - y^2}$

45. $\mathbf{F} = \langle yx, 1, x \rangle$, S is the portion of $z = 2 - x - y$ above the square $0 \leq x \leq 1, 0 \leq y \leq 1$ (**n** upward)

46. $\mathbf{F} = \langle y, 3, z \rangle$, S is the portion of $z = x^2 + y^2$ above the triangle with vertices $(0, 0)$, $(0, 1)$, $(1, 1)$ (**n** downward)

47. $\mathbf{F} = \langle y, 0, 2 \rangle$, S is the boundary of the region bounded above by $z = \sqrt{8 - x^2 - y^2}$ and below by $z = \sqrt{x^2 + y^2}$ (**n** outward)

48. $\mathbf{F} = \langle 3, z, y \rangle$, S is the boundary of the region between $z = 8 - 2x - y$ and $z = \sqrt{x^2 + y^2}$ and inside $x^2 + y^2 = 1$ (**n** outward)

In exercises 49–52, find the mass and center of mass of the region.

49. The portion of the plane $3x + 2y + z = 6$ inside the cylinder $x^2 + y^2 = 4$, $\rho(x, y, z) = x^2 + 1$

50. The portion of the plane $x + 2y + z = 4$ above the region bounded by $y = x^2$ and $y = 1$, $\rho(x, y, z) = y$

51. The hemisphere $z = \sqrt{1 - x^2 - y^2}$, $\rho(x, y, z) = 1 + x$

52. The portion of the paraboloid $z = x^2 + y^2$ inside the cylinder $x^2 + y^2 = 4$, $\rho(x, y, z) = z$

53. State the formula converting a surface integral into a double integral for a projection into the yz-plane.

54. State the formula converting a surface integral into a double integral for a projection into the xz-plane.

In exercises 55–62, use the formulas of exercises 53 and 54 to evaluate the surface integral.

55. $\iint_S z\, dS$, where S is the portion of $x^2 + y^2 = 1$ with $x \geq 0$ and z between $z = 1$ and $z = 2$

56. $\iint_S yz\,dS$, where S is the portion of $x^2 + y^2 = 1$ with $x \geq 0$ and z between $z = 1$ and $z = 4 - y$

57. $\iint_S (y^2 + z^2)\,dS$, where S is the portion of the paraboloid $x = 9 - y^2 - z^2$ in front of the yz-plane

58. $\iint_S (y^2 + z^2)\,dS$, where S is the hemisphere $x = \sqrt{4 - y^2 - z^2}$

59. $\iint_S x^2\,dS$, where S is the portion of the paraboloid $y = x^2 + z^2$ to the left of the plane $y = 1$

60. $\iint_S (x^2 + z^2)\,dS$, where S is the hemisphere $y = \sqrt{4 - x^2 - z^2}$

61. $\iint_S 4dS$, where S is the portion of $y = 1 - x^2$ with $y \geq 0$ and between $z = 0$ and $z = 2$

62. $\iint_S (x^2 + z^2)\,dS$, where S is the portion of $y = \sqrt{4 - x^2}$ between $z = 1$ and $z = 4$

63. Explain the following result geometrically. The flux integral of $\mathbf{F}(x, y, z) = \langle x, y, z\rangle$ across the cone $z = \sqrt{x^2 + y^2}$ is 0.

64. In geometric terms, determine whether the flux integral of $\mathbf{F}(x, y, z) = \langle x, y, z\rangle$ across the hemisphere $z = \sqrt{1 - x^2 - y^2}$ is 0.

65. For the cone $z = c\sqrt{x^2 + y^2}$ (where $c > 0$), show that in spherical coordinates $\tan\phi = \frac{1}{c}$. Then show that parametric equations are $x = \dfrac{u\cos v}{\sqrt{c^2+1}}$, $y = \dfrac{u\sin v}{\sqrt{c^2+1}}$ and $z = \dfrac{cu}{\sqrt{c^2+1}}$.

66. Find the surface area of the portion of $z = c\sqrt{x^2 + y^2}$ below $z = 1$, using the parametric equations in exercise 65.

67. Find the flux of $\langle x, y, z\rangle$ across the portion of $z = c\sqrt{x^2 + y^2}$ below $z = 1$. Explain in physical terms why this answer makes sense.

68. Find the flux of $\langle x, y, z\rangle$ across the entire cone $z^2 = c^2(x^2 + y^2)$.

69. Find the flux of $\langle x, y, 0\rangle$ across the portion of $z = c\sqrt{x^2 + y^2}$ below $z = 1$.

70. Find the limit as c approaches 0 of the flux in exercise 69. Explain in physical terms why this answer makes sense.

EXPLORATORY EXERCISES

1. If $x = 3\sin u\cos v$, $y = 3\cos u$ and $z = 3\sin u\sin v$, show that $x^2 + y^2 + z^2 = 9$. Explain why this equation doesn't guarantee that the parametric surface defined is the entire sphere, but it does guarantee that all points on the surface are also on the sphere. In this case, the parametric surface is the entire sphere. To verify this in graphical terms, sketch a picture showing geometric interpretations of the "spherical coordinates" u and v. To see what problems can occur, sketch the surface defined by $x = 3\sin\dfrac{u^2}{u^2+1}\cos v$, $y = 3\cos\dfrac{u^2}{u^2+1}$ and $z = 3\sin\dfrac{u^2}{u^2+1}\sin v$. Explain why you do not get the entire sphere. To see a more subtle example of the same problem, sketch the surface $x = \cos u\cosh v$, $y = \sinh v$, $z = \sin u\cosh v$. Use identities to show that $x^2 - y^2 + z^2 = 1$ and identify the surface. Then sketch the surface $x = \cos u\cosh v$, $y = \cos u\sinh v$, $z = \sin u$ and use identities to show that $x^2 - y^2 + z^2 = 1$. Explain why the second surface is not the entire hyperboloid. Explain in words and pictures exactly what the second surface is.

14.7 THE DIVERGENCE THEOREM

Recall that at the end of section 14.5, we had rewritten Green's Theorem in terms of the divergence of a two-dimensional vector field. We had found there (see equation 5.3) that

$$\oint_C \mathbf{F}\cdot\mathbf{n}\,ds = \iint_R \nabla\cdot\mathbf{F}(x, y)\,dA.$$

Here, R is a region in the xy-plane enclosed by a piecewise-smooth, positively oriented, simple closed curve C. Further, $\mathbf{F}(x, y) = \langle M(x, y), N(x, y), 0\rangle$, where $M(x, y)$ and $N(x, y)$ are continuous and have continuous first partial derivatives in some open region D in the xy-plane, with $R \subset D$.

We can extend this two-dimensional result to three dimensions in exactly the way you might expect. That is, for a solid region $Q \subset \mathbb{R}^3$ bounded by the surface ∂Q, we have

$$\iint_{\partial Q} \mathbf{F} \cdot \mathbf{n}\, dS = \iiint_{Q} \nabla \cdot \mathbf{F}(x, y, z)\, dV.$$

This result (referred to as the **Divergence Theorem** or **Gauss' Theorem**) has great significance in a variety of settings. If **F** represents the velocity field of a fluid in motion, the Divergence Theorem says that the total flux of the velocity field across the boundary of the solid is equal to the triple integral of the divergence of the velocity field over the solid.

In Figure 14.48, the velocity field **F** of a fluid is shown superimposed on a solid Q bounded by the closed surface ∂Q. Observe that there are two ways to compute the rate of change of the amount of fluid inside of Q. One way is to calculate the fluid flow into or out of Q across its boundary, which is given by the flux integral $\iint_{\partial Q} \mathbf{F} \cdot \mathbf{n}\, dS$. On the other hand, instead of focusing on the boundary, we can consider the accumulation or dispersal of fluid at each point in Q. As we'll see, this is given by $\nabla \cdot \mathbf{F}$, whose value at a given point measures the extent to which that point acts as a source or sink of fluid. To obtain the total change in the amount of the fluid in Q, we "add up" all of the values of $\nabla \cdot \mathbf{F}$ in Q, giving us the triple integral of $\nabla \cdot \mathbf{F}$ over Q. Since the flux integral and the triple integral both give the net rate of change of the amount of fluid in Q, they must be equal.

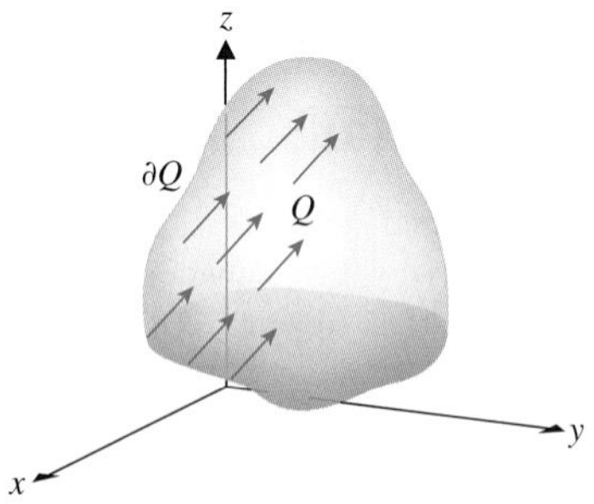

FIGURE 14.48
Flow of fluid across ∂Q

We now state and prove the result.

> **THEOREM 7.1** (Divergence Theorem)
>
> Suppose that $Q \subset \mathbb{R}^3$ is bounded by the closed surface ∂Q and that $\mathbf{n}(x, y, z)$ denotes the exterior unit normal vector to ∂Q. Then, if the components of $\mathbf{F}(x, y, z)$ have continuous first partial derivatives in Q, we have
>
> $$\iint_{\partial Q} \mathbf{F} \cdot \mathbf{n}\, dS = \iiint_{Q} \nabla \cdot \mathbf{F}(x, y, z)\, dV.$$

Although we have stated the theorem in the general case, we prove the result only for the case where the solid Q is fairly simple. A proof for the general case can be found in a more advanced text.

PROOF

For $\mathbf{F}(x, y, z) = \langle M(x, y, z), N(x, y, z), P(x, y, z)\rangle$, the divergence of **F** is

$$\nabla \cdot \mathbf{F}(x, y, z) = \frac{\partial M}{\partial x} + \frac{\partial N}{\partial y} + \frac{\partial P}{\partial z}.$$

We then have that

$$\iiint_{Q} \nabla \cdot \mathbf{F}(x, y, z)\, dV = \iiint_{Q} \frac{\partial M}{\partial x}\, dV + \iiint_{Q} \frac{\partial N}{\partial y}\, dV + \iiint_{Q} \frac{\partial P}{\partial z}\, dV. \tag{7.1}$$

Further, we can write the flux integral as

$$\iint_{\partial Q} \mathbf{F} \cdot \mathbf{n}\, dS = \iint_{\partial Q} M(x, y, z)\mathbf{i} \cdot \mathbf{n}\, dS + \iint_{\partial Q} N(x, y, z)\mathbf{j} \cdot \mathbf{n}\, dS$$
$$+ \iint_{\partial Q} P(x, y, z)\mathbf{k} \cdot \mathbf{n}\, dS. \tag{7.2}$$

Looking carefully at (7.1) and (7.2), observe that the theorem will follow if we can show that

$$\iiint_Q \frac{\partial M}{\partial x}\, dV = \iint_{\partial Q} M(x, y, z)\mathbf{i} \cdot \mathbf{n}\, dS, \tag{7.3}$$

$$\iiint_Q \frac{\partial N}{\partial y}\, dV = \iint_{\partial Q} N(x, y, z)\mathbf{j} \cdot \mathbf{n}\, dS \tag{7.4}$$

and

$$\iiint_Q \frac{\partial P}{\partial z}\, dV = \iint_{\partial Q} P(x, y, z)\mathbf{k} \cdot \mathbf{n}\, dS. \tag{7.5}$$

As you might imagine, the proofs of (7.3), (7.4) and (7.5) are all virtually identical (and all fairly long). Consequently, we prove only one of these three equations here. In order to prove (7.5), we assume that we can describe the solid Q as follows:

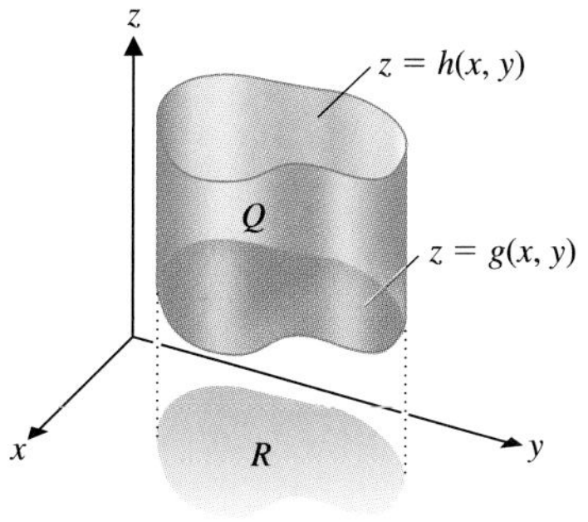

FIGURE 14.49a
The solid Q

$$Q = \{(x, y, z) | g(x, y) \le z \le h(x, y), \text{ for } (x, y) \in R\},$$

where R is some region in the xy-plane, as illustrated in Figure 14.49a. (We can prove (7.3) and (7.4) by making corresponding assumptions regarding Q.) Now, notice from Figure 14.49a that there are three distinct surfaces that make up the boundary of Q. In Figure 14.49b, we have labeled these surfaces S_1 (the bottom surface), S_2 (the top surface) and S_3 (the lateral surface), where we have also indicated exterior normal vectors to each of the surfaces.

Notice that on the lateral surface S_3, the $\mathbf{k}$ component of the exterior unit normal $\mathbf{n}$ is zero and so, the flux integral of $P(x, y, z)\mathbf{k}$ over S_3 is zero. This gives us

$$\iint_{\partial Q} P(x, y, z)\mathbf{k} \cdot \mathbf{n}\, dS = \iint_{S_1} P(x, y, z)\mathbf{k} \cdot \mathbf{n}\, dS + \iint_{S_2} P(x, y, z)\mathbf{k} \cdot \mathbf{n}\, dS. \tag{7.6}$$

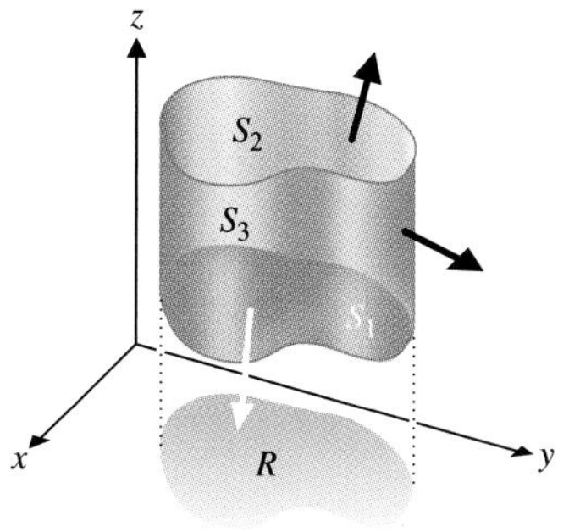

FIGURE 14.49b
The surfaces S_1, S_2 and S_3 and several exterior normal vectors

In order to prove the result, we need to rewrite the two integrals on the right side of (7.6) as double integrals over the region R in the xy-plane. First, you must notice that on the surface S_1 (the bottom surface), the exterior unit normal $\mathbf{n}$ points downward (i.e., it has a negative $\mathbf{k}$ component). Now, S_1 is defined by

$$S_1 = \{(x, y, z) | z = g(x, y), \text{ for } (x, y) \in R\}.$$

If we define $k_1(x, y, z) = z - g(x, y)$, then the exterior unit normal on S_1 is given by

$$\mathbf{n} = \frac{-\nabla k_1}{\|\nabla k_1\|} = \frac{g_x(x, y)\mathbf{i} + g_y(x, y)\mathbf{j} - \mathbf{k}}{\sqrt{[g_x(x, y)]^2 + [g_y(x, y)]^2 + 1}}$$

and

$$\mathbf{k} \cdot \mathbf{n} = \frac{-1}{\sqrt{[g_x(x, y)]^2 + [g_y(x, y)]^2 + 1}},$$

since the unit vectors **i**, **j** and **k** are all mutually orthogonal. We now have

$$\iint_{S_1} P(x,y,z)\mathbf{k}\cdot\mathbf{n}\,dS = -\iint_{S_1} \frac{P(x,y,z)}{\sqrt{[g_x(x,y)]^2+[g_y(x,y)]^2+1}}\,dS$$
$$= -\iint_R \frac{P(x,y,g(x,y))}{\sqrt{[g_x(x,y)]^2+[g_y(x,y)]^2+1}} \cdot\sqrt{[g_x(x,y)]^2+[g_y(x,y)]^2+1}\,dA$$
$$= -\iint_R P(x,y,g(x,y))\,dA, \tag{7.7}$$

thanks to the two square roots canceling out one another. In a similar way, notice that on S_2 (the top surface), the exterior unit normal **n** points upward (i.e., it has a positive **k** component). Since S_2 corresponds to the portion of the surface $z = h(x, y)$ for $(x, y) \in R$, if we take $k_2(x, y) = z - h(x, y)$, we have that on S_2,

$$\mathbf{n} = \frac{\nabla k_2}{\|\nabla k_2\|} = \frac{-h_x(x,y)\mathbf{i} - h_y(x,y)\mathbf{j} + \mathbf{k}}{\sqrt{[h_x(x,y)]^2+[h_y(x,y)]^2+1}}$$

and so,
$$\mathbf{k}\cdot\mathbf{n} = \frac{1}{\sqrt{[h_x(x,y)]^2+[h_y(x,y)]^2+1}}.$$

We now have

$$\iint_{S_2} P(x,y,z)\,\mathbf{k}\cdot\mathbf{n}\,dS = \iint_{S_2} \frac{P(x,y,z)}{\sqrt{[h_x(x,y)]^2+[h_y(x,y)]^2+1}}\,dS$$
$$= \iint_R \frac{P(x,y,h(x,y))}{\sqrt{[h_x(x,y)]^2+[h_y(x,y)]^2+1}} \cdot\sqrt{[h_x(x,y)]^2+[h_y(x,y)]^2+1}\,dA$$
$$= \iint_R P(x,y,h(x,y))\,dA. \tag{7.8}$$

Putting together (7.6), (7.7) and (7.8) gives us

$$\iint_{\partial Q} P(x,y,z)\mathbf{k}\cdot\mathbf{n}\,dS = \iint_{S_1} P(x,y,z)\,\mathbf{k}\cdot\mathbf{n}\,dS + \iint_{S_2} P(x,y,z)\mathbf{k}\cdot\mathbf{n}\,dS$$
$$= \iint_R P(x,y,h(x,y))\,dA - \iint_R P(x,y,g(x,y))\,dA$$
$$= \iint_R [P(x,y,h(x,y)) - P(x,y,g(x,y))]\,dA$$
$$= \iint_R \int_{g(x,y)}^{h(x,y)} \frac{\partial P}{\partial z}\,dz\,dA \qquad \text{By the Fundamental Theorem of Calculus}$$
$$= \iiint_Q \frac{\partial P}{\partial z}\,dV,$$

which proves (7.5). With appropriate assumptions on Q, we can similarly prove (7.3) and (7.4). This proves the theorem for the special case where the solid Q can be described as indicated. ■

EXAMPLE 7.1 Applying the Divergence Theorem

Let Q be the solid bounded by the paraboloid $z = 4 - x^2 - y^2$ and the xy-plane. Find the flux of the vector field $\mathbf{F}(x, y, z) = \langle x^3, y^3, z^3 \rangle$ over the surface ∂Q.

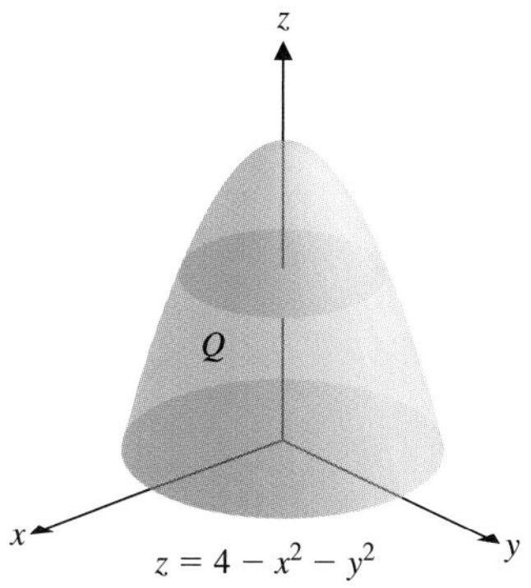

FIGURE 14.50
The solid Q

Solution We show a sketch of the solid in Figure 14.50. Notice that to compute the flux directly, we must consider the two different portions of ∂Q (the surface of the paraboloid and its base in the xy-plane) separately. Alternatively, observe that the divergence of $\mathbf{F}$ is given by

$$\nabla \cdot \mathbf{F}(x, y, z) = \nabla \cdot \langle x^3, y^3, z^3 \rangle = 3x^2 + 3y^2 + 3z^2.$$

From the Divergence Theorem, we now have that the flux of $\mathbf{F}$ over ∂Q is given by

$$\iint_{\partial Q} \mathbf{F} \cdot \mathbf{n}\, dS = \iiint_Q \nabla \cdot \mathbf{F}(x, y, z)\, dV = \iiint_Q (3x^2 + 3y^2 + 3z^2)\, dV.$$

If we rewrite the triple integral in cylindrical coordinates, we get

$$\begin{aligned}\iint_{\partial Q} \mathbf{F} \cdot \mathbf{n}\, dS &= \iiint_Q (3x^2 + 3y^2 + 3z^2)\, dV \\ &= 3\int_0^{2\pi}\int_0^2\int_0^{4-r^2} (r^2 + z^2)\, r\, dz\, dr\, d\theta \\ &= 3\int_0^{2\pi}\int_0^2 \left(r^2 z + \frac{z^3}{3}\right)\Bigg|_{z=0}^{z=4-r^2} r\, dr\, d\theta \\ &= 3\int_0^{2\pi}\int_0^2 \left[r^3(4-r^2) + \frac{1}{3}(4-r^2)^3 r\right] dr\, d\theta \\ &= 96\pi,\end{aligned}$$

where we have left the details of the final integrations as a straightforward exercise. ■

Notice that in example 7.1, we used the Divergence Theorem to replace a very messy surface integral calculation by a comparatively simple triple integral. In example 7.2, we prove a general result regarding the flux of a certain vector field over any surface, something we would be unable to do without the Divergence Theorem.

EXAMPLE 7.2 Proving a General Result with the Divergence Theorem

Prove that the flux of the vector field $\mathbf{F}(x, y, z) = \langle 3y \cos z, x^2 e^z, x \sin y \rangle$ is zero over any closed surface ∂Q enclosing a solid region Q.

Solution Notice that in this case, the divergence of $\mathbf{F}$ is

$$\begin{aligned}\nabla \cdot \mathbf{F}(x, y, z) &= \nabla \cdot \langle 3y \cos z, x^2 e^z, x \sin y \rangle \\ &= \frac{\partial}{\partial x}(3y \cos z) + \frac{\partial}{\partial y}(x^2 e^z) + \frac{\partial}{\partial z}(x \sin y) = 0.\end{aligned}$$

From the Divergence Theorem, we then have that the flux of **F** over ∂Q is given by

$$\iint_{\partial Q} \mathbf{F} \cdot \mathbf{n}\, dS = \iiint_Q \nabla \cdot \mathbf{F}(x, y, z)\, dV$$

$$= \iiint_Q 0\, dV = 0,$$

for any solid region $Q \subset \mathbb{R}^3$. ■

In section 4.4, we saw that for a function $f(x)$ of a single variable, if f is continuous on the interval $[a, b]$ then the average value of f on $[a, b]$ is given by

$$f_{\text{ave}} = \frac{1}{b-a} \int_a^b f(x)\, dx.$$

Similarly, when $f(x, y, z)$ is a continuous function on the region $Q \subset \mathbb{R}^3$ (bounded by the surface ∂Q), the **average value** of f on Q is given by

$$f_{\text{ave}} = \frac{1}{V} \iiint_Q f(x, y, z)\, dV,$$

where V is the volume of Q. Further, by continuity, there must be a point $P(a, b, c) \in Q$ at which f equals its average value, that is, where

$$f(P) = \frac{1}{V} \iiint_Q f(x, y, z)\, dV.$$

This says that if $\mathbf{F}(x, y, z)$ has continuous first partial derivatives on Q, then div **F** is continuous on Q and so, there is a point $P(a, b, c) \in Q$ for which

$$(\nabla \cdot \mathbf{F})|_P = \frac{1}{V} \iiint_Q \nabla \cdot \mathbf{F}(x, y, z)\, dV$$

$$= \frac{1}{V} \iint_{\partial Q} \mathbf{F}(x, y, z) \cdot \mathbf{n}\, dS,$$

by the Divergence Theorem. Finally, observe that since the surface integral represents the flux of **F** over the surface ∂Q, then $(\nabla \cdot \mathbf{F})|_P$ represents the flux per unit volume over ∂Q.

In particular, for any point $P_0(x_0, y_0, z_0)$ in the interior of Q (i.e., in Q, but not on ∂Q), let S_a be the sphere of radius a, centered at P_0, where a is sufficiently small so that S_a lies completely inside Q. From the preceding discussion, there must be some point P_a in the interior of S_a for which

$$(\nabla \cdot \mathbf{F})|_{P_a} = \frac{1}{V_a} \iint_{S_a} \mathbf{F}(x, y, z) \cdot \mathbf{n}\, dS,$$

where V_a is the volume of the sphere ($V_a = \frac{4}{3}\pi a^3$). Finally, taking the limit as $a \to 0$, we have by the continuity of $\nabla \cdot \mathbf{F}$ that

$$(\nabla \cdot \mathbf{F})\big|_{P_0} = \lim_{a \to 0} \frac{1}{V_a} \iint_{S_a} \mathbf{F}(x, y, z) \cdot \mathbf{n}\, dS$$

or

$$\operatorname{div} \mathbf{F}(P_0) = \lim_{a \to 0} \frac{1}{V_a} \iint_{S_a} \mathbf{F}(x, y, z) \cdot \mathbf{n}\, dS. \tag{7.9}$$

In other words, the divergence of a vector field at a point P_0 is the limiting value of the flux per unit volume over a sphere centered at P_0, as the radius of the sphere tends to zero.

In the case where $\mathbf{F}$ represents the velocity field for a fluid in motion, (7.9) provides us with an important interpretation of the divergence of a vector field. In this case, if $\operatorname{div} \mathbf{F}(P_0) > 0$, then the flux per unit volume at P_0 is positive. From (7.9), this means that for a sphere S_a of sufficiently small radius centered at P_0, the net (outward) flux through the surface of S_a is positive. For an incompressible fluid (such as a liquid), this says that more fluid is passing out through the surface of S_a than is passing in through the surface, which can happen only if there is a source somewhere in S_a, where additional fluid is coming into the flow. Likewise, if $\operatorname{div} \mathbf{F}(P_0) < 0$, there must be a sphere S_a for which the net (outward) flux through the surface of S_a is negative. This says that more fluid is passing in through the surface than is flowing out. Once again, for an incompressible fluid, this can occur only if there is a sink somewhere in S_a, where fluid is leaving the flow. For this reason, in incompressible fluid flow, a point where $\operatorname{div} \mathbf{F}(P) > 0$ is called a **source** and a point where $\operatorname{div} \mathbf{F}(P) < 0$ is called a **sink.** Notice that for an incompressible fluid flow with no sources or sinks, we must have that $\operatorname{div} \mathbf{F}(P) = 0$ throughout the flow.

EXAMPLE 7.3 Finding the Flux of an Inverse Square Field

Show that the flux of an inverse square field over every closed surface enclosing the origin is a constant.

Solution Suppose that S is a closed surface forming the boundary of the solid region Q, where the origin lies in the interior of Q and suppose that $\mathbf{F}$ is an inverse square field. That is,

$$\mathbf{F}(x, y, z) = \frac{c}{\|\mathbf{r}\|^3}\mathbf{r},$$

where $\mathbf{r} = \langle x, y, z \rangle$, $\|\mathbf{r}\| = \sqrt{x^2 + y^2 + z^2}$ and c is a constant. Before you rush to apply the Divergence Theorem, notice that $\mathbf{F}$ is *not* continuous in Q, since $\mathbf{F}$ is undefined at the origin and so, we cannot apply the theorem in Q. Notice, though, that if we could somehow exclude the origin from the region, then we could apply the theorem. A very common method of doing this is to "punch out" a sphere S_a of radius a centered at the origin, where a is sufficiently small that S_a is completely contained in the interior of Q. (See Figure 14.51.) That is, if we define Q_a to be the set of all points inside Q, but outside of S_a (so that Q_a corresponds to Q, where the sphere S_a has been "punched out"), we can now apply the Divergence Theorem on Q_a. Before we do that, notice that

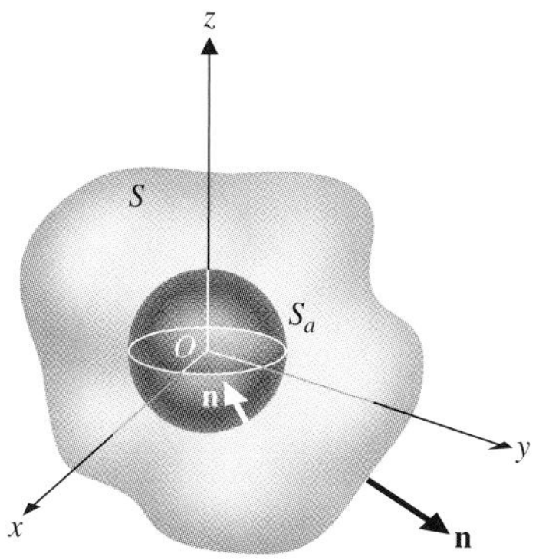

FIGURE 14.51
The region Q_a

the boundary of Q_a consists of the two surfaces S and S_a. We now have

$$\iiint_{Q_a} \nabla \cdot \mathbf{F}\, dV = \iint_S \mathbf{F} \cdot \mathbf{n}\, dS + \iint_{S_a} \mathbf{F} \cdot \mathbf{n}\, dS.$$

We leave it as an exercise to show that for any inverse square field, $\nabla \cdot \mathbf{F} = 0$. This now gives us

$$\iint_S \mathbf{F} \cdot \mathbf{n}\, dS = -\iint_{S_a} \mathbf{F} \cdot \mathbf{n}\, dS. \tag{7.10}$$

Since the integral on the right side of (7.10) is taken over a sphere centered at the origin, we should be able to easily calculate it. You need to be careful, though, to note that the exterior normals here point *out* of Q_a and so, the normal on the right side of (7.10) must point *toward* the origin. That is,

$$\mathbf{n} = -\frac{1}{\|\mathbf{r}\|}\mathbf{r} = -\frac{1}{a}\mathbf{r},$$

since $\|\mathbf{r}\| = a$ on S_a. We now have from (7.10) that

$$\begin{aligned}
\iint_S \mathbf{F} \cdot \mathbf{n}\, dS &= -\iint_{S_a} \frac{c}{a^3}\mathbf{r} \cdot \left(-\frac{1}{a}\mathbf{r}\right) dS \\
&= \frac{c}{a^4}\iint_{S_a} \mathbf{r} \cdot \mathbf{r}\, dS \\
&= \frac{c}{a^4}\iint_{S_a} \|\mathbf{r}\|^2\, dS \\
&= \frac{c}{a^2}\iint_{S_a} dS = \frac{c}{a^2}(4\pi a^2) = 4\pi c,
\end{aligned}$$

Since $\|\mathbf{r}\| = a$

since $\iint_{S_a} dS$ simply gives the surface area of the sphere, $4\pi a^2$. Notice that this says that over any closed surface enclosing the origin, the flux of an inverse square field is a constant: $4\pi c$. ■

The principle derived in example 7.3 is called **Gauss' Law** for inverse square fields and has many important applications, notably in the theory of electricity and magnetism. The method we used to derive Gauss' Law, where we punched out a sphere surrounding the discontinuity of the integrand, is a common technique used in applying the Divergence Theorem to a variety of important cases where the integrand is discontinuous. In particular, such applications to discontinuous vector fields are quite important in the field of differential equations.

We close the section with a straightforward application of the Divergence Theorem to show that the flux of a magnetic field across a closed surface is always zero.

EXAMPLE 7.4 Finding the Flux of a Magnetic Field

Use the Divergence Theorem and Maxwell's equation $\nabla \cdot \mathbf{B} = 0$ to show that $\iint_S \mathbf{B} \cdot \mathbf{n}\, dS = 0$ for any closed surface S.

Solution Applying the Divergence Theorem to $\iint_S \mathbf{B}\cdot\mathbf{n}\,dS$ and using $\nabla\cdot\mathbf{B}=0$, we have

$$\iint_S \mathbf{B}\cdot\mathbf{n}\,dS = \iiint_Q \nabla\cdot\mathbf{B}\,dV = 0.$$

Observe that this result says that the flux of a magnetic field over any closed surface is zero. ■

EXERCISES 14.7

WRITING EXERCISES

1. If $\mathbf{F}$ is the velocity field of a fluid, explain what $\mathbf{F}\cdot\mathbf{n}$ represents and then what $\iint_{\partial Q} \mathbf{F}\cdot\mathbf{n}\,dS$ represents.

2. If $\mathbf{F}$ is the velocity field of a fluid, explain what $\nabla\cdot\mathbf{F}$ represents and then what $\iiint_Q \nabla\cdot\mathbf{F}\,dV$ represents.

3. Use your answers to exercises 1 and 2 to explain in physical terms why the Divergence Theorem makes sense.

4. For fluid flowing through a pipe, give one example each of a source and a sink.

In exercises 1–4, verify the Divergence Theorem by computing both integrals.

1. $\mathbf{F}=\langle 2xz, y^2, -xz\rangle$, Q is the cube $0\le x\le 1, 0\le y\le 1, 0\le z\le 1$

2. $\mathbf{F}=\langle x, y, z\rangle$, Q is the ball $x^2+y^2+z^2\le 1$

3. $\mathbf{F}=\langle xz, zy, 2z^2\rangle$, Q is bounded by $z=1-x^2-y^2$ and $z=0$

4. $\mathbf{F}=\langle x^2, 2y, -x^2\rangle$, Q is the tetrahedron bounded by $x+2y+z=4$ and the coordinate planes

In exercises 5–16, use the Divergence Theorem to compute $\iint_{\partial Q} \mathbf{F}\cdot\mathbf{n}\,dS$.

5. Q is bounded by $x+y+2z=2$ (first octant) and the coordinate planes, $\mathbf{F}=\langle 2x-y^2, 4xz-2y, xy^3\rangle$.

6. Q is bounded by $4x+2y-z=4$ $(z\le 0)$ and the coordinate planes, $\mathbf{F}=\langle x^2-y^2z, x\sin z, 4y^2\rangle$.

7. Q is the cube $-1\le x\le 1, -1\le y\le 1, -1\le z\le 1$, $\mathbf{F}=\langle 4y^2, 3z-\cos x, z^3-x\rangle$.

8. Q is the rectangular box $0\le x\le 2, 1\le y\le 2, -1\le z\le 2$, $\mathbf{F}=\langle y^3-2x, e^{xz}, 4z\rangle$.

9. Q is bounded by $z=x^2+y^2$ and $z=4$, $\mathbf{F}=\langle x^3, y^3-z, xy^2\rangle$.

10. Q is bounded by $z=\sqrt{x^2+y^2}$ and $z=4$, $\mathbf{F}=\langle y^3, x+z^2, z+y^2\rangle$.

11. Q is bounded by $z=4-x^2-y^2, z=1$ and $z=0$, $\mathbf{F}=\langle z^3, x^2y, y^2z\rangle$.

12. Q is bounded by $z=\sqrt{x^2+y^2}, z=1$ and $z=2$, $\mathbf{F}=\langle x^3, x^2z^2, 3y^2z\rangle$.

13. Q is bounded by $x^2+y^2=1, z=0$ and $z=1$, $\mathbf{F}=\langle x-y^3, x^2\sin z, 3z\rangle$.

14. Q is bounded by $x^2+y^2=4, z=1$ and $z=8-y$, $\mathbf{F}=\langle y^2z, 2y-e^z, \sin x\rangle$.

15. Q is bounded by $z=\sqrt{1-x^2-y^2}$ and $z=0$, $\mathbf{F}=\langle x^3, y^3, z^3\rangle$.

16. Q is bounded by $z=-\sqrt{4-x^2-y^2}$ and $z=0$, $\mathbf{F}=\langle x^3, y^3, z^3\rangle$.

In exercises 17–28, find the flux of F over ∂Q.

17. Q is bounded by $z=\sqrt{x^2+y^2}$ and $z=\sqrt{2-x^2-y^2}$, $\mathbf{F}=\langle x^2, z^2-x, y^3\rangle$.

18. Q is bounded by $z=\sqrt{x^2+y^2}$ and $z=\sqrt{8-x^2-y^2}$, $\mathbf{F}=\langle 3xz^2, y^3, 3zx^2\rangle$.

19. Q is bounded by $z=\sqrt{x^2+y^2}, x^2+y^2=1$ and $z=0$, $\mathbf{F}=\langle y^2, x^2z, z^2\rangle$.

20. Q is bounded by $z=x^2+y^2$ and $z=8-x^2-y^2$, $\mathbf{F}=\langle 3y^2, 4x^3, 2z-x^2\rangle$.

21. Q is bounded by $x^2+z^2=1, y=0$ and $y=1$, $\mathbf{F}=\langle z-y^3, 2y-\sin z, x^2-z\rangle$.

22. Q is bounded by $y^2 + z^2 = 4$, $x = 1$ and $x = 8 - y$, $\mathbf{F} = \langle x^2 z, 2y - e^z, \sin x \rangle$.

23. Q is bounded by $x = y^2 + z^2$ and $x = 4$, $\mathbf{F} = \langle x^3, y^3 - z, z^3 - y^2 \rangle$.

24. Q is bounded by $y = 4 - x^2 - z^2$ and the xz-plane, $\mathbf{F} = \langle z^2 x, x^2 y, y^2 x \rangle$.

25. Q is bounded by $3x + 2y + z = 6$ and the coordinate planes, $\mathbf{F} = \langle y^2 x, 4x^2 \sin z, 3 \rangle$.

26. Q is bounded by $x + 2y + 3z = 12$ and the coordinate planes, $\mathbf{F} = \langle x^2 y, 3x, 4y - x^2 \rangle$.

27. Q is bounded by $z = 1 - x^2$, $z = -3$, $y = -2$ and $y = 2$, $\mathbf{F} = \langle x^2, y^3, x^3 y^2 \rangle$.

28. Q is bounded by $z = 1 - x^2$, $z = 0$, $y = 0$ and $x + y = 4$, $\mathbf{F} = \langle y^3, x^2 - z, z^2 \rangle$.

29. Coulomb's law for an electrostatic field applied to a point charge q at the origin gives us $\mathbf{E}(\mathbf{r}) = q\dfrac{\mathbf{r}}{r^3}$, where $r = \|\mathbf{r}\|$. Let Q be bounded by the sphere $x^2 + y^2 + z^2 = a^2$ for some constant $a > 0$. Show that the flux of $\mathbf{E}$ over ∂Q equals $4\pi q$. Discuss the fact that the flux does not depend on the value of a.

30. Show that for any inverse square field (see example 7.3), the divergence is 0.

31. Prove Green's first identity in three dimensions (see exercise 43 in section 14.5 for Green's first identity in two dimensions):

$$\iiint_Q f\nabla^2 g \, dV = \iint_{\partial Q} f(\nabla g) \cdot \mathbf{n} \, dS - \iiint_Q (\nabla f \cdot \nabla g) \, dV.$$

(Hint: Use the Divergence Theorem applied to $\mathbf{F} = f\nabla g$.)

32. Prove Green's second identity in three dimensions (see exercise 44 in section 14.5 for Green's second identity in two dimensions):

$$\iiint_Q (f\nabla^2 g - g\nabla^2 f) \, dV = \iint_{\partial Q} (f\nabla g - g\nabla f) \cdot \mathbf{n} \, dS.$$

(Hint: Use Green's first identity from exercise 31.)

Exercises 33–36 use Gauss' Law $\nabla \cdot \mathbf{E} = \dfrac{\rho}{\epsilon_0}$ for an electric field $\mathbf{E}$, charge density ρ and permittivity ϵ_0.

33. If S is a closed surface, show that the total charge q enclosed by S satisfies $q = \epsilon_0 \iint_S \mathbf{E} \cdot \mathbf{n} \, dS$.

34. Let $\mathbf{E}$ be the electric field for an infinite line charge on the z-axis. Assume that $\mathbf{E}$ has the form $\mathbf{E} = c\hat{\mathbf{r}} = c\dfrac{\langle x, y, 0 \rangle}{x^2 + y^2}$, for some constant c and let the charge density ρ (with respect to length on the z-axis) be a constant.

  a. If S is a portion of the cylinder $x^2 + y^2 = 1$ with height h, argue that $q = \rho h$.

  b. Use the results of exercise 33 and part (a) to find c in terms of ρ and ϵ_0.

35. Let $\mathbf{E}$ be the electric field for an infinite plane of charge density ρ. Assume that $\mathbf{E}$ has the form $\mathbf{E} = \left\langle 0, 0, \dfrac{cz}{|z|} \right\rangle$, for some constant $c > 0$.

  a. If S is a portion of the cylinder $x^2 + y^2 = 1$ with height h extending above and below the xy-plane, argue that $q = 2\pi\rho$.

  b. Use the technique of exercise 34 to determine the constant c.

36. The integral form of Gauss' Law is $\iint_S \mathbf{E} \cdot \mathbf{n} \, dS = \dfrac{q}{\epsilon_0}$, where $\mathbf{E}$ is an electric field, q is the total charge enclosed by S and ϵ_0 is the permittivity constant. Use equation (7.1) to derive the differential form of Gauss' Law: $\nabla \cdot \mathbf{E} = \dfrac{\rho}{\epsilon_0}$, where ρ is the charge density.

EXPLORATORY EXERCISES

1. In this exercise, we develop the **continuity equation,** one of the most important results in vector calculus. Suppose that a fluid has density ρ (a scalar function of space and time) and velocity $\mathbf{v}$. Argue that the rate of change of the mass m of the fluid contained in a region Q can be written as $\dfrac{dm}{dt} = \iiint_Q \dfrac{\partial \rho}{\partial t} \, dV$. Next, explain why the only way that the mass can change is for fluid to cross the boundary of Q (∂Q). Argue that $\dfrac{dm}{dt} = -\iint_{\partial Q} (\rho \mathbf{v}) \cdot \mathbf{n} \, dS$. In particular, explain why the minus sign in front of the surface integral is needed. Use the Divergence Theorem to rewrite this expression as a triple integral over Q. Explain why the two triple integrals must be equal. Since the integration is taken over arbitrary solids Q, the integrands must be equal to each other. Conclude that the continuity equation holds:

$$\nabla \cdot (\rho \mathbf{v}) + \frac{\partial \rho}{\partial t} = 0.$$

14.8 STOKES' THEOREM

Recall that, after introducing the curl in section 14.5, we observed that for a piecewise, smooth, positively oriented, simple closed curve C in the xy-plane enclosing the region R, we could rewrite Green's Theorem in the vector form

$$\oint_C \mathbf{F}\cdot d\mathbf{r} = \iint_R (\nabla \times \mathbf{F})\cdot \mathbf{k}\, dA, \tag{8.1}$$

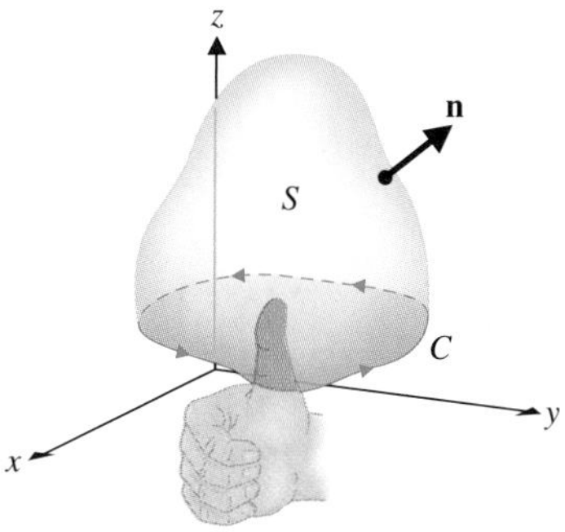

FIGURE 14.52a
Positive orientation

where $\mathbf{F}(x, y)$ is a vector field of the form $\mathbf{F}(x, y) = \langle M(x, y), N(x, y), 0\rangle$. In this section, we generalize this result to the case of a vector field defined on a surface in three dimensions. Suppose that S is an oriented surface with unit normal vector $\mathbf{n}$. If S is bounded by the simple closed curve C, we determine the orientation of C using a right-hand rule like the one used to determine the direction of a cross product of two vectors. Align the thumb of your right hand so that it points in the direction of one of the unit normals to S. Then if you curl your fingers, they will indicate the **positive orientation** on C, as indicated in Figure 14.52a. If the orientation of C is opposite that indicated by the curling of the fingers on your right hand, as shown in Figure 14.52b, we say that C has **negative orientation.** The vector form of Green's Theorem in (8.1) generalizes as follows.

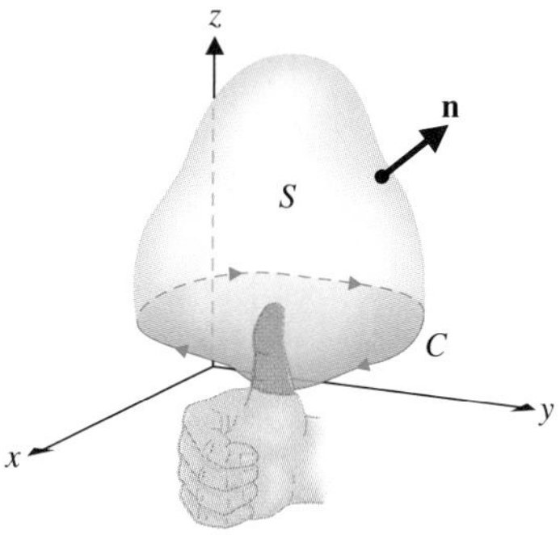

FIGURE 14.52b
Negative orientation

THEOREM 8.1 (Stokes' Theorem)

Suppose that S is an oriented, piecewise-smooth surface with unit normal vector $\mathbf{n}$, bounded by the simple closed, piecewise-smooth boundary curve ∂S having positive orientation. Let $\mathbf{F}(x, y, z)$ be a vector field whose components have continuous first partial derivatives in some open region containing S. Then,

$$\int_{\partial S} \mathbf{F}(x, y, z)\cdot d\mathbf{r} = \iint_S (\nabla \times \mathbf{F})\cdot \mathbf{n}\, dS. \tag{8.2}$$

Notice right away that the vector form of Green's Theorem (8.1) is a special case of (8.2), as follows. If S is simply a region in the xy-plane, then a unit normal to the surface at every point on S is the vector $\mathbf{n} = \mathbf{k}$. Further, $dS = dA$ (i.e., the surface area of the plane region is simply the area) and (8.2) simplifies to (8.1). The proof of Stokes' Theorem for the special case considered below hinges on Green's Theorem and the chain rule.

One important interpretation of Stokes' Theorem arises in the case where $\mathbf{F}$ represents a force field. Note that in this case, the integral on the left side of (8.2) corresponds to the work done by the force field $\mathbf{F}$ as the point of application moves along the boundary of S. Likewise, the right side of (8.2) represents the net flux of the curl of $\mathbf{F}$ over the surface S. A general proof of Stokes' Theorem can be found in more advanced texts. We prove it here only for a special case of the surface S.

PROOF (Special Case)

We consider here the special case where S is a surface of the form

$$S = \{(x, y, z) | z = f(x, y), \text{ for } (x, y) \in R\}\text{-}$$

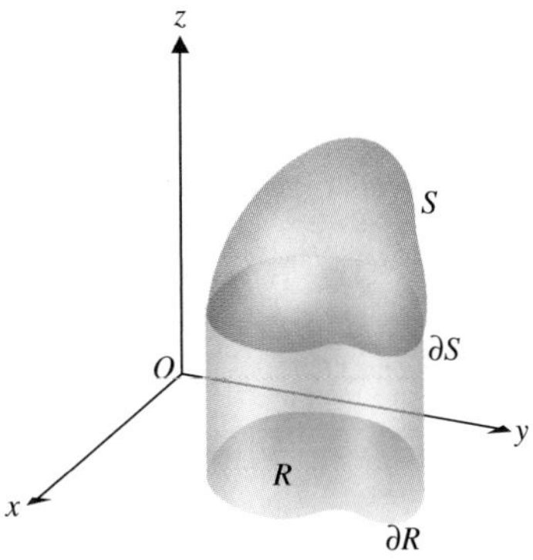

FIGURE 14.53
The surface S and its projection R onto the xy-plane

where R is a region in the xy-plane with piecewise-smooth boundary ∂R, where $f(x, y)$ has continuous first partial derivatives and for which ∂R is the projection of the boundary of the surface ∂S onto the xy-plane. (See Figure 14.53.)

Let $\mathbf{F}(x, y, z) = \langle M(x, y, z), N(x, y, z), P(x, y, z)\rangle$. We then have

$$\nabla \times \mathbf{F} = \begin{vmatrix} \mathbf{i} & \mathbf{j} & \mathbf{k} \\ \frac{\partial}{\partial x} & \frac{\partial}{\partial y} & \frac{\partial}{\partial z} \\ M & N & P \end{vmatrix}$$
$$= \left(\frac{\partial P}{\partial y} - \frac{\partial N}{\partial z}\right)\mathbf{i} + \left(\frac{\partial M}{\partial z} - \frac{\partial P}{\partial x}\right)\mathbf{j} + \left(\frac{\partial N}{\partial x} - \frac{\partial M}{\partial y}\right)\mathbf{k}.$$

Note that a normal vector at any point on S is given by

$$\mathbf{m} = \langle -f_x(x, y), -f_y(x, y), 1\rangle$$

and so, we orient S with the unit normal vector

$$\mathbf{n} = \frac{\langle -f_x(x, y), -f_y(x, y), 1\rangle}{\sqrt{[f_x(x, y)]^2 + [f_y(x, y)]^2 + 1}}.$$

Since $dS = \sqrt{[f_x(x, y)]^2 + [f_y(x, y)]^2 + 1}\, dA$, we now have

$$\iint_S (\nabla \times \mathbf{F}) \cdot \mathbf{n}\, dS$$
$$= \iint_R \left[-\left(\frac{\partial P}{\partial y} - \frac{\partial N}{\partial z}\right) f_x - \left(\frac{\partial M}{\partial z} - \frac{\partial P}{\partial x}\right) f_y + \left(\frac{\partial N}{\partial x} - \frac{\partial M}{\partial y}\right)\right]_{z=f(x,y)} dA.$$

Equation (8.2) is now equivalent to

$$\int_{\partial S} M\,dx + N\,dy + P\,dz$$
$$= \iint_R \left[-\left(\frac{\partial P}{\partial y} - \frac{\partial N}{\partial z}\right) f_x - \left(\frac{\partial M}{\partial z} - \frac{\partial P}{\partial x}\right) f_y + \left(\frac{\partial N}{\partial x} - \frac{\partial M}{\partial y}\right)\right]_{z=f(x,y)} dA. \quad (8.3)$$

We will now show that

$$\int_{\partial S} M(x, y, z)\,dx = -\iint_R \left(\frac{\partial M}{\partial y} + \frac{\partial M}{\partial z} f_y\right)_{z=f(x,y)} dA. \quad (8.4)$$

Suppose that the boundary of R is described parametrically by

$$\partial R = \{(x, y) | x = x(t), y = y(t), a \le t \le b\}.$$

Then, the boundary of S is described parametrically by

$$\partial S = \{(x, y, z) | x = x(t), y = y(t), z = f(x(t), y(t)), a \le t \le b\}$$

and we have

$$\int_{\partial S} M(x, y, z)\,dx = \int_a^b M(x(t), y(t), f(x(t), y(t)))\,x'(t)\,dt.$$

Now, notice that for $m(x, y) = M(x, y, f(x, y))$, this gives us

$$\int_{\partial S} M(x, y, z)\,dx = \int_a^b m(x(t), y(t))\,x'(t)\,dt = \int_{\partial R} m(x, y)\,dx. \quad (8.5)$$

HISTORICAL NOTES

George Gabriel Stokes (1819–1903) English mathematician who published important results in the field of hydrodynamics. Many of Stokes' results, including what are now known as the Navier-Stokes equations, were independently developed but duplicated previously published European results. (Due to the lasting bitterness over the Newton-Leibniz calculus dispute, the communication of results between Europe and England was minimal.) Stokes had a long list of publications, including a paper in which he named and explained fluorescence.

From Green's Theorem, we know that

$$\int_{\partial R} m(x, y)\, dx = -\iint_R \frac{\partial m}{\partial y} dA. \tag{8.6}$$

However, from the chain rule,

$$\frac{\partial m}{\partial y} = \frac{\partial}{\partial y} M(x, y, f(x, y)) = \left(\frac{\partial M}{\partial y} + \frac{\partial M}{\partial z} f_y\right)_{z=f(x,y)}.$$

Putting this together with (8.5) and (8.6) gives us

$$\int_{\partial S} M(x, y, z)\, dx = -\iint_R \frac{\partial m}{\partial y}\, dA = -\iint_R \left(\frac{\partial M}{\partial y} + \frac{\partial M}{\partial z} f_y\right)_{z=f(x,y)} dA,$$

which is (8.4). Similarly, you can show that

$$\int_{\partial S} N(x, y, z)\, dy = \iint_R \left(\frac{\partial N}{\partial x} + \frac{\partial N}{\partial z} f_x\right)_{z=f(x,y)} dA \tag{8.7}$$

and

$$\int_{\partial S} P(x, y, z)\, dz = \iint_R \left(\frac{\partial P}{\partial x} f_y - \frac{\partial P}{\partial y} f_x\right)_{z=f(x,y)} dA. \tag{8.8}$$

Putting together (8.4), (8.7) and (8.8) now gives us (8.3), which proves Stokes' Theorem for this special case of the surface. ■

EXAMPLE 8.1 Using Stokes' Theorem to Evaluate a Line Integral

Evaluate $\int_C \mathbf{F} \cdot d\mathbf{r}$, for $\mathbf{F}(x, y, z) = \langle -y, x^2, z^3 \rangle$, where C is the intersection of the circular cylinder $x^2 + y^2 = 4$ and the plane $x + z = 3$, oriented so that it is traversed counterclockwise when viewed from high up on the positive z-axis.

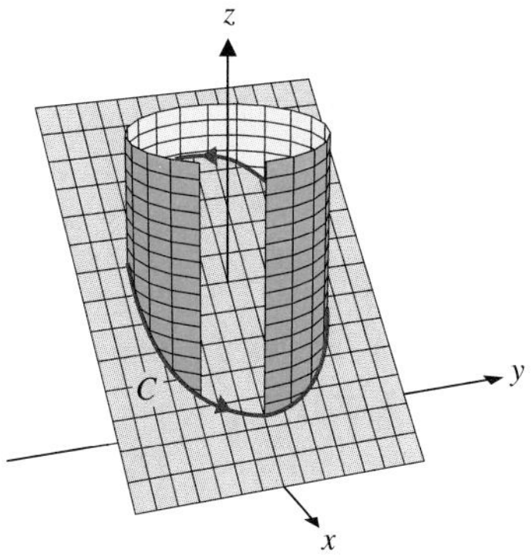

FIGURE 14.54
Intersection of the plane and the cylinder producing the curve C

Solution First, notice that C is an ellipse, as indicated in Figure 14.54. Unfortunately, C is rather difficult to parameterize, which makes the direct evaluation of the line integral somewhat difficult. Instead, we can use Stokes' Theorem to evaluate the integral. First, we calculate the curl of $\mathbf{F}$:

$$\nabla \times \mathbf{F} = \begin{vmatrix} \mathbf{i} & \mathbf{j} & \mathbf{k} \\ \frac{\partial}{\partial x} & \frac{\partial}{\partial y} & \frac{\partial}{\partial z} \\ -y & x^2 & z^3 \end{vmatrix} = (2x + 1)\,\mathbf{k}.$$

Notice that on the surface S, consisting of the portion of the plane $x + z = 3$ enclosed by C, we have the unit normal vector

$$\mathbf{n} = \frac{1}{\sqrt{2}}\langle 1, 0, 1 \rangle.$$

From Stokes' Theorem, we now have

$$\int_C \mathbf{F} \cdot d\mathbf{r} = \iint_S (\nabla \times \mathbf{F}) \cdot \mathbf{n}\, dS = \iint_R \underbrace{\frac{1}{\sqrt{2}}(2x + 1)}_{(\nabla \times \mathbf{F}) \cdot \mathbf{n}} \underbrace{\sqrt{2}\, dA}_{dS},$$

where R is the disk of radius 2, centered at the origin (i.e., the projection of S onto the xy-plane). Introducing polar coordinates, we have

$$\begin{aligned}\int_C \mathbf{F}\cdot d\mathbf{r} &= \iint_R (2x+1)\,dA = \int_0^{2\pi}\int_0^2 (2r\cos\theta+1)\,r\,dr\,d\theta \\ &= \int_0^{2\pi}\int_0^2 (2r^2\cos\theta + r)\,dr\,d\theta \\ &= \int_0^{2\pi}\left(2\frac{r^3}{3}\cos\theta + \frac{r^2}{2}\right)\Big|_{r=0}^{r=2}\,d\theta \\ &= \int_0^{2\pi}\left(\frac{16}{3}\cos\theta + 2\right)d\theta = 4\pi,\end{aligned}$$

where we have left the final details of the calculation to you. ■

EXAMPLE 8.2 Using Stokes' Theorem to Evaluate a Surface Integral

Evaluate $\iint_S (\nabla\times\mathbf{F})\cdot\mathbf{n}\,dS$, where $\mathbf{F}(x,y,z) = \langle e^{z^2}, 4z-y, 8x\sin y\rangle$ and where S is the portion of the paraboloid $z = 4-x^2-y^2$ above the xy-plane, oriented so that the unit normal vectors point to the outside of the paraboloid, as indicated in Figure 14.55.

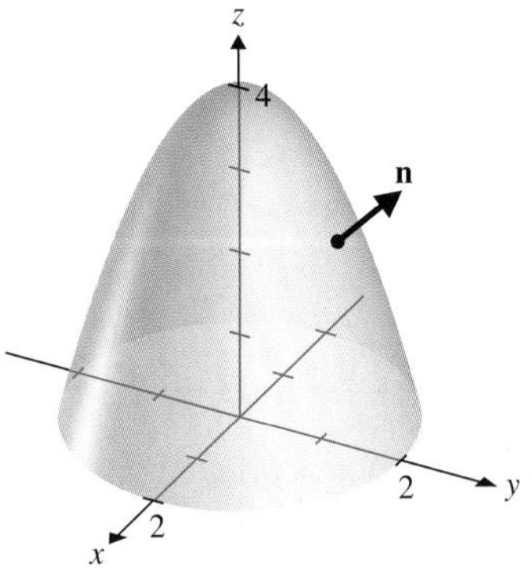

FIGURE 14.55
$z = 4 - x^2 - y^2$

Solution Notice that the boundary curve is simply the circle $x^2+y^2=4$ lying in the xy-plane. By Stokes' Theorem, we then have

$$\begin{aligned}\iint_S (\nabla\times\mathbf{F})\cdot\mathbf{n}\,dS &= \int_{\partial S}\mathbf{F}(x,y,z)\cdot d\mathbf{r} \\ &= \int_{\partial S} e^{z^2}dx + (4z-y)\,dy + 8x\sin y\,dz.\end{aligned}$$

Now, we can parameterize ∂S by $x = 2\cos t$, $y = 2\sin t$, $z=0$, $0\le t\le 2\pi$. This says that on ∂S, we have $dx = -2\sin t$, $dy = 2\cos t$ and $dz = 0$. In view of this, we have

$$\begin{aligned}\iint_S (\nabla\times\mathbf{F})\cdot\mathbf{n}\,dS &= \int_{\partial S} e^{z^2}dx + (4z-y)\,dy + 8x\sin y\,dz \\ &= \int_0^{2\pi}\{e^0(-2\sin t) + [4(0) - 2\sin t](2\cos t)\}\,dt = 0,\end{aligned}$$

where we leave the (straightforward) details of the calculation to you. ■

In example 8.3, we consider the same surface integral as in example 8.2, but over a different surface. Although the surfaces are different, they have the same boundary curve, so that they must have the same value.

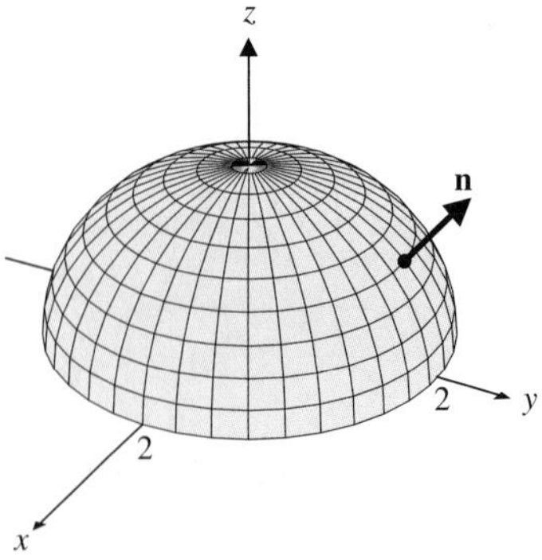

FIGURE 14.56
$z = \sqrt{4-x^2-y^2}$

EXAMPLE 8.3 Using Stokes' Theorem to Evaluate a Surface Integral

Evaluate $\iint_S (\nabla\times\mathbf{F})\cdot\mathbf{n}\,dS$, where $\mathbf{F}(x,y,z) = \langle e^{z^2}, 4z-y, 8x\sin y\rangle$ and where S is the hemisphere $z = \sqrt{4-x^2-y^2}$, oriented so that the unit normal vectors point to the outside of the hemisphere, as indicated in Figure 14.56.

Solution Notice that although this is not the same surface as in example 8.2, the two surfaces have the same boundary curve, the circle $x^2 + y^2 = 4$ lying in the xy-plane, and the same orientation. Just as in example 8.2, we then have

$$\iint_S (\nabla \times \mathbf{F}) \cdot \mathbf{n}\, dS = 0.$$ ■

Much as we used the Divergence Theorem in section 14.7 to give an interpretation of the meaning of the divergence of a vector field, we can use Stokes' Theorem to give some meaning to the curl of a vector field. Suppose once again that $\mathbf{F}(x, y, z)$ represents the velocity field for a fluid in motion and let C be an oriented closed curve in the domain of $\mathbf{F}$, traced out by the endpoint of the vector-valued function $\mathbf{r}(t)$ for $a \le t \le b$. Notice that the closer the direction of $\mathbf{F}$ is to the direction of $\frac{d\mathbf{r}}{dt}$, the larger its component is in the direction of $\frac{d\mathbf{r}}{dt}$. (See Figure 14.57.) In other words, the closer the direction of $\mathbf{F}$ is to the direction of $\frac{d\mathbf{r}}{dt}$, the larger $\mathbf{F} \cdot \frac{d\mathbf{r}}{dt}$ will be. Now, recall that $\frac{d\mathbf{r}}{dt}$ points in the direction of the unit tangent vector along C. Then, since

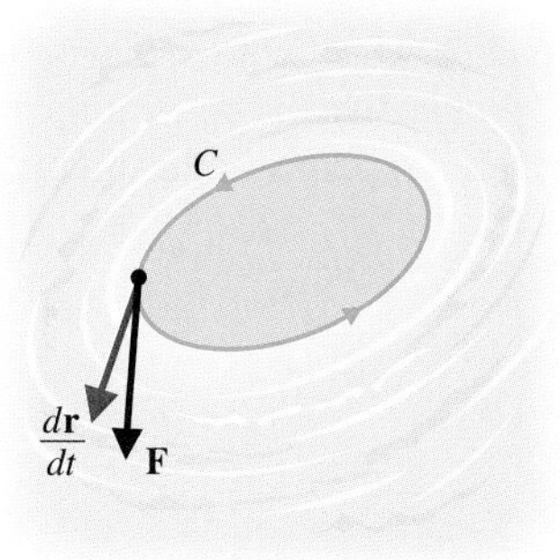

FIGURE 14.57
The surface S in a fluid flow

$$\int_C \mathbf{F} \cdot d\mathbf{r} = \int_a^b \mathbf{F} \cdot \frac{d\mathbf{r}}{dt} dt,$$

it follows that the closer the direction of $\mathbf{F}$ is to the direction of $\frac{d\mathbf{r}}{dt}$ along C, the larger $\int_C \mathbf{F} \cdot d\mathbf{r}$ will be. This says that $\int_C \mathbf{F} \cdot d\mathbf{r}$ measures the tendency of the fluid to flow around or *circulate* around C. For this reason, we refer to $\int_C \mathbf{F} \cdot d\mathbf{r}$ as the **circulation** of $\mathbf{F}$ around C.

For any point (x_0, y_0, z_0) in the fluid flow, let S_a be a disk of radius a centered at (x_0, y_0, z_0), with unit normal vector $\mathbf{n}$, as indicated in Figure 14.58 and let C_a be the (positively oriented) boundary of S_a. Then, by Stokes' Theorem, we have

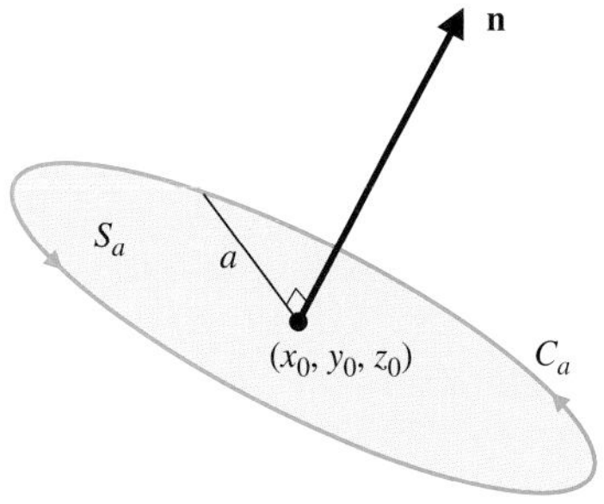

FIGURE 14.58
The disk S_a

$$\int_{C_a} \mathbf{F} \cdot d\mathbf{r} = \iint_{S_a} (\nabla \times \mathbf{F}) \cdot \mathbf{n}\, dS. \tag{8.9}$$

Notice that the average value of a function f on the surface S_a is given by

$$f_{\text{ave}} = \frac{1}{\pi a^2} \iint_{S_a} f(x, y, z)\, dS.$$

Further, if f is continuous on S_a, there must be some point P_a on S_a at which f equals its average value, that is, where

$$f(P_a) = \frac{1}{\pi a^2} \iint_{S_a} f(x, y, z)\, dS.$$

In particular, if the velocity field $\mathbf{F}$ has continuous first partial derivatives throughout S_a, then it follows from equation (8.9) that for some point P_a on S_a,

$$(\nabla \times \mathbf{F})(P_a) \cdot \mathbf{n} = \frac{1}{\pi a^2} \iint_{S_a} (\nabla \times \mathbf{F}) \cdot \mathbf{n}\, dS = \frac{1}{\pi a^2} \int_{C_a} \mathbf{F} \cdot d\mathbf{r}. \tag{8.10}$$

TODAY IN MATHEMATICS

Cathleen Synge Morawetz (1923–) a Canadian mathematician whose work on transonic flows greatly influenced the design of air foils. With similar methods, she made fundamental contributions to a variety of physically important questions about wave propagation. Her mother was a mathematics teacher for a few years and her father, John Lighton Synge, was Ireland's most distinguished mathematician of the twentieth century. However, her mathematics career was almost derailed by World War II and a culture in which she felt that, "It really was considered very bad form for a woman to be overly ambitious." Fortunately, she overcame both social and mathematical obstacles with a work ethic that would not let a problem go until it was solved. She has served as Director of the Courant Institute and President of the American Mathematical Society, and is a winner of the United States' National Medal of Science.

Notice that the expression on the far right of (8.10) corresponds to the circulation of **F** around C_a per unit area. Taking the limit as $a \to 0$, we have by the continuity of curl **F** that

$$(\nabla \times \mathbf{F})(x_0, y_0, z_0) \cdot \mathbf{n} = \lim_{a \to 0} \frac{1}{\pi a^2} \int_{C_a} \mathbf{F} \cdot d\mathbf{r}. \tag{8.11}$$

Read equation (8.11) very carefully. Notice that it says that at any given point, the component of curl **F** in the direction of **n** is the limiting value of the circulation per unit area around circles of radius a centered at that point (and normal to **n**), as the radius a tends to zero. In this sense, $(\nabla \times \mathbf{F}) \cdot \mathbf{n}$ measures the tendency of the fluid to rotate about an axis aligned with the vector **n**. You can visualize this by thinking of a small paddle wheel with axis parallel to **n**, which is immersed in the fluid flow. (See Figure 14.59.) Notice that the circulation per unit area is greatest (so that the paddle wheel moves fastest) when **n** points in the direction of $\nabla \times \mathbf{F}$.

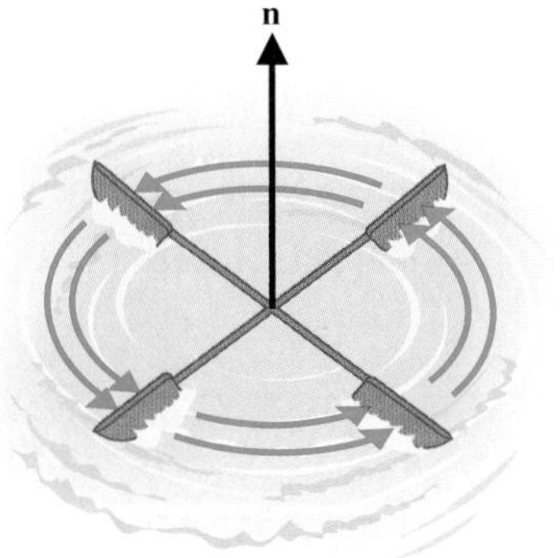

FIGURE 14.59
Paddle wheel

If $\nabla \times \mathbf{F} = \mathbf{0}$ at every point in a fluid flow, we say that the flow is **irrotational,** since the circulation about every point is zero. In particular, notice that if the velocity field **F** is a constant vector throughout the fluid flow, then

$$\text{curl } \mathbf{F} = \nabla \times \mathbf{F} = \mathbf{0},$$

everywhere in the fluid flow and so, the flow is irrotational. Physically, this says that there are no eddies in such a flow.

Notice, too, that by Stokes' Theorem, if curl $\mathbf{F} = \mathbf{0}$ at every point in some open region D, then we must have that for every simple closed curve C that is the boundary of an oriented surface contained in D,

$$\oint_C \mathbf{F} \cdot d\mathbf{r} = 0.$$

In other words, the circulation is zero around every such curve C lying in the region D. It turns out that by suitably restricting the type of regions $D \subset \mathbb{R}^3$ we consider, we can show that the circulation is zero around every simple closed curve contained in D. (The converse of this is also true. That is, if $\oint_C \mathbf{F} \cdot d\mathbf{r} = 0$, for every simple closed curve C contained in the region D, then we must have that curl $\mathbf{F} = \mathbf{0}$ at every point in D.) To obtain this result, we consider regions in space that are simply-connected. Recall that in the plane a region is said to be simply-connected whenever every closed curve contained in the region encloses only points in the region (that is, the region contains no holes). In three dimensions, the situation is slightly more complicated. A region D in $\mathbb{R}^3$ is called **simply-connected** whenever every simple closed curve C lying in D can be continuously shrunk to a point without crossing the boundary of D. For instance, notice that the interior of a sphere or a rectangular box is simply-connected, but a region with a hole drilled through it is not simply-connected. Be careful not to confuse connected with simply-connected. Recall that a connected region is one where every two points contained in the region can be connected with a path that is completely contained in the region. We illustrate connected and simply-connected two-dimensional regions in Figures 14.60a to 14.60c. We can now state the complete theorem.

FIGURE 14.60a
Connected and simply-connected

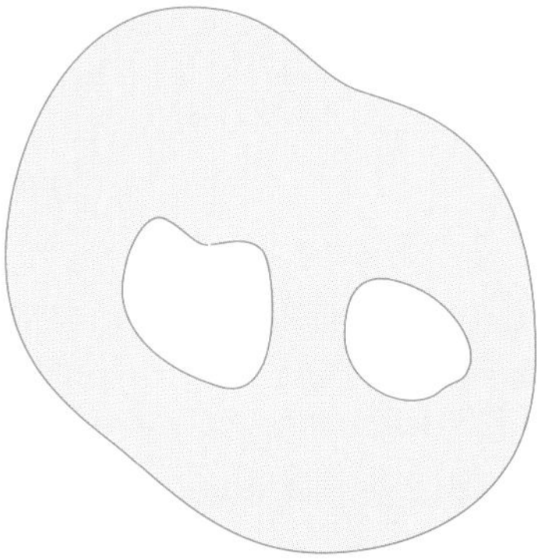

FIGURE 14.60b
Connected but not simply-connected

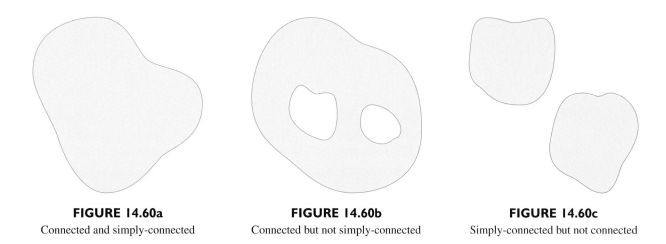

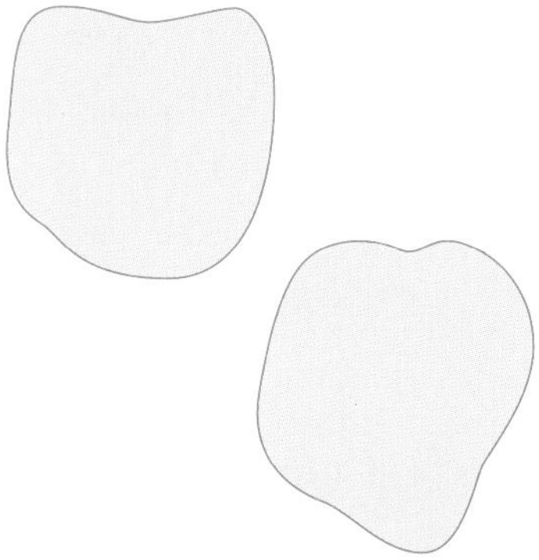

FIGURE 14.60c
Simply-connected but not connected

THEOREM 8.2

Suppose that $\mathbf{F}(x, y, z)$ is a vector field whose components have continuous first partial derivatives throughout the simply-connected open region $D \subset \mathbb{R}^3$. Then, curl $\mathbf{F} = \mathbf{0}$ in D if and only if $\oint_C \mathbf{F} \cdot d\mathbf{r} = 0$, for every simple closed curve C contained in the region D.

PROOF (Necessity)

We have already suggested that when curl $\mathbf{F} = \mathbf{0}$ in an open, simply-connected region D, it can be shown that $\oint_C \mathbf{F} \cdot d\mathbf{r} = 0$, for every simple closed curve C contained in the region D (although the proof is beyond the level of this text). Conversely, suppose now that $\oint_C \mathbf{F} \cdot d\mathbf{r} = 0$ for every simple closed curve C contained in the region D and assume that curl $\mathbf{F} \neq \mathbf{0}$ at some point $(x_0, y_0, z_0) \in D$. Since the components of $\mathbf{F}$ have continuous first partial derivatives, curl $\mathbf{F}$ must be continuous in D and so, there must be a sphere of radius $a_0 > 0$, contained in D and centered at (x_0, y_0, z_0), throughout whose interior S, curl $\mathbf{F} \neq \mathbf{0}$ and curl $\mathbf{F}\,(x, y, z) \cdot \operatorname{curl} \mathbf{F}(x_0, y_0, z_0) > 0$. (Note that this is possible since curl $\mathbf{F}$ is continuous and curl $\mathbf{F}\,(x_0, y_0, z_0) \cdot \operatorname{curl} \mathbf{F}(x_0, y_0, z_0) > 0$.) Let S_a be the disk of radius $a < a_0$ centered at (x_0, y_0, z_0) and oriented by the unit normal vector $\mathbf{n}$ having the same direction as curl $\mathbf{F}\,(x_0, y_0, z_0)$. Notice that since $a < a_0$, S_a will be completely contained in S as illustrated in Figure 14.61. If C_a is the boundary of S_a, then we have by Stokes' Theorem that

$$\int_{C_a} \mathbf{F} \cdot d\mathbf{r} = \iint_{S_a} (\nabla \times \mathbf{F}) \cdot \mathbf{n}\, dS > 0,$$

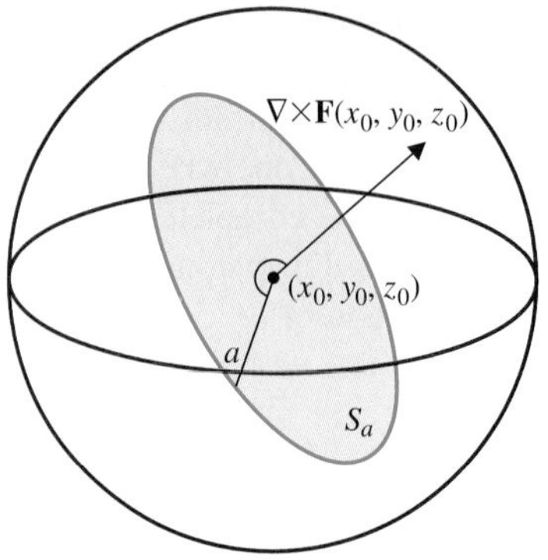

FIGURE 14.61
The disk S_a

since $\mathbf{n}$ was chosen to be parallel to $\nabla \times \mathbf{F}(x_0, y_0, z_0)$. This contradicts the assumption that $\oint_C \mathbf{F} \cdot d\mathbf{r} = 0$, for every simple closed curve C contained in the region D. It now follows that curl $\mathbf{F} = \mathbf{0}$ throughout D. ■

Recall that we had observed earlier that a vector field is conservative in a given region if and only if $\oint_C \mathbf{F} \cdot d\mathbf{r} = 0$, for every simple closed curve C contained in the region. Theorem 8.2 has then established the following results.

THEOREM 8.3

Suppose that $\mathbf{F}(x, y, z)$ has continuous first partial derivatives in a simply-connected region D. Then, the following statements are equivalent.

(i) $\mathbf{F}$ is conservative in D. That is, for some scalar function $f(x, y, z)$, $\mathbf{F} = \nabla f$;
(ii) $\int_C \mathbf{F} \cdot d\mathbf{r}$ is independent of path in D;
(iii) $\mathbf{F}$ is irrotational (i.e., curl $\mathbf{F} = \mathbf{0}$) in D; and
(iv) $\oint_C \mathbf{F} \cdot d\mathbf{r} = 0$, for every simple closed curve C contained in D.

We close this section with a simple application of Stokes' Theorem.

EXAMPLE 8.4 Finding the Flux of a Magnetic Field

Use Stokes' Theorem and Maxwell's equation $\nabla \cdot \mathbf{B} = 0$ to show that the flux of a magnetic field $\mathbf{B}$ across a surface S satisfying the hypotheses of Stokes' Theorem equals the circulation of $\mathbf{A}$ around ∂S, where $\mathbf{B} = \nabla \times \mathbf{A}$.

Solution The flux of $\mathbf{B}$ across S is given by $\iint_S \mathbf{B} \cdot \mathbf{n}\, dS$. Since $\nabla \cdot \mathbf{B} = 0$, it follows from exercise 66 in section 14.5 that there exists a vector field $\mathbf{A}$ such that $\mathbf{B} = \nabla \times \mathbf{A}$. We can now rewrite the flux of $\mathbf{B}$ across S as $\iint_S (\nabla \times \mathbf{A}) \cdot \mathbf{n}\, dS$. Applying Stokes' Theorem gives us

$$\iint_S \mathbf{B} \cdot \mathbf{n}\, dS = \iint_S (\nabla \times \mathbf{A}) \cdot \mathbf{n}\, dS = \oint_{\partial S} \mathbf{A} \cdot d\mathbf{r}.$$

You should recognize the line integral on the right side as the circulation of $\mathbf{A}$ around ∂S, as desired. ■

EXERCISES 14.8

WRITING EXERCISES

1. Describe circumstances (e.g., example 8.1) in which the surface integral of Stokes' Theorem will be simpler than the line integral.

2. Describe circumstances (e.g., example 8.2) in which the line integral of Stokes' Theorem will be simpler than the surface integral.

3. The surfaces in example 8.2 and 8.3 have the same boundary curve C. Explain why all surfaces above the xy-plane with the boundary C will share the same value of

$$\iint_S (\nabla \times \mathbf{F}) \cdot \mathbf{n}\, dS.$$

What would change if the surface were below the xy-plane?

4. Explain why part (iv) of Theorem 8.3 follows immediately from part (ii). Explain why parts (ii) and (iii) follow immediately from part (i).

In exercises 1–4, verify Stokes' Theorem by computing both integrals.

1. S is the portion of $z = 4 - x^2 - y^2$ above the xy-plane, $\mathbf{F} = \langle zx, 2y, z^3 \rangle$.

2. S is the portion of $z = 1 - x^2 - y^2$ above the xy-plane, $\mathbf{F} = \langle x^2 z, xy, xz^2 \rangle$.

3. S is the portion of $z = \sqrt{4 - x^2 - y^2}$ above the xy-plane, $\mathbf{F} = \langle 2x - y, yz^2, y^2 z \rangle$.

4. S is the portion of $z = \sqrt{1 - x^2 - y^2}$ above the xy-plane, $\mathbf{F} = \langle 2x, z^2 - x, xz^2 \rangle$.

In exercises 5–14, use Stokes' Theorem to compute $\iint_S (\nabla \times \mathbf{F}) \cdot \mathbf{n}\, dS$.

5. S is the portion of the tetrahedron bounded by $x + y + 2z = 2$ and the coordinate planes with $z > 0$, $\mathbf{n}$ upward, $\mathbf{F} = \langle zy^4 - y^2, y - x^3, z^2 \rangle$.

6. S is the portion of the tetrahedron bounded by $x + y + 4z = 8$ and the coordinate planes with $z > 0$, $\mathbf{n}$ upward, $\mathbf{F} = \langle y^2, y + 2x, z^2 \rangle$.

7. S is the portion of $z = 1 - x^2 - y^2$ above the xy-plane with $\mathbf{n}$ upward, $\mathbf{F} = \left\langle zx^2, ze^{xy^2} - x, x \ln y^2 \right\rangle$.

8. S is the portion of $z = \sqrt{4 - x^2 - y^2}$ above the xy-plane with $\mathbf{n}$ upward, $\mathbf{F} = \left\langle zx^2, ze^{xy^2} - x, x \ln y^2 \right\rangle$.

9. S is the portion of the tetrahedron in exercise 5 with $y > 0$, $\mathbf{n}$ to the right, $\mathbf{F} = \langle zy^4 - y^2, y - x^3, z^2 \rangle$.

10. S is the portion of $y = x^2 + z^2$ with $y \le 2$, $\mathbf{n}$ to the left, $\mathbf{F} = \left\langle xy, 4xe^{z^2}, yz + 1 \right\rangle$.

11. S is the portion of the unit cube $0 \le x \le 1, 0 \le y \le 1, 0 \le z \le 1$ with $z < 1$, $\mathbf{n}$ upward, $\mathbf{F} = \langle xyz, 4x^2 y^3 - z, 8 \cos xz^2 \rangle$.

12. S is the portion of the unit cube $0 \le x \le 1, 0 \le y \le 1, 0 \le z \le 1$ with $z < 1$, $\mathbf{n}$ downward, $\mathbf{F} = \langle xyz, 4x^2 y^3 - z, 8 \cos xz^2 \rangle$.

13. S is the portion of the cone $z = \sqrt{x^2 + y^2}$ below the sphere $x^2 + y^2 + z^2 = 2$, $\mathbf{n}$ downward, $\mathbf{F} = \left\langle x^2 + y^2, ze^{x^2+y^2}, e^{x^2+z^2} \right\rangle$.

14. S is the portion of the cone $z = \sqrt{x^2 + y^2}$ inside the cylinder $x^2 + y^2 = 2$, $\mathbf{n}$ downward, $\mathbf{F} = \left\langle zx, x^2 + y^2, z^2 - y^2 \right\rangle$.

In exercises 15–24, use Stokes' Theorem to evaluate $\int_C \mathbf{F} \cdot d\mathbf{r}$.

15. C is the boundary of the portion of the paraboloid $y = 4 - x^2 - z^2$ with $y > 0$, $\mathbf{n}$ to the right, $\mathbf{F} = \langle x^2 z, 3 \cos y, 4z^3 \rangle$.

16. C is the boundary of the portion of the paraboloid $x = y^2 + z^2$ with $x \le 4$, $\mathbf{n}$ to the back, $\mathbf{F} = \langle yz, y - 4, 2xy \rangle$.

17. C is the boundary of the portion of $z = 4 - x^2 - y^2$ above the xy-plane, oriented upward, $\mathbf{F} = \left\langle x^2 e^x - y, \sqrt{y^2 + 1}, z^3 \right\rangle$.

18. C is the boundary of the portion of $z = x^2 + y^2$ below $z = 4$, oriented downward, $\mathbf{F} = \langle x^2, y^4 - x, z^2 \sin z \rangle$.

19. C is the intersection of $z = x^2 + y^2$ and $z = 8 - y$, oriented clockwise from above, $\mathbf{F} = \left\langle 2x^2, 4y^2, e^{8z^2} \right\rangle$.

20. C is the intersection of $x^2 + y^2 = 1$ and $z = x - y$, oriented clockwise from above, $\mathbf{F} = \langle \cos x^2, \sin y^2, \tan z^2 \rangle$.

21. C is the triangle from $(0, 1, 0)$ to $(0, 0, 4)$ to $(2, 0, 0)$, $\mathbf{F} = \langle x^2 + 2xy^3 z, 3x^2 y^2 z - y, x^2 y^3 \rangle$.

22. C is the square from $(0, 2, 2)$ to $(2, 2, 2)$ to $(2, 2, 0)$ to $(0, 2, 0)$, $\mathbf{F} = \langle x^2, y^3 + x, 3y^2 \cos z \rangle$.

23. C is the intersection of $z = 4 - x^2 - y^2$ and $x^2 + z^2 = 1$ with $y > 0$, oriented clockwise as viewed from the right, $\mathbf{F} = \left\langle x^2 + 3y, \cos y^2, z^3 \right\rangle$.

24. C is the intersection of $z = x^2 + y^2 - 4$ and $z = y - 1$, oriented clockwise as viewed from above, $\mathbf{F} = \langle \sin x^2, y^3, z \ln z - x \rangle$.

25. Show that $\oint_C (f\,\nabla f)\cdot d\mathbf{r} = 0$ for any simple closed curve C and differentiable function f.

26. Show that $\oint_C (f\,\nabla g + g\nabla f)\cdot d\mathbf{r} = 0$ for any simple closed curve C and differentiable functions f and g.

27. Let $\mathbf{F}(x, y) = \langle M(x, y), N(x, y)\rangle$ be a vector field whose components M and N have continuous first partial derivatives in all of $\mathbb{R}^2$. Show that $\nabla\cdot\mathbf{F} = 0$ if and only if $\int_C \mathbf{F}\cdot\mathbf{n}\,ds = 0$ for all simple closed curves C. (Hint: Use a vector form of Green's Theorem.)

28. Under the assumptions of exercise 27, show that $\int_C \mathbf{F}\cdot\mathbf{n}\,ds = 0$ for all simple closed curves C if and only if $\int_C \mathbf{F}\cdot\mathbf{n}\,ds$ is path-independent.

29. Under the assumptions of exercise 27, show that $\nabla\cdot\mathbf{F} = 0$ if and only if $\mathbf{F}$ has a **stream function** $g(x, y)$ such that $M(x, y) = g_y(x, y)$ and $N(x, y) = -g_x(x, y)$.

30. Combine the results of exercises 27–29 to state a two-variable theorem analogous to Theorem 8.3.

31. If S_1 and S_2 are surfaces that satisfy the hypotheses of Stokes' Theorem and that share the same boundary curve, under what circumstances can you conclude that

$$\iint_{S_1} (\nabla\times\mathbf{F})\cdot\mathbf{n}\,dS = \iint_{S_2} (\nabla\times\mathbf{F})\cdot\mathbf{n}\,dS?$$

32. Give an example where the two surface integrals of exercise 31 are not equal.

33. Use Stokes' Theorem to verify that

$$\oint_C (f\nabla g)\cdot d\mathbf{r} = \iint_S (\nabla f\times\nabla g)\cdot\mathbf{n}\,dS,$$

where C is the positively oriented boundary of the surface S.

34. Use Stokes' Theorem to verify that $\oint_C (f\nabla g + g\nabla f)\cdot d\mathbf{r} = 0$, where C is the positively oriented boundary of some surface S.

EXPLORATORY EXERCISES

1. The **circulation** of a vector field $\mathbf{F}$ around the curve C is defined by $\int_C \mathbf{F}\cdot d\mathbf{r}$. Show that the curl $\nabla\times\mathbf{F}(0, 0, 0)$ is in the same direction as the normal to the plane in which the circulation per unit area around the origin is a maximum as the area around the origin goes to 0. Relate this to the interpretation of the curl given in section 14.5.

2. The Fundamental Theorem of Calculus can be viewed as relating the values of the function on the boundary of a region (interval) to the sum of the derivative values of the function within the region. Explain what this statement means and then explain why the same statement can be applied to Theorem 3.2, Green's Theorem, the Divergence Theorem and Stokes' Theorem. In each case, carefully state what the "region" is, what its boundary is and what type derivative is involved.

14.9 APPLICATIONS OF VECTOR CALCULUS

Through the past eight sections, we have developed a powerful set of tools for analyzing vector quantities. You can now compute flux integrals and line integrals for work and circulation, and you have the Divergence Theorem and Stokes' Theorem to relate these quantities to one another. To this point in the text, we have emphasized the conceptual and computational aspects of vector analysis. In this section, we present a small selection of applications from fluid mechanics and electricity and magnetism. As you work through the examples in this section, notice that we are using vector calculus to derive general results that can be applied to any specific vector field that you may run across in an application.

Our first example is similar to example 7.4, which concerns magnetic fields. Here, we also apply Stokes' Theorem to derive a second result.

EXAMPLE 9.1 Finding the Flux of a Velocity Field

Suppose that the velocity field $\mathbf{v}$ of a fluid has a vector potential $\mathbf{w}$, that is, $\mathbf{v} = \nabla\times\mathbf{w}$. Show that $\mathbf{v}$ is incompressible and that the flux of $\mathbf{v}$ across any closed surface is 0. Also, show that if a closed surface S is partitioned into surfaces S_1 and S_2 (that is, $S = S_1\cup S_2$ and $S_1\cap S_2 = \emptyset$), then the flux of $\mathbf{v}$ across S_1 is the additive inverse of the flux of $\mathbf{v}$ across S_2.

Solution To show that $\mathbf{v}$ is incompressible, compute $\nabla \cdot \mathbf{v} = \nabla \cdot (\nabla \times \mathbf{w}) = 0$, since the divergence of the curl of *any* vector field is zero. Next, suppose that the closed surface S is the boundary of the solid Q. Then from the Divergence Theorem, we have

$$\iint_S \mathbf{v} \cdot \mathbf{n}\, dS = \iiint_Q \nabla \cdot \mathbf{v}\, dV = \iiint_Q 0\, dV = 0.$$

Finally, since $S = S_1 \cup S_2$ and $S_1 \cap S_2 = \emptyset$, we have

$$\iint_{S_1} \mathbf{v} \cdot \mathbf{n}\, dS + \iint_{S_2} \mathbf{v} \cdot \mathbf{n}\, dS = \iint_S \mathbf{v} \cdot \mathbf{n}\, dS = 0,$$

so that

$$\iint_{S_1} \mathbf{v} \cdot \mathbf{n}\, dS = -\iint_{S_2} \mathbf{v} \cdot \mathbf{n}\, dS.$$

The general result shown in example 9.1 also has practical implications for computing integrals. One use of this result is given in example 9.2.

EXAMPLE 9.2 Computing a Surface Integral Using the Complement of the Surface

Find the flux of the vector field $\nabla \times \mathbf{F}$ across S, where $\mathbf{F} = \langle e^{x^2} - 2xy, \sin y^2, 3yz - 2x \rangle$ and S is the portion of the cube $0 \le x \le 1, 0 \le y \le 1, 0 \le z \le 1$ above the xy-plane.

Solution We have several options for computing $\iint_S (\nabla \times \mathbf{F}) \cdot \mathbf{n}\, dS$. Notice that the surface S consists of five faces of the cube, so five separate surface integrals would be required to compute it directly. We could use Stokes' Theorem and rewrite it as $\oint_C \mathbf{F} \cdot d\mathbf{r}$, where C is the square boundary of the open face of S. However, this would require four line integrals involving a complicated vector field $\mathbf{F}$. Example 9.1 gives us a third option: the flux over the entire cube is zero, so that the flux over S is the additive inverse of the flux over the missing side of the cube. Notice that the (outward) normal vector for this side is $\mathbf{n} = -\mathbf{k}$, and the curl of $\mathbf{F}$ is given by

$$\nabla \times \mathbf{F} = \begin{vmatrix} \mathbf{i} & \mathbf{j} & \mathbf{k} \\ \dfrac{\partial}{\partial x} & \dfrac{\partial}{\partial y} & \dfrac{\partial}{\partial z} \\ e^{x^2} - 2xy & \sin y^2 & 3yz - 2x \end{vmatrix}$$
$$= \mathbf{i}(3z - 0) - \mathbf{j}(-2 - 0) + \mathbf{k}(0 + 2x) = 3z\mathbf{i} + 2\mathbf{j} + 2x\mathbf{k}.$$

So,

$$(\nabla \times \mathbf{F}) \cdot \mathbf{n} = (3z\mathbf{i} + 2\mathbf{j} + 2x\mathbf{k}) \cdot (-\mathbf{k}) = -2x$$

and $dS = dA$. Taking S_2 as the bottom face of the cube, we now have that the flux is given by

$$\iint_S (\nabla \times \mathbf{F}) \cdot \mathbf{n}\, dS = -\iint_{S_2} (\nabla \times \mathbf{F}) \cdot \mathbf{n}\, dS = -\iint_R -2x\, dA = \int_0^1 \int_0^1 2x\, dx\, dy = 1.$$

One very important use of the Divergence Theorem and Stokes' Theorem is in deriving certain fundamental equations in physics and engineering. The technique we use here to derive the **heat equation** is typical of the use of these theorems. In this technique, we start with two different descriptions of the same quantity and use the vector calculus to draw conclusions about the functions involved.

For the heat equation, we analyze the amount of heat per unit time leaving a solid Q. Recall from example 6.7 that the net heat flow out of Q is given by $\iint_S (-k\nabla T)\cdot \mathbf{n}\, dS$, where S is a closed surface bounding Q, T is the temperature function, $\mathbf{n}$ is the outward unit normal and k is a constant (called the heat conductivity). Alternatively, physics tells us that the total heat within Q equals $\iiint_Q \rho\sigma T\, dV$, where ρ is the (constant) density and σ is the **specific heat** of the solid. From this, it follows that the heat flow out of Q is given by $-\frac{d}{dt}\left[\iiint_Q \rho\sigma T\, dV\right]$. Notice that the negative sign is needed to give us the heat flow *out of* the region Q. If the temperature function T has a continuous partial derivative with respect to t, we can bring the derivative inside the integral and write this as $-\iiint_Q \rho\sigma \frac{\partial T}{\partial t} dV$. Equating these two expressions for the heat flow out of Q, we have

$$\iint_S (-k\nabla T)\cdot \mathbf{n}\, dS = -\iiint_Q \rho\sigma \frac{\partial T}{\partial t}\, dV. \tag{9.1}$$

EXAMPLE 9.3 Deriving the Heat Equation

Use the Divergence Theorem and equation (9.1) to derive the heat equation $\frac{\partial T}{\partial t} = \alpha^2 \nabla^2 T$, where $\alpha^2 = \frac{k}{\rho\sigma}$ and $\nabla^2 T = \nabla\cdot(\nabla T)$ is the Laplacian of T.

Solution Applying the Divergence Theorem to the left-hand side of equation (9.1), we have

$$\iiint_Q \nabla\cdot(-k\nabla T)\, dV = -\iiint_Q \rho\sigma \frac{\partial T}{\partial t}\, dV.$$

Combining the preceding two integrals, we get

$$\begin{aligned} 0 &= \iiint_Q -k\nabla\cdot(\nabla T)\, dV + \iiint_Q \rho\sigma \frac{\partial T}{\partial t}\, dV \\ &= \iiint_Q \left(-k\nabla^2 T + \rho\sigma \frac{\partial T}{\partial t}\right) dV. \end{aligned} \tag{9.2}$$

Observe that the only way for the integral in (9.2) to be zero for *every* solid Q is for the integrand to be zero. (Think about this carefully; you can let Q be a small sphere around any point you like.) That is,

$$0 = -k\nabla^2 T + \rho\sigma \frac{\partial T}{\partial t}$$

or

$$\rho\sigma \frac{\partial T}{\partial t} = k\nabla^2 T.$$

Finally, dividing both sides by $\rho\sigma$ gives us

$$\frac{\partial T}{\partial t} = \frac{k}{\rho\sigma}\nabla^2 T = \alpha^2 \nabla^2 T,$$

as desired. ■

We next derive a fundamental result in the study of fluid dynamics, diffusion theory and electricity and magnetism. We consider a fluid that has density function ρ (in general, ρ is a scalar function of space and time). We also assume that the fluid has velocity field $\mathbf{v}$ and that there are no sources or sinks. Since the total mass of fluid contained in a given region Q is given by the triple integral $m = \iiint_Q \rho(x, y, z, t)\, dV$, the rate of change of the mass is given by

$$\frac{dm}{dt} = \frac{d}{dt}\left[\iiint_Q \rho(x, y, z, t)\, dV\right] = \iiint_Q \frac{\partial \rho}{\partial t}(x, y, z, t)\, dV, \tag{9.3}$$

assuming that the density function has a continuous partial derivative with respect to t, so that we can bring the derivative inside the integral. Now, look at the same problem in a different way. In the absence of sources or sinks, the only way for the mass inside Q to change is for fluid to cross the boundary ∂Q. That is, the rate of change of mass is the additive inverse of the flux of the velocity field across the boundary of Q. (You will be asked in the exercises to explain the negative sign. Think about why it needs to be there!) So, we also have

$$\frac{dm}{dt} = -\iint_{\partial Q} (\rho\mathbf{v}) \cdot \mathbf{n}\, dS. \tag{9.4}$$

Given these alternative representations of the rate of change of mass, we derive the ***continuity equation*** in example 9.4.

EXAMPLE 9.4 Deriving the Continuity Equation

Use the Divergence Theorem and equations (9.3) and (9.4) to derive the ***continuity equation:*** $\nabla \cdot (\rho\mathbf{v}) + \dfrac{\partial \rho}{\partial t} = 0.$

Solution We start with equal expressions for the rate of change of mass in a generic solid Q, given in (9.3) and (9.4). We have

$$\iiint_Q \frac{\partial \rho}{\partial t}(x, y, z, t)\, dV = -\iint_{\partial Q} (\rho\mathbf{v}) \cdot \mathbf{n}\, dS.$$

Applying the Divergence Theorem to the right-hand side gives us

$$\iiint_Q \frac{\partial \rho}{\partial t}(x, y, z, t)\, dV = -\iiint_Q \nabla \cdot (\rho\mathbf{v})\, dV.$$

Combining the two integrals, we have

$$0 = \iiint_Q \nabla \cdot (\rho \mathbf{v})\, dV + \iiint_Q \frac{\partial \rho}{\partial t}(x, y, z, t)\, dV$$
$$= \iiint_Q \left[\nabla \cdot (\rho \mathbf{v}) + \frac{\partial \rho}{\partial t} \right] dV.$$

Since this equation must hold for *all* solids Q, the integrand must be zero. That is,

$$\nabla \cdot (\rho \mathbf{v}) + \frac{\partial \rho}{\partial t} = 0,$$

which is the continuity equation, as desired. ■

Bernoulli's Theorem is often used to explain the lift force of a curved airplane wing. This result relates the speed and pressure in a steady fluid flow. (Here, steady means that the fluid's velocity, pressure etc., do not change with time.) The starting point for our derivation is Euler's equation for steady flow. In this case, a fluid moves with velocity $\mathbf{u}$ and vorticity $\mathbf{w}$ through a medium with density ρ and the speed is given by $u = \|\mathbf{u}\|$. We consider the case where there is an external force, such as gravity, with a potential function ϕ and where the fluid pressure is given by the scalar function p. Since the flow is steady, all quantities are functions of position (x, y, z), but not time. In this case, **Euler's equation** states that

$$\mathbf{w} \times \mathbf{u} + \frac{1}{2}\nabla u^2 = -\frac{1}{\rho}\nabla p - \nabla\phi. \tag{9.5}$$

Bernoulli's Theorem then says that $\frac{1}{2}u^2 + \phi + \frac{p}{\rho}$ is constant along flow lines. A more precise formula is given in the derivation in example 9.5.

EXAMPLE 9.5 Deriving Bernoulli's Theorem

Use Euler's equation (9.5) to derive Bernoulli's Theorem.

Solution Recall that the flow lines are tangent to the velocity field. So, to compute the component of a vector function along a flow line, you start by finding the dot product of the function with velocity. In this case, we take Euler's equation and find the dot product of each term with $\mathbf{u}$. We get

$$\mathbf{u} \cdot (\mathbf{w} \times \mathbf{u}) + \mathbf{u} \cdot \left(\frac{1}{2}\nabla u^2\right) = -\mathbf{u} \cdot \left(\frac{1}{\rho}\nabla p\right) - \mathbf{u} \cdot \nabla\phi$$

or

$$\mathbf{u} \cdot (\mathbf{w} \times \mathbf{u}) + \mathbf{u} \cdot \left(\frac{1}{2}\nabla u^2\right) + \mathbf{u} \cdot (\nabla\phi) + \mathbf{u} \cdot \frac{1}{\rho}\nabla p = 0.$$

Notice that $\mathbf{u} \cdot (\mathbf{w} \times \mathbf{u}) = 0$, since the cross product $\mathbf{w} \times \mathbf{u}$ is perpendicular to $\mathbf{u}$. All three remaining terms are gradients, so factoring out the scalar functions involved, we have

$$\mathbf{u} \cdot \nabla \left(\frac{1}{2}u^2 + \phi + \frac{p}{\rho}\right) = 0.$$

This says that the component of $\nabla\left(\frac{1}{2}u^2 + \phi + \frac{p}{\rho}\right)$ along $\mathbf{u}$ is zero. So, the directional derivative of $\frac{1}{2}u^2 + \phi + \frac{p}{\rho}$ is zero in the direction of the tangent to the flow lines. This gives us Bernoulli's Theorem, that $\frac{1}{2}u^2 + \phi + \frac{p}{\rho}$ is constant along flow lines. ■

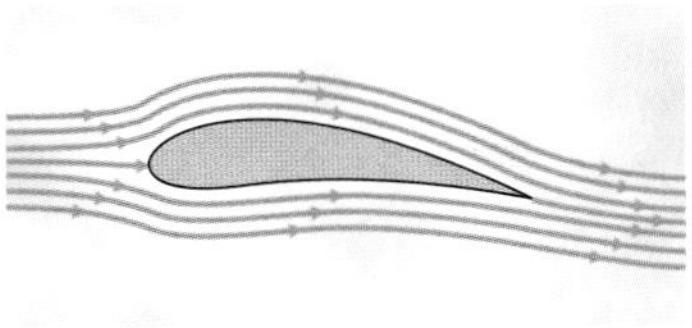

FIGURE 14.62
Cross section of a wing

Consider now what Bernoulli's Theorem means in the case of steady airflow around an airplane wing. Since the wing is curved on top (see Figure 14.62), the air flowing across the top must have a greater speed. From Bernoulli's Theorem, the quantity $\frac{1}{2}u^2 + \phi + \frac{p}{\rho}$ is constant along flow lines, so an increase in speed must be compensated for by a decrease in pressure. Due to the lower pressure on top, the wing experiences a lift force. Of course, airflow around an airplane wing is more complicated than this. The interaction of the air with the wing itself (the boundary layer) is quite complicated and determines many of the flight characteristics of a wing. Still, Bernoulli's Theorem gives us some insight into why a curved wing produces a lift force.

Maxwell's equations are a set of four equations relating the fundamental vector fields of electricity and magnetism. From these equations, you can derive many more important relationships. Taken together, Maxwell's equations give a concise statement of the fundamentals of electricity and magnetism. The equations can be written in different ways, depending on whether the integral or differential form is given and whether magnetic or polarizable media are included. Also, you may find that different texts refer to these equations by different names. Listed below are Maxwell's equations in differential form in the absence of magnetic or polarizable media.

MAXWELL'S EQUATIONS

$$\nabla \cdot \mathbf{E} = \frac{\rho}{\epsilon_0} \quad \text{(Gauss' Law for electricity)}$$

$$\nabla \cdot \mathbf{B} = 0 \quad \text{(Gauss' Law for magnetism)}$$

$$\nabla \times \mathbf{E} = -\frac{\partial \mathbf{B}}{\partial t} \quad \text{(Faraday's Law of induction)}$$

$$\nabla \times \mathbf{B} = \frac{1}{\epsilon_0 c^2}\mathbf{J} + \frac{1}{c^2}\frac{\partial \mathbf{E}}{\partial t} \quad \text{(Ampere's Law)}$$

In these equations, **E** represents an electrostatic field, **B** is the corresponding magnetic field, ϵ_0 is the permittivity, ρ is the charge density, c is the speed of light and **J** is the current density. In example 9.6, we derive a simplified version of the differential form of Ampere's law. The hypothesis in this case is a common form for Ampere's law: the line integral of a magnetic field around a closed path is proportional to the current enclosed by the path.

EXAMPLE 9.6 Deriving Ampere's Law

In the case where **E** is constant and I represents current, use the relationship $\oint_C \mathbf{B} \cdot d\mathbf{r} = \frac{1}{\epsilon_0 c^2} I$ to derive Ampere's law: $\nabla \times \mathbf{B} = \frac{1}{\epsilon_0 c^2}\mathbf{J}$.

Solution Let S be any capping surface for C, that is, any positively oriented two-sided surface bounded by C. (See Figure 14.63.) The enclosed current I is then related to the current density by $I = \iint_S \mathbf{J} \cdot \mathbf{n}\, dS$. By Stokes' Theorem, we can rewrite the line integral of **B** as

$$\oint_C \mathbf{B} \cdot d\mathbf{r} = \iint_S (\nabla \times \mathbf{B}) \cdot \mathbf{n}\, dS.$$

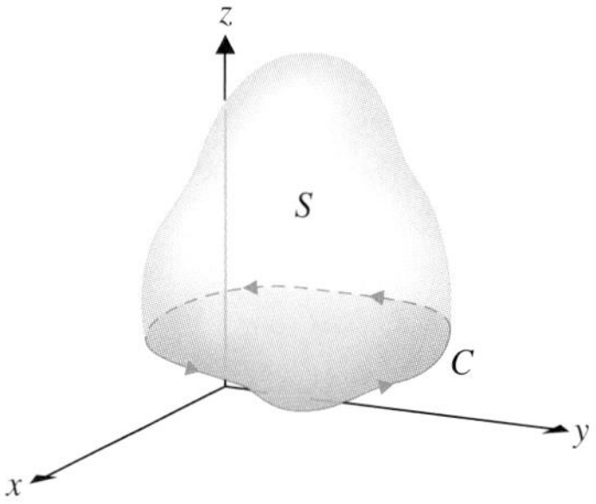

FIGURE 14.63
Positive orientation

Equating the two expressions, we now have

$$\iint_S (\nabla \times \mathbf{B}) \cdot \mathbf{n}\, dS = \frac{1}{\epsilon_0 c^2} \iint_S \mathbf{J} \cdot \mathbf{n}\, dS,$$

from which it follows that

$$\iint_S \left(\nabla \times \mathbf{B} - \frac{1}{\epsilon_0 c^2} \mathbf{J} \right) \cdot \mathbf{n}\, dS = 0.$$

Since this holds for all capping surfaces S, it must be that $\nabla \times \mathbf{B} - \frac{1}{\epsilon_0 c^2}\mathbf{J} = \mathbf{0}$ or $\nabla \times \mathbf{B} = \frac{1}{\epsilon_0 c^2}\mathbf{J}$, as desired. ■

In our final example, we illustrate one of the uses of Faraday's law. In an AC generator, the turning of a coil in a magnetic field produces a voltage. In terms of the electric field $\mathbf{E}$, the voltage generated is given by $\oint_C \mathbf{E} \cdot d\mathbf{r}$, where C is a closed curve. As we see in example 9.7, Faraday's law relates this to the magnetic flux function $\phi = \iint_S \mathbf{B} \cdot \mathbf{n}\, dS$.

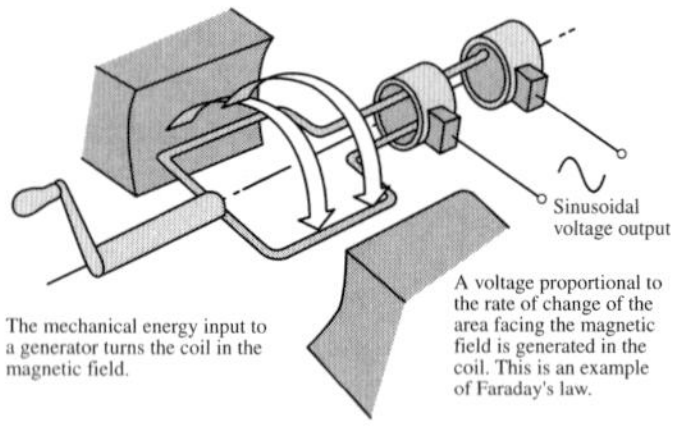
Sinusoidal
voltage output
A voltage proportional to the rate of change of the area facing the magnetic field is generated in the coil. This is an example of Faraday's law.
The mechanical energy input to a generator turns the coil in the magnetic field.

EXAMPLE 9.7 Using Faraday's Law to Analyze the Output of a Generator

An AC generator produces a voltage of $120 \sin(120\pi t)$ volts. Determine the magnetic flux ϕ.

Solution The voltage is given by

$$\oint_C \mathbf{E} \cdot d\mathbf{r} = 120 \sin(120\pi t).$$

Applying Stokes' Theorem to the left-hand side, we have

$$\iint_S (\nabla \times \mathbf{E}) \cdot \mathbf{n}\, dS = \oint_C \mathbf{E} \cdot d\mathbf{r} = 120 \sin(120\pi t).$$

Applying Faraday's law to the left-hand side, we have

$$\iint_S \left(-\frac{\partial \mathbf{B}}{\partial t} \right) \cdot \mathbf{n}\, dS = \iint_S (\nabla \times \mathbf{E}) \cdot \mathbf{n}\, dS = 120 \sin(120\pi t).$$

Assuming that the integrand is continuous and we can bring the derivative outside, we get

$$-\frac{d}{dt} \iint_S (\mathbf{B} \cdot \mathbf{n})\, dS = 120 \sin(120\pi t).$$

Writing this in terms of the magnetic flux ϕ, we have

$$-\frac{d}{dt}\phi = 120 \sin(120\pi t)$$

or

$$\phi'(t) = -120 \sin(120\pi t).$$

Integrating this gives us

$$\phi(t) = \frac{1}{\pi} \cos(120\pi t) + c,$$

for some constant c. ■

EXERCISES 14.9

WRITING EXERCISES

1. Give an example of a fluid with velocity field with zero flux as in example 9.1.
2. Give an example of a fluid with velocity field with nonzero flux.
3. In the derivation of the continuity equation, explain why it is important to assume no sources or sinks.
4. From Bernoulli's Theorem, if all other things are equal and the density ρ increases, in what way does velocity change?

---

1. Rework example 9.2 by computing $\oint_C \mathbf{F} \cdot d\mathbf{r}$.
2. Rework example 9.2 by directly computing $\iint_S (\nabla \times \mathbf{F}) \cdot \mathbf{n}\, dS$.

In exercises 3–8, use Gauss' Law for electricity and the relationship $q = \iiint_Q \rho\, dV$.

3. For $\mathbf{E} = \langle yz, xz, xy \rangle$, find the total charge in the hemisphere $z = \sqrt{R^2 - x^2 - y^2}$.
4. For $\mathbf{E} = \langle 2xy, y^2, 5x \rangle$, find the total charge in the hemisphere $z = \sqrt{R^2 - x^2 - y^2}$.
5. For $\mathbf{E} = \langle 4x - y, 2y + z, 3xy \rangle$, find the total charge in the hemisphere $z = \sqrt{R^2 - x^2 - y^2}$.
6. For $\mathbf{E} = \langle 2xz^2, 2yx^2, 2zy^2 \rangle$, find the total charge in the hemisphere $z = \sqrt{R^2 - x^2 - y^2}$.
7. For $\mathbf{E} = \langle 2xy, y^2, 5xy \rangle$, find the total charge in the cone $z = \sqrt{x^2 + y^2}$ below $z = 4$.
8. For $\mathbf{E} = \langle 4x - y, 2y + z, 3xy \rangle$, find the total charge in the solid bounded by $z = R - x^2 - y^2$ and $z = 0$.
9. Faraday showed that $\oint_C \mathbf{E} \cdot d\mathbf{r} = -\frac{d\phi}{dt}$, where $\phi = \iint_S \mathbf{B} \cdot \mathbf{n}\, dS$, for any capping surface S (that is, any positively oriented open surface with boundary C). Use this to show that $\nabla \times \mathbf{E} = -\frac{\partial \mathbf{B}}{\partial t}$. What mathematical assumption must be made?
10. If an electric field $\mathbf{E}$ is conservative with potential function $-\phi$, use Gauss' Law of electricity to show that **Poisson's equation** must hold: $\nabla^2 \phi = -\frac{\rho}{\epsilon_0}$.
11. Use Maxwell's equation and $\mathbf{J} = \rho \mathbf{v}$ to derive the continuity equation. (Hint: Start by computing $\nabla \cdot \mathbf{J}$.) What mathematical assumption must be made?
12. For a magnetic field $\mathbf{B}$, Maxwell's equation $\nabla \cdot \mathbf{B} = 0$ implies that $\mathbf{B} = \nabla \times \mathbf{A}$ for some vector field $\mathbf{A}$. Show that the flux of $\mathbf{B}$ across an open surface S equals the circulation of $\mathbf{A}$ around the closed curve C, where C is the positively oriented boundary of S.
13. Let I be the current crossing an open surface S, so that $I = \iint_S \mathbf{J} \cdot \mathbf{n}\, dS$. Given that $I = \oint_C \mathbf{B} \cdot d\mathbf{r}$ (where C is the positively oriented boundary of S), show that $\mathbf{J} = \nabla \times \mathbf{B}$.
14. Using the same notation as in exercise 13, start with $I = \iint_S \mathbf{J} \cdot \mathbf{n}\, dS$ and $\mathbf{J} = \nabla \times \mathbf{B}$ and show that $I = \oint_C \mathbf{B} \cdot d\mathbf{r}$.

In exercises 15–18, use the electrostatic force $\mathbf{E} = \frac{q}{4\pi\epsilon_0 r^3}\mathbf{r}$ for a charge q at the origin, where $\mathbf{r} = \langle x, y, z \rangle$ and $r = \sqrt{x^2 + y^2 + z^2}$.

15. If S is a closed surface not enclosing the origin, show that $\iint_S \mathbf{E} \cdot \mathbf{n}\, dS = 0$.
16. If S is the sphere $x^2 + y^2 + z^2 = 1$, show that $\iint_S \mathbf{E} \cdot \mathbf{n}\, dS = \frac{q}{\epsilon_0}$.
17. If S is the sphere $x^2 + y^2 + z^2 = R^2$, show directly that $\iint_S \mathbf{E} \cdot \mathbf{n}\, dS = \frac{q}{\epsilon_0}$.
18. Use exercises 15 and 16 to show that $\iint_S \mathbf{E} \cdot \mathbf{n}\, dS = \frac{q}{\epsilon_0}$, for any closed surface S enclosing the origin.
19. Assume that $\iint_S \mathbf{D} \cdot \mathbf{n}\, dS = q$, for any closed surface S, where $\mathbf{D} = \epsilon_0 \mathbf{E}$ is the electric flux density and q is the charge enclosed by S. Show that $\nabla \cdot \mathbf{D} = Q$, where Q is the charge density satisfying $q = \iiint_R Q\, dV$.
20. The moment of inertia about the z-axis of a solid Q with constant density ρ is $I_z = \iiint_Q (x^2 + y^2)\rho\, dV$. Express this as a surface integral.
21. Let u be a scalar function with continuous second partial derivatives. Define the **normal derivative** $\frac{\partial u}{\partial n} = \nabla u \cdot \mathbf{n}$. Show that $\iint_S \frac{\partial u}{\partial n}\, dS = \iiint_Q \nabla^2 u\, dV$.
22. Suppose that u is a harmonic function (that is, $\nabla^2 u = 0$). Show that $\iint_S \frac{\partial u}{\partial n}\, dS = 0$.
23. If the heat conductivity k is not constant, our derivation of the heat equation is no longer valid. If $k = k(x, y, z)$, show that the heat equation becomes $k\nabla^2 T + \nabla k \cdot \nabla T = \sigma\rho \frac{\partial T}{\partial t}$.

24. If h has continuous partial derivatives and S is a closed surface enclosing a solid Q, show that $\iint_S (h\nabla h) \cdot \mathbf{n}\, dS = \iiint_Q (h\nabla^2 h + \nabla h \cdot \nabla h)\, dV$.

25. Suppose that f and g are both harmonic (that is, $\nabla^2 f = \nabla^2 g = 0$) and $f = g$ on a closed surface S, where S encloses a solid Q. Use the result of exercise 24, with $h = f - g$, to show that $f = g$ in Q.

Review Exercises

WRITING EXERCISES

The following list includes terms that are defined and theorems that are stated in this chapter. For each term or theorem, (1) give a precise definition or statement, (2) state in general terms what it means and (3) describe the types of problems with which it is associated.

Vector field	Velocity field	Flow lines
Gradient field	Potential function	Conservative field
Curl	Divergence	Laplacian
Line integral	Work line integral	Path independence
Green's Theorem	Surface integral	Flux integral
Divergence Theorem	Stokes' Theorem	Heat equation
Continuity equation	Bernoulli's Theorem	Maxwell's equations

TRUE OR FALSE

State whether each statement is true or false and briefly explain why. If the statement is false, try to "fix it" by modifying the given statement to a new statement that is true.

1. The graph of a vector field shows vectors $\mathbf{F}(x, y)$ for all points (x, y).
2. The antiderivative of a vector field is called the potential function.
3. If the flow lines of $\mathbf{F}(x, y)$ are straight lines, then $\nabla \times \mathbf{F}(x, y) = \mathbf{0}$.
4. $\mathbf{F}$ is conservative if and only if $\nabla \times \mathbf{F} = \mathbf{0}$.
5. The line integral $\int_C f\, ds$ equals the amount of work done by f along C.
6. If $\nabla \times \mathbf{F} = \mathbf{0}$, then the work done by $\mathbf{F}$ along any path is 0.
7. If the curve C is split into pieces C_1 and C_2, then $\int_{C_1} \mathbf{F} \cdot d\mathbf{r} = -\int_{C_2} \mathbf{F} \cdot d\mathbf{r}$.
8. Green's Theorem cannot be applied to a region with a hole.
9. When using Green's Theorem, positive orientation means counterclockwise.
10. When converting a surface integral to a double integral, you must replace z with a function of x and y.
11. A flux integral is always positive.
12. The Divergence Theorem applies only to three-dimensional solids without holes.
13. By Stokes' Theorem, the flux of $\nabla \times \mathbf{F}$ across two nonclosed surfaces sharing the same boundary is the same.

In exercises 1 and 2, sketch several vectors in the velocity field by hand and verify your sketch with a CAS.

1. $\langle x, -y\rangle$ **2.** $\langle 0, 2y\rangle$

3. Match the vector fields with their graphs. $\mathbf{F}_1(x, y) = \langle \sin x, y\rangle$, $\mathbf{F}_2(x, y) = \langle \sin y, x\rangle$, $\mathbf{F}_3(x, y) = \langle y^2, 2x\rangle$, $\mathbf{F}_4(x, y) = \langle 3, x^2\rangle$

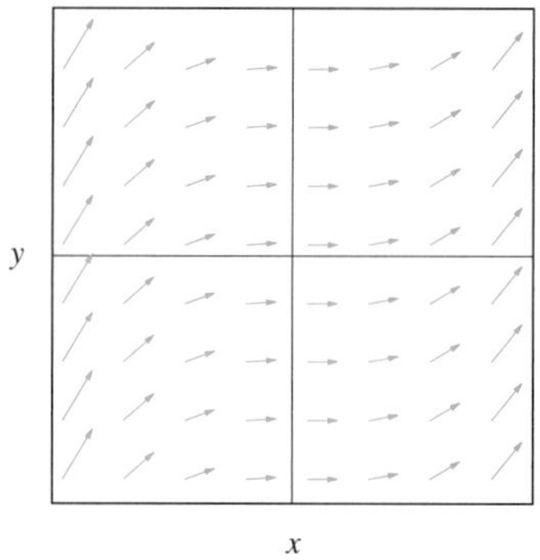

GRAPH A

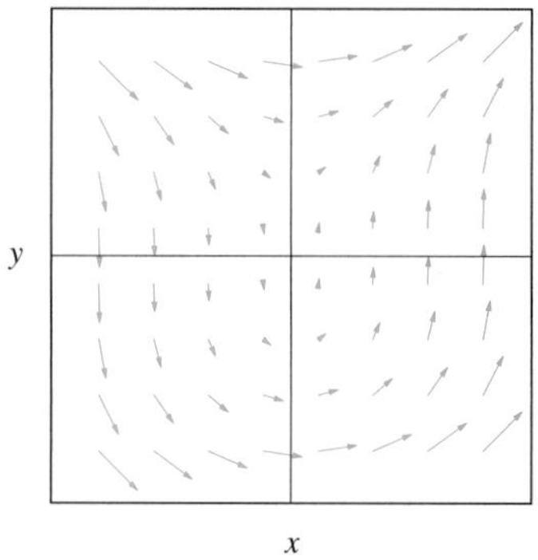

GRAPH B

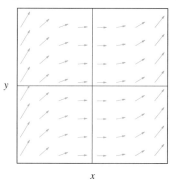

Review Exercises

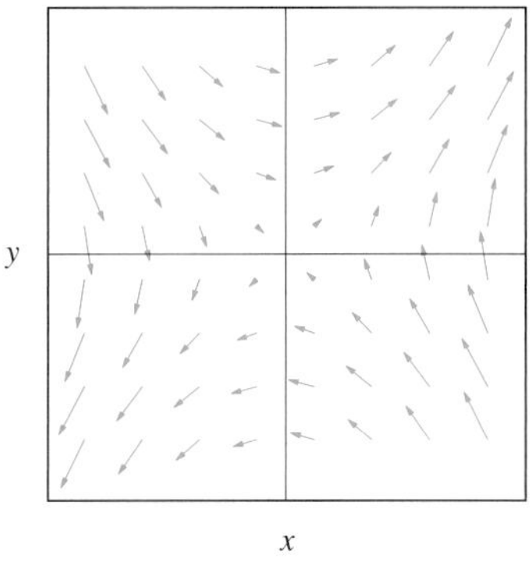

GRAPH C

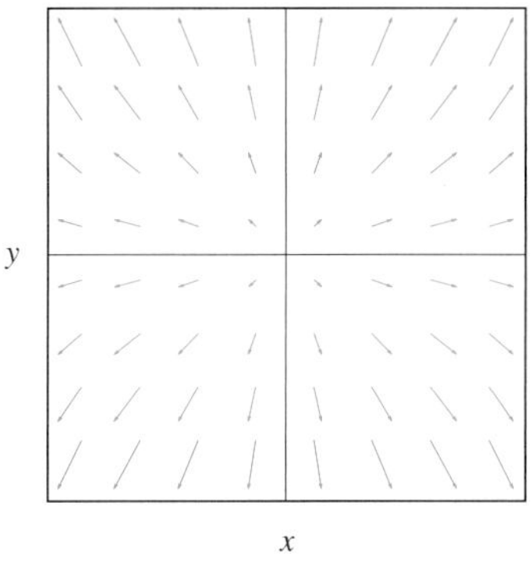

GRAPH D

4. Find the gradient field corresponding to f. Use a CAS to graph it.

   a. $f(x, y) = \ln \sqrt{x^2 + y^2}$ b. $f(x, y) = e^{-x^2-y^2}$

In exercises 5–8, determine whether or not the vector field is conservative. If it is, find a potential function.

5. $\langle y - 2xy^2, x - 2yx^2 + 1 \rangle$ 6. $\langle y^2 + 2e^{2y}, 2xy + 4xe^{2y} \rangle$

7. $\langle 2xy - 1, x^2 + 2xy \rangle$ 8. $\langle y \cos xy - y, x \cos xy - x \rangle$

In exercises 9 and 10, find equations for the flow lines.

9. $\left\langle y, \frac{2x}{y} \right\rangle$ 10. $\left\langle \frac{3}{x}, y \right\rangle$

In exercises 11 and 12, use the notation $\mathbf{r} = \langle x, y \rangle$ and $r = \|\mathbf{r}\| = \sqrt{x^2 + y^2}$.

11. Show that $\nabla (\ln r) = \frac{\mathbf{r}}{r^2}$. 12. Show that $\nabla \left(\frac{1}{r}\right) = -\frac{\mathbf{r}}{r^3}$.

In exercises 13–18, evaluate the line integral.

13. $\int_C 3y\,dx$, where C is the line segment from (2, 3) to (4, 3)

14. $\int_C (x^2 + y^2)\,ds$, where C is the half-circle $x^2 + y^2 = 16$ from (4, 0) to (−4, 0) with $y \geq 0$

15. $\int_C \sqrt{x^2 + y^2}\,ds$, where C is the circle $x^2 + y^2 = 9$, oriented clockwise

16. $\int_C (x - y)\,ds$, where C is the portion of $y = x^3$ from (1, 1) to (−1, −1)

17. $\int_C 2x\,dx$, where C is the upper half-circle from (2, 0) to (−2, 0), followed by the line segment to (2, 0)

18. $\int_C 3y^2 dy$, where C is the portion of $y = x^2$ from (−1, 1) to (1, 1), followed by the line segment to (−1, 1)

In exercises 19–22, compute the work done by the force F along the curve C.

19. $\mathbf{F}(x, y) = \langle x, -y \rangle$, C is the circle $x^2 + y^2 = 4$ oriented counterclockwise

20. $\mathbf{F}(x, y) = \langle y, -x \rangle$, C is the circle $x^2 + y^2 = 4$ oriented counterclockwise

21. $\mathbf{F}(x, y) = \langle 2, 3x \rangle$, C is the quarter-circle from (2, 0) to (0, 2), followed by the line segment to (0, 0)

22. $\mathbf{F}(x, y) = \langle y, -x \rangle$, C is the square from (−2, 0) to (2, 0) to (2, 4) to (−2, 4) to (−2, 0)

In exercises 23 and 24, use the graph to determine whether the work done is positive, negative or zero.

23.

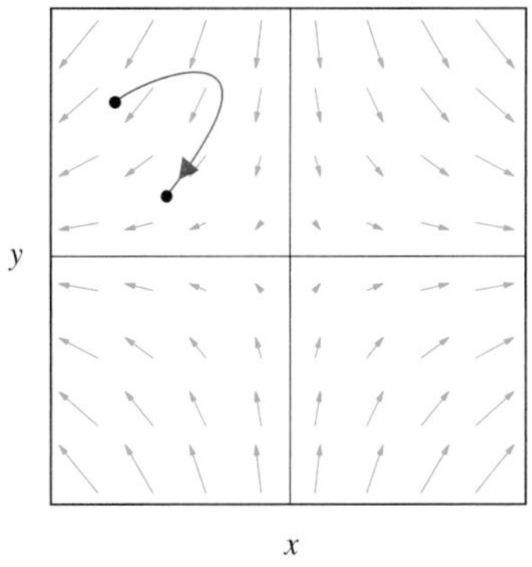

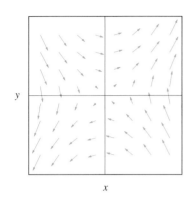

Review Exercises

24.

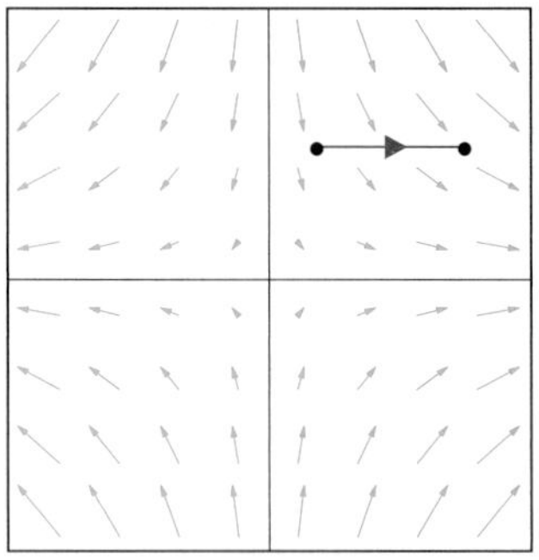

In exercises 25 and 26, find the mass of the indicated object.

25. A spring in the shape of $\langle \cos 3t, \sin 3t, 4t \rangle, 0 \le t \le 2\pi$, $\rho(x, y, z) = 4$

26. The portion of $z = x^2 + y^2$ under $z = 4$ with $\rho(x, y) = 12$

In exercises 27 and 28, show that the integral is independent of path and evaluate the integral.

27. $\int_C (3x^2y - x)\,dx + x^3 dy$, where C runs from $(2, -1)$ to $(4, 1)$

28. $\int_C (y^2 - x^2)\,dx + (2xy + 1)\,dy$, where C runs from $(3, 2)$ to $(1, 3)$

In exercises 29–32, evaluate $\int_C \mathbf{F} \cdot d\mathbf{r}$.

29. $\mathbf{F}(x, y) = \langle 2xy + y\sin x + e^{x+y}, e^{x+y} - \cos x + x^2 \rangle$, C is the quarter-circle from $(0, 3)$ to $(3, 0)$

30. $\mathbf{F}(x, y) = \langle 2y + y^3 + \frac{1}{2}\sqrt{y/x}, 3xy^2 + \frac{1}{2}\sqrt{x/y} \rangle$, C is the top half-circle from $(1, 3)$ to $(3, 3)$

31. $\mathbf{F}(x, y, z) = \langle 2xy, x^2 - y, 2z \rangle$, C runs from $(1, 3, 2)$ to $(2, 1, -3)$

32. $\mathbf{F}(x, y, z) = \langle yz - x, xz - y, xy - z \rangle$, C runs from $(2, 0, 0)$ to $(0, 1, -1)$

In exercises 33 and 34, use the graph to determine whether or not the vector field is conservative.

33.

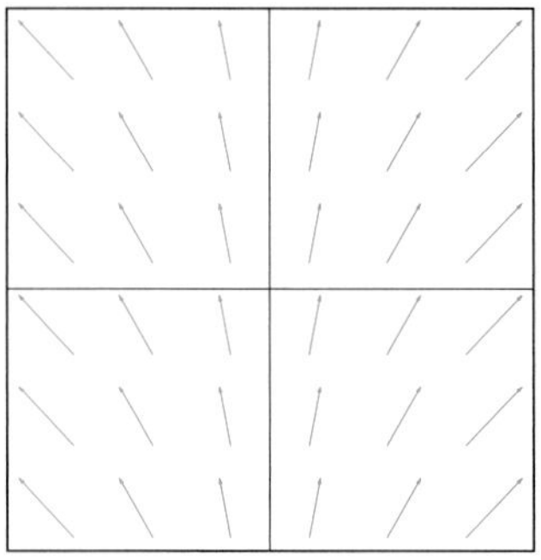

34.

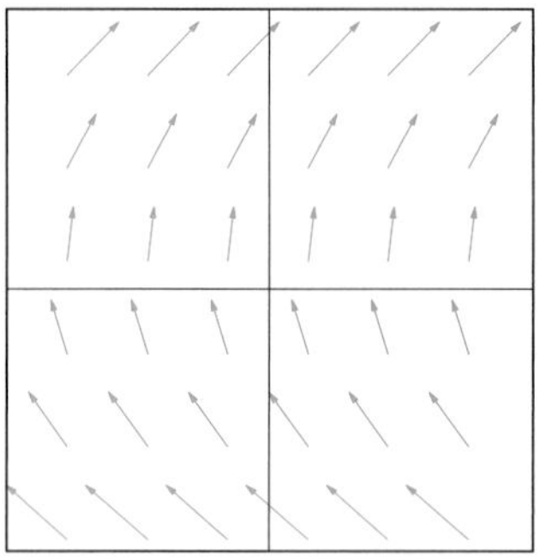

In exercises 35–40, use Green's Theorem to evaluate the indicated line integral.

35. $\oint_C \mathbf{F} \cdot d\mathbf{r}$, where $\mathbf{F} = \langle x^3 - y, x + y^3 \rangle$ and C is formed by $y = x^2$ and $y = x$, oriented positively.

36. $\oint_C \mathbf{F} \cdot d\mathbf{r}$, where $\mathbf{F} = \langle y^2 + 3x^2y, xy + x^3 \rangle$ and C is formed by $y = x^2$ and $y = 2x$, oriented positively.

37. $\oint_C \tan x^2 dx + x^2 dy$, where C is the triangle from $(0, 0)$ to $(1, 1)$ to $(2, 0)$ to $(0, 0)$.

38. $\oint_C x^2 y\,dx + \ln\sqrt{1 + y^2}\,dy$, where C is the triangle from $(0, 0)$ to $(2, 2)$ to $(0, 2)$ to $(0, 0)$.

39. $\oint_C \mathbf{F} \cdot d\mathbf{r}$, where $\mathbf{F} = \langle 3x^2, 4y^3 - z, z^2 \rangle$ and C is formed by $z = y^2$ and $z = 4$, oriented positively in the yz-plane.

40. $\oint_C \mathbf{F} \cdot d\mathbf{r}$, where $\mathbf{F} = \langle 4y^2, 3x^2, 8z \rangle$ and C is $x^2 + y^2 = 4$, oriented positively in the plane $z = 3$.

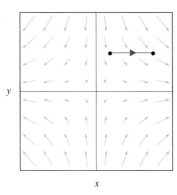

Review Exercises

In exercises 41 and 42, use a line integral to compute the area of the given region.

41. The ellipse $4x^2 + 9y^2 = 36$

42. The region bounded by $y = \sin x$ and the x-axis for $0 \le x \le \pi$

In exercises 43–46, find the curl and divergence of the given vector field.

43. $x^3\mathbf{i} - y^3\mathbf{j}$

44. $y^3\mathbf{i} - x^3\mathbf{j}$

45. $\langle 2x, 2yz^2, 2y^2z \rangle$

46. $\langle 2xy, x^2 - 3y^2z^2, 1 - 2zy^3 \rangle$

In exercises 47–50, determine whether the given vector field is conservative and/or incompressible.

47. $\langle 2x - y^2, z^2 - 2xy, xy^2 \rangle$

48. $\langle y^2z, x^2 - 3z^2y, z^3 - y \rangle$

49. $\langle 4x - y, 3 - x, 2 - 4z \rangle$

50. $\langle 4, 2xy^3, z^4 - x \rangle$

In exercises 51 and 52, conjecture whether the divergence at point P is positive, negative or zero.

51.

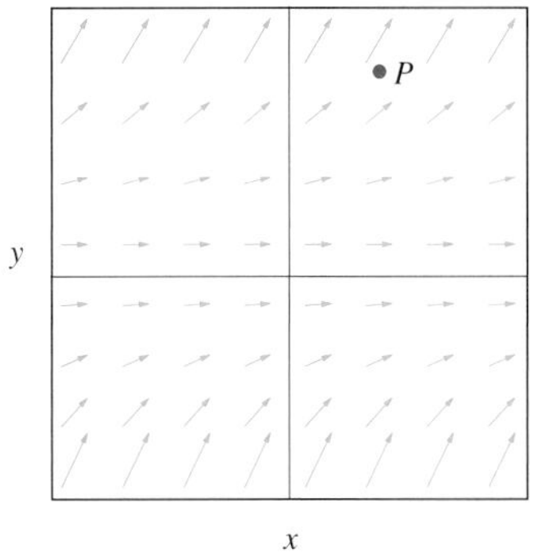

52.

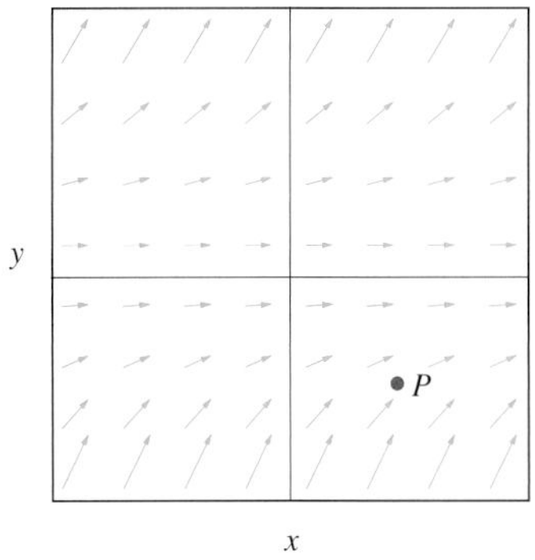

In exercises 53 and 54, sketch a graph of the parametric surface.

53. $x = u^2,\ y = v^2,\ z = u + 2v$

54. $x = (3 + 2\cos u)\cos v,\ y = (3 + 2\cos u)\sin v,\ z = 2\cos v$

55. Match the parametric equations with the surfaces.

a. $x = u^2,\ y = u + v,\ z = v^2$

b. $x = u^2,\ y = u + v,\ z = v$

c. $x = u,\ y = u + v,\ z = v^2$

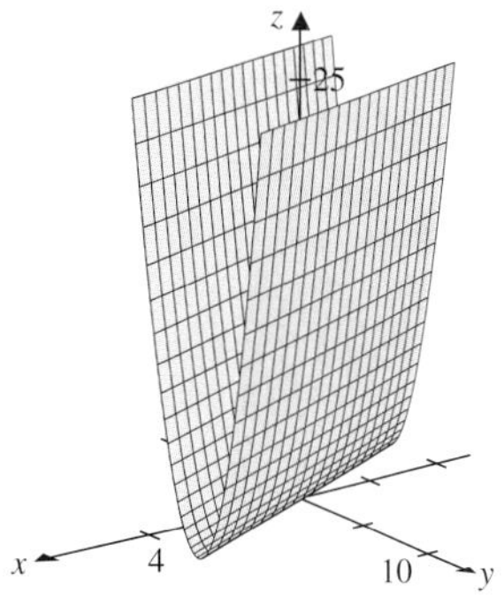

SURFACE A

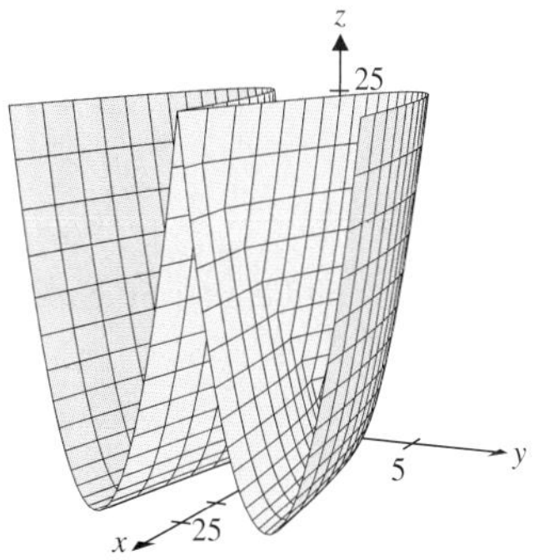

SURFACE B

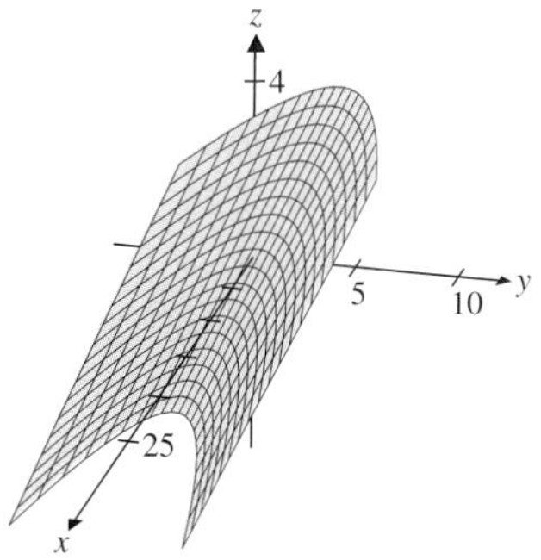

SURFACE C

Review Exercises

56. Find a parametric representation of $x^2 + y^2 + z^2 = 9$.

In exercises 57 and 58, find the surface area.

57. The portion of the paraboloid $z = x^2 + y^2$ between the cylinders $x^2 + y^2 = 1$ and $x^2 + y^2 = 4$

58. The portion of the paraboloid $z = 9 - x^2 - y^2$ between the cylinders $x^2 + y^2 = 1$ and $x^2 + y^2 = 4$

In exercises 59–64, evaluate the surface integral $\iint_S f(x, y, z)\,dS$.

59. $\iint_S (x - y)\,dS$, where S is the portion of the plane $3x + 2y + z = 12$ in the first octant

60. $\iint_S (x^2 + y^2)\,dS$ where S is the portion of $y = 4 - x^2$ above the xy-plane, $y \geq 0$ and below $z = 2$

61. $\iint_S (4x + y + 3z)\,dS$, where S is the portion of the plane $4x + y + 3z = 12$ inside $x^2 + y^2 = 1$

62. $\iint_S (x - z)\,dS$, where S is the portion of the cylinder $x^2 + z^2 = 1$ above the xy-plane between $y = 1$ and $y = 2$

63. $\iint_S yz\,dS$, where S is the portion of the cone $y = \sqrt{x^2 + z^2}$ to the left of $y = 3$

64. $\iint_S (x^2 + z^2)\,dS$, where S is the portion of the paraboloid $x = y^2 + z^2$ behind the plane $x = 4$

In exercises 65 and 66, find the mass and center of mass of the solid.

65. The portion of the paraboloid $z = x^2 + y^2$ below the plane $z = 4$, $\rho(x, y, z) = 2$

66. The portion of the cone $z = \sqrt{x^2 + y^2}$ below the plane $z = 4$, $\rho(x, y, z) = z$

In exercises 67–70, use the Divergence Theorem to compute $\iint_{\partial Q} \mathbf{F} \cdot \mathbf{n}\,dS$.

67. Q is bounded by $x + 2y + z = 4$ (first octant) and the coordinate planes, $\mathbf{F} = \langle y^2 z, y^2 - \sin z, 4y^2 \rangle$.

68. Q is the cube $-1 \leq x \leq 1$, $-1 \leq y \leq 1$, $-1 \leq z \leq 1$, $\mathbf{F} = \langle 4x, 3z, 4y^2 - x \rangle$.

69. Q is bounded by $z = 1 - y^2$, $z = 0$, $x = 0$ and $x + z = 4$, $\mathbf{F} = \langle 2xy, z^3 + 7yx, 4xy^2 \rangle$.

70. Q is bounded by $z = \sqrt{4 - x^2}$, $z = 0$, $y = 0$ and $y + z = 6$, $\mathbf{F} = \langle y^2, 4yz, 2xy \rangle$.

In exercises 71 and 72, find the flux of F over ∂Q.

71. Q is bounded by $z = \sqrt{x^2 + y^2}$, $x^2 + y^2 = 4$ and $z = 0$, $\mathbf{F} = \langle xz, yz, x^2 - z \rangle$.

72. Q is bounded by $z = x^2 + y^2$ and $z = 2 - x^2 - y^2$, $\mathbf{F} = \langle 4x, x^2 - 2y, 3z + x^2 \rangle$.

In exercises 73–76, use Stokes' Theorem, if appropriate, to compute $\iint_S (\nabla \times \mathbf{F}) \cdot \mathbf{n}\,dS$.

73. S is the portion of the tetrahedron bounded by $x + y + 2z = 2$ and the coordinate planes in front of the yz-plane, $\mathbf{F} = \langle zy^4 - y^2, y - x^3, z^2 \rangle$.

74. S is the portion of $z = x^2 + y^2$ below $z = 4$, $\mathbf{F} = \langle z^2 - x, 2y, z^3 xy \rangle$.

75. S is the portion of the cone $z = \sqrt{x^2 + y^2}$ below $x + 2y + 3z = 24$, $\mathbf{F} = \langle 4x^2, 2ye^{2y}, \sqrt{z^2 + 1} \rangle$.

76. S is the portion of the paraboloid $y = x^2 + 4z^2$ to the left of $y = 8 - z$, $\mathbf{F} = \langle xe^{3x}, 4y^{2/3}, z^2 + 2 \rangle$.

In exercises 77 and 78, use Stokes' Theorem to evaluate $\int_C \mathbf{F} \cdot d\mathbf{r}$.

77. C is the triangle from (0, 1, 0) to (1, 0, 0) to (0, 0, 20), $\mathbf{F} = \langle 2xy \cos z, y^2 + x^2 \cos z, z - x^2 y \sin z \rangle$.

78. C is the square from (0, 0, 2) to (1, 0, 2) to (1, 1, 2) to (0, 1, 2), $\mathbf{F} = \langle x^3 + yz, y^2, z^2 \rangle$.

EXPLORATORY EXERCISES

1. In exploratory exercise 2 of section 14.1, we developed a technique for finding equations for flow lines of certain vector fields. The field $\langle 2, 1 + 2xy \rangle$ from example 1.5 is such a vector field, but the calculus is more difficult. First, show that the differential equation is $y' - xy = \frac{1}{2}$ and show that an integrating factor is $e^{-x^2/2}$. The flow lines come from equations of the form $y = e^{x^2/2} \int \frac{1}{2} e^{-x^2/2} dx + ce^{x^2/2}$. Unfortunately, there is no elementary function equal to $\int \frac{1}{2} e^{-x^2/2} dx$. It can help to write this in the form $y = e^{x^2/2} \int_0^x \frac{1}{2} e^{-u^2/2} du + ce^{x^2/2}$. In this form, show that $c = y(0)$. In example 1.5, the curve passing through (0, 1) is $y = e^{x^2/2} \int_0^x \frac{1}{2} e^{-u^2/2}\,du + e^{x^2/2}$. Graph this function and compare it to the path shown in Figure 14.7b. Find an equation for and plot the curve through (0, −1). To find the curve through (1, 1), change the limits of integration and rewrite the solution. Plot this curve and compare to Figure 14.7b.

CHAPTER

15

Second-Order Differential Equations

In a classic television commercial from years ago, the great jazz vocalist Ella Fitzgerald broke a wineglass by singing a particular high-pitched note. The phenomenon that makes this possible is called *resonance,* which is one of the topics in this chapter. Resonance results from the fact that the crystalline structures of certain solids have natural frequencies of vibration. An external force of the same frequency will "resonate" with the object and create a huge increase in energy. For instance, if the frequency of a musical note matches the natural vibration of a crystal wineglass, the glass will vibrate with increasing amplitude until it shatters.

Resonance causes such a dramatic increase in energy that engineers pay special attention to its presence. Buildings are designed to eliminate the chance of destructive resonance. Electronic devices are built to limit some forms of resonance (for example, vibrations in a CD player) and utilize others (for example, stochastic resonance to amplify the desirable portions of a signal).

One commonly known instance of resonance is that caused by soldiers marching in step across a footbridge. Should the frequency of their steps match the natural frequency of the bridge, the resulting resonance can cause the bridge to start moving violently up and down. To avoid this, soldiers are instructed to march out of step when crossing a bridge.

A related incident is the Tacoma Narrows Bridge disaster of 1940. You have likely seen video clips of this large suspension bridge twisting and undulating more and more until it finally tears itself apart. This disaster was initially thought to be the result of resonance, but the cause has come under renewed scrutiny in recent years. While engineers are still not in complete agreement as to its cause, it has been shown that the bridge was not a victim of resonance, but failed due to some related design flaw. Some of the stability issues that had a role in this disaster will be explored in this chapter.

In this chapter, we extend our study of differential equations to those of second order. We develop the basic theory and explore a small number of important applications.

15.1 SECOND-ORDER EQUATIONS WITH CONSTANT COEFFICIENTS

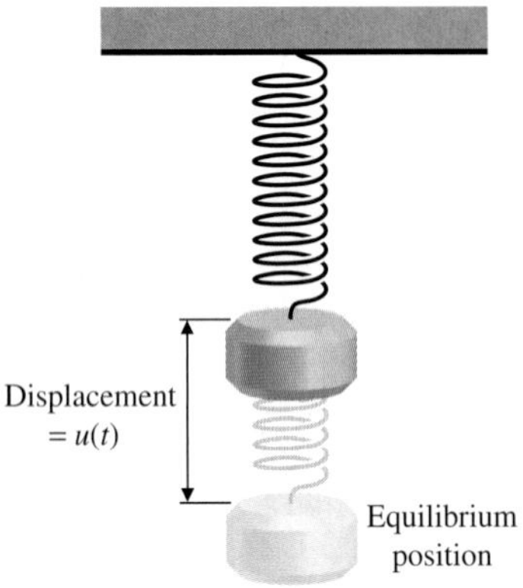

FIGURE 15.1
Spring-mass system

Today's sophisticated technology often requires very precise motion control to maintain acceptable performance. For instance, portable CD players must absorb bumps and twists without skipping. Similarly, a video camera should record a steady image even when the hand holding it is shaking. In this section and section 15.2, we begin to explore the mathematics behind such mechanical vibrations.

A simple version of this problem is easy to visualize. In Figure 15.1, we show a mass hanging from a spring that is suspended from the ceiling. We call the natural length of the spring l. Observe that hanging the mass from the spring will stretch the spring a distance Δl from its natural length. We measure the displacement $u(t)$ of the mass from this **equilibrium position.** Further, we consider downward (where the spring is stretched beyond its equilibrium position) to be a positive displacement and consider upward (where the spring is compressed from its equilibrium position) to be the negative direction. So, the mass in Figure 15.1 has been displaced from its natural length by a total of $u(t) + \Delta l$.

To describe the motion of a spring-mass system, we begin with Newton's second law of motion: $F = ma$. Note that there are three primary forces acting on the mass. First, gravity pulls the mass downward, with force mg. Next, the spring exerts a restoring force when it is stretched or compressed. If the spring is compressed to less than its natural length, the spring exerts a downward force. If the spring is stretched beyond its natural length, the spring exerts an upward force. So, the spring force has the opposite sign from the total displacement from its natural length. Hooke's law states that this force is proportional to the displacement from the spring's natural length. (That is, the more you stretch or compress the spring, the harder the spring resists.) Putting this together, the spring force is given by

$$\text{Spring force} = -k(u + \Delta l),$$

for some positive constant k (called the **spring constant**), determined by the stiffness of the spring. The third force acting on the mass is the **damping force** that resists the motion, due to friction such as air resistance. (A familiar device for adding damping to a mechanical system is the shock absorber in your car.) The damping force depends on velocity: the faster an object moves, the more damping there is. A simple model of the damping force is then

$$\text{Damping force} = -cv,$$

where $v = u'$ is the velocity of the mass and c is a positive constant.

Combining these three forces, Newton's second law gives the following:

$$mu''(t) = ma = F = mg - k[u(t) + \Delta l] - cu'(t)$$

or

$$mu''(t) + cu'(t) + ku(t) = mg - k\Delta l. \tag{1.1}$$

We can simplify this equation with a simple observation. If the mass is not in motion, then $u(t) = 0$, for all t. In this case, $u'(t) = u''(t) = 0$ for all t and equation (1.1) reduces to

$$0 = mg - k\Delta l.$$

REMARK 1.1

The spring constant k is given by

$$k = \frac{mg}{\Delta l}$$

(weight divided by displacement from natural length).

While we can use this to solve for the spring constant k in terms of the mass and Δl, this also simplifies (1.1) to

$$mu''(t) + cu'(t) + ku(t) = 0. \tag{1.2}$$

Equation (1.2) is a **second-order** differential equation, since it includes a second derivative. Solving equations such as (1.2) gives us insight into spring motion as well as many other diverse phenomena.

Before trying to solve this general equation, we first solve a few simple examples of second-order equations. The simplest second-order equation is $y''(t) = 0$. Integrating this once gives us

$$y'(t) = c_1,$$

for some constant c_1. Integrating again yields

$$y(t) = c_1 t + c_2,$$

where c_2 is another arbitrary constant. We refer to this as the **general solution** of the differential equation, meaning that *every* solution of the equation can be written in this form. It should not be surprising that the general solution of a second-order differential equation should involve two arbitrary constants, since it requires two integrations to undo the two derivatives. A slightly more complicated equation is

$$y'' - y = 0. \tag{1.3}$$

We can discover the solution of this, if we first rewrite the equation as

$$y'' = y.$$

Think about it this way: we are looking for a function whose second derivative is itself. One such function is $y = e^t$. It's not hard to see that a second solution is $y = e^{-t}$. It turns out that every possible solution of the equation can be written as a combination of these two solutions, so that the general solution of (1.3) is

General solution of $y'' = y$

$$\boxed{y = c_1 e^t + c_2 e^{-t},}$$

for constants c_1 and c_2.

More generally, we want to solve

$$ay''(t) + by'(t) + cy(t) = 0, \tag{1.4}$$

where a, b and c are constants. Notice that equation (1.4) is the same as equation (1.2), except for the name of the dependent variable. In a full course on differential equations, you will see that if you can find two solutions $y_1(t)$ and $y_2(t)$, neither of which is a constant multiple of the other, then *all solutions* can be written in the form

$$y(t) = c_1 y_1(t) + c_2 y_2(t),$$

for some constants c_1 and c_2. The question remains as to how to find these two solutions. As we've already seen, the answer starts with making an educated guess.

Notice that equation (1.4) asks us to find a function whose first and second derivatives are similar enough that the combination $ay''(t) + by'(t) + cy(t)$ adds up to zero. As we already saw with equation (1.3), one candidate for such a function is the exponential function e^{rt}. So, we look for some (constant) value(s) of r for which $y = e^{rt}$ is a solution of (1.4).

Observe that if $y(t) = e^{rt}$, then $y'(t) = re^{rt}$ and $y''(t) = r^2 e^{rt}$. Substituting into (1.4), we get

$$ar^2 e^{rt} + bre^{rt} + ce^{rt} = 0$$

and factoring out the common e^{rt}, we have

$$(ar^2 + br + c)e^{rt} = 0.$$

Since $e^{rt} > 0$, this can happen only if

$$ar^2 + br + c = 0. \tag{1.5}$$

Equation (1.5) is called the **characteristic equation,** whose solution(s) can be found by the quadratic formula to be

$$r_1 = \frac{-b + \sqrt{b^2 - 4ac}}{2a} \quad \text{and} \quad r_2 = \frac{-b - \sqrt{b^2 - 4ac}}{2a}.$$

So, there are three possibilities for solutions of (1.5) : (1) r_1 and r_2 are distinct real solutions (if $b^2 - 4ac > 0$), (2) $r_1 = r_2$ is a (repeated) real solution (if $b^2 - 4ac = 0$) or (3) r_1 and r_2 are complex solutions (if $b^2 - 4ac < 0$). All three of these cases lead to different solutions of the differential equation (1.4), which we must consider separately.

Case 1: Distinct Real Roots

If r_1 and r_2 are distinct real solutions of (1.5), then $y_1 = e^{r_1 t}$ and $y_2 = e^{r_2 t}$ are two solutions of (1.4) and $y(t) = c_1 e^{r_1 t} + c_2 e^{r_2 t}$ is the general solution of (1.4). We illustrate this in example 1.1.

EXAMPLE 1.1 Finding General Solutions

Find the general solution of (a) $y'' - y' - 6y = 0$ and (b) $y'' + 4y' - 2y = 0$.

Solution In each case, we solve the characteristic equation and interpret the solution(s).

For part (a), the characteristic equation is

$$0 = r^2 - r - 6 = (r - 3)(r + 2).$$

So, there are two distinct real solutions of the characteristic equation: $r_1 = 3$ and $r_2 = -2$. The general solution is then

$$y(t) = c_1 e^{3t} + c_2 e^{-2t}.$$

For part (b), the characteristic equation is

$$0 = r^2 + 4r - 2.$$

Since the polynomial does not easily factor, we use the quadratic formula to get

$$r = \frac{-4 \pm \sqrt{16 + 8}}{2} = -2 \pm \sqrt{6}.$$

We again have two distinct real solutions and so, the general solution of the differential equation is

$$y(t) = c_1 e^{(-2+\sqrt{6})t} + c_2 e^{(-2-\sqrt{6})t}. \quad \blacksquare$$

Case 2: Repeated Root

If $r_1 = r_2$ (repeated root of the characteristic equation), then we have found only one solution of (1.4): $y_1 = e^{r_1 t}$. We leave it as an exercise to show that a second solution in this case is $y_2 = te^{r_1 t}$. The general solution of (1.4) is then $y(t) = c_1 e^{r_1 t} + c_2 t e^{r_1 t}$. We illustrate this case in example 1.2.

EXAMPLE 1.2 Finding General Solutions (Repeated Root)

Find the general solution of $y'' - 6y' + 9y = 0$.

Solution Here, the characteristic equation is

$$0 = r^2 - 6r + 9 = (r - 3)^2.$$

So, here we have the repeated root $r = 3$. The general solution is then

$$y(t) = c_1 e^{3t} + c_2 t e^{3t}. \blacksquare$$

Case 3: Complex Roots

If r_1 and r_2 are complex roots of the characteristic equation, we can write these as $r_1 = u + vi$ and $r_2 = u - vi$, where i is the imaginary number $i = \sqrt{-1}$. The question is how to interpret a complex exponential like $e^{(u+vi)t}$. The answer lies with Euler's formula, which says that $e^{i\theta} = \cos\theta + i \sin\theta$. The solution corresponding to $r = u + vi$ is then

$$e^{(u+vi)t} = e^{ut+vti} = e^{ut} e^{vti} = e^{ut}(\cos vt + i \sin vt).$$

It can be shown that both the real and the imaginary parts of this solution (that is, both $y_1 = e^{ut} \cos vt$ and $y_2 = e^{ut} \sin vt$) are solutions of the differential equation. So, in this case, the general solution of (1.4) is

$$y(t) = c_1 e^{ut} \cos vt + c_2 e^{ut} \sin vt. \quad (1.6)$$

In example 1.3, we see how to use this solution.

EXAMPLE 1.3 Finding General Solutions (Complex Roots)

Find the general solution of the equations (a) $y'' + 2y' + 5y = 0$ and (b) $y'' + 4y = 0$.

Solution For part (a), the characteristic equation is

$$0 = r^2 + 2r + 5.$$

Since this does not factor, we use the quadratic formula to obtain

$$r = \frac{-2 \pm \sqrt{4 - 20}}{2} = -1 \pm 2i.$$

From (1.6) the general solution is

$$y(t) = c_1 e^{-t} \cos 2t + c_2 e^{-t} \sin 2t.$$

For part (b), there is no y'-term and so, the characteristic equation is simply

$$r^2 + 4 = 0.$$

This gives us $r^2 = -4$, so that $r = \pm\sqrt{-4} = \pm 2i$. From (1.6) the general solution is then

$$y(t) = c_1 \cos 2t + c_2 \sin 2t. \blacksquare$$

We can now find the general solution of any equation of the form (1.4). Notice that the general solution of a second-order differential equation always involves two arbitrary constants. In order to determine the value of these constants, we specify two initial conditions, most often $y(0)$ and $y'(0)$ (corresponding to the initial position and initial velocity of the mass, in the case of a spring-mass system). A second-order differential equation plus two initial conditions is called an **initial value problem.** Example 1.4 illustrates how to apply these conditions to the general solution of a differential equation.

EXAMPLE 1.4 Solving an Initial Value Problem

Find the solution of the initial value problem $y'' + 4y' + 3y = 0$, $y(0) = 2$, $y'(0) = 0$.

Solution Here, the characteristic equation is

$$0 = r^2 + 4r + 3 = (r + 3)(r + 1),$$

so that $r = -3$ and $r = -1$. The general solution is then

$$y(t) = c_1 e^{-3t} + c_2 e^{-t},$$

so that

$$y'(t) = -3c_1 e^{-3t} - c_2 e^{-t}.$$

The two initial conditions then give us

$$2 = y(0) = c_1 + c_2 \tag{1.7}$$

and

$$0 = y'(0) = -3c_1 - c_2. \tag{1.8}$$

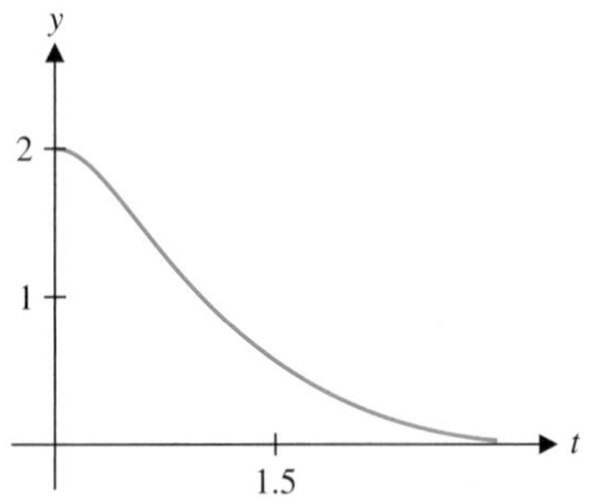

FIGURE 15.2
$y = -e^{-3t} + 3e^{-t}$

We solve the two equations (1.7) and (1.8) for c_1 and c_2, as follows. From (1.8), $c_2 = -3c_1$. Substituting this into (1.7) gives us

$$2 = c_1 + c_2 = c_1 - 3c_1 = -2c_1,$$

so that $c_1 = -1$. Then $c_2 = -3c_1 = 3$. The solution of the initial value problem is then

$$y(t) = -e^{-3t} + 3e^{-t}.$$

A graph of this solution is shown in Figure 15.2. ■

In example 1.5, the differential equation has no y'-term. Physically, this corresponds to the case of a spring-mass system with no damping.

EXAMPLE 1.5 Solving an Initial Value Problem

Find the solution of the initial value problem $y'' + 9y = 0$, $y(0) = 4$, $y'(0) = -6$.

Solution Here, the characteristic equation is $r^2 + 9 = 0$, so that $r^2 = -9$ and $r = \pm\sqrt{-9} = \pm 3i$. The general solution is then

$$y(t) = c_1 \cos 3t + c_2 \sin 3t$$

and

$$y'(t) = -3c_1 \sin 3t + 3c_2 \cos 3t.$$

From the initial conditions, we now have

$$4 = y(0) = c_1$$

and

$$-6 = y'(0) = 3c_2.$$

So, $c_2 = -2$ and the solution of the initial value problem is

$$y(t) = 4\cos 3t - 2\sin 3t.$$

A graph is shown in Figure 15.3. ■

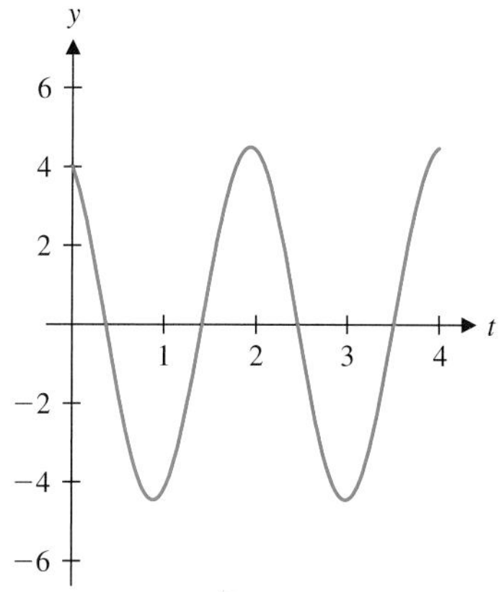

FIGURE 15.3
$y = 4\cos 3t - 2\sin 3t$

We now have the mathematical machinery needed to analyze a simple spring-mass system. In example 1.6, pay careful attention to the amount of work we do in setting up the differential equation. Remember: you can't get the right solution if you don't have the right equation!

EXAMPLE 1.6 Spring-Mass System with No Damping

A spring is stretched 6 inches by an 8-pound weight. The mass is then pulled down an additional 4 inches and released. Neglect damping. Find an equation for the position of the mass at any time t and graph the position function.

Solution The general equation describing the spring-mass system is $mu'' + cu' + ku = 0$. We need to identify the mass m, damping constant c and spring constant k. Since we are neglecting damping, we have $c = 0$. The mass m is related to the weight W by $W = mg$, where g is the gravitational constant. Since the weight is 8 pounds, we have $8 = m(32)$ or $m = \frac{8}{32} = \frac{1}{4}$ (the units of mass here are slugs). The spring constant k is determined from the equation $mg = k\Delta l$. Here, the mass stretches the spring 6 inches, which we must convert to $\frac{1}{2}$ foot. So, $\Delta l = \frac{1}{2}$ and $8 = k\left(\frac{1}{2}\right)$, leaving us with $k = 16$. The equation of motion is then

$$\frac{1}{4}u'' + 0u' + 16u = 0$$

or

$$u'' + 64u = 0.$$

REMARK 1.2

In the English system of units, with pounds (weight), feet and seconds,

$$g \approx 32\ \text{ft/s}^2.$$

In the metric system with kg (mass), meters and seconds,

$$g \approx 9.8\ \text{m/s}^2.$$

Here, the characteristic equation is $r^2 + 64 = 0$, so that $r = \pm\sqrt{-64} = \pm 8i$ and the general solution is

$$u(t) = c_1 \cos 8t + c_2 \sin 8t. \tag{1.9}$$

To determine the values of c_1 and c_2, we need the initial values $u(0)$ and $u'(0)$. Read the problem carefully and notice that the spring is released after it is pulled down 4 inches. This says that the initial position is 4 inches or $\frac{1}{3}$ foot down (the positive direction) and so, $u(0) = \frac{1}{3}$. Further, since the weight is pulled down and simply released, its initial velocity is zero, $u'(0) = 0$. We then have the initial conditions

$$u(0) = \frac{1}{3} \quad \text{and} \quad u'(0) = 0.$$

From (1.9), we have $u'(t) = -8c_1 \sin 8t + 8c_2 \cos 8t$. The initial conditions now give us

$$\frac{1}{3} = u(0) = c_1(1) + c_2(0) = c_1$$

and

$$0 = u'(0) = -8c_1(0) + 8c_2(1) = 8c_2,$$

so that $c_1 = \frac{1}{3}$ and $c_2 = 0$. The solution of the initial value problem is now

$$u(t) = \frac{1}{3}\cos 8t.$$

The graph of this function in Figure 15.4 shows the smooth up and down motion of an undamped spring (called **simple harmonic motion**). ■

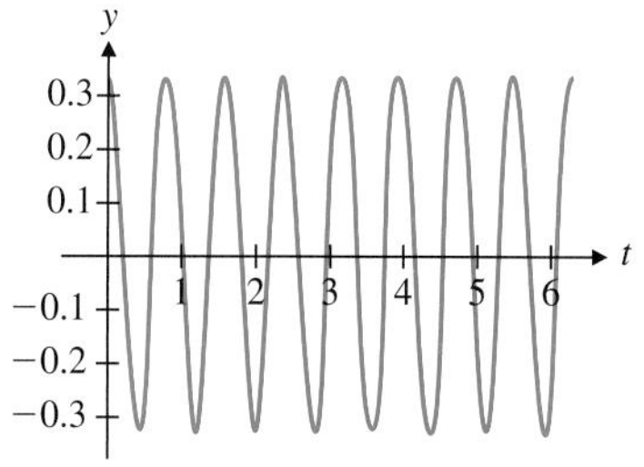

FIGURE 15.4
$u(t) = \frac{1}{3}\cos 8t$

For a real spring-mass system, there is always some damping, so that the idealized perpetual motion of example 1.6 must be modified somewhat. In example 1.7, again take note of the steps required to obtain the equation of motion.

EXAMPLE 1.7 Spring-Mass System with Damping

A spring is stretched 5 cm by a 2-kg mass. The mass is set in motion from its equilibrium position with an upward velocity of 2 m/s. The damping constant equals $c = 4$. Find an equation for the position of the mass at any time t and graph the position function.

Solution The equation of motion is $mu'' + cu' + ku = 0$, where the damping constant is $c = 4$ and the mass is $m = 2$ kg. With these units, $g \approx 9.8$ m/s^2, so that the weight is $W = mg = 2(9.8) = 19.6$. The displacement of the mass is 5 cm or 0.05 meter. The spring constant is then $k = \frac{W}{\Delta l} = \frac{19.6}{0.05} = 392$ and the equation of motion is

$$2u'' + 4u' + 392u = 0$$

or

$$u'' + 2u' + 196u = 0.$$

Here, the characteristic equation is $r^2 + 2r + 196 = 0$. From the quadratic formula, we have $r = \dfrac{-2 \pm \sqrt{4 - 784}}{2} = -1 \pm \sqrt{195}i$. The general solution is then

$$u(t) = c_1 e^{-t} \cos\sqrt{195}t + c_2 e^{-t} \sin\sqrt{195}t,$$

so that

$$u'(t) = -c_1 e^{-t} \cos\sqrt{195}t - c_1\sqrt{195}e^{-t} \sin\sqrt{195}t - c_2 e^{-t} \sin\sqrt{195}t + c_2\sqrt{195}e^{-t} \cos\sqrt{195}t.$$

Since the mass is set in motion from its equilibrium position, we have $u(0) = 0$ and since it's set in motion with an upward velocity of 2 m/s, we have $u'(0) = -2$. (Keep in mind that upward motion corresponds to negative displacement.) These initial conditions now give us

$$0 = u(0) = c_1$$

and

$$-2 = u'(0) = -c_1 + c_2\sqrt{195} = c_2\sqrt{195},$$

since $c_1 = 0$. So, $c_2 = \frac{-2}{\sqrt{195}}$ and the displacement of the mass at any given time is given by

$$u(t) = -\frac{2}{\sqrt{195}} e^{-t} \sin\sqrt{195}t.$$

The graph of this solution in Figure 15.5 shows a spring whose oscillations rapidly die out. ■

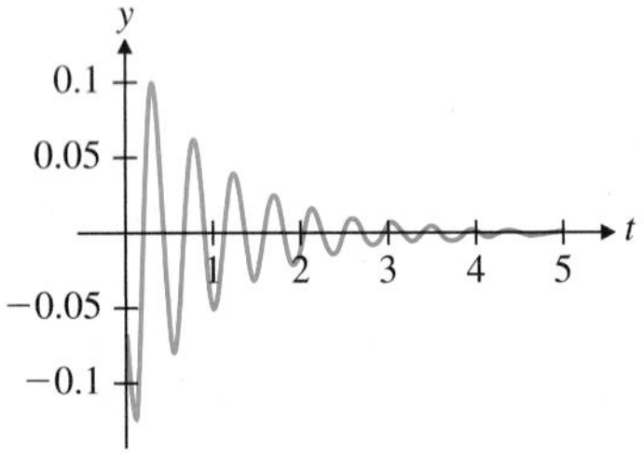

FIGURE 15.5
$u(t) = -\frac{2}{\sqrt{195}} e^{-t} \sin\sqrt{195}t$

BEYOND FORMULAS

A major difference between solving second-order equations and solving first-order equations is that for second-order equations of the form $ay''(t) + by'(t) + cy(t) = 0$, you need to find two different solutions y_1 and y_2 (neither one of which is a constant multiple of the other). The form of these solutions depends on the type of the solutions to the characteristic equation, but in all cases the general solution is given by $c_1 y_1(t) + c_2 y_2(t)$, where the values of c_1 and c_2 are determined from two initial conditions.

EXERCISES 15.1

WRITING EXERCISES

1. Briefly discuss the role that theory plays in this section. In particular, if we didn't know that two different solutions were enough, would our method of guessing exponential solutions lead to solutions?

2. Briefly describe why our method of guessing exponential solutions would not work on equations with nonconstant coefficients. (You may want to work with a specific example like $y'' + 2ty' + 3y = 0$.)

3. It can be shown that e^{2t} and $2e^{2t}$ are both solutions of $y'' - 3y' + 2y = 0$. Explain why these can't be used as the two functions in the general solution. That is, you can't write all solutions in the form $c_1e^{2t} + c_2 2e^{2t}$.

4. Discuss Figures 15.4 and 15.5 in physical terms. In particular, discuss the significance of the y-intercept and the increasing/decreasing properties of the graph in terms of the motion of the spring. Further, relate the motion of the spring to the forces acting on the spring.

In exercises 1–12, find the general solution of the differential equation.

1. $y'' - 2y' - 8y = 0$
2. $y'' - 2y' - 6y = 0$
3. $y'' - 4y' + 4y = 0$
4. $y'' + 2y' + 6y = 0$
5. $y'' - 2y' + 5y = 0$
6. $y'' + 6y' + 9y = 0$
7. $y'' - 2y' = 0$
8. $y'' - 6y = 0$
9. $y'' - 2y' - 6y = 0$
10. $y'' + y' + 3y = 0$
11. $y'' - \sqrt{5}y' + y = 0$
12. $y'' - \sqrt{3}y' + y = 0$

In exercises 13–20, solve the initial value problem.

13. $y'' + 4y = 0,\ y(0) = 2,\ y'(0) = -3$
14. $y'' + 2y' + 10y = 0,\ y(0) = 1,\ y'(0) = 0$
15. $y'' - 3y' + 2y = 0,\ y(0) = 0,\ y'(0) = 1$
16. $y'' + y' - 2y = 0,\ y(0) = 3,\ y'(0) = 0$
17. $y'' - 2y' + 5y = 0,\ y(0) = 2,\ y'(0) = 0$
18. $y'' - 4y' + 4y = 0,\ y(0) = 2,\ y'(0) = 1$
19. $y'' - 2y' + y = 0,\ y(0) = -1,\ y'(0) = 2$
20. $y'' + 3y' = 0,\ y(0) = 4,\ y'(0) = 0$

21. Show that $c_1 \cos kt + c_2 \sin kt = A \sin(kt + \delta)$, where $A = \sqrt{c_1^2 + c_2^2}$ and $\tan \delta = \frac{c_1}{c_2}$. We call A the **amplitude** and δ the **phase shift.** Use this identity to find the amplitude and phase shift of the solution of $y'' + 9y = 0$, $y(0) = 3$ and $y'(0) = -6$.

In exercises 22–24, solve the initial value problem and use the result of exercise 21 to find the amplitude and phase shift of the solution.

22. $y'' + 4y = 0,\ y(0) = 1,\ y'(0) = -2$
23. $y'' + 20y = 0,\ y(0) = -2,\ y'(0) = 2$
24. $y'' + 12y = 0,\ y(0) = -1,\ y'(0) = -2$

25. A spring is stretched 6 inches by a 12-pound weight. The weight is then pulled down an additional 8 inches and released. Neglect damping. Find an equation for the position of the spring at any time t and graph the position function.

26. A spring is stretched 20 cm by a 4-kg mass. The weight is released with a downward velocity of 2 m/s. Neglect damping. Find an equation for the position of the spring at any time t and graph the position function.

27. A spring is stretched 10 cm by a 4-kg mass. The weight is pulled down an additional 20 cm and released with an upward velocity of 4 m/s. Neglect damping. Find an equation for the position of the spring at any time t and graph the position function. Find the amplitude and phase shift of the motion.

28. A spring is stretched 2 inches by a 6-pound weight. The weight is then pulled down an additional 4 inches and released with a downward velocity of 4 ft/s. Neglect damping. Find an equation for the position of the spring at any time t and graph the position function. Find the amplitude and phase shift of the motion.

29. A spring is stretched 4 inches by a 16-pound weight. The damping constant equals 10. The weight is then pushed up 6 inches and released. Find an equation for the position of the spring at any time t and graph the position function.

30. A spring is stretched 8 inches by a 32-pound weight. The damping constant equals 0.4. The weight is released with a downward velocity of 3 ft/s. Find an equation for the position of the spring at any time t and graph the position function.

31. A spring is stretched 25 cm by a 4-kg mass. The weight is pushed up $\frac{1}{2}$ meter and released. The damping constant equals $c = 2$. Find an equation for the position of the spring at any time t and graph the position function.

32. A spring is stretched 10 cm by a 5-kg mass. The weight is released with a downward velocity of 2 m/s. The damping constant equals $c = 5$. Find an equation for the position of the spring at any time t and graph the position function.

33. Show that in the case of a repeated root $r = r_1$ to the characteristic equation, the function $y = te^{r_1 t}$ is a second solution of the differential equation $ay'' + by' + cy = 0$.

34. Show that in the case of complex roots $r = u \pm vi$ to the characteristic equation, the functions $y = e^{ut} \cos vt$ and $y = e^{ut} \sin vt$ are solutions of the differential equation $ay'' + by' + cy = 0$.

35. For the equation $u'' + cu' + 16u = 0$, compare solutions with $c = 7$, $c = 8$ and $c = 9$. The first case is called **underdamped,** the second case is called **critically damped** and the last case is called **overdamped.** Briefly explain why these terms are appropriate.

36. For the general equation $mu'' + cu' + ku = 0$, show that critical damping occurs with $c = 2\sqrt{km}$. Without solving any equations, briefly describe what the graph of solutions look like with $c < 2\sqrt{km}$, compared to $c > 2\sqrt{km}$.

37. A spring is stretched 3 inches by a 16-pound weight. Use exercise 36 to find the critical damping value.

38. Show that for both the critically damped case and the overdamped case, the mass can pass through its equilibrium position at most once. (Hint: Show that $u(t) = 0$ has at most one solution.)

39. If you are designing a screen door, you can control the damping by changing the viscosity of the fluid in the cylinder in which the closure rod is embedded. Discuss whether overdamping or underdamping would be more appropriate.

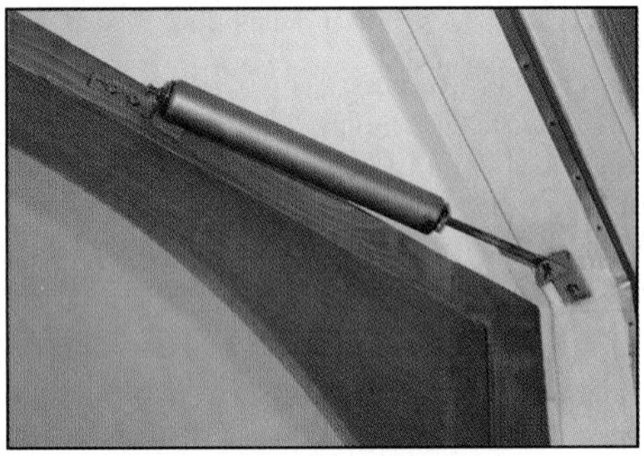

40. Show that e^t and e^{-t} are solutions of the equation $y'' - y = 0$, and conclude that a general solution is given by $y = c_1 e^t + c_2 e^{-t}$. Then show that $\sinh t$ and $\cosh t$ are solutions of $y'' - y = 0$, and conclude that a general solution is given by $y = c_1 \sinh t + c_2 \cosh t$. Discuss whether or not these two general solutions are equivalent.

41. As in exercise 40, show that $y = c_1 \sinh at + c_2 \cosh at$ is a general solution of $y'' - a^2 y = 0$, for any constant $a > 0$. Compare this to the general solution of $y'' + a^2 y = 0$.

42. For the general equation $ay'' + by' + cy = 0$, if the roots of the characteristic equation are complex and $b > 0$, show that the solution $y(t) \to 0$ as $t \to \infty$.

43. For the general equation $ay'' + by' + cy = 0$, if $ac > 0$, $b > 0$ and the roots of the characteristic equation are real numbers $r_1 < r_2$, show that both roots are negative and thus, the solution $y(t) \to 0$ as $t \to \infty$.

44. For the general equation $ay'' + by' + cy = 0$, suppose that there is a repeated root $r_1 < 0$ of the characteristic equation. Show that $\lim_{t\to\infty} te^{r_1 t} = 0$ and thus, the solution $y(t) \to 0$ as $t \to \infty$.

45. Use the results of exercises 42–44 to show that if a, b and c are all positive, then the solution $y(t)$ of $ay'' + by' + cy = 0$ goes to 0 as $t \to \infty$.

46. Interpret the result of exercise 45 in terms of the spring equation $mu'' + cu' + ku = 0$. In particular, if there is nonzero damping, then what is the eventual motion of the spring?

EXPLORATORY EXERCISES

1. In this exercise, you will explore solutions of a different type of second-order equation. An **Euler equation** has the form $x^2 y'' + axy' + by = 0$ for constants a and b. Notice that this equation requires that x times the first derivative and x^2 times the second derivative be similar to the original function. Explain why a reasonable guess is $y = x^r$. Substitute this into the equation and (similar to our derivation of the characteristic equation) show that r must satisfy the equation

$$r^2 + (a - 1)r + b = 0.$$

Use this to find the general solution of (a) $x^2 y'' + 4xy' + 2y = 0$ and (b) $x^2 y'' - 3xy' + 3y = 0$. Discuss the main difference in the graphs of solutions to (a) and (b). Can you say anything definite about the graph of a solution of (c) $x^2 y'' + 2xy' - 6y = 0$ near $x = 0$? There remains the issue of what to do with complex and repeated roots. Show that if you get complex roots $r = u \pm vi$, then $y = x^u \cos(v \ln x)$ and $y = x^u \sin(v \ln x)$ are solutions for $x > 0$. Use this information to find the general solution of (d) $x^2 y'' + xy' + y = 0$. Use the form of the solutions corresponding to complex roots to guess the second solution in the repeated roots case. Find the general solution of (e) $x^2 y'' + 5xy' + 4y = 0$.

2. In this exercise, you will explore solutions of **higher-order** differential equations. For a third-order equation with constant coefficients such as (a) $y''' - 3y'' - y' + 3y = 0$, make a reasonable guess of the form of the solution, write down the characteristic equation and solve the equation (which factors). Use this idea to find the general solution of (b) $y''' + y'' + 3y' - 5y = 0$ and (c) $y''' - y'' - y' + y = 0$. Oddly enough, an equation like (d) $y''' - y = 0$ causes more problems than (a)–(c). How many solutions of the characteristic equation do you find? Show that $y = te^t$ is *not* a solution. Show that $y = e^{t/2} \cos \frac{\sqrt{3}}{2} t$ and $y = e^{t/2} \sin \frac{\sqrt{3}}{2} t$ are two additional solutions. Identify the two r-values to which these solutions correspond. Show that these r-values are in fact solutions of the characteristic equation. Conclude that a more thorough understanding of solutions of complex equations is necessary to fully master third-order equations. To end on a more positive note, find the general solutions of the fourth-order equation (e) $y^{(4)} - y = 0$ and the fifth-order equation (f) $y^{(5)} - 3y^{(4)} - 5y''' + 15y'' + 4y' - 12y = 0$.

15.2 NONHOMOGENEOUS EQUATIONS: UNDETERMINED COEFFICIENTS

Imagine yourself trying to videotape an important event. You might be more concerned with keeping a steady hand than with understanding the mathematics of motion control, but mathematics plays a vital role in helping you produce a professional-looking tape. In section 15.1, we modeled mechanical vibrations when the motion is started by an initial displacement or velocity. In this section, we extend that model to cases where an external force such as a shaky hand continues to affect the system.

The starting place for our model again is Newton's second law of motion: $F = ma$. We now add an external force to the spring force and damping force considered in section 15.1. If the external force is $F(t)$ and $u(t)$ gives the displacement from equilibrium, as defined before, we have

$$mu''(t) = -ku(t) - cu'(t) + F(t)$$

or

$$mu''(t) + cu'(t) + ku(t) = F(t). \tag{2.1}$$

The only change from the spring model in section 15.1 is that the right-hand side of the equation is no longer zero. Equations of the form (2.1) with zero on the right-hand side are called **homogeneous.** In the case where $F(t) \neq 0$, we call the equation **nonhomogeneous.**

Our goal is to find the general solution of such equations (that is, the form of all solutions). We can do this by first finding one **particular solution** $u_p(t)$ of the nonhomogeneous equation (2.1). Notice that if $u(t)$ is any other solution of (2.1), then we have that

$$\begin{aligned} m(u-u_p)'' + c(u-u_p)' + k(u-u_p) &= (mu'' + cu' + ku) - (mu_p'' + cu_p' + ku_p) \\ &= F(t) - F(t) = 0. \end{aligned}$$

That is, the function $u - u_p$ is a solution of the homogeneous equation $mu'' + cu' + ku = 0$ solved in section 15.1. So, if the general solution of the homogeneous equation is $c_1u_1 + c_2u_2$, then $u - u_p = c_1u_1 + c_2u_2$, for some constants c_1 and c_2 and

$$u = c_1u_1 + c_2u_2 + u_p.$$

We summarize this in Theorem 2.1.

THEOREM 2.1

Let $u = c_1u_1 + c_2u_2$ be the general solution of $mu'' + cu' + ku = 0$ and let u_p be any solution of $mu'' + cu' + ku = F(t)$. Then the general solution of $mu'' + cu' + ku = F(t)$ is given by

$$u = c_1u_1 + c_2u_2 + u_p.$$

We illustrate this result with example 2.1.

TODAY IN MATHEMATICS

Shigefumi Mori (1951–)
A Japanese mathematician who earned the Fields Medal in 1990. A colleague wrote, "The most profound and exciting development in algebraic geometry during the last decade or so was the Minimal Model Program or Mori's Program. . . . Shigefumi Mori initiated the program with a decisively new and powerful technique, guided the general research direction with some good collaborators along the way and finally finished up the program by himself overcoming the last difficulty. . . . Mori's theorems were stunning and beautiful by the totally new features unimaginable by those who had been working, probably very hard too, only in the traditional world of algebraic . . . surfaces."

EXAMPLE 2.1 Solving a Nonhomogeneous Equation

Find the general solution of $u'' + 4u' + 3u = 30e^{2t}$, given that $u_p = 2e^{2t}$ is a solution.

Solution One of the two pieces of the general solution is given to us: we have $u_p = 2e^{2t}$. The other piece is the solution of the homogeneous equation

$u'' + 4u' + 3u = 0$. Here, the characteristic equation is

$$0 = r^2 + 4r + 3 = (r+3)(r+1),$$

so that $r = -3$ or $r = -1$. The general solution of the homogeneous equation is then $c_1 e^{-3t} + c_2 e^{-t}$, so that the general solution of the nonhomogeneous equation is

$$u(t) = c_1 e^{-3t} + c_2 e^{-t} + 2e^{2t}. \blacksquare$$

While Theorem 2.1 shows us how to piece together the solution of a nonhomogeneous equation from a particular solution and the general solution of the corresponding homogeneous equation, we still do not know how to find a particular solution. The method presented here, called the **method of undetermined coefficients,** works for equations with constant coefficients and where the nonhomogeneous term is not too complicated. The method relies on our ability to make an educated guess about the form of a particular solution. We begin by illustrating this technique for example 2.1.

If $u'' + 4u' + 3u = 30e^{2t}$, then the most likely candidate for the form of $u(t)$ is a constant multiple of e^{2t}. (How else would u'', $4u'$ and $3u$ all add up to $30e^{2t}$?) Be sure that you understand the logic here, because we will be using it in the examples to come. So, an educated guess is $u_p(t) = Ae^{2t}$, for some constant A. Substituting this into the differential equation, we try to solve for A. (If it turns out to be impossible to solve for A, then we have simply made a bad guess.) Here, we have $u_p' = 2Ae^{2t}$ and $u_p'' = 4Ae^{2t}$ and so, requiring u_p to be a solution of the nonhomogeneous equation gives as

$$\begin{aligned} 30e^{2t} &= u_p'' + 4u_p' + 3u_p \\ &= 4Ae^{2t} + 4(2Ae^{2t}) + 3(Ae^{2t}) \\ &= 15Ae^{2t}. \end{aligned}$$

So, $15A = 30$ or $A = 2$. A particular solution is then $u_p(t) = 2e^{2t}$, as desired.

We learn more about making good guesses in examples 2.2 through 2.4.

EXAMPLE 2.2 Solving a Nonhomogeneous Equation

Find the general solution of $u'' + 2u' - 3u = -30 \sin 3t$.

Solution First, we solve the corresponding homogeneous equation: $u'' + 2u' - 3u = 0$. The characteristic equation here is

$$0 = r^2 + 2r - 3 = (r+3)(r-1),$$

so that $r = -3$ or $r = 1$. This gives us $u = c_1 e^{-3t} + c_2 e^t$ as the general solution of the homogeneous equation. Next, we guess the form of a particular solution. Since the right-hand side is a constant multiple of sin 3t, a reasonable guess might seem to be $u_p = A \sin 3t$. However, it turns out that this is too specific a guess, since when we compute derivatives to substitute into the equation, we will also get $\cos 3t$ terms. This suggests the slightly more general guess $u_p = A \sin 3t + B \cos 3t$. Substituting this into the equation, we get

$$u_p = A \sin 3t + B \cos 3t$$
$$u_p' = 3A \cos 3t - 3B \sin 3t$$
$$u_p'' = -9A \sin 3t - 9B \cos 3t$$

$$\begin{aligned} -30 \sin 3t &= -9A \sin 3t - 9B \cos 3t + 2(3A \cos 3t - 3B \sin 3t) \\ &\quad - 3(A \sin 3t + B \cos 3t) \\ &= (-12A - 6B) \sin 3t + (6A - 12B) \cos 3t. \end{aligned}$$

Equating the corresponding coefficients of the sine and cosine terms (imagine the additional term $0\cos 3t$ on the left-hand side), we have

$$-12A - 6B = -30$$

or

$$2A + B = 5$$

and

$$6A - 12B = 0$$

or

$$A = 2B.$$

Substituting into the first equation, we get $2(2B) + B = 5$ or $B = 1$. Then, $A = 2B = 2$ and a particular solution is $u_p = 2\sin 3t + \cos 3t$. We now put together all of the pieces to obtain the general solution of the original equation:

$$u(t) = c_1 e^{-3t} + c_2 e^t + 2\sin 3t + \cos 3t. \blacksquare$$

Observe that while the calculations get a bit messy, the process of making a guess is not exceptionally challenging. To keep the details of calculation from getting in the way of the ideas, we next focus on making good guesses. In general, you start with the function $F(t)$ and then add terms corresponding to each derivative. For instance, in example 2.2, we started with $A\sin 3t$ and then added a term corresponding to its derivative: $B\cos 3t$. We do not need to add other terms, because all other derivatives are simply constant multiples of either $\sin 3t$ or $\cos 3t$. However, suppose that $F(t) = 7t^5$. The initial guess would include At^5 and the derivatives Bt^4, Ct^3 and so on. To save letters, you can use subscripts and write the initial guess as

$$A_5t^5 + A_4t^4 + A_3t^3 + A_2t^2 + A_1t + A_0.$$

There is one exception to the preceding rule. If any term in the initial guess is also a solution of the homogeneous equation, you must multiply the initial guess by a sufficiently high power of t so that nothing in the modified guess is a solution of the homogeneous equation. (In the case of second-order equations, this means multiplying the initial guess by either t or t^2.) To see why, consider $u'' + 2u' - 3u = 4e^{-3t}$. The initial guess Ae^{-3t} won't work, as seen in example 2.3.

EXAMPLE 2.3 Modifying an Initial Guess

Show that $u(t) = Ae^{-3t}$ is not a solution of $u'' + 2u' - 3u = 4e^{-3t}$, for any value of A, but that there is a solution of the form $u(t) = Ate^{-3t}$.

Solution For $u(t) = Ae^{-3t}$, we have $u'(t) = -3Ae^{-3t}$ and $u''(t) = 9Ae^{-3t}$ and so,

$$u'' + 2u' - 3u = 9Ae^{-3t} + 2(-3Ae^{-3t}) - 3Ae^{-3t} = 0.$$

That is, $u(t) = Ae^{-3t}$ is a solution of the homogeneous equation for every choice of A. As a result, Ae^{-3t} is not a solution of the nonhomogeneous equation for any choice of A. However, if we multiply our initial guess by t, we have $u(t) = Ate^{-3t}$, $u'(t) = Ae^{-3t} + At(-3e^{-3t})$ and $u''(t) = -3Ae^{-3t} - 3Ae^{-3t} + At(9e^{-3t}) = -6Ae^{-3t} + 9Ate^{-3t}$. Substituting into the equation, we have

$$\begin{aligned} 4e^{-3t} &= u'' + 2u' - 3u = -6Ae^{-3t} + 9Ate^{-3t} + 2(Ae^{-3t} - 3Ate^{-3t}) - 3Ate^{-3t} \\ &= -4Ae^{-3t}. \end{aligned}$$

So, $-4A = 4$ and $A = -1$. A particular solution of the nonhomogeneous equation is then $u_p(t) = -te^{-3t}$. ■

A summary of rules for making good guesses is given in the accompanying table. Notice that sine and cosine terms always go together, and all polynomials are complete with terms from t^n all the way down to t and a constant.

The form of u_p for $au'' + bu' + cu = F(t)$

$F(t)$	*Initial Guess*	*Modify Initial Guess if $ar^2 + br + c = 0$ for*
$e^{r_1 t}$	$Ae^{r_1 t}$	$r = r_1$
$\cos kt$ or $\sin kt$	$A\cos kt + B\sin kt$	$r = \pm ki$
t^n	$C_n t^n + C_{n-1}t^{n-1} + \cdots + C_1 t + C_0$	$r = 0$
$e^{ut}\cos vt$ or $e^{ut}\sin vt$	$e^{ut}(A\cos vt + B\sin vt)$	$r = u \pm vi$
$t^n e^{r_1 t}$	$(C_n t^n + C_{n-1}t^{n-1} + \cdots + C_1 t + C_0)e^{r_1 t}$	$r = r_1$

We illustrate the process of making good guesses in example 2.4.

EXAMPLE 2.4 Finding the Form of Particular Solutions

Determine the form for a particular solution of the following equations: (a) $y'' + 4y' = t^4 + 3t^2 + 2e^{-4t}\sin t + 3e^{-4t}$ and (b) $y'' + 4y = 3t^2\sin 2t + 3te^{2t}$.

Solution For part (a), the characteristic equation is $0 = r^2 + 4r = r(r+4)$, so that $r = 0$ and $r = -4$. The solution of the homogeneous equation is then $y = c_1 + c_2 e^{-4t}$. Looking at the right-hand side, we have a sum of three types of terms: a polynomial, an exponential/sine combination and an exponential. From the table, our initial guess is

$$y_p = (C_4 t^4 + C_3 t^3 + C_2 t^2 + C_1 t + C_0) + e^{-4t}(A\cos t + B\sin t) + De^{-4t}.$$

However, referring back to the solution of the homogeneous equation, note that the constant C_0 and the exponential De^{-4t} are solutions of the homogeneous equation. (Also note that the exponential/sine term is not a solution of the homogeneous equation.) Multiplying the first and third terms by t, the correct form of a solution to the nonhomogeneous equation is

$$y_p = t(C_4 t^4 + C_3 t^3 + C_2 t^2 + C_1 t + C_0) + e^{-4t}(A\cos t + B\sin t) + Dte^{-4t}.$$

Now that we have the correct form of a solution, it's a straightforward (though tedious) matter to find the value of all the constants.

For part (b), the characteristic equation is $r^2 + 4 = 0$, so that $r = \pm 2i$ and the solution of the homogeneous equation is then $y = c_1\cos 2t + c_2\sin 2t$. Here, the right-hand side consists of two terms: the product of a polynomial and a sine function and the product of a polynomial and an exponential function. Multiplying guesses from the table, we make our initial guess

$$y_p = (A_2 t^2 + A_1 t + A_0)\sin 2t + (B_2 t^2 + B_1 t + B_0)\cos 2t + (C_1 t + C_0)e^{2t}.$$

Observe that both $A_0\sin 2t$ and $B_0\cos 2t$ are solutions of the homogeneous equation. So, we must multiply the first two terms by t to obtain the modified guess:

$$y_p = t(A_2 t^2 + A_1 t + A_0)\sin 2t + t(B_2 t^2 + B_1 t + B_0)\cos 2t + (C_1 t + C_0)e^{2t}. \blacksquare$$

NOTES

The letters used in writing the forms of the solutions are completely arbitrary.

We now return to the study of mechanical vibrations. Recall that the movement of a spring-mass system with an external force $F(t)$ is modeled by $mu'' + cu' + ku = F(t)$.

EXAMPLE 2.5 The Motion of a Spring Subject to an External Force

A mass of 0.2 kg stretches a spring by 10 cm. The damping constant is $c = 0.4$. External vibrations create a force of $F(t) = 0.2 \sin 4t$ newtons, setting the spring in motion from its equilibrium position. Find an equation for the position of the spring at any time t.

Solution We are given $m = 0.2$ and $c = 0.4$. Recall that the spring constant k satisfies the equation $mg = k\Delta l$, where $\Delta l = 10\text{ cm} = 0.1$ m. (Notice that since the mass is given in kg, $g = 9.8\text{ m/s}^2$ and we need the value of Δl in meters.) This enables us to solve for k, as follows:

$$k = \frac{mg}{\Delta l} = \frac{(0.2)(9.8)}{0.1} = 19.6.$$

The equation of motion is then

$$0.2u'' + 0.4u' + 19.6u = 0.2 \sin 4t$$

or

$$u'' + 2u' + 98u = \sin 4t.$$

This gives us the characteristic equation $r^2 + 2r + 98 = 0$, which has solutions $r = \frac{-2 \pm \sqrt{4 - 392}}{2} = -1 \pm \sqrt{97}i$. The solution of the homogeneous equation is then $u(t) = c_1 e^{-t} \cos \sqrt{97}t + c_2 e^{-t} \sin \sqrt{97}t$. A particular solution has the form $u_p = A \sin 4t + B \cos 4t$. Substituting this into the equation, we get

$$\begin{aligned} \sin 4t &= u'' + 2u' + 98u \\ &= -16A \sin 4t - 16B \cos 4t + 2(4A \cos 4t - 4B \sin 4t) \\ &\quad + 98(A \sin 4t + B \cos 4t) \\ &= (82A - 8B) \sin 4t + (8A + 82B) \cos 4t. \end{aligned}$$

Then, $82A - 8B = 1$ and $8A + 82B = 0$. The solution is $A = \frac{41}{3394}$ and $B = \frac{-2}{1697}$ and so,

$$u(t) = c_1 e^{-t} \cos \sqrt{97}t + c_2 e^{-t} \sin \sqrt{97}t + \frac{41}{3394} \sin 4t - \frac{2}{1697} \cos 4t.$$

The initial conditions are $u(0) = 0$ and $u'(0) = 0$. With $t = 0$ and $u = 0$, we get $0 = c_1 - \frac{2}{1697}$ or $c_1 = \frac{2}{1697}$. Computing the derivative $u'(t)$ and substituting in $t = 0$ and $u' = 0$, we get $0 = -c_1 + \sqrt{97}c_2 + \frac{82}{1697}$ or $c_2 = \frac{-80}{1697\sqrt{97}}$. The general solution of the nonhomogeneous equation is then

$$\begin{aligned} u(t) = {} & \frac{2}{1697} e^{-t} \cos \sqrt{97}t - \frac{80}{1697\sqrt{97}} e^{-t} \sin \sqrt{97}t \\ & + \frac{41}{3394} \sin 4t - \frac{2}{1697} \cos 4t. \end{aligned}$$

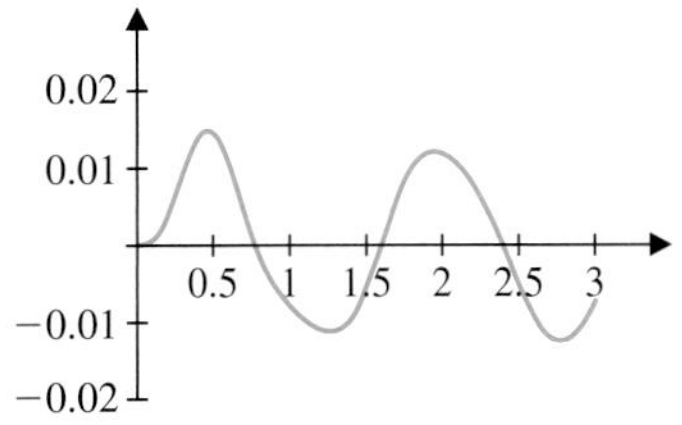

FIGURE 15.6
Spring motion with an external force

A graph is shown in Figure 15.6. ■

Notice in Figure 15.6 that after a very brief time, the motion appears to be simple harmonic motion. We can verify this by a quick analysis of our solution in example 2.5.

Recall that the solution comes in two pieces, the particular solution

$$u_p(t) = \frac{41}{3394}\sin 4t - \frac{2}{1697}\cos 4t$$

and the solution of the homogeneous equation

$$c_1 e^{-t}\cos\sqrt{97}t + c_2 e^{-t}\sin\sqrt{97}t.$$

As t increases, the presence of the exponential e^{-t} causes the homogeneous solution to approach 0, regardless of the value of the constants c_1 and c_2. So, for any initial conditions, the solution will eventually be dominated by the particular solution, which is a simple oscillation. For this reason, the solution of the homogeneous equation is called the **transient solution** and the particular solution is called the **steady-state solution.** This is true of many, but not all equations. (Can you think of cases where the homogeneous solution does not tend to 0 as t increases?) If we are interested only in the steady-state solution, we can avoid much of the work in example 2.5 and simply solve for the particular solution.

EXAMPLE 2.6 Finding a Steady-State Solution

For $u'' + 3u' + 2u = 20\cos 2t$, find the steady-state solution.

Solution For the homogeneous solution, the characteristic equation is $0 = r^2 + 3r + 2 = (r+1)(r+2)$ and so, the solutions are $r = -2$ and $r = -1$. The solution of the homogeneous equation is then $u(t) = c_1 e^{-2t} + c_2 e^{-t}$. Since this tends to 0 as $t \to \infty$, we ignore it. A particular solution has the form $u_p(t) = A\cos 2t + B\sin 2t$. Substituting into the equation, we have

$$\begin{aligned} 20\cos 2t &= u'' + 3u' + 2u \\ &= -4A\cos 2t - 4B\sin 2t + 3(-2A\sin 2t + 2B\cos 2t) \\ &\quad + 2(A\cos 2t + B\sin 2t) \\ &= (-2A + 6B)\cos 2t + (-6A - 2B)\sin 2t. \end{aligned}$$

So, we must have

$$-2A + 6B = 20$$

and

$$-6A - 2B = 0.$$

From the second equation, $B = -3A$. Substituting into the first equation, we have $-2A - 18A = 20$ or $A = -1$. Then, $B = -3A = 3$ and the steady-state solution is

$$u_p(t) = -\cos 2t + 3\sin 2t.$$

A graph is shown in Figure 15.7. ■

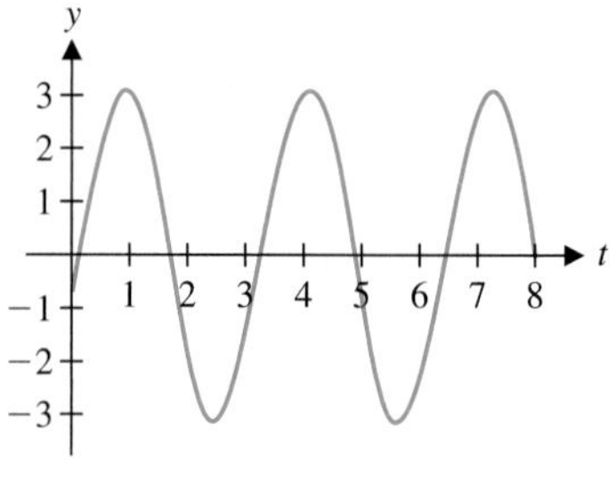

FIGURE 15.7
Steady-state solution

There are several possibilities for the steady-state motion of a mechanical system. Two interesting cases, called **resonance** and **beats,** are introduced here. In their pure forms, both occur only when there is no damping and the external force is a sine or cosine. In these cases, then, the equation of motion is

$$mu'' + ku = F(t).$$

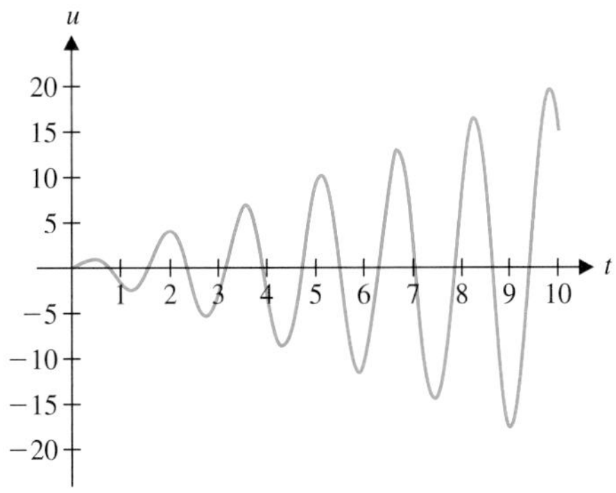

FIGURE 15.8
Resonance: $u = 2t \sin 4t$

The characteristic equation for the homogeneous equation is $mr^2 + k = 0$, which has solutions $r = \pm\sqrt{\frac{k}{m}}i$ and the solution of the homogeneous equation is

$$u(t) = c_1 \cos \omega t + c_2 \sin \omega t,$$

where $\omega = \sqrt{\frac{k}{m}}$ is called the **natural frequency** of the system.

Resonance occurs in a mechanical system when the external force is a sine or cosine whose frequency exactly matches the natural frequency of the system. For example, suppose that $F(t) = \sin \omega t$. Then our initial guess $u_p(t) = A \sin \omega t + B \cos \omega t$ matches the homogeneous solution and must be modified to the guess $u_p(t) = t(A \sin \omega t + B \cos \omega t)$. The graph of such a function would oscillate, but the presence of the factor t would cause the oscillations to grow larger and larger without bound. The graph of $u = 2t \sin 4t$ in Figure 15.8 illustrates this behavior.

Physically, resonance can cause impressive disasters. A singer hitting a note (thus producing an external force) at exactly the natural frequency of a wineglass can shatter it. Soldiers marching in step across a bridge at exactly the natural frequency of the bridge can create large oscillations in the bridge that can cause it to collapse.

The phenomenon of beats occurs when the forcing frequency is close to (but not equal to) the natural frequency. For example, for $u'' + 4u = 2\sin(2.1t)$ with $u(0) = u'(0) = 0$, the homogeneous solution is $u(t) = c_1 \sin 2t + c_2 \cos 2t$ and so, the forcing frequency of 2.1 is close to the natural frequency of 2. We leave it as an exercise to show that the solution is

$$u(t) = 5.1219 \sin(2t) - 4.878 \sin(2.1t)$$

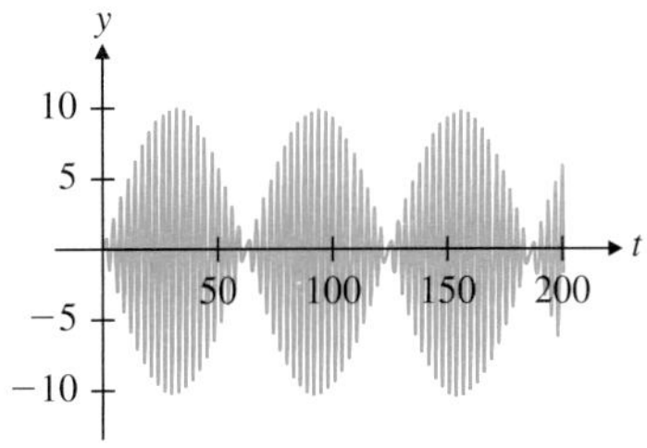

FIGURE 15.9
Beats

The graph in Figure 15.9 illustrates the beats phenomenon of periodically increasing and decreasing amplitudes.

This can be heard when tuning a piano. If a note is slightly off, its frequency is close to the frequency of the external tuning fork and you will hear the amplitude variation illustrated in Figure 15.9.

EXAMPLE 2.7 Resonance and Beats

For the system $u'' + 5u = 3 \sin \omega t$, find the natural frequency, the value of ω that produces resonance and a value of ω that produces beats.

Solution The characteristic equation for the homogeneous equation is $r^2 + 5 = 0$, with solutions $r = \pm\sqrt{5}i$. The natural frequency is then $\sqrt{5}$ and this is the value of ω that produces resonance. Values close to $\sqrt{5}$ (such as $\omega = 2$) produce beats. ■

BEYOND FORMULAS

Be sure that you understand the difference between the equations solved in section 15.2 and those solved in section 15.1. For the nonhomogeneous equations explored in this section, you must first find the homogeneous solution $c_1y_1(t) + c_2y_2(t)$ as in section 15.1 and then find a particular solution $y_p(t)$. Always keep in mind that the overall structure of the general solution of a nonhomogeneous equation is $y(t) = c_1y_1(t) + c_2y_2(t) + y_p(t)$. Your task is then to fill in the details one at a time.

EXERCISES 15.2

WRITING EXERCISES

1. In many cases, a guess for the form of a particular solution may seem logical but turn out to be a bad guess. Identify the criterion for whether a guess is ultimately good or bad. (See example 2.3.)
2. In example 2.4 part (a), the initial guess e^{-4t} is multiplied by t but $e^{-4t}\cos t$ is not. Explain why these terms are treated differently by comparing the r-values in a characteristic equation with solution e^{-4t} to one with solution $e^{-4t}\cos t$.
3. Soldiers are taught to break step when marching across a bridge. Briefly explain why this is a good idea.
4. Is there any danger to a party of people dancing on a strong balcony? Would it help if some of the people had a bad sense of rhythm?

In exercises 1–4, find the general solution of the equation given the particular solution.

1. $u'' + 2u' + 5u = 15e^{-2t}$, $u_p(t) = 3e^{-2t}$
2. $u'' + 2u' - 8u = 14e^{3t}$, $u_p(t) = 2e^{3t}$
3. $u'' + 4u' + 4u = 4t^2$, $u_p(t) = t^2 - 2t + \frac{3}{2}$
4. $u'' + 4u = 6\sin t$, $u_p(t) = 2\sin t$

In exercises 5–10, find the general solution of the equation.

5. $u'' + 2u' + 10u = 26e^{-3t}$
6. $u'' - 2u' + 5u = 10e^{2t}$
7. $u'' + 2u' + u = 25\sin t$
8. $u'' + 4u = 24\cos 4t$
9. $u'' - 4u = 2t^3$
10. $u'' + u' - 6u = 18t^2$

In exercises 11–18, determine the form of a particular solution of the equation.

11. $u'' + 2u' + 10u = 2e^{-t} + 3e^{-t}\cos 3t + 2\sin 3t$
12. $u'' - 2u' + 5u = e^t \sin 2t - t^2e^t$
13. $u'' + 2u' = 5t^3 - 2t + 4e^{2t}$
14. $u'' + 4u = 2t\cos 2t - t^2 \sin t$
15. $u'' + 9u = e^t \cos 3t - 2t\sin 3t$
16. $u'' - 4u = t^3e^{2t} + t^2e^{-2t}$
17. $u'' + 4u' + 4u = t^2e^{-2t} + 2te^{-2t}\sin t$
18. $u'' + 2u' + u = t^2 - 4 + 2e^{-t}$
19. A mass of 0.1 kg stretches a spring by 2 mm. The damping constant is $c = 0.2$. External vibrations create a force of $F(t) = 0.1\cos 4t$ newtons, setting the spring in motion from its equilibrium position with zero initial velocity. Find an equation for the position of the spring at any time t.
20. A mass of 0.4 kg stretches a spring by 2 mm. The damping constant is $c = 0.4$. External vibrations create a force of $F(t) = 0.8\sin 3t$ newtons, setting the spring in motion from its equilibrium position with zero initial velocity. Find an equation for the position of the spring at any time t.
21. A mass weighing 0.4 lb stretches a spring by 3 inches. The damping constant is $c = 0.4$. External vibrations create a force of $F(t) = 0.2e^{-t/2}$ lb. The spring is set in motion from its equilibrium position with a downward velocity of 1 ft/s. Find an equation for the position of the spring at any time t.
22. A mass weighing 0.1 lb stretches a spring by 2 inches. The damping constant is $c = 0.2$. External vibrations create a force of $F(t) = 0.2e^{-t/4}$ lb. The spring is set in motion by pulling it down 4 inches and releasing it. Find an equation for the position of the spring at any time t.

Exercises 23–28 refer to amplitude and phase shift. (See exercise 21 in section 15.1.)

23. For $u'' + 2u' + 6u = 15\cos 3t$, find the steady-state solution and identify its amplitude and phase shift.

24. For $u'' + 3u' + u = 5\sin 2t$, find the steady-state solution and identify its amplitude and phase shift.

25. For $u'' + 4u' + 8u = 15\cos t + 10\sin t$, find the steady-state solution and identify its amplitude and phase shift.

26. For $u'' + u' + 6u = 12\cos t + 8\sin t$, find the steady-state solution and identify its amplitude and phase shift.

27. A mass weighing 2 lb stretches a spring by 6 inches. The damping constant is $c = 0.4$. External vibrations create a force of $F(t) = 2\sin 2t$ lb. Find the steady-state solution and identify its amplitude and phase shift.

28. A mass of 0.5 kg stretches a spring by 20 cm. The damping constant is $c = 1$. External vibrations create a force of $F(t) = 3\cos 2t$ N. Find the steady-state solution and identify its amplitude and phase shift.

29. For the system $u'' + 3u = 4\sin\omega t$, find the natural frequency, the value of ω that produces resonance and a value of ω that produces beats.

30. For the system $u'' + 10u = 2\cos\omega t$, find the natural frequency, the value of ω that produces resonance and a value of ω that produces beats.

31. A mass weighing 0.4 lb stretches a spring by 3 inches. Ignore damping. External vibrations create a force of $F(t) = 2\sin\omega t$ lb. Find the natural frequency, the value of ω that produces resonance and a value of ω that produces beats.

32. A mass of 0.4 kg stretches a spring by 3 cm. Ignore damping. External vibrations create a force of $F(t) = 2\sin\omega t$ N. Find the natural frequency, the value of ω that produces resonance and a value of ω that produces beats.

33. In this exercise, we compare solutions where resonance is present and solutions of the same system with a small amount of damping. Start by finding the solution of $y'' + 9y = 12\cos 3t$, $y(0) = 1$, $y'(0) = 0$. Then solve the initial value problem $y'' + 0.1y' + 9y = 12\cos 3t$, $y(0) = 1$, $y'(0) = 0$. Graph both solutions on the same set of axes, and estimate a range of t-values for which the solutions stay close.

34. Repeat exercise 33 for $y'' + 0.01y' + 9y = 12\cos 3t$, $y(0) = 1$, $y'(0) = 0$.

35. For $u'' + 4u = \sin\omega t$, explain why the form of the particular solutions is simply $A\sin\omega t$, for $\omega^2 \neq 4$.

36. For $u'' + 4u = 2t^3$, identify a simplified form of the particular solution.

37. Find the solution of $u'' + 4u = 2\sin(2.1t)$, with $u(0) = u'(0) = 0$.

38. Find the solution of $u'' + 4u = 2\sin 2t$ with $u(0) = u'(0) = 0$. Compare the graphs of the solutions to exercises 37 and 38.

39. For $u'' + 4u = \sin\omega t$, $u(0) = u'(0) = 0$, find the solution as a function of ω. Compare the graphs of the solutions for $\omega = 0.5$, $\omega = 0.9$ and $\omega = 1$.

40. For $u'' + 4u = \sin\omega t$, $u(0) = u'(0) = 1$, find the solution as a function of ω. Compare the graphs of the solutions for $\omega = 0.5$, $\omega = 0.9$ and $\omega = 1$. Discuss the effects of the initial conditions by comparing the graphs in exercises 39 and 40.

41. For $u'' + 0.1u' + 4u = \sin\omega t$, find the amplitude of the steady-state solution as a function of ω.

42. For the spring problem in exercise 41, what happens to the steady-state amplitude as ω approaches 0? Explain why this makes sense.

EXPLORATORY EXERCISES

1. A washing machine whose tub spins with rotational speed ω generates a downward force of $f_0 \sin\omega t$ for some constant f_0. If the machine rests on a spring and damping mechanism (see diagram), the vertical motion of the machine satisfies the familiar equation $u'' + cu' + ku = f_0\sin\omega t$. Explain what happens to the motion as c and k are increased. We now add one layer to the design problem for the machine. The forces absorbed by the spring and damper are transmitted to the floor. That is, $F(t) = cu' + ku$ is the force of the machine on the floor. We would like this to be small. Explain in physical terms why this force increases if c and k increase. So the design of the machine must balance vertical movement versus force transmitted to the floor. Consider the following argument for an equation for $F(t)$.

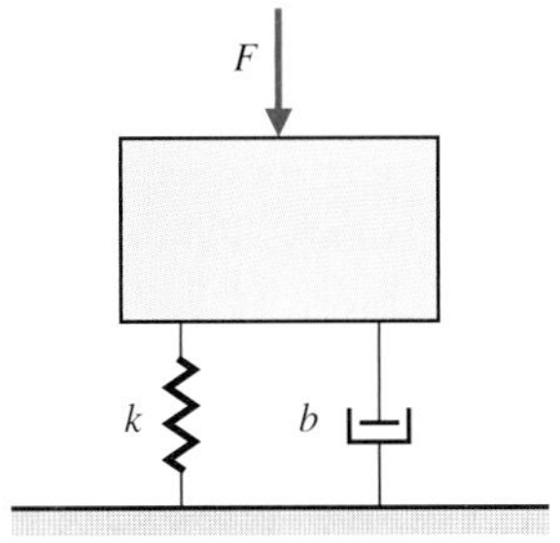

Machine schematic

Let the symbol D stand for derivative. Then we can write $F = cu' + ku = (cD + k)u$. Solving for u, we get $u = \dfrac{F}{cD + k}$. Now, writing the equation $u'' + cu' + ku = f_0\sin\omega t$ as $(D^2 + cD + k)u = f_0\sin\omega t$, we solve for u and get $u = \dfrac{f_0\sin\omega t}{D^2 + cD + k}$. Setting the two expressions for u equal to each other, we have

$$\frac{F}{cD + k} = \frac{f_0\sin\omega t}{D^2 + cD + k}.$$

Multiplying this out, we have
$(D^2 + cD + k)F = (cD + k)f_0 \sin \omega t.$
Show that this gives the correct answer: that is, $F(t)$ satisfies the equation
$F'' + cF' + kF = c(f_0 \sin \omega t)' + k(f_0 \sin \omega t).$

2. Spring devices are used in a variety of mechanisms, including the railroad car coupler shown in the photo. The coupler allows the railroad cars a certain amount of slack but applies a restoring force if the cars get too close or too far apart. If y measures the displacement of the coupler back and forth, then $y'' = F(y)$, where $F(y)$ is the force produced by the coupler. A simple model is $F(y) = \begin{cases} -y - d & \text{if } y \le -d \\ 0 & \text{if } -d \le y \le d. \\ -y + d & \text{if } y \ge d \end{cases}$ This models a restoring force with a dead zone in the middle. Suppose the initial conditions are $y(0) = 0$ and $y'(0) = 1$. That is, the coupler is centered at $y = 0$ and has a positive velocity. At $y = 0$, the coupler is in the dead zone with no forces. Solve $y'' = 0$ with the initial conditions and show that $y(t) = t$ for $0 \le t \le d$. At this point, the coupler leaves the dead zone and we now have $y'' = -y + d$. Explain why initial conditions for this part of the solution are $y(d) = d$ and $y'(d) = 1$. Solve this problem and determine the time at which the coupler reenters the dead zone. Continue in this fashion to construct the solution piece by piece. Describe in words the pattern that emerges. Then, explain in which sense this model ignores damping. Revise the function $F(y)$ to include damping.

15.3 APPLICATIONS OF SECOND-ORDER EQUATIONS

In sections 15.1 and 15.2, we developed models of spring-mass systems with and without external forces. Surprisingly, the charge in a simple electrical circuit can be modeled with the same equation as for the motion of a spring-mass system. An *RLC*-circuit consists of resistors, capacitors, inductors and a voltage source. The net resistance R (measured in ohms), the capacitance C (in farads) and the inductance L (in henrys) are all positive. For now, we will assume that there is no impressed voltage. If $Q(t)$ (coulombs) is the total charge on the capacitor at time t and $I(t)$ is the current, then $I = Q'(t)$. The basic laws of electricity tell us that

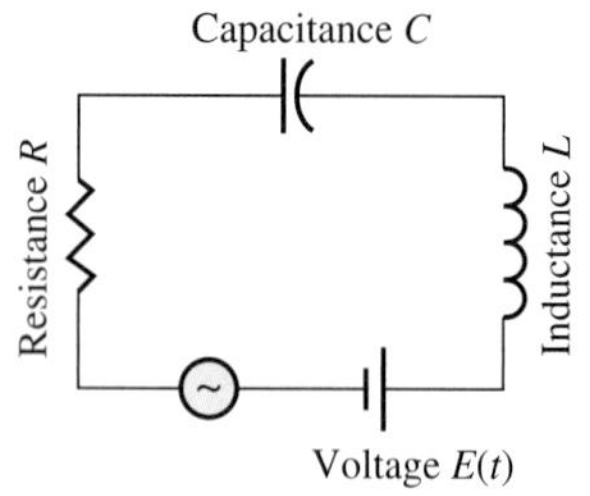

the voltage drop across the resistors is IR,
the voltage drop across the capacitor is $\dfrac{Q}{C}$
and the voltage drop across the inductor is $LI'(t)$.

These voltage drops must sum to the impressed voltage. If there is none, then

$$LI'(t) + RI(t) + \frac{1}{C}Q(t) = 0$$

or

$$LQ''(t) + RQ'(t) + \frac{1}{C}Q(t) = 0. \tag{3.1}$$

Observe that this is the same as equation (1.2), except for the names of the constants. Example 3.1 works the same as the examples from section 15.1.

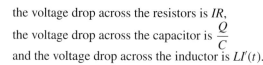

EXAMPLE 3.1 Finding the Charge in an Electrical Circuit

A series circuit has an inductor of 0.2 henry, a resistor of 300 ohms and a capacitor of 10^{-5} farad. The initial charge on the capacitor is 10^{-6} coulomb and there is no initial current. Find the charge on the capacitor and the current at any time t.

Solution From (3.1), with $L = 0.2$, $R = 300$ and $C = 10^{-5}$, the equation for the charge is

$$0.2Q''(t) + 300Q'(t) + 100{,}000Q(t) = 0$$

or

$$Q''(t) + 1500Q'(t) + 500{,}000Q(t) = 0.$$

The characteristic equation is then

$$0 = r^2 + 1500r + 500{,}000 = (r + 500)(r + 1000),$$

so that the roots are $r = -500$ and $r = -1000$. The general solution is then

$$Q(t) = c_1e^{-500t} + c_2e^{-1000t}. \tag{3.2}$$

The initial conditions are $Q(0) = 10^{-6}$ and $Q'(0) = 0$ [since $Q'(t)$ gives the current]. This gives us

$$10^{-6} = Q(0) = c_1 + c_2$$

and

$$0 = Q'(0) = -500c_1 - 1000c_2,$$

from which we obtain $c_1 = -2c_2$. The first equation now gives us $c_1 = 2 \times 10^{-6}$. The charge function is then

$$Q(t) = 10^{-6}(2e^{-500t} - e^{-1000t}).$$

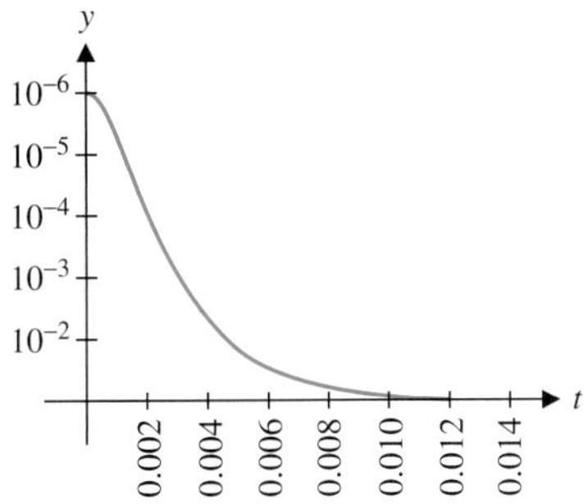

FIGURE 15.10
$Q(t) = 10^{-6}(2e^{-500t} - e^{-1000t})$

The graph in Figure 15.10 shows a rapidly declining charge. The current function is simply the derivative of the charge function. That is,

$$I(t) = -10^{-3}(e^{-500t} - e^{-1000t}). \blacksquare$$

If an impressed voltage $E(t)$ from a power supply is added to the circuit of example 3.1, equation (3.1) is replaced by the nonhomogeneous equation

$$LQ''(t) + RQ'(t) + \frac{1}{C}Q(t) = E(t). \tag{3.3}$$

Notice that here, the impressed voltage plays a role equivalent to the external force in a spring-mass system. We can use the techniques of section 15.2 to solve such an equation, as we illustrate in example 3.2.

EXAMPLE 3.2 Finding the Charge in a Circuit with an Impressed Voltage

Suppose that the circuit of example 3.1 is attached to an alternating current power supply with the impressed voltage $E(t) = 170\sin(120\pi t)$ volts. Find the steady-state charge on the capacitor and the steady-state current.

Solution From (3.3) using the values of example 3.1, we obtain the equation for the charge:

$$0.2Q''(t) + 300Q'(t) + 100{,}000Q(t) = 170\sin(120\pi t)$$

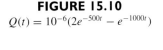

or
$$Q''(t) + 1500Q'(t) + 500{,}000Q(t) = 850\sin(120\pi t). \tag{3.4}$$

As in example 3.1, the roots of the characteristic equation are $r = -500$ and $r = -1000$, so that the solution of the homogeneous equation is $c_1e^{-500t} + c_2e^{-1000t}$. Since this part of the solution tends to 0 as t increases, the steady-state solution is simply the particular solution, which here has the form

$$Q_p(t) = A\sin(120\pi t) + B\cos(120\pi t).$$

This gives us

$$Q_p'(t) = 120\pi A\cos(120\pi t) - 120\pi B\sin(120\pi t)$$

and
$$Q_p''(t) = -14{,}400\pi^2 A\sin(120\pi t) - 14{,}400\pi^2 B\cos(120\pi t).$$

Substituting these into (3.4) gives us

$$\begin{aligned}850\sin(120\pi t) &= [-14{,}400\pi^2 A\sin(120\pi t) - 14{,}400\pi^2 B\cos(120\pi t)]\\ &\quad + 1500\,[120\pi A\cos(120\pi t) - 120\pi B\sin(120\pi t)]\\ &\quad + 500{,}000\,[A\sin(120\pi t) + B\cos(120\pi t)]\\ &= [(500{,}000 - 14{,}400\pi^2)A - 180{,}000\pi B]\sin(120\pi t)\\ &\quad + [(500{,}000 - 14{,}400\pi^2)B + 180{,}000A]\cos(120\pi t).\end{aligned}$$

Matching up the coefficients of $\sin(120\pi t)$ and $\cos(120\pi t)$ gives us

$$(500{,}000 - 14{,}400\pi^2)A - 180{,}000\pi B = 850$$

and
$$(500{,}000 - 14{,}400\pi^2)B + 180{,}000\pi A = 0.$$

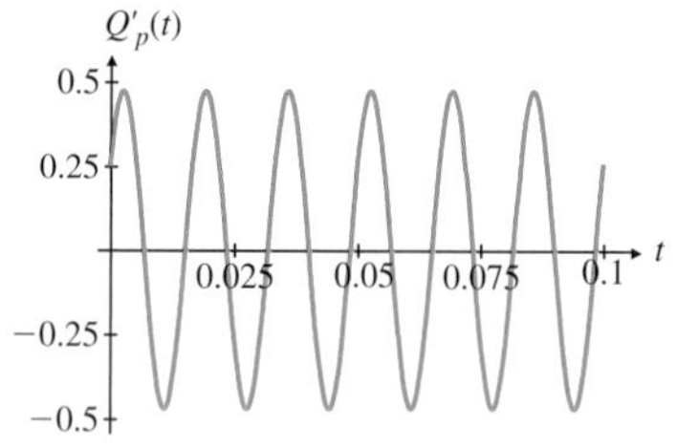

FIGURE 15.11
Steady-state current

Solving for A and B, we get the approximate values $A \approx 0.000679$ and $B \approx -0.00107$. The steady-state charge is then approximately

$$Q_p(t) \approx 0.000679\sin(120\pi t) - 0.00107\cos(120\pi t),$$

which gives us a steady-state current of

$$Q_p'(t) \approx 0.2561\cos(120\pi t) + 0.4046\sin(120\pi t).$$

We show a graph of this in Figure 15.11.

Notice that this is an alternating current with amplitude $\sqrt{(0.2561)^2 + (0.4046)^2} \approx 0.4788$ and the same 60 hertz (cycles per second) as the power supply. The large resistance in this circuit has greatly reduced the current. ■

The properties of electrical circuits are sometimes summarized in a frequency response curve, as constructed in example 3.3. The numbers are simplified so that we can illustrate a basic principle behind radio reception. For convenience, we use the fact that the solution of the system of equations

$$\begin{aligned}c_1A + c_2B &= d_1\\ c_3A + c_4B &= d_2\end{aligned}$$

can be written in the form

$$A = \frac{c_4d_1 - c_2d_2}{c_1c_4 - c_2c_3} \quad\text{and}\quad B = \frac{c_1d_2 - c_3d_1}{c_1c_4 - c_2c_3}, \tag{3.5}$$

provided $c_1c_4 - c_2c_3 \neq 0$.

EXAMPLE 3.3 A Frequency Response Curve

For a circuit whose charge satisfies $u'' + 8u' + 2532u = \sin \omega t$, find the amplitude of the steady-state solution as a function f of the external frequency ω and plot the resulting **frequency response curve** $y = f(\omega)$. Explain why this circuit could be useful for tuning in a radio station.

Solution We leave it as an exercise to show that the solution of the homogeneous equation tends to 0 as t increases. The steady-state solution is then the particular solution $u_p(t) = A \sin \omega t + B \cos \omega t$. Here, we have

$$u_p'(t) = A\omega \cos \omega t - B\omega \sin \omega t$$

and

$$u_p''(t) = -A\omega^2 \sin \omega t - B\omega^2 \cos \omega t.$$

Substituting into the equation, we have

$$\begin{aligned} \sin \omega t &= (-A\omega^2 \sin \omega t - B\omega^2 \cos \omega t) + 8(A\omega \cos \omega t - B\omega \sin \omega t) \\ &\quad + 2532(A \sin \omega t + B \cos \omega t) \\ &= [(2532 - \omega^2)A - 8B\omega] \sin \omega t + [8A\omega + (2532 - \omega^2)B] \cos \omega t. \end{aligned}$$

Equating the coefficients of the sine and cosine terms gives us the system of equations

$$(2532 - \omega^2)A - 8\omega B = 1$$

and

$$8\omega A + (2532 - \omega^2)B = 0.$$

From (3.5), the solution is

$$A = \frac{2532 - \omega^2}{(2532 - \omega^2)^2 + 64\omega^2} \quad \text{and} \quad B = \frac{-8\omega}{(2532 - \omega^2)^2 + 64\omega^2}.$$

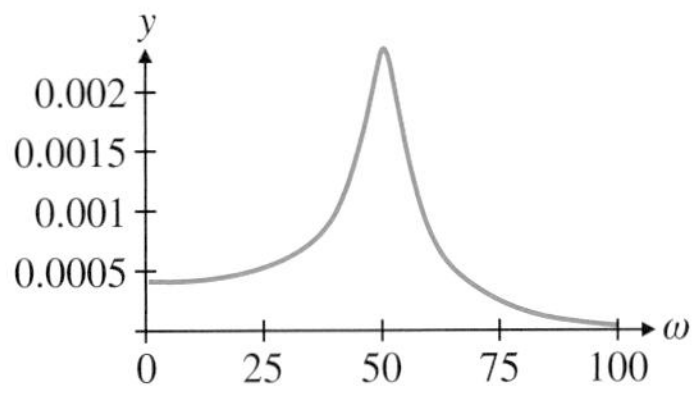

FIGURE 15.12
Frequency response curve $y = f(\omega)$

Without simplifying this, we can write the steady-state solution as $u_p(t) = A \sin \omega t + B \cos \omega t$ as $u_p(t) = \sqrt{A^2 + B^2} \sin(\omega t - \delta)$, for some constant δ, so that the amplitude of the steady-state solution is $\sqrt{A^2 + B^2}$. Notice that since A and B have the same denominator, it factors out of the square root and leaves us with

$$\begin{aligned} \sqrt{A^2 + B^2} &= \frac{1}{(2532 - \omega^2)^2 + 64\omega^2} \sqrt{(2532 - \omega^2)^2 + (-8\omega)^2} \\ &= \frac{1}{\sqrt{(2532 - \omega^2)^2 + 64\omega^2}}. \end{aligned}$$

The frequency response curve is the graph of this function, as shown in Figure 15.12. Notice that the graph has a sharp peak at about $\omega = 50$. Thinking of the right-hand side of the original equation, $\sin \omega t$, as a radio signal, we see that this circuit would "hear" the frequency $\omega = 50$ much better than any other frequency and could thus tune in on a radio station broadcasting at frequency 50. ■

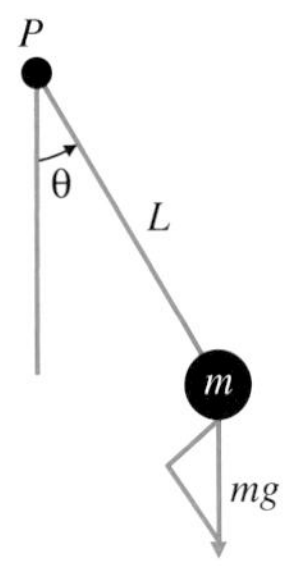

FIGURE 15.13
A simple pendulum

Another basic physical example with a surprising number of applications is the pendulum. In the sketch in Figure 15.13, a weight of mass m is attached to the end of a massless rod of length L that rotates about a pivot point P in two dimensions.

We first model the undamped pendulum, where the only force is due to gravity and the pendulum bob moves along a circular path centered at the pivot point. We can track its position s on the circle by measuring the angle θ from the vertical, where counterclockwise is positive. Since $s = L\theta$, the acceleration is $s'' = L\theta''$. The only force is gravity, which has magnitude mg in the downward direction. The component of gravity along the direction of motion is then $-mg\sin\theta$. Newton's second law of motion $F = ma$ gives us

$$mL\theta''(t) = -mg\sin\theta(t) \quad \text{or} \quad \theta''(t) + \frac{g}{L}\sin\theta(t) = 0. \tag{3.6}$$

Notice that (3.6) is *not* an equation of the form solved in sections 15.1 and 15.2, because of the term $\sin\theta(t)$. However, if we simplify (3.6) by replacing $\sin\theta(t)$ by $\theta(t)$, then (3.6) can be solved quite easily. This replacement is often justified with the statement. "For small angles θ, $\sin\theta$ is approximately equal to θ." As calculus students, you can say more. From the Maclaurin series

$$\sin\theta = \theta - \frac{\theta^3}{3!} + \frac{\theta^5}{5!} + \cdots,$$

it follows that the approximation $\sin\theta \approx \theta$ has an error bounded by $|\theta|^3/6$. So, if $|\theta|^3/6$ is small enough to safely neglect, then you can replace (3.6) with

$$\theta''(t) + \frac{g}{L}\theta(t) = 0. \tag{3.7}$$

This equation is easy to solve, as we see in example 3.4.

EXAMPLE 3.4 The Undamped Pendulum

A pendulum of length 5 cm satisfies equation (3.7). The bob is released from rest from a starting angle $\theta = 0.2$. Find an equation for the position at any time t and find the amplitude and period of the motion.

Solution Taking $g = 9.8$ m/s^2, we convert the length to $L = 0.05$ m. Then (3.7) becomes

$$\theta''(t) + 196\,\theta(t) = 0.$$

The characteristic equation is then $r^2 + 196 = 0$, so that $r = \pm 14i$ and the general solution is

$$\theta(t) = c_1 \sin 14t + c_2 \cos 14t,$$

so that

$$\theta'(t) = 14c_1 \cos 14t - 14c_2 \sin 14t.$$

Since the bob is *released* from rest, it has no initial velocity and so, the initial conditions are $\theta(0) = 0.2$ and $\theta'(0) = 0$. From these, we have that $0.2 = \theta(0) = c_2$ and $0 = \theta'(0) = 14c_1$, so that $c_1 = 0$. The solution is then

$$\theta(t) = 0.2\cos 14t,$$

which has amplitude 0.2 and period $\frac{2\pi}{14} = \frac{\pi}{7}$. A graph of the solution is shown in Figure 15.14. ■

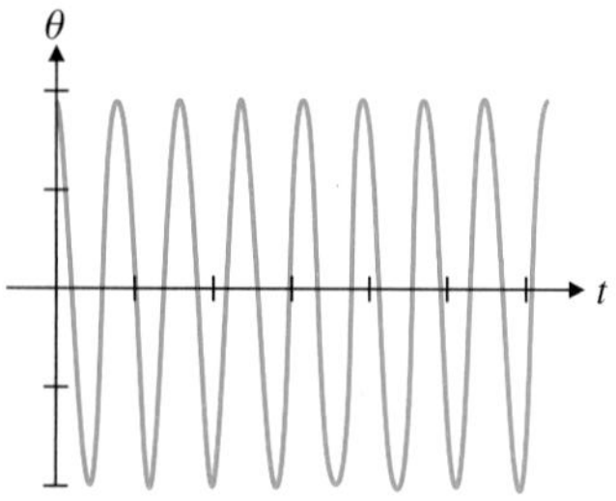

FIGURE 15.14
$\theta(t) = 0.2\cos 14t$

In the exercises, you will show that the period of any solution of (3.7) is $2\pi\sqrt{\frac{L}{g}}$, which gives an approximation of the period of the undamped pendulum. Notice that the period is independent of the mass but depends on the length L.

Observe that the pendulum of example 3.4 oscillates forever. Of course, the motion of a real pendulum dies out due to damping from friction at the pivot and air resistance. The simplest model of the force due to damping effects represents the damping as proportional to the velocity, or $k\theta'(t)$ for some constant k. Retaining the approximation $\sin\theta \approx \theta$ yields the following model for the damped pendulum:

$$\theta''(t) + k\theta'(t) + \frac{g}{L}\theta(t) = 0,$$

for some constant $k > 0$. If we further allow the pendulum to be driven by some external force $F(t)$, we have the more general model

$$\theta''(t) + k\theta'(t) + \frac{g}{L}\theta(t) = \frac{1}{m}F(t). \tag{3.8}$$

Several areas of current biological research involve situations where one periodic quantity serves as input into some other system that is naturally periodic. The effect of sunlight on circadian rhythms and the response of the heart to electrical signals from the sinoatrial node are examples of this phenomenon. In example 3.5, we explore what happens when a small amount of damping is present.

EXAMPLE 3.5 A Damped Forced Pendulum

For a pendulum of weight 2 pounds, length 6 inches, damping constant $k = 0.1$ and forcing function $F(t) = 0.5\sin 4t$, find the amplitude and period of the steady-state motion.

Solution Using $g = 32\text{ ft/s}^2$, we have $L = \frac{1}{2}$ ft and $m = \frac{2}{32}$ slug since weight $= mg$. Equation (3.8) then becomes

$$\theta''(t) + 0.1\theta'(t) + 64\,\theta(t) = 8\sin 4t.$$

We leave it as an exercise to show that the solution of the homogeneous equation approaches 0 as t increases. The steady-state solution is then the particular solution

$$\theta_p(t) = A\sin 4t + B\cos 4t.$$

This gives us
$$\theta_p'(t) = 4A\cos 4t - 4B\sin 4t$$

and
$$\theta_p''(t) = -16A\sin 4t - 16B\cos 4t.$$

Substituting into the differential equation, we have

$$\begin{aligned} 8\sin 4t &= (-16A\sin 4t - 16B\cos 4t) + 0.1(4A\cos 4t - 4B\sin 4t) \\ &\quad + 64\,(A\sin 4t + B\cos 4t) \\ &= (48A - 0.4B)\sin 4t + (0.4A + 48B)\cos 4t. \end{aligned}$$

It follows that

$$48A - 0.4B = 8 \quad \text{and} \quad 0.4A + 48B = 0.$$

The solution of this system is $A = \frac{384}{2304.16} \approx 0.166655$ and $B = -\frac{3.2}{2304.16} \approx -0.001389$. The steady-state solution can now be rewritten as

$$\theta_p(t) = A\sin 4t + B\cos 4t = \sqrt{A^2 + B^2}\sin(4t - \delta) \approx 0.16666\sin(4t - \delta),$$

so that the amplitude is approximately 0.16666 and the period is $\frac{2\pi}{4} = \frac{\pi}{2}$. ■

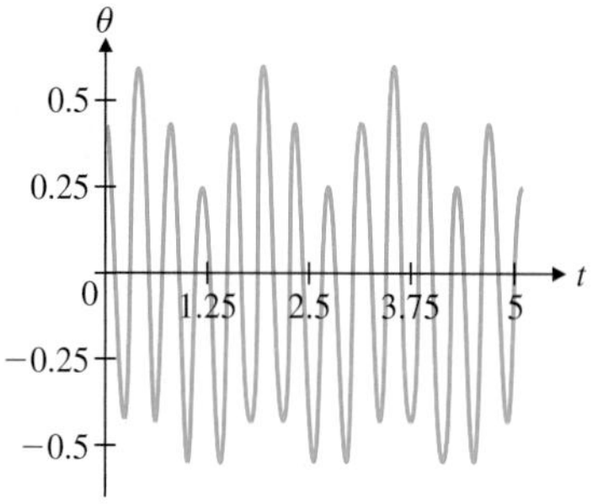

FIGURE 15.15a
$0 \le t \le 5$

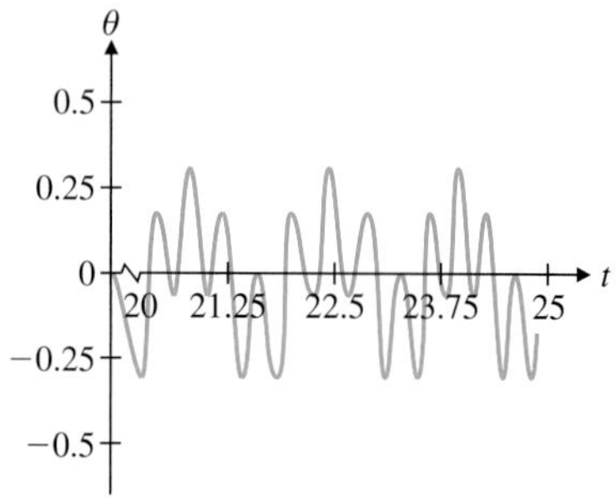

FIGURE 15.15b
$20 \le t \le 25$

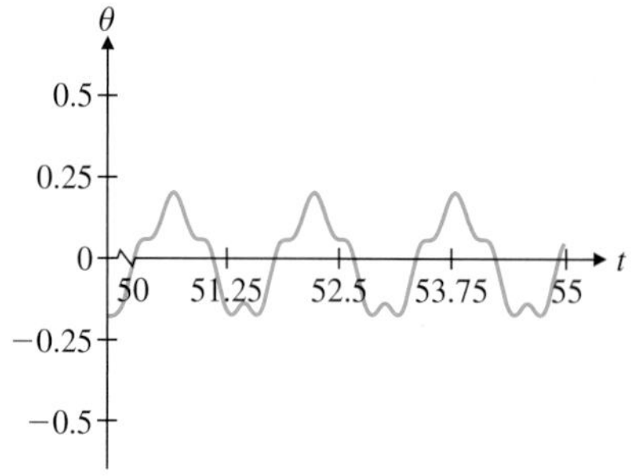

FIGURE 15.15c
$50 \le t \le 55$

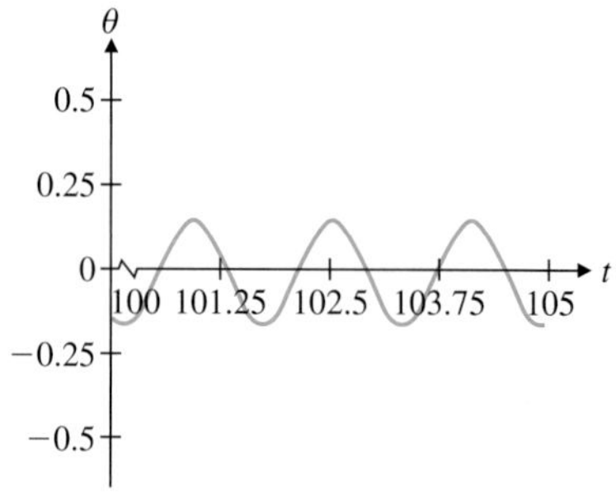

FIGURE 15.15d
$100 \le t \le 105$

Notice that the period of the steady-state solution in example 3.5 matches the period of the forcing function $8 \sin 4t$ and not the natural period of the undamped pendulum, $2\pi\sqrt{\frac{L}{g}} = \frac{\pi}{4}$. Keep in mind that the steady-state solution gives the behavior of the solution for very large t. For small t, the motion of the pendulum in example 3.5 can be erratic. For initial conditions $\theta(0) = 0.5$ and $\theta'(0) = 0$, the solution for $0 \le t \le 5$ is shown in Figure 15.15a, while Figures 15.15b–15.15d show the solutions for larger values of t. Notice that the solution seems to go through different stages until settling down to the steady-state solution around $t = 100$.

EXERCISES 15.3

WRITING EXERCISES

1. The correspondence between mechanical vibrations and electrical circuits is surprising. To start to understand the correspondence, develop an analogy between the roles of a resistor in a circuit and damping in spring motion. Continue by drawing an analogy between the roles of the spring force and the capacitor in storing and releasing energy.

2. In example 3.3, explain why the sharper the peak is on the frequency response curve, the clearer the radio reception would be.

3. For most objects, the magnitude of air drag is proportional to the *square* of the speed of the object. Explain why we would not want to use that assumption in equation (3.8).

4. To understand why the forced pendulum behaves erratically, consider the case where a child is on a swing and you push the swing. If the swing is coming back at you, does your push increase or decrease the child's speed? If the swing is moving forward away from you, does your push increase or decrease the child's speed? If you push every three seconds and the

swing is not on a three-second cycle, describe how your pushing would affect the movement of the swing.

---

1. A series circuit has an inductor of 0.4 henry, a resistor of 200 ohms and a capacitor of 10^{-4} farad. The initial charge on the capacitor is 10^{-5} coulomb and there is no initial current. Find the charge on the capacitor and the current at any time t.

2. A series circuit has an inductor of 0.4 henry, no resistance and a capacitor of 10^{-4} farad. The initial charge on the capacitor is 10^{-5} coulomb and there is no initial current. Find the charge on the capacitor and the current at any time t. Find the amplitude and phase shift of the charge function. (See exercise 21 in section 15.1.)

3. A series circuit has an inductor of 0.2 henry, no resistance and a capacitor of 10^{-5} farad. The initial charge on the capacitor is 10^{-6} coulomb and there is no initial current. Find the charge on the capacitor and the current at any time t. Find the amplitude and phase shift of the charge function. (See exercise 21 in section 15.1.)

4. A series circuit has an inductor of 0.6 henry, a resistor of 400 ohms and a capacitor of 2×10^{-4} farad. The initial charge on the capacitor is 10^{-6} coulomb and there is no initial current. Find the charge on the capacitor and the current at any time t.

5. A series circuit has an inductor of 0.5 henry, a resistor of 2 ohms and a capacitor of 0.05 farad. The initial charge on the capacitor is zero and the initial current is 1 A. A voltage source of $E(t) = 3\cos 2t$ volts is analogous to an external force. Find the charge on the capacitor and the current at any time t.

6. A series circuit has an inductor of 0.2 henry, a resistor of 20 ohms and a capacitor of 0.1 farad. The initial charge on the capacitor is zero and there is no initial current. A voltage source of $E(t) = 0.4\cos 4t$ volts is analogous to an external force. Find the charge on the capacitor and the current at any time t.

7. A series circuit has an inductor of 1 henry, a resistor of 10 ohms and a capacitor of 0.5 farad. A voltage source of $E(t) = 0.1\cos 2t$ volts is analogous to an external force. Find the steady-state solution and identify its amplitude and phase shift. (See exercise 21 in section 15.1.)

8. A series circuit has an inductor of 0.2 henry, a resistor of 40 ohms and a capacitor of 0.05 farad. A voltage source of $E(t) = 0.2\ \sin 4t$ volts is analogous to an external force. Find the steady-state solution and identify its amplitude and phase shift. (See exercise 21 in section 15.1.)

Exercises 9–16 involve frequency response curves and Bode plots.

9. Suppose that the charge in a circuit satisfies the equation $x''(t) + 2x'(t) + 5x(t) = A_1 \sin \omega t$ for constants A_1 and ω. Find the steady-state solution and rewrite it in the form $A_2 \sin(\omega t + \delta)$, where $A_2 = \dfrac{A_1}{\sqrt{(5-\omega^2)^2 + 4\omega^2}}$. The ratio $\dfrac{A_2}{A_1}$ is called the **gain** of the circuit. Notice that it is independent of the actual value of A_1.

10. Graph the gain function $g(\omega) = \dfrac{1}{\sqrt{(5-\omega^2)^2 + 4\omega^2}}$ from exercise 9 as a function of $\omega > 0$. This is called a **frequency response curve.** Find $\omega > 0$ to maximize the gain by minimizing the function $f(\omega) = (5-\omega^2)^2 + 4\omega^2$. This value of ω is called the **resonant frequency** of the circuit. Also graph the **Bode plot** for this circuit, which is the graph of $20\log_{10} g$ as a function of $\log_{10}\omega$. (In this case, the units of $20\log_{10} g$ are decibels.)

11. The charge in a circuit satisfies the equation $x''(t) + 0.4x'(t) + 4x(t) = A\sin\omega t$. Find the gain function and the value of $\omega > 0$ that maximizes the gain, and graph the Bode plot of $20\log_{10} g$ as a function of $\log_{10}\omega$.

12. The charge in a circuit satisfies the equation $x''(t) + 0.4x'(t) + 5x(t) = A\sin\omega t$. Find the gain function and the value of $\omega > 0$ that maximizes the gain, and graph the Bode plot of $20\log_{10} g$ as a function of $\log_{10}\omega$.

13. The charge in a circuit satisfies the equation $x''(t) + 0.2x'(t) + 4x(t) = A\sin\omega t$. Find the gain function and the value of $\omega > 0$ that maximizes the gain, and graph the Bode plot of $20\log_{10} g$ as a function of $\log_{10}\omega$.

14. Based on your answers to exercises 11–13, which of the constants b, c and A affect the gain in the circuit described by $x''(t) + bx'(t) + cx(t) = A\sin\omega t$?

15. The motion of the arm of a seismometer is modeled by $y'' + by' + cy = \omega^2\cos\omega t$, where the horizontal shift of the ground during the earthquake is proportional to $\cos\omega t$. (See *Multimedia ODE Architect* for details.) If $b = 1$ and $c = 4$, find the gain function and the value of $\omega > 0$ that maximizes the gain.

16. The amplitude A of the motion of the seismometer in exercise 15 and the distance D of the seismometer from the epicenter of the earthquake determine the Richter measurement M through the formula $M = \log_{10} A + 2.56\log_{10} D - 1.67$. Use the result of exercise 15 to prove that A depends on the frequency of the horizontal motion as well as the actual horizontal distance moved. Explain in terms of the motion of the ground during an earthquake why the frequency affects the amount of damage done.

17. In exercise 10, we sketched the Bode plot of the gain as a function of frequency. The other Bode plot, of phase shift as a function of frequency, is considered here. First, recall that in the general relationship $a\sin\omega t + b\cos\omega t = B\sin(\omega t + \theta)$, we have $a = B\cos\theta$ and $b = B\sin\theta$. We can "solve" for θ as $\cos^{-1}\left(\dfrac{a}{B}\right)$ or $\sin^{-1}\left(\dfrac{b}{B}\right)$ or $\tan^{-1}\left(\dfrac{b}{a}\right)$. In exercise 10, we have $a = \dfrac{(5-\omega^2)A}{(5-\omega^2)^2 + 4\omega^2}$ and $b = \dfrac{-2\omega A}{(5-\omega^2)^2 + 4\omega^2}$, so

that $B = \sqrt{a^2+b^2} = \dfrac{A}{(5-\omega^2)^2+4\omega^2}$. For the frequencies $\omega > 0$, this tells us that $\sin\theta < 0$, so that θ is in quadrant III or IV. Explain why the functions $\sin^{-1}\left(\dfrac{b}{B}\right)$ and $\tan^{-1}\left(\dfrac{b}{a}\right)$ are not convenient for this range of angles. However, $-\cos^{-1}\left(\dfrac{a}{B}\right)$ gives the correct quadrants. Show that $\theta = -\cos^{-1}\left(\dfrac{5-\omega^2}{\sqrt{(5-\omega^2)^2+(2\omega)^2}}\right)$ and sketch the Bode plot.

18. Sketch the plot of phase shift versus frequency for exercise 11.

19. Show that if a, b and c are all positive numbers, then the solutions of $ay'' + by' + cy = 0$ approach 0 as $t \to \infty$.

20. For the electrical charge equation $LQ''(t) + RQ'(t) + \frac{1}{C}Q(t) = 0$, if there is nonzero resistance, what is the eventual charge on the capacitor?

21. Show that the gain in the general circuit described by $ax''(t)+bx'(t)+cx(t) = A\sin\omega t$ equals $\dfrac{1}{\sqrt{(c-a\omega^2)^2+(b\omega)^2}}$.

22. Show that in exercise 21 the general resonant frequency equals $\sqrt{\dfrac{2ac-b^2}{2a^2}}$.

23. A pendulum of length 10 cm satisfies equation (3.7). The bob is released from a starting angle $\theta = 0.2$. Find an equation for the position at any time and find the amplitude and period of the motion. Compare your solution to that of example 3.4. What effect does a change in length have?

24. Repeat exercise 23 with a starting angle of $\theta = 0.4$. What effect does doubling the starting angle have?

25. A pendulum of length 10 cm satisfies equation (3.7). The bob is released from a starting angle $\theta = 0$ with an initial angular velocity of $\theta' = 0.1$. Find an equation for the position at any time and find the amplitude and period of the motion.

26. Repeat exercise 25 with initial angular velocity $\theta' = 0.2$. What effect does doubling the initial angular velocity have?

27. For a pendulum of weight 6 pounds, length 8 inches, damping constant $k = 0.2$ and forcing function $F(t) = \cos 3t$, find the amplitude and period of the steady-state motion.

28. For a pendulum of weight 6 pounds, length 8 inches, damping constant $k = 0.2$ and forcing function $F(t) = \cos 6t$, find the amplitude and period of the steady-state motion. Compare your solution to that of exercise 27. Does the frequency of the forcing function affect the amplitude of the motion?

29. In example 3.3, find the general homogeneous solution and show that it approaches 0 as $t \to \infty$.

30. In example 3.5, find the general homogeneous solution and show that it approaches 0 as $t \to \infty$.

31. Use Taylor's Theorem to prove that the error in the approximation $\sin\theta \approx \theta$ is bounded by $|\theta|^3/6$.

32. To keep the error in the approximation $\sin\theta \approx \theta$ less than 0.01, how small does θ need to be?

33. Show that the solution of $\theta'' + \frac{g}{L}\theta = 0$ has period $2\pi\sqrt{\frac{L}{g}}$. Galileo deduced that the square of the period varies directly with the length. Is this consistent with a period of $2\pi\sqrt{\frac{L}{g}}$?

34. Galileo believed that the period of a pendulum is independent of the weight of the bob. Determine whether the model (3.7) is consistent with this prediction.

35. Galileo further believed that the period of a pendulum is independent of its amplitude. Use exercise 24 to determine whether the model (3.7) supports this conjecture.

36. Taking into account damping, Galileo found that a pendulum will eventually come to rest, with lighter ones coming to rest faster than heavy ones. Show that this is implied by (3.8) in that for pendulums of identical length and damping constant c (note that c is different from k) but different masses, the pendulum with the smaller mass will come to rest faster.

37. The gun of a tank is attached to a system with springs and dampers such that the displacement $y(t)$ of the gun after being fired at time 0 is

$$y'' + 2\alpha y' + \alpha^2 y = 0,$$

for some constant α. Initial conditions are $y(0) = 0$ and $y'(0) = 100$. Estimate α such that the quantity $y^2 + (y')^2$ is less than 0.01 at $t = 1$. This enables the gun to be fired again rapidly.

38. Let $G(t)$ be the concentration of glucose in the bloodstream and $g(t) = G(t) - G_0$ the difference between the glucose level and the ideal concentration G_0. Braun derives an equation of the form

$$g''(t) + 2\alpha g'(t) + \omega^2 g(t) = 0$$

for the concentration t hours after a glucose injection. It turns out that if the natural period $\dfrac{2\pi}{\omega}$ of the solution is less than 4 hours, the patient is not likely to be diabetic, whereas $\dfrac{2\pi}{\omega} > 4$ is an indicator of mild diabetes. Using $\alpha = 1$ and initial conditions $g(0) = 10$ and $g'(0) = 0$, compare the graphs of glucose levels for a healthy patient with $\omega = 2$ and a diabetic patient with $\omega = 1$.

39. If $0 < \alpha < \omega$, show that the solution in exercise 38 is a damped exponential. Show that the time between zeros of the solution is greater than $\dfrac{\pi}{\omega}$. Use this result to determine whether the following patient would be suspected of diabetes. The optimal glucose level is 75 mg glucose/100 ml blood. The glucose levels are 90 mg glucose/100 ml blood one hour after an injection, 70 mg glucose/100 ml blood two hours after the injection and 78 mg glucose/100 ml blood three hours after the injection.

40. Show that the data in exercise 39 are inconsistent with the case $0 < \omega < \alpha$.

41. Consider an *RLC*-circuit with capacitance C and charge $Q(t)$ at time t. The energy in the circuit at time t is given by $u(t) = \frac{[Q(t)]^2}{2C}$. Show that the charge in a general *RLC*-circuit has the form

$$Q(t) = e^{-(R/L)t/2}|Q_0 \cos \omega t + c_2 \sin \omega t|,$$

where $Q_0 = Q(0)$ and $\omega = \frac{1}{2L}\sqrt{R^2 - 4L/C}$. The relative energy loss from time $t = 0$ to time $t = \frac{2\pi}{\omega}$ is given by $U_{loss} = \frac{u(2\pi/\omega) - u(0)}{u(0)}$ and the **inductance quality factor** is defined by $\frac{2\pi}{U_{loss}}$. Using a Taylor polynomial approximation of e^x, show that the inductance quality factor is approximately $\omega \frac{L}{R}$.

EXPLORATORY EXERCISES

1. In quantum mechanics, the possible locations of a particle are described by its **wave function** $\Psi(x)$. The wave function satisfies **Schrödinger's wave equation**

$$\frac{\hbar}{2m}\Psi''(x) + V(x)\Psi(x) = E\Psi(x).$$

Here, $\hbar$ is Planck's constant, m is mass, $V(x)$ is the potential function for external forces and E is the particle's energy. In the case of a bound particle with an infinite square well of width $2a$, the potential function is $V(x) = 0$ for $-a \le x \le a$. We will show that the particle's energy is quantized by solving the **boundary value problem** consisting of the differential equation $\frac{\hbar}{2m}\Psi''(x) + v(x)\Psi(x) = E\Psi(x)$ plus the boundary conditions $\Psi(-a) = 0$ and $\Psi(a) = 0$. The theory of boundary value problems is different from that of the initial value problems in this chapter, which typically have unique solutions. In fact, in this exercise we specifically want more than one solution. Start with the differential equation and show that for $V(x) = 0$; the general solution is $\Psi(x) = c_1 \cos kx + c_2 \sin kx$, where $k = \sqrt{2mE}/\hbar$. Then set up the equations $\Psi(-a) = 0$ and $\Psi(a) = 0$. Both equations are true if $c_1 = c_2 = 0$, but in this case the solution would be $\Psi(x) = 0$. To find **nontrivial solutions** (that is, nonzero solutions), find all values of k such that $\cos ka = 0$ or $\sin ka = 0$. Then, solve for the energy E in terms of a, m and $\hbar$. These are the only allowable energy levels for the particle. Finally, determine what happens to the energy levels as a increases without bound.

2. Imagine a hole drilled through the center of the Earth. What would happen to a ball dropped in the hole? Galileo conjectured that the ball would undergo **simple harmonic motion,** which is the periodic motion of an undamped spring or pendulum. This solution requires no friction and a nonrotating Earth. The force due to gravity of two objects r units apart is $\frac{Gm_1m_2}{r^2}$, where G is the universal gravitation constant and m_1 and m_2 are the masses of the objects. Let R be the radius of the Earth and y the displacement from the center of the Earth.

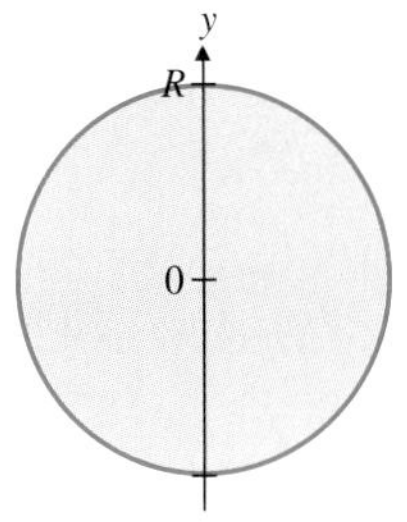

For a ball at position y with $|y| \le R$, the ball is attracted to the center of the Earth as if the Earth were a single particle located at the origin with mass ρv, where ρ is the density of the Earth and v is the volume of the sphere of radius $|y|$. (This assumes a constant density and a spherical Earth.) If M is the mass of the Earth, show that if you neglect damping, the position of the ball satisfies the equation $y'' + \frac{GM}{R^3}y = 0$. Use $g = \frac{GM}{R^2}$ to simplify this. Find the motion of the ball. Does the period depend on the starting position? Compare the motions of balls dropped simultaneously from the Earth's surface and halfway to the center of the Earth. Explore the motion of a ball thrown from the surface of the Earth at $y = R$ with initial velocity $-R/100$.

15.4 POWER SERIES SOLUTIONS OF DIFFERENTIAL EQUATIONS

So far in this chapter, we have seen how to solve only those second-order equations with constant coefficients, such as

$$y'' - 6y' + 9y = 0.$$

What if the coefficients aren't constant? For instance, suppose you wanted to solve the equation

$$y'' + 2xy' + 2y = 0.$$

We leave it as an exercise to show that substituting $y = e^{rx}$ in this case does not lead to a solution. However, in many cases such as this, we can find a solution by assuming that the solution can be written as a power series, such as

$$y = \sum_{n=0}^{\infty} a_n x^n.$$

The idea is to substitute this series into the differential equation and then use the resulting equation to determine the coefficients, $a_0, a_1, a_2 \ldots, a_n$. Before we see how to do this in general, we illustrate this for a simple equation whose solution is already known, to demonstrate that we arrive at the same solution using either method.

EXAMPLE 4.1 Power Series Solution of a Differential Equation

Use a power series to determine the general solution of

$$y'' + y = 0.$$

Solution First, observe that this equation has constant coefficients and its general solution is

$$y = c_1 \sin x + c_2 \cos x,$$

where c_1 and c_2 are constants.

We now look for a solution of the equation in the form of the power series

$$y = a_0 + a_1 x + a_2 x^2 + a_3 x^3 + \cdots = \sum_{n=0}^{\infty} a_n x^n.$$

To substitute this into the equation, we first need to obtain representations for y' and y''. Assuming that the power series is convergent and has a positive radius of convergence, recall that we can differentiate term-by-term to obtain the derivatives

$$y' = a_1 + 2a_2 x + 3a_3 x^2 + \cdots = \sum_{n=1}^{\infty} n a_n x^{n-1}$$

and

$$y'' = 2a_2 + 6a_3 x + \cdots = \sum_{n=2}^{\infty} n(n-1) a_n x^{n-2}.$$

Substituting these power series into the differential equation, we get

$$0 = y'' + y = \sum_{n=2}^{\infty} n(n-1) a_n x^{n-2} + \sum_{n=0}^{\infty} a_n x^n. \tag{4.1}$$

REMARK 4.1

Notice that when we change $\sum_{n=2}^{\infty} n(n-1) a_n x^{n-2}$ to $\sum_{n=0}^{\infty} (n+2)(n+1) a_{n+2} x^n$, the index in the sequence increases by 2 (for example, a_n becomes a_{n+2}), while the initial value of the index decreases by 2.

The immediate objective here is to combine the two series in (4.1) into one power series. Since the powers in the one series are of the form x^{n-2} and in the other series are of the form x^n, we will first need to rewrite one of the two series. Notice that we have

$$y'' = \sum_{n=2}^{\infty} n(n-1) a_n x^{n-2} = 2a_2 + 3 \cdot 2a_3 x + 4 \cdot 3a_4 x^2 + \cdots$$

$$= \sum_{n=0}^{\infty} (n+2)(n+1) a_{n+2} x^n.$$

Substituting this into equation (4.1) gives us

$$0 = y'' + y = \sum_{n=0}^{\infty}(n+2)(n+1)a_{n+2}x^n + \sum_{n=0}^{\infty} a_n x^n$$

$$= \sum_{n=0}^{\infty}[(n+2)(n+1)a_{n+2} + a_n]x^n. \tag{4.2}$$

Read equation (4.2) carefully; it says that the power series on the right converges to the constant function $f(x) = 0$. In view of this, all of the coefficients must be zero. That is,

$$0 = (n+2)(n+1)a_{n+2} + a_n,$$

for $n = 0, 1, 2, \ldots$. We solve this for the coefficient with the *largest* index, to obtain

$$a_{n+2} = \frac{-a_n}{(n+2)(n+1)}, \tag{4.3}$$

for $n = 0, 1, 2, \ldots$. Equation (4.3) is called the **recurrence relation,** which we use to determine all of the coefficients of the series solution. The general idea is to write out (4.3) for a number of specific values of n and then try to recognize a pattern that the coefficients follow. From (4.3), we have for the even-indexed coefficients that

$$a_2 = \frac{-a_0}{2\cdot 1} = \frac{-1}{2!}a_0,$$

$$a_4 = \frac{-a_2}{4\cdot 3} = \frac{1}{4\cdot 3\cdot 2\cdot 1}a_0 = \frac{1}{4!}a_0,$$

$$a_6 = \frac{-a_4}{6\cdot 5} = \frac{-1}{6!}a_0,$$

$$a_8 = \frac{-a_6}{8\cdot 7} = \frac{1}{8!}a_0$$

and so on. (Try to write down a_{10} by recognizing the pattern, without referring to the recurrence relation.) Since we can write each even-indexed coefficient as a_{2n}, for some n, we can now write down a simple formula that works for any of these coefficients. We have

$$a_{2n} = \frac{(-1)^n}{(2n)!}a_0,$$

for $n = 0, 1, 2, \ldots$. Similarly, using (4.3), we have that the odd-indexed coefficients are

$$a_3 = \frac{-a_1}{3\cdot 2} = \frac{-1}{3!}a_1,$$

$$a_5 = \frac{-a_3}{5\cdot 4} = \frac{1}{5!}a_1,$$

$$a_7 = \frac{-a_5}{7\cdot 6} = \frac{-1}{7!}a_1,$$

$$a_9 = \frac{-a_7}{9\cdot 8} = \frac{1}{9!}a_1$$

and so on. Since we can write each odd-indexed coefficient as a_{2n+1} (or alternatively as a_{2n-1}), for some n, note that we have the following simple formula for the odd-indexed coefficients:

$$a_{2n+1} = \frac{(-1)^n}{(2n+1)!}a_1,$$

for $n = 0, 1, 2, \ldots$. Since we have now written every coefficient in terms of either a_0 or a_1, we can rewrite the solution by separating the a_0 terms from the a_1 terms. We have

$$\begin{aligned} y &= \sum_{n=0}^{\infty} a_n x^n = a_0 + a_1 x + a_2 x^2 + a_3 x^3 + \cdots \\ &= a_0 \left(1 - \frac{1}{2!}x^2 + \frac{1}{4!}x^4 + \cdots\right) + a_1 \left(x - \frac{1}{3!}x^3 + \frac{1}{5!}x^5 + \cdots\right) \\ &= a_0 \underbrace{\sum_{n=0}^{\infty} \frac{(-1)^n}{(2n)!} x^{2n}}_{y_1(x)} + a_1 \underbrace{\sum_{n=0}^{\infty} \frac{(-1)^n}{(2n+1)!} x^{2n+1}}_{y_2(x)} \\ &= a_0 y_1(x) + a_1 y_2(x), \end{aligned} \tag{4.4}$$

where $y_1(x)$ and $y_2(x)$ are two solutions of the differential equation (assuming the series converge). At this point, you should be able to easily check that both of the indicated power series converge absolutely for all x, by using the Ratio Test. Beyond this, you might also recognize that the series solutions $y_1(x)$ and $y_2(x)$ that we obtained are, in fact, the Maclaurin series expansions of $\cos x$ and $\sin x$, respectively. In light of this, (4.4) is an equivalent solution to that found by using the methods of section 15.1. ■

The method used to solve the differential equation in example 4.1 is certainly far more complicated than the methods we used in section 15.1 for solving the same equation. However, this new method can be used to solve a wider range of differential equations than those solvable using our earlier methods. We now return to the equation mentioned in the introduction to this section.

EXAMPLE 4.2 Solving a Differential Equation with Variable Coefficients

Find the general solution of the differential equation

$$y'' + 2xy' + 2y = 0.$$

Solution First, observe that since the coefficient of y' is not constant, we have little choice but to look for a series solution of the equation. As in example 4.1, we begin by assuming that we may write the solution as a power series,

$$y = \sum_{n=0}^{\infty} a_n x^n.$$

As before, we have

$$y' = \sum_{n=1}^{\infty} n a_n x^{n-1}$$

and

$$y'' = \sum_{n=2}^{\infty} n(n-1) a_n x^{n-2}.$$

Substituting these three power series into the equation, we get

$$\begin{aligned} 0 = y'' + 2xy' + 2y &= \sum_{n=2}^{\infty} n(n-1)a_n x^{n-2} + 2x \sum_{n=1}^{\infty} n a_n x^{n-1} + 2 \sum_{n=0}^{\infty} a_n x^n \\ &= \sum_{n=2}^{\infty} n(n-1)a_n x^{n-2} + \sum_{n=1}^{\infty} 2n a_n x^n + \sum_{n=0}^{\infty} 2a_n x^n, \end{aligned} \tag{4.5}$$

where in the middle term, we moved the x into the series and combined powers of x. In order to combine the three series, we must only rewrite the first series so that its general term is a multiple of x^n, instead of x^{n-2}. As we did in example 4.1, we write

$$\sum_{n=2}^{\infty} n(n-1)a_n x^{n-2} = \sum_{n=0}^{\infty}(n+2)(n+1)a_{n+2}x^n$$

and so, from (4.5), we have

$$\begin{aligned} 0 &= \sum_{n=2}^{\infty} n(n-1)a_n x^{n-2} + \sum_{n=1}^{\infty} 2na_n x^n + \sum_{n=0}^{\infty} 2a_n x^n \\ &= \sum_{n=0}^{\infty}(n+2)(n+1)a_{n+2}x^n + \sum_{n=0}^{\infty} 2na_n x^n + \sum_{n=0}^{\infty} 2a_n x^n \\ &= \sum_{n=0}^{\infty}[(n+2)(n+1)a_{n+2} + 2na_n + 2a_n]x^n \\ &= \sum_{n=0}^{\infty}[(n+2)(n+1)a_{n+2} + 2(n+1)a_n]x^n. \qquad (4.6) \end{aligned}$$

To get this, we used the fact that $\sum_{n=1}^{\infty} 2na_n x^n = \sum_{n=0}^{\infty} 2na_n x^n$. (Notice that the first term in the series on the right is zero!) Reading equation (4.6) carefully, note that we again have a power series converging to the zero function, from which it follows that all of the coefficients must be zero:

$$0 = (n+2)(n+1)a_{n+2} + 2(n+1)a_n,$$

for $n = 0, 1, 2, \ldots$. Again solving for the coefficient with the largest index, we get the recurrence relation

REMARK 4.2

> Always solve for the coefficient with the largest index.

$$a_{n+2} = -\frac{2(n+1)a_n}{(n+2)(n+1)}$$

or

$$a_{n+2} = -\frac{2a_n}{n+2}.$$

Much like we saw in example 4.1, the recurrence relation tells us that all of the even-indexed coefficients are related to a_0 and all of the odd-indexed coefficients are related to a_1. In order to try to recognize the pattern, we write out a number of terms, using the recurrence relation. We have

$$\begin{aligned} a_2 &= -\frac{2}{2}a_0 = -a_0, \\ a_4 &= -\frac{2}{4}a_2 = \frac{1}{2}a_0, \\ a_6 &= -\frac{2}{6}a_4 = -\frac{1}{3!}a_0, \\ a_8 &= -\frac{2}{8}a_6 = \frac{1}{4!}a_0 \end{aligned}$$

and so on. At this point, you should recognize the pattern for these coefficients. (If not, write out a few more terms.) Note that we can write the even-indexed coefficients as

$$a_{2n} = \frac{(-1)^n}{n!}a_0,$$

for $n = 0, 1, 2, \ldots$. Be sure to match this formula against those coefficients calculated above to see that they match. Continuing with the odd-indexed coefficients, we have from the recurrence relation that

$$a_3 = -\frac{2}{3}a_1,$$

$$a_5 = -\frac{2}{5}a_3 = \frac{2^2}{5\cdot 3}a_1,$$

$$a_7 = -\frac{2}{7}a_5 = -\frac{2^3}{7\cdot 5\cdot 3}a_1,$$

$$a_9 = -\frac{2}{9}a_7 = \frac{2^4}{9\cdot 7\cdot 5\cdot 3}a_1$$

and so on. While you might recognize the pattern here, it's hard to write down this pattern succinctly. Observe that the products in the denominators are not quite factorials. Rather, each is the product of the first so many odd numbers. The solution to this is to write this as a factorial, but then cancel out all of the even integers in the product. In particular, note that

$$\frac{1}{9\cdot 7\cdot 5\cdot 3} = \frac{\overbrace{8}^{2\cdot 4}\cdot\overbrace{6}^{2\cdot 3}\cdot\overbrace{4}^{2\cdot 2}\cdot\overbrace{2}^{2\cdot 1}}{9!} = \frac{2^4\cdot 4!}{9!},$$

so that a_9 becomes

$$a_9 = \frac{2^4}{9\cdot 7\cdot 5\cdot 3}a_1 = \frac{2^4\cdot 2^4\cdot 4!}{9!}a_1 = \frac{2^{2\cdot 4}\cdot 4!}{9!}a_1.$$

More generally, we now have

$$a_{2n+1} = \frac{(-1)^n 2^{2n} n!}{(2n+1)!}a_1,$$

for $n = 0, 1, 2 \ldots$.

Now that we have expressions for all of the coefficients, we can write the solution of the differential equation as

$$\begin{aligned} y &= \sum_{n=0}^{\infty} a_n x^n = \sum_{n=0}^{\infty}(a_{2n}x^{2n} + a_{2n+1}x^{2n+1}) \\ &= a_0\underbrace{\sum_{n=0}^{\infty}\frac{(-1)^n}{n!}x^{2n}}_{y_1(x)} + a_1\underbrace{\sum_{n=0}^{\infty}\frac{(-1)^n 2^{2n} n!}{(2n+1)!}x^{2n+1}}_{y_2(x)} \\ &= a_0 y_1(x) + a_1 y_2(x), \end{aligned}$$

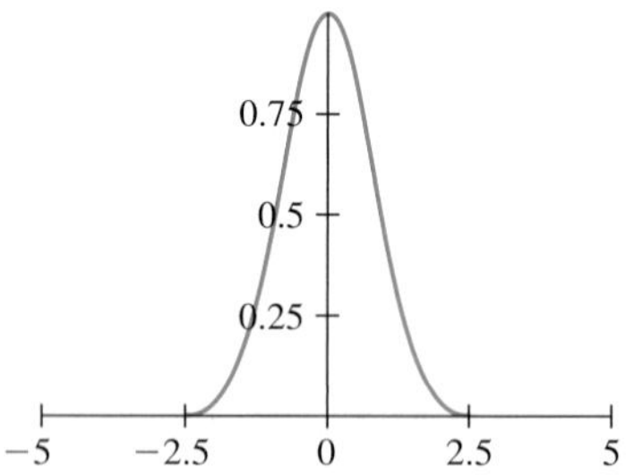

FIGURE 15.16a
$y = y_1(x) = e^{-x^2}$

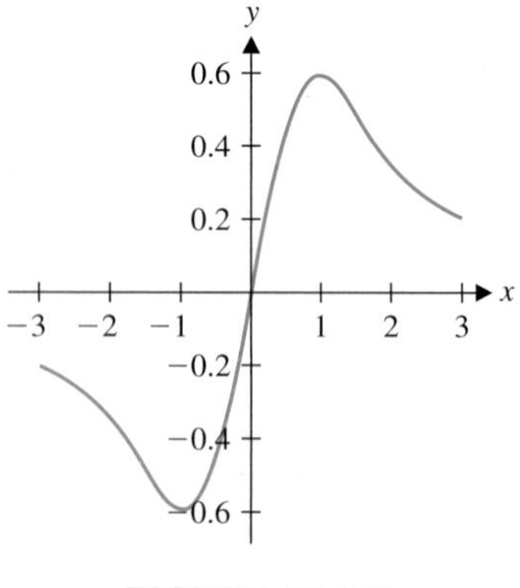

FIGURE 15.16b
10-term approximation to $y = y_2(x)$

where y_1 and y_2 are two power series solutions of the differential equation. We leave it as an exercise to use the Ratio Test to show that both of these series converge absolutely for all x. You might recognize $y_1(x)$ as the Maclaurin series expansion for e^{-x^2}, but in practice recognizing series solutions as power series of familiar functions is rather unlikely. To give you an idea of the behavior of these functions, we draw a graph of $y_1(x)$ in Figure 15.16a and of $y_2(x)$ in Figure 15.16b. We obtained the graph of $y_2(x)$ by plotting the partial sums of the series. ■

From examples 4.1 and 4.2, you might get the idea that if you look for a series solution, you can always recognize the pattern of the coefficients and write the pattern down succinctly. Unfortunately, the pattern is most often difficult to see and even more difficult to write down compactly. Still, series solutions are a valuable means of solving a differential equation. In the worst case, you can always compute a number of the coefficients of the series from the recurrence relation and then use the first so many terms of the series as an approximation to the actual solution.

In example 4.3, we illustrate the more common case where the coefficients are a bit more challenging to find.

EXAMPLE 4.3 A Series Solution Where the Coefficients Are Harder to Find

Use a power series to find the general solution of **Airy's equation**

$$y'' - xy = 0.$$

Solution As before, we assume that we may write the solution as a power series

$$y = \sum_{n=0}^{\infty} a_n x^n.$$

Again, we have

$$y' = \sum_{n=1}^{\infty} n a_n x^{n-1}$$

and

$$y'' = \sum_{n=2}^{\infty} n(n-1) a_n x^{n-2}.$$

Subsituting these power series into the equation, we get

$$\begin{aligned} 0 = y'' - xy &= \sum_{n=2}^{\infty} n(n-1) a_n x^{n-2} - x \sum_{n=0}^{\infty} a_n x^n \\ &= \sum_{n=2}^{\infty} n(n-1) a_n x^{n-2} - \sum_{n=0}^{\infty} a_n x^{n+1}. \end{aligned}$$

In order to combine the two preceding series, we must rewrite one or both series so that they both have the same power of x. For simplicity, we rewrite the first series only. We have

$$\begin{aligned} 0 &= \sum_{n=2}^{\infty} n(n-1) a_n x^{n-2} - \sum_{n=0}^{\infty} a_n x^{n+1} \\ &= \sum_{n=-1}^{\infty} (n+3)(n+2) a_{n+3} x^{n+1} - \sum_{n=0}^{\infty} a_n x^{n+1} \\ &= (2)(1)a_2 + \sum_{n=0}^{\infty} (n+3)(n+2) a_{n+3} x^{n+1} - \sum_{n=0}^{\infty} a_n x^{n+1} \\ &= 2a_2 + \sum_{n=0}^{\infty} [(n+3)(n+2) a_{n+3} - a_n] x^{n+1}, \end{aligned}$$

where we wrote out the first term of the first series and then combined the two series, once both had an index that started with $n = 0$. Again, this is a power series expansion of the zero function and so, all of the coefficients must be zero. That is,

$$0 = 2a_2 \tag{4.7}$$

and
$$0 = (n+3)(n+2)a_{n+3} - a_n, \tag{4.8}$$

for $n = 0, 1, 2, \ldots$. Equation (4.7) says that $a_2 = 0$ and (4.8) gives us the recurrence relation

$$a_{n+3} = \frac{1}{(n+3)(n+2)}a_n, \tag{4.9}$$

for $n = 0, 1, 2, \ldots$. Notice that here, instead of having all of the even-indexed coefficients related to a_0 and all of the odd-indexed coefficients related to a_1, (4.9) tells us that every *third* coefficient is related. In particular, notice that since $a_2 = 0$, (4.9) now says that

$$a_5 = \frac{1}{5 \cdot 4}a_2 = 0,$$
$$a_8 = \frac{1}{8 \cdot 7}a_5 = 0$$

and so on. So, every third coefficient starting with a_2 is zero. But, how do we concisely write down something like this? Think about the notation a_{2n} and a_{2n+1} that we have used previously. You can view a_{2n} as a representation of every second coefficient starting with a_0. Likewise, a_{2n+1} represents every second coefficient starting with a_1. In the present case, if we want to write down every third coefficient starting with a_2, we write a_{3n+2}. We can now observe that

$$a_{3n+2} = 0,$$

for $n = 0, 1, 2, \ldots$. Continuing on with the remaining coefficients, we have from (4.9) that

$$a_3 = \frac{1}{3 \cdot 2}a_0,$$
$$a_6 = \frac{1}{6 \cdot 5}a_3 = \frac{1}{6 \cdot 5 \cdot 3 \cdot 2}a_0,$$
$$a_9 = \frac{1}{9 \cdot 8}a_6 = \frac{1}{9 \cdot 8 \cdot 6 \cdot 5 \cdot 3 \cdot 2}a_0$$

and so on. Hopefully, you see the pattern that's developing for these coefficients. The trouble here is that it's not as easy to write down this pattern as it was in the first two examples. Notice that the denominator in the expression for a_9 is almost 9!, but with every third factor in the product deleted. Since we don't have a way of succinctly writing this down, we write the coefficients by indicating the pattern, as follows:

$$a_{3n} = \frac{(3n-2)(3n-5)\cdots 7 \cdot 4 \cdot 1}{(3n)!}a_0,$$

where this is not intended as a literal formula, as explicit substitution of $n = 0$ or $n = 1$ would result in negative values. Rather, this is an indication of the general pattern. Similarly, the recurrence relation gives us

$$a_4 = \frac{1}{4 \cdot 3}a_1,$$
$$a_7 = \frac{1}{7 \cdot 6}a_4 = \frac{1}{7 \cdot 6 \cdot 4 \cdot 3}a_1,$$
$$a_{10} = \frac{1}{10 \cdot 9}a_7 = \frac{1}{10 \cdot 9 \cdot 7 \cdot 6 \cdot 4 \cdot 3}a_1$$

and so on. More generally, we can establish the pattern:

$$a_{3n+1} = \frac{(3n-1)(3n-4)\cdots 8\cdot 5\cdot 2}{(3n+1)!}a_1,$$

where again, this is not intended as a literal formula.

Now that we have found all of the coefficients, we can write down the solution, by separately writing out every third term of the series, as follows:

$$\begin{aligned} y &= \sum_{n=0}^{\infty} a_n x^n = \sum_{n=0}^{\infty} \left(a_{3n}x^{3n} + a_{3n+1}x^{3n+1} + a_{3n+2}x^{3n+2}\right) \\ &= a_0 \underbrace{\sum_{n=0}^{\infty} \frac{(3n-2)(3n-5)\cdots 7\cdot 4\cdot 1}{(3n)!}x^{3n}}_{y_1(x)} + a_1 \underbrace{\sum_{n=0}^{\infty} \frac{(3n-1)(3n-4)\cdots 8\cdot 5\cdot 2}{(3n+1)!}x^{3n+1}}_{y_2(x)} \\ &= a_0 y_1(x) + a_1 y_2(x). \end{aligned}$$

We leave it as an exercise to use the Ratio Test to show that the power series defining y_1 and y_2 are absolutely convergent for all x. ■

You may have noticed that in all three of our examples, we assumed that there was a solution of the form

$$y = \sum_{n=0}^{\infty} a_n x^n = a_0 + a_1 x + a_2 x^2 + \cdots,$$

only to arrive at the general solution

$$y = a_0 y_1(x) + a_1 y_2(x),$$

where y_1 and y_2 were power series solutions of the equation. This is in fact not coincidental. One can show that (at least for certain equations) this is always the case. One clue as to why this might be so lies in the following.

Suppose that we want to solve the initial value problem consisting of a second-order differential equation and the initial conditions $y(0) = A$ and $y'(0) = B$. Taking $y(x) = \sum_{n=0}^{\infty} a_n x^n$ gives us

$$y'(x) = \sum_{n=0}^{\infty} n a_n x^{n-1} = a_1 + 2a_2 x + 3a_3 x^2 + \cdots.$$

So, imposing the initial conditions, we have

$$A = y(0) = a_0 + a_1(0) + a_2(0)^2 + \cdots = a_0$$

and

$$B = y'(0) = a_1 + 2a_2(0) + 3a_3(0)^2 + \cdots = a_1.$$

So, irrespective of the particular equation we're solving, we always have $y(0) = a_0$ and $y'(0) = a_1$.

You might ask what you'd do if the initial conditions were imposed at some point other than at $x = 0$, say at $x = x_0$. In this case, we look for a power series solution of the form

$$y = \sum_{n=0}^{\infty} a_n (x - x_0)^n.$$

It's easy to show that in this case, we still have $y(x_0) = a_0$ and $y'(x_0) = a_1$.

In the exercises, we explore finding series solutions about a variety of different points.

BEYOND FORMULAS

This section connects two important threads of calculus: solutions of differential equations and infinite series. In Chapter 8, we expressed known functions like $\sin x$ and $\cos x$ as power series. Here, we just extend that idea to unknown solutions of differential equations. For second-order homogeneous equations, keep in mind that the general solution has the format $c_1 y_1(x) + c_2 y_2(x)$. You should think about the problems in this section as following this strategy: write the solution as a power series, substitute into the differential equation and find relationships between the coefficients of the power series, remembering that two of the coefficients will be left as arbitrary constants.

EXERCISES 15.4

WRITING EXERCISES

1. After substituting a power series representation into a differential equation, the next step is always to rewrite one or more of the series, so that all series have the same exponent. (Typically, we want x^n.) Explain why this is an important step. For example, what would we be unable to do if the exponents were not the same?

2. The recurrence relation is typically solved for the coefficient with the largest index. Explain why this is an important step.

3. Explain why you can't solve equations with nonconstant coefficients, such as

$$y'' + 2xy' + 2y = 0,$$

by looking for a solution in the form $y = e^{rx}$.

4. The differential equations solved in this section are actually of a special type, where we find power series solutions centered at what is called an **ordinary point.** For the equation $x^2y'' + y' + 2y = 0$, the point $x = 0$ is not an ordinary point. Discuss what goes wrong here if you look for a power series solution of the form $\sum_{n=0}^{\infty} a_n x^n$.

In exercises 1–8, find the recurrence relation and general power series solution of the form $\sum_{n=0}^{\infty} a_n x^n$.

1. $y'' + 2xy' + 4y = 0$
2. $y'' + 4xy' + 8y = 0$
3. $y'' - xy' - y = 0$
4. $y'' - xy' - 2y = 0$
5. $y'' - xy' = 0$
6. $y'' + 2xy = 0$
7. $y'' - x^2y' = 0$
8. $y'' + xy' - 2y = 0$

9. Find a series solution of $y'' + (1-x)y' - y = 0$ in the form $y = \sum_{n=0}^{\infty} a_n (x-1)^n$.

10. Find a series solution of $y'' + y' + (x-2)y = 0$ in the form $\sum_{n=0}^{\infty} a_n (x-2)^n$.

11. Find a series solution of Airy's equation $y'' - xy = 0$ in the form $\sum_{n=0}^{\infty} a_n (x-1)^n$. [Hint: First rewrite the equation in the form $y'' - (x-1)y - y = 0$.]

12. Find a series solution of Airy's equation $y'' - xy = 0$ in the form $\sum_{n=0}^{\infty} a_n (x-2)^n$.

13. Solve the initial value problem $y'' + 2xy' + 4y = 0$, $y(0) = 5$, $y'(0) = -7$. (See exercise 1.)

14. Solve the initial value problem $y'' + 4xy' + 8y = 0$, $y(0) = 2$, $y'(0) = \pi$. (See exercise 2.)

15. Solve the initial value problem $y'' + (1-x)y' - y = 0$, $y(1) = -3$, $y'(1) = 12$. (See exercise 9.)

16. Solve the initial value problem $y'' + y' + (x-2)y = 0$, $y(2) = 1$, $y'(2) = -1$. (See exercise 10.)

17. Determine the radius of convergence of the power series solutions about $x_0 = 0$ of $y'' - xy' - y = 0$. (See exercise 3.)

18. Determine the radius of convergence of the power series solutions about $x_0 = 0$ of $y'' - xy' - 2y = 0$. (See exercise 4.)

19. Determine the radius of convergence of the power series solutions about $x_0 = 1$ of $y'' + (1-x)y' - y = 0$. (See exercise 9.)

20. Determine the radius of convergence of the power series solutions about $x_0 = 1$ of $y'' - xy = 0$ (See exercise 11.)

21. Find a series solution of the form $y = \sum_{n=0}^{\infty} a_n x^n$ to the equation $x^2y'' + xy' + x^2y = 0$ (Bessel's equation of order 0).

22. Find a series solution of the form $y = \sum_{n=0}^{\infty} a_n x^n$ to the equation $x^2y'' + xy' + (x^2 - 1)y = 0$ (Bessel's equation of order 1).

23. Determine the radius of convergence of the series solution found in example 4.3.

24. Determine the radius of convergence of the series solution found in problem 12.

25. For the initial value problem $y'' + 2xy' - xy = 0$, $y(0) = 2$, $y'(0) = -5$, substitute in $x = 0$ and show that $y''(0) = 0$. Then take $y'' = -2xy' + xy$ and show that $y''' = -2xy'' + (x - 2)y' + y$. Conclude that $y'''(0) = 12$. Then compute $y^{(4)}(x)$ and find $y^{(4)}(0)$. Finally, compute $y^{(5)}(x)$ and find $y^{(5)}(0)$. Write out the fifth-degree Taylor polynomial for the solution, $P_5(x) = y(0) + y'(0)x + y''(0)\frac{x^2}{2} + y'''(0)\frac{x^3}{3!} + y^{(4)}(0)\frac{x^4}{4!} + y^{(5)}(0)\frac{x^5}{5!}$.

26. Use the technique of exercise 25 to find the fifth-degree Taylor polynomial for the solution of the initial value problem $y'' + x^2y' - (\cos x)y = 0$, $y(0) = 3$, $y'(0) = 2$.

27. Use the technique of exercise 25 to find the fifth-degree Taylor polynomial for the solution of the initial value problem $y'' + e^x y' - (\sin x)y = 0$, $y(0) = -2$, $y'(0) = 1$.

28. Use the technique of exercise 25 to find the fifth-degree Taylor polynomial for the solution of the initial value problem $y'' + y' - (e^x)y = 0$, $y(0) = 2$, $y'(0) = 0$.

29. Use the technique of exercise 25 to find the fifth-degree Taylor polynomial for the solution of the initial value problem $y'' + xy' + (\sin x)y = 0$, $y(\pi) = 0$, $y'(\pi) = 4$.

30. Use the technique of exercise 25 to find the fifth-degree Taylor polynomial for the solution of the initial value problem $y'' + (\cos x)y' + xy = 0$, $y(\frac{\pi}{2}) = 3$, $y'(\frac{\pi}{2}) = 0$.

EXPLORATORY EXERCISES

1. The equation $y'' - 2xy' + 2ky = 0$ for some integer $k \geq 0$ is known as **Hermite's equation.** Following our procedure for finding series solutions in powers of x, show that, in fact, one of the series solutions is simply a polynomial of degree k. For this polynomial solution, choose the arbitrary constant such that the leading term of the polynomial is $2^k x^k$. The polynomial is called the **Hermite polynomial** $H_k(x)$. Find the Hermite polynomials $H_0(x), H_1(x), \ldots, H_5(x)$.

2. The Chebyshev polynomials are polynomial solutions of the equation $(1 - x^2)y'' - xy' + k^2y = 0$, for some integer $k \geq 0$. Find polynomial solutions for $k = 0, 1, 2$ and 3.

Review Exercises

WRITING EXERCISES

The following list includes terms that are defined and theorems that are stated in this chapter. For each term or theorem, (1) give a precise definition or statement, (2) state in general terms what it means and (3) describe the types of problems with which it is associated.

Nonhomogeneous equation	Method of undetermined coefficients	Resonance
Second-order differential equation	Damping	Recurrence relation

TRUE OR FALSE

State whether each statement is true or false and briefly explain why. If the statement is false, try to "fix it" by modifying the given statement to a new statement that is true.

1. The form of the solution of $ay'' + by' + cy = 0$ depends on the value of $b^2 - 4ac$.

2. The current in an electrical circuit satisfies the same differential equation as the displacement function for a mass attached to a spring.

3. The particular solution of a nonhomogeneous equation $mu'' + ku' + cu = F$ has the same form as the forcing function F.

4. Resonance cannot occur if there is damping.

5. A recurrence relation can always be solved to find the solution of a differential equation.

In exercises 1–6, find the general solution of the differential equation.

1. $y'' + y' - 12y = 0$

2. $y'' + 4y' + 4y = 0$

3. $y'' + y' + 3y = 0$

4. $y'' + 3y' - 8y = 0$

Review Exercises

5. $y'' - y' - 6y = e^{3t} + t^2 + 1$
6. $y'' - 4y = 2e^{2t} + 16\cos 2t$

In exercises 7–10, solve the initial value problem.

7. $y'' + 2y' - 8y = 0,\ y(0) = 5,\ y'(0) = -2$
8. $y'' + 2y' + 5y = 0,\ y(0) = 2,\ y'(0) = 0$
9. $y'' + 4y = 3\cos t,\ y(0) = 1,\ y'(0) = 2$
10. $y'' - 4y = 2e^{2t} + 16\cos 2t,\ y(0) = 0,\ y'(0) = 1/2$
11. A spring is stretched 4 inches by a 4-pound weight. The weight is then pulled down an additional 2 inches and released. Neglect damping. Find an equation for the position of the weight at any time t and graph the position function.
12. In exercise 11, if an external force of $4\cos\omega t$ pounds is applied to the weight, find the value of ω that would produce resonance. If instead $\omega = 10$, find and graph the position of the weight.
13. A series circuit has an inductor of 0.2 henry, a resistor of 160 ohms and a capacitor of 10^{-2} farad. The initial charge on the capacitor is 10^{-4} coulomb and there is no initial current. Find the charge on the capacitor and the current at any time t.
14. In exercise 13, if the resistor is removed and an impressed voltage of $2\sin\omega t$ volts is applied, find the value of ω that produces resonance. In this case, what would happen to the circuit?

In exercises 15 and 16, determine the form of a particular solution.

15. $u'' + 2u' + 5u = 2e^{-t}\sin 2t + 4t^3 - 2\cos 2t$
16. $u'' + 2u' - 3u = (3t^2 + 1)e^t - e^{-3t}\cos 2t$
17. A spring is stretched 4 inches by a 4-pound weight. The weight is then pulled down an additional 2 inches and set in motion with a downward velocity of 2 ft/s. A damping force equal to $0.4u'$ slows the motion of the spring. An external force of magnitude 2 sin $2t$ pounds is applied. Completely set up the initial value problem and then find the steady-state motion of the spring.
18. A spring is stretched 2 inches by an 8-pound weight. The weight is then pushed up 3 inches and set in motion with an upward velocity of 1 ft/s. A damping force equal to $0.2u'$ slows the motion of the spring. An external force of magnitude $2\cos 3t$ pounds is applied. Completely set up the initial value problem and then find the steady-state motion of the spring.

In exercises 19 and 20, find the recurrence relation and a general power series solution of the form $\sum_{n=0}^{\infty} a_n x^n$.

19. $y'' - 2xy' - 4y = 0$
20. $y'' + (x - 1)y' = 0$

In exercises 21 and 22, find the recurrence relation and a general power series solution of the form $\sum_{n=0}^{\infty} a_n (x-1)^n$.

21. $y'' - 2xy' - 4y = 0$
22. $y'' + (x - 1)y' = 0$

In exercises 23 and 24, solve the initial value problem.

23. $y'' - 2xy' - 4y = 0,\ y(0) = 4,\ y'(0) = 2$
24. $y'' - 2xy' - 4y = 0,\ y(1) = 2,\ y'(1) = 4$

EXPLORATORY EXERCISES

1. A pendulum that is free to rotate through 360 degrees has two equilibrium points. One is hanging straight down and the other is pointing straight up. The $\theta = \pi$ equilibrium is unstable and is classified as a saddle point. This means that for *most but not all* initial conditions, solutions that start near $\theta = \pi$ will get farther away. Explain why with initial conditions $\theta(0) = \pi$ and $\theta'(0) = 0$, the solution is exactly $\theta(t) = \pi$. However, explain why initial conditions $\theta(0) = 3.1$ and $\theta'(0) = 0$ would have a solution that gets farther from $\theta = \pi$. For the model $\theta''(t) + \frac{g}{L}\theta(t) = 0$, show that if $v = \pi\sqrt{\frac{g}{L}}$, then the initial conditions $\theta(0) = 0$ and $\theta'(0) = v$ produce a solution that reaches the state $\theta = \pi$ and $\theta' = 0$. Physically, explain why the pendulum would remain at $\theta = \pi$ and then explain why the solution of our model does not get "stuck" at $\theta = \pi$. Explain why for any starting angle θ, there exist two initial angular velocities that will balance the pendulum at $\theta = \pi$. The undamped pendulum model $\theta''(t) + \frac{g}{L}\sin\theta(t) = 0$ is equivalent to the system of equations (with $y_1 = \theta$ and $y_2 = \theta'$)
$$y_1' = y_2,$$
$$y_2' = -\frac{g}{L}\sin y_1.$$
Use a CAS to sketch the phase portrait of this system of equations near the equilibrium point $(\pi, 0)$. Explain why the phase portrait shows an unstable equilibrium point with a small set of initial conditions that lead to the equilibrium point.
2. In exploratory exercise 2 of section 15.3, we investigate the motion of a ball dropped in a hole drilled through a non-rotating Earth. Here, we investigate the motion taking into account the Earth's rotation. (See Andrew Simoson's article in the June 2004 *Mathematics Magazine*.) We describe the motion in polar coordinates with respect to a fixed plane through the equator. Define unit vectors $\mathbf{u}_r = \langle\cos\theta, \sin\theta\rangle$ and

Review Exercises

$\mathbf{u}_\theta = \langle -\sin\theta, \cos\theta\rangle$. If the ball has position vector $r\mathbf{u}_r$, show that its acceleration is given by

$$[r\theta''(t) + 2r'(t)\theta'(t)]\mathbf{u}_\theta + \{r''(t) - r(t)[\theta'(t)]^2\}\mathbf{u}_r.$$

Since gravity acts in the radial direction only, $r\theta''(t) + 2r'(t)\theta'(t) = 0$. Show that this implies that $r^2(t)\theta'(t) = k$ for some constant k. (This is the law of conservation of angular momentum.) If the acceleration due to gravity is $f(r)\mathbf{u}_r$ for some function f, show that

$$r''(t) - \frac{k^2}{r^3} = f(r).$$

Initial conditions are $r(0) = R, r'(0) = 0, \theta(0) = 0$ and $\theta'(0) = \frac{2\pi}{Q}$. Here, Q is the period of one revolution of the Earth and we assume that the ball inherits the initial angular velocity from the rotation of the Earth. For the gravitational force $f(r) = -c^2 r$, show that a solution is

$$r(t) = \sqrt{R^2\cos^2 ct + \frac{k^2}{c^2R^2}\sin^2 ct}$$

or

$$r(\theta) = \frac{1}{\sqrt{\frac{1}{R^2}\cos^2\theta + \frac{c^2R^2}{k^2}\sin^2\theta}}.$$

Show that this converts to

$$\frac{x^2}{R^2} + \frac{y^2}{(k/Rc)^2} = 1$$

and describe the path of the ball.

Appendix A

PROOFS OF SELECTED THEOREMS

In this appendix, we provide the proofs of selected theorems from the body of the text. These are results that were not proved in the body of the text for one reason or another.

The first several results require the formal (ε-δ) definition of limit, which was not discussed until section 1.6. Given what we've done in section 1.6, we are now in a position to prove these results. The first of these results concerns our routine rules for calculating limits and appeared as Theorem 3.1 in section 1.3.

THEOREM A.1

Suppose that $\lim_{x \to a} f(x)$ and $\lim_{x \to a} g(x)$ both exist and let c be any constant. The following then apply:

(i) $\lim_{x \to a} [c \cdot f(x)] = c \cdot \lim_{x \to a} f(x)$,

(ii) $\lim_{x \to a} [f(x) \pm g(x)] = \lim_{x \to a} f(x) \pm \lim_{x \to a} g(x)$,

(iii) $\lim_{x \to a} [f(x) \cdot g(x)] = \left[\lim_{x \to a} f(x)\right]\left[\lim_{x \to a} g(x)\right]$ and

(iv) $\lim_{x \to a} \dfrac{f(x)}{g(x)} = \dfrac{\lim_{x \to a} f(x)}{\lim_{x \to a} g(x)}$ $\left(\text{if } \lim_{x \to a} g(x) \neq 0\right)$.

PROOF

(i) Given $\lim_{x \to a} f(x) = L_1$, we know by the precise definition of limit that given any number $\varepsilon_1 > 0$, there is a number $\delta_1 > 0$ for which

$$|f(x) - L_1| < \varepsilon_1, \text{ whenever } 0 < |x - a| < \delta_1. \tag{A.1}$$

In order to show that $\lim_{x \to a} [cf(x)] = c \lim_{x \to a} f(x)$, we need to be able to make $cf(x)$ as close to cL_1 as desired. We have

$$|cf(x) - cL_1| = |c||f(x) - L_1|.$$

We already know that we can make $|f(x) - L_1|$ as small as desired. Specifically, given any number $\varepsilon_1 > 0$, there is a number $\delta_1 > 0$ for which

$$|cf(x) - cL_1| = |c||f(x) - L_1| < |c|\varepsilon_1, \text{ whenever } 0 < |x - a| < \delta_1.$$

Taking $\varepsilon_1 = \frac{\varepsilon}{|c|}$ and $\delta = \delta_1$, we get

$$|cf(x) - cL_1| < |c|\varepsilon_1 = |c|\frac{\varepsilon}{|c|} = \varepsilon, \text{ whenever } 0 < |x - a| < \delta.$$

This says that $\lim_{x \to a}[cf(x)] = cL_1$, as desired.

(ii) Likewise, given $\lim_{x \to a} g(x) = L_2$, we know that given any number $\varepsilon_2 > 0$, there is a number $\delta_2 > 0$ for which

$$|g(x) - L_2| < \varepsilon_2, \text{ whenever } 0 < |x - a| < \delta_2. \tag{A.2}$$

Now, in order to verify that

$$\lim_{x \to a}[f(x) + g(x)] = L_1 + L_2,$$

we must show that, given any number $\varepsilon > 0$, there is a number $\delta > 0$ such that

$$|[f(x) + g(x)] - (L_1 + L_2)| < \varepsilon, \text{ whenever } 0 < |x - a| < \delta.$$

Notice that
$$\begin{aligned} |[f(x) + g(x)] - (L_1 + L_2)| &= |[f(x) - L_1] + [g(x) - L_2]| \\ &\leq |f(x) - L_1| + |g(x) - L_2|, \end{aligned} \tag{A.3}$$

by the triangle inequality. Of course, both terms on the right-hand side of (A.3) can be made arbitrarily small, from (A.1) and (A.2). In particular, if we take $\varepsilon_1 = \varepsilon_2 = \frac{\varepsilon}{2}$, then as long as

$$0 < |x - a| < \delta_1 \quad \textit{and} \quad 0 < |x - a| < \delta_2,$$

we get from (A.1), (A.2) and (A.3) that

$$\begin{aligned} |[f(x) + g(x)] - (L_1 + L_2)| &\leq |f(x) - L_1| + |g(x) - L_2| \\ &< \frac{\varepsilon}{2} + \frac{\varepsilon}{2} = \varepsilon, \end{aligned}$$

as desired. This will occur if we take

$$0 < |x - a| < \delta = \min\{\delta_1, \delta_2\},$$

where taking $\delta = \min\{\delta_1, \delta_2\}$ simply means to pick δ to be the smaller of δ_1 and δ_2. (Recognize that if $0 < |x - a| < \delta = \min\{\delta_1, \delta_2\}$, then $0 < |x - a| < \delta_1$ *and* $0 < |x - a| < \delta_2$.)

(iii) In this case, we need to show that for any given $\varepsilon > 0$, we can find a $\delta > 0$, such that

$$|f(x)g(x) - L_1L_2| < \varepsilon, \text{ whenever } 0 < |x - a| < \delta.$$

The object, then, is to make $|f(x)g(x) - L_1L_2|$ as small as needed. Notice that we have

$$\begin{aligned} |f(x)g(x) - L_1L_2| &= |f(x)g(x) - g(x)L_1 + g(x)L_1 - L_1L_2| \\ &= |[f(x) - L_1]g(x) + L_1[g(x) - L_2]| \\ &\leq |f(x) - L_1||g(x)| + |L_1||g(x) - L_2|, \end{aligned} \tag{A.4}$$

by the triangle inequality. Now, notice that we can make $|f(x) - L_1|$ and $|g(x) - L_2|$ as small as we like. If we make both of the terms in (A.4) less than $\frac{\varepsilon}{2}$, then the sum will be less than ε, as desired. In particular, we know that there is a number $\delta_2 > 0$, such that

$$|g(x) - L_2| < \frac{\varepsilon}{2|L_1|}, \text{ whenever } 0 < |x - a| < \delta_2,$$

assuming $L_1 \neq 0$, so that

$$|L_1||g(x) - L_2| < |L_1|\frac{\varepsilon}{2|L_1|} = \frac{\varepsilon}{2}.$$

If $L_1 = 0$, then

$$|L_1||g(x) - L_2| = 0 < \frac{\varepsilon}{2}.$$

So, no matter the value of L_1, we have that

$$|L_1||g(x) - L_2| < \frac{\varepsilon}{2}, \text{ whenever } 0 < |x - a| < \delta_2. \tag{A.5}$$

Notice that the first term in (A.4) is slightly more complicated, as we must also estimate the size of $|g(x)|$. Notice that

$$|g(x)| = |g(x) - L_2 + L_2| \leq |g(x) - L_2| + |L_2|. \tag{A.6}$$

Since $\lim_{x \to a} g(x) = L_2$, there is a number $\delta_3 > 0$, such that

$$|g(x) - L_2| < 1, \text{ whenever } 0 < |x - a| < \delta_3.$$

From (A.6), we now have that

$$|g(x)| \leq |g(x) - L_2| + |L_2| < 1 + |L_2|.$$

Returning to the first term in (A.4), we have that for $0 < |x - a| < \delta_3$,

$$|f(x) - L_1||g(x)| < |f(x) - L_1|(1 + |L_2|). \tag{A.7}$$

Now, since $\lim_{x \to a} f(x) = L_1$, given any $\varepsilon > 0$, there is a number $\delta_1 > 0$ such that for $0 < |x - a| < \delta_1$,

$$|f(x) - L_1| < \frac{\varepsilon}{2(1 + |L_2|)}.$$

From (A.7), we then have

$$\begin{aligned} |f(x) - L_1||g(x)| &< |f(x) - L_1|(1 + |L_2|) \\ &< \frac{\varepsilon}{2(1 + |L_2|)}(1 + |L_2|) \\ &= \frac{\varepsilon}{2}, \end{aligned}$$

whenever $0 < |x - a| < \delta_1$ *and* $0 < |x - a| < \delta_3$. Together with (A.4) and (A.5), this tells us that for $\delta = \min\{\delta_1, \delta_2, \delta_3\}$, if $0 < |x - a| < \delta$, then

$$\begin{aligned} |f(x)g(x) - L_1 L_2| &\leq |f(x) - L_1||g(x)| + |L_1||g(x) - L_2| \\ &< \frac{\varepsilon}{2} + \frac{\varepsilon}{2} = \varepsilon, \end{aligned}$$

which proves (iii).

(iv) We first show that

$$\lim_{x \to a} \frac{1}{g(x)} = \frac{1}{L_2},$$

for $L_2 \neq 0$. Notice that in this case, we need to show that we can make $\left|\frac{1}{g(x)} - \frac{1}{L_2}\right|$ as small as possible. We have

$$\left|\frac{1}{g(x)} - \frac{1}{L_2}\right| = \left|\frac{L_2 - g(x)}{L_2 g(x)}\right|. \tag{A.8}$$

Of course, since $\lim_{x \to a} g(x) = L_2$, we can make the numerator of the fraction on the right-hand side as small as needed. We must also consider what happens to the denominator, though. Recall that given any $\varepsilon_2 > 0$, there is a $\delta_2 > 0$ such that

$$|g(x) - L_2| < \varepsilon_2, \text{ whenever } 0 < |x - a| < \delta_2.$$

In particular, for $\varepsilon_2 = \frac{|L_2|}{2}$, this says that

$$|g(x) - L_2| < \frac{|L_2|}{2}.$$

Notice that by the triangle inequality, we can now say that

$$|L_2| = |L_2 - g(x) + g(x)| \le |L_2 - g(x)| + |g(x)| < \frac{|L_2|}{2} + |g(x)|.$$

Subtracting $\frac{|L_2|}{2}$ from both sides now gives us

$$\frac{|L_2|}{2} < |g(x)|,$$

so that

$$\frac{2}{|L_2|} > \frac{1}{|g(x)|}.$$

From (A.8), we now have that for $0 < |x - a| < \delta_2$,

$$\left| \frac{1}{g(x)} - \frac{1}{L_2} \right| = \left| \frac{L_2 - g(x)}{L_2 g(x)} \right| < \frac{2|L_2 - g(x)|}{L_2^2}. \tag{A.9}$$

Further, given any $\varepsilon > 0$, there is a $\delta_3 > 0$ so that

$$|L_2 - g(x)| < \frac{\varepsilon L_2^2}{2}, \text{ whenever } 0 < |x - a| < \delta_3.$$

From (A.9), we now have that for $\delta = \min\{\delta_2, \delta_3\}$,

$$\left| \frac{1}{g(x)} - \frac{1}{L_2} \right| < \frac{2|L_2 - g(x)|}{L_2^2} < \varepsilon,$$

whenever $0 < |x - a| < \delta$, as desired. We have now established that

$$\lim_{x \to a} \frac{1}{g(x)} = \frac{1}{L_2}.$$

From (iii), we now have that

$$\lim_{x \to a} \frac{f(x)}{g(x)} = \lim_{x \to a} \left[f(x) \frac{1}{g(x)} \right] = \left[\lim_{x \to a} f(x) \right] \left[\lim_{x \to a} \frac{1}{g(x)} \right]$$

$$= L_1 \left(\frac{1}{L_2} \right) = \frac{L_1}{L_2},$$

which proves the last part of the theorem. ■

The following result appeared as Theorem 3.3 in section 1.3.

THEOREM A.2

Suppose that $\lim_{x\to a} f(x) = L$ and n is any positive integer. Then,

$$\lim_{x\to a} \sqrt[n]{f(x)} = \sqrt[n]{\lim_{x\to a} f(x)} = \sqrt[n]{L},$$

where for n even, we assume that $L > 0$.

PROOF

Since $\lim_{x\to a} f(x) = L$, we know that given any number $\varepsilon_1 > 0$, there is a number $\delta_1 > 0$, so that

$$|f(x) - L| < \varepsilon_1, \text{ whenever } 0 < |x - a| < \delta_1.$$

To show that $\lim_{x\to a} \sqrt[n]{f(x)} = \sqrt[n]{L}$, we need to show that given any $\varepsilon > 0$, there is a $\delta > 0$ such that

$$\left|\sqrt[n]{f(x)} - \sqrt[n]{L}\right| < \varepsilon, \text{ whenever } 0 < |x - a| < \delta.$$

Notice that this is equivalent to having

$$\sqrt[n]{L} - \varepsilon < \sqrt[n]{f(x)} < \sqrt[n]{L} + \varepsilon.$$

Now, raising all sides to the nth power, we have

$$\left(\sqrt[n]{L} - \varepsilon\right)^n < f(x) < \left(\sqrt[n]{L} + \varepsilon\right)^n.$$

Subtracting L from all terms gives us

$$\left(\sqrt[n]{L} - \varepsilon\right)^n - L < f(x) - L < \left(\sqrt[n]{L} + \varepsilon\right)^n - L.$$

Since ε is taken to be small, we now assume that $\varepsilon < \sqrt[n]{L}$. Observe that in this case, $0 < \sqrt[n]{L} - \varepsilon < \sqrt[n]{L}$. Let $\varepsilon_1 = \min\{\left(\sqrt[n]{L} + \varepsilon\right)^n - L, L - \left(\sqrt[n]{L} - \varepsilon\right)^n\} > 0$. Then, since $\lim_{x\to a} f(x) = L$, we know that there is a number $\delta > 0$, so that

$$-\varepsilon_1 < f(x) - L < \varepsilon_1, \text{ whenever } 0 < |x - a| < \delta.$$

It then follows that

$$(\sqrt[n]{L} - \varepsilon)^n - L \le -\varepsilon_1 < f(x) - L < \varepsilon_1 \le (\sqrt[n]{L} + \varepsilon)^n - L,$$

whenever $0 < |x - a| < \delta$. Reversing the above sequence of steps gives us $\lim_{x\to a} \sqrt[n]{f(x)} = \sqrt[n]{L}$, as desired. ■

The following result appeared in section 1.3, as Theorem 3.5.

THEOREM A.3 (Squeeze Theorem)

Suppose that

$$f(x) \le g(x) \le h(x), \tag{A.10}$$

for all x in some interval (c, d), except possibly at the point $a \in (c, d)$ and that

$$\lim_{x\to a} f(x) = \lim_{x\to a} h(x) = L,$$

for some number L. Then, it follows that

$$\lim_{x\to a} g(x) = L, \text{ also.}$$

PROOF

To show that $\lim_{x\to a} g(x) = L$, we must prove that given any $\varepsilon > 0$, there is a $\delta > 0$, such that

$$|g(x) - L| < \varepsilon, \text{ whenever } 0 < |x - a| < \delta.$$

Since $\lim_{x\to a} f(x) = L$, we have that given any $\varepsilon > 0$, there is a $\delta_1 > 0$, such that

$$|f(x) - L| < \varepsilon, \text{ whenever } 0 < |x - a| < \delta_1.$$

Likewise, since $\lim_{x\to a} h(x) = L$, we have that given any $\varepsilon > 0$, there is a $\delta_2 > 0$, such that

$$|h(x) - L| < \varepsilon, \text{ whenever } 0 < |x - a| < \delta_2.$$

Now, choose $\delta = \min\{\delta_1, \delta_2\}$. Then, if $0 < |x - a| < \delta$, it follows that $0 < |x - a| < \delta_1$ *and* $0 < |x - a| < \delta_2$, so that

$$|f(x) - L| < \varepsilon \quad \textit{and} \quad |h(x) - L| < \varepsilon.$$

Equivalently, we can say that

$$L - \varepsilon < f(x) < L + \varepsilon \quad \text{and} \quad L - \varepsilon < h(x) < L + \varepsilon. \tag{A.11}$$

It now follows from (A.10) and (A.11) that if $0 < |x - a| < \delta$, then

$$L - \varepsilon < f(x) \le g(x) \le h(x) < L + \varepsilon,$$

which gives us

$$L - \varepsilon < g(x) < L + \varepsilon$$

or $|g(x) - L| < \varepsilon$ and it follows that $\lim_{x\to a} g(x) = L$, as desired. ■

The following result appeared as Theorem 4.3 in section 1.4.

THEOREM A.4

Suppose $\lim_{x\to a} g(x) = L$ and f is continuous at L. Then,

$$\lim_{x\to a} f(g(x)) = f\left(\lim_{x\to a} g(x)\right) = f(L).$$

PROOF

To prove the result, we must show that given any number $\varepsilon > 0$, there is a number $\delta > 0$ for which

$$|f(g(x)) - f(L)| < \varepsilon, \text{ whenever } 0 < |x - a| < \delta.$$

Since f is continuous at L, we know that $\lim_{t \to L} f(t) = f(L)$. Consequently, given any $\varepsilon > 0$, there is a $\delta_1 > 0$ for which

$$|f(t) - f(L)| < \varepsilon, \text{ whenever } 0 < |t - L| < \delta_1.$$

Further, since $\lim_{x \to a} g(x) = L$, we can make $g(x)$ as close to L as desired, simply by making x sufficiently close to a. In particular, there must be a number $\delta > 0$ for which $|g(x) - L| < \delta_1$ whenever $0 < |x - a| < \delta$. It now follows that if $0 < |x - a| < \delta$, then $|g(x) - L| < \delta_1$, so that

$$|f(g(x)) - f(L)| < \varepsilon,$$

as desired. ■

The following result appeared as Theorem 5.1 in section 1.5.

THEOREM A.5

For any rational number $t > 0$,

$$\lim_{x \to \pm\infty} \frac{1}{x^t} = 0,$$

where for the case where $x \to -\infty$, we assume that $t = \frac{p}{q}$, where q is odd.

PROOF

We first prove that $\lim_{x \to \infty} \frac{1}{x^t} = 0$. To do so, we must show that given any number $\varepsilon > 0$, there is an $M > 0$ for which $\left|\frac{1}{x^t} - 0\right| < \varepsilon$, whenever $x > M$. Since $x \to \infty$, we can take x to be positive, so that

$$\left|\frac{1}{x^t} - 0\right| = \frac{1}{x^t} < \varepsilon,$$

which is equivalent to

$$\frac{1}{x} < \varepsilon^{1/t}$$

or

$$\frac{1}{\varepsilon^{1/t}} < x.$$

Notice that taking M to be any number greater than $\frac{1}{\varepsilon^{1/t}}$, we will have $\left|\frac{1}{x^t} - 0\right| < \varepsilon$ whenever $x > M$, as desired.

For the case $\lim_{x\to-\infty}\frac{1}{x^t}=0$, we must show that given any number $\varepsilon>0$, there is an $N<0$ for which $\left|\frac{1}{x^t}-0\right|<\varepsilon$, whenever $x<N$. Since $x\to-\infty$, we can take x to be negative, so that

$$\left|\frac{1}{x^t}-0\right|=\frac{1}{|x^t|}<\varepsilon,$$

which is equivalent to

$$\frac{1}{|x|}<\varepsilon^{1/t}$$

or

$$\frac{1}{\varepsilon^{1/t}}<|x|=-x,$$

since $x<0$. Multiplying both sides of the inequality by -1, we get

$$-\frac{1}{\varepsilon^{1/t}}>x.$$

Notice that taking N to be any number less than $-\frac{1}{\varepsilon^{1/t}}$, we will have $\left|\frac{1}{x^t}-0\right|<\varepsilon$, whenever $x<N$, as desired. ■

In section 3.2, we proved l'Hôpital's Rule only for a special case. Here, we present a general proof for the $\frac{0}{0}$ case. First, we need the following generalization of the Mean Value Theorem.

THEOREM A.6 (Generalized Mean Value Theorem)

Suppose that f and g are continuous on the interval $[a,b]$ and differentiable on the interval (a,b) and that $g'(x)\neq 0$, for all x on (a,b). Then, there is a number $z\in(a,b)$, such that

$$\frac{f(b)-f(a)}{g(b)-g(a)}=\frac{f'(z)}{g'(z)}.$$

Notice that the Mean Value Theorem (Theorem 9.4 in section 2.9) is simply the special case of Theorem A.6 where $g(x)=x$.

PROOF

First, observe that since $g'(x)\neq 0$, for all x on (a,b), we must have that $g(b)-g(a)\neq 0$. This follows from Rolle's Theorem (Theorem 9.1 in section 2.9), since if $g(a)=g(b)$, there would be some number $c\in(a,b)$ for which $g'(c)=0$. Now, define

$$h(x)=[f(b)-f(a)]g(x)-[g(b)-g(a)]f(x).$$

Notice that h is continuous on $[a,b]$ and differentiable on (a,b), since both f and g are continuous on $[a,b]$ and differentiable on (a,b). Further, we have

$$\begin{aligned}h(a)&=[f(b)-f(a)]g(a)-[g(b)-g(a)]f(a)\\&=f(b)g(a)-g(b)f(a)\end{aligned}$$

and
$$h(b) = [f(b) - f(a)]g(b) - [g(b) - g(a)]f(b)$$
$$= g(a)f(b) - f(a)g(b),$$

so that $h(a) = h(b)$. In view of this, Rolle's Theorem says that there must be a number $z \in (a, b)$ for which

$$0 = h'(z) = [f(b) - f(a)]g'(z) - [g(b) - g(a)]f'(z)$$

or
$$\frac{f(b) - f(a)}{g(b) - g(a)} = \frac{f'(z)}{g'(z)},$$

as desired. ■

We can now give a general proof of l'Hôpital's Rule for the $\frac{0}{0}$ case. The proof of the $\frac{\infty}{\infty}$ case can be found in a more advanced text.

THEOREM A.7 (l'Hôpital's Rule)

Suppose that f and g are differentiable on the interval (a, b), except possibly at some fixed point $c \in (a, b)$ and that $g'(x) \neq 0$, on (a, b), except possibly at $x = c$. Suppose further that $\lim_{x \to c} \frac{f(x)}{g(x)}$ has the indeterminate form $\frac{0}{0}$ or $\frac{\infty}{\infty}$ and that $\lim_{x \to c} \frac{f'(x)}{g'(x)} = L$ (or $\pm\infty$). Then,

$$\lim_{x \to c} \frac{f(x)}{g(x)} = \lim_{x \to c} \frac{f'(x)}{g'(x)}.$$

PROOF

($\frac{0}{0}$ case) In this case, we have that $\lim_{x \to c} f(x) = \lim_{x \to c} g(x) = 0$. Define

$$F(x) = \begin{cases} f(x) & \text{if } x \neq c \\ 0 & \text{if } x = c \end{cases} \quad \text{and} \quad G(x) = \begin{cases} g(x) & \text{if } x \neq c \\ 0 & \text{if } x = c \end{cases}.$$

Notice that
$$\lim_{x \to c} F(x) = \lim_{x \to c} f(x) = 0 = F(c)$$

and
$$\lim_{x \to c} G(x) = \lim_{x \to c} g(x) = 0 = G(c),$$

so that both F and G are continuous on all of (a, b). Further, observe that for $x \neq c$, $F'(x) = f'(x)$ and $G'(x) = g'(x)$ and so, both F and G are differentiable on each of the intervals (a, c) and (c, b). We first consider the interval (c, b). Notice that F and G are continuous on $[c, b]$ and differentiable on (c, b) and so, by the Generalized Mean Value Theorem, for any $x \in (c, b)$, we have that there is some number z, with $c < z < x$, for which

$$\frac{F'(z)}{G'(z)} = \frac{F(x) - F(c)}{G(x) - G(c)} = \frac{F(x)}{G(x)} = \frac{f(x)}{g(x)},$$

where we have used the fact that $F(c) = G(c) = 0$. Notice that as $x \to c^+$, $z \to c^+$, also, since $c < z < x$. Taking the limit as $x \to c^+$, we now have

$$\lim_{x \to c^+} \frac{f(x)}{g(x)} = \lim_{z \to c^+} \frac{F'(z)}{G'(z)} = \lim_{z \to c^+} \frac{f'(z)}{g'(z)} = L.$$

Similarly, by focusing on the interval (a, c), we can show that $\lim_{x \to c^-} \frac{f(x)}{g(x)} = L$, which proves that $\lim_{x \to c} \frac{f(x)}{g(x)} = L$ (since both one-sided limits agree). ■

The following theorem corresponds to Theorem 6.1 in section 8.6.

THEOREM A.8

Given any power series, $\sum_{k=0}^{\infty} b_k (x-c)^k$, there are exactly three possibilities:

(i) the series converges for *all* $x \in (-\infty, \infty)$ and the radius of convergence is $r = \infty$;
(ii) the series converges *only* for $x = c$ (and diverges for all other values of x) and the radius of convergence is $r = 0$ or
(iii) the series converges for $x \in (c - r, c + r)$ and diverges for $x < c - r$ and for $x > c + r$, for some number r with $0 < r < \infty$.

In order to prove Theorem A.8, we first introduce and prove two simpler results.

THEOREM A.9

(i) If the power series $\sum_{k=0}^{\infty} b_k x^k$ converges for $x = a \neq 0$, then it also converges for all x with $|x| < |a|$.
(ii) If the power series $\sum_{k=0}^{\infty} b_k x^k$ diverges for $x = d$, then it also diverges for all x with $|x| > |d|$.

PROOF

(i) Suppose that $\sum_{k=0}^{\infty} b_k a^k$ converges. Then, by Theorem 2.2 in section 8.2, $\lim_{k \to \infty} b_k a^k = 0$. For this to occur, we must be able to make $|b_k a^k|$ as small as desired, just by making k sufficiently large. In particular, there must be a number $N > 0$, such that $|b_k a^k| < 1$, for all $k > N$. So, for $k > N$, we must have

$$|b_k x^k| = \left| b_k a^k \left(\frac{x^k}{a^k} \right) \right| = |b_k a^k| \left| \frac{x}{a} \right|^k < \left| \frac{x}{a} \right|^k .$$

If $|x| < |a|$, then $\left|\frac{x}{a}\right| < 1$ and so, $\sum_{k=0}^{\infty} \left|\frac{x}{a}\right|^k$ is a convergent geometric series. It then follows from the Comparison Test (Theorem 3.3 in section 8.3) that $\sum_{k=0}^{\infty} |b_k x^k|$ converges and hence, $\sum_{k=0}^{\infty} b_k x^k$ converges absolutely.

(ii) Suppose that $\sum_{k=0}^{\infty} b_k d^k$ diverges. Notice that if x is any number with $|x| > |d|$, then $\sum_{k=0}^{\infty} b_k x^k$ must diverge, since if it converged, we would have by part (i) that $\sum_{k=0}^{\infty} b_k d^k$ would also converge, which contradicts our assumption. ■

Next, we state and prove a slightly simpler version of Theorem A.8.

THEOREM A.10

Given any power series, $\sum_{k=0}^{\infty} b_k x^k$, there are exactly three possibilities:

(i) the series converges for *all* $x \in (-\infty, \infty)$ and the radius of convergence is $r = \infty$;
(ii) the series converges *only* for $x = 0$ (and diverges for all other values of x) and the radius of convergence is $r = 0$ or
(iii) the series converges for $x \in (-r, r)$ and diverges for $x < -r$ and $x > r$, for some number r with $0 < r < \infty$.

PROOF

If neither (i) nor (ii) is true, then there must be nonzero numbers a and d such that the series converges for $x = a$ and diverges for $x = d$. From Theorem A.9, observe that $\sum_{k=0}^{\infty} b_k x^k$ diverges for all values of x with $|x| > |d|$. Define the set S to be the set of all values of x for which the series converges. Since the series converges for $x = a$, S is nonempty. Further, $|d|$ is an upper bound on S, since the series diverges for all values of x with $|x| > |d|$. By the Completeness Axiom (see section 8.1), S must have a least upper bound r. So, if $|x| > r$, then $\sum_{k=0}^{\infty} b_k x^k$ diverges. Further, if $|x| < r$, then $|x|$ is not an upper bound for S and there must be a number t in S with $|x| < t$. Since $t \in S$, $\sum_{k=0}^{\infty} b_k t^k$ converges and by Theorem A.9, $\sum_{k=0}^{\infty} b_k x^k$ converges since $|x| < t \leq |t|$. This proves the result. ■

We can now prove the original result (Theorem A.8).

PROOF OF THEOREM A.8

Let $t = x - c$ and the power series $\sum_{k=0}^{\infty} b_k (x - c)^k$ becomes simply $\sum_{k=0}^{\infty} b_k t^k$. By Theorem A.10, we know that either the series converges for all t (i.e., for all x) or only for $t = 0$ (i.e., only for $x = c$) or there is a number $r > 0$ such that the series converges for $|t| < r$ (i.e., for $|x - c| < r$) and diverges for $|t| > r$ (i.e., for $|x - c| > r$). This proves the original result. ■

Appendix B

ANSWERS TO ODD-NUMBERED EXERCISES

CHAPTER 10

Exercises 10.1, page 794

1.

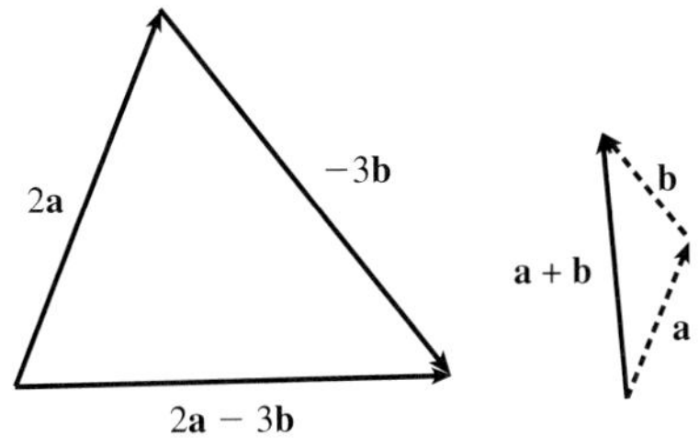

3. $\langle 5, 3\rangle, \langle -4, 6\rangle, \langle 6, 12\rangle, \sqrt{290}$

5. $4\mathbf{i}+\mathbf{j}, -5\mathbf{i}+4\mathbf{j}, 3\mathbf{i}+6\mathbf{j}, 5\sqrt{10}$

7.

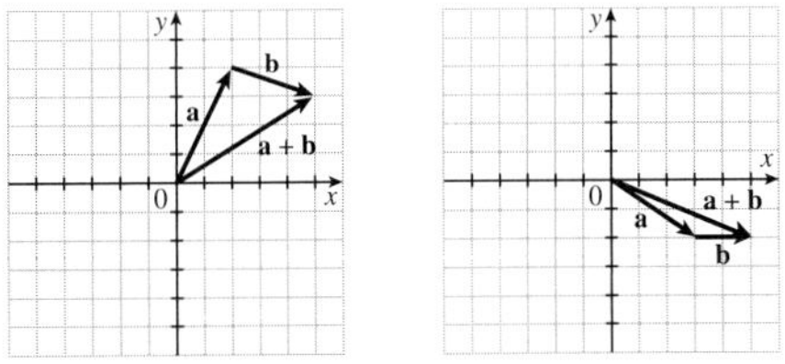

9. parallel **11.** not **13.** parallel **15.** $\langle 3, 1\rangle$

17. $\langle 2, -3\rangle$ **19.** (a) $\left\langle \frac{4}{5}, -\frac{3}{5}\right\rangle$ (b) $5\left\langle \frac{4}{5}, -\frac{3}{5}\right\rangle$

21. (a) $\frac{1}{\sqrt{5}}\mathbf{i} - \frac{2}{\sqrt{5}}\mathbf{j}$ (b) $2\sqrt{5}\left\langle \frac{1}{\sqrt{5}}, -\frac{2}{\sqrt{5}}\right\rangle$

23. (a) $\left\langle \frac{3}{\sqrt{10}}, \frac{1}{\sqrt{10}}\right\rangle$ (b) $\sqrt{10}\left\langle \frac{3}{\sqrt{10}}, \frac{1}{\sqrt{10}}\right\rangle$

25. $\frac{9}{5}\mathbf{i} + \frac{12}{5}\mathbf{j}$ **27.** $\langle 2\sqrt{29}, 5\sqrt{29}\rangle$ **29.** $\langle 4, 0\rangle$

31. 10 pounds down, 20 pounds to the right

33. 190 pounds up, 30 pounds to the right

35. $\langle 13, 17\rangle$; right and up

37. $\langle -80\sqrt{14}, 20\rangle$ or 3.8° north of west

39. $\langle 20, 20\sqrt{399}\rangle$ or 2.9° east of north **41.** 10 feet

43. $20\sqrt{101}$ pounds at 5.7° above the horizontal

45. speed, $\sqrt{17} \approx 4.123$ ft/s; angle, $\tan^{-1} 4 \approx 75.964°$ **47.** 7, 1, 5

51. $\|\mathbf{a}+\mathbf{b}\| = \sqrt{58} < \sqrt{13} + \sqrt{17} = \|\mathbf{a}\| + \|\mathbf{b}\|$

53. $\mathbf{a} = c\mathbf{b} (c > 0)$; $\mathbf{a} \perp \mathbf{b}$; $\|\mathbf{a}+\mathbf{b}\|^2 > \|\mathbf{a}\|^2 + \|\mathbf{b}\|^2$ when $\mathbf{a} = c\mathbf{b}$ for $c > 0$ or when the angle between **a** and **b** in the triangle formed by **a**, **b** and $\mathbf{a}+\mathbf{b}$ is obtuse, $\|\mathbf{a}+\mathbf{b}\|^2 < \|\mathbf{a}\|^2 + \|\mathbf{b}\|^2$ when $\mathbf{a} = c\mathbf{b}$ for $c < 0$ or when the angle between **a** and **b** in the triangle formed by **a**, **b** and $\mathbf{a}+\mathbf{b}$ is acute, $\|\mathbf{a}+\mathbf{b}\|^2 = \|\mathbf{a}\|^2 + \|\mathbf{b}\|^2$ when $\mathbf{a} \perp \mathbf{b}$.

Exercises 10.2, page 802

1.

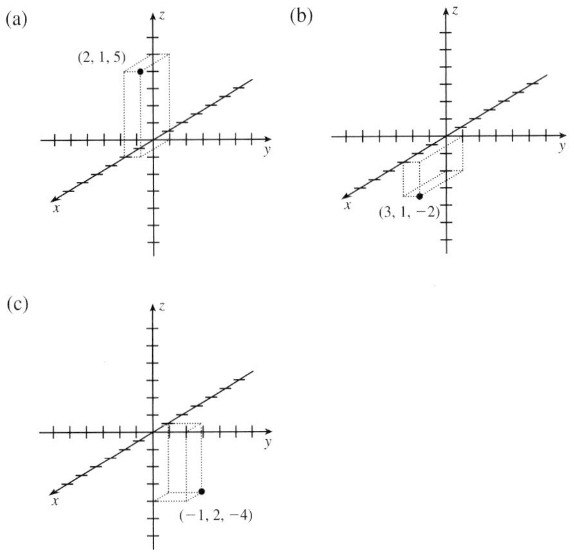

3.

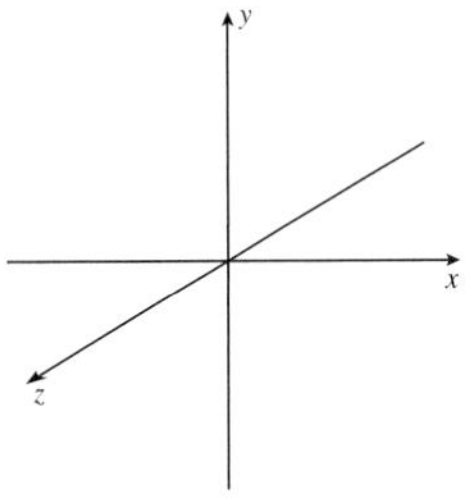

5. 5 **7.** 3 **9.** $\langle 3, 4, -2\rangle, \langle -1, -8, -2\rangle, 2\sqrt{66}$

11. $8\mathbf{i}+4\mathbf{k}, -12\mathbf{i}-4\mathbf{j}+4\mathbf{k}, 2\sqrt{186}$

13. (a) $\pm\frac{1}{\sqrt{14}}\langle 3, 1, 2\rangle$ (b) $\sqrt{14}\left\langle \frac{3}{\sqrt{14}}, \frac{1}{\sqrt{14}}, \frac{2}{\sqrt{14}}\right\rangle$

15. (a) $\pm\frac{1}{3}(2\mathbf{i}-\mathbf{j}+2\mathbf{k})$ (b) $3\left(\frac{2}{3}\mathbf{i} - \frac{1}{3}\mathbf{j} + \frac{2}{3}\mathbf{k}\right)$

17. (a) $\pm\frac{1}{\sqrt{2}}\langle 1, 0, -1\rangle$ (b) $2\sqrt{2}\left\langle \frac{1}{\sqrt{2}}, 0, -\frac{1}{\sqrt{2}}\right\rangle$

19. $\langle 4, 4, -2\rangle$ **21.** $\frac{4}{\sqrt{14}}(2\mathbf{i} - \mathbf{j} + 3\mathbf{k})$

23. $(x-3)^2 + (y-1)^2 + (z-4)^2 = 4$

25. $(x-\pi)^2 + (y-1)^2 + (z+3)^2 = 5$

27. sphere, center $(1, 0, -2)$, radius 2

29. sphere, center $(1, 0, 2)$, radius $\sqrt{5}$

31. plane parallel to xz-plane **33.** plane parallel to xy-plane

35. $y = 0$ **37.** $x = 0$ **43.** $\langle 2, -1, 1\rangle$, $\langle 4, -2, 2\rangle$, yes

45. not equilateral **47.** they form a right triangle

49. not a square

51. net force is 149 pounds in the direction $\langle -1, 5, -14\rangle$; force required to balance is $\langle 10, -50, 140\rangle$ pounds

53. direction is $\langle 41, 38, 20\rangle$ and speed is 593.72 mph

55. $\langle 4, -1, 7, 7\rangle$ **57.** $\langle 0, -8, 10, 1, -9, -1\rangle$ **59.** $\sqrt{31}$

61. 7 **63.** $\sqrt{2} - 1$ **65.** $\sqrt{n} - 1 > 2$ if $n \geq 10$

Exercises 10.3, page 811

1. 10 **3.** 10 **5.** 1 **7.** $\cos^{-1}\frac{1}{\sqrt{26}} \approx 1.37$

9. $\cos^{-1}\frac{-8}{\sqrt{234}} \approx 2.12$ **11.** yes **13.** yes

15. possible answer: $\langle 1, 2\rangle$ **17.** possible answer: $\mathbf{j} + 2\mathbf{k}$

19. $2, \langle \frac{6}{5}, \frac{8}{5}\rangle$ **21.** $2, \frac{2}{3}\langle 1, 2, 2\rangle$ **23.** $-\frac{8}{5}, -\frac{8}{25}\langle 0, -3, 4\rangle$

25. 105,600 foot-pounds **29.** 920 foot-pounds

31. (a) false (b) true (c) true (d) false (e) false

33. $\mathbf{a} = c\mathbf{b}$ **37.** 15 **39.** $\sum_{k=1}^{n} \frac{1}{k^3} \leq \frac{\pi^3}{6\sqrt{15}}$

47. $\cos^{-1}\left(\frac{1}{3}\right) \approx 109.5°$

51. $-\frac{200}{3\sqrt{14}} \simeq -17.8$ newtons **53.** $-2000 \sin 10° \approx -347$ pounds

55. 92 ft/s $\approx$ 63 mph **57.** total revenue

61. $\mathbf{v} \cdot \mathbf{n} = 0$, $\text{comp}_{\mathbf{v}} w = -w \sin\theta$, $\text{comp}_{\mathbf{n}} w = -w \cos\theta$

63. 45° **65.** \$190,000; monthly revenue

Exercises 10.4, page 825

1. 1 **3.** 4 **5.** $\langle 4, -3, -2\rangle$ **7.** $\langle -9, -4, 1\rangle$

9. $\langle 4, -2, 8\rangle$ **11.** $\pm\frac{1}{\sqrt{69}}\langle 8, 1, -2\rangle$ **13.** $\pm\frac{1}{\sqrt{46}}\langle -3, -6, 1\rangle$

15. $\pm\frac{1}{\sqrt{154}}\langle -1, -3, 12\rangle$ **17.** $\sin^{-1}\frac{7}{\sqrt{85}} \approx 0.86$

19. $\sin^{-1}\frac{13}{\sqrt{170}} \approx 1.49$ **21.** $\sqrt{\frac{7}{2}} \approx 1.87$ **23.** $\sqrt{\frac{61}{5}} \approx 3.49$

25. $\frac{20\sqrt{2}}{3} \approx 9.4$ foot-pounds **27.** 10 foot-pounds

29. (a) up (b) up and to the right

31. (a) down, left (b) up and to the left **33.** ball rises

35. ball drops **37.** no effect **39.** ball rises **41.** false

43. false **45.** true **47.** 5 **49.** $\frac{11\sqrt{3}}{2}$ **51.** 10

53. $\mathbf{0}$ **55.** $-\mathbf{i}$ **57.** $-3\mathbf{j}$ **59.** yes **61.** no

65. Figure A; 12

Exercises 10.5, page 835

1. (a) $x = 1 + 2t$, $y = 2 - t$, $z = -3 + 4t$
(b) $\frac{x-1}{2} = \frac{y-2}{-1} = \frac{z+3}{4}$

3. (a) $x = 2 + 2t$, $y = 1 - t$, $z = 3 + t$
(b) $\frac{x-2}{2} = \frac{y-1}{-1} = \frac{z-3}{1}$

5. (a) $x = 1 - 3t$, $y = 4$, $z = 1 + t$ (b) $\frac{x-1}{-3} = \frac{z-1}{1}$, $y = 4$

7. (a) $x = 2 - 4t$, $y = -t$, $z = 1 + 2t$ (b) $\frac{x-2}{-4} = \frac{y}{-1} = \frac{z-1}{2}$

9. (a) $x = 1 + 2t$, $y = 2 - t$, $z = -1 + 3t$
(b) $\frac{x-1}{2} = \frac{y-2}{-1} = \frac{z+1}{3}$

11. $\cos^{-1}\frac{-13}{\sqrt{234}} \approx 2.59$ **13.** perpendicular **15.** parallel

17. intersect **19.** parallel

21. $2(x-1) - (y-3) + 5(z-2) = 0$

23. $2(x-2) - 7y - 3(z-3) = 0$

25. $2(x+2) + 6(y-2) - 3z = 0$ **27.** $-2x + 4(y+2) = 0$

29. $(x-1) - (y-2) + (z-1) = 0$

31.

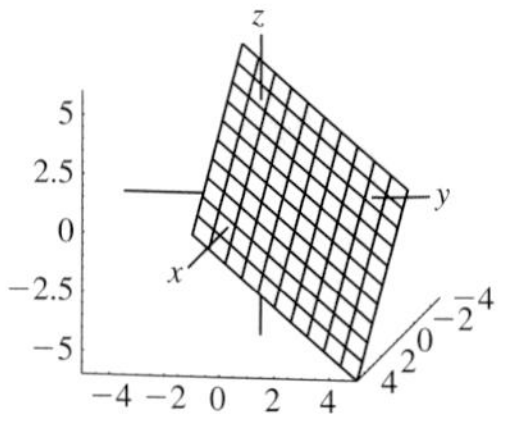

33.

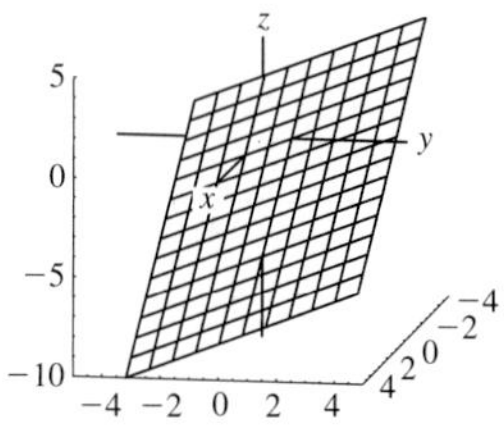

35.

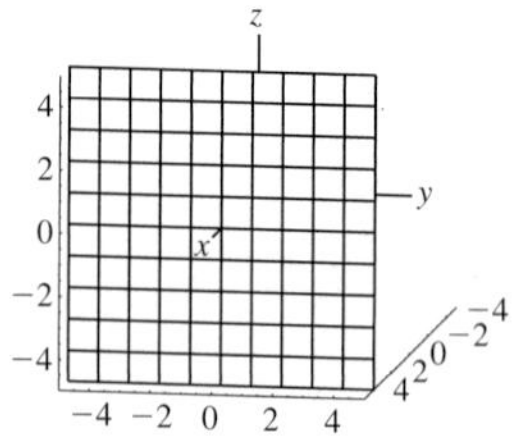

37.

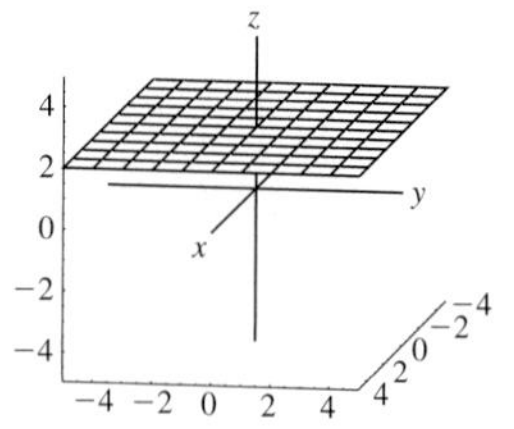

39.

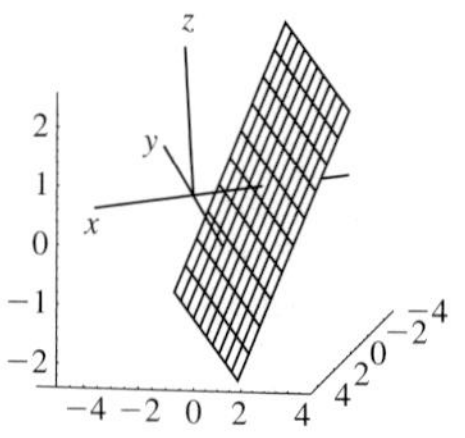

41. $x = t$, $y = \frac{5}{3}t - \frac{4}{3}$, $z = \frac{1}{3}t - \frac{8}{3}$

43. $x = 4t + 11$, $y = -3t - 8$, $z = t$

45. $\frac{2}{3}$ **47.** $\frac{2}{\sqrt{3}}$ **49.** $\frac{3}{\sqrt{6}}$ **53.** $-4(x-4) + 2(z-3) = 0$

55. true (if the planes coincide, they intersect *and* are parallel)

57. false

59. false (true if we take all lines perpendicular to a given line through a given point on the line)

61. true **63.** yes **65.** no

67. intersect at $(3, 4, 4)$, collide if $s = 1$ when $t = 1$

Exercises 10.6, page 847

1.

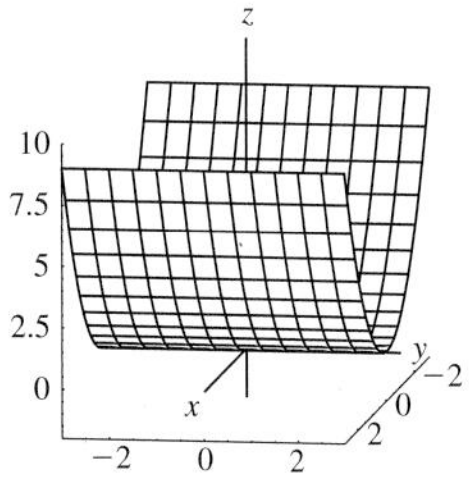

cylinder

3.

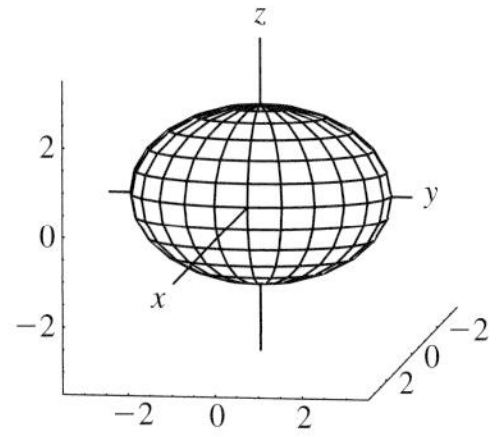

ellipsoid

5.

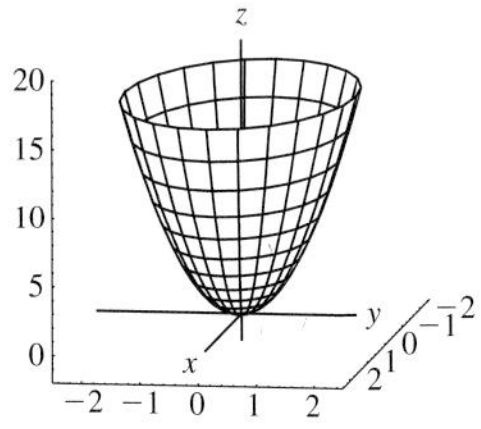

circular paraboloid

7.

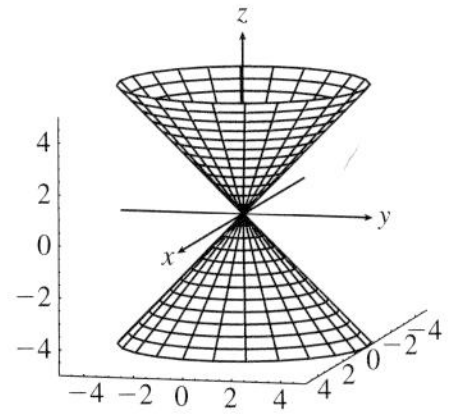

elliptic cone

9.

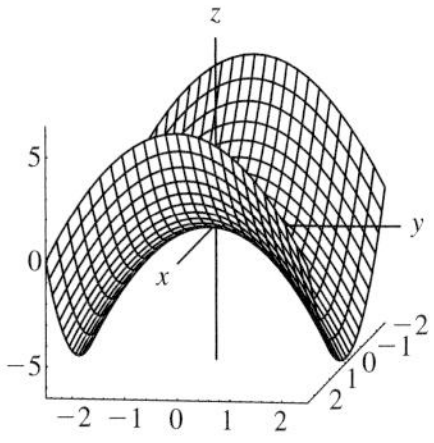

hyperbolic paraboloid

11.

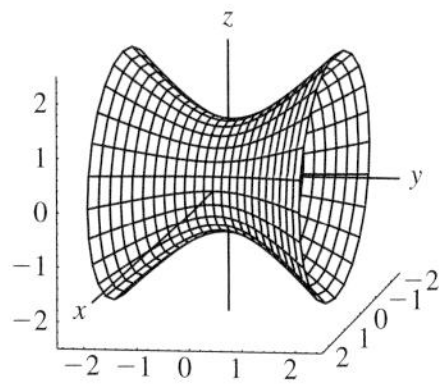

hyperboloid of 1 sheet

13.

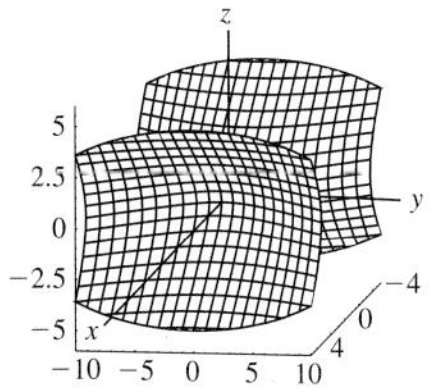

hyperboloid of 2 sheets

15.

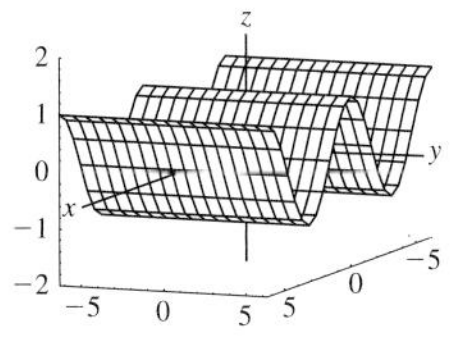

cylinder

17.

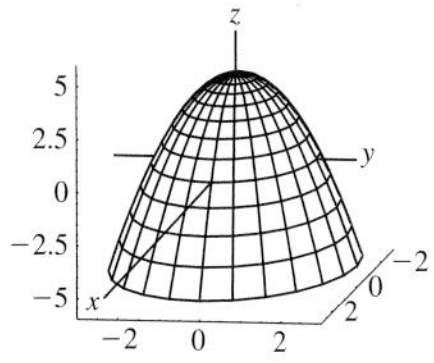

circular paraboloid

19.

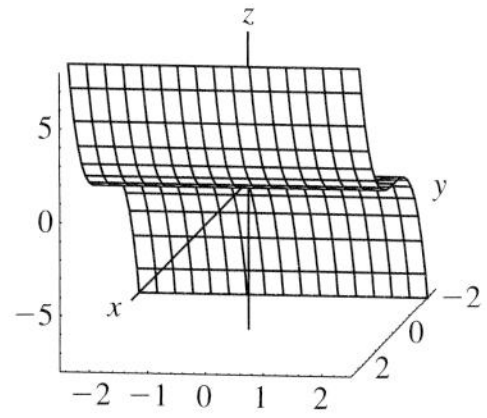

cylinder

21.

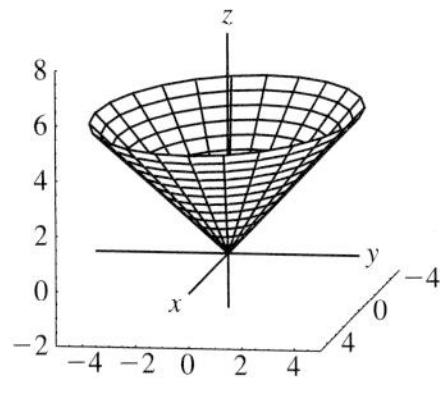

circular cone, $z \geq 0$

23.

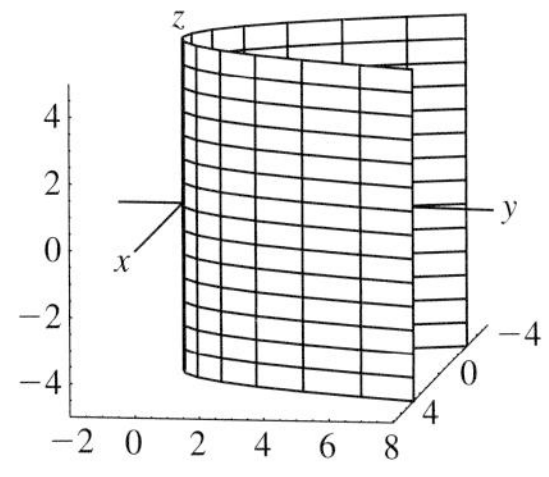

cylinder

25.

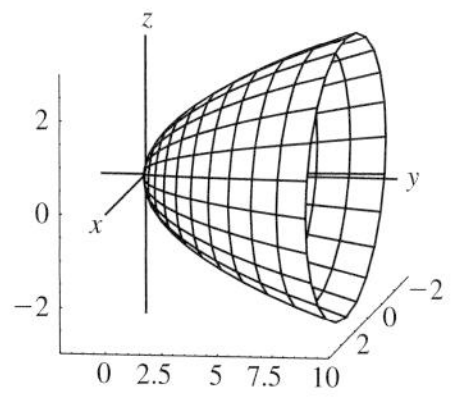

circular paraboloid

27.

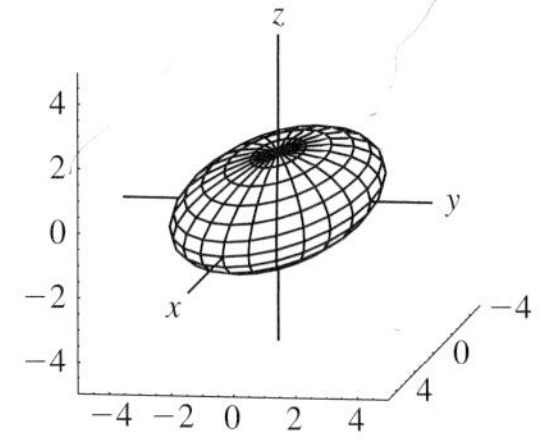

ellipsoid

29.

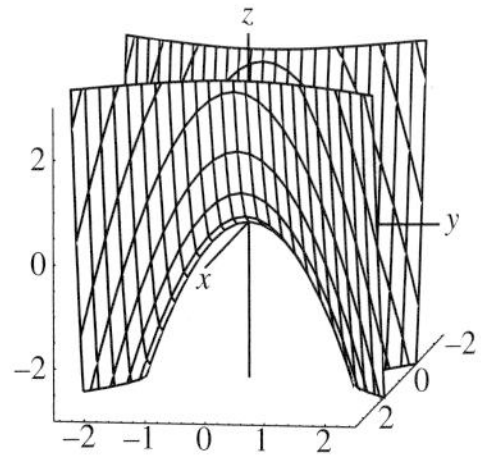

hyperbolic paraboloid

31.

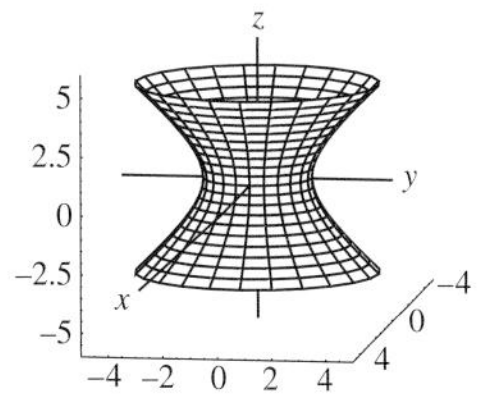

hyperboloid of 1 sheet

33.

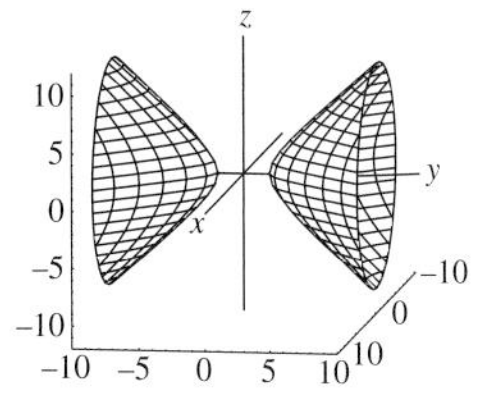

hyperboloid of 2 sheets

35.

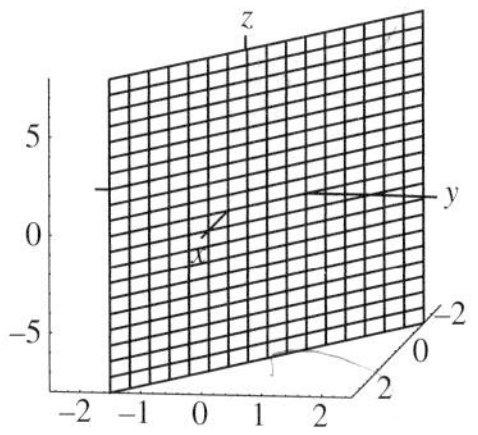

cylinder (plane)

37.

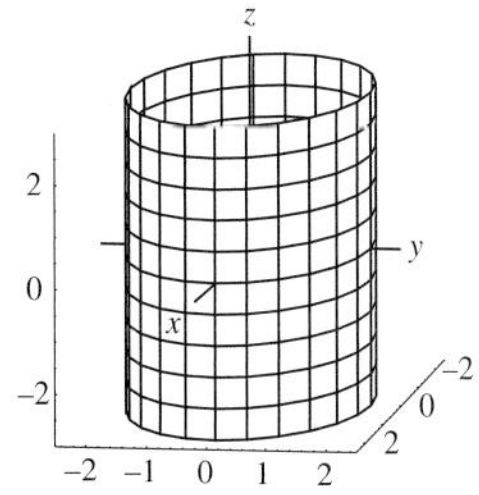

cylinder

39.

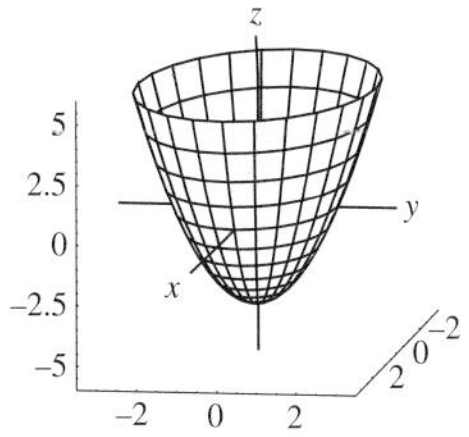

circular paraboloid

41.

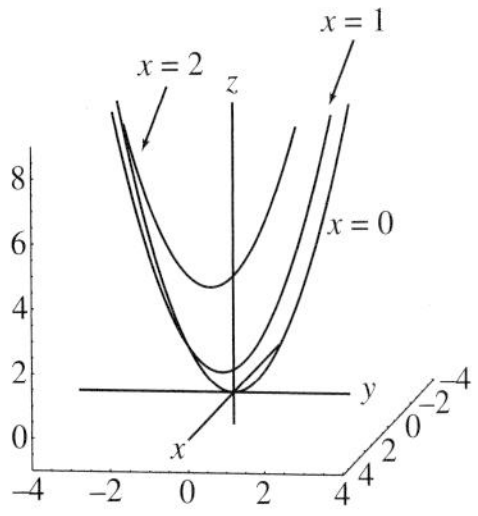

43.

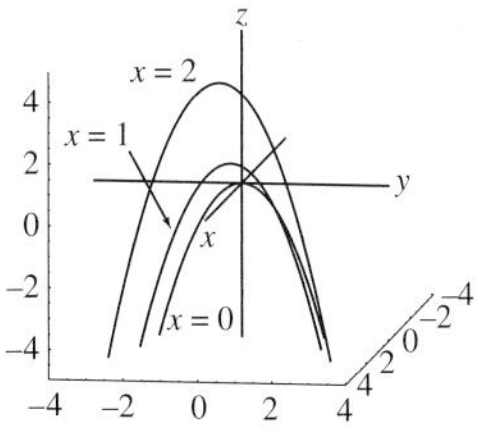

53. exercise 3: $x = \sin s \cos t$, $y = 3 \sin s \sin t$, $z = 2 \cos s$;
exercise 5: $x = \frac{1}{2} s \cos t$, $y = \frac{1}{2} s \sin t$, $z = s^2$;
exercise 7: $x = \frac{1}{2} s \cos t$, $y = s \sin t$, $z = s$

55. possible answer: $x = s \cos t$, $y = s \sin t$, $z = 4 - s^2$

Chapter 10 Review Exercises, page 849

1. $\langle -1, 3\rangle, \langle 4, 0\rangle, 5$ **3.** $6\mathbf{i} + 5\mathbf{j}, -16\mathbf{i} + 12\mathbf{j} + 8\mathbf{k}, 2\sqrt{94}$

5. neither **7.** parallel **9.** $\langle -1, -2, 3\rangle$ **11.** $\left\langle \frac{1}{\sqrt{5}}, \frac{2}{\sqrt{5}} \right\rangle$

13. $\frac{1}{3\sqrt{3}}(5\mathbf{i} + \mathbf{j} - \mathbf{k})$ **15.** $\left\langle -\frac{3}{5}, 0, \frac{4}{5} \right\rangle$ **17.** $\sqrt{46}$

19. $\frac{2}{\sqrt{3}}(\mathbf{i} - \mathbf{j} + \mathbf{k})$ **21.** $\langle 20\sqrt{609}, 80\rangle$ or 9.2° north of east

23. $x^2 + (y+2)^2 + z^2 = 36$ **25.** 0 **27.** -8

29. $\cos^{-1} \frac{1}{\sqrt{84}} \approx 1.46$ **31.** $\frac{1}{\sqrt{6}}, \frac{1}{6}(\mathbf{i} + 2\mathbf{j} + \mathbf{k})$

33. $\langle -2, 1, 4\rangle$ **35.** $-4\mathbf{i} + 4\mathbf{j} - 8\mathbf{k}$ **37.** $\pm\frac{1}{\sqrt{21}}\langle -2, 1, 4\rangle$

39. 1700 foot-pounds **41.** 3 **43.** $\sqrt{41}$

45. $\frac{25}{2}$ foot-pounds

47. (a) $x = 2 - 2t, y = -1 + 3t, z = -3$
(b) $\frac{x-2}{-2} = \frac{y+1}{3}, z = -3$

49. (a) $x = 2 + 2t, y = -1 + \frac{1}{2}t, z = 1 - 3t$
(b) $\frac{x-2}{2} = 2(y+1) = \frac{z-1}{-3}$

51. $\cos^{-1} \frac{5}{\sqrt{30}} \approx 0.42$ **53.** skew

55. $4(x+5) + y - 2(z-1) = 0$

57. $4(x-2) - (y-1) + 2(z-3) = 0$

59.

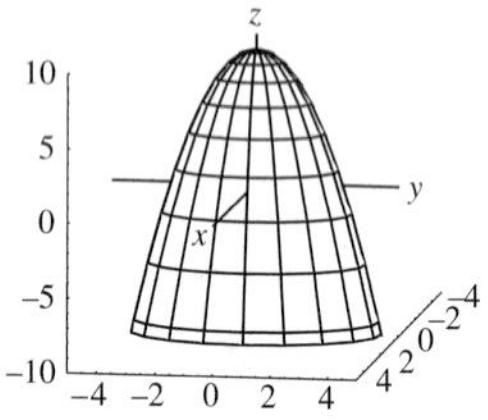

elliptic paraboloid

61.

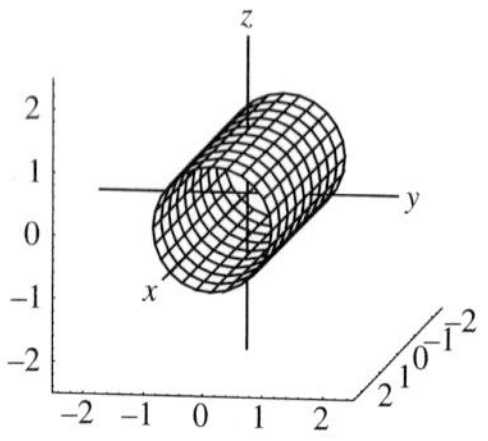

cylinder

63.

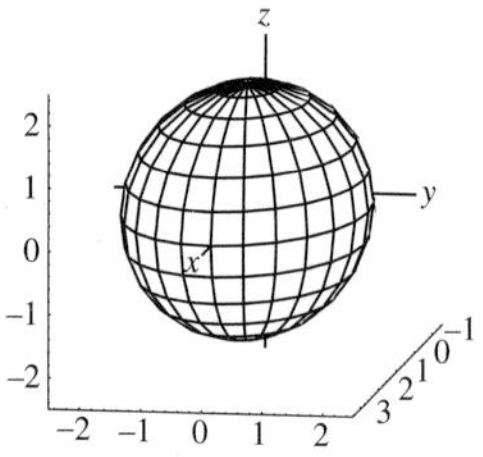

sphere

65.

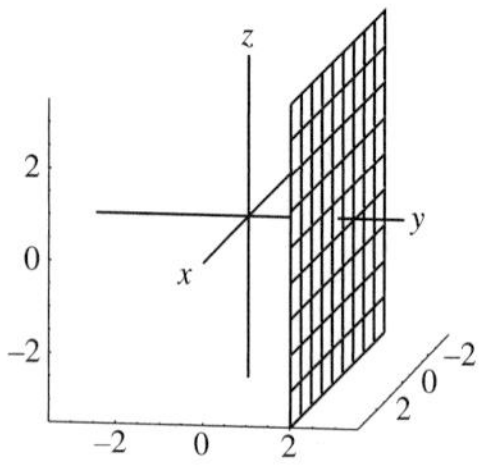

plane

67.

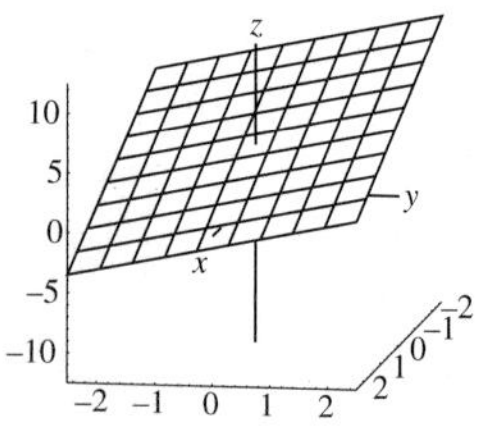

plane

69.

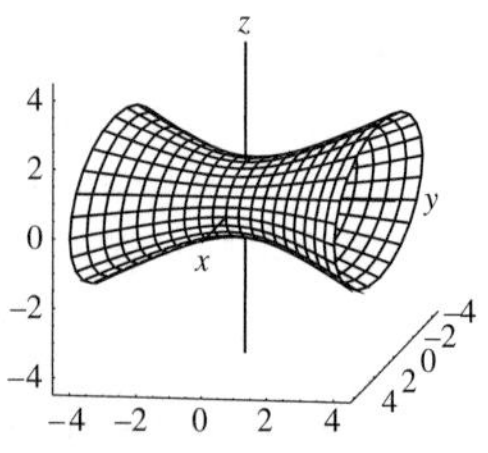

hyperboloid of 1 sheet

71.

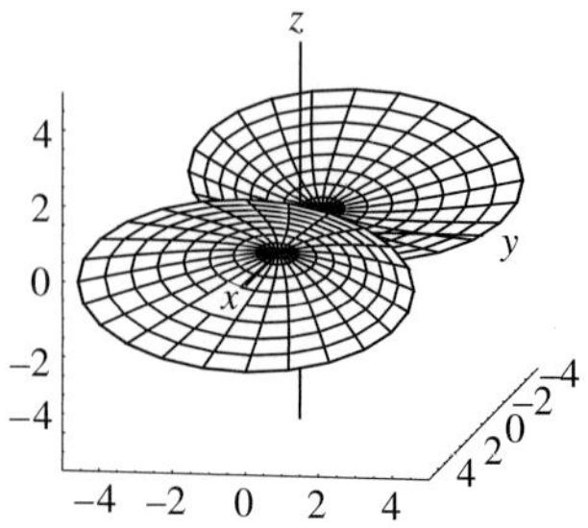

hyperboloid of two sheets

CHAPTER 11

Exercises 11.1, page 864

1.

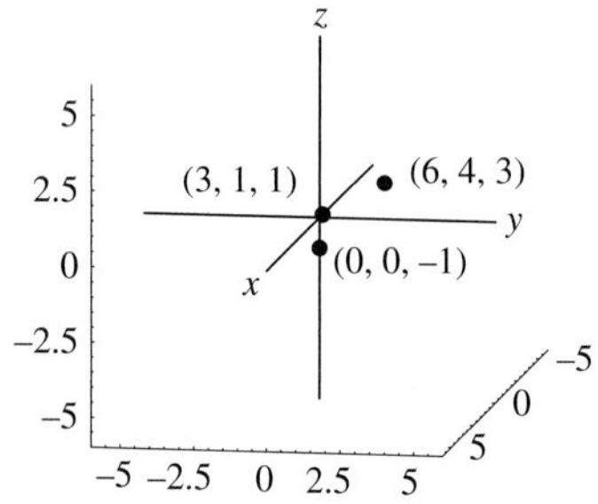

3.

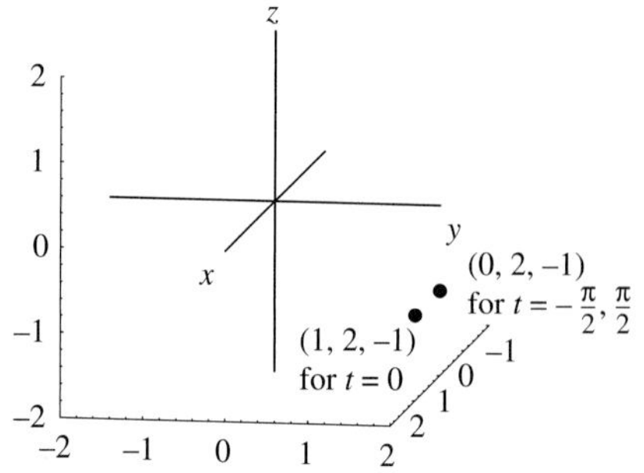

5.

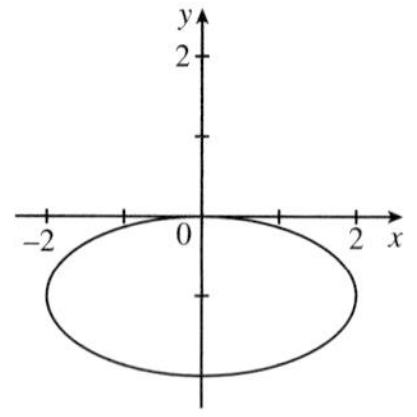

7.

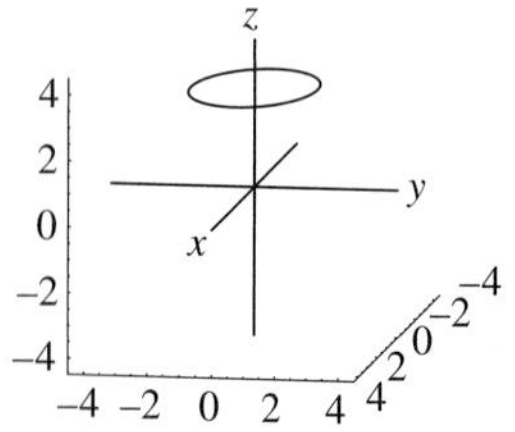

9.

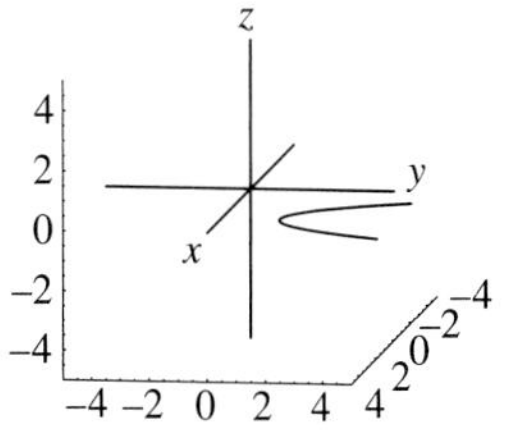

11.

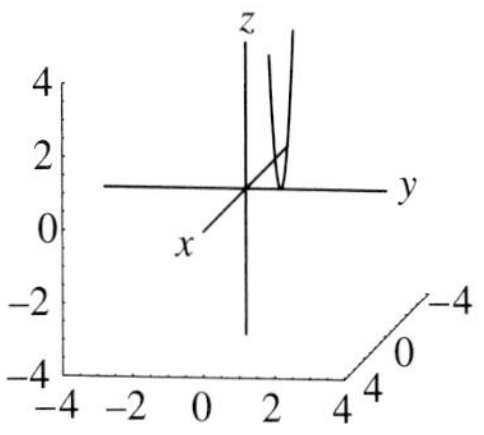

13.

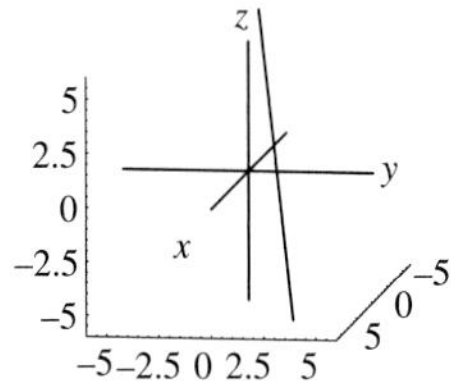

15.

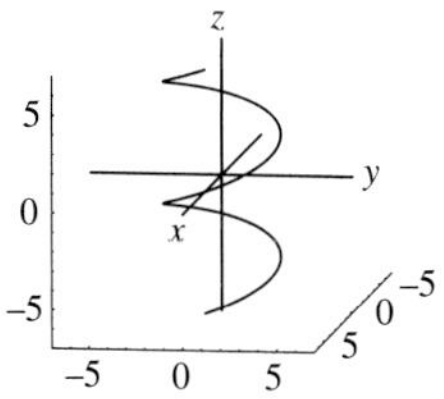

17.

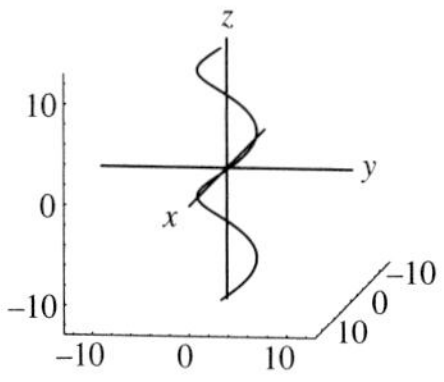

19.

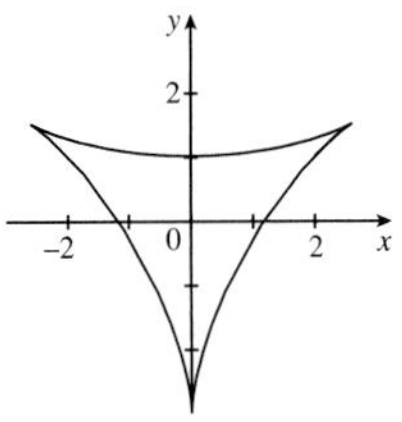

21.

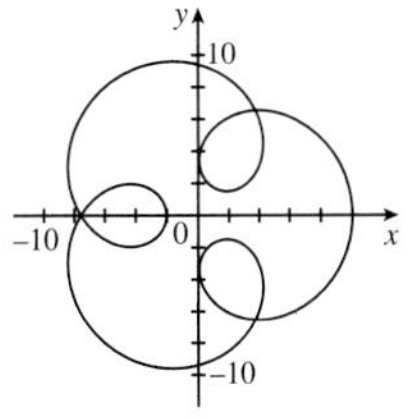

23.

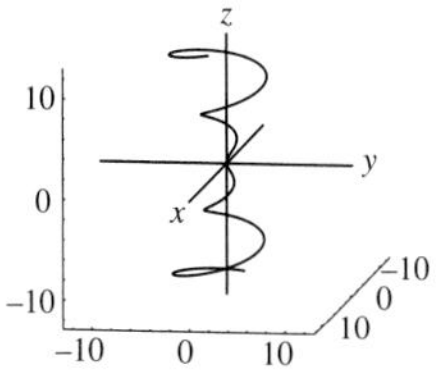

25.

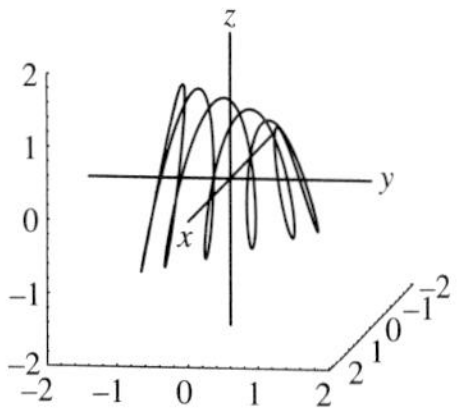

27.

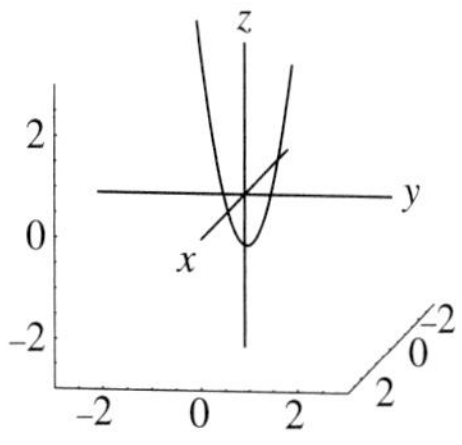

29.

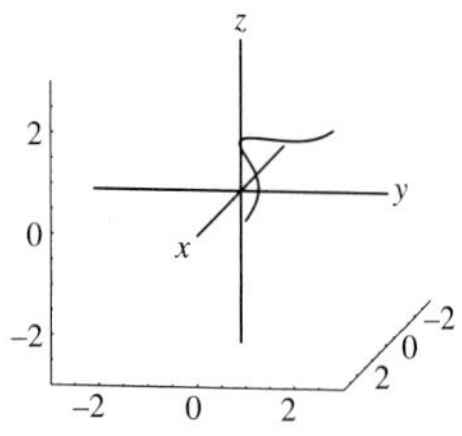

31. (a) F (b) C (c) E (d) A (e) B (f) D

33. 10.54 **35.** 21.56 **37.** 9.57

39. $\cos 2t = \cos^2 t - \sin^2 t$

43. same except for domains: $-\infty < x < \infty, -1 \le x \le 1, 0 \le x$

45. $x = 2\cos t,\ y = 2\sin t,\ z = 2,\ 0 \le t \le 2\pi$

47. $x = 3\cos t,\ y = 3\sin t,\ z = 2 - 3\sin t,\ 0 \le t \le 2\pi$

49. $\sqrt{144\pi^2 + 100} \approx 39.00$ feet

Exercises 11.2, page 876

1. $\langle -1, 1, 0\rangle$ **3.** $\langle 1, 1, -1\rangle$ **5.** does not exist

7. $t \neq 1$ **9.** $t \neq \frac{n\pi}{2}$, n odd **11.** $t \geq 0$

13. $\left\langle 4t^3, \frac{1}{2\sqrt{t+1}}, -\frac{6}{t^3}\right\rangle$ **15.** $\langle \cos t, 2t\cos t^2, -\sin t\rangle$

17. $\langle 2te^{t^2}, 2t, 2\sec 2t\tan 2t\rangle$

19.

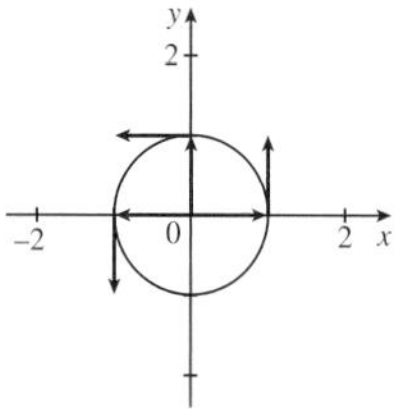

21.

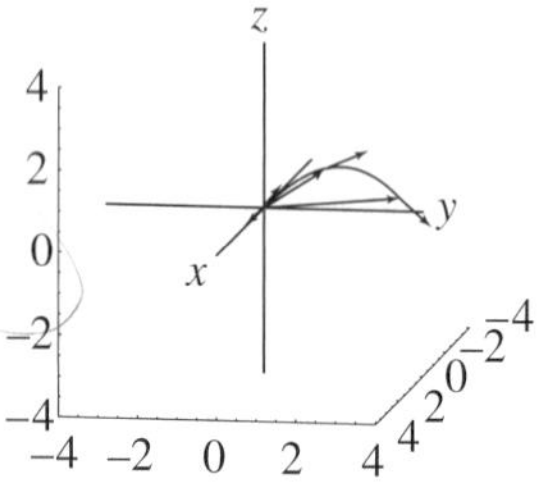

23. $\left\langle \frac{3}{2}t^2 - t, \frac{2}{3}t^{3/2}\right\rangle + \mathbf{c}$ **25.** $\left\langle \frac{1}{3}\sin 3t, -\cos t, \frac{1}{4}e^{4t}\right\rangle + \mathbf{c}$

27. $\left\langle \frac{1}{2}e^{t^2}, 3\sin t - 3t\cos t, \frac{3}{2}\ln(t^2+1)\right\rangle + \mathbf{c}$ **29.** $\left\langle -\frac{2}{3}, \frac{3}{2}\right\rangle$

31. $\langle 4\ln 3, 1 - e^{-2}, e^2 + 1\rangle$ **33.** all t **35.** $t = 0$

39. $t = 0$ **41.** $t = \frac{n\pi}{4}$, n odd **47.** false

51.

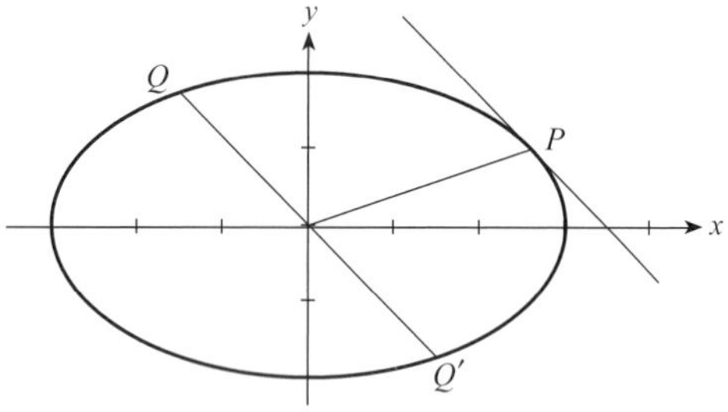

$a = 3, b = 2, t = \frac{\pi}{6}$

53. $\mathbf{f}'(t)\cdot[\mathbf{g}(t)\times\mathbf{h}(t)] + \mathbf{f}(t)\cdot[\mathbf{g}'(t)\times\mathbf{h}(t) + \mathbf{g}(t)\times\mathbf{h}'(t)]$

Exercises 11.3, page 886

1. $\langle -10\sin 2t, 10\cos 2t\rangle$, $\langle -20\cos 2t, -20\sin 2t\rangle$

3. $\langle 25, -32t + 15\rangle$, $\langle 0, -32\rangle$

5. $\langle (4 - 8t)e^{-2t}, -4e^{-2t}, -32t\rangle$, $\langle (-16 + 16t)e^{-2t}, 8e^{-2t}, -32\rangle$

7. $\langle 10t + 3, -16t^2 + 4t + 8\rangle$ **9.** $\langle 5t, -16t^2 + 16\rangle$

11. $\langle 10t, -3e^{-t} - 3, -16t^2 + 4t + 20\rangle$

13. $\left\langle \frac{1}{6}t^3 + 12t + 5, -4t, -8t^2 + 2\right\rangle$ **15.** $-160\langle \cos 2t, \sin 2t\rangle$

17. $-960\langle \cos 4t, \sin 4t\rangle$ **19.** $\langle -120\cos 2t, -200\sin 2t\rangle$

21. $\langle 60, 0\rangle$ **23.** $\frac{1875}{16} \approx 117$ feet, $\frac{625\sqrt{3}}{4} \approx 271$ feet, 100 ft/s

25. 210 feet, $400 + 40\sqrt{105} \approx 810$ feet, $8\sqrt{410} \approx 162$ ft/s

27. 810 feet, $1600 + 720\sqrt{5} \approx 3210$ feet, $8\sqrt{1610} \approx 321$ ft/s

29. approximately quadruples **33.** $\left\langle 60\sqrt{3}t, 3 + 60t - 16t^2\right\rangle$, no-hits wall 5.7′ up **35.** $\langle 130t, 6 - 16t^2\rangle$, 2.59 feet

37. $\langle 120t, 8 - 16t^2\rangle$, out **39.** 3.86 seconds **41.** $\langle 271, 117, 0\rangle$

43. $a = 100, b = -1, c = 10$ **45.** 56.57 ft/s **47.** 1275.5 m

49. 5 rad/s **51.** $\frac{225}{2\pi}$ rad/s^2

53. the additional rotation increases speed by a factor of $\sqrt{2}$

59. $w = \frac{\pi}{43082}$, $b \approx 42,168$ km

61. $x_0 = \frac{x_1 t_2 - x_2 t_1}{t_2 - t_1}$, $y_0 = \frac{y_1 t_2 - y_2 t_1}{t_2 - t_1}$ **63.** 0.2473 mi/s

Exercises 11.4, page 895

1. $x = 2\cos\left(\frac{s}{2}\right)$, $y = 2\sin\left(\frac{s}{2}\right)$, $0 \leq s \leq 4\pi$

3. $x = \frac{3}{5}s$, $y = \frac{4}{5}s$, $0 \leq s \leq 5$

5. $\langle 1, 0\rangle$, $\frac{1}{\sqrt{13}}\langle 3, -2\rangle$, $\frac{1}{\sqrt{13}}\langle 3, 2\rangle$

7. $\langle 0, 1\rangle$, $\langle 1, 0\rangle$, $\langle -1, 0\rangle$

9. $\frac{1}{\sqrt{13}}\langle 3, 0, 2\rangle$, $\frac{1}{\sqrt{13}}\langle 3, 0, 2\rangle$, $\frac{1}{\sqrt{13}}\langle 3, 0, 2\rangle$

11.

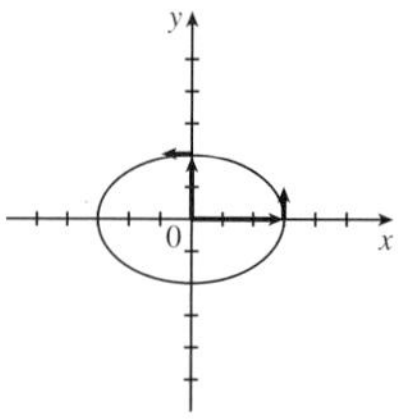

13.

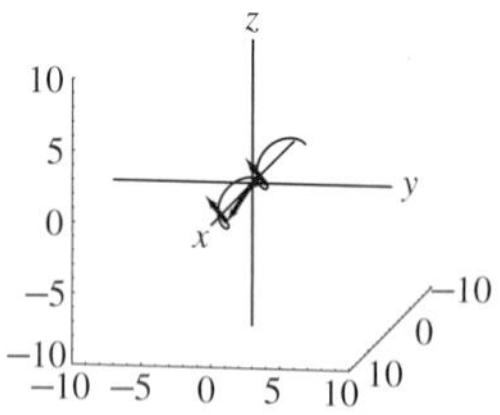

15. $2^{-3/2} \approx 0.3536$ **17.** 0 **19.** $(6)(37^{-3/2}) \approx 0.0267$

21. 1 **23.** smaller **25.** $\frac{8}{25}, \frac{8}{25}$ **27.** $1, \frac{1}{27}$

29. max at $\langle 0, \pm 3\rangle$, min at $\langle \pm 2, 0\rangle$ **31.** max at $\langle 0, -3\rangle$, no min

33. 0 **35.** 0 **37.** curve straightens **39.** false

41. true **45.** $\frac{2}{3}, 10$ **47.** $\frac{1}{\sqrt{45}}, \frac{1}{\sqrt{45}}e^{-2}$ **49.** $\frac{25}{52} > \frac{1}{10}$

Exercises 11.5, page 908

1. $\langle 1, 0\rangle$ and $\langle 0, 1\rangle$, $\frac{1}{\sqrt{5}}\langle 1, 2\rangle$ and $\frac{1}{\sqrt{5}}\langle -2, 1\rangle$

3. $\langle 0, 1\rangle$ and $\langle -1, 0\rangle$, $\langle -1, 0\rangle$ and $\langle 0, -1\rangle$

5. $\frac{1}{\sqrt{5}}\langle 0, 1, 2\rangle$, and $\langle -1, 0, 0\rangle$, $\frac{1}{\sqrt{5}}\langle 0, 1, -2\rangle$ and $\langle 1, 0, 0\rangle$

7. $\frac{1}{\sqrt{2}}\langle 1, 0, 1\rangle$ and $\langle 0, 1, 0\rangle$, $\frac{1}{\sqrt{6}}\langle 1, 2, 1\rangle$ and $\frac{1}{\sqrt{3}}\langle -1, 1, -1\rangle$

9. $x^2 + \left(y - \frac{1}{2}\right)^2 = \frac{1}{4}$ **11.** $x^2 + y^2 = 1$

13. $a_T = -\frac{64}{\sqrt{5}}$ and $a_N = \frac{32}{\sqrt{5}}$, $a_T = \frac{64}{\sqrt{5}}$ and $a_N = \frac{32}{\sqrt{5}}$

15. $a_T = 0$ and $a_N = \sqrt{20}$, $a_T = \frac{2\pi}{\sqrt{16+\pi^2}}$ and $a_N = 4\sqrt{\frac{20+\pi^2}{16+\pi^2}}$

17. neither; increasing **19.** $a_T = 0$ and $a_N = a$

21. $\frac{1}{\sqrt{5}}\langle 2, -1, 0\rangle$, $\frac{1}{\sqrt{5}}\langle 2, -1, 0\rangle$

23. $\frac{1}{\sqrt{1+16\pi^2}}\langle 0, -1, 4\pi\rangle$, $\frac{1}{\sqrt{1+16\pi^2}}\langle 0, 1, 4\pi\rangle$

25. true **27.** true

29. $10{,}000\pi^2\langle -\cos \pi t, -\sin \pi t\rangle$

31. $40{,}000\pi^2\langle -\cos 2\pi t, -\sin 2\pi t\rangle$ **33.** doubles

35. $1 < \frac{3\sqrt{3}}{2}$, $|-1| > \left|-\frac{1}{\sqrt{2}}\right|$

Exercises 11.6, page 915

1.

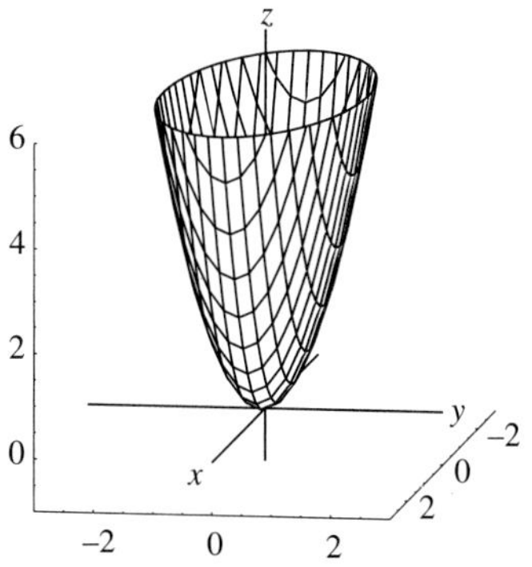

3.

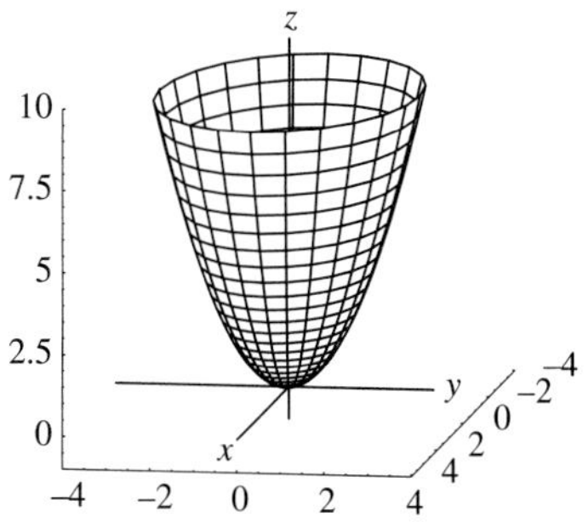

5.

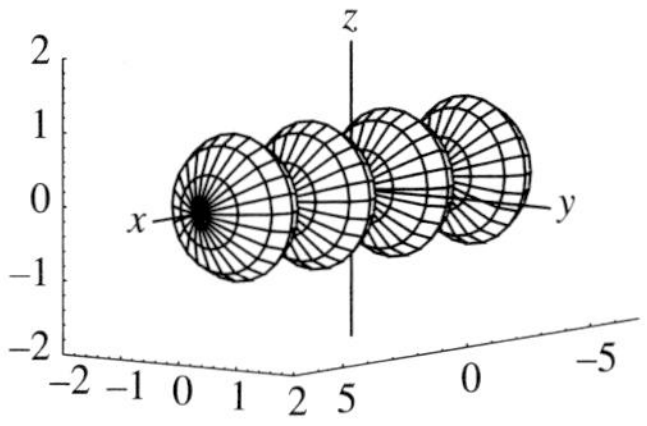

7.

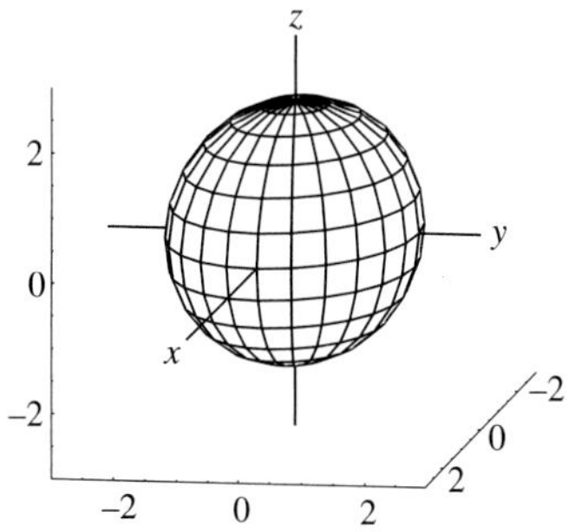

9. half of a hyperbolic paraboloid (y is restricted to positive values)

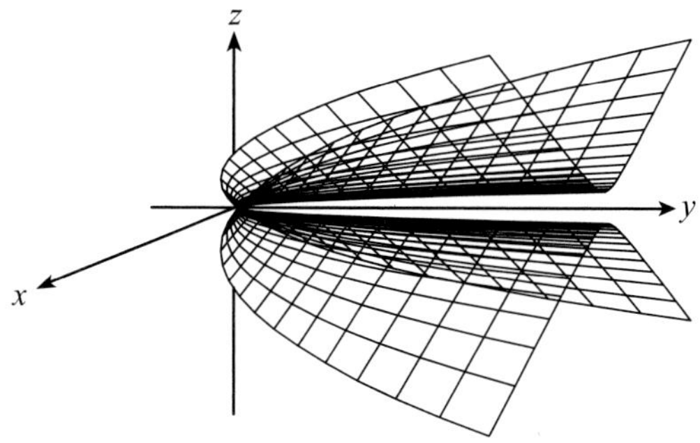

11. one sheet of a hyperboloid of two sheets

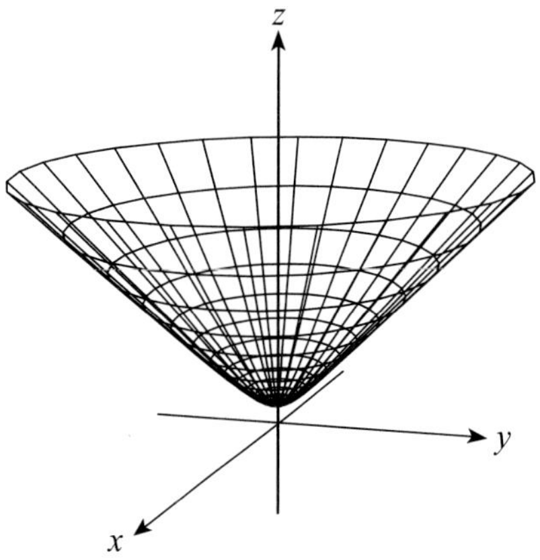

13. $x = x$, $y = y$, $z = 3x + 4y$

15. $x = \cos u \cosh v$, $y = \sin u \cosh v$, $z = \sinh v$, $0 \le u \le 2\pi$, $-\infty < v < \infty$

17. $x = 2\cos\theta$, $y = 2\sin\theta$, $z = z$, $0 \le \theta \le 2\pi$, $0 \le z \le 2$

19. $x = r\cos\theta$, $y = r\sin\theta$, $z = 4 - r^2$, $0 \le \theta \le 2\pi$, $0 \le r \le 2$

21. $x = \pm\cosh u$, $y = \sinh u \cos v$, $z = \sinh u \sin v$, $-\infty < u < \infty$, $0 \le v \le 2\pi$

23. (a) A (b) C (c) B **27.** a sphere of radius 3

29. half of a circular cone with the z-axis as its axis (the half with $z \ge 0$)

31. a half-plane with the z-axis as its boundary

33. a half-cone (except for the case $\phi = \frac{\pi}{2}$, which is the xy-plane)

35. $x = 3\cos\theta \sin\phi$, $y = 3\sin\theta\sin\phi$, $z = 3\cos\phi$ with $0 \le \theta \le 2\pi$, $0 \le \phi \le \frac{\pi}{2}$

37. $x = \rho \cos\theta \sin\frac{\pi}{4}$, $y = \rho \sin\theta \sin\frac{\pi}{4}$, $z = \rho \cos\frac{\pi}{4}$ with $0 \le \theta \le 2\pi$, $0 \le \rho$

39. $x = \rho \cos\theta \sin\frac{\pi}{4}$, $y = \rho \sin\theta \sin\frac{\pi}{4}$, $z = \rho \cos\frac{\pi}{4}$ with $0 \le \theta \le 2\pi$, $0 \le \rho \le 2$ and $x = 2\cos\theta \sin\phi$, $y = 2\sin\theta \sin\phi$, $z = 2\cos\phi$ with $0 \le \theta \le 2\pi$, $0 \le \phi \le \frac{\pi}{4}$

41. $u = v = 0$ gives $(2, -1, 3)$;
$u = 1$ and $v = 0$ gives $(3, 1, 0)$;
$u = 0$ and $v = 1$ gives $(4, -2, 5)$

43. a normal is $\mathbf{v}_1 \times \mathbf{v}_2$

45. $\mathbf{r} = \langle 3, 1, 1\rangle + \langle -1, -2, 2\rangle u + \langle 1, 1, 0\rangle v$

49. $x = r\cos\theta$, $y = r\sin\theta$, $z = z$ with $0 \le r \le 2$, $0 \le \theta \le \frac{\pi}{4}$, $0 \le z \le 1$

53. $x = r\cos\theta$, $y = r\sin\theta$, $z = \sin r$ with $r \ge 0$, $0 \le \theta \le 2\pi$

Chapter 11 Review Exercises, page 917

1.

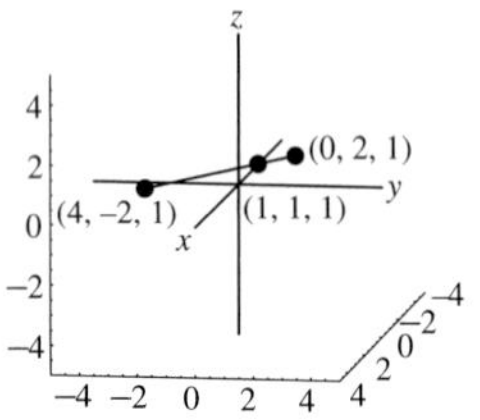

3.

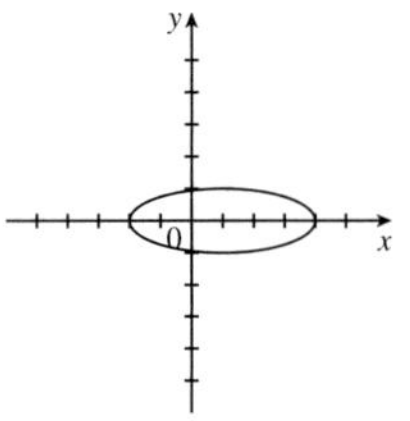

5.

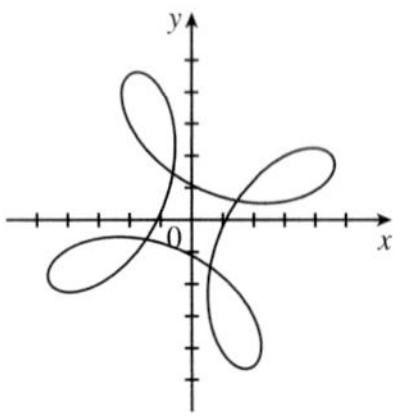

7.

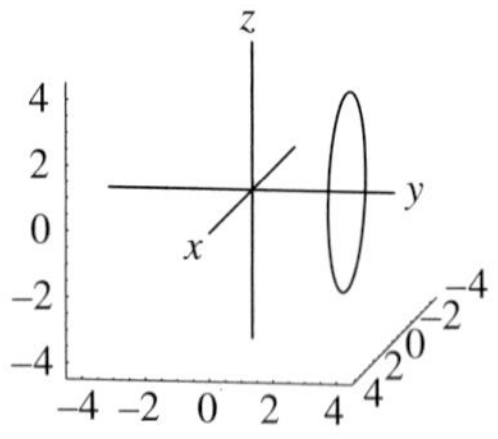

9.

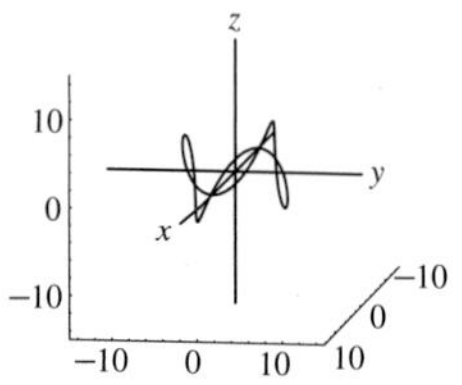

11.

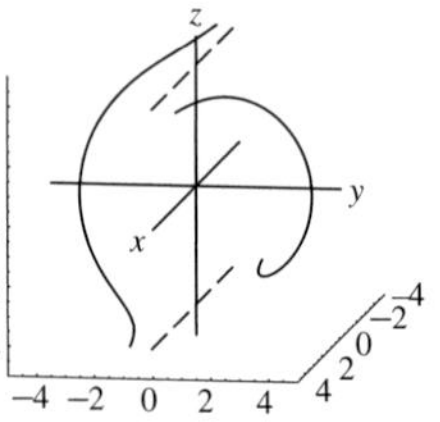

13. (a) B (b) C (c) A (d) F (e) D (f) E **15.** $2\pi\sqrt{37}$

17. $\langle 0, e^2, -1\rangle$ **19.** $t \ne 0$ **21.** $\left\langle \frac{t}{\sqrt{t^2+1}}, 4\cos 4t, \frac{1}{t}\right\rangle$

23. $\left\langle -\frac{1}{4}e^{-4t}, -t^{-2}, 2t^2 - t\right\rangle + \mathbf{c}$ **25.** $\langle 0, 2, 2\rangle$

27. $\langle -8\sin 2t, 8\cos 2t, 4\rangle$, $\langle -16\cos 2t, -16\sin 2t, 0\rangle$

29. $\langle t^2 + 4t + 2, -16t^2 + 1\rangle$ **31.** $\langle 4t + 2, -16t^2 + 3t + 6\rangle$

33. $\langle 0, -128\rangle$ **35.** $25(2 - \sqrt{3}) \approx 6.70$ feet, 100 feet, 80 ft/s

37. $\frac{1}{\sqrt{2}}\langle -1, 1, 0\rangle$, $\frac{1}{\sqrt{e^{-4}+1}}\langle -e^{-2}, 1, 0\rangle$ **39.** $\frac{1}{2}, \frac{4}{3\sqrt{3}}$

41. 0, 0 **43.** $\frac{1}{\sqrt{2}}\langle 0, 1, 1\rangle$, $\langle -1, 0, 0\rangle$

45. $a_T = 0, a_N = 2; a_T = \sqrt{2}, a_N = \sqrt{2}$

47. $345{,}600\langle -\cos 6t, -\sin 6t\rangle$

49. (assuming $0 \le u \le 2\pi$ and $0 \le r$)

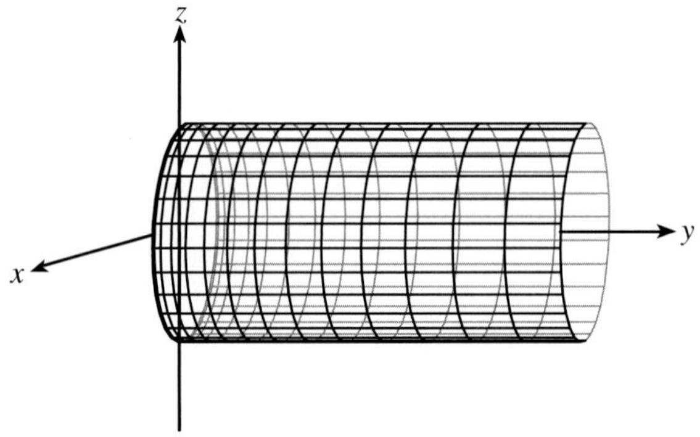

51.

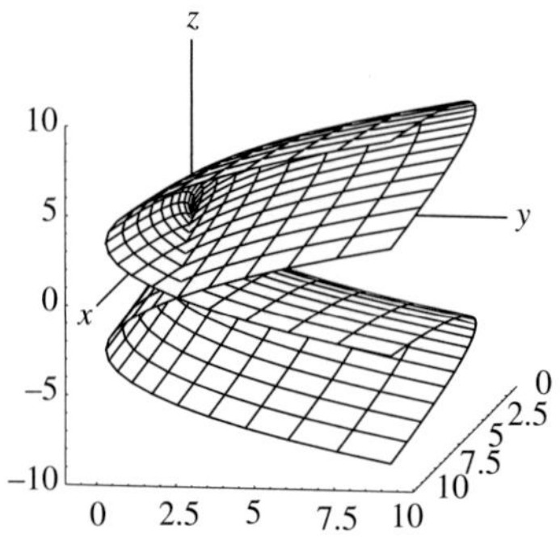

53. (a) B (b) C (c) A

CHAPTER 12

Exercises 12.1, page 930

1. $y \neq -x$ **3.** $x + y + 2 > 0$ **5.** $x^2 + y^2 + z^2 < 4$

7. $f \geq 0$ **9.** $f \geq -1$ **11.** 3, 3 **13.** $-\frac{1}{5}, -5$

15. (a) 312 (b) 333 (c) 350 (d) about 19 feet

17.

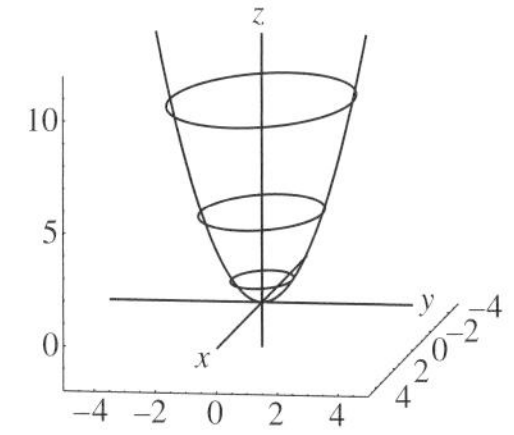

19.

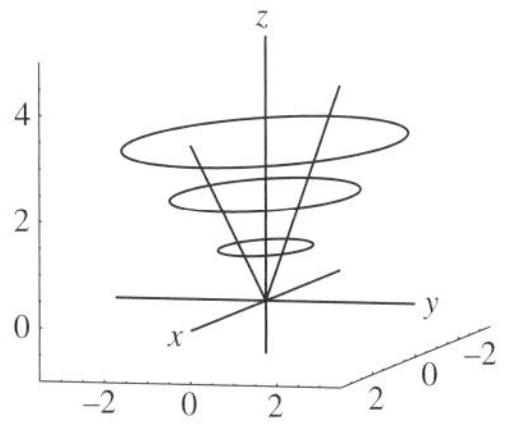

For 21–29, one view is shown.

21.

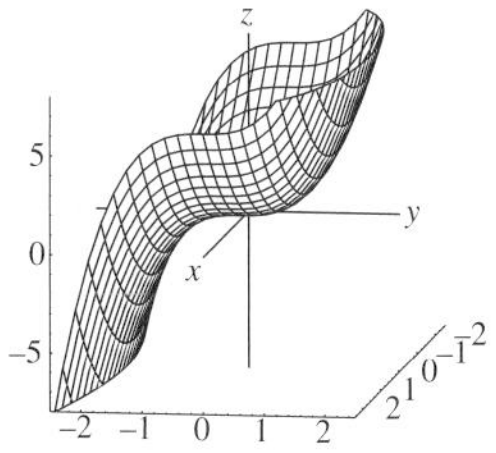

23.

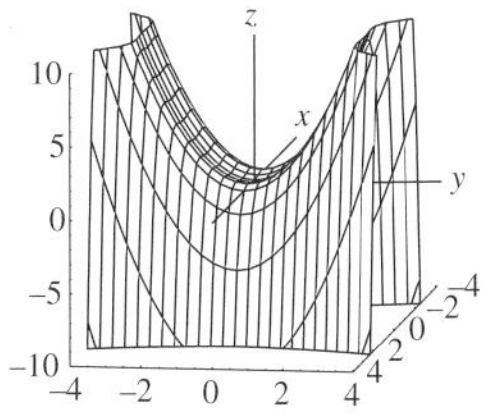

25.

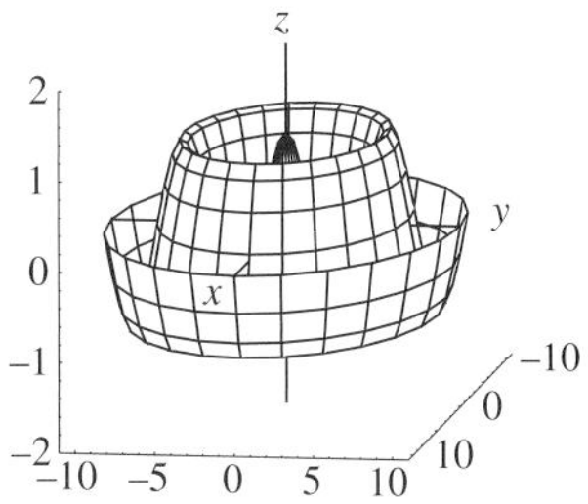

27.

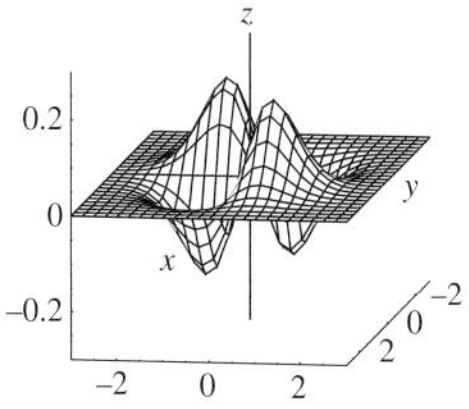

29.

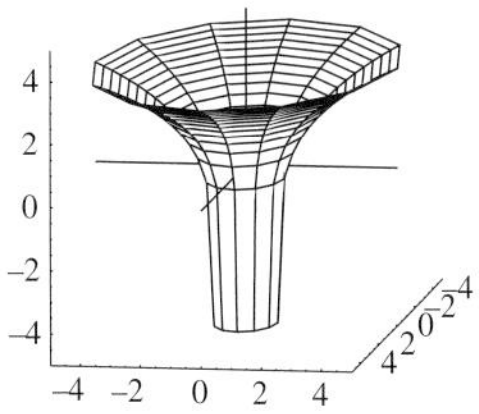

31. A: 480, B: 470, C: about 475 **33.** center, toward upper left

35. 77.4, 80.4, 82.8, 2.7°

37.

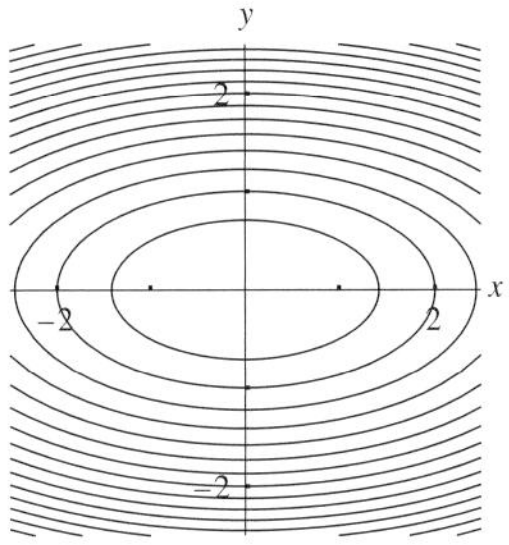

39.

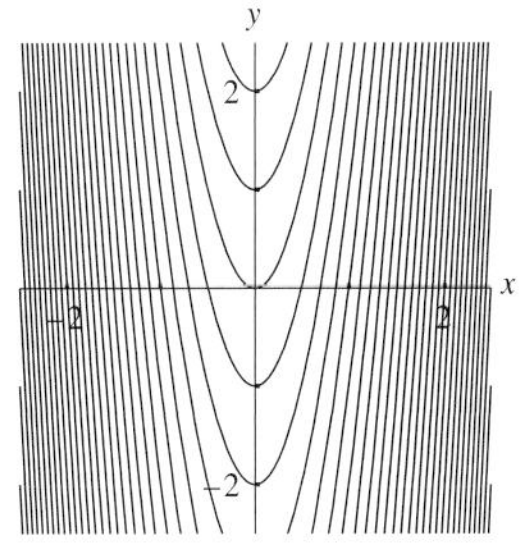

41.

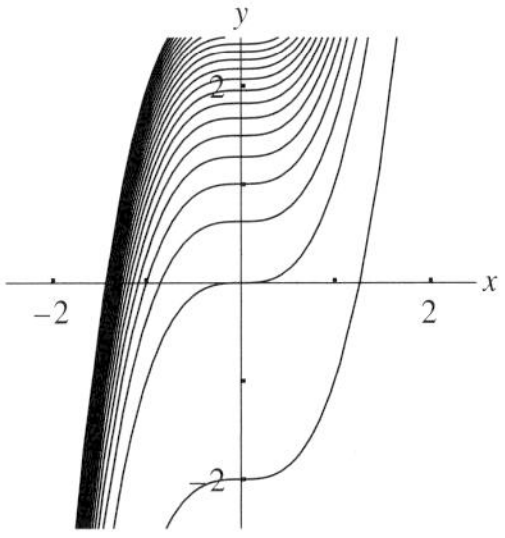

43.

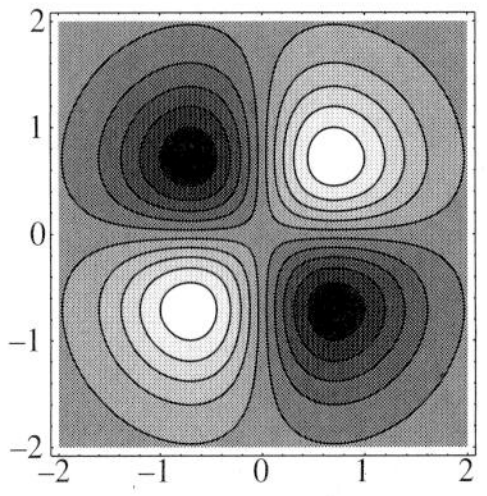

45.

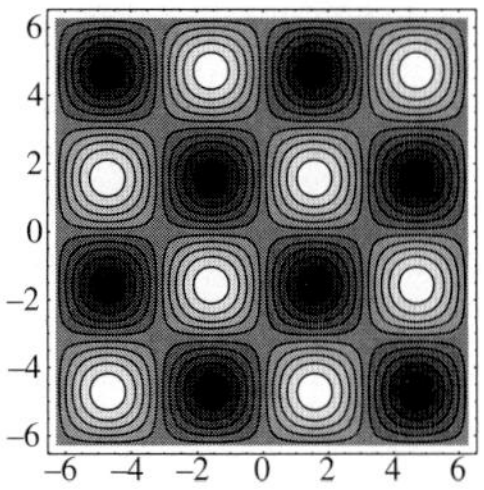

47. (a) A (b) D (c) C (d) B

49. (a) B (b) D (c) A (d) F (e) C (f) E

51.

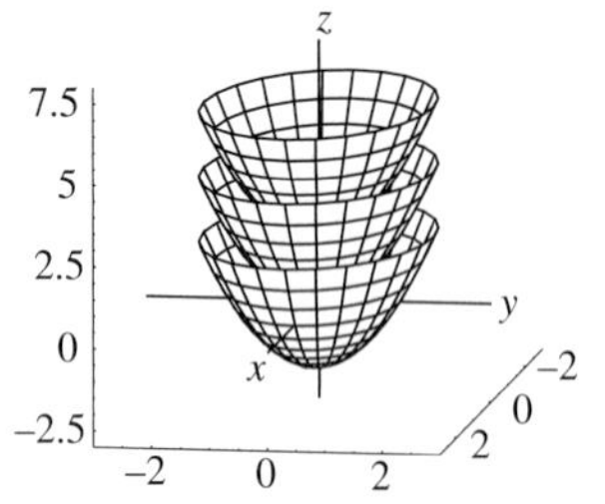

53. (a) B (b) A **55.** no height is visible

59. any point on the line $(x, \sqrt{3}x, 0)$

61. upper left, restaurants, roads

63. inside smallest oval, power is increasing away from frame

65. max = 3.942, min = −0.57, HS **67.** 60 mph, impossible

Exercises 12.2, page 947

1. 3 **3.** $-\frac{1}{2}$ **5.** 2

7. Along $x = 0$, $L_1 = 0$; along $y = 0$, $L_2 = 3$, therefore L does not exist.

9. Along $x = 0$, $L_1 = 0$; along $y = x$, $L_2 = 2$, therefore L does not exist.

11. Along $x = 0$, $L_1 = 0$; along $y = x^2$, $L_2 = 1$, therefore L does not exist.

13. Along $x = 0$, $L_1 = 0$; along $y^3 = x$, $L_2 = \frac{1}{2}$, therefore L does not exist.

15. Along $x = 0$, $L_1 = 0$; along $y = x$, $L_2 = \frac{1}{2}$, therefore L does not exist.

17. Along $x = 1$, $L_1 = 0$; along $y = x + 1$, $L_2 = \frac{1}{2}$, therefore L does not exist.

19. Along $x = 0$, $L_1 = 0$; along $x^2 = y^2 + z^2$, $L_2 = \frac{3}{2}$, therefore L does not exist.

21. Along $x = 0$, $L_1 = 0$; along $x = y = z$, $L_2 = \frac{1}{3}$, therefore L does not exist.

23. 0 **25.** 0 **27.** 2 **29.** 0 **35.** $x^2 + y^2 \le 9$

37. $x^2 - y < 3$ **39.** $x^2 + y^2 + z^2 \ge 4$

41. (0, 2) and all other points for which $x \ne 0$ **43.** all (x, y)

45. $\frac{1}{2}$ **47.** true **49.** false **55.** 1 **57.** 0

Exercises 12.3, page 957

1. $f_x = 3x^2 - 4y^2$, $f_y = -8xy + 4y^3$

3. $f_x = 2x \sin xy + x^2 y \cos xy$, $f_y = x^3 \cos xy - 9y^2$

5. $f_x = \frac{4}{y} e^{x/y} + \frac{y}{x^2}$, $f_y = -\frac{4x}{y^2} e^{x/y} - \frac{1}{x}$

7. $f_x = 3 \sin y + 12x^2 y^2 z$, $f_y = 3x \cos y + 8x^3 yz$, $f_z = 4x^3 y^2$

9. $\frac{\partial^2 f}{\partial x^2} = 6x$, $\frac{\partial^2 f}{\partial y^2} = -8x$, $\frac{\partial^2 f}{\partial y \partial x} = -8y$

11. $f_{xx} = 12x^2 - 6y^3$, $f_{xy} = -18xy^2$, $f_{xyy} = -36xy$

13. $f_{xx} = 6xy^2$, $f_{yz} = -\cos yz + yz \sin yz$, $f_{xyz} = 0$

15. $f_{ww} = 2xy - z^2 e^{wz}$, $f_{wxy} = 2w$, $f_{wwxyz} = 0$

17.

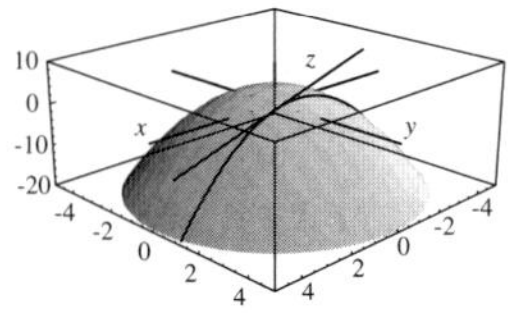

19.

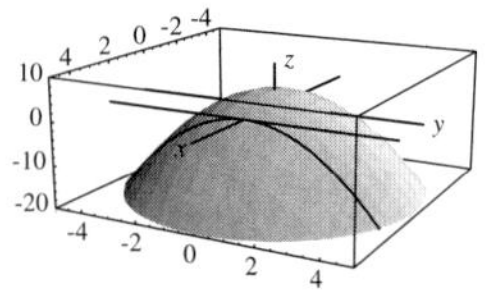

21. $\frac{nRV^3}{PV^3 - n^2 aV + 2n^3 ab}$ Hint: Hold pressure constant.

23. about $\frac{12}{V}$, assuming V is much greater than 0.004

27. h **29.** (0, 0, 0) = min

31. $\left(\frac{\pi}{2} + m\pi, \frac{\pi}{2} + n\pi, 1\right)$ = max for m, n odd or m, n even; $\left(\frac{\pi}{2} + m\pi, \frac{\pi}{2} + n\pi, -1\right)$ = min for m odd and n even or m even and n odd; $(m\pi, n\pi, 0)$ neither max nor min

33. 4, 2 **35.** $1, -\frac{2}{3}$ **37.** 1.4, −2.4 **39.** 2.2, 0.0195

43. $\frac{\partial^2 f}{\partial x^2} = -n^2\pi^2 \sin n\pi x \cos n\pi ct$, $\frac{\partial^2 f}{\partial t^2} = -c^2 n^2 \pi^2 \sin n\pi x \cos n\pi ct$

45. $-\frac{5}{1+I} V$, $-\frac{0.5}{1 + 0.1(1-T)} V$, inflation

47. $\cos x \cos t$, $-\sin x \sin t$ **51.** $\left(\frac{R}{R_1}\right)^2$

53. 400, $\frac{1}{4}$, decrease by 27

55. $\frac{\partial P}{\partial L} = 0.75 L^{-0.25} K^{0.25}$, $\frac{\partial P}{\partial K} = 0.25 L^{0.75} K^{-0.75}$

59. Production increases by $\frac{5}{3}$ **63.** $-2y^3 \sin(xy^2)$

65. concavity of intersection of $z = f(x, y)$ with $y = y_0$ at $x = x_0$

Exercises 12.4, page 970

1. (a) $4(x-2) + 2(y-1) = z - 4$; $x = 2 + 4t$, $y = 1 + 2t$, $z = 4 - t$
(b) $4(y-2) = z - 3$; $x = 0$, $y = 2 + 4t$, $z = 3 - t$

3. (a) $-(x - 0) = z - 0$ or $x + z = 0$; $x = -t$, $y = \pi$, $z = -t$
(b) $z + 1 = 0$; $x = \frac{\pi}{2}$, $y = \pi$, $z = -1 - t$

5. (a) $-\frac{3}{5}(x+3)+\frac{4}{5}(y-4)=z-5; x=-3-\frac{3}{5}t, y=4+\frac{4}{5}t,$
$z=5-t$

(b) $\frac{4}{5}(x-8)-\frac{3}{5}(y+6)=z-10; x=8+\frac{4}{5}t, y=-6-\frac{3}{5}t,$
$z=10-t$

7. (a) $L(x,y)=x$ **(b)** $L(x,y)=-y$

9. (a) $L(x,y)=x+6y-3$ **(b)** $L(x,y)=x$

11. (a) $L(w,x,y,z)=-12w+4x+12y+2z-37$
(b) $L(w,x,y,z)=2w-1$

13. values of L are 3, 3.1, 3.1; values of f are 3.002, 3.1, 3.102

15. values of L are 0, −0.1, −0.1; values of f are 0, −0.1, −0.099

17. 1.5552 ± 0.6307 **19.** 3.85 **21.** 4.03

23. $2y\Delta x+(2x+2y)\Delta y+(2\Delta y)\Delta x+(\Delta y)\Delta y$

25. $2x\Delta x+2y\Delta y+(\Delta x)\Delta x+(\Delta y)\Delta y$ **27.** yes

29. $(ye^x+\cos x)\,dx+e^x\,dy$

31. $f_x(0,0)=f_y(0,0)=0$

35. $6+4x+2y$ **37.** $3+x-\frac{2}{3}y$

39. $-9+1.4(t-10)-2.4(s-10); -13.4$ **45.** $4x+4y-z=8$

47. $x=1$

Exercises 12.5, page 981

3. $(2t+t^2+1-\cos e^t)e^t$

5. $\frac{\partial g}{\partial u}=512u^6(3u^2-v\cos u)(u^3-v\sin u)+1536u^5(u^3-v\sin u)^2;$
$\frac{\partial g}{\partial v}=512u^6\sin u(v\sin u-u^3)$

7. $g'(t)=\frac{\partial f}{\partial x}x'(t)+\frac{\partial f}{\partial y}y'(t)+\frac{\partial f}{\partial z}z'(t)$

9. $\frac{\partial g}{\partial u}=\frac{\partial f}{\partial x}\frac{\partial x}{\partial u}+\frac{\partial f}{\partial y}\frac{\partial y}{\partial u}, \frac{\partial g}{\partial v}=\frac{\partial f}{\partial x}\frac{\partial x}{\partial v}+\frac{\partial f}{\partial y}\frac{\partial y}{\partial v},$
$\frac{\partial g}{\partial w}=\frac{\partial f}{\partial x}\frac{\partial x}{\partial w}+\frac{\partial f}{\partial y}\frac{\partial y}{\partial w}$

11. −0.6271 **13.** 0.0587

17. $\frac{\partial^2 f}{\partial x^2}[x'(t)]^2+2\frac{\partial^2 f}{\partial y\partial x}x'(t)y'(t)+\frac{\partial^2 f}{\partial y^2}[y'(t)]^2$
$+\frac{\partial f}{\partial x}x''(t)+\frac{\partial f}{\partial y}y''(t)$

19. $\frac{\partial^2 f}{\partial x^2}\left(\frac{\partial x}{\partial u}\right)^2+\frac{\partial^2 f}{\partial y\partial x}\frac{\partial y}{\partial u}\frac{\partial x}{\partial u}+\frac{\partial f}{\partial x}\frac{\partial^2 x}{\partial u^2}$
$+\frac{\partial^2 f}{\partial x\partial y}\frac{\partial x}{\partial u}\frac{\partial y}{\partial u}+\frac{\partial^2 f}{\partial y^2}\left(\frac{\partial y}{\partial u}\right)^2+\frac{\partial f}{\partial y}\frac{\partial^2 y}{\partial u^2}$

21. $\frac{\partial z}{\partial x}=\frac{-6xz}{3x^2+6z^2-3y}, \frac{\partial z}{\partial y}=\frac{3z}{3x^2+6z^2-3y}$

23. $\frac{\partial z}{\partial x}=\frac{3yze^{xyz}-4z^2+\cos y}{8xz-3xye^{xyz}}, \frac{\partial z}{\partial y}=\frac{3xze^{xyz}-x\sin y}{8xz-3xye^{xyz}}$

33. $f(\Delta x,\Delta y)=\Delta x-\frac{1}{6}\Delta x^3-\frac{1}{2}\Delta x\Delta y^2$

35. $f(\Delta x,\Delta y)=\Delta x\Delta y$

37. $f(\Delta x,\Delta y)=1+2\Delta x+\Delta y+2\Delta x^2+2\Delta x\Delta y+\frac{1}{2}\Delta y^2+$
$\frac{4}{3}\Delta x^3+2\Delta x^2\Delta y+\Delta x\Delta y^2+\frac{1}{6}\Delta y^3$

39. $R(1+\Delta c, 1+\Delta h)=1+0.55\Delta c+0.45\Delta h$

43. 2 points; halved

45. $\frac{dg}{dt}=vu^{v-1}\frac{du}{dt}+(\ln u)u^v\frac{dv}{dt};$
$\left[\frac{6t^2}{2t+1}+6t\ln(2t+1)\right](2t+1)^{3t^2}$

Exercises 12.6, page 992

1. $\langle 2x+4y^2, 8xy-5y^4\rangle$

3. $\langle e^{xy^2}+xy^2e^{xy^2}, 2x^2ye^{xy^2}-2y\sin y^2\rangle$

5. $\langle -8e^{-8}-2, -16e^{-8}\rangle$ **7.** $\langle 0,0,-1\rangle$

9. $\langle -4, 12e^{4\pi}, 3e^{4\pi}, 3\pi e^{4\pi}\rangle$ **11.** $2+6\sqrt{3}$ **13.** $\frac{17}{5\sqrt{13}}$

15. 0 **17.** $-\frac{6}{\sqrt{5}}$ **19.** $-\frac{12}{\sqrt{5}}$ **21.** $-\frac{3}{\sqrt{29}}$

23. $\frac{-2}{\sqrt{30}}$ **25.** $\frac{37}{10}$ **27.** $\langle 4,-3\rangle, \langle -4,3\rangle, 5, -5$

29. $\langle 16,-4\rangle, \langle -16,4\rangle, \sqrt{272}, -\sqrt{272}$ **31.** $\langle 1,0\rangle, \langle -1,0\rangle, 1, -1$

33. $\left\langle \frac{3}{2}, -\frac{1}{8}\right\rangle, \left\langle -\frac{3}{2}, \frac{1}{8}\right\rangle, \frac{\sqrt{145}}{8}, -\frac{\sqrt{145}}{8}$

35. $\langle 16,4,24\rangle, \langle -16,-4,-24\rangle, \sqrt{848}, -\sqrt{848}$ **37.** parallel

39. $\langle \cos(x+y), \cos(x+y)\rangle = \cos(x+y)\langle 1,1\rangle; \langle 2,-1\rangle$

41. $2(x-1)+3(y+1)-z=0,$
The equation of the normal line is $\begin{cases} x=1+2t \\ y=-1+3t \\ z=-t \end{cases}$

43. $-2(x+1)+4(y-2)+2(z-1)=0,$
The equation of the normal line is $\begin{cases} x=-1-2t \\ y=2+4t \\ z=1+2t \end{cases}$

45. $(0,0,0), (1,1,-1), (-1,-1,-1)$

47.

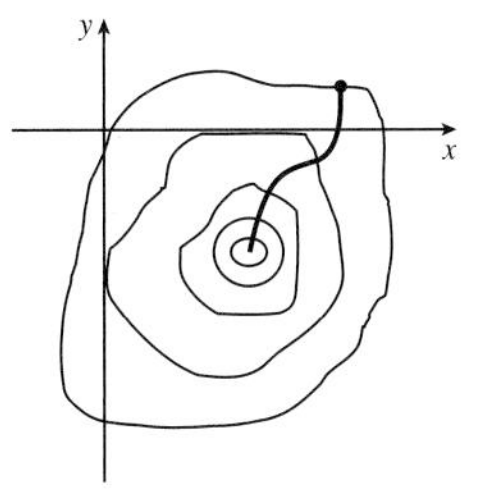

49.

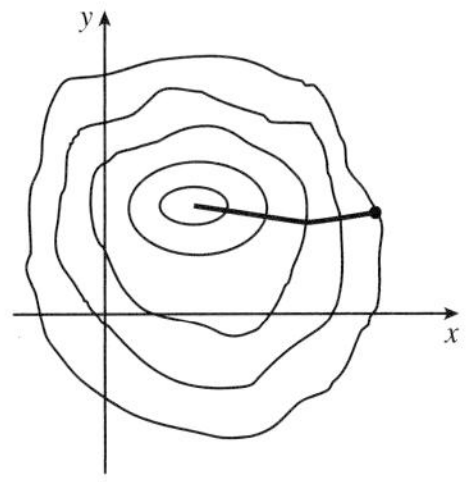

51. possible answer: $\left\langle -\frac{4}{3}, -2\right\rangle$ **53.** $\langle 2,-2\rangle$

55. $\langle -\tan 10^\circ, \tan 6^\circ\rangle \approx \langle -0.176, 0.105\rangle, 11.6^\circ$ **57.** $\langle 8,4\rangle$

61. y-coordinate of position should decrease, x-component of velocity should increase

63. If the shark moves toward higher charge, it moves in the direction $\langle 12,-20,5\rangle$.

65. (a) true (b) true **67.** $\langle 0.8, 0.3, -0.004\rangle$

Exercises 12.7, page 1008

1. (0, 0) saddle **3.** (0, 0) saddle, (1, 1) local min

5. (0, 1) local min, (±2, −1) saddle **7.** (0, 0) max

9. (0, 0) saddle, (1, 1) and (−1, −1) local min

11. $\left(\frac{1}{\sqrt{2}}, 0\right)$ local max, $\left(-\frac{1}{\sqrt{2}}, 0\right)$ local min

13. (0, 0) saddle, $\pm\left(\sqrt{\frac{1}{2}}, \sqrt{\frac{1}{2}}\right)$ local max, $\pm\left(\sqrt{\frac{1}{2}}, -\sqrt{\frac{1}{2}}\right)$ local min

15. (2.82, 0.18) local min, (−2.84, −0.18) saddle, (0.51, 0.99) saddle

17. $(\pm 1, 0)$ local max, $\left(0, -\sqrt{\frac{3}{2}}\right)$ local max, $\left(0, \sqrt{\frac{3}{2}}\right)$ local min, (0, 0) saddle, $\left(\pm\frac{\sqrt{19}}{3\sqrt{3}}, -\frac{2}{3}\right)$ saddle

21. $1.37x - 2.80$ **23.** 9176 **25.** 247, 104

27. (a) 1.29 **(b)** 2.75

29. (−0.3210, −0.5185), (−0.1835, −0.4269)

31. (0.9044, 0.8087), (3.2924, −0.3853)

33. (0, 0) is a saddle point **35.** $f(2, 0) = 4$, $f(2, 2) = -2$

37. $f(3, 0) = 9$, $f(0, 0) = 0$ **39.** 4 ft by 4 ft by 4 ft

41. $a = b = c = \frac{2}{3}s$

43. (a) $f(x, 0) = f(0, y) = 0$; minima
(b) $f(x, 0) = f(0, y) = 0$; minima

51. false **53.** false

55. extrema at $\left(\pm\frac{\pi}{2}, \pm\frac{\pi}{2}\right)$, saddles at $(\pm n\pi, \pm m\pi)$

57. extrema at $\pm(1, 1)$, saddle (0, 0)

59. extrema $(\pm 0.1, 0.1)$, saddle (0, 0)

61. $d(x, y) = \sqrt{(x-3)^2 + (y+2)^2 + (3 - x^2 - y^2)^2}$, (1.55, −1.03)

63. (1.6, 0.8, −2.4) **65.** (1, 0), $f(1, 0) < f(-10, 0)$

Exercises 12.8, page 1020

1. $x = \frac{6}{5}, y = -\frac{2}{5}$ **3.** $x = 2, y = -1$

5. $x = 1, y = 1$ **7.** $x = 1, y = 1$

9. $\max = f(2, 2) = f(-2, -2) = 16$,
$\min = f(2, -2) = f(-2, 2) = -16$

11. $\max = f(\pm\sqrt{2}, 1) = 8$, $\min = f(\pm\sqrt{2}, -1) = -8$

13. $\max = f(1, 1) = e$, $\min = f(-1, 1) = -e$

15. $\max = f(\pm\sqrt{2}, 1) = 2e$, $\min = f(0, \pm\sqrt{3}) = 0$

17. $\max = f(2, 2) = f(-2, -2) = 16$
$\min = f(2, -2) = f(-2, 2) = -16$

19. $\max = f(\pm\sqrt{2}, 1) = 8$, $\min = f(\pm\sqrt{2}, -1) = -8$

21. $u = \frac{128}{3}$, $z = 195$ feet **25.** $P(20, 80, 20) = 660$

27. $P\left(\sqrt{\frac{8801}{22}}, 4\sqrt{\frac{8801}{22}}, \sqrt{\frac{8801}{22}}\right) \approx 660.0374989$; $660 + \lambda = 660.0375$

29. $x = y$ **31.** $f\left(-\frac{\sqrt{2}}{2}, \frac{\sqrt{2}}{2}\right) = \sqrt{2}$

33. $\alpha = \beta = \theta = \frac{\pi}{6}$; $f\left(\frac{\pi}{6}, \frac{\pi}{6}, \frac{\pi}{6}\right) = \frac{1}{8}$

37. $L = 50$, $K = 10$

39. $C(L, K) = C(64, 8) = 2400$ **41.** $c = d = 12$

43. $f(4, -2, 2) = 24$ **45.** $f(1, 1, 2) = 2$

47. (a) $(\pm 1, 0, 0)$ (b) $(0, \pm 1, \pm 1)$

51. (0.5898, 0, 0.3478)

Chapter 12 Review Exercises, page 1022

1.

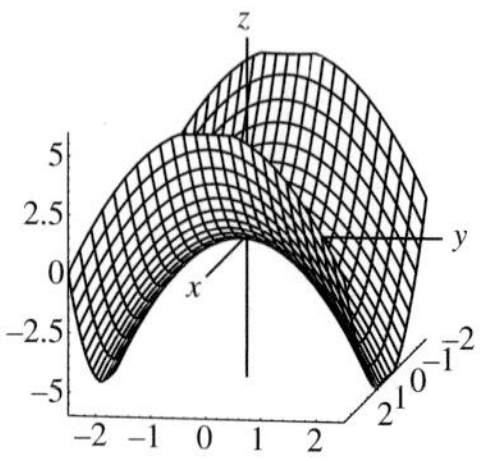

3.

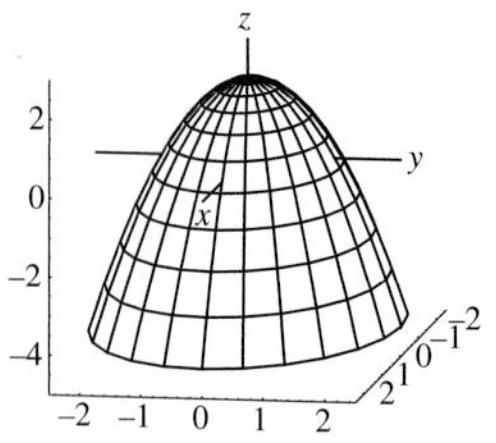

5.

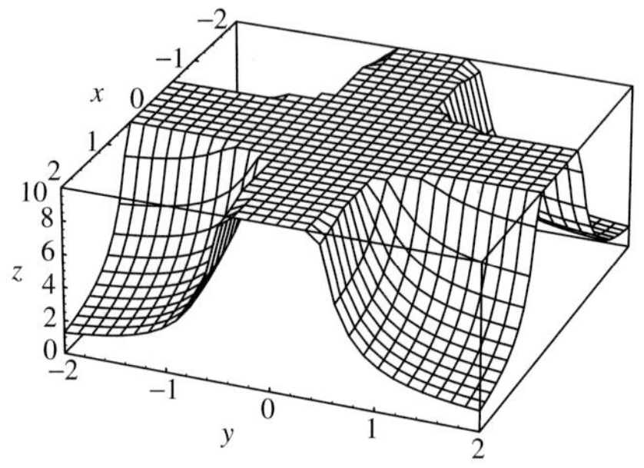

7.

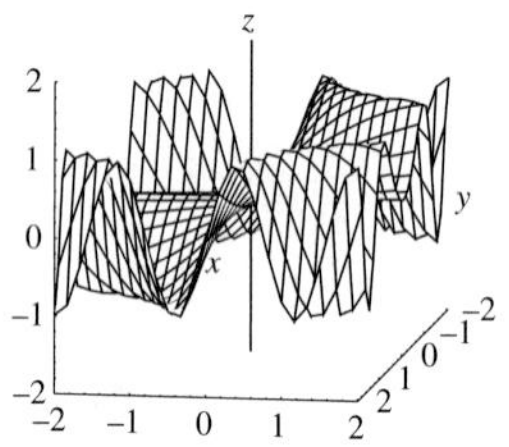

9.

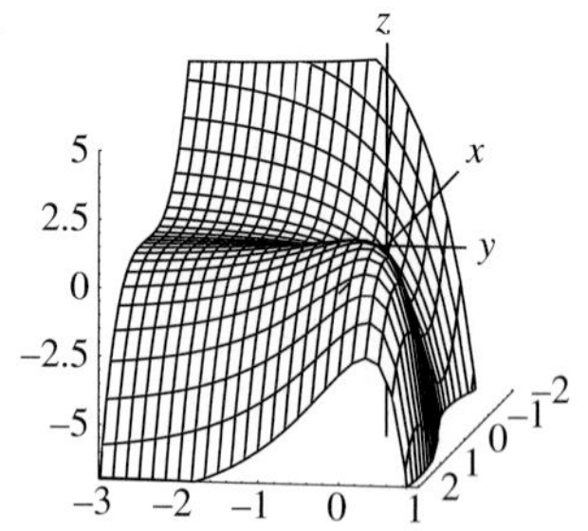

11. (a) D (b) B (c) C (d) A (e) F (f) E

13. (a) C (b) A (c) D (d) B

15. Along $y = 0$, $L_1 = 0$; along $y = x^2$, $L_2 = \frac{3}{2}$; therefore L does not exist.

17. Along $x = 0$, $L_1 = 1$; along $y = x$, $L_2 = \frac{2}{3}$; therefore L does not exist.

19. 0 **21.** $x \neq 0$

23. $f_x = \frac{4}{y} + (1+xy)e^{xy}$, $f_y = -\frac{4x}{y^2} + x^2e^{xy}$

25. $f_x = 6xy\cos y - \frac{1}{2\sqrt{x}}$, $f_y = 3x^2\cos y - 3x^2y\sin y$

29. $-0.04, 0.06$ **31.** $45 - 10(x+2) + 9(y-5)$

33. $(x-\pi) + 2\left(y - \frac{\pi}{2}\right)$

35. $f_{xx} = 24x^2y + 6y^2$, $f_{yy} = 6x^2$, $f_{xy} = 8x^3 + 12xy$

37. $3(y+1) - z = 0$ **39.** $4x + 4(y-2) + 2(z-1) = 0$

41. $8e^{8t}\sin t + (e^{8t} + 2\sin t)\cos t$

43. $g'(t) = \frac{\partial f}{\partial x}x'(t) + \frac{\partial f}{\partial y}y'(t) + \frac{\partial f}{\partial z}z'(t) + \frac{\partial f}{\partial w}w'(t)$

45. $\frac{\partial z}{\partial x} = -\frac{x+y}{z}$, $\frac{\partial z}{\partial y} = -\frac{x+y}{z}$ **47.** $\left\langle -\frac{1}{2}, 12\pi - \frac{1}{2}\right\rangle$

49. -4 **51.** $-\frac{7}{\sqrt{5}}$ **53.** $\pm\frac{1}{\sqrt{145}}\langle 9, -8\rangle, \pm 4\sqrt{145}$

55. $\pm\langle 1, 0\rangle, \pm 4$ **57.** $\langle 16, 2\rangle$

59. $(0, 0)$ relative min, $(2, \pm 8)$ saddles

61. $\left(\frac{4}{3}, \frac{4}{3}\right)$ relative max, $(0, 0)$ saddle **63.** 212, 112

65. $f(4, 0) = 512$, $f(0, 0) = 0$ **67.** $f(1, 2) = 5$, $f(-1, -2) = -5$

69. $f\left(\sqrt{\frac{1}{2}}, \sqrt{\frac{1}{2}}\right) = f\left(-\sqrt{\frac{1}{2}}, -\sqrt{\frac{1}{2}}\right) = \frac{1}{2}$, $f\left(\sqrt{\frac{1}{2}}, -\sqrt{\frac{1}{2}}\right) = f\left(-\sqrt{\frac{1}{2}}, \sqrt{\frac{1}{2}}\right) = -\frac{1}{2}$

71. $(1, 1)$

CHAPTER 13

Exercises 13.1, page 1043

1. 6 **3.** $\frac{13}{2}$ **5.** 40 **7.** $\frac{16}{3}$ **9.** $\frac{19}{2} - \frac{1}{2}e^6$

11.

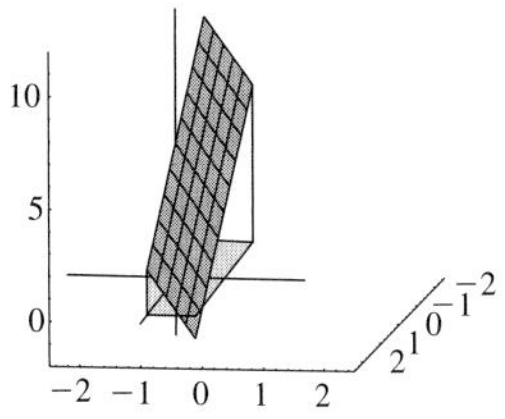

13.

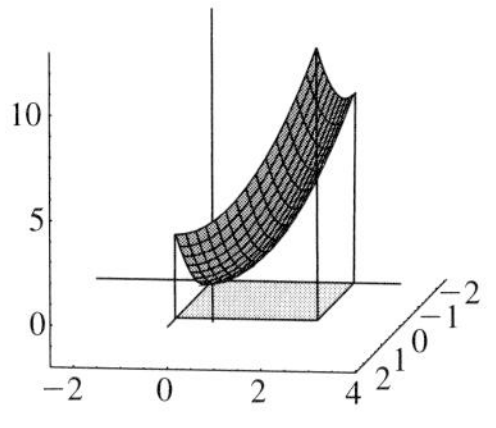

15. 2 **17.** $\frac{62}{21}$ **19.** $e^4 - 1$ **21.** $2(\ln 2)(\sin 1)$

23. $\int_0^1\int_0^{2x} x^2\,dy\,dx = \frac{1}{2}$; $\int_0^2\int_0^{y/2} x^2\,dx\,dy = \frac{1}{6}$

25. $\int_0^3\int_1^4 (x^2+y^2)\,dy\,dx = 90$

27. $\int_{-1}^1\int_{x^2}^1 (x^2+y^2)\,dy\,dx = \frac{88}{105}$

29. $\int_{-2}^2\int_0^{4-y^2} (6-x-y)\,dx\,dy = \frac{704}{15}$

31. $\int_0^2\int_0^x y^2\,dy\,dx = \frac{4}{3}$

33. -1.5945 **35.** 1.6697 **37.** $\int_0^2\int_{y/2}^1 f(x, y)\,dx\,dy$

39. $\int_0^4\int_0^{x/2} f(x, y)\,dy\,dx$ **41.** $\int_1^4\int_0^{\ln y} f(x, y)\,dx\,dy$

43. $\int_0^2\int_0^y 2e^{y^2}\,dx\,dy = e^4 - 1$ **45.** $\int_0^1\int_0^x 3xe^{x^3}\,dy\,dx = e - 1$

49.

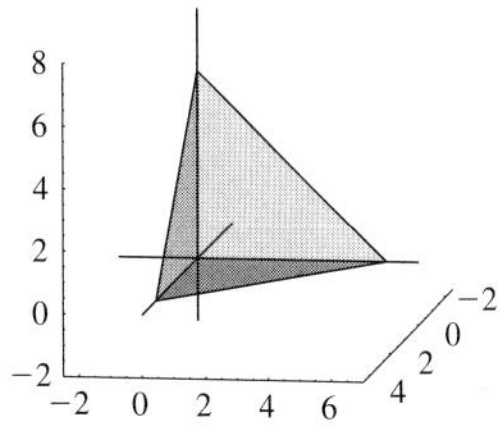

51.

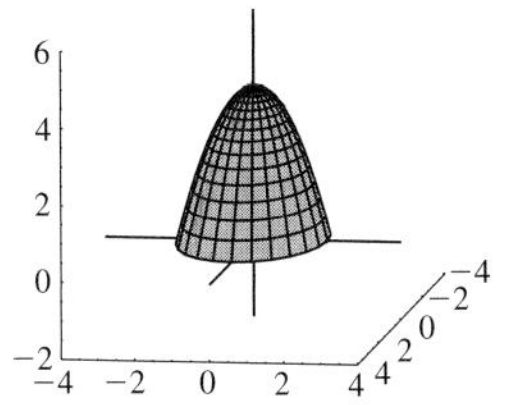

53. different domains **55.** $\frac{2\pi}{3}$ **59.** 2.375 **61.** 1.025

63. 0.93125 **65.** 2.10469 **67.** 4

69. $f(b, d) - f(b, c) - f(a, d) + f(a, c)$; 4

71. $\ln 5 - \frac{1}{2}\ln 17 + 4(\tan^{-1}4 - \tan^{-1}2)$ **73.** $\frac{14}{3}$

Exercises 13.2, page 1055

1. $\int_{-2}^2\int_{x^2}^{8-x^2} 1\,dy\,dx = \frac{64}{3}$ **3.** $\int_0^2\int_{y/2}^{3-y} 1\,dx\,dy = 3$

5. $\int_0^1\int_{x^2}^{\sqrt{x}} 1\,dy\,dx = \frac{1}{3}$ **7.** 6 **9.** $\frac{40}{3}$ **11.** $\frac{1}{2}$

13. $\frac{5}{12}$ **15.** $\frac{10{,}816}{105}$ **17.** $\frac{279}{20}$

19. $\int_0^2\int_0^{4-x^2}\sqrt{x^2+y^2}\,dy\,dx = 10.275$

21. $\int_0^4\int_0^{2-x/2} e^{xy}\,dy\,dx = 9.003$

23. $m = \frac{1}{3}, \bar{x} = \frac{3}{5}, \bar{y} = \frac{12}{35}$ **25.** $m = \frac{12}{5}, \bar{x} = \frac{41}{63}, \bar{y} = 0$

27. $m = 4, \bar{x} = \frac{16}{15}, \bar{y} = \frac{8}{3}$

29. In exercise 26, $\rho(x, y)$ is not x-axis symmetric.

31. $\rho(-x, y) = \rho(x, y)$ **33.** 1164

35. $1200(1 - e^{-2/3}) \approx 583.899$ **37.** $m = \frac{32}{3}, I_y = \frac{128}{15}, I_x = \frac{512}{7}$

39. $I_y = \frac{68}{3}, \frac{5}{3}$; second spin rate 13.6 times faster

41. 100.531, 508.938 **43.** $\frac{12}{5}$ **45.** same

47. 3.792 **49.** 50.113

51. (a) total rainfall in region
(b) average rainfall per unit area in region

53. $\bar{y} = \frac{1}{3}$ **55.** $\int_0^a\int_0^{b-(b/a)x}\left(c - \frac{c}{a}x - \frac{c}{b}y\right)dy\,dx = \frac{abc}{6}$

59. $c = \frac{1}{32}$ **61.** $\int_0^2\int_0^{\min\{y,x^2\}} f(x, y)\,dy\,dx = \int_0^1\int_0^{x^2} f(x, y)\,dy\,dx + \int_1^2\int_0^x f(x, y)\,dy\,dx$

63. $c = \frac{3}{16}$; $\frac{17\sqrt{17}-25}{64} \approx 0.705$ **65.** $\bar{y} = \frac{4}{3\pi}[27 - 16\sqrt{2}] \approx 1.8558 < 2$

Exercises 13.3, page 1064

1. 11π **3.** $\frac{\pi}{12}$ **5.** $\frac{\pi}{9} + \frac{\sqrt{3}}{6}$ **7.** 18π **9.** $\pi - \pi e^{-4}$

11. 0 **13.** $\frac{81\pi}{2}$ **15.** $\frac{16}{3}$ **17.** $\frac{81\pi}{2}$ **19.** $\frac{16\pi}{3}$

21. $\frac{\pi}{12}[61 - 15\sqrt{15}]$ **23.** 36 **25.** $\frac{\pi}{2}$

27. $\int_0^{2\pi}\int_0^2 r^2 dr d\theta = \frac{16\pi}{3}$

29. $\int_{-\pi/2}^{\pi/2}\int_0^2 re^{-r^2} dr\, d\theta = \frac{\pi}{2}(1 - e^{-4})$

31. $\int_{\pi/4}^{\pi/2}\int_0^{2\sqrt{2}} r^4 dr\, d\theta = \frac{32\pi\sqrt{2}}{5}$ **33.** $1 - e^{-1/16} \approx 0.06$

35. $\frac{1}{20}(e^{-225/16} - e^{-16}) \approx 0.000000033$ **37.** $\frac{31\pi}{320}$

39. $\bar{x} = 0, \bar{y} = \frac{2}{3}$ **41.** $20,000\pi(1 - e^{-1}) \approx 39,717$

43. $\frac{\pi R^4}{4}$ **45.** $V = 2\int_0^{2\pi}\int_0^a r\sqrt{a^2 - r^2} dr\, d\theta = \frac{4\pi a^3}{3}$

47. $2\int_{-\pi/2}^{\pi/2}\int_0^{2\cos\theta} \sqrt{9 - r^2} r\, dr\, d\theta \approx 17.164$

49. $\int_0^{\pi}\int_0^{\frac{4\sin\theta}{1+\sin^2\theta}} \left[\sqrt{4 - r^2} - (2 - r\sin\theta)\right] r\, dr\, d\theta$

51. cone: $V = \frac{\pi}{3}k^3$; paraboloid: $V = \frac{\pi}{2}k^2$ **53.** 1.2859

Exercises 13.4, page 1071

1. $\frac{1}{12}(69^{3/2} - 5^{3/2}) \approx 46.831$ **3.** $\frac{\pi}{6}(17^{3/2} - 1) \approx 36.177$

5. $4\sqrt{2}\pi$ **7.** $6\sqrt{11}$ **9.** $4\sqrt{6}$ **11.** 8π **13.** 583.7692

15. 31.3823 **17.** 37.174 **19.** 12.045 **21.** $\sqrt{2}A$

23. $\frac{A}{|\cos\theta|}$ **25.** $4L$

27. $k = \frac{3}{4} + \frac{1}{16}(4 + 68\sqrt{17})^{2/3} \approx 3.453$

31. $\pi[\sqrt{2} - \ln(\sqrt{2} - 1)] \approx 7.212$

Exercises 13.5, page 1082

1. 16 **3.** $-\frac{2}{3}$ **5.** $\frac{4}{15}$ **7.** $\frac{171}{5}$ **9.** 0

11. 0 **13.** 64 **15.** symmetry; yes; no

17. $\int_0^2\int_{-1}^1\int_{x^2}^1 dz\, dx\, dy = \frac{8}{3}$

19. $\int_{-1}^1\int_0^{1-y^2}\int_{2-z/2}^4 dx\, dz\, dy = \frac{44}{15}$

21. $\int_{-\sqrt{10}}^{\sqrt{10}}\int_{-6}^{4-x^2}\int_0^{y+6} dz\, dy\, dx = \frac{160\sqrt{10}}{3}$

23. $\int_{-1}^1\int_{x^2}^1\int_0^{3-x} dy\, dz\, dx = 4$ **25.** $\frac{4}{3}$ **27.** 8π

29. $m = 32\pi, \bar{x} = \bar{y} = 0, \bar{z} = \frac{8}{3}$

31. $m = 138, \bar{x} = \frac{186}{115}, \bar{y} = \frac{56}{115}, \bar{z} = \frac{168}{115}$

33. right side is heavier in #30

35. $\int_0^1\int_0^{2-2y}\int_0^{2-x-2y} 4yz\, dz\, dx\, dy$

$= \int_0^1\int_0^{2-2y}\int_0^{2-2y-z} 4yz\, dx\, dz\, dy$

$= \int_0^2\int_0^{2-x}\int_0^{1-x/2-z/2} 4yz\, dy\, dz\, dx$

37. $\int_0^2\int_0^{4-2y}\int_0^{4-2y-x} dz\, dx\, dy$

39. $\int_0^1\int_0^{\sqrt{1-x^2}}\int_0^{\sqrt{1-x^2-z^2}} dy\, dz\, dx$

41. $\int_0^2\int_{x^2}^4\int_0^{\sqrt{y-x^2}} dz\, dy\, dx$

43. total amount is 0.01563 gram; average density is 1.356×10^{-5} gram per cubic foot

45. $c = \frac{3}{2}$ **47.** $k = 2 - 2^{2/3} \approx 0.413$

49. $\left[\int_a^b f(x)\, dx\right]\left[\int_c^d g(y)\, dy\right]\left[\int_r^s h(z)\, dz\right]$; only if a, b, c, d, r, and s are all constants

51. The volume of the parallelepiped is abc; the volume of the tetrahedron is $abc/6$.

Exercises 13.6, page 1090

1. $r = 4$ **3.** $r = 4\cos\theta$ **5.** $z = r^2$ **7.** $\theta = \tan^{-1}(2)$

9. $\int_0^{2\pi}\int_0^2\int_r^{\sqrt{8-r^2}} rf(r\cos\theta, r\sin\theta, z)\, dz\, dr\, d\theta$

11. $\int_0^{2\pi}\int_0^3\int_0^{9-r^2} rf(r\cos\theta, r\sin\theta, z)\, dz\, dr\, d\theta$

13. $\int_0^{2\pi}\int_{\sqrt{3}}^{\sqrt{8}}\int_{r^2-1}^8 rf(r\cos\theta, r\sin\theta, z)\, dz\, dr\, d\theta$

15. $\int_0^{2\pi}\int_0^2\int_0^{4-r^2} rf(r\cos\theta, y, r\sin\theta)\, dy\, dr\, d\theta$

17. $\int_0^{2\pi}\int_0^1\int_{r^2}^{2-r^2} rf(x, r\cos\theta, r\sin\theta)\, dx\, dr\, d\theta$

19. $\int_2^3\int_0^{2\pi}\int_0^z rf(r\cos\theta, r\sin\theta, z)\, dr\, d\theta\, dz$

21. $\int_0^{2\pi}\int_0^2\int_1^2 re^{r^2} dz\, dr\, d\theta = \pi(e^4 - 1)$

23. $\int_0^2\int_0^{3-3z/2}\int_0^{6-2y-3z} (x + z)\, dx\, dy\, dz = 12$

25. $\int_0^{2\pi}\int_0^{\sqrt{2}}\int_r^{\sqrt{4-r^2}} zr\, dz\, dr\, d\theta = 2\pi$

27. $\int_0^2\int_0^{4-2y}\int_0^{4-x-2y} (x + y)\, dz\, dx\, dy = 8$

29. $\int_0^{2\pi}\int_{\sqrt{3}}^2\int_{-\sqrt{4-r^2}}^0 e^z r\, dz\, dr\, d\theta = \pi\left(\frac{4}{e} - 1\right)$

31. $\int_0^{\pi}\int_0^{2\sin\theta}\int_0^r 2r^2\cos\theta\, dz\, dr\, d\theta = 0$

33. $\int_0^{2\pi}\int_0^1\int_0^r 3z^2 r\, dz\, dr\, d\theta = \frac{2\pi}{5}$

35. $\int_0^{\pi}\int_0^2\int_r^{\sqrt{8-r^2}} 2r\, dz\, dr d\theta = \frac{32\pi}{3}(\sqrt{2} - 1)$

37. $\int_{\pi}^{2\pi}\int_0^3\int_0^{r^2} r^3 dy\, dr\, d\theta = \frac{243\pi}{2}$

39.

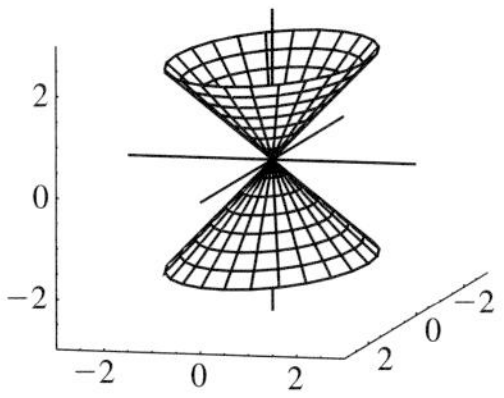

41.

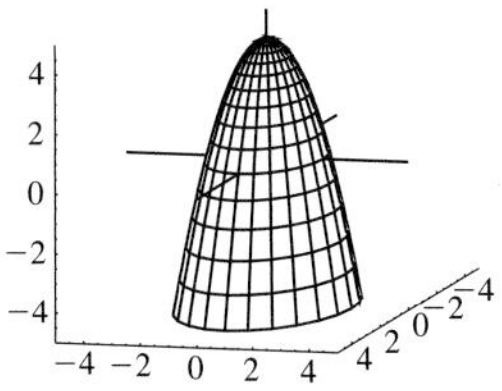

43.

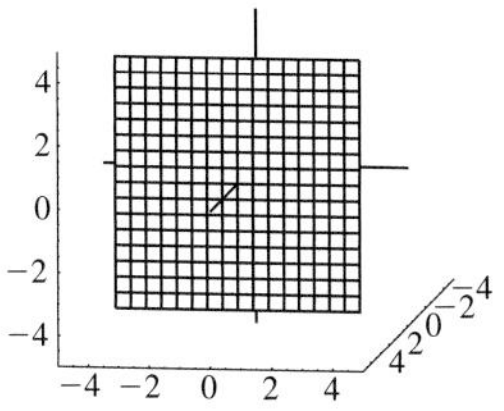

45.

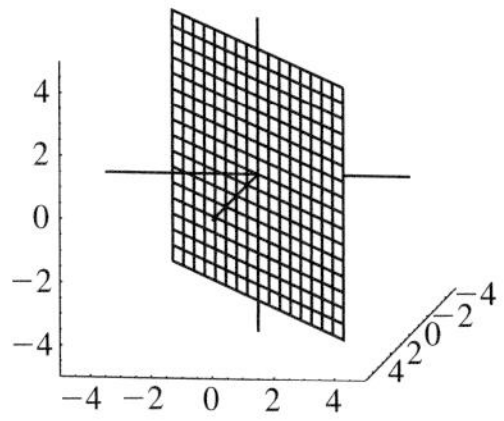

47. $m = \frac{128\pi}{3}, \bar{x} = \bar{y} = 0, \bar{z} = \frac{16}{5}$

49. $m = 10\pi, \bar{x} = 0, \bar{y} = \frac{4}{5}, \bar{z} = \frac{38}{15}$ **51.** $\langle \cos\theta, \sin\theta, 0 \rangle$

53. $\frac{d\hat{\mathbf{r}}}{dt} = \hat{\boldsymbol{\theta}}\frac{d\theta}{dt}, \frac{d\hat{\boldsymbol{\theta}}}{dt} = -\hat{\mathbf{r}}\frac{d\theta}{dt}$ **55.** $c = 2\sqrt{2}$

57. $c = \sqrt{2}$ **59.** $\mathbf{v} = -2\sqrt{2}\,\hat{\mathbf{r}} = \sqrt{2}\int_{-3\pi/4}^{\pi/4} \hat{\boldsymbol{\theta}} d\theta$

Exercises 13.7, page 1098

1. $(0, 0, 4)$ **3.** $(4, 0, 0)$ **5.** $(\sqrt{2}, 0, \sqrt{2})$

7. $\left(\frac{\sqrt{2}}{4}, \frac{\sqrt{6}}{4}, \frac{\sqrt{6}}{2}\right)$ **9.** $\rho = 3$

11. $\theta = \frac{\pi}{4}$ or $\frac{5\pi}{4}$

13. $\rho\cos\phi = 2$ **15.** $\phi = \frac{\pi}{6}$

17. $x^2 + y^2 + z^2 = 2$

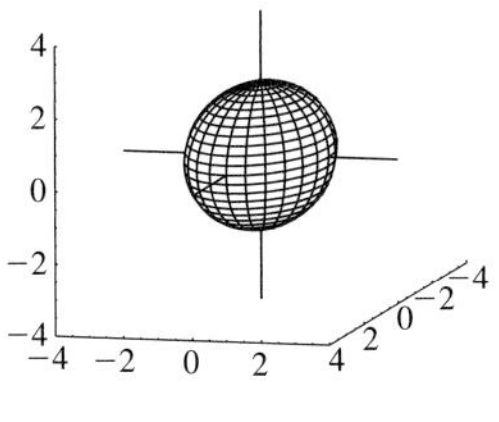

19. $z = \sqrt{x^2 + y^2}$

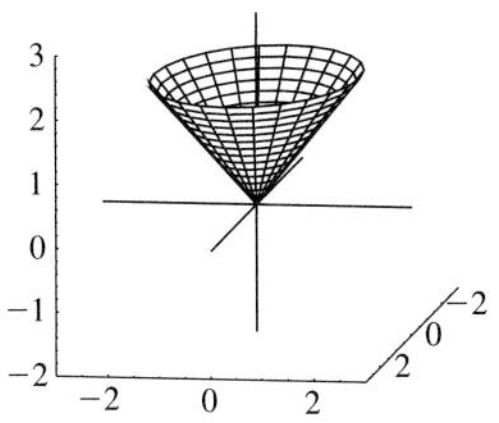

21. $y = 0; x \geq 0$

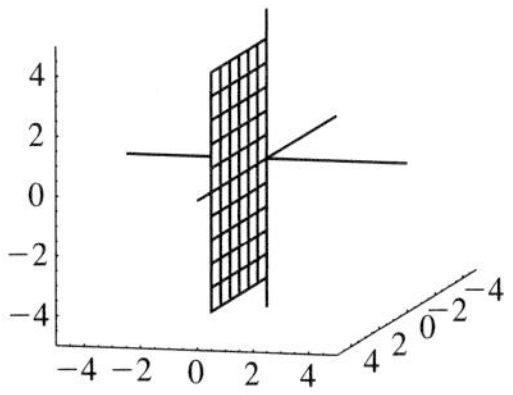

23. $z = \frac{1}{\sqrt{3}}\sqrt{x^2 + y^2}$

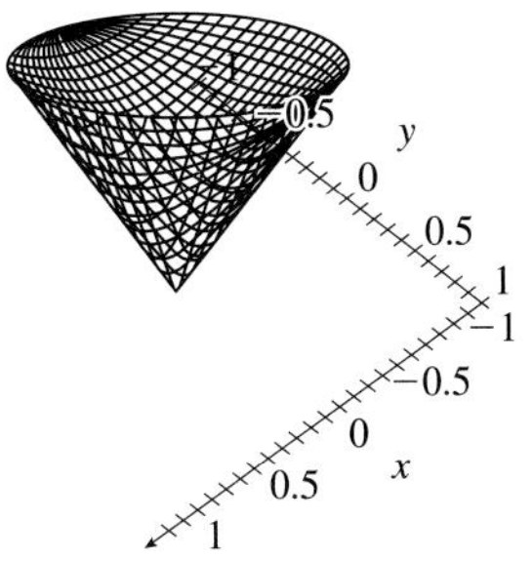

25.

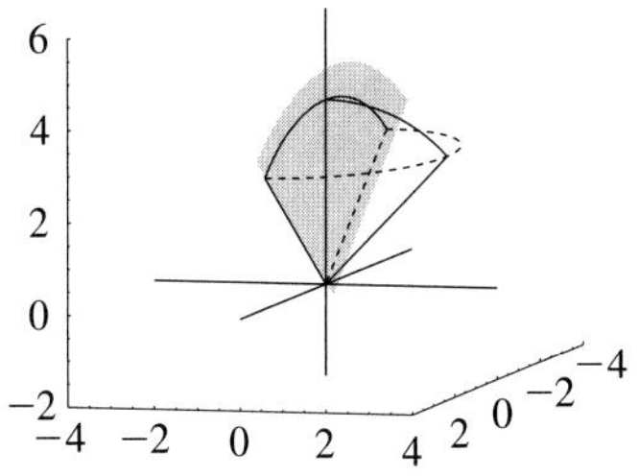

27.

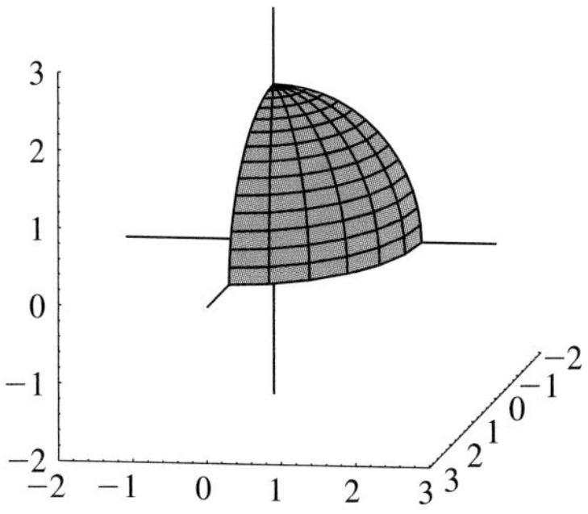

29.

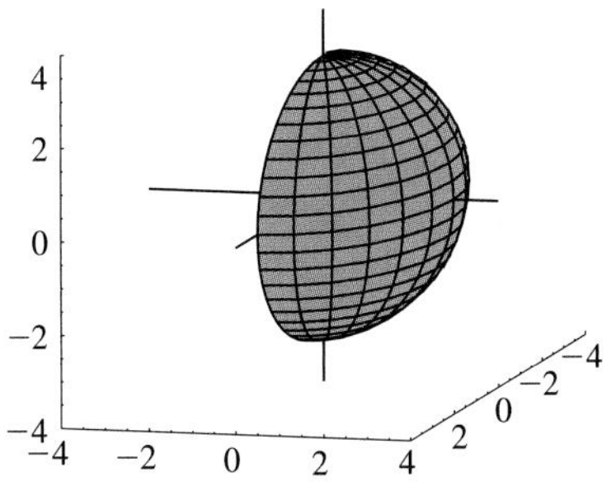

31. 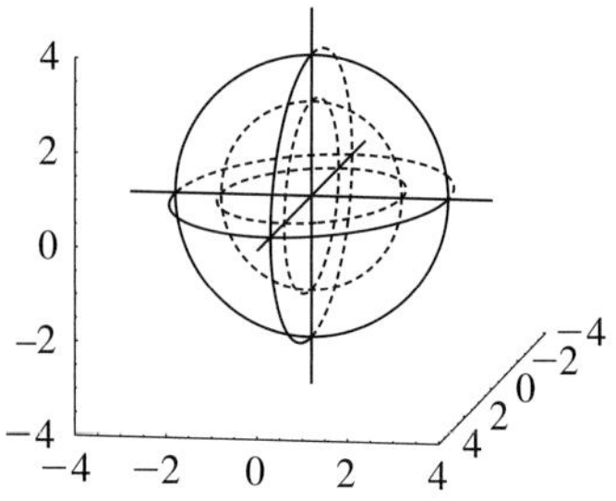

33. $\int_0^{2\pi}\int_0^{\pi/2}\int_0^2 e^{\rho^3}\rho^2\sin\phi\,d\rho\,d\phi\,d\theta = \frac{2}{3}\pi(e^8-1)$

35. $\int_0^{2\pi}\int_{\pi/4}^{3\pi/4}\int_{\csc\phi}^{\sqrt{2}} \rho^2\sin\phi\,d\rho\,d\phi\,d\theta \approx 14.2381$

37. $\int_0^1\int_1^2\int_3^4 (x^2+y^2+z^2)\,dz\,dy\,dx = 15$

39. $\int_0^{2\pi}\int_0^2\int_0^{4-r^2} r^3\,dz\,dr\,d\theta = \frac{32\pi}{3}$

41. $\int_0^{2\pi}\int_0^{\pi/4}\int_0^{\sqrt{2}} \rho^3\sin\phi\,d\rho\,d\phi\,d\theta = (2-\sqrt{2})\pi$

43. $\int_0^{2\pi}\int_0^{\pi/4}\int_0^{4\cos\phi} \rho^2\sin\phi\,d\rho\,d\phi\,d\theta = 8\pi$

45. $\int_0^{2\pi}\int_2^4\int_0^{z/\sqrt{2}} r\,dr\,dz\,d\theta = \frac{28\pi}{3}$

47. $\int_{-1}^1\int_{-1}^1\int_0^{\sqrt{x^2+y^2}} dz\,dy\,dx \approx 3.061$

49. $\int_0^{\pi/2}\int_0^{\pi/4}\int_0^2 \rho^2\sin\phi\,d\rho\,d\phi\,d\theta = \frac{4-2\sqrt{2}}{3}\pi$

51. $\int_0^{2\pi}\int_0^2\int_0^r r\,dz\,dr\,d\theta = \frac{16\pi}{3}$

53. $\int_{-\pi/2}^{\pi/2}\int_0^{\pi}\int_0^1 \rho^3\sin\phi\,d\rho\,d\phi\,d\theta = \frac{\pi}{2}$

55. $\int_0^{\pi}\int_0^{\pi/4}\int_0^{\sqrt{8}} \rho^5\sin\phi\,d\rho\,d\phi\,d\theta = \frac{256-128\sqrt{2}}{3}\pi$

57. $\bar{x}=\bar{y}=0, \bar{z}=\frac{3}{4}+\frac{3\sqrt{2}}{8}$

59. $\langle\cos\theta\sin\phi, \sin\theta\sin\phi, \cos\phi\rangle$ **61.** $c=2$ **63.** $c=2$

65. 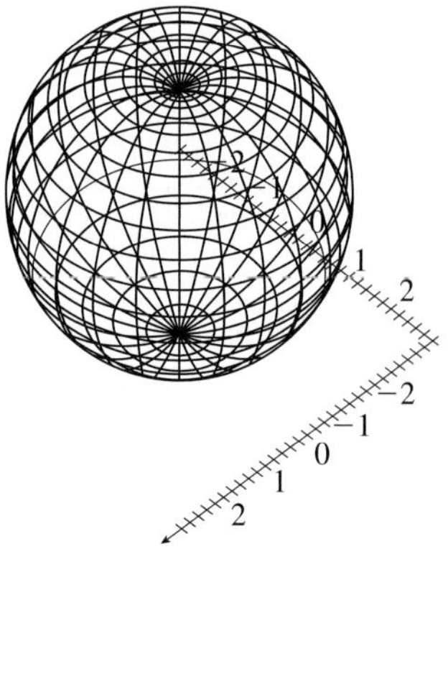

67. 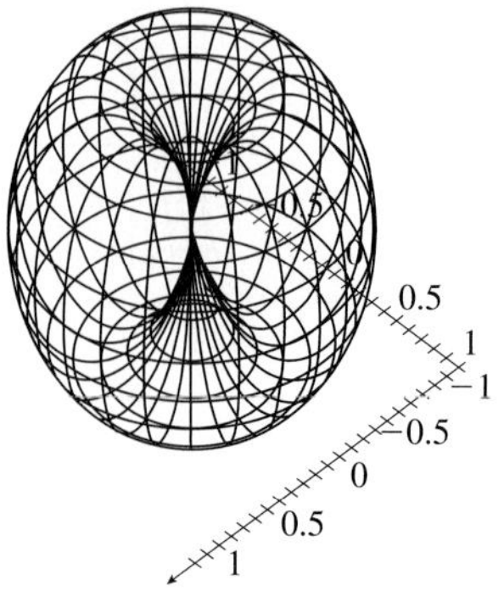

Exercises 13.8, page 1109

1. $x=\frac{1}{6}(v-u), y=\frac{1}{3}(u+2v), 2\le u\le 5, 1\le v\le 3$

3. $x=\frac{1}{4}(u-v), y=\frac{1}{4}(u+3v), 1\le u\le 3, -3\le v\le -1$

5. $x=r\cos\theta, y=r\sin\theta, 1\le r\le 2, 0\le\theta\le\frac{\pi}{2}$

7. $x=r\cos\theta, y=r\sin\theta, 2\le r\le 3, \frac{\pi}{4}\le\theta\le\frac{3\pi}{4}$

9. $x=\sqrt{\frac{1}{2}(v-u)}, y=\frac{1}{2}(u+v), 0\le u\le 2, 2\le v\le 4$

11. $x=\ln\left(\frac{1}{2}(v-u)\right), y=\frac{1}{2}(u+v), 0\le u\le 1, 3\le v\le 5$

13. $\frac{7}{2}$ **15.** $\frac{13}{3}$ **17.** $\frac{7}{3}$ **19.** $\frac{\ln 3}{6}\left(e^5-e^2\right)$

21. $\frac{9}{4}$ **23.** $-2u$ **25.** 2

27. $x=u-w, y=\frac{1}{2}(-u+v+w), z=\frac{1}{2}(u-v+w)$,
$1\le u\le 2, 0\le v\le 1, 2\le w\le 4$

29. 1 **33.** $\frac{\pi^2}{8}$

Chapter 13 Review Exercises, page 1111

1. 18 **3.** 207 **5.** $\pi(e^{-1}-e^{-4})$ **7.** $\frac{2}{3}$ **9.** 0

11. −19.92 **13.** $\frac{4}{3}$ **15.** 16π **17.** $\frac{128}{3}$ **19.** $\frac{64\pi}{3}$

21. $\frac{\pi}{3}(16-8\sqrt{2})$ **23.** $\frac{11\pi}{2}$ **25.** $\int_0^4\int_{\sqrt{y}}^2 f(x,y)\,dx\,dy$

27. $\int_{-\pi/2}^{\pi/2}\int_0^2 2r^2\cos\theta\,dr\,d\theta = \frac{32}{3}$ **29.** $m=\frac{16}{3}, \bar{x}=\frac{3}{2}, \bar{y}=\frac{9}{4}$

31. $m=\frac{64}{15}, \bar{x}=0, \bar{y}=\frac{23}{28}, \bar{z}=\frac{5}{14}$ **33.** $\int_0^1\int_{\sqrt{y}}^{2-y} dx\,dy = \frac{5}{6}$

35. $\frac{1}{2}$ **37.** $2\sqrt{21}$ **39.** $\frac{13\pi}{3}$ **41.** $16\pi\sqrt{2}$

43. $\int_0^2\int_{-1}^1\int_{-1}^1 z(x+y)\,dz\,dy\,dx = 0$

45. $\int_0^{2\pi}\int_0^{\pi/4}\int_0^2 \rho^3\sin\phi\,d\rho\,d\phi\,d\theta = \pi(8-4\sqrt{2})$

47. $\int_0^2\int_x^2\int_0^{6-x-y} f(x,y,z)\,dz\,dy\,dx$

49. $\int_0^{2\pi}\int_0^{\pi/2}\int_0^{2} f(\rho\sin\phi\cos\theta,\ \rho\sin\phi\sin\theta,\ \rho\cos\phi)\rho^2\sin\phi\,d\rho\,d\phi\,d\theta$

51. $\int_{\pi/4}^{\pi/2}\int_0^{\sqrt{2}}\int_0^{r} e^z r\,dz\,dr\,d\theta = \dfrac{\pi e^{\sqrt{2}}(\sqrt{2}-1)}{4}$

53. $\int_0^{\pi}\int_0^{\pi/4}\int_0^{\sqrt{2}} \rho^3\sin\phi\,d\rho\,d\phi\,d\theta = \pi\left(1-\dfrac{\sqrt{2}}{2}\right)$

55. (a) $r\sin\theta = 3$ (b) $\rho\sin\phi\sin\theta = 3$

57. (a) $r^2 + z^2 = 4$ (b) $\rho = 2$

59. (a) $z = r$ (b) $\phi = \dfrac{\pi}{4}$

61.

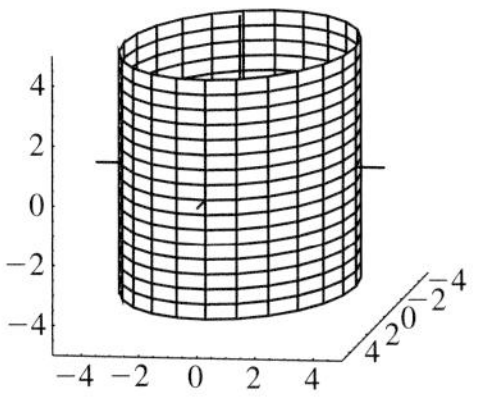

63.

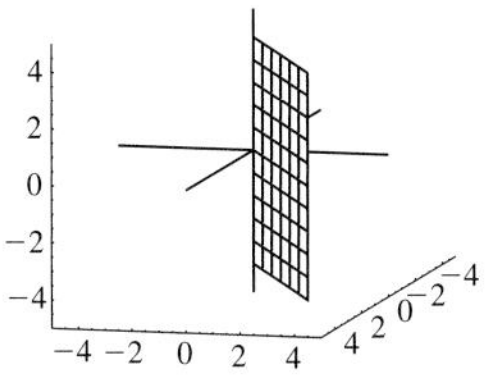

65.

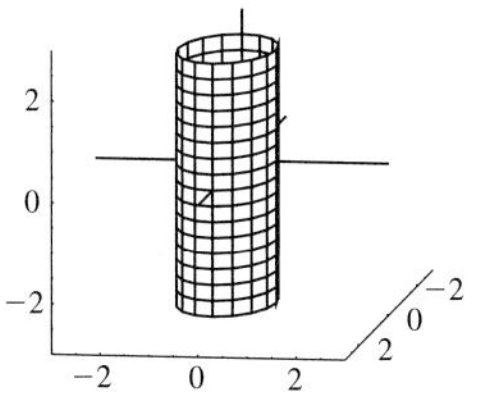

67. $x = \dfrac{1}{4}(v-u),\ y = \dfrac{1}{2}(u+v),\quad 1 \le u \le 1,\ 2 \le v \le 4$

69. $\dfrac{1}{2}(e - e^{-1})$ **71.** $4uv^2 - 4u^2$

CHAPTER 14

Exercises 14.1, page 1125

1.

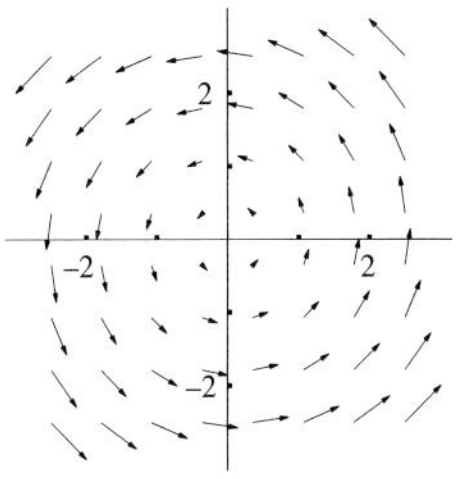

3.

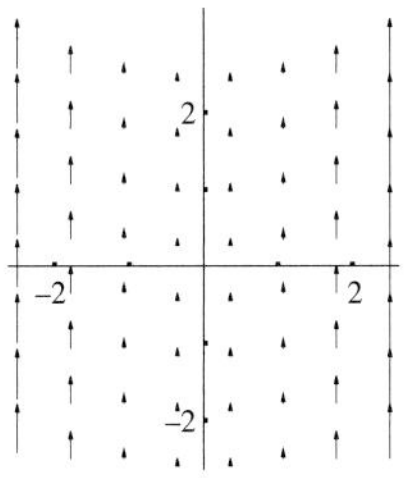

5.

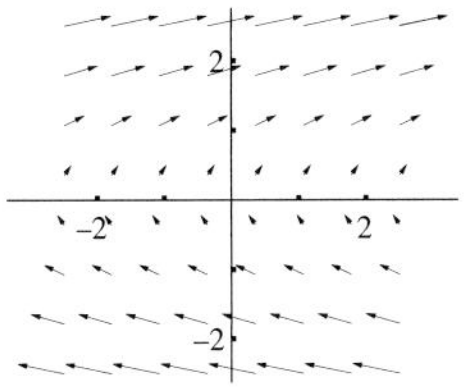

7.

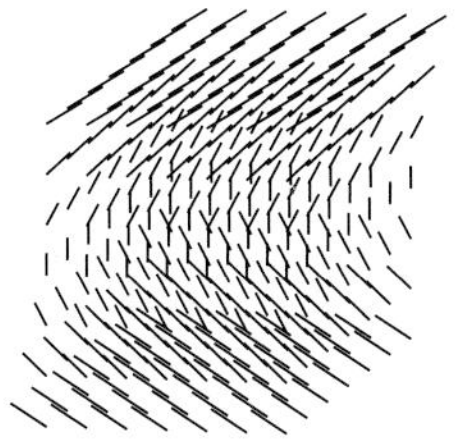

9.

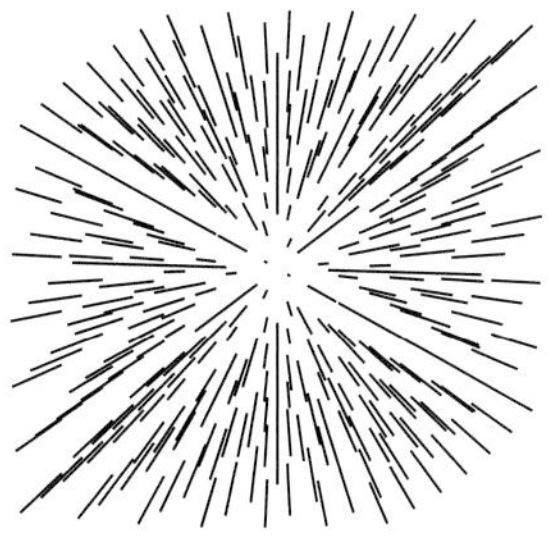

11. $\mathbf{F_1} = \text{D}, \mathbf{F_2} = \text{B}, \mathbf{F_3} = \text{A}, \mathbf{F_4} = \text{C}$ **13.** $\langle 2x, 2y\rangle$

15. $\dfrac{\langle x, y\rangle}{\sqrt{x^2+y^2}}$ **17.** $\langle e^{-y}, -xe^{-y}\rangle$ **19.** $\dfrac{\langle x, y, z\rangle}{\sqrt{x^2+y^2+z^2}}$

21. $\langle 2xy, x^2+z, y\rangle$ **23.** $f(x, y) = xy + c$ **25.** not

27. $f(x, y) = \frac{1}{2}x^2 - x^2y + \frac{1}{3}y^3 + c$

29. $f(x, y) = -\cos xy + c$

31. $f(x, y, z) = 2x^2 - xz + \frac{3}{2}y^2 + yz + c$ **33.** not

35. $y = \frac{1}{2}\sin x + c$ **37.** $y^2 = x^3 + c$

39. $(y+1)e^{-y} = -\frac{1}{2}x^2 + c$ **41.** $y^2 + 1 = ce^{2x}$

43. $f(x, y, z) = \int_0^x f(u)\,du + \int_0^y g(u)\,du + \int_0^z h(u)\,du + c$

47. $3r\mathbf{r}$

51.

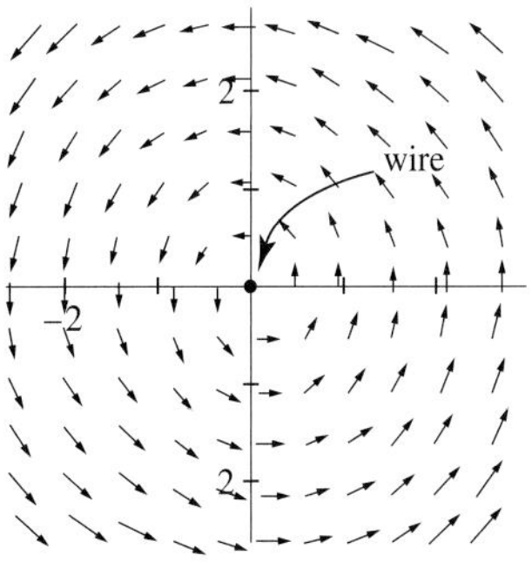

53. $\dfrac{3\langle x, y\rangle}{\sqrt{x^2+y^2}}$ (undefined at the origin)

55. $\sqrt{x^2+y^2+z^2}\,\langle -x, -y, -z\rangle$

57. $\dfrac{q\langle x+1, y\rangle}{[(x+1)^2+y^2]^{3/2}} + \dfrac{q\langle x-1, y\rangle}{[(x-1)^2+y^2]^{3/2}} - \dfrac{q\langle x, y\rangle}{[(x^2+y^2)]^{3/2}}$

59. from hot to cold (assuming $k > 0$)

Exercises 14.2, page 1140

1. $4\sqrt{13}$ **3.** $\frac{21}{2}\sqrt{17}$ **5.** 4 **7.** 12 **9.** −4 **11.** 3π

13. 25.41 **15.** −4 **17.** 14 **19.** $\frac{9}{2}$ **21.** $6\sqrt{6}$

23. −4 **25.** 31 **27.** 0 **29.** $\frac{8}{3}$ **31.** 26 **33.** 0

35. $4\pi - \frac{19}{3}$ **37.** positive **39.** zero **41.** negative

43. 18.67 **45.** $\bar{x} = 2.227, \bar{y} = 5.324$ **47.** 99.41

49. 359.9 **51.** $\dfrac{\pi^3}{3}\sqrt{5}$ **55.** 4π **57.** $\dfrac{32}{3}$ **59.** 12

61. (a) 22.1 (b) 15.35 (c) 3.65

Exercises 14.3, page 1151

1. $f(x, y) = x^2y - x + c$

3. $f(x, y) = \frac{x}{y} - x^2 + \frac{1}{2}y^2 + c$

5. not **7.** $f(x, y) = e^{xy} + \sin y + c$

9. $f(x, y, z) = xz^2 + x^2y + y - 3z + c$

11. $f(x, y, z) = xy^2z^2 + \frac{1}{2}x^2 + \frac{1}{2}y^2 + c$

13. $f(x, y) = x^2y - y$; 8

15. $f(x, y) = e^{xy} - y^2$; -16

17. $f(x, y, z) = xz^2 + x^2y$; -38 **19.** $\frac{152}{3}$ **21.** 18

23. $\sqrt{30} - \sqrt{14}$ **25.** -2 **27.** $10 - e^{18}$ **29.** 0

31. yes **33.** no **35.** no **41.** false

43. true **45.** $\tan^{-1}\left(\frac{y}{x}\right) + c,\ x \neq 0$; 0

47. (a) simply-connected (b) not simply-connected

49. a potential for **F** is $\frac{-kq}{\sqrt{x^2 + y^2 + z^2}}$

51. (*a*) $RT_2 \ln\left(\frac{P_2}{P_1}\right) - R(T_2 - T_1)$ (*b*) $RT_1 \ln\left(\frac{P_2}{P_1}\right) - R(T_2 - T_1)$

53 0; 0

Exercises 14.4, page 1162

1. π **3.** 16 **5.** -54 **7.** $\frac{32}{3}$

9. 6π **11.** $\frac{1}{3}$ **13.** $\frac{4}{3} + \frac{1}{2}e^2 + \frac{3}{2}e^{-2}$

15. $\frac{32}{5}$ **17.** 4 **19.** 0 **21.** 8π

23. $\frac{3}{8}\pi$ **25.** $\frac{32}{3}$ **29.** $\bar{x} = 0, \bar{y} = \frac{4}{7}$ **33.** 0

35. 0 **37.** $\{(x, y) \in \mathbb{R}^2 | (x, y) \neq (0, 0)\}$; no **39.** yes

Exercises 14.5, page 1171

1. $\langle 0, 0, -3y\rangle, -x$ **3.** $\langle -3, 2x, 0\rangle, 2z$

5. $\langle -y, -2x, -x\rangle, y + z$

7. $\langle xe^y + 1, -e^y, 0\rangle, 2x + 1$

9. $\langle -x \sin y, 3y - \cos y, 2x - 3z\rangle, 0$

11. $\langle 2yz - 2z, 2x, 0\rangle, 2z + 1 + y^2$ **13.** conservative

15. incompressible **17.** conservative **19.** neither

21. incompressible **23.** conservative **25.** conservative

27. (a) scalar (b) undefined (c) undefined (d) vector (e) vector

31. positive **33.** negative **35.** negative **53.** $\frac{2}{\sqrt{x^2 + y^2 + z^2}}$

55. (a) equal (b) less than

57. $\mathbf{F} = x\mathbf{i} + y^3\mathbf{j} + z\mathbf{k}, \nabla \cdot \mathbf{F} = 3y^2 + 2$

61. $g(x, y) = xy^2 - \frac{x^2}{2}$

63. $\nabla \times \mathbf{F} = \left\langle 0, 0, \frac{-2x}{(1 + x^2)^2}\right\rangle$; rotational, axis of rotation perpendicular to the xy-plane

Exercises 14.6, page 1185

1. $x = x, y = y, z = 3x + 4y$

3. $x = \cos u \cosh v, y = \sin u \cosh v, z = \sinh v, 0 \leq u \leq 2\pi, -\infty < v < \infty$

5. $x = 2\cos\theta, y = 2\sin\theta, z = z, 0 \leq \theta \leq 2\pi, 0 \leq z \leq 2$

7. $x = r\cos\theta, y = r\sin\theta, z = 4 - r^2, 0 \leq \theta \leq 2\pi, 0 \leq r \leq 2$

9.

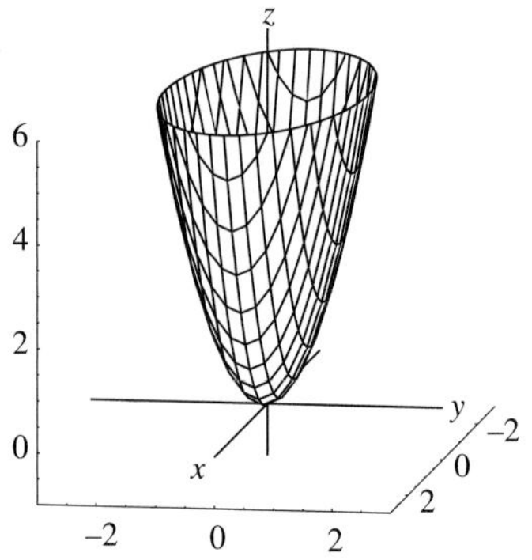

11.

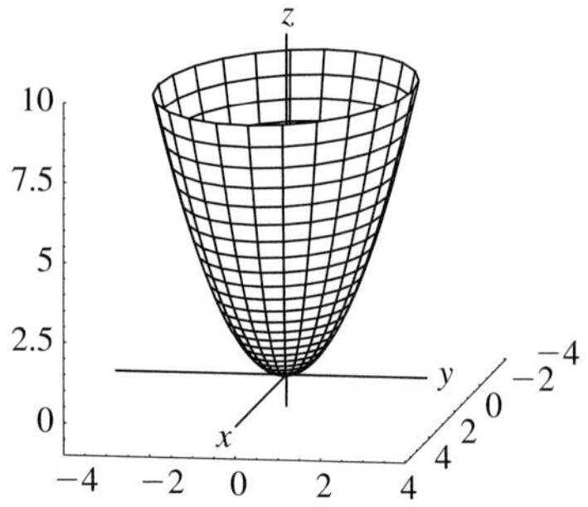

13.

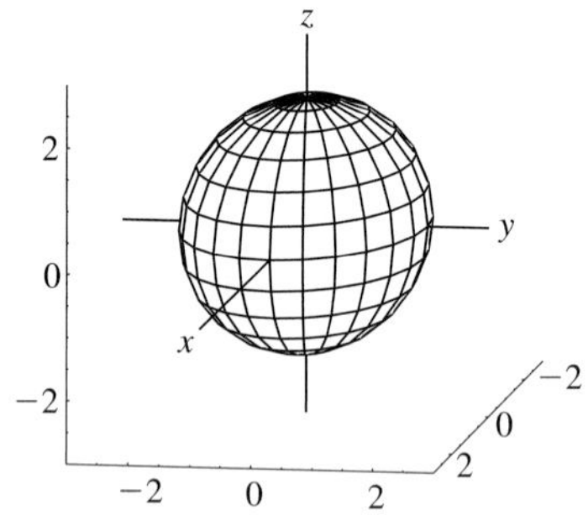

15.

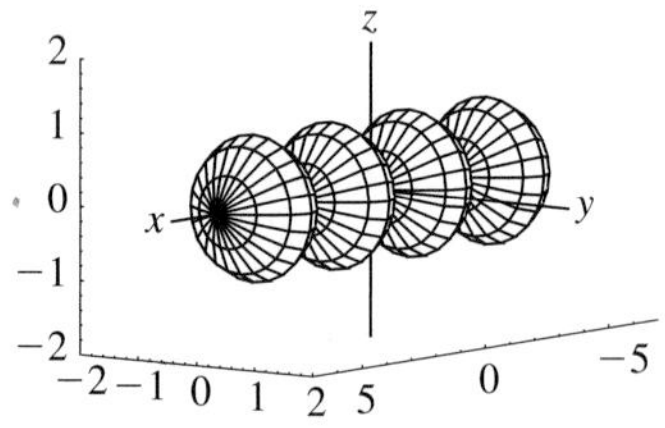

17. (a) A (b) C (c) B **19.** $16\pi\sqrt{2}$ **21.** $2\pi\sqrt{14}$

23. $\frac{\sqrt{2}}{2}$ **25.** 4π **27.** $\int_1^3 \int_1^2 (2x^2 + 3xy)\sqrt{14}\, dx\, dy = \frac{82\sqrt{14}}{3}$

29. 486π (the integrand is the constant 27 on a hemisphere of radius 3)

31. $\int_0^{2\pi} \int_{\sqrt{2}}^{\sqrt{3}} (2r^3 - 4r)\sqrt{4r^2 + 1}\, dr\, d\theta = \frac{\pi}{10}[81 - 13\sqrt{13}]$

33. $2\int_0^{2\pi} \int_0^4 \sqrt{2}r^3 dr\, d\theta = 256\sqrt{2}\pi$

35. 0, by symmetry **37.** 24π **39.** -18π **41.** $\frac{5}{2}$

43. $\frac{9\pi}{2}$ **45.** $\frac{7}{4}$ **47.** 0 **49.** $m = 8\sqrt{14}\pi, \bar{x} = \bar{y} = 0, \bar{z} = 6$

51. $m = 2\pi, \bar{x} = \frac{1}{3}, \bar{y} = 0, \bar{z} = \frac{1}{2}$

53. $\iint_S g(x, y, z)\, dS = \iint_R g(f(y, z), y, z)\sqrt{(f_y)^2 + (f_z)^2 + 1}\, dA$
where S is given by $x = f(y, z)$ for (y, z) in region R in $\mathbb{R}^2$

55. $\frac{3\pi}{2}$ **57.** 198.8π **59.** 0.47π

61. 23.66 **63.** flow lines don't cross boundary

67. 0; flow is along the surface **69.** $\frac{2\pi}{3c^2}$

Exercises 14.7, page 1195

1. $\frac{3}{2}$ **3.** π **5.** 0 **7.** 8 **9.** 32π **11.** $\frac{37\pi}{6}$

13. 4π **15.** $\frac{6\pi}{5}$ **17.** 0 **19.** $\frac{\pi}{2}$ **21.** π **23.** 224π

25. $\frac{27}{5}$ **27.** $\frac{512}{3}$ **35.** $c = \frac{\rho}{2\varepsilon_0}$

Exercises 14.8, page 1205

1. 0 **3.** 4π **5.** $-\frac{4}{3}$ **7.** $-\pi$ **9.** 0 **11.** 1/2

13. 0 **15.** 4π **17.** 4π **19.** 0 **21.** 0

23. 3π **31.** Both boundary curves have the same orientation.

33. Use $\nabla \times (f\nabla g) = (\nabla f) \times (\nabla g)$.

Exercises 14.9, page 1213

1. 1 **3.** 0 **5.** $4\pi\varepsilon_0 R^3$ **7.** 0

15. For the given field, $\nabla \cdot \mathbf{E} = 0$ in any region not containing the origin.

17. $\mathbf{E} \cdot \mathbf{n} = \frac{q}{4\pi\varepsilon_0 R^2}$ on the sphere and the area of the sphere is $4\pi R^2$

21. $\iint_S \nabla u \cdot \mathbf{n}\, dS = \iiint_Q \nabla \cdot (\nabla u)\, dV = \iiint_Q \nabla^2 u\, dV$ where Q is the region enclosed by S

25. Use exercise 24 to show that $\nabla(f - g) = 0$ everywhere inside S; since $f - g = 0$ on S, $f = g$ inside S as well.

Chapter 14 Review Exercises, page 1214

1.

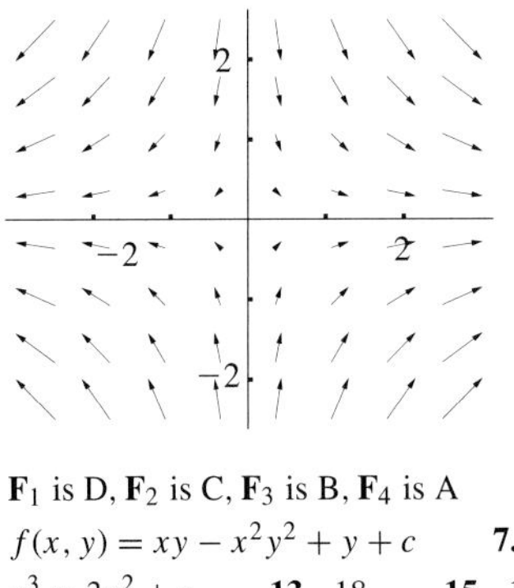

3. $\mathbf{F}_1$ is D, $\mathbf{F}_2$ is C, $\mathbf{F}_3$ is B, $\mathbf{F}_4$ is A

5. $f(x, y) = xy - x^2y^2 + y + c$ **7.** not conservative

9. $y^3 = 3x^2 + c$ **13.** 18 **15.** 18π **17.** 0 **19.** 0

21. $3\pi - 4$ **23.** zero **25.** 40π **27.** 66 **29.** 3

31. 10 **33.** conservative **35.** $\frac{1}{3}$ **37.** -2 **39.** $\frac{32}{3}$

41. 6π **43.** $\langle 0, 0, 0\rangle, 3x^2 - 3y^2$ **45.** $\langle 0, 0, 0\rangle, 2 + 2z^2 + 2y^2$

47. neither **49.** both **51.** positive

53.

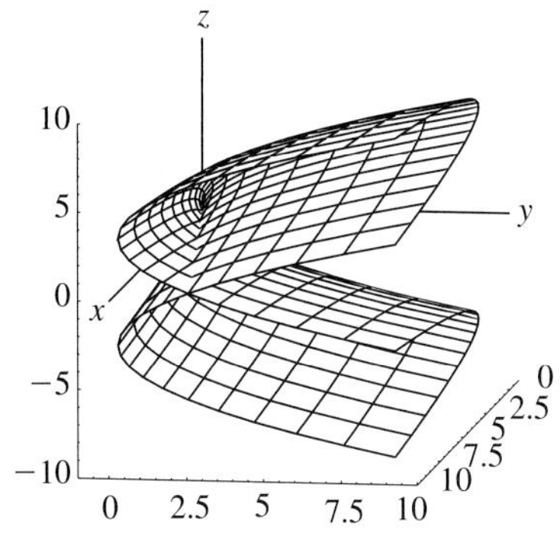

55. (a) B (b) C (c) A

57. $\frac{\pi}{6}(17^{3/2} - 5^{3/2})$ **59.** $-8\sqrt{14}$

61. $4\pi\sqrt{26}$ **63.** 0

65. $m = \frac{\pi}{3}(17\sqrt{17} - 1), \bar{x} = \bar{y} = 0, \bar{z} = \frac{1 + 391\sqrt{17}}{10(17\sqrt{17} - 1)}$

67. $\frac{16}{3}$ **69.** $\frac{304}{5}$ **71.** $\frac{8\pi}{3}$ **73.** 0

75. 0 **77.** 0

CHAPTER 15

Exercises 15.1, page 1227

1. $y(t) = c_1e^{4t} + c_2e^{-2t}$

3. $y(t) = c_1e^{2t} + c_2te^{2t}$

5. $y(t) = e^t(c_1 \cos 2t + c_2 \sin 2t)$

7. $y(t) = c_1 + c_2e^{2t}$

9. $y(t) = c_1e^{(1+\sqrt{7})t} + c_2e^{(1-\sqrt{7})t}$

11. $y(t) = c_1e^{\left(\frac{\sqrt{5}+1}{2}\right)t} + c_2e^{\left(\frac{\sqrt{5}-1}{2}\right)t}$

13. $y(t) = -\frac{3}{2}\sin 2t + 2\cos 2t$ **15.** $y(t) = -e^t + e^{2t}$

17. $y(t) = e^t(2\cos 2t - \sin 2t)$ **19.** $y(t) = (3t - 1)e^t$

21. $A = \sqrt{13}, \delta \approx -0.983$ **23.** $A = \frac{\sqrt{105}}{5}, \delta \approx -1.351$

25. $u(t) = \frac{2}{3}\cos 8t$

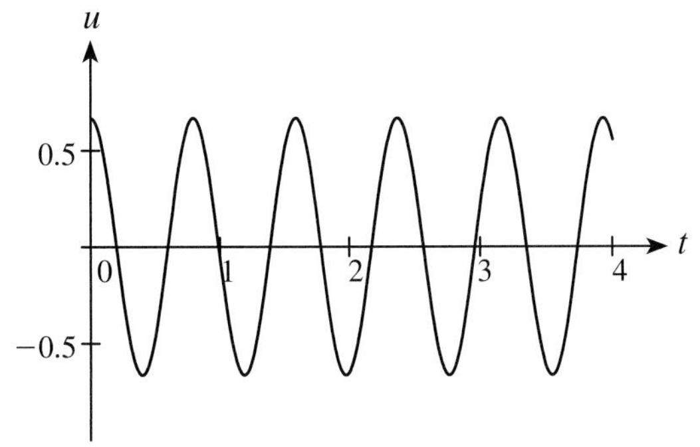

27. $u(t) = \frac{1}{5}\cos(7\sqrt{2}t) - \frac{2\sqrt{2}}{7}\sin(7\sqrt{2}t)$; $A = \frac{\sqrt{249}}{35}$, $\delta \approx -0.460$

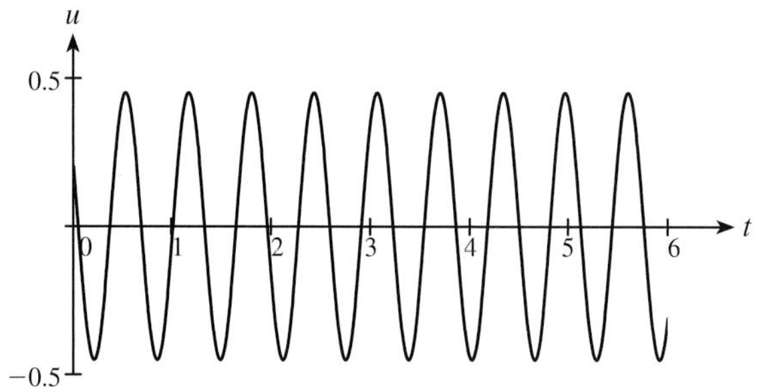

29. $u(t) = e^{-12t} - \frac{3}{2}e^{-8t}$

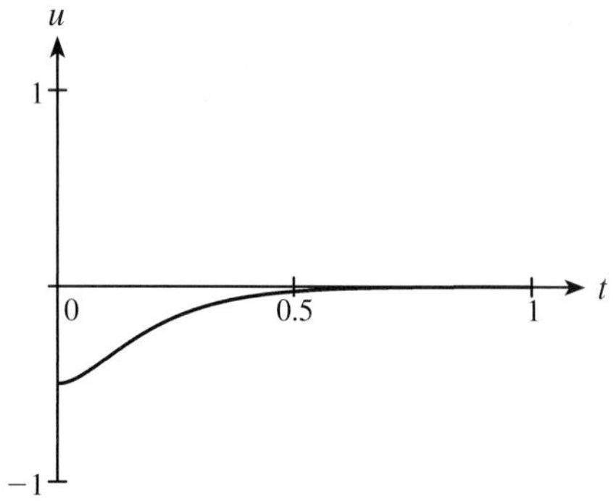

31. $u(t) = e^{-(1/4)t}\left(-\frac{1}{2}\cos\left(\frac{\sqrt{15655}}{20}t\right) - \frac{\sqrt{15655}}{6262}\sin\left(\frac{\sqrt{15655}}{20}t\right)\right)$

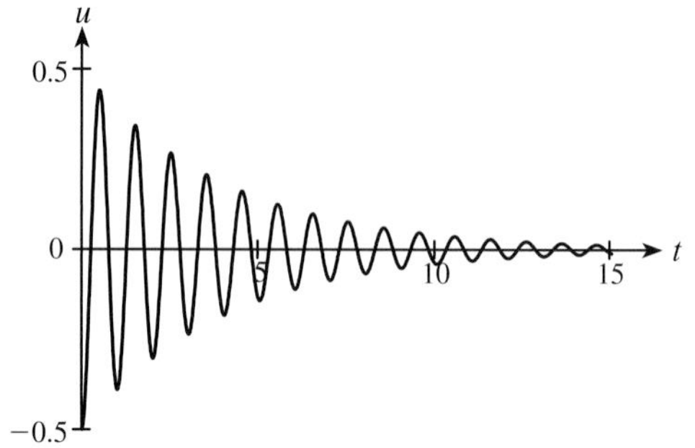

37. $c = 8\sqrt{2}$

41. The solution to $y'' + a^2y = 0$ has ordinary sine and cosine functions instead of hyperbolic sine and cosine.

Exercises 15.2, page 1236

1. $u(t) = e^{-t}(c_1\cos 2t + c_2\sin 2t) + 3e^{-2t}$

3. $u(t) = e^{-2t}(c_1 + c_2t) + t^2 - 2t + \frac{3}{2}$

5. $u(t) = e^{-t}(c_1\cos 3t + c_2\sin 3t) + 2e^{-3t}$

7. $u(t) = e^{-t}(c_1 + c_2t) - \frac{25}{2}\cos t$

9. $u(t) = c_1e^{2t} + c_2e^{-2t} - \frac{1}{2}t^3 - \frac{3}{4}t$

11. $u(t) = Ae^{-t} + Bte^{-t}\cos 3t + Cte^{-t}\sin 3t + D\cos 3t + E\sin 3t$

13. $u(t) = t\,(C_3t^3 + C_2t^2 + C_1t + C_0) + Ae^{2t}$

15. $u(t) = Ae^t\cos 3t + Be^t\sin 3t + (Ct^2 + Dt)\cos 3t + (Et^2 + Ft)\sin 3t$

17. $u(t) = t^2e^{-2t}(At^2 + Bt + C) + e^{-2t}(Dt + E)\cos t + e^{-2t}(Ft + G)\sin t$

19. $u(t) = e^{-t}\left(-\frac{1221}{5963380}\cos(\sqrt{4899}\,t) - \frac{1229\sqrt{4899}}{29214598620}\sin(\sqrt{4899}\,t)\right) + \left(\frac{1221}{5963380}\cos 4t + \frac{1}{2981690}\sin 4t\right)$

21. $u(t) = \frac{-1024 - 543\sqrt{2}}{14368}e^{(-16+8\sqrt{2})t} + \frac{-1024 + 543\sqrt{2}}{14368}e^{(-16-8\sqrt{2})t} + \frac{64}{449}e^{-t/2}$

23. $u(t) = -\cos 3t + 2\sin 3t$; amplitude $= \sqrt{5}$; phase shift ≈ -0.464

25. $u(t) = \cos t + 2\sin t$; amplitude $= \sqrt{5}$; phase shift ≈ 0.464

27. $u(t) = -\frac{640}{5881}\cos 2t + \frac{3000}{5881}\sin 2t$; amplitude ≈ 0.522, phase shift ≈ -0.210

29. natural frequency and value of ω that produces resonance, $\sqrt{3}$; beats, for example, at $\omega = 1.9$

31. natural frequency and value of ω that produces resonance, $8\sqrt{2}$; beats, for example, at $\omega = 11$

33. undamped: $y(t) = \cos 3t + 2t\sin 3t$;
damped: $y(t) =$
$e^{-t/20}\left(\cos\left(\frac{\sqrt{3599}}{20}t\right) - \frac{2399}{\sqrt{3599}}\sin\left(\frac{\sqrt{3599}}{20}t\right)\right) + 40\sin 3t$

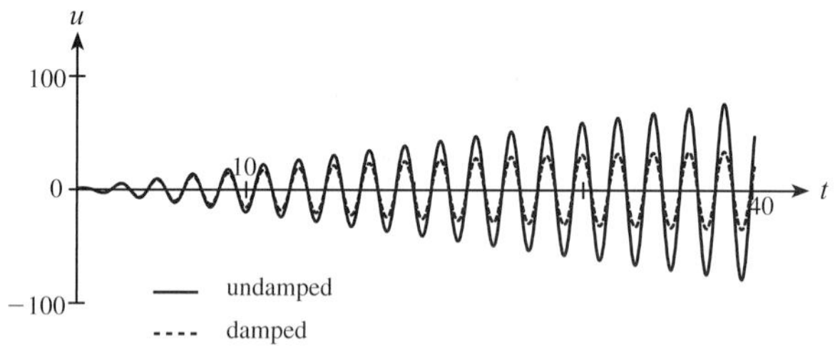

37. $u(t) = 5.12194\sin 2t - 4.87804\sin 2.1t$

39. $u(t) = \frac{1}{\omega^2 - 4}\cos 2t - \frac{1}{\omega^2 - 4}\sin\omega t$

41. $\frac{1}{\sqrt{\omega^4 - 7.99\,\omega^2 + 16}}$

Exercises 15.3, page 1245

1. $Q(t) = -1.4549 \times 10^{-6}e^{-443.65t} + 1.1455 \times 10^{-5}e^{-56.35t}$;
$I(t) = 6.4547 \times 10^{-4}e^{-443.65t} - 6.4549 \times 10^{-4}e^{-56.35t}$

3. $Q(t) = 10^{-6}\cos 707.11t$; $I(t) = -7.0711 \times 10^{-4}\sin 707.11t$; amplitude is 10^{-6}, phase shift is $\frac{\pi}{2}$

5. $Q(t) = e^{-2t}\left(-\frac{27}{170}\cos 6t + \frac{26}{255}\sin 6t\right) + \frac{27}{170}\cos 2t + \frac{3}{85}\sin 2t$;
$I(t) = e^{-2t}\left(\frac{79}{85}\cos 6t + \frac{191}{255}\sin 6t\right) + \frac{6}{85}\cos 2t - \frac{27}{85}\sin 2t$

7. $u(t) = -\frac{1}{2020}\cos 2t + \frac{1}{202}\sin 2t$; amplitude $\approx 4.975 \times 10^{-3}$; phase shift ≈ -0.100

9. Choose $\delta = \tan^{-1}\left(\dfrac{\omega^2 - 5}{2\omega}\right)$ and A_2 as given in the exercise.

11. $g(\omega) = \dfrac{1}{\sqrt{\omega^4 - \dfrac{196}{25}\omega^2 + 16}}$; for maximum gain, $\omega = \dfrac{7\sqrt{2}}{5}$

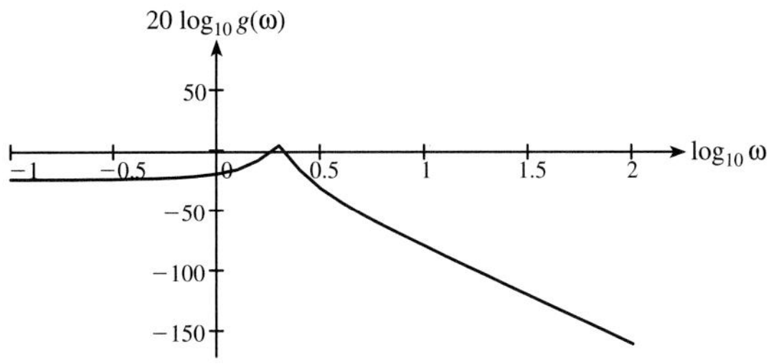

13. $g(\omega) = \dfrac{1}{\sqrt{\omega^4 - \dfrac{199}{25}\omega^2 + 16}}$; for maximum gain, $\omega = \sqrt{\dfrac{199}{50}}$

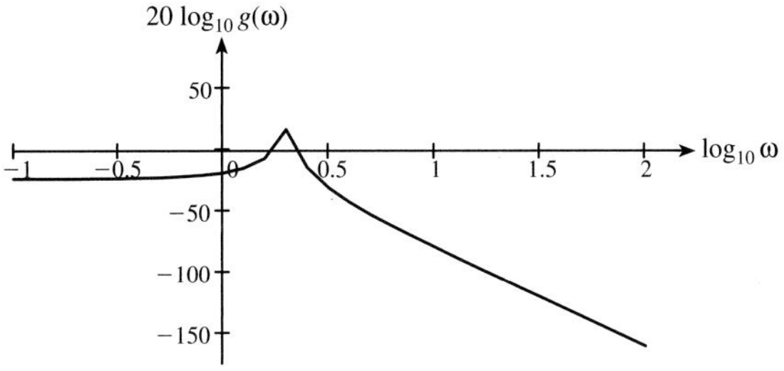

15. $g(\omega) = \dfrac{\omega^2}{\sqrt{\omega^4 - 7\omega^2 + 16}}$; for maximum gain, $\omega = \dfrac{4\sqrt{14}}{7}$

23. $\theta(t) = 0.2\cos(7\sqrt{2}t)$; amplitude $= 0.2$; period $= \dfrac{\pi\sqrt{2}}{7}$; if the length of the pendulum is doubled, the amplitude remains the same but the period increases by a factor of $\sqrt{2}$

25. $\theta(t) = \dfrac{1}{70\sqrt{2}}\sin(7\sqrt{2}t)$; amplitude $= \dfrac{1}{70\sqrt{2}}$; period $= \dfrac{\pi\sqrt{2}}{7}$

27. amplitude ≈ 0.1367 rad; period $= \dfrac{2\pi}{3}$

29. $y = e^{-4t}(c_1\cos\sqrt{2516}t + c_2\sin\sqrt{2516}t)$

33. yes **35.** yes **37.** $\alpha > 8.99$ **39.** $\dfrac{2\pi}{w} < 4$; no

Exercises 15.4, page 1256

1. $a_{n+2} = \dfrac{-2}{n+1}a_n$; $a_0\displaystyle\sum_{k=0}^{\infty}\frac{(-1)^k 2^{2k}k!}{(2k)!}x^{2k} + a_1\sum_{k=0}^{\infty}\frac{(-1)^k}{k!}x^{2k+1}$

3. $a_{n+2} = \dfrac{1}{n+2}a_n$; $a_0\displaystyle\sum_{k=0}^{\infty}\frac{x^{2k}}{k!2^k} + a_1\sum_{k=0}^{\infty}\frac{2^k k! x^{2k+1}}{(2k+1)!}$

5. $a_{n+2} = \dfrac{n}{(n+2)(n+1)}a_n$ for
$n = 1, 3, 5, \ldots$; $a_0 + a_1\displaystyle\sum_{k=0}^{\infty}\frac{x^{2k+1}}{(2k+1)k!2^k}$

7. $a_{n+3} = \dfrac{n}{(n+2)(n+3)}a_n$ for
$n = 1, 4, 7, \ldots$; $a_0 + a_1\displaystyle\sum_{k=0}^{\infty}\frac{x^{3k+1}}{(3k+1)k!3^k}$

9. $a_0\displaystyle\sum_{k=0}^{\infty}\frac{(x-1)^{2k}}{2^k k!} + a_1\sum_{k=0}^{\infty}\frac{2^k k!(x-1)^{2k+1}}{(2k+1)!}$

11. $\displaystyle\sum_{n=0}^{\infty} a_n(x-1)^n$ where a_0 and a_1 are chosen freely, $a_2 = \dfrac{a_0}{2}$, and
$a_{n+2} = \dfrac{a_{n-1} + a_n}{(n+1)(n+2)}$ for $n \geq 1$

13. $5\displaystyle\sum_{k=0}^{\infty}\frac{(-1)^k x^{2k}}{k!} + 7\sum_{k=0}^{\infty}\frac{(-1)^{k+1}2^{2k}k!x^{2k+1}}{(2k+1)!}$

15. $-3\displaystyle\sum_{k=0}^{\infty}\frac{(x-1)^{2k}}{2^k k!} + 12\sum_{k=0}^{\infty}\frac{2^k k!(x-1)^{2k+1}}{(2k+1)!}$

17. ∞ **19.** ∞ **21.** $\displaystyle\sum_{k=0}^{\infty}\frac{c(-1)^k x^{2k}}{2^{2k}(k!)^2}$ **23.** ∞

27. $P_5(x) = -2 + x - \dfrac{x^2}{2} - \dfrac{x^3}{3} + \dfrac{5x^4}{24} + \dfrac{x^5}{60}$

29. $P_5(x) = 4(x-\pi) - 2\pi(x-\pi)^2 + \left(\dfrac{2\pi^2 - 2}{3}\right)(x-\pi)^3$
$- \dfrac{\pi^3 - 3\pi - 2}{6}(x-\pi)^4 + \dfrac{4\pi^4 - 24\pi^2 - 20\pi + 12}{120}(x-\pi)^5$

Chapter 15 Review Exercises, page 1257

1. $y(t) = c_1 e^{-4t} + c_2 e^{3t}$

3. $y(t) = e^{-t/2}\left(c_1\cos\left(\dfrac{\sqrt{11}}{2}t\right) + c_2\sin\left(\dfrac{\sqrt{11}}{2}t\right)\right)$

5. $y(t) = c_1 e^{-2t} + c_2 e^{3t} + \dfrac{1}{5}te^{3t} - \dfrac{1}{6}t^2 + \dfrac{1}{18}t - \dfrac{25}{108}$

7. $y(t) = 2e^{-4t} + 3e^{2t}$ **9.** $y(t) = \sin 2t + \cos t$

11. $u(t) = \dfrac{1}{6}\cos(4\sqrt{6}t)$

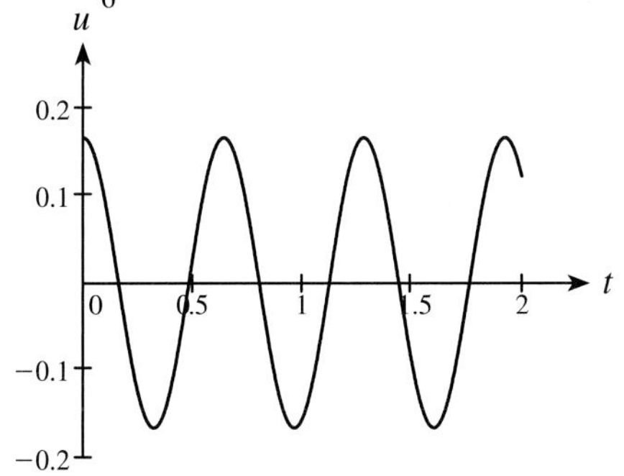

13. $Q(t) = \dfrac{-499 + 20{,}000\sqrt{1595}}{10{,}000}e^{(-400+10\sqrt{1595})t}$
$+ \dfrac{1 - 40\sqrt{1595}}{20}e^{(-400-10\sqrt{1595})t}$
$I(t) = 63801e^{(-400-10\sqrt{1595})t} - 49.9279e^{(-400+10\sqrt{1595})t}$

15. $u(t) = te^{-t}(A\cos 2t + B\sin 2t) + C_3t^3 + C_2t^2 + C_1t + C_0$
$+ D\cos 2t + E\sin 2t$

17. $u(t) = -\dfrac{160}{13289}\cos 2t + \dfrac{2300}{13289}\sin 2t$

19. $a_{n+2} = \dfrac{2}{n+1}a_n$; $a_0\displaystyle\sum_{k=0}^{\infty}\frac{2^{2k}k!x^{2k}}{(2k)!} + a_1\sum_{k=0}^{\infty}\frac{x^{2k+1}}{k!}$

21. $\displaystyle\sum_{n=0}^{\infty} a_n x^n$, where a_0 and a_1 are freely chosen and
$a_{n+2} = 2\left(\dfrac{a_n}{n+1} + \dfrac{a_{n+1}}{n+2}\right)$ for $n \geq 0$

23. $4\displaystyle\sum_{k=0}^{\infty}\frac{2^{2k}k!x^{2k}}{(2k)!} + 2\sum_{k=0}^{\infty}\frac{x^{2k+1}}{k!}$

Credits

CHAPTER 10

Page 785 bottom: © Sam Sharpe/CORBIS; **785 middle:** © Bill Pugliano/Getty Images; **785 top:** © George Tiedemann/NewSport/CORBIS; **796:** © The Granger Collection; **813:** © Sam Sharpe/CORBIS; **818:** © Culver Pictures, Inc.; **848 bottom:** © Jeremy Hoare/Photo Disc; **848 top:** © Neil Beer/Photo Disc

CHAPTER 11

Page 853 top: © The RoboCup Federation; **853 bottom, all:** Courtesy of Roland Minton; **882:** © Dimitri Iundt/CORBIS Sygma; **886:** © Christopher J. Morris/CORBIS; **899:** Courtesy of Edward Witten; **903:** © The Granger Collection

CHAPTER 12

Page 919 bottom: © Bettmann/CORBIS; **919 top:** © Duomo/CORBIS; **928, fig. 12.12a and b:** WW2010 Project, Department of Atmospheric Sciences, University of Illinois at Urbana-Champaign; **929, fig. 12.12c and d:** Jet Propulsion Laboratory, California Institute of Technology; **931:** NOAA; **936 right and left:** Courtesy of Roland Minton; **1013:** © The Granger Collection

CHAPTER 13

Page 1035: University of St. Andrews; **1057:** © Gray Mortimore/Getty Images; **1104:** © The Granger Collection; **1088:** Enrico Bombieri: Courtesy of the Institute for Advanced Study, Princeton, N.J.H. Landshoff, photographer

CHAPTER 14

Page 1115 top: © Dale E. Boyer/Courtesy NASA/Photo Researchers, Inc.; **1115 middle:** © David Reed/CORBIS; **1115 bottom:** Courtesy of Volkwagen of America Inc.; **fig 14.5:** JPL/California Institute of Technology/NASA; **1129:** NOAA; **1156:** Courtesy of the University of Nottingham; **1184:** University of St. Andrews; **1200:** © The Granger Collection; **1204:** Courtesy of Cathleen Synge Morawetz; Photographer: Hamilton

CHAPTER 15

Page 1221 top: © Steve Bronstein/Getty Images; **1221 bottom:** © AP/Wide World Photos; **1230:** © Mark Gibson Photography; **1240:** © Jonathan Nourok/Photo Edit

Index

A

B

C

D

E

F

G

H

I

J

K

L

M

N

O

P

Q

R

S

T

U

V

W

Table of Integrals

Forms Involving $a + bu$

1. $\displaystyle\int \frac{1}{a+bu}\,du = \frac{1}{b}\ln|a+bu| + c$

2. $\displaystyle\int \frac{u}{a+bu}\,du = \frac{1}{b^2}(a+bu - a\ln|a+bu|) + c$

3. $\displaystyle\int \frac{u^2}{a+bu}\,du = \frac{1}{2b^3}[(a+bu)^2 - 4a(a+bu) + 2a^2\ln|a+bu|] + c$

4. $\displaystyle\int \frac{1}{u(a+bu)}\,du = \frac{1}{a}\ln\left|\frac{u}{a+bu}\right| + c$

5. $\displaystyle\int \frac{1}{u^2(a+bu)}\,du = \frac{b}{a^2}\ln\left|\frac{a+bu}{u}\right| - \frac{1}{au} + c$

Forms Involving $(a + bu)^2$

6. $\displaystyle\int \frac{1}{(a+bu)^2}\,du = \frac{-1}{b(a+bu)} + c$

7. $\displaystyle\int \frac{u}{(a+bu)^2}\,du = \frac{1}{b^2}\left(\frac{a}{a+bu} + \ln|a+bu|\right) + c$

8. $\displaystyle\int \frac{u^2}{(a+bu)^2}\,du = \frac{1}{b^3}\left(a+bu - \frac{a^2}{a+bu} - 2a\ln|a+bu|\right) + c$

9. $\displaystyle\int \frac{1}{u(a+bu)^2}\,du = \frac{1}{a(a+bu)} + \frac{1}{a^2}\ln\left|\frac{u}{a+bu}\right| + c$

10. $\displaystyle\int \frac{1}{u^2(a+bu)^2}\,du = \frac{2b}{a^3}\ln\left|\frac{a+bu}{u}\right| - \frac{a+2bu}{a^2u(a+bu)} + c$

Forms Involving $\sqrt{a + bu}$

11. $\displaystyle\int u\sqrt{a+bu}\,du = \frac{2}{15b^2}(3bu-2a)(a+bu)^{3/2} + c$

12. $\displaystyle\int u^2\sqrt{a+bu}\,du = \frac{2}{105b^3}(15b^2u^2 - 12abu + 8a^2)(a+bu)^{3/2} + c$

13. $\displaystyle\int u^n\sqrt{a+bu}\,du = \frac{2}{b(2n+3)}u^n(a+bu)^{3/2} - \frac{2na}{b(2n+3)}\int u^{n-1}\sqrt{a+bu}\,du$

14. $\displaystyle\int \frac{\sqrt{a+bu}}{u}du = 2\sqrt{a+bu} + a\int \frac{1}{u\sqrt{a+bu}}\,du$

15. $\displaystyle\int \frac{\sqrt{a+bu}}{u^n}\,du = \frac{-1}{a(n-1)}\frac{(a+bu)^{3/2}}{u^{n-1}} - \frac{(2n-5)b}{2a(n-1)}\int \frac{\sqrt{a+bu}}{u^{n-1}}\,du,\ n \neq 1$

16a. $\displaystyle\int \frac{1}{u\sqrt{a+bu}}\,du = \frac{1}{\sqrt{a}}\ln\left|\frac{\sqrt{a+bu}-\sqrt{a}}{\sqrt{a+bu}+\sqrt{a}}\right| + c,\ a>0$

16b. $\displaystyle\int \frac{1}{u\sqrt{a+bu}}\,du = \frac{2}{\sqrt{-a}}\tan^{-1}\sqrt{\frac{a+bu}{-a}} + c,\ a<0$

17. $\displaystyle\int \frac{1}{u^n\sqrt{a+bu}}\,du = \frac{-1}{a(n-1)}\frac{\sqrt{a+bu}}{u^{n-1}} - \frac{(2n-3)b}{2a(n-1)}\int \frac{1}{u^{n-1}\sqrt{a+bu}}\,du,\ n\neq 1$

18. $\displaystyle\int \frac{u}{\sqrt{a+bu}}\,du = \frac{2}{3b^2}(bu-2a)\sqrt{a+bu} + c$

19. $\displaystyle\int \frac{u^2}{\sqrt{a+bu}}\,du = \frac{2}{15b^3}(3b^2u^2-4abu+8a^2)\sqrt{a+bu} + c$

20. $\displaystyle\int \frac{u^n}{\sqrt{a+bu}}\,du = \frac{2}{(2n+1)b}u^n\sqrt{a+bu} - \frac{2na}{(2n+1)b}\int \frac{u^{n-1}}{\sqrt{a+bu}}\,du$

Forms Involving $\sqrt{a^2+u^2}$, $a>0$

21. $\displaystyle\int \sqrt{a^2+u^2}\,du = \tfrac{1}{2}u\sqrt{a^2+u^2} + \tfrac{1}{2}a^2\ln\left|u+\sqrt{a^2+u^2}\right| + c$

22. $\displaystyle\int u^2\sqrt{a^2+u^2}\,du = \tfrac{1}{8}u(a^2+2u^2)\sqrt{a^2+u^2} - \tfrac{1}{8}a^4\ln\left|u+\sqrt{a^2+u^2}\right| + c$

23. $\displaystyle\int \frac{\sqrt{a^2+u^2}}{u}\,du = \sqrt{a^2+u^2} - a\ln\left|\frac{a+\sqrt{a^2+u^2}}{u}\right| + c$

24. $\displaystyle\int \frac{\sqrt{a^2+u^2}}{u^2}\,du = \ln\left|u+\sqrt{a^2+u^2}\right| - \frac{\sqrt{a^2+u^2}}{u} + c$

25. $\displaystyle\int \frac{1}{\sqrt{a^2+u^2}}\,du = \ln\left|u+\sqrt{a^2+u^2}\right| + c$

26. $\displaystyle\int \frac{u^2}{\sqrt{a^2+u^2}}\,du = \frac{1}{2}u\sqrt{a^2+u^2} - \frac{1}{2}a^2\ln\left|u+\sqrt{a^2+u^2}\right| + c$

27. $\displaystyle\int \frac{1}{u\sqrt{a^2+u^2}}\,du = \frac{1}{a}\ln\left|\frac{u}{a+\sqrt{a^2+u^2}}\right| + c$

28. $\displaystyle\int \frac{1}{u^2\sqrt{a^2+u^2}}\,du = -\frac{\sqrt{a^2+u^2}}{a^2u} + c$

Forms Involving $\sqrt{a^2-u^2}$, $a>0$

29. $\displaystyle\int \sqrt{a^2-u^2}\,du = \tfrac{1}{2}u\sqrt{a^2-u^2} + \tfrac{1}{2}a^2\sin^{-1}\frac{u}{a} + c$

30. $\displaystyle\int u^2\sqrt{a^2-u^2}\,du = \tfrac{1}{8}u(2u^2-a^2)\sqrt{a^2-u^2} + \tfrac{1}{8}a^4\sin^{-1}\frac{u}{a} + c$

31. $\displaystyle\int \frac{\sqrt{a^2-u^2}}{u}\,du = \sqrt{a^2-u^2} - a\ln\left|\frac{a+\sqrt{a^2-u^2}}{u}\right| + c$

32. $\displaystyle\int \frac{\sqrt{a^2-u^2}}{u^2}\,du = -\frac{\sqrt{a^2-u^2}}{u} - \sin^{-1}\frac{u}{a} + c$

33. $\displaystyle\int \frac{1}{\sqrt{a^2-u^2}}\,du = \sin^{-1}\frac{u}{a} + c$

34. $\displaystyle\int \frac{1}{u\sqrt{a^2-u^2}}\,du = -\frac{1}{a}\ln\left|\frac{a+\sqrt{a^2-u^2}}{u}\right| + c$

35. $\displaystyle\int \frac{u^2}{\sqrt{a^2-u^2}}\,du = -\frac{1}{2}u\sqrt{a^2-u^2} + \frac{1}{2}a^2\sin^{-1}\frac{u}{a} + c$

36. $\displaystyle\int \frac{1}{u^2\sqrt{a^2-u^2}}\,du = -\frac{\sqrt{a^2-u^2}}{a^2u} + c$

Forms Involving $\sqrt{u^2-a^2}$, $\quad a > 0$

37. $\displaystyle\int \sqrt{u^2-a^2}\,du = \tfrac{1}{2}u\sqrt{u^2-a^2} - \tfrac{1}{2}a^2\ln\left|u+\sqrt{u^2-a^2}\right| + c$

38. $\displaystyle\int u^2\sqrt{u^2-a^2}\,du = \tfrac{1}{8}u(2u^2-a^2)\sqrt{u^2-a^2} - \tfrac{1}{8}a^4\ln\left|u+\sqrt{u^2-a^2}\right| + c$

39. $\displaystyle\int \frac{\sqrt{u^2-a^2}}{u}\,du = \sqrt{u^2-a^2} - a\sec^{-1}\frac{|u|}{a} + c$

40. $\displaystyle\int \frac{\sqrt{u^2-a^2}}{u^2}\,du = \ln\left|u+\sqrt{u^2-a^2}\right| - \frac{\sqrt{u^2-a^2}}{u} + c$

41. $\displaystyle\int \frac{1}{\sqrt{u^2-a^2}}\,du = \ln\left|u+\sqrt{u^2-a^2}\right| + c$

42. $\displaystyle\int \frac{u^2}{\sqrt{u^2-a^2}}\,du = \frac{1}{2}u\sqrt{u^2-a^2} + \frac{1}{2}a^2\ln\left|u+\sqrt{u^2-a^2}\right| + c$

43. $\displaystyle\int \frac{1}{u\sqrt{u^2-a^2}}\,du = \frac{1}{a}\sec^{-1}\frac{|u|}{a} + c$

44. $\displaystyle\int \frac{1}{u^2\sqrt{u^2-a^2}}\,du = \frac{\sqrt{u^2-a^2}}{a^2u} + c$

Forms Involving $\sqrt{2au-u^2}$

45. $\displaystyle\int \sqrt{2au-u^2}\,du = \frac{1}{2}(u-a)\sqrt{2au-u^2} + \frac{1}{2}a^2\cos^{-1}\left(\frac{a-u}{a}\right) + c$

46. $\displaystyle\int u\sqrt{2au-u^2}\,du = \frac{1}{6}(2u^2-au-3a^2)\sqrt{2au-u^2} + \frac{1}{2}a^3\cos^{-1}\left(\frac{a-u}{a}\right) + c$

47. $\displaystyle\int \frac{\sqrt{2au-u^2}}{u}\,du = \sqrt{2au-u^2} + a\cos^{-1}\left(\frac{a-u}{a}\right) + c$

48. $\displaystyle\int \frac{\sqrt{2au-u^2}}{u^2}\,du = -\frac{2\sqrt{2au-u^2}}{u} - \cos^{-1}\left(\frac{a-u}{a}\right) + c$

49. $\displaystyle\int \frac{1}{\sqrt{2au-u^2}}\,du = \cos^{-1}\left(\frac{a-u}{a}\right) + c$

50. $\displaystyle\int \frac{u}{\sqrt{2au-u^2}}\,du = -\sqrt{2au-u^2} + a\cos^{-1}\left(\frac{a-u}{a}\right) + c$

51. $\displaystyle\int \frac{u^2}{\sqrt{2au-u^2}}\,du = -\frac{1}{2}(u+3a)\sqrt{2au-u^2} + \frac{3}{2}a^2\cos^{-1}\left(\frac{a-u}{a}\right) + c$

52. $\displaystyle\int \frac{1}{u\sqrt{2au-u^2}}\,du = -\frac{\sqrt{2au-u^2}}{au} + c$

Forms Involving sin *u* OR cos *u*

53. $\displaystyle\int \sin u\,du = -\cos u + c$

54. $\displaystyle\int \cos u\,du = \sin u + c$

55. $\displaystyle\int \sin^2 u\,du = \tfrac{1}{2}u - \tfrac{1}{2}\sin u\cos u + c$

56. $\displaystyle\int \cos^2 u\,du = \tfrac{1}{2}u + \tfrac{1}{2}\sin u\cos u + c$

57. $\displaystyle\int \sin^3 u\,du = -\tfrac{2}{3}\cos u - \tfrac{1}{3}\sin^2 u\cos u + c$

58. $\displaystyle\int \cos^3 u\,du = \tfrac{2}{3}\sin u + \tfrac{1}{3}\sin u\cos^2 u + c$

59. $\displaystyle\int \sin^n u\,du = -\frac{1}{n}\sin^{n-1}u\cos u + \frac{n-1}{n}\int \sin^{n-2}u\,du$

60. $\displaystyle\int \cos^n u\,du = \frac{1}{n}\cos^{n-1}u\sin u + \frac{n-1}{n}\int \cos^{n-2}u\,du$

61. $\displaystyle\int u\sin u\,du = \sin u - u\cos u + c$

62. $\displaystyle\int u\cos u\,du = \cos u + u\sin u + c$

63. $\displaystyle\int u^n\sin u\,du = -u^n\cos u + n\int u^{n-1}\cos u\,du$

64. $\displaystyle\int u^n\cos u\,du = u^n\sin u - n\int u^{n-1}\sin u\,du$

65. $\displaystyle\int \frac{1}{1+\sin u}\,du = \tan u - \sec u + c$

66. $\displaystyle\int \frac{1}{1-\sin u}\,du = \tan u + \sec u + c$

67. $\displaystyle\int \frac{1}{1+\cos u}\,du = -\cot u + \csc u + c$

68. $\displaystyle\int \frac{1}{1-\cos u}\,du = -\cot u - \csc u + c$

69. $\displaystyle\int \sin(mu)\sin(nu)\,du = \frac{\sin(m-n)u}{2(m-n)} - \frac{\sin(m+n)u}{2(m+n)} + c$

70. $\displaystyle\int \cos(mu)\cos(nu)\,du = \frac{\sin(m-n)u}{2(m-n)} + \frac{\sin(m+n)u}{2(m+n)} + c$

71. $\int \sin(mu)\cos(nu)\,du = \dfrac{\cos(n-m)u}{2(n-m)} - \dfrac{\cos(m+n)u}{2(m+n)} + c$

72. $\int \sin^m u \cos^n u\,du = -\dfrac{\sin^{m-1} u \cos^{n+1} u}{m+n} + \dfrac{m-1}{m+n}\int \sin^{m-2} u \cos^n u\,du$

Forms Involving Other Trigonometric Functions

73. $\int \tan u\,du = -\ln|\cos u| + c = \ln|\sec u| + c$

74. $\int \cot u\,du = \ln|\sin u| + c$

75. $\int \sec u\,du = \ln|\sec u + \tan u| + c$

76. $\int \csc u\,du = \ln|\csc u - \cot u| + c$

77. $\int \tan^2 u\,du = \tan u - u + c$

78. $\int \cot^2 u\,du = -\cot u - u + c$

79. $\int \sec^2 u\,du = \tan u + c$

80. $\int \csc^2 u\,du = -\cot u + c$

81. $\int \tan^3 u\,du = \frac{1}{2}\tan^2 u + \ln|\cos u| + c$

82. $\int \cot^3 u\,du = -\frac{1}{2}\cot^2 u - \ln|\sin u| + c$

83. $\int \sec^3 u\,du = \frac{1}{2}\sec u \tan u + \frac{1}{2}\ln|\sec u + \tan u| + c$

84. $\int \csc^3 u\,du = -\frac{1}{2}\csc u \cot u + \frac{1}{2}\ln|\csc u - \cot u| + c$

85. $\int \tan^n u\,du = \dfrac{1}{n-1}\tan^{n-1} u - \int \tan^{n-2} u\,du,\ n \neq 1$

86. $\int \cot^n u\,du = -\dfrac{1}{n-1}\cot^{n-1} u - \int \cot^{n-2} u\,du,\ n \neq 1$

87. $\int \sec^n u\,du = \dfrac{1}{n-1}\sec^{n-2} u \tan u + \dfrac{n-2}{n-1}\int \sec^{n-2} u\,du,\ n \neq 1$

88. $\int \csc^n u\,du = -\dfrac{1}{n-1}\csc^{n-2} u \cot u + \dfrac{n-2}{n-1}\int \csc^{n-2} u\,du,\ n \neq 1$

89. $\int \dfrac{1}{1 \pm \tan u}\,du = \frac{1}{2}u \pm \ln|\cos u \pm \sin u| + c$

90. $\int \dfrac{1}{1 \pm \cot u}\,du = \frac{1}{2}u \mp \ln|\sin u \pm \cos u| + c$

91. $\displaystyle\int \frac{1}{1 \pm \sec u}\, du = u + \cot u \mp \csc u + c$

92. $\displaystyle\int \frac{1}{1 \pm \csc u}\, du = u - \tan u \pm \sec u + c$

Forms Involving Inverse Trigonometric Functions

93. $\displaystyle\int \sin^{-1} u\, du = u \sin^{-1} u + \sqrt{1 - u^2} + c$

94. $\displaystyle\int \cos^{-1} u\, du = u \cos^{-1} u - \sqrt{1 - u^2} + c$

95. $\displaystyle\int \tan^{-1} u\, du = u \tan^{-1} u - \ln \sqrt{1 + u^2} + c$

96. $\displaystyle\int \cot^{-1} u\, du = u \cot^{-1} u + \ln \sqrt{1 + u^2} + c$

97. $\displaystyle\int \sec^{-1} u\, du = u \sec^{-1} u - \ln |u + \sqrt{u^2 - 1}| + c$

98. $\displaystyle\int \csc^{-1} u\, du = u \csc^{-1} u + \ln |u + \sqrt{u^2 - 1}| + c$

99. $\displaystyle\int u \sin^{-1} u\, du = \tfrac{1}{4}(2u^2 - 1) \sin^{-1} u + \tfrac{1}{4} u\sqrt{1 - u^2} + c$

100. $\displaystyle\int u \cos^{-1} u\, du = \tfrac{1}{4}(2u^2 - 1) \cos^{-1} u - \tfrac{1}{4} u\sqrt{1 - u^2} + c$

Forms Involving e^u

101. $\displaystyle\int e^{au}\, du = \frac{1}{a} e^{au} + c$

102. $\displaystyle\int u e^{au}\, du = \left(\frac{1}{a} u - \frac{1}{a^2}\right) e^{au} + c$

103. $\displaystyle\int u^2 e^{au}\, du = \left(\frac{1}{a} u^2 - \frac{2}{a^2} u + \frac{2}{a^3}\right) e^{au} + c$

104. $\displaystyle\int u^n e^{au}\, du = \frac{1}{a} u^n e^{au} - \frac{n}{a} \int u^{n-1} e^{au}\, du$

105. $\displaystyle\int e^{au} \sin bu\, du = \frac{1}{a^2 + b^2}(a \sin bu - b \cos bu) e^{au} + c$

106. $\displaystyle\int e^{au} \cos bu\, du = \frac{1}{a^2 + b^2}(a \cos bu + b \sin bu) e^{au} + c$

Forms Involving $\ln u$

107. $\displaystyle\int \ln u\, du = u \ln u - u + c$

108. $\displaystyle\int u \ln u\, du = \tfrac{1}{2} u^2 \ln u - \tfrac{1}{4} u^2 + c$